T0140577

Advances in Intelligent Systems and Computing

Volume 736

Series editor

Janusz Kacprzyk, Polish Academy of Sciences, Warsaw, Poland
e-mail: kacprzyk@ibspan.waw.pl

The series "Advances in Intelligent Systems and Computing" contains publications on theory, applications, and design methods of Intelligent Systems and Intelligent Computing. Virtually all disciplines such as engineering, natural sciences, computer and information science, ICT, economics, business, e-commerce, environment, healthcare, life science are covered. The list of topics spans all the areas of modern intelligent systems and computing.

The publications within "Advances in Intelligent Systems and Computing" are primarily textbooks and proceedings of important conferences, symposia and congresses. They cover significant recent developments in the field, both of a foundational and applicable character. An important characteristic feature of the series is the short publication time and world-wide distribution. This permits a rapid and broad dissemination of research results.

More information about this series at http://www.springer.com/series/11156

Ajith Abraham · Pranab Kr. Muhuri
Azah Kamilah Muda · Niketa Gandhi
Editors

Intelligent Systems Design and Applications

17th International Conference on Intelligent Systems Design and Applications (ISDA 2017) Held in Delhi, India, December 14–16, 2017

 Springer

Editors
Ajith Abraham
Machine Intelligence Research Labs
Auburn, WA
USA

Pranab Kr. Muhuri
Department of Computer Science
South Asian University
Chanakyapuri, Delhi
India

Azah Kamilah Muda
Faculty of Information and Communication
    Technology
Universiti Teknikal Malaysia Melaka
Durian Tunggal, Melaka
Malaysia

Niketa Gandhi
Machine Intelligence Research Labs
Auburn, WA
USA

ISSN 2194-5357   ISSN 2194-5365 (electronic)
Advances in Intelligent Systems and Computing
ISBN 978-3-319-76347-7   ISBN 978-3-319-76348-4 (eBook)
https://doi.org/10.1007/978-3-319-76348-4

Library of Congress Control Number: 2018935895

Printed on acid-free paper

This Springer imprint is published by the registered company Springer International Publishing AG
part of Springer Nature
The registered company address is: Gewerbestrasse 11, 6330 Cham, Switzerland

Preface

Welcome to the Proceedings of the 17th International Conference on Intelligent Systems Design and Applications (ISDA17), which was held in South Asian University, Delhi, India, during December 14–16, 2017. ISDA 2017 is jointly organized by the Machine Intelligence Research Labs (MIR Labs), USA, and South Asian University, Delhi, India.

ISDA 2017 brings together researchers, engineers, developers, and practitioners from academia and industry working in all interdisciplinary areas of intelligent systems and system engineering to share their experiences, and to exchange and cross-fertilize their ideas. The aim of ISDA 2017 is to serve as a forum for the dissemination of state-of-the-art research and development of intelligent systems, intelligent technologies, and applications. ISDA 2017 was organized in conjunction with the 7th World Congress on Information and Communication Technologies (WICT 2017).

The themes of the contributions and scientific sessions range from theories to applications, reflecting a wide spectrum of the coverage of intelligent systems and computational intelligence areas. ISDA 2017 received submissions from over 30 countries, and each paper was reviewed by at least 5 reviewers in a standard peer-review process. Based on the recommendation by 5 independent referees, finally about 100 papers were accepted for publication in the proceedings published by Springer, Verlag.

Many people have collaborated and worked hard to produce the successful ISDA 2017 conference. First, we would like to thank all the authors for submitting their papers to the conference, for their presentations and discussions during the conference. Our thanks go to Program Committee members and reviewers, who carried out the most difficult work by carefully evaluating the submitted papers. Our special thanks to Patricia Melin, Tijuana Institute of Technology, Tijuana, Mexico, and Alexander Gelbukh, Instituto Politécnico Nacional, Mexico City, Mexico, for the exciting plenary talks.

We express our sincere thanks to special session chairs and organizing committee chairs for helping us to formulate a rich technical program.

Ajith Abraham
Pranab Kr. Muhuri
ISDA 2017 - General Chairs

ISDA 2017 Organization

General Chairs

Ajith Abraham   Machine Intelligence Research Labs, USA
Pranab Kr. Muhuri   South Asian University, Delhi, India

Program Committee Co-chairs

Simone Ludwig   North Dakota State University, USA
Aswani Kumar   VIT University, Vellore, India
Punam Bedi   University of Delhi, India
Millie Pant   Indian Institute of Technology Roorkee, India
Antonio J. Tallón-Ballesteros   University of Seville, Spain

Advisory Board

Albert Zomaya   The University of Sydney, Australia
Andre Ponce de Leon   University of Sao Paulo at Sao Carlos, Brazil
   F. de Carvalho
Bruno Apolloni   University of Milano, Italy
Hideyuki Takagi   Kyushu University, Japan
Imre J. Rudas   Óbuda University, Hungary
Janusz Kacprzyk   Polish Academy of Sciences, Poland
Javier Montero   Complutense University of Madrid, Spain
Krzysztof Cios   Virginia Commonwealth University, USA
Marina Gavrilova   University of Calgary, Canada
Mario Koeppen   Kyushu Institute of Technology, Japan
Mohammad Ishak Desa   Universiti Teknikal Malaysia Melaka, Malaysia
Patrick Siarry   Université Paris-Est Créteil, France
Ronald Yager   Iona College, USA

Salah Al-Sharhan	Gulf University of Science and Technology, Kuwait
Sebastian Ventura	University of Cordoba, Spain
Vincenzo Piuri	Università degli Studi di Milano, Italy

Publication Chairs

| Azah Kamilah Muda | UTeM, Malaysia |
| Niketa Gandhi | Machine Intelligence Research Labs, USA |

Local Organizing Committee

| Q. M. Danish Lohani | South Asian University, India |
| | danishlohani@cs.sau.ac.in |

Local Organizing Committee Members

Amit K. Shukla	South Asian University, Delhi, India
Ashraf Zubair	South Asian University, Delhi, India
Manvendra Janmaijaya	South Asian University, Delhi, India
Amit Rauniyar	South Asian University, Delhi, India
Rahul Nath	South Asian University, Delhi, India
Sandeep Kumar	South Asian University, Delhi, India
Taniya Seth	South Asian University, Delhi, India
Deepika Malhotra	South Asian University, Delhi, India

Web Service

| Kun Ma | University of Jinan, China |

International Program Committee

Ajith Abraham	Machine Intelligence Research Labs, USA
Akila Muthuramalingam	KPR Institute of Engineering and Technology, India
Alberto Cano	University of Córdoba, Spain
Amiya Tripathy	Don Bosco Institute of Technology, Mumbai, India
Andrzej Skowron	Warsaw University of Technology, Poland
Anna Jordanous	University of Kent, UK
Antonio J. Tallón Ballesteros	Universidad de Sevilla, Spain

Aswani Cherukuri	Vellore Institute of Technology, India
Bharanidharan Shanmugam	Universiti Teknologi Malaysia, Malaysia
Bin Li	University of Science and Technology of China, China
Carlos Pereira	Instituto Superior de Engenharia de Coimbra, Portugal
Cerasela Crisan	"Vasile Alecsandri" University of Bacau, Romania
César Hervás Martínez	University of Córdoba, Spain
Chao Chun Chen	Southern Taiwan University of Science and Technology, Taiwan
Chin-Shiuh Shieh	National Kaohsiung University of Applied Sciences, Taiwan
Daniela Zaharie	West University of Timisoara, Romania
Diaf Moussa	Université Mouloud Mammeri, Algeria
Dilip Pratihar	Indian Institute of Technology Roorkee, India
Eduardo Solteiro Pires	University of Trás-os-Montes and Alto Douro, Portugal
Efrén Mezura Montes	Universidad Veracruzana, Mexico
Eiji Uchino	Yamaguchi University, Japan
Elizabeth Goldbarg	Universidade Federal do Rio Grande do Norte, Brazil
Enrique Dominguez	Universidad de Málaga, Spain
Fabrício Olivetti de França	Universidade Federal do ABC, Brazil
Fedja Netjasov	University of Belgrade, Serbia
José Francisco Martínez Trinidad	National Institute of Astrophysics, Optics and Electronics, Puebla, Mexico
Gagandeep Kaur	JIIT, Noida, India
Georg Peters	Munich University of Applied Sciences, Germany
Hector Benitez-Perez	Universidad Nacional Autónoma de México, Mexico
Heder Bernardino	Universidade Federal de Juiz de For a, Brazil
Hema Banati	University of Delhi, India
Hiroshi Dozono	Saga University, Japan
Ilhem Kallel	École Nationale d'Ingénieurs de Sfax, Tunisia
Isabel Barbancho	Universidad de Málaga, Spain
Isabel S. Jesus	Instituto Superior de Engenharia do Porto, Portugal
Janos Botzheim	Tokyo Metropolitan University, Japan
Jerzy Grzymala Busse	University of Kansas, USA
Jolanta Mizera-Pietraszko	Opole University, Poland
Kelemen Arpad	University of Maryland, USA
Keun Ho Ryu	Chungbuk National University, South Korea
Konstantinos Parsopoulos	University of Ioannina, Greece

Korhan Karabulut — Yaşar Üniversitesi, Turkey
Kyriakos Kritikos — Foundation for Research and Technology (FORTH) Hellas, Greece
Laurence Amaral — Universidade Federal de Uberlândia, Brazil
Lee Chang Yong — Kongju National University, South Korea
Leocadio G. Casado — University of Almería, Spain
Leticia Hernando — The University of the Basque Country, Spain
Lin Wang — Jinan University, China
Lubna Gabralla — Sudan University of Science and Technology, Sudan
Ludwig Simone — North Dakota State University, USA
Luigi Troiano — University of Sannio, Italy
Matthias Becker — Leibniz Universität Hannover, Germany
Mauricio Ayala Rincon — Universidade de Brasilia, Brazil
Mdrafiul Hassan — King Fahd University of Petroleum & Minerals, Dhahran, KSA
Millie Pant — Indian Institute of Technology Roorkee, India
Mohammad Shojafar — Sapienza University of Rome, Italy
Mrutyunjaya Panda — Gandhi Institute for Technological Advancement, India
Nebojsa Bacanin — Megatrend Univerzitet, Serbia
Neetu Sardana — JIIT, Noida, India
Niketa Gandhi — Machine Intelligence Research Labs, USA
Olfa Jemai — Université de Sfax, Tunisia
Oscar Castillo — Tijuana Institute of Technology, Tijuana
Oscar Gabriel Reyes Pupo — The University of Central Oklahoma, USA
Patrick Siarry — Université de Paris, France
Paulo Carrasco — Universidade do Algarve, Portugal
Paulo Moura Oliveira — University of Trás-os-Montes and Alto Douro, Portugal
Pranab Muhuri — South Asian University, Delhi, India
Ramzan Muhammad — Maulana Mukhtar Ahmad Nadvi Technical Campus, India
Shikha Mehta — JIIT, Noida, India
Shing Chiang Tan — Multimedia University, Malaysia
Shu Fen Tu — Chinese Culture University, China
Siddhivinayak Kulkarni — University of Ballarat, Australia
Tarun Sharma — Amity University, Rajasthan
Terry Gafron — Bio Inspired Technologies, USA
Thomas Hanne — University of Applied Sciences Northwestern Switzerland, Switzerland
Usue Mori — University of the Basque Country, Spain
Varun Kumar Ojha — Swiss Federal Institute of Technology, Switzerland

Additional Reviewers

Kaushik Das Sharma	University of Calcutta, India
Safia Djemame	Ferhat Abbas University, Algeria
Yi-Fei Pu	Sichuan University, China
Md Sarwar Haque	King Fahd University of Petroleum & Minerals Dammam, Saudi Arabia
Denis Felipe	Federal University of Rio Grande do Norte, Brazil
Sílvia M. D. M. Maia	Federal University of Rio Grande do Norte, Brazil
Lucas Daniel M. S. Pinheiro	Universidade Federal do Rio Grande do Norte, Brazil
Hector-Gabriel Acosta-Mesa	Universidad Veracruzana, Mexico
Edgar-Alfredo Portilla-Flores	Instituto Politécnico Nacional, Mexico
Md Sarwar Haque	King Fahd University of Petroleum & Minerals, Saudi Arabia
Adelaide Cerveira	INESC TEC and UTAD, Portugal
Joslaine Cristina Jeske de Freitas	Universidade Federal de Goiás, Brazil
Eliana Pantaleão	Universidade Federal de Uberlândia, Brasil
Ariane Alves Almeida	University of Brasília, Brazil
Lucas Angelo Silveira	University of Brasília, Brazil
Daniele Nantes-Sobrinho	University of Brasília, Brazil
Daniel Saad Nogueira Nunes	University of Brasília, Brazil
Sumit Kumar Banshal	South Asian University, Delhi, India
Rajesh Piryani	South Asian University, Delhi, India
Sandeep Kumar	South Asian University, Delhi, India
Amit Kumar Shukla	South Asian University, Delhi, India
Thatiana C. N. Souza	Federal University Rural Semi-Arid, Brazil
Shadrack Maina Mambo	Kenyatta University, Nairobi, Kenya
Nawel Drira	Ecole Nationale d'Electronique et des Télécommunications de Sfax, Tunisia
Esteban José Palomo	University of Malaga, Spain

Contents

Enhancing Job Opportunities in Rural India Through Constrained Cognitive Learning Process: Reforming Basic Education

Shivangi Nigam, Abhishek Bajpai[✉], and Bineet Gupta

Shri Ramswaroop Memorial University, Lucknow, UP, India
abhishek.srmu@gmail.com

Abstract. Technological advancements in cognitive learning suggest significant changes in methods of teaching and learning process. A Constrained Cognitive Learning (CCL) model links various forms of cognitive learning methods with a restrictive domain. The main objective of this research study is to propose a CCL scheme that integrates cognitive learning theories and instructional prescriptions to achieve an effective learning environment for the basic education system in rural India. It improves both knowledge acquisition and employment in optimized way. Furthermore, our objective is that, the proposed research contributes in promoting the dialogue between professional learners, academic researchers and practitioners that increasingly brings empirical educational and research orientation into the contemporary educational environment across the rural India. Our focus is to plan such a cognitive learning environment so that the learner not only acquire knowledge but also improve their cognitive abilities to apply their knowledge for the employment and extend their knowledge depth to move towards research oriented innovative skills.

Keywords: Constrained Cognitive Learning (CCL) · Learning paradigm
Behaviorism · Cognitivism · Constructivism · Self regulated learning

1 Introduction

"Our profession has always stood at the interface of person and environment-a tenuous place to be theoretically, since no theory has to date effectively spanned this bridge".

Cooper (Educationist), 1979

Learning is a continuous pattern of perceiving various goings-on and attaining knowledge from them. The learning process leads to change in behaviors', increase in the intellect levels and the ways to analyze and process the perceived information. The analysis of the learning process is significant for student-centered learning environment to enhance the learning abilities of the students. The students vary greatly in their potential to grasp any new information. Their attitude of apprehension is very low as they do not try to engage themselves towards the understanding of the new subject. The Learning theories paramount the learning propensities by analyzing the learning patterns and unraveling the ways of perception, perseverance and exploitation of the knowledge.

© Springer International Publishing AG, part of Springer Nature 2018
A. Abraham et al. (Eds.): ISDA 2017, AISC 736, pp. 1–9, 2018.
https://doi.org/10.1007/978-3-319-76348-4_1

As per the quote of Jean Piaget:

"The goal of education is not to increase the amount of knowledge but to create the possibilities for a child to invent and discover, to create men who are capable of doing new things."

With this objective, Learning Theories have proposed various models for an enhanced learning process. The Learning theories have been broadly categorized in 3 models as Behaviorism, Cognitivism and Constructivism. The behaviorism, fabricated by John Watson, is the study of learner's behaviour. The conditioning of the learner is done via reinforcement (rewards) or punishment which is perceptible from the behaviour of the learner Greeno (1996). The Cognitivism stemmed from Gestalt psychology, explains the learning as processing of information gained as past experiences. This theory focuses on thinking caliber of an individual with a previously owned knowledge regarding the subject matter. The Behaviorist expostulate learning process as the incremental pattern of change in behaviour according to the conditioning of individual i.e. the conditioning is exhibited by the behaviour of the learner. It prepares students for performing in a knlearning is the mental escalation instead of a behavioral reform. The persistent knowledge acquisition upshots the mental structure which aids the individual in resolving novel issues. The psychoown state of affairs. The individuals fail to enact in any inconclusive environment. They lack the problem solving and creative thinking capabilities. The Cognitivist theory juxtaposes the Behaviorist theory by propounding that logist, Jean Piaget gave a constructive approach to learning as Constructivism. It accentuates the students to intensively partake in acquiring knowledge for themselves. Where Cognitivism is based on previous knowledge, Constructivism focuses on striving to achieve a novel information with the help of previous knowledge thus providing better perceivance and exploitation. The learner is free to choose his goal which is not the case in Behaviorism/Cognitivism where it is confined to a predetermined set of goals Hannafin (2010).

The Curricular and instructional design which strongly affect educational practice and advancements extend the scope of learning paradigm of instructional design and management Hill (1997). Cognitive system of learning can be analyzed from several points of view in a dialogue between different parties cross and host. Their informal conversation provides the background for displaying examples and different styles of learning, teaching, testing, and group dynamics situations. The applicability of cognitive system of teaching and learning is emphasized in two ways. First, not all learners have same level of understanding that match the traditional lecture delivery and laboratory format of teaching. Second, certain teachers lecture delivery style is more effective with certain types of students because they have matching cognitive styles and levels. Efforts should be made, therefore, to ascertain cognitive styles, to match students and teachers with compatible styles, and to develop individualized level of materials appropriateness for specific cognitive styles of learning.

2   Background

The cognitive learning has been the concern of various research studies in the recent past. The study by Van Merriënboer and Ayres (2005) proposes a student centered cognitive learning in web based environments. The research analyses the current pattern of student learning from web based multimedia Iiyoshi et al. (2005). The research focus is on exposing students for a better learning experience as proved by various researches. Although the internet technologies may be fruitful for deep learning of the learners, there has to be some curtailment for the exploration otherwise it can mislead the learners from the subject. Thus the web based learning needs to be refined for better learning results. The research in Pintrich (1999) draws the distinction between the self regulated learning and motivational learning. The research study briefs about various categories of self regulated learning as cognitive learning strategies, self-regulatory strategies to control cognition, and resource management strategies. The study aims to scrutinize the role of self regulated learning techniques in student centered classroom environment. The study by Ohlsson (2016) restrains the cognitive learning process to provide the learners a supervised learning environment. The research commingles the cognitive mechanisms by the user specific domains thus providing aid for Intelligent tutoring systems (ITS). The work in Roberson and Merriam (2005) explores the correlation among epistemological beliefs and conceptual change learning (CCL). The study reveals that prior knowledge and learning abilities of an individual are significant in conceptual learning process. Turning over to the rural population which has least exposure to the technological learning, various studies have been done to study the factors affecting the learning patterns of students as well as the elderly/adults. The research by Qian and Alvermann (1995) takes into consideration the adults a rural town of America. The study revolves around the self-directed learning (SDL) also termed as personal learning of the elderly in a rural domain. The results depict a positive response of adults towards making the most of learning resources such as computers, mobile phones etc. thus rendering self directed learning activities. An application of cognitive learning is presented by Rahman et al. (2008) as providing assistance to the health workers in rural areas of Pakistan. The study examined pregnant women over a period of 12 months. The cognitive learning of health works in a resource deficient environment proved a successful intervention.

3 Constrained Cognitive Learning Scheme

3.1 Issues

The Basic Education in India, predominantly, in rural India needs to be brushed up as it will eventually enhance India's economic development Roediger (2013). According to the Annual Status of Education Report (ASER) in 2012, a larger fragment (96.5%) of children of age group 6–14 years are enrolled in some school. Although the quantity of this percentage has perked up, but the make-up of the knowledge measure is still not competent. This deterioration is the repercussion of various factors as teacher quality, classroom quality, curriculum concerns and other dismissive factors as motivation to study etc. The teacher based learning have shown appalling pattern of learning stature. There is a need to develop a supervised learning methodology so as to develop better learning scheme specifically designed for rural environments.

3.2   Research Questions

The research probes the repercussions of bringing in the restrictive cognitive learning into the current learning paradigm. The study aims to give the way outs for following questions:

Q1. What are the factors in dominance by introducing Constrained Cognitive Learning (CCL) in curriculum of rural area students?
Q2. How can the CCL paradigm up-skill the teachers in rural areas?
Q3. What determinants can motivate the students for CCL?
Q4. What is the stumbling block the pavement for quality learning?

3.3   Methodology

The Constrained Cognitive Learning (CCL) scheme proposes learning in 2 phases Up-skill phase and Learning phase in recurrence depicted by Fig. 1.

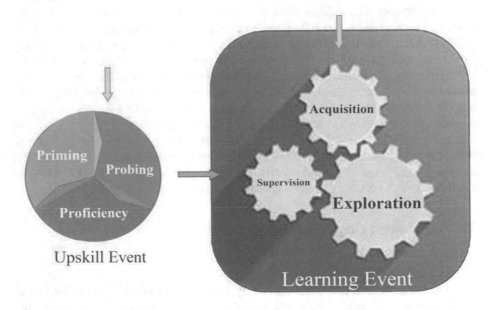

Fig. 1. Constrained cognitive learning scheme

The up-skill phase involve perk up the teachers in the existing environ. This phase recapitulate for 3 steps: Priming, Probing and Proficiency. The first step, Priming of teachers is done to enhance their current skills as per the student apriori knowledge. It is important to consider the student stature before imparting knowledge to them of a new subject. It makes them have better understanding of the new subject. The teachers should ruminate over their level of apprehension. The second step, Probing of the teachers is done to observe them if they are using any supplementary methods for imparting a better learning to the students. The reflection of positive probe results depict that this phase is successful. The third step, Proficiency of the teachers is a significant concern. The teacher's proficiency is decided as per the result of probe. It should be very clear that the teachers are provided the domains in which they are proficient in. It has a substantial effect on the learning. The up-skill phase helps the supervisors to have finer practical knowledge thus evolving out their exceptional aspects.

The Learning Event involves the student and his environment. Before irrupting into the learning phase, the student is supposed to have a basic understanding of the subject. The learning event is a process involving 3 intrinsic units in chorus. In the Acquisition part, the student strives to acquire knowledge for a problem solving. The students are given elementary knowledge of the subject. With some understanding regarding a subject, the students are exposed to the practical problems of the subject area. The learner's are expected to knock themselves out and explores the problem for a solution. The exploration of student is guided by a supervisor to fabricate the domain of his anatomy. The supervisor is required to keep the domain of research restricted so as to guide the student for a correct path of knowledge.

4 Implementation: Data Collection and Analysis

4.1 Data Collection

The study involves participants of a school in a rural area near Lucknow, India. A change of schedule of the teachers and children was done for a month. A prior text of student's stature was done to be compared later on with the end results. The initial tests include quiz, interviews and brief problem solving sessions. The students were observed for 1 week prior to the refashion of their curriculum. In the mean time, the up-skill session of teachers was programmed as per the above mentioned scheme. The teachers were dispensed with various development stratagems such as video lectures, presentations etc. They were enlightened with various teaching schemes and were appraised to disseminate teaching with a practical aspect. They were asked to put in for the students more towards practical learning. This scheme also objected to perceive to best of the skills of the teachers and thus making them involve in subjects of their proficiencies.

The program commenced with a reprogrammed curriculum and a schedule with some sessions of problem solving in a practical environments. For instance, the students were taken to fields to learn about the crops and they free to explore the environ. The CCL scheme decreased the theory sessions and replaced it with more of practical sessions. They were also exposed to various technologies in the area of their subjects. This was done to make students more jobs oriented as they can have an idea of what are their special aspects and they can pursue their higher education in the same to achieve their goals. The students were probed once every week for their performance. The inquest involved comprehensive problem solving and quizzes sessions. Table 1 shows the standard deviation of students as well as supervisors computed on the basis of questionnaire conducted at various intervals of the program.

Table 1. Standard deviation of supervisors and students based on questionnaire

Participant	Pre-CCL	1st iteration	2nd iteration	3rd iteration	4th iteration
Supervisors	0.60	0.74	0.99	0.76	0.71
Students	0.71	0.85	0.78	0.71	0.60

Table 2 presents a comparison of the parameters to distinguish the performance of students before and after. The parameters value of the conceptual parameters is the standard deviation of the aspects before and after the program commencement. The parameters of practical learning have been to measure of responsiveness (Resp.) and satisfaction (Satisf.) of the students on a ranking scale of 1 to 5.

Table 2. Comparison of various practical and conceptual aspects of CCL scheme.

Concept		Before		After	
Conceptual knowledge					
Objectivity of subject		0.92		0.88	
Curriculum design		0.61		0.72	
Teaching practices (Visual/Theoretical)		0.76		0.79	
Practical learning					
	Resp.	Satisf.	Resp.	Satisf.	
Problem elaboration	2.2	1.2	1.3	2.1	
Logical attitude	1.5	1.6	1.6	3.4	
Solution exploration	2.8	2.3	2.3	3.8	
Performance	1.2	3.1	3.1	4.6	
Grasping speed	1.5	3.2	3.2	3.5	

4.2 Analysis

The CCL scheme focuses on introducing supervision in previous cognitive learning scheme. The up-skilling of the supervisors has improved the overall results of the program. The average probing result of the teachers has improved significantly.

The development programs have been rated as positive by the faculties as they themselves have shown better performance also due to the restriction introduced as the proficiency part of up-skill phase. The higher the standard deviation, the parameters depict a wider distribution of values. Figure 2 depicts growing graph from the 6–8 year students to the 14–16 year students. This demonstrates that the students from the age group 10–16 year have a good effect of CCL scheme. The exploration skills of the students exhibit tremendous amelioration in their problem solving skills with significant increase in standard deviation of the overall performance.

The above research findings were verified by the Chi square tests to determine the fitness of various variables among each others. The results of the tests were found fruitful as the chi values achieved are lower. The small values of chi square signify that the results have proved to be as expected. The comparison of chi values achieved for Responsiveness and Satisfaction of the students can be depicted in the Fig. 2.

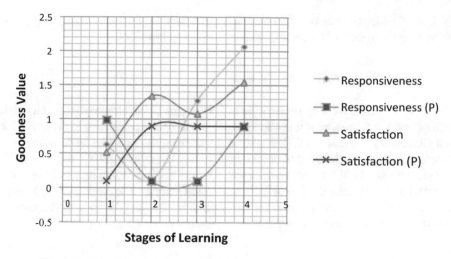

Fig. 2. Goodness of the values of responsiveness and satisfaction of students.

4.3　Research Findings

The study postulated four questions to find out the correct formulation of the CCL scheme for the basic education of rural India. The factors of Dominance in the investigation were found as the teaching practices, technology exposure and curriculum design (Fig. 3).

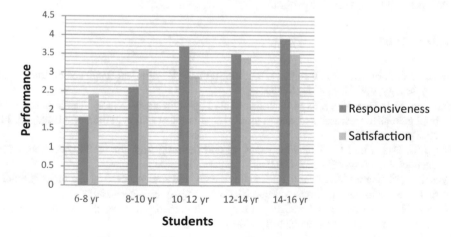

Fig. 3. Analysis of student's responsiveness and satisfaction of performance

The results show that they are as per the predictions made by the CCL scheme. The scheme also finds out ways to ameliorate the teaching practices as video lectures, presentations to promote visual learning process. Other empirical practices include workshops, group discussion sessions, seminars, and technology exposure. These practices promote students for a qualitative learning process. The dismissive factors towards

CCL are the lack of resources like electricity, internet connectivity, inadequate curriculum design and lack of technology usage as learning practices.

5   Future Research Directions

The initial introduction of cognitive style of learning as an instructional parameter at the college level; however, now it is also applicable and become effective to the secondary and elementary levels as well. The certain dimensions of cognitive learning practices depend a lot upon interactions of host and receivers. The effects of CCL scheme will be to a greater extent if the supervisors are aware of the student capabilities and can supervise the learners to enhance their skills in a effectual manner. The current CCL paradigm can be extended to introduce an Ingenious Learning System (ILS) which can adapt to the current learner stature and supervises the learner accordingly.

6   Conclusion

The CCL scheme is program proposed for the basic education system in rural India. The main objective is to combine practical aspects of the learning into the current curriculum and also improving the curriculum to make it a conceptual one. The up-skill and learning phase has a stupendous improvement in the problem solving of students and also the proficiencies of the teachers. This research proved efficacious to accomplish the ultimate goal of qualitative learning program.

References

Greeno, J.G., Collins, A.M., Resnick, L.B.: Cognition and learning. In: Handbook of Educational Psychology, vol. 77, pp. 15–46 (1996)

Hannafin, M.J., Hannafin, K.M.: Cognition and student-centered, web-based learning: issues and implications for research and theory. In: Learning and Instruction in the Digital Age, pp. 11–23. Springer (2010)

Hill, J.R., Hannafin, M.J.: Cognitive strategies and learning from the world wide web. Educ. Technol. Res. Dev. 45(4), 37–64 (1997)

Iiyoshi, T., Hannafin, M.J., Wang, F.: Cognitive tools and student- centred learning: rethinking tools, functions and applications. Educ. Media Int. 42(4), 281–296 (2005)

Ohlsson, S.: Constraint-based modeling: from cognitive theory to computer tutoring–and back again. Int. J. Artif. Intell. Educ. 26(1), 457–473 (2016)

Pintrich, P.R.: The role of motivation in promoting and sustaining self-regulated learning. Int. J. Educ. Res. 31(6), 459–470 (1999)

Qian, G., Alvermann, D.: Role of epistemological beliefs and learned helpless-ness in secondary school students' learning science concepts from text. J. Educ. Psychol. 87(2), 282 (1995)

Rahman, A., Malik, A., Sikander, S., Roberts, C., Creed, F.: Cognitive behaviour therapy-based intervention by community health workers for mothers with depression and their infants in rural Pakistan: a cluster-randomised controlled trial. Lancet 372(9642), 902–909 (2008)

Roberson Jr., D.N., Merriam, S.B.: The self-directed learning process of older, rural adults. Adult Educ. Q. 55(4), 269–287 (2005)

Roediger III, H.L.: Applying cognitive psychology to education: Translational educational science. Psychol. Sci. Publ. Interest **14**(1), 1–3 (2013)

Van Merriënboer, J.J., Ayres, P.: Research on cognitive load theory and its design implications for e-learning. Educ. Technol. Res. Dev. **53**(3), 5–13 (2005)

UML2ADA for Early Verification of Concurrency Inside the UML2.0 Atomic Components

Taoufik Sakka Rouis[1(✉)], Mohamed Tahar Bhiri[2], Mourad Kmimech[3], and
Layth Sliman[4]

[1] Cristal Laboratory, National School of Computer Sciences, University of Manouba, 2010
Manouba, Tunisia
srtaoufik@yahoo.fr
[2] Miracl Laboratory, ISIMS, Technological Pole of Sfax, 3021 Sakiet Ezzit, Sfax, Tunisia
tahar_bhiri@yahoo.fr
[3] UR-OASIS Laboratory, ENIT, University of Tunis El Manar, Tunis, Tunisia
mkmimech@gmail.com
[4] Efrei - École d'Ingénieur Généraliste en Informatique, 94800 Villejuif, France
layth.sliman@efrei.fr

Abstract. In recent years, the Unified Modeling Language (UML) has emerged
as a de facto industrial standard for modeling Component-Based Software (CBS).
However, in order to ensure the safety and vivacity of UML CBS, many
approaches have been proposed to verify the concurrency between interconnected
components. But, rare are the works that tackle concurrency verification inside
atomic components. In this paper, our purpose was the verification of the concurrency inside UML2.0 atomic components endowed with behavioral specifications
described by protocol state machines (PSM). To achieve this, we propose to
translate the UML2.0/PSM source component to an Ada concurrent program.
Using an Ada formal analysis tool such as FLAVERS or INCA tools, we could
detect the potential behavioral concurrency properties such as the deadlock of an
Ada concurrent program.

Keywords: UML2.0 · Component · Protocol state machine
Concurrency verification · Ada concurrent program

1 Introduction

In recent years, the Unified Modeling Language (UML) has emerged as a de facto
industrial standard for describing both the structure and behavior of software systems.
It is simple, intuitive and easy to understand. Furthermore, its graphical notations have
been extensively applied to design systems in various application domains, from small
embedded-software systems to huge distributed systems. In addition, the UML2.0
standard introduces a number of new concepts and refines a number of existing ones for
modeling Component-Based Software (CBS). Indeed, the UML2.0 component model
introduces the major architectural elements such as component, connector and port. The
introduction of these concepts and the distinction between provided and required

© Springer International Publishing AG, part of Springer Nature 2018
A. Abraham et al. (Eds.): ISDA 2017, AISC 736, pp. 10–20, 2018.
https://doi.org/10.1007/978-3-319-76348-4_2

interfaces allow an interesting set of notations for the modeling of component-based software. However, in this CBS engineering, software construction is based on the reutilization by an easy and coherent assembly of components. A coherent component assembly requires that the atomic components be correct. Hence, an atomic component is said correct if and only if its partial behaviors associated with the interfaces offered by this component are ensured by the global behavior of the said component.

In order to ensure the safety and vivacity of UML CBS, many approaches have been proposed to verify the concurrency between interconnected components [1]. However, rare are the works that tackle concurrency verification between a component interfaces and the component itself. In this study, an approach is proposed to verify the behavioral concurrency between an UML2.0 component and its interfaces. The behaviors associated with interfaces are called partial behaviors, and the behavior associated with the atomic component is called global behavior. The sought after objective in this paper is to give an answer to the checking problem of the partial behaviors compatibility (interface per interface) with the component global behavior. Thus, we provided an approach allowing the formal checking of the UML2.0 atomic components behavioral consistency. The behavioral aspect of this component and its interfaces is described by the Protocol State Machine (PSM), where PSM is a refinement of the (generic) behavioral State Machine (SM), imposing a restriction on its transitions and requiring that no activities are associated with neither transitions nor states.

To achieve this verification approach, the models transformation is used to translate the UML2.0/PSM atomic components into Ada. The choice of this concurrent language is justified by the presence of different analysis tools related to the detection of the dynamic and specific problems of an Ada program. For example, Naumovich et al. apply in [2] INCA and FLAVERS, two static concurrency analysis tools used for proving the behavioral properties of an Ada concurrent program, where INCA is a flow equation-based tool, and FLAVERS a data flow analysis-based tool. In addition, José et al. propose in [3] an Ada source code analyzer called CodePeer that detects run-time and logic errors. Using control-flow, data-flow, and other advanced static analysis techniques, CodePeer detects errors that would otherwise only be found through labor-intensive debugging. Currently, the newest version of the Ada programming language – Ada 2012 – provides a language of contracts for the dynamic verification of functional properties. Pedro et al. [4] propose an Ada framework called RMF4Ada that enriched this Ada 2012 specification with behavioral runtime verification. This RMF4Ada framework combines aspects of runtime monitoring, formal languages, and software architecture methods to provide the infrastructure that is needed to equip an Ada program with runtime verification functionality.

The rest of the paper is organized as follows: Sect. 2 presents the main related work; Sect. 3 presents the main formalisms used by the recommended approach, namely the UML2.0/PSM component model and the Ada concurrent language; Sect. 4 deals with our systematic rules allowing the translation of UML2.0/PSM source model to the Ada concurrent program; Sect. 5 exhibits a validation of our approach. Finally, Sect. 6 provides a conclusion and possible future work.

2 Related Work

In the component-based software engineering, software construction is based on the assembly of coherent components. However, the most of existing component models do not offer mechanisms to check their component's properties. A comparison of the most known components models based on their ability to specify and verify the component's properties is proposed in [5]. Table 1 summarizes this comparison, which is based on the Meyer's property classification [6].

Table 1. Components models classification [5]

Property	Component model	Checking Tool	Comments
Syntactic	UML2.0/OCL Fractal/CCLJ, Acme/Armani ArchJava	Evaluator of predicate logic	OCL is adopted on OO model
Semantic	Kmelia → B UML2.0 → B	B proves	Gap between the source and target models
Behavioral	Fractal/LTS	CADP	PSM can describe only one direction of communication
	Darwin/FSP	LTSA	
	Pi_ADL	ArchWare	
	Wright/CSP	FDR	Wright proposes a standard contracts
	UML2.0/PSM		
Qualitative	AADL/proper (property concept)	Plugin OSATE	AADL proposes specific qualities such as safety and security
	Acme/Armani (property comcept)	AcmeStudio	It offers a powerful evaluator of predicate logic

Based on this Table 1, we can conclude that no component model covers the various properties types. For example, Wright, Darwin and Pi-ADL allow the description and verification of behavioral properties, while AADL and Acme models allow the description and verification of qualitative properties. The UML2.0 does not describe all aspects of software based on components. It needs other formalisms more or less integrated into UML2.0 as PSM to specify the behavioral aspects.

Seeing that UML is a semi-formal language, it comprises several notations with no rigorous semantics. Therefore, it is not possible to apply a rigorous analysis on a UML model. To deal with this defect, several works propose to open the UML models on a formal language such as LTS, B and CSP. To achieve this, these works offer more or less systematic translations of source model to the target model. In this section, only the works related to the behavioral verification of the UML2.0 components are mentioned. The works described in [7, 8] suggest translating UML2.0 state machines into an LTS specification verifiable by LOTOS tool. The Wr2fdr and FDR2 tools [9] are successfully used in [10] to check the behavioral properties of a UML2.0/PoSM atomic component, where PoSM [10] is an extension of UML2.0 Protocol State Machine (PSM). The author of [11] offers an approach based on a formal method to check the interaction behavioral conformity between UML2.0 components. This approach consists in abstracting PSM diagrams into a similar semantic domain, i.e., automata which are combined using various synchronization operators. These automata can be verified by automated tools (i.e., model checkers) such as the UPPAAL tool. The authors of [12] propose an approach based on MDE to detect conformability in component models. It focuses on verifying the internal interface specification with the maps describing the execution of tasks between the UML2.0 components. Firstly, each PSM associated with an interface is formalized by a B abstract machine. Then, the UCM (Use Case Maps) describing a scenario execution in the component interaction is formalized by a B implementation machine. Consistency between UCM and PSMs can, thus, be automatically validated by B provers. The CSP process-algebraic formalism is suited to modeling patterns of behavioral interaction. The fact that CSP is supported by model-checkers such as FDR gives birth to works that open UML on CSP to verify the behavioral consistency of UML diagrams. For example, the works presented in [13–15] bind, respectively, the state machine diagrams, the sequence diagrams and the activity diagrams to CSP.

3   Retained Models

3.1   UML2.0/PSM Component Model

The UML 2.0 standard provides a number of diagrams capable of describing the structural and behavioral aspects of component-based software. For example, the UML 2.0 component model is especially well suited to depict the structural systems based on components. In the UML 2.0 component model, components are interconnected via their interfaces. Messages exchanged between interacting components are often specified as operations of interfaces. Thus, each UML 2.0 component specifies a set of interfaces it supports and the set of interfaces it calls in another component.

To describe the behavior view of a UML2.0 component or of its interfaces (or ports), UML2.0 proposes a new behavioral diagram called Protocol State Machine (PSM). The latter constitutes a specialization of the behavioral State Machines. Consequently, PSM is similar to State Machines and can use compound transitions, sub-state machines, composite states and concurrent regions, but cannot have deep or shallow history pseudo-states. However, a state in a PSM cannot have actions nor activities, so entry, exit and do features no longer exist in PSM. Typically, a PSM is associated with an interface whose concrete usage in a component determines whether the events captured by the PSM are received (in case of a provided interface) or sent (in case of a required interface).

3.2   Ada Concurrent Language

Ada [16] is a programming language highly recommended for the development of distributed software systems thanks to its careful and safe concurrency mechanism. Indeed, Ada defines a powerful fundamental concurrent unit called task. Hence, an Ada task is a unit of modularization comprising a specification and a body. Similar to a type declaration, the task specification defines the task interface. This task specification can have entries for synchronization. Each task entry can have one or more accept statements within the task body. If the control flow of the task reaches an accept statement, the task is blocked until the corresponding entry is called by another task (similarly, a calling task is blocked until the called task reaches the corresponding accept statement).

Using an Ada formal analysis tool such as FLAVERS [20] or INCA [2], we can detect the potential behavioral properties such as the deadlock of an Ada concurrent program. Indeed, there are two static concurrency analysis tools used for proving the behavioral properties of an Ada concurrent program, where INCA is a flow equation-based tool, and FLAVERS a data flow analysis-based tool.

4   Translation of UML2.0/PSM into ADA

In this section, a set of rules allowing the translation of a UML2.0/PSM component to an Ada concurrent program is proposed. This allows the verification of behavioral properties supported by Ada tools.

4.1   Static View Translation

In UML 2.0, a component is a modular part of a system that encapsulates its contents and whose manifestation is replaceable within its environment. This generalization permits the component concept to be used to describe component-based software. A UML2.0 component can be instantiated as a classifier and can represent an internal structure as a structured classifier. As an encapsulated classifier, a UML2.0 component can have behavioral descriptions, an internal structure, ports and interfaces, and instantiation.

Regarding the static aspect, we propose to translate a UML2.0 component by an Ada concurrent program in which each component's interface (or port) is specified by an Ada task. The latter declares the interface's methods by a set of Ada entry. The implementation of the main program body and these task bodies are explained in the next section.

Table 2 illustrates these translation rules using a UML2.0 compressing proxy component. This example is to some extent identical to an example originally used by [17]. The compressing component proposes tree interfaces: Stream, TxtCompressing and GifCompressing. The compressing proxy distinguishes textual and graphical data received at its Stream interface and forwards data for textual compression via TxtCompressing interface and graphical compression via GifCompressing interface. This UML2.0 component is translated to an Ada concurrent program that proposes three tasks.

Table 2. Translation of UML2.0 component into Ada

UML2.0 Component	Ada Code
| 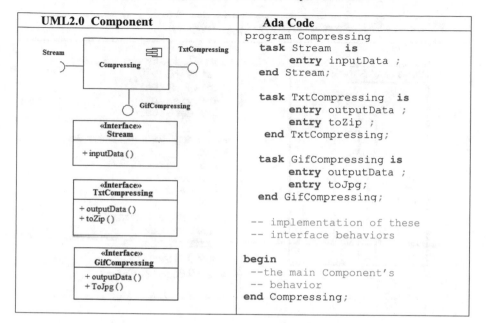 | ```
program Compressing
    task Stream   is
        entry inputData ;
    end Stream;

    task TxtCompressing  is
        entry outputData ;
        entry toZip ;
    end TxtCompressing;

    task GifCompressing is
        entry outputData ;
        entry toJpg;
    end GifCompressing;

--  implementation of these
--  interface behaviors

begin
--the main Component's
--  behavior
end Compressing;
``` |

4.2   Behavioral View Transition

In a UML2.0 protocol state machine (PSM), we can have several types of states. Indeed, one can find simple states, choice states and composite states. For each type of state, a translation rule will be proposed.

- In a PSM that describes a component's partial behavior, a simple transition (oper) can be translated by an acceptation of the entry (oper) declared in the corresponding task.
- In the PSM that describes the component's global behavior, a simple transition in the form (InterfaceA.Oper) can be translated by a call of this task's entry in the main program (taskA.Oper;).
- A loop transition between two states or more can be translated by an Ada loop.
- Unlike a simple state, a composite state is made up of a set of sub-machines that can be executed at the same time. These sub-machines can be translated to a set of Ada concurrent tasks.
- In the case of a state with multiple outgoing transitions, delayed transitions by this state are translated with an Ada select clause if the choice between target states is non-deterministic –transition without pre-condition-. Also, the choice can be implemented in Ada with a simple conditional structure such as: if or case.

Table 3 illustrates the use of these behavioral translation rules on a simple PSM machine. This PSM modeled the behavioral aspect of a UML2.0 interface called Video-Stream which is a simple interface of a digital Video-camera component.
```

Table 3. Translation of a PSM machine into Ada

PSM of the VideoStream Interface	Ada Code

(diagram: VideoStream {Protocol} state machine with play, stop, forward transitions)	```task VideoStream is```
	 ``` entry play;```
	 ``` entry forward;```
	 ``` entry stop;```
	 ```end VideoStream;```
	
	 ```task body VideoStream is```
	 ``` begin```
	 ``` loop```
	 ``` select```
	 ``` accept play ;```
	 ``` or```
	 ``` accept forward ;```
	 ``` end select;```
	 ``` accept stop ;```
	 ``` end loop;```
	
	 ``` end VideoStream;```

5   Validation on a Video-Camera Component

In order to illustrate our translation rules on an example, we choose to use a simple video-camera component as a case study. This component is inspired from [18]. This Video-camera component proposes two interfaces. The first called Memorization and proposes three services: store, remove and stop. However, the second, called Video-Stream, provides four services: play, forward, rewind and stop.

Figure 1 shows the UML2.0 modeling of the static view of this component. The behaviors of the component's interfaces and the global behavior of the component are specified in Fig. 2 by three PSM machines.

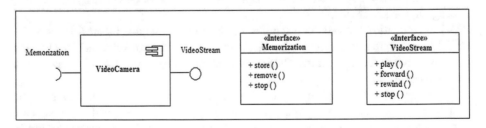

Fig. 1.   UML2.0 VideoCamera component

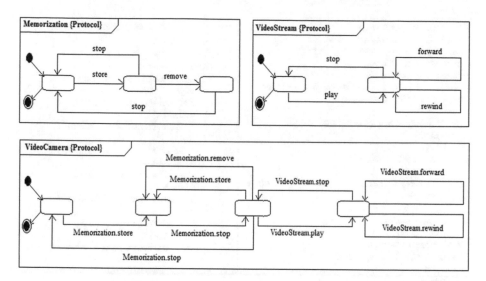

Fig. 2. PSMs of VideoCamera component

Using the translation rules presented in the previous sections, we obtained the Ada concurrent program (see Fig. 3) corresponding to the UML2.0/PSM VideoCamera component source model.

```
Program VideoCamera                --the main Component's
  task Memorization   is           -- behavior
   entry store, remove, stop;       begin
  end Memorization   ;                loop
  task VideoStream  is                  Memorization. store;
   entry play,forward,rewind,stop;      loop
  end VideoStream  ;                      Memorization. stop;
                                          select
  task body Memorization is                 Memorization. remove ;
   begin                                  or
    loop                                     Memorization. store;
      accept store ;                       end select ;
      select                             Memorization. stop;
       accept remove ; accept stop ;     select
      or                                   Memorization. stop ;
       accept stop ;                     or
      end select;                          VideoStream. play ;
    end loop;                              select
  end Memorization;                          VideoStream. stop ;
  task body VideoStream is               or
   begin                                    loop
    loop                                      select
      accept play ;                            VideoStream. forward ;
      select                                 or
       accept stop ;                          VideoStream. rewind ;
      or                                     end select ;
       loop                                end loop ;
        select                           end select ;
         accept forward ;              end loop ;
        or                           end VideoCamera;
         accept rewind ;
        end select ;
       end loop ;
      end select ; end loop ;
    end VideoStream;
```

Fig. 3. Ada implementation of the camera-video component

Using an Ada analysis tool such as SPIN, SMV, FLAVERS and INCA [19], we can detect several errors in our UML2.0/PSM component source model. For example, using the Control Flow Graph (CFG), Flavers allows the detection of the potential deadlock in Ada concurrent program. In particular, using the FLAVERS toolset we detect that our UML2.0/PSM VideoCamera component has a deadlock problem between its global behavior and its interface behaviors. Indeed, its global behavior allows the call of two successive rendezvous of the Memorization.stop entry; however, the behavior of this Memorization interface cannot accept these rendezvous. To cover this error, we can replace the global PSM of the component by another compatible with its interfaces.

6 Conclusion

In this paper, an approach for verifying the coherence of UML2.0/PSM atomic components has been proposed. To achieve this, systematic translation rules allowing the translation of UML2.0/PSM atomic component to an Ada concurrent program have been suggested. The choice of this language is justified by the presence of many Ada tools able to detect several error types in an Ada concurrent program. Vis-à-vis the properties specification to be verified, we can choose the adequate Ada analysis tool. For example, the SMV and SPIN tools promote the property-oriented state, while INCA and FLAVERS promote the property- oriented trace. These Ada tools are four complementary approaches permitting the static analysis of Ada concurrent programs.

Currently, we are extending this work by an automation of these translation rules using a model transformation language.

References

1. Han, J.: A comprehensive interface definition framework for software components. In: Asia Pacific Software Engineering Conference, pp. 110–117. IEEE Computer Society (1998)
2. Naumovich, G., Avrunin, G.S., Clarke, L.A., Osterweil, L.J.: Applying static analysis to software architectures. In: ACM SIGSOFT 1997 Softw. Eng. Notes 22(6), pp. 77–93 (1997)
3. Ruiz, J.F., Comar, C., Moy, Y.: Source code as the key artifact in requirement-based development: the case of ada 2012. In: Ada-Europe 2012, pp. 49–59 (2012)
4. Pedro, A.M., Pereira D., Pinho, L.M., Pinto, J.S.: Towards a runtime verification framework for the ada programming language. In: Ada-Europe, pp. 58–73 (2014)
5. Sakka Rouis, T., Bhiri, M.T., Kmimech, M., Moussa, F.: A contractual approach for the verification of UML2.0 software architectures. Appareats Int. J. Comput. Appl. Technol. 57(1) (2018)
6. Meyer, B.: Applying design by contract. IEEE Comput. 25, 40–51 (1992)
7. Luong, H.V., Courbis, A.L., Lambolais, T., Phan, T.: IDCM: un outil d'analyse de composants et d'architectures dédié à la construction incrémentale. 11èmes Journées Francophones sur les Approches Formelles dans l'Assistance au Développement de Logiciels, Grenoble, France, pp. 50–53, January 2012
8. Lambolais, T., Courbis, A.L., Luong, H.V.: Raffinement de modèles comportementaux UML, vérification des relations d'implantation et d'extension sur les machines d'états, AFADL 2009, France, 14 p. (2009)
9. Sakka Rouis T., Bhiri M.T., Kmimech, M.: Behavioral verification of UML2.0/PoSM components. In: Proceeding in 15th International SoMeT, Larnaca, Cyprus, pp. 246–257, 12–14 September 2016
10. Sakka Rouis, T., Bhiri, M.T., Kmimech, M., Moussa, F.: Wr2Fdr tool maintenance for models checking. In: Proceeding in 16th International SoMeT Conference. Kitakyushu, Japan, pp. 425–440 (2017)
11. Hammal, Y.: Towards checking protocol conformance of active components. Int. J. Softw. Eng. Appl. 5(2) (2011)
12. Thuan, T.N., Anh, T.V.V., Ha, N.V.: Consistency between UCM and PSMs in component models. In: IEEE International Conference on Research, Innovation and Vision for the Future in Computing and Communication Technologies, Ho Chi Minh City (2008)
13. Ng, M.Y., Butler, M.: Towards formalizing UML state diagrams in CSP. In: SEFM, pp. 138–147. IEEE (2003)
14. Jacobs, J., Simpson, A.C.: On a process algebraic representation of sequence diagrams. In: SaFoMe 2014. LNCS, vol. 8938, pp. 71–85 (2014)
15. Dong, X., Philbert, N., Zongtian, L., Wei, L.: Towards formalizing UML activity diagrams in CSP. In: ISCSCT, pp. 450–453. IEEE (2008)

16. Taft, S.T., Duff, R.A., Brukardt, R.L., Ploedereder, E., Leroy, P.: Ada 2005 Reference Manual. Language and Standard Libraries: International Standard ISO/IEC 8652/1995(E) with Technical Corrigendum 1 and Amendment 1. Springer, New York (2007)
17. Bernardo, M., Ciancarini, P., Donatiello, L.: Architecting families of software systems with process algebras. ACM Trans. Softw. Eng. Methodol. 11(4), 386–426 (2002)
18. Bhiri, M.T., Sakka Rouis, T., Kmimech, M.: Checking non-functional properties of UML2.0 components assembly. In: IEEE WETICE Conference, Tunisia, pp. 278–283 (2013)
19. Dwyer, M.B., Pasarean, S.C., Corbett J.C.: Translating ADA programs for Model checking: A tutorial. Technical Report 1998–12, Kansas State University
20. Cobleigh, J.M., Clarke, L.A., Osterweil, L.J.: FLAVERS: a finite state verification technique for software systems. IBM Syst. J. 41(1), 140–165 (2002)

A New Approach for the Diagnosis of Parkinson's Disease Using a Similarity Feature Extractor

João W. M. de Souza, Jefferson S. Almeida,
and Pedro Pedrosa Rebouças Filho[✉]

Laboratório de Processamento de Imagens e Simulação Computacional,
Instituto Federal de Educação, Ciência e Tecnologia do Ceará, Fortaleza, CE, Brazil
{wellmendes,jeffersonsilva}@lapisco.ifce.edu.br, pedrosarf@ifce.edu.br

Abstract. Parkinson's disease affects millions of people worldwide. Nowadays there are several ways to help diagnose this disease. Among which we can highlight handwriting exams. One of the main contributions of the computational field to help diagnose this disease is the feature extraction of handwriting exams. This paper proposed a similarity extraction approach which was applied to the exam template and the handwritten trace of the patient. The similarity metrics used in this work were: structural similarity, mean squared error and peak signal-to-noise ratio. The proposed approach was evaluated with variations in obtaining the exam template and the handwritten trace generated by the patient. Each of these variations was used together with the Nave Bayes, OPF, and SVM classifiers. In conclusion, the proposed approach demonstrated that it was better than the other approach found in the literature, and is therefore a potential aid in the detection and monitoring of Parkinson's disease.

Keywords: Parkinson's disease · Similarity extractor
Medical diagnosis

1 Introduction

According to the World Health Organization, neurological disorders are diseases of the nervous system. About 20% of people over 60 suffer from some kind of neurological disorder [20,21]. Parkinson's disease (PD), first described by James Parkinson [8], is a degenerative disease of the central nervous system that is both chronic and progressive [20]. The Parkinson's Disease Foundation claims that this disease affects about 7–10 million people worldwide and 4% of people with PD are diagnosed before the age of 50. There is no cure for PD, but an early diagnosis helps in treating the patient's illness.

There are numerous studies in the computational field that are related to PD such as the extraction of different features from diagnostic exams [2,6] that can demonstrate the existence of the illness as well as its degree of severity in PD patients [1], among others [12,13].

© Springer International Publishing AG, part of Springer Nature 2018
A. Abraham et al. (Eds.): ISDA 2017, AISC 736, pp. 21–31, 2018.
https://doi.org/10.1007/978-3-319-76348-4_3

Most of the studies related to PD are based on the signals obtained in different ways on diagnostic exams [14]. However, there are few studies based on handwriting exams. An example of diagnosis based on such exams is the quality control of the patients hand writing movements [2,3].

Handwriting exams have advantages as they are ease to obtain. These exams can be performed on paper [10], as well as using more sophisticated methods such as a digitizer [2,3], an electronic pen [17], or even a smartphone [4]. The exams performed on paper are practical, but the feature extraction becomes complicated since the exams have some printing error and the information in the exam is not so clear. Handwriting exams are very diverse and use spirals, ellipses, connected syllables, connected words, and many other ways to test the patients ability to trace these forms.

This paper proposes a similarity feature extraction approach between the drawing of the handwriting examination template and the drawing generated by the patient in the exam. The similarity metrics are used in this paper as attributes to measure this similarity since these parameters are used as classification attributes for the Nave Bayes, Optimal-Path Forest (OPF) and Support Vector Machine (SVM) classifiers. This approach arose since these metrics are able to quantify differences between images, and thus they are promising for this type of application. Another advantage related to this type of extraction is the non-use of parameters, which makes the classification more robust.

2   Diagnosis of Parkinson's Disease Through Handwriting Exams

Some papers have contributed to the diagnosis of PD using handwriting exams. Here, we can cite the works of Drotár, who used the examination time as a parameter for the medical diagnosis [2,3]. The work by Surangsrirat et al. applied polar coordinates to interpret features for diagnosis [15]. There are also works based on the difference between the patient's tracings and the template, as proposed by Pereira et al. [10].

The approach by Pereira et al. [10] is easy to obtain as the exam is performed on paper. The patient performs this exam by attempting to trace the template correctly. Another tool of Pereiras work was a set of images composed of handwriting exams named HandPD dataset [10].

The HandPD dataset [9] consists of the handwriting exams of two groups: Control Group (CG) and Patient Group (PG), where the latter is composed of people affected by PD. The exams consist of handwriting in a spiral and meander formats. These exams were collected at the Medical School of Botucatu, State University of São Paulo - Brazil.

This set of images [9] consists of 736 images divided into two groups: the CG containing 72 images, and the PG containing 296 images. The exams were obtained from 92 individuals, of which 18 were healthy individuals (CG) and 74 were patients (PG) (Fig. 1).

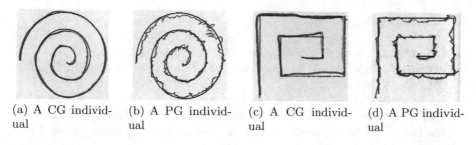

(a) A CG individ-
ual

(b) A PG individ-
ual

(c) A CG individ-
ual

(d) A PG individ-
ual

Fig. 1. Some examples of handwriting exams in the spirals and meanders format [10].

The attributes used to extract information from the images of the Pereira *et al.* approach [10] are based on the difference between the exam template (ET) and the handwritten trace (HT). Equations 1, 2, 3, 4 and the list below present the attributes used to describe the HandPD dataset exams [10].

– Root mean square (RMS) of the difference between HT and ET radius:

$$RMS = \sqrt{\frac{1}{n}\sum_{i=0}^{n}\left(r_{HT}^{i} - r_{ET}^{i}\right)^2};\qquad(1)$$

where n is the number of sample points drawn for each HT and ET skeleton, and r_{HT}^i and r_{ET}^i denote the HT and ET radius, which is basically the length of the straight line that connects the *i-th* sampled point, respectively, to the center of the spiral or meander.
– Maximum difference between HT and ET radius:

$$\Delta_{max} = \underset{i}{\mathrm{argmax}}\left\{\left|r_{HT}^i - r_{ET}^i\right|\right\};\qquad(2)$$

– Minimum difference between HT and ET radius:

$$\Delta_{min} = \underset{i}{\mathrm{argmin}}\left\{\left|r_{HT}^i - r_{ET}^i\right|\right\};\qquad(3)$$

– Standard deviation of the difference between HT and ET radius;
– Mean relative tremor (MRT) [10] is a quantitative evaluation to measure the "amount of tremor" of a given individual's HT:

$$MRT = \frac{1}{n-d}\sum_{i=d}^{n}\left\{\left|r_{HT}^i - r_{HT}^{i-d+1}\right|\right\};\qquad(4)$$

where d is the displacement of the sample points used to compute the radius difference.
– Maximum ET radius;
– Minimum ET radius;
– Standard deviation of HT radius;
– The number of times the difference between HT and ET radius changes from negative to positive, or the opposite.

3   Methodology

In this section, the similarity feature extractor approach is presented in this study using the handwriting exam to diagnose PD. Figure 2 shows the flowchart of the proposed approach.

An important step in the handwriting exam is the process used to obtain the two images, ET and HT that are highlighted in Fig. 2 subsection 1. These images are obtained by segmentation of the handwriting exam using digital image processing techniques.

The ET and HT segmentations are divided into two steps. The first step, in both cases, consists of smoothing the image using a Median filter (5 × 5), to eliminate any noise obtained in the acquisition of these exams.

An erosion (9 × 9 ellipse structure) was applied to ensure that there was no discontinuity in the ET segmentation and then a threshold was defined empirically. To make sure of a successful segmentation, a second erosion was applied with the same structuring element, thus providing the segmented ET image.

The HT segmentation is carried out by converting the handwriting exam to a grayscale, in order to use the Otsu threshold. Following, we apply a difference operation between the grayscale image and the ET. Figure 3 illustrates this process.

Images generated in segmentation, as well as in the handwriting exam, are converted to the grayscale for the next step in the proposed methodology. This conversion is shown in Fig. 2.

After segmentation, a Gaussian filter with $\sigma = 1$ was applied which blurred these images to eliminate the discontinuity with pixel agglutination. Several types of settings were necessary to define the parameters for the filter. The variation of these settings depends on the masks and the number of times (iterations) the filter is used. The mask sizes used were (3 × 3), (5 × 5), (7 × 7), (9 × 9) and (11 × 11), and the filters were applied between 1 and 5 times. The mask sizes and the number of times the filter was applied were empirically defined for further analysis of the results. The process chosen was applied to the grayscale images obtained after the segmentation process. This process is illustrated in Fig. 2 subsection 2.

The feature extraction of the images obtained after the handwriting exam segmentation was performed through similarity metrics and the application of the similarity feature extractor occurs between two images. Thus, we proposed three combinations of the three images, which are: handwriting exam and handwritten trace, handwriting exam and exam template, handwritten trace and exam template. The similarity degree is the quality of these two images, that is, the comparison between the quality of the patients trace and the exam template. The feature extraction process is presented in Fig. 2 subsection 2.

The metrics adopted are widely used in quality analysis between images [11]. Machine learning methods use these similarity metrics as attributes (Fig. 2 subsection 3). A brief description of the 3 metrics adopted is presented in the Eqs. 5, 6 and 7:

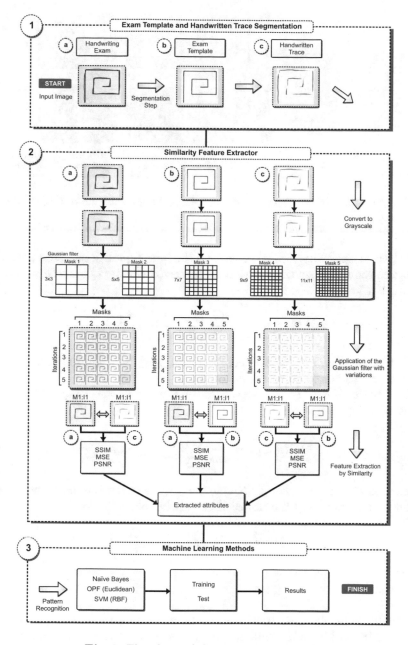

Fig. 2. Flowchart of the proposed approach.

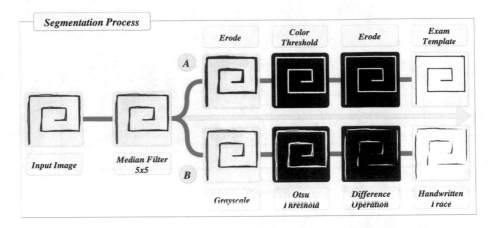

Fig. 3. An example of the segmentation process: (a) Segmentation of ET; (b) Segmentation of HT.

– Structural similarity (SSIM), measures the image quality based on perception that considers image degradation as perceived change in structural information [19], i.e.:

$$SSIM(x,y) = \frac{(2\mu_y\mu_x + c_1)(2\sigma_{xy} + c_2)}{(\mu_x^2 + \mu_y^2 + c_1)(\sigma_x^2 + \sigma_y^2 + c_2)} \tag{5}$$

where μ_x and μ_y are the average for x and y; σ_x and σ_y are the variance for x and y; σ_{xy} is related to the covariance.

– Mean squared error (MSE) measures the average of the squares of the errors between the images, i.e.:

$$MSE(x,y) = \frac{1}{mn}\sum_{x=0}^{m-1}\sum_{y=0}^{n-1}[I(x,y)-J(x,y)]^2 \tag{6}$$

– Peak signal-to-noise ratio (PSNR) is the ratio between the maximum possible power of a signal and the power of corrupting noise [5], i.e.:

$$PSNR = 10 \cdot log_{10}\left(\frac{MAX_I^2}{MSE}\right) \tag{7}$$

4   Results

The results were obtained from the machine learning methods applied to the features extracted from the similarity measures used in this work. In this paper, we used the handwriting exam model proposed by Pereira *et al.* [10]. These exams show the patients ability to trace the exam template.

The information obtained by the similarity feature extractor was separated into three groups: differences in the handwriting exam format; variation of the

Gaussian filter; and the combination of the segmented images. These groups were evaluated using the following machine learning methods: Nave Bayes [16], OPF [7] with Euclidean distance and SVM [18] with Radial basis function in order to have a direct comparison with the Pereira *et al.* approach [10].

The metric used to evaluate the classifications of the machine learning methods was Accuracy. The Accuracy refers to the closeness of a measured value to a standard or known value and is defined in the confusion matrix generated after the application of the machine learning methods.

We evaluated three experiments, which presented the accuracies achieved by the machine learning methods on the previously separated groups. All three

Table 1. Results obtained using handwriting exams in the format of meanders, spirals and both with a variation on masks and iterations of a Gaussian filter.

Meander

	Mask 3x3 a⇔c	a⇔b	c⇔b	Mask 5x5 a⇔c	a⇔b	c⇔b	Mask 7x7 a⇔c	a⇔b	c⇔b	Mask 9x9 a⇔c	a⇔b	c⇔b	Mask 11x11 a⇔c	a⇔b	c⇔b
Iter1 BAYES	70.65%	47.66%	71.79%	75.11%	51.20%	72.07%	**78.59%**	55.11%	69.62%	76.68%	57.43%	69.79%	76.95%	60.38%	70.00%
Iter1 OPF	77.83%	69.78%	68.53%	75.71%	68.15%	73.97%	75.76%	69.40%	74.78%	77.17%	68.59%	76.20%	77.66%	71.52%	**76.79%**
Iter1 SVM	79.84%	79.62%	80.54%	81.14%	80.00%	80.49%	79.95%	80.05%	81.74%	80.49%	79.62%	80.87%	82.34%	80.00%	80.54%
Iter2 BAYES	75.87%	52.28%	73.21%	77.55%	52.01%	68.04%	74.35%	58.26%	69.02%	76.96%	61.96%	75.76%	73.53%	60.00%	74.73%
Iter2 OPF	77.61%	70.98%	72.12%	73.75%	66.58%	75.05%	**80.49%**	72.55%	75.87%	76.14%	70.82%	73.97%	74.84%	68.91%	73.26%
Iter2 SVM	80.05%	80.11%	80.22%	80.82%	**80.43%**	80.43%	80.98%	79.78%	81.90%	81.47%	80.16%	81.79%	82.07%	80.05%	**84.35%**
Iter3 BAYES	78.32%	53.70%	70.22%	77.01%	56.74%	69.84%	75.71%	61.30%	75.71%	77.07%	60.82%	72.01%	72.55%	61.96%	71.96%
Iter3 OPF	75.27%	66.41%	75.16%	76.20%	68.26%	75.82%	75.87%	71.58%	75.82%	75.60%	69.62%	74.02%	73.21%	70.71%	76.30%
Iter3 SVM	80.98%	79.84%	81.30%	81.79%	80.33%	81.30%	81.68%	80.16%	82.50%	80.82%	80.16%	81.25%	**83.21%**	80.16%	81.58%
Iter4 BAYES	77.28%	56.96%	69.78%	76.03%	58.32%	69.95%	75.71%	57.12%	**77.45%**	72.50%	63.91%	68.64%	69.46%	61.58%	59.13%
Iter4 OPF	75.05%	67.99%	74.67%	77.55%	72.50%	75.33%	76.09%	71.90%	75.60%	72.88%	**73.32%**	73.26%	74.78%	69.51%	73.80%
Iter4 SVM	80.65%	79.95%	81.36%	81.90%	80.33%	81.14%	81.36%	80.43%	81.96%	82.45%	80.33%	81.30%	82.07%	80.38%	80.98%
Iter5 BAYES	78.37%	52.66%	69.35%	73.42%	57.50%	74.18%	74.29%	62.34%	71.58%	69.57%	65.49%	63.48%	61.41%	**65.98%**	73.04%
Iter5 OPF	79.24%	68.26%	74.89%	76.30%	71.90%	74.57%	77.72%	69.89%	75.16%	73.15%	72.39%	73.64%	73.64%	68.37%	73.75%
Iter5 SVM	81.36%	79.89%	82.39%	82.17%	80.05%	81.09%	80.98%	80.16%	81.47%	82.12%	80.38%	82.07%	81.52%	79.89%	81.30%

Spiral

	Mask 3x3 a⇔c	a⇔b	c⇔b	Mask 5x5 a⇔c	a⇔b	c⇔b	Mask 7x7 a⇔c	a⇔b	c⇔b	Mask 9x9 a⇔c	a⇔b	c⇔b	Mask 11x11 a⇔c	a⇔b	c⇔b
Iter1 BAYES	69.73%	55.11%	70.76%	78.91%	57.34%	71.63%	82.61%	56.47%	72.17%	84.13%	58.15%	73.32%	84.46%	57.39%	73.86%
Iter1 OPF	75.49%	70.98%	70.65%	76.96%	70.87%	74.29%	78.53%	71.30%	74.35%	81.68%	69.67%	73.70%	82.34%	67.50%	**77.61%**
Iter1 SVM	81.14%	79.51%	79.40%	81.85%	79.84%	79.35%	84.46%	79.89%	80.49%	86.85%	79.73%	80.05%	**88.91%**	80.33%	80.05%
Iter2 BAYES	78.64%	56.96%	69.84%	81.74%	56.74%	69.78%	84.95%	59.51%	68.91%	83.86%	65.87%	73.10%	81.41%	63.26%	74.40%
Iter2 OPF	76.79%	**71.52%**	73.80%	81.58%	71.47%	74.08%	82.61%	68.91%	75.33%	82.55%	71.25%	76.25%	77.88%	70.11%	73.53%
Iter2 SVM	81.36%	79.51%	80.05%	85.65%	79.18%	80.27%	86.74%	79.89%	80.11%	85.92%	80.43%	79.08%	83.80%	80.22%	78.48%
Iter3 BAYES	80.87%	56.36%	67.28%	**86.14%**	60.11%	70.49%	84.13%	59.73%	75.16%	79.46%	63.32%	72.28%	77.50%	61.96%	70.54%
Iter3 OPF	76.68%	69.57%	72.55%	**83.91%**	69.02%	75.33%	80.65%	71.47%	75.92%	82.23%	69.40%	75.38%	77.83%	70.11%	74.78%
Iter3 SVM	81.90%	79.73%	79.67%	87.61%	80.11%	80.11%	87.12%	79.95%	79.84%	83.10%	**80.60%**	79.46%	82.17%	80.27%	79.95%
Iter4 BAYES	82.01%	55.05%	72.07%	84.73%	54.18%	73.37%	80.27%	58.80%	72.50%	76.74%	63.37%	74.13%	65.22%	59.95%	72.23%
Iter4 OPF	81.90%	69.62%	74.57%	83.42%	70.16%	75.76%	76.68%	70.65%	73.42%	78.80%	71.47%	75.22%	72.98%	70.65%	75.65%
Iter4 SVM	84.51%	79.35%	80.22%	87.23%	80.11%	79.29%	84.35%	80.16%	78.91%	83.21%	80.33%	80.49%	79.29%	80.27%	80.00%
Iter5 BAYES	84.51%	56.79%	72.61%	85.54%	59.57%	75.38%	80.43%	61.85%	74.08%	69.40%	**68.04%**	68.91%	69.35%	63.37%	**79.84%**
Iter5 OPF	81.96%	69.46%	73.70%	81.47%	70.54%	74.84%	76.14%	70.00%	73.37%	75.54%	70.65%	73.37%	72.28%	68.86%	76.63%
Iter5 SVM	85.82%	80.00%	80.16%	88.53%	79.40%	79.29%	82.61%	80.33%	79.35%	82.12%	80.27%	**81.85%**	77.45%	80.11%	80.71%

Meander and Spiral

	Mask 3x3 a⇔c	a⇔b	c⇔b	Mask 5x5 a⇔c	a⇔b	c⇔b	Mask 7x7 a⇔c	a⇔b	c⇔b	Mask 9x9 a⇔c	a⇔b	c⇔b	Mask 11x11 a⇔c	a⇔b	c⇔b
Iter1 BAYES	69.28%	59.10%	66.68%	71.42%	59.52%	66.50%	74.62%	59.77%	66.51%	75.14%	59.73%	66.95%	75.34%	59.09%	66.67%
Iter1 OPF	69.35%	59.39%	63.40%	64.09%	58.04%	62.69%	76.51%	57.98%	64.29%	69.39%	57.47%	63.93%	70.78%	60.17%	63.71%
Iter1 SVM	70.04%	54.98%	67.08%	71.44%	55.54%	66.32%	74.92%	55.93%	66.88%	75.21%	57.39%	65.83%	76.79%	57.98%	66.22%
Iter2 BAYES	72.13%	59.31%	66.88%	75.16%	60.10%	66.28%	**76.90%**	59.25%	66.11%	76.53%	60.19%	67.42%	76.71%	**61.01%**	67.19%
Iter2 OPF	63.70%	56.33%	62.04%	67.44%	59.00%	64.61%	69.60%	59.44%	64.00%	72.15%	58.92%	63.97%	72.91%	60.89%	61.11%
Iter2 SVM	71.24%	54.49%	**67.72%**	74.26%	57.34%	65.34%	76.71%	58.12%	66.18%	**78.18%**	57.99%	66.78%	77.76%	58.17%	65.01%
Iter3 BAYES	73.21%	59.70%	67.01%	76.52%	60.30%	66.84%	76.03%	59.95%	68.35%	76.50%	60.03%	65.55%	76.42%	60.73%	64.88%
Iter3 OPF	65.43%	56.70%	63.79%	68.66%	59.75%	64.57%	70.40%	61.52%	63.77%	**73.94%**	58.80%	62.62%	70.20%	**61.44%**	61.84%
Iter3 SVM	73.93%	55.87%	66.48%	76.21%	58.49%	66.89%	77.97%	58.30%	64.92%	77.02%	59.19%	64.90%	77.20%	58.65%	63.56%
Iter4 BAYES	75.02%	59.77%	67.10%	76.02%	60.05%	66.42%	76.17%	59.80%	66.08%	75.88%	60.28%	66.71%	74.46%	60.32%	62.91%
Iter4 OPF	67.59%	59.22%	**64.65%**	70.90%	58.93%	64.52%	72.26%	60.91%	61.51%	70.19%	58.93%	61.84%	67.78%	60.09%	60.81%
Iter4 SVM	75.82%	56.69%	66.13%	76.95%	58.39%	66.40%	77.81%	59.83%	64.92%	77.03%	**61.33%**	66.07%	74.90%	59.83%	64.37%
Iter5 BAYES	76.31%	59.94%	**68.42%**	75.28%	60.04%	66.81%	76.52%	60.29%	66.81%	73.73%	59.73%	66.13%	71.52%	60.99%	62.79%
Iter5 OPF	68.93%	59.28%	64.52%	70.47%	60.89%	64.31%	73.01%	61.32%	62.63%	70.61%	58.90%	61.93%	66.62%	58.07%	62.95%
Iter5 SVM	74.98%	57.43%	65.77%	77.45%	59.31%	64.16%	78.12%	59.29%	64.12%	76.07%	58.39%	65.68%	73.31%	59.67%	63.20%

experiments were evaluated with 75% for training and 25% for classification. A cross-validation model with 20 runnings was used for the reliability of the results.

The general results are presented in Table 1. Table 1 shows the subdivisions of the proposed configurations in the Gaussian filter application. There is also a division in the combinations of the images that had their features extracted by similarity. The best results were highlighted in bold.

Table 2 presents the best results obtained in Table 1, for the classifiers and the combinations between the images obtained in the segmentation. Table 2 also presents the lowest accuracy rates of the experiment with meander and spiral formats together. The inferiority about the others experiments are determined by the most significant difference between the spiral and meander formats.

Another important factor, observed in the best results obtained with the classifiers in Table 2, was the predominance of the best results using the combination of handwriting exam and the handwritten trace.

Table 2. Best results from the best classifiers and combinations.

		Meander	Spiral	Meander/Spiral
		Acc (%)	Acc (%)	Acc (%)
Naïve Bayes	a⇔c	**78.59 ± 2.94**	**86.14 ± 3.86**	**76.90 ± 1.48**
	a⇔b	65.98 ± 9.78	68.04 ± 6.52	61.01 ± 1.42
	c⇔b	77.45 ± 3.64	79.84 ± 3.03	68.42 ± 1.72
OPF	a⇔c	**80.49 ± 3.65**	**83.91 ± 2.52**	**73.94 ± 1.89**
	a⇔b	73.32 ± 4.41	71.52 ± 3.70	61.44 ± 1.96
	c⇔b	76.79 ± 4.23	77.61 ± 3.02	64.65 ± 2.27
SVM	a⇔c	83.21 ± 2.32	**88.91 ± 2.63**	**78.18 ± 1.37**
	a⇔b	80.43 ± 2.01	80.60 ± 2.92	61.33 ± 2.12
	c⇔b	**84.35 ± 3.36**	81.85 ± 2.75	67.72 ± 2.26

Table 3 presents a comparison between the best results obtained in this paper and the best results achieved in the work by Pereira *et al.* [10]. Table 3 shows each classifier with subdivisions for the experiments using handwriting exams in the formats of meanders, spirals and their combination. The highest accuracy results are highlighted in green color and the lowest accuracy results in red, for each of the classifiers used in this paper. Also, the highest and the lowest accuracy results of both approaches and experiments are highlighted in bold.

The proposed approach in this paper obtained superior results in all of the experiments compared to the results obtained by Pereira *et al.* [10]. The highest accuracy result in this work was 88.91% using the SVM classifier with handwriting in a spiral format for the combination of the handwriting exam and the handwritten trace. The Gaussian filter mask size was (11 × 11) and was applied

Table 3. Comparison between the best results of this paper and the best results of the Pereira *et al.* [10].

Classifier	Database	Feature extractor	Accuracy (%)
Naïve Bayes	Meander	**Proposed approach**	78.59 ± 2.94
		Pereira *et al.* approach [10]	59.20 ± 4.78
	Spiral	**Proposed approach**	86.14 ± 3.86
		Pereira *et al.* approach [10]	64.23 ± 7.11
	Meander/Spiral	**Proposed approach**	76.90 ± 1.48
		Pereira *et al.* approach [10]	**45.79 ± 4.15**
OPF	Meander	**Proposed approach**	80.49 ± 3.65
		Pereira *et al.* approach [10]	57.54 ± 6.35
	Spiral	**Proposed approach**	83.91 ± 2.52
		Pereira *et al.* approach [10]	52.48 ± 5.32
	Meander/Spiral	**Proposed approach**	73.94 ± 1.89
		Pereira *et al.* approach [10]	55.86 ± 3.63
SVM	Meander	**Proposed approach**	84.35 ± 3.36
		Pereira *et al.* approach [10]	66.72 ± 5.33
	Spiral	**Proposed approach**	**88.91 ± 2.63**
		Pereira *et al.* approach [10]	50.16 ± 1.71
	Meander/Spiral	**Proposed approach**	78.18 ± 1.37
		Pereira *et al.* approach [10]	58.61 ± 2.84

only once. The lowest accuracy among the best results presented in Table 3 was 45.79% for the Pereira *et al.* approach [10].

5   Conclusion

The proposed approach is based on feature extraction of the similarity measures between the exam template and the handwritten trace. The SVM classifier with the RBF kernel and handwriting in the spiral format proved to be more promising for similarity analysis. This configuration got a hit rate of 88.91% along with the combination of the handwriting exam and the handwritten trace in feature extraction. The proposed approach was 22.19% superior to the best results achieved by Pereira *et al.* [10].

We conclude that this approach is promising for the use of a system to aid in the diagnosis of PD. Another advantage of this proposal is the lack of parameters in the analysis by similarity, facilitating its application. These results encourage us to propose future works related to handwriting feature extractions.

References

1. Bouadjenek, N., Nemmour, H., Chibani, Y.: Robust soft-biometrics prediction from off-line handwriting analysis. Appl. Soft Comput. **46**, 980–990 (2016)
2. Drotár, P., Mekyska, J., Rektorová, I., Masarová, L., Smékal, Z., Faundez-Zanuy, M.: Evaluation of handwriting kinematics and pressure for differential diagnosis of Parkinson's disease. Artif. Intell. Med. **67**, 39–46 (2016)
3. Drotár, P., Mekyska, J., Smékal, Z., Rektorová, I., Masarová, L., Faundez-Zanuy, M.: Contribution of different handwriting modalities to differential diagnosis of Parkinson's disease. In: 2015 IEEE International Symposium on Medical Measurements and Applications, pp. 344–348 (2015)
4. Graça, R., e Castro, R.S., Cevada, J.: Parkdetect: Early diagnosing Parkinson's disease. In: 2014 IEEE International Symposium on Medical Measurements and Applications, pp. 1–6 (2014)
5. Huynh-Thu, Q., Ghanbari, M.: Scope of validity of PSNR in image/video quality assessment. Electron. Lett. **44**(13), 800–801 (2008)
6. de Ipiña, K.L., Iturrate, M., Calvo, P.M., Beitia, B., Garcia-Melero, J., Bergareche, A., De la Riva, P., Marti-Masso, J.F., Faundez-Zanuy, M., Sesa-Nogueras, E., Roure, J., Solé-Casals, J.: Selection of entropy based features for the analysis of the archimedes' spiral applied to essential tremor. In: 2015 4th International Work Conference on Bioinspired Intelligence (IWOBI), pp. 157–162 (2015)
7. Papa, J.P., Falcão, A.X., Suzuki, C.T.N.: Supervised pattern classification based on optimum-path forest. Int. J. Imaging Syst. Technol. **19**(2), 120–131 (2009)
8. Parkinson, J.: An Essay on the Shaking Palsy. Whittingham and Rowland, London (1817)
9. Pereira, C.R., Pereira, D.R., da Silva, F.A., Hook, C., Weber, S.A.T., Pereira, L.A.M., Papa, J.P.: A step towards the automated diagnosis of Parkinson's disease: analyzing handwriting movements. In: 2015 IEEE 28th International Symposium on Computer-Based Medical Systems, pp. 171–176 (2015)
10. Pereira, C.R., Pereira, D.R., Silva, F.A., Masieiro, J.P., Weber, S.A.T., Hook, C., Papa, J.P.: A new computer vision-based approach to aid the diagnosis of Parkinson's disease. Comput. Methods Programs Biomed. **136**, 79–88 (2016)
11. Rebouças Filho, P.P., Moreira, F.D.L., de Lima Xavier, F.G., Gomes, S.L., dos Santos, J.C., Freitas, F.N.C., Freitas, R.G.: New analysis method application in metallographic images through the construction of mosaics via speeded up robust features and scale invariant feature transform. Materials **8**(7), 3864–3882 (2015)
12. Schiffer, A., Nevado-Holgado, A.J., Johnen, A., Schönberger, A.R., Fink, G.R., Schubotz, R.I.: Intact action segmentation in Parkinson's disease: hypothesis testing using a novel computational approach. Neuropsychologia **78**, 29–40 (2015)
13. Sengoku, R., Matsushima, S., Bono, K., Sakuta, K., Yamazaki, M., Miyagawa, S., Komatsu, T., Mitsumura, H., Kono, Y., Kamiyama, T., Ito, K., Mochio, S., Iguchi, Y.: Olfactory function combined with morphology distinguishes Parkinson's disease. Parkinsonism & Relat. Disord. **21**(7), 771–777 (2015)
14. Shah, V.V., Goyal, S., Palanthandalam-Madapusi, H.J.: A perspective on the use of high-frequency stimulation in deep brain stimulation for Parkinson's disease. In: 2016 Indian Control Conference (ICC), pp. 19–24 (2016)
15. Surangsrirat, D., Intarapanich, A., Thanawattano, C., Bhidayasiri, R., Petchrutchatachart, S., Anan, C.: Tremor assessment using spiral analysis in time-frequency domain. In: 2013 Proceedings of IEEE Southeastcon, pp. 1–6, April 2013

16. Theodoridis, S., Koutroumbas, K.: Pattern Recognition, 4th edn. Academic Press, New York (2009)
17. Ünlü, A., Brause, R., Krakow, K.: Handwriting Analysis for Diagnosis and Prognosis of Parkinson's Disease, pp. 441–450. Springer, Heidelberg (2006)
18. Vapnik, V.N.: Statistical Learning Theory. Wiley, New York (2009)
19. Wang, Z., Bovik, A.C., Sheikh, H.R., Simoncelli, E.P.: Image quality assessment: from error visibility to structural similarity. IEEE Trans. Image Process. **13**(4), 600–612 (2004)
20. WHO: Neurological Disorders: Public Health Challenges. World Health Organization (2006)
21. WHO: Mental health and older adults, April 2016. http://www.who.int/mediacentre/factsheets/fs381/en/

A Novel Restart Strategy for Solving Complex Multi-modal Optimization Problems Using Real-Coded Genetic Algorithm

Amit Kumar Das and Dilip Kumar Pratihar[(✉)]

Department of Mechanical Engineering, Indian Institute of Technology Kharagpur,
Kharagpur 721302, India
amit.besus@gmail.com, dkpra@mech.iitkgp.ernet.in

Abstract. Genetic algorithm (GA) is one of the most popular and robust stochastic optimization tools used in various fields of research and industrial applications. It had been applied for solving many global optimization problems for the last few decades. However, it has a poor theoretical assurance to reach the globally optimal solutions, while solving the complex multi-modal problems. Restart strategy plays an important role in overcoming this limitation of a GA to a certain extent. Although there are a few restart methods available in the literature, these are not adequate. In this paper, a novel restart strategy is proposed for solving complex multi-modal optimization problems using a real-coded genetic algorithm (RCGA). To show the superiority of the proposed scheme, ten complex multi-modal test functions have been selected from the CEC 2005 benchmark functions and its results are compared with that of the other strategies.

Keywords: Restart strategy · Real-coded genetic algorithm
CEC 2005 benchmark test functions · Multi-modal optimization problems

1 Introduction

Optimization is a method of searching the best feasible solution out of several possibilities. The optimization problems can be classified broadly into two categories. The first case is that, where the user is satisfied with the obtained locally optimal solution. However, in the second case, the search algorithm will try to locate the globally optimal point out of the several locally optimal solutions. Now, the global optimization technique uses mainly two types of approaches, namely deterministic method and stochastic method [1]. Deterministic approaches have the assurance to obtain globally optimal solutions to the given accuracy for certain types of problems. However, they may not be suitable for tackling the non-smoothed or ill-conditioned problems. Stochastic approaches can be applied to all kinds of problems, but they have probabilistic promise to reach the globally optimal solution [2]. It is a well-accepted fact that whenever the size of variable search space is expanded exponentially and other exact approaches fail to produce the optimal solution, the stochastic method can be used effectively to search for the optimal or near-optimal solution.

© Springer International Publishing AG, part of Springer Nature 2018
A. Abraham et al. (Eds.): ISDA 2017, AISC 736, pp. 32–41, 2018.
https://doi.org/10.1007/978-3-319-76348-4_4

Among several stochastic approaches, some popular algorithms like genetic algorithm [3], particle swarm optimization [4], differential evolution [5], ant colony optimization [6], cuckoo search [7], bat algorithm [8] etc. are in use. Genetic algorithm (GA), which was introduced by J. Holland [3], is one of the most popular global optimization techniques. It has been used successfully for solving different optimization problems in various fields of engineering, computational sciences, bio-medicine, etc. [9–11]. However, as like others, GA has also a probabilistic assurance to find the globally optimal solution. This limitation of a GA is more visible in case of solving complex multi-modal optimization problems. Although there does not exist any such algorithm, which can solve all types of global optimization problems with certainty in finite time [12], it is always desirable for a GA to have a high probabilistic guarantee in search of the globally best result. Now, to enhance the global search capability of a GA, one way might be to increase its population size. However, a GA's search process would be computationally expensive due to a very large size of population. Another method could be to multi-start the GA. This is nothing but to restart the algorithm under some user-defined conditions and consider the best result, achieved among all the generations, as the globally optimal solution. There are a few restart strategies for GA available in the literature. However, there is a chance of further improvement of these strategies. In this study, a novel restart strategy is proposed for a real-coded genetic algorithm (RCGA) to solve complex multi-modal optimization problems. The performance of the RCGA with the proposed restart strategy is evaluated through ten multi-modal problems, selected from the CEC 2005 benchmark test functions [13], and its results are compared with that of the other strategies.

2 Literature Survey

For global optimization, a few restart strategies for GA are available in the literature. Ghannadian et al. [14] proposed a concept of the random restart to GA. In their work, the mutation operator is substituted by a random restart scheme. Whenever the search of a GA is seen to be stagnated, random restart introduces a set of new initial population and the next evolution starts with these new solutions. Beligiannis et al. [15] introduced a technique for restarting of the classical GA, where an insertion operator is used when the restarting criteria are satisfied. This insertion operator is nothing but the act of incorporating a certain constant percentage of the genomes from the last generation to the new initial population of the GA during restarting of the algorithm. For example, if the size of the population is 50 and the constant percent is taken as 20%, then 40 new solutions are created randomly and ten randomly chosen solutions from the last generation are merged together and the next evolution begins with this mixed population. The algorithm is restarted after every predefined number of generations. Hughes et al. [16] introduced a recentering-restarting evolutionary algorithm for the epidemic network to evolve and their method could be applied for TSP implementation, ordered gene problems, etc. Dao et al. [17] suggested an improved structure of a GA, where the algorithm restarts, if the best solution obtained so far is not improved till a certain number of consecutive generations. During restart, the new population consists of a number of elite chromosomes (which is defined by the user), and the rest are randomly created solutions. Suksut et al. [18] introduced a support vector machine with restarting GA for classifying imbalanced data. In their study, a GA restarts when the fitness value of the new generation becomes less than that of the old population.

These restart strategies were not sufficient for solving complex multi-modal optimization problems using an RCGA. So, in this paper, a novel restart strategy is proposed and the performance of the same is demonstrated through ten multi-modal optimization case studies. Section 3 gives a detailed description of the developed restart strategy. Section 4 deals with the results and discussion, and the conclusions are presented in Sect. 5.

3 Developed Restart Strategy

To describe the developed restart strategy, we take the help of two terms, namely locally best solution (l_best) and globally best solution (g_best). The developed strategy consists of four different restart conditions. If any one of the conditions is found to be satisfied, the algorithm gets restarted. During restart, the next evolution begins with the N number of randomly created new solutions, where N is the population size of the GA. It is to be noted that no elite solution transfer takes place at the time of restart. The said conditions are as follows:

1st condition: At first, we assume an accuracy level of the problem, let say, $1e-14$. For a particular generation, if the change in the fitness value of l_best is seen to be less than the stated accuracy level, then it is considered that there is no improvement of the l_best solution in that generation. Now, if the total number of such consecutive generations, say m, exceeds a predefined threshold generation value (say th_gen), then the GA gets restarted. The initial value of m is set equal to zero at the starting of an evolution and whenever there is an improvement in the l_best, m again takes the value of zero. So, the GA restarts, $if\ m > th\_gen$.

The task of deciding the value of th_gen is difficult, as it depends on the nature of the objective function's space. If the said parameter is set at a very low value, then it might happen that the algorithm gets restarted very frequently and the optimal points may not be reached. On the contrary, a too large value of th_gen can make the global search of a GA less efficient and time consuming. The thumb rule is as follows: if the complexity of the optimization problem is on the higher side, then the value of th_gen should be set at a lower value and vice-versa. A user can determine the value of th_gen using Eq. (1):

$$th\_gen = d_1 \times max\_gen,\tag{1}$$

where max_gen represents the maximum number of generations of an RCGA and d_1 is set equal to 0.03 through some trial experiments with the test functions studied in this paper.

2nd condition: In an evolution, the algorithm is supposed to count the number of generations (say, c_gen), where the value of m is found to be greater than zero. This means that if the value of m is seen to be greater than zero in a certain generation, then the parameter (c_gen) is going to be incremented by one. When c_gen exceeds the value of threshold count parameter (say, th_count) and the value of m is also found to be greater than zero at that moment, then the RCGA prepares for a new start. This condition of restart can be written as follows:

If($c\_gen > th\_count$ && $m > 0$), there is a restart of the algorithm.

Similar to m, the value of c_gen is initialized to zero in every evolution. For setting the value of th_count, the same thumb rule, as in case of th_gen, is applicable. Nevertheless, the value of th_count can be estimated using Eq. (2), as given below.

$$th\_count = d_2 \times max\_gen, \tag{2}$$

where d_2 is set equal to 0.1 through some trial experiments with the test functions.

3rd condition: Another situation of the restart for RCGA occurs, when the value of c_gen exceeds a pre-fixed parameter (say, th_count_1) and the objective function value of l_best is seen to be inferior to the same of g_best. This condition of restart for a minimization problem can be expressed as follows:

If($c\_gen > th\_count\_1$ && $l\_best > g\_best$), there is a restart of the algorithm.

The probable value of the parameter th_count_1, can be evaluated using Eq. (3):

$$th\_count\_1 = d_3 \times th\_count, \tag{3}$$

where d_3 has been assigned a value of 0.3 through some trial experiments with the test functions considered in this study.

4th condition: This condition is designed to avoid revisiting either the local or global optimum basin. It is obvious that if an RCGA can detect and keep away from the already visited local basins, then the probability of reaching the globally optimal solution will be more, as the search process will be extended over more number of unvisited regions in the variable space. For the purpose of detecting, whenever the algorithm restarts after satisfying any of the above-mentioned first three conditions, the obtained locally best solution (l_best) is memorized. In the next evolution, for every generation, the Euclidean distances are calculated between the presently available l_best solution and the previously memorized locally best points. If any one of the Euclidean distances is found to be less than a pre-allocated threshold value (say, th_val) and the l_best is seen to be inferior to the g_best in terms of the objective function value, then the next restart occurs. Here, one important point is to be noted that if the algorithm gets restarted under this 4th condition, then the locally best solution is not going to be captured for that evolution, as it is a revisited one. Also, another significant point to be noted is that, this condition is not going to be satisfied for the first evolution of the RCGA.

The selection of the best value for the parameter th_val is very difficult. However, a method for calculating this value from the problem information can be stated as follows:

1st Step: At first, the ranges of all the variables are calculated. The range of a variable is nothing but the absolute difference between the given boundaries for that variable. Next, the maximum value among all the ranges is found out and it is denoted as max_range.

2nd Step: Another parameter, say s_range, is calculated. This is done after multiplying all the ranges by a factor of d_4 and then, by taking the sum of these values.

$$s\_range = \sum_{i=1}^{k} (d_4 \times range_i),\tag{4}$$

where k is the total number of variables of the optimization problem and $range_i$ represents the range for i^{th} variable. Here, a suitable value of d_4 is obtained through some preliminary experiments with the test functions studied in this paper and it is found to be equal to 0.04.

3rd Step: In this step, the minimum value in between the s_range and $(max\_range \times d_5)$ is determined. This obtained value is considered as the threshold value (th_val) of the problem, which is calculated as follows:

$$th\_val = minimum(max\_range \times d_5, s\_range),\tag{5}$$

where d_5 is seen to be equal to 0.4 through some trial experiments.

From the above description of the restart strategy, it can be comprehended easily that the last two conditions (i.e., 3rd and 4th) are never going to be satisfied during the first evolution of an RCGA, as l_best and g_best solutions are found to be same for that period. The novelty of the proposed strategy lies with the developed restart conditions, specially from 2nd to 4th. These are totally new types of conditions used to restart an RCGA and simultaneously, they are very efficient, robust and easy to implement. Nevertheless, it is important to note that the values of d_1 through d_5 of the Eqs. (1) through (5), respectively, are dependent on the nature of test functions.

4   Results and Discussion

For the purpose of evaluating the performance of the proposed restart strategy for the RCGA, ten complex multi-modal test functions have been chosen from the CEC 2005 benchmark test functions [13]. These functions were designed from some classical optimization functions by rotating, shifting or hybridizing. These operations increased the complexity of these functions. Table 1 shows the list of these selected test functions with their variable boundaries and global minimum objective function values (f_{min}).

The experiment is carried out with a standard elitism-based RCGA, equipped with tournament selection [19], simulated binary crossover operator [20] and polynomial mutation operator [21]. The RCGA has been run with four different strategies, such as

- Proposed Restart Strategy of this paper (say, St-1)
- Restart strategy suggested by Beligiannis et al. [15] (say, St-2)
- Restart strategy introduced by Dao et al. [17] (say, St-3)
- Without Restart (say, St-4)

Table 1. A list of ten complex multi-modal problems from CEC'05 benchmark test functions [13].

Functions	Hybrid Composition function	Bound	f_{min}
F01	Hybrid Composition function 1	$[-5, 5]$	120
F02	Rotated Hybrid Composition function 1	$[-5, 5]$	120
F03	Rotated Hybrid Composition function 1 with Noise in fitness	$[-5, 5]$	120
F04	Rotated Hybrid Composition function 2	$[-5, 5]$	10
F05	Rotated Hybrid Composition function 2 with a Narrow basin for the Global optimum	$[-5, 5]$	10
F06	Rotated Hybrid Composition function 2 with the Global optimum on the bounds	$[-5, 5]$	10
F07	Rotated Hybrid Composition function 3	$[-5, 5]$	360
F08	Rotated Hybrid Composition function 3 with High Conditioned Number Matrix	$[-5, 5]$	360
F09	Non-Continuous Rotated Hybrid Composition function 3	$[-5, 5]$	360
F10	Rotated Hybrid Composition function 4	$[-5, 5]$	260

4.1 Parameters' Settings

For the purpose of fair comparison, all the common parameters of RCGA, such as crossover probability $(p_c = 1.0)$, mutation probability $(p_m = 0.02)$, user index parameter for crossover $(\eta_c = 2.0)$, user index parameter for mutation $(\eta_m = 10)$, population size $(N = 50)$, number of variables $(k = 10)$, maximum number of fitness evaluation (taken as 1,00,000) and maximum number of generations $(max\_gen = 2000)$ are kept fixed. The stopping criterion is set as the maximum number of fitness evaluation. Special parameters for St-1 like, th_gen, th_count, th_count_1 and th_val, are calculated using the Eqs. (1), (2), (3) and (5), respectively, and accuracy level of the problems is considered as 1e−14. Table 2 displays the values of these parameters.

Table 2. Special parameters' values for St-1

Special parameters	Values
th_gen	60
th_count	200
th_count_1	60
th_val	4

In St-2, after every 60 generations, the algorithm gets restarted and during a new evolution, randomly chosen 20% of the population of the last generation is inserted in the randomly created $(0.8 \times N)$ number of new solutions. In case of St-3, if there is no change in the best solution up to a consecutive number of 60 generations, the RCGA will go for a restart. During the restart, the best found solution (g_best) is going to be transferred into the next evolution and the rest (N−1) solutions are generated at random.

Each problem has been evaluated for 50 times with the same initial population. We have used the same seed in each run to serve this purpose. After each run, the gained globally best objective function value is captured and its absolute deviation from the actual globally best objective function value is calculated. Next, the average of the obtained 50 deviations is found out. This type of experiment is carried out for all the ten problems using four different strategies, as mentioned earlier. Table 3 presents a comparison of the average deviations for these selected test functions (best results are shown in bold).

Table 3. Comparison of the average deviations obtained using RCGA with four different strategies

Function	St-1	St-2	St-3	St-4
F01	**1.308E−02**	8.564E+00	1.682E+01	1.702E+01
F02	**1.233E+02**	1.298E+02	1.320E+02	1.334E+02
F03	1.298E+02	1.301E+02	1.354E+02	**1.297E+02**
F04	**7.639E+02**	9.551E+02	8.787E+02	9.595E+02
F05	**7.610E+02**	9.493E+02	9.393E+02	9.652E+02
F06	**7.608E+02**	9.589E+02	9.351E+02	9.713E+02
F07	**4.682E+02**	9.735E+02	9.657E+02	9.895E+02
F08	**7.619E+02**	8.278E+02	7.901E+02	8.303E+02
F09	**5.353E+02**	1.097E+03	8.912E+02	1.104E+03
F10	**2.000E+02**	3.887E+02	3.345E+02	4.698E+02

From the comparison of these results, it is clear that the proposed restart strategy (St-1) has outperformed the other three strategies for nine problems out of ten (that is, except in case of F03). Moreover, in case of F03, the obtained value of deviation yielded by St-1 is found to be comparable with the best value gained using St-4. Figure 1 depicts the variations in absolute deviations of the obtained g_best for 50 runs of the four strategies. From this figure, it is clear that the RCGA with the proposed restart strategy, is able to yield the better results for most of the times compared to the other three strategies. The superiority of the proposed strategy lies with the construction of the efficient restart conditions. This strategy is capable of detecting and avoiding the already visited locally or globally optimal points. It enhances the global search capability of an RCGA within the given maximum number of fitness evaluations. Nevertheless, as there is no elite solution transfer takes place during a restart, a new evolution is not influenced by the previously found best solutions. This makes each search stage independent of the others.

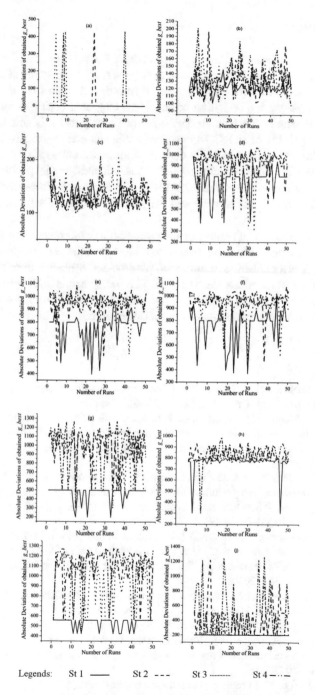

Legends: St 1 ——   St 2 - - -   St 3 ··········   St 4 —··—

Fig. 1. Absolute deviations of obtained globally optimal function values in 50 runs: (a) F01, (b) F02, (c) F03, (d) F04, (e) F05, (f) F06, (g) F07, (h) F08, (i) F09 and (j) F10.

5   Conclusions

In this study, a novel restart strategy of an RCGA for solving complex multi-modal optimization problems has been proposed and presented. Four types of conditions for restarting have been embedded in the scheme introduced. This strategy is capable of detecting and avoiding the revisiting of any local or global optimum basin. Also, the proposed restart scheme has been designed in such a way that it is able to increase the global search capability of an RCGA and due to this fact, the probabilistic guarantee of an RCGA of finding out the globally optimal solution is enhanced by a considerable amount. From the CEC 2005 benchmark functions, ten multi-modal problems have been selected and experiments are carried out to test the performance of the developed strategy of restarting. The results of the proposed scheme are compared with that of the other three schemes.

The proposed strategy of restarting has yielded the better results compared to other three schemes for nine test functions out of ten. However, the performance of the developed strategy is decided by a number of parameters, whose numerical values are dependent on the test functions to be studied. The performance of the proposed scheme of restarting will be tested in future on other CEC 2005 benchmark test functions and some practical problems related multi-modal optimization.

References

1. Liberti, L., Kucherenko, S.: Comparison of deterministic and stochastic approaches to global optimization. Int. Trans. Oper. Res. **12**(3), 263–285 (2005)
2. Moles, C.G., Mendes, P., Banga, J.R.: Parameter estimation in biochemical pathways: a comparison of global optimization methods. Genome Res. **13**(11), 2467–2474 (2003)
3. Holland, J.H.: Adaptation in Natural and Artificial Systems. University of Michigan Press, Ann Arbor (1975)
4. Eberhart, R., Kennedy, J.: A new optimizer using particle swarm theory. In: Proceedings of the Sixth International Symposium on Micro Machine and Human Science, pp. 39–43 (1995)
5. Das, S., Suganthan, P.N.: Differential evolution: a survey of the state-of-the-art. IEEE Trans. Evol. Comput. **15**(1), 4–31 (2011)
6. Dorigo, M., Maniezzo, V., Colorni, A.: Ant system: optimization by a colony of cooperating agents. IEEE Trans. Syst. Man Cybern. Part B (Cybern.) **26**(1), 29–41 (1996)
7. Yang, X.-S., Deb, S.: Engineering optimisation by cuckoo search. Int. J. Math. Model. Numer. Optim. **1**(4), 330–343 (2010)
8. Yang, X.-S.: A new metaheuristic bat-inspired algorithm. In: González, J.R., Pelta, D.A., Cruz, C., Terrazas, G., Krasnogor, N. (eds.) Nature Inspired Cooperative Strategies for Optimization (NICSO 2010), pp. 65–74. Springer, Heidelberg (2010)
9. Wang, Y., Huang, J., Dong, W.S., Yan, J.C., Tian, C.H., Li, M., Mo, W.T.: Two-stage based ensemble optimization framework for large-scale global optimization. Eur. J. Oper. Res. **228**(2), 308–320 (2013)
10. Ng, C.-K., Li, D.: Test problem generator for unconstrained global optimization. Comput. Oper. Res. **51**(Suppl. C), 338–349 (2014)
11. dos Santos Coelho, L., Ayala, H.V.H., Mariani, V.C.: A self-adaptive chaotic differential evolution algorithm using gamma distribution for unconstrained global optimization. Appl. Math. Comput. **234**(Suppl. C), 452–459 (2014)

12. Boender, C.G.E., Romeijin, H.E.: Stochastic methods. In: Horst, R., Pardalos, P.M. (eds.) Handbook of Global Optimization. Kluwer Academic Publishers, Boston (1995)
13. Suganthan, P.N., Hansen, N., Liang, J.J., Deb, K., Chen, Y., Auger, A., Tiwari, S.: Problem definitions and evaluation criteria for the CEC 2005 special session on real-parameter optimization. Technical report, Nanyang Technological University, Singapore, May 2005 and KanGAL Report 2005, IIT Kanpur, India (2005)
14. Ghannadian, F., Alford, C., Shonkwiler, R.: Application of random restart to genetic algorithms. Inf. Sci. **95**(1), 81–102 (1996)
15. Beligiannis, G.N., Tsirogiannis, G.A., Pintelas, P.E.: Restartings: a technique to improve classic genetic algorithms' performance. In: International Conference on Computational Intelligence 2004, pp. 404–407 (2004)
16. Hughes, J.A., Houghten, S., Ashlock, D.: Recentering and restarting a genetic algorithm using a generative representation for an ordered gene problem. Int. J. Hybrid Intell. Syst. **11**(4), 257–271 (2014)
17. Dao, S.D., Abhary, K., Marian, R.: An improved structure of genetic algorithms for global optimisation. Prog. Artif. Intell. **5**(3), 155–163 (2016)
18. Suksut, K., Kerdprasop, K., Kerdprasop, N.: Support vector machine with restarting genetic algorithm for classifying imbalanced data. Int. J. Futur. Comput. Commun. **6**(3), 92 (2017)
19. Goldberg, D.E., Deb, K.: A comparative analysis of selection schemes used in genetic algorithms. Found. Genet. Algorithms **1**, 69–93 (1991)
20. Agrawal, R.B., Deb, K.: Simulated binary crossover for continuous search space. Complex Syst. **9**(2), 115–148 (1995)
21. Deb, K., Goyal, M.: A combined genetic adaptive search (GeneAS) for engineering design. Comput. Sci. inf. **26**, 30–45 (1996)

Evaluating SPL Quality with Metrics

Jihen Maazoun[1][(✉)], Nadia Bouassida[1], and Hanêne Ben-Abdallah[2]

[1] Mir@cl Laboratory, University of Sfax, Sfax, Tunisia
jihenmaazoun@gmail.com   nadia.Bouassida@isimsf.rnu.tn
[2] King Abdulaziz University, Jeddah, Saudi Arabia
HBenAbdallah@kau.edu.sa

Abstract. A Software Product Line (SPL) is a set of systems that share a group of manageable features and satisfy the specific needs of a particular domain. The features of an SPL can be used in variable combinations to derive product variants in the SPL domain. Because SPLs promote product development through reuse, it is vital to have a means to measure their quality in terms of quality attributes like complexity, reusability,… In this paper, we propose a set of metrics to evaluate the quality of an SPL at three levels: the feature model, design and code. We adapted a set of metrics for software quality and defined new metrics to deal with the inherent characteristics of SPLs, specifically the feature model and the traceability between features, design and code. Furthermore, to assist in interpreting the quality of a given SPL, we conducted an empirical study over ten open source SPLs to identify thresholds for the proposed metrics.

Keywords: SPL · Feature model · Metrics · SPL quality

1 Introduction

The advantages of Software Product Lines (SPL) [2] as good quality solutions have been widely accepted in the research and industrial communities. In fact, reusing SPLs accelerates the development process and reduces the cost of software development. A Software Product Line (SPL) [2] is a set of systems that share a group of manageable, distinctive quality characteristics called *features* [1] and it is based on variability management. An SPL encodes variability by describing all possible variation points (*i.e.,* information about how to customize the SPL to derive a product), and all the variants (*i.e.,* information about the available options for each variation point). To represent the variability in an SPL, a *feature model* (FM) is often used. This indicates the features and the constraints (and, or, require, etc.) relating the features to one another. It is used to configure new products by instantiating the features while respecting the constraints. In other words, a feature model provides for development by reuse of software product lines.

Given that SPLs are dedicated to promote product development through reuse, it is therefore vital to have a means to measure their quality in terms of quality attributes/ characteristics like complexity, reusability,… Indeed, several researchers have been interested in SPL quality (*e.g.,* [6, 10, 14]); however, until now there is no recognized

© Springer International Publishing AG, part of Springer Nature 2018
A. Abraham et al. (Eds.): ISDA 2017, AISC 736, pp. 42–51, 2018.
https://doi.org/10.1007/978-3-319-76348-4_5

and adopted technique for measuring SPL quality. The herein proposed approach differs from existing works by adopting a comprehensive view of the quality of an SPL: It considers the SPL quality as a combination of the quality of its feature model, the quality of its design, the quality of its code and the quality of the traceability between the feature, design and code. For example, if a feature is associated to a very large number of classes, while all the other features contain a small number of classes, then this feature is most likely complex and its reuse would be difficult. As such, our approach proposes to consider SPL quality measurement at three levels: FM, design and code, to which it adds the correlation between the three artifacts types by measuring the traceability amongst them.

In this paper, SPL quality is assessed quantitatively by measuring understandability, modularity, traceability, and reusability. To do so, our approach uses several structural design metrics used in OO software [3] to which we propose new metrics to deal with the inherent characteristics of SPLs, specifically the feature model and the traceability between features, design and code. Besides the definition of a metric suite for SPLs, we find it important to have appropriate thresholds to assist a designer in analyzing the quality of a particular SPL. To this end, we relied on an empirical study we conducted on ten open source SPLs from the Feature House [18].

The remainder of this paper is organized as follows. Section 2 overviews currently proposed approaches for SPL quality evaluation. Section 3 presents a metric suite for the evaluation of SPL quality and it illustrates the calculus of these metrics through an example. Section 4 shows the empirical study we conducted to identify thresholds for the metrics. Section 5 presents an evaluation of our metrics and, finally, Sect. 6 summarizes the paper and outlines our future work.

2 Overview of SPL Quality Evaluation Approaches

Several works examined SPL quality and proposed metrics adapted to this field (e.g., [3–5, 7, 10, 14, 15]). Each of these propositions was interested in measuring a certain quality characteristic (like complexity, reusability …).

Some works were interested in metrics for SPL variability (e.g. [12, 14]). As an example, Lopez-Herrejon and Trujillo [12] argue that SPLs have to express variability and that SPLs are structured thanks to variation points, which justifies the proposition of metrics pertinent to this quality characteristic. They adapted the work of McCabe [13] on Cyclomatic Complexity to SPL variation points. That is, they measure the complexity of SPLs by measuring the complexity of their variation points. Also measuring SPL variability, Oliviera et al. [11] proposed metrics that analyze variability modelled through specific stereotypes in UML artefacts. The proposed metrics tightly depend on the profile of Oliviera [14] and they cannot be calculated on SPLs whose design is not modeled with this UML profile.

Zhang et al. [17] proposed an approach based on a Bayesian Belief Network (BBN) also to assess the SPL variability. The BBN is used to explicitly model the impact of variants (especially design decisions) on system quality attributes. The feature model is used to capture functional requirements and the BBN model to capture the impact of functional variants on quality attributes. Zhang et al. [17] treat the impact of variants on quality attributes without considering the quality of SPL design and source code.

Besides the variability characteristic of quality, other works were interested in the quality of feature models (e.g. [8, 9]). Benavides et al. [8] proposed metrics to assess the reusability of a feature model by using Constraint Satisfaction Programming. Their metrics aim to evaluate the influences on reuse and number of derivable solutions by measuring the number of features and the types of relations among them. On the other hand, Bagheri and Gasevic [4] explored the graphical aspect of feature models to propose a set of structural metrics for SPLs. Their proposed metrics are relative to size, structural complexity and length. They are shown to be valid from a measurement theoretical perspective (construct validity). These metrics are adopted in our approach and will be presented with more details in Sect. 3.1.

Besides the above quality characteristics, many works focused on evaluating the reusability characteristic of SPLs (e.g. [6]) For example, Torkamani [6] proposed new metrics for estimating SPL reusability by considering the weighted values of assets which are elements of SPL. In order to determine the weight value, assets and artifacts should be converted to a common measurement unit such as "Line of code".

In summary, there is a lack of consensus on the metrics appropriate for assessing SPL quality characteristics. Most of the existing works are initial investigations in the area. Each investigation tackled quality assessment from different aspect, *e.g.,* reusability, variability, complexity. In addition, none of the proposed investigations tackled the traceability among the various SPL assets; this quality characteristic is vital for maintainability. Covering a wide range of quality characteristics within a standard quality model motivated our work in proposing new metrics for SPL.

3   Metrics for Evaluating SPL Quality

In the context of software quality, models have been applied to link measures of software artefacts with external, high-level, quality characteristics. Quality models use a hierarchical decomposition of the quality concept into quality *Factors* (also called characteristics), Sub-factors (also called sub-characteristics), and *Criteria*. The evaluation of software begins with measuring each quality criteria with numerical values from metrics. In this paper, we focus on the following four quality factors for SPLs: (1) *Reusability,* (2) *Understandability, (3) Modularity* and (4) *Complexity*.

In the following subsections, we overview those metrics we reused from the literature and we present our new metrics.

3.1   Useful Existing Syntactic Metrics

We retained from the metrics suite of [4] the following set of metrics:

- **Cyclomatic complexity (CC):** Counts the number of distinct cycles that can be found in the feature model. Since feature models are in the form of trees, no cycles can exist in a feature model; however, integrity constraints between features can cause cycles.

Then, CC counts the number of "exclude" and "require" without considering "OR", "XOR" and "AND" constraints.

- **Ratio of variability (RoV):** Counts the ratio of the average branching factor of the parent features in the feature model. In other words, the average number of children of the nodes in the feature model tree.
- **Coefficient of connectivity density (CoC):** Counts the ratio of the number of edges over the number of nodes in the graph.
- **Number of features (NF):** Counts the number of nodes in a feature model.
- **Number of top feature (NTop):** Counts the number of features that are first direct descendants of the feature model root.
- **Number of leaf feature (NLeaf):** Counts the number of features with no children or further specializations.

3.2 SPL New Metrics

We believe that the available metrics for estimating the quality of the feature model, code and design of an SPL are insufficient. In our work, our main objective is to identify a set of measures for the feature model, that is meaningfully when associated with external quality of software product line design. Our metrics are classified into understandability, modularity and reuse metrics.

Table 1. SPL Understandability metrics

Metrics		Description
Number of mandatory features	NB_mandatory	Counts the number of mandatory features in a feature model
Number of optional features	NB_optional	Counts the number of optional features in a feature model
Number of constraints	NB_OR	Counts the number of OR constraints in a feature model
	NB_XOR	Counts the number of XOR constraints in a feature model
	NB_Exclude	Counts the number of exclude constraints in a feature model
	NB_Require	Counts the number of Require constraints in a feature model
Degree of traceability	FNOP	Counts the number of packages associated to a feature
	FNOC	Counts the number of classes associated to a feature
	FNOM	Counts the number of methods associated to a feature
	FNOA	Counts the number of attributes associated to a feature
	FNOAs	Counts the number of associations between classes associated to a feature

Understandability Metrics. Understandability metrics measure the system in terms of the attributes, methods, classes, packages, mandatory, optional features and constraints. They are associated to complexity. In general, higher values of these metrics mean an increase in memory footprint, lower performance, and higher complexity.

In Table 1, we provide a description of understandability metrics adapted to SPLs. The first goal is to measure the size of optional and mandatory features and the number of constraints in a feature model. The second goal is to measure the degree of traceability between the feature model, the design and code. As presented in Sect. 3.2, a feature can be composed of several elements like package(s), Class(es), attribute(s), method(s). To measure the degree of traceability, we present the following set of metrics.

Modularity Metrics. In an object oriented design, modularity metrics measure the interdependencies of different classes. A design with a large number of inter class dependencies (high coupling) is weak and fragile. In an object oriented design, modularity is used to measure the number of classes to which a class is coupled. Two classes are coupled if a method belonging to a certain class uses a method or an attribute of another class. By analogy, the coupling in an SPL (presented in Table 2) measures the relation between the elements of features (packages, classes, methods and attributes). An excessive coupling between features would be at the expense of modularity and it prevents reuse. In fact, promoting encapsulation necessitates the minimization of the coupling between features. Moreover, the more a feature is combined with other features, the more any change of this feature influences other features. As such, high coupling impedes maintainability and reusability.

Table 2. SPL modularity metrics

Metrics	Description
NCE	Counts the number of common elements (packages, classes, methods, attributes) between two features
NFC	Counts the number of features combined in a class

The NFC metric measures the relationship between features. In fact, as illustrated in Fig. 1, a class "AddMediaToAlbum" stereotyped "Media" and belonging to feature

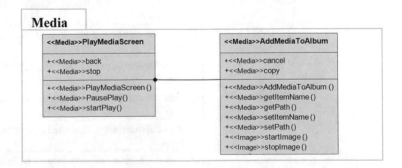

Fig. 1. An example illustrating the NFC metric

"Media" may contain two methods ("startImage()", "StopImage()") stereotyped "Image" and belonging to feature "Image". Then, the number of features combined in the class "AddMediaToAlbum" is equal to 2. If the number of NFC increases, the maintenance becomes more difficult.

The NCE metric measures the number of elements used by different features. As an example, Fig. 2 presents two features named "File" and "Save" containing the same methods "getReadFile()" and "setReadFile()". In this figure, the NCE is equal to 2. High values of NCE lead to a feature model that is poor in encapsulation, reusability, and maintainability.

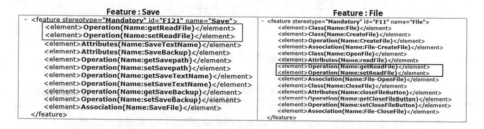

Fig. 2. An example illustrating the NCE metric

Reuse Metrics. In this sub-section, we propose new reuse metrics inspired from [15] and adapted to SPLs. Some of these metrics are related to feature reuse like *Level of Reusing a Feature model (LRF)* and *Reuse Frequency of a Feature (RFF)* and others are related to SPL reuse like *Feature model Reuse Percentage (FRP)*. The overall Reuse Level (RL) is defined as the ratio of reused items to the total number of items in a product [16]. The LRF metric is adpted from [16]. It is defined as the ratio between the number of reused features of a given product and the total number of features in the feature model as shown in Eq. (1).

$$LRF_j = \frac{NF\_product_j}{NF} \tag{1}$$

where:

NF is the number of features in the SPL.
$NF\_Product_j$: number of reused features in the product j.

Often, a lower level item is used more than once in a higher level item. Counting the multiple uses of a lower level item is not done for Reuse Level. Hence a new term, Reuse Frequency, was defined which is a count of the number of references to a lower level item from a higher level item [16]. The *RFFi* metric is inspired from [16] and it is defined as the number of apparitions of the feature *i* in the products divided by the number of products NP (Eq. (2)).

$$RFF_i = \frac{\sum_{j=1}^{NP} Pj\backslash i}{NP} \tag{2}$$

where: NP is the number of products reusing the SPL and Pj\i is equal to 1 if the feature
i appears in the product Pj, otherwise it is equal to 0. If the RFF_i of a feature i is equal
to 1, then this feature is mandatory, otherwise it is optional.

The RFF of the feature treat "Malware" equals the sum of the number of reusing the
SPL and that reuse this feature divided by the number of products.

Finally, we define the FRP metric which measures how much the feature model is
qualified for reuse. This metric is defined as the ratio between the number of optional
features in a feature model and the total number of features (NF) in an SPL (Eq. (3)).

$$FRP = \frac{NB\_optional}{NF} * 100 \tag{3}$$

4   Descriptive Analysis of the Metrics Values

It is essential to make an empirical study on existing feature models in order to approximate the threshold values for the metrics. Although these are not fixed and definitive
thresholds, but they can give an idea about the quality, i.e., if the values of a feature
model's metrics are similar to the values obtained in our empirical study, then it is within
the good-quality standards. The feature models that were used in our experimental evaluation were all taken from Feature House[1]. We selected ten feature models from this
repository.

To approximate the threshold values for the metrics, Sect. 4 presents the analysis
results. We have too few systems, thus a valid statistical analysis could not be performed
for such measures. Consequently, we propose a descriptive statistics (min, max, mean)
for the different metrics.

From applying these metrics on different feature models, we observed the following
points:

- The average of NCE is in the interval [0, 4] (the min is equal to 0 and the max is
  equal to 4). Evidently, in order to have a reusable and easy to maintain SPL, we
  consider that the average of NCE should be minimized and do not exceed 4.
- If the average NFC does not exceed 5 in an SPL, this means that the considered SPL
  is sufficiently modular.
- If the number of optional features Nb_optional is relatively high, i.e., higher than the
  mean, then the degree of reusability is high.
- The CC depends on the metrics NB_require and NB_exclude. High values of
  NB_require and NB_exclude exceeding the mean make the feature model more
  complex.
- The mean of the FRP is 50.43%, consequently a feature model is considered reusable
  if the FRP is superior to 50.43%.

---

[1] http://www.infosun.fim.uni-passau.de/spl/apel/fh/.

5 Evaluation

The overall objective of this section is to judge the performance of our metrics, presented in Sect. 3, when evaluating the quality of feature models. Note that in order to have a fair comparison with other existing works and to be able to evaluate really the efficiency of the different existing approaches, we must evaluate all the approaches on the same corpus. We believe that it is necessary to have a benchmark for software product line quality and this would be an interesting research axis.

Consequently, we conducted a comparison between our proposed metrics values obtained for a list of SPLs and evaluations of the quality of the same SPLs that were handled by experts. More specifically, we presented a list of questions about the quality of the SPLs to six experts that work in the domain of software engineering and particularly in SPL engineering. They were asked about complexity, number of interdependencies between features, size and reusability of the three SPLs present in the literature. In fact, we asked the experts to judge if the number of packages associated to the features is reasonable, if the number of classes associated to the features is reasonable, … and then, we calculated the average FNOP, the average FNOC, the average of FNOM, the average FNOA and the average FNOAs and we found that the calculated metrics and their interpretations are similar to the estimation of the experts.

The advantage of the three considered FMs is that they correspond to different domains and therefore exhibit various constraints and relations among their features (FM1: Antivirus domain, FM2: Gophone messaging domain and FM3: Flight reservation domain).

Note that experts, responded to our list of questions about the quality of the SPLs, they judged that FM2 has an equitable size and degree of traceability between feature model, design and code source while FM1 is rather complex and its reusability degree is very low, finally FM3 has a high value of interdependency between features. The metrics chosen for analysis can be divided into 4 categories: size, modularity, complexity, and reuse metrics. We remark that the evaluated quality obtained by experts is very close to the quality as judged by experts. Concerning the size value and the degree of traceability (FNOP, FNOC, FNOA, FNOM, FNOAs) are equitable. Their average values vary between 1 and 5. This makes the feature model less complex. We notice that for FM1 the coupling is high: NCE is equal to 5 (exceeds the threshold presented in Sect. 4) and NFC equals 6 (exceeds the threshold presented in Sect. 4). On the other hand, the value of FRP is equal to 36.88%. Thus, we deduce that the reusability of FM1 is very low. For FM2, the values of the calculated metrics imply that the FM has a good quality. For FM3, we deduce that the NCE is high. The coupling is high and on the other hand, the reusability is not very high, since the value of FRP is equal to 56.46%.

To conclude our preliminary empirical study shows that our proposed metrics are conform to those obtained by experts.

Threats to Validity: Threats to validity are conditions that limit our ability to generalize the results of our preliminary empirical study. An important threat concerns the choice of experts that judge the quality of different SPL. They all were from the computer science discipline while feature models were selected keeping in mind various domains of the real life. However, further experiments involving industrial professional subjects are needed to ensure the external validity of our experiments. Also, we have difficulty in including professionals in our study to judge the quality of our SPL.

Although, we have select different SPL which have high quality according to the judgment of expert while it is possible that these SPL have a poor quality (e.g., poor understandability) which would make the derived threshold biased towards poor-quality system.

6 Conclusion

In this paper, we first reviewed existing works for the evaluation of SPLs quality and we argued that there are insufficient measuring techniques in the field of software product lines. Secondly, we proposed a metrics suite adapted to the evaluation of four SPL quality factors, namely *understandability*, *modularity*, *reusability* and *complexity*.

Our metrics suite has the advantage of measuring the quality of the feature model, the design and the traceability between them. In addition, we performed an empirical study on existing SPLs in order to identify some thresholds/guidelines for the proposed metrics.

In our future works, we plan to apply the proposed metrics to larger real case studies; this could provide insights and would allow the validation of the thresholds of the metrics. Moreover, we are currently examining the definition of well-formedness rules for SPLs and also bad SPL practices.

References

1. Kang, K., Cohen, S., Hess, J., Novak, W., Peterson, A.: Feature-Oriented Domain Analysis (FODA) feasibility study. Technical report CMU/SEI-90-TR-21, Software Engineering Institute, Carnegie Mellon University (1990)
2. Clements, P., Northrop, L.: Software Product Lines: Practices and Patterns. SEI Series in Software Engineering. Addison-Wesley, Boston (2001)
3. Chidamber, S., Kemerer, C.: A metrics suite for object oriented design. IEEE Trans. Softw. Eng. **20**, 476–493 (1994)
4. Bagheri, E., Gasevic, D.: Assessing the maintainability of software product line feature models using structural metrics. Softw. Qual. J. **19**(3), 579–612 (2011)
5. Trendowicz, A., Punter, T.: Quality modeling for software product lines. In: Workshop on Quantitative Approaches in Object-Oriented Software Engineering, QAOOSE 2003 (2003)
6. Torkamani, M.-A.: Metric suite to evaluate reusability of software product line. Int. J. Electr. Comput. Eng. (IJECE) **4**(2), 285–294 (2014)
7. Zubrow, D., Chastek, G.: Measures for Software Product Lines. Technical report, Carnegie Mellon University (2003)
8. Benavides, D., Trinidad, P., Ruiz-cortés, A.: Automated reasoning on feature models. In: 17th International Conference on Advanced Information Systems Engineering, CAISE. LNCS (2005)
9. Benavides, D., Segura, S., Ruiz-Cortés, A.: Automated analysis of feature models 20 years later: a literature review. Inf. Syst. **35**, 615–636 (2010)

10. Berger, C., Rendel, H., Rumpe, B.: Measuring the ability to form a product line from existing products. In: Variability Modelling of Software-Intensive Systems (VaMos) (2010)
11. Oliveira Jr. E.A., de Souza Gimenes, I.M., Maldonado, J.C.: A metric suite to support software product line architecture evaluation. In: CLEI 2008, pp. 489–498 (2008)
12. Lopez-Herrejon, R.-E., Trujillo, S.: How complex is my Product Line? The case for Variation Point Metrics. VaMoS, pp. 97–100 (2008)
13. McCabe, T.J.: A complexity measure. IEEE Trans. Softw. Eng. **2**, 308–320 (1976)
14. Oliveira Junior E. A.; Gimenes, I.M. S.; Maldonado, J. C.; Masiero, P. C. and Barroca, L.: Systematic evaluation of software product line architectures. J. Univ. Comput. Sci. **19**(1), 25–52 (2013)
15. Frakes, W., Anguswamy, R., Sarpotdar, S.: Reuse ratio metrics RL and RF. In: 11th International Conference on Software Reuse, Falls Church, VA, USA (2009)
16. Frakes, W.B.: An empirical framework for software reuse research. In: Proceedings of the Third Workshop on Methods and Tools for Reuse, Syracuse University CASE Center Technical Report, no. 9014, 5 p. (1990)
17. Zhang, H., Jarzabek, S., Yang, B.: Quality prediction and assessment for product lines. In: Eder, J., Missikoff, M. (eds.) Proceedings of the 15th International Conference on Advanced Information Systems Engineering, CAiSE. LNCS, vol. 268, pp. 681–695. Springer (2003)
18. http://www.infosun.fim.uni-passau.de/spl/apel/fh/#download

Using Sentence Similarity Measure for Plagiarism Detection of Arabic Documents

Wafa Wali$^{(\boxtimes)}$, Bilel Gargouri, and Abdelmajid Ben Hamadou

MIRACL Laboratory, Sfax University, Sfax, Tunisia
{wafa.wali,bilel.gargouri}@fsegs.rnu.tn,
abdelmajid.benhamadou@isimsf.rnu.tn

Abstract. Plagiarism detection it is a challenging task, particularly in natural language texts. Some plagiarism detection tools have been developed for diverse natural languages, especially English. In this paper, we propose, a new plagiarism detection system devoted to Arabic text documents. This system is based on an algorithm that uses a semantic sentence similarity measure. Indeed, the sentence similarity measure aggregates in a linear function between three components: the lexical-based LS including the common words, the semantic-based SS using the synonymy relationships, and the syntactico-semantic- based SSS semantic arguments properties notably semantic argument and thematic role. It measures the semantic similarity between words that play the same syntactic role. Concerning the word-based semantic similarity, an information content-based measure is used to estimate the SS degree between words by exploiting the LMF Arabic standardized dictionary ElMadar. The performance of the proposed system was confirmed through experiments with student thesis reports that promising capabilities in identifying literal and some types of intelligent plagiarism. We also demonstrate its advantages over other plagiarism detection tools, including Aplag.

Keywords: Plagiarism · Sentence similarity · Arabic language
Lexical Markup Framework · Semantic information
Syntactico-semantic information

1   Introduction

The problem of plagiarism has existed for a long time but with the advance of information technology this problem becomes worse by the fact there are many electronic versions of published materials available to everyone. In general, plagiarism refers to copying others' publications or intellectual works without quoting the sources. In order to prevent this problem, some researches were conducted on plagiarism detection in the last years, particularly for English. Language-independent tools exist as well, but are considered restrictive as they usually do take into account the specific language features. However, detecting

© Springer International Publishing AG, part of Springer Nature 2018
A. Abraham et al. (Eds.): ISDA 2017, AISC 736, pp. 52–62, 2018.
https://doi.org/10.1007/978-3-319-76348-4_6

plagiarism in Arabic documents is particularly a challenging task because of the complex linguistic structure of Arabic. To the best of our knowledge, the only works in this field are those of Menai et al. [14] and Jadalla et al. [11]. All of them addressed the plagiarism detection using content based methods. However, semantic sentence similarity has not been yet applied to Arabic texts.

In this research, we present a plagiarism detection tool to compare Arabic documents to identify potential similarities. The tool is based on a new algorithm that uses semantic sentence similarity. Indeed, we have proposed a novel method to compute semantic similarity between sentences [17–19] using LMF standardized Arabic dictionary ElMadar [13]. The proposed method presents a model for the estimation of the semantic similarity between sentences based, firstly, on the semantic similarity of their words and secondly on the syntactico-semantic similarity notably the common semantic arguments properties such as semantic class and thematic role. The present method is the first which exploits an SS (Syntactico-Semantic) knowledge-based semantic similarity measure in the quantification of word semantic similarity. Moreover, we have evaluated the performance of developed tool in terms of precision and recall on a data set of student thesis reports, and showed its capability of identifying direct and sophisticated copying, such as sentence reordering and synonym substitution. We also demonstrate its advantages over other plagiarism detection tools, including Aplag [14].

The remaining part of this paper is organized as follows. Section 2 presents the related work about plagiarism detection methods. Section 3 presents an overview of Arabic language characteristics and challenges. Section 4 details our approach about plagiarism detection and describes a semantic similarity measure of sentences. Section 5 describes the experiments and discusses their results. The paper ends with a conclusion and a future work we are planning to undertake to extend this work.

2   Related Work

In general, there is an extensive literature on plagiarism detection of documents [1,4,8,16] but there are very few publications relating to the Arabic detection plagiarism [5,11,14].

This section reviews some related work studies in order to explore the strengths and limitations of the previous methods, and identify the particular difficulties in detecting the similarity between Arabic documents.

Related work studies can roughly be classified into two major categories: intrinsic and external. Intrinsic automatic plagiarism detection methods uncover the changes in the writing style, and use them as an evidence of a plagiarism act. In contrast, external automatic plagiarism detection methods do not rely on internal evidence, but rather they are based on detecting the similarities between the suspicious document and other documents from a reference corpus.

For external methods, Jadalla et al. [11] proposed Iqtebas 1.0, which is a primary solid and complete piece of work for plagiarism detection in Arabic documents. In fact, it is similar to a search engine. The goal of the Iqtbas 1.0 is to calculate the originality, value of the examined document, by computing the distance between each sentence in the text and the closest sentence in the suspected documents.

On their part, Farahat et al. [2] proposed the ZPLAG system, which is a prototype for detecting plagiarism in Arabic documents where some hidden plagiarism forms, such as change of sentence structure and replacement of synonym can be detected. Indeed, the aim of this system is to enable to submit students' assignments to teachers in classrooms. The teacher, in turn, can retrieve the students' assignments in one of his/her classes and view a report that highlights the plagiarized parts in each submitted assignment. The results showed that the ZPLAG system is very appropriate for Arabic scripts

In another study, Menai presented APlag [14], which is a prototype of a plagiarism detector for Arabic documents in which some hidden forms of plagiarism, such as sentence structure change and synonym replacement can be detected. In fact, the author described main components of this prototype, in particular, heuristic algorithms at different logical levels (document, paragraph, and sentence) to pass up redundant comparisons.

However, in another study, Alzahrani et al. [3] produced an Arabic plagiarized detection (APD) tool which combines the fuzzy and semantic similarity models derived from a lexical database. Indeed, they retrieved a list of candidate documents for each suspicious document using Jaccard coefficient, and then made a sentence-wise detailed comparison between the suspicious and the associated candidate documents using the fuzzy similarity model. Their preliminary results indicated that fuzzy semantic-based similarity model can be used to detect plagiarism in Arabic documents.

Furthermore, Khan et al. [12] developed a web-based plagiarism detection system to detect plagiarism in Arabic written documents. The proposed plagiarism detection framework consists of two main components, one global and the other local. The former is heuristics-based, in which a potentially plagiarized given document is used to construct a set of representative queries by using different best performing heuristics. These queries are then submitted to Google via Google's search API to retrieve candidate source documents from the Web. Next, the latter carries out detailed similarity computations to detect if the given document was plagiarized from the documents retrieved from the Web or not. The global component is thoroughly evaluated, whereas the local component is partially implemented.

On the other hand, Bensalem et al. [14] developed a system which uses various stylistic features to account for intrinsic plagiarism. Unfortunately, this system was evaluated on a small corpus therefore it is difficult to quantify its effectiveness.

3   Characteristics of Arabic Language

Arabic belongs to the Semitic language group. The main characteristics of Modern Standard Arabic (MSA) [6, 7] are the following:

- One must add to this alphabet eight other forms: The hamza with six forms (أ, إ, آ, أُ, إي, أو), the T marbouta (ة) and the "alif" maksour (ى) as well as the ligation of the letters ل (L) and ا A, which is لا (called lamalif).
- A special feature of the Arabic language is that the letters change shape depending on their location in the word.
- An advantage of the Arabic alphabet compared to other alphabets, according to which the letters can be written in uppercase or lowercase, is that it is not capitalizable.
- The Arabic script is cursive, the words are written at once, whether printed or handwritten characters.
- A certain difficulty is that the short vowels are replaced in Arabic by diacritical marks الشكل often omitted in the texts, making it difficult to determine the meaning of the word, if not impossible mainly when the word is isolated from its context.

Example: كَتَبَ and كُتُبٌ (respectively pronounced kataba and kotobUN) are written exactly in the same way, but respectively mean "he wrote" and "books".

- Arabic is a templatic morphology language based on having roots and patterns (templates) to identify derivational lemmas in the language. An Arabic word may be composed of a root plus affixes and clitics. The affixes include inflectional markers for tense, gender, and/or number, whereas verbs, which derive from a three-consonant root, have ten patterns. For example, if the root (k t b) is combined with the patterns 1a2a3a, the result will be "كتب - kataba" (in English to write); if combined with 1A2a3a, the result will be "كاتب - kAtaba", (in English to correspond), where the numbers (1,2,3) designate the root radicals.
- The construction of the sentence is relatively free.

Example: الولد ذهب إلى المدرسة (went the child to the school), ذهب الولد إلى المدرسة (the child went to the school), إلى المدرسة الولد ذهب (to the school the child went), ذهب إلى المدرسة الولد (went to the school the child)

4   Our Plagiarism Detection Approach

Plagiarism can be detected by establishing the "content similarity" between documents. The proposed approach identified "DP" as a document plagiarized from a source document "DS", if DP contains (words in) sentences with high degrees of similarity to (words in) sentences in DS. In reality, plagiarism detection is not as simple as matching sentences with sentences, since sentences in DS may not be copied entirely into DP. Indeed, establishing which sentences of DS have been plagiarized is quite broad in scope. For this reason, our proposal considers a measure that computes the degree of sentence similarity between S1, that belong to DS, and S2, that belong to DP taking advantage of LMF standardized Arabic dictionaries [13], notably the syntactic-semantic knowledge that they contain.

The suggested measure is discussed in the following subsections.

4.1   Sentence Similarity Measure

The proposed approach computes the degree of similarity of any two sentences which takes into account lexical, semantic and syntactico semantic information. The proposed measure computes the sentence semantic similarity via the LMF standardized Arabic dictionary [13]. Indeed, the LMF standardized Arabic dictionary diversifies lexical knowledge at the morphological, syntactic and semantic levels. Furthermore, it is finely structured, which facilitates access to information. These lexical pieces of information are interconnected. For example, the senses can be related by semantic relationships such as synonymy or antonymy. Moreover, the LMF standardized Arabic dictionary identifies the syntactico-semantic relationship. The originality of our measure is that it underlines the synonymy relations, semantic class and thematic role extracted from LMF Arabic dictionary to compute the sentence similarities.

Our semantic sentence similarity measure is based on three similarity levels, namely the lexical, the semantic and the syntactico-semantic ones using the LMF standardized Arabic dictionary [13].

4.1.1   Lexical Similarity

Before, analyzing potential plagiarized documents, we first remove all the stop ones and reduce all the non-stop words in a document D to their stems, using the morphological analyzer MADAMIRA [15]. In addition, as part of the pre-processing step, short sentences (i.e., semantically uninformative with regard to plagiarism detection) are removed from D due to the high probability that independent authors can create similar short sentences rather than long, similar ones. In fact, the longer two sentences are, the less likely they are similar.

At the lexical level, we compare the lexical units constituting the sentences in order to extract the similar words.

To calculate the degree of lexical similarity, which we call LS(S1,S2), using the Jaccard coefficient [10], we used the following formula:

$$LS(S1, S2) = \frac{CW(S1, S2)}{WS1 + WS2 - CW(S1, S2)} \tag{1}$$

where
CW(S1,S2) denotes the common word number between sentences S1 and S2,
WS1 denotes the word number of sentences S1,
and WS2 denotes the word number of sentences S2.

4.1.2   Semantic Similarity

The semantic similarity score is based to semantic vectors where the entry number of a semantic vector is equal to that the of distinct words in sentences S1 and S2. In fact, we intend to present all the distinct words from both sentences in a list called "T".

For example, if we have sentences:

S1: الوقت المتد من الفجر الى غروب الشمس in English "The time from dawn to sunset"

S2: زمن مقداره من طلوع البشمس الى غروبها in English "The time from sunrise to sunset"

then   we   will   have   preprocessing   step   T = وقت,ممتد,من,فجر,الى,غروب, طلوع,مقدار,زمن,شمس

The process of deriving a semantic vector for S1 and S2 is explained as follows: each entry of the semantic vector SV_i is determined by the semantic similarity of the corresponding word in "T" to a word in the sentence.

The semantic similarity between words can be determined from one of two cases taking S1 as an example:

Case1: if Wi (the word at position i in "T") appears in the sentence S1, then SV_i is set to 1.

Case2: if Wi is not contained in S1, a semantic similarity score is computed between Wi and each word in S2, using the LMF standardized Arabic dictionary.

Indeed, the LMF Arabic dictionary model defines many types of semantic relationships (e.g., synonymy, antonymy, etc.) between the senses of two or several lexical entries by means of the Sense Relation class.

Given the words W1 and W2, the semantic similarity SW(W1,W2) is deduced from the analysis relations between word meanings as follows: words are linked to a semantic relation in the LMF standard (in this paper, we are interested in the synonymy relation) and with relation pointers to other synsets. In fact, our objective is to find the synonymy set of each word so as to detect the common synonyms between two words.

Then, once the two sets of synonyms for each word have been collected, we calculate the degree of similarity between them using the Jaccard coefficient [10]:

$$SW(W1, W2) = \frac{CS}{SW1 + SW2 - CS} \tag{2}$$

where
CS is the common synonymy number between words W1 and W2,
SW1 is the synonymy number of word W1,
and SW2 is the synonymy number of word W2.

Once the semantic vectors for each sentence is computed, we calculate the semantic similarity between sentences SS(S1,S2) using the cosine similarity as indicated in the following formula:

$$SS(S1, S2) = \frac{\sum_{i=0}^{n} V1i * V2i}{\sqrt{\sum_{i=0}^{n} V1i^2} \sqrt{\sum_{i=0}^{n} V2i^2}} \tag{3}$$

where
V1i and V2i are the components of vectors V1 and V2 respectively to S1 and S2.

4.1.3   Syntactico-Semantic Similarity

In establishing the degree of syntactico-semantic similarity between sentences, we firstly to define the syntactic structure for each sentence from Stanford analyzer [9]. Indeed, each lexical entry sense is related to syntactic dependencies via Syntactic Behavior class where each behavior is linked to a semantic predicate in the Predicative Representation class. The semantic predicate specified the semantic argument properties, notably the semantic class and the thematic role.

Hence from the semantic knowledge, especially the semantic class and the thematic role, we computed the syntactico-semantic similarity between sentences SSS(S1,S2) using the Jaccard index [10].

The SSS(S1,S2) is defined as follows:

$$SSS(S1, S2) = \frac{CSA}{SA1 + SA2 - CSA} \tag{4}$$

where
CSA is the common semantic argument number between sentences S1 and S2,
SA1 denotes the semantic number of arguments of sentence S1,
and SA2 denotes the semantic number of arguments of sentence S2.

4.2   Sentence to Document Similarity

Having determined the similarity score of each sentence S in a plagiarized document DP with respect to each sentence in a source document DS, the plagiarism detection approach identifies the highest degree of similarity of S with the sentences in DS, which increases the probability of S to have the same content as one of the sentences in DS, as indicated in the following formula:

$$Sim(S, DS) = \max(\forall_{Sj \in Ds} Sim(S, DS)) \\ such \, that \quad sim(S, DS) >= 0.8 \tag{5}$$

where
Sim(S,DS) returns the highest Sim of S with respect to the sentences in DS, if S is not created by merging two or more sentences in DS; otherwise, the combined similarity of the sentences in DS that are merged to yield P is assigned to be the sentence-to-document value of S with respect to DS.

4.3   Document Similarity

After identifying similar sentences in a source document DS related to the sentences in a (potential) plagiarized document DP, we determine the overall percentage of plagiarism of DP with respect to DS as follows:

$$Sim(Dp, Ds) = \frac{\sum_{i=1}^{n} Sim(Si, DS)}{n} \tag{6}$$

where
n is the sentences number in DP,
Sim(DP,DS) is the degree of resemblance of DP with respect to DS.

5   Experimental Results

In this section, we introduce the data sets used for conducting various empirical studies on automatic plagiarism detection and then present several evaluation measures for analyzing the performance of our proposed approach measure in terms of accuracy and recall in detecting plagiarized documents.

5.1   The Data Set

In assessing the performance of our plagiarism detection measure, we manually built a corpus of 300 student thesis reports with different sizes and topics downloaded from the doctoral school of the Law faculty of Sfax-Tunisia. The thesis reports are related to information protection in new technologies. Some statistics about this corpus are provided in Table 1.

Table 1. Corpus statistics.

Number of documents	300
Number of all sentences	36 520
Number of plagiarized sentences	18 548
Number of words	219 101
Number of tokens	167 899

5.2   Performance Evaluation

Two measures were used to evaluate the performance of automatic plagiarism detection measure: accuracy (Eq. 7) and recall (Eq. 8). The results are shown in Table 2.

$$Accuracy = \frac{Number\ of\ plagiarized\ sentence\ identified}{Total\ number\ of\ sentence\ identified} \tag{7}$$

Table 2. Evaluation results.

Accuracy	0.97
Recall	0.94
F-measure	0.954

$$Recall = \frac{Number\ of\ plagiarized\ sentence\ identified}{Total\ number\ of\ plagiarized\ sentences} \qquad (8)$$

The performance of our proposal is dependent on morphological analyzer MADAMIRA [15], the synonyms and semantic properties, notably semantic class and thematic role, retrieved from LMF standardised Arabic dictionary ElMadar [13]. According to comparative evaluation study of Arabic morphological analyzers and stemmers, MADAMIRA [15] achieves the highest accuracy. There, we do not expect to increase the performance of our automatic plagiarism detection measure by using other stemmers. However, using other semantic properties databases might impact the performance of our proposal.

Aplag [14] is used as baseline for our proposed measure.

The performance results of Aplag are returned in terms of synonymy replacement (SR): percentage of matched words the tool was able to find for the tested document. For this reason, the SR is also estimated for our suggested measure. In fact, Fig. 1 shows the mean of the synonymy replacement, given by our proposed measure and Aplag for each data set.

Aplag could not to detect any semantic property however, its performance is close to our measure in detecting synonymy replacement: the Mean(SR)= 90% for APlag and Mean(SR)= 93% for our proposal. Nevertheless, the Aplag' results for detecting similarities in our data sets are not competitive. This indicates that semantic properties, especially the semantic class and the thematic role are important for detecting the similarity between sentences and hence, for plagiarism detection.

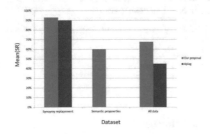

Fig. 1. Mean synonymy replacement for Aplag and our proposal

6   Conclusion

We have proposed a sentence similarity-based plagiarism-detection tool which relies on semantic and syntactico-semantic knowledge to identify the sentence-to-sentence similarity values that determine the degree of resemblance of any two documents to detect the plagiarized one, if it exists. Our proposal is designed to detect plagiarized text documents profiting from the LMF standardized Arabic dictionary where the semantic and syntactico-semantic knowledge, notably the semantic class and the thematic role are well-defined.

Actually, we have conducted an empirical study which shows that our proposal achieves high accuracy, i.e., 94%, in detecting (non-)plagiarized documents. Furthermore, we have compared the performance of our approach to the existing plagiarism-detection tool Aplag introduced in [14], which shows that our proposal thus outperforms the document- based plagiarism-detection approach using the constructed dataset.

Moreover, these experiments were carried out using a small corpus; As a result, there is a need to build large corpora that can confirm the above finding and conduct further studies.

In the future, we intend to integrate the different components of the suggested approach to build one sentence similarity based plagiarism detection system. We will be thoroughly investigating the performance of different similarity measures before incorporating them into the final similarity computation model.

References

1. Abdi, A., Idris, N., Alguliyev, R.M., Aliguliyev, R.M.: PDLK: plagiarism detection using linguistic knowledge. Expert Syst. Appl. **42**(22), 8936–8946 (2015)
2. Riad, A.M., Farahat, A.S., Zaher, M.A.: Studying different methods for plagiarism detection. Int. J. Comput. Sci. Eng. (IJCSE) **2**(5), 147–154 (2013)
3. Alzahrani, S.M., Salim, N., Abraham, A.: Understanding plagiarism linguistic patterns, textual features, and detection methods. IEEE Trans. Syst. Man Cybern. Part C (Appl. Rev.) **42**(2), 133–149 (2012)
4. Barrón-Cedeño, A., Vila, M., Martí, M.A., Rosso, P.: Plagiarism meets paraphrasing: insights for the next generation in automatic plagiarism detection. Comput. Linguist. **39**(4), 917–947 (2013)
5. Bensalem, I., Rosso, P., Chikhi, S.: Intrinsic plagiarism detection using n-gram classes. In: EMNLP, pp. 1459–1464 (2014)
6. Darwish, K., Magdy, W. et al.: Arabic information retrieval. Found. Trends® Inf. Retr. **7**(4), 239–342 (2014)
7. Farghaly, A., Shaalan, K.: Arabic natural language processing: challenges and solutions. ACM Trans. Asian Lang. Inf. Process. (TALIP) **8**(4), 14 (2009)
8. Franco-Salvador, M., Rosso, P., Montes-y Gómez, M.: A systematic study of knowledge graph analysis for cross-language plagiarism detection. Inf. Process. Manag. **52**(4), 550–570 (2016)
9. Green, S., Manning, C.D.: Better Arabic parsing: baselines, evaluations, and analysis. In: Proceedings of the 23rd International Conference on Computational Linguistics, COLING 2010, pp. 394–402. Association for Computational Linguistics, Stroudsburg (2010)

10. Jaccard, P.: Etude comparative de la distribution florale dans une portion des Alpes et du Jura. Impr. Corbaz, Paris (1901)
11. Jadalla, A., Elnagar, A.: A plagiarism detection system for Arabic text-based documents, pp. 145–153. Springer, Heidelberg (2012)
12. Khan, I.H., Siddiqui, M.A., Mansoor, K.: A framework for plagiarism detection in Arabic documents. Comput. Sci. Inf. Technol. 01–09 (2015)
13. Khemakhem, A., Gargouri, A., Hamadou, A.B., Francopoulou, G.: ISO standard modeling of a large Arabic dictionary. Nat. Lang. Eng. **22**, 849–879 (2016)
14. Menai, M.E.B.: Detection of plagiarism in Arabic documents. Int. J. Inf. Technol. Comput. Sci. (IJITCS) **4**(10), 80 (2012)
15. Pasha, A., Al-Badrashiny, M., Diab, M.T., El Kholy, A., Eskander, R., Habash, N., Pooleery, M., Rambow, O., Roth, R.: MADAMIRA: a fast, comprehensive tool for morphological analysis and disambiguation of Arabic. In: LREC, vol. 14, pp. 1094–1101 (2014)
16. Velásquez, J.D., Covacevich, Y., Molina, F., Marrese-Taylor, E., Rodríguez, C., Bravo-Marquez, F.: Docode 3.0 (document copy detector): a system for plagiarism detection by applying an information fusion process from multiple documental data sources. Inf. Fusion **27**, 64–75 (2016)
17. Wali, W., Gargouri, B., Hamadou, A.B.: Supervised learning to measure the semantic similarity between Arabic sentences. In: Computational Collective Intelligence, pp. 158–167. Springer, Cham (2015)
18. Wali, W., Gargouri, B., Hamadou, A.B.: Enhancing the sentence similarity measure by semantic and syntactico-semantic knowledge. Vietnam J. Comput. Sci. **4**(1), 51–60 (2017)
19. Wali, W., Gargouri, B., Hamadou, A.B.: Using standardized lexical semantic knowledge to measure similarity. In: Knowledge Science, Engineering and Management, pp. 93–104. Springer, Cham (2014)

Computer Aided Recognition and Classification of Coats of Arms

Frantisek Vidensky[✉] and Frantisek Zboril Jr.

FIT, IT4Innovations Centre of Excellence, Brno University of Technology,
Brno, Czech Republic
{ividensky, zborilf}@fit.vutbr.cz

Abstract. This paper describes the design and development of a system for detection and recognition of coat of arms and its heraldic parts (components). It introduces the methods by which individual features can be implemented. Most of the heraldic parts are segmented using a convolution neural networks and the rest of them are segmented using active contour model. The Histogram of the gradient method was chosen for coats of arms detection in an image. For training and functionality verification we used our own data that was created as a part of our research. The resulting system can serve as an auxiliary tool used in heraldry and other sciences related to history.

Keywords: Image segmentation · Neural networks · Heraldry

1 Introduction

Each aristocratic family inherited their own coat of arms by which they represent themselves. Although the age of aristocratic families has already passed in most countries, we can still find its remains. For example, in the heart of Europe, on the territory of Bohemia, Moravia and Silesia, we can still find hundreds of castles and chateaus and other dwelling of aristocratic families. In these dwellings, we can find coats of arms above gates, on walls, on furniture, or on items of ordinary use. Coats of arms can also be found in the literature that deals with them.

In recent years, computer vision methods have been increasingly used for various purposes. For example, in the industry, computer vision is used to control products in industry, modern cars can detect pedestrians and or other obstacles and can read traffic signs. Until now, as we believe, no one has developed an application for recognizing coats of arms. Even though there is a science that deals with them, this science is called a heraldry.

This is our first work that deals with the recognition of coats of arms. During the research, we tried different methods for detecting and segmenting the objects. The best result for detection was achieved with the histogram of oriented gradients (HOG) method [1].

During segmentation, the classic method failed (edge based methods, Hough transform, clustering, watershed). We decided to try to segment the components of coat of arms with known shape using the semantic segmentation, which is a technique that classifies each pixel in the image according to its semantic meaning. Semantic

© Springer International Publishing AG, part of Springer Nature 2018
A. Abraham et al. (Eds.): ISDA 2017, AISC 736, pp. 63–73, 2018.
https://doi.org/10.1007/978-3-319-76348-4_7

segmentation can be performed with multiple methods, for example, using semantic texton forests [2], but more recently, convolutional neural networks [3, 4] have been used for this technique. The other components are segmented with active contours [5].

The paper is structured as follows. First, we briefly introduce the heraldry in Sect. 2. Then we outline structure of the system for coat of arms recognition in Sect. 3. In Sect. 4 we describe the implementation of the system. Section 5 contains evaluations and experiments with the system. In Sect. 6 we describe the experiments. We conclude with explanation of our further work in this area.

2 Heraldry

Heraldry is one of Auxiliary sciences of history and deals with coats of arms. It does not only deal with family coats of arms but also states coats of arms, urban coats of arms, ecclesiastical coats of arms and others. Heraldry have not always developed in the same way. For example, English heraldry, French heraldry and Italian heraldry are different from each other. In our work, we focus on Czech heraldry, which evolved in Czech lands (Bohemia, Moravia, and Czech Silesia).

Typical family coat of arms consists of the components shown in Fig. 1(a). They are described below, because it's not a well-known thing (the text is based on [6]). The shield is the only required and most important (coat of arms is displayed here) component of the coat of arms. The shield may have different shape and its choice only depended on the heraldic artist.

This choice often depends on the period or country but the heraldic artist always could choose his or her favorite shape. The basic shapes of the shields can be seen in Fig. 1(b). The shield can be drawn straight or tilted and we can put a helmet, coronet or crown on it.

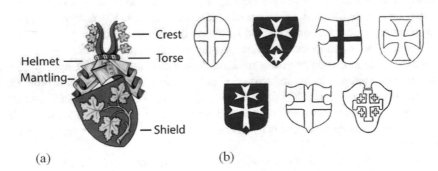

(a)   (b)

Fig. 1. (a) Split coat of arms into individual components. Extracted from [7]. (b) Basic shapes of shield from the left: Norman, Heater, Tilting, Iberian, French, Polish and Italian. Extracted from [6].

The helmet can only be placed on the shield. There are several types of helmets which have evolved over time. These types can be seen in Fig. 2. Each helmet has mantling. It is drapery attached to the helmet and used to protect the helmet from the sun, rain or chopping by the sword. Previously, they were drawn very simply (as you can see in Fig. 1(a)) but later they started being drawn split and more glamorous.

The crest is an item attached to the top of the helmet and the mantling. The appearance of the crest may vary, but it is usually chosen to be in harmony with the coat of arms. Originally crests wore knights in tournaments, but it was impractical in battles.

The join between the helmet and the crest was not nice to look at. For this reason, torse and ducal coronet were created for masking this join. The torse was created by twisting of the fabrics from the mantling and the ducal coronet has the shape of a three-leaf ring.

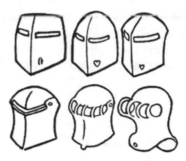

Fig. 2. Types of helmets. From the left to right are $3\times$ great helmet, tilting helmet and $2\times$ barred helmet. Extracted from [6].

3   Design of Decomposition System of Family Coats of Arms

Decomposition system must be able to find a coat of arms at any position in the image. Furthermore, it must be able to decompose the coat of arms into particular components and to classify all of them. We divided the system into the following parts:

Shield detector. In the first step, areas where coats of arms should be located, are searched. It is enough to detect only the shield, which is the only mandatory component of the coat of arms. The detector can find areas in which the shield is not located, but later these areas will be discarded. It is better to explore the areas where the shield is not located than skipping it incorrectly.

Segmentation and classification of the shield, helmets, torses, ducal. These components are precisely defined and quite easy to detect in an image. The second part of the
system divides the located area into heraldry components and classifies them. If it was
not possible to locate the shield, is know that the shield detector has found a wrong area
and the search in this part of the image is terminated.

Segmentation of crest and mantling This part of the system is only used if there is a
helmet in the segmentation map. Because the system does not know exact shape of
these components it is difficult to segment them. However, their position related to
other components is known. Detecting areas where the coat of arms is located should
be done by cutting a small area around the shield. From the observation of the various
coat of arms drawings, we found that 2.1 multiple of the height of the shield is
sufficient to ensure that other components are in the cut-out.

After receiving a segmentation map with components with a distinguishable shape,
the inappropriate areas will be removed. The main criterion for removing the cut-out
should be that it will not contain any or only an insignificant area with a shield.

In the next stage the system will search the remaining components in the extended
area according to the extended segmentation map from the previous step. It is enough
to enlarge the area to double size to assure that the remaining parts will be found in the
search area. Finally, the system draws the segment map to the selected.

4   Implementation

This section describes the implementation of the individual parts of the system
described in the previous section.

4.1   Detector of Shields

The HOG method [1] requires a detection window divided into blocks that are further
divided into cells. Furthermore, it needs to determine the number of bins in which the
gradient orientation is divided. The authors of [1] used a 64×128 pixel detection
window to detect pedestrians. For detection of shields a square shape detection window
is sufficient with a size of 64×64 pixels. To select the size of blocks, cells, and bin
counts, we made experiments described in Sect. 6.

For this purpose, we programmed in Emgu CV[1] an application that trains the
Sup-port vector machines (SVM) [9] classifier according to the given dataset. In HOG,
a linear kernel SVM is used.

Block sizes were defined as 64 divisors belonging to the interval of integers <8, 32>
and the size of cells belongs to the interval of integers <4, 16>, because the cell must be
smaller than the block size. For the bin count, we have determined the possible values
of divisors as 180 from the interval of integers <6, 36>.

---

[1] http://www.emgu.com/.

The result can be seen in Fig. 3.

(a)   (b)

Fig. 3. (a) Detected shields on drawings extracted from [7, 13]. Manually marked required area is green, found areas by detector are red. (b) Detected shields in photos extracted from [14].

4.2 Segmentation of Components with Distinguishable Shape

During reimplementation of the convolutional neural network for semantic segmentation, we proceeded according to paper [4]. The authors of the paper transposed the 16-layer VGG convolutional neural network [10] into a fully convolutional network. Subsequently they used upsampling (increasing the sampling rate) with factor 32 to the output of this network for semantic segmentation and they called the network FCN-32s. Then they made two improvements with using factors 16 and 8 which they used on multiple places and modified the architecture of the network. These networks are called FCN-16s and FCN-8s.

We achieved the best results with the FCN-8s network which we reimplemented using the Tensorflow[2] and TS-Slim[3] libraries. For training the network we were able to collect a training set containing Heater, Tilting, Iberian, French and Renaissance shields. Shape of the helmets is represented in the dataset by a great, tilting and barred helmet.

4.3 Segmentation of Crest and Mantling

We can perform the segmentation of the last components only if we have successfully segmented the shield and the helmet. Without knowing the position of these two components, it is not possible to continue.

As mentioned above, torse or ducal coronet can be inserted between the helmet and the crest, but the crest can also be placed directly on the helmet. First of all, it is necessary to examine the area directly above the helmet, whether there is a torse or ducal coronet. After successful search, it is necessary to remember the center of the lower edge as a dividing point. If the torse or ducal coronet has been successfully found, the centre of the helmet's top edge is taken as the dividing point.

---

[2] https://www.tensorflow.org/.

[3] https://github.com/tensorflow/models/blob/master/.../inception/inception/slim/README.md.

The last step was to find the contour of the entire coat of arms and remove the shield and the helmet from it. The rest of the contour is divided into two parts at the dividing point. In this way, contour of crest and contour of mantling are obtained. The result can be seen in Fig. 4(b).

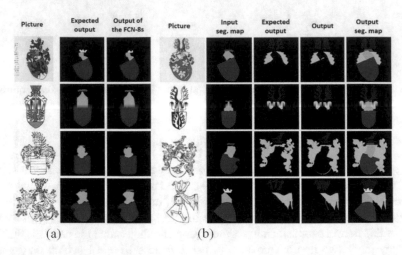

Fig. 4. (a) Segmentation results using the FCN-8s network. Images extracted from [7, 8, 13]; (b) Results of segmentation of crests, mantlings, and display of the result segmentation map. Images extracted from [7, 8, 13].

5   Verification of Functionality

This section describes verification of functionality of each system parts proposed in Sect. 3 and whose implementation is described in Sect. 4.

5.1   Verification of Functionality of the Shields Detector

To evaluate shield detectors, we chose the Intersection over Union metric (IoU) [11]. To use this metric, it is necessary to mark in the test pictures the area in which the search shield is located and this area is compared with the area marked by the detector. IoU between box A and box B is calculated according to the formula:

$$IoU = \frac{A \cap B}{A \cup B} \tag{1}$$

The result of the metric is a value from the interval <0, 1>. The smallest acceptable value is 0.5.

To measure the success of the segmentation we used two metrics. The pixel accuracy metric [4] shows how well the entire picture was segmented:

$$pixel\,accuracy = \sum_i n_{ii} \Big/ \sum_i t_i \qquad (2)$$

where n_{ij} is the number of pixels of the class i predicted in class j and $t_i = \sum j\, n_{ij}$ is the total number of pixels in the class i.

As the second metric, we used a different version of Intersection over Union, which we would label as IU in line with [4] article authors. This metric shows how well individual objects have been segmented and according to [12] we calculate it as:

$$IU = \frac{TP}{FP + TP + FN} \qquad (3)$$

where TP stands for true positive, FP false positive, and FN false negative. This metric is counted separately for each class, and pixels belonging to another class are understood as part of the background.

Both metrics give results in interval <0, 1>. We, however, recalculate the results in accordance to [4] into the interval <0, 100> and round it to one decimal place. To evaluate the best trained HOG detector, we performed tests for 29 combinations of settings described in Sect. 4.1. Table 1 lists the results for the top 10 detectors.

In Table 1 we can see high false positive values. Most of these false positive detections are located in the area of the crest of mantling, because their rounded parts can look like rounded parts of shield. False positive values could be reduced by a large expansion of the negative data set.

Table 1. Results of the top 10 detectors for different sizes (blocks, cells, bin counts), true positives, false positives, the success of finding shields, and the time of the calculation.

#	Detector	TP	FP	Success [%]	Time [s]
1	(32, 4, 6)	271	1543	90.33	181.54
2	(16, 4, 6)	263	1372	87.67	96.67
3	(32, 8, 6)	261	577	87.00	24.44
4	(32, 8, 9)	260	844	86.67	24.95
5	(8, 4, 20)	260	1959	86.67	99.42
6	(32, 8, 20)	257	960	85.67	28.23
7	(8, 4, 6)	257	1848	85.67	70.50
8	(16, 4, 9)	256	1614	85.33	116.03
9	(32, 8, 30)	255	638	85.00	30.95
10	(32, 8, 15)	255	765	85.00	26.88

5.2   Segmentation of the Components with Distinguishable Shape Using the FCN-8s Network

For verification of functionality, we created a test set in which all the shapes of shields, types of helmets, ducal coronets and torses from all the artist whose drawings were used in the training set are represented. The results can be seen in the Fig. 4(a). In the dataset, the coat of arms is not always complete, but the height of the cut is approximately 2.1 × the height of the shield.

Table 2 shows the results for segmentation using the FCN-8s network. The worst results were achieved in the case of segmentation of ducal coronets. The reason can be imperfection of both the training and testing set, because both the ducal coronet and the torse were very complicated and insufficiently manually segmented,

Table 2. Results of individual metrics for segmentation of distinguishable shape.

Metric	Result
Pixel accuracy	98,5
Shield (IU)	95,3
Helmet (IU)	87,5
Torse (IU)	63,3
Ducal coronet (IU)	51,7

5.3   Segmentation of the Crest and Mantling Using the Active Contours

We have collected coats of arms with different shapes of mantlings. We manually created an input segmentation map and the required output for segmentation of crests and mantlings, which we compared with the output according to the same metrics as in the previous part of the system. The results are listed in Table 3 and shown in Fig. 4(b).

We can evaluate the results of segmentation of crests and mantlings as good. The reduction of the results occurs mainly when the contour of the entire coat of arms is obtained, because in some cases is emerging a connection between the crest and the mantling.

Table 3. Results of individual metrics for segmentation of crests and mantlings.

Metric	Result
Pixel accuracy	96,3
Crest (IU)	77,2
Mantling (IU)	77,9

6 Experiments

All parts of the system were interconnected in such a way that the detector is used to locate areas where shields could be found in different resolutions of the input image. Subsequently, two cut-outs were created around each area. A small cut-out in which the system searched for a component using the FCN-8s network and we got the first segmentation map. This cut-out was then expanded to the size of a large cut-out and we tried to find the crest and the mantling in large cut-out. Upon successful finding, all the components are drawn into the segmentation map, and it will eventually be drawn into the cut-out of the input image in which the coat of arms is located.

In Fig. 5, the first column contains input images and marked areas where shields should be located. Other columns include the output of the FCN-8s network from a small cut-out, the resulting segmentation map obtained from a large cut-out and finally the output of the system. As we can see for the cut-out from the book that was used for the training data set, the results of the experiment went well (the first coat of arms). Also for the second coat of arms, whose author uses a similar style to the authors in the training set. On the third coat of arms we see that the shield was inaccurately segmented, but all components were still found and properly segmented. In the last one the ducal coronet is not segmented and the shape of the shield is segmented inaccurately.

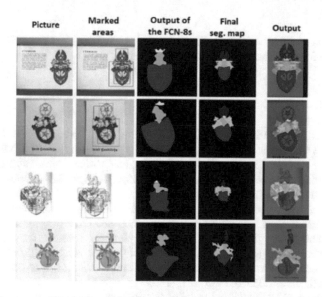

Fig. 5. Partial outputs of the system. The first coat of arms comes from the drawing artist who is in the training set and is extracted from [7]. Other coats come from drawers who were not in the training set. The second coat of arms comes from the Landštejn postcard, the third from [15] and the last from [14].

We also tested the system on a set of photos containing real coats of arms. Here, however, the segmentation of individual components completely failed. One of the reasons for this may be that the drawings of each coat of arms are bounded by a distinct black edge (it has a high gradient, and it is likely that the convolutional network has been taught to search it). The edge of the coat embossed in the wall is not so visible. In some places, components of coat of arms merge with each other or with the background.

7   Conclusion and Future Work

This paper describes the design and implementation of the system that can detect the coat of arms in the entrance image, make its decomposition to individual components and classify it. For each part of the system, their functionality was tested and experiments were performed with the resulting system. The system works well for book cuts that have been used for a training set or drawings similar to them. In the other artists, the results of the worse and with the images of real coats of arms the system completely fail. The system can be expanded in many ways. We can find and classify other possible components, identify the owner of the coat of arms or analyze the ordinary (all signs that can be placed on the shield) drawn on the shield and eventually perform blazoning (a formal description of the coat of arms according to heraldic rules [6]) of the coat of arms.

Acknowledgements. This work was supported by the BUT project FIT-S-17-4014 and the IT4IXS: IT4Innovations Excellence in Science project (LQ1602).

References

1. Dalal, N., Triggs, B.: Histograms of oriented gradients for human detection. In: Proceedings of the IEEE Conference on Computer Vision and Pattern Recognition (CVPR) (2005)
2. Cipolla, R., Battiatob, S., Farinella, G.M.: Computer Vision: Detection, Recognition and Reconstruction. Studies in Computational Intelligence, vol. 285. Springer, New York (2010). ISBN 978-3-642-12847-9
3. Hyeonwooh, N., Seunghoon, H., Bohyung, H.: Learning deconvolution network for semantic segmentation. arXiv preprint arXiv:1505.04366 (2015)
4. Long, J., Shellhamer, E. Darrell, T.: Fully convolutional networks for semantic segmentation. CoRR, volume abs/1605.06211 (2016). http://arxiv.org/abs/1605.06211
5. Kass, M.: Snakes: active contour models. Int. J. Comput. Vis. **1**(4), 321–331 (1998). https://doi.org/10.1007/bf00133570. ISSN 1573-1405
6. Schwarzenberg, K.F.: Heraldika: heraldika, čili, Přehled její theorie se zřetelem k Čechám na vývojovém základě. In: Vyšehrad 2, vol. 3. Vyšehrad, Prague (2007). Historica (Vyšehrad). ISBN 978-80-7021-827-3
7. Janáček, J., Louda, J.: České erby. Albatros, Prague (1974). Oko (Albatros)
8. Mysliveček, M.: Erbovník, aneb, Kniha o znacích i osudech rodů žijících v Čechách a na Moravě podle starých pramenů a dávných ne vždy věrných svědectví. Horizont, Praha (1993). ISBN 80-7012-070-3

9. Introduction to Support Vector Machines. OpenCV (2016). http://docs.opencv.org/2.4/doc/tutorials/ml/introduction_to_svm/introduction_to_svm.html. Accessed 30 Dec 2016

10. Simonyan, K., Zisserman, A.: Very deep convolutional networks for large-scale image recognition. CoRR, volume abs/1409.1556 (2014). http://arxiv.org/abs/1409.1556

11. Rosebrock, A.: Intersection over Union (IoU) for object detection. https://www.pyimagesearch.com/2016/11/07/intersection-over-union-iou-for-object-detection/. Accessed 30 Dec 2016

12. Rahman, A., Wang, Y.: Optimizing intersection-over-union in deep neural networks for image segmentation. In: International Symposium on Visual Computing, pp. 234–244. Springer, Cham (2016)

13. Halada, J.: Lexikon české šlechty II: erby, fakta, osobnosti, sídla a zajímavosti. Illustrated by František Doubek. Akropolis, Prague (1993). ISBN 80-85770-04-0

14. Sedláček, A., Růžek, V.: Atlasy erbů a pečetí české a moravské středověké šlechty: vol. 2. Academia, Prague (2001). ISBN 80-200-0935-3

15. Vavřínek, K.: Almanach českých šlechtických a rytířských rodů 2027. Vavřínek Zdeněk (2016)

Mining Gene Expression Data: Patterns Extraction for Gene Regulatory Networks

Manel Gouider$^{(\boxtimes)}$, Ines Hamdi, and Henda Ben Ghezala

RIADI Laboratory, ENSI, Campus of Manouba, 2010 Manouba, Tunisia
manelguider@yahoo.fr, nshamdi@gmail.com,
henda.benghezala@ensi.rnu.tn

Abstract. Gene interaction modeling is a fundamental step in the understanding of cellular functions. The high throughput technologies (microarrays, …) generate a large volume of gene expression data. However, gene expression data mining is a very complex process, it becomes necessary to analyze these data to discover new knowledge about genes and their interactions in purpose to model the Gene Regulatory Network GRN.

In this paper, we compare some patterns extraction approaches used in the literature to infer Gene Regulatory Networks and we propose to use gradual patterns of the form *(when A increases, B decreases)* to extract knowledge about genes. Furthermore, we rely on GO Gene Ontology as a knowledge source to semantically annotate genes and to add information that can be useful in the process of knowledge extraction.

Keywords: Genetic interactions · Knowledge extraction · GRN
Gene expression data · GO · Gradual patterns

1 Introduction

Several biological databases [1] are developed to archive the great mass of relevant and interesting data to make information available to the biologist, so it becomes necessary to discover and extract new knowledge from these sources. Furthermore, with the emergence of microarrays technologies that generate a large amount of genetic expression data, the extraction of knowledge about genes becomes a very important and complex task. The modeling of gene interactions is a considerable scientific interest to biologists as it is a fundamental step in the understanding of cellular function. The goal of the gene regulatory network is to understand the functioning of genes and how these genes are repressed or activated. Besides, the extraction of interactions between genes for an organism is a major challenge for bioinformaticiens. The techniques of Data mining are effective in extracting new knowledge about genes but the large volume of data generated by the large-scale microarray technologies makes difficult the task of discovering knowledge. Gradual patterns having this form (when A increases, B decreases) can express co-variation of genes expression data. We propose to use this type of patterns to extract knowledge from microarrays.

© Springer International Publishing AG, part of Springer Nature 2018
A. Abraham et al. (Eds.): ISDA 2017, AISC 736, pp. 74–82, 2018.
https://doi.org/10.1007/978-3-319-76348-4_8

In this paper, we are interested in a particularly important problematic: Knowledge extraction about genetic interactions facilitating the modeling of dynamics of gene regulatory networks GRN. The GRN is a network of transcription factor (TF) proteins and their target genes, where each edge represents the TF regulatory activity on a target gene. This article is organized as follow: The Sect. 2 describes some approaches of gene expression data analysis (clustering of genes, patterns extraction and GRN inference). In Sect. 3, we present our adopted method for extracting genetic interactions and finally we conclude.

2   Mining Gene Expression Data

The technologies of microarrays are powerful tools for functional genomics. These techniques provide the measurements of expression levels of thousands of genes simultaneously in few different experimental conditions (samples) or time points. The microarrays are used for the study of Gene regulatory network GRN or transcriptional regulation network. The microarrays are represented in the form of gene expression data matrix (G × C), the lines represent the genes and the columns represent the experimental conditions. An analysis of such type of numerical data, using data mining techniques, can give useful knowledge about genes, their functions and their interactions.

2.1   Clustering of Genes

The classification aims to group the genes under a predefined number of groups. Some supervised methods such as the neuronal network or the SVM are used for genes classification. Whereas the clustering, the unsupervised classification, aims to determine clusters of co-expressed genes whose expression varies in a similar way in different experiments. Several algorithms are used in the literature such as: Hierarchical clustering, K-means, Self-organizing maps, Self-organizing trees and Evolutionary algorithms, see [2]. This traditional clustering groups the genes over all samples or experimental conditions, but it doesn't give good results with large data sets and in this case a gene can belong only to one cluster, whereas in the reality, the genes are not to be co-expressed over all samples. In addition, a gene can participate to many biological processes and have many functions, then the biclustering try to resolve these limits. It aims to group simultaneously genes and experimental conditions, to get co-expressed genes likely to be co-regulated. Unlike the clustering methodology, a gene can belong to many biclusters. The biclustering is an important technique to extract regulatory modules. We can cite some biclustering algorithms [3]: BIMAX, FABIC, ISA, QUBIC, SAMBA...

2.2   Patterns Extraction

Patterns extraction is the major task in data mining, it aims to extract and discover from large datasets the patterns that frequently occur.

Extraction of Frequent Patterns and Association Rules
Agrawal [4] introduced the notion of association rules to extract associations between frequent items in a transaction database. An example of a rule: {milk → cereal}, customers who purchases milk is likely to purchase cereal. This method applies to binary attributes (0 or 1) to indicate the presence or absence of the item (purchased product) in the transaction. The search of association rules is done in two steps: search of frequent patterns then extraction of associations rules into these patterns. We can call a frequent itemset if its support *supp* (its frequency in the database) greater or equal to the minsup (threshold) defined by the user. An association rule denoted by $X \rightarrow Y$, a valid rule is a rule with support and confidence greater or equal to the minsup and minconf thresholds respectively. These measures (support and the confidence) are the two criteria for evaluating the quality of rules. The support estimates the importance of rule, whereas the confidence estimates the precision of rule.

Definition 1 (support of rule): *The support of rule denotes the frequency of the rule within n transactions. supp(X → Y) = supp(X ∪ Y)/n.*

Definition 2 (Confidence of rule): *The confidence of rule denotes the percentage of transactions containing X which contain also Y. It is an estimation of conditional probability P(Y|X). conf(X → Y) = supp(X ∪Y)/ supp(X).*

In addition to the tow traditional criteria (support and confidence) for evaluating rules, there are many others quality measures [5]. This popular method is largely used for gene expression data analysis [6–8] to extract new knowledge about genes sharing the same regulatory mechanism. *Apriori* is the most popular algorithm of extraction frequent itemsets. Unlike the clustering and biclustering which can't extract the relationships between genes in the same group, this method groups together the co-expressed genes and discovers their relationships. Carmona-Saez et al. [7] integrate the expression data with gene annotation to extract association rules and discover the association between data based in the co-occurrence. The extracted rules have the following form: *{cell cycle → [+]condition1, [+]condition2, [+]condition3, [−]condition6}.* This rule expresses that a large number of genes which are annotated by this term *'cell cycle'* are overexpressed under conditions 1, 2 and 3 whereas they are under-expressed in condition 6. This symbolic patterns extraction method produces a huge number of rules and a large proportion of this generated rules are redundant and it is generally applied like the other symbolic methods on a binary context. And to resolve these limits many works focused on the reduction of the number of rules and on the optimization of their quality [9]. In addition, other works use the threshold method [8] as discretization step to convert the gene expression level to discrete values: 1 for over expressed and 0 for under expressed. The difficulty of this method lies in the preprocessing step of gene expression levels discretization, knowing that the gene level is discretized without taking into account the other genes levels.

Extraction of Sequential Patterns. This type of patterns introduces the notion of order in searching association between the itemsets: each itemset of a sequence represents a set of events that appear at the same time while the different itemsets of a sequence are associated with different times. For example this sequence <(A, B) (C)> means that we have A and B simultaneously then later C. The extraction of sequential patterns consists in searching the maximum sequences (not included in another sequences) whose support is greater or equal to a minimum threshold minsup defined by the user. Salle et al. [10], extract sequential discriminant patterns from DNA chips by demonstrating the order relationship according to gene expression data; the patterns extracted can be in this form <(GeneA) (GeneB) (GeneC)> , this is a frequent sequence in a particular class (for example in young people class). In [11], an approach using novel sequential pattern mining to extract regulatory modules (co-regulated genes and their co-regulating conditions) from microarray data has been adopted. GSP, PrefixSpan, SPAM, SPADE are some examples of algorithms of sequential patterns extraction. The CloSpan is a closed sequential patterns extraction algorithm, it comes to reduce the number of extracted frequent patterns.

Extraction of Gradual Patterns.   Another type of itemsets having this form (the more A, the more B) is called Gradual patterns, this method have been used in the command and the control systems, but recently it becomes useful in data mining, especially when it applicable on numerical and large data like biological data, sensor readings or data streams. Many algorithms extract automatically gradual patterns [12–16]. A gradual rule that can be extracted from gene expression data has this following form (*the more gene G1 is expressed, the less gene G2 is expressed*), it can be evaluated using the support and the confidence measures.

These gradual rules are the extension of association rules but they differ by the type of data to be analyzed and by the type of correlation to be extracted. They are applicable on numerical or fuzzy data and they extract the co-variation of attributes. Contrary to the classic patterns, whose correlation is expressed for each object by checking the absence or the presence, the variations of gradual patterns are measured by taking into account all the data. Hullermeier [17] used regression based approach. In 2007, Berzal et al. [12] used for the first time the gradual patterns for data mining and adopted an approach based on classical association patterns using Apriori. GRITE [13] is an efficient algorithm but can only extract gradual patterns and rules from datasets that contains hundreds of attributes and can't be applicable on the gene expression data having thousands of lines. Mining these rules is a difficult task especially with real databases that have a large size. To solve this problem, some parallel approaches comes to exploit the multicores processors of computer in order to reduce the computation runtimes such as the parallel PGP-mc approach [15]. Trong et al. [14] mined for the first time, the gradual patterns from the gene expression data using PGLCM, a parallel algorithm for mining closed gradual itemsets, that represents the parallelized version of GLCM algorithm. Paraminer [16] is a generic, efficient and parallel mining algorithm of different pattern (frequent patterns, frequent relational graphs, and gradual itemsets). It comes to overcome the limits of other mining algorithms especially for the gradual pattern mining problem, it can compete the ad-hoc algorithms and it is available as an open source software.

2.3 GRN Inference

The GRN inference can be defined in system biology as the process of the GRN identification, it can be considered as data mining task. Several approaches are used to infer gene regulation networks from the expression data [18, 19], the most popular are:

- *Information-theoretic approaches:* based on the pairwise distances or similarities: The Correlation Coefficient and the mutual information. The correlation: is measured between each pair of genes, then by using a threshold the significant relationships can be defined to reconstruct the network. So This approach requires the definition of a threshold, in addition it does not make it possible to distinguish the direct relation of the indirect one between two genes. To overcome this problem, some works use mutual information to evaluate the degree of dependence between genes. The *REVEAL algorithm (REVerse Engineering Algorithm)* and *the ARACNe (Algorithm for the Reconstruction of Accurate Cellular NEtworks)* are based on mutual information to infer networks. The Conditional mutual information can overcome this limit but it fails to specify the direction of regulation in a network.
- *Boolean networks:* these networks are simply and easy applied but they provide only a qualitative description of system. The gene can be either active or inactive. The gene interactions are modeled using Boolean logic functions.
- *Bayesian networks*: it is a probabilistic approach for modeling gene regulatory network, it allows us to have an idea about the most probable state of gene y, if a gene x is up. The feedbacks loops can't be modeled by the Bayesian networks but the dynamic Bayesian networks comes to overcome this issue.
- *Differential equation models:* deterministic approach that provide a quantitative description of the network

It is possible to combine two methods to overcome the limits of previous approaches, for example in [19] the authors combine Bayesian networks and differential equations to infer gene regulatory networks. Huang et al. [6] proposed two algorithms for learning gene regulatory networks from microarray expression data: modified Bayesian learning algorithm and modified algorithm of association rules. In [20] Liu et al. use the conditional mutual information to construct a GRN then this GRN is decomposed into local networks and the Bayesian network is employed. The DNA chip or microarray measures the expression levels of several genes in a few conditions, while for network inference, it is preferred to have few relevant genes under many conditions.

3 Genetic Interactions Extraction Method

To define the GRN, we need to extract knowledge about genetic interactions from microarray available in databases such as: Gene Expression Omnibus GEO and ArrayExpress. This technology provides dynamic information about thousands of genes in a genome. Hence, it is necessary to do preprocessing analysis or pretreatment step, as shown in Fig. 1, to prepare and reduce the data for data mining method. The extraction of frequent patterns and sequential patterns are applied to binary data but the gradual patterns are applied to quantitative data without any discretization step. While we need to work on quantitative data to extract knowledge about the spatio-temporal dynamics of genetic interactions, we propose to use gradual patterns in our extraction approach of genetic interactions. So, our method which is based on the gradual patterns takes into account the qualitative aspect of the data, focusing on the co-variation of values. This method will reduce the complexity of the data because it directly kicks out all the not-differential expressed genes. The knowledge extracted in the form of rules is more comprehensible by the user, and provide an explicit representation of the regulation links. The GRN is represented by a list of rules and the interactions between genes are represented by rules of this form: gi* → gj*, * in {≥, ≤}.

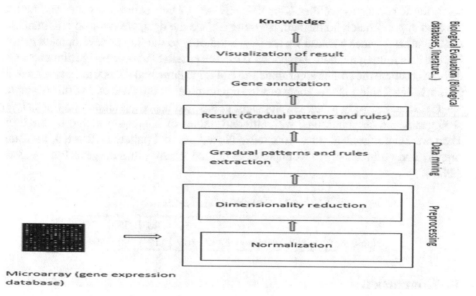

Fig. 1. Gradual rules from gene expression data - Extraction process

Our approach described in Fig. 1 integrates expression data and gene annotation (GO/TFBS enrichment analysis). In the annotation phase we use gene ontology annotation and transcription factor binding sites annotation TFBS to analyze the co-regulation of genes and to have more information about GRNs. The transcription factor is defined as a protein that binds to specific sites in the DNA to active or inhibit the expression of a gene. We can also integrate the expression data with chip-seq data (Next Generation Sequencing) to better understand the mechanisms of gene regulation. The in vivo binding data help to identify which genes regulated by which transcription factors so in our case this technology is useful to identify the direct links or direct relationships between proteins and target genes.

4   Analysis of Dataset

We are interested to the GRN of flower development of a model plant Arabidopsis Thaliana. The genes involved in the development of the flower have spatio-temporal expression profiles. So, we used the gene expression data of this plant, GSE5632 Dataset: 'AtGenExpress: Developmental series (flowers and pollen)', available in GEO database. This dataset contains 22 810 genes and 66 samples. We have done the pretreatment step to prepare and reduce the data for the data mining step: a GCRMA normalization and reduction of data are performed using R packages. The GCRMA normalization outputs log2-transformed expression values and corrects for non-specific binding. We start by filtering out uninformative data. We eliminate the duplicated genes because in some cases a gene is matched by the multiple probes due to the updates in the genome annotation. Furthermore, we removed the genes with low variance and non-expressed genes using the R package genefilter. This filtering step of non-expressed genes can reduce the data approximately to the half. To more reduce the data, we can also eliminate the photosynthesis genes because we are interested only to the flower development genes For the data mining method, we choose to use the parallel Paraminer algorithm because it is can compete the ad-hoc algorithms and can mine large real dataset to extract gradual patterns. The Table 1 shows some results of paraminer in our dataset. For this example, we are interested only to the development genes, we tested paraminer on our dataset (100 genes and 66 conditions) with different values of V (minsup) fixed by the user. Then we extract gradual rules from these closed gradual patterns extracted, by using support and confidence. An example of extracted gradual rule is (gene PI+ → gene AP3+).

Table 1. Gradual patterns extraction

V	30	28	25	20
Number of gradual patterns extracted	103	113	180	1046

5   Conclusion

In this paper, we have discussed some data mining approaches used to discover genetic knowledge then we have proposed to use the gradual rules to extract the co-variation and interactions between genes. Unlike others existing data mining approaches of extraction genetic knowledge, we focus on the qualitative and quantitative aspect of gene expression data to extract the correlation between genes by using gradual patterns. This is the first time that the method of gradual patterns extracts genetic knowledge to infer gene regulatory networks GRNs. Future works aim at modeling the GRN using the rules describing the relations between Arabidopsis thaliana genes, involved in the development of the flower. This approach allows us, not only to know the co-regulated genes but also to know the variations of the transcription factors and the target genes. Hence, this approach can give the biologists insights about the mechanistic functions of genes.

References

1. Gouider, M., Hamdi, I., Ben Ghezala, H.: A review: data mining and text mining tools in biological domain. In: Proceedings of the IBIMA 2016, 28th International Business Information Management Association (IBIMA), Sevilla, Spain, 9–10 November 2016, pp. 2737–2746 (2016). ISBN: 978-0-9860419-8-3
2. Jiang, D., Tang, C., Zhang, A.: Cluster analysis for gene expression data: a survey. IEEE Trans. Knowl. Data Eng. **16**(11), 1370–1386 (2004)
3. Ayadi, W., Elloumi, M.: Biclustering of microarray data. In: Algorithms in Computational Molecular Biology: Techniques, Approaches and Applications. Wiley Book Series on Bioinformatics: Computational Techniques and Engineering, pp. 651–664 (2011)
4. Agrawal, R., Imielinski, T., Swami, A.: Mining association rules between sets of items in large databases. In: Proceedings of the 1993 ACMSIGMOD International Conference on Management of Data, pp. 207–216. ACM Press, Washington, DC (1993)
5. Datta, S., Bose, S.: Mining and ranking association rules in support, confidence, correlation, and dissociation framework. In: Proceedings of the 4th International Conference on Frontiers in Intelligent Computing: Theory and Applications (FICTA) 2015, pp. 141–152. Springer, India (2016)
6. Huang, Z., Li, J., Su, H., et al.: Large-scale regulatory network analysis from microarray data: modified Bayesian network learning and association rule mining. Decis. Support Syst. **43**(4), 1207–1225 (2007)
7. Carmona-Saez, P., Chagoyen, M., Rodriguez, A., et al.: Integrated analysis of gene expression by association rules discovery. BMC Bioinform. **7**(1), 1 (2006)
8. Alagukumar, S., Lawrance, R.: A Selective analysis of microarray data using association rule mining. Procedia Comput. Sci. **47**, 3–12 (2015)
9. Jourdan, L.: Métaheuristiques pour l'extraction de connaissances: Application à la génomique. Doctoral dissertation, Université des Sciences et Technologie de Lille-Lille I (2003)
10. Salle, P., Bringay, S., Teisseire, M.: Motifs Séquentiels Discriminants pour les puces ADN. In: InforSID'09: 27ème Congrès Informatique des organisations et systèmes d'information et de décision, pp. 397–412, May 2009
11. Kim, M., Shin, H., Chung, T.S., et al.: Extracting regulatory modules from gene expression data by sequential pattern mining. BMC Genomics **12**(3), S5 (2011)
12. Berzal, F., Cubero, J.C., Sanchez, D., Vila, M.A., Serrano, J.M.: An alternative approach to discover gradual dependencies. Int. J. Uncertain. Fuzziness Knowl. Based Syst. (IJUFKS) **15**(5), 559–570 (2007)
13. Di Jorio, L., Laurent, A., Teisseire, M.: Mining frequent gradual itemsets from large databases. In: International Conference on Intelligent Data Analysis, IDA 2009 (2009)
14. Do, T.D.T., Laurent, A., Termier, A.: PGLCM: efficient parallel mining of closed frequent gradual itemsets. In: International Conference on Data Mining (ICDM), pp. 138–147 (2010)
15. Laurent, A., Negrevergne, B., Sicard, N. et al.: PGP-mc:: Towards a multicore parallel approach for mining gradual patterns. In: Database Systems for Advanced Applications, pp. 78–84. Springer, Heidelberg (2010)
16. Negrevergne, B., Termier, A., Rousset, M.-C., et al.: Paraminer: a generic pattern mining algorithm for multi-core architectures. Data Mining Knowl. Discov. **28**(3), 593–633 (2014)
17. Hullermeier, E.: Association rules for expressing gradual dependencies. In: Proceedings of the 6th European Conference on Principles of Data Mining and Knowledge Discovery, PKDD 2002, pp. 200–211. Springer (2002)

18. Bansal, M., Belcastro, V., Ambesi-Impiombato, A., et al.: How to infer gene networks from expression profiles. Mol. Syst. Biol. **3**(1), 78 (2007)
19. Kaderali, L., Radde, N.: Inferring gene regulatory networks from expression data. In: Computational Intelligence in Bioinformatics, pp. 33–74. Springer, Heidelberg (2008)
20. Liu, F., Zhang, S.W., Guo, W.F., Wei, Z.G., Chen, L.: Inference of gene regulatory network based on local bayesian networks. PLoS Comput. Biol. **12**(8), e1005024 (2016)

Exploring Location and Ranking
for Academic Venue Recommendation

Nour Mhirsi and Imen Boukhris[✉]

LARODEC, Institut Supérieur de Gestion, Université de Tunis, Tunis, Tunisia
nour.mhirsi@gmail.com, imen.boukhris@hotmail.com

Abstract. Publishing scientific results is extremely important for each researcher. The concrete challenge is how to select the right academic venue that corresponds to researcher's current interest and without missing the deadline at the same time. Due to the huge number of academic venues especially in the field of computer science, it is difficult for researchers to choose a conference or a journal to submit their works. A lot of time is wasted asking about the conference topics, its host country, its ranking, its submission deadline, etc. To tackle this problem, this paper proposes a recommendation approach that suggests personalized upcoming academic venues to computer scientists that fit their current research area and also their interests in terms of venue location and ranking. The target researcher and his community current preferences are taken into consideration. Experiments demonstrate the effectiveness of our proposed rating and recommendation method and show that it outperforms the baseline venue recommendations in terms of accuracy and ranking quality.

Keywords: Personalized academic venue recommendation
Current preference analysis · Venue location · Venue ranking

1 Introduction

Recommender systems are helping users to deal with the flood of information. They are tools that automatically filter a large set of items in order to identify those that are most relevant to a user's interest [9].

Academic events (e.g., journals, conferences, symposiums, workshops, and seminars) play an important role in scientific communities since they provide higher visibility and greater impact. In the field of computer science, the number of academic events and venues has considerably increased in recent years. According to DBLP statistics[1], the number of published papers in conferences and workshops increased from 103,554 in 2000 to 1,971,559 in 2017. Besides, when a researcher reaches results, choosing the most appropriate conference or journal to get them published is crucial. Moreover, the review cycle can be time consuming and, if the paper is rejected because it is not a good fit, valuable

---

[1] http://dblp.uni-trier.de/statistics/recordsindblp.html.

© Springer International Publishing AG, part of Springer Nature 2018
A. Abraham et al. (Eds.): ISDA 2017, AISC 736, pp. 83–91, 2018.
https://doi.org/10.1007/978-3-319-76348-4_9

time can be lost. Generally, researchers learn of scholarly venues related to their research interests from limited sources: by word of mouth from laboratory members, departmental colleagues, and members of other scholarly communities; by conducting online searches (e.g., confsearch[2]) and reviewing the research articles returned by these searches; from venue rankings; or from publishers reputations. In the past, these approaches worked satisfactorily because relatively few related venues existed for any particular field. Today, however, researchers become acquainted with newly available and specialized venues only by spending considerable time browsing and evaluating. In this work, we present a personalized recommender system that aside the focus on the scientist's current interest in research topics, it take into account context information namely, the venue ranking and the location as relevant factors when providing suggestions. Whereas, say, an Asian researcher will prefer to publish in academic venue that will be hosted in Australia than in Africa for different reasons such as travel conditions or budget. Besides, a researcher that usually published papers in ranked A* conferences will not be interested in unranked conferences. To tackle these issues, this paper proposes a recommender system that suggests academic venues fitting the current interests of computer scientists. It is based on the venues attended by the target researcher's, his co-citers and co-authors. Moreover, preferences about academic venues location and rating are considered to make recommendation. The rest of the paper is organized as follows: Sect. 2, we discuss related work. In Sect. 3, we detail our approach for measuring an implicit rating for academic venues by monitoring researchers behavior. In Sect. 4, we expose the experimental results conducted on real world datasets. In Sect. 5, we present results and we discuss it. Section 6, concludes the paper and discuss future work in this area.

2   Academic Venue Recommendation: Related Work

In recent years, despite its importance, only few studies focused on methods for recommending scholarly events and venues. These latter can be divided into two categories according to context type: recommendation based on authors and recommendation based on papers content.

2.1   Recommendation Based on Authors

Authors in [1] developed an approach that recommended academic events based on a researcher's event participation history. A preliminary work was proposed in [2], where a hybrid recommender system suggesting conferences in computer science is proposed. Besides, authors in [3] clustered users on social networks and used the number of papers published in a venue by a researcher to derive the researcher's rating for that venue.

---

[2]   http://www.confsearch.org/confsearch/.

2.2 Recommendation Based on Papers Content

Approaches based on paper content such as topic and writing style, title and/or abstract were proposed in [4,5]. In [6], an approach based on personal bibliographies and citations was suggested. Authors in [7] developed a recommender system based on user changing interest while techniques to automatically compare and recommend conferences were implemented in [10].

In fact, there are several recommendation approaches but the most popular are content based and collaborative filtering (CF). In the case of academic venue suggestion, recommendations based on authors correspond to CF approaches while those based on papers content correspond to content based approaches. According to the literature, CF is considered to be the most popular and the widely used approach in this area [11–13]. Besides, with using CF solutions, we get more diverse and serendipitous recommendation [8]. This have motived us to propose a new approach that enhance the quality of recommended personalized upcoming academic venues in computer science according to the authors current interest. Indeed, the proposed personalized recommender system takes as input implicit ratings based on author's current preferences.

3 Personalized Recommender System for Computer Scientists

3.1 Overview

Unlike for movies where users are looking for unseen movies, a researcher may attend the same conference several times. This makes recommending academic venues different from the traditional recommendation process. Accordingly in this paper, a new approach that collects implicitly ratings from papers already published by researchers and then, suggests a list of appropriate upcoming venues whose submission deadlines are not outdated to a computer science researcher is proposed. Both conference location and ranking are taken into consideration to make recommendation by analysing target researcher preference location from attended venues.

It consists of an item based collaborative filtering recommender system that improves the quality of the serendipity list in how far recommendations may positively surprise researchers.

3.2 Current Personal Academic Venue Rating (CPAVR)

In order to predict implicitly researcher u rating to a specific venue v, we proposed a measure so-called Current Personal Academic Venue Rating (CPAVR).

We define $CPAVR_{u,v}$ for researcher u to venue v as a weighted sum as follows:

$$CPAVR_{u,v} = \sum_{i=1}^{5} NP_i \times (1 - \alpha) + L + R \tag{1}$$

Where:

- NP_i ($i = 1, \ldots, 5$) denotes the number of papers published by researcher u in the venue v in (current year-i).
  Note that only papers published in the five last years are considered. Indeed, we assumed that papers published in this period reflect the researcher current research area.
- The discount rate $(1 - \alpha)$ with ($\alpha \in [0, 1]$) as explained in Table 1 increases the impact of newly published papers. The main idea is to promote venues that the target researcher and his community have recently published in and recommend them to him.

Table 1. Discount operation

Year	α
Current year-5	0.8
Current year-4	0.6
Current year-3	0.4
Current year-2	0.2
Current year-1	0

- The weight L denotes the preference of researcher u to continent j.
  We consider the six following continents:

$$j \in \{Asia, \ North \ America, \ South \ America, \ Africa, \ Europe, \ Australia\}$$

It is defined as:
$$L = \frac{NPP_j + 1}{T + 6} \tag{2}$$

NPP_j reflects the number of papers that researcher u has published in venues hosted in countries of a continent j, and T denotes the total number of papers published by researcher u.
Here, we used *Laplace estimator* (*Add − One Smooting*) to estimate the weight L assuming that each unseen continent actually occurred once.

- The weight R denotes the preference of researcher u to venues rank k.
  We have five ranks defined as follows:

$$k \in \{A^*, \ A, \ B, \ C, \ unranked\}$$

It is defined as:
$$R = \frac{NPP_k + 1}{T + 5} \tag{3}$$

NPP_k denotes the number of papers that researcher u has published in venues rank k and T denotes the total number of papers published by researcher u.
As we did before with L, we also used *Laplace estimator* (*Add − One Smooting*) to estimate the weight R.

3.3   CPAVR Recommender Engine

The traditional collaborative based engine recommends items that are not yet being rated by the target user. Thus, if the target user likes a given book or a movie, the system will recommend him novel movies that are similar to the ones he has liked in the past. In the academic context by following the same process suggested venues will be those that he did not never attended before. This is a non-realistic and non-personalized recommendation. In fact, if a researcher has published in ISDA conference in 2016, he will be interested to attend the ISDA conference in 2017. Note that he may not be aware about ISDA 2017 since deadlines may considerably change.

To overcome the limitation of traditional recommender systems, we propose a personalized academic venue recommender engine as depicted in Fig. 1.

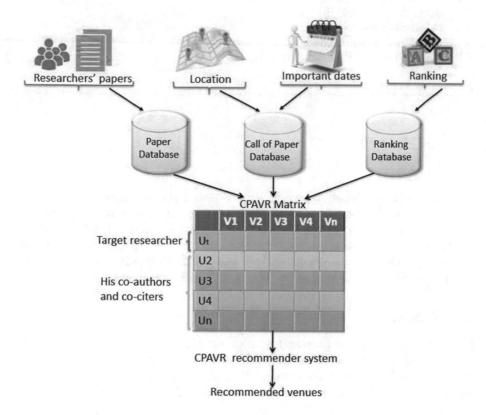

Fig. 1. CPAVR recommender engine

The recommendation process takes as input a matrix composed by the implicit preferences of the target researcher, denoted by U_t, and his co-authors and co-citers, denoted by U_c ($c \in [1, n], c \neq t$). Each entry in the matrix, denoted by $CPAVR_{u,v}$, represents a researcher's current personalized rating in a particular academic venue v.

As explained in Algorithm 1, L and R are firstly calculated for each researcher based on his current preferences. Then, after selecting NP_i, $CPAVR_{u,v}$ is calculated and CPAVR matrix is established. After that, the similarity between target user and his community members is determined. Finally, the top-N recommendation venues for the target researcher U_t is suggested in a descending order.

Algorithm 1. Personalized CPAVR recommendation algorithm

Input: Set of researchers (U_t and U_c ($c \in [1,n], c \neq t$)) and set of venues ($v \in V$)
Output: The list of recommended venues for the target researcher U_t
1: **For** U_t **and** U_c ($c \in [1,n], c \neq t$)
2:     select NP_i ($\forall i \in [1,5]$)
3:     calculate L using Eq. 2
4:     calculate R using Eq. 3
5:     calculate each $CPAVR_{u,v}$ using Eq. 1
6: **End For**
7:Top-N recommendations for U_t

---

4   Experiments

4.1   Datasets

To evaluate the proposed approach, experiments were conducted on the DBLP citation dataset[3] which consists of bibliographic data augmented by the citation relations between papers. It contains information about 2,084,055 academic papers. Information about computer science venues (venue's title, acronym, publisher, venue ranking) are extracted from the CORE Conference Portal[4]. We can distinguish five ranking categories (A*, A, B, C and unranked).

4.2   Baseline

In order to evaluate our proposed recommendation approach, we compared the method described in Sect. 3 with a baseline. Following the experimental process of [7], we used a Boolean recommendation as a baseline and compared it with recommendation for academic venues based on CPAVR implicit ratings.

4.3   Evaluation Metrics

Precision and Recall. Precision and recall measures are two popular metrics to evaluate the performance of a RS. Precision is used to evaluate how many

---

[3] http://arnetminer.org/DBLP_Citation.
[4] http://103.1.187.206/core/.

selected venues are relevant and Recall is used to evaluate how many relevant venues are selected. Those latters are defined as follows:

$$Precision = \frac{Relevant\_venues\_recommended}{Venues\_recommended} \tag{4}$$

$$Recall = \frac{Relevant\_venues\_recommended}{Relevant\_venues} \tag{5}$$

where:

- *Relevant_venues_recommended* corresponds to the number of venues in the recommended list in which a researcher participates in the next year.
- *Venues_recommended* corresponds to the number of venues recommended.
- *Relevant_venues* corresponds to the number of venues which a researcher takes part in the next year.

Normalized Discounted Cumulative Gain (NDCG). NDCG measures the performance of a recommendation system based on the graded relevance of the recommended entities. It varies from 0 to 1, with 1 representing the ideal ranking of the entities.

$$DCG_p = \sum_{v=1}^{p} \frac{2^{rel_v} - 1}{\log_2(1 + v)} \tag{6}$$

where:

- p is the maximum number of entities that can be recommended.
- rel_v is the graded relevance of the result at position v.

We measured the normalized discounted cumulative gain $NDCG_p$ as follows:

$$NDCG_p = \frac{DCG_p}{IDCG_p} \tag{7}$$

$IDCG_p$ corresponds to the maximum possible (ideal) DCG for a given set of queries, documents, and relevances.

5   Results and Discussion

In this section, we will compare the baseline boolean recommendation and the recommendations obtained from our proposed engine.

We used 70% of the data as a training set and 30% as a test set. In order to highlight the extent to which personalization can improve recommendations, we will compare results of boolean recommendation with our personalized recommendation based on CPAVR implicit rating. We calculated precision@5, recall@5 and NDCG@5 (the top 5 most similar researchers to the target researcher) and we also calculated precision@10, recall@10 and NDCG@10.

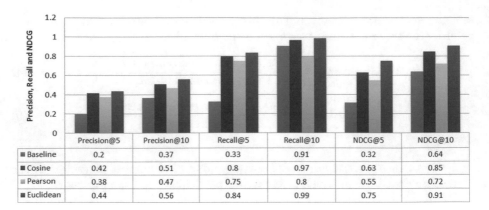

	Precision@5	Precision@10	Recall@5	Recall@10	NDCG@5	NDCG@10
■ Baseline	0.2	0.37	0.33	0.91	0.32	0.64
■ Cosine	0.42	0.51	0.8	0.97	0.63	0.85
▥ Pearson	0.38	0.47	0.75	0.8	0.55	0.72
■ Euclidean	0.44	0.56	0.84	0.99	0.75	0.91

Fig. 2. Comparison of the proposed algorithm with different similarities

Figure 2 illustrates the comparison of the proposed algorithm with different similarities. We can conclude that Euclidean distance achieved better results than Cosine and Pearson by improving precision@5 by 20%, recall@5 by 50% and NDCG@5 by 40% and also by improving precision@10 by 20%, recall@10 by 8% and NDCG@10 by 30%. Further, we can conclude that our proposed approach give a good results and that the CPAVR implicit ratings significantly improve Precision, Recall, and NDCG.

Besides, we compared our method with academic venue recommendation method based on authors [3] which clusters researchers according to their social relationships (the co-authorship network). As shown in Table 2, experimental results on the DBLP database demonstrate the importance of CPAVR implicit ratings and the effectiveness of our approach.

Table 2. Comparaison with existing recommendation method in term of precision

	Clustering − based [3]	CF based on CPAVR
Precision	0.35	0.51

As noticeable, our proposed method performs better than the clustering-based method with an improve of 16% in terms of precision.

Then, considering weights as venue ranking and location influence researcher's preference analysis and improves the quality of recommendations.

6   Conclusion

Obtaining recommendations from trusted sources is a critical component of the natural process of human decision making. As the academic venues are increasing in computer science field, selecting a relevant venue to the author is a challenging issue and researchers ask for the personalization of recommendations. In this paper, we have presented a personalized academic venue recommender system based on user's current interest with a focus on current academic venue location and ranking. Experimental results demonstrate the significance of our approach. As future work, we plan to integrate other weights to get more personalized recommendations.

References

1. Klamma, R., Cuong, P.M., Cao, Y.: You never walk alone: recommending academic events based on social network analysis. In: Proceedings of the First International Conference on Complex Science, Complex 2009, pp. 657–670 (2009)
2. Boukhris, I., Ayachi, R.: A novel personalized academic venue hybrid recommender. In: IEEE 15th International Symposium on Computational Intelligence and Informatics, CINTI, pp. 465–470 (2014)
3. Pham, M.C., Cao, Y., Klamma, R., Jarke, M.: A clustering approach for collaborative filtering recommendation using social network analysis. J. Univ. Comput. Sci. **17**(4), 583–604 (2011)
4. Yang, Z., Davison, B.D.: Venue recommendation: submitting your paper with style. In: Proceedings of the 11th International Conference on Machine Learning and Applications, ICMLA 2012, pp. 681–686 (2012)
5. Kang, N., Doornenbal, M.A., Schijvenaars, R.J.A.: Elsevier journal finder: recommending journals for your paper. In: Proceedings of the 9th ACM Conference on Recommender Systems, RecSys 2015, pp. 261–264 (2015)
6. Kucuktunc, O., Saule, E., Kaya, K., Catalyurek, U.V.: The advisor: a webservice for academic recommendation. In: Proceedings of the 13th ACM IEEE - CS Joint Conference on Digital Libraries, JCDL 2013, pp. 433–434 (2013)
7. Alhoori, H., Furuta, R.: Recommendation of scholarly venues based on dynamic user interests. J. Informetrics **11**(2), 553–563 (2017)
8. Ricci, F., Rokach, L., Shapira, B., Kantor, P.B.: Recommender Systems Handbook (2011)
9. Luong, H., Huynh, T., Gauch, S., Hoang, K.: Exploiting social networks for publication venue recommendations. In: International Conference on Knowledge Discovery and Information Retrieval, KDIR 2012, pp. 239–245 (2012)
10. García, G.M., Nunes, B.P., Lopes, G.R., Casanova, M.A., Paes Leme, L.A.P.: Techniques for comparing and recommending conferences. J. Brazil. Comput. Soc. (2017)
11. Bobadilla, J., Ortega, F., Hernando, A., Gutierrez, A.: Recommender systems survey. Knowl.-Based Syst. **46**, 109–132 (2013)
12. Park, Y., Park, S., Jung, W., Lee, S.G.: Reversed CF: a fast collaborative filtering algorithm using a k-nearest neighbor graph. Expert Syst. Appl. **42**(8), 4022–4028 (2015)
13. Su, X., Khoshgoftaar, T.M.: A survey of collaborative filtering techniques. Adv. Artif. Intell. **2009**, 1–19 (2009)

Designing Compound MAPE Patterns for Self-adaptive Systems

Marwa Hachicha[✉], Riadh Ben Halima, and Ahmed Hadj Kacem

ReDCAD, University of Sfax, B.P. 1173, 3038 Sfax, Tunisia
marwahachicha@gmail.com

Abstract. Self-adaptive systems are able to change their own behavior whenever the software or hardware is not accomplishing what it was intended to do. In this context, the MAPE (Monitoring, Analysis, Planning, Execution) control loop model has been identified as crucial element for realizing self-adaptation in software systems. Complex self-adaptive systems often exhibit several architectural patterns in their design which leads to the need of architectural pattern composition. In this paper, we focus on modeling and composing MAPE patterns for decentralized control in self-adaptive systems. We illustrate our approach using a case study example of the fall-detection ambient assisting living system for elderly people.

1 Introduction

Modern advanced systems are required to continuously perceive important structural and dynamic changes in their contexts, and autonomously react to such changes. Therefore, there is a need to have self-adaptive systems that can adapt their behavior in response to their perception. MAPE control loops [3] have been identified as crucial elements for realizing self-adaptation. When systems are large, complex, and heterogeneous, a centralized architecture that assigns the whole system control to a single block will inherently cause problems such as the single point of failure. Consequently, a decentralized approach based on multiple MAPE loops for self-adaptive systems is required.

Architecting and engineering software systems can benefit from the concept of pattern. Patterns provide a general reusable solution to a commonly occurring problem within a given context. In this context, Weyns et al. [6] proposed a set of patterns for describing multiple interacting MAPE loops for self-adaptive systems. They proposed two classes of patterns: the first class is based on a fully decentralized approach and comprises two patterns: coordinated control and information sharing patterns. The second class is based on a hierarchical distribution approach where higher level MAPE components control the subordinate MAPE control loops. It comprises three patterns: master/slave, regional planning and hierarchical control patterns. In our work, we have adopted these patterns to manage the adaptation in self-adaptive systems.

A. Abraham et al. (Eds.): ISDA 2017, AISC 736, pp. 92–101, 2018.
https://doi.org/10.1007/978-3-319-76348-4_10

In real world self-adaptive systems, recurring problems are complex and their solutions can be represented by patterns in complex forms that require the combination and the reuse of other existing architectural patterns. The MAPE patterns for decentralized control in self-adaptive systems proposed by Weyns et al. [6] are typically defined imprecisely using informal and simple graphical notation. On the other hand, these patterns do not fully enumerate all the possible decentralization patterns, so these patterns could potentially be combined in any number of ways [6].

In this paper, we propose an approach to compose the MAPE patterns for self-adaptive systems which is illustrated with a fall-detection application for elderly people. It consists in describing the compound pattern with patterns that include it and resulting from the application of merging operators and several composition rules. These composite patterns are useful when the systems requirements present features that can be found in different patterns that, brought together better describe the system, than using only one of them. In this paper, we focus on modeling structural features of the compound MAPE pattern for self adaptive systems.

The remainder of this paper is organized as follows. In Sect. 2, we present the different types of pattern composition. Then, we describe our proposed approach. Section 4 presents our case study example to illustrate and validate our approach. In Sect. 5, we describe the MAPE patterns that form the compound MAPE pattern. In Sect. 6, we present the resulting hybrid compound MAPE pattern after concretizing merging operators. Finally, the last section concludes the paper and gives future work directions.

2   Types of Pattern Composition

There are three types of pattern composition. Suppose that we have two patterns $P1$ and $P2$, the different types of composition of $P1$ with $P2$ are the followings:

– Connection of patterns: A compound pattern $P12$ can be formed by connecting two patterns $P1$ and $P2$ [4].
– Blend of patterns: A compound pattern $P12$ can be formed by fusing one or multiple components from Pattern $P1$ with another component from pattern $P2$ [4,7].
– Inclusion of patterns: When building a solution for the problem addressed by pattern $P1$, pattern $P2$ can be used to solve a sub-problem in $P1$ [4,7]. So, pattern $P2$ becomes the internal structure of pattern $P1$.

In this work, we apply two types of pattern composition: connection and blend of patterns.

3   Approach Overview

We propose an approach to compose the MAPE patterns for self-adaptive systems including two steps as shown in Fig. 1. The first step consists in describing

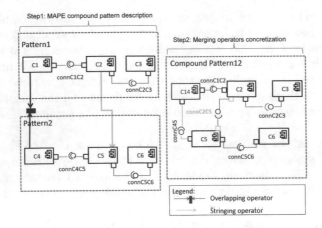

Fig. 1. Composition approach overview

the compound MAPE pattern as a set of unit patterns linked together by merging operators. The second step consists in concretizing merging operators in order to form the compound MAPE pattern. Such the work of [4,5] we propose using two types of merging operators: stringing and overlapping operators. The stringing operator links one component from pattern *P1* to another component from pattern *P2* as shown in the right corner side of Fig. 2. Therefore, a new connector is added to the pattern model. The overlapping operator merges two components from pattern *P1* and pattern *P2*. So, the two components became one component after their fusion as shown in the left corner side of Fig. 2.

The composition approach is based on a composition strategy. The composition is in pairs and incremental. In the first increment, we combined two patterns (*P1* with *P2*). In the next increment, the following pattern (*P3*) is combined with the obtained compound pattern (*P12*). In the last increment, the obtained compound pattern from the previous increment is combined with the final pattern (*Pn*). As presented in Fig. 1, to obtain the compound pattern *Pattern12* we perform two steps. In the first step we describe the compound MAPE pattern using two merging operators. An overlapping operator links the two components *C1* and *C4*. A stringing operator links the source component *C2* with the target component *C5*. In the second step, the merging operators described in the first step are concretized. Therefore, the components (*C1* and *C4*) are merged with the overlapping operator to the component *C14*. Additionally, the two connections (*connC1C2* and *connC4C5*) are preserved in the compound pattern. The stringing operator is concretized by adding a new connection (*connC2C5*) to the compound pattern *Pattern12* to describe the communication between the two components: *C2* and *C5*. Also, a new connection port is added to the components *C2* and *C5*.

We note that the overlapping operator rules have more priority than the stringing operator rules. In other words, we apply the overlapping rules and then the stringing operator rules.

Our composition approach applied to MAPE patterns for self-adaptive systems produces a new compound pattern that preserves the direction of the communication between the different MAPE components and the order of the MAPE control loop process. To design the MAPE patterns for self-adaptive system, we use our proposed UML profile that extends UML component diagram for modeling MAPE patterns. Fore more details about our proposed UML profile, we refer readers to our previous work in [1, 2].

In the two following sub-sections, we present the merging operators and their corresponding composition rules that we propose.

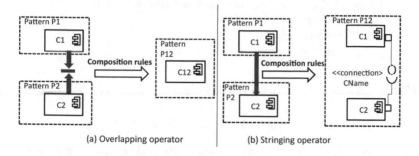

(a) Overlapping operator   (b) Stringing operator

Fig. 2. Overlapping and stringing operators description

3.1   Stringing Operator Transformation

The stringing operator [4] is concretized in the compound MAPE pattern if source and target components are in the first or in the second class as shown in the algorithm in below. The first class comprises the connected MAPE components having the same type: they can be either two monitors, two analyzers, two planners, two executors, two managed elements, two managers, etc.

The second class comprises the connected MAPE components having different types: they can be a monitor and an analyzer, an analyzer and a planner, a planner and an executor, etc.

The stringing operator is transformed to a new connection (line 6, line 11) to relate source and target components. Consequently, new component ports are also added to the source (line 4, line 9) and target (line 5, line 10) components in the compound pattern [4]. We classify the different connections between MAPE components into two main classes. The first class comprises connections between the components having the same type such as: *MonitoringInf* which represents connections between two monitor components, *AnalysisInf* which represents the connection between two analyzer components, etc.

The second class comprises connections between MAPE components having different types such as: *Monitoringdata* which represents communication between probes to monitor components in order to collect sensed data from the local context, *Symptom* represents communication between the monitor and analyzer components, *RFC* represents the communication between analyzer to planner components in order to send request for change if there is a possible degradation, etc.

1: Input=source, target, conc1, conc2
2: conc1 ∈ { symptoms, ...} and conc2 ∈ { MonitoringInf, ..}
3: **if** source=target **then**
4:     AddNewPort(source)
5:     AddNewPort(target)
6:     Addconnection(source, target, conc1)
7: **else**
8:     **if** source ≠ target and source.CanConnect(target) **then**
9:         AddNewPort(source)
10:        AddNewPort(target)
11:        Addconnection(source, target, conc2)
12:    **end if**
13: **end if**

3.2   Overlapping Operator Transformation

The overlapping operator [4] is concretized if merged components have the same type as shown in the algorithm in below. They can be either two monitors, two analyzers, two planners, two executors, two probes, two effectors, etc. The concretization of the overlapping operator is transformed to a new component having a new name (line 4 and line 5) which preserve all the features of the source and the target components. The new component obtained after fusion must preserves all components ports (line 6) from the source and the target components. Also, via these components ports all connections (line 7) of the source and target components must be preserved.

1: Input=source, target, newName
2: output=mergedComponent
3: **if** source=target **then**
4:     mergedComponent ⟵ fusion(source,target)
5:     mergedComponent.Name=newName
6:     mergedComponent.Ports ⟵ source.Ports ∪ target.Ports
7:     mergedComponent.connections ⟵ source.connections ∪ target.connections
8: **end if**

4   Case Study Example: The Fall Detection Ambient Assisting Living System

Automatic fall detection system based on a set of interconnected sensors and devices has the potential of improving the health care situation of elderly people. We propose a self-adaptive fall detection system composed of sensors that detect the movement of elderly people and measure the battery level of devices. This intelligent system can perform several actions when a fall is detected (No movement) such as triggering alarm and increasing the monitoring frequency to allow caregivers having more information about the criticality of the accident. Moreover, the self-adaptive fall detection system can change the routing protocol of the sensor node or triggering an alarm when the battery level reaches a low level.

The master/slave pattern proposed in [6] can be applied in our case study. It contains a base station that analyzes and plans adaptation and also communicates with the major blocks (Managed Element and slaves) in the system. However, centralizing adaptation in the base station can create a single point of failure for the system. If the base station fails, then the whole system will fail. Consequently, a decentralized approach is required. In this context, the coordinated control pattern could be applied to overcome this problem. In this case, the managed system (fall detection system) can be divided into several managed elements. Besides, each MAPE control loop can coordinate its operation with the corresponding peers of other loops. However, the coordinated control pattern has some disadvantages. The cost of reaching a consensus about a suitable adaptation may be high since it may lead to sub-optimal adaptation decisions and actions, regarding the overall system viewpoint. Therefore, we propose combining both of the master/slave and coordinated control patterns, to have an optimal solution. The composition of these two patterns leads to the definition of an hybrid MAPE pattern where the control of the adaptation of the managed element can be distributed or performed by a central component named the Master having a global supervision.

5 MAPE Patterns Description

In this section, we present the first step of our approach.

5.1 Master/Slave Pattern Modeling

The Master/Slave pattern organizes the adaptation logic as a hierarchy between one centralized master component and multiple slave components [6]. Consequently, the entities that make up the architecture of a master/slave pattern are *Master* and *Slave*. They represent a sub-class of UML package meta-model. There is a single instance of the master element containing a P and an A component, and there can be an arbitrary number of instances of slave elements with an M and an E component. *Monitor*, *Analyzer*, *Planner* and *Executor* extend the UML component meta-model. The slave has one or several *Probe* and *Effector*. The *Probe* makes measurements about a *ContextElement* with the purpose of sensing a specific variable of interest during run-time. The *Effector* is configured to affect the changes needed to alter the target system behavior according to the adaptation needs. We noticed that each MAPE component contains three sub-components, namely a receiver, a processor and a sender.

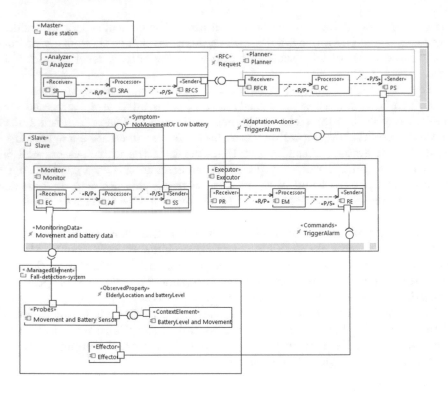

Fig. 3. Modeling of the master/slave MAPE pattern

- The *Sender* sub-component publishes symptoms, request for change (RFC), plans, etc.
- The *Processor* sub-component aggregates and filters the events, analyzes the data, processes the plans, etc.
- The *Receiver* sub-component receives information and monitoring data.

The *Receiver* and *Processor* are connected through the *R/P* dependency. The *Processor* and *Sender* are connected through the *P/S* dependency. We present the model describing the master/slave pattern applied to the fall detection system in Fig. 3.

5.2   Coordinated Control Pattern Modeling

In the coordinated control pattern, a MAPE control loop is associated with each part of the managed system being under its direct control [6].

The main entities that constitute the architecture of a coordinated control pattern are the *Manager* and the *Managed Element*. They represent a sub-class of UML package meta-model. The managed element contains effector and probe components to perceive and affect the environment. The manager contains all the MAPE components. Each MAPE component can interact with its peers to share particular information or coordinate adaptation actions. For example, the analyzer components exchange information to make decisions about the need for an adaptation. Figure 4 presents the coordinated control pattern applied to our case study.

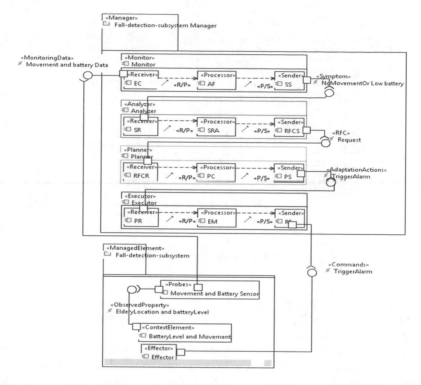

Fig. 4. Modeling of the coordinated control MAPE pattern

6   Merging Operators Concretization

To obtain the hybrid compound pattern, master/slave and coordinated control patterns are connected together through merging operators (stringing and overlapping operators). More specifically, by applying the composition approach, we add three merging operators to our pattern model. Figure 5 describes the hybrid compound pattern obtained after the enforcement of merging operators and the application of composition rules. The enforcement of merging operators is based on the following actions:

1. The component *ManagedElement* of the coordinated control pattern (Fall-detection-subsystem) is merged using the overlapping operator with the component *ManagedElement* of the master/slave pattern (Fall-detection-system)

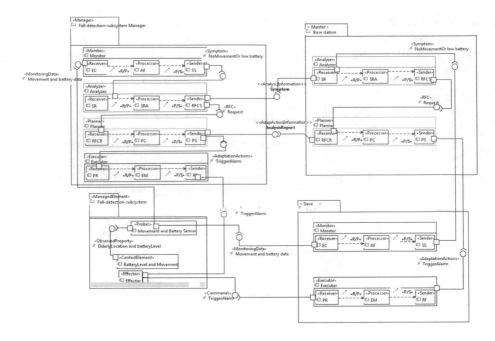

Fig. 5. Modeling of the Hybrid MAPE pattern

to obtain the component *Fall-detection-subsystem*. Since these two components have the same type and also the same internal structure, the overlapping
operator can be applied.

2. The analyzer component of the manager element in the coordinated control
   pattern is related to the analyzer component of the master element in the
   master/slave pattern. A stringing operator whose source is the analyzer in the
   coordinated control pattern and its target is the analyzer in the master/slave
   pattern, is added. The concretization of the stringing operator leads to the
   addition of new component ports and a new connection (*Symptom*) to form
   communication between the two analyzer components as shown in Fig. 5.

3. The planner component of the manager element in the coordinated control
   pattern is related with the planner component of the master element in the
   master/slave pattern. A stringing operator whose source is the planner in the
   coordinated control pattern and its target is the planner in the master/slave
   pattern, is added. The enforcement of the stringing operator leads to the
   addition of new component ports and a new connection (*AnalysisReport*) to
   build communication between the two planner components as shown in Fig. 5.
   The application of the two stringing operators allows the analyzer or planner
   components of the manager element in the coordinated control pattern to
   exchange information with the analyzer or planner component of the master
   element in the master/slave pattern in order to choose the suitable adaptation
   actions based on a global view of the system. So, coordinating their operations
   with the master avoid making sub-optimal adaptation.

7 Conclusion and Future Work

Through this paper we proposed an approach to describe and compose MAPE patterns for self-adaptive systems. We obtained the resulting compound MAPE pattern by the enforcement of merging operators. We illustrate our composition approach using the fall detection system case study example to specify the hybrid MAPE pattern. One of our future research studies is to formally specify the compound MAPE patterns to verify some properties using the Event-b method.

References

1. Hachicha, M., Dammak, E., Halima, R.B., Kacem, A.H.: A correct by construction approach for modeling and formalizing self-adaptive systems. In: 2016 17th IEEE/ACIS International Conference on Software Engineering, Artificial Intelligence, Networking and Parallel/Distributed Computing (SNPD), pp. 379–384, May 2016
2. Hachicha, M., Halima, R.B., Kacem, A.H.: Modeling, specifying and verifying self-adaptive systems instantiating MAPE patterns. Int. J. Comput. Appl. Technol., 57 (2018, to appear)
3. Kephart, J.O., Chess, D.M.: The vision of autonomic computing. Computer **36**(1), 41–50 (2003)
4. That, M.T.T., Sadou, S., Oquendo, F., Borne, I.: Composition-centered architectural pattern description language. In: Proceedings of the 7th European Conference on Software Architecture, ECSA 2013, pp. 1–16. Springer, Heidelberg (2013)
5. Tounsi, I., Kacem, M.H., Kacem, A.H., Drira, K.: An approach for SOA design patterns composition. In: 2015 IEEE 8th International Conference on Service-Oriented Computing and Applications (SOCA), pp. 219–226, October 2015
6. Weyns, D., Schmerl, B., Grassi, V., Malek, S., Mirandola, R., Prehofer, C., Wuttke, J., Andersson, J., Giese, H., Goschka, K.M.: On patterns for decentralized control in self-adaptive systems. In: Software Engineering for Self-adaptive Systems II: International Seminar, Dagstuhl Castle, Germany, Revised Selected and Invited Papers, 24–29 October 2010, pp. 76–107. Springer, Heidelberg (2013)
7. Zimmer, W.: Relationships between design patterns. In: Pattern Languages of Program Design, pp. 345–364. ACM Press/Addison-Wesley Publishing Co., New York (1995). http://dl.acm.org/citation.cfm?id=218662.218687

CRF+LG: A Hybrid Approach for the Portuguese Named Entity Recognition

Juliana P. C. Pirovani[✉] and Elias de Oliveira

Programa de Pós-Graduação em Informática,
Universidade Federal do Espírito Santo (UFES), Vitória, ES 29060-970, Brazil
juliana.campos@ufes.br, elias@lcad.inf.ufes.br

Abstract. Named Entity Recognition is an important and challenging task of Information Extraction. Conditional Random Fields (CRF) is a probabilistic method for structured prediction, which can be used in the Named Entity Recognition task. This paper presents the use of Conditional Random Fields for Named Entity Recognition in Portuguese texts considering an additional feature informed by a Local Grammar. Local grammars are handmade rules to identify named entities within the text. Moreover, we also present a study about the boundaries of CRF's performance when using a result coming from any other classifier as an additional feature. Two well-known collections in Portuguese were used as training and test sets respectively. The results obtained outperform results of state-of-the-art systems reported in the literature for the Portuguese.

Keywords: Named Entity Recognition · Conditional Random Fields
Local Grammar

1 Introduction

The automatic identification and classification of entities in texts, known as Named Entity Recognition (NER), is an important task in Information Extraction (IE). According to [1], identifying entities such as persons, organizations, places and time expressions is a fundamental step of preprocessing tasks in IE for several other applications (e.g., relation and event extraction).

NER is not a simple task. Several categories of named entities (NEs) are written similarly and they appear in similar contexts. For example, persons and places names begin with a capitalized letter, as well as temporal expressions and values contain numbers. In addition, the same NE can be classified into different categories depending on the surrounding context and some entities do not appear even in large training sets. Dictionaries are not always useful since, for example, a list of person names is growing every day [2].

The Message Understanding Conference (MUC-6) [3] included the NER task for the first time over English text, in 1995, carrying out a joint assessment of

© Springer International Publishing AG, part of Springer Nature 2018
A. Abraham et al. (Eds.): ISDA 2017, AISC 736, pp. 102–113, 2018.
https://doi.org/10.1007/978-3-319-76348-4_11

the area. Several similar assessments have emerged later such as the CoNLL [4] and the HAREM [5,6]. HAREM was a joint assessment for the Portuguese and the annotated corpora used in the First and Second HAREM[1], known as the Golden Collections (GC), have served as a golden standard reference for NER systems in Portuguese.

HAREM differs from other similar joint assessments in two aspects [6]: the classification of an NE depends only on its use in context and more than one classification can be assigned to an NE. Moreover, the HAREM classifies 10 categories of NEs (Person, Place, Organization, Value, Time, Event, Abstraction, Work, Thing and Other) that have a total of 34 types and 17 subtypes, while the MUC classifies 5 and CoNLL classifies only 4. Thus, the HAREM presents a more demanding task and, therefore, the performance values obtained using its reference data sets are still lower compared to the others [5].

NER systems can be developed using the following approaches: linguistics [7], machine learning [8] or hybrid [9]. In the linguistics approach, rules in which NEs can appear are manually identified and constructed, that is, the rules leads to a detection of NEs. In the machine learning approach, systems learn to identify and classify NEs from a training corpus. Describing the rules of the linguistic approach requires human expertise and manual efforts. However, systems based on machine learning depend on many annotated corpora for training purposes and do not consider human expertise to capture the rules that do not appear in the examples of the annotated corpus used for training. Thus, the human expert needs to be involved to supply these rules. The hybrid approach combines both previous methods.

This work seeks to explore the potential of both linguistics and machine learning approaches by constructing a hybrid system for NER in Portuguese. The presented strategy, CRF+LG, combines a labeling obtained by a Conditional Random Fields (CRF) [10], with a term classification obtained from Local Grammars (LGs) [11]. LGs are one means of representing rules of the linguistics approach. We apply LGs to perform the pre-labeling by capturing general evidence of NEs in texts and the CRF performs sequential labeling using this pre-labeling. The pre-labeling is sent to the CRF together with other features of the word as its grammatical class and can be seen as a suggestion for the CRF.

This paper is organized in 5 sections. Section 2 presents a literature review while the Sect. 3 presents the methodology used in this work. The results of CRF+LG and a study about the boundaries of CRF's performance combined with another classifier are presented in Sect. 4. Section 5 presents conclusions and future works.

2   The Literature Review

The LG formalism for identifying person names in Portuguese was used in [7]. The authors built an LG for the book Senhora (a novel by José de Alencar)

---

[1] http://www.linguateca.pt/HAREM/.

from linguistic studies on the text of this book. During the linguistic study, they observed the context to the right and left of the person names in the book. Later, they applied this exact LG to articles in a local newspaper of the Esprito Santo called A Tribuna. The goal was to observe the appropriation of an LG built from one context to another. While this is possible, some adaptations were necessary such as including a rule to identify specificative appositives and a rule to identify compound names in a dictionary. The study confirmed that the automatic identification of names is corpus dependent.

Some approaches (e.g., [8,12]) have used machine learning techniques such as Hidden Markov Models (HMM), Transformation-Based Learning (TBL), Support Vector Machines (SVM), Naive Bayes and Decision Table for NER over Portuguese texts. However, the systems presented in [13,14] achieved the best results for the 10 categories of the HAREM to date using Conditional Random Fields (CRF) and CharWNN Deep Neural Network (DNN).

The CRF for Portuguese NER was used by [13] to identify and classify the 10 categories of HAREM NEs. The BILOU [15] notation, the corpus annotated with part-of-speech tags (POS-tagging), the HAREM-defined categories and a feature vector are used as input for the training phase. In the testing phase, the HAREM-defined categories are removed. The HAREM corpora was used for training and testing. The NERP-CRF system proposed by [13] achieved the best Precision compared to systems of the Second HAREM when using the GC of the First and Second HAREM as training and test sets respectively.

NERP-CRF was one of the four tools (Language Tasks, PALAVRAS, Freeling and NERP-CRF) used to recognize NEs in Portuguese texts compared in [16]. They performed experiments over the HAREM corpora while considering only person, place and organization categories since these are the only ones present in all chosen tools. The results reveal the advantages of different tools according to different classes of NEs. For instance, Language Tasks and PALAVRAS achieved better performance for the person class. NERP-CRF showed higher degrees of Precision and the best performance for the Organization class.

A language-independent system was proposed by [14]. The system is based on the CharWNN Deep Neural Network (DNN) [17], which uses word-level and character-level representations to perform sequential classification. The approach was compared to the ETL_{CMT} system, an ensemble method based on Entropy Guided Transformation Learning (ETL), using the GC of the First HAREM as training set and the MiniHAREM as the test set. CharWNN outperforms the ETL_{CMT} system in both total (10 categories of HAREM) and selective (categories Person, Place, Organization, Time and Value) scenarios.

This paper aims to perform the NER for the 10 categories of the HAREM using CRF as it was carried out in [13]; however, the preprocessing of the texts was performed differently and an initial information about the label of each word was obtained by LGs and added to the feature set sent to the CRF training phase.

This work also differs from those presented in [18], that performs a combination of K-Nearest Neighbors (KNN) and CRF for English NER in tweets. We combined a rules-based approach and we do not used gazetteers or dictionaries.

To the best of our knowledge, there is not yet a work that combines LG and CRF for NER in Portuguese.

3   The Methodology

In this work, the GC of the First and Second HAREM were used as training and test sets respectively. Both GC have 129 texts written in Portuguese of various genres such as technical, political, journalistic, emails and interviews.

Figure 1 presents an overview of the methodology used for training.

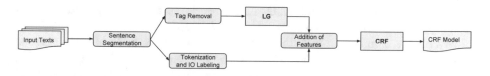

Fig. 1. Overview of the methodology used for training

Initially, each input file is splitted into sentences. The sentence segmentation was performed by the tool Unitex[2], a free software that allows the preprocessing of texts, creation and application of LGs, among other NLP tasks. Unitex uses LGs to describe the different ways that indicate the end of a sentence. For this work, the LG that performs sentence segmentation in Unitex was changed to not split the sentences in a colon (:) and a semicolon (;), thus maintaining the complete sentence for the supervised learning of the CRF. This flexibility is a strength of this tool.

The training files contain the annotations (tags) of the NEs. So it is necessary to remove them for LG application. After removing the tags of the already segmented files, the LG is applied over these texts and the NEs tags identified by LG are placed in.

On the other hand, the segmented files are tokenized using the OpenNLP[3] library. This library performs common NLP tasks. In order to represent the NER as a sequence labeling problem, a label must be assigned to each text token. Several notations can be used to delimit NEs and identify tokens in text, but the IO notation [15] was chosen because it presented better results in previous tests performed during this work. In addition, [19] presented better results with the IO notation compared to BILOU in the GC of the Second HAREM.

The IO notation is used as follows: all tokens which are part of the NE are then labeled with I (Inside) and all other tokens with O (Outside). In this case, the class of the NE is also mentioned in label I. For example, IO notation for the sentence *Meu pai é Gabriel Raimundo da Silva* is (Meu O) (pai O) (é O) (Gabriel I-PERSON) (Raimundo I-PERSON) (da I-PERSON) (Silva I-PERSON) (. O).

---

[2] http://unitexgramlab.org/.

[3] http://opennlp.apache.org/.

Next, several features are added for each token in the files, including the NE label previously assigned by LG. These features are used during supervised learning of the CRF prediction model.

The methodology used for testing is similar, but the input files do not have the NEs tags. In addition to the files containing the tokens and features, the CRF receives the previously trained model to predict a label for each token.

3.1   Local Grammars (LG)

Local Grammars [11] are one means of representing the contextual rules of the linguistics approach. An LG created in Unitex is represented as a set of one or more graphs. The LG built manually in this work consists of 10 graphs, one for each of the NEs categories considered by HAREM.

We observed in the training file in which context each type of NE appeared and which words could somehow indicate the existence of NE to construct each graph. We observed that, for example, words with the first letter capitalized preceded by the preposition *em* (in) were labeled as Place. Moreover, we observed that NEs of the Abstraction category are preceded by words such as *denominada* (denominated), *chamava* (called), *professor de* (teacher of), etc.

Thus, the graphs created capture some simple heuristics to the recognition of NEs in the training set. An example of rule in the graph created for the Event category is presented in Fig. 2.

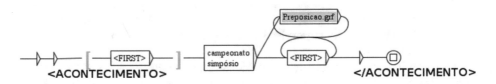

Fig. 2. Example of rule in the graph that recognizes the Event category

This graph recognizes *Campeonato* (Championship) or *Simpósio* (Symposium) followed by words with the first letter capitalized, as identified by the code <FIRST> in Unitex dictionaries. <FIRST> between [and] before the node containing *Campeonato* and *Simpósio* indicates that these words also need to have the first letter capitalized to be recognized. Among words with the first letter capitalized, prepositions may appear whose recognition has been previously detailed in graph Preposicao.grf included as subgraph. Examples of sentences recognized by the graph include *Campeonato Distrital de Seniores* and *Simpósio da Vida*. The identified events will appear between the tags <ACONTECIMENTO> (<EVENT>) and </ACONTECIMENTO> in the concordance file containing the list of occurrences identified.

3.2   Conditional Random Fields (CRF)

Conditional Random Fields (CRF) is a probabilistic method for structured prediction proposed by [10], which has been successfully used in several NLP tasks,
including NER. It is used to model the dependencies among random variables
that can be represented as an undirected graph.

Let $X = (x_1, x_2, ..., x_n)$ be a sequence of words in a text, we want to determine
the best sequence of labels $Y = (y_1, y_2, ..., y_n)$ for these words, corresponding to
the categories of NEs. The CRF models a conditional distribution $p(Y|X)$ that
represents the probability of obtaining the output Y given the input X.

In this work, we used a linear-chain CRF that models the output variables Y
as a sequence. According to [20], a linear-chain CRF is a conditional distribution
that takes the form shown in Eq. 1:

$$p(y|x) = \frac{1}{Z(x)} \prod_{t=1}^{T} \exp \left\{ \sum_{k=1}^{K} \theta_k f_k(y_t, y_{t-1}, \mathbf{x}_t) \right\} \tag{1}$$

where Z(x) is a normalization function given by Eq. 2:

$$Z(x) = \sum_{y} \prod_{t=1}^{T} \exp \left\{ \sum_{k=1}^{K} \theta_k f_k(y_t, y_{t-1}, \mathbf{x}_t) \right\} \tag{2}$$

$F = \{f_k(y_t, y_{t-1}, \mathbf{x}_t)\}_{k=1}^{K}$ is a set of feature functions that must be defined
according to the problem. An example is a function which takes the value 1 when
the word is written in uppercase, its label is Organization and the previous label
is Other and 0 otherwise. $\theta = \{\theta_k\}$ is a vector of weights that must be estimated
from the training set. The weights depend on each feature function and the more
discriminating the function, the higher its computed weight will be.

A standard way of estimating the vector of weights is maximum likelihood
and a way of predicting the most probable assignment Y* for a given input X is
the Viterbi algorithm [20]. The MALLET[4] toolkit was used in this work to infer
the CRF model and then apply this model to label the test set.

3.3   Addition of Features

In this step, several features are added to the tokenized file containing the word
and its label. The features used were the same proposed by [13], in addition
to that feature corresponding to the label assigned by LG. The feature set is
presented in Table 1.

The POS-Tagging of a word corresponds to its grammatical class and it was
also assigned by the OpenNLP library. When a word does not have one of the
previous words ($p - 1$ or $p - 2$) or posterior ($p + 1$ or $p + 2$), the corresponding
feature values are "null". Table 2 presents an example of a vector of features.

---

[4] http://mallet.cs.umass.edu/.

Table 1. Feature set assigned to each token

Features	Description
word	current word (position p)
tag	POS-Tagging of the word corresponding to its grammatical class
cap	if the word is composed of only capital letters, only lowercase or mixed
ini	if the word starts with uppercase, lowercase or symbols
simb	if the word is composed of symbols, digits or letters
prevW, prevT, prevCap	word, tag and cap for the word in position $p - 1$
prev2W, prev2T, prev2Cap	word, tag and cap for the word in position $p - 2$
nextW, nextT, nextCap	word, tag and cap for the word in position $p + 1$
next2W, next2T, next2Cap	word, tag and cap for the word in position $p + 2$
tip	label assigned by LG to the word

Table 2. Example of vector of features to the Gabriel token in sentence *Meu pai é Gabriel Raimundo da Silva*

Token	Vector of features	IO Notation
Gabriel	word=Gabriel tag=prop cap=maxmin ini=cap simb=alpha prevW= prevT=v-fin prevCap=min nextW=Raimundo nextT=n nextCap=maxmin prev2W=pai prev2T=n prev2Cap=min next2W=da next2T=v-pcp next2Cap=min tip=I-PERSON	I-PERSON

3.4   Evaluation

The metrics commonly used to evaluate NER systems are Precision (P), Recall (R) and F-Measure (F) as presented in Eqs. 3, 4 and 5.

$$Precision = \frac{\text{Total of NEs correctly identified}}{\text{Total of NEs identified}} \qquad (3)$$

$$Recall = \frac{\text{Total of NEs correctly identified}}{\text{Total of NEs actually existing}} \qquad (4)$$

$$F - Measure = \frac{2 \times \text{Precision} \times \text{Recall}}{\text{Precision} + \text{Recall}} \qquad (5)$$

The Precision metric measures the total number of correctly identified NEs. A high rate of Precision indicates that few NEs have been misidentified (false positives). The Recall metric denotes the number of hits for all existing NEs in the corpus. The higher the Recall value, the fewer NEs are in the text that have not been identified through the devised procedure (false negatives). F-Measure denotes a combination of both.

In this work, these metrics were computed using the evaluation scripts from the Second HAREM[5] for two tasks: identification that evaluates only if the recognized string is actually an NE, that is, it checks the boundaries of the NE; or classification that in addition to checking the boundaries of the NE also checks if the category is correct.

The scripts from the Second HAREM use a formula to compute what is correct (Total of NEs correctly identified). This formula [6] considers the identification of the NEs, classification of category, type and subtype according to the evaluation mode chosen by the user and penalizes wrong classifications.

4   Results and Discussion

LG and CRF techniques were applied individually to evaluate their effects in relation to the combined strategy CRF+LG. Table 3 shows the results. The Recall value obtained by LG individually was lower because LG captures only some general heuristics for NER. As expected, the gain obtained by CRF+LG in comparison to the CRF was slightly higher in the classification task. This gain (1.27% in F-measure) may seem small, but it corresponds to an extra 45 NEs correctly classified. The Wilcoxon test [21] confirmed for the 10 ENs categories tested the significant difference between CRF and CRF+LG ($\alpha = 0.05$).

The performance of CRF+LG was compared to the performance of NERP–CRF [13]. The .xml file annotated by NERP-CRF was obtained as indicated in [16][6]. We only modified the identifiers (ID) of each NE by adding a unique number at the end since this is necessary to compute the metrics correctly using the Second HAREM scripts.

Table 3. Performances evaluation of LG, CRF, CRF+LG and NERP-CRF using the Second HAREM scripts

Techniques	Identification			Classification		
	P (%)	R (%)	F (%)	P (%)	R (%)	F (%)
LG	71.27	28.49	40.70	64.80	20.70	31.38
CRF	79.05	66.07	71.98	64.95	43.41	52.04
CRF+LG	**79.84**	**66.75**	**72.71**	**66.52**	**44.47**	**53.31**
NERP-CRF	74.81	54.84	63.29	62.11	36.40	45.90

Table 3 shows that the results obtained outperform the NERP-CRF results by approximately 12% for the Recall metric in the identification task and more

---

[5] http://www.linguateca.pt/HAREM/.

[6] http://www.inf.pucrs.br/linatural/recursos_para_reconhecimento_de_entidades_nomeadas/NERP_CRF.xml.

than 8% in the classification task. For the F-measure metric, CRF+LG outperforms the NERP-CRF results in more than 9% and 7% in the identification and classification tasks respectively, representing considerable gain.

Note that only with the use of CRF we already outperformed the results reported in NERP-CRF. This is an interesting result obtained possibly because of the different tools used for the preprocessing of the texts and different notation used for labeling. For example, analyzing the .xml file annotated by NERP-CRF, we observed that most of the names preceded by the treatment pronoun $D.$ were not recognized or the pronoun was not included as part of the NE as it should be. The sentence segmentation tool used may have inserted segmentation after $D.$ and caused this error. CRF+LG recognized all these names correctly.

Santos and Guimaraes [14] presented results using the GC of the First HAREM as training set and the MiniHAREM as the test set. As [14] did not present the results for the GC of the Second HAREM, CRF+LG was rerun using the GC that they used for training and testing. The CoNLL-2002[7] evaluation script which evaluates the classification task was also used as done by them to compute the metrics. The results are presented in Table 4. Note that CRF+LG achieved a gain of approximately 2% in each metric evaluated.

Table 4. Comparison with CharWNN (result from [14] for selective scenario) using the CoNLL-2002 evaluation script

Systems	P (%)	R (%)	F (%)
CharWNN	65.21	52.27	58.03
CRF+LG	**67.09**	**54.85**	**60.36**

The results presented in Table 4 were obtained for a selective scenario (categories Person, Place, Organization, Time and Value) because the results presented by [14] for the 10 categories of the HAREM were obtained using word-level embeddings previously trained with three other corpus (Portuguese Wikipedia, CETENFolha and CETEMPublico) to perform this unsupervised pre-training. Therefore the comparison with this result would be unfair since the CRF+LG uses only the GC of the First HAREM for the CRF training phase and LG construction. Hence, just for the sake of comparison, the GC of the First HAREM has approximately 78667 words while only the CETEMPublico, one of the three corpus used by [14], has about 180 million words.

4.1   Study of the Lower Bound and Upper Bound for CRF

In order to study the boundaries of the CRF using as an additional feature the result of another classifier, we performed the following experiments keeping the same training files and test (GCs of the First and Second HAREM):

---

[7] http://www.cnts.ua.ac.be/conll2002/ner/bin/conlleval.txt.

- Lower bound: a random tip among the 11 possible (10 categories of the HAREM plus label O) was inserted into the vectors of features rather than the label assigned by LG. The experiment was repeated 100 times and the mean was calculated for each metric (the standard deviation obtained was less than 0.01 for each metric).
- Bound: a wrong tip (contrary to the correct category of the word) was inserted into the vectors of features rather than the label assigned by LG. If the label of the word was I-X where X is one of the 10 categories of the HAREM, the tip assigned was O; if the label was O, the tip assigned was I-X where the category X was chosen randomly among the 10 categories of the HAREM. This experiment was also repeated 100 times and the mean was calculated for each metric (the standard deviation obtained was less than or equal to 0.01 for each metric).
- Upper bound: the correct label was inserted as tip into the vectors of features rather than the label assigned by LG.

The results obtained are presented in Table 5. As expected, the CRF with random tip (lower bound) obtained results very close (slightly lower) to the CRF without a tip. This indicates that the weight estimated to the random tip feature function during training was very low since it is non-discriminant.

Table 5. Comparison with results for lower bound and upper bound

Experiments	Identification			Classification		
	P (%)	R (%)	F (%)	P (%)	R (%)	F (%)
CRF	79.05	66.07	71.98	64.95	43.41	52.04
CRF+LG	79.85	66.75	72.72	66.51	44.47	53.30
Lower bound	*78.61*	*65.56*	*71.49*	*64.48*	*43.01*	*51.60*
Bound	92.32	87.75	89.98	73.08	55.63	63.17
Upper bound	**95.67**	**94.54**	**95.10**	**95.65**	**75.71**	**84.52**

At first, we thought that the CRF with wrong tip (bound) would be a lower bound since the tip was completely the opposite of the label. However, the CRF identified and learned a pattern. When the tip is I-X, the CRF learned that the label is O, distinguishing between what is not NE from what it is. When the tip is O, the CRF learned that the word is an EN (I-X), but does not know which one and needs the other features to predict and may make some mistake.

Obviously, the results for the upper bound are the best since the tip feature will be more discriminating and will have a higher weight. The results presented in Table 5 for lower bound and upper bound represent the bounds for the performance of the CRF combined with another classifier in the manner proposed in this paper (inserting the result of the classifier as an additional feature for the CRF) and under the conditions of the experiments (using the GCs of the

HAREM as training and test and the 18 features proposed). The upper bound results are very important because it allows us to foresee the maximum gain that we can achieve by combining the CRF with other better classifiers.

5 Conclusions

This paper presented the Named Entity Recognition in Portuguese texts using Conditional Random Fields and Local Grammars. The classification obtained from LG was sent as an additional feature for the learning process of the CRF prediction model. The CRF model assigns the final label of the NEs. Our approach is a good way to take into account the human expertise for capturing the rules that do not appear in examples of the annotated corpus used for training by the CRF.

The results obtained were higher than those results obtained by other systems performing under equivalent conditions. Using our approach, CRF+LG, Precision and Recall metrics were increased by 1.5% and 1% in the classification task, respectively. At first glance, these values may seem small, but they correspond to an extra 45 NEs now correctly classified in comparison to the CRF without the suggestion given by the LG. It is important to mention that these are the results for a small corpus and the gains can become more expressive when using a larger corpus for training. Furthermore, a deeper linguistic study was not carried out to construct our LG.

We observed that using CRF with some different preprocessing decisions we already outperformed the results reported in NERP-CRF. This suggests a future work to investigate the impact of some preprocessing decisions on the performance of the CRF. The upper bound of the proposed strategy also suggests testing other classifiers to inform new features for the CRF learning process rather than an LG in order to get results closer to the upper bound indicated in the experiments.

References

1. Jiang, J.: Information extraction from text. In: Mining Text Data, pp. 11–41. Springer, Boston (2012)
2. Manning, C.D., Schütze, H.: Foundations of Statistical Natural Language Processing. MIT Press, Cambridge (1999)
3. Grishman, R., Sundheim, B.: Message understanding conference-6: a brief history. In: COLING, vol. 96, pp. 466–471 (1996)
4. Tjong Kim Sang, E.F., De Meulder, F.: Introduction to the CoNLL-2003 shared task: language-independent named entity recognition. In: Daelemans, W., Osborne, M. (eds.) Proceedings of CoNLL-2003, Edmonton, Canada, pp. 142–147 (2003)
5. Santos, D., Cardoso, N.: Reconhecimento de entidades mencionadas em português: Documentação e actas do HAREM, a primeira avaliação conjunta na área. Linguateca (2007). http://www.linguateca.pt/aval_conjunta/LivroHAREM/Livro-SantosCardoso2007.pdf. ISBN 978-989-20-0731-1

6. Mota, C., Santos, D.: Desafios na avaliação conjunta do reconhecimento de entidades mencionadas: O Segundo HAREM. Linguateca (2008). ISBN 978-989-20-1656-6

7. Pirovani, J.P.C., Oliveira, E.: Extração de Nomes de Pessoas em Textos em Português: uma Abordagem Usando Gramáticas Locais. In: Computer on the Beach 2015. SBC, Florianópolis, March 2015

8. Pellucci, P.R.S., de Paula, R.R., de Oliveira Silva, W.B., Ladeira, A.P.: Utilização de técnicas de aprendizado de máquina no reconhecimento de entidades nomeadas no português. e-Xacta 4(1), 73–81 (2011)

9. Oudah, M., Shaalan, K.F.: A pipeline Arabic named entity recognition using a hybrid approach. In: COLING, pp. 2159–2176 (2012)

10. Lafferty, J., McCallum, A., Pereira, F.: Conditional random fields: probabilistic models for segmenting and labeling sequence data. In: Proceedings of the Eighteenth International Conference on Machine Learning, ICML 2001, vol. 1, pp. 282–289 (2001)

11. Gross, M.: The construction of local grammars. In: Roche, E., Schabs, Y. (eds.) Finite-State Language Processing, Language, Speech, and Communication, pp. 329–354. The MIT Press, Cambridge (1997)

12. Milidiú, R.L., Duarte, J.C., Cavalcante, R.: Machine learning algorithms for Portuguese named entity recognition. Inteligencia Artif. 11(36), 67–75 (2007). Revista Iberoamericana de Inteligencia Artificial

13. do Amaral, D.O.F.: O reconhecimento de entidades nomeadas por meio de conditional random fields para a língua portuguesa. Master's thesis, Pontifícia Universidade Católica do Rio Grande do Sul, Porto Alegre, Brazil (2013)

14. dos Santos, C.N., Guimaraes, V.: Boosting named entity recognition with neural character embeddings. In: Proceedings of the Fifth Named Entities Workshop, ACL 2015, pp. 25–33 (2015)

15. Konkol, M., Konopík, M.: Segment representations in named entity recognition. In: International Conference on Text, Speech, and Dialogue, pp. 61–70. Springer (2015)

16. Amaral, D.O., Fonseca, E.B., Lopes, L., Vieira, R.: Comparative analysis of Portuguese named entities recognition tools. In: Proceedings of the Ninth International Conference on Language Resources and Evaluation (LREC 2014), pp. 2554–2558 (2014)

17. dos Santos, C.N., Zadrozny, B.: Learning character-level representations for part-of-speech tagging. In: ICML, pp. 1818–1826 (2014)

18. Liu, X., Zhang, S., Wei, F., Zhou, M.: Recognizing named entities in tweets. In: Proceedings of the 49th Annual Meeting of the Association for Computational Linguistics: Human Language Technologies, vol. 1, pp. 359–367. Association for Computational Linguistics (2011)

19. do Amaral, D.O.F., Buffet, M., Vieira, R.: Comparative analysis between notations to classify named entities using conditional random fields (2015)

20. Sutton, C., McCallum, A.: An introduction to conditional random fields. Found. Trends® Mach. Learn. 4(4), 267–373 (2012)

21. Bussab, W.d.O., Morettin, P.A.: Estatística básica. Saraiva (2010)

A Secure and Efficient Temporal Features Based Framework for Cloud Using MapReduce

P. Srinivasa Rao[1] and P. E. S. N. Krishna Prasad[2(✉)]

[1] MVGR College of Engineering, Vizianagaram, India
psr.sri@gmail.com
[2] Prasad V. Potluri Siddhartha Institute of Technology, Vijayawada, India
surya125@gmail.com

Abstract. A new data mining method called temporal pattern identification in cloud is developed using MapReduce: a power full feature of hadoop with temporal features. In the paper a new approach called temporal features based authentication approach has been introduced where a user will be verified by 2 phases. In phase1 the user facial features will be stored at HIB (Hadoop Image Bundle) and in second phase the user credentials will be checked by using symmetric encryption technique. The hadoop cluster can be exported to cloud based on user demand such as IAAS, SAAS, and PAAS. This cloud model is useful to deploy applications that can be use to transmit massive data over cloud environment. It also allows different kinds of optimization techniques and functionalities. The framework can also be used to give optimized security to the massive data stored by user at cloud environment. The article make use of temporal patters of user who entered into cloud by updating it in a logfile. Experimentation conducted by hiding text, image and both in video file and the performance of the cluster is monitored by using efficient monitoring tool called ganglia to provide auto-scaling functionality of the cluster.

Keywords: Bigdata · Hadoop · MapReduce · Cloud computing
Temporal patterns

1 Introduction

In many real-world applications where data has been increasing enormously such as library reader analysis, patient disease analysis, stock fluctuation, it is important to study the behavioral patterns of dynamically changed patters of the user through the temporal patters. In this paper, an analysis has been conducted on temporal categorical data such as Electronic health records, traffic incident logs and weblogs, usability study logs in cloud environment. Specifically, in this analysis experimentation has been done on weblog data based on the usability of the customer who registered at his respective cloud.

Cloud computing environment is more preferred aspect in the present big data world due to the massive data maintenance, storage and processing cost [14]. These clusters will be deployed on to a Cloud computing environment which is one of the

© Springer International Publishing AG, part of Springer Nature 2018
A. Abraham et al. (Eds.): ISDA 2017, AISC 736, pp. 114–123, 2018.
https://doi.org/10.1007/978-3-319-76348-4_12

latest emerging technologies, used widely in IT and other important organizations to reduce the capital expenditure and to increase the scalability resources, this provides services at different levels like Saas, Paas, Iaas. Adding networking functionalities to the cloud is discussed in [1, 3] which addressed the challenges over the security systems. The network over cloud resources address about Network Virtualization [11], that connects multiple systems and data centers over a simple physical topology and divides the network environment into two different providers like Infrastructure providers and Traditional ISP to manage the Infrastructure layer over virtual network topology. Cloud computing focuses on many functionalities and privacy issues and mainly deals with security to analyze security issues [12] that arises in the cloud environment and also addresses the security issues at ISP level with authorization and authentication methodology. Security and privacy issues play major role in cloud environment [2] with service level agreements and identity management. Cloud computing also follows many methodologies over communication channel to find out the geo-location verification of cloud data over the network [4], by using the global networking positioning system to locate various nodes and a link of cloud resources that are connected on Internet. Cloud computing follows the Two – Phase Validation (2PV) commit protocol to measure the accuracy and precision of the cloud data and to balance the cloud transactions in a secure way [16], by undergoing proof of authorization mechanism using certificate authority thereby protecting the cloud data and their transactions. Cloud computing performs key management operation for encryption based cloud storage system [7] over the virtual network environment and follows policy enforcement techniques to protect cloud data and their resources. An open source framework called Hadoop is placed in a cloud environment towards optimizing and provisioning the resources [5] in the cloud and to manage the data access system [18] in the cloud by enforcing the security parameters and policies. A hadoop Cluster is placed in the cloud at Application level, which consists of two phases such as Map and Reduce. The files stored in the cloud framework allow the users to do sorting and searching techniques. To protect the data of user in a secured way models will be deployed in the framework.

The article presents an efficient framework for handling massive data and also identifies the user behavioral patterns when the users access it in a dynamic cloud environment by considering two stage security mechanisms. To handle massive data, an efficient big data processing tool called hadoop with Map Reduce functionality is customized by considering a methodology called data hiding technique [13].

1.1 Data Hiding with Hadoop

Digital knowledge concealing refers to the method of concealing secondary knowledge in host knowledge. Knowledge is hidden while not distorting the initial transmission content to a plain level and transferred while not requiring further channel information measure. Knowledge is hid within the pictures that are the thought of steganography which will provide security to the information. The concept will be integrated with hadoop MapReduce so that the processing can be done very fast when compared to the traditional existing models.

1.2   Contribution and Plan of the Paper

The article planning to address security connected problems with the information that has been changed between Hadoop machines in Distributed atmosphere. The remainder of this paper is organized as follows. In Sect. 2, the connected work is mentioned. In Sect. 3, projected system model and design were conferred. In Sect. 4 the Methodology is conferred. In Sect. 5 experimentation results are conferred.

Finally, Sect. 6 includes Conclusion and Future scope.

2   Related Work

Schoo et al. [12] experimented a secure framework with virtualization infrastructure where the user level authentication is verified by using various techniques. Streitberger and Ruppel [18] demonstrated cloud security methodologies by considering various challenges in the environment. Ries et al. [17], had given a report on the placement of cloud computing atmosphere and discussed the locating techniques over web to verify the cloud locating resources. Kambatla et al. [5], introduced a Hadoop open supply implementation techniques to scale back the duty in extracting the data over the information analysis and explained the 3 phases within the Hadoop that are Map, Copy and cut back, that are utilized in the cloud. Munir and Palaniappan [6], generated a report on the transactions that are concerned within the cloud computing atmosphere, wherever they cope with the proof of authorization and mentioned the approaches concerned in accessing the cloud resources and explained regarding two-phase validation commit protocol as an answer for cloud security. Senhadji et al. [15] experimented on grid to extract association rules with load equalization strategy.

Taking into consideration all existing problems, the paper goes to deal with Privacy Protection through digital video in Hadoop Distributed System by exploitation bilateral algorithms like Filekey, DES and AES, to attenuate the delay by an alternate approach. The paper proposes a completely unique methodology where the security is verified at two stages of the model implemented by using Hadoop MapReduce framework.

3   System Model

As shown in the Fig. 1 the user who wants to enter into the cloud will be authenticated at stage 1 with his signature via facial features, one time password which is having a validity period and a captcha. Once the user is authorized will be allowed to submit the data to Namenode in the Hadoop Cluster.

3.1   Internal Operation

As shown in the Fig. 2 at any Namenode the user can put a request to either upload or download the data. Based on the credentials acceptance of the user the temporal patterns of the log file will be updated.

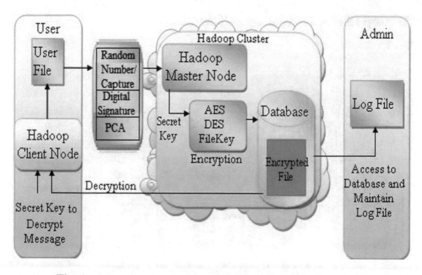

Fig. 1. Internal system model of developed Hadoop framework

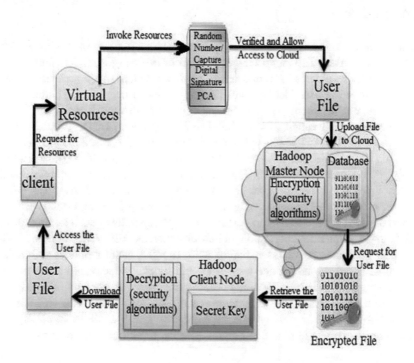

Fig. 2. Overall architecture of signature based cloud environment encrypted data transfer.

3.2   Hadoop Cluster Internal Operations

See Fig. 3.

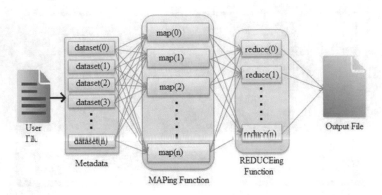

Fig. 3. Hadoop cluster internal function

4   Methodology

A cloud framework is designed with 2-phase authentication approach where user will be authenticated using ELGamal Digital Signature [9] using a random number and a captcha in order to protect the cloud resources from unauthorized access. This digital signature verifies the user identity and provides a tag to determine whether a user is authorized or not. In second phase, the user is authenticated using PCA (Principal Component Analysis) which will be stored in HIB (Hadoop Image Bundle) [10]. For each of these phases Tag value will be updated in log file of Namenode in hadoop cluster. Tag defines the user level of authorization, for example, Tag "0" = Authorized and Tag "1" = Unauthorized. This Tag level varies up on service user tasks inside the cloud resources usage. When a service user gets access to cloud resources with Tag "0", cloud monitors and maintain a service list using ganglia based on the service user tasks and allows the service user to perform multiple operations such as encrypting the user data using symmetric encryption techniques such as DES, AES, File key. When a service user gets access to cloud resources with Tag "1", cloud monitors and maintains a service list based on the service user tasks and blocks the service user at one level, when the Tag value is increasing continuously, the user with corresponding IP will be blocked from the service. As per the literature the symmetric encryption is more advantageous than one key encryption explained by kitu File whilte et al. [8].

Our user authentication and temporal pattern identification contains following steps:

1. The User U passes his user information M_u and parameters (r_u, s_u) to the Namenode (verifier), where $S_u = r_u^{s_u} mod\, P$.
2. After receiving these parameters the verifier checks the database log file for his authenticity if it is correct then it prepares a random number (secrete key) and send it to client (Datanode).

3. The user first uses this secret key to compute the Diffie-Hellman secret key and send it to Namenode as Ack. The name node uses this secret key to compute Diffie Hellman shared secrete key.
4. After receiving ack from name node the user facial features will be captured automatically and are compared with the features stored in HIB (Hadoop Image Bundle) with PCA Technique.
5. If this verification is successful, the certificate owner (Datanode) is authenticated by the verifier (Namenode) and a one-time session key is shared between them. This shared key can be useful for transmission of data or for further communication.

5 Experimentation

5.1 Environment Setup

The experimentation was carried out by installing Hadoop 2.7.1 on 4 node cluster each with configuration of 2.5-GHz CPUs, 2 GB of RAM, and 500 GB of storage space. Once the cluster is established to know the cluster status ganglia 3.6.0 monitoring tool was also installed.

5.2 Implementation

5.2.1 Mapper

The algorithm for checking the user credentials is given in Table 1 where the user will select the input data set which he wanted to upload into cloud. If the credential are matched with his facial features, one time password and Captcha then stage1 is successful. If any user whose credentials are not matched, the temporal patters will be updated in log file associated with Namenode in the framework. A typical Map Reduce algorithm for distinguishing such temporal patterns is listed in Table 1.

Table 1. TPMap & Reduce Algorithm

Customized input and output path Hadoop have to be set Initially

Step 1: Select input DataSet
Step 2: If User Login Successful then goto step 3
else go to step 8.
Step 3: Check captcha and random number issued by CA if it is matched, then verify MapKey else goto Step 8.
Step 4: Check the Features with HIB. If Match not found update tag to to step 2.
Step 5: Embedded File can be uploaded by User.
Step 6: Encrypt the data with the proposed Methodolgy.
Step 7: Update the Logfile.
Step 8: Logout

MapKey Algorithm

The Mapkey algorithm plays a key role in improving the efficiency of the framework developed. Once the user wants to upload or download the data if the credentials are

Table 2. Shows running time for different log file size

File type	File format	Size (GB)	Number of nodes	Time (sec)
Text	Original file	10	4	600.8745
			8	500.3254
	Compression low 8		4	500.6789
	High 6		8	400.5973
			4	458.1020
			8	300.9848
Image	Original file	12	4	800.5672
			8	600.8972
	Compression low 9		4	500.9876
	High 7		8	500.8732
			4	500.4321
			8	400.1234
Text or image in video	Original file	15	4	900.3214
			8	700.8765
	Compression low 12		4	800.5672
	High 9		8	600.8972
			4	500.9876
			8	500.8732

matched, by using various messages digest algorithms the Mapkey will return a valid key with which user can login into the cloud, otherwise the temporal patterns will be updated in log file.

MapKey Algorithm

1. Read File from any Data node
2. Generate Random Number by using any methodology to encrypt the file to generate Key
3. Return MapKey value
4. Stop

As shown in the above, MapKey algorithm will generate a random key, when a user uploads the file in a cloud, the security algorithms are applied over the user file.

The following Table 2 shows time required to process the files uploaded by user with a random key where the data file is of size 10 GB.

Sample Log File

Once the log file is created by Hadoop MapReduce framework, the temporal patters of the user can be generated as shown in Fig. 4.

The framework created in the article can be deployed in the real cloud environment once the Hadoop cluster is ready to use then the performance of it can be monitored by using Ganglia which is an open source software tool. Figure 5, shows monitoring of 8 Node cluster by using Ganglia.

Table 3. Shows authorized vs unauthorized users

IP Address	URL's	Digital sign verify	Login time	Logout time	Tag	User type	Tag session level	Allow/Deny
192.168.10.2	http://www.youtube.com	Yes	09:00	09:18	0	Authorized	3	Allow
192.68.10.12	http://www.google.com	No	21:00	21:36	1	Unauthorized	6	Block
192.168.5.12	http://www.gutenberg.com	No	02:45	03:16	1	Unauthorized	8	Block
192.168.5.22	http://www.pratibha.net	Yes	09:00	09:09	0	Authorized	2	Allow
192.168.9.45	http://www.gitam.edu.in	Yes	06:30	06:39	0	Authorized	3	Allow
192.168.9.14	http://www.mvgrce.edu.in	Yes	10:44	11:07	0	Authorized	3	Allow
192.168.6.25	http://www.sharekhan.com	Yes	12:35	01:20	1	Unauthorized	4	Block

---

192.168.5.154,http://www.youtube.com/watch?v=tQ2wJmFAvlE&feature=watch-now-button&wide=1

192.168.5.65,http://apache.techartifact.com/mirror/hadoop/core

192.168.5.178,http://www.gutenberg.org/ebooks/4300

192.168.5.185,http://en.community.dell.com/support-

Fig. 4. Log file of different users using cloud at different time intervals

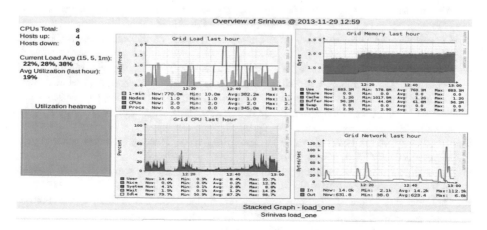

Fig. 5. Monitoring of 8 node cluster.

6 Conclusion and Future Work

In the article a new framework was developed to store, process and transmit massive data in a secure way by incorporating a two level security mechanism by identifying the temporal patterns of user who will log into the framework. The novel approach of temporal pattern identification will be improved in the coming works by integrating other models with image API library deployed in hadoop to identify user facial features over a period of time so that fraudulent people can be identified even though they change either their facial features or geographical location identity.

References

1. Wong, C.K., Lam, S.S.: Digital signatures for flows and multicasts. IEEE/ACM Trans. Netw. **7**(4), 502–513 (1999)
2. Basescu, C., Leordeanu, C., Costan, A., Carpen-Amarie, A., Antoniu, G.: Managing data access on clouds: a generic framework for enforcing security policies. In: 25th International Conference on Advanced Information Networking and Applications, vol. 10, pp. 459–466 (2011)
3. Koslovski, G., Yeow, W.-L., Westphal, C., Huu, T.T., Montagnat, J., Vicat-Blanc, P.: Reliability support in virtual infrastructures. In: IEEE International Conference on Cloud Computing Technology and Science, vol. 10 (2010)
4. Yu, J., et al.: Towards dynamic resource provisioning for traffic mining service cloud. White paper (2011)
5. Kambatla, K., Pathak, A., Pucha, H.: Towards optimizing Hadoop provisioning in the cloud. White paper (2011)
6. Munir, K., Palaniappan, S.: Framework for secure cloud computing. Int. J. Cloud Comput. Serv. Archit. (IJCCSA) **3**(2), pp. 21–35 (2013)
7. Hamlen, K., Kantarcioglu, M., Khan, L., Thuraisingham, B.: Security issues for cloud computing. Int. J. Inf. Secur. Priv. **4**(2), 39–51 (2010)
8. kitu File whilte, et al.: Symmetric vs asymmetric encryption (2010)
9. Harn, L., et al.: Generalized digital certificate for user authentication and key establishment for secure communications. IEEE (2011)
10. Almeer, M.H., et al.: Cloud Hadoop map reduce for remote sensing image analysis. J. Emerg. Trends Comput. Inf. Sci. **3**(4), 637–644 (2012)
11. Chowdhur, N.M.M.K., Boutaba, R.: A Survey of Network Virtualization, vol. 54, pp. 862–876. Elsevier (2010)
12. Schoo, P., Fusenig, V., Souza, V., Melo, M., Murray, P., Debar, H., Medhioub, H., Zeghlache, D.: Challenges for cloud networking security. In: 2nd International ICST Conference on Mobile Networks and Management, vol. 137, 22–24 September 2010
13. Dutta, P., et al.: Data hiding in audio signal: a review. Int. J. Database Theor. Appl. **2**(2), 1–8 (2009)
14. Rao, P.S., et al.: A novel approach for identifying cloud Hadoop temporal patterns using MapReduce. IJITCS **4**, 37–42 (2014)
15. Senhadji, S., et al.: Association rule mining and load balancing strategy in grid systems. IAJIT **11**(4), 338–344 (2013)
16. De Capitani di Vimercati, S., Foresti, S., Jajodia, S., Paraboschi, S., Pelosi, G., Samarati, P.: Encryption-based policy enforcement for cloud storage. White paper (2012)

17. Ries, T., Fusenig, V., Vilbois, C., Engel, T.: Verification of data location in cloud networking. In: Fourth IEEE International Conference on Utility and Cloud Computing, vol. 72, pp. 439–444 (2011)
18. Streitberger, W., Ruppel, A.: Cloud computing security – protection goals, taxonomy, market review. Institute for Secure Information Technology SIT, Technical report (2010)

A Comparison of Machine Learning Methods to Identify Broken Bar Failures in Induction Motors Using Statistical Moments

Navar de Medeiros Mendonça e Nascimento, Cláudio Marques de Sá Medeiros, and Pedro Pedrosa Rebouças Filho[✉]

Programa de Pós-Graduao em Energias Renováveis, Instituto Federal de Educação, Ciência e Tecnologia do Ceará, Fortaleza, Brazil
navarmedeiros@ppger.ifce.edu.br, {claudiosa,pedrosarf}@ifce.edu.br

Abstract. Induction motors are reported as the horse power in industries. Due to its importance, researchers studied how to predict its faults in order to improve reliability. Condition health monitoring plays an important role in this field, since it is possible to predict failures by analyzing its operational data. This paper proposes the usage of vibration signals, combined with Higher-Order Statistics (HOS) and machine learning methods to detect broken bars in a squirrel-cage three-phase induction motor. The Support Vector Machines (SVM), Multi-Layer Perceptron (MLP), Optimum-Path Forest and Naive-Bayes were used and have achieved promising results: high classification rate with SVM, high sensitivity rate with MLP and fast training convergence with OPF.

Keywords: Higher-Order Statistics · Induction motor
Patter recognition · Vibration signal

1 Introduction

It is estimated there are more than 300 million electric motors all over the world, consuming 7400 TWh, which represents 40% of global electricity production [1]. Among all rotational machines the induction motor is quite representative, it is presented in 90% of industrial applications [2]. Brazil highlights it, because 70% of industrial electric consumption is due to electrical motors [1].

The Three-phase Induction Motor (TIM) is split in two types: the squirrel-cage rotor and the wound rotor [3]. However, Francisco [2] pointed the squirrel-cage rotor as the most used type of motor, because it can be easily constructed and still be robust, guaranteeing it as reliable and cheap machine.

Although the TIM advantages, it is not a failure proof equipment and has its limitations. In general, the problems are associated with electrical and mechanical factors, such as: mechanical overload, overheating, misalignment [4]. An research presented by Bonnet [5] compile 40 years data from inductions motors

© Springer International Publishing AG, part of Springer Nature 2018
A. Abraham et al. (Eds.): ISDA 2017, AISC 736, pp. 124–133, 2018.
https://doi.org/10.1007/978-3-319-76348-4_13

installed in a petroleum and chemical industry, and exhibits the most expressive failure causes: 51% were related to rotor bearing, 16% to stator winding, while the broken rotor bars represents 5%.

That has risen an concern to failure analysis in electrical motors, and were presented relevant methods in literature such as: motor current signature analysis and vibration analysis [6]. Both methods are applied for fault detection in induction motor's bearing [7], in stator's winding [8]. There are also researches focusing on identifying broken bars in motor rotor [9]. Those methods mainly use frequency domain techniques, such as Fourier and Wavelet Transforms [10].

However, the Higher-Order Statistics (HOS) is a promising technique for describe time domain signals [11]. Researchers have directed their attention for HOS as an auxiliary tool for fault detections on rotational machines [10,12]. The work presented in this paper is a different approach than the model made by Martinez et al. [9,13], wherein they provided theoretical equations for vibration patterns in electrical motors, during abrupt broken rotor bars. Because we use non deterministic methods to describe machine health state. Additionally, Vibration Signals (VS) have been widely used for failure analyses in rotational machines [12], because there are unique features in these signals, associated with operational condition, such as: misalignment, imbalance, poor lubrication, overload and mechanical fatigue.

The relation between broken rotor bars and mechanical vibration on TIM was observed by [9], and years ahead its model is presented and validated by Martinez *et al.* [13]. Assuming the set of bars around the rotor is a multiphase circuit, Martinez *et al.* [13] express the produced magnetic field on the rotor as a function of the number of bars which composes it, and torque as a function of the magnetic field. Furthermore, the current in the rotor circuit with broken bars is asymmetric, resulting in a higher current density in the bars close to the broken one [13], and this induces an equally unbalanced magnetic field. In this case the torque produced in the rotor axis triggers the appearance of axial vibration [13].

The aforementioned mentioned studies have contributed to understanding the relation between broken rotor bar and its vibrations on an electrical machine, because they have provided analytical models to identify failures, through vibration intensity, on TIM. But, the approach proposed by Martinez et al. [9,13] is more complex than the work proposed on this paper, as we use processing signal and machine learning techniques as a mean to identify broken rotor bar in three-phase induction motors.

A possible use of machine learning is to automatically identify regularities and irregularities in data, through the usage of algorithms, that might learn the behavior of the system and perform classification tasks [14]. For the work presented in this paper we have chosen different methods to perform identification of broken bars in electrical motors: (i) The Naive-Bayes classifier due to its probabilistic approach; (ii) the classic neural network Multi-layer Perceptron (MLP); (iii) a classifier based on kernel, the Support Vector Machines (SVM) and (iv) the Optimum-Path Forest (OPF) [15]. The contribution of our research is the combined usage of (i) HOS and (ii) machine learning techniques, to classify failure data from electrical motors, through (iii) vibration signals obtained from an experimental set-up.

This paper is organized in the following order: The methodology is presented in Sect. 2, wherein we explain steps to acquire vibration signals, data pre-processing and the metrics chosen for models evaluations; In Sect. 3 we discussed the results and in Sect. 4 we provide our conclusions and recommendations for future works.

2   Methodology

In this section it is explained each one of the steps we have followed to achieve our goal. The general steps are shown in Fig. 1, which starts on (i) acquiring raw data from our laboratory set-up. The following step consists in (ii) pre-processing raw data, over the HOS, and creating a dataset to feed (iii) the last step, where several classifiers will be used.

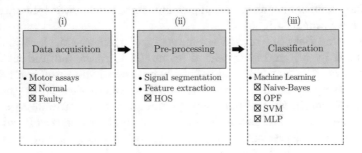

Fig. 1. General steps in methodology

2.1   Raw Data Acquisition

We have chosen a three-phase squirrel-cage induction motor, from WEG, 1 HP, 220 V, powered by a CFW-08 frequency converter as a the electrical machine in this experiment. The vibration model exhibit by Martinez, Belahcen and Muetze [9] shows a straight relation between speed and vibration intensity, thus the motor is operated under 30, 35, 40, 45, 50, 55, and 60 Hz frequencies. A Eddy current brake was used to simulate different loading operations: unload, 50% and 100% load in motor axis. The set-up is exhibit in Fig. 2. In order to simulate normal and faulty conditions we used two similar motors, which one of them has 3 adjacent broken rotor bars.

To acquire the vibration data it was chosen a three-axis MEMs accelerometer, attached to the motor frame [16], and three analogical channels, with 14 bits resolution, of a National Instruments data acquisition module, USB-6009.

Fig. 2. Experimental set-up with TIM and Eddy current

The combination of 3 different load operation with the range of 7 frequencies, commanded by converter, results in 21 tests for each motor (i.e. normal and faulty), totalizing 42 assays. Each experiment is performed while the machine on steady-state, during 10 s, with a 10 kHz sampling rate, and the outcome is 100000 data points. According Nyquist-Shannon sampling theorem, the maximum frequency observed by our sensor is 5 kHz, which is under the accelerometer resonance frequency, 6 kHz [17]. We adopted x as the longitudinal axis along motor axis, y as the transversal and z as lateral axis. The signals collected here are name raw data and the next step consists in pre-processing this data.

2.2   Pre-processing Vibration Data

We use the kurtosis as originally proposed by Dwyer [18], as a statistical tool to indicate non-gaussian components in signals, and later reformulated by Capdevielle and Lacoume [19] under the High-Order Statistics basis. To identify skewness we use just as Martin and Hanorvar [7] and Samanta and Al-Balushi [20]. Reports of its practical usage have been made in [21].

In order to increase the number of samples, each one of the 42 collected signals are split in 5 equal parts, which will be treated as independents signals. Therefore, are extracted 5 values of Kurtosis, Skewness and Variance of each vibration signal along three-axis. Thus, 21 assays for normal conditions and 21 assays for faulty (i.e. broken rotor bars) will provide a dataset with 210 samples, 9 features and 2 classes, this processes is exhibit in Fig. 3. The next and final step consists in fed data into several classifiers to predict two different operational conditions.

2.3   Motor Operational Condition Classification

Having the raw data pre-processed and set-up, the dataset will serve as input for the following methods: (i) Naive-Bayes; (ii) OPF, (iii) SVM and (iv) MLP. For all classifiers we have randomly separated 80% of sample for training and 20% for testing, with equal number of samples per classes.

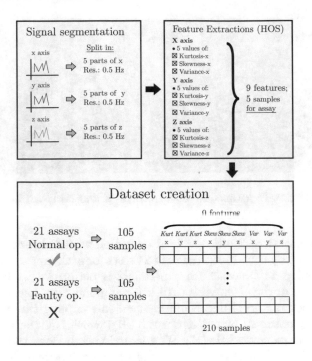

Fig. 3. The process to create a reliable dataset

Since Naive-Bayes is a particularity of Fisher's Linear Discriminant, a covariance matrix established as diagonal, containing only the variances of each classes (i.e. relied on the Naive-Bayes assumption that, there is no correlation between classes). For OPF we have chosen three types of distances: Euclidean (OPF-E); Gaussian (OPF-G) and Manhattan (OPF-M). The SVM is trained using four types of kernels: Linear (SVM-L); Polynomial (SVM-P); RBF (SVM-R).

The trainings performed on MLP is based on backpropagation algorithm, using stochastic gradient descent, having 1 output neuron (i.e. since is a binary classification) and varying the number of hidden neurons within the range 1 to 20, with the hyperbolic tangent as activation function. The model has shown convergence after 200 epochs and the two best configurations after 50 independent trainings are: 5 hidden neurons (MLP-1) and 8 hidden neurons (MLP-2). All networks have only 1 hidden layer.

The result of each training is a confusion matrix, which accounts hits and miss of each classifier. For each algorithm are made 10 independent trainings and the result are exhibit in Sect. 3.

2.4 Evaluation Metrics

The two classes involved in our problem are: (i) Normal, which represents the TIM expected operation condition or no-failure and (ii) Faulty, which is when the TIM experience 3 broken rotor bars. We can extract the following information from confusion matrix:

- **True Positive (TP)**: indicates the number of time Faulty conditions were correctly classified.
- **True Negative (TN)**: suggest the number of times Normal conditions were correctly classified.
- **False Positive (FP)**: imply the number of times Normal conditions were misclassified as Faulty.
- **False Negative (FN)**: designate the number of times Faulty conditions were misclassified as Normal.

Given the confusion matrix information the evaluation metrics are: (i) Accuracy (Acc), which relates the total amount of correct classifications among Normal and Faulty conditions; (ii) Specificity (Spe), which indicates the amount of Normal conditions correctly classified; (iii) Sensitivity (Sen) denotes the number of Faulty which were correctly classified and the (iv) F-Score (FS) comes to reaffirm the sensitivity result, exhibiting the methods robustness to identify broken rotor bars.

3   Results and Discussion

The results obtained from all methods will be presented in this section, according to the methodology proposed. In Fig. 4 are shown the accuracy distribution over 10 trainings. Aside from outliers, OPF-M, SVM-L, SVM-P, SVM-R and MLP-2 were quite similar, because all of them reached accuracies near to 100%, while Naive-Bayes, OPF-E, OPF-G and MLP-1 had a bit more dispersion. However none of the methods have reached rates below 92%.

The average results, over all classes and 10 trainings, are presented in Table 1, wherein SVM-P and SVM-R have achieved the maximum accuracy, sensitivity, specificity and F-score values, demonstrating itself to be an effective approach to identify broken bar in three-phase induction motors. However, other methods

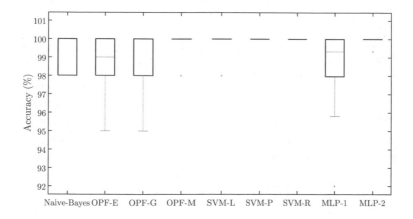

Fig. 4. Accuracy distribution obtained from all classifiers

have gotten near to SVM-P/R, like MLP-2, SVM-L, OPF-M and Naive-Bayes. The sensitivity analyses, through Table 1, exhibits that MLP-2 has a higher true positive rate, which means it is able to identified 100% of Faulty conditions cases. Besides the simplicity (i.e. less number of parameters to be adjusted during training), Naive-Bayes was able to classify 100% of Normal and Faulty conditions correctly in its best training. Therefore, MLP-1 obtained the worst average accuracy and a higher variance, which indicates a poor capacity for detection either the motor is operating in normal or faulty condition.

Table 2 shows the results by classes, demonstrating where each one of the methods were better or worse. All methods were able to identify each class correctly, but Naive-Bayes, SVM-L, SVM-P, SVM-R and MLP-2 have demonstrated to be more sensitive to identify changes in vibration patterns, caused by broken bars in squirrel cage rotors. MLP-2 was both, sensitive and specific, to identify normal and faulty vibration motor patterns. The three OPF methods have reached similar results, but the Manhattan distance provided lower variance, resulting in more stable classifications. The Training Time (TrT), Testing Time (TsT) for all methods are shown in Table 3, alongside with its accuracy. As expected, MLP-1 and MLP-2 took much more time than other methods, because of its recursive training processed, based on backpropagation and gradient descent. They were followed by SVM-P, that indicates, regards its accuracy, the polynomial kernel might not be suitable. The OPF-E/G and M had much faster training them MLP and SVM, due to its graph reorganizing theory and fast distance computing, compared to backpropagation (i.e. MLP) and inverse Kernel Matrix (i.e. SVM). But, the faster method, among those, were Naive-Bayes, essentially because its batch learning mode through a simplified covariance matrix. On the other hand, MLP was faster than other methods, because it only requires linear algebra operation since is properly trained.

Table 1. Overall results obtained by all classifiers

Overall				
Classifier	Acc (%)	Sen (%)	Spe (%)	FS (%)
Naive-Bayes	99.29 ± 1.09	99.29 ± 1.09	99.29 ± 1.09	99.29 ± 1.09
OPF-E	98.57 ± 1.57	98.57 ± 1.57	98.57 ± 1.57	98.57 ± 1.57
OPF-G	98.57 ± 1.57	98.57 ± 1.57	98.57 ± 1.57	98.57 ± 1.57
OPF-M	99.52 ± 0.95	99.52 ± 0.95	99.52 ± 0.95	99.52 ± 0.95
SVM-L	99.52 ± 0.95	99.52 ± 0.95	99.52 ± 0.95	99.52 ± 0.95
SVM-P	100 ± 0.00	100 ± 0.00	100 ± 0.00	100 ± 0.00
SVM-R	100 ± 0.00	100 ± 0.00	100 ± 0.00	100 ± 0.00
MLP-1	98.18 ± 8.49	98.32 ± 14.14	98.37 ± 9.76	97.36 ± 14.51
MLP-2	99.81 ± 0.74	100 ± 0.00	99.62 ± 1.49	99.81 ± 0.72

Table 2. Classes (i.e. normal and faulty) results, obtained from all classifiers

Per class					
Classifier	Class	Acc (%)	Sen (%)	Spe (%)	FS (%)
Naive Bayes	Normal	99.29 ± 1.09	98.57 ± 2.18	100 ± 0.00	99.27 ± 1.11
	Faulty	99.29 ± 1.09	100 ± 0.00	98.57 ± 2.18	99.30 ± 1.06
OPF-E	Normal	98.57 ± 1.58	100 ± 0.00	97.14 ± 3.15	98.62 ± 1.52
	Faulty	98.57 ± 1.58	97.14 ± 3.15	100 ± 0.00	98.52 ± 1.64
OPF-G	Normal	98.57 ± 1.58	99.52 ± 1.42	97.62 ± 3.19	98.60 ± 1.52
	Faulty	98.57 ± 1.58	97.62 ± 3.19	99.52 ± 1.42	98.54 ± 1.63
OPF-M	Normal	99.52 ± 0.95	99.52 ± 1.42	99.52 ± 1.42	99.52 ± 0.95
	Faulty	99.52 ± 0.95	99.52 ± 1.42	99.52 ± 1.42	99.52 ± 0.95
SVM-L	Normal	99.52 ± 0.95	99.05 ± 1.90	100 ± 0.00	99.51 ± 0.97
	Faulty	99.52 ± 0.95	100 ± 0.00	99.05 + 1.90	99.53 ± 0.03
SVM-P	Normal	100 ± 0.00	100 ± 0.00	100 ± 0.00	100 ± 0.00
	Faulty	100 ± 0.00	100 ± 0.00	100 ± 0.00	100 ± 0.00
SVM-R	Normal	100 ± 0.00	100 ± 0.00	100 ± 0.00	100 ± 0.00
	Faulty	100 ± 0.00	100 ± 0.00	100 ± 0.00	100 ± 0.00
MLP-1	Normal	98.37 ± 9.76	98.37 ± 9.76	98 ± 14.14	97.36 ± 14.51
	Faulty	98.15 ± 14.14	98 ± 14.14	98.37 ± 9.76	97.36 ± 14.51
MLP-2	Normal	99.62 ± 1.49	99.62 ± 1.49	100 ± 0.00	99.81 ± 0.72
	Faulty	100 ± 0.00	100 ± 0.00	99.62 ± 1.49	99.81 ± 0.72

Table 3. Training and testing time, followed by accuracy, for all classifiers.

Geral			
Classifier	Acc (%)	TrT (ms)	TsT (ms)
Naive-Bayes	99.29 ± 1.09	0.567 ± 1.200	0.233 ± 0.228
OPF-E	98.57 ± 1.57	2.852 ± 0.510	0.276 ± 0.269
OPF-G	98.57 ± 1.57	3.673 ± 0.436	0.1 ± 0.09
OPF-M	99.52 ± 0.95	2.900 ± 0.434	0.1 ± 0.06
SVM-L	99.52 ± 0.95	22.259 ± 3.767	0.017 ± 0.016
SVM-P	100 ± 0.00	185.592 ± 56.317	0.057 ± 0.056
SVM-R	100 ± 0.00	115.707 ± 3.856	0.197 ± 0.192
MLP-1	99.29 ± 1.09	17580.430 ± 521	0.0017 ± 0.016
MLP-2	99.29 ± 1.09	19580.382 ± 721	0.0018 ± 0.016

4   Conclusions

The Higher-Order Statistics has demonstrated to be an effective and powerful tool to identify pattern in vibration data from three-phase induction motors. It does not require much processing, since it is as a time-domain tool, thus HOS might be well suitable for embedded applications.

From the tests conducted with MLP it is concluded that as number of hidden neurons past 10, the accuracy decays due to overfitting. And, MLP-2 (i.e. 8 hidden units) was far better than MLP-1 (i.e. 5 hidden units) in all metrics, specially because it has shown less dispersion in classification, indicating to be a more stable classifier.

The highlight goes to SVM with kernels Polynomial and RBF, because they classify correctly all samples in all trainings. However, it was not considerate the number of support vectors in this work, but will be taken into account for future works with embedded systems.

But, since Naive-Bayes had similar results to SVM and MLP the 9-dimensional problem (i.e. three axis vibration plus HOS as feature extraction) might be linearly-separable. That is ratified by SVM with linear kernel (SVM-L), and simplicity Naive-Bayes should be taken into consideration to the chosen classifier for the purpose of broken bars detection in induction motors.

Since there were not applications using electrical machine alongside with OPF we have evaluated its performance and reaffirm what [22] suggest: that OPF is faster then MLP and SVM and possible to achieve similar or better results.

References

1. Abinee: Eficiência Energética. ABINEE TEC (2012)
2. Francisco, A.M.S.: Motores de Indução Trifásicos (2006)
3. Chapman, S.J.: Fundamentos de Máquinas Elétricas. Bookman (2013)
4. Bonnett, A.H., Soukup, G.C.: Cause and analysis of stator and rotor failures in 3-phase squirrel cage induction motors. In: Conference Record of 1991 Annual Pulp and Paper Industry Technical Conference, pp. 22–42 (1991)
5. Bonnett, A.H.: Root cause failure analysis for AC induction motors in the petroleum and chemical industry. In: 2010 Record of Conference Papers Industry Applications Society 57th Annual Petroleum and Chemical Industry Conference (PCIC), pp. 1–13 (2010)
6. Torabizadeh, M., Noshadi, A.: Artificial neural network-based fault diagnostics of an electric motor using vibration monitoring. In: Proceedings 2011 International Conference on Transportation, Mechanical, and Electrical Engineering (TMEE), pp. 1512–1516 (2011)
7. Martin, H.R., Honarvar, F.: Application of statistical moments to bearing failure detection. Appl. Acoust. 44(1), 67–77 (1995)
8. Jin, C., Ompusunggu, A.P., Liu, Z., Ardakani, H.D., Petré, F., Lee, J.: Envelope analysis on vibration signals for stator winding fault early detection in 3-phase induction motors. Int. J. Progn. Health Manag. 2153–2648 (2015)
9. Martinez, J., Arkkio, A., Belahcen, A.: Broken bar indicators for cage induction motors and their relationship with the number of consecutive broken bars. IET Electr. Power Appl. 7(8), 633–642 (2013)

10. Keskes, H., Braham, A., Lachiri, Z.: Broken rotor bar diagnosis in induction machines through stationary wavelet packet transform and multiclass wavelet SVM. Electr. Power Syst. Res. **97**, 151–157 (2013)
11. Mendel, J.M.: Tutorial on higher-order statistics (spectra) in signal processing and system theory: theoretical results and some applications. Proc. IEEE **79**(3), 278–305 (1991)
12. Doguer, T., Strackeljan, J.: Vibration analysis using time domain methods for the detection of small roller bearing defects, pp. 23–25. Mechanik (2009)
13. Martinez, J., Belahcen, A., Muetze, A.: Analysis of the vibration magnitude of an induction motor with different numbers of broken bars. IEEE Trans. Ind. Appl. **9994**(c), 1 (2017)
14. Bishop, C.M.: Pattern Recognit. Mach. Learn. **53**(9) (2013)
15. Papa, J.P., Falca, C.T.N., Suzuki, A.X.: Supervised pattern classification based on optimum-path forest. J. Imaging Syst. Technol. **2**, 120–131 (2009)
16. Ramalho, G.L.B., Pereira, A.H., Rebouças Filho, P.P., de Sá Medeiros, C.M.: Deteção De Falhas Em Motores Elétricos Através Da Classificação De Padrões De Vibração Utilizando Uma Rede Neural Elm. Holos **4**, 185 (2014)
17. Ramalho, G.L.B., Rebouças Filho, P.P., Júnior, C.R.S., Dias, S V · Detecção de talhas através de características do sinal de vibração e rede SOFM. In: XI Simpósio Brasileiro de Automação Inteligente, 2013, Fortaleza-CE.Simpósio Brasileiro de Automação Inteligente 2013 (SBAI 2013) (2013)
18. Dwyer, R.: Detection of non-Gaussian signals by frequency domain Kurtosis estimation. In: IEEE International Conference on Acoustics, Speech, and Signal Processing (ICASSP 1983), vol. 8, pp. 607–610 (1983)
19. Capdevielle, C.S.V., Lacoume, J.-L.: Blind separation of wide-band sources: application to rotating machine signals. In: Proceedings of the Eighth European Signal Processing Conference, vol. 3, pp. 2085–2088 (1996)
20. Samanta, B., Al-Balushi, K.R.: Artificial neural network based fault diagnostics of rolling element bearings using time-domain features. Mech. Syst. Signal Process. **17**(2), 317–328 (2003)
21. Lin, F.-J., Tan, K.-H., Fang, D.-Y., Lee, Y.-D.: Intelligent controlled three-phase squirrel-cage induction generator system using wavelet fuzzy neural network for wind power. IET Renew. Power Gener. **7**(5), 552–564 (2013)
22. Papa, J.P., Falcao, A.: Optimum-path forest: a novel and powerful framework for supervised graph-based pattern recognition techniques, Institute of Computing University of Campinas, pp. 41–48 (2010)

Canonical Correlation-Based Feature Fusion Approach for Scene Classification

J. Arunnehru$^{(\boxtimes)}$, A. Yashwanth, and Shaik Shammer

Department of Computer Science and Engineering, SRM University,
Vadapalani Campus, Chennai, Tamilnadu, India
arunnehru.aucse@gmail.com, aineeryyashwanth@gmail.com

Abstract. Vision-based scene recognition and analysis is an emerging field and actively conceded in computer vision and robotics area. Classifying the complex scenes in a real-time environment is a challenging task to solve. In this paper, an indoor and outdoor scene recognition approach by linear combination (fusion) of global descriptor (GIST) and Local Energy based Shape Histogram (LESH) descriptor with Canonical Correlation Analysis (CCA) is proposed. The experiments have been carried out using publicly available 15-dataset and the fused features are modeled by Random forest and K-Nearest Neighbor for classification. In the experimental results, K-NN exhibits the good performance in our proposed approach with an average accuracy rate of 81.62%, which outperforms the random forest classifier.

Keywords: Scene recognition · Feature extraction · Feature fusion
K-Nearest Neighbor · Random forest

1 Introduction

Nowadays, indoor and outdoor scene recognition algorithms accessible for hand held assistance to guide visually impaired inside and outside of the unknown places like roads, railway station, airports, park, restaurant, playground and shopping mall. In addition, there are a lot photos being taken by different photographers at different places all over the world. The collection of digital photography resulted in enormous increase in image collection. Redeeming a picture or an image from such huge pool of databases consumes a lot of searching time. In order to make this process easier, scene classification has been implemented. From the recent research works, scene recognition has achieved immense success using image processing techniques and methods. Currently, with the emergent prominence of dual-camera with Kinect sensors mobile phone, researchers are motivated to handle the conventional image recognition task related to computer vision approach and it depends on the ability to distinguish environment scenes, and classifications. Many approaches for scene recognition have been proposed to extract local and global shape descriptions that train scene formation such as color and texture.

© Springer International Publishing AG, part of Springer Nature 2018
A. Abraham et al. (Eds.): ISDA 2017, AISC 736, pp. 134–143, 2018.
https://doi.org/10.1007/978-3-319-76348-4_14

In this paper, we classified the image scenes into indoor and outdoor scenes. The indoor/outdoor detection remains an open research field when we are aiming for seamless positioning and context-aware sensing. But there are some challenges in this field. The main challenge is that our algorithm must be efficient in semi-indoor/outdoor environments like balcony of a building, lawn in a house, theme parks etc. Another challenge is to extract the discriminated features from complex scene images. The dominant features which we extract from the image should be precise and able to give optimal result for classification. In this paper, proposed scene recognition system works under, strong light changing conditions and varieties in the vision, in addition to clamor and impediments as such our visual acknowledgment framework. In the recognition process, consistent data about the object exploitation factor is much needed one.

2   Related Work

The indoor and outdoor scene recognition is a challenging problem in computer vision. Most of the scene recognition models [1,2] computes well on outdoor scenes and perform inadequately in the indoor environment. Scene recognition and classification [3] or scene categorization has been broadly carried out in various environments. Authors, Szummer and Picard [4] demonstrated that the classification performance can be enhanced by measuring the features on sub blocks, categorizing these sub blocks, and then merging these results in a way similar to stacking for scene classification. Luo and Savakis [5], recommended the use of a Bayesian network for collecting low-level features such as color, texture and semantic features for classification of images into indoor and outdoor scenes. Result shows that the classification efficiency has been significantly increased when semantic features are used in the classification. Kim et al. [6] proposed an approach for indoor/outdoor scene classification using edge and color orientation histogram (ECOH) with SVM classifier. Li et al. [7] integrated information from mid-level in order to obtain superior scene description. The low-level descriptions obtained from the pixel which is useful and their performance can be improved by computing the mid-level region information. Experimental results show that it is possible to obtain proportionate results of low level features with the help of mid-level features. By combining with low-level features, the classification results get improved. Chen et al. [8] suggested a method for indoor scene understanding by RGB-Depth images. Shahriari and Bergevin [9] proposed a two-stage Convolutional Neural Networks (CNN) for outdoor-indoor scene classification. In addition, Ren et al. [10] presented a learned feature related framework for indoor and outdoor scene labeling using CNN and their proposed Region Consistency Activation (RCA) algorithm. Experiments had shown that proposed method frequently produces better accuracy and visual consistency.

The rest of the paper is organized as follows. Section 3 presents the proposed approach, Sect. 4 discuss the feature extraction and fusion techniques. Section 5 presents the Experimental results. Finally, Sect. 6 concludes the paper.

3   Proposed Approach

This paper deals with scene recognition and classification that aims to understand texture patterns from static images of various environments. The workflow
of the proposed approach is shown in Fig. 1. Initially, input images are preprocessed by the median filter to eliminate noise ratio for fine feature extraction and
classification. The proposed approach is evaluated using 15 class scene dataset
[3,11,12]. In feature extraction phase, GIST descriptor (global features) is calculated by Convolving the image by applying 32 Gabor filters on 4 different scales
and 8 orientations, which gives 32 feature maps of an input image having same
size. Further it is divided into 16 sub regions and the averaged values of the each
region is concatenated of all 32 feature maps, gives $16 \times 32 = 512$ dimensional
feature vector. GIST features produces a high level representation of an input
image by gradient information for different regions.

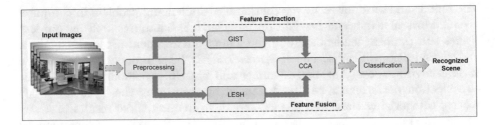

Fig. 1. Overview of the proposed system.

Secondly, LESH descriptors (shape histogram) are extracted from 64 different sub-regions of the input scene image by applying Gaussian filter orientations
on each region (8 bin local histogram), which gives $64 \times 8 = 512$ dimensional
feature vector. The extracted GIST (512) and LESH (512) features are combined
with Canonical Correlation Analysis (CCA) method for compact representation
of highly discriminated features which gives single 512 dimensional feature vector and the fused features are modeled by Random forest and KNN for scene
recognition.

4   Feature Extraction

In image processing, feature extraction is a unique compact representation by
reducing the dimension. Transforming the input image into the set of distinct
features is called feature extraction. The following sub sections, we discussed
about GIST and LESH feature extraction methods and CCA feature fusion
technique are presented.

4.1 GIST Descriptor

The GIST [11] descriptor was originally developed and used for image classification tasks. The GIST descriptor were calculated from an input image by convolving it with 32 Gabor filters at 8 orientations and 4 scales, which produces 32 feature maps of the similar resolution as the original input image, as global vectors providing us with a holistic representation with dominant spatial structure of a scene. Then, image is divided into a 4 × 4 grid, each feature map is further divided into 16 regions and each region values are averaged in order to obtain 512 dimensional feature vector (32 feature maps each comprising of 16 (4 × 4) points) and the corresponding illustration is shown in Fig. 2.

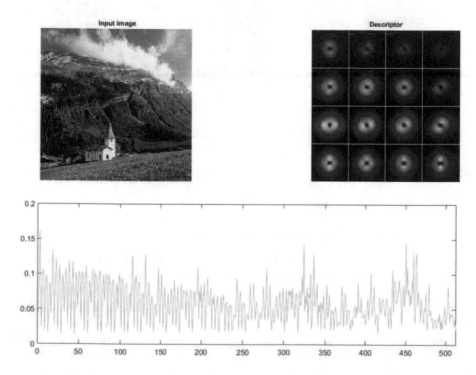

Fig. 2. Extracted GIST descriptor from 'mountain' image.

4.2 Local Energy-Based Shape Histogram (LESH)

The LESH [13] descriptors were utilized to represent the shapes to built local energy model for feature perception. The energy response varies on different orientation with respect to the primary shape which signifies the local energy. It comprises information like corners and edges, to create a local histogram and combining the local energy along with filter orientation in order to obtain 512 dimensional feature vectors and the corresponding illustration is shown in Fig. 3. The local energy histogram h is obtained by the following Eqs. (1) and (2):

$$h_{r,b} = \sum w_r \times E \times \delta(L - b) \tag{1}$$

Where, b represents the current bin value, L is the orientation label map, E is the local energy. w is a function of Gaussian weighing centered at region r.

$$w_r = \frac{1}{\sqrt{2\pi\sigma}} e^{\frac{\left[(x-r_{xo})^2 + (y-r_{yo})^2\right]}{\sigma^2}} \tag{2}$$

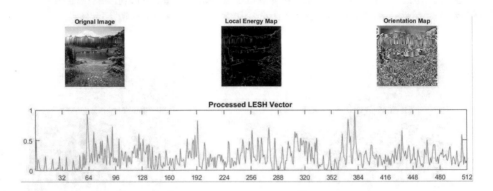

Fig. 3. Extracted LESH feature vectors (energy and orientation maps) from 'mountain' image.

4.3   Canonical Correlation Analysis (CCA)

Canonical correlation analysis (CCA) [14] is a method for finding the relationships among two multivariate sets (vectors). Initially, it calculates the input canonical coefficients of the $n \times D_1$ and $n \times D_2$ data matrices (features) M and N, where M and N should contains the similar number of observations (rows) but not limited to columns and finds the correlations between the two matrice. Then CCA will gives linear combinations of the M_i and N_j which have maximum correlation with each other. For example, sample feature points $(A_1, B_1), ...(A_{d_1}, B_{d_2})$, where A_{d_1} is d_1 dimensional feature and B_{d_2} is d_2 dimensional feature. Then find the directions u and v so that $u^T A$ and $v^T B$ is maximized. Finally the canonical correlation vector obtained by the linear dependence of two multivariate vectors and calculated by the following equation.

$$\rho = \max_{u,v} \frac{Cov\left[u^T A, \ v^T B\right]}{\sqrt{Var\left[u^T A\right]} \ Var\left[v^T B\right]} \tag{3}$$

5   Experimental Results

The experiments conducted on scene dataset consists of 15 categories collected from [3, 11, 12]. This dataset contains a broad range of indoor and outdoor scene environments at resolution of 250×300 with an average of 350 images per class and categories are class_bedroom, class_suburb, class_industrial, class_kitchen, class_living room, class_coast, class_forest, class_highway, class_inside city, class_mountain, class_open country, class_street, class_tall building, class_office and class_store. The sample images from 15-scene dataset are illustrated in Fig. 4 and two supervised classifiers are utilized for scene recognition, namely random forest [15] and K-Nearest Neighbor [16].

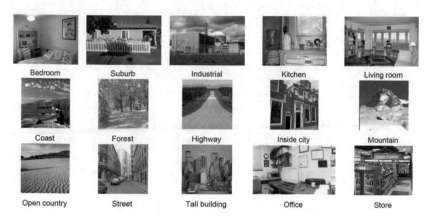

Fig. 4. Sample images from 15-scene dataset

The performance of the proposed approach is evaluated using 5 - fold cross-validation approach and evaluation metrices for the recognition system is calculated by

$$accuracy\ (A) = \frac{tp + tn}{tp + fp + tn + fn}, \qquad precision\ (P) = \frac{tp}{tp + fp}$$

$$recall\ (R) = \frac{tp}{tp + fn}, \qquad F\text{-}measure\ (F) = 2\frac{P \times R}{P + R}$$

where tp is the true positive (predicted as positive), tn is the true negative (predicted as negative), fp is the false positive (predicted incorrectly as positive), fn is the false negative (predicted incorrectly as negative) which are calculated from the confusion matrix.

The average recognition accuracies obtained from random forest and K-NN classifier are 73.62% and 81.62% respectively and the corresponding confusion matrix is shown in Figs. 5 and 6. The each scene category instance is represented by the rows and the scene category predicted by the classifier is represented by the columns. In random forest, the scenes like suburb, coast, forest, highway, tall building, open country, store and inside city are classified well with accuracy greater than 70%. From this, office, kitchen, living room and bedroom classes are imprecisely confused, where these classes intuitively seem hard to differentiate

	Bedroom	Suburb	Industrial	Kitchen	Living room	Coast	Forest	Highway	Inside city	Mountain	Open country	Street	Tall building	Office	Store
Bedroom	0.61	0.02	0.05	0.02	0.07	0.07	0	0	0.03	0	0.02	0.03	0.01	0.06	0
Suburb	0.01	0.79	0.03	0	0.01	0.08	0	0	0.03	0	0.02	0.02	0.01	0.01	0.01
Industrial	0.01	0.03	0.65	0.02	0.02	0.06	0	0.01	0.03	0.02	0.05	0	0.06	0.01	0.02
Kitchen	0.04	0.06	0.05	0.59	0.06	0.03	0	0	0.03	0	0.02	0.01	0.03	0.06	0.05
Living room	0.04	0.03	0.04	0.05	0.65	0.02	0	0.01	0.01	0.02	0.02	0.03	0.01	0.01	0.06
Coast	0	0	0	0	0.01	0.83	0	0.05	0	0.01	0.09	0	0	0	0
Forest	0	0	0	0	0	0.03	0.79	0	0.02	0.02	0.04	0.02	0.04	0	0.04
Highway	0	0	0	0	0	0.03	0	0.00	0	0.03	0.05	0.01	0	0	0.01
Inside city	0	0.03	0.02	0.02	0.01	0.02	0	0.01	0.75	0.01	0.02	0.04	0.03	0	0.03
Mountain	0.01	0.01	0.03	0	0.01	0.05	0.02	0.02	0.03	0.69	0.11	0.01	0.02	0	0.02
Open country	0	0	0.01	0	0	0.08	0.02	0.02	0.04	0.04	0.77	0.01	0	0	0.01
Street	0	0.03	0.01	0	0	0	0	0.03	0.09	0	0	0.8	0.02	0	0
Tall building	0	0	0.04	0	0.01	0.02	0.01	0.01	0.03	0.02	0.01	0.01	0.84	0	0.01
Office	0.05	0.04	0.05	0.03	0.03	0.13	0	0	0.02	0.01	0.01	0	0.02	0.61	0.01
Store	0.01	0.01	0.05	0.02	0.02	0.01	0.02	0	0.04	0.01	0	0	0.01	0.01	0.79

Fig. 5. Confusion matrix for the 15-scene dataset on random forest classifier.

	Bedroom	Suburb	Industrial	Kitchen	Living room	Coast	Forest	Highway	Inside city	Mountain	Open country	Street	Tall building	Office	Store
Bedroom	0.75	0.03	0.06	0.02	0.04	0.01	0.01	0.01	0.01	0	0	0	0.02	0.02	0
Suburb	0.02	0.75	0.07	0.03	0.02	0.03	0	0.02	0.01	0.01	0.02	0.01	0.01	0.01	0.02
Industrial	0.02	0.02	0.77	0.02	0.02	0.03	0	0	0.02	0.01	0.01	0.01	0.05	0	0.02
Kitchen	0.05	0.06	0.04	0.72	0.03	0.01	0.01	0.01	0.02	0.02	0	0.02	0.02	0.01	0.02
Living room	0.02	0.01	0.04	0.06	0.81	0.01	0	0	0	0	0.01	0.01	0	0.01	0.01
Coast	0	0.01	0.02	0.01	0.02	0.86	0	0.03	0.01	0.01	0.02	0	0	0	0
Forest	0	0	0	0	0	0.01	0.93	0	0.01	0.02	0.01	0	0	0	0
Highway	0.01	0.02	0	0.01	0.01	0.03	0	0.9	0.01	0.01	0	0	0	0	0
Inside city	0.02	0.01	0.02	0.01	0.01	0.01	0.02	0	0.81	0.01	0.02	0.02	0.01	0	0.01
Mountain	0.01	0.01	0.03	0.01	0.01	0.01	0.03	0.01	0	0.77	0.07	0	0.03	0	0.01
Open country	0.01	0.01	0.02	0	0	0.05	0.02	0.02	0	0.05	0.82	0	0	0	0
Street	0	0.01	0.04	0.01	0.01	0.01	0.02	0.01	0.07	0.01	0.01	0.77	0.03	0	0.01
Tall building	0.01	0	0.02	0.01	0.01	0.01	0.01	0	0.01	0.02	0.01	0	0.89	0	0
Office	0.04	0.03	0.03	0.03	0.03	0.03	0	0	0.01	0.01	0	0	0.01	0.78	0.01
Store	0.01	0.01	0.05	0.02	0.02	0.01	0.02	0	0.04	0.01	0	0	0.01	0.01	0.79

Fig. 6. Confusion matrix for the 15-scene dataset on KNN classifier.

and it needs further attention. In K-NN, the scenes like living room, coast, forest highway, inside city open country and tall building classified with accuracy greater than 75% to 90%. From this, kitchen class is mis-predicted as bedroom and suburb, additionally, street class is confused as inside street classes, where these classes intuitively seem hard to differentiate and it needs further attention.

Table 1. Performance measure of the 15-scene dataset on random forest and KNN classifier.

Class	Random forest			K-NN		
	Precision	Recall	F-measure	Precision	Recall	F-measure
class_bedroom	0.773481	0.614035	0.684597	0.744589	0.754386	0.749455
class_suburb	0.670996	0.786802	0.724299	0.706731	0.746193	0.725926
class_industrial	0.674342	0.644654	0.659164	0.677686	0.773585	0.722467
class_kitchen	0.686747	0.584615	0.631579	0.696517	0.717949	0.707071
class_living room	0.738956	0.650177	0.691729	0.807018	0.812721	0.809859
class_coast	0.654064	0.833735	0.733051	0.822171	0.857831	0.839623
class_forest	0.911111	0.785942	0.843911	0.856305	0.932907	0.892966
class_highway	0.812883	0.854839	0.833333	0.891026	0.896774	0.893891
class_inside city	0.672872	0.750742	0.709677	0.824773	0.810089	0.817365
class_mountain	0.815152	0.686224	0.745152	0.831492	0.767857	0.798408
class_open country	0.658482	0.770235	0.709988	0.835106	0.819843	0.827404
class_street	0.768908	0.802632	0.785408	0.884422	0.77193	0.824356
class_tall building	0.810606	0.83812	0.824134	0.859296	0.89295	0.8758
class_office	0.741935	0.605263	0.666667	0.89697	0.778947	0.833803
class_store	0.737762	0.674121	0.704508	0.878571	0.785942	0.82968
Average	**0.741886**	**0.725476**	**0.729813**	**0.814178**	**0.807994**	**0.809872**

Table 1 shows the statistical measures are computed from the confusion matrix obtained from the random forest and K-NN classifier. However, random forest classifier produces nominal results on 15-scene dataset. In this case, the average performance reported by overall Precision (P) = 74.18%, Recall (R) = 72.54% and F-measure = 72.98% (trade-off between precision and recall). In KNN, the overall average performance of Precision (P) = 81.41%, Recall (R) = 80.79% and F-measure = 80.98% results are consistently better than the random forest, is almost 8% improved on the KNN classifier.

6   Conclusion

In this paper, a novel approach was introduced for indoor and outdoor scene recognition. The experiments are conducted on scene dataset considering 15 different indoor and outdoor environments addressed various challenges. The proposed features extraction was obtained by fusing the global descriptor (GIST) and Local Energy based Shape Histogram (LESH) descriptor with Canonical Correlation Analysis (CCA) and modeled using random forest and K-NN classifiers. Experimental results show an overall accuracy of random forest and K-NN as 73.62% and 81.62% respectively. The performance result indicates that K-NN outperforms random forest. It is observed from the experiments that the system could not distinguish kitchen, bedroom, suburb and street with high accuracy and is further interest.

References

1. Quattoni, A., Torralba, A.: Recognizing indoor scenes. In: IEEE Conference on Computer Vision and Pattern Recognition (CVPR 2009), pp. 413–420. IEEE (2009)
2. Niu, Z., Hua, G., Gao, X., Tian, Q.: Context aware topic model for scene recognition. In: IEEE Conference on Computer Vision and Pattern Recognition (CVPR), pp. 2743–2750. IEEE (2012)
3. Fei-Fei, L., Perona, P.: A Bayesian hierarchical model for learning natural scene categories. In: IEEE Computer Society Conference on Computer Vision and Pattern Recognition (CVPR 2005), vol. 2, pp. 524–531. IEEE (2005)
4. Szummer, M., Picard, R.W.: Indoor-outdoor image classification. In: Proceedings of the IEEE International Workshop on Content-Based Access of Image and Video Database, pp. 42–51. IEEE (1998)
5. Luo, J., Savakis, A.: Indoor vs outdoor classification of consumer photographs using low-level and semantic features. In: Proceedings of the International Conference on Image Processing, vol. 2, pp. 745–748. IEEE (2001)
6. Kim, W., Park, J., Kim, C.: A novel method for efficient indoor-outdoor image classification. J. Signal Process. Syst. **61**(3), 251–258 (2010)
7. Li, Q., Wu, J., Tu, Z.: Harvesting mid-level visual concepts from large-scale internet images. In: Proceedings of the IEEE Conference on Computer Vision and Pattern Recognition, pp. 851–858 (2013)
8. Chen, Y., Pan, D., Pan, Y., Liu, S., Gu, A., Wang, M.: Indoor scene understanding via monocular RGB-D images. Inf. Sci. **320**, 361–371 (2015)
9. Shahriari, M., Bergevin, R.: A two-stage outdoor-indoor scene classification framework: experimental study for the outdoor stage. In: International Conference on Digital Image Computing: Techniques and Applications (DICTA), pp. 1–8. IEEE (2016)
10. Ren, Y., Chen, C., Li, S., Kuo, C.C.J.: GAL: a global-attributes assisted labeling system for outdoor scenes. J. Vis. Commun. Image Represent. **42**, 192–206 (2017)
11. Oliva, A., Torralba, A.: Modeling the shape of the scene: a holistic representation of the spatial envelope. Int. J. Comput. Vis. **42**(3), 145–175 (2001)
12. Lazebnik, S., Schmid, C., Ponce, J.: Beyond bags of features: spatial pyramid matching for recognizing natural scene categories. In: IEEE Computer Society Conference on Computer Vision and Pattern Recognition, vol. 2, pp. 2169–2178. IEEE (2006)
13. Wajid, S.K., Hussain, A.: Local energy-based shape histogram feature extraction technique for breast cancer diagnosis. Expert Syst. Appl. **42**(20), 6990–6999 (2015)
14. Thompson, B.: Canonical correlation analysis. In: Encyclopedia of Statistics in Behavioral Science (2005)

15. Liaw, A., Wiener, M., et al.: Classification and regression by randomForest. R News **2**(3), 18–22 (2002)
16. Wu, X., Kumar, V., Quinlan, J.R., Ghosh, J., Yang, Q., Motoda, H., McLachlan, G.J., Ng, A., Liu, B., Philip, S.Y., et al.: Top 10 algorithms in data mining. Knowl. Inf. Syst. **14**(1), 1–37 (2008)

A Mixed-Integer Linear Programming Model and a Simulated Annealing Algorithm for the Long-Term Preventive Maintenance Scheduling Problem

Roberto D. Aquino(✉), Jonatas B. C. Chagas, and Marcone J. F. Souza

Departamento de Computação, Universidade Federal de Ouro Preto,
Ouro Preto, Brazil
roberto.dias@vale.com, {jonatas.chagas,marcone}@iceb.ufop.br

Abstract. This paper addresses a problem arising in the long-term maintenance programming of an iron ore processing plant of a company in Brazil. The problem is a complex maintenance programming where we have to assign the equipment preventive programming orders to the available work teams over a 52 week planning. We first developed a general mixed integer programming model which was not able for solving real instances using the CPLEX optimizer. Therefore, we also proposed a heuristic approach, based on the Simulated Annealing meta-heuristic, that was able to handle the instances.

Keywords: Long-term maintenance programming · Scheduling
Simulated Annealing · Combinatorial optimization · Heuristic

1 Introduction

This work has its focus on the problem of assigning the equipment preventive maintenance orders to the work teams over a specific time period. The goal is to maximize the number of orders performed and minimize the number of work teams necessary to execute them.

In real scenarios, we have to take into account specific constraints related to the maintenance order programming. Most of these constraints are directly related to the scheduling problems, widely studied in the literature. The work team (a group of workers that have to work together) cannot perform more than one maintenance order (task, for short) at the same time. Moreover, each equipment cannot have more than one task being executed at the same time. Each task also has a time window limit that must be obeyed. Besides above-mentioned constraints, each task has to be performed by a team that has the ability to perform it. Furthermore, there exists usually more than one team that is able to execute the same task.

At the studied company, the long-term maintenance programming is made by the engineering team using a specific maintenance software. It is a general

© Springer International Publishing AG, part of Springer Nature 2018
A. Abraham et al. (Eds.): ISDA 2017, AISC 736, pp. 144–153, 2018.
https://doi.org/10.1007/978-3-319-76348-4_15

maintenance software that manages all preventive and corrective maintenance activities such as work order, inventory, and material management. It also has the data of all maintenance procedures performed and all the maintenance plans. The maintenance plan for each equipment includes all the preventive tasks that have to be performed, the frequency that each task has to be performed, the expected time to complete the task, the team which is able to perform the task and the materials and tools that are necessary. From this database, it is built 52 weeks preventive maintenance plan, which consists of assigning the tasks to the teams. The problem is that the maintenance plan is unfeasible and it has to be rescheduled. To do it, the maintenance team works in extra hours, but even so, some tasks are not allocated.

For solving this problem, we developed a general mixed integer programming model and also a heuristic approach, based on the Simulated Annealing meta-heuristic.

The remaining of this paper is organized as follows. In Sect. 2, we give a brief review on existing papers on the related subjects. Section 3 shows the mathematical formulation for the addressed problem, which formally describes the problem. Section 4 presents a heuristic approach, based on the Simulated Annealing meta-heuristic, for solving the problem. The results are discussed in Sect. 5 and the conclusions are presented in Sect. 6.

2   Related Papers

Maintenance optimization problems is not a young research field. Sharma et al. [1] made an extensive review in this area. They reviewed 104 articles starting from the early 1960s. According to these authors, maintenance optimization could have several optimization criteria such as maintenance cost rate, profitability, plant utilization, performance efficiency and worker safety. On our work, the maintenance cost is what we want to minimize. Although there are some articles on optimization of maintenance cost, they are all focused on the maintenance strategy such as Reliability Centered Maintenance (RCM), Total Productive Maintenance (TPM) and Plant Asset Management (PAM). All these techniques have several trade-offs that have to be balanced to give an optimal solution. Our work is based on an already maintenance program where all the maintenance orders, frequency, and duration are known. Our focus is then scheduled these maintenance orders.

The authors from [2] made a review on maintenance performance measurement published in 67 journals between 1969 and 2009. They showed that cost is the subject that had the higher number occurrences. As in [1], the cost was also focused on maintenance strategy.

Yamayee and Sidenblad [3] proposed a mathematical formulation for optimal preventive maintenance scheduling problem and used dynamic programming as a framework to solve it. The problem was to schedule 21 maintenance orders with different capacities and different costs.

Yao et al. [4] proposed a mixed integer programming model for the short-term preventive maintenance scheduling for 29 maintenance orders distributed on 11 different tools associated with a time window. Scheduling several tasks to the same tool is equivalent to scheduling to the same equipment in our problem. The objective is to maximize the overall tool availability and minimize unavailability during the periods when a significant amount of work is expected.

Saraiva et al. [5] proposed a Simulated Annealing (SA) approach to schedule the preventive maintenance of thermal generators. The goal is to minimize the yield loss related to the downtime of the generators. The case study had 29 different generators and each had one maintenance order. As in our problem, each maintenance order has a time window.

The main difference of our problem from the ones that we found in the literature is the objective function. Here we want to maximize the number of orders executed and try to minimize the number of teams necessary to execute them. The greatest challenge is the huge size of it. In our problem, we have more than 30,000 maintenance orders to be executed.

3   Problem Definition and Mathematical Model

The Long-term Preventive Maintenance Scheduling Problem (LTPMSP) can be formally described as follows. Let $E = \{1, 2, \cdots, Q\}$ be the set of Q industrial equipment's which must be realized preventive maintenance. Let also $\mathcal{T} = \{1, 2, \cdots, N\}$ be the set of N preventive maintenance, and let $\mathcal{W} = \{1, 2, \cdots, M\}$ be the set of M work teams available and responsible to realize them. The processing time of the preventive maintenance i is denoted by P_i. Each preventive maintenance $i \in \mathcal{T}$ is associated with a single equipment E_i and just one work team can perform it, but two or more different type of maintenance may be associated with the same equipment.

In addition, each preventive maintenance has a time window, i.e., an interval $[e_i, l_i]$, where e_i and l_i correspond, respectively, to the earliest and the latest time available to perform the preventive maintenance i. Each preventive maintenance $i \in \mathcal{T}$ requires a specific ability to be performed, being that $\mathcal{W}_i \subseteq \mathcal{W}$ indicates the set of work teams with the ability to perform it. If the preventive maintenance i is not performed, there is a cost C_i that must be paid. Each work team $k \in \mathcal{W}$ can process at most one preventive maintenance at a time and has availability to work by h_k continuous hours. The set $\mathcal{T}_k \subseteq \mathcal{T}$ indicates the preventive maintenance that can be executed by the work team k.

The aim of the LTPMSP is to determine a scheduling plan to perform as many preventive maintenance as possible in order to minimize the required number of work teams.

In order to formulate the problem, we define five sets of decision variables, which are described as follows:

- x_{ij}^k: binary variable that gets 1 if maintenance i is performed immediately before maintenance j by the work team k; 0, otherwise;
- y_{ik}: binary variable that gets 1 if maintenance i is performed by the work team k; 0, otherwise;
- z_k: binary variable that gets 1 if work team k is used; 0, otherwise;

- c_{ik}: completion time of the maintenance i when it is performed by the work team k;
- r_{ij}: binary variable that gets 1 if maintenance i is performed before maintenance j and 0, otherwise.

With these variables we can describe the LTPMSP by the Mixed-Integer Linear Programming (MILP) formulation expressed by Eqs. (1)–(17):

$$\min \sum_{k \in \mathcal{W}} z_k + \sum_{i \in \mathcal{T}} C_i \left(1 - \sum_{k \in \mathcal{W}_i} y_{ik} \right) \tag{1}$$

$$\sum_{k \in \mathcal{W}_i} y_{ik} \leq 1 \qquad\qquad i \in \mathcal{T} \quad (2)$$

$$\sum_{i \in \mathcal{T}_k \cup \{0\} \setminus \{j\}} x_{ij}^k = y_{jk} \qquad\qquad j \in \mathcal{T}, k \in \mathcal{W}_j \quad (3)$$

$$\sum_{j \in \mathcal{T}_k} x_{0j}^k = z_k \qquad\qquad k \in \mathcal{W} \quad (4)$$

$$\sum_{i \in \mathcal{T}_k \cup \{0\} \setminus \{l\}} x_{il}^k = \sum_{j \in \mathcal{T}_k \cup \{0\} \setminus \{l\}} x_{lj}^k \qquad\qquad k \in \mathcal{W}, l \in \mathcal{T}_k \quad (5)$$

$$c_{0k} = 0 \qquad\qquad k \in \mathcal{W} \quad (6)$$

$$c_{jk} \geq c_{ik} + P_j - M_{ij}'(1 - x_{ij}^k) \qquad\qquad k \in \mathcal{W}, i \in \mathcal{T}_k \cup \{0\}, j \in \mathcal{T}_k \quad (7)$$

$$c_{ik} \geq (e_i + P_i) y_{ik} \qquad\qquad k \in \mathcal{W}, i \in \mathcal{T}_k \quad (8)$$

$$c_{ik} \leq l_i \qquad\qquad k \in \mathcal{W}, i \in \mathcal{T}_k \quad (9)$$

$$c_{jk'} \geq c_{ik} + P_j - M_{ij}'(1 - r_{ij}) \qquad\qquad \begin{array}{c} k \in \mathcal{W}, k' \in \mathcal{W}, i \in \mathcal{T}_k, j \in \mathcal{T}_{k'}, \\ | \, k \neq k', i < j, E_i = E_j \end{array} \quad (10)$$

$$c_{jk'} \leq c_{ik} - P_i + M_{ij}'' r_{ij} \qquad\qquad \begin{array}{c} k \in \mathcal{W}, k' \in \mathcal{W}, i \in \mathcal{T}_k, j \in \mathcal{T}_{k'}, \\ | \, k \neq k', i < j, E_i = E_j \end{array} \quad (11)$$

$$c_{ik} \leq h_k \qquad\qquad k \in \mathcal{W}, i \in \mathcal{T}_k \quad (12)$$

$$x_{ij}^k \in \{0,1\} \qquad\qquad k \in \mathcal{W}, i \in \mathcal{T}_k \cup \{0\}, j \in \mathcal{T}_k \cup \{0\} \quad (13)$$

$$y_{ik} \in \{0,1\} \qquad\qquad k \in \mathcal{W}, i \in \mathcal{T}_k \cup \{0\} \quad (14)$$

$$z_k \in \{0,1\} \qquad\qquad k \in \mathcal{W} \quad (15)$$

$$c_{ik} \geq 0 \qquad\qquad k \in \mathcal{W}, i \in \mathcal{T}_k \quad (16)$$

$$r_{ij} \in \{0,1\} \qquad\qquad i \in \mathcal{T}, j \in \mathcal{T} \, | \, i < j, E_i = E_j \quad (17)$$

We define a fictitious maintenance 0 that precedes immediately the first maintenance and follows immediately the last maintenance performed by each work team. The completion time of this maintenance is zero for all work teams, as imposed by the constraint (6).

The objective function (1) minimizes the total number of work teams, while maximizes the number of maintenance performed through the minimization of the penalties. Constraint (2) ensures that each maintenance is performed by at most one work team. Constraint (3) guarantees if a work team k performs a maintenance j, that maintenance must be contained in the maintenance schedule of work team k. By constraint (4), if at least one maintenance is assigned to the work team k, this work team is used. Constraint (5) ensures the continuity of the maintenance schedule of each work team. Constraint (7) calculates the completion time of all preventive maintenance. The constraints (8) and (9) force that all maintenance is performed in their respective time windows. The constraints (10) and (11) ensure that two or more maintenance orders are not performed at the same time on the same equipment. Note that these last two constraints can be applied only between different work crews because there will be no overlap of execution of maintenance by the same work team, it is ensured by the predecessor constraints. The constraint (12) ensures that the number of working hours of work teams is not exceeded. And, finally, the constraints (13) to (17) define the scope and domain of the decision variables.

The constants M'_{ij} and M''_{ij} in (7), (10) and (11) can be any sufficiently large number that is greater or equal to $l_i + P_j$ and $l_j + P_i$, respectively.

4   Heuristic Approach

Due to the huge size of real instances, the MILP model was not even able to run, running out of memory in a short time. For small instances, the MILP model was able to give an optimal solution for some instances but not for others with a time limit of one hour.

In order to solve real (large) instances and find high-quality solutions with low computational time, in this section we describe an algorithm based on the Simulated Annealing meta-heuristic for the LTPMSP.

4.1   Solution Representation and Evaluation Function

A solution to the problem is represented by a permutation $\pi = \langle \pi_1, \pi_2, ..., \pi_N \rangle$ of the N maintenance orders. The evaluation of the quality of a solution π is done as follows. Consider a procedure where the tasks are sequentially allocated to the first available time slot that has a duration greater or equal to the task duration. For each task, the available time slot is built checking its time window and the non-allocated time of the teams. When all the allocation conditions are met, the task is allocated at the beginning of the available time slot; otherwise, a penalty is incurred.

To illustrate, consider an instance of 6 maintenance orders involving 3 types of equipment, which should be executed by 3 teams. The maintenance orders 1, 2, 3 and 6 can be executed by the work teams 1 and 2, and the maintenance orders 4 and 5 by work team 2. The maintenance orders 1, 3 and 5 must be executed at equipment type 1, while the maintenance orders 2 and 6 at equipment type 2 and maintenance order 3 at equipment type 3. The time windows of the maintenance orders start at the following time: 0, 2, 3, 2, 3, 4; and end in the following time:

4, 7, 9, 6, 8, 7. The processing times of the tasks are the following: 1, 2, 3, 1, 2, 1. The penalty for not performing a maintenance order is the following: 20, 30, 40, 20, 30, 40. Figure 1 shows an allocation to these permutation $\pi = \langle 1, 2, 3, 4, 5, 6 \rangle$. The time window of each maintenance order is represented by a horizontal bar, and the time and duration are represented by the filled part.

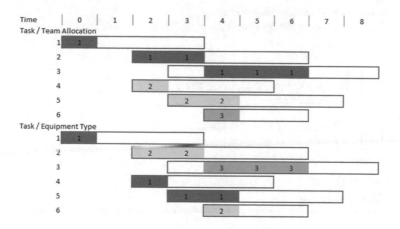

Fig. 1. Allocation example.

4.2   Initial Solution and Neighborhood Structure

The initial solution is obtained choosing any random permutation π of N maintenance orders.

In order to explore the solution space, we define a simple neighborhood structure that consists of exchanging two positions of the permutation π. Figure 2 shows a solution s and one of its neighbors s' for a instance with 6 maintenance orders. All neighbors of a solution s are represented by $N(s)$.

$$s = \langle\ 1,\ \mathbf{2},\ 3,\ 4,\ \mathbf{5},\ 6\ \rangle \qquad\qquad s' = \langle\ 1,\ \mathbf{5},\ 3,\ 4,\ \mathbf{2},\ 6\ \rangle$$

Fig. 2. A solution s and a neighbour s'.

4.3   Simulated Annealing

Proposed by Kirkpatrick et al. in [6], Simulated Annealing (SA) is a probabilistic meta-heuristic, which is inspired by the annealing process of steel thermodynamic optimization. Algorithm 1 describes the pseudo-code of SA for solving the LTMPSP.

Algorithm 1. Simulated Annealing (SA)

---

1    $s \leftarrow$ initial solution
2    $bestSolution \leftarrow s$
3    $t \leftarrow$ maxTemperature
4 **while** $t > minTemperature$ **do**
5      $iter \leftarrow 0$
6   **while** $iter < maxIterations$ **do**
7          $s' \leftarrow$ select a random solution $s' \in N(s)$
8          $\Delta \leftarrow f(s') - f(s)$
9   **if** $\Delta < 0$ **then**
10            $s \leftarrow s'$
11   **if** $f(s') < f(bestSolution)$ **then**
12              $bestSolution \leftarrow s'$
13   **else if** $rand(0,1) < e^{-\Delta/t}$ **then**
14            $s \leftarrow s'$
15          $iter \leftarrow iter + 1$
16      $t \leftarrow t \times (1 - \alpha)$
17 **return** $bestSolution$

The algorithm's initial solution is generated randomly and it is considered as the best solution so far. While the minimum temperature and the maximum number of iterations are not achieved, the algorithm randomly selects a solution nearby the current solution and accepts it according to a probability function given by the method, selecting it as the best solution so far if it is better than the last best solution. After *maxIterations* iterations, the temperature is updated by a cooling rate $\alpha \in [0,1]$ (line 16 of Algorithm 1). The algorithm ends when the minimum temperature is reached.

5   Computational Experiments

Our proposed MILP formulation was coded in C++ using the Concert Technology Library of CPLEX 12.5 Academic Version, with default settings, except for the runtime that was limited to 1 h. The SA algorithm was coded in C++ and executed sequentially. All experiments were performed on an Intel Core i5-4440 CPU @ 3.10 GHz x 4 computer, 8 GB RAM, Ubuntu 14.04 LTS 64 bits.

The real instance of the studied one iron ore processing plant consists of 33,484 preventive maintenance orders involving 1,032 equipment types to be allocated to 145 work teams. As already mentioned, the MILP formulation is not able to solve real-size instances. In order to compare the SA algorithm to the MILP, we have created 60 different instances, which are sub-instances of the real instance. As reported in Table 1, these instances differ from each other by the number of preventive maintenance orders, the number of equipment types and the number of work teams.

5.1   Parameter Calibration

The proposed Simulated Annealing algorithm has 3 parameters: maximum temperature (*maxTemperature*), minimum temperature (*minTemperature*), maximum number of iterations (*maxIterations*) and cooling rate (α). To make a fare calibration of these parameters we used an automated algorithm called Irace (Iterated Racing for Automatic Algorithm Configuration) [7]. This algorithm was designed to give the most appropriate parameters for an optimization algorithm and a set of instances. Irace runs as a package of the R software. The R is a free environment for statistical computing and graphics.

Experiments were conducted using a sample of 20 training instances. Different values were tested for each parameter, defined as follows: *maxTemperature* $\in \{10, 20, 50, 100, 500, 1000\}$; *minTemperature* $\in \{0.1, 1.0, 5.0\}$; *maxIterations* $\in \{1N, 2N, 5N, 10N\}$; $\alpha \in \{0.001, 0.002, 0.005, 0.01, 0.02, 0.05, 0.1\}$.

After running Irace we got the following parameters: *maxTemperature* $= 500$; *minTemperature* $- 0.1$; *maxIterations* $= 2N$; $\alpha = 0.002$.

5.2   Experimental Results

We now present the results obtained by our approaches for the LTPMSP. Notice that since there is no literature regarding this problem, we are only going to compare the results of our approaches.

In Table 1, the three first columns describe the instances, where the columns Q, N and M inform, respectively, the number of equipment types, the number of preventive maintenance orders and the number work teams available. The results obtained by the MILP formulation are described in the columns *Obj*, *#T*, *#P*, *Gap* and $t(s)$. The column *Obj* shows the solution value obtained at the end of the computation, *#T* informs the number of work teams used to perform the *#P* preventive maintenance orders, *Gap* shows the relative gap between upper bound (*Obj*) and *LB* computed as (*Obj* - *LB*)/*Obj*, where *LB* is the lower bound value at the end of the computation, and the column $t(s)$ informs the processing time in seconds spent by the MILP formulation. In the last six columns, we present the SA results. The column *Obj* shows the best solution value found after 10 executions of the SA algorithm, while the column σ informs the standard deviation of the all 10 solution's value. The columns *#T*, *#P* have the same meaning of the columns for the MILP formulation. The column $\frac{SA}{MILP}$ shows the ration between the solutions obtained by the MILP formulation and the SA algorithm. Finally, the column $t(s)$ shows the processing time in seconds required by the SA.

The results show that, within a 1 h time limit, our MILP formulation found the optimal solution (column Gap $= 0.00$) for 21 (35%) instances. It is also noticed that for 56 (93%) instances, the SA algorithm found solutions whose values are equal or better to those obtained by the MILP formulation.

According to the column σ, we can note that the SA algorithm presented good convergence since the standard deviation of the 10 solutions obtained is small (0.0 for most instances).

Table 1. Comparative analysis of the MILP formulation and the SA algorithm.

Instances			MILP					SA					
Q	N	M	Obj	$\#T$	$\#P$	Gap	$t(s)$	Obj	$\#T$	$\#P$	$\frac{SA}{MILP}$	σ	$t(s)$
2	20	2	434	2	18	0.00	0.3	434	2	18	1.00	0.0	5.0
		3	219	3	19	0.00	7.2	219	3	19	1.00	0.0	5.4
		4	219	3	19	0.00	82.0	219	3	19	1.00	0.0	5.7
	30	2	434	2	28	0.00	2.0	434	2	28	1.00	0.0	13.6
		3	219	3	29	0.00	10.7	219	3	29	1.00	0.0	15.3
		4	219	3	29	0.00	333.4	219	3	29	1.00	0.0	15.3
	40	2	434	2	38	0.00	57.2	650	2	37	1.50	0.0	27.1
		3	219	3	39	0.00	186.3	219	3	39	1.00	0.0	28.0
		4	220	4	39	0.01	1 h	219	3	39	1.00	0.0	29.6
	60	2	434	2	58	0.00	67.7	434	2	58	1.00	91.1	76.8
		3	219	3	59	0.00	1358.2	219	3	59	1.00	0.0	78.6
		4	220	4	59	1.00	1 h	219	3	59	1.00	0.0	80.0
	80	2	866	2	76	0.50	1 h	1082	2	75	1.25	0.0	139.3
		3	219	3	79	0.99	1 h	219	3	79	1.00	0.0	142.8
		4	436	4	78	1.00	1 h	219	3	79	0.50	0.0	150.2
3	20	3	579	3	17	0.00	2.1	579	3	17	1.00	0.0	4.9
		4	220	4	19	0.00	14.7	220	4	19	1.00	0.0	5.3
		5	220	4	19	0.00	129.9	220	4	19	1.00	0.0	5.5
	30	3	579	3	27	0.00	6.3	579	3	27	1.00	0.0	11.8
		4	220	4	29	0.00	38.3	220	4	29	1.00	0.0	12.4
		5	220	4	29	0.00	316.8	220	4	29	1.00	0.0	12.6
	40	3	579	3	37	0.00	807.3	579	3	37	1.00	69.6	25.6
		4	220	4	39	0.00	238.5	220	4	39	1.00	0.0	27.3
		5	221	5	39	0.01	1 h	220	4	39	1.00	0.0	27.9
	60	3	1515	3	51	0.76	1 h	1659	3	50	1.10	60.7	65.5
		4	220	4	59	0.00	1619.3	220	4	59	1.00	0.0	78.5
		5	221	5	59	0.01	1 h	220	4	59	1.00	0.0	85.5
	80	3	3891	3	59	0.32	1 h	1803	3	69	0.46	74.4	116.0
		4	2524	4	68	0.13	1 h	220	4	79	0.09	0.0	136.7
		5	2525	5	68	0.26	1 h	220	4	79	0.09	0.0	140.4
4	20	3	723	3	16	0.00	530.6	723	3	16	1.00	0.0	4.8
		4	220	4	19	0.00	1862.6	220	4	19	1.00	0.0	5.5
		5	220	4	19	0.01	1 h	220	4	19	1.00	0.0	5.7
	30	3	723	3	26	0.40	1 h	723	3	26	1.00	0.0	11.6
		4	220	4	29	0.01	1 h	220	4	29	1.00	0.0	12.5
		5	221	5	29	0.02	1 h	220	4	29	1.00	0.0	13.0
	40	3	867	3	35	1.00	1 h	867	3	35	1.00	69.6	22.4
		4	220	4	39	0.01	1 h	220	4	39	1.00	0.0	24.8
		5	221	5	39	0.02	1 h	220	4	39	1.00	0.0	25.3
	60	3	2163	3	47	1.00	1 h	2307	3	46	1.07	0.0	62.2
		4	652	4	57	0.73	1 h	220	4	59	0.34	0.0	75.0
		5	221	5	59	0.61	1 h	220	4	59	1.00	0.0	78.0
	80	3	3171	3	62	1.00	1 h	3099	3	58	0.98	60.7	98.8
		4	220	4	79	0.01	1 h	220	4	79	1.00	0.0	129.9
		5	221	5	79	0.99	1 h	220	4	79	1.00	0.0	131.9
5	20	4	724	4	16	0.00	62.9	724	4	16	1.00	0.0	4.8
		5	221	5	19	0.00	277.8	221	5	19	1.00	0.0	5.4
		6	221	5	19	0.01	1 h	221	5	19	1.00	0.0	5.6
	30	4	724	4	26	0.20	1 h	724	4	26	1.00	0.0	10.4
		5	221	5	29	0.01	1 h	221	5	29	1.00	0.0	11.3
		6	221	5	29	0.01	1 h	221	5	29	1.00	0.0	11.7
	40	4	868	4	35	1.00	1 h	868	4	35	1.00	0.0	20.2
		5	221	5	39	0.01	1 h	221	5	39	1.00	0.0	21.8
		6	222	6	39	0.02	1 h	221	5	39	1.00	0.0	22.6
	60	4	1588	4	50	1.00	1 h	1588	4	50	1.00	0.0	55.5
		5	221	5	59	0.02	1 h	221	5	59	1.00	0.0	63.7
		6	222	6	59	0.02	1 h	221	5	59	1.00	0.0	65.1
	80	4	3388	4	61	0.92	1 h	3244	4	61	0.96	0.0	91.9
		5	509	5	78	0.43	1 h	221	5	78	0.43	0.0	127.3
		6	510	6	78	0.43	1 h	221	5	77	0.43	0.0	126.1

Regarding the computational time and the solution's quality, we can see that the SA algorithm is more efficient while compared to MILP formulation. Just for smaller instances, the MILP was faster than SA algorithm, and by the column $\frac{SA}{MILP}$ we can note that the relation between the two methods is less than or equal to 1.00 for most the instances, which indicates the SA algorithm obtained better or equal solutions, respectively.

For the real instance, our SA algorithm would take approximately 102 days to run, using the parameter values described in Sect. 5.1. To get a result in a reasonable time, we changed the *maxTemperature* to 100, *minTemperature* to 1 and α to 0.01. With these parameters 90% of the maintenance orders were allocated.

6 Conclusions

In this paper, we examined different optimization approach proposed in the literature and built our own MILP model based on the real objective function and restrictions of a preventive maintenance of an existing iron ore processing plant. This MILP model is very straightforward and helps to understand the problem mathematically. Although its easiness to understand, it could not solve bigger instances. In order to get a reasonable solution, we proposed a constructive solution and a Simulated Annealing approach that was able to handle the real instance with good results.

Acknowledgements. The authors thank FAPEMIG, CNPq and UFOP for supporting this research.

References

1. Sharma, A., Yadava, G.S., Deshmukh, S.G.: A literature review and future perspectives on maintenance optimization. J. Qual. Maint. Eng. **17**(1), 5–25 (2011)
2. Simões, J.M., Gomes, C.F., Yasin, M.M.: A literature review of maintenance performance measurement: a conceptual framework and directions for future research. J. Qual. Maint. Eng. **17**(2), 116–137 (2011)
3. Yamayee, Z., Sidenblad, K., Yoshimura, M.: A computationally efficient optimal maintenance scheduling method. IEEE Trans. Power Appar. Syst. **102**(2), 330–338 (1983)
4. Yao, X., Fernández-Gaucherand, E., Fu, M.C., Marcus, S.I.: Optimal preventive maintenance scheduling in semiconductor manufacturing. IEEE Trans. Semicond. Manuf. **17**(3), 345–356 (2004)
5. Saraiva, J.T., Pereira, M.L., Mendes, V.T., Sousa, J.C.: A simulated annealing based approach to solve the generator maintenance scheduling problem. Electr. Power Syst. Res. **81**(7), 1283–1291 (2011)
6. Kirkpatrick, S., Gelatt, C.D., Vecchi, M.P.: Optimization by simulated annealing. Science **220**(4598), 671–680 (1983)
7. López-Ibáñez, M., Dubois-Lacoste, J., Cáceres, L.P., Birattari, M., Stützle, T.: The irace package: iterated racing for automatic algorithm configuration. Oper. Res. Perspect. **3**, 43–58 (2016)

Interval Valued Feature Selection
for Classification of Logo Images

D. S. Guru and N. Vinay Kumar[(✉)]

Department of Studies in Computer Science, University of Mysore,
Manasagangotri, Mysore 570006, Karnataka, India
dsg@compsci.uni-mysore.ac.in, vinaykumar.natraj@gmail.com

Abstract. A model for classification of logo images through a symbolic feature selection is proposed in this paper. The proposed model extracts three global features viz., color, texture, and shape from logo images. These features are then fused to emphasize the superiority of feature level fusion strategy. Due to the existence of large variations across the samples in each class, the samples are clustered and represented in the form of symbolic interval valued data during training. The symbolic feature selection is then adopted to show the efficacy of feature sub-setting in classifying the logo images. During testing, a query logo image is classified as one of the members of three classes with only few discriminable set of features using a suitable symbolic classifier. For experimentation purpose, a huge corpus of 5044 color logo images has been used. The proposed model is validated using suitable validity measures viz., f-measure, precision, recall, accuracy, and time. The results with the comparative analysis show the superiority of the symbolic feature selection method with that of without feature selection in terms of time and average f-measure.

Keywords: Logo image · K-Means clustering · Symbolic feature selection
Symbolic classification

1 Introduction

In the current digital era, with the proliferation of social networking sites, personal logging and digital forums, the size of image data being generated in the web is myriad. Managing such a huge amount of image data is more or less testing and at the same time it is an interesting research problem. Working in this direction, we can find a couple of readily available tools in the web such as Google image search engine, ArcGIS search engine, Bing search engine, Flickr search engine, etc. These tools are basically designed to perform classification, retrieval, and detection of images based on their visual characteristics. In this work, we attend to a problem of classification of logo images. A logo is a representation or a symbol which exemplifies the functionalities and responsibilities of an organization.

The freshness and exclusivity of a logo needs to be tested once it is designed for any organization. Otherwise many invaders can re-design the logos which appear very similar to the prevailing logos and perhaps affect the status of the particular organization.

© Springer International Publishing AG, part of Springer Nature 2018
A. Abraham et al. (Eds.): ISDA 2017, AISC 736, pp. 154–165, 2018.
https://doi.org/10.1007/978-3-319-76348-4_16

To evade such trade contravention or replication, a system to test a newly intended logo for its novelty is requisite. To test for the novelty, the system has to authenticate the newly intended logo by comparing with the prevailing logos. Since the quantity of logos obtainable for assessment is extremely big, either a swift approach for assessment or any other choice needs to be investigated. One such choice is to recognize the category of logos to which the newly intended logo belongs and then authenticating it by comparing against only those logos of the corresponding class. Hence, the course of classification reduces the search space of a logo authentication system to a larger extent. With this motivational background, a problem correlated to classification of logos based on their visual appearance is addressed.

In literature, a couple of works are found on logo image detection, retrieval and classification. In [1], a logo and trademark image detection model has been proposed which works based on wavelet co-occurrence histogram. Nourbakhsh et al. [2], proposed a logo image retrieval model based on polar image representation. Romberg et al. [3] proposes a work which majorly concentrates on detection and classification of logos from natural scene images. Kalantidis et al. [4], also proposed a model for detection and classification of logos from natural scene images based on Delaunay triangulation representation. Romberg and Lienhart [5] proposed a model based on bundle min-hashing technique which detects and classifies the logos extracted from natural scene images. In [6], a logo classification model has been designed to classify a logo image as either tainted or non-tainted logo image. University of Maryland (UMD) logo image database is used for experimentation. Sun and Chen [7] designed a logo classification prototype which classifies a logo image captured through high resolution mobile cameras. Arafat et al. [8] presented a comparative analysis of different invariant schemes associated with the logo image classification. In [9], the authors emphasize on the logo image classification using reasonably very large UoMLogo Database through features fusion. They have considered the fusion of global features for classification. They categorized logo images into three classes namely- logo images consists of both texts and symbols, logo images consists of only texts and logo images consists of only symbols. Further in [10], a logo image classification model based on symbolic interval valued representation of appearance based features for the UoMLogo Database is addressed. Due to the presence of large intra class variations, they preserved such intra class variations through interval valued representation of data which improved the classification rate compared to [9], which does not preserve intra-class variations.

It is suggested from the literature that the symbolic interval valued representation and classification [10] outperforms the existing works. Hence, in this work, the symbolic representation of logo images is recommended for classification. But, during classification, the common problem one can face is the curse of dimensionality [11]. Therefore, a symbolic feature selection method [12] which suits the interval valued data and is better among all the contemporary symbolic feature selection methods [13–16] is adopted. Perhaps, the major difference between this paper and [10] is the number of features used for classification of logo images. For classification of logo images, a suitable symbolic classifier which operates on interval valued data during training and a crisp (single valued) data during testing is needed. Hence, a symbolic classifier [17] which serves the purpose is recommended. The symbolic classifier [17] is recommended over contemporary symbolic classifiers [18, 19], due to the fact that these classifiers operate only on interval valued data both at training and testing stages.

The proposed classification model uses symbolic feature selection method for classification of logo images. During training, the pre-processing and features fusion are performed on training set of logo images. Due to the existence of large variations across the samples in each class, further the logo image samples are clustered and represented in the form of interval valued data. The symbolic feature selection is then adopted to show the efficacy of feature sub-setting in classifying the logo images. During testing, a query logo image is classified as one of the members of the three classes with only few discriminable set of features using suitable symbolic classifier.

The following are the major contributions of this paper.

1. Exploring the practical feasibility of the symbolic (interval valued) feature selection method for logo image classification.
2. Successful attempt in reducing the column and row dimension of the feature matrix during classification.
3. Extensive experimentation has been done to test the efficacy of the proposed model.

The organization of remaining parts of the paper is as follows. The details on the proposed model which explains the logo image classification through symbolic feature selection is presented in Sect. 2. In Sect. 3, the details of experimentation and results are given. The comparative analysis is given in Sect. 4 and finally Sect. 5 concludes the paper.

2   Proposed Model

The proposed model majorly comprised two phases namely, training phase and testing phase. The different steps followed during training and testing phases are shown in Fig. 1.

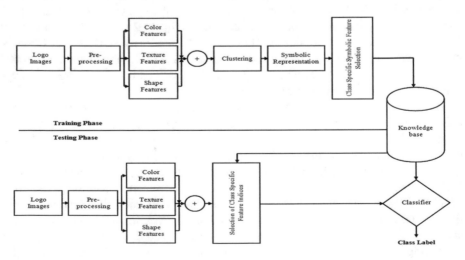

Fig. 1.   General architecture of the proposed model.

2.1   Training Phase

The different procedures followed during the training phase are pre-processing, feature extraction, features fusion, clustering, symbolic representation, followed by symbolic feature selection.

In pre-processing step, the logo images are uniformly resized from M × N dimension to m × n dimension to ease the computational burden. Further, the images are converted from RGB scale to gray scale.

For discrimination purpose, the three types of global features are extracted from logo images viz., color, texture, and shape [9]. The feature extraction step results with the 48-color, 8-shape, and 4-texture features respectively. These three global features are tested with different combinations viz., single/double/triple set of features. Among these combinations, the fusion of three features (i.e., color+texture+shape) appears to be more effective and efficient compared to other feature combinations [9]. Thus, in this work, feature fusion is preferred over individual (or paired) features for classification.

Though, the three features obtained are from three different domains, these features need to be normalized before fusion. Thus, the features are first normalized using Eq. (1) and the normalized features are then fused using simple vector concatenation rule.

$$F_l = \frac{x_{w,l}}{\max(x_{all,l})} \tag{1}$$

where, l = 1, 2, …, No. of Features; w = 1, 2, …, No. of Samples.

Due to the presence of large variations among logo images within a class. The images are clustered using K-Means partitional clustering algorithm [20]. The partitional clustering algorithm is chosen over the hierarchical clustering because the type of data which are used in this work is not of type taxonomical data [20]. The value of K is fixed empirically.

After clustering, the samples within each cluster may (not) vary and it is difficult to preserve such intra-cluster variations and hence symbolic interval valued representation is recommended [10]. The notion of K-Means clustering and symbolic interval valued representation is illustrated in Fig. 2.

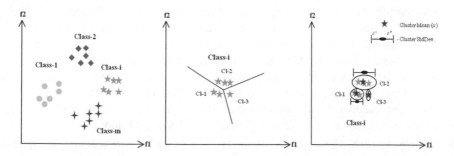

Fig. 2. Illustration of K-Means clustering and symbolic interval representation. (a) Samples are spread over m-classes in 2-D feature space, (b) Clustered samples of ith class with K = 3, (c) Intra-cluster variations are preserved using mean-standard deviation interval representation.

Consider a set of samples $X_i = \{x_1, x_2, \ldots, x_{n_i}\}'(x_h \in \mathbb{R}^d)$, belongs to i^{th} *Class* $(n_i = No.\ of\ samples\ in\ i^{th}\ class; i = 1, 2, \ldots, m; m = Number\ of\ Classes)$. If K-Means clustering algorithm is applied on these samples then K (henceforth termed as K_s) number of clusters are formed with each cluster consisting of $u_j (j = 1, 2, \ldots, K_s)$ number of samples $(u_1 + u_2 + \ldots + u_{K_s} = n_i)$. The intra-cluster variations present among the samples in each cluster are preserved using the mean-standard deviation interval representation [17]. The mean and standard deviation computed for the samples of each cluster is given by (2) and (3).

$$\mu_{j_i}^l - \frac{1}{u_j} \sum_{h=1}^{u_j} r_h^l \tag{2}$$

$$\sigma_{j_i}^l = \sqrt{\frac{1}{(u_j - 1)} \sum_{h=1}^{u_j} \left(x_h^l - \mu_{j_i}^l\right)^2} \tag{3}$$

Where, $\mu_{j_i}^l$ and $\sigma_{j_i}^l$ are respectively the mean and standard deviation values of lth feature belongs to j^{th} cluster corresponding to i^{th} class.

After computing the mean and the standard deviation values of each cluster belonging to a particular class, the difference between them represents a lower limit and the sum between them represents an upper limit of an interval. Finally, from each class, K_s number of such cluster interval valued representatives are formed. Thus, an interval valued feature matrix of dimension $(K_s xm)xd$ is created. The cluster representative associated with j^{th} cluster and i^{th} class is formulated as follows:

$$CR_j^i = \left\{ \left[\left(\mu_{j_i}^1 - \sigma_{j_i}^1\right), \left(\mu_{j_i}^1 + \sigma_{j_i}^1\right)\right], \left[\left(\mu_{j_i}^2 - \sigma_{j_i}^2\right), \left(\mu_{j_i}^2 + \sigma_{j_i}^2\right)\right], \ldots, \left[\left(\mu_{j_i}^d - \sigma_{j_i}^d\right), \left(\mu_{j_i}^d + \sigma_{j_i}^d\right)\right] \right\}$$

$$CR_j^i = \left\{ [f_1^-, f_1^+], [f_2^-, f_2^+], \ldots, [f_d^-, f_d^+] \right\}$$

$$where,\ f_l^- = \left\{ \left(\mu_{j_i}^l - \sigma_{j_i}^l\right) \right\} and f_l^+ = \left\{ \left(\mu_{j_i}^l + \sigma_{j_i}^l\right) \right\}$$

Symbolic Feature Selection. The supervised interval feature matrix (of dimension $(K_s xm)xd$) obtained in the previous step is the input to the feature selection method. The method initially transforms the supervised feature matrix (Fig. 3(a)) and soon divides the transformed matrix into several (equal to number of classes) interval feature sub-matrices (Fig. 3(b)). The transformed feature sub-matrices are then fed into modified interval K-means clustering algorithm [12]. Thus, results with the K_f clusters from each sub-matrix. Further, for each cluster, a cluster representative is chosen based on the feature's maximum affinity towards the remaining features present in the cluster. Such feature is considered as the cluster representative and thus be a feature get selected. The affinity between two interval features is computed using symbolic similarity kernel (SSK) proposed in [12]. Finally, the method produces with K_f features (representatives) corresponding to each class.

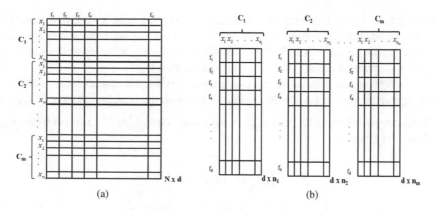

Fig. 3. (a) Interval feature matrix, (b) Transformed class-specific interval feature sub-matrices.

If $K_f^{(a)}$ and $K_f^{(b)}$ features respectively are selected from any two classes say *class-a* and *class-b*, then the selected $K_f^{(a)}$ features from *class-a* need to/not be same as the selected $K_f^{(b)}$ features from *class-b* $(a = 1, 2, \ldots, m; b = 1, 2, \ldots, m)$.

After the class specific feature selection process, the dimension of the original supervised interval matrix thus becomes $(K_s x m) x K_f$ and thus called reduced supervised interval matrix. A sample in the reduced supervised interval feature matrix is given by:

$$CRFS_j^i = \left\{ \left[\left(\mu_{j_i}^1 - \sigma_{j_i}^1 \right), \left(\mu_{j_i}^1 + \sigma_{j_i}^1 \right) \right], \left[\left(\mu_{j_i}^2 - \sigma_{j_i}^2 \right), \left(\mu_{j_i}^2 + \sigma_{j_i}^2 \right) \right], \ldots, \left[\left(\mu_{j_i}^{K_f} - \sigma_{j_i}^{K_f} \right), \left(\mu_{j_i}^{K_f} + \sigma_{j_i}^{K_f} \right) \right] \right\}$$

$$CRFS_j^i = \left\{ \left[f_1^-, f_1^+ \right], \left[f_2^-, f_2^+ \right], \ldots, \left[f_{K_f}^-, f_{K_f}^+ \right] \right\}$$

$$\text{where, } f_l^- = \left\{ \left(\mu_{j_i}^l - \sigma_{j_i}^l \right) \right\} \text{ and } f_l^+ = \left\{ \left(\mu_{j_i}^l + \sigma_{j_i}^l \right) \right\}$$

Further, the reduced matrix is preserved in the knowledgebase and is used in classification process.

2.2 Testing Phase

During testing, the same procedures which are followed in training phase are followed till the features fusion. Further, the class specific feature indices are selected from the knowledgebase and a logo image is classified as a member of any one of the m classes using a symbolic classifier proposed in [17].

Classification. Let us consider a test sample $T_q = \{t^1, t^2, \ldots, t^d\}$, contains d features. Now, the test sample T_q needs to be classified as a member of any one of the m classes. Hence, the adopted class specific feature selection model emphasizes on the selection of class specific feature indices from the knowledgebase. Thus, the test sample T_q becomes $S_q^i = \{s^1, s^2, \ldots, s^{K_f}\}$ after the selection of feature indices corresponding to a ith class (the dimension of K_f remains constant in each class). Further, the similarity is computed between the test sample and all reference (interval) samples of ith class (the feature dimension of test sample and reference samples are equal). Here, feature level similarity is computed for every test sample. Hence, the similarity between a test crisp (single valued) feature and a reference interval valued feature can be inferred as: the similarity value becomes 1, if the test crisp value lies between the lower limit and upper limit of an reference interval valued feature, otherwise 0.

If S_q^i is said to be a member of any one of the m classes, then the acceptance count $AC_q^{j_i}$ takes high value with respect to the ith class with which the reference sample belongs to.

The acceptance count $AC_q^{j_i}$ for a test sample associated with j^{th} cluster corresponding to i^{th} class is given by:

$$AC_q^{j_i} = \sum_{j=1}^{K_s} \sum_{l=1}^{K_f} Sim\left(S_q^i, CRFS_j^i\right)$$

$$Where, \; Sim\left(S_q^i, CRFS_j^i\right) = \begin{cases} 1 & if \; s^l \geq f_l^- \; and \; s^l \leq f_l^+ \\ 0 & Otherwise \end{cases} \; and \tag{4}$$

$$i = 1, 2, \ldots, m; j = 1, 2, \ldots, K_s; \; and \; l = 1, 2, \ldots, K_f$$

3   Experimentation and Results

3.1   Dataset

For experimentation purpose, a huge corpus of color logo images has been used in this work. The corpus consists of 5044 color logo images collected from different web sources of the internet [9]. The logo images are spread across three different classes (viz., both text and symbol, only text, and only symbol), where each class consists of 3171, 1246, and 627 images respectively.

3.2   Experimental Setup

In this section, details on the experimental setup are given. Initially, the logo image dataset is divided into training and testing sets. The percentage of training-testing sets of data are varied from 20%–80% to 80%–20% (in steps of 10%) respectively. During training phase, the logo images undergo various steps viz., pre-processing, feature extraction, feature fusion, clustering and symbolic representation as explained in Sect. 2.1. Further, the symbolic feature selection is applied on the interval feature matrix as explained in Subsect. 2.1 and selected features are preserved in the knowledgebase. During testing phase, an unknown crisp (single valued) test sample is classified as a member of the known three logo image classes using the symbolic classifier as described in Subsect. 2.2.

In this work, the K-means clustering is adopted for clustering both samples as well as features and hence the value K is treated as K_s and K_f in the former and latter cases respectively. The value of parameter K_f is varied from 2 to d − 1 (one less than the actual dimension). Similarly, the value of the parameter K_s is varied from 2 to γ (where, γ is the convergent parameter). Here, γ is fixed to 10, as the clustering algorithm fails to group the samples above this value.

3.3 Results

The performance of the proposed classification system is evaluated based on precision, recall, F-Measure and accuracy computed from a confusion matrix. In addition to these measures, it is also evaluated with respect to time (in seconds). The details of the said measures are given in [10].

With respect to each train-test percentage, the experimentation is conducted for 20 trials. In each trial, the feature subset which gives maximum f-measure for different values of K_s (2 to 10) is obtained and from all 20 trials, 180 (20 × 9) such f-measure along with associated precision, recall, time, K_f and K_s are obtained. Using this, the minimum, maximum and average values of f-measure along with accuracy, precision, recall, time, K_f and K_s are tabulated. Similarly, the same procedure is followed for all remaining train-test percentage of samples.

Table 1 shows the minimum, maximum, and average values associated with accuracy, precision, recall, and f-measure obtained under varying train test percentage of samples. The first and second values within a pair of bracket in Table 1 respectively denote the number of features selected through feature clustering (K_f) and number of clusters formed while clustering the samples (K_s). The last row in the Table 1 shows the best performance results obtained based on the average f-measure.

Table 1. Minimum, maximum, and average values of accuracy, precision, recall, and f-measure obtained under varied training and testing percentage of samples (with feature selection).

Train-Test %	Accuracy			Precision			Recall			F-Measure		
	Min	Max	Avg	Min	Max	Avg	Min	Max	Avg	Min	Max	Avg
20–80	74.19 (40,6)	74.34 (54,5)	74.19 (44,9)	89.08 (40,6)	89.43 (54,5)	88.51 (44,9)	53.71 (40,6)	54.11 (54,5)	54.18 (44,9)	67.01 (40,6)	67.42 (54,5)	67.17 (44,9)
30–70	74.19 (43,8)	75.49 (33,5)	74.34 (45,8)	85.51 (43,8)	85.87 (33,5)	87.94 (45,8)	54.30 (43,8)	56.29 (33,5)	54.23 (45,8)	66.42 (43,8)	68.01 (33,5)	67.07 (45,8)
40–60	74.25 (31,10)	74.21 (46,7)	74.17 (42,7)	84.62 (31,10)	89.48 (46,7)	85.23 (42,7)	54.53 (31,10)	53.81 (46,7)	54.95 (42,7)	66.32 (31,10)	67.21 (46,7)	66.75 (42,7)
50–50	74.18 (53,3)	74.77 (38,9)	73.71 (45,7)	82.55 (53,3)	87.84 (38,9)	85.35 (45,7)	55.11 (53,3)	54.85 (38,9)	55.05 (45,7)	66.09 (53,3)	67.53 (38,9)	66.83 (45,7)
60–40	74.16 (58,6)	74.65 (53,8)	74.46 (43,7)	82.45 (58,6)	86.61 (53,8)	85.40 (43,7)	54.74 (58,6)	55.16 (53,8)	54.99 (43,7)	65.80 (58,6)	67.40 (53,8)	66.89 (43,7)
70–30	74.07 (47,7)	75.60 (38,3)	74.51 (40,7)	83.09 (47,7)	84.19 (38,3)	84.71 (40,7)	54.80 (47,7)	56.54 (38,3)	55.37 (40,7)	66.04 (47,7)	67.65 (38,3)	66.92 (40,7)
80–20	**73.91 (58,5)**	**75.20 (40,6)**	**74.55 (39,7)**	**83.66 (58,5)**	**84.92 (40,6)**	**84.90 (39,7)**	**54.84 (58,5)**	**56.29 (40,6)**	**55.75 (39,7)**	**66.25 (58,5)**	**67.70 (40,6)**	**67.22 (39,7)**
Best	**73.91 (58,5)**	**75.20 (40,6)**	**74.55 (39,7)**	**83.66 (58,5)**	**84.92 (40,6)**	**84.90 (39,7)**	**54.84 (58,5)**	**56.29 (40,6)**	**55.75 (39,7)**	**66.25 (58,5)**	**67.70 (40,6)**	**67.22 (39,7)**

From Table 1, it is very clear that the best classification results with the reduced set of features and samples are obtained for 80–20% of training-testing samples. Thus, accomplishing the classification with a compact representation of a feature matrix (reducing the column and row dimensions of a feature matrix).

Figure 4 shows the distribution of K_f and K_s obtained from 20 trials corresponding to each train-test percentage. In the Fig. 4(a) and (b), each bin describes a cumulative frequency corresponding to the (K_f) and (K_s) spread across different training and testing percentage.

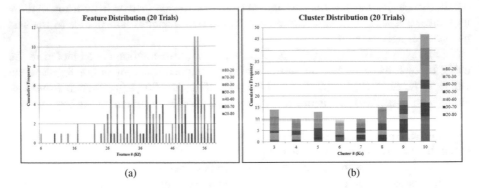

(a)   (b)

Fig. 4. Distributions of K_f and K_s obtained from 20 trials (a) feature clusters distribution (K_f) (i.e., features selected), (b) sample clusters distribution (K_s).

4   Comparative Analysis

To test the efficacy of the proposed model, a comparative analysis is given against the existing model which does not uses any symbolic feature selection models during classification [10]. The comparative analysis is made in-terms average f-measure and also in terms of time. Figure 5(a) and (b) show the results of the proposed classification model (with $K_s = 7$ and $K_f = 39$) against the results of classification model which does not use any symbolic feature selection methods in-terms average f-measure and time respectively.

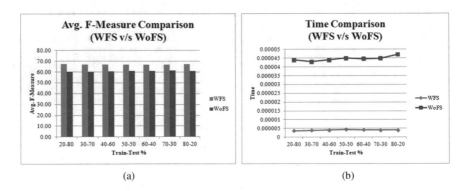

(a)   (b)

Fig. 5. Comparative analysis of with feature selection method v/s without feature selection method in classifying the logo images- in terms of (a) average F-measure and (b) time.

From Fig. 5, it is very clear that the proposed classification model with feature selection outperforms the existing model which does not use any symbolic feature selection

during classification in terms of both average f-measure and time. For better visualization on classification of logo images, the confusion matrices obtained from the existing
model and the proposed model are shown in Tables 2 and 3 respectively.

Table 2. Confusion matrix obtained from [10] (with $K_s = 4$ and $K_f = 60$ (70%–30%))

	Both	Text	Symbol
Both (951)	818	86	47
Text (419)	154	194	25
Symbol (188)	93	24	71

Table 3. Confusion matrix obtained from the proposed model (with $K_s = 7$ and $K_f = 39$ (80%–20%))

	Both	Text	Symbol
Both (634)	545	15	74
Text (249)	81	141	27
Symbol (125)	55	2	68

Figure 6 illustrates the classification of sample logo images. Figure 6(a) shows the
correct and in-correct classification of logo images with all features are in consideration
and Fig. 6(b) shows the correct and in-correct classification of logo images with subset
of features are in consideration. From the correctly classified logo images, it is guaranteed that the subset of features is good enough to classify a logo image.

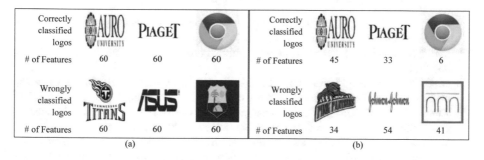

Fig. 6. Illustration of correctly and wrongly classified logo images (a) with using symbolic feature
selection method, (b) without using symbolic feature selection method.

5   Conclusion

In this paper, the efficacy of symbolic feature selection in logo image classification has
been demonstrated. The proposed model preserves the large variations present across
the samples in each class through clustering and are represented in the form of interval
valued data. The symbolic feature selection is further adopted to classify the logo images.
The proposed model not only reduces the row dimension but also reduces the column
dimension of a feature matrix. Thus, performing classification through compact

representation. For experimentation purpose, a huge corpus of 5044 color logo images has been used. The proposed model has been validated with validity measures viz., precision, recall, f-measure and time. The comparative analysis show the superiority of the classification through symbolic feature selection method with that of classification without feature selection in terms of average f-measure and time.

Acknowledgement. The second author would like to acknowledge the Department of Science Technology, INDIA, for their financial support through DST-INSPIRE fellowship.

References

1. Hesson, A., Androutsos, D.: Logo and trademark detection in images using color wavelet co-occurrence histograms. In: IEEE International Conference on Acoustics, Speech and Signal Processing, ICASSP 2008, pp.1233–1236 (2008)
2. Nourbakhsh, F., Karatzas, D., Valveny, E.: A polar-based logo representation based on topological and colour features. In: Proceedings of the 9th IAPR Workshop on Document Analysis Systems, pp. 341–348. ACM Press (2010)
3. Romberg, S., Pueyo, L.G., Lienhart, R., Zwol, R.V.: Scalable logo recognition in real-world images. In: ACM International Conference on Multimedia Retrieval 2011, ICMR 2011 (2011)
4. Kalantidis, Y., Pueyo, L.G., Trevisiol, M., Zwol, R.V., Avrithis, Y.: Scalable triangulation-based logo recognition. In: ACM International Conference on Multimedia Retrieval, ICMR 2011 (2011)
5. Romberg, S., Lienhart, R.: Bundle min-hashing for logo recognition. In: Proceedings of the 3rd ACM International Conference on Multimedia Retrieval, ICMR 2013, pp. 113–120 (2013)
6. Neumann, J., Samet, H., Soffer, A.: Integration of local and global shape analysis for logo classification. Pattern Recogn. Lett. **23**, 1449–1457 (2002)
7. Sun, S.K., Chen, Z.: Logo recognition by mobile phone cameras. J. Inf. Sci. Eng. **27**, 545–559 (2011)
8. Arafat, Y.S., Saleem, M., Hussain, A.S.: Comparative analysis of invariant schemes for logo classification. In: International Conference on Emerging Technologies, pp. 256–261 (2009)
9. Kumar, N.V., Kantha, P.V., Govindaraju, K.N., Guru, D.S.: Fusion of features for classification of logos. Procedia Comput. Sci. **85**, 370–379 (2016)
10. Guru, D.S., Kumar, N.V.: Symbolic representation and classification of logos. In: Proceedings of International Conference on Computer Vision and Image Processing (CVIP-2016). AISC Series, vol. 459, pp 555–569. Springer, Singapore (2016)
11. Duda, O.R., Hart, E.P., Stork, G.D.: Pattern Classification, 2nd edn. Wiley-Interscience, New York (2000)
12. Guru, D.S., Kumar, N.V.: Class specific feature selection for interval valued data through interval K-Means clustering. In: RTIP2R 2016. CCIS, vol. 709, pp. 228–239. Springer, Singapore (2017)
13. Ichino, M.: Feature selection for symbolic data classification. In: New Approaches in Classification and Data Analysis, pp. 423–429. Springer, Heidelberg (1994)
14. Kiranagi, B.B., Guru D.S., Ichino M.: Exploitation of multivalued type proximity for symbolic feature selection. In: Proceedings of the Internal Conference on Computing: Theory and Applications, pp. 320–324. IEEE (2007)
15. Hedjazi, L., Martin, A.J., Lann, M.V.L.: Similarity-margin based feature selection for symbolic interval data. Pattern Recogn. Lett. **32**, 578–585 (2011)

16. Guru, D.S., Kumar, N.V.: Novel feature ranking criteria for interval valued feature selection. In: International Conference on Advances in Computing, Communications and Informatics (ICACCI), pp. 149–155. IEEE (2016)

17. Guru, D.S., Prakash, H.N.: Online signature verification and recognition: an approach based on symbolic representation. IEEE Trans. Pattern Anal. Mach. Int. **31**(6), 1059–1073 (2009)

18. Silva, A.P.D., Brito, P.: Linear discriminant analysis for interval data. Comput. Stat. **21**, 289–308 (2006)

19. Barros, A.P., Carvalho, F.A.T., Neto, E.A.L.: A pattern classifier for interval-valued data based on multinomial logistic regression model. In: IEEE International Conference on Systems, Man, and Cybernetics, pp. 541–546 (2012)

20. Jain, A.K., Dubes, R.C.: Algorithms for Clustering Data. Prentice-Hall Inc., Upper Saddle River (1988)

An Hierarchical Framework for Classroom Events Classification

D. S. Guru, N. Vinay Kumar$^{(\boxtimes)}$, K. N. Mahalakshmi Gupta,
S. D. Nandini, H. N. Rajini, and G. Namratha Urs

Department of Studies in Computer Science, University of Mysore,
Manasagangotri, Mysore 570006, Karnataka, India
dsg@compsci.uni-mysore.ac.in,
vinaykumar.natraj@gmail.com,
mahalakshmigupta5@gmail.com, nandini.sd60@gmail.com,
rajininagrajl@gmail.com, namrathaurs.13@gmail.com

Abstract. In this paper, a model for classroom events classification is proposed. Major classroom events which are considered in this work are drowsiness of a student, group discussion, steady and alert, and noisy classroom. These events are classified using a two level classification model. It makes use of simple threshold based classifiers for classification. In the first level, classes such as noisy classroom and drowsiness are separated from that of remaining classes based on global threshold. The global threshold is computed based on correlation coefficients, computed across the intensity values of the video frames. The correlation scores obtained from each video are used for classification. During second level, a partially labeled video is classified as a member of any of the said four classes based on the local threshold computed from each class of videos. Local threshold is computed based on the global characteristics extracted from the videos. For classification purpose, the events which are considered here are strictly mutually exclusive events. Due to the lack of classroom events video datasets, the dataset has been created consisting of 96 videos spread across 4 different classes. The proposed model is validated using suitable validity measures viz., accuracy, precision, recall, and f-measure. The results show that the proposed model performs better in classifying the said events.

Keywords: Classroom events · Statistical features · Threshold based classifier

1 Introduction

Nowadays due to the increase in the demand of security and monitoring systems across the globe, designing and building of such systems have become very challenging task across the community of researchers. The system basically handles with the videos which are normally present in an unstructured format. The task of converting such unstructured format into structured and further processing of structured data has become very tedious job for the researchers in the field of video analytics. In this regard many researchers have taken up this issue seriously and have contributed for the conversion of unstructured video data into a structured one [1–3]. In addition, researchers also contributed to the field of

© Springer International Publishing AG, part of Springer Nature 2018
A. Abraham et al. (Eds.): ISDA 2017, AISC 736, pp. 166–179, 2018.
https://doi.org/10.1007/978-3-319-76348-4_17

video analytics in general, events classification/detection/retrieval in particular with respect to specific type of application [4, 5].

In this paper, one such video analytics problem has been addressed. The addressed problem completely relied on events classification. There are several events found in our day to day activities which have been considered as the problems of video analytics. Some of them are, abnormal events classification in traffic circles [6, 7], abnormal events classification in crowds [8, 6], and sports events classification [9–11]. Most of the problems have been well explored in the field of events classification, but the classroom events classification has not been explored in the literature of events classification. Hence, classroom events videos are considered here for designing the classification model. Basically, classroom events are the activities performed in the classrooms either by teachers or by students or by both of them. The classroom events include student-teacher interaction, teacher's teaching, students' drowsiness, group discussion, steady and alert, and noisy classroom.

Nowadays, these events have been monitored through surveillance cameras in most of the educational institutions and organizations. But, it requires manual intervention for monitoring the classroom events which results with a dedicated man power for analyzing the classroom videos. To avoid this, an automated system is necessary to monitor the classroom events automatically. With this motivation, the classroom events classification model has been proposed which takes care of automatic classification of classroom events without any manual intervention. From the literature, it is proven that there are no works found on the classroom events classification.

In this paper, basically four classroom events, which are more concentrated on students' activities, are considered for classification viz., students' drowsiness, group discussion, steady and alert, and noisy classroom. The proposed model makes use of statistical features for distinguishing the four different events. It is built up with two-level classification hierarchy for classifying the videos. The model initially makes global threshold based classifier for classification of a video into global two groups and then based on the behavior of objects present in the video, again using local thresholds the videos are classified specifically.

The major contributions of this work include the following:

1. Proposal of classroom events classification model.
2. Creation of classroom events dataset.
3. Extensive experimentation to corroborate the efficacy of the proposed model.

The organization of the paper is as follows. Section 2 describes the conceptualization of the proposed work. The details on the proposed model are given in Sect. 3. Experimentation and results are brought out in Sect. 4. Section 5 concludes the paper.

2 Conceptualization

Basically, the classroom events are captured using video cameras due to its natural behaviour of continuous sequence of actions performed in classrooms. In this work, majorly, four classroom events viz., student's drowsiness (C_1), group discussion (C_2), steady and alert (C_3), and noisy classroom (C_4) are considered.

If these events are studied further, an inference can be achieved such a way that the statistical features are proven to be fit for the videos classification. For example, the events viz., noisy classroom and students' drowsiness have much variation in the pixel intensities across the frames compared to the other classroom events viz., group discussion and steady and alert. In case of latter cases, the behaviour of students is almost idle and there will be no such variations found in pixel intensities across the frames. This motivated us to preserve such characteristics using statistical feature-correlation coefficient for grouping four different classes into two groups. Using the correlation co-efficient, a threshold has been fixed such that below the threshold the videos are said to be from students' drowsiness or noisy classroom and above the threshold the videos are said to be from steady and alert or group discussion. This ends up with conceptualization of grouping the videos into two global groups viz., group-A and group-B.

The videos in group-A are basically the videos with maximum variation across the frames. Here, motion of students changes rapidly from frame to frame with respect to C_4 while it is minimum with respect to C_1. This motivates us to extract the shape features of the objects in video frames and then preserve the variations across the shape features to distinguish between these two classes. Similarly, the videos in group-B are basically the videos with minimum variations or no variations across the frames. In this regard, only key frames are considered for further processing. The key frames are the frames which give maximum information in terms of distinguishing capability of the video. In the class of group discussion (C_2) and steady and alert (C_3), the students are usually seated in a group and in a sequential manner respectively. By keeping this in mind, the centroids of every student object are extracted and further the distances between these centroids are calculated. So that the distance among the centroids is more with respect to group discussion class and is less with respect to the steady and alert class of classroom events videos.

3 Proposed Model

The proposed model consists of mainly two phases viz., training phase and testing phase. The former phase includes pre-processing, correlation computation, global threshold fixation, feature extraction, and local threshold fixation. Similarly, the latter phase includes the similar steps of training phase but in addition to these steps a two level classification model is introduced for classification. The architecture of the proposed model is visualized in Fig. 1 and the details of each step are given in the subsequent sections.

3.1 Training Phase

Initially, the video dataset has been divided into training and testing datasets. The training video dataset is used to train the model. The training dataset is further divided into learning and validation sets. Finally, the testing dataset is used for testing the performance of the proposed model. The different steps of training phase are explained in subsequent sections.

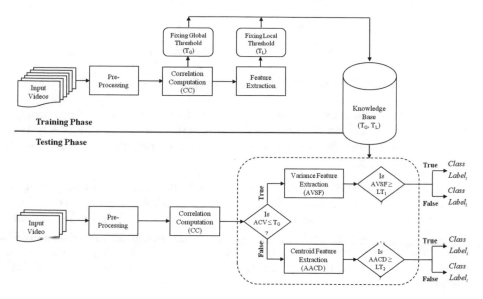

Fig. 1. Architecture of the proposed model.

Pre-processing. In this work, the model is specially designed to work on frames of the videos and hence initially, the frames are extracted from the given video. These frames are then resized from $M \times N$ to standard $m \times n$ dimensions to decrease the burden of the computational complexity. Handling with the grayscale video frames is much simpler than the RGB video frames and hence the resized frames are converted from RGB to grayscale video frames.

Correlation Computation. Based on the conceptualization discussed in Sect. 2, the correlation is computed across the video frames. Let us consider the input video V consisting of d number of frames. Each video frame is then reshaped into a column vector of dimension $((m \times n) \times 1)$, such that the correlation computation across the frames becomes much easier. The average correlation value computed among the video frames for a video is given by:

$$ACV = \frac{1}{d} \sum_{j=2}^{d} CC(F_1, F_j)$$

$$CC(F_1, F_j) = \frac{Cov(F_1, F_j)}{\sqrt{Var(F_1) * Var(F_j)}}$$

$$where, Cov(F_1, F_j) = \frac{1}{(m*n) - 1} \sum_{a=1}^{(m*n)} (x_a - \mu_u) * (y_a - \mu_v)$$

$$Here, \mu_I = \frac{1}{(m*n)} \sum_{I=1}^{(m*n)} X_I; \ var(F_I) = \frac{1}{(m*n)-1} \sum_{I=1}^{(m*n)} (X_I - \mu_I)^2$$

$$\forall u \in F_1 \ and \ v \in F_j; \ j = 2, \ldots d; \ I = 1, j; \ X = x, y$$

If the average correlation value (ACV) is high then the probability of video belongs to the class of steady-alert or group discussion is high. If the score is relatively low, then the probability that the video belongs to the class of noisy classroom or student's drowsiness will be high.

Global Threshold Fixation. The dataset is divided into training and testing dataset. The training set is further used to fix up the global threshold which is later used for classification during testing stage. Hence, the 2-fold cross validation [12] is recommended to fix up the global threshold which is useful in separating the four classroom events into two major groups viz., group-A and group-B. The former group consists of C_1 and C_4 and the latter group consists of C_2 and C_3. The computation of the global threshold is given by:

$$T_G^{Max} = Avg\left(Max\left(Max(ACV_{1k}^l), Max(ACV_{4k}^l)\right), Max\left(Max(ACV_{1b}^v), Max(ACV_{4b}^v)\right)\right)$$

$$T_G^{Min} = Avg\left(Min\left(Min(ACV_{2k}^l), Min(ACV_{3k}^l)\right), Min\left(Min(ACV_{2b}^v), Min(ACV_{3b}^v)\right)\right)$$

$$T_G = Avg\left(T_G^{Max}, T_G^{Min}\right)$$

Where, ACV_{ak}^l is the average correlation value obtained from training set $\in$ class C_a and ACV_{ab}^v is the average correlation value obtained from validation set $\in$ class C_a

$$\forall a = 1, 2, 3, 4; k = 1, 2, \ldots, No. of \ learning \ samples;$$

$$b = 1, 2, \ldots, validation \ samples$$

Feature Extraction. In this sub-section, we present different discriminative features extracted from the video frames grouped into two groups. The extracted features will help in classifying the event video as any one of the two classes in the respective groups. Based on the conceptualization discussed in Sect. 2, different features are extracted with respect to two different groups. The shape and centroid based features are extracted with respect to group-A and group-B of video frames respectively.

The Zernike moments shape features viz., amplitude and phase angle [13] are extracted from each video frame belonging to the video of group-A. Let us consider a video V which belongs to group-A consisting d number of frames. The Zernike features for a frame in a video is given by

$$Zernike \ Features(F_i) = [\omega(F_i), \psi(F_i)]$$

Where, $\omega(F_i)$ *and* $\psi(F_i)$ are respectively the amplitude and the phase angle of the frame F_i calculated from [13] corresponding to video V, $i = 1, 2, \ldots, d$.

Similarly, the Zernike shape features are extracted for all the frames of the corresponding video. Hence, there are d number of features and these features are then aggregated by calculating its variance across the frames for both features of the video and is given by:

$$Var(V_\omega) = \frac{1}{d-1} \sum_{i=1}^{d} (\omega(F_i) - \mu_\omega)^2$$

$$Var(V_\psi) = \frac{1}{d-1} \sum_{i=1}^{d} (\psi(F_i) - \mu_\psi)^2$$

Where, V_ω and V_ψ respectively are the variances of amplitude and phase angle computed across the d frames corresponding to the video V which belongs to group-A.

Now, the average of variances corresponding to two shape features is computed to represent the video in the scalar form and is given by:

$$AVSF(V) = \left(\frac{Var(V_\omega) + Var(V_\psi)}{2} \right)$$

This, $AvgVar(V)$ is used for classifying the video into any one of the particular class.

Further, in classifying the videos grouped into group-B, the key frame is selected. The key frame is selected based on the following:

$$KF_V = \left\lfloor \frac{1+d}{2} \right\rfloor$$

Where, d = No. of frames.

The key frame KF_V is selected as the mid frame since the videos belonging to the group-B have less variations with maximum information in middle frames. In addition to this the videos of group-B does not have much variations across the frames. In this regard, the middle frame is selected as the key-frame.

The edge components are extracted from gray scale key frame. From these edge components, the connected components are identified. Let there be c number of connected components present in a frame. Then the centroids corresponding to each connected component are computed. Based on the conceptualization discussed in Sect. 2, the distance is computed among the different centroids present in each key frame.

$$DM_{yz} = ED(c_y, c_z)$$

$$\forall y = 1, 2, \ldots, C; \ z = 1, 2, \ldots, C; \ C = No.of \ Connected \ Components$$

where, $ED(c_y, c_z)$ is the Euclidean distance computed between two centroids.

The average distance corresponding to each centroid computed from the distance matrix is given by:

$$CD_y = Avg(DM_{y*})$$

The average of average centroids distance $(AACD^V)$ corresponding to each video V belonging to group-B is given by:

$$AACD^V = Avg(CD_y)$$

Now, the $AACD$ is used as the features for deciding the video of group-B into any one of the particular class. If the value of the $AACD$ is relatively high, then the video is said to belong to class group discussion otherwise, the video is said belong to be steady and alert class.

Local Threshold Computation. In this sub-section, the details of the local threshold computation useful in classifying the video into any one of the particular class which are inherited from the group-A and group-B video set is given. Two local thresholds LT_1 and LT_2 are computed corresponding to the two different groups.

Let us consider the local threshold computed for group-A set of videos. Here, the 2-fold cross validation [12] is recommended to fix up the local threshold-1 which is useful in separating the two classroom events into either noisy classroom or students drowsiness classes. The computation of the local threshold-1 is given by:

$$LT_1^{Max} = Avg(Max(AVSF_{1k}^l), Max(AVSF_{1b}^v))$$

$$LT_1^{Min} = Avg(Min(AVSF_{4k}^l), Min(AVSF_{4b}^v))$$

$$LT_1 = Avg(LT_1^{Max}, LT_1^{Min})$$

$$\forall k = 1, 2, \ldots, No.of\ learning\ samples;\ b = 1, 2, \ldots, validation\ samples$$

Where, $AVSF_{ak}^l$ is the average variance of shape features value obtained from the learning set of videos $\in$ group-A and $AVSF_{ab}^v$ is the average variance of shape features value obtained from validation set of videos $\in$ group-A. $\forall a = 1, 4$.

Let us now consider the local threshold-2 computed for group-B set of videos. Here, the 2-fold cross validation [12] is recommended to fix up the local threshold-2 which is useful in separating the two classroom events into either group discussion or steady and alert classes. The computation of the local threshold-2 is given by:

$$LT_2^{Max} = Avg(Max(AACD_{3k}^l), Max(AACD_{3b}^v))$$

$$LT_2^{Min} = Avg(Min(AACD_{2k}^l), Min(AACD_{2b}^v))$$

$$LT_2 = Avg\left(LT_1^{Max}, LT_1^{Min}\right)$$

$$\forall k = 1, 2, \ldots, No.\,of\ learning\ samples;\ b = 1, 2, \ldots, validation\ samples$$

Where, $AACD_{ak}^{l}$ is the average of average Centroid distance value obtained from learning set of videos $\in$ group-B and $AACD_{ab}^{v}$ is the average of average Centroid distance value obtained from validation set of videos $\in$ group-B. $\forall a = 2, 3$.

3.2   Testing Phase

In this section, the details of the proposed two-level classification model in classifying the classroom events videos is given. Initially, for a video, pre-processing is done as explained in first sub-section of Sect. 3.1, before the video is classified as any one of the two groups using the global threshold. Further, features are extracted from the frames video and then video is classified as any one of the four classes using either local threshold-1 or local threshold-2. The details of two level classification model are given in subsequent sub-sections.

Level-1 Classification. In this sub-section, a video is classified into any of the two groups based on the global threshold comparison. The classifier used for first level classification is termed a Level-1 Classifier ($L1C$). It is also called as threshold based classifier [14].

Let us consider a query video taken from the test dataset. Pre-process it completely as explained Sect. 3.1. Then, the average correlation value is computed for the video as explained in Sect. 3.2. If the video is said to be of group-A viz., students' drowsiness (C_1) or noisy classroom (C_4), then if the $ACV(V_q)$ is greater than or equal to the global threshold (T_G), otherwise, the video belongs to the group-B viz., group discussion (C_2) or steady and alert (C_3).

$$L1C(V_q, C) = \begin{cases} group\,A & if\ ACV(V_q) \leq T_G \\ group\,B & Otherwise \end{cases}$$

In this way, at the end of level-1 classification, all the test sample videos are categorized into two sub-groups viz., group-A and group-B.

Level-2 Classification. In this sub-section, the classifier used classification is termed as Level-2 Classifier ($L2C$). Let us consider the same query video chosen either from group-A or group-B depending on the results obtained from $L1C$ classifier. If the query video is from group-A, then $AVSF$ is computed for the video and if the value of $AVSF$ is greater than or equal to the LT_1 then it is classified as the member of C_4 otherwise it is classified as the member of C_1.

$$L2C(V_q, C) = \begin{cases} C_4 & if\ AVSF(V_q) \geq LT_1 \\ C_1 & Otherwise \end{cases}$$

If the query video is from group-B, then $AACD$ is computed for the video and if the value of $AACD$ is greater than or equal to the LT_2 then it is classified as the member of C_2 otherwise it is classified as the member of C_3.

$$L2C(V_q, C) = \begin{cases} C_2 & \text{if } AACD(V_q) \geq LT_2 \\ C_3 & \text{Otherwise} \end{cases}$$

In this way, at the end of level-2 classification, all the test sample videos are categorized as the member of any one class of the given four classes.

4   Experimentation

4.1   Dataset

Due to the non availability of the classroom events dataset in the literature, in this work, an attempt is made for the creation of the classroom events dataset and it is named as UoMCS_Classroom_Events_Dataset. The dataset is created by considering only four classroom events viz., students' drowsiness (C_1), group discussion (C_2), steady and alert (C_3), and noisy classroom (C_4).

The dataset consists of 96 classroom events videos spread across four different classes, where each class consists of 24 video samples captured using mobile cameras.

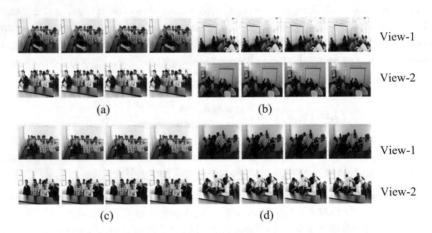

Fig. 2. Sample video frames extracted from UoMCS_Classroom_Events_Dataset (a) Students Drowsiness, (b) Group Discussion, (c) Steady & Alert and (d) Noisy Classroom. The top row represents video frames captured at view-1 and the bottom row represents the video frames captured at view-2.

Out of 24 video samples, first 12 videos are captured at left-view (view-1) and the remaining 12 videos are captured at right view (view-2) of the classroom. For the sake of simplification during experimentation, each video is trimmed to 600 frames/video. The sample video frames extracted from the dataset are shown in Fig. 2.

4.2 Experimental Setup

The experimentation is conducted on the newly created classroom events datasets which comprised of four different classes for classification. Basically, the dataset is divided into training and testing dataset. The training dataset is used for training the model and the testing dataset is used for testing the validity of the proposed classification model. During training, the frames are extracted from the video and then converted to gray scale frames. These frames are then resized from 1280×720 dimensions to standard 256×256 dimensions for the sake of computational simplicity. The resized video frames are subjected to compute the correlation across the video frames to fix the global threshold T_G as explained in third sub-section of Sect. 3.1. Further, the shape and centroid based features are extracted from the video frames. Using these features the two local thresholds are fixed as explained in fifth sub-section of Sect. 3.1. Hence, during training phase, three parameters $(T_G, LT_1, and LT_2)$ are fixed using 2-fold cross validation technique.

During testing phase, a query video being classified as a member of anyone of the classes-will undergo pre-processing step followed by computation of *ACV* from the pre-processed video frames. The details of the former and latter process are explained in first and second sub-sections of Sect. 3.1 respectively. Further, using *L1* classifier, the video is grouped into either group-A or group-B as given in third sub-section of Sect. 3.1. If the video is labeled as group-A then the *AVSF* is computed for the video frames as explained in fourth sub-section of Sect. 3.1 and then the video is classified as either class-1 or class-4 using *L2C* classifier. Similarly, if the video is labeled as group-B, then the *AACD* value is computed as explained in fourth sub-section of Sect. 3.1 and classified as either class-2 or class-4 using *L2C* classifier.

The dataset is divided into 20% (40%, 60%, and 80%) of training and 80% (60%, 40%, and 20%) of testing set for experimentation. The training set is further divided into 50% for learning set and remaining 50% for validation set respectively.

4.3 Results

The performance of the proposed classification model is evaluated based on suitable validity measures viz., classification accuracy, precision, recall, and F-Measure computed from the confusion matrix.

Let us consider a confusion matrix CM_{ij}, generated during classification of the classroom event videos at testing stage. From this confusion matrix, the accuracy, the precision, the recall, and the F-Measure are all computed to measure the efficacy of the proposed model. The overall accuracy of a system is given by:

$$Accuracy = \frac{No.of\ Correctly\ classified\ Samples}{Total\ number\ of\ Samples} * 100$$

The precision and recall can be computed in two ways. The one way is to compute with respect to each class and the other is with respect to overall classification system. The class wise precision and class wise recall computed from the confusion matrix are given by:

$$P_i = \frac{No.of\ Correctly\ classified\ Samples}{No\ of\ Samples\ classified\ as\ a\ member\ of\ a\ class} * 100$$

$$R_i = \frac{No.of\ Correctly\ classified\ Samples}{Expected\ number\ of\ Samples\ to\ be\ classified\ as\ a\ member\ of\ a\ class} * 100$$

Where, i = 1, 2, 3, 4.

The system precision and system recall computed from the class wise precision and class wise recall is given by:

$$Precision = \frac{\sum_{i=1}^{4} P_i}{4}$$

$$Recall = \frac{\sum_{i=1}^{4} R_i}{4}$$

The F-measure computed from the precision and recall is given by:

$$F - Measure = \frac{2 * Precision * Recall}{Precision + Recall} * 100$$

The classification results obtained for various training and testing percentages of video samples are shown in Fig. 3(a)–(d). These are measured in terms of class-wise precision, class-wise recall and class-wise F-Measure. In addition to this the system's precision, recall, F-measure and accuracy are tabulated in Table 2.

From the above graphs, it is very clear that the performance of the proposed model works better in terms of class-wise precision, recall, and F-measure for all training and testing percentages of video samples. It is also observed from the graphs that the results corresponding to class-1 is bit lower compared to the remaining classes. It is because, the behavior of the class-1 events are more overlapping with the behavior of class-3 events. The confusion matrices in Table 1 show the misclassified samples from class-1 to class-3.

From Table 2, it is very clear that the proposed classification model performs well in classifying the classroom events under varying training and testing percentages of video samples.

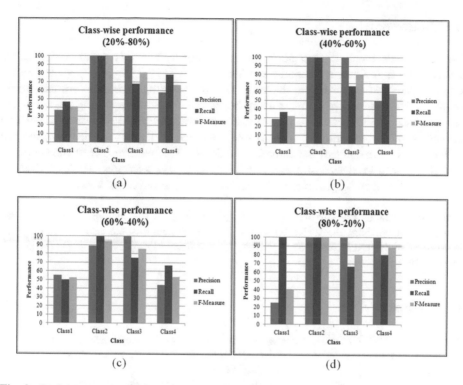

Fig. 3. Performance evaluation of the proposed model evaluated in terms of class-wise precision, recall, and F-measure for various training and testing percentages of samples, (a) 20%–40%, (b) 40%–60%, (c) 60%–40%, and (d) 80%–20%.

Table 1. Confusion matrices obtained during classification for various training and testing percentages of video samples

20%–80%				40%–60%				60%–40%				80%–20%			
	C1	C2	C3	C4	C1	C2	C3	C4	C1	C2	C3	C4	C1	C2	C3	C4
C1	7	0	9	3	4	0	7	3	5	0	3	1	1	0	2	1
C2	0	19	0	0	0	14	0	0	0	8	0	1	0	4	0	0
C3	0	0	19	0	0	0	14	0	0	0	9	0	0	0	4	0
C4	8	0	0	11	7	0	0	7	5	0	0	4	0	0	0	4

Table 2. Performance evaluation of the proposed model evaluated in terms of system's accuracy, precision, recall, and F-measure for various training and testing percentages of samples

Train% – Test%	Performance measures			
	Accuracy	Precision	Recall	F-Measure
20–40	73.68421	73.68421	73.27381	72.17355
40–60	69.64286	69.64286	68.25758	67.58333
60–40	72.22222	72.22222	72.91667	71.44921
80–20	81.25	81.25	86.66667	77.222222

5 Conclusion

The classroom events classification model is proposed in this paper which works basically for four different categories of classroom events. Initially, the proposed model extracts statistical features from the video frames. Later, these features are used to fix up the global and local parameters using 2-fold cross validation. During classification, the proposed model makes use of two different threshold based classifiers to perform classification. For experimentation, the classroom events dataset has been created due to the lack of the standard benchmarking datasets. The performance of the proposed model has been validated using the suitable performance measures. The experimental results are found to be positive in classifying the classroom event videos.

Acknowledgement. The authors would like to acknowledge the support rendered by IISc Bangalore for providing the VADS resources. The second author would also acknowledge the Department of Science & Technology, INDIA, for their financial support through DST-INSPIRE fellowship.

References

1. Zhang, L., Li, S.Z., Yuan, X., Xiang, S.: Real-time object classification in video surveillance based on appearance learning. In: 2007 IEEE Conference on Computer Vision and Pattern Recognition, CVPR 2007, pp. 1–8. IEEE (2007)
2. Guru, D.S., Manjunath, S., Kiranagi, B.B.: SVARS: Symbolic video archival and retrieval system. In: Bangalore Compute Conference, vol. 4, pp. 1–9 (2010)
3. Vijayakumar, V., Nedunchezhian, R.: A study on video data mining. Int. J. Multimed. Inf. Retrieval 1(3), 153–172 (2012)
4. Guru, D.S., Dallalzadeh, E., Manjunath, S.: A symbolic approach for classification of moving vehicles in traffic videos. ICPRAM 2, 351–356 (2012)
5. Bhaumik, H., Bhattacharyya, S., Nath, M.D., Chakraborty, S.: Hybrid soft computing approaches to content based video retrieval: a brief review. Appl. Soft Comput. 46, 1008–1029 (2016)
6. Kotikalapudi, U.K.: Abnormal event detection in video, M. Tech Thesis. Indian Institute of Science, Bangalore (2007)
7. Cui, L., Li, K., Chen, J., Li, Z.: Abnormal event detection in traffic video surveillance based on local features. In: Image and Signal Processing (CISP), pp. 362–366 (2011)
8. Zhong, H., Shi, J., Visontai, M.: Detecting unusual activity in video. In: Computer Vision and Pattern Recognition, pp. II-819-II-826 (2004)
9. Gong, Y., Han, M., Hua, W., Xu, W.: Maximum entropy model based baseball highlight detection and classification. Comput. Vis. Image Underst. 96(2), 181–199 (2004)
10. Fleischman, M., Roy, D.: Temporal feature induction for baseball highlight classification. In: Proceedings of ACM Multimedia, pp. 333–336 (2007)
11. Harikrishna, N., Sanjeev, S., Sriram, D.S.: Automatic summarization of cricket video events using genetic algorithm. In: Proceedings of the 12th Annual Conference Companion on Genetic and Evolutionary Computation, Oregon, USA, pp. 2051–2054 (2010)

12. Burman, P.: A comparative study of ordinary cross-validation, v-fold cross-validation and the repeated learning-testing methods. Biometrika **76**, 503–514 (1989)
13. Tahmasbi, A., Saki, F., Shokouhi, S.B.: Classification of benign and malignant masses based on zernike moments. Comput. Biol. Med. **41**(8), 726–735 (2011)
14. Duda, O.R., Hart, E.P., Stork, G.D.: Pattern Classification, 2nd edn. Wiley-Interscience, New York (2000)

Hand Gesture Recognition System Based
on Local Binary Pattern Approach
for Mobile Devices

Houssem Lahiani[1,3,4(✉)], Monji Kherallah[2], and Mahmoud Neji[3,4]

[1] National School of Electronics and Telecommunications, University of Sfax, Sfax, Tunisia
lahianihoussem@gmail.com
[2] Faculty of Sciences, University of Sfax, Sfax, Tunisia
monji.kherallah@gmail.com
[3] Faculty of Economics and Management, University of Sfax, Sfax, Tunisia
mahmoud.neji@gmail.com
[4] Multimedia Information Systems and Advanced Computing Laboratory, Sfax, Tunisia

Abstract. Since the appearance of mobile devices, gesture recognition is being a challenging task in the field of computer vision. In this paper, a simple and fast algorithm for static hand gesture recognition for mobile device is described. The hand pose is recognized by using gentle AdaBoost learning algorithm and Local Binary Pattern features. The system is developed on an Android OS platform. The method used consists of two steps: a real-time gesture captured by a smartphone's camera and the recognition of the hand gestures. It presents a system based on a real-time hand posture recognition algorithm for mobile devices. The aim of this work is to allow the mobile device interpreting the sign made by the user without the need to touch the screen. In this system, the device is able to perform all necessary steps to recognize hand posture without the need to connect to any distant device.

Keywords: Hand posture recognition · Android · LBP · AdaBoost
Human-machine interaction

1 Introduction

Nowadays, smartphones are being at the heart of technological innovation. After the success of touch screens, accelerometers and gyroscopes, researchers aims to improve ever more the user experience. An innovation that will undoubtedly attract the attention is the vision based hand gesture recognition integrated in mobile devices. As gestures constitute a natural and intuitive way for communication, vision based hand gesture recognition has been widely used in the Human Computer Interaction field, and it has become necessary to integrate it into the applications of today's mobile devices. However, because the mobile device has computational limits, and because the gesture itself has diversity and visual discomfort qualitative, vision-based hand gesture recognition is being a challenging task. Indeed, a hand posture captured by the camera of the mobile device could be interpreted to recognize sign language or to be analyzed to

© Springer International Publishing AG, part of Springer Nature 2018
A. Abraham et al. (Eds.): ISDA 2017, AISC 736, pp. 180–190, 2018.
https://doi.org/10.1007/978-3-319-76348-4_18

control the smartphone without touching the screen. Regarding our system, we use the camera of the mobile device as a direct source of video acquisition, which allows us to process image in real time. In this work, the system is made to run on Android OS platform. The choice of the Android OS platform is made for many reasons. Among the non-technical reason, most of mobile device users use Android OS [1]. Thus, According to the independent web analytics company StatsCounter, Android is the most popular operating system in the world and reached a very big milestone in March 2017, surpassing Windows OS as the world's most popular operating system in terms of total Internet usage on the Portable, desktop, tablet, and mobile handset [2]. Moreover, for technical reasons, because Android platform allows integrating libraries for image processing. For this system, we integrate Open CV library (Open Computer Vision library) [3], a free graphics library for image processing. In addition, according to a comparative study for image processing between Android and iOS [4], better perform- ance was achieved with Android platform. The mainly challenge of hand posture recog- nition for mobile devices is that mobile devices are computationally limited and some vision based tasks are computationally expensive. Several approaches for hand gesture recognition for mobile devices had been proposed. Among those approaches, we find glove-based approach [5, 6], which is cumbersome for users. We find also vision-based approaches which are based on skin color of the hand (Lahiani et al.) [7–10]. However, even those methods are suitable for mobile device since they are based on color, which is computationally inexpensive; they are still non-robust against various kinds of lighting. Moreover, Shape based approaches could achieve good results for rigid objects, but not for articulated objects; in addition, most of them are computationally expensive and not suitable for mobile devices. The proposed system is based on a cascade archi- tecture using AdaBoost algorithm. Viola and Jones firstly introduced this architecture for face recognition [11] and after that, many researchers have used it for object detection because it allows robust and fast detection and due to its robustness against illumination changes. We propose a robust and real-time system for hand posture recognition, which is based on the boosting architecture (gentle AdaBoost) and Local Binary Pattern (LBP) features.

2   Related Works

We have studied many systems made by different researchers in hand gesture recognition field using mobile devices. Due to their availability and simplicity, smartphones have motivated researchers to use them in hand gesture recognition. Many approaches have been developed in this field. The vision-based approach uses the camera of the smart- phone to capture the image of the hand performing the sign. However, this approach encounters challenges such as variations in luminance and the difference in skin color of users and complex backgrounds, which risks producing a low accuracy compared to sensor-based approaches. Thus, we propose a cascade-based approach that is efficient against illumination changes. LBP features and boosting algorithm has been widely used in object detection and recognition because most of its performance depends on the training process and not on the lightning condition.

H. Lahiani et al. proposed a real time hand gesture recognition system based on skin color segmentation and SVM classifier. The mobile device does all steps and recognition rate of this system was about 93%.

H. Lahiani et al. proposed a Static hand pose estimation system to control the smartphone, the system is based on Viola-Jones Algorithm to detect the palm and SVM classifier to recognize postures. The mobile device does all steps and recognition rate of this system was about 91%.

In [12] Hays et al. Made a real time system to interpret American Sign Language by a consumer-level mobile device. They used YCrCb color space for Skin detection. The used classification algorithm is based on Locality Preserving Projections (LPP) as manifold learning along with a multi-class Support Vector Machine (SVM).

In [13] Jin et al. makes a Mobile Application of American Sign Language Translation. They used the Canny edge detection to detect the hand and SURF algorithm for the feature extraction step. They used Support Vector Machine for classification. Obtained results show that the system can recognize 16 different American Sign Language gestures with an overall accuracy of 97.13%.

In [14] Kamat et al. made an Android Application for hearing impaired people. The developed application was named MonVoix, which constitute a French remark for "my voice", and it act as a boon for the deaf and dump people by eliminating the requisite of a human interpreter. The application uses a smartphone camera to capture hand gestures and convert the image file to the corresponding message and audio using RGB color space for skin detection and template matching for classification.

In [15] Setiawardhana et al. makes an Android-based application that interpret sign language submitted by deaf people into written language. Translation process begins from the hand detection using Viola-Jones Haar Filters and translation of hand gestures with the K-NN classification. Obtained results was 100% if distance <50 cm and 25% if distance = 75 cm.

In [16] Prasuhn et al. make a static hand gesture recognition system for the American Sign Language using mobile device. They make a system based on (HOG) Histogram of Oriented Gradients features due to its invariance against rotation. The sensitivity of HOG is treated by using a database of Histogram of Oriented Gradients descriptors corresponding to generated hand images.

3   Theoretical Background

3.1   Hand Gesture Recognition

Hand Gesture Recognition is a computer science field that have to interpret gestures made by human being using mathematical algorithms. It represents a way for computers to understand the intentions of humans, creating an interface between computer and humans. An input device like smartphone in our case, detect frames to read those gestures. They are then interpreted using algorithms based on either artificial intelligence techniques or statistical analysis.

3.2 Local Binary Patterns: LBP

Local binary patterns are features used in computer vision to recognize textures or to detect objects in digital images. This descriptor was mentioned for the first time in 1993 to measure the local contrast of an image [17] but actually popularized three years later by Ojala et al. to analyze the textures [18]. The general principle is to compare the luminance level of a pixel with the levels of its neighbors. Depending on the scale of the neighborhood used, certain areas of interest such as corners or edges can be detected by this descriptor. LBP represents grey scale invariant local texture operator with low computational complexity and powerful discrimination and which make it suitable for integration in mobile device applications. It is calculated on a region of 3 × 3 pixels. Every pixel of an image is labeled by thresholding its 3 × 3 neighborhood pixels with the center value and considering the result as a binary number.

After that, histogram of labels can be used as a texture descriptor. In Fig. 1 an LBP descriptor is shown. When the classifier is under training, each image in the database containing the hand gesture is divided into small squares from which LBP histograms are calculated as a feature. LBP of all pixels of an image is computed and it replaces pixel intensity to give a transformed image. The 'histogram of all pixels of the transformed image is calculated on a circular region of radius R_{region} and considered as 'texture descriptor.

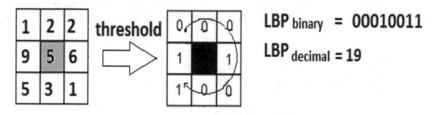

Fig. 1. Calculating a local binary pattern (LBP)

3.3 AdaBoost (Adaptive Boosting)

Boosted cascades have become popular due to a more than satisfactory detection of a large number of objects. Success is driven by the effectiveness and ability of this approach to recognize objects in real time and under different conditions and on low-cost ARM architectures like smartphones and embedded devices. One of the most used algorithms in boosting is called AdaBoost, an abbreviation for adaptive boosting. AdaBoost relies on the iterative selection of weak classifier according to a distribution of the learning examples. Each example is weighted according to its difficulty with the current classifier.

AdaBoost tries several weak classifiers over several cycles, selecting the best weak classifier in each cycle and combining the best classifier to create a strong classifier.

The proposed system is based on a cascade architecture using AdaBoost algorithm. Viola and Jones firstly introduced this architecture. The Viola-Jones Algorithm has four stages: Haar Feature Selection, creation of an Integral Image, AdaBoost Training and

Cascading Classifiers. In our case we use LBP features instead of Haar-like feature. LBP is more suitable for embedded and mobile device because it is faster even it is less (10–20%) accurate than Haar, and it could be remedied by using more positive samples in the training step. With LBP, all calculations are done in integers. Haar uses floats instead of integer, which is a killer for mobile devices. In addition, in training step, LBP is faster than Haar. In our system, we used gentle AdaBoost. The gentle AdaBoost is a variant of the powerful boosting learning technique [19]. The different AdaBoost algorithms differ in the weights update scheme. In [20] according to Lienhart et al. the gentle AdaBoost is the most successful learning method tested for face detection applications. Gentle AdaBoost doesn't require the calculation of log-ratios that can be numerically unstable. The Experimental results on benchmark data demonstrates that the conservative Gentle AdaBoost shows similar performance to Real AdaBoost and Logit AdaBoost, and in majority of cases outperforms these other two variants [21].

4   System Architecture

In this paper, we present a hand posture recognition system running on an android OS platform. The captured hand gesture by the camera of the device constitute the input of our system. We used techniques that are computationally inexpensive to compensate the weak processing capability of smartphones. The proposed algorithm consists of four main steps: Hand detection, features extraction, gesture classification and recognition (Fig. 2).

Fig. 2. System design

4.1   Hand Detection

This step is to extract Hand from the image. This step is more delicate when the acquired image contains several objects like hand or a non-uniform background that creates a texture disturbing the correct segmentation of the hand. This step is dependent on the quality of the acquired images.

4.2   Feature Extraction: LBP Features

LBP is a simple but highly effective descriptive technique for the classification of objects within the computer vision. Due to its high discriminatory capacity, it is a usual approach for solving a multitude of problems. One of its most important characteristics is the robustness of its invariant to light variations.

Consider first how to compute the LBP descriptor. First, we convert the input image to grayscale because LBP works on grayscale images. For every pixel in the grayscale image, a neighborhood is assigned around the current pixel, and then the LBP value is computed for the pixel using the neighborhood. After computing the LBP value of this pixel, the corresponding pixel location is updated in the LBP mask with the LBP value computed. To compute an LBP value for a pixel in grayscale image, the central pixel value id compared with neighboring pixel values. The process can start from any neighboring pixel, then clockwise or counterclockwise could be transversed, but the same order for all pixels must be used. Since there are eight neighboring pixels for each pixel, so eight comparisons must be made. The whole process is illustrated in Fig. 3.

Greyscale image   LBP image

Fig. 3. Grayscale image and corresponding LBP image

To classify texture in the LBP approach, an histogram is made to collect the occurrences of the LBP codes in an image. After that, the classification is performed by calculating simple histogram similarities. Nevertheless, keeping in mind that a similar approach to the representation of the hand posture image results in a loss of spatial information and, subsequently, one must codify texture information while also retaining their locations. One way to reach this purpose is to use LBP texture descriptors to create multiple local descriptions of the hand pose and combine them into a global description. Such local descriptions have gained interest in recent years, which is understandable given the limits of holistic representations. These methods based on local features are more robust against variations of pose or lighting than holistic methods.

Once LBP mask is calculated, the LBP histogram will be calculated. The values of the LBP mask range from 0 to 255, so the LBP descriptor will be 1×256. The LBP histogram will be normalized. The whole process is illustrated in Fig. 4.

Fig. 4. Calulating LBP histogram

4.3 Gesture Classification: AdaBoost

The LBP-based features cited above are used in the framework of AdaBoost learning for detecting Hand postures. In this work, Gentle AdaBoost is proposed as a classifier. A set of positive and negative samples were used for training purposes. During the training process, only weak classifiers made out of LBP features that would be able to improve the prediction are selected. For every weak classifier AdaBoost chose an acceptance threshold, but a single weak classifier is not capable of classifying desired hand gesture with low error rate. In every iteration, weights are normalized and the best weak classifier is selected in function of the weighted error value. In the next stage, the weight values are updated and it could be decided if an example is correctly classified or not. After having traversed all the iterations, a set of weak classifiers characterized by a specified error rate is selected and we obtain a resulted strong-trained classifier. The classification is done iteratively and the number of learning examples affects the effectiveness of the classification process [22].

4.4 Training Data

In order to train classifier we need samples, which means that we need a lot of images that show the hand postures we want to detect (they are called positive sample) and even more images without the hand gesture (negative sample). For training purposes, we used The NUS hand posture datasets I [23]. This dataset contain 240 images containing hand postures. It has 10 classes of postures and 24 sample images per class that have been captured by varying the position and size of the hand within the image frame.

Training an accurate classifier could take a lot of time and a huge number of samples. In our case, for each hand posture we choose the 24 sample images representing the hand gestures as positive samples. Due to their limited number, we enhanced the existing database by adding 50 images to each class of gestures. We cropped positive images to obtain only the needed area, which represents the hand posture. The ratios of the cropped images should not differ that much. We used the opencv_createsamples utility to increase the number of positive samples to 1000 samples for each class of gestures. Indeed, a large set of positive images is created from the set of positive data having the background from the set of negative images. If we want to train a highly accurate classifier, we must have many negative images that look exactly like the positive ones, except that they do not contain the hand posture we want to detect. We used more than 4000 negative images containing backgrounds, walls, faces, etc. To recognize those gestures, we assign to each one of them a letter or word (Fig. 5).

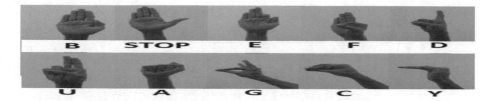

Fig. 5. Grammar to interpret gestures

5   Experimental Results

In order to test the system and to know how well the recognition rate is, we tested hand poses made by different persons in different lightning condition and different backgrounds. For each gesture, ten frames were captured in different backgrounds and luminance condition. The recognition result depends on the distance. The Fig. 6 presents some gestures recognized correctly made in different lightening conditions and with different backgrounds. We can see the name that corresponds to each sign on the top of the green box in the capture.

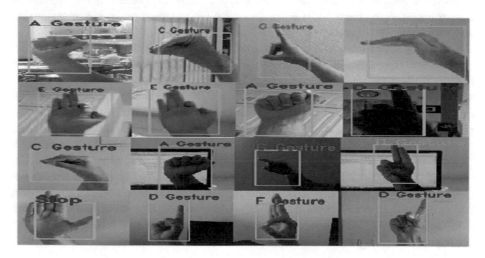

Fig. 6. Different poses in different luminance conditions and backgrounds

We tested the system by inviting 50 different people to make the gestures in different lightning condition and different backgrounds in front of the camera of the device. Results are shown respectively in Table 1.

Table 1. Recognition rate of different hand poses.

Distance <= 75 cm			
Sign	Number of capture	Recognized	Recognition rate
B	50	45	90%
STOP	50	45	90%
E	50	43	86%
F	50	44	88%
D	50	44	88%
U	50	45	90%
A	50	45	90%
G	50	42	84%
C	50	44	88%
Y	50	43	86%
Average			88%

At the end, we can say that detection of different gestures would be optimal if done at a distance less than or equals to 75 cm. The system was deployed onto different smartphone models. The average execution time of detecting poses was about 60 ms in a smartphone which has 1.5 GB RAM and a 1.4 GHz Octo-core processor and it was about 72 ms in a smartphone equipped with a 1.2 GHz Quad-Core processor and 1 GB RAM. The performance of the system essentially relies on the training data and the quality of positive samples. In other terms, if the «training» step is well done, the probability that the hand gesture could be classified in the right category is high. At the end we can say that the recognition system based on LBP features and AdaBoost achieved a satisfactory result since it achieve a good recognition rate and at the same time it remedy the problems caused by the changes in brightness for the system cited at the related work section. As well as it achieve the detection fastly compared to systems based on the detection of contours, which are greedy in terms of consumption of critical resource of the device. The LBP approach demonstrated that it is suitable for mobile device as it is computationally inexpensive since it does all the calculations in integers contrarily to other approaches that uses floats, which is a killer for embedded and mobile. A video that demonstrate the performance of the system is available on youtube: https://www.youtube.com/watch?v=FJC5H83dcWE.

6   Conclusion

In this paper, we described a simple and fast algorithm for a hand gesture recognition issues for mobile device.

Here, we have used Open CV library for Android for image processing, and Tegra Android Development Pack (TADP) to develop the android application.

At first, the NUS hand poses database was used for training purposes. We have choose to work with LBP approach that demonstrate a good performance in the training time compared to other approaches that needs many days instead of some hours [24]. Indeed, the detection of gestures was fast and accurate even if the number of positive samples used for training for each gesture was not so numerous. The LBP approach demonstrated that it is suitable for mobile device as it is computationally inexpensive. In a future work, to improve the training step, we will try also to enhance the experimentation by using PR curve and by inviting more and more people to test our application.

References

1. Manjoo, F.: A Murky road ahead for android, despite market dominance. The New York Times (2015). ISSN 0362-4331. Accessed 27 May 2015
2. Statcounter company web site. http://gs.statcounter.com/press/android-overtakes-windows-for-first-time. Accessed 3 Apr 2017
3. OpenCV. http://opencv.org/platforms/android.html
4. Cobârzan, C., Hudelist, M.A., Schoeffmann, K., Primus, M.J.: Mobile image analysis: android vs. iOS. In: 21st International Conference on MultiMedia Modelling (MMM), pp. 99–110 (2015)
5. Seymour, M., Tšoeu, M.: A mobile application for South African sign language (SASL) recognition. In: IEEE AFRICON 2015, pp. 281–285 (2015)
6. Xie, C., Luan, S., Wang, H., Zhang, B.: Gesture recognition benchmark based on mobile phone. In: Chinese Conference on Biometric Recognition (CCBR). Lecture Notes in Computer Science, vol. 9967, pp. 432–440 (2016)
7. Lahiani, H., Elleuch, M., Kherallah, M.: Real time hand gesture recognition system for android devices. In: 15th International Conference on Intelligent Systems Design and Applications (ISDA), pp. 592–597 (2015)
8. Lahiani, H., Elleuch, M., Kherallah, M.: Real time static hand gesture recognition system for mobile devices. J. Inf. Assur. Secur. 11, 67–76 (2016). ISSN 1554-1010
9. Lahiani, H., Kherallah, M., Neji, M.: Hand pose estimation system based on Viola-Jones algorithm for android devices. In: 13th ACS/IEEE International Conference on Computer Systems and Applications (AICCSA) (2016)
10. Lahiani, H., Kherallah, M., Neji, M.: Vision based hand gesture recognition for mobile devices: a review. In: 16th International Conference on Hybrid Intelligent Systems (HIS 2016). Advances in Intelligent Systems and Computing, pp. 308–318 (2016)
11. Viola, P., Jones, M.: Rapid objet detection using a boosted cascade of simple features. In: Proceedings of the IEEE Conference on Computer Vision and Pattern Recognition, pp. 511–518 (2001)
12. Hays, P., Ptucha, R., Melton, R.: Mobile device to cloud co-processing of ASL finger spelling to text conversion. In: 2013 IEEE Western New York Image Processing Workshop (WNYIPW), pp. 39–43 (2013)
13. Jin, C., Omar, Z., Jaward, M.H.: A mobile application of American sign language translation via image processing algorithms. In: 2016 IEEE Region 10 Symposium (TENSYMP), pp. 104–109 (2016)
14. Kamat, R., Danoji, A., Dhage, A., Puranik, P., Sengupta, S.: Monvoix-an android application for hearing impaired people. J. Commun. Technol. Electron. Comput. Sci. 8, 24–28 (2016)
15. Setiawardhana, Hakkun, R.Y., Baharuddin, A.: Sign language learning based on android for deaf and speech impaired people. In: 2015 International Electronics Symposium (IES), pp. 114–117 (2015)

16. Prasuhn, L., Oyamada, Y., Mochizuki, Y., Ishikawa, H.: A HOG-based hand gesture recognition system on a mobile device. In: IEEE International Conference on Image Processing (ICIP), pp. 3973–3977 (2014)
17. Harwood, D., Ojala, T., Pietikäinen, M., Kelman, S., Davis, S.: Texture classification by center-symmetric auto-correlation, using Kullback discrimination of distributions. Technical report, Computer Vision Laboratory, Center for Automation Research, University of Maryland, College Park, Maryland, CAR-TR-678 (1993)
18. Ojala, T., Pietikïnen, M., Harwood, D.: A comparative study of texture measures a with classification based on feature distributions. Pattern Recogn. **29**, 51–59 (1996)
19. Freund, Y., Schapire, R.E.: Experiments with a new boosting algorithm. In: Proceedings of the 13th International Conference in Machine Learning, pp. 148–156 (1996)
20. Lienhart, R., Maydt, J.: An extended set of haar-like features for rapid object detection. In: Proceedings of the IEEE Conference on Image Processing (ICIP 2002), New York, USA, pp. 155–162, September 2002
21. Ferreira, A., Figueiredo, M.: Boosting algorithms: a review of methods, theory, and applications. In: Ensemble Machine Learning: Methods and Applications. Springer, New York, pp. 35–85 (2012)
22. Frejlichowski, D., Gosciewska, K., Forczmanski, P., Nowosielski, A., Hofman, R.: Applying image features and AdaBoost classification for vehicle detection in the 'SM4Public' system. In: Image Processing and Communications Challenges 7. Advances in Intelligent Systems and Computing, vol. 389, pp. 81–88 (2015)
23. The NUS hand posture datasets I. https://www.ece.nus.edu.sg/stfpage/elepv/NUS-HandSet/
24. Lahiani, H., Kherallah, M., Neji, M.: Hand pose estimation system based on a cascade approach for mobile devices. In: 17th International Conference on Intelligent Systems Design and Applications (ISDA) (2017)

An Efficient Real-Time Approach for Detection of Parkinson's Disease

Joyjit Chatterjee$^{(\boxtimes)}$, Ayush Saxena, Garima Vyas, and Anu Mehra

Amity School of Engineering and Technology, Amity University, Noida 201303, U.P., India
joyjit.c.in@ieee.org, ayushsaxena709@gmail.com,
{gvyas,amehra}@amity.edu

Abstract. Parkinson's disease is one of the most complex neurological disorders which have affected mankind since ages. Recent studies in the field of Biomedical Engineering have shown that by analyzing the Verbal Response of any human being, it is highly feasible to predict the odds of having the deadly disease. A simple analysis of an utterance of "ahh" sound by a person can help to analyze the person's state of neurological health from a layman's perspective. The paper initially utilizes the SVM (Support Vector Machine) Learning algorithm to predict the odds of having the Parkinson's disease from a variety of audio samples consisting of healthy and unhealthy population. The cepstral features are used to develop a Real-Time Program for user-friendly application which asks the user to utter "ahh" for as long and as boldly as possible and finally displays whether the user has Parkinson's Disease or not. The Real-Time Program can prove to be a helpful tool for the people as well as the medical community in general, assisting in early diagnosis of the Parkinson's disease.

Keywords: SVM · MFCC · Length · Energy · Volume · Zero Crossing Rate
Real-time machine learning · Parkinson's disease

1 Introduction

Parkinson's disease is a type of disorder that affects the central nervous system [1]. The symptoms of this disease are shown slowly over a period of time. The main symptoms that indicate the Parkinson's disease are rigidity, shaking, and slowness with movement. The diminished sensory response is one of the major symptoms which can be noticed in early stages. As the disease progresses, neurons are lost and cure becomes unfeasible. Approximately 15% population with Parkinson's Disease has family history of the disease. This shows that it is not completely a genetic disease, but can affect any individual at any point of his life, though, it mainly occurs in the later part of life [4]. This disease also affects the verbal and non verbal movements of the human body. Behavior and mood of the patient also changes as the effect of the intensity of the disease increases. Sleep disorder is another critical indicator of this disease. The person suffers from both motor movement and non motor movement. Motor symptoms are - tremor of arms and legs, muscle stiffness, slowed movement etc. and non motor movements include cognition, psychiatric, autonomous systems.

© Springer International Publishing AG, part of Springer Nature 2018
A. Abraham et al. (Eds.): ISDA 2017, AISC 736, pp. 191–200, 2018.
https://doi.org/10.1007/978-3-319-76348-4_19

Audio processing helps in detecting the disease at minimal cost as compared to the original biomedical systems through which the disease is detected. Presently, there is no blood test which can detect Parkinson's disease, making early intervention almost close to null. The state of the art technique of simply using the voice as an early indicator of the disease can help sort out things in a simpler and a more effective manner. In audio processing, the voice of the human subject can directly be used for further processing and calculation of features. For the purpose of this paper, the features that are used are MFCC, Volume, Zero Crossing Rate, and Short Term Energy. The classifier used is SVM to differentiate between a healthy and an unhealthy subject, typically on the basis of the "ahh" sound which the person utters [2]. The ongoing research at the National Parkinson's Foundation carried out by Max Little has shown that the "ahh" sound utterance can be analyzed for mainly 3 clusters of symptoms in voice, namely the voice tremors, breathiness and weakness of voice and the manner in which tongue, jaws and lips of the subject fluctuate during speech. Little's Algorithm has proved to have over 83% accuracy in detecting the disease in both, people with and without the syndrome.

The present work is significantly different from the previous research [11] in this area as it makes use of certain new features, other than just duration of the signal, which the previous algorithm proposes. The flow of the paper is organized as follows: Sect. 2 gives a brief overview of the database used; Sect. 3 outlines the methodology of the work done, including the brief description of the features extracted and their significant differences in the Parkinson's and Healthy subject's cases. Section 4 illustrates the results obtained after SVM Classification and the Real Time Implementation and finally, Sect. 5 concludes the paper.

2 Database

The database is used from the Open Source University of California, Irvine - Machine Learning Repository [3]. A total of 58 audio samples have been used, of which 36 are those of subjects having Parkinson's Disease and 22 are those of Normal Human Subjects [10]. The sampling rate for the wav files is 22,050 Hz (mono). The "ahh" sound utterance is the essence of the entire database used in the paper as already outlined in the introduction. Ahh sound is used because ahh word clearly provides the required results.

3 Methodology

Figure 1 shows the flow of algorithms adopted to diagnose Parkinson disease from the speech samples using audio signal processing techniques.

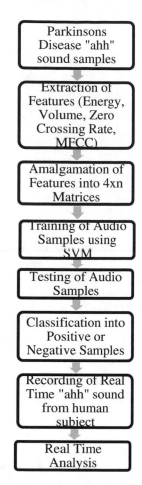

Fig. 1. Flowchart of methodology adopted

3.1 Feature Extraction

The features which are robust in nature and are suitable for machine learning are extracted from the frames of the audio clips. The most discriminatory features are selected by employing box plots on each feature set of the healthy and patient speech samples. It is observed from the experiments that Short Term Energy, Volume, MFCC and Zero Crossing Rate are the most discriminatory features [6]. Figure 2(a)–(d) shows the box plots for these four features.

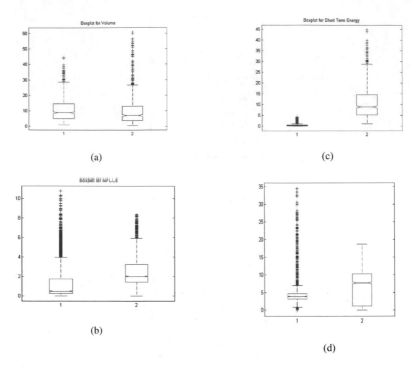

Fig. 2. (a) Boxplot for volume (1-Healthy, 2-Patient); (b) Boxplot for MFCCs (1-Healthy, 2-Patient); (c) Boxplot for short term energy (1-Healthy, 2-Patient); (d) Boxplot for ZCR (1-Healthy, 2-Patient)

3.1.1   Short Term Energy (STE)

It is the average of the energy that is linked with the short term duration of any speech signal. The Eq. 1 explains the STE as follows:

$$E_t = \sum_{n=-\infty}^{+\infty} s^2(n) \tag{1}$$

Where E_t : Short Term Energy

$s(n)$: Amplitude of Signal

The Fig. 3(a) and (b) compares the STE for the healthy and patient speech respectively.

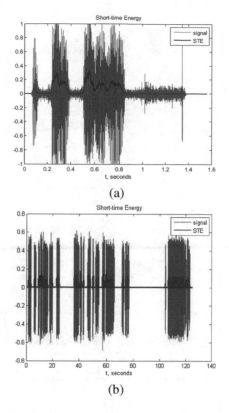

Fig. 3. (a) STE of healthy person; (b) STE of Parkinson affected person

It can clearly be visualized that for a healthy subject, the STE reaches the peak values up to 0.4 while for the patient's speech the STE has highest peak at 0.1. In addition to this, the amplitude level of the healthy speaker is fluctuating rapidly in comparison to the patient speaker which indicates that the patient has very low energy content in the voiced region [5].

3.1.2   Volume

The volume of the speech signal has been used as it is a paramount indicator of the degree of loudness or softness of the voice, where a too soft voice is an indicator of weak neurological functioning [7]. Figure 4(a) and (b) compares the absolute values of the Volume for a healthy and patient subject respectively.

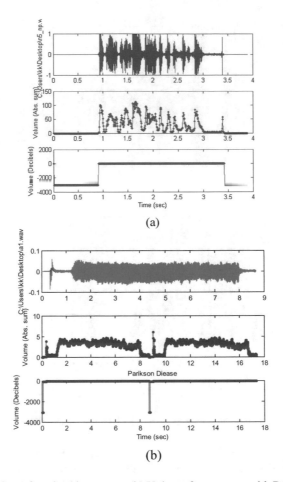

Fig. 4. (a) Volume for a healthy person; (b) Volume for a person with Parkinson's disease

The volume values are typically high for a Normal speakers and are close to the 0–100 levels in the absolute scale [12]. While, for the patient, the volumes levels are very low and near to 5 in the absolute scale. Hence, this feature is discriminatory in nature and is suitable for machine learning.

3.1.3   Zero Crossing Rate

Zero Crossing Rate (ZCR) shows the rate at which a signal typically varies in the discrete time domain [8]. Figure 5(a) and (b) compare the Short Term ZCR for Healthy and Patient Speech respectively. As can be visualized, the amplitude of the ahh signal for a normal person varies just between −1 and +1, and the Short Term Zero Crossing Rate (STZCR) varies from 0 to about 0.22. While, for the patient, the speech signal is merely concentrated in amplitude ranges from −0.8 to +0.8, showing lesser peak amplitude, while the STZCR reaches peak values of more than 0.6 at about 120 s, signifying rapid fluctuation in the signal intensity [9].

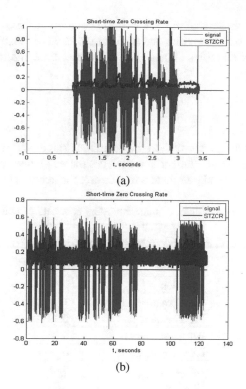

(a)

(b)

Fig. 5. (a) Zero Crossing Rate for normal person; (b) Short time Zero Crossing Rate for Parkinson's affected patient

3.1.4 Mel Frequency Cepstral Coefficients (MFCCs)

MFCC is a good choice for feature because it accurately represents the envelope of the short time power spectrum of the speech. MFCC spectrum for a healthy person has wide frame indices varying from 1 to about 2000. While, for the patient, the frame indices are in a narrow range, ranging from 0 to about 350 and are mainly concentrated from frame indices 150 to about 180. This shows that the "ahh" sound of Parkinson's Disease affected patient is mainly concentrated in the lower values of the frequency spectrum.

3.2 SVM for Classification

The SVM is implemented to segregate patient speech samples and healthy speech samples. In this experiment, 40 audio files (20 healthy, 20 patient) are used for training and 18 audio files (9 healthy, 9 patient) are used to test the performance of a classifier.

3.3 Real Time Implementation

To test the intelligibility of the designed system on real time, the speech samples of "ahh" sound are recorded by the in-build microphone of the laptop. The four features are extracted and amalgamated to form a single complex feature set [14]. This feature matrix is then fed into the classifier and is classified as either healthy or patient speech. The SVM accurately classifies almost 80% of the real time audio signals into healthy and patient speech.

4   Results

Several experiments are performed on the designed system and the results are encouraging. The system is quite efficient and delivers the accuracy up to 85.71%. Table 1 shows the various statistical parameters to measure the accuracy of the system.

Table 1. Values of some important parameters obtained

Measure	Value
Sensitivity	1
Specificity	0
Precision	0.5
Negative predictive value	1
False positive rate	0.38
False discovery rate	0.14
False negative rate	0
Accuracy	0.85
F1 score	0.92

Every single audio clip in the database is labeled as healthy or patient speech. The same is explained by glyph plot in Fig. 6. The symbol "." (Dot) represents the Parkinson speakers sample and symbol "—" shows the healthy speakers. Figure 7 show the real time classification process in MATLAB. First, the user is asked to press key number 1

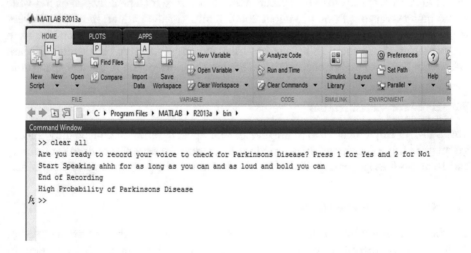

Fig. 6. Real time classification system for patient speech sample

to start recording his voice and say "ahh" for as long as he can. Then, the SVM Classifier runs in the background and classifies the person into either healthy or patient.

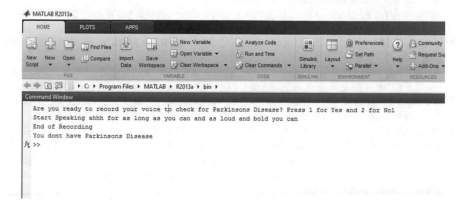

Fig. 7. Real time classification system for healthy speaker

5 Conclusion

The paper provides for an interesting and innovative algorithm to facilitate real time analysis of the "ahh" speech signal utterance by a human subject and detect whether or not the human subject has odds of having the Parkinson's disease. Through the use of some critical features, including the MFCC, Energy, Zero Crossing Rate and Volume, the paper paves way for testing a given set of audio samples by employing the SVM Classifier. The accuracy of the classifier comes to 85.71% which is quite good, as not much research has been done in the area of identification of Parkinson's disease. Further, the real time analysis of the speech signal provides for an easy way for a layman to detect Parkinson's disease in the human subject. The real time analysis works quite well, on few tested subjects, though it needs to be further tested and analyzed medically for viable results. The paper can pave way for a real time web framework which can be used by any user to record his/her "ahh" sound utterance from a Laptop [12] and identify the chances of having the deadly neurological disease.

References

1. Sankur, B., Güler, E.C., Kahya, Y.P.: Multiresolution biological transient extraction applied to respiratory crackles. Comput. Biol. Med. **26**, 25–39 (1996)
2. Nemerovskii, L.I.: Analysis of a pulmophonogram for assessing local pulmonary ventilation (1980)
3. Lax, A., et al.: Exploiting nonlinear recurrence and fractal scaling properties for voice disorder detection. Biomed. Eng. OnLine **6**(1), 23 (2007)
4. Loudon, R.G., Murphy, R.L.: Parkinson's sounds. Am. Rev. Respir. (1984)
5. Kraman, S.S., Bohadana, A.B.: Transmission to the chest of sound introduced at the mouth. J. Appl. Physiol. **66**(1), 278–281 (1989)

6. Venugopal, K.R., Vibhudendra Simha, G.G., Joshi, S., Shenoy, D., Rrashmi, P.L., Patnaik, L.M.: Classification of non Parkinson's disease and Parkinson's disease by using machine learning and neural network methods (2008)
7. Ramezani, H., Khaki, H., Erzin, E., Akan, O.: Speech features for monitoring of Parkinson's disease symptoms (2017)
8. Keloth, S.M., Arjunan, S.P., Kumar, D.: Computing the variations in the self-similar properties other various intervals in Parkinson disease patients
9. Kostas, M., Georgios, T., Gatsios, R.D., Antonini, A., Konitsiotis, S., Koutsouris, D.D.: Predicting rapid progression of Parkinson's Disease at baseline patients evaluation (2017)
10. Gümüşçü, A., Karadağ, K., İbrahim, M.E.T., Aydılek, B.: Genetic algorithm based feature selection on diagnosis of Parkinson disease via vocal analysis (2016)
11. Sengupta, N., Sahidullah, M., Saha, G.: Lung sound classification using cepstral-based statistical features. Comput. Biol. Med., US National Library of Medicine, NIH, August 2016
12. Fernandez-Granero, M.A., Sanchez-Morillo, D., Leon-Jimenez, A.: Computerised analysis of telemonitored respiratory sounds for predicting acute exacerbations of COPD. Sensors **15**(10), 26978–26996 (2015)

Dual Image Encryption Technique: Using Logistic Map and Noise

Muskaan Kalra$^{(\boxtimes)}$, Hemant Kumar Dua, and Reena Singh

Department of Computer Science,
Bharati Vidyapeeth's College of Engineering, Delhi, India
muskteer_dps@yahoo.com, hemant.dua56@gmail.com,
reena.singh@bharatividyapeeth.edu

Abstract. Any web based computer system is susceptible to attacks from hackers who intrude a system to obtain information for illegal use as well as for sabotaging a company's business operations. This is where we need encryption techniques and image security. Image security includes visual authentication and access control based user images identification. In this paper, we propose a secured image encryption technique using Chaos-Logistic Maps. The algorithm proposed uses Logistic Map to shuffle the image and the noise and then superimposes them. Two shares of this resulting image are then created and XORed to make the image ready for transmission. The performance analysis using histograms and correlation coefficient show that the algorithm has high security, and strong robustness. The algorithm uses a logistic map and XOR gates to achieve a multi-chaos encryption effect.

Keywords: Encryption technique · Logistic map · Noise · Visual cryptography
XOR gates

1 Introduction

Security of data on the internet is one of the main concerns in today's emerging world. Cryptography has been used since the ancient times in the form of hieroglyphics. A number of unusual symbols to obscure the meaning of the inscriptions were used back in about 1900 BC. Scytale [1] was a device used in 500 BC to send and receive secret messages.

Various war driver cryptography methods were used in WWI and WWII. In modern times, we use one-time pads that make use of a key using a modular addition (XOR) to combine plaintext elements with key elements. Pseudo random number generators are used to create apparently random numbers from a designated key. Symmetric key encryption involves the use of a private key. A person with access to this key only can open the encrypted information. Asymmetric key encryption makes use of a different key for encryption and a different key for decryption.

At present, the image information security has turned into a world-wide problem which has made increasing number of researchers interested in studying robust and secure image cryptography schemes to protect important images from leakage. A lot of research efforts have been done to apply Physics and Mathematics concepts into real

© Springer International Publishing AG, part of Springer Nature 2018
A. Abraham et al. (Eds.): ISDA 2017, AISC 736, pp. 201–208, 2018.
https://doi.org/10.1007/978-3-319-76348-4_20

world engineering applications. In 1992, Bourbakis and Alexopoulos [2] proposed an image encryption scheme which makes use of the SCAN language to encrypt as well as compress an image simultaneously. Fridrich [3] showed the construction of asymmetric block technique for encryption based on two-dimensional standard baker map.

Chaos theory is getting a lot of attention since the last two decades. Chaos is a branch of mathematics that studies complex systems. Because we can never know all the initial conditions of a complex system in sufficient detail, we cannot hope to predict the ultimate fate of a complex system making the chaotic system highly unpredictable. Chaotic maps are maps capable of exhibiting chaotic behavior and usually occur in the study of dynamical systems. The non-linear behavior of chaotic maps, make it difficult to decrypt the image without the presence of secret key which would thereby protect secret data.

2 Literature Survey

In year 1992, Bourbakis and Alexopoulos [2] proposed the first digital image encryption based on SCAN language. After this many researchers came up with the idea of chaos based encryption and decryption systems.

Guan et al. [4] in 2005, proposed a new encryption scheme in which two processes-shuffling the positions of pixels and changing their grey values are combined to confuse the relationship between the cipher-image and the plain-image. Pareek et al. [5] in 2006, proposed a new scheme in which an external secret key and two chaotic logistic maps are employed. Zhu and Li [6] in 2010, proposed a new algorithm that gave rise to nine chaotic patterns by using a single key from which six were used to shuffle the image pixels while the other three were used to confuse and diffuse image pixels. In 2012, Mishra and Mankar [7] proposed a hybrid message scheme using non-linear feedback shift register and 1-D logistic map as chaotic map. Liu and Miao [8] in 2016, proposed a new scheme in which parameters for logistic map were varied and a dynamic algorithm, was used to encrypt the image. In 2013, Saraereh et al. [9] proposed to improve an already existing algorithm (NCA) by modifying logistic map. Borujeni and Ehsani [10] proposed two modified versions of logistic map - First Modification of Logistic Map (FML) and Second Modification of Logistic Map (SML) in which symmetry and transformation have been changed.

3 Background Study

(i) Butterfly Effect

Chaos theory can be better explained with Edward Lorenz's 'butterfly effect' theory [11]. The theory shows that the smallest of differences are producing the largest effects. The idea is that the flapping of a butterfly's wings in one continent could cause a tornado in another continent. According to the butterfly effect, if the butterfly had never flapped its wings, the tornado would not have happened. This shows that the initial conditions are extremely important, and they have a major impact on the outcome of things. Another important aspect of chaos theory is unpredictability. A small error or change in the initial conditions can change the outcome.

Chaotic maps are equations that generate different outcomes on every value that is put as an input. Since recent years they are used in crypto systems which consist of multiple rounds of substitution and diffusion. In the substitution phase, another image is substituted on top of the input image. In the diffusion phase, the pixels are shuffled to make encryption secure.

(ii) Logistic Map

A sequence of numbers that are chosen at random are useful in many different kinds of applications like simulation, sampling, numerical analysis, decision making, recreation etc. Producing a sample which is truly random is not possible.

With the advent of fast computers, a lot of information about chaotic systems has come into the picture over the last two decades. One of them is the logistic map that can be used to demonstrate complex, dynamic phenomena occurring in chaos theory.

Logistic map is a quadratic map that is capable of very complicated behavior. The Logistic equation is as follows:

$$x_{n+1} = rx_n(1 - x_n) \tag{1}$$

Where r (sometimes also denoted μ) and is any real value in the interval [0, 4] as obtained from the bifurcation diagram in Fig. 1. The iteration of this formula gives back values between [0, 1]. For r values less than 3.5 the iteration quickly stabilizes into one value. But at higher values of r the solutions start bifurcating, stabilizing in 2 values then 4, 8, 16... until it reaches total chaos. This is called period doubling. For a r value of 4, the behavior is totally random.

A global representation of the various regimes that are encountered as any control parameter is varied. This is done with the help of bifurcation diagrams - Fig. 1.

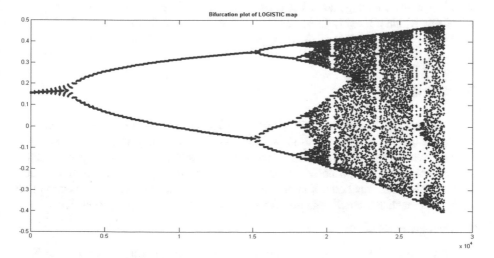

Fig. 1. Bifurcation diagram of Logistic Map obtained by MATLAB

4 Proposed Research Scheme

In our paper we used Logistic maps and XOR gates for image encryption. We initially take as input the original image, noise and any random value of r in the interval $[0, 4]$ say, $r = 3.777$.

1. The Logistic equation cannot be applied directly to the image.
2. It is applied separately to the red, blue and green pixels of the image.
3. The image is thus first broken down into its three RGB components.
4. Then the Logistic value is applied to each pixel of each component.
5. After applying the Logistic equation to each of the pixels, the value of each pixel obtained which is very low.
6. Therefore, a Scale = 10,000 to make the value higher and thereby making computation easier.
7. For changing the value of each pixel, the original value of each pixel is added to the value obtained from the Logistic equation.
8. Then the value is divided by 256. This gives us the value of each pixel which will be used in the encryption process.
9. This is done in order to embed a part of the original image into the Logistic value so as to make it possible to retrieve the image at the receiving end.
10. Noise is an additional feature that has been added to this research paper to make the cryptography technique two-fold secure. The same process is repeated for the image that is used as noise.
11. The value of each pixel in the image to be sent are generated.
12. The value of each pixel in the noise image are also generated.
13. Now, the desired image is taken and Logistic equation is applied to each pixel.
14. Two shares of the image are then generated.
15. To make the shares the Logistic value is used i.e. if the logistic value corresponding to a pixel is even, the pixel is placed in share 1 else in share 2.
16. Finally, the shares generated and the noise and XORed to generate two separate shares which are ready to be sent through the transmission channel.
17. We receive two shares of the encrypted image as the output (Fig. 2).

5 Results

A good encryption system should not be affected by statistical, brute-force and exhaustive attacks. In this section we discuss security analysis such as statistical analysis, sensitivity analysis with respect to the key and plaintext, key space analysis etc. to prove that the proposed crypto system is secure against the most common attacks. This scheme is better than the previous ones because it uses a dual encryption method rather than a single one and thereby makes the system more secure.

(i) **Histogram Analysis**

   To prove the robustness of the proposed image encryption procedure, we have performed statistical analysis by doing an analysis of histograms [12].

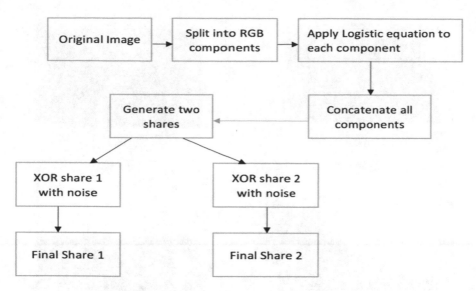

Fig. 2. Block diagram of proposed encryption technique

As we can observe from Fig. 3, the histograms for the original image and the encrypted image are significantly different. Also, the histograms of the encrypted image are uniform and hence resistant to statistical attacks.

(ii) **Correlation Coefficient and Sensitivity Analysis**

A prominent feature of chaos based systems is its sensitivity to initial parameters. On changing the value of the key slightly, the encryption pattern changes. To demonstrate that on changing the key value i.e. r slightly, the value of correlation coefficient changes, we have used 2 key values $r_1 = 3.777$ and $r_2 = 3.999$ and the encrypted images obtained using both these values for two images - lady and bird are shown in Fig. 4.

The encrypted images on both values look similar but on analysing the correlation coefficient between both of them, we have found the correlation coefficient to be extremely low indicating that the similarity between the encrypted images is very less as shown in Table 1.

This is a clear indication that encryption using different values will yield to different outputs thereby shows that the encryption scheme proposed is sensitive to even small changes in the value of r.

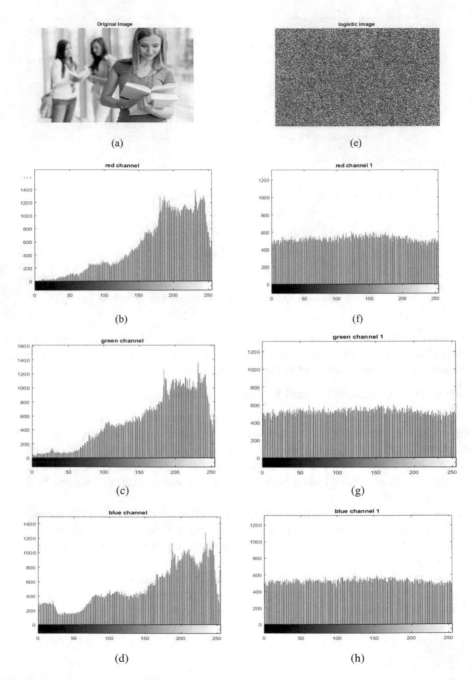

Fig. 3. (a) shows the original image. (b), (c) and (d) respectively, show the histograms of red, green and blue channels of the original image shown in (a). (e) shows the encrypted image of the original image shown in (a) using the logistic map. (f), (g) and (h) respectively, show the histograms of red, green and blue channels of the encrypted image shown in (e)

Fig. 4. (a) and (b) show the original image of the lady and the bird respectively. (c) and (d) show the encrypted image of a lady at r = 3.777 and r = 3.999 respectively. (e) and (f) show the encrypted image of a bird at r = 3.777 and r = 3.999 respectively.

Table 1. Correlation coefficient between original and encrypted image (RGB values)

Image	R	G	B
Lady	0.006	0.0014	0.0020
Bird	0.0044	0.0024	0.0047
Flower	0.0005	0.0056	0.0056
Lemon	0.0012	0.0016	0.0034

6 Future Work

We can use different chaotic maps to carry out the encryption process. A map with more degree of randomness than Logistic map can be employed. Maps like - Arnold Cat's Map, Tent Map, etc. can be used and their properties can be identified and utilised to make the encryption scheme more secure and fast. The decryption process for the proposed scheme can also be devised by using the inverse Logistic equation.

# 7	Conclusion

In this paper we have proposed a method for encryption of images using Logistic Map and increasing the security by creating a dual security system by using noise and superimposing it on the encrypted image. This method has three steps - applying Logistic equation, concatenation of shares and XOR of the shares with noise. The algorithm proposed is efficient in generating a good cipher as is indicated by the low correlation between the original and encrypted image. The proposed technique is also at a low risk of statistical attacks as can be seen by the uniformity of histograms of the encrypted shares. The sensitivity, correlation and time analysis has also been performed which indicate that the proposed scheme is good. The correlation coefficient shows difference in encryption at different values of the key indicating that the scheme is secure. The correlation before and after superimposing with noise has been found and it decreases after adding noise indicating that the dual security method is more successful and efficient.

References

1. Count on - Transposition Ciphers, counton.org (2006). http://www.counton.org/explorer/codebreaking/transposition-ciphers.php. Accessed 13 Oct 17
2. Bourbakis, N., Alexopoulos, C.: Picture data encryption using scan patterns. Pattern Recogn. 25(6), 567–581 (1992). https://doi.org/10.1016/0031-3203(92)90074-S. ISSN 0031-3203
3. Fridrich, J.: Symmetric ciphers based on two-dimensional chaotic maps. Int. J. Bifurc. Chaos 8(6), 1259–1284 (1998)
4. Guan, Z.-H., Huang, F., Guan, W.: Chaos-based image encryption algorithm. Phys. Lett. A 346, 153–157 (2005). https://doi.org/10.1016/j.physleta.2005.08.006
5. Pareek, N., Patidar, V., Sud, K.K.: Image encryption using chaotic logistic map. Image Vis. Comput. 24, 926–934 (2006). https://doi.org/10.1016/j.imavis.2006.02.021
6. Zhu, A., Li, L.: Improving for chaotic image encryption algorithm based on logistic map. In: 2010 the 2nd Conference on Environmental Science and Information Application Technology, Wuhan, pp. 211–214 (2010). https://doi.org/10.1109/esiat.2010.5568374
7. Mishra, M., Mankar, V.H.: Chaotic encryption scheme using 1-D chaotic map. Int. J. Commun. Netw. Syst. Sci. 4, 452–455 (2011). https://doi.org/10.4236/ijcns.2011.47054
8. Liu, L., Miao, S.: A new image encryption algorithm based on logistic chaotic map with varying parameter. SpringerPlus 5, 289 (2016). PMC. Web, 14 Nov 2017
9. Saraereh, O.A., Aisafasfeh, Q., Arfoa, A.: Improving a new logistic map as a new chaotic algorithm for image encryption. Mod. Appl. Sci. 7(12) (2013)
10. Borujeni, S.E., Ehsani, M.S.: Modified logistic maps for cryptographic application. Appl. Math. 6, 773–782 (2015)
11. Borwein, J.J., Rose, M.: 19 November 2012. https://theconversation.com/explainer-what-is-chaos-theory-10620
12. Shodhganga: a reservoir of Indian theses. http://hdl.handle.net/10603/25696

A Memetic Algorithm for the Network Construction Problem with Due Dates

Jonatas B. C. Chagas[1]([⊠]), André G. Santos[2], and Marcone J. F. Souza[1]

[1] Departamento de Computação, Universidade Federal de Ouro Preto,
Ouro Preto, Brazil
{jonatas.chagas,marcone}@iceb.ufop.br
[2] Departamento de Informática, Universidade Federal de Viçosa, Viçosa, Brazil
andre@dpi.ufv.br

Abstract. In this work, we present an effective memetic algorithm for a transportation network reconstruction problem. The problem addressed arises when the connections of a transportation network have been destroyed by a disaster, and then need to be rebuilt by a construction crew in order to minimize the damage caused in the post-disaster phase. Each vertex of the network has a due date that indicates its self-sufficiency, i.e., a time that this vertex may remain isolated from the network without causing more damages. The objective of the problem is to reconnect all the vertices of the network in order to minimize the maximum lateness in the recovery of the vertices. The computational results show that our memetic algorithm is able to find solutions with same or higher quality in short computation time for most instances when compared to the methods already present in the literature.

Keywords: Network construction · Post-disaster
Scheduling · Combinatorial optimization · Heuristic
Memetic algorithm

1 Introduction

The problem addressed in this work arises in emergency situations, when a transportation network needs to be rebuilt after it has been destroyed by a disaster. We know that, unfortunately, terrorist attacks and natural disasters such as floods, earthquakes, hurricanes and storms have been marking human existence throughout history, always causing enormous ecological and economic damage to mankind, as well as making thousands of victims every year [1, 2]. It is important to emphasize that generally the number of victims increases in the post-disaster phase due to the difficulty of accessing the affected areas, since the roads connecting those areas are usually also affected by disasters. Therefore, an effective rescue planning is crucial to reduce the impact of catastrophes [3].

The literature presents several works that deal with different problems of network reconstruction (see Çelik [4]) and some of them are briefly presented here.

© Springer International Publishing AG, part of Springer Nature 2018
A. Abraham et al. (Eds.): ISDA 2017, AISC 736, pp. 209–220, 2018.
https://doi.org/10.1007/978-3-319-76348-4_21

Feng and Wan [5] developed a multi-objective model involving maximizing the performance of emergency rehabilitation, minimizing the risk of rescuers, and maximizing lives saved, where roads are repaired by multiple work groups in order to maximize the effectiveness of the rescue. Hu and Sheu [6] have also developed a multi-objective model that considers conflicts of logistical operations, environmental protection, and specially psychological recovery experienced by local residents who were waiting for medical treatment and removal of debris. Berktaş et al. [3] proposed mathematical models and heuristic algorithms for two different problems, where the objective of the first one is to minimize the total time to reach all critical regions of the network and the objective of the second problem is to minimize the weighted sum of the times to reach the regions, using weights associated with each region to indicate their priority.

The problems described and studied by Averbakh and Pereira [7], referred to Network Construction Problems with Due Dates (NCPDD), consider that a transportation network needs to be rebuilt in order to reconnect all their vertices, that is, restore the network so that all vertices can be reached from some other. The restoration is performed by a single construction crew, which is initially located at one specific vertex of the network (depot) and is able to work only one connection at a time. The instant when a vertex is reached for the first time by the construction crew is called the recovery time of this vertex. Moreover, for each vertex a due date is associated, which informs the limit of self-sustainability of the vertex, that is, the maximum time that the construction crew must take to build a connection to that vertex to avoid greater damage.

Averbakh and Pereira [7] defined three problems with the aim to find a construction planning of the edges in order to optimize some scheduling objectives. The first problem, referred to as problem L, consists in minimizing the maximum lateness of the vertices. The objective of the second problem (problem T) is to minimize the number of tardy vertices, and the last problem, named as problem F, is a decision problem requiring to determine the existence of a feasible solution that does not violate any due dates, or to show that no such solution exists. The author then presented a Mixed Integer Linear Programming (MILP) formulation and a branch-and-bound (B&B) algorithm for each problem (see Averbakh and Pereira [7]). The experimental results showed that the branch-and-bound algorithm is significantly more efficient than MILP formulation, specially for the problems L and T.

In this paper we describe a Memetic Algorithm (MA) proposed to find high quality solutions for the problem L. Our algorithm can also be applied, with few changes, to solve the problems T and F described by Averbakh and Pereira [7]. However, due to the difficulty of comparing the results presented in their paper for these problems, we presented only a comparative analysis with respect to the results obtained for the problem L, as such results were informed in details by the authors.

To the best of our knowledge, only one work has addressed the NCPDD-L by heuristic methods. Chagas and Santos [8] proposed a simple algorithm based on the Simulated Annealing metaheuristic that obtained moderate results for the problem, which were overcome by the algorithm proposed here. For this reason, we present a comparative analysis only between the results obtained by our MA and the B&B algorithm proposed by Averbakh and Pereira [7].

The structure of the rest of this paper is as follows. In Sect. 2, we present a formalization of the NCPDD-L, as was presented by Averbakh and Pereira [7]. In Sect. 3, we describe the proposed heuristic method based on memetic algorithms. Finally, results are reported in Sect. 4 and in Sect. 5 we present our conclusions.

2   Problem Definition

As was formally introduced by Averbakh and Pereira [7], the Network Construction Problems with Due Dates (NCPDD) can be represented by an undirected graph $G = (V, E)$, where the set of vertices $V = \{1, 2, ..., n\}$ represents the points to be recovered by a single construction crew, and E is the set of candidates edges to reconnect the vertices. For each edge $\{i, j\} \in E$ is associated a value C_{ij} that indicates its length.

The construction crew initially is located at the depot (vertex 1) and starts working at time 0. The construction crew works only one connection (edge) at a time and is able to build a connection with a constant speed of 1 unit size per unit of time. This speed is substantially low when compared to the travel speed on already constructed edges. It is so low that the speed of travel of the construction crew, within the network already restored, is considered infinite, i.e., at any time the construction team can move to any point within the network already restored with zero travel time.

The time c_j when the construction crew reaches the vertex j for the first time is called the recovery time of the vertex j, i.e., the recovery time c_j represents the time that the vertex j becomes connected to the depot (vertex 1).

For each vertex $j > 1$ is associated a value d_j of deadline that informs the maximum time that the construction crew can take to rebuilt a connection until the vertex j. The depot (vertex 1) has no due date value.

The value $c_j - d_j$, if greater than zero, is the lateness of vertex j, and the objective of the NCPDD-L is to determine a construction schedule so that the largest lateness should be minimized, i,e., $\min\{max_{j=2}^{n}(c_j - d_j)\}$. Notice that the objective includes "negative" lateness (which is in fact earliness), so that in the case that all vertices can be recovered before deadline, we want to maximize the earliness.

3   Memetic Algorithm Proposed

Genetic Algorithms (GAs) are computational techniques frequently used to solve complex problems through mechanisms that mimic the processes of Darwinian Evolution [9]. The functioning mechanism of GAs is based on the evolution of a population of individuals, which represent solutions to the problem treated. In order to explore the solution space of the problem, a simulation of the evolution of species is made through selection, reproduction and mutation of individuals.

Throughout the simulation, characteristics of individuals with greater fitness tend to survive, thus guiding the algorithm to explore more promising regions of the solutions space. GAs became popular from the works of Holland in 1973 [10] and 1975 [11], and, since then, they have been successfully applied to many complex optimization problems such as scheduling problems.

In 1989, a new class of hybrid algorithms, known as Memetic Algorithms (MAs), was introduced by Moscato [12]. These hybrid algorithms combine GAs with local search methods. Just like GAs, MAs are also based on the biological principles of selection, reproduction and mutation. The difference is the use of cultural evolution (meme) in MAs, a process in which cultural information is transmitted by communication between individuals and not by the recombination mechanism.

The local search procedure is applied in order to find local optimal solutions possibly not found when only evolution mechanism is applied. According to Gong et. al [13], MAs have been demonstrated to be more effective than traditional evolutionary algorithms for some problems, specially in the combinatorial optimization field.

In the remainder of this section we describe the proposed memetic algorithm, detailing the form of representation and evaluation of the solution, the crossover and mutation operators, as well as the hybridization and steps of the algorithm.

3.1   Solution Representation and Fitness Evaluation

As mentioned by Averbakh and Pereira in [7], problem NCPDD-L can be solved in polynomial time $O(n \log n)$ when network G is a tree. In order to take advantage of this information, we firstly considered to represent the solutions of the problem as trees, thus the best reconstruction planning of each tree would be decided in polynomial time. However, due to the difficulty and instability of manipulating trees in evolutionary algorithms (crossover and mutation generally produce completely different solutions from the original ones), we decided to represent a solution of the problem as a permutation $\pi = \langle \pi_1, \pi_2, ..., \pi_{n-1} \rangle$ of the $n - 1$ vertices to be recovered and a preliminary procedure is applied to build a tree from this permutation.

The procedure BUILD-TREE, shown in Fig. 1, describes formally how a tree is constructed from a permutation π. The edges belonging to the shortest path $\mathcal{P}_i$ between the vertex π_i and some vertex that has already been recovered, set $\mathcal{R}$, are inserted into the tree $\mathcal{T}$. Initially, the set $\mathcal{R}$ contains only the vertex 1 (depot), and for each new vertex π_i recovered, all edges belonging to the path $\mathcal{P}_i$ are inserted into the tree and consequently all vertices belonging to $\mathcal{P}_i$ are inserted to $\mathcal{R}$. Note that the path $\mathcal{P}_i$ may contain vertices that have not yet been analyzed, i.e., it is possible that $\pi_j \in \mathcal{P}_i \mid j > i$.

procedure BUILD-TREE (π, n)
    $\mathcal{T} \leftarrow \{\}$
    $\mathcal{R} \leftarrow \{1\}$ $\triangleright$ depot
    for $i \leftarrow 1$ **to** $n-1$ **do**
        if $\pi_i \notin \mathcal{R}$ **then**
            $\mathcal{P}_i \leftarrow$ shortest path between vertex π_i and some vertex in $\mathcal{R}$
            for each $\{u,v\} \in \mathcal{P}_i$ **do**
                $\mathcal{T} \leftarrow \mathcal{T} \cup \{u,v\}$
            for each $v \in \mathcal{P}_i$ **do**
                $\mathcal{R} \leftarrow \mathcal{R} \cup v$
    return $\mathcal{T}$

Fig. 1. BUILD-TREE.

3.2 Crossover Operators

The crossover operators are responsible for the creation of new solutions, which allow the memetic algorithm to better explore the search space and find better solutions. These new solutions, called offspring solutions, are created by exchanging information between two or more solutions, called parent solutions, of the current population.

In this work, we consider two simple crossover operators widely used. Figure 2 shows in details their operating mechanisms when applied to permutations representing solutions for a network of 9 vertices. The first crossover operator C_1 is the single-point crossover and it operates on two parent solutions p_1 and p_2. After randomly selecting a crossover-point on the permutation π, as exemplified in Fig. 2(a), the first segment of p_1 and the second segment of p_2 form the child solution f_1', and the first segment of p_2 and the second segment of p_1 form the child solution f_2'. The second crossover operator C_2 also operates on two parent solutions p_1 and p_2, but it uses two crossover-points, which are randomly selected on the permutation. The part delimited by these crossover-points is replicated to the respective children solutions f_1' and f_2', and the other parts are exchanged, as shown in the Fig. 2(b).

$p_1 = \langle\ 4,6,8,9,\ |\ 2,3,5,7\ \rangle$ $\rightarrow$ $f_1' = \langle\ 4,6,8,9,\ 9,6,3,2\ \rangle$ $\rightarrow$ $f_1 = \langle\ 4,6,8,9,5,7,3,2\ \rangle$
$p_2 = \langle\ 5,8,7,4,\ |\ 9,6,3,2\ \rangle$ $f_2' = \langle\ 5,8,7,4,\ 2,3,5,7\ \rangle$ $f_2 = \langle\ 5,8,7,4,2,3,9,6\ \rangle$

(a) Single point crossover operator (C_1)

$p_1 = \langle\ 4,6,8,\ |\ 9,2,3,\ |\ 5,7\ \rangle$ $\rightarrow$ $f_1' = \langle\ 5,8,7,\ 9,2,3,\ 3,2\ \rangle$ $\rightarrow$ $f_1 = \langle\ 5,8,7,9,2,3,4,6\ \rangle$
$p_2 = \langle\ 5,8,7,\ |\ 4,9,6,\ |\ 3,2\ \rangle$ $f_2' = \langle\ 4,6,8,\ 4,9,6,\ 5,7\ \rangle$ $f_2 = \langle\ 3,2,8,4,9,6,5,7\ \rangle$

(b) Two point crossover operator (C_2)

Fig. 2. Crossover operators.

Note that both operators may generate infeasible solutions, because the offspring solutions may contain repeated elements (highlighted in bold) and missed ones. In order to turn the offspring solutions feasible, a repair operation is applied in solutions f_1' and f_2'. This repair operator consists in replacing the duplicates with the missing ones generating the feasible solutions f_1 and f_2.

3.3   Mutation Operators

In order to introduce diversity to the population of solutions, we define three different mutation operators. The first one, which we call by M_1, consists of removing a randomly chosen vertex π_i from the permutation and inserting it in a randomly chosen position j. Referred to as M_2, the second mutation operator randomly selects two vertices π_i and π_j and exchanges their positions. The third one (M_3) operates on three different vertices π_i, π_j and π_k, performing a shuffling of their positions. The mechanisms of the three mutation operators are exemplified in Fig. 3. Note that none of the three mutation operators produces infeasible solutions.

$$s = \langle\, 4, 6, 8,9,2,3,5, \uparrow 7\, \rangle \qquad s' = \langle\, 4,8,9,2,3,5, 6, 7\, \rangle$$
(a) Insert mutation operator (M_1)

$$s = \langle\, 4,6, 8, 9,2, 3, 5,7\, \rangle \qquad s' = \langle\, 4,6, 3, 9,2, 8, 5,7\, \rangle$$
(b) Exchange mutation operator (M_2)

$$s = \langle\, 4, 6, 8,9, 2, 3,5, 7\, \rangle \qquad s' = \langle\, 4, 7, 8,9, 2, 3,5, 6\, \rangle$$
(c) Shuffle mutation operator (M_3)

Fig. 3. Mutations operators.

3.4   Framework

Algorithm 1 describes in details the steps of the Memetic Algorithm (MA) here proposed for the Network Construction Problem with Due Dates when the maximum lateness must be minimized.

Initially, Algorithm 1 generates a population of random solutions (random permutations). While time limit (10 s) is not reached, the algorithm performs cycles of evolution, creating through crossover and mutations new solutions from the current population. After each evolutionary cycle, a heuristic based on Reduced Variable Neighborhood Search (RVNS) is applied in 10% of the solutions of the current population. The RVNS is a variant of VNS, proposed by Mladenović and Hansen [14]. The procedure RVNS, shown in Fig. 4, describes the heuristic proposed to improve some solutions in order to guide the algorithm in the next evolutionary cycles to a part of the search space with possible higher quality solutions.

Algorithm 1. Memetic Algorithm (MA) (μ, P_c, P_m)

1: Time $\leftarrow 0$;
2: Pop $\leftarrow$ Randomly generates a population of μ solutions;
3: **while** Time < 10 seconds **do**
4:     Parents $\leftarrow$ Selects $\mu/2$ pairs of solutions from Pop using binary tournament;
5:     Children $\leftarrow \{ \}$
6:     **for each** $(p_1, p_2) \in$ Parents **do**
7:         **if** $rand\,(0,1) \leq P_c$ **then**
8:             Randomly selects a crossover operator O_c from $\{C_1, C_2\}$;
9:             Generates two new solutions f_1 and f_2 from p_1 and p_2 using O_c;
10:        **else**
11:            $f_1 \leftarrow p_1$; $f_2 \leftarrow p_2$;
12:        **if** $rand\,(0,1) \leq P_m$ **then**
13:            Randomly selects a mutation operator O_m from $\{M_1, M_2, M_3\}$;
14:            Apply the mutation operator O_m in the solution f_1;
15:        **if** $rand\,(0,1) \leq P_m$ **then**
16:            Randomly selects a mutation operator O_m from $\{M_1, M_2, M_3\}$;
17:            Apply the mutation operator O_m in the solution f_2;
18:        Children $\leftarrow$ Children $\cup \{f_1, f_2\}$;
19:    Pop $\leftarrow$ Children;
20:    H $\leftarrow$ Randomly selects $\mu/10$ solutions from Pop;
21:    Pop $\leftarrow$ Pop $\setminus$ H;
22:    **for each** $s \in$ H **do**
23:        $s' \leftarrow$ RVNS(s, n);   ▷ where n is the number of vertices
24:        Pop $\leftarrow$ Pop $\cup s'$;
25:    Update Time;
26: **return** best solution found throughout the process;

procedure RVNS (s, n)
    $iter \leftarrow 0$
    while $iter < n$ **do**
        $k \leftarrow 1$
        repeat
            $s' \leftarrow$ generates a solution from s using M_k mutation operator
            if $f(s') < f(s)$ **then**
                $s \leftarrow s', k \leftarrow 1$
            else $k \leftarrow k + 1$
        until $k > 3$
        $iter \leftarrow iter + 1$
    return s

Fig. 4. Reduced Variable Neighborhood Search (RVNS).

4   Computational Experiments

Our algorithm was coded in C/C++, compiled with GNU version 4.8.4, and executed sequentially, on an Intel Core i7-6700HQ CPU @ 2.60 × 8 computer with 8 GB of RAM running Ubuntu 16.04 LTS 64-bits.

In order to validate our algorithm, we use two sets of instances defined by Averbakh and Pereira [7]. The first one consists of 640 randomly generated instances and the second set consists of 80 instances whose data were obtained from a real earthquake that occurred in February 2010 in Chile. Further details on the generation of those instances can be obtained in the work of Averbakh and Pereira [7].

4.1   Parameter Tuning

We use the I/F-Race implementation provided in the irace package [15] in order to find the best parameter values for our algorithm, which has three parameters: population size (μ), crossover rate (P_c), and mutation rate (P_m). We tested the following values: $\mu \in \{20, 50, 100, 500\}$, $P_c \in \{0.70, 0.75, 0.80, 0.85, 0.90, 0.95\}$, $P_m \in \{0.01, 0.02, 0.05, 0.10, 0.15, 0.20\}$. After calibration the following parameter values were determined: $\mu = 500$, $P_c = 0.90$, and $P_m = 0.15$.

4.2   Experimental Results

The results given by the literature (B&B algorithm from Averbakh and Pereira [7]) and by the proposed algorithm are presented in Tables 1 and 2. Table 1 shows the results for the 640 instances randomly generated, while Table 2 shows the results for the 80 instances related to the earthquake that occurred in 2010 in Chile.

Averbakh and Pereira [7] showed that their B&B algorithm proved to be efficient for most of the instances. The authors limited the execution time of their algorithm to 600 s, and within this time limit the algorithm B&B was able to find the optimal solution for 506 of the 640 random instances, and the optimal solution for all 80 real instances.

The first two columns of Table 1 describe the random instances and the remaining columns show the results obtained from the literature and by our memetic algorithm. The columns n and RDD inform, respectively, the number of vertices and the range of due dates used in generating of the instances (see Averbakh and Pereira [7]). For each combination of n and RDD values there are 20 different instances. The B&B algorithm information was extracted from Averbakh and Pereira [7] and is described in the columns Obj, $\#Opt$, $Time$ (s) and $\#Nodes$ that inform, respectively, for each set of 20 instances, the average solution value, the number of optimal solutions found, the average execution time, and the average number (in millions) of nodes explored in the B&B search tree. The results of the MA are shown in the last five columns. The algorithm was run 5 times for each instance. The columns Obj and σ report respectively, for each set of 20 instances, the average of the best solutions and the standard deviation after 5 runs, while columns $\#Worse$, $\#Equal$ and $\#Better$

Table 1. Comparative analysis of B&B and MA algorithms on the random instances.

n	RDD	B&B				MA				
		Obj	#Opt	Time (s)	#Nodes	Obj	#Worse	#Equal	#Better	σ
25	0.2	1838.7	20	0.0	0.00	1838.7	0	20	0	0.00
	0.4	1549.2	20	0.0	0.01	1549.2	0	20	0	0.00
	0.6	1352.3	20	0.0	0.01	1352.3	0	20	0	0.00
	0.8	1582.2	20	0.0	0.02	1582.2	0	20	0	0.00
30	0.2	2010.9	20	0.1	0.01	2010.9	0	20	0	0.00
	0.4	1709.2	20	0.0	0.05	1709.2	0	20	0	0.00
	0.6	1627.0	20	0.3	0.11	1627.0	0	20	0	0.00
	0.8	1862.2	20	0.2	0.10	1862.2	0	20	0	0.00
35	0.2	2324.6	20	0.8	0.12	2324.6	0	20	0	0.00
	0.4	1893.0	20	2.6	0.41	1893.0	0	20	0	0.00
	0.6	1773.0	20	3.1	0.51	1773.0	0	20	0	0.00
	0.8	1801.7	20	2.0	0.34	1801.7	0	20	0	0.00
40	0.2	2384.9	20	1.6	0.32	2384.9	0	20	0	0.00
	0.4	2049.1	20	18.1	2.03	2049.1	0	20	0	0.00
	0.6	1902.0	20	47.4	5.54	1902.0	0	20	0	0.00
	0.8	2156.9	20	45.0	5.57	2156.9	0	20	0	0.00
45	0.2	2596.7	20	10.9	1.17	2596.7	0	20	0	0.00
	0.4	2103.3	18	152.9	16.23	2101.6	0	19	1	0.00
	0.6	1976.0	15	245.1	26.07	1975.0	0	19	1	0.00
	0.8	2126.9	20	170.8	18.41	2126.9	0	20	0	0.00
50	0.2	2804.5	20	7.8	0.81	2804.5	0	20	0	0.40
	0.4	2149.9	18	113.8	10.50	2150.7	1	19	0	0.00
	0.6	2097.5	6	483.2	17.05	2078.5	0	15	5	0.00
	0.8	2226.9	8	389.8	38.46	2190.4	0	14	6	0.30
55	0.2	2885.8	20	48.7	4.68	2885.8	0	20	0	1.80
	0.4	2461.7	9	382.3	34.27	2454.1	0	17	3	1.20
	0.6	2314.6	3	533.0	46.09	2305.2	0	13	7	1.10
	0.8	2234.2	4	515.3	45.28	2195.8	0	9	11	0.00
60	0.2	2992.2	19	110.7	9.38	2992.6	2	18	0	0.90
	0.4	2483.9	5	467.4	38.64	2474.9	3	13	4	1.00
	0.6	2468.8	1	577.3	46.74	2384.9	0	4	16	3.40
	0.8	2277.8	0	600.0	50.64	2188.8	0	6	14	0.20
Average		2125.6	$\frac{506}{640}$	154.1	13.11	2116.4	$\frac{6}{640}$	$\frac{566}{640}$	$\frac{68}{640}$	0.32

inform, respectively, the number of instances for which the solution found by MA algorithm was worse, equal and better than the ones found by the B&B algorithm.

The results of Table 1 show that both the B&B algorithm and the proposed MA algorithm found the optimal solution for all 320 instances up to 40 vertices ($n \leq 40$). Notice that, for a given n, the B&B algorithm behaves better for the smaller values of RDD, since the smaller this value the more homogeneous are the due dates of the vertices, facilitating the pruning of the B&B, which can be verified in the #Nodes column. For the instances with $n \geq 45$, the proposed MA obtained solutions equal or better than the B&B in less execution time for almost all the groups of instances. For the larger instances ($n \geq 50$), the MA was able to obtain several better solutions (column #Better), specially for higher values of RDD.

As seen from Table 1, the MA found worse solutions than the B&B algorithm for only 6 instances. In addition, our MA was able to find solutions that outperformed 68 solutions of the B&B algorithm, and for 566 instances both algorithms found solutions with same objective value. We may also conclude, analyzing the column σ, that MA has a good convergence, as the standard deviation is zero or close to zero for all instances groups. Therefore, only one or very few executions would be enough.

Regarding the average quality of solutions and the average computational times (last row of Table 1) of all 640 instances, the MA was more effective and more efficient than the B&B algorithm. It is easily verified that the processing time of the B&B algorithm increases with increasing number of vertices, exceeding 500 s for some instances. While our MA, even having executed for only 10 s, provided equal or higher quality solutions than B&B for most of the instances.

Table 2 provides the results obtained for the Chilean instances. All columns have the same meaning as the columns in the Table 1. We can see that both the B&B algorithm and the MA were able to find the optimal solution for all 80 instances. Moreover, MA has found the optimal solution in all runs, as the standard deviation (column σ) is 0 in all cases. However, even being a network of 53 vertices, the B&B algorithm was more efficient when compared to MA. We believe that the structure of the Chilean network has facilitated the pruning of the B&B algorithm, increasing its performance.

Table 2. Comparative analysis of algorithms B&B and MA on the Chilean instances.

n	RDD	B&B			MA				
		Obj	#Opt	Time (s)	Obj	#Worse	#Equal	#Better	σ
53	0.2	435392.5	20	0.0	435392.5	0	20	0	0.00
	0.4	3892019.3	20	1.8	3892019.3	0	20	0	0.00
	0.6	7348237.6	20	4.5	7348237.6	0	20	0	0.00
	0.8	10842480.6	20	2.9	10842480.6	0	20	0	0.00

5   Conclusions and Future Work

In this paper we proposed a Memetic Algorithm for the Network Construction Problem with Due Dates. The proposed approach was tested using instances available in the literature, and the results found show a significant superiority of the proposed Memetic Algorithm when compared to the B&B of the literature, mainly, for the large instances.

In the future, we would like to use a self-tuning technique to dynamically update the parameters of the Memetic Algorithm. We also intend to explore other problems of network construction already present in the literature through evolutionary algorithms, such as the Memetic Algorithm proposed here.

Acknowledgments. The authors thank Coordenação de Aperfeiçoamento de Pessoal de Nível Superior (CAPES) and Fundação de Amparo à Pesquisa do Estado de Minas Gerais (FAPEMIG) for the financial support of this project.

References

1. Abadie, A., Gardeazabal, J.: Terrorism and the world economy. Eur. Econ. Rev. **52**(1), 1–27 (2008)
2. Guha-Sapir, D., Hoyois, P.: Estimating populations affected by disasters: A review of methodological issues and research gaps. Centre for Research on the Epidemiology of Disasters (CRED), Institute of Health and Society (IRSS), University Catholique de Louvain, Brussels (2015)
3. Berktaş, N., Kara, B.Y., Karaan, O.E.: Solution methodologies for debris removal in disaster response. EURO J. Comput. Optim. **4**(3–4), 403–445 (2016)
4. Çelik, M.: Network restoration and recovery in humanitarian operations: framework, literature review, and research directions. Surv. Oper. Res. Manage. Sci. **21**(2), 47–61 (2016)
5. Feng, C.M., Wang, T.C.: Highway emergency rehabilitation scheduling in post-earthquake 72 hours. J. 5th East. Asia Soc. Transp. Stud. **5**, 3276–3285 (2003)
6. Hu, Z.H., Sheu, J.B.: Post-disaster debris reverse logistics management under psychological cost minimization. Transp. Res. Part B: Methodol. **55**, 118–141 (2013)
7. Averbakh, I., Pereira, J.: Network construction problems with due dates. Eur. J. Oper. Res. **244**(3), 715–729 (2015)
8. Chagas, J.B.C., Santos, A.G.: Abordagem Heurística para o Problema de Reconstrução de Redes de Transporte com Prazos de Recuperação. In: XLVIII Simpósio Brasileiro de Pesquisa Operacional (SBPO), pp. 1696–1707, Vitória, Espírito Santo, Brazil (2016)
9. Darwin, C., Bynum, W.F.: The Origin of Species by Means of Natural Selection: Or, the Preservation of Favored Races in the Struggle for Life. AL Burt, New York (2009)
10. Holland, J.H.: Genetic algorithms and the optimal allocation of trials. SIAM J. Comput. **2**(2), 88–105 (1973)
11. Holland, J.H.: Adaptation in Natural and Artificial Systems: An Introductory Analysis with Applications to Biology, Control, and Artificial Intelligence. MIT press, Cambridge, MA (1992)
12. Moscato, P.: On evolution, search, optimization, genetic algorithms and martial arts: Towards memetic algorithms. Technical Report, Caltech Concurrent Computation Program, C3P Report (1989)
13. Gong, M., Cai, Q., Li, Y., Ma, J.: An improved memetic algorithm for community detection in complex networks. In: IEEE Congress on Evolutionary Computation (CEC), pp. 1–8. IEEE Press, Brisbane, Australia (2012)

14. Mladenović, N., Hansen, P.: Variable neighborhood search. Comput. Oper. Res. **24**(11), 1097–1100 (1997)
15. López-Ibáñez, M., Dubois-Lacoste, J., Cáceres, L.P., Birattari, M., Stützle, T.: The irace package: Iterated racing for automatic algorithm configuration. Oper. Res. Perspect. **3**, 43–58 (2016)

Incremental Real Time Support Vector Machines

Fahmi Ben Rejab$^{(\boxtimes)}$ and Kaouther Nouira

BESTMOD, Institut Supérieur de Gestion de Tunis, Université de Tunis,
41 Avenue de la Liberté, 2000 Le Bardo, Tunisia
fahmi.benrejab@gmail.com, kaouther.nouira@planet.tn

Abstract. This paper investigates the problem of handling large data stream and adding new attributes over time. We propose a new app-roach that employs the dynamic learning when classifying dynamic datasets. Our proposal consists of the incremental real time support vec-tor machines (I-RTSVM) which is an improved version of the support vector machines (SVM) and LASVM. On one hand, the I-RTSVM han-dles large databases and uses the model produced by the LASVM to train data. It updates this model to be appropriate to new observations in test phase without re-training. On the other hand, the I-RTSVM presents a dynamic approach that adds attributes over time. It uses the final model of classification and updates it with new attributes without re-training from the beginning. Experiments are illustrated using real-world UCI databases and by applying different evaluation criteria. Results of com-parison between the I-RTSVM and other approaches mainly the SVM and LASVM shows the efficiency of our proposal.

Keywords: Incremental learning · SVM · Data stream

1 Introduction

Support Vector Machines (SVM) [16] is one from the most popular classifica-tion techniques. Applied to different fields, the SVM has provided successful results [3,4,10] compared to other techniques [2]. The SVM is based on the use of support vectors. However, the number of training vectors used by the SVM proportionally increases with the increase of the memory capacity. This is why, the main problem of the SVM is suffering from the high computing require-ments. Numerous works have been proposed to overcome this issue. We can mention the LASVM [5,6] and ISVM [8] which are two incremental and mod-ified versions of SVM. They are improved versions of the SVM and they deal with large datasets in training phase. Besides, in [10,13], authors have used the parallelization machines to decrease the execution time needed by the standard SVM.

Despite these proposed approaches, the problem of data scalability still exist in classification. In fact, data are considerably growing such as data generated

© Springer International Publishing AG, part of Springer Nature 2018
A. Abraham et al. (Eds.): ISDA 2017, AISC 736, pp. 221–230, 2018.
https://doi.org/10.1007/978-3-319-76348-4_22

from the web, monitor devices, and the flow of information, etc. However, handling large data and conserving all important information is essential and can help when making decisions. Many techniques of classification cannot deal with large database. There is a loss of information when just a part of datasets are analyzed and interpreted. However, ignoring some information in data mining or medicine fields for example can lead to inaccurate results and wrong decisions. These latter can, in some cases, produce many issues.

In addition, most of classification techniques are using the same model produced in training phase to classify new instances of test phase. However, it is necessarily to update the model and consider all new information which can help to make better and more accurate decisions. The standard SVM needs a high memory requirement and faces some problems to update its model in test phase. Avoiding the re-training of initial observations is needed not only when updating the model in test phase but also, when adding new attributes over time. The incremental learning decreases the memory requirement and the execution time.

In this paper, we propose a new method that takes profits of both SVM and LASVM techniques and avoids their limitations. Our proposal called I-RTSVM which means the incremental real time SVM is a modified and improved version of SVM that uses the model generated by the LASVM in training phase. This model is used by the I-RTSVM as an input to train initial data. Then, this model takes into consideration of new observations in test phase. Another advantage of the I-RTSVM is the possibility to add new attributes over time. The use of the incremental learning by the I-RTSVM saves time and memory and avoids the re-training by using only the last model produced by the classifier.

The rest of the paper is organized as follows: Sect. 2 reviews support vector machines technique followed by the LASVM technique. Section 3 describes our contribution through the incremental real time SVM i.e., I-RTSVM. Section 4 illustrates and analyzes the experimental results. Section 5 concludes the paper.

2   The Support Vector Machines and the LASVM

In this section, we present an overview of the support vector machines (SVM) and its modified version LASVM by detailing their main concepts.

2.1   Support Vector Machines

Support vector machines (SVM), proposed by Vapnik [16], is a supervised technique used for classification and regression problems. SVM is considered as one from the most desirable approaches to bi-classes datasets. It was widely used in various researches including [2,4,10] where has produced important results in batch mode. Generally, the SVM performs linear as well as non-linear separable data. As follows, we present a brief overview explaining how the SVM handles these two types of data.

1. ***Linearly separable data:*** Data can be separable by using an optimal hyperplane. This algorithm make a separation of data into two classes. It is possible

to define two more hyperplanes, H_1 and H_2 parallel to the separating hyperplane. The distance between them represents the SVM's margin. Thus, the main aim is to find the optimal hyperplane which maximizes the margin while is equidistant from both H_1 and H_2.

The setting of w (a vector normal) and b (a scalar) parameters requires the resolve of a convex quadratic programming (QP) problem.

$$\min \frac{1}{2}||w||^2 + C, \ subject \ to \ y_i(x_i w + b) \geq 1 \ for \ i = 1 \ldots m. \qquad (1)$$

with the decision rule given by: $f_{w,b}(x) = sign(w^T x_i + b)$.

Where w presents the weight vector, b is the bias (or $-b$ is the threshold), x_i is an observation and P and N are respectively positive and negative data (the class of x_i), y_i is the class of the observation x_i, m is the number of observations and R^d is the number of dimension.

2. **Non-linearly separable data:** SVM is also able to perform non-linear data classification. It separates classes that cannot be separated with a linear classifier. In this case, its inputs (the original observations) are mapped into a feature space (high dimensional may be infinite) using non-linear functions called feature functions ϕ. In the new space, a linear classifier separates the two classes. The data mapping is defined using the function Φ defined by $R^d \rightarrow R^D (D >> d)$, with R^D is Hilbert space. The detection of the hyperplane requires the resolve of the optimization problem defined using the slack variable ξ_i.

$$\min \frac{1}{2}||w||^2 + C \sum_{i=1}^{m} \xi_i \ subject \ to \ y_i(x_i w + b) \geq 1 - \xi_i, \xi_i \geq 1 \ for \ i = 1 \ldots m. \ (2)$$

The SVM technique provides advantages by handling both linear and non-linear data. However, it cannot perform the classification of incremental data when new information are provided over time [15]. To overcome this limitation, two improved and incremental versions of SVM have been proposed namely the LASVM [9] and the ISVM [8]. The LASVM is reviewed in the following section.

2.2   Online and Active SVM

The online and active SVM denoted by LASVM and proposed in [5,6] is a popular incremental algorithm for classification. It handles large training sets and uses incremental learning by adding data over time [7].

The LASVM is an incremental approach. It is considered as an improved version of SVM. LASVM was applied in various works such as [11,17]. The main advantages of the LASVM algorithm is being faster than the SVM and needs less memory capacity. Besides, it can handle noisy datasets and avoid the over-fitting problem by discarding the useless support vectors [12]. Actually, the LASVM algorithm is based on two main procedures mainly the process and reprocess procedures.

1. **The process procedure:** by selecting data as new training instances. After that, it is necessarily to find the support vectors that correspond to these data. These vectors will form with the new instances.
2. **The reprocess procedure:** is based on removing support vectors with $\alpha = 0$. Then, the update of the α weight is made. It is necessarily to verify if there is any support vector with $\alpha = 0$ in order to remove it. Finally, the update of the gradient g and the bias b are made.

3   Incremental and Real Time SVM: The I-RTSVM

3.1   I-RTSVM Description

Dealing with incremental learning is very important, since in real-world situations data are rapidly growing [1]. There is often a need to update the obtained model (classifier) or even completely scrap it and retrain to get accurate result.

Our proposal (I-RTSVM) emphasizes on two main aims. The first one is relative to the incremental aspect in test phase. It consists of having new data validated by an expert in test phase and the classification model should be updated. The second aim of the I-RTSVM is about the incremental strategy applied when adding new attributes. To this end, the I-RTSVM uses as input the model generated by the LASVM technique in the training phase. After that, it classifies new observations in test phase. The I-RTSVM updates the model through the validation of the expert. Besides, adding new attributes is possible without training data from the beginning. The structure of the I-RTSVM is described in Fig. 1.

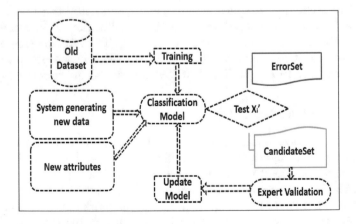

Fig. 1. The structure of the I-RTSVM

In real-world situations, there are several applications where the attribute set evolves over time. We will incorporate a new attribute into the existing model without retraining from the beginning. However, within standard classification approaches, all observations have to be present with a known attribute set, before the classification process. For a dynamic dataset, we need to process the whole dataset, the new and the old attributes. So, we need a new method that can consider the incremental learning and data stream. To this end, our novel approach studies how to update the initial model by the new information. By then, there are two steps: building the initial model from a given dataset. Updating the current model when we have any extensions each time interval. Besides, by the new method, we can simulate the expert task.

3.2   I-RTSVM Algorithm

Let $T = (x_i, y_i), i = \{1 \dots l\}$ where $x_i \in R^n$ and $y \in \{1, -1\}$ be the training set to be used to build a classification model. Each observation is initially characterized by A attributes. We mentioned above, the initialization (the first step) consists in applying LASVM algorithm to our dataset T to build an initial model.

Our method makes also an extension of attribute set, with one new attribute, any object X_i becomes X_i'. When extending dataset, with new observations, we check if these examples can be a support vector or an error vector. After that, we try to incorporate this new attribute and add new observations in the current dataset, initially obtained by applying LASVM, and updating the models' parameters (which are b and w) using our new approach I-RTSVM. Starting with the current model, we will study in which conditions an extended observation X_i' can be a support vector (added to a candidateSet) or an error vector (ErrorSet). The incremental real time SVM (I-RTSVM) algorithm is presented as flow.

INPUTS

1. TrainingSet (X_i, Y_i): X_i.
2. The weight W and the bias b, obtained in training phase.
3. Used parameters: the cost C, KernelType, and KernelParameters.
4. Set of new annotated observations denote by S.

OUTPUTS

1. A new TrainingSet: $(X_i, Y_i, i = 1 \dots l)$.
2. Updated values of the weight W and the bias b.

I-RTSVM Algorithm
Begin
    $M(b, w) \leftarrow TrainingLASVM(T)$ *(the initial model)*
    IF there is a new attribute $A_{+}1$
        $X_i' \leftarrow X_i(A_{+}1)$ // X_i' has $A + 1$ attributes.
    ELSE
        $X_i' \leftarrow X_i$

END IF
FOR each new observation X_i'
    IF (X_i', Y_i') is a candidate
        AddCandidate (X_i', Y_i') at the CandidateSet.
    ELSE   AddErrorSet(X_i')
    END IF
FOR each time interval
        Print CandidateSet
        Update CandidateSet from Expert Decision
        PROCESS()
        REPROCESS()
        $M'(b, w) \leftarrow UpdateModel(M, parameters)$
End For
End For
End

Based on the pseudo code of I-RTSVM algorithm, it is obvious that it takes as input the initial training set T. After that, it builds the original model of classification using LASVM. The I-RTSVM starts classifying the new observations in test phase.

For each newly added observation, with a possibility to add a new attribute, the I-RTSVM makes the classification task using the same model of LASVM. Then, the I-RTSVM assigns to the vector of the new observation a label equals to the value $(+1)$ if it belongs to the first class or (-1) otherwise. In back office, I-RTSVM checks the nature of the new vector, if it is a candidate (i.e. can be a support vector), the I-RTSVM adds it to the CandiateSet.

During the classification of new data, the expert have to validate the CandidateSet for each predetermined time interval. The validation consists of proving or modifying the label assigned by the I-RTSVM.

After the validation of the CandidateSet by confirming or modifying the assigned class, I-RTSVM calls *UpdateModel* $(M, parameters)$ to update the original model parameters. This step has many advantages mainly reducing the execution time of training by re-training the initial model with only the CandidateSet (and not using the whole dataset).

I-RTSVM can learn new parameters from new attributes and keeping historic parameters that are already learned from previous data available. Suppose the former training data contains M attributes, denoted as $M \in R^m$, newly joined training data contains $(M + N)$ attributes with N new attributes, and suppose the structural parameters of classifier trained from newly joined training data with $(M + N)$ attributes are w and b. In order to make use of the historical parameters w_M and b_M, w can be denoted as $w = [w_M, w_N]$ with w_M representing the parameter which are trained from previous M attributes and w_N representing the parameter which are trained from the rest of new N attributes. After solving the set of linear equations of the standard SVM the corresponding label y is given by: $y(x) = sign(w_M^T \phi_1(x_M) + w_N^T \phi_1(x_N) + b)$.

By applying two different implicit functions to the already existing M and N attributes, we reformulate the decision rule.

4   Experiments

4.1   The Framework

We have tested and applied our new algorithm to two-classes real-world datasets taken from UCI machine learning repository [14]. A brief description of these databases is given in Table 1.

Table 1. Description of the used UCI databases

Databases	#Instances	Attributes
Breast Cancer Wisconsin (BCW)	683	10
Bank (B)	4521	12
Banknote Authentication (DBA)	1372	5
Connectionist Bench (CB)	208	60

4.2   Evaluation Criteria

Three evaluation criteria are used to test and evaluate the I-RTSVM compared to other approaches. These criteria are described as follows:

1. The Percent of Correct Classification (PCC): shows the effectiveness i.e. the quality of the proposed incremental approach:
   $PCC = nbr\ of\ well\ classified\ instances/total\ nbr\ of\ classified\ instances * 100.$
2. The execution time: presented through the time needed to build and get the final model. This criterion shows the efficiency of the algorithm i.e. speed of the convergence.
3. Size: defined as the number of support vectors needed to build the model. Less the number of support vectors is, better are the results.

4.3   Results and Discussion

In this section, we report and detail the results of our new proposal, using the evaluation criteria, compared to the standard SVM and the online training method the LASVM. The comparisons between the three methods (i.e. I-RTSVM, SVM, and LASVM) using as evaluation criteria the PCC, the execution time and the size are shown in Table 2.

Looking at Table 2 and for BCW, B, and CB datasets, we can remark that the incremental real-time SVM results are very close from the SVM results in classification precision. For the other datasets, our algorithm i.e., the I-RTSVM provides a higher result in terms of PCC than the SVM. We can also remark that our I-RTSVM provides similar results that the LASVM method, since it uses the same model in training phase.

We can see that the PCC for both incremental as well as non-incremental approaches are competitive. Thus, these interesting results confirm that our method is well appropriate within such dynamic environment. We can also remark that the proposed I-RTSVM and LASVM methods are considerably faster than the standard algorithm of SVM and their training time is linear with the dataset size. For example, for Bank dataset (B), the execution time of the I-RTSVM and LASVM (2804 s) is lower than the execution time of SVM (4010 s).

Table 2. Training phase: the comparison of I-RTSVM, SVM, and LASVM based on the PCC, the execution time, and the size

Databases	PCC			Execution time			Size		
	SVM	LASVM	I-RTSVM	SVM	LASVM	I-RTSVM	SVM	LASVM	I-RTSVM
BCW	89,12	88,56	88,56	94	76	76	150	118	118
B	89,09	89,08	89,08	4010	2804	2804	1284	1009	1009
DBA	88,75	88,32	88,32	202	176	176	181	143	143
CB	72,41	75,82	75,82	31	27	27	67	53	53

From Table 2, the SVM needs much more time to train the data compared to our proposal and LASVM. Hence, the SVM also needs higher number of support vectors than others methods i.e., LASVM and I-RTSVM to get its model.

We can conclude that the I-RTSVM provides final results faster than the SVM. It can be explained by the need of the SVM to load the whole training set in the memory before starting the training task. However, the LASVM and I-RTSVM can update the model parameters by the new information without re-training from the beginning.

The following Table 3, shows the obtained results relative to the test phase when we are adding new features over time.

Table 3. Test phase: the comparison of I-RTSVM, SVM, and LASVM based on the PCC, the execution time, and the size

Databases	PCC			Execution time			Size		
	SVM	LASVM	I-RTSVM	SVM	LASVM	I-RTSVM	SVM	LASVM	I-RTSVM
BCW	74,14	74,12	78,71	143	114	48	221	147	95
B	97,71	97,51	97,76	4553	3201	1745	1784	1352	112
DBA	89,56	89,51	92,38	274	188	53	265	174	88
CB	84,12	86,49	88,72	49	36	9	94	81	85

From Table 3, we can notice that the training time of the standard SVM for 4521 training samples takes important time to get the final results in test phase. Since it retains more unnecessary support vectors, the online SVM (LASVM) runs slower than the proposed method for large data sets. We remark that the new method I-RTSVM shows its performance in test phase which confirms the necessity to update the initial model by new observations.

Based on the second evaluation criterion, we can notice that our method I-RTSVM has less support vectors than the other two methods. Both online and standard SVM training algorithms increase the number of support vectors as the size of the data set increases. However, our method keeps almost a constant number of support vectors. This can be interpreted as the I-RTSVM updates its classifier based on the candidateSet collected in test phase.

More generally, experiments prove that our new algorithm can effectively minimize the training sample set while providing a high classification precision and minimizing the execution time. The improvement made through our algorithm and shown using these criteria are the result of the ability of the I-RTSVM to obtain the classifier model without re-training the whole data from the beginning. In fact, the I-RTSVM only uses the candidateset (observation can probably be a support vector) in contrast to the SVM and LASVM which re-train the whole data to get a new model.

5   Conclusion

In this paper, we have highlighted and solved a problem of classification by proposing a new method consisting of the incremental real time support vector machines (I-RTSVM). This issue is adding new attributes when classifying observations in test phase. The I-RTSVM has avoided this problems by using a model generated by the LASVM technique and then, making several improvements and updates on it in order to fit new data. Besides, it uses the incremental aspect by adding new attributes over time without re-classifying initial instances. Our proposal is tested using data bases from UCI machine learning repository. Results of I-RTSVM prove the improvement made by this new method compared to the LASVM and SVM. When dealing with classification task with a large amount of new additional features, our experiments validate that the proposed algorithm have higher classification precision while saving training time.

References

1. Anqi, B., Shitong, W.: Incremental enhanced α-expansion move for large data: a probability regularization perspective. Int. J. Mach. Learn. Cybernet. **8**, 1615–1631 (2017)
2. Ben Rejab, F., Nouira, K., Trabelsi, A.: Support vector machines versus multi-layer perceptrons for reducing false alarms in intensive care units. Int. J. Comput. Appl. Found. Comput. Sci. **49**, 41–47 (2012)

3. Ben Rejab, F., Nouira, K., Trabelsi, A.: On the use of the incremental support vector machines for monitoring systems in intensive care unit. In: TAEECE 2013, pp. 266–270 (2013)
4. Ben Rejab, F., Nouira, K., Trabelsi, A.: Health monitoring systems using machine learning techniques. In: Intelligent Systems for Science and Information, pp. 423–440 (2014)
5. Bordes, A., Bottou, L.: The Huller: a simple and efficient online SVM. In: Machine Learning: ECML 2005. Lecture Notes in Artificial Intelligence, LNAI, vol. 3720, pp. 505–512. Springer (2005)
6. Bordes, A., Ertekin, S., Weston, J., Bottou, J.: Fast kernel classifiers with online and active learning. J. Mach. Learn. Res. 6, 1579–1619 (2005)
7. Bottou, L., Curtis, Frank E., Nocedal, J.: Optimization methods for large-scale machine learning. International Archives of the Photogrammetry, ArXiv e-prints (2016)
8. Cauwenberghs, G., Poggio, T.: Incremental and decremental support vector machine learning. In: Advances in Neural Information Processing Systems (NIPS*2000), vol. 13, pp. 409–415 (2000)
9. Cortes, C., Vapnik, V.: Support vector networks. Mach. Learn. 20, 273–297 (1995)
10. Chang, E.Y.: PSVM: parallelizing support vector machines on distributed computers, pp. 213–230 (2011)
11. Frasconi, P., Passerini, A.: Predicting the geometry of metal binding sites from protein sequence, vol. 9, pp. 203–213 (2012)
12. Ghaemi, Z., Farnaghi, M., Alimohammadi, A.: Hadoop-based distribution system for online prediction of air pollution based on support vector machine. In: International Archives of the Photogrammetry, Remote Sensing and Spatial Information Sciences, vol. XL-1/W5, pp. 215–219 (2015)
13. Graf, H.P., Cosatto, E., Bottou, L., Durdanovic, I., Vapnik, V.: Parallel support vector machines: the cascade SVM. In: NIPS, pp. 521–528 (2005)
14. Lichman, M.: UCI Machine Learning Repository. University of California, School of Information and Computer Science, Irvine, CA (2013). http://archive.ics.uci.edu/ml
15. Liu, X., Zhang, G., Zhan, Y., Zhu, E.: An incremental feature learning algorithm based on least square support vector machine. Front. Algorithmics 5059, 330–338 (2008)
16. Vapnik, V.: Statistical Learning Theory, pp. 1–736. Wiley, New York (1998)
17. Wang, Z., Vucetic, S.: Online training on a budget of support vector machines using twin prototypes. Stat. Anal. Data Min. 3, 149–169 (2010)

Content-Based Classification Approach for Video-Spam Identification

Palak Agarwal$^{(\boxtimes)}$, Mahak Sharma, and Gagandeep Kaur

Department of CSE & IT, JIIT, Noida, Uttar Pradesh, India
Palakagarwal124@gmail.com, mahaksharma3016@gmail.com,
gagandeep.kaur@jiit.ac.in

Abstract. In this paper the authors have worked on YouTube comment spamming. The work has been carried out on a large and labeled dataset of text-comments. Filtration and pre-processing was done to speed up the detection, elimination of redundancies as well as to increase the accuracy. Spam flags on each set of text-comments were used to check the accuracy in implementation of classification techniques. An improved algorithm has also been proposed based on term frequencies. The results were compared based on accuracy-score and F-score considering the spam flag corresponding to each comment. Further, the accuracy of SVM model was compared with respect to size of dataset, pre-processing of data as well as with XGBoost.

Keywords: Spam/Ham · XGBoost · TF-IDF · RCA · SVM · LDA
Video · Security

1 Introduction

As reported by Wall Street Journal the number of videos in different categories being viewed by YouTube users is growing exponentially. There is ten times increase in it's users base since year 2012 and at present more than 1 billion hours per day [1]. However a sudden increase in the amount of fake comments on videos being uploaded on YouTube has resulted in big companies like Google and Facebook to take extreme steps by withdrawing their advertisements from YouTube. Since YouTube's revenues are primarily from these advertisements therefore this decision has caused high revenue losses to it [2]. It is under these circumstances that both the research based and industry based communities are working to find better solutions for spam filtering in large, and video-spam comment detection along with spam user detection in specific.

By definition, Spam means sending of unwanted digital messages to legitimate users of the Internet services, like gmail, YouTube, Facebook, Twitter, Google etc. Spam detection, identification and prevention is therefore not a simple task and involves not only search for more accurate and fast identification techniques but also upgradation of old existing ones to handle new breed of spams. Spam methods have been classified into different classes based on: (1) how spam content is delivered to the user, like via text messages, audio messages and video messages (YouTube videos); (2) characteristics of the spammer and (3) motive of the spammer behind sending or posting spam messages. Many academicians and researchers have contributed to the

© Springer International Publishing AG, part of Springer Nature 2018
A. Abraham et al. (Eds.): ISDA 2017, AISC 736, pp. 231–242, 2018.
https://doi.org/10.1007/978-3-319-76348-4_23

cause of spam filtering and identification. The classification techniques in use for spam have therefore been categorized as link-based spam classification techniques, hidden-spam classification techniques and content-based spam classification techniques [9]. Content-based spam classification are based on methods that build models by applying feature classifications like word counts, repetition or other language based characters. The technical contribution of this paper is to propose the more improved and efficient set of algorithms relevant with the real scenario as well as to depict the best algorithm with respect to various constraints.

2 Related Work

Wattenhofer et al. in [3] have employed unreduced data of YouTube for measuring subscription graph, comment graph and video content corpus. The authors have based their research on three main ideas: the explicit social graph illustrating subscriptions, implicit social graph rendering commenting activities and aggregated metrics of user uploaded content. Their paper illustrated that YouTube deviates from traditionally Online Social Network (OSN) characteristics. Nevertheless, the datasets collected were only fractions of entire corpus and algorithms used were highly imbalanced in nature. Chaudhary and Sureka in [4] have investigated the user's comment activity based on implementation of usage-based features. They have shown that methods of time, difference, similar comment are effective for spammer detection. They involved Microsoft SQL Server Data Mining Tools (SSDT) for a heuristic classification of videos as Spam or Legitimate. However, the SSDT Tool involves creation of OLAP mining structure/model which is not appropriate when working for relational dataset. Jin et al. in [5] have given a scalable online social media detection system using the implementation of General Activity Detection (GAD) clustering, in integration with active learning algorithm to cover real-time detection challenges. Although, the authors' claims to have laid down the improved efficiencies and scalabilities compared to previous approaches, they have failed to provide substantial results to support their work. O'Calloghan et al. in [6] scrutinized the networks retrieved from the user's comments on videos to evaluate recurring campaigns. The authors worked on primarily two strategies i.e. small number of spam user accounts to comment on large number of videos and using large number of accounts to comment on one or two videos. They showed the effectiveness of dynamic network analysis. Though, a small period (only 72 h) was considered for complete analysis of such a fast networking social media - YouTube. Abdulhamid et al. in [7] on the other hand have reviewed currently available methods, challenges and future research opportunities on the mobile SMS Spam detection, and also discussed about various filtering and mitigation methods. Spirin and Han in [8] have categorized algorithms into three categories namely content-based methods, link based methods and methods based on non-traditional data. They did various studies considering the broad spectrum of technologies but didn't provide any relevant and explanatory accuracy results. Ghiam and Pour have surveyed and classified spam detection techniques in [9] mainly into three broad categories including Link Based, Hiding and Content-Based techniques and discussed about the related detection methods. However, there can be more Spam detection methods introduced till now. They have failed to give preventive frameworks for email, image or any kind of web spam in the findings of paper. Radulescu et al. in [10] detected Spam comments using various indicators such as discontinuous text flow, inadequate and vulgar language or identification of the unrelated topics to a specific context/domain and further used machine learning algorithms for topic detection. However, using only 1024 comments related to a particular domain is not sufficient for spam analysis and detection work.

After studying these works we realized that there can be many additional features (described in next section) of spam comments, not considered here can also impact spam detection. Our proposed detection methodology has been discussed next in Sect. 3.

3   Proposed Methodology

In our research we have shown that comments of the video can be classified as spam or ham based on the unique and distinctive features exercised independently by spammers and legitimate users. In this section system architecture has been discussed first, followed by system component details and classifiers used in our methodology.

3.1   System Architecture

As shown in Figs. 1 and 2, the proposed architecture consisted of different phases based on clustering, self-designed relatable comment algorithm for supervised learning, term-frequency and inverse-document-frequency algorithms (TF-IDF) for significant term weighting and machine learning algorithms for accuracy determination.

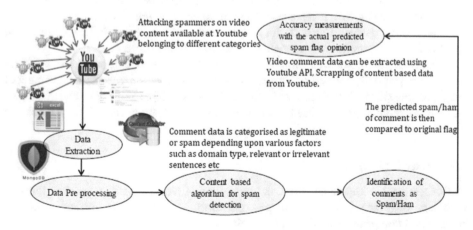

Fig. 1. System architecture

Firstly, large raw dataset from YouTube was crawled and was filtered and preprocessed before applying the algorithms. The filtration process was used to do dimensionality reduction by scrapping out extra features of the comments, namely number of

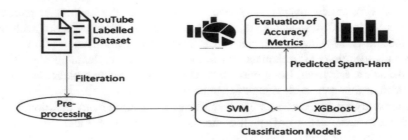

Fig. 2. System architecture for machine learning algorithms

likes, posting time, replies, uploaded by, number of subscribers etc. Preprocessing also involved removal of unwanted and disrupting characters like stopwords, punctuation marks, Latin characters, null values, white spaces and any other remaining inappropriate content in the comment. As a next step, the raw dataset was converted to target data used as input for content-based algorithm to determine the term's clusters. Term clusters were formed out of every text comment for all the unique terms present in the target dataset. In cognitive filtering, set of terms or descriptors are used to represent items in the dataset. The proposed algorithm established comment-term relationships with respect to unique existence of terms corresponding to each comment.

The proposed model accepts labeled data with each classified comment which in the real-life scenario is done by YouTube users by marking the spam flag relative to each comment as spam/ham in user's consideration. In the next step, the outlier comments were predicted as spam and remaining ones were sent as input to Relatable Comment Algorithm (RCA). Similarly, preprocessed target dataset was sent as input to TF-IDF algorithm. It converted the target dataset into a sparse matrix to determine term weight and calculate the inverse-document frequency of terms with respect to each comment in the dataset. For both the proposed algorithms, a threshold value was set independently to predict the comment as a spam comment or a ham.

Thereafter, the combined output of both algorithms was compared to classifiers which were primarily marked by the users as spam and ham to ascertain the accuracy of predicted classifiers in the proposed model. As shown in Fig. 2 SVM and XGBoost were used having preprocessed data input to classification model to predict the classifiers.

3.2   System Components

Various system components used in the system involved data extraction through crawling, raw dataset filtering & preprocessing, algorithms approaches, accuracy metrics and comparative analysis of models.

Data Collection, Filtering and Preprocessing

Since no relevant dataset consisting of spam flag was available with any of the data dispensing websites, it was extracted using the crawlers for YouTube. Our unprocessed raw dataset consisted of 5575 comments for single YouTube video titled Mobile Messaging. The gathered data was filtered for retrieving the required information out of a corpus. The significant columns involved were: (1) identical video-id for single video extraction; (2) user-id, to determine each comment uniquely; (3) text, to consider users reaction to video; (4) spam flag, acted as a predefined test condition for spam/ham tag given by user's. Table 1 illustrates the instance of the filtered dataset.

Table 1. Demonstrating the filtered unprocessed dataset

User-Id	Comment	Flag
H2MI6YaBW_w	You have won ?1,000 cash or a ?2,000 prize! To claim, call 0950000000327	Spam
O42s5Ff9cCs	Finally the match heading towards draw as your prediction	Ham

Preprocessing of the data was done using following steps:

Removal of Latin characters: The filtered dataset involved non-ascii characters which are undesired because they reduce the efficiency of models and therefore removed. Figure 3 shows output snippet after removal of Latin characters.

Input		Output	
WINNER!! As a valued network customer you have been selected to receivea å£900 prize reward! To claim call 09061701461. Claim code KL341. Valid 12 hours only.	Spam	WINNER !! As valued network customer you have been selected to selected receivea 900 prize reward!To claim call 09061701461. Claim code KL341. Valid 12 hours only.	spam
Fine if thatåÕs the way u feel. ThatåÕs the way its gota b	ham	Fine if that's the way u feel. Thats the way its gota b	ham

Fig. 3. Output after Latin characters removal

Removal of punctuations: The punctuations marks in English involve a set consisting of 14 elements, like ! , : ; . ? " " '() - _ {}, etc. Our corpus involved all of them which was a hurdle and hence removed. Figure 4 shows a snippet of output after removal.

> WINNER   As a valued network customer you have been selected to receivea   900 prize reward   To claim call 09061701461. Claim code KL341. Valid 12 hours only.

Fig. 4. Output after punctuations removal

Conversion to Lowercase Letters: The idea to convert the comprehensive text into lowercase was to free from the entanglement of case-sensitive nature in various programming platforms. This led the whole text to a common background which eased the comparison and tokenization. Figure 5 shows the snippet after conversion of text to lowercase.

Removing Stopwords: Stopwords like *the, is, an, at, which, on* etc., are a group of words that occur frequently in the text but are not relevant for ranking. Because of their recurring nature these can acutely affect the outcomes. For an instance, if the word "THE" is recurring in a comment then due to its increased the weight and more accurately increased repetition can impact the outcome of classifier identification. Stopwords were removed from the text. Figure 6 shows the snippet of output received after removing stopwords.

> winner   as a valued network customer you have been selected to receive   900 prize reward   to claim call 09061701461. Claim code kl341 valid 12 hours only.

Fig. 5. Output after lowercase conversion

> winner   as valued network customer selected re-ceivea   900 prize reward   claim call 09061701461 claim code kl341 valid 12 hours

Fig. 6. Snippet of output after removing stopwords

Stemming: The requirement that the words like call, calling, called and so on should be treated as same, driven by the fact that they all possess the same weight in reference to the term frequency and term clusters. It constitutes of hashing of last characters considering the similar initial characters. Considering all such words to be similar, improved the accuracy of the sparse matrix generated and decreased the number of clusters. Figure 7 shows the snippet of output received after stemming.

> winner   as value network customer select receive 900 prize reward   claim call 09061701461 claim code kl341 valid 12 hour

Fig. 7. Snippet of output after stemming

Removal of Null Characters and Comments: The disadvantage to the procedure of pre-processing is some add-on white spaces within the text, as shown in Fig. 8. The spaces are the result of the removal of Latin characters, punctuation marks etc. The null values need to be excluded to work on the real-time and relevant dataset.

Finally, the output file of preprocessing has a reduced size with more relevant data. It involved relevant and appropriate terms which further improved the efficiency of our algorithm. The entire process of pre-processing has been step-wise illustrated in Table 2.

winner as value network customer select receive 900prize reward claim call 09061701461 claim code kl341 valid 12 hour

Fig. 8. Snippet of output after removing null values

Table 2. Demonstration of step by step pre-processing

Input	Examples	Reduced size(KB)
Removal of non-ascii characters	Å, £, ≅, α, å, Õ etc.	15
Removal of punctuations	(!) (,) (:) (;) (.) (?) (") (') () (-) (_) ({})	3
Lowercase letter conversion	a, b, c, d, e, f, g, h, i, j, k, l, m, n, o, p, q, r, s, t, u, v, w, x, y, z	0
Stopwords removal	the, is, an, at, which, on, was, with, or, etc.	136
Stemming	It should identify the string "stems", "stemmer", "stemming", "stemmed" as based on "stem". A stemming algorithm reduces such words to nouns	143

3.3 Classifiers

Data Mining involved the predicting spam/ham results using relatable comment algorithm as well as TF-IDF. Under machine learning, SVM and XGBoost were used and numerous experiments were conducted.

Relatable Comment Algorithm (RCA)

In RCA at first, each unique term with respect to each comment was required to create a main cluster node connecting to various comments containing the cluster term. The further processing involved the creation of sparse matrix to determine term comment relationship with value 0 and 1 for each occurrence of the term in a respective comment. The constructed matrix was sent as input to get a relatable term and comment index relationship. The comments were then grouped based on their unique index to get the unique terms involved in it. To increase and efficiently predict the comments

we computed normalized values, $(Normalized\ Value = X_i/|X|, \quad |X| = \sqrt{\sum_{i=1}^{n} X_i^2},$

where n denotes length of dataset). Since the spam comments are required to have neither too less number of terms and neither too more, a relevant number was set as a threshold. $Value_{THRESH} = X_{MIN} + rand[X_{MAX} - X_{MIN}]$, where $Value_{THRESH}$ is threshold, X_{MIN}, X_{MAX} are min and max values of idf count and $rand = [0, 1]$. The described algorithm requires less computation due to involved concepts of network clustering.

Term Frequency-Inverse Document Frequency (TF-IDF)

TF-IDF as proposed in [11] is used to determine the significance of tokenizers based on its frequency. The algorithm involved constituting a vocabulary box out of the comments text which sets the significant score for each term and thereby helps in determining the relevant comments with respect to the kind of video. It then creates an IDF vector which determines that the less the value of this vector, the more relevant term is. Then the values of IDF were normalized and the threshold was set based on the equations given before. The threshold was set to get 20% of terms and hence the comments as spam. Thereafter, each pre-processed comment was tokenized and compared with the terms determined as spam by TF-IDF analysis. The proposed concept was further used to ascertain the useful and most relevant comments and hence the most relevant user.

INPUT:-	12. *for* all comments (C_i) in input data
Initial data : unprocessed set of comments	13. create vector using CountVectorizer
for YouTube video ($C_i \in$ Comments, $i \in$ N)	14. create Array and get feature-
Input data: processed set of comments for	names(term)
YouTube video ($C_i \in$ Comments $i \in$ Z+)	15. mat[ ] = sparse matrix of Array
n: no. of comments in initial data	16. flag = ClassifyFun (feed mat[] for
x: no. of comments in input data	analyzing)
spamflag: column in data files.	17. accuracy = Compare (flag,spamflag).
OUTPUT:-	accuracy
Classified comments & classified users.	18. EVALUATE compare(flag,spamflag)
1. LOAD initial data	for accuracy
2. *for* all comments (C_I) in initial data	19. END when i = x+1
3. C_i .remove non-ascii characters	
4. C_i .remove punctuations	1. **ClassifyFun(input)**
5. C_i .to LowerCase	2. create dataframe [term,C_i]
6. C_i .remove stopwords	3. termcount = dataframe.groupby [C_I].
7. C_i .stemming	count
8. C_i .remove null values	4. if (termcount < threshold)
9. input data <- C[i]	5. return spam
10. END when i = n+1 or initial data = { }	6. else
11. LOAD input data	7. return legitimate

Algorithm 1: Relatable Comment Classification

Support Vector Machine (SVM)

SVM in [11] was used to implement both classification and regression analysis. It sets the partition for different classes of data by considering different features for classes to be identified. The data points were marked at different coordinates based on the value of features.

After plotting the points, it sets a hyper plane to separate the two classes uniquely. Since our classification of comments involved spam/ham identification, is adopted spam and ham as two different classes while considering the features namely, length of comment, repetition of words, presence of URL, irrelevant or inadequate statements.

Extreme Gradient Boosting (XGBoost)
This algorithm in [11] has been an improved form of gradient boosting framework. To implement XGBoost, the entire corpus was converted to a sparse matrix which was then processed by different CPU cores simultaneously to construct a tree. The tree was constructed considering the features of small length of spam comments, number of relevant terms, repetition of terms etc.

Latent Dirichlet Allocation (LDA)
To examine the dataset, this algorithm as in [11] was used to retrieve the topics in order to regulate the type of comments. The identification of most useful topics further assisted in determining the accuracy of spam flag specifically the ham comments and thereby the legitimate users. After the implementation of various algorithms, the XGBoost outperformed other classifiers for which detailed explanations are described in results Sect. 4.

4 Results and Discussions

We analyzed the entire dataset of comments to evaluate the accuracy between predicted and true spam/ham comments. The combined observations for cluster analysis and RCA depicted in Table 3 are both with respect to pre-processing and unprocessed data to depict the improved accuracy after pre-processing. The comparison of accuracies of data mining algorithms depicts that TF-IDF outperformed other data-mining classifiers. RCA being a cluster dependent algorithm depicted itself to be less accurate. Table 4 reveals the computation of threshold values. All the comments which gratified the threshold values were depicted as spam.

Table 3. Accuracy for different Algorithms

Algorithms	Accuracy (Without Processing)	Accuracy (With Pre-processing)
RCA	84.5%	85%
TF-IDF	85%	87%
SVM	94%	96%
XGBoost	95%	97%

We further observed the accuracy of the predictions by proposed classifier models. XGBoost proved to be most effective. The accuracy evaluated out to be the best in case of pre-processed data. The accuracy of 95% outperformed all other classification models which were further improved to 97% by pre-processing the dataset. Figures 9 and 10 depicts the accuracy model. Further, it was observed that on decreasing the size of the dataset to half, accuracy of SVM and XGBoost decreased. SVM proved to be more accurate compared to XGBoost on decreasing the size of the dataset. The decreased efficiency has been shown in Fig. 11.

Table 4. Threshold values

Models	Threshold values
RCA (For spam)	<0.0210507331
RCA (For useful)	>0.0220507331
TF-IDF (For spam)	>0.0109896487
TF-IDF (For useful)	<0.0108896487

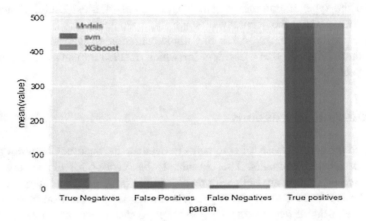

Fig. 9. Comparison of classifier performance

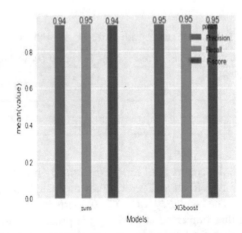

Fig. 10. Comparison of classifier performance based on precision, recall and F-score

Thereafter, manually inspecting the comments and users of numerous popular videos, we can conclude that spamming is prevalent in YouTube and the proposed rules are efficient, reliable and relevant in detection.

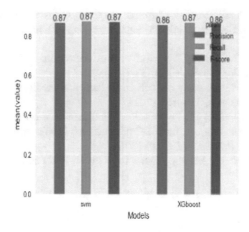

Fig. 11. Decreased accuracy with halved dataset.

5 Conclusion

This paper describes the methods to differentiate legitimate users from spam users through various classification model techniques. We considered some features of spam users to predict them and performed a comparative study on our predictions. Future works may involve application of more approaches on larger datasets to further determine the changes in curve.

References

1. Balakrishnan, A.: Google claims YouTube is 10x as Popular as Netflix or Facebook Video, and Approaching TV (2017). https://www.cnbc.com/2017/02/27/youtube-viewers-reportedly-watch-1-billion-hours-of-videos-a-day--us-tv-viewers-watch-125-billion-and-dropping.html
2. Google's Bad Week: YouTube Loses Millions as Advertising Row Reaches US. https://www.theguardian.com/technology/2017/mar/25/google-youtube-advertising-extremist-content-att-verizon
3. Wattenhofer, M., Wattenhofer, R., Zhu, Z.: The YouTube social network. In: Proceedings of the Sixth International AAAI Conference on Weblogs and Social Media (2012)
4. Chaudhary, V., Sureka, A.: Contextual feature based one-class classifier approach for detecting video response spam on Youtube. In: Eleventh Annual International Conference on Privacy, Security and Trust (PST), pp. 195–204, July 2013
5. Jin, X., Lin, C.X., Luo, J., Han, J.: Social spam guard: a data mining-based spam detection system for social media networks. In: Proceedings of the Very Large Data Bases, pp. 1458–1461 (2011)
6. O'Callaghan, D., Harrigan, M., Carthy, J., Cunningham, P.: Network analysis of recurring youtube spam campaigns. In: Proceedings of the 6th International Conference on Weblogs and Social Media (ICWSM 2012), Dublin, Ireland, pp. 531–534

7. Abdulhamid, S.M., Latiff, M.S.A., Chiroma, H., Osho, O., Abdul-Salaam, G., Abubakar, A. I., Herawan, T.: A review on mobile SMS spam filtering techniques. IEEE Access **5**, 15650–15666 (2017)

8. Spirin, N., Han, J.: Survey on web spam detection: principles and algorithms. ACM SIGKDD Explor. Newsl. **13**(2), 50–64 (2012)

9. Ghiam, S., Pour, A.N.: A survey on web spam detection methods: taxonomy. Int. J. Netw. Secur. Appl. (IJNSA) **4**(5), 119–134 (2012)

10. Rădulescu, C., Dinsoreanu, M., Potolea, R.: Identification of spam comments using natural language processing techniques. In: IEEE International Conference on Intelligent Computer Communication and Processing (ICCP) (2014)

11. Lesmeister, C.: Mastering Machine Learning with R. Packt Publishing Ltd., Birmingham (2017)

Kinematic Analysis and Simulation of a 6 DOF Robot in a Web-Based Platform Using CAD File Import

Ujjal Dey$^{(\boxtimes)}$ and Kumar Cheruvu Siva

Department of Mechanical Engineering, Indian Institute of Technology (IIT),
Kharagpur 721302, India
ujjal.dey@iitkgp.ac.in, kumar@mech.iitkgp.ernet.in

Abstract. The current trend of simulator-based analysis, especially in the area of robotics had emerged broadly. This kind of simulation gives initial familiarization of the system, which is very useful for introductory level courses in robotics as well as research-based work. Simulation developed through any commercial software's requires its installation on the user's system for any animation or analysis. Therefore, an open source platform for this type of robot motion analysis had a much impact due to its light version, better graphics, and web-based running capability. This paper describes an efficient and very straightforward approach of building a 6 degree of freedom KGP50 robot simulation model in a web interface using WebGL technology. Here a component is first designed in SolidWorks and then it is imported directly into the WebGL-based platform utilizing a library of Three.js. The forward kinematics analysis of our KGP50 robot is presented through this simulator, which gives the idea of the whole framework and also an exploration of the KGP50 robot.

Keywords: KGP50 · WebGL · Web · Three.js · CAD model

1 Introduction

Analysis of any sort of robot motion through simulation requires creating a virtual model of its representation. In general, a 3D model is build using some CAD designing software and its simulation is observed through MATLAB/Simulink/Simscape based environment [1–5]. Some other tools like RoboAnalyzer, RoboDK are also available for this type of kinematic study and programming of robots [6]. This procedure requires a user to have that software installed in the system to run the simulation, which in many cases may put a constraint for frequent access. Hence, web-based interfaces are developed to visualize 3D simulations with interactive application enabled, which is free from any supporting software or plugins. For better enrichment of the display of 3D-content WebGL based platform are emerging. Web Graphics Library, (WebGL) is a JavaScript Application Programming Interface (API) for rendering 2D or 3D graphics within any HTML5 supported web browser [7]. It uses the HTML5 canvas element and is accessed using Document Object Model (DOM) interfaces.

© Springer International Publishing AG, part of Springer Nature 2018
A. Abraham et al. (Eds.): ISDA 2017, AISC 736, pp. 243–250, 2018.
https://doi.org/10.1007/978-3-319-76348-4_24

WebGL provides creating 3D geometries with specific commands from which a virtual model is made, and its simulation is performed. For complicated 3D components method of CAD file import is used for building robot models. A 3D CAD model is first designed in software's like Solidworks, CATIA, Creo, etc. and then extracted triangular vertex coordinate data from the file is implied into the program. This type of WebGL technology method has been implemented with various approaches for standard industrial robot motion simulations [8]. It seems to be a fine approach, but the direct implication of WebGL technology is very tiresome, especially when a model with multiple components is to be rendered. For viewing a 3D element one need to describe the entire vertex coordinates extracted from the CAD file through some application. Therefore, using a library reduces the steps and makes the process much easier and faster. Currently, many JavaScript libraries like Three.js, SceneJS, BabylonJS uses WebGL and provide many high-level features for 3D object visualization. Three.js is a high-level utility library which makes WebGL program much easier and more straightforward. Since it uses JavaScript coding language, it can be combined with other libraries, which are capable of giving interactive applications. It also has different features of camera control, light control, and animations. Implementation of this technology for various web-based experimentations is going on in many fields of study [9, 10]. This paper describes the simulation and forward kinematic analysis of the KGP50 robot and also its workspace visualization in a web-based platform. The virtual model of the robot is created through the direct import of CAD file in STL format. Here WebGL technology is explored via Three.js library for building the simulator, which gives the idea of the whole framework of importing a particular CAD file, and its assemblage with other components.

2   The Methodology of Importing a CAD Model

Three.js is an open source JavaScript 3D library which uses WebGL Technology [11, 12]. It is used to create and display 3D scene through WebGL renderer module. For viewing a particular object, one needs to set up a Canvas within which a scene is created and position of the camera is specified. Then that scene is then rendered through THREE.WebGLRenderer() function. Three.js has some predefined commands for creating some basic 3D geometries, which is appropriately assembled to model some simple robotic configuration [13]. Since KGP50 has an intricate design, therefore methodology of CAD file import is needed for accurate modeling and better visualization. A particular part is first designed in SolidWorks and saved as stereolithography (STL) format. Then after preparing the primary interface of the program, it allows a user to visualize and manipulate that STL geometry within HTML canvas. In order to view the 3D model, the path of CAD file location and STL loader library must be specified properly within the program. The flow diagram of the framework for viewing a CAD model within the scene using Three.js library is given in Fig. 1. Apart from STL format Three.js also support other extensions of the 3D object, which can be loaded through that particular loader library. One can add different material and texture to make it more visually compelling.

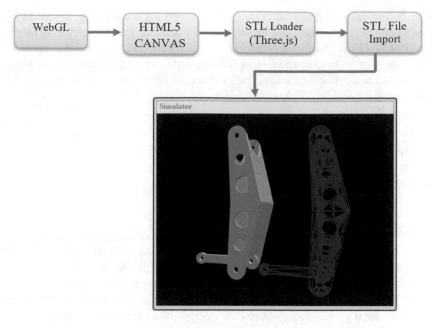

Fig. 1. Sequential steps for viewing a CAD model within canvas

3   KGP50 Robot

KGP50 is a prototype industrial robot with all features of a manipulator along with modular controllers and control techniques. It has six degrees of freedom with high payload capability and parallel linkage structure with the recirculating screwball

Fig. 2. KGP50 model

mechanism in joints 2 and 3. It is used as a testbed for cutting-edge technologies of human-computer interaction and intelligent systems. The robot is powered by digital servo control of AC motors synchronized and precision-controlled for coordinated motions using a real-time digital control station, working on a Digital Signal Processing-based motion controller. The actual model of the KGP50 robot is shown in Fig. 2.

Specifications:

- 6-Axis, Continuous path control
- 50 kg Payload
- 1.5 m reach
- 1.5 m/s maximum speed
- 0.1 mm repeatability

3.1 Kinematics

Kinematics is the science of motion that treats the subject without regard to the forces that cause it. In robot kinematics, two major aspects of the study are the forward and inverse kinematics analysis of any manipulator. The problem of forward kinematics deals with the determination of the position and orientation of the end-effector of the manipulator for a given set of joint angles. Here forward kinematics analysis is carried out by assigning D-H parameters to all links. To assign the D-H parameters kinematic diagram of the robot is the first setup, which is presented in Fig. 3. D-H parameters of the robotic arm are given in Table 1. The transformation matrix relation between the end effector and the base frame attached to the robot base is expressed as:

$$^0T_6 = {}^0T_1\,{}^1T_2\,{}^2T_3\,{}^3T_4\,{}^4T_5\,{}^5T_6 \tag{1}$$

Each homogeneous transformation matrix is expressed as a product of four basic transformations associated with joints i and j (l-link length, α-link twist, d-link offset, and θ-joint angle) and I is a 4×4 identity matrix. The general form of each transformation matrix of the ith frame with respect to $i-1^{th}$ frame is given by [14]:

$$^{i-1}T_i = \begin{bmatrix} C\theta_i & -S\theta_i & 0 & a_{i-1} \\ S\theta_i C\alpha_{i-1} & C\theta_i C\alpha_{i-1} & -S\alpha_{i-1} & -S\alpha_{i-1}d_i \\ S\theta_i S\alpha_{i-1} & C\theta_i S\alpha_{i-1} & C\alpha_{i-1} & C\alpha_{i-1}d_i \\ 0 & 0 & 0 & 1 \end{bmatrix}$$

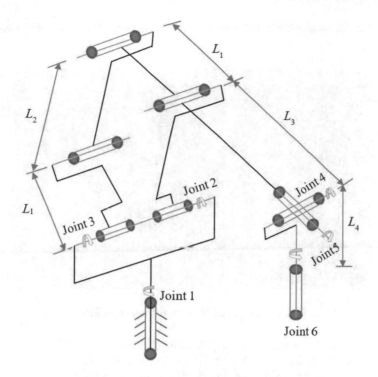

Fig. 3. Kinematic diagram

Table 1. D-H parameters of KGP50

Link i	α_{i-1}	a_{i-1} (mm)	d_i (mm)	θ_i
1	0	0	0	θ_1
2	−90	0	0	θ_2
3	0	L_2	0	θ_3
4	−90	0	L_3	θ_4
5	90	0	0	θ_5
6	−90	0	L_4	θ_6

4   Assembly and Simulation of KGP50 Robot

To perform the kinematic analysis, the whole model is constructed by importing all the components of KGP50 and assembling them sequentially. As described earlier, once an STL file is introduced within the HTML5 canvas it is considered as geometry, and that geometry can be manipulated as desired. It can be translated and rotated any amount as desired within the canvas for assembling. For accurate positioning and orientation of the geometry, the transformation matrix can be applied to that geometry. The assembled virtual model of KGP50 is shown in Fig. 4. The whole model can be visually enhanced by adding different materials and texture, which will look almost similar to our actual robot.

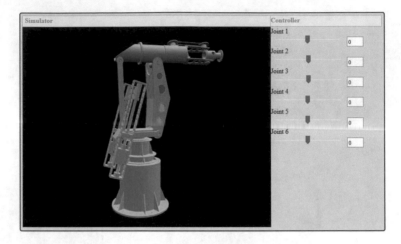

Fig. 4. Assembled 3D virtual model of KGP50

After the model is ready, separate programs are developed to observe it's all joint motions separately and perform forward kinematic analysis. An Input panel is created for submission of joint angles data. Specifying the joint angular values final transformation matrix is displayed, which shows the position and orientation of the tool frame with respect to the base frame. In robot kinematics, one other important aspect of the study is the workspace analysis of robot, which indicates the volume of space that the end-effector of the manipulator can reach. Therefore a charting library of plotly.js is used for graphical representation of manipulator position information. For each submission of joint angular values, a 3D scatter graph is generated which indicates end-effector position and from which workspace of the robot can be visualized. Now switching the format of the chart from point to line graph the motion trajectory of the robot can be seen. Figure 5 shows two different configurations of the KGP50 robot in the forward

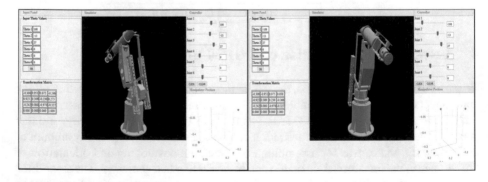

Fig. 5. Screenshot of the interface for two distinct configurations of the robot

kinematics analysis. The whole program in compacted within an HTML file, which can be run in any WebGL compatible web browser for analysis.

5 Conclusion

This paper illustrates the utilization of Three.js library for developing the 3D model of the KGP50 robot which is used for its forward kinematic analysis in a web-based virtual environment. It uses the technology of WebGL, which provides a very efficient platform for 3D object visualization. Using the library of Three.js helps in the direct rendering of the 3D object in Cartesian space. Any 3D model can be easily built up using this CAD file import procedure, which removes commercial software's dependencies. The same environment can be further used for Inverse kinematics analysis by just adding the mathematical calculation in a JavaScript file format. Here multiple robots motion simulation can also be incorporated into the same environment, which is a unique feature in this type of methodology. This kind of web-based platform is beneficial for virtual laboratories, which enhance the process of E-learning/education. One major prospect of this kind of open-source program is that the source code is freely available to others and it can be modified as desired to improvise the design. Since the whole analysis is carried out considering actual dimensions and kinematic parameters, it is almost similar to real system operation. It can be further programmed to connect this simulation model and real-time control system of the actual robot for remote operation.

Acknowledgement. The authors gratefully acknowledge the financial support from MHRD, New Delhi and Department of Mechanical Engineering, Indian Institute of Technology, Kharagpur for providing the platform to carry out the work.

References

1. Alshamasin, M.S., Ionescu, F., Al-Kasasbeh, R.T.: Kinematic modeling and simulation of a scara robot by using solid dynamics and verification by Matlab/Simulink. Eur. J. Sci. Res. **37**(3), 388–405 (2009)
2. Gousmi, M., Ouali, M., Fernini, B., Meghatria, M.H.: Kinematic modelling and simulation of a 2-R robot using SolidWorks and verification by MATLAB/Simulink. Int. J. Adv. Robot. Syst. **9**(6), 245 (2012)
3. Zodey, S., Pradhan, S.K.: Matlab toolbox for kinematic analysis and simulation of dexterous robotic grippers. Procedia Eng. **97**, 1886–1895 (2014)
4. Corke, P.: A robotics toolbox for MATLAB. IEEE Robot. Autom. Mag. **3**(1), 24–32 (1996)
5. Fedák, V., Ďurovský, F., Üveges, R.: Analysis of robotic system motion in SimMechanics and MATLAB GUI environment. In: Bennett, K. (ed.) MATLAB Applications for the Practical Engineer, 3rd edn. InTech (2014)
6. Bahuguna, J., Chittawadigi, R.G., Saha, S.K.: Teaching and learning of robot kinematics using RoboAnalyzer software. In: Proceedings of First International Conference on Advances in Robotics, Pune, pp. 1–6. ACM (2013)
7. Tavares, G.: WebGL fundamentals. HTML5 Rocks (2012)

8. Lianzhong, L., Kun, Z., Yang, X.: A cloud-based framework for robot simulation using WebGL. In: Proceedings of Sixth International Conference on Intelligent Systems Design and Engineering Applications (ISDEA), Marrakesh, pp. 5–8. IEEE (2015)
9. Severa, O., Goubej, M., Konigsmarkova, J.: Unified framework for generation of 3D web visualization for mechatronic systems. J. Phys. Conf. Ser. **659**(1), 012053 (2015)
10. Tudjarov, B., Botzheim, J., Kubota, N.: Facilitation of cognitive robotics by web-based computational intelligent models. In: Micro-NanoMechatronics and Human Science (MHS), pp. 144–148. IEEE (2012)
11. Dirksen, J.: Learning Three.js: The JavaScript 3D Library for WebGL. Packt Publishing Ltd., Birmingham (2013)
12. Parisi, T.: Programming 3D Applications with HTML5 and WebGL: 3D Animation and Visualization for Web Pages. O'Reilly Media, Inc., Sebastopol (2014)
13. Dey, U., Jana, P.K., Kumar, C.S.: Modeling and kinematic analysis of industrial robots in WebGL interface. In: Proceedings of Eighth International Conference on Technology for Education (T4E), Mumbai, pp. 256–257. IEEE (2016)
14. Craig, J.J.: Introduction to Robotics: Mechanics and Control. Pearson Prentice Hall, Upper Saddle River (2005)

Large Scale Deep Network Architecture of CNN for Unconstraint Visual Activity Analytics

Naresh Kumar[1,2(✉)]

[1] Computer Society of India (CSI), Delhi, India
[2] Department of Mathematics, Indian Institute of Technology Roorkee, Roorkee 247667, India
atrindma@iitr.ac.in

Abstract. Handling the issues of massive datasets for information retrieval, feature learning, is expected one of the most challenging problems in machine learning and computer vision research. The issues in this work, have been focused to maintain the data scalability problems for machine learning classifiers in social media activity analysis. The research highlights the machine learning perform-ance techniques which can provide promising results against the large and unstructured complex data of social media activities. This work has been focused on the biologically inspired processing techniques by neural network and intro-duces the extension of this network to resolve the problems of complex data pertaining to human activity analysis. It is presented various architectures of CNN and several phases of visual data processing for detection and recognition prob-lems. Some selected techniques are highlighted that create the interest for deep network learning in various domains of research under the consideration of complex data handlings. It has been introduced activation functions and sequence pooling methodology for fast training of convolutional network with massive data of unstructured human activity recognition. Overall, it is highlighted that fast training aspects of the network against large scale and complex data, can be improved by choosing activation function and pooling methodology at fully connected layers of the neural network. Moreover, the sounding techniques of deep learning and data analytics are highly applicable for human health, medicine, robotics, education and industrial applications.

Keywords: Artificial neural network · Convolutional neural network
Deep learning · Human activity recognition · Large scale data

1 Introduction

Human activity recognition is one of the sounding challenge of several visual analytics research problems by many decades. The interest towards rich features statistics and computation of features changes among the video frames, attracts the researcher for

N. Kumar—IEEE Member.

human action recognition [18–22, 24]. Machine learning techniques are highly applicable for the problems discretized by real life scenario in which huge amount of various kinds of data is processed. Highly complex and unstructured data captured from unconstraint environment is entertained in machine learning research. The research work for large scale visual analytics include natural language processing (NPL), robotic vision, computational medicine and social media analytics. The processing of such data may ranges from small mobile devices like phones and tablets to highly scaled distributed systems like the clusters of thousands of high computational devices. Detection and recognition of visual contents related to human or any real life scene, in videos and images can be a common interest. The processing complexity in such work is nicely resolved by deep analytics of the network. Moreover, all these issues may belong to several domains like visual surveillance, human activity recognition and health and medicine. Video is a sequence of frames in which frame boundary, shot and scene change play a vital role for several multimedia analytics problems.

1.1 Motivation to Research Problem

With the evolution of big data handling techniques, the research of human action recognition can be motivated by promising resources to get through the unconstraint environmental issues. The goal of activity analysis can be projected to video surveillance and intruder detection, psychovisual analysis, weather forecasting and many fields related to motion estimations. The requirement of economic model to train global features extracted from these fields of activity analysis is highly demanded. This research is closely applicable to almost every part of real life ranging from daily activity to real time security and geographical changes. The motivation of the research problem encourages the evolution of deep network by enhancing the features of traditional neural network as presented in Figs. 1 and 2.

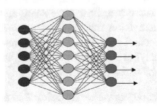

Fig. 1. Traditional neural network

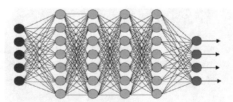

Fig. 2. Deep learning neural network

1.2 Challenges to Process the Data

Massively fast growing data belonging to several activities is the biggest hurdle to retrieve particular information from unconstraint environment. The processing of such a complex and massive visual data challenges to the computational devices and traditional machine learning techniques. Big data is generally conceptualized by 5 v's of volume, velocity, variety, veracity and values. All these are referred for high dimensional features which are hard to process without highly computational devices and rich source of machine learning algorithms to train such data.

2 Related Work

Large variability and complexity of fast varying actions performed in unconstraint video data mark the video description and action recognition a big data handling problem. In this text, it has been outlined the literature for human activity recognition which is simply an estimation of change with time for actors localization in video frames. This is accomplished based on tracking feature updates of the scene. Human activity recognition from unstructured data in video of unconstraint environment [18–24] is a challenging problem of global features computation. Such features of massive activities comprise big data analytics [1] as considered the extracting information from pool of petabytes and supposed to solve by deep learning analytics.

For unsupervised and large scale feature learning on deep network [2] Dean et al. introduced downpour stochastic gradient decent (SGD) methods and sandblaster optimization that provide data parallelism and model processing. The spatial and temporal data is processed by long term r-CNN [4] for video description and activity recognition, which is basically a doubly deep training network. The sequence learning of Long Short Term Memory (LSTM) network [4] is promoted by TS-LSTM [5, 6] by introducing temporal convolution and segmentation fusion. The extended dynamic temporal features of Human Activity Recognition (HAR) [8] are promoted by [6] in which HAR is considered as a ranking problem and the rank and score pooling functions are introduced by DSIFT and HOG/HOF descriptors. Furthermore, SIFT with CNN [25] is incorporated for fast information retrieval. This ranking model is tested on HMBD and ASLAN datasets with 30.24% and 38.97% accuracy respectively. According to Song et al. [7] HAR data pertains multimodal spatial temporal resolution which inspires to develop Hierarchical Sequence Summarization (HSS) model and outperformed to the state of the art by 99.59% and 75.56% accuracy at ArmGesture and Canal9 datasets respectively. Detailed experimental study of deep residual network [9] BoosResNet outperformed traditional backpropagation. Long short-term memory (LSTM) [10] provides compromising results for long training sequence while it is demonstrated faster and exact detection [11] of the object without estimation the features of bounding box of the objects. The accuracy enhancement of single shot detection (SSD) follows feed-forward CNN which gives 73.4% precision on PASCAL VOC as compared to You Look Only Once (YOLO) with 64.3% precision while the detection rate was 45 FPS. Other versions of SSD are SSD512 and SDD300 which are 3-times faster than faster r-CNN [12, 25] and YOLO by detecting 59 FPS in video data of action recognition. Outperforming the faster r-CNN with ResNet and SSD, YOLO9000 [16] and its faster version YOLOv2 achieved 76.8% at 67 FPS and 78.6% at 40 FPS on PASCAL VOC dataset. Faster r-CNN is basically a fully convolutional network based Region Proposal Network (RPN) which combined fast r-CNN [29] and RPN. This network adopts the attention based technology developed in YOLO [13] and achieved accuracy up to 75% with faster r-CNN [12]. To beat object detection challenges in benchmark PASCAL VOC dataset, CNN is combined with region proposal and termed as r-CNN [14]. Restricted Boltzmann Machine (RBM) [17] ensures that the thresholding function rectified linear unit (ReLU) [32] increase the efficiency of object detection. Today ReLU and its versions are giving promising accurate results at fully connected layer of CNN. To process large scale data video analytics, an architecture of fast learning convolutional network (Caffe) [15] which can process 40 million images on a single Titan X in 24 h, developed at Berkeley Vision and Learning Center (BVLC) as a parallel computing open source. To improve the performance of neural network several activation functions like sigmoid, Tan hyperbolic, Elue, ReLU and steps functions [31, 33, 35] have been widely used. In many learning algorithms, Parametric Algebraic Activation (PAA), which a family of S-shaped curve, has been ensured number of epochs and error reduction is highly significant for deep network training. Again, in case of over-fitting of large training data, dropout and ReLU [32] have improve the error rate by using Hessian-free optimizer software. Like ReLU, leaky ReLU and parameterized ReLU, exponential linear unit (ELU) [34] speed up Deep Neural Network (DNN) learning for higher accuracy with low computational complexity and significant generalization performance.

Human actions datasets and state of the art [19] are presented in detail to highlight activity recognition as a challenging problem. Several benchmark human action datasets like Hollywood2, UCF101 and HMDB51 has proved HAR is a challenging problem due to frame level structures of action appearance and multi-view of diversified activities a single a video frame. For this problem in [20], Seng et al. presented a novel stratified pooling for CNN training (SP-CNN) which maps frame level features to video level features in fully connected CNN layers. A wild class of dataset THUMOS for human actions recognition [22] is proved a challenge for classification and detection of actions and provides large annotation of the objects and events in visual categorization. Deep learning for HAR, 3D CNN architecture is presented [23] on TRECVID dataset. The tedious efforts looks for unsupervised classifications because for large network, it has used SVM at fully connected layer of CNN. This work is fully centered for human action recognition from large scale unstructured and complex visual data. Rahmani et al. [18] developed Robust Nonlinear Knowledge Transfer Modal (R-NKTM) for unsupervised human action recognition in which they computed 2D dense trajectory from 3D video sequence of human model.

3   Convolutional Networks Architecture

This research is exploiting deep structure of neural network in the context of fast computation prospective and the highlights the evolution of several CNNs to resolve the problems that are challenges for traditional neural network. In this reference, the vital technology includes convolutional neural network (CNN) [26–28, 30], restricted Boltzmann machine (RBM) and recursive neural network (RNN). In this section a brief architecture of CNN is represented to incorporate its contribution for human activity analysis and their four basic building blocks convolution, non-linearity unit, sampling and fully connected layer are introduced for features classification.

3.1 LeNet (1990)

Yann LeCun introduced very first CNN to create the interest for deep learning, which is earlier used fir Optical Character Recognition (OCR) and digit recognition research. The basic LeNet architecture is given in Fig. 3 which has cascading pooling and non-linearity introduced by rectified linear unit (ReLU) [33, 35] before the fully connected layers. In Fig. 3, as in put figure showing high probability of being boat (0.94) among rest of categories at fully connected layer.

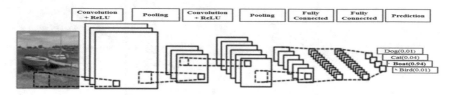

Fig. 3. LeNet: a first convolution network

3.1.1 Non-linearity Introduction

This is a kind of thresholding achievements in traditional neural network which accomplished by sigmoid or activation function. To incorporate with real life problem which are generally non-linear, activation function of same nature is adopted to get threshold for CNN. Non-linearity is introduced by basic thresholding functions which include logistic function, rectified linear unit (ReLU) [31, 33], exponential linear unit (ELU) [34], sink and Gaussian function. All these functions perform pixel wise operations. As evidences from literature the modified ReLU in Eq. (2), leaky ReLU, Parametric ReLU (PReLU), randomize leaky ReLU (RReLU) are outperforming.

3.1.2 Convolution Filters

Convolution by a 3 * 3 kernel with image produces the features accordingly the kernel features are selected. The basic kernel features are Gaussian, sobel, Laplacian, box filters and median and average filter. The selection of more number of filters give better training for our network if it is taken care of processing overheads. Convolution step includes three important things to decide before processing. First is depth of convolution that corresponds to the number of filters used and second is strides which means how many pixels are decided to jump in a single convolution and finally zero padding which allows to convolve with boundary pixels of the input image.

$$f(u) = \begin{cases} u, & u \geq 0 \\ 0, & u < 0 \end{cases} \tag{1}$$

$$f(y) = \begin{cases} y, & y \geq 0 \\ ay, & y < 0 \end{cases} \tag{2}$$

$$f(z) = \begin{cases} z, & z \geq 0 \\ \left(\dfrac{z}{a}\right), & z < 0 \end{cases} \tag{3}$$

The parameter 'a' is uniformly distributed random number which affects the performance of CNN used in parametric, leaky (PReLU) (2) and RReLU (3) and corresponding graph is represented by Fig. 4.

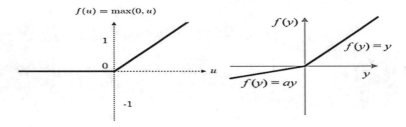

Fig. 4. Rectified Linear Unit (ReLU) and Parametric Rectified Linear Unit (PReLU)

3.1.3 Pooling in CNN Layers

Introduction of non-linearity to previous linear convolution operation increase the dimension, although unstructured data of human activities is highly dimensional to create processing hurdles. These issues are resolved in the third layer of CNN by a spatial operation called pooling on the feature maps. Pooling is generally termed as down sampling and can have several classes. Some basic pooling are max, average, sum but high features pooling techniques include temporal, spatial pyramid, subspace pooling and content adaptive spatial pooling.

3.1.4 Fully Connected Layer

In the previous three layers, useful and reduced dimensional features are achieved which can be easily classified in last fully connected (FC) layer of CNN. In FC layer all the neuron are connected to every neurons of previous and next layer. Any general classification techniques used to classify the high level features obtained from convolutional and nonlinearity layers.

3.2 AlexNet (2012)

The basic details of every CNN borrows from LeNet, only difference varies with number of convolution filters chosen to dimensionally reduce the non-linearity by various pooling strategies. This network [3] is trained on 2 NVidia GTX 580, for purely super-vised classification over 1.2 million sample images of large scale dataset of ImageNet during five to six days. This CNN has five convolution layers in which softmax activation is used to train over 60 million parameters and 6.5 lakhs neurons.

3.3   ZF Net (2013)

Tweaking the architecture proposed in AlexNet CNN, Zeiler and Fergus introduce ZF Net to visualize the function inside the intermediate layers of the network and classifi-cation methodology at the convolution layers. This convonet model proves state of the art as a generalization at the dataset Caltech 101 and Caltech-256. The is network is rained on GTX 580 for twelve days to develop feature to pixel maps as opposite to the convolution layer. The activation and error function are performed by ReLU and cross-entropy loss function. The novel visualization benchmark is established at ImageNet [3] for classification in which regularization is achieved by Dropout strategy.

3.4   GoogleNet (2014)

Earlier network suffer from the massiveness of big data, in which it is hard to maintain the huge number of parameters. Google Net reduces up to 4 million as it is 60 million in AlexNet. This achievement is possible due to the development of inception model in Fig. 5 which set the state of the art for classification and detection in large scale bench-mark ImageNet dataset for visual recognition in 2014. Hebbian principle used to achieve the optimization control and a 22 layers Google Net is the successful outcome regarding detection and classification. This reduce the layers in CNN as a result get free from over-fitting and these layers are referred as global average pooling.

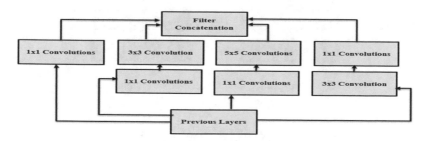

Fig. 5. GoogleNet inception layer architecture

3.5   VGGNet (2014)

This network is highly sounding in the literature of visual recognition. The visual recog-nition performance of this network is directly characterized by the depth of convolution network. This promotes the researchers to increase the depth of CNN layers by 16–19 to reduce the number of parameters used in the network. Three to four week are taken to train on four Titan Black GPUs.

This model is built on Caffe toolbox and used jittering as data augmentation techniques. Training data is optimized by stochastic batch gradient scheme and every convolutional layer follows ReLU for non-linearity introduction. From the experimental report [9] Table 1, it has been observed that 19-layer network has 7.5% error on validation set and 7.3% error on test set on top-5 classification layer.

Table 1. VGGNet classification performance

Model	Top-5 classification error on ILSVRC	
	Validation set	Test set
16 layer	7 5%	7,4%
19-layer	7.5%	7.3%
Model fusion	7.1%	7.0%

3.6   ResNet (2015)

The development of residual network (ResNet) was awarded to Kaiming H. in ILSVRC 2015. Deeper neural networks outperforms for big and complex datasets but training such a big data is a tedious job. In this network the layers are formulated as a learning residual functions [9] which optimizes and gains accuracy at the higher depth scales. The residual learning model is given by Fig. 6 and its architecture is presented in Fig. 7.

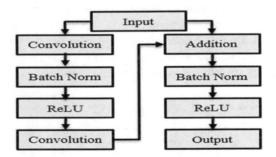

Fig. 6. Inspection module for ResNet architecture

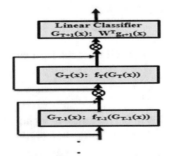

Fig. 7. ResNet architecture

It is reported that batch normalization is also fail while adding the extra layer to reduce training and validation errors. This problem is resolved in ResNet by adding bypass to sum up with convolution layers. This network is trained on 8 GPUs during 21 days by the processing 152 layers (Table 2).

Table 2. ResNet classification performance

Model	Error %	Time (ms)
VGG-A	28.9	380
ResNet-34	26.7	230
BN-Inception	26.1	191
ResNet-50	24.0	403
ResNet-101	22.1	652
Inception-v3	20.9	492

4   Limitations of Deep Learning Networks

Deep learning is good tool to deal with highly growing data of several discipline but remains a black box because of billions of complex layers and huge number parameters in nodes for each layer. This makes the elementary computation more tedious. The propensity of machine learning methods assisted in case of over fitting of unstructured data. Learning complex model from large scale data needs a great deal with computational cost. Lacking high domain expertize creates time consuming overhead for feature engineering and data analysis. The drawbacks in deep network are generally considered in terms of confidence reduction, source and target misclassification.

5   Conclusion and Future Directions

This work puts novel efforts to produce human action recognition as central theme and highly sounds the issues related to big data of visual analytics for human detection and recognition problems. For this deep learning architecture and its components are incorporated as an outstanding resources. The vital problem is focused on high cost hardware requirement, challenging demand of global features and computational algorithms to deal with massively unstructured data. So, this problem is closely considered as a multi-discipline of high performance computing of data analytics related to multimedia activity recognition and information processing.

The novelty of this work can be summarized as a multidiscipline of research that correlates visual activity recognition from large scale unconstraint visual data with big data issues and portraits deep learning architecture as a high performance computing resources. The hardware computing resources like NVidia GTX are not economical. On the other hand, some fast network training schemes are also highlighted which can incorporate to deal with the economy of high cost processing devices. In the background of this work, cost effective processing sounds as a future challenge as it is demanded several GPUs and CPUs to resolve the problems of multimedia visual detection and classification.

References

1. Gantz, J., Reinsel, D.: Extracting Value from Chaos. EMC, Hopkinton (2011)
2. Dean, J., Corrado, G., Monga, R., Chen, K., Devin, M., Mao, M., Ng, A.Y.: Large scale distributed deep networks. In: Advances in Neural Information Processing Systems, pp. 1223–1231 (2012)
3. Krizhevsky, A., Sutskever, I., Hinton, G.E.: ImageNet classification with deep convolutional neural networks. In: Advances in Neural Information Processing Systems, pp. 1097–1105 (2012)
4. Donahue, J., Anne Hendricks, L., Guadarrama, S., Rohrbach, M., Venugopalan, S., Saenko, K., Darrell, T.: Long-term recurrent convolutional networks for visual recognition and description. In: Proceedings of the IEEE Conference on Computer Vision and Pattern Recognition, pp. 2625–2634 (2015)
5. Ma, C.Y., Chen, M.H., Kira, Z., AlRegib, G.: TS-LSTM and temporal-inception: exploiting spatiotemporal dynamics for activity recognition. arXiv preprint arXiv:1703.10667 (2017)
6. Fernando, B., Gavves, E., Oramas, J.M., Ghodrati, A., Tuytelaars, T.: Modeling video evolution for action recognition. In: Proceedings of the IEEE Conference on Computer Vision and Pattern Recognition, pp. 5378–5387 (2015)
7. Song, Y., Morency, L.P., Davis, R.: Action recognition by hierarchical sequence summarization. In: Proceedings of the IEEE Conference on Computer Vision and Pattern Recognition, pp. 3562–3569 (2013)
8. Can, E.F., Manmatha, R.: Formulating action recognition as a ranking problem. In: Proceedings of the IEEE Conference on Computer Vision and Pattern Recognition Workshops, pp. 251–256 (2013)
9. Huang, F., Ash, J., Langford, J., Schapire, R.: Learning deep ResNet blocks sequentially using boosting theory. arXiv preprint arXiv:1706.04964 (2017)
10. Hochreiter, S., Schmidhuber, J.: Long short-term memory. Neural Comput. 9(8), 1735–1780 (1997)
11. Liu, W., Anguelov, D., Erhan, D., Szegedy, C., Reed, S., Fu, C.Y., Berg, A.C.: SSD: single shot multibox detector. In: European Conference on Computer Vision, pp. 21–37. Springer, Cham (2016)
12. Ren, S., He, K., Girshick, R., Sun, J.: Faster r-CNN: towards real-time object detection with region proposal networks. IEEE Trans. Pattern Anal. Mach. Intell. 39(6), 1137–1149 (2017)
13. Redmon, J., Divvala, S., Girshick, R., Farhadi, A.: You only look once: unified, real-time object detection. In: Proceedings of the IEEE Conference on Computer Vision and Pattern Recognition, pp. 779–788 (2016)
14. Girshick, R., Donahue, J., Darrell, T., Malik, J.: Region-based convolutional networks for accurate object detection and segmentation. IEEE Trans. Pattern Anal. Mach. Intell. 38(1), 142–158 (2016)
15. Jia, Y., Shelhamer, E., Donahue, J., Karayev, S., Long, J., Girshick, R., Darrell, T.: Caffe: convolutional architecture for fast feature embedding. In: Proceedings of the 22nd ACM International Conference on Multimedia, pp. 675–678. ACM (2014)
16. Redmon, J., Farhadi, A.: YOLO9000: better, faster, stronger. arXiv preprint arXiv: 1612.08242 (2016)

17. Nair, V., Hinton, G.E.: Rectified linear units improve restricted Boltzmann machines. In: Proceedings of the 27th International Conference on Machine Learning (ICML-2010), pp. 807–814 (2010)

18. Rahmani, H., Mian, A., Shah, M.: Learning a deep model for human action recognition from novel viewpoints. IEEE Trans. Pattern Anal. Mach. Intell. **40**(3), 667–681 (2017)

19. Liu, A.A., Xu, N., Nie, W.Z., Su, Y.T., Wong, Y., Kankanhalli, M.: Benchmarking a multimodal and multiview and interactive dataset for human action recognition. IEEE Trans. Cybern. **47**(7), 1781–1794 (2017)

20. Yu, S., Cheng, Y., Su, S., Cai, G., Li, S.: Stratified pooling based deep convolutional neural networks for human action recognition. Multimed. Tools Appl. **76**(11), 13367–13382 (2017)

21. Liu, A.A., Su, Y.T., Nie, W.Z., Kankanhalli, M.: Hierarchical clustering multi-task learning for joint human action grouping and recognition. IEEE Trans. Pattern Anal. Mach. Intell. **39**(1), 102–114 (2017)

22. Idrees, H., Zamir, A.R., Jiang, Y.G., Gorban, A., Laptev, I., Sukthankar, R., Shah, M.: The THUMOS challenge on action recognition for videos "in the wild". Comput. Vis. Image Underst. **155**, 1–23 (2017)

23. Ji, S., Xu, W., Yang, M., Yu, K.: 3D convolutional neural networks for human action recognition. IEEE Trans. Pattern Anal. Mach. Intell. **35**(1), 221–231 (2013)

24. Xia, L., Chen, C.C., Aggarwal, J.K.: View invariant human action recognition using histograms of 3D joints. In: IEEE Computer Society Conference on Computer Vision and Pattern Recognition Workshops (CVPRW), pp. 20–27. IEEE, June 2012

25. He, K., Gkioxari, G., Dollár, P., Girshick, R.: Mask r-CNN. arXiv preprint arXiv:1703.06870 (2017)

26. Zheng, L., Yang, Y., Tian, Q.: SIFT meets CNN: a decade survey of instance retrieval. IEEE Trans. Pattern Anal. Mach. Intell. **14**, 1–20 (2017)

27. Razavian, A.S., Azizpour, H., Sullivan, J., Carlsson, S.: CNN features off-the-shelf: an astounding baseline for recognition. In: Proceedings of the IEEE Conference on Computer Vision and Pattern Recognition Workshops, pp. 806–813 (2014)

28. Chua, L.O., Roska, T.: The CNN paradigm. IEEE Trans. Circuits Syst. I Fundam. Theor. Appl. **40**(3), 147–156 (1993)

29. Girshick, R.: Fast r-CNN. In: Proceedings of the IEEE International Conference on Computer Vision, pp. 1440–1448 (2015)

30. Karpathy, A., Toderici, G., Shetty, S., Leung, T., Sukthankar, R., Fei-Fei, L.: Large-scale video classification with convolutional neural networks. In: Proceedings of the IEEE Conference on Computer Vision and Pattern Recognition, pp. 1725–1732 (2014)

31. Naresh Babu, K.V., Edla, D.R.: New algebraic activation function for multi-layered feed forward neural networks. IETE J. Res. **63**(1), 71–79 (2017)

32. Dahl, G.E., Sainath, T.N., Hinton, G.E.: Improving deep neural networks for LVCSR using rectified linear units and dropout. In: IEEE International Conference on Acoustics, Speech and Signal Processing (ICASSP), pp. 8609–8613. IEEE (2013)

33. Xu, B., Wang, N., Chen, T., Li, M.: Empirical evaluation of rectified activations in convolutional network. arXiv preprint arXiv:1505.00853 (2015)

34. Clevert, D.A., Unterthiner, T., Hochreiter, S.: Fast and accurate deep network learning by exponential linear units (ELUs). arXiv preprint arXiv:1511.07289 (2015)

35. Jin, X., Xu, C., Feng, J., Wei, Y., Xiong, J., Yan, S.: Deep learning with s-shaped rectified linear activation units. In: AAAI, pp. 1737–1743 (2016)

An Automated Support Tool to Compute State Redundancy Semantic Metric

Dalila Amara[✉], Ezzeddine Fatnassi, and Latifa Rabai

SMART Laboratory, Institut Supérieur de Gestion de Tunis, Université de Tunis, Tunis, Tunisia
dalilaa.amara@gmail.com

Abstract. Semantic metrics are quantitative measures of software quality characteristics based on semantic information extracted from the different phases of the software process. The empirical validation of these metrics is necessary required to consider them as quality indicators; which can't be achieved only through their automatic computing based on the appropriate software tools. However, some semantic metrics are only based on theoretical formulation and require further empirical studies and experiments to validate and exploit them. This paper will take into consideration one of the theoretical metrics to be automatically calculated using various basic programs. The experimental results show that automatical computing of this metric is beneficial and fruitful in two sides. On one side, it has an efficient role in computing semantic metrics from the program functional attitude. On the other side, this step is essential to empirically validate this metric as a software quality indicator.

Keywords: Semantic metrics · State redundancy metric
Semantic metrics tools

1 Introduction

Software measurement is the process of collecting information about an entity (program code, completed project) in order to describe it in a quantitative way [1]. One of the useful methods to measure software quality is software metrics which are generally classified into two syntactic and semantic categories. Syntactic metrics are related to the syntax' code to measure internal attributes i.e. cohesion and complexity [2, 3]. Concerning semantic metrics, they are independent on the code structure which makes it possible to calculate them from earlier phases (requirement and design). Additionally, a deep understanding of the software functionality is compulsory needed. That's why their computation is tough and hard [4, 5].

There exist many semantic metrics in the literature. Most of them as those proposed by Etzkorn and Delugach [6] and Stein et al. [4] are focusing on measuring internal characteristics like cohesion and complexity. However, few of them such the suite proposed by Mili et al. [7] insists on measuring external attributes like reliability. To the best of our knowledge, there is no automated tool in the literature to compute this metric. Consequently, this gap in the literature limits the validation of the state redundancy metric.

© Springer International Publishing AG, part of Springer Nature 2018
A. Abraham et al. (Eds.): ISDA 2017, AISC 736, pp. 262–272, 2018.
https://doi.org/10.1007/978-3-319-76348-4_26

As a remedy to the previous problem, we will take into consideration in this paper, one of the Mili et al. metrics [7] suite: State Redundancy to suggest an automated tool to compute it which will be useful in our future works.

This paper is organized as follows. In Sect. 2, we present a literature review of the most proposed semantic metrics and the way of their computation in order to identify which of them is still theoretically computed. Section 3 describes the state redundancy metric, its objective and the way of its computation. In Sect. 4, we provide our contribution and the results of the different experiments. Conclusion and perspectives will be drown in Sect. 5.

2   Related Work and Motivation

In this section, we will emphasize on proposed semantic metrics as well as different tools proposed for their calculation.

As noted above, there exist different semantic metrics. We started with Voas and Miller [8] who proposed a semantic metric called Domain Range Ratio (DDR) in order to minimize hidden functions in testing phase. It was followed by two major suites which are suggested by Etzkorn and Delugach [6] and Stein et al. [4, 5] in order to measure object oriented (OO) properties like cohesion, coupling and complexity. Additionally, conceptual cohesion of classes (C3) suite is proposed by Marcus and Poshyvanyk [9]. Recently, another suite is added to the previous ones in which Mili et al. [7, 10] objective is to measure a program's ability to detect errors and avoid failure.

Proposed metrics certainly need the appropriate tools to collect them. By contrast to syntactic metrics which are computed from the structure of the code, semantic ones require knowledge based program understanding and natural language processing (NLP) techniques to analyze the program functionality ignoring its syntax [11].

NLP techniques aim at understanding the human speech through lexical, syntactic and semantic analysis. They might be used in the different software development phases. Part Of Speech (POS) technique [12], parsers like Sleator and Temperley [12], Standford parser [13] and OpenNLP parser [14] are examples of NLP techniques.

The program understanding approach aims at understanding the functionality of each software component. It consists on extracting the identifiers and function names from the source code based on knowledge-based NL-understanding systems that aid in program understanding and metrics analysis [12]. The most common are the following:

DESign Information Recovery Environment (DESIRE) system: stores domain knowledge in the form of concepts within a semantic net. Natural language tokens, such as identifiers and keywords from comments, are compared to low level domain concepts [12].

Program Analysis Tool for Reuse (PATRicia) system aims at analyzing available comments and identifiers in the code source. It involves a module called the Conceptual Hierarchy for Reuse Including Semantics (CHRiS) in order to understand the program and extract information. This module supplies us with a report comporting concepts and keywords used to compute semantic metric [15].

SemMet is based on the Etzkorn's PATRicia system [12]. The knowledge-base format of semMet includes an interface layer consisting of keywords tagged with parts of speech and a conceptual graph layer.

To sum up, most of the proposed semantic metrics, their objective, their use through the development process as well as their computation tools are identified in Table 1.

Table 1. Semantic metrics and correspondent tools.

Semantic metrics	Objective	Metric' use through development phases	Correspondent tool
DRR metric [8]	Identify errors in the code [8]	Proposed to be applied in the specification phase [8]	Dynamic Error Flow Analysis (DEFA) system [8]
Etzkorn and Delugash suite [6]	Measures OO attributes i.e. complexity and cohesion [6, 12]	Proposed to be used in the code level. Then adopted to be computed from the design phase [6, 12]	PATRicia system [6, 12]
Stein et al. suite [4, 5]	measure different quality attributes i.e. complexity, cohesion, etc. [4]	Proposed and used in the code level then adopted to be computed from the design phase [4]	SemMet tool [4, 5]
Marcus and Poshyvanyk [9]	Proposed to measure the conceptual cohesion of classes [9, 16]	Proposed firstly to be used in the code phase [9, 16]	A tool support for C++ software projects in MS Visual Studio .NET is implemented [9, 16]
Mili et al. [7]	Measuring the program's ability to detect errors at run-time and avoid failures based on program's specification [10]	Proposed to be used in the code level [7]	An automated tool still required

The Table 1 shows that the last metric suite [7] is incomplete since an automated support is missed. This suite is important because it belongs to the first suites that take into account the measurement of external attributes. This concept motivates us to propose an automated tool in order to compute this suite.

As the suite consists of four major metrics which are state redundancy, functional redundancy, non determinacy and non-injectivity, we will focus in this study only on the automatic computation of the state redundancy metric. The others will be analyzed in the future papers.

3   The State Redundancy Semantic Metric

In this section, we will focus on the state redundancy metric by presenting its objective and formulation.

3.1   Definition and Objective

The state redundancy reflects the excess data in the representation of a state and proposed to be used for software error detection. The idea comes from the fact that valid program states are not always represented by all elements in a space S. For example, the age of an employee is generally declared as an integer variable type. However, only a restrict range i.e. between 0 and 120 is really needed [7, 10]. The main idea of this metric may be summarized in Fig. 1.

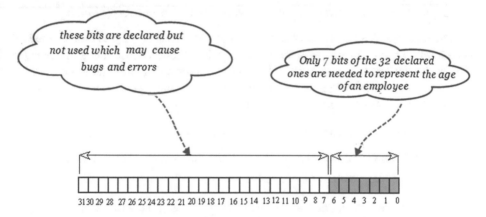

Fig. 1. The range of bits really required to represent the age of an employee

3.2   Formulation

According to Mili et al. [7], the state redundancy of a program is the interval defined by the state redundancy of its initial state (the set of values that the declared program variables may take), and the state redundancy of its final state (the set of states that the program may be in). It is defined for a program g by:

$$K(g) = [k(\sigma 1) \dots k(\sigma F)], \tag{1}$$

where:

$K(\sigma 1)$: State redundancy of the initial state

$$K(\sigma 1) = H(S) - H(\sigma 1),$$

where:
$H(S)$: declared state space of the program g

H(σ1): state space of σ1, σ1 is the initial state of the program

$$k(\sigma) = H(S) - H(\sigma)$$

Assumption: variables used in the program are 32 bits of width.

3.3 Detailed Steps

Based on the theoretical definition of Mili et al. suite [7, 10], we state that the state redundancy calculation process may be done in four basic steps

The first step consists on computing the state space H(S) of the program which is the maximum value (size in bits) that the declared program variables may take [7]. It is theoretically defined as:

$$H(S) = \text{number of declared variables} * \text{word size in the memory}(32 \text{ bits}) \qquad (2)$$

The second step aims at computing the initial state space of the program denoted by H(σ1). It is the actual range of states or simply the exactly needed range of states (in bits) at the starting of the program as it is shown in formula (3). In the same step, the final state space H(σf) which is the number of bits required to store the result of the problem execution has to be computed using formula (4).

$$H(\sigma 1) = \text{number of declared variables} * \text{the range value of these variables} \qquad (3)$$

$$H(\sigma f) = \text{number of variables as the result of the program execution}$$
$$* \text{range value of these variables} \qquad (4)$$

In the third step, the initial state redundancy of the program K(σ1) which is the gap between the minimal bandwidth required to store the program state H(S) and the actual bandwidth reserved to that effect might be deduced using formula (5). K(σ1) can simply be defined as the number of bits which are partially used from the declared ones. Similarly, it is possible in this step to compute the final state space of the program K(σf) which is the maximum bandwidth of relationships that hold between program variables as a result of the execution of the program as indicated by formula (6).

$$K(\sigma 1) = H(S) - H(\sigma 1) \qquad (5)$$

$$K(\sigma f) = H(S) - H(\sigma f) \qquad (6)$$

In the final step, it's possible to deduce the state redundancy metric of a program g denoted by K(g) based on formulas (5) and (6) as shown in formula (7).

$$K(g) = [K(\sigma 1) \cdots K(\sigma f)] \qquad (7)$$

Computing the state redundancy metric from the program functionality attitude makes it a semantic measure. Moreover, this interval is helpful to identify the number of bits losses which are declared but not used by the program. They will result in bugs in the program execution which in turn will lead to errors.

4   Automatic Derivation of State Redundancy (SR) Metric

To compute the state redundancy metric, we resort to Java language and the Eclipse development environment (version: Neon.3 Release (4.6.3)). The used experiment consists on using four basic programs to check the possibility and the correctness to calculate this metric. The selected programs are the Greatest Common Divisor (GCD), the Selection Sort program (SSort), the Palindrome program and the maximum value program (MaxValue). There are different motivations behind these programs. One motivation is that the two first GCD and SSort programs are used by Mili et al. [7] to theoretically compute the SR metric. We use the same programs to automatically compute this metric in order to verify the obtained results. One other motivation is that these programs represent the first steps to validate this metric. Moreover, they support the usual fundamental data types: integer, Boolean, Character and string types which are needed to show the possibility to compute this metric for different data types.

4.1   Computing State Redundancy for GCD Program

The previous steps which are theoretically defined will be exploited to automatically calculate the SR metric for the GCD program. Consequently, our effort to shift the computing from theoretical to automatic is shown in the following four steps:

To begin with, we compute the state space H(S) of the GCD based on the Integer.**MAX_VALUE** function which gives us the maximum value an integer can have ($2^{31} - 1$.) as presented in lines 11 and 12 of our program (Fig. 2). It is the maximum value (size in bits) that the declared program variables x and y may take.

```
11      int values= Integer.MAX_VALUE;
12      int statespace=2*sizeOfBits(values);

16      int initialsizex= sizeOfBits(x);
17      int initialsizey= sizeOfBits(y);
18      int initialstatespace=initialsizex+initialsizey;

37      int initialstateredundancy = statespace-initialstatespace;
38      int finalstateredundancy = statespace-finalstatespace;
```

Fig. 2.   Part of the GCD program for State Redundancy computing

Secondly, we determine the initial state space of GCD H(σ1). To achieve this goal, we resort to **sizeOfBits** (int value) function which provides us with the exact number of bits effectively used from the 32 word size in the memory as indicated from line 16 to

18 (Fig. 2). The same function is used to compute the final state space $H(\sigma f)$ of the GCD program.

Third, we compute the initial and the final state redundancy of the program by applying previous formulas (5) and (6) as it is shown in lines 37 and 38 (Fig. 2).

Eventually, we deduce the state redundancy semantic metric K(GCD) by applying the formula (7).

Results

To test our proposed automatic computation tool, we consider two integers x = 8 and y = 20. The obtained result is presented in Fig. 3:

```
state space H(S):62
the initial state space H(σ1):9
the final state space H(σf):6

the state redundancy K(GCD) is:[53Bits...56Bits]
```

Fig. 3. State Redundancy metric for GCD program

Through this interval, we can identify the inappropriate use of the declared bits since we have between 55 bits and 58 bits which are left from the declared 32 bits. As noted above, these unused bits may cause bugs and errors.

4.2   Computing State Redundancy for SSort Program

Another example we consider in this study is the SSort program. The automatic computation of the SR metric for this program will be also based on the four basic steps revealed previously.

First, we compute the state space of the program H(S). Since in our example, we use an integer array, we start by computing the maximum value an integer can have ($2^{31} -$ 1.) based on the **Integer.Maxvalue** function. As we use a table of more than one integer, we need to exploit the **sizeOfBits (int)** function that gives the size in bits of an integer multiplied by the size of the table as indicated in lines 10 and 11 of Fig. 4.

```
10  int values= Integer.MAX_VALUE; //the declared state space H(s)
11  int statespace=13*sizeOfBits(values);

17     int Initialactualsize=0;
18
19  for (int i = 0; i < table.length; i++) {
20      Initialactualsize=Initialactualsize+sizeOfBits(table[i]);
21  }
22
23  Initialactualsize=Initialactualsize+sizeOfBits(c)+sizeOfBits(d)+sizeOfBits(p)+sizeOfBits(swap);

58  int borninf = statespace -(Finalactualsize);
59  int bornsup=statespace-(Initialactualsize);
```

Fig. 4. Part of the SSort program for State Redundancy computing

Second, we compute the initial state redundancy by applying the formula (3) as indicated from line 17 to 23 (Fig. 4). Similarly, we compute the final state space by applying the formula (4).

Third, once we obtain the sorted table, we can deduce the initial and the final state redundancy by applying respectively the formulas (5) and (6) as indicated in lines 58 and 59 (Fig. 4).

Finally, we deduce the state redundancy semantic metric K(SSort) by applying the formula (7).

Results

The SR of the sort function is presented in Fig. 5.

```
the state space is 403
the actual sate space is 76
the final sate space is 97
the state redundancy is the interval :[306...327]
```

Fig. 5. State Redundancy metric for SSort program

As presented above, the obtained interval shows the number of bits which are declared but does not used by the program. They can cause bugs and errors.

4.3 Computing State Redundancy for MaxValue Program

The SR of the MaxValue program is computed on the same way as the previous programs. Firstly we compute the state space of the program H(S) as it is indicated in lines 11 and 12 (Fig. 6). Secondly, the initial state space is computed as indicated in line 14 (Fig. 6). Similarly, we compute the final state space by applying the formula (4). Finally the state redundancy metric is deduced by applying the formulas (5)–(7).

```
11   int a = 0,b = 0,max = 0;
12   int maxSizeInBits=32*3;

14   int initialState=maxSizeInBits-(sizeOfBits(a)+sizeOfBits(b)+sizeOfBits(max));
```

Fig. 6. Part of the MaxValue program for State Redundancy computing

Results

We consider for example two values which are 15 and 12. The maximum value is 15 and the SR is presented as follows (Fig. 7).

```
the  maximum value is 15
The state redunduncy interval is [84,96]
```

Fig. 7. State Redundancy metric for MaxValue program

The interval shown in Fig. 7 gives the number of unneeded bits which may cause bugs and errors.

4.4 Computing State Redundancy for the Palindrome Program

Java palindrome program is used to check if a string is a palindrome or not (if it remains unchanged when reversed or not). We choose this example to show that the state redundancy metric may be computed for different programs which manipulate different data types.

To compute this metric for the Palindrome program, we proceed as the same manner used in the previous examples. Hence we start by computing the state space of the program as indicated in lines 12 and 13 (Fig. 8). After that, we compute the initial and the final states from which we deduce the final interval as indicated in lines 38 and 39 (Fig. 8).

```
12   BigInteger maximum= new BigInteger(String.valueOf(Integer.MAX_VALUE));
13   maximum=maximum.multiply(new BigInteger(String.valueOf(2)));


38   BigInteger BornSup= maximum.subtract(initBig);
39   BigInteger BornInf= maximum.subtract(finBig );
```

Fig. 8. Part of the Palindrome program for State Redundancy computing

Results

The state redundancy of the Palindrome program is presented in Fig. 9.

```
Enter a string to check if it is a palindrome
abccba
Entered string is a palindrome.
The interval is  [4294966398 , 4294966654]

Enter a string to check if it is a palindrome
abcd
Entered string is not a palindrome.
The interval is  [4294966526 , 4294966654]
```

Fig. 9. State Redundancy metric for the Palindrome program

The obtained interval shows that there exist a number of bits which are declared but does not used by the program. They can cause bugs and errors.

5 Conclusion and Perspectives

Although semantic metrics aren't a new concept, we notice that only a few studies are focusing on exploiting them as software quality indicators. The survey we performed showed that there exist some newly proposed semantic metrics which aims at measuring external attributes like reliability. However, these metrics are still incomplete since they are only theoretically presented.

As a remedy, we proposed in this paper an automated tool to compute the state redundancy metric which is proposed as a measure of error detection. We started by presenting the theoretical way to compute it, then, we exploited different experiments which consist of four basic java programs to compute this metric in an automated way.

Three major benefits are driven from the automatic calculation of this metric. The most important merit is that validating one metric as a software quality measure requires its automatic calculation. Equally important, thanks to this tool we can exploit the functional attitude of the program to compute semantic metrics. The last advantage but not the least is that identifying the state redundancy interval of bits enhances developers to optimize their programs in order to avoid bugs and errors.

Although these benefits, we could enhance the quality of this work by adapting it to other bigger java projects. In fact, we are currently working on computing this metric for open source projects like Velocity and Camel systems in order to validate it as a measure of the defect density attribute.

References

1. Fenton, N., Pfleeger, S.L.: Software Metrics: A Rigorous & Practical Approach, 2nd edn. International Thomson Computer Press, London (1997)
2. Chidamber, S.R., Kemerer, C.F.: A metrics suite for object oriented design. IEEE Trans. Softw. Eng. **20**(6), 476–493 (1994)
3. Li, W.: Another for object-oriented programming metric suite. J. Syst. Softw. **44**(2), 155–162 (1998)
4. Stein, C., Etzkorn, L., Cox, G., Farrington, P., Gholston, S., Utley, D., Fortune, J.: A new suite of metrics for object-oriented software. In: Proceedings of the 1st International Workshop on Software Audits and Metrics, Portugal, pp. 49–58 (2004)
5. Stein, C., Etzkorn, L., Gholston, S., Farrington, P., Utley, D., Cox, G., Fortune, J.: Semantic metrics: metrics based on semantic aspects of software. Appl. Artif. Intell. **23**(1), 44–77 (2009)
6. Etzkorm, L., Delugach, H.: Towards a semantic metrics suite for object-oriented design. In: 34th International Conference on Technology of Object-Oriented Languages and Systems, USA, pp. 71–80 (2000)
7. Mili, A., Jaoua, A., Frias, A.: Semantic metrics for software products. Innov. Syst. Softw. Eng. **10**(3), 203–217 (2014)
8. Voas, J.M., Miller, K.: Semantic metrics for software testability. J. Syst. Softw. **20**(3), 207–216 (1993)
9. Marcus, A., Poshyvanyk, D.: The conceptual cohesion of classes. In: Proceedings of the 21st International Conference on Software Maintenance, Budapest, pp. 133–142 (2005)
10. Mili, A., Tchier, F.: Software Testing: Concepts and Operations, 2nd edn. Wiley, New Jersey (2015)
11. Cox, G.W., Gholston, S.E., Utley, D.R., Etzkorn, L.H., Gall, C.S., Farrington, P.A., Fortune, J.L.: Empirical validation of the RCDC and RCDE semantic complexity metrics for object-oriented software. J. Comput. Inf. Technol. (CIT) **15**(2), 151–160 (2007)

12. Etzkorn, LH.: A metrics-based approach to the automated identification of object-oriented reusable software components. Doctoral Dissertation, University of Alabama in Huntsville (1997)
13. Wang, Y.: Semantic information extraction for software requirements using semantic role labeling. In: IEEE International Conference on Progress in Informatics and Computing (PIC), pp. 332–337 (2015)
14. Ibrahim, M., Ahmad, R.: Class diagram extraction from textual requirements using natural language processing (NLP) techniques. In: Proceedings of Second International Conference on Computer Research and Development, pp. 200–204. IEEE (2010)
15. Etzkorn, L., Gholston, S., Hughes, W.E.: A semantic entropy metric. J. Softw. Maint. Evol. Res. Pract. **14**(5), 293–310 (2002)
16. Marcus, A.: Using the conceptual cohesion of classes for fault prediction in object- oriented systems. IEEE Trans. Softw. Eng. **34**(2), 287–300 (2008)

Computing Theory Prime Implicates
in Modal Logic

Manoj K. Raut$^{(\boxtimes)}$, Tushar V. Kokane, and Rishabh Agarwal

Dhirubhai Ambani Institute of Information and Communication Technology,
Gandhinagar 382007, India
manoj_raut@daiict.ac.in, tushar.kokane15@gmail.com,
agarwalrishabh98@gmail.com

Abstract. The algorithm to compute theory prime implicates, a generalization of prime implicates, in propositional logic has been suggested in [9]. As a preliminary result, in this paper we have extended that algorithm to compute theory prime implicates of a modal knowledge base X with respect to another modal knowledge base $\Box Y$ using [1], where Y is a propositional knowledge base and $X \models Y$ in modal system $\mathcal{T}$ and we have also proved its correctness. We have also proved that it is an equivalence preserving knowledge compilation and the size of theory prime implicates of X with respect to $\Box Y$ is less than the size of the prime implicates of $X \wedge \Box Y$. We have also extended the query answering algorithm in modal logic.

Keywords: Modal logic · Theory prime implicates
Knowledge compilation

1 Introduction

Propositional entailment problem is a fundamental issue in artificial intelligence due to its high complexity. Determining whether a query logically follows from a given knowledge base is intractable [4] in general as every known algorithm runs in time exponential in the size of the given knowledge base. To overcome such computational intractability, the propositional entailment problem is split into two phases such as off-line and on-line. In the off-line phase, the original knowledge base X is compiled into a new knowledge base $X^{'}$ and in on-line phase queries are actually answered from the new knowledge base in time polynomial in their size. In such type of compilation most of the computational overhead shifted into the off-line phase, is amortized over large number of on-line query answering. The off-line computation is called *knowledge compilation*.

Several approaches of knowledge compilation in propositional logic, first order logic and modal logic has been suggested so far in literature [5,7,8,10,11]. The first kind of approach consists of an equivalence preserving knowledge compilation. In such an approach, the knowledge base X is compiled into another equivalent knowledge base $\Pi(X)$, called the prime implicates of X with respect

© Springer International Publishing AG, part of Springer Nature 2018
A. Abraham et al. (Eds.): ISDA 2017, AISC 736, pp. 273–282, 2018.
https://doi.org/10.1007/978-3-319-76348-4_27

to which queries are answered from $\Pi(X)$ in polynomial time. In another approach to equivalence preserving compilation in propositional logic, Marquis suggested the computation of theory prime implicates [9] of a knowledge base X with respect to another knowledge base Y, so that queries can be answered from theory prime implicates in polynomial time.

Most of the work in knowledge compilation have been restricted to propositional logic and first order logic in spite of an increasing intrest in modal logic. Due to lack of expressive power in propositional logic and the undecidability of first order logic, modal logic is required as a knowledge representation language in many problems. Modal logic gives a trade-off between expressivity and complexity as they are more expressive than propositional logic and computationally better behaved than first order logic. An algorithm to compute the set of prime implicates of modal logic $\mathcal{K}$ and $\mathcal{K}_n$ have been proposed in [1] and [2] respectively.

In [9], the notion of prime implicates is generalized to theory prime implicates in propositional logic where the size of theory prime implicate compilation of a knowledge base is always exponentially smaller than the size of its prime implicate compilation. Moreover, query answering from theory prime implicate compilation can be performed in time polynomial in their size. In this paper we extend this concept from propositional to modal logic using the algorithm in [1]. So here we compute the theory prime implicates of a modal knowledge base X with respect to another *restricted* modal knowledge base $\Box Y$, i.e., $\Theta(X, \Box Y)$ where Y is a propositional knowledge base such that $X \models Y$. It can be noted that if $Y = \emptyset$ then $\Theta(X, \Box Y)$ becomes $\Pi(X)$. This paper is a revised version of [12].

The paper is organized as follows. In Sect. 2 We give basic results in modal logic. In Sect. 3 we propose basic definitions of prime implicates, theory prime implicates and we describe the properties of theory prime implicates, the algorithm for computing theory prime implicates and query answering in modal logic. Section 4 concludes the paper.

2   Preliminaries

Let us now discuss the basics of modal logic $\mathcal{K}$ from [3]. The alphabet of modal formulas is $Var \cup \{\neg, \vee, \Diamond, (,)\}$. Var is a countable set of variables denoted by $p, q, r, \ldots$. The connectives $\neg$ and $\vee$ are negation and disjunction. $\Diamond$ is the modal operator 'possible'. The modal formulas MF are defined inductively as follows. Variables are modal formulas. If A and B are modal formulas then $\neg A, A \vee B, \Diamond A$ are modal formulas. For the sake of convenience, we introduce the connectives $A \to B \equiv \neg A \vee B, A \leftrightarrow B \equiv (A \to B) \wedge (B \to A)$. The 'necessary' operator $\Box$ is defined as $\Box A \equiv \neg \Diamond \neg A$. We avoid using parentheses whenever possible. The length of a formula A, written as $|A|$, is defined as the number of occurrences of propositional variables, logical connectives, and modal operators in A. For example, $|(\neg p \vee \neg q)| = 5$ and $|\Box(\neg p \wedge \neg q)| = 6$.

Definition 1. *The semantics of modal logic $\mathcal{K}$ is defined using Kripke models. A Kripke model M is a triple $\langle W, R, v \rangle$ where W is a nonempty set (of worlds),*

R is a binary relation on W called the accessibility relation, so if $(w, w') \in R$ then we say w' is accessible from w, and $v : W \to 2^{Var}$ is a valuation function, which assigns to each world $w \in W$ a subset $v(w)$ of Var such that p is true at a world w iff $p \in v(w)$.

Definition 2. *Given any Kripke model $M = \langle W, R, v \rangle$, a world $w \in W$, and a formula $\phi \in MF$, the truth of ϕ at w of M denoted by $M, w \models \phi$, is defined inductively as follows:*

- $M, w \models p$ *where* $p \in Var$ *iff* $p \in v(w)$,
- $M, w \models \neg \phi$ *iff* $M, w \not\models \phi$,
- $M, w \models \phi \wedge \psi$ *iff* $M, w \models \phi$ *and* $M, w \models \psi$,
- $M, w \models \phi \vee \psi$ *iff* $M, w \models \phi$ *or* $M, w \models \psi$,
- $M, w \models \Box \phi$ *iff for all* $w' \in W$ *with* wRw' *we have* $M, w' \models \phi$,
- $M, w \models \Diamond \phi$ *iff for some* $w' \in W$ *with* wRw' *we have* $M, w' \models \phi$.

We say that a formula ϕ is satisfiable if there exists a model M and a world w such that $M, w \models \phi$ and say ϕ is valid denoted by $\models \phi$ if $M, w \models \phi$ for all M and w. A formula ϕ is unsatisfiable written as $\phi \models \bot$ if there exists no M and w for which $M, w \models \phi$. A formula ψ is a logical consequence of a formula ϕ written as $\phi \models \psi$ if $M, w \models \phi$ implies $M, w \models \psi$ for every model M and world $w \in W$.

There are two types of logical consequences given in [3] in modal logic which are:

1. a formula ψ is a global consequence of ϕ if whenever $M, w \models \phi$ for every world w of a model M, then $M, w \models \psi$ for every world w of M.
2. a formula ψ is a local consequence of ϕ if $M, w \models \phi$ implies $M, w \models \psi$ for every model M and world w.

Eventhough both consequences exist, in this paper we will only study local consequences and whenever $\phi \models \psi$ we mean ψ is a local consequence of ϕ.

Two formulas ϕ and ψ are equivalent written as $\phi \equiv \psi$ or $\models \phi \leftrightarrow \psi$ if both $\phi \models \psi$ and $\psi \models \phi$. A formula ϕ is said to be logically stronger than ψ or ψ is said to be weaker than ϕ if $\phi \models \psi$ and $\psi \not\models \phi$. We can always strengthen a premise and weaken a consequence as $\phi \models \psi$ implies $\phi \wedge \chi \models \psi$ and $\phi \models \psi \vee \chi$ for any formula χ.

It can be noted that in Definitions 1 and 2, if we take R to be a reflexive relation then system $\mathcal{K}$ becomes system $\mathcal{T}$. There are some results in this paper which holds in system $\mathcal{T}$ only. As any theorem of $\mathcal{K}$ is a theorem in $\mathcal{T}$ so every result holding in $\mathcal{K}$ also holds in $\mathcal{T}$.

The definitions of literals, clauses, terms and formulas in modal logic $\mathcal{T}$ known as definition $D4$ in [1] are given below.

Definition 3. *The literals L, clauses C, terms T, and formulas F are defined as follows:*

$$L ::= a \mid \neg a \mid \Box F \mid \Diamond F$$
$$C ::= L \mid C \vee C$$
$$T ::= L \mid T \wedge T$$
$$F ::= a \mid \neg a \mid F \wedge F \mid F \vee F \mid \Box F \mid \Diamond F$$

A formula is said to be in conjunctive normal form (CNF) if it is a conjunction of clauses and it is in disjunctive normal form (DNF) if it is a disjunction of terms. The transformation of a formula to CNF or DNF is exponential in both time and space. The number of clauses in a CNF formula ϕ is denoted as $\text{nb\_cl}(\phi)$.

We now present some basic properties of logical consequences and equivalences in $\mathcal{K}$ which will be used in the proofs of some theorems in our paper.

Lemma 1. *Let ϕ and ψ be modal formulas. Then the following three statements are equivalent.*

(i) $\phi \equiv \psi$
(ii) $\Diamond \phi \equiv \Diamond \psi$
(iii) $\Box \phi \equiv \Box \psi$

Proof. Refer to [12].

We now extend the definition of $\models$ with respect to a formula Z written as $\models_Z$.

Definition 4. *Let X_1, X_2, Z be modal formulas. We define $\models_Z$ over $MF \times MF$ (as the extension of $\models$) by $X_1 \models_Z X_2$ iff $X_1 \wedge Z \models X_2$. When $X_1 \models_Z X_2$ holds then we say that X_2 is a Z-logical consequence of X_1 or X_1 Z-entails X_2 or X_1 entails X_2 with respect to Z. We define the equivalence relation $\equiv_Z$ over MF by $X_1 \equiv_Z X_2$ iff $X_1 \models_Z X_2$ and $X_2 \models_Z X_1$. When $X_1 \equiv_Z X_2$ holds we say X_1 and X_2 are Z-equivalent.*

We now present the following lemmas which will be used in the proofs of Theorems 4 and 5 later.

Lemma 2. *Let ψ, χ, and Y be any modal formulas. Then the following three statements are equivalent with respect to $\models_Y$.*

(i) $\psi \models_Y \chi$
(ii) $\models_Y \neg \psi \vee \chi$
(iii) $\psi \wedge \neg \chi \models_Y \bot$

Proof. Refer to [12].

Lemma 3. *Let ψ and χ be modal formulas, and Y be any propositional formula. Then the following three statements are equivalent in modal system $\mathcal{K}$.*

(i) $\psi \models_Y \chi$.
(ii) $\Diamond\psi \models_{\Box Y} \Diamond\chi$.
(iii) $\Box\psi \models_{\Box Y} \Box\chi$.

Proof. Refer to [12].

The following lemma says that a formula is unsatisfiable with respect to $\Box Y$ if and only if one of the following seven conditions hold. The following Lemma which holds in $\mathcal{K}$ is used in the proof of Theorems 4 and 5.

Lemma 4. *Let* $\beta_1, \beta_2, \ldots, \beta_m, \gamma_1, \gamma_2, \ldots, \gamma_n, \phi_1, \phi_2, \ldots, \phi_q, \xi_1, \xi_2, \ldots, \xi_r$ *be modal formulas and* $\alpha_1, \alpha_2, \ldots, \alpha_l, \psi_1, \psi_2, \ldots, \psi_p, Y$ *be propositional formulas. Then* $(\vee_{i=1}^l \alpha_i) \wedge (\vee_{j=1}^m \Diamond\beta_j) \wedge (\vee_{k=1}^n \Box\gamma_k) \wedge ((\vee_{i=1}^l \alpha_i) \vee (\vee_{j=1}^m \Diamond\beta_j)) \wedge ((\vee_{i=1}^l \alpha_i) \vee (\vee_{k=1}^n \Box\gamma_k)) \wedge ((\vee_{j=1}^m \Diamond\beta_j) \vee (\vee_{k=1}^n \Box\gamma_k)) \wedge ((\vee_{i=1}^l \alpha_i) \vee (\vee_{j=1}^m \Diamond\beta_j) \vee (\vee_{k=1}^n \Box\gamma_k)) \wedge \psi_1 \wedge \ldots \wedge \psi_p \wedge \Box\phi_1 \wedge \ldots \wedge \Box\phi_q \wedge \Diamond\xi_1 \wedge \ldots \wedge \Diamond\xi_r \models_{\Box Y} \bot$ *if and only if*

1. $(\vee_{i=1}^l \alpha_i) \wedge \psi_1 \wedge \ldots \wedge \psi_p \models_Y \bot$ *or*
2. $(\vee_{j=1}^m \beta_j) \wedge \phi_1 \wedge \ldots \wedge \phi_q \models_Y \bot$ *or*
3. $(\vee_{k=1}^n \gamma_k) \wedge \xi_u \wedge \phi_1 \wedge \ldots \wedge \phi_q \models_Y \bot$ *for* $1 \le u \le r$ *or*
4. $((\vee_{i=1}^l \alpha_i) \vee (\vee_{j=1}^m \beta_j)) \wedge \phi_1 \wedge \ldots \wedge \phi_q \models_Y \bot$ *or*
5. $((\vee_{i=1}^l \alpha_i) \vee (\vee_{k=1}^n \gamma_k)) \wedge \xi_u \wedge \phi_1 \wedge \ldots \wedge \phi_q \models_Y \bot$ *for* $1 \le u \le r$ *or*
6. $((\vee_{j=1}^m \beta_j) \vee (\vee_{k=1}^n \gamma_k)) \wedge \phi_1 \wedge \ldots \wedge \phi_q \models_Y \bot$ *or*
7. $((\vee_{i=1}^l \alpha_i) \vee (\vee_{j=1}^m \beta_j) \vee (\vee_{k=1}^n \gamma_k)) \wedge \phi_1 \wedge \ldots \wedge \phi_q \models_Y \bot$.

Proof. Refer to [12].

3   Theory Prime Implicates

Now we give the definitions of prime implicates and prime implicants of a knowledge base X with respect to $\models$ in modal logic.

Definition 5. *A clause C is said to be an implicate of a formula X if $X \models C$. A clause C is a prime implicate of X if C is an implicate of X and there is no other implicate C' of X such that $C' \models C$. The set of prime implicates of X is denoted by $\Pi(X)$.*

Definition 6. *A term C is said to be an implicant of a formula X if $C \models X$. A term C is said to be a prime implicant of X if C is an implicant of X and there is no other implicant C' of X such that $C \models C'$.*

Definition 7. *A clause $C' \in X$ is a minimal element of X if for all $C \in X$, $C \models C'$ implies $C \equiv C'$. Similarly, a clause $C' \in X$ is a minimal element of X with respect to a propositional formula Y if for all $C \in X$, $C \models_Y C'$ implies $C \equiv_Y C'$. We denote the minimal element of X with respect to a formula Y by $min(X, Y)$.*

So we note that prime implicates (or prime implicants) of a knowledge base X are minimal elements with respect to $\models$ among the implicates (or implicants) of X respectively.

We now extend the definition of prime implicate to theory prime implicate with respect to $\models_Y$ as follows.

Definition 8. *Let X and Y be any modal formulas. A clause C is a theory implicate of X with respect to Y iff $X \models_Y C$. A clause C is a theory prime implicate of X with respect to Y iff C is a theory implicate of X with respect to Y and there is no theory implicate C' of X with respect to Y such that $C' \models_Y C$. We denote $\Theta(X, Y)$ as the set of theory prime implicates of X with respect to Y.*

We note that the set of theory prime implicates of X with respect to Y, i.e., $\Theta(X, Y)$, is the minimal elements with respect to $\models_Y$ among the set of theory implicates of X with respect to Y.

As a preliminary result, in this paper we discuss the properties and the computation of $\Theta(X, \Box Y)$, i.e., the theory prime implicates of a modal formula X with respect to a restricted modal formula $\Box Y$ where Y is a propositional formula. We consider computing $\Theta(X, Z)$, i.e., the theory prime implicates of X with respect to an arbitrary modal formula Z as a future research work.

3.1   Properties of Theory Prime Implicates

Below we list some of the properties of theory prime implicates.

Lemma 5. *Let X be a modal formula and Y be any propositional formula. Then $\Theta(X, \Box Y) \subseteq \Pi(X \wedge \Box Y)$.*

Proof. Refer to [12].

Lemma 6. *If $C_1, C_2 \in \Pi(X \wedge \Box Y)$ and $C_1 \models_{\Box Y} C_2$ then $C_2 \notin \Theta(X, \Box Y)$.*

Proof. Refer to [12].

So we conclude from Lemmas 5 and 6 that the set of theory prime implicates of X with respect to $\Box Y$ can be defined from the set of prime implicates of $X \wedge \Box Y$ as follows:

Theorem 1. $\Theta(X, \Box Y) = min(\Pi(X \wedge \Box Y), \models_{\Box Y})$

The above theorem is used in proving the correctness of computation of theory prime implicate algorithm $MODALTPI$.

The following theorem says that the set of theory prime implicates of X with respect to $\Box Y$ captures all the theory implicates of X with respect to $\Box Y$. It is useful in proving the correctness of query answering algorithm QA later.

Theorem 2. *Let X and Y be modal formulas and C be a clause. Then $X \models_{\Box Y} C$ holds if and only if there is a theory prime implicate C' of X with respect to $\Box Y$ such that $C' \models_{\Box Y} C$ holds.*

Proof. Refer to [12].

The following theorem is a metalogical property of prime implicates.

Lemma 7. *Let X and X' be formulae in $\mathcal{T}$. Then $X \equiv X'$ if and only if $\Pi(X) \equiv \Pi(X')$.*

Proof. It is easy to prove.

The following theorem is a metalogical property of theory prime implicates.

Theorem 3. *Suppose X, X', are formulae in $\mathcal{T}$ and Y, Y' be any propositional formulae. If $X \equiv_{\Box Y} X'$ and $Y \equiv Y'$ then $\Theta(X', \Box Y') = \Theta(X, \Box Y)$.*

Proof. Refer to [12].

When we prove the equivalence preserving knowledge compilation in Theorem 7 we use the following lemma.

Lemma 8. *Let X be a modal formula and Y be any propositional formula. Then $X \equiv_{\Box Y} \Theta(X, \Box Y)$.*

Proof. Refer to [12].

The following result which holds in $\mathcal{T}$ shows that weakening the consequence Y does not increases the number of clauses of $\Theta(X, \Box Y)$.

Theorem 4. *Let X be a modal formula and Y, Y' be any propositional formulas such that $X \models Y$ and $Y \models Y'$. For every $\pi' \in \Theta(X, \Box Y')$ there exists a $\pi \in \Theta(X, \Box Y)$ such that $\pi' \equiv_{\Box Y'} \pi$. Consequently, $nb\_cl(\Theta(X, \Box Y')) \leq nb\_cl(\Theta(X, \Box Y))$.*

Proof. Refer to [12].

The following result which holds in $\mathcal{T}$ shows that the size of the theory prime implicates of X with respect to $\Box Y$, i.e., $\Theta(X, \Box Y)$, is always smaller than the size of the prime implicates of $X \wedge \Box Y$, i.e., $\Pi(X \wedge \Box Y)$ which is an advantage to our compilation.

Theorem 5. *Let X be a modal formula and Y be any propositional formula. For every $\pi' \in \Theta(X, \Box Y)$ there exists a $\pi \in \Pi(X \wedge \Box Y)$ such that $\pi' \equiv_{\Box Y} \pi$. Consequently, $nb\_cl(\Theta(X, \Box Y)) \leq nb\_cl(\Pi(X \wedge \Box Y))$.*

Proof. Refer to [12].

Remark 1. Like Theorem 5 we can also prove that for every $\pi' \in \Theta(X, \Box Y)$ there exists a $\pi \in \Pi(X \wedge Y)$ such that $\pi' \equiv_{\Box Y} \pi$. Consequently, $nb\_cl(\Theta(X, \Box Y)) \leq nb\_cl(\Pi(X \wedge Y))$.

3.2   Algorithm for Computing Theory Prime Implicates

Let us now present the algorithm for computation of theory prime implicates. The following algorithm is based on the Bienvenu's algorithm [1]. First, our algorithm computes the theory implicates of X with respect to $\Box Y$ using the algorithm in [1] which is called as the set $CANDIDATES$. Here Y is a propositional formula such that $X \models Y$. The assumption $X \models Y$ is considered in Definition 9 below. Then it removes logically entailed clauses with respect to $\models_{\Box Y}$ to get theory prime implicates of X with respect to $\Box Y$.

Algorithm $MODALTPI(X, \Box Y)$

Input: Two formulas X and Y where X is a modal formula and Y is any
          propositional formula
Output: Set of theory prime implicates of X with respect to $\Box Y$
begin
    Compute CANDIDATES for $X \wedge \Box Y$
    Remove π_j from CANDIDATES if $\pi_i \models_{\Box Y} \pi_j$ for some π_i in CANDIDATES
    Return CANDIDATES$(=\Theta(X, \Box Y))$
end

Theorem 6. *The algorithm MODALTPI terminates.*

Proof. Refer to [12].

The correctness of the above algorithm $MODALTPI$ follows from Theorems 1 and 6.

3.3   Theory Prime Implicate Compilation

A formula is said to be tractable if its satisfiability can be solved by a computer algorithm in polynomial time.

Definition 9. *Let X be a modal formula and Y be any propositional formula such that $X \models Y$ and $\Box Y$ be tractable. The theory prime implicate compilation of X with respect to $\Box Y$ is defined as $\Omega_{\Box Y}(X) = \Theta(X, \Box Y) \cup \Box Y$.*

Theorem 7. $\Omega_{\Box Y}(X) \equiv X$.

Proof. Refer to [12].

Algorithm $QA(\Omega_{\Box Y}(X), Q)$

Input: The theory prime implicate compilation $\Omega_{\Box Y}(X)$ and a clausal query Q
Output: either $X \models Q$ or $X \not\models Q$
begin
    if $\pi' \models_{\Box Y} Q$ for any $\pi' \in \Theta(X, \Box Y) \cup \Box Y$
        then return $X \models Q$

```
        else
            return X ⊭ Q
    end
```

The correctness of the above algorithm follows from Theorems 2 and 6. Let us now see how query answering can be performed in polynomial time.

Theorem 8. *Let X be a modal formula and Y be any propositional formula such that $X \models Y$ and $\Box Y$ is tractable. So checking whether $\pi' \models_{\Box Y} Q$ holds in algorithm QA can be done in time $O(|\Box Y \cup Q|^m)$ where m is the number of literals in π' and answering a query in algorithm QA can be performed in time $O(|\Theta(X, \Box Y) \cup \Box Y| * |\Box Y \cup Q|^m))$.*

Proof. Refer to [12].

Remark 2. The set of theory prime implicates, i.e., $\Theta(X, \Box Y)$ is easily exponential with respect to $|X \wedge \Box Y|$ by [7] but if we have exponential number of queries then obviously each query can be answered from $\Theta(X, \Box Y)$ in polynomial time in their size. As in query answering algorithm QA, we check whether $\pi' \models_{\Box Y} Q$ for every $\pi' \in \Theta(X, \Box Y) \cup \Box Y$ and moreover query answering with respect to $\Theta(X, \Box Y)$ is polynomial so if $\Box Y$ becomes tractable then query answering can be done in polynomial time. As $\Box Y$ is a modal Horn clause and by [6] modal Horn clauses are tractable in $S5$ so $\Box Y$ is tractable. So query answering from $\Omega_{\Box Y}(X)$ can be done in polynomial time.

We have proved theorems in $\mathcal{K}$ and in $\mathcal{T}$. Proofs of Theorems 4 and 5 have used formulas in $\mathcal{T}$ as can be seen from the proof. As all the theorems of $\mathcal{K}$ are theorems of $\mathcal{T}$, so all the results proved above hold in $\mathcal{T}$.

Example 1. Consider a formula $X = (p_1 \vee p_2) \wedge \Diamond\Box\neg p_3 \wedge \Box\Diamond p_2$. We take $Y = p_1 \vee p_2 [= X \setminus (\Diamond\Box\neg p_3 \wedge \Box\Diamond p_2)]$. Clearly $X \models Y$. $\Box Y = \Box(p_1 \vee p_2)$ and $\Box Y$ is tractable in system $S5$. So $X \wedge \Box Y = (p_1 \vee p_2) \wedge \Diamond\Box\neg p_3 \wedge \Box\Diamond p_2 \wedge \Box(p_1 \vee p_2)$. So we have $CANDIDATES = \{p_1 \vee p_2, p_1 \vee \Box(\Diamond p_2 \wedge (p_1 \vee p_2)), p_1 \vee \Diamond(\Box\neg p_3 \wedge \Diamond p_2 \wedge (p_1 \vee p_2)), \Box(\Diamond p_2 \wedge (p_1 \vee p_2)) \vee p_2, \Box(\Diamond p_2 \wedge (p_1 \vee p_2)), \Box(\Diamond p_2 \wedge (p_1 \vee p_2)) \vee \Diamond(\Box\neg p_3 \wedge \Diamond p_2 \wedge (p_1 \vee p_2)), \Diamond(\Box\neg p_3 \wedge \Diamond p_2 \wedge (p_1 \vee p_2)) \vee p_2, \Diamond(\Box\neg p_3 \wedge \Diamond p_2 \wedge (p_1 \vee p_2)) \vee \Box(\Diamond p_2 \wedge (p_1 \vee p_2)), \Diamond(\Box\neg p_3 \wedge \Diamond p_2 \wedge (p_1 \vee p_2))\}$. After removing logically entailed clauses with respect to $\Box Y$, the set of theory prime implicates of X with respect to $\Box Y$, i.e., $\Theta(X, \Box Y) = \{p_1 \vee p_2, \Box(\Diamond p_2 \wedge (p_1 \vee p_2)), \Diamond(\Box\neg p_3 \wedge \Diamond p_2 \wedge (p_1 \vee p_2))\}$.

4   Conclusion

In this paper the definitions and results of theory prime implicates in propositional logic [9] is extended to modal logic $\mathcal{T}$ and the algorithm for computing theory prime implicates in propositional logic is also extended to modal logic according to [1] and its correctness has been proved. Another algorithm for query answering in [9] from $\Omega_{\Box Y}(X)$ is also extended to modal logic. Due to Lemma 3, Theorems 4 and 5 we had to compute theory prime implicates of a modal formula X with respect to another restricted modal formula $\Box Y$ in the algorithm $MODALTPI$ instead of theory prime implicates of a modal formula X with respect to an arbitrary modal formula Z. So, if Y is empty, then $\Theta(X, \Box Y) = \Pi(X)$. As a future work, we want to compute theory prime implicates of a knowledge base X with respect to another arbitrary modal knowledge base Z instead of the knowledge base $\Box Y$ for a proposition knowledge base Y assumed here. By Theorem 7 we have shown that the theory prime implicate compilation $\Omega_{\Box Y}(X)$ is equivalent to X so queries will be answered from $\Omega_{\Box Y}(X)$ in polynomial time by Theorem 8. Our algorithm $MODALTPI$ is based on Bienvenu's algorithm [1] which relies on distribution property whereas Marquis's theory prime implicate algorithm [9] is based on prime implicate generation algorithm of Kean and Tsiknis [7] and of de Kleer [8] which rely on resolution. As another direction of research we want to compute theory prime implicates of a modal knowledge base X with respect to another arbitrary modal knowledge base Z using global consequence relation.

Acknowledgement. Authors thank the National Board for Higher Mathematics, Department of Atomic Energy, Mumbai, India for financial support under grant reference number 2/48(16)/2014/NBHM(R.P.)/R&D II/1392.

References

1. Bienvenu, M.: Prime implicates and prime implicants: from propositional to modal logic. J. Artif. Intell. Res. (JAIR) **36**, 71–128 (2009)
2. Bienvenu, M.: Consequence finding in modal logic. Ph.D. thesis, Université Paul Sabatier, 7 May 2009
3. Blackburn, P., de Rijke, M., Venema, Y.: Modal Logic. Cambridge University Press, Cambridge (2002)
4. Cook, S.A.: The complexity of theorem-proving procedures. In: Proceedings of the 3rd ACM Symposium on the Theory of Computing, pp 151–158. ACM Press (1971)
5. Cadoli, M., Donini, F.M.: A survey on knowledge compilation. AI Commun. Eur. J. AI **10**, 137–150 (1998)
6. Cerro, L.F.D., Penttonen, M.: A note on the complexity of the satisfiability of modal Horn clauses. J. Logic Program. **4**, 1–10 (1987)
7. Kean, A., Tsiknis, G.: An incremental method for generating prime implicants/implicates. J. Symb. Comput. **9**(2), 185–206 (1990)
8. de Kleer, J.: An improved incremental algorithm for generating prime implicates. In: Proceedings of the Tenth National Conference on Artificial Intelligence, AAAI 1992, pp. 780–785. AAAI Press (1992)
9. Marquis, P.: Knowledge compilation using theory prime implicates. In: Proceedings of International Joint conference on Artificial Intelligence, IJCAI 1995, pp. 837–843 (1995)
10. Raut, M.K., Singh, A.: Prime implicates of first order formulas. IJCSA **1**(1), 1–11 (2004)
11. Raut, M.K.: An incremental algorithm for computing prime implicates in modal logic. In: Proceedings of the 11th Annual Conference on Theory and Applications of Models of Computation (TAMC), LNCS 8402, pp. 188–202 (2014)
12. Raut, M.K.: Computing Theory Prime Implicates in Modal logic. https://arxiv.org/abs/1512.08366

Fault Tolerance in Real-Time Systems: A Review

Egemen Ertugrul and Ozgur Koray Sahingoz(✉)

Computer Engineering Department, Istanbul Kultur University, 34158 Istanbul, Turkey
egemenertugrul@yandex.com, o.sahingoz@iku.edu.tr

Abstract. Real-time systems are safety/mission critical computing systems which behave deterministically, and gives correct reactions to the inputs (or changes in the physical environment) in a timely manner. As in all computing systems, there is always a possibility of the presence of some faults in real time systems. Due to the its time critical missions a fault-tolerance mechanism should be constructed. Fault tolerance can be achieved by hardware, software or time redundancy and especially in safety-critical applications. There are strict time and cost constraints, which should be satisfied. This leaves us to the situation where constraints should be satisfied and at the same time, faults should be tolerated. In this paper, the basic concepts, terminology, history, features and techniques of fault tolerance approach on real-time systems, are detailed and related works are reviewed for composing a good resource for the researchers.

Keywords: Real-time systems · Safety-critical applications · Fault tolerance

1 Introduction

In a real-time system (RTS), the correct behavior depends both on the logical result of the calculation, and the physical time of execution. The later one emphasizes that late execution means incorrect calculation due to the requirement of time constraints which can also be identified as deadlines. For example, many embedded systems can be identified as RTSs such as air traffic control systems, robotic systems, monitoring services (e.g. health devices), cruise control system, the avionics of a plane and etc. If an internet based airline reservation system is slow, that is an annoying thing. However, the system which controls the plane is slow for responses, that will cause dangerous results.

A real-time operating system (RTOS) is a key component for building a real-time system. Additionally, some hardware and software components are also used in it. RTOSs emphasize predictability, efficiency with features that support time-constraints. Violating time constraints in RTOS leads to system failure. These systems are usually found on embedded systems, which are considered as special purpose computer systems, designed to perform one or a few dedicated functions with real-time computing [1, 2].

Nowadays, the consumer electronics market was surrounded with lots of this type micro electronic devices with unseen performances. These devices are take a great place in our daily lives such as smart phones, tablets, laptops, etc. Also they are widely used different industrial domains. Due to the size and resource constraints, these micro

© Springer International Publishing AG, part of Springer Nature 2018
A. Abraham et al. (Eds.): ISDA 2017, AISC 736, pp. 283–293, 2018.
https://doi.org/10.1007/978-3-319-76348-4_28

technologies are affected by some software and hardware errors which cause malfunctioning of the system. If a software type simple errors occur in a smart phone, it can be easily handled by restarting it. However, what should be done if such an error occurs in a crucial component of real time system such as an airplane or a nuclear reactor? Can restarting be a good solution? In the history there are lots of bad examples. In one of them, Air France Flight 447 crashed into the ocean after a sensor malfunction caused a series of system errors. The pilots stalled the aircraft while responding to outdated instrument readings. All 12 crew and 216 passengers were dead. The research area which tries to solve this type of problem is called as Fault Tolerance. Fault tolerance mechanism is a tool for enabling the system to continue its execution even if there exist some faults. This can be accomplished by using some error detection mechanism or and subsequent system recovery tools especially in safety-critical applications such as avionics [3].

In this paper it is aimed to give some specifications/definitions of fault tolerance mechanisms, explaining important techniques which are built on a real time (operating) system. By using this review paper, it will be easy for the researchers who study on this area to find necessary background information. The rest of the paper is organized as follows. In the next section the basic concepts and terminologies are detailed. Some recent related works on this topic are explained and RTOS Features with Fault Tolerance Techniques are described in Sect. 3 and Sect. 4 relatively. Finally, the conclusion and future works are depicted.

2   Basic Concepts and Terminology

Real-time systems are critical systems, where tasks have to be completed before reaching their deadline. Real-time systems need to produce valid and correct results in timely manner. Validness is achieved when correct results are produced on time. A typical real-time system consists of a controlling system, a controlled system, and the environment (as shown in Fig. 1).

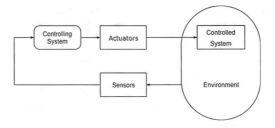

Fig. 1. A typical real-time system [4]

One important criterion in RTSs is the deadline which means a certain time instance, when the results are expected to be produced before it. As depicted in Fig. 2, RTSs can be classified into three groups according to their deadlines:

- **Soft:** Consequences are relatively more tolerable if a deadline is missed. Value for a result after its deadline is not zero (has utility).

- **Firm:** Consequences are relatively tolerable if deadline is missed. The usefulness of a result is zero, but the system is not considered to have failed after task's deadline.
- **Hard:** Consequences of missing a deadline can be catastrophic. The system is considered to have failed if the system fails to meet the deadline.

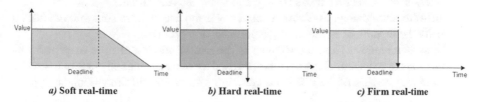

| *a)* Soft real-time | *b)* Hard real-time | *c)* Firm real-time |

Fig. 2. Classification of real-time systems based on deadline strictness

If the system timing requirements are soft, a system that has the best effort to satisfy the timing constraints and also has a minimum quality loss while violating timing constraints should be designed. If the system has hard real-time constraints, the designer has to spend a lot of time to guaranty system safety and predictability and also guaranty that all timing constraints (deadlines) are met. Especially in safety critical systems, if the occurrence of a failure is not catched, it will cause some catastrophic effects.

2.1 Fault Tolerance

Fault tolerance is a very important criterion in RTSs. Therefore, before entering its details, it will be helpful to explain some important concepts:

- **Error:** A human action that produces an incorrect result. The mistakes made by programmer is known as an 'Error'. This could happen because of the following reasons: some confusion in understanding the requirement of the software; some miscalculation of the values; or/and misinterpretation of any value, etc.
- **Fault:** A manifestation of an 'Error' in software. Faults are also known colloquially as defaults or bugs. There are three types of faults: Permanent in which faults don't die away by time and remain until the system is recovered; Intermittent in which faults occurs at irregular intervals and Transient in which faults die away after some time.
- **Defect:** The departure of a quality characteristic from its specified value that results in a product not satisfying its normal usage requirements. The 'Error' introduced by programmer inside the code is known as a 'Defect'. This can happen because of some software errors.
- **Failure:** Failure is a deviation of the software from its intended purpose. If under certain circumstances these 'Defects' get executed by the tester during the testing, then it results in the 'Failure' which is known as software 'Failure'.
- **Redundancy:** It is one of the most used techniques in fault-tolerance. A redundant (spare) part of hardware or software can help the system run even in the presence of faults; it is an almost crucial aspect of a fault-tolerant system. There are four kinds of redundancy available for fault-tolerance [2, 4]:

1. Hardware redundancy is the addition of extra hardware to the system, such as spare processors that are used if one of the running processors fail.
2. Information redundancy: Information redundancy is the addition of extra information to data, to allow error detection and correction. This is typically error-detecting codes, error-correcting codes (ECC), and self-checking circuits.
3. Software redundancy: Software redundancy is the use of extra software modules to verify the results, or to use multiple versions of a program.
4. Time redundancy: Time redundancy is the use of additional time to perform the functions of a system. This time might be used to re-execute a faulty task or to execute a different version of the task (thus combining software and time redundancy).

When the weight of the system, power consumption and price constraints are taken into account, software redundancy is more commonly used to increase fault tolerance, than hardware redundancy. Hardware redundancy is avoided as far as possible.

Redundancy techniques are typically used on an autopilot system on-board a large-sized passenger aircraft. This kind of passenger aircrafts usually consists of a central autopilot system, along with two backups, which can be considered as redundant hardware. If central autopilot system is broken down, the backups will kick in. However, hardware redundancy is not sufficient only on its own, other techniques should also be implemented, such as software redundancy.

Every process in the autopilot system is copied onto other computers. The system gathers the results from all computers that have completed the process. Finally, the system votes on the results and decides on which result to take into account. This is an example of N-Version programming (covered in the next sections) as well as software redundancy.

3   Related Works

Fault tolerance techniques are implemented in some distributed systems [5, 6] and also in many real time systems. There are numerous works done on improving the fault tolerance and reliability while maintaining the overall quality of a real-time system. Transient faults are counted as one of the challenges to solve in real-time computing. Triple modular redundancy or standby-sparing techniques can be deployed in such systems where transient faults may occur. In [7], it's been stated that check pointing with rollback-recovery is a relatively cost-effective technique against transient faults, compared to the aforementioned techniques; a two-state check pointing scheme to achieve low-energy fault tolerance for hard real-time systems has been proposed.

Networked control systems (NCS) are distributed systems, which are also considered as hard real-time systems. [8] has proposed a model-based coupling approach to deal with coupling imperfections occurring in NCSs or real-time co-simulation problems. With this approach, time varying delays and data losses can be compensated.

In real-time cloud computing, failures are normal and expected, while fault detection and system recovery are two issues that should be efficiently solved. [9] proposed an optimized fault tolerance approach where a model is designed to tolerate faults based on the reliability of each compute node (virtual machine) and can be replaced if the performance is not optimal.

In embedded systems, a platform-wide software verification for safety is considered expensive. In [10], a design approach to deploy safe-by-design embedded systems was proposed. The main idea is to detect the faults by monitoring the application timing and perform a full platform restart in case of fault-recovery. Such ability is enabled by the short restart time of embedded systems. SAFEbus R, developed by Honeywell, can achieve Byzantine fault tolerance with on the order of a microsecond of added latency [11]. This system is accepted by ARINC 659 specifications (Avionics Application Standard Software Interface).

Byzantine Fault Tolerance is a technique deployed on fault-tolerant computer systems, in order to eliminate the "Byzantine Generals' Problem", which is the situation where a battle zone to attack a city is illustrated. Generals of armies separated in the battle zone have to communicate with each other to attack or retreat the city at the same time. If the communication fails somehow, this might result in some armies attacking and some armies retreating; this is a failure and an unwanted result (Fig. 3).

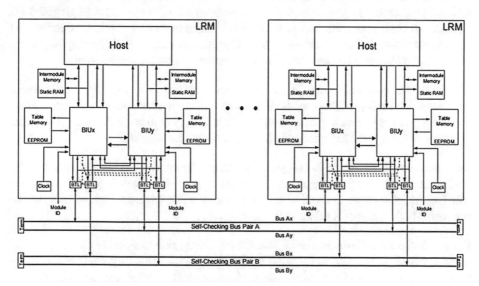

Fig. 3. Two Hosts Connected Via SAFEbus [9]

According to [10], "Honey Honeywell's SAFEbus (ARINC 659) uses selfchecking pair (SCP) buses and SCP BIUs to ensure that Byzantine faults are not propagated (i.e. a Byzantine input will not cause the halves of a pair to disagree)." and "SAFEbus is the only standard bus topology that can tolerate a Byzantine fault, made possible by SAFEbus's full coverage, fault-tolerant hardware". SAFEbus is deployed on Boeing 777 Aircraft Information Management System (IMS).

4   RTOS Features and Fault Tolerance Techniques

Fault tolerance in real-time (operating) systems is an extremely important aspect, as discussed earlier. This section is about features of real time operating systems and the way fault tolerance techniques are applied to those features.

4.1   Memory Management

Memory management is the first feature of RTOSs to examine. The data in memory, affect the way programs are run; non-integrated/faulty data in memory would eventually lead to system errors and failure. Thus, the way memory management is done in RTOSs plays a key role to prevent failures in the first place [1].

Fault Tolerance Techniques in Memory Management: Real-time operating systems are expected to be predictable and flexible if needed. The use of dynamic storage allocation (DSA) leads to uncertainty in RTOSs, because of unconstrained response time and the fragmentation problem. A DSA algorithm called Two-Level Segregated Fit provides explicit allocation and deallocation of memory blocks with a bounded and acceptable timing behavior; usage of this algorithm is encouraged. Bitmapping technique is an alternative for achieving fault tolerance in memory management.

Memory Management Unit: In some real-time systems, the memory management unit (MMU) is disabled. This causes all processes run in the same address space. Hence, a badly written code or a bug in a code, would lead to system crash. In order to prevent failures caused by pointers and bugs, it is advised to enable MMU.

Redundancy in Memory Management: Redundancy is a technique that can be applied also in memory management. Whenever a process is loaded, the OS can duplicate the data and states in more than one place/memory. Memory redundancy can be supported both software level and hardware level.

Error-Correcting Code Memory (ECCM):   Error-correcting code memory (ECC memory) is an instrument to improve operating systems reliability from the memory protection perspective. Error-correcting code memory is a perfect example of information redundancy. An ECCM can significantly improve the fault-tolerance of the system.

4.2   Kernel Considerations

The kernel of a fault-tolerant real-time operating system should provide a mechanism that whenever an error occurs, a notification is sent to an agent. This agent, called supervisor has to perform some types of error recovery operations [1].

Fault Tolerance Techniques in Kernel The kernel of a RTOS should support event logging mechanism to for improved detection of errors, software watchdog capability to be notified of a task that was supposed to run but did not (software redundancy), protection against improper system calls and prevention of spread of faults to the kernel, thus avoiding system crashes.

4.3   Process and Thread Management

Process management is responsible for process creation, process termination, scheduling, dispatching, context switching and other related activities. The definition and activation of processes are important roles of RTOSs. If a RTOS has activated a process once, it should also release it once or periodically, on-time and before deadline. If a RTOS does not implement process management, a new task may not be able to use the processor or other system resources as a result of malicious or careless execution of other tasks; this situation may result in miss of deadline.

4.4   Scheduling

(Task) Scheduling is a very critical process for a RTSs. It is the process to decide on when and which processor the given tasks should be executed. The system, or the CPU, should determine which task to run, switch between given tasks without making them past their deadlines, while guaranteeing the recovery of tasks that do not generate correct outputs due to faults in the system.

Scheduler is the most important task in RTOS and it should be protected against failures. In case of failure of the scheduler, the system crashes since the other system tasks are not scheduled and released correctly. Scheduling algorithms can be mainly classified into two major models:

Preemptive algorithms assume that any task can be interrupted during its execution, while *non-preemptive algorithms* do not allow a running task to be interrupted; while both of these algorithms have advantages and disadvantages, this paper covers the preemptive algorithms.

Rate Monotonic algorithm is a fixed-priority, preemptive scheduling algorithm, which tasks' priorities are defined in advance and tasks with smaller period have higher priority. This algorithm is widely used in industry because of its simplicity, flexibility, ease of implementation and modification.

In the Earliest Deadline First algorithm deadline is a dynamic-priority, and it is a preemptive scheduling algorithm, which tasks' priorities are defined dynamically in run-time and tasks with closer deadline have higher priority.

Least Laxation First is a dynamic-priority scheduling algorithm in which the task with the least slack time has higher priority.

Fault Tolerance Techniques in Scheduling Task can be used to significantly improve fault-tolerance in a RTOS. Fault-tolerant RTOSs should be able to recover processors from transient and permanent faults. Five different techniques and approaches are detailed in the following part for fault-tolerant scheduling in real-time systems [4, 12–14].

N-copy Programming is one of the most popular fault tolerance techniques that can be used for both fixed-priority and dynamic-priority scheduling algorithms as depicted in Fig. 4 [13]. N-copy programming is a very straightforward and simple approach. N number of copies of a scheduler (n ≥ 3) are run concurrently in different address spaces.

The output data is obtained by using a voting scheme. N-copy programming can be used to tolerate transient faults. The best example for N-copy programming is Triple Modular Redundancy. Again, in TMR, the output is decided by a majority-voting system, with the processes from three different modules.

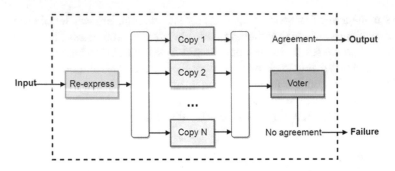

Fig. 4. Basic structure of N-copy programming

N-Version Programming is a technique that is based on the principle of design diversity; where a task is coded by different teams of programmers, in multiple versions. This technique can tolerate not only software but also hardware faults.

Recovery Blocks is an approach that again uses multiple alternates to perform the same function as depicted in Fig. 5 [13]. While one task is considered primary, to others are considered backup (alternate) tasks.

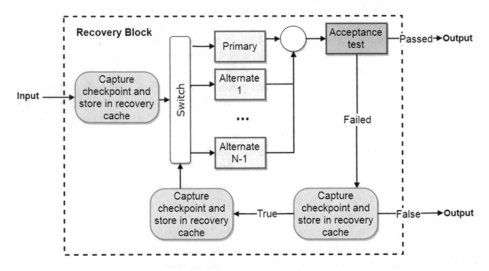

Fig. 5. The basic structure and flow of a recovery block

The system rolls back to the checkpoint if the acceptance test is failed, candidates are not exhausted and the deadline is not exceeded, else the system fails. Primary-Backup approach is an example of Recovery Blocks, where every task has one backup (copy).

This approach is implemented for both nonpreemptive and preemptive situations. Imprecise Computations Imprecise Computation model provides scheduling flexibility by trading off the quality of the results in order to meet the task deadlines.

Backup Overloading Algorithm is an algorithm that involves primary backup approach in multi-processor real-time systems, as there can be other variations with different approaches, such as with dynamic grouping. It uses backup overloading and backup deallocation strategies. Backup overloading, allows the backup copies of different tasks to overlap in time on the same processor. Backup deallocation, reclaim the time allocated to a backup copy in the case of fault-free operation of its primary. This algorithm assumes that there is at most one failure in the system.

An example of backup overloading algorithm shown in Fig. 6. Backups 1, 3 are scheduled on Processor 2 and they overlap on execution, while their primaries, Primary 1 and 3 are scheduled on different processors (processor 1 and 3 respectively).

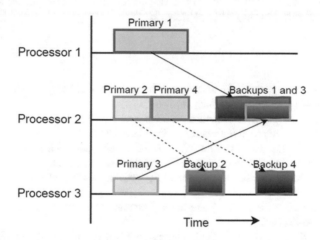

Fig. 6. Backup overloading [14]

4.5 I/O Management

Input/Output Management, is another important feature in RTOS, where possibility of faults and fault tolerance techniques should be taken into account. RTOS should be able to manage the order of I/O accesses, without interrupting and preventing the tasks to meet their timing constraints [1]. Fault Tolerance Techniques in I/O Management Replication is a common technique that can be used, by replicating I/O devices. Backup devices can continue the unfinished process of the primary I/O device. Robustness is also an important aspect of I/O management to achieve fault-tolerance.

4.6 Programming Languages

The programming languages of a real-time operating system are determined by the purpose and special use of that system. Some certain characteristics of a programming language are looked for RTS: well-defined language semantics, the strong type checking, and structuring mechanisms. Generally, programming languages of RTS are determined in three models: synchronous, scheduled and timed.

5   Conclusion

Real-time systems are closely related to embedded systems and safety-critical applications. Obtaining fault tolerance in such systems is a crucial step during both design, development and testing phases. Implementing fault tolerance techniques can highly increase the reliability of a real-time system and reduce costs in the long term by preventing any catastrophic results. In this paper, firstly, some basic concepts and examples were given. Then the features of RTSs were mentioned in categories, as they are inseparable attributes to achieve fault tolerance, then fault tolerance techniques accordingly. Various fault tolerance techniques done on real-time computing and operating systems were covered in related work section.

References

1. Ramezani, R., Sedaghat, Y.: An overview of fault tolerance techniques for realtime operating systems. In: ICCKE 2013, Mashhad, pp. 1–6 (2013)
2. Persya, C., Nair, G.: Fault tolerant realtime systems. In: International Conference on Managing Next Generation Software Application, MNGSA 2008, Coimbatore (2008)
3. Imai, S., Blasch, E., Galli, A., Zhu, W., Lee, F., Varela, C.A.: Airplane flight safety using error-tolerant data stream processing. IEEE Aerosp. Electron. Syst. Mag. **32**(4), 4–17 (2017)
4. Al-Omari, R.M.S: Controlling schedulability-reliability trade-offs in real-time systems (2001)
5. Sahingoz, O.K., Sonmez, A.C.: Agent-based fault tolerant distributed event system. Comput. Inf. **26**, 489–506 (2007)
6. Sahingoz, O.K., Sonmez, A.C.: Fault tolerance mechanism of agent-based distributed event system. In: 6th International Conference Computational Science, ICCS 2006, Reading, UK, 28–31 May, pp. 192–199 (2006)
7. Salehi, M., Tavana, M.K., Rehman, S., Shafique, M., Ejlali, A., Henkel, J.: Two-state checkpointing for energy-efficient fault tolerance in hard real-time systems. IEEE Trans. Very Large Scale Integr. Syst. **24**(7), 2426–2437 (2016)
8. Tranninger, M., Haid, T., Stettinger, G., Benedikt, M., Horn, M.: Fault-tolerant coupling of real-time systems: a case study. In: 3rd Conference on Control and Fault-Tolerant Systems (SysTol), Barcelona, pp. 756–762 (2016)
9. Mohammed, B., Kiran, M., Awan, I.U., Maiyama, K.M.: Optimising fault tolerance in real-time cloud computing IaaS environment. In: IEEE 4th International Conference on Future Internet of Things and Cloud (FiCloud), Vienna, pp. 363–370 (2016)
10. Abdi, F., Mancuso, R., Tabish, R., Caccamo, M.: Restart-Based Fault-Tolerance: System Design and Schedulability Analysis. CoRR (2017)
11. Driscoll, K., Hall, B., Sivencrona, H., Zumsteg, P.: Byzantine fault tolerance, from theory to reality. In: LNCS, pp. 235–248 (2003)
12. Murthy, C.: Resource Management in Real-Time Systems and Networks. The MIT Press, Cambridge (2016)

13. Hiller, M.: Software Fault-Tolerance Techniques from a Real-Time Systems Point of View, 1st edn. Chalmers University of Technology, Gothenburg (1998)
14. Al-Omari, R.M.S., Govindarasu, M., Arun, S.: An efficient backup-overloading for fault-tolerant scheduling of real-time tasks. In: LNCS, pp. 1291–1295 (2000)

Gauss-Newton Representation Based Algorithm for Magnetic Resonance Brain Image Classification

Lingraj Dora[1], Sanjay Agrawal[2(✉)], and Rutuparna Panda[2]

[1] Department of EEE, VSSUT, Burla 768018, India
lingraj02uce157ster@gmail.com
[2] Department of Electronics and Telecommunication Engineering, VSSUT,
Burla 768018, India
agrawals_72@yahoo.com, r_ppanda@yahoo.co.in

Abstract. Brain tumor is a harmful disease worldwide. Every year, a majority of adults as well as children dies due to brain tumor. Early detection of the tumor can enhance the survival rate. Many brain image classification schemes are reported in the literature for early detection of tumors. Thus, it has become a challenging problem in the field of medical image analysis. In this paper, a novel hybrid method is proposed that uses the Gauss-Newton representation based algorithm (GNRBA) with feature selection approach. The proposed method is threefold. Firstly, discrete wavelet transform (DWT) is used as a pre-processing step to extract the features from the brain images. Secondly, principal component analysis (PCA) is used to address the dimensionality problem. Finally, the extracted features in the lower dimensional space are utilized by GNRBA for classification. To show the robustness of the proposed method, real human brain magnetic resonance (MR) images are used to experiment. It is witnessed from the results that the performance of the proposed method is superior as compared to the existing brain image classification methods.

Keywords: Discrete wavelet transform
Principal component analysis
Gauss-Newton representation based algorithm

1 Introduction

Brain tumor is a critical disease among the children and adults worldwide. It is a mass of abnormal cells developed inside or around the brain, due to uncontrolled growth of the abnormal cells. A tumor can be Benign or Malignant. Benign tumors are not life threatening, as they do not spread into nearby tissues or other parts of the body. Malignant tumors are considered as cancerous as they can invade different parts of the body [1]. In this paper, our primary focus is on classification of MR brain image as normal or pathological brain

© Springer International Publishing AG, part of Springer Nature 2018
A. Abraham et al. (Eds.): ISDA 2017, AISC 736, pp. 294–304, 2018.
https://doi.org/10.1007/978-3-319-76348-4_29

(i.e. brain with benign or malignant tumor). According to American cancer society's statistics for 2017, in the United States around 23,800 malignant tumors may be diagnosed and 16,700 may die from brain tumors [2]. Early detection of brain tumors could be a preventive measure to lower the rate of death due to brain tumors. It also helps in analyzing the brain tumors for further studies. Neuroimaging technique like MR imaging (MRI) is used as a standard clinical assessment for the early diagnosis of brain tumor. However, to identify the tumor (Benign or Malignant) as well as for further treatment planning a domain expert is required. Image processing techniques can aid in analyzing the brain tumors. They show a respectable classification accuracy which can assist the domain experts in diagnosing diseases. Hence, they are widely used in the medical field for disease diagnosis and classification [3,4].

In the literature, a wide range of methods are available for the pathological brain image classification. Most of the approaches in the pathological brain image classification involve feature extraction followed by classification method. A hybrid method is proposed in [5] for classifying the brain images as normal or abnormal. The method uses DWT for feature extraction. PCA is then used to reduce the dimension of the extracted features. Two different classifiers are used for testing, namely: artificial neural network (ANN) and k-nearest neighbor (k-NN). Ain et al. [6] proposed a method that uses first order and second order texture feature methods for extracting features from the brain MR images. Ensemble based support vector machine (SVM) is used for classification of images as normal or pathological brain. After successful classification of pathological brain, fuzzy c-means (FCM) is used to segment the tumor region. The method is able to classify normal or tumorous brain. However, classification of the type of tumor (Benign or Malignant) is not included in the method. A support system is proposed in [7] that utilizes subband texture fractal features for classification of meningioma brain from histopathological images. The method achieved a classification accuracy of 94.12%. Recently, machine learning techniques are gaining attention in the medical field. They have the capability to obtain the information from past knowledge and learning. In [8], a hybrid intelligent system is proposed for brain tumor classification. The method uses gray level co-occurrence matrix (GLCM) and gray level different matrix to extract features from the input image. Tolerance rough set Firefly based Quick Reduct (TRSFFQR) technique is used to select the relevant features. Three different types of classifiers such as Naive Bayes, J48 and IBk were used for classification. Othman and Basri [9] proposed a probabilistic neural network (PNN) for tumor classification of MR brain images. The authors used PCA for feature extraction followed by PNN for classification.

Nowadays, computer aided detection (CAD) systems are developing rapidly to improve the diagnosis capability of the doctors. The CAD systems should have the abilities comparable to the doctors in terms of learning and disease diagnosis. For this reason, CAD systems utilize pattern recognition techniques with machine learning techniques for accurate diagnosis. In [10], a CAD system is proposed to segment and classify brain tumor. The method uses content based active contour to segment the region of interest (ROI). Genetic algorithm (GA) is used to extract the optimal intensity and texture features from the segmented ROI. Two hybrid machine learning techniques, namely GA-SVM and GA-ANN are used to classify the optimal features. A detailed survey highlighting the state-of-the-art CAD systems for brain tumor detection and classification is presented in [4]. In addition, the authors in [4] also proposed a CAD system for brain image classification as normal or abnormal. The system uses feedback pulse coupled neural network (FPCNN) to segment the tumor region from the brain MR images. DWT is then used to extract the tumor features. To avoid the dimensionality problem, PCA is used to reduce the dimension of the extracted features. Feedforward back propagation neural network (BPNN) is used to perform classification. In the literature, a variety of feature extraction methods are available. However, shape, size and texture of tumor tissue makes the feature extraction a complicated task. Most of the feature extraction techniques use DWT. That's why, we have used DWT for feature extraction in this paper.

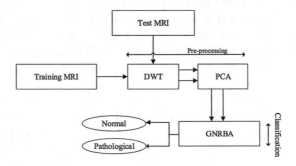

Fig. 1. Block diagram of the proposed method.

In this paper, we have proposed a novel hybrid method to classify an unknown test MR brain image as normal or pathological, by using labelled training MR brain images. The proposed method involves three major steps, as depicted in Fig. 1. Firstly, it uses DWT as a pre-processing step to extract the features from both the training and test images. It allows us to capture the features at different resolutions because of its multi-resolution analytic property. Secondly, PCA is used to curb the curse of dimensionality and computational complexity. In addition, PCA selects a set of discriminating features from the training images for performing accurate classification. In this way, the proposed method transforms the high dimensional image into a reduced feature space. Finally, GNRBA is used for classification. In the reduced feature space, GNRBA uses the test image to select the most significant training images by using Euclidean distance. Each training image is labelled as normal or pathological, which represents its class. GNRBA is based on sparse representation technique, which intelligently evaluates the contribution of each class on the test image. The test image is classified to the class having maximum contribution.

GNRBA is successfully implemented for breast cancer classification problem with improved results [11]. This has motivated us to implement GNRBA for the brain image classification problem. To the best of our knowledge, it has not been used in such application before. It is observed that results obtained using our proposed method are superior to other existing methods. The rest of the paper is organized as follows: Sect. 2 explains our proposed method. Section 3 presents the results and discussions and finally Sect. 4 draws the conclusion.

2   Proposed Method

2.1   Discrete Wavelet Transform

The wavelet transform has been proven to be an impressive tool for feature extraction [12,13]. It allows us to capture both the spatial and frequency domain information. Among, different types of wavelets, Harr wavelet is the popular, simple and the most preferred wavelet used in several applications [4,5,14]. The DWT is a well-accepted technique used to transform an image from the spatial domain to frequency domain. For a 2D image, the 2D DWT is implemented to each dimension separately. It decomposes an image into different sub-bands with their respective coefficients. It is based on cascaded filter bank scheme, which involves low pass and high pass filters to ensure certain fixed constraints. The DWT decomposes a brain MR image into four different sub-bands images, namely: LL, LH, HH and HL, corresponding to each scale. The LL sub-band is considered as the approximation component of the image. The other three sub-bands (LH, HH and HL) are considered as the detailed components of the image. For feature extraction, only the LL sub-band is used for the DWT decomposition at different levels. The last level of the resulting LL sub-band is considered as the feature vector. In this paper, we have used the LL sub-band for decomposition via Haar wavelet at three different levels to extract features from brain MR images. We have utilized symmetric padding technique [15] in the DWT to calculate the boundary values.

2.2   Principal Component Analysis

The feature vector extracted from the original image using the DWT is usually a high dimensional vector. A high dimensional feature vector often increases the response time as well as the memory storage. In addition, they also limit the classification accuracy due to the curse of dimensionality. A method to avoid the above problems, is to reduce the dimension of the feature vector. PCA is a well-known technique used to project the high dimensional data into a reduced sub-space involving less computational complexity. It allows us to transform a high dimensional correlated data into a lower dimensional space with uncorrelated data, while preserving the variance. The new low dimensional space contains a group of uncorrelated data that are called principal components. A detailed description about the PCA can be found in the literature [16]. In this paper, we have used PCA to get a low dimensional representation of the feature vectors. Let $D = [d_1, d_2, \ldots, d_N]$ represents the DWT feature vectors extracted from N training images. All the N training images belong to either of the two classes (normal or pathological). PCA transforms the original n-dimensional feature vector into m-dimensional feature vector such that, $m \ll n$. using the linear transformation given as:

$$Y = W^T D \tag{1}$$

where, $W \in \Re^{n \times m}$ is a matrix having orthogonal columns and $Y \in \Re^m$ is the new feature vectors in the low dimensional space.

2.3 GNRBA

In this section, the GNRBA used to classify an unknown test brain MR image is discussed. The GNRBA is based on sparse representation technique [11]. It represents a test feature vector as a weighted linear combination of the training feature vectors. All the training feature vectors belong to either of the two classes, i.e. normal or pathological. The objective is to evaluate the contribution of each class on the test feature vector. The test feature vector is then classified to the class having maximum contribution. Let $Y = [y_1, y_2, \ldots, y_N]$ be the feature vectors of the N training images. All the feature vectors of the training images may not contain the necessary information about the test feature vector. GNRBA uses the Euclidean distance measure to select the training feature vectors that are similar (nearest) to the test feature vector, given as:

$$d_i = \|z - y_i\|_2 \tag{2}$$

where, z represents a test feature vector. A small d_i indicates that a training feature vector y_i is nearest to the test feature vector z for i $= 1, 2, \ldots,$ N. A large d_i indicates that the vector y_i is an unsuitable feature vector. Thus, (2) is used to select a subset of K training feature vectors, i.e. $X = [x_1, x_2, \ldots, x_K]$, that are similar to test feature vector. At the same time, the class labels (normal or pathological) of the selected feature vectors are also preserved. GNRBA represents a test feature vector as the weighted linear combination of the K selected training feature vectors, given as:

$$z = w_1 x_1 + w_2 x_2 + \ldots + w_K x_K \tag{3}$$

where, x_i is the selected training feature vector and w_i is the weighting coefficient assigned to each feature vector x_i for i $= 1, 2, \ldots, K$. Each feature vector belongs to either of the two classes (normal or pathological). The contribution of each feature vector towards their class is decided by the value of the weighting coefficient. GNRBA uses Gauss-Newton representation to find the optimal values of the weighting coefficients w_i, such that it consistently reflects the contribution of training feature vectors, given as:

$$w_{next} = w_{now} + \Delta w \tag{4}$$

where, $\Delta w = (X^T X + \lambda I)^{-1} X^T \Delta f$, I is the identity matrix and λ is the regularization parameter, which is equal to 0.01, as suggested in [17]. It represents the difference between the desired output value and the actual output value when the input is X as defined in [18]. The stopping criteria to obtain an optimum w can be the maximum number of iterations or until the function converges. Once the optimum w is obtained from (4), it can be used to show the contribution of each class in representing the test feature vector. However, as the selected training feature vectors might be from different classes, the weighted sum of the training feature vectors from a class is used to represent the test feature vector, given as:

$$X_k = \sum_{i=1}^{N_k} w_i x_i \tag{5}$$

where, N_k is the number of training feature vectors from k^{th} class; w_i is the ith coefficient value of the ith training feature vector X_i and X_k indicates the contribution of kth class. The sum in (5) is used to classify a test feature vector z as:

$$C_k = \|z - X_k\|_2 \tag{6}$$

In (6), a smaller distance C_k between the test feature vector z and X_k indicates a maximum contribution of the kth class to represent the test feature vector. Hence, the test feature vector is classified into the class k.

3 Results and Discussions

We have developed our proposed hybrid method in MATLAB on a MAC with Intel core i5. To justify the robustness of the proposed hybrid method, it is validated using different performance indices like classification accuracy (CA), sensitivity, specificity, area under receiver operating characteristics (AUC) curves [19,20]. In addition, real clinical images are used to show the effectiveness of the proposed method for clinical use. In this paper, the experiments are carried out using real human brain MR images from the Harvard Medical School Website (http://www.med.harvard.edu/aanlib/home.html). All the images in the database are T2-weighted MRI having a dimension of 256×256 pixels. We have randomly selected 10 images of each brain type for the experiment. The database contains only one type of normal brain and 11 types of pathological brains ($11 \times 10 = 110$ pathological images). Thus, a total of 120 brain MR images are used for the experiment. The details of training-testing partition is depicted in Table 1. In addition, we have also used a 10-fold cross validation technique [21] to show the robustness of the proposed method. In this technique, the whole database is divided into 10 equal blocks. For training 90% of the blocks are used and the remaining 10% are used for testing. The process is repeated 10 times with 10% block for testing exchanged in all runs. We have performed the experiment to demonstrate the dependence of the proposed hybrid method on the value of K (size of the subset). The variation of CA with K is shown in Fig. 2. From Fig. 2, it is observed that small values of K yield a higher CA. When the

Table 1. Details of training-testing partition

Training-testing partition	Number of images used for training		Number of images used for testing	
	Normal	Pathological	Normal	Pathological
50-50	5	55	5	55
60-40	6	66	4	44
70-30	7	77	3	33

value of K increases, some unsuitable training feature vectors are also included in the training subset which brings in a negative effect on the performance of the proposed method.

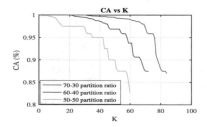

Fig. 2. Variation of CA with respect to the value of K.

The CA, sensitivity and specificity of the proposed hybrid method on different training-testing partition ratio is illustrated in Table 2. However, such training-testing partition ratio results are not available with the other methods used for comparison.

A comparison of the CA, sensitivity and specificity of the proposed hybrid method with the different brain MR image classification methods is illustrated in Table 3. For comparison we have used other methods of extraction of alternative characteristics to the PCA such as GLCM and TRSFFQR technique.

Table 2. CA, sensitivity, specificity of the proposed hybrid method

Training-testing partition		CA	Sensitivity	Specificity
50-50		99.89%	100%	100%
60-40		100%	100%	100%
70-30		100%	100%	100%
10-fold cross validation	mean	99.78%	99.51%	99.13%
	max	100%	100%	100%
	std. dev	0.0341	0.0214	0.0544

Additionally, classifiers namely: ANN, k-NN, SVM, BPNN and Naive Bayes, J48 and IBk are also considered for the comparison. The criterion used to select the comparison methods is inspired by the review literature suggested in [4]. From Table 3, it is witnessed that our proposed hybrid method performs better in terms of CA, sensitivity and specificity as compared to the different brain MR image classification methods. The reason behind this improvement may be due to (1) selection of a subset of training feature vectors which allows us to eliminate the redundant ones and (2) use of GNRBA for classification, which is efficiently able to identify the correct class of the test feature vector. The results demonstrate that the proposed hybrid method can be used as a supportive tool to assist the experts in the clinical diagnosis. The cells with no value indicates unavailability of data from the concerned reference.

Table 3. Comparison of the proposed hybrid method with different brain MR image classification methods

Name of the method	CA(%)	Sensitivity(%)	Specificity(%)
DWT+PCA+ANN	97	95.9	96
DWT+PCA+k-NN	98	96	97
FPCNN+DWT+PCA+BPNN	99	100	92.8
Texture feature+Ensemble based SVM	99.78	-	-
TRSFFQR	93.5	92.5	-
Proposed hybrid method	100	100	100

Sometimes, CA alone can't be used to predict the performance of a method because it does not provide detailed information. The performance index like confusion matrix allows us to capture the detailed information from a method. For a binary classifier, it is a matrix consisting of two rows and two columns. Each element of the matrix represents one of the four possible outcomes, i.e. true positive (TP), true negative (TN), false positive (FP) and false negative (FN). The elements along the diagonal represents correct classification. The elements along the off-diagonal represents the amount of misclassification. A comparison of the confusion matrix of the proposed hybrid method with the different brain image classification methods is shown in Table 4. From Table 4, it is observed that TP and TN is 100% for the proposed hybrid method. The result demonstrates the superiority of the proposed hybrid method as compared to the other methods.

Table 5 presents the comparison of the proposed hybrid method with different methods, in terms of the AUC values. The AUC values are calculated by using the trapezoidal rule. From Table 5, it is noticed that the AUC value of the proposed hybrid method is higher as compared to all other methods in all training-testing partition ratio. The result reveals the superiority of the proposed hybrid method.

Table 4. Comparison of the confusion matrix

Name of the method	Actual result	Predicted result	
		Normal	Pathological
DWT+PCA+ANN	Normal	58	0
	Pathological	2	9
DWT+PCA+k-NN	Normal	60	0
	Pathological	1	9
FPCNN+DWT+PCA+BPNN	Normal	87	0
	Pathological	1	13
TRSFFQR	Normal	43	1
	Pathological	16	140
Proposed hybrid method	Normal	3	0
	Pathological	0	33

Table 5. Comparison of the AUC Values

Name of the method	AUC
TRSFFQR	0.9320
FPCNN+DWT+PCA+BPNN	0.9800
Proposed hybrid method	0.9892 (50-50)
	0.9994 (60-40)
	0.9997 (70-30)
	0.9977 (10-fold cross validation)

4   Conclusion

Feature selection and classification is a challenging task in the field of medical image analysis. It could assist experts for making clinical diagnosis. In this paper, we have proposed a novel hybrid method for pathological brain MR image classification. The method utilizes the strengths and benefits of DWT and GNRBA. From the results, it is witnessed that the proposed hybrid method outperforms the existing methods. It is noticed that the proposed hybrid method is better as compared to all other methods in terms of CA, sensitivity, specificity, confusion matrix and AUC value. The superiority of the method reveals that it could assist the experts in clinical diagnosis or can be used for providing the second opinion for final diagnosis.

The performance of the proposed hybrid method can be tested with multimodal images in the future. In addition, the use of different feature extraction methods using our approach can also be investigated.

Acknowledgment. This work is supported by seed fund grant provided under TEQIP-II, Veer Surendra Sai University of Technology, Burla.

References

1. Selvanayaki, K., Karnan, M.: CAD system for automatic detection of brain tumor through magnetic resonance image-a review. Int. J. Eng. Sci. Technol. **2**(10), 5890–5901 (2010)
2. American Cancer Society. https://www.cancer.org/cancer/brain-spinal-cord-tumors-adults/about/key-statistics.html
3. Kharrat, A., Benamrane, N., Messaoud, M.B., Abid, M.: Detection of brain tumor in medical images. In: 3rd International Conference on Signals, Circuits and Systems (SCS), Medenine, Tunisia, pp. 1–6 (2009)
4. El-Dahshan, E.S.A., Mohsen, H.M., Revett, K., Salem, A.B.M.: Computer-aided diagnosis of human brain tumor through MRI: a survey and a new algorithm. Expert Syst. Appl. **41**(11), 5526–5545 (2014)
5. El-Dahshan, E.S.A., Hosny, T., Salem, A.B.M.: Hybrid intelligent techniques for MRI brain images classification. Digit. Signal Proc. **20**(2), 433–441 (2010)
6. Ain, Q., Jaffar, M.A., Choi, T.S.: Fuzzy anisotropic diffusion based segmentation and texture based ensemble classification of brain tumor. Appl. Soft Comput. **21**, 330–340 (2014)
7. Al-Kadi, O.S.: A multiresolution clinical decision support system based on fractal model design for classification of histological brain tumours. Comput. Med. Imaging Graph. **41**, 67–79 (2015)
8. Jothi, G., Inbarani, H.H.: Hybrid tolerance rough set-firefly based supervised feature selection for MRI brain tumor image classification. Appl. Soft Comput. **46**, 639–651 (2016)
9. Othman, M.F., Basri, M.A.M.: Probabilistic neural network for brain tumor classification. In: 2nd International Conference on Intelligent Systems, Modelling and Simulation (ISMS), Kuala Lumpur, Malaysia, pp. 136–138 (2011)
10. Sachdeva, J., Kumar, V., Gupta, I., Khandelwal, N., Ahuja, C.K.: A package-SFERCB-Segmentation, feature extraction, reduction and classification analysis by both SVM and ANN for brain tumors. Appl. Soft Comput. **47**, 151–167 (2016)
11. Dora, L., Agrawal, S., Panda, P., Abraham, A.: Optimal breast cancer classification using Gauss-Newton representation based algorithm. Expert Syst. Appl. **85**, 134–145 (2017)
12. Daubechies, I.: Ten Lectures on Wavelets. Society for Industrial and Applied Mathematics, Philadelphia (1992)
13. Hiremath, P.S., Shivashankar, S., Pujari, J.: Wavelet based features for color texture classification with application to CBIR. Int. J. Comput. Sci. Netw. Secur. **6**(9A), 124–133 (2006)
14. Zhang, Y., Wang, S., Wu, L.: A novel method for magnetic resonance brain image classification based on adaptive chaotic PSO. Prog. Electromagn. Res. **109**, 325–343 (2010)
15. Messina, A.: Refinements of damage detection methods based on wavelet analysis of dynamical shapes. Int. J. Solids Struct. **45**(14), 4068–4097 (2008)
16. Belhumeur, P.N., Hespanha, J.P., Kriegman, D.J.: Eigenfaces vs. fisherfaces: recognition using class specific linear projection. IEEE Trans. Pattern Anal. Mach. Intell. **19**(7), 711–720 (1997)
17. Jang, J.S.R., Sun, C.T., Mizutani, E.: Neuro-Fuzzy and Soft Computing: A Computational Approach to Learning and Machine Intelligence. Prentice-Hall, Englewood Cliffs (1997)

18. Gill, P.E., Murray, W., Wright, M.H.: Practical Optimization, University of Michigan. Academic Press, USA (1981)
19. Fawcett, T.: An introduction to ROC analysis. Pattern Recognit. Lett. **27**(8), 861–874 (2006)
20. Sokolova, M., Lapalme, G.: A systematic analysis of performance measures for classification tasks. Inf. Process. Manag. **45**(4), 427–437 (2009)
21. Hastie, T., Tibshirani, R., Friedman, J.: The Elements of Statistical Learning, 2nd edn. Springer Science & Business Media, New York (2009)

Evaluating Different Similarity Measures for Automatic Biomedical Text Summarization

Mozhgan Nasr Azadani and Nasser Ghadiri[✉]

Department of Electrical and Computer Engineering, Isfahan University of Technology,
84156-83111 Isfahan, Iran
mozhgan.nasr@ec.iut.ac.ir, nghadiri@cc.iut.ac.ir

Abstract. Automatic biomedical text summarization is maturing and can provide a solution for biomedical researchers to access the information they need efficiently. Biomedical summarization approaches often rely on the similarity measure to model the source document, mainly when they employ redundancy removal or graph structures. In this paper, we examine the impact of the similarity measure on the performance of the summarization methods. We model the document as a weighted graph. Various similarity measures are used to build different graphs based on biomedical concepts, semantic types and a combination of them. We next use the graphs to generate and evaluate the automatic summaries. The results suggest that the selection of the similarity measure has a substantial effect on the quality of the summaries ($\approx$37% improvement in ROUGE-2 metric, and $\approx$29% in ROUGE-SU4). The results also demonstrate that exploiting both biomedical concepts and semantic types yields slightly better performance.

Keywords: Biomedical similarity measure · Graph-based text summarization ·
Sentence similarity

1 Introduction

Information overload in biomedical domain is becoming an unprecedented and severe problem for biomedical researchers and clinicians. The number of biomedical works available online has increased exponentially. More than 27 million journal citations for biomedical literature have been indexed in MEDLINE bibliographic database [1]. Moreover, biomedical information is now accessible from various resources such as Electronic Health Record (EHR) systems, online clinical reports or multimedia documents [2]. Notably, this information overload has undermined everyday work of biomedical researchers such as interpreting experimental results, generating research hypotheses or broadening their knowledge [3]. The reason is that the information overload has made it extremely difficult to find the relevant information. Aware of this situation, considerable efforts have been made aiming at helping manage and alleviate this problem.

Automatic text summarization of biomedical literature can assist clinical researchers and information seekers to deal with the information overload challenge and efficiently identify and process the information most pertinent to their needs. This task is

© Springer International Publishing AG, part of Springer Nature 2018
A. Abraham et al. (Eds.): ISDA 2017, AISC 736, pp. 305–314, 2018.
https://doi.org/10.1007/978-3-319-76348-4_30

accomplished by generating a condensed representation of the source document that consists of the most significant parts of its content [4]. Consequently, there have been substantial advances in biomedical text summarization. Most recent text summarization systems in biomedical domain model the input document regardless of the exact words used [5], focusing on the concept-level analysis of the text using biomedical knowledge resources. Moreover, the performance of the summarization systems heavily depends on the similarity measures they use. Some systems use simple methods that benefit from concept frequency, redundancy removal, or cue phrases [6, 7] or more advanced ones such as graph-based methods [8] where the existence or weights of an edge rely on how similar graph nodes are.

The primary objective of this work is to assess the impact of different similarity measures employed to represent the document on the quality of the system-generated summaries. Similarity measures are so crucial for many summarization systems that we hypothesize the use of appropriate similarity measure can be advantageous when we summarize scientific biomedical articles. To this aim, we propose a framework to represent a source document as an undirected weighted graph. Different similarity measures are taken into account to construct various graphs. Afterward, we compare the automatic summaries generated by each graph to evaluate the impact of the similarity measure. The evaluation results demonstrate that selection of the similarity measures to be used to construct the graphs has a significant impact on the performance of the system in terms of the Recall-Oriented Understudy for Gisting Evaluation metrics (ROUGE) [9] ($\approx$37% improvement in ROUGE-2).

The rest of the paper is organized as follows. In Sect. 2, we represent a background of biomedical text summarization as well as a description of different similarity measures. We then discuss our summarization framework in Sect. 3. The evaluation methodology is also explained in this section. The experimental results are presented in Sect. 4. Eventually, we conclude and outline future work in Sect. 5.

2   Background

Automatic text summarization involves distilling the most significant information of a source document to generate a shortened representation. Although many efforts have been made toward developing automatic summarization system commenced in the late 1950s [10, 11], paying attention to biomedical domain singularities has recently interested the researchers. Unlike domain-independent systems, biomedical methods take advantage of domain-specific information and concept-level analysis of the text by using domain knowledge resources such as the Unified Medical Language System (UMLS) [12]. UMLS is a biomedical repository that covers over 1 million biomedical concepts and their corresponding semantic types. It has been employed by various biomedical methods [6, 7, 13–15].

We can classify automatic biomedical summarizers regarding different dimensions. Concerning the output of the summarization, most of the biomedical summarizers are typically an *extractive* summarizer, i.e., a summarizer which identifies, selects, and puts together the most significant and informative sentences of the source document.

However, *abstractive* biomedical summarization has not been neglected [16–18], in which a summarizer generates summary making use of new sentences not seen in the original document. Regarding the goal of the produced summaries, we can classify them into *generic* and *query-focused* summaries, depending on whether or not their contained information provides an answer to a user query [5]. We can also categorize the summarization approaches into *single-document* and *multi-document* based on the number of the input documents to the system.

Various biomedical summarization approaches have been proposed that exploit different summarization techniques and methods. For instance, Reeve et al. [19] managed to combine two different methods, called BioChain [13] and FreqDist [6]. The latter was used to remove redundancy while the former was responsible for finding thematic sentences using concept chaining. In another work, Plaza et al. [8] constructed a semantic graph exploiting UMLS concepts and their semantic relations to represent the text. Their method proceeded by clustering the concepts using a degree-based algorithm. In a similar method, Menendez et al. [15] demonstrated the usefulness of the genetic clustering algorithm for biomedical summarization. Their summarizer took advantage of a genetic graph clustering algorithm to identify and cover the existing topics within a source document.

Recently, Plaza [20] evaluated the use of different knowledge sources for modeling the source document in biomedical text summarization. She investigated the various combinations of UMLS ontologies to retrieve biomedical concepts. It was demonstrated that using a proper knowledge source can improve the performance of the produced summaries. In another work [21], different positional strategies were compared for sentence selection in biomedical summarization. The results showed that taking account of a proper positional strategy can yield better summarization performance. However, to the best of the author's knowledge, the impact of different similarity measures has never been investigated for biomedical summarization.

3   Summarization Framework

In this section, we first discuss the similarity measures used in modeling the source document and then present our graph-based summarization framework based on which the similarity measures are evaluated. Afterward, we explain the evaluation method.

3.1   Similarity Measures

Similarity measures are crucial in text summarization if the method is contingent upon finding how similar two data objects are, particularly for weighting graph nodes or redundancy removal. Different similarity measures can be taken into account when summarizing a document. In this study, we investigate four different similarity measures including cosine similarity [22], Jaccard similarity [22], positional similarity, and TextRank [23] each of which can be calculated using different relations and aspects of data objects (see Sect. 3.2). Imagine we want to compute the similarity between two sentences S_i and S_j that each one consists of a set of terms $S_i = \{t_1, t_2, \ldots, t_n\}$. Using this representation, the similarity measures mentioned above are denoted as:

- Cosine similarity:

$$Similarity_C(S_i, S_j) = \frac{S_i \cdot S_j}{\|S_i\| \|S_j\|} \tag{1}$$

- Jaccard similarity:

$$Similarity_J(S_i, S_j) = \frac{|S_i \cap S_j|}{|S_i \cup S_j|} \tag{2}$$

- TextRank comparer:

$$Similarity_T(S_i, S_j) = \frac{|S_i \cap S_j|}{\log(|S_i|) + \log(|S_j|)} \tag{3}$$

- Positional similarity:

$$Similarity_P(S_i, S_j) = \frac{1}{|pos(S_i) - pos(S_j)|} \tag{4}$$

3.2 Graph-Based Summarizer

We used a single graph-based summarization system for evaluation. Our framework is based on constructing a concept-based model of the input document, exploiting the UMLS, and taking advantage of an agglomerative hierarchical clustering to discover different subthemes of the source document. Figure 1 illustrates the architecture of our proposed summarizer including four main steps of preprocessing, graph creating, graph clustering, and sentence selection, each of which is briefly explained in the following.

- In the first step, we omit unnecessary parts of the content for summary generation such as *references* or *conflicts of interest* parts. Next, we map the source document to biomedical concepts and their corresponding semantic types existing in the UMLS Metathesaurus using MetaMap tool [24]. MetaMap is a widely-used program for mapping the contents of a document to biomedical concepts employing the national language processing approach. Once the text sentences and their noun phrases are identified, MetaMap generates and ranks all potential candidate concepts and selects the one with the highest score. It is worth mentioning that those UMLS concepts whose semantic types are very generic are removed, for they have been assumed excessively broad. Some of these semantic types include *intellectual product*, *quantitative concept, temporal concept,* and *spatial concept* [8]. For example, considering the following sample sentence:

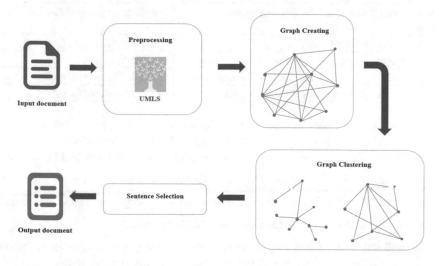

Fig. 1. The architecture of the graph-based summarization method

"For example, schizophrenia is treated with antipsychotics, bipolar disorder with mood stabilizers and antipsychotics, major depression with antidepressants and attention-deficit/hyperactivity disorder with psychostimulants" [25].

Its identified UMLS biomedical concepts and their corresponding semantic types are shown in Table 1.

- In the second step, we model the source document as an undirected weighted graph, where nodes are the sentences of the document, and the links represent how similar these sentences are. Different similarity measures can be defined and used in this step showing various aspects of the sentences. In this work, we use four different comparers mentioned in Sect. 3.1 to define sentence similarity based on extracted biomedical concepts and semantic types. We also suggest using the combination of both biomedical concepts and their corresponding semantic types. As a result, we measure a single similarity function like cosine similarity in three different ways. The term-level comparison is excluded, for it does not either take biomedical domain singularities into account or produce a satisfactory performance. Regardless of the used similarity measure, there will be a link between every two sentence pairs that share common information. The outcome representative graph is somehow connected where edges weights show the strength of the connections between different sentence pairs. For example, consider the following sentence from the sample document and the one mentioned in the previous step. *"Schizophrenia genome-wide association studies data support a polygenic basis for the disorder and estimate that common single nucleotide polymorphisms explain approximately one-third of the total variation in liability to schizophrenia"* [25]. In this example, there are two semantic types (mobd and phsu) and one concept (schizophrenia) shared

Table 1. Extracted UMLS concepts for the sample sentence.

Concept (semantic type)	
Schizophrenia (mental or behavioral dysfunction (mobd))	Stabilizer (organ or tissue function)
Major depressive disorder (mobd)	Antidepressive agent (phsu)
Treat (therapeutic or preventive procedure)	Bipolar disorder (mobd)
Antipsychotic agents (pharmacologic substance (phsu))	Example (conceptual entity)
Attention-deficit/hyperactivity disorder (mobd)	Psychostimulants (phsu)

between the two sentences one of which contains 10 concepts and 5 semantic types and the other one has 7 concepts and 9 semantic types. Therefore, according to Eq. (2), their Jaccard similarity based on concepts, semantic types and the combination of them is $\frac{1}{16}, \frac{2}{12}$, and the average of them, respectively. Consequently, in the graph constructed using concepts and Jaccard similarity, the weight of the edge between the nodes representing these two sentences will be $\frac{1}{16}$.

- The third step aims to divide the vertices of the created graph into groups to discover and cover different subthemes of the document. To this end, we employ clustering technique which refers to grouping several data objects into various clusters in a way that the object similarity is maximized within a cluster and minimized between the clusters. Our summarization system benefits from an agglomerative hierarchical clustering algorithm. The aim is to make sets of sentences closely related in meaning. First, each sentence is initially considered as a single cluster. Afterward, we define a similarity matrix $M_{k \times k}$ based on the sentence similarities obtained in the previous step, where k is the total number of vertices in the graph. The matrix consists of the weights of the links in the graph showing how similar the sentence pairs in the input document are. It is calculated as follows:

$$M_{ij} = \begin{cases} Similarity(S_i, S_j) & i < j \\ 0, & otherwise \end{cases} \tag{5}$$

After calculating the matrix, we identify the most similar clusters each time and merge them into a new cluster. The similarity between two clusters C_m and C_n is defined as the average of the similarity values between two sentences each of which belongs to one of the clusters, and it is denoted as:

$$Similarity(C_m, C_n) = \frac{\sum_{i,j | S_i \in C_m \wedge S_j \in C_n} M_{ij}}{N} \tag{6}$$

where N is the number of similarity values added in the numerator. The clustering algorithm stops only if the number of the clusters reaches to a predefined number. The pseudocode of our agglomerative hierarchical clustering algorithm is shown in Algorithm 1.

- The last step consists of selecting the most informative sentences for the summary. Our sentence selection strategy is to score the sentences based on their similarity and choose the most similar ones from all existing clusters constructed in the previous step. This way, we can cover all subthemes of the document. The number of the selected sentences from each cluster is proportional to their size.

Algorithm 1. Agglomerative hierarchical clustering

Input: G – a graph representing document D, N –number of clusters, M – similarity matrix
Output: SC = {C₁,C₂,...,Cₙ} – a set of clusters

SC = Ø;
for all nodes in G **do**
    "Create a new cluster in SC";
End for
while |SC| is greater than N **do**
    for all cluster pairs **do**
        "Find the similarity between clusters"
    end for
    "Merge the most two similar clusters"
end while
return SC;

3.3 Evaluation Design

To assess the impact of different similarity measures on the final automatic summaries, we compare the generated summaries produced by varying the similarity function under a single graph-based summarization method. It is noteworthy that all parameters of the summarization system are kept unchanged so that we could only vary the similarity measure based on which the representative graph is constructed. In the following, we describe the evaluation method and corpus as well as the used evaluation metrics.

The system-generated summaries also known as peer are commonly compared with human-made summaries, called model summaries, regarding shared information content. A peer is assumed to be better if more information is shared between the peer and the model summary. Although this comparison can be made either manually or automatically, it is most commonly done automatically, since manual evaluation is time-consuming and it is difficult for human judges to reach a consensus. Therefore, we exploit ROUGE package [9] to evaluate the generated summaries. ROUGE is a widely-used evaluation method for text summarization which evaluates the quality of an automatic summary by estimating the shared content between it and a model summary. While estimating, ROUGE takes account of n-gram co-occurrence and term sequence matching. In this research, we employ four different ROUGE metrics including ROUGE-1, ROUGE-2, ROUGE-W-1.2, and ROUGE-SU4.

As mentioned earlier in this subsection, the automatic summaries are commonly compared with model or references summaries. To the authors' knowledge, there are no biomedical documents with their model summaries available to be used in summary evaluation. In this case, the most common approach is to use biomedical articles as the input of the summarization system and their abstracts as model summaries. In this work, we use a collection of 300 biomedical articles randomly selected from the open access corpus of BioMed Central [26]. Moreover, we consider the abstract of the selected articles as their model summaries.

To evaluate the automatically generated summaries, we exploit a 30% compression rate. That means only 30% of all sentences existing in the input document are included in the final summary. This choice is based on a well-accepted standard of compression rate ranging from 15% to 35% [27].

4   Experimental Results

In this section, we present the experimental results. We compared different ROUGE scores obtained from our summarization framework when different similarity measures are used in graph creating process. We construct the graphs employing cosine similarity, Jaccard similarity, TextRank comparer and positional similarity each of which is based on biomedical concepts (C), semantic types (S) and a combination of both (C+S). Table 2 represents different ROUGE scores (ROUGE-2, ROUGE-W-1.2, and ROUGE-SU4) for all similarity measures and relations.

Table 2. ROUGE scores for summaries generated by graph-based summarizer using different similarity measures and relations. The best scores per measure are shown in bold.

Similarity measures	ROUGE-2	ROUGE-W-1-2	ROUGE-SU4
Cosine sim (C)	0.3255	0.0767	0.3813
Cosine sim (S)	0.3260	0.0773	0.3848
Cosine sim (C+S)	**0.3330**	**0.0782**	**0.3989**
Jaccard sim (C)	0.3187	0.0752	0.3752
Jaccard sim (S)	0.3187	0.0759	0.3777
Jaccard sim (C+S)	**0.3319**	**0.0775**	**0.3848**
TextRank sim (C)	0.3340	0.0775	0.3879
TextRank sim (S)	0.3228	**0.0783**	0.3797
TextRank sim (C+S)	**0.3349**	0.0780	**0.3887**
Positional sim	0.2444	0.0617	0.3089

As can be seen in Table 2, the selection of the appropriate similarity measure is of utmost importance while summarizing the biomedical scientific articles. The results illustrated an approximately 37% improvement in ROUGE-2, 27% improvement in ROUGE-1-W-2, and 29% improvement in ROUGE-SU4. Moreover, concerning the relationships used to measure the similarity between the sentences, it may be observed that using both the biomedical concepts and semantic types together produces relatively better results. The reason might be the fact that taking both concepts and semantic types into account conveys more semantics and meaning rather than using them separately. As an example, for cosine and Jaccard similarity, the results demonstrated a roughly 5% improvement in ROUGE-SU4 and 4% improvement in ROUGE-2, respectively.

5   Conclusion

Automatic biomedical text summarization is a useful technique aims at helping the biomedical researchers to manage information overload challenge. In this paper, we have investigated the effect of exploiting various similarity measures and relations used to model biomedical documents on the summarization performance. To this aim, we present a graph-based biomedical text summarization framework. The framework represents the source document as a weighted graph where the nodes are the sentences, and the weights of the edges are based on the similarity relation between the sentences. We evaluated the summarizer performance on different representative graphs that consider different similarity measures. Overall, the results of the evaluation revealed that the selection of similarity measure as well as the relations based on which the similarity is computed had a significant impact on the quality of the generated summaries. We also found that taking account of both biomedical concepts and their corresponding semantic types while computing the similarity provides slightly better summaries.

As future work, we will focus on investigating the impact of combining different similarity measures not only on the quality of the generated summaries but also on the accuracy of clusters used in text summarization and clustering.

References

1. MEDLINE. https://www.nlm.nih.gov/bsd/pmresources.html. Accessed 7 Oct 2017
2. Afantenos, S., Karkaletsis, V., Stamatopoulos, P.: Summarization from medical documents: a survey. J. Artif. Intel. Med. **33**, 157–177 (2005)
3. Fleuren, W.W.M., Alkema, W.: Application of text mining in the biomedical domain. J. Meth. **74**, 97–106 (2015)
4. Jones, K.S.: Automatic summarising: the state of the art. J. Inf. Process. Manage. **43**, 1449–1481 (2007)
5. Mishra, R., Bian, J., Fiszman, M., Weir, C.R., Jonnalagadda, S., Mostafa, J., et al.: Text summarization in the biomedical domain: a systematic review of recent research. J. Biomed. Inform. **52**, 457–467 (2014)
6. Reeve, L.H., Han, H., Nagori, S., Yang, J.C., Schwimmer, T.A.: Concept frequency distribution in biomedical text summarization. In: Proceedings of the 15th ACM International Conference on Information and Knowledge Management, pp. 604–611 (2006)
7. Sarkar, K.: Using domain knowledge for text summarization in medical domain. Int. J. Recent Trends Eng. **1**, 200–205 (2009)
8. Plaza, L., Díaz, A., Gervás, P.: A semantic graph-based approach to biomedical summarisation. J. Artif. Intell. Med. **53**, 1–14 (2011)
9. Lin, C.-Y.: Rouge: a package for automatic evaluation of summaries. In: Proceedings of Workshop on Text Summarization Branches Out, Workshop of ACL (2004)
10. Gambhir, M., Gupta, V.: Recent automatic text summarization techniques: a survey. J. Artif. Intell. Rev. **47**, 1–66 (2017)
11. Yao, J.-G., Wan, X., Xiao, J.: Recent advances in document summarization. J. Knowl. Inf. Syst. **53**, 297–336 (2017)
12. Nelson, S.J., Powell, T., Humphreys, B.L.: The Unified Medical Language System (UMLS) project, Encyclopedia of library (2002)

13. Reeve, L.H., Han, H., Brooks, A.D.: BioChain: lexical chaining methods for biomedical text summarization. In: Proceedings of the 2006 ACM Symposium on Applied Computing, pp. 180–184. ACM (2006)

14. Yoo, I., Hu, X., Song, I.-Y.: A coherent graph-based semantic clustering and summarization approach for biomedical literature and a new summarization evaluation method. J. BMC Bioinform. **8**, S4 (2007)

15. Menendez, H.D., Plaza, L., Camacho, D.: A genetic graph-based clustering approach to biomedical summarization. In: Proceedings of the 3rd International Conference on Web Intelligence, Mining and Semantics, pp. 1–8. ACM (2013)

16. Fiszman, M., Demner-Fushman, D., Kilicoglu, H., Rindflesch, T.C.: Automatic summarization of MEDLINE citations for evidence-based medical treatment: a topic-oriented evaluation. J. Biomed. Inform. **42**, 801–813 (2009)

17. Zhang, H., Fiszman, M., Shin, D., Wilkowski, B., Rindflesch, T.C.: Clustering cliques for graph-based summarization of the biomedical research literature. J. BMC Bioinform. **14**, 182 (2013)

18. Zhang, H., Fiszman, M., Shin, D., Miller, C.M., Rosemblat, G., Rindflesch, T.C.: Degree centrality for semantic abstraction summarization of therapeutic studies. J. Biomed. Inform. **44**, 830–838 (2011)

19. Reeve, L.H., Han, H., Brooks, A.D.: The use of domain-specific concepts in biomedical text summarization. J. Inf. Process. Manage. **43**, 1765–1776 (2007)

20. Plaza, L.: Comparing different knowledge sources for the automatic summarization of biomedical literature. J. Biomed. Inform. **52**, 319–328 (2014)

21. Plaza, L., Carrillo-de-Albornoz, J.: Evaluating the use of different positional strategies for sentence selection in biomedical literature summarization. J. BMC Bioinform. **14**, 71 (2013)

22. Han, J., Kamber, M., Pei, J.: Data Mining: Concepts and Techniques, 3rd edn. Morgan Kaufmann Publishers, Burlington (2011)

23. Mihalcea, R., Tarau, P.: TextRank: bringing order into text. In: Proceeding of the 2004 Conference on Empirical Methods in Natural Language Processing, pp. 404–411 (2004)

24. National Library of Medicine. MetaMap portal. https://mmtx.nlm.nih.gov/. Accessed 25 May 2017

25. Doherty, J.L., Owen, M.J.: Genomic insights into the overlap between psychiatric disorders: implications for research and clinical practice. J. Genome Med. **6**, 29 (2014)

26. BioMed Central. https://old.biomedcentral.com/about/datamining. Accessed 15 Mar 2017

27. Mitkov, R.: The Oxford Handbook of Computational Linguistics. Oxford University Press, Oxford (2003)

Fingerprint Image Enhancement Using Steerable Filter in Wavelet Domain

K. S. Jeyalakshmi[1] and T. Kathirvalavakumar[2(✉)]

[1] Department of Computer Science,
N.M.S.S.Vellaichamy Nadar College (Autonomous), Madurai 625 019, India
jeyal2007@gmail.com
[2] Research center in Computer Science, V.H.N.S.N. College (Autonomous),
Virudhunagar 626 001, India
kathirvalavakumar@yahoo.com

Abstract. The proposed work is to enhance the features of the fingerprint image using steerable filter in wavelet domain to increase the accuracy and speed of Automatic fingerprint identification system. The proposed method uses steerable filter and wavelet. The steerable filter allows filtering process adaptively to any orientation and determining analytically the filter output as a function of orientation and the wavelet domain speeds up the computation process. The steerable filter is applied on each local blocks of approximation image of wavelet transform for tuning up the fingerprint image features and then smoothing the resultant which leads to enhanced image. Experiments are conducted on FVC databases and results show that enhancement process reveals clear visualization of fingerprint images.

Keywords: Fingerprint enhancement · Wavelet transform
Steerable filter · Orientation field · Principal component analysis
Multi-scale pyramid decomposition

1 Introduction

Most widely agreeable and trustable biometric technique is fingerprint though there are various biometrics available for identifying a person because of its uniqueness, reliability and ease of use. In the hectic world, poor quality (wet or dry) images are captured without any care, for the automatic personal identification. These poor quality fingerprint images lead to degrade the accuracy of Automatic fingerprint identification system (AFIS). Its accuracy can be increased if an enhancement process is used prior to the feature extraction process.

Hong et al. [9] have used Gabor filter as a low pass filter to eliminate noise and retain features of fingerprint but produced blocking artifacts in the enhanced fingerprint image. Hsieh et al. [10] have proved that the multi-resolution analysis and wavelet transform improve the fingerprint image visualization and continuity of ridge structures. Yang et al. [11] have developed the modified Gabor filter

© Springer International Publishing AG, part of Springer Nature 2018
A. Abraham et al. (Eds.): ISDA 2017, AISC 736, pp. 315–325, 2018.
https://doi.org/10.1007/978-3-319-76348-4_31

using the traditional Gabor filter which preserves fingerprint image structure and achieves image enhancement consistency than traditional Gabor filter. Blotta and Moler [1] have used differential hysteresis processing for enhancing the defective fingerprint images. Zhang et al. [22] have proposed a fingerprint enhancement using wavelet transform combined with Gabor filter. Cheng and Tian [3] have introduced the scale space theory in the computer vision to enhance the fingerprint. Anisotropic filter combined with directional median filter was used by Wu et al. [18] for fingerprint image enhancement. In signal processing, short time fourier transform (STFT) analysis is used to analyze non-stationary signals. An extension of STFT's application to 2-D fingerprint images was applied by Chikkerur et al. [4] for fingerprint enhancement. Cavusoglu and Gorgunoglu [2] have proposed a fingerprint enhancement algorithm that uses the direction for designing a mask of parabolic coefficients and leads to fast filtering. Wang et al. [16] have proposed an algorithm that enhances fingerprints using log-Gabor filter. Gottschlich [8] has introduced a filter named curved Gabor filters that locally adapt their shape with respect to the orientation. Ye et al. [21] have used 2-D empirical mode decomposition in their proposed fingerprint enhancement algorithm.

Even though lots of algorithms were proposed for fingerprint image enhancement, improvement in speed and accuracy are still needed. In the proposed method, segmentation is done for identifying region of interest in a input fingerprint image and then orientation field of fingerprint image is calculated based on Principle Component Analysis and multi-scale pyramid decomposition those have both robust to noise and accuracy in finding local orientation. Next the fingerprint image is transformed into wavelet domain since it has the faster computation speed. Steerable filter is then applied on the local blocks of approximation image for having clarity in the image features and enhanced image is obtained by replacing the intensity of each pixel in the small window with the average intensity of its neighboring pixels to remove blocking artifacts. Rest of the paper is organized as follows: Fingerprint image enhancement procedure is elaborated in Sects. 2 and 3 discusses the experimental results.

2   Fingerprint Image Enhancement

2.1   Segmentation

The ridge & valley area, the area of interest, are segmented from irretrievable non ridge and non valley area. The method of Wu et al. [17] is applied to generate a mask on resized fingerprint image using morphological operations 'erode' and 'open' for separating ridge and valley area from the background.

2.2   Orientation Field Estimation

Flow of ridges over the point has the direction called orientation which lies in $[0,\pi)$ [12]. Local ridge orientation has a significant role in fingerprint singular point detection, classification and enhancement. Calculation of local ridge orientation using Feng and Milanfar [5] follows:

Step 1: Initialize orientation image and energy rate R as 0.

Step 2: Construct n layers of gradient pyramid.

Step 3. Do the following task from n^{th} layer to 1st layer.

(A) Divide the gradient image of each layer into local blocks with 1 pixel overlapping.

(B) Estimate dominant orientation of each block as follows.

    (a) Group the gradient of each block into a matrix G of size N × 2 where N is a number of pixels in a block.

    (b) Compute SVD for the matrix G, $G = USV^T$ [5] where U is order of N × N; S is order of N × 2; and V is order of 2 × 2, in which the first column v_1 represents the dominant orientation of the gradient field.

    (c) Rotate v_1 by 90° to obtain the dominant orientation.

    (d) Find energy rate $R = \frac{s_1 - s_2}{s_1 - s_2}$ where s_1 and s_2 are singular values of S.

    (e) If energy rate R of current layer is greater than energy rate of its parent layer then energy rate R and propagation weight are updated.

    (f) Up-sample the orientation image and energy rate R by 2.

2.3 Homogeneous Zones Division

Removing noise and enhancing fingerprint image features by filter uses local orientation of the fingerprint [22]. This orientation is to be homogenized in order to retain originality and to avoid artifacts in the enhanced image. Divide the orientation image into n homogeneous zones.
H^k be the label of the orientation of kth homogeneous zone.

$$H^k = \frac{(k-1)}{n}\pi \quad if(\frac{(k-1)}{n}\pi - w_0 \le O(i,j) < \frac{(k-1)}{n}\pi + w_0) \qquad (1)$$

where $w_0 = \frac{\pi}{2n}$ and the orientation of block(i,j) is represented by O(i,j). Assign a label for each homogeneous zone as follows.

$$Homogeneous\ zone\ H^1 = 0 \quad if(-w_0 \le O(i,j) < w_0) \qquad (2)$$

$$Homogeneous\ zone\ H^2 = \frac{\pi}{n} \quad if(\frac{\pi}{n} - w_0 \le O(i,j) < \frac{\pi}{n} + w_0) \qquad (3)$$

$$Homogeneous\ zone\ H^3 = \frac{2\pi}{n} \quad if(\frac{2\pi}{n} - w_0 \le O(i,j) < \frac{2\pi}{n} + w_0) \qquad (4)$$

$$\vdots$$

$$Homogeneous\ zone\ H^n = \frac{(n-1)\pi}{n} \quad if(\frac{(n-1)\pi}{n} - w_0 \le O(i,j) < \frac{(n-1)\pi}{n} + w_0) \qquad (5)$$

2.4   Wavelet Coefficient Adjustment

By applying discrete wavelet transform with level one, approximation image and three detail images are obtained by decomposing the fingerprint image. Generally gain is applied on all the wavelet coefficients by using a non mapping function. If the gain is applied to all the wavelet coefficients then enhancement will be affected because noise is also enhanced. Therefore gain is applied only to the approximation image coefficients and not to the noise [13]. In the proposed work, steerable filter is applied on local blocks of approximation image with its orientation to perform wavelet coefficient adjustment.

2.5   Steerable Filtering

Steerable filter is a filter of arbitrary orientation synthesized as a linear combination of a set of basis filters. It can change the shape of a filter for any orientation [6]. Basis filters are designed with second order derivatives since they have a stronger response to fine details and produce double response when steps change in gray level that leads to image enhancement [7]. In the proposed work, Gaussian distribution function G (6) is used and adopted the filter design used by Freeman and Adelson [6].

$$Gaussian\ distribution\ function\ G(x,y) = \frac{e^{\frac{-(x^2+y^2)}{2\sigma^2}}}{\sqrt{2\pi}} \quad (6)$$

The degree of (6) is 2 and at least 3 basis filters are required for the function G (minimum number of basis filters required = degree of function + 1 = 2 + 1 = 3 [6]. These basis filters are G_2^x the second order derivative of G w.r.to x, G_2^y the second order derivative of G w.r.to y and $G_2^x y$ the second order derivative of G w.r.to x, y. The first order derivative of Gaussian function w.r.to x is

$$G_1^x = \frac{\partial G}{\partial x} = \frac{\partial}{\partial x}\left(\frac{e^{\frac{-(x^2+y^2)}{2\sigma^2}}}{\sqrt{2\pi}}\right) = -\frac{xe^{\frac{-(x^2+y^2)}{2\sigma^2}}}{\sigma^3\sqrt{2\pi}} = -\frac{x}{\sigma^2}G(x,y) \quad (7)$$

The basis filter G_2^x is

$$G_2^x = \frac{\partial^2 G}{\partial x^2} = \frac{\partial}{\partial x}\left(-\frac{x}{\sigma^2}G(x,y)\right) = \frac{(x^2-\sigma^2)e^{\frac{-(x^2+y^2)}{2\sigma^2}}}{\sigma^5\sqrt{2\pi}} = \frac{(x^2-\sigma^2)}{\sigma^4}G(x,y) \quad (8)$$

The basis filter G_2^{xy} is

$$G_2^{xy} = \frac{\partial^2 G}{\partial y\partial x} = \partial\left(-\frac{x}{\sigma^2}G(x,y)\right)\partial y = \frac{xye^{\frac{-(x^2+y^2)}{2\sigma^2}}}{\sigma^5\sqrt{2\pi}} = \frac{xy}{\sigma^4}G(x,y) \quad (9)$$

The first order derivative of Gaussian function w.r.to y is

$$G_1^y = \frac{\partial G}{\partial y} = \frac{\partial}{\partial y}\left(\frac{e^{\frac{-(x^2+y^2)}{2\sigma^2}}}{\sqrt{2\pi}}\right) = -\frac{ye^{\frac{-(x^2+y^2)}{2\sigma^2}}}{\sigma^3\sqrt{2\pi}} = -\frac{y}{\sigma^2}G(x,y) \qquad (10)$$

The basis filter G_2^y is

$$G_2^y = \frac{\partial^2 G}{\partial y^2} = \frac{\partial}{\partial y}\left(-\frac{y}{\sigma^2}G(x,y)\right) = \frac{(y^2-\sigma^2)e^{\frac{-(x^2+y^2)}{2\sigma^2}}}{\sigma^5\sqrt{2\pi}} = \frac{(y^2-\sigma^2)}{\sigma^4}G(x,y) \qquad (11)$$

G_2^x, G_2^{xy} and G_2^y are basis filters for G_2^θ where θ is a filter angle. A linear combination of G_2^x, G_2^{xy} and G_2^y synthesize G_2^θ, the steerable filter, with an arbitrary orientation θ as [6]:

$$G_2^\theta = \cos^2\theta\, G_2^x - 2\cos\theta\sin\theta\, G_2^{xy} + \sin^2\theta\, G_2^y \qquad (12)$$

An image I convolved with basis filters G_2^x, G_2^{xy} and G_2^y of G_2^θ yield the filter responses R_2^x, R_2^{xy} and R_2^y. Then linear combinations of the filter responses at an arbitrary orientation θ synthesize an enhanced image R_2^θ

$$R_2^x = G_2^x * I \qquad (13)$$

$$R_2^{xy} = G_2^x y * I \qquad (14)$$

$$R_2^y = G_2^y * I \qquad (15)$$

Then,

$$R_2^\theta = \cos^2\theta\, R_2^x - 2\cos\theta\sin\theta\, R_2^{xy} + \sin^2\theta\, R_2^y \qquad (16)$$

Steerable filter with the homogeneous orientation is applied on each block of approximation image to enhance fingerprint image. Non-overlapping blocks of size of 8×8 pixels are obtained by dividing approximation image. (16) is applied on each block of the approximation image with the orientation θ the dominant angle of that block to get the enhanced image. Pixel averaging is done to remove blocking artifacts in the enhancement process by replacing every pixel's intensity value with the average of intensity values of all neighboring pixels in a window of size n × n. In this process pixel considered for averaging is to be a centre of the window. Necessary cell padding is to be done for making the pixel as centre of the window.

3   Results and Discussion

The proposed algorithm is implemented on the benchmark fingerprint databases set B of FVC 2000 [23], FVC 2002 [24], and FVC 2004 [25]. There are totally 960 fingerprint images. Each database has 8 different rolling of 10 fingers. The characteristics of each database are given in Table 1. The proposed algorithm is implemented in MATLAB 2013a for all the databases.

Table 1. Characteristics of fingerprint databases

FVC database	Sensor type	Image size	# Images	Resolution
2000 DB1_B	Low-cost optical sensor	300×300	80	500 dpi
2000 DB2_B	Low-cost capacitive sensor	256×364	80	500 dpi
2000 DB3_B	Optical sensor	448×478	80	500 dpi
2000 DB4_B	Synthetic generator	240×320	80	About 500 dpi
2002 DB1_B	Optical	388×374	80	500 dpi
2002 DB2_B	Optical	296×560	80	569 dpi
2002 DB3_B	Capacitive	300×300	80	500 dpi
2002 DB4_B	Synthetic	288×374	80	500 dpi
2004 DB1_B	Optical sensor	640×480	80	500 dpi
2004 DB2_B	Optical sensor	328×364	80	500 dpi
2004 DB3_B	Thermal sweeping Sensor	300×480	80	500 dpi
2004 DB4_B	Synthetic	288×384	80	About 500 dpi

The input fingerprint images are resized for uniformity and segmented to identify the region of interest. Figure 1 shows the fingerprint images before and after segmentation. Dominant orientation of each non-overlapping block is estimated (Fig. 2) and the estimated orientation image (Fig. 2) is divided into homogeneous zones as in Fig. 3. DWT with level one is applied on the input image. The resultant image composed of approximation image and the detail images are shown in Fig. 4(a) (b), (c) and (d) respectively. The basis filters of steerable filter are shown in Fig. 5. The sample input blocks and its corresponding filtered blocks obtained after applying the steerable filter G_2^θ is shown in Fig. 6. Each non-overlapping block of approximated image was tuned by steerable filter with its orientation followed by pixel averaging which leads to enhanced fingerprint image as shown in Fig. 7a and its thinned image is shown in Fig. 7b. Figure 8 shows the extracted minutiae before and after enhancement. The performance of the fingerprint image enhancement is assessed by minutiae based method. In this, D is a set of minutiae identified from the enhanced fingerprint image using Chikkerur and Govindaraju [4] and M is a set of minutiae marked by human expert. For evaluating the quality of identified minutiae, the following terms are used Ratha et al. [14]:

- True minutiae (M_T): A detected minutiae D_i which matches with the minutiae identified by human expert. ($D_i = M_i$)
- False minutiae (M_F): A detected minutiae D_i which does not match with the minutiae identified by human expert. ($D_i \neq M_i$)
- Missed minutiae (M_M): A minutiae identified by human expert Mi which was not detected. ($M_i \notin D_i$)

Fig. 1. Input image (Left) and Segmented image (Right)

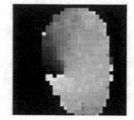

Fig. 2. Orientation image   **Fig. 3.** Homogeneously divided image

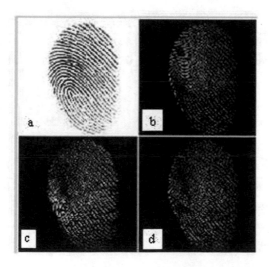

Fig. 4. Sub-images after DWT

Fig. 5. Basis filters

- Type-Exchanged minutiae (M_E): Type of detected minutiae Di which matched wrongly with the type of minutiae identified by human expert. (Type(D_i) ≠ Type(M_i) i.e. bifurcation as ending and ending as bifurcation). The true minutiae ratio (TMR), false minutiae ratio (FMR), missed minutiae ratio (MMR) and type-exchanged minutiae ratio (EMR) are calculated as the ratio of number of M_T, M_F, M_M and M_E divided by total number of minutiae respectively [4] for the randomly selected eight fingerprint images from each database and average of its values are tabulated in Table 2. The result shows that the enhanced image obtained by the proposed method found no FMR and no EMR whereas the image without enhancement results with FMR and EMR. Similarly better evaluation is found for TMR and MMR for the enhanced image.

Table 2. Average performance of original and enhanced image

Image	TMR (%)	FMR (%)	MMR(%)	EMR (%)
Before enhancement	40.8	39.5	14.3	5.4
After enhancement	97.15	0	2.85	0

Table 3. Comparison of average TMR, FMR, MMR and EMR of original and enhanced image

Image	TMR (%)	FMR (%)	MMR(%)	EMR (%)
Original image	40.8	39.5	14.3	5.4
Chikkerur et. al. [4]	92.5	3.3	2.11	2.09
Yang et al. [20]	95.4	2.33	1.14	1.13
Proposed method	**97.15**	**0**	**2.85**	**0**

Table 4. Comparison of computational time

Method	Time
Hong et al. [9]	2.2 s
Hsieh et al. [10]	2.3 s
Chikkerur et al. [4]	4.3 s
Proposed method	**1.6 s**

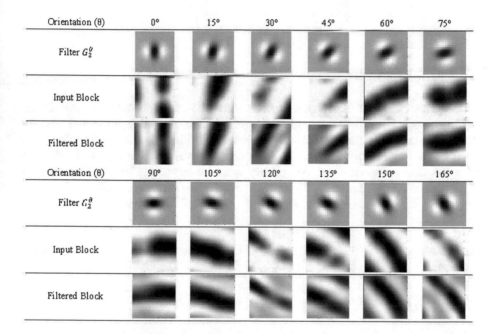

Fig. 6. Input block, its corresponding filter at orientation (θ) and its corresponding filtered block

Fig. 7. a. Enhanced image b. Thinned image

Fig. 8. Extracted minutiae before (left) and after Enhancement (right)

4   Conclusion

Fingerprint image enhancement process has a significant improvement in the performance of AFIS. The proposed method uses steerable filter and wavelet domain to enhance fingerprint image. Steerable filter allows one to adaptively steer a filter to any orientation. Wavelet transform reduces image size, influence of noise, and computational complexity and hence speeds up the consequence processes of AFIS. Hence AFIS performance is also improved in time and accuracy. The proposed method yields good quality of fingerprint images in lesser time.

Acknowledgement. This work is funded by University Grants Commission Major Research Project (MRP: F.No. 42-144/2013(SR)), New Delhi, INDIA

References

1. Blotta, E., Moler, E.: Fingerprint image enhancement by differential hysteresis processing. Forensic Sci. Int. **141**(2), 109–113 (2004)
2. Cavusoglu, A., Gorgunoglu, S.: A fast fingerprint image enhancement algorithm using a parabolic mask. Comput. Electr. Eng. **34**(3), 250–256 (2008)
3. Cheng, J., Tian, J.: Fingerprint enhancement with dyadic scale-space. Pattern Recognit. Lett. **25**(11), 1273–1284 (2004)
4. Chikkerur, S., Cartwright, A.N., Govindaraju, V.: Fingerprint enhancement using STFT analysis. Pattern Recognit. **40**(1), 198–211 (2007)
5. Feng, X.G., Milanfar, P.: Multiscale principal components analysis for image local orientation estimation. IEEE Signals Syst. Comput. **1**, 478–482 (2002)
6. Freeman, W.T., Adelson, E.H.: The design and use of steerable filter. IEEE Trans. Pattern Anal. Mach. Intell. **13**(9), 891–906 (1991)
7. Gonzalez, R.C., Woods, R.E.: Digital Image Processing, 3rd edn, p. 717. Pearson Education India, New Jersey (2009). Image Analysis
8. Gottschlich, C.: Curved-region-based ridge frequency estimation and curved Gabor filters for fingerprint image enhancement. IEEE Trans. Image Process. **21**(4), 2220–2227 (2012)

9. Hong, L., Wan, Y., Jain, A.: Fingerprint image enhancement: algorithm and performance evaluation. IEEE Trans. Pattern Anal. Mach. Intell. **20**(8), 777–789 (1998)
10. Hsieh, C.-T., Lai, E., Wang, Y.-C.: An effective algorithm for fingerprint image enhancement based on wavelet transform. Pattern Recognit. **36**(2), 303–312 (2003)
11. Yang, J., Liu, L., Jiang, T., Fan, Y.: A modified Gabor filter design method for fingerprint image enhancement. Pattern Recognit. Lett. **24**(12), 1805–1817 (2003)
12. Maltoni, D., Maio, D., Jain, A., Prabhakar, S.: Handbook of Fingerprint Recognition, p. 103. Springer Science & Business Media, London (2009)
13. Paul, A.M., Lourde, R.M.: A study on image enhancement techniques for fingerprint identification. In: IEEE International Conference on Video and Signal Based Surveillance, p. 16 (2006)
14. Ratha, N.K., Chen, S., Jain, A.K.: Adaptive flow orientation-based feature extraction in fingerprint image. Pattern Recognit. **28**(11), 1657 1672 (1995)
15. Shi, Z., Govindaraju, V.: A chaincode based scheme for fingerprint feature extraction. Pattern Recogni. Lett. **27**(5), 462–468 (2006)
16 Wang, W., Li, J., Huang, F., Feng, H.: Design and implementation of Log-Gabor filter in fingerprint image enhancement. Pattern Recognit. Lett. **29**(3), 301–308 (2008)
17. Wu, C., Tulyakov, S., Govindaraju, V.: Robust point-based feature fingerprint segmentation algorithm. In: Lee, S.-W., Li, S.Z. (eds.) International Conference on Biometrics. LNCS, vol. 4642, pp. 1095–1103. Springer, Heidelberg (2007)
18. Wu, C., Shi, Z., Govindaraju, V.: Fingerprint image enhancement method using directional median filter. In: Proceedings of SPIE, vol. 5404, p. 67 (2004)
19. Yang, J., Liu, L., Jiang, T., Fan, Y.: An effective algorithm for fingerprint image enhancement based on wavelet transform. Pattern Recognit. **36**(2), 303–312 (2003)
20. Yang, J., Xiong, N., Vasilakos, A.V.: Two-stage enhancement scheme for low quality fingerprint images by learning from the images. IEEE Trans. Hum. Mach. Syst. **43**(2), 235–248 (2013)
21. Ye, Q., Xiang, M., Cui, Z.: Fingerprint image enhancement algorithm based on two dimension EMD and Gabor filter. Procedia Eng. **29**, 1840–1844 (2012)
22. Zhang, W., Tang, Y.Y., You, X.: Fingerprint enhancement using wavelet transform combined with Gabor filter. Int. J. Pattern Recognit. Artif. Intell. **18**(8), 1391–1406 (2004)
23. FVC2000 Set B Databases. http://bias.csr.unibo.it/fvc2000/download.asp
24. FVC2002 Set B Databases. http://bias.csr.unibo.it/fvc2002/download.asp
25. FVC2004 Set B Databases. http://bias.csr.unibo.it/fvc2004/download.asp

Privacy Preserving Hu's Moments
in Encrypted Domain

G. Preethi$^{(\boxtimes)}$ and Aswani Kumar Cherukuri

School of Information Technology and Engineering, VIT University,
Vellore, Tamil Nadu, India
preetgitty@gmail.com, cherukuri@acm.org

Abstract. Privacy preserving image processing is an active area of
research that focuses on ensuring security of sensitive images stored
in an untrusted environment like cloud. Hu introduced the concept of
moment invariants that are widely employed in pattern recognition. The
moment invariants are used to represent the global shape features of an
image that are insensitive to basic geometric transformations like rotation, scaling and translation. In view of this fact, this paper addresses
the problem of moment invariants computation in an encrypted domain.
A secure Hu's moments computation is proposed based on a fully homomorphic encryption scheme. This method may be employed for feature
extraction without revealing sensitive image information in an untrusted
environment.

Keywords: Privacy · Homomorphic encryption · Feature extraction
Geometric moment · Central moment · Normalized central moment
Hu's moments

1 Introduction

Digital images are semantically rich sources of information that are being stored
and managed on a large scale by remote third parties. The conventional image
encryption techniques used to address the user's privacy demands do not allow
any processing on the encrypted images. Hence the extensive computational
services provided by platforms like Cloud cannot be effectively utilized. Some
important security based applications like secure querying on encrypted data
have been proposed [9]. Image processing in encrypted domain is an active area of
research that focuses on achieving image processing applications without exposing the sensitive images to malicious tampering and unauthorized access. Homomorphic encryption schemes are cryptographic tools that allow computations
on encrypted data. Significant partial [1,2] and fully homomorphic encryption
schemes [3–5] have been proposed over the years with enhanced functionalities on encrypted data. But their application to image processing techniques
is little explored. Local feature descriptors like SIFT [10] and SURF [11] have
been implemented on encrypted images using homomorphic encryption and used

© Springer International Publishing AG, part of Springer Nature 2018
A. Abraham et al. (Eds.): ISDA 2017, AISC 736, pp. 326–336, 2018.
https://doi.org/10.1007/978-3-319-76348-4_32

for privacy preserving image recognition and retrieval. However global feature descriptors like moment invariants have not been addressed so far to the best of our knowledge.

The concept of moment invariants was first introduced by Hu [12] in 1962. He derived seven invariants based on the theory of algebraic invariants. The analysis of Hu's moments in [13] shows that the fluctuation in rotation and scaling invariance decreases with increase in spatial resolution under a minimum threshold. Hu's moments have been applied in geometric invariant feature extraction and pattern recognition [14–16]. In this paper, we have proposed a secure Hu's moments computation based on a fully homomorphic encryption [7] scheme.

The proposed privacy preserving Hu's moments computation is described as a client-server model. The server computes the Hu's moment values on the homomorphically encrypted image of the client. The client on decryption retrieves the original image's moment values. In this paper, the computation of the basic geometric moment, central moments and normalized central moments that are needed for the computation of the seven moment invariants in the encrypted domain has also been realized.

The remainder of this paper is organized as follows: Sect. 2 is the preliminaries on Hu's moments and Fan and Vercauteren cryptosystem. In Sect. 3, the overall design and description of the proposed privacy preserving Hu's moments on encrypted images is presented. Experimental analysis of the proposed model is provided in Sect. 4. Conclusion and future directions are given in Sect. 5.

2 Preliminaries

2.1 Moment Invariants

Moment invariants are widely used global descriptors in pattern recognition and classification for describing image features that are insensitive to transformations (geometric and radiometric). Regular moment invariants are a class of contour-based feature descriptors introduced by Hu [12]. The seven invariant values were derived from normalized central moments upto third order.

I. Geometric Moments
A two dimensional $(p+q)^{th}$ order geometric moment (also known as raw moment) m_{pq} of a MXN image I is defined as

$$m_{pq} = \sum_{x=0}^{x=M-1} \sum_{y=0}^{y=N-1} x^p y^q I(x,y) \tag{1}$$

where (x,y) is the pixel position I and $p, q = 0, 1, 2, \ldots$.

II. Central Moments
The central moment μ_{pq} is generated by shifting the center of the geometric moment in (1) to the centroid $(\bar{x}, \bar{y})$ of $I(x,y)$ as follows:

$$\mu_{pq} = \sum_{x=0}^{x=M-1} \sum_{y=0}^{y=N-1} (x - \bar{x})^p (y - \bar{y})^q I(x,y) \tag{2}$$

where $\bar{x} = m_{10}/m_{00}$ and $\bar{y} = m_{01}/m_{00}$. The central moment values are invariant to image translation.

III. Normalized Central Moments

Central moments in (2) can be used to generate the scale invariant normalized central moments as follows:

$$\eta_{pq} = \frac{\mu_{pq}}{\mu_{00}{}^{\gamma}} \tag{3}$$

where $\gamma = (p+q+2)/2$ and $p+q = 2, 3, \ldots.$

IV. Hu's Moments

Using the normalized central moments in (3), seven moment invariants derived by Hu are as follows:

$$\phi_1 = \eta_{20} + \eta_{02}$$
$$\phi_2 = (\eta_{20} - \eta_{02})^2 + 4\eta_{11}{}^2$$
$$\phi_3 = (\eta_{30} - 3\eta_{12})^2 + (3\eta_{21} - \eta_{03})^2$$
$$\phi_4 = (\eta_{30} + \eta_{12})^2 + (\eta_{21} + \eta_{03})^2$$
$$\phi_5 = (\eta_{30} - 3\eta_{12})(\eta_{30} + \eta_{12})[(\eta_{30} + \eta_{12})^2 - 3(\eta_{21} + \eta_{03})^2]$$
$$\quad + (3\eta_{21} - \eta_{03})(\eta_{21} + \eta_{03})[3(eta_{30} + \eta_{12})^2 - (\eta_{21} + \eta_{03})^2]$$
$$\phi_6 = (\eta_{20} - \eta_{02})[(\eta_{30} + \eta_{12})^2 - (\eta_{21} + \eta_{03})^2] + 4\eta_{11}(\eta_{30} + \eta_{12})(\eta_{21} + \eta_{03})$$
$$\phi_7 = (3\eta_{21} - \eta_{03})(\eta_{30} + \eta_{12})[(\eta_{30} + \eta_{12})^2 - 3(\eta_{21} + \eta_{03})^2]$$
$$\quad - (\eta_{30} - 3\eta_{12})(\eta_{21} + \eta_{03})[3(\eta_{30} + \eta_{12})^2 - (\eta_{21} + \eta_{03})^2]$$

2.2   V. Fan and Vercauteren Cryptosystem

For the proposed privacy preserving Hu's moments computation, the fully homomorphic Fan and Vercauteren cryptosystem [7] is chosen for encrypting the image as the system supports homomorphic addition and homomorphic multiplication. It is a more practical fully homomorphic encryption scheme and is an improvisation of the fully homomorphic scheme proposed in [6]. This section briefs on the key generation, encryption, decryption and homomorphic operations of the Fan and Vercauteren cryptosystem. The notations and symbols for this cryptosystem are used as in [7]. The readers may refer to [7] for an elaborate study. In this cryptosystem, the plaintext space is taken as $\mathbb{R}_t$ for some $t > 1$. Let $\Delta = \lfloor (q'/t) \rfloor$ and $r_t(q')$ denote $q' \bmod t$ such that $q' = \Delta.t + r_t(q')$. It is to be noted that neither q nor t have to be prime nor that t and q have to be coprime.

KeyGen(1^{λ}): For the security parameter λ, the secret key and public key are generated as follows:

A random element s is sampled from a distribution $\chi = \chi(\lambda)$ over $\mathbb{R}$. Set the secret key $sk = s$.

To generate the public key pk, sample elements $a \leftarrow \mathbb{R}_{q'}$ and $e \leftarrow \chi$ and compute $pk = ([-(a \cdot s + e)]_{q'}, a)$.

Encrypt(pk, m): To encrypt a message $m \in \mathbb{R}_t$, let $p'_0 = pk[0]$, $p'_1 = pk[1]$, sample $u, e_1, e_2 \leftarrow \chi$ and compute ciphertext ct as follows:

$$ct = ([p'_0 \cdot u + e_1 + \Delta \cdot m]_{q'}, [p'_1 \cdot u + e_2]_{q'}) \tag{4}$$

Let $ct_0 = [p_0 \cdot u + e_1 + \Delta \cdot m]_{q'}$ and $ct_1 = [p'_1 \cdot u + e_2]_{q'}$

Decrypt(sk, ct): The decryption function to retrieve the original message m is given as follows:

$$m = \left[\left\lfloor \frac{t \cdot [c_0 + c_1 \cdot s]_{q'}}{q'} \right\rceil \right]_t \tag{5}$$

The elements of the ciphertext ct in this scheme is interpreted as coefficients of the polynomial $ct(x)$. Evaluating this polynomial in s, we obtain

$$\lfloor ct(s) \rceil_{q'} = \Delta \cdot m + v$$

from which the original message m can be recovered easily. Using this interpretation, the homomorphic addition and multiplication can be obtained.

Homo.Add(ct_1, ct_2): Given two ciphertexts ct_1 and ct_2, decryption of $ct_1 + ct_2$ will give the addition of their corresponding plaintexts $m_1 + m_2$ as follows:

$$[ct_1(s) + ct_2(s)]_{q'} = [[ct_1[0] + ct_2[0]]_{q'}, [ct_1[1] + ct_2[1]]_{q'}]$$

The noise in the resulting ciphertext grows additively upto a maximum of t.

Homo.Mul(ct_1, ct_2, rlk): Given two ciphertexts ct_1 and ct_2, multiplying $ct_1 * ct_2$ will increase the number of elements in the resultant ciphertexts. Hence a concept called "relinearization" is introduced to reduce the degree 2 ciphertext to degree 1 ciphertext using a relinearization key rlk. There are two ways of generating rlk for homomorphic multiplication.

rlk.version1(sk, T): For $i = 0 \cdots \ell = \lfloor log_T(q') \rfloor$, sample $a \leftarrow \mathbb{R}_{q'}, e_i \leftarrow \chi$ and generate

$$rlk = \left[\left(\left[-(a_i \cdot s + e_i) + T^i \cdot s^2 \right]_{q'}, a_i \right) : i \in [0 \cdots \ell] \right]$$

rlk.version2(sk, p): Sample $a \leftarrow \mathbb{R}_{p' \cdot q'}, e_i \leftarrow \chi'$ and generate

$$rlk = ([-(a \cdot s + e) + p' \cdot s^2]_{q'}, a)$$

Step 1 (Basic Multiplication): Compute

$$c_0 = \left[\left\lfloor \frac{t \cdot (ct_1[0] \cdot ct_2[0])}{q'} \right\rceil \right]_{q'}$$

$$c_1 = \left[\left\lfloor \frac{t \cdot (ct_1[0] \cdot ct_2[1] + ct_1[1] \cdot ct_2[0])}{q'} \right\rceil \right]_{q'}$$

$$c_2 = \left[\left\lfloor \frac{t \cdot (ct_1[1] \cdot ct_2[1])}{q'} \right\rceil \right]_{q'}$$

Step 2a (Relinearization Version 1): Write c_2 in base T, $c_2 = \sum_{i=0}^{\ell} c_2^i T^i$ with $c_2^i \in \mathbb{R}_T$ and set

$$(c_0', c_1') = \left(\left[c_0 + \sum_{i=0}^{\ell} rlk[i][0] \cdot c_2^i \right]_{q'}, \left[c_1 + \sum_{i=0}^{\ell} rlk[i][1] \cdot c_2^i \right]_{q'} \right)$$

Step 2b (Relinearization Version 2): Compute

$$(c_{2,0}, c_{2,1}) = \left(\left[\left\lfloor \frac{c_2 \cdot rlk[0]}{p'} \right\rceil \right]_{q'}, \left[\left\lfloor \frac{c_2 \cdot rlk[1]}{p'} \right\rceil \right]_{q'} \right)$$

The above mentioned scheme is somewhat homomorphic. The authors have shown that this scheme can be converted into a fully homomorphic scheme by using a bootstrapping procedure using a modulus switching technique. The detailed analysis of this scheme can be referred to in [7].

3   Proposed Model

In this section the proposed model of a privacy preserving Hu's moments is described. Incorporating Hu's moments on encrypted domain is not a straight-forward since computation of centroids and normalized central moments involves division operation which is not directly possible on encrypted images. Certain modifications have been introduced to enable privacy preserving computation of Hu's moments. In Fig. 1., the overview of the proposed model is illustrated. As shown in Fig. 1., the client initially encrypts the images using Fan and Vercauteren encryption scheme and sends the encrypted image to the server. On the

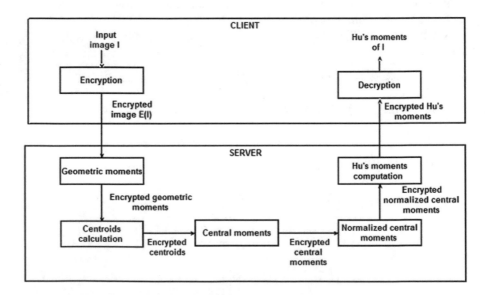

Fig. 1. Overview of the proposed model

server, the Hu's moments are computed over the encrypted images. The resultant encrypted features are sent back to the client that decrypts the result to get the original image's Hu's moments.

I. Image Encryption

The first step in the proposed model is the encryption of the input image by the client. Since the Fan and Vercauteren encryption operates only on integer values, the input image is preprocessed into a binary image. The client encrypts the two dimensional binary input image I pixel-wise and sends the encrypted image E_I to the server for further processing. Given an input image I of size MXN, the pixel-wise encryption is given as follows: For $x = 0 \cdots M - 1$ and $y = 0 \cdots N - 1$,

$$E_I(x,y) = Encrypt(pk, I(x,y)) \qquad (6)$$

Here (x, y) is the pixel position. Computation of Hu's moments require the normalized central moments upto order 3. In the following section, the step wise computation of geometric moments, central moments and normalized central moments is presented.

II. Encrypted Geometric Moments

The $(p + q)^{th}$ order encrypted geometric moment (also known as raw moment) Em_{pq} of a MXN encrypted image E_I can be computed using the homomorphic addition and multiplication properties of Fan and Vercauteren cryptosystem. The algorithm for computing the nth power of a ciphertext followed by the computation of geometric moments on encrypted image is illustrated in Algorithms 1 and 2 respectively.

Algorithm 1. Computation of nth power of ciphertext

1: **procedure** $E_{pow}(a, b)$
2:     **if** $b == 0$ **then**
3:         **return** $Encrypt(pk, 1)$
4:     **end if**
5:     $apow \leftarrow Encrypt(pk, 1)$
6:     $temp \leftarrow a$
7:     **while** $b > 0$ **do**
8:         $apow \leftarrow apow * temp$
9:         $b \leftarrow b - 1$
10:    **end while**
11:    **return** $apow$
12: **end procedure**

Algorithm 2. Computation of Encrypted Geometric moments

1: **procedure** $EncGeometricMoment(E_I, p, q)$
2:     $Em_{pq} \leftarrow Encrypt(pk, 0)$
3:     **for** $x \leftarrow 1 \cdots M$ **do**
4:         **for** $y \leftarrow 1 \cdots N$ **do**
5:             $xpow \leftarrow E_{pow}(Encrypt(pk, x), p)$
6:             $ypow \leftarrow E_{pow}(Encrypt(pk, y), q)$
7:             $Em_{pq} \leftarrow Em_{pq} + (xpow * ypow * E_I(x, y))$
8:         **end for**
9:     **end for**
10:    **return** Em_{pq}
11: **end procedure**

III. Encrypted Central Moments

To compute central moments, centroid values should be calculated. In this proposed model, the division operation required to compute the centroids is achieved by a simple privacy preserving division protocol between client and server. The values exchanged between client and server involves only encrypted data without exposing original data or sharing the decryption key. The privacy preserving approximate division protocol is illustrated in Algorithm 3.

Algorithm 3. Privacy preserving Approximate Division Protocol

    procedure PP_{div}(Enum, $Eden$)
        $sign \leftarrow 1$
3:        $result \leftarrow 1$
        $m1 \leftarrow Enum * Eden$
        if $Decrypt(sk, m1) < 0$ **then**
6:            $sign \leftarrow -1$
        end if
        $result1 \leftarrow 0$
9:        $Enum \leftarrow Enum - Eden$
        while $Decrypt(sk, Enum) >= 0$ **do**
            $Enum \leftarrow Enum - Eden$
12:            $result1 \leftarrow result1 + 1$
        end while
        $result \leftarrow sign * result1$
15:    **return** $Encrypt(pk, result)$
    end procedure

IV. Encrypted Normalized Central Moments

Applying the formula for the normalized central moments in Eq. 3 on ciphertexts using the PP_{div} protocol yields a resultant that decrypts to 0 for most of p and q values. This results in all the moment values to 0 which is not a correct result. In order to obtain more precise moment values, the normalized central moments on encrypted central moment values are approximated as follows:

$$E\eta_{pq} = PP_{div}((E\mu_{pq} * Encrypt(pk, 10^{exp})), E_{pow}(E\mu_{pq}, \gamma))$$

where $\gamma = \lfloor (p + q + 2)/2 \rfloor$ and $p + q = 2, 3, \ldots$. Here the power value exp can be determined based on the number of decimal point values needed to calculate more meaningful moments. Setting the value of $exp = 4$ or 5 can lead to more meaningful moment values. The computation of normalized central moment values is shown in Algorithm 5.

V. Encrypted Hu's Moments

The resulting normalized central moment from Algorithm 5 can be used to generate the unencrypted's moments as follows:

$$E\phi_1 = E\eta_{20} + E\eta_{02}$$

$$E\phi_2 = E_{pow}((E\eta_{20} - E\eta_{02}), 2) + Encrypt(pk, 4) * E_{pow}((E\eta_{11}), 2)$$

$$E\phi_3 = E_{pow}((E\eta_{30} - Encrypt(pk, 3) * E\eta_{12}), 2)$$
$$+ E_{pow}((Encrypt(pk, 3) * E\eta_{21}) - E\eta_{03}), 2)$$

$$E\phi_4 = E_{pow}(E\eta_{30} + E\eta_{12}, 2) + E_{pow}(E\eta_{21} + E\eta_{03}, 2)$$

$$E\phi_5 = (E\eta_{30} - (Encrypt(pk, 3) * E\eta_{12}))(E\eta_{30} + E\eta_{12})$$
$$[E_{pow}((E\eta_{30} + E\eta_{12}), 2) - (Encrypt(pk, 3) * E_{pow}((\eta_{21} + \eta_{03}), 2)]$$
$$+ ((Encrypt(pk, 3) * E\eta_{21}) - E\eta_{03})(E\eta_{21} + E\eta_{03})$$
$$[3(\eta_{30} + \eta_{12})^2 - (\eta_{21} + \eta_{03})^2]$$

$$E\phi_6 = (E\eta_{20} - E\eta_{02})[E_{pow}((E\eta_{30} + E\eta_{12}), 2) - E_{pow}((\eta_{21} + \eta_{03}), 2)]$$
$$+ Encrypt(pk, 4) * E\eta_{11}(E\eta_{30} + E\eta_{12})(E\eta_{21} + E\eta_{03})$$

$$E\phi_7 = ((Encrypt(pk, 3) * E\eta_{21}) - E\eta_{03})(E\eta_{30} + E\eta_{12})$$
$$[E_{pow}((E\eta_{30} + E\eta_{12}), 2) - (Encrypt(pk, 3) * E_{pow}((E\eta_{21}) + E\eta_{03}), 2)]$$
$$- (E\eta_{30} - (Encrypt(pk, 3) * E\eta_{12}))(E\eta_{21} + E\eta_{03})$$
$$[(Encrypt(pk, 3) * E_{pow}((E\eta_{30} + E\eta_{12}), 2)) - E_{pow}((E\eta_{21} + E\eta_{03}), 2)]$$

Obtaining Original Moments:
The above encrypted Hu's moment values form the encrypted feature vector $E\phi = [E\phi_1, E\phi_2, E\phi_3, E\phi_4, E\phi_5, E\phi_6, E\phi_7]$. The server sends the vector $E\phi$ along with the encrypted exp value to the client $(E\phi, Encrypt(pk, n))$.

The client at the receiving end computes $\dfrac{Decrypt(sk, E\phi_i)}{Decrypt(sk, exp)}$ to obtain the final Hu's moment values. As the underlying cryptosystem is provably secure based on the hardness of the RLWE problem, the proposed privacy preserving Hu's moment computation is secure from a third party adversary. The proposed scheme may be further extended into a secure recognition scheme by implementing a homomorphic comparison between moment invariant feature vectors of two

---

Algorithm 4. Computation of Encrypted Central moments

    procedure $EncCentralMoment(E_I, p, q)$
        $\bar{E}_x \leftarrow PP_{div}(Em_{10}, Em_{00})$
        $\bar{E}_y \leftarrow PP_{div}(Em_{01}, Em_{00})$
4:     $E\mu_{pq} \leftarrow Encrypt(pk, 0)$
        for $x \leftarrow 1 \cdots M$ **do**
            for $y \leftarrow 1 \cdots N$ **do**
                $xpow \leftarrow E_{pow}((Encrypt(pk, x) - \bar{E}_x), p)$
8:            $yow \leftarrow E_{pow}((Encrypt(pk, y) - \bar{E}_y), q)$
                $E\mu_{pq} \leftarrow E\mu_{pq} + (xpow * ypow * E_I(x, y))$
            end for
        end for
12:    **return** $E\mu_{pq}$
    end procedure

encrypted images. This may be achieved using secure comparison protocols for homomorphic encryption available in the literature.

4   Experimental Analysis

The proposed secure Hu's moments is implemented on a sample image to validate its performance. We have taken the a binary face-1 image from MPEG7 CE shape database as the sample image. The sample image is resized into 128×128 dimensions and pixel-wise encrypted into a 128×128 encrypted image matrix over which the proposed scheme is executed. An open source HomomorphicEncryption R package [8] was used to implement the proposed proposed model. The sample image was rotated at various angles (every $45°$) and encrypted. On the encrypted image, the proposed privacy preserving Hu's moments were implemented. The resulting encrypted Hu's moment values on decryption were same as the original moment values. Homomorphic evaluation gave correct decryption for the values $d = 8192, \sigma = 512, qmodulus = 1024$ and $t = 268435456$. Table 1 lists out the moment invariant values generated for the sample image under various rotation angles. From Table 1, it is shown that the obtained Hu's moments have only slight variation during rotation. Since the underlying cryptosystem is an fully homomorphic encryption scheme, the obtained values at the decryption end are divided by 10^{exp} to get the desired precision of the moments. Compared to the partial homomorphic schemes like Paillier, the proposed model that is based on a somewhat practical homomorphic encryption scheme achieves secure computation against chosen ciphertext attack and does not disclose potential information about the original image. However the interactions between client and server cause computational complexity to be high. The future direction of this work is aimed at reducing the computational complexity and exploring other statistical image analysis methods for secure computation (Fig. 2).

Fig. 2. Sample face image

Algorithm 5. Computation of Encrypted Normalized Central moments

   procedure $EncNormalizedCentralMoment(E\mu_{pq},p,q,exp)$

$$\gamma \leftarrow \lfloor \frac{p+q+2}{2} \rceil$$

     $temp1 \leftarrow Encrypt(pk, 10^{exp})$

     $temp2 \leftarrow E_{pow}(E\mu_{00}, \gamma)$

5:    $temp3 \leftarrow Encrypt(pk, 1)$

     Select $rand \in [1, q')$

     $temp4 \leftarrow E\mu_{pq} * Encrypt(pk, rand)$

     if $Decrypt(sk, temp4) < 0$ **then**

        $temp3 \leftarrow E\mu_{pq} * Encrypt(pk, -1)$

10:      $E\eta_{pq} = PP_{div}(temp3 * temp1, temp2)$

        $E\eta_{pq} \leftarrow (E\eta_{pq} * Encrypt(pk, -1)) + Encrypt(pk, -1)$

     end if

     if $Decrypt(sk, temp4) > 0$ **then**

        $E\eta_{pq} \leftarrow PP_{div}(E\mu_{pq} * temp1, temp2)$

15:     **end if**

     return $E\eta_{pq}$

   end procedure

Table 1. Hu's moment values for the sample face image at different angles

Moments	Original image	45	90	135	180	225	270	315	360
ϕ_1	0.1700	0.1690	0.1700	0.1700	0.1700	0.1700	0.1700	0.1690	0.1700
ϕ_2	3.1360	3.1370	1.2544	1.2464	3.1360	3.1400	1.2544	1.2441	3.1360
ϕ_3	2.4888	1.2014	5.2479	2.5006	2.9864	5.2638	2.2597	2.5313	2.4888
ϕ_4	3.4743	189.5840	6.7411	2.9883	6.5288	6.4880	3.2574	3.1451	3.4743
ϕ_5	−3.2123	7.8095	1.2677	−2.4905	4.4062	1.1985	8.8369	−2.7778	−3.2123
ϕ_6	1.9425	1.0316	7.5478	5.2637	−8.3189	3.6346	2.2907	2.6847	1.9425
ϕ_7	3.4402	4.5691	−2.3154	6.8554	−7.9810	−3.3538	1.2090	−3.9856	3.4402

5   Conclusion

In this paper, computation of privacy preserving Hu's moments in an encrypted domain is proposed. The proposed model is based on the fully homomorphic encryption scheme enabling both addition and multiplication on encrypted images. The modifications to compute operations like division and power computation for deriving the moment values on the encrypted images have also been discussed. Our proposed scheme may be extended to securely compute more efficient moments like complex moments, orthogonal and affine moments. We believe that our system is the first step towards exploring moment based image feature extraction in an encrypted domain.

References

1. ElGamal, T.: A public key cryptosystem and a signature scheme based on discrete logarithms. IEEE Trans. Inf. Theor. **31**, 469–472 (1985)
2. Paillier, P.: Public-key cryptosystems based on composite degree residuosity classes. In: Proceedings of EUROCRYPT-99, pp. 223–238. Springer, Heidelberg (1999)
3. Gentry, C.: A fully homomorphic encryption scheme. Ph.D thesis. Stanford University, September 2009. http://crypto.stanford.edu/craig
4. Van Dijk, M., Gentry, C., Halevi, S., Vaikuntanathan, V.: Fully homomorphic encryption over the integers. In: Proceedings of Eurocrypt-10, LNCS, vol. 6110, pp. 24–43. Springer, Heidelberg (2010)
5. Brakerski, Z., Vaikuntanathan, V.: Efficient fully homomorphic encryption from (standard) LWE. In: FOCS 2011 Proceedings of the 2011 IEEE 52nd Annual Symposium on Foundations of Computer Science, pp. 97–106 (2011)
6. Brakerski, Z.: Fully homomorphic encryption without modulus switching from classical GapSVP. In: Advances in Cryptology CRYPTO 2012. LNCS, vol. 7417, pp. 868–886. Springer, Heidelberg (2012)
7. Fan, J., Vercauteren, F.: Somewhat Practical Fully Homomorphic Encryption, Cryptology ePrint Archive, Report 2012/144 (2012). http://eprint.iacr.org/2012/144
8. Aslett, L.J.M., Esperan, P.M., Holmes, C.C.: A Review of Homomorphic Encryption and Software Tools for Encrypted Statistical Machine Learning, Technical Report. University of Oxford (2015)
9. Baby, T., Cherukuri, A.K.: On query execution over encrypted data. Secur. Commun. Netw. **8**(2), 321–331 (2015)
10. Hsu, C.Y., Lu, C.S., Pei, S.C.: Image feature extraction in encrypted domain with privacy-preserving SIFT. IEEE Trans. Image Process. **21**, 4593–4607 (2012). https://doi.org/10.1109/TIP.2012.2204272
11. Bai, Y., Zhuo, L., Cheng, B. and Peng, Y. F.: Surf feature extraction in encrypted domain. In: IEEE International Conference on Multimedia and Expo (2014). https://doi.org/10.1109/ICME.2014.6890170
12. Hu, M.: Visual pattern recognition by moment invariants. IRE Trans. Inf. Theor. **IT–08**, 179–187 (1962)
13. Huang, Z., Leng, J.: Analysis of Hu's moment invariants on image scaling and rotation. In: 2nd International Conference on Computer Engineering and Technology, pp. 476–480 (2010). https://doi.org/10.1109/ICCET.2010.5485542
14. Urooj, S., Singh, S.P.: Geometric invariant feature extraction of medical images using Hu's invariants. In: 3rd International Conference on Computing for Sustainable Global Development (INDIACom), New Delhi, pp. 1560–1562 (2016)
15. Isnanto, R.R., Zahra, A.A., Julietta, P.: Pattern recognition on herbs leaves using region-based invariants feature extraction. In: 3rd International Conference on Information Technology, Computer, and Electrical Engineering (ICITACEE), Semarang, pp. 455–459 (2016). https://doi.org/10.1109/ICITACEE.2016.7892491
16. Zhang, Y.: Pathological brain detection based on wavelet entropy and Hu moment invariants. Bio-Med. Mater. Eng. **26**, 1283–1290 (2015)
17. Xu, D., Li, H.: Geometric moment invariants. Pattern Recognit. **41**(1), 240–249 (2008)

Ensemble of Feature Selection Methods for Text Classification: An Analytical Study

D. S. Guru, Mahamad Suhil, S. K. Pavithra[✉], and G. R. Priya

Department of Studies in Computer Science, University of Mysore,
Manasagangotri, Mysore 570006, India
dsg@compsci.uni-mysore.ac.in, mahamad45@yahoo.co.in,
skpavithra163@gmail.com, priya.gr08@gmail.com

Abstract. In this paper, alternative models for ensembling of feature selection methods for text classification have been studied. An analytical study on three different models with various rank aggregation techniques has been made. The three models proposed for ensembling of feature selection are homogeneous ensemble, heterogeneous ensemble and hybrid ensemble. In homogeneous ensemble, the training feature matrix is randomly partitioned into multiple equal sized training matrices. A common feature evaluation function (FEF) is applied on all the smaller training matrices so as to obtain multiple ranks for each feature. Then a final score for each feature is computed by applying a suitable rank aggregation method. In heterogeneous ensemble, instead of partitioning the training matrix, multiple FEFs are applied onto the same training matrix to obtain multiple rankings for every feature. Then a final score for each feature is computed by applying a suitable rank aggregation method. Hybrid ensembling combines the ranks obtained by multiple homogeneous ensembling through multiple FEFs. It has been experimentally proven on two benchmarking text collections that, in most of the cases the proposed ensembling methods achieve better performance than that of any one of the feature selection methods when applied individually.

Keywords: Feature selection · Ranking aggregation
Feature evaluation function · Text classification · Document term matrix

1 Introduction

For the last two decades, automatic content based classification of electronic text documents from huge collections is gaining a vital role in the development of information systems due to exponential growth of the electronic data over the internet and www (Sebastiani 2002; Aggarwal and Zhai 2012). The wide variety of applications of text classification have also made the researchers to explore alternative and effective ways of analyzing the text data. High dimensionality is an important issue in the design of an effective text classification system which is generally handled through feature selection or feature transformation. Feature selection for text classification is the task of reducing dimensionality through the selection of a small subset of features which possess a very high degree of discriminating ability in separating documents of

© Springer International Publishing AG, part of Springer Nature 2018
A. Abraham et al. (Eds.): ISDA 2017, AISC 736, pp. 337–349, 2018.
https://doi.org/10.1007/978-3-319-76348-4_33

different classes. Feature Selection is a major task in the design of text classification system due to the presence of (i) a large number of features (ii) redundant features and (iii) noisy features when the documents are represented in a feature space. The primary goal of feature selection is to improve classification with reduced computational burden. Thus, feature selection helps in building a simple yet effective classification model using a best subset of features (Yang and Pederson 1997; Li et al. 2009).

Feature selection approaches can be subdivided into three categories (i) filter based approaches (ii) wrapper based approaches and (iii) embedded approaches. However, in the literature of text classification the majority of methods developed for feature selection are filter based due to their simplicity and scalability to handle large collections of text data (Bharti and Singh 2015).

A filter based feature selection method uses a feature evaluation function (FEF) to rank the features for their ability in discriminating different classes. Though there are a number of FEFs available for text classification, none of them is capable performing consistently well on verities of text datasets. This is because; (i) text data from different sources carry different complexities and (ii) each FEF tries to capture the importance of a feature by extracting the characteristics of different classes in its own way (Pinheiro et al. 2012). Thus selection of an appropriate FEF to be used during feature selection for a given text collection is a challenging task. One solution to this problem is to combine different feature selection methods in order to obtain a best subset of features. In this direction, we study alternative models for ensembling of feature selection methods to enhance the performance of a text classification system.

The remaining part of the paper is organized as follows. Section 2 provides a brief survey of feature selection methods for text classification along with the alternative ways for ensembling of feature selection methods. Section 3 presents three alternative ways of performing ensemble of feature selection methods along with the rank aggregation techniques. Results of experimentation and analysis is given in Sect. 4. Conclusions and future avenues are presented in Sect. 5.

2 Literature Survey and Motivation

Until the last decade, only a few FEFs such as mutual information, chi-square statistic, and information gain etc., along with their variants have been predominantly used for text classification (Yang and Pedersen 1997; Sebastiani 2002). Recently, a number of novel feature selection algorithms can be traced in literature based on Bayesian principle (Feng et al. 2012; Sarkar et al. 2014; Fenga et al. 2015; Jiang et al. 2016; Zhang et al. 2016), clustering of features (Bharti and Singh 2015), global information gain (Shang et al. 2013), adaptive keyword (Tasci and Gungor 2013), global ranking (Pinheiro et al. 2012; Pinheiro et al. 2015) and term frequency (Wang et al. 2014; Azam and Yao 2012). Further, there are some methods based on combination of feature selection methods and feature transformation techniques such as genetic algorithm (Bharti and Singh 2015; Ghareb et al. 2016) and ant colony optimization (Meena et al. 2012; Dadaneh et al. 2016; Moradi and Gholampour 2016; Uysal 2016).

Given a set of documents for classification, the choice of an appropriate FEF for feature selection is still a difficult task as there is no single FEF which achieves better classification performance for all types of datasets. Hence, instead of working with a subset of features decided by a single FEF, ranks from multiple FEFs can be combined to assign a final rank to each future. This helps in identifying a best subset of features by analyzing different properties of the documents. Recent studies have shown that combining feature selection methods can improve the performance of a classifier by identifying features that are weak as an individual but strong as a group and by determining the features that have high discriminating power (Brahim and Limam 2015; Haque et al. 2016). It is based on the assumption that combining the output of multiple expert decisions is better than single decision.

Ensemble of feature selection methods has been a very less studied topic in text classification. A hybrid feature selection method with two stages has been proposed by Gunal (2012). In the first stage features are selected using a feature selection method. Next, the features selected by a feature selection method are combined together in the second stage through a wrapper method (Gunal 2012). Another method of reducing the dimensionality by combining features through unionization has been proposed by Jalilvanda and Salim (2017) and evaluated for sentiment classification data. A feature relevancy criterion is used to compute the relation of a feature according to their target class.

Ensemble feature selection can be formed in many ways. The two commonly used ensembles for are homogeneous ensemble and heterogeneous ensemble. Homogeneous ensemble uses a same FEF with different training data through distribution of the training dataset over several samplings while the heterogeneous ensemble uses different FEFs with the same training data (Seijo-Pordo et al. 2017; Brahim and Limam 2015; Kumar and Kumar 2012).

In spite of the extensive number of FEFs available for feature selection on text data, there is no significant work investigating the efficacy combining the features selected by a variety of FEFs under different conditions. With this backdrop, in this paper an analytical study of alternative ways of ensembling feature selection methods resulted by multiple FEFs has been made with the following objectives.

- To explore different ways for ensemble of various FEFs based feature selection methods for text classification.
- To identify a best ranking aggregation for combining the ranks of the features.
- Extensive experimentation on benchmarking text datasets to observe the effect of different parameters.

3 Proposed Method

The proposed text classification system involves two major stages viz., training and testing. A collection of N documents from K different classes is used for training a classifier. Initially, the documents undergo various preprocessing such as tokenization, stop word removal and stemming. The preprocessed documents are then used to form a term frequency based document-term matrix (DTM) of size $N \times P$ where P is the number of unique words present across all the N documents after preprocessing which is generally known as bag of words (BoW). The DTM is given as an input to an ensemble feature selection method for ranking of terms which will be discussed in the subsequent subsections. Then a reduced DTM will be created by considering the columns corresponding to only the top d ranked terms. Thus, the reduced DTM is used to train a classifier for the purpose of labeling of unknown documents. During testing, given an unlabeled document d_q, it will be represented in the form of a d-dimensional feature vector of term frequencies using the same d terms selected during training. Then, the class of the document d_q will be identified by the use of the trained classifier.

The major objective of this paper is to design an efficient ensemble feature selection method for selecting a best subset of features which helps in discriminating the documents of different classes effectively, To this end, three different ways for ensembling of feature selection methods have been recommended viz., homogeneous ensemble, heterogeneous ensemble, and hybrid ensemble. The three methods which have been studied are available in literature but have not been explored for the purpose of text classification (Seijo-Pordo et al. 2017; Brahim and Limam 2015). We also have modified the core ideology of the ensemble algorithms to suit our requirements. In the following subsections we present the three ensembling methods in detail.

3.1   Homogenous Ensemble

The training document-term matrix DTM is partitioned into m equal sized partitions by distributing the documents of each class into the M partitions equally through random sampling. Each matrix is called as a sampled training matrix (STM) and thus the STMs corresponding to the m partitions are denoted by STM_1, STM_2, ..., STM_m. Feature selection is performed by applying an FEF uniformly on all the STM's to rank the features. Thus the ith feature f_i gets m ranks due to the application of the FEF on to the m different STM's individually. Then the final rank to the feature f_i is computed through aggregation of the m ranks using a rank aggregation method. Finally, the features are sorted according to their final ranks. Figure 1 shows the architecture of the proposed homogeneous ensemble for feature selection.

3.2   Heterogeneous Ensemble

The given training feature matrix DTM is given as input to the n different of feature selection methods using n different FEFs. Each FEF evaluates the importance of each feature by analyzing the DTM in its own way. Therefore, the ith feature f_i gets n ranks due to the application of n different FEFs on to the DTM. Similar to the homogeneous ensemble, in this case also a rank aggregation mechanism is used to assign a final rank to the feature f_i by using the n ranks. Finally, the features are sorted according to their final ranks. Figure 2 shows a general architecture of the proposed heterogeneous ensemble for feature selection for text classification.

3.3   Hybrid Ensemble

The homogeneous and heterogeneous ensemble methods for feature selection can be combined to further select a most discriminative subset of features. Given are the n different FEFs for feature ranking and m sampled training matrices STM_1,

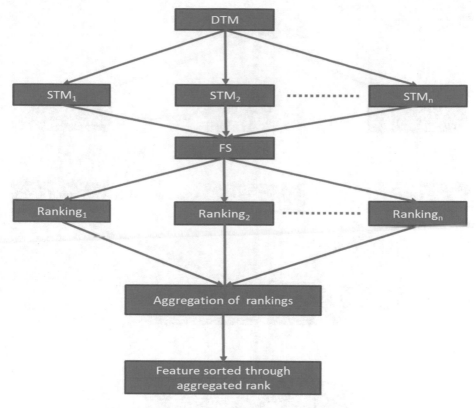

Fig. 1. Typical architecture of a homogeneous ensemble based feature selection

$STM_2, \ldots, STM_m$ of the training feature matrix DTM, the idea of hybrid ensemble is as follows. Initially, for each feature evaluation function, FEF_i (where $i = 1, \ldots, n$), homogeneous ensemble is applied on the DTM to arrive at n different intermediate rankings of the features say $IR_1, IR_2, \ldots, IR_n$. Further, the intermediate rankings are aggregated again using a suitable rank aggregation technique. This approach takes into account the strengths and weaknesses of both homogeneous and heterogeneous ensembles. Figure 3 shows a general architecture of the proposed hybrid ensemble for feature selection for text classification.

3.4   Ranking Aggregation

During ensemble feature selection, the intermediate ranks assigned to the features from different STMs in case of homogeneous, from different FEFs in case of heterogeneous and from different homogeneous ensembles in case of hybrid ensemble should be aggregated to form a final subset of features to be used to represent the original training document-term matrix DTM in a lower dimensional space. In literature, we can find many ways of aggregating multiple subsets of features (Kolde et al. 2012; Abeel et al. 2010; Brahim and Limam 2015). In the proposed work, we make use of union and

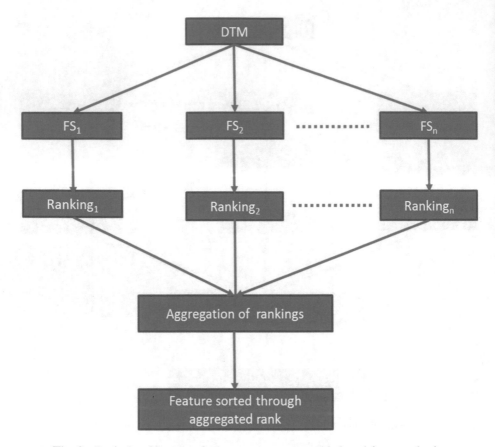

Fig. 2. Typical architecture of a heterogeneous ensemble based feature selection

intersection based ranking aggregation to select a final subset of features. In case of union based aggregation, the union of top d features from all the intermediate rankings are considered as the final subset of features. In contrast to this in intersection based aggregation, intersection of the top d features from all the intermediate rankings are considered to be the final subset of features.

4 Experimentation and Results

In this section, we present the details of experimentation and results obtained by the proposed ensemble feature selection techniques on two benchmarking text collections viz., 20Newsgroups dataset (http://qwone.com/~jason/20Newsgroups/) and Reuters21578 dataset. In case of Reuters21578, a subset formed by the ten categories with higher number of documents has been considered as in (Pinheiro et al. 2015; Chen et al. 2009; Shang et al. 2007; Chang et al. 2008). A comparative study of the performance of the proposed ensembles with that of the individual feature selection methods has been made using the classification accuracy.

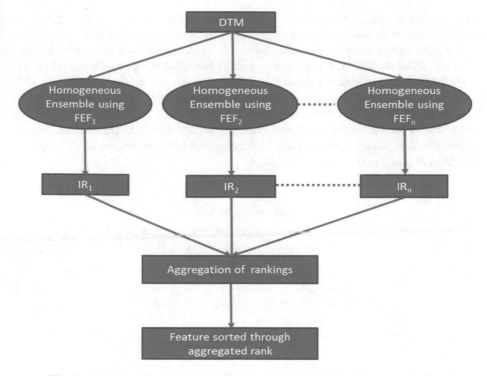

Fig. 3. Typical architecture of a homogeneous ensemble based feature selection

4.1 Experimental Setup

A detailed study of the effectiveness of the proposed ensemble techniques has been made with the help of three FEFs viz., chi-square statistic (CHI), information gain (IG) and weighted log likelihood ratio (WLLR) (Forman 2003; Debole and Sebastiani 2003; Nigam et al. 2000). These FEFs have been selected as they are based on different metrics which ensure diversity in the final ensemble and they are widely used for text classification.

A dataset is initially partitioned into two subsets with 60:40 ratio for training and testing the proposed classification system respectively. Experiments were conducted by varying the number of features, d, from 10 to 4000 to study the nature of the classification accuracy. Results of individual ranking and ensemble ranking approaches were evaluated using the KNN classifier. The parameter K for KNN classifier has been empirically fixed to 5 through various random experiments.

4.2 Results Using Homogeneous Ensemble

Figure 4 shows the accuracy of classification obtained by the proposed homogeneous ensemble on 20 newsgroup dataset with CHI, IG and WLLR feature selection methods. Both union and intersection based ranking aggregation have been used. It can be observed that with the increase in number of features up to 150, the proposed ensemble has attained a consistent increase in performance. After 200 number of features the performance of different ensemble have been slightly deteriorated. But there are some more cross overs among the individual feature selection and ensemble feature selection. The overall observation is that the homogeneous ensemble is superior to any one of the individual feature selection methods in majority of the cases. The ensemble of CHI with union based rank aggregation has outperformed all other homogeneous ensemble combinations. When the number of features is more than 450 accuracy of homogeneous ensemble is similar to the individual feature selection method. The performance of union based ranking aggregation has been relatively better when compared to that of the intersection based ranking aggregation.

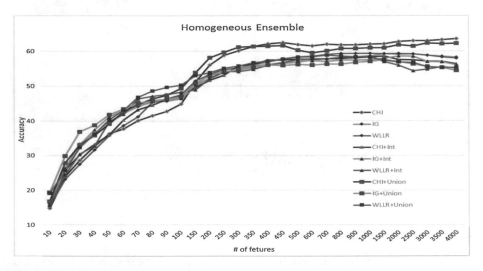

Fig. 4. Classification accuracy of the homogeneous ensemble feature selection on 20News-groups dataset using different FEFs and ranking aggregations

Figure 5 shows the accuracy of classification obtained by the proposed homogeneous ensemble on Reuters21578 dataset with CHI, IG and WLLR feature selection methods. Similar observations as in the case of 20newsgroups dataset can be arrived from the Fig. 5. That is the performance of the homogeneous ensemble based feature selection has attained better performance than of the individual feature selection with number of features up to 150. But there have been some cross overs among performances obtained by the individual feature selection and ensemble feature selection when the number of features are greater than 150. The ensemble of IG with union based rank aggregation has outperformed all other homogeneous ensemble combinations. Similar to the 20Newsgroups dataset, the performance obtained by union based ranking aggregation has been relatively better when compared to that of the intersection based ranking aggregation.

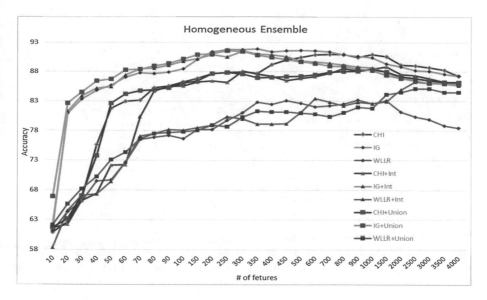

Fig. 5. Classification accuracy of the homogeneous ensemble feature selection on Reuters21578 dataset using different FEFs and ranking aggregations

4.3   Results Using Heterogeneous Ensemble

Figures 6 and 7 show the accuracy of classification obtained by the heterogeneous ensemble on 20Newsgroup and Reuters21578 datasets respectively for different combinations of FEFs. Figures 6(a) and 7(a) show the results using union based ranking aggregation while the results using intersection based ranking aggregations are given in Figs. 6(b) and 7(b) respectively. It can be observed that with the increase in

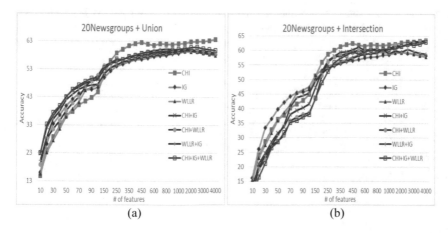

Fig. 6. Classification accuracy of the heterogeneous ensemble feature selection on 20Newsgroups dataset using different FEFs with (a) Union and (b) Intersection based ranking aggregations

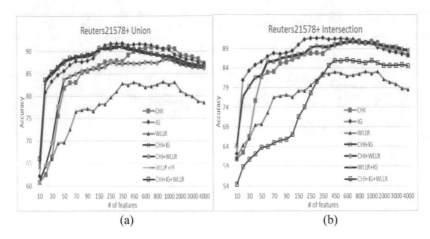

Fig. 7. Classification accuracy of the heterogeneous ensemble feature selection on Reuters21578 dataset using different FEFs with (a) Union and (b) Intersection based ranking aggregations

number of features the heterogeneous ensemble has attained a consistent increase in performance up to 150 number of features. Similar to the homogeneous ensemble feature selection, here too the accuracy of the ensemble feature selection has been superior to any one of the individual feature selection methods in many cases especially for the union based ranking aggregation. This ascertains that the proposed ensemble is superior to any one of the individual feature selection methods. Further, the accuracy obtained by union based ranking aggregation is better than that of the intersection based ranking aggregation. The combination of all the three FEFs has given better results than the other combinations.

Table 1. The feature selection methods (either individual or an ensemble) for which best accuracies are obtained for different number of features (Abbreviations used in table: I = IG, C = CHI, W = WLLR, U = Union, In = Intersection, D1 = 20Newsgroups, D2 = Reuters21578, Hm = Homogeneous, Ht = Heterogeneous)

Ensemble	Dataset	Number of features											
		10	20	30	40	50	60	70	80	90	100	150	200
Hm	D1	I+U	I+U	I+U	I+U	I+U	I+U	W+U	W+U	W+U	W+U	C+U	C+U
	D2	I+U	I+U	I+U	I+U	I+U	I+U	I+In	I+U	I+U	I+U	I+U	I+U
Ht + U	D1	C+I+W	C+I+W	C+I+W	C+I+W	C+I	C+I+W	C+I+W	C+I+W	C+I+W	C+W	C+I+W	C
	D2	C+I+W	C+I+W	C+I+W	C+I	C+I	C+I+W	C+I'	C+I+W	C+I	C+I	C+I+W	I
Ht + In	D1	I	I	I	I	I	I	I	I	W	W	W	C
	D2	C+I	I	I	I	I	I	I	I	I	I	I	I

It can be observed from the figures from Figs. 4, 5, 6 and 7 that beyond 200 number of features the classification accuracies of the ensemble feature selection methods are sometimes lesser than that of some of the individual feature selection methods. However, in majority of the cases, the ensemble methods are outperforming any other individual feature selection methods when the number of features are from 10 to roughly 200. To make this analysis more clear, the details of a feature selection method which has attained best accuracy for number of features from 10 to 200 are shown in Table 1. It can be observed that in all other cases except the case of heterogeneous ensemble with intersection as rank aggregation, ensemble approaches have outperformed than any other individual feature selection methods.

4.4 Results Using Hybrid Ensemble

Figure 8(a) and (b) show the accuracy of classification obtained by the hybrid ensemble on 20Newsgroup and Reuters21578 datasets respectively using union, intersection and average based ranking aggregations. In average based ranking aggregation, a final rank is assigned to each feature as the average of the ranks given to the feature by multiple FEFs using homogeneous ensemble as shown in Fig. 3. In addition, the results of the hybrid ensemble have been compared against the best results obtained among all the previous experiments viz., individual FEFs, homogeneous ensemble and heterogeneous ensemble based classification. It can be observed from the figure that the proposed hybrid ensemble feature selection outperforms the best results obtained in previous experiments. Further, the union and average based ranking aggregations are recommended instead of intersection based ranking aggregation.

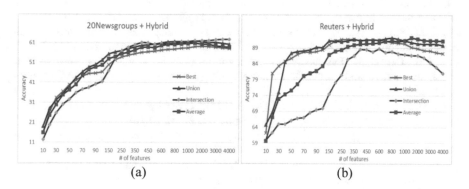

Fig. 8. Classification accuracy of the hybrid ensemble feature selection on (a) 20Newsgroups and (b) Reuters21578 datasets in comparison with the best results of homogeneous and heterogeneous ensemble

5 Conclusion

In this paper, an analytical study has been made on the three alternative models viz., homogeneous ensemble, heterogeneous ensemble and hybrid ensemble for ensembling of feature selection methods driven by various FEFs for text classification. Further, to combine the ranks obtained by different ensemble feature selection methods, rank aggregation techniques have also been explored. To study the performance of the ensemble feature selection methods, chi-square statistic, information gain and weighted log likelihood ratio have been used as FEFs. It has been demonstrated through extensive experiments on the two benchmarking text 20Newsgroups and Reuters21578 that the ensembling methods achieve better performance than any one of the feature selection methods when applied individually. Further, the study can be extended by experimenting with other FEFs, and ranking aggregation algorithms to decide up on a best subset of features for a given dataset. The ensemble based feature selection is a most efficient way of reducing the dimension of feature matrix for text classification, if parallelism can be achieved in applying multiple FEFs to rank the features.

References

Abeel, T., Helleputte, T., Van de Peer, Y., Dupont, C., Saeys, Y.: Robust biomarker identification for cancer diagnosis with ensemble feature selection methods. Bioinformatics 26(3), 392–398 (2010)

Aggarwal, C.C., Zhai, C.X.: Mining Text Data. Springer, Boston (2012). ISBN 978-1-4614-3222-7

Azam, N., Yao, J.: Comparison of term frequency and document frequency based feature selection metrics in text categorization. Expert Syst. Appl. 39, 4760–4768 (2012)

Bharti, K.K., Singh, P.K.: Hybrid dimension reduction by integrating feature selection with feature extraction method for text clustering. Expert Syst. Appl. 42, 3105–3114 (2015)

Brahim, A.B., Limam, M.: Ensemble feature selection for high dimensional data: a new method and a comparative study. Adv. Data Anal. Classif. (2015). https://doi.org/10.1007/s11634-017-0285-y

Dadaneh, B.Z., Markid, H.Y., Zakerolhosseini, A.: Unsupervised probabilistic feature selection using ant colony optimization. Expert Syst. Appl. 53, 27–42 (2016)

Feng, G., Guo, J., Jing, B.Y., Hao, L.: A Bayesian feature selection paradigm for text classification. Inf. Process. Manag. 48, 283–302 (2012)

Fenga, G., Guoa, J., Jing, B.Y., Sunb, T.: Feature subset selection using Naive Bayes for text classification. Pattern Recogn. Lett. 65, 109–115 (2015)

Ghareb, S., Bakar, A.A., Hamdan, A.R.: Hybrid feature selection based on enhanced genetic algorithm for text categorization. Expert Syst. Appl. 49, 31–47 (2016)

Gunal, S.: Hybrid feature selection for text classification. Turk. J. Electr. Eng. Comp. Sci. 20(2), 1296–1311 (2012)

Jalilvanda, A., Salim, N.: Feature unionization: a novel approach for dimension reduction. Appl. Soft Comput. 52, 1253–1261 (2017)

Jiang, L., Li, C., Wang, S., Zhang, L.: Deep feature weighting for Naive Bayes and its application to text classification. Eng. Appl. Artif. Intell. 52, 26–39 (2016)

Kolde, R., Laur, S., Adler, P., Vilo, J.: Robust rank aggregation for gene list integration and meta-analysis. Bioinformatics 28(4), 573–580 (2012)

Kumar, G., Kumar, K.: The use of artificial-intelligence-based ensembles for intrusion detection: a review. Appl. Comput. Intell. Soft Comput. **2012**, 1–20 (2012)

Li, Y.H., Jain, A.K.: Classification of text documents. Comput. J. **41**(8), 537–546 (1998)

Meena, M.J., Chandran, K.R., Karthik, A., Samuel, A.V.: An enhanced ACO algorithm to select features for text categorization and its parallelization. Expert Syst. Appl. **39**, 5861–5871 (2012)

Moradi, P., Gholampour, M.: A hybrid particle swarm optimization for feature subset selection by integrating a novel local search strategy. Appl. Soft Comput. **43**, 117–130 (2016)

Seijo-Pardo, B., Porto-Díaz, I., Bolón-Canedo, V., Alonso-Betanzos, A.: Ensemble feature selection: homogeneous & heterogeneous approach. Knowl. Based Syst. **118**, 124–139 (2017)

Pinheiro, R.H.W., Cavalcanti, G.D.C., Ren, T.I.: Data-driven global-ranking local feature selection methods for text categorization. Expert Syst. Appl. **42**, 1941–1949 (2015)

Pinheiro, R.H.W., Cavalcanti, G.D.C., Correa, R.F., Ren, T.I.: A global-ranking local feature selection method for text categorization. Expert Syst. Appl. **39**, 12851–12857 (2012)

Sarkar, S.D, Goswami, S., Agarwal, A., Aktar, J.: A novel feature selection technique for text classification using Naive Bayes. Int. Sch. Res. Not. **2014**, 1–10 (2014)

Sebastiani, F.: Machine learning in automated text categorization. ACM Comput. Surv. **34**(1), 1–47 (2002)

Shang, C., Li, M., Feng, S., Jiang, Q., Fan, J.: Feature selection via maximizing global information gain for text classification. Knowl. Based Syst. **54**, 298–309 (2013)

Tasci, S., Gungor, T.: Comparison of text feature selection policies and using an adaptive framework. Expert Syst. Appl. **40**, 4871–4886 (2013)

Uysal, A.K.: An improved global feature selection scheme for text classification. Expert Syst. Appl. **43**, 82–92 (2016)

Wang, D., Zhang, H., Li, R., Lv, W., Wang, D.: t-Test feature selection approach based on term frequency for text categorization. Pattern Recogn. Lett. **45**, 1–10 (2014)

Yang, Y., Pedersen, J.O.: A comparative study on feature selection in text categorization. In: Proceedings of the 14th International Conference on Machine Learning, vol. 97, pp. 412–420 (1997)

Zhang, L., Jiang, L., Li, C., Kong, G.: Two feature weighting approaches for Naive Bayes text classifiers. Knowl. Based Syst. **100**, 137–144 (2016)

Correlation Scaled Principal Component Regression

Krishna Kumar Singh[✉], Amit Patel, and Chiranjeevi Sadu

RGUKT IIIT Nuzvid, Krishna 521202, AP, India
krishnasingh@rgukt.in, amtptl93@gmail.com, chiranjeevi@gmail.com

Abstract. Multiple Regression is a form of model for prediction purposes. With large number of predictor variables, the multiple regression becomes complex. It may underfit on higher number of dimension (variables) reduction. Most of the regression techniques are either correlation based or principal components based. The correlation based method becomes ineffective if the data contains a large amount of multicollinearity, and the principal component approach also becomes ineffective if response variables depends on variables with lesser variance. In this paper, we propose a Correlation Scaled Principal Component Regression (CSPCR) method which constructs orthogonal predictor variables having scaled by corresponding correlation with the response variable. That is, the construction of such predictors is done by multiplying the predictors with corresponding correlation with the response variable and then PCR is applied on a varying number of principal components. It allows higher reduction in the number of predictors, compared to other standard methods like Principal Component Regression (PCR) and Least Squares Regression (LSR). The computational results show that it gives a higher coefficient of determination than PCR, and simple correlation based regression (CBR).

Keywords: Multiple regression · Principal components · Correlations
Multicollinearity

1 Introduction

Regression analysis is very useful in forecasting and prediction and in the field of machine learning as well. It is also used to understand which the predictor variables are related to the dependent (response) variable, and to explore the forms of their relationships [1]. In some circumstances, It is used to find out causal relationships between the independent and dependent variables. Sometimes this can lead to spurious or false relationships [2], for example, correlation does not imply causation. In classical multiple linear regression analysis, problems will occur if the regressors are either multicollinear or if the number of regressors is larger than the number of observations [5].

The earliest form of regression was the least squares method which was published by Legendre in 1805 and by Gauss in 1809 [2]. Gauss published his further work of the theory of least squares in 1821, including a technique of the Gauss–Markov theorem.

© Springer International Publishing AG, part of Springer Nature 2018
A. Abraham et al. (Eds.): ISDA 2017, AISC 736, pp. 350–356, 2018.
https://doi.org/10.1007/978-3-319-76348-4_34

The majority of analyses of multidimensional systems are multiple linear regression (MLR), principal component regression (PCR), and partial least squares (PLS) [5, 7, 8]. In principal component regression (PCR) [6] the first k principal components (PCs) of the predictors X are obtained and used as regressors. The main idea behind PCR is to compute the principal components and then use the first few of these components as predictors in a linear regression model and fitting is done using the classical least squares procedure. If all PCs are used in the regression model, the response variable will be predicted with the same accuracy as with the least square approach. Although PCR can deal with multicollinearity, but it does not directly infer the correlation between the predictor variables and the response variable.

The usual way to construct latent predictors is to obtain the first k principal components (PCs) of the predictors' variables X. This approach is called principal component regression (PCR). It is observed in [4] that in some circumstances where the response variable is depending on the predictor variables with lesser variance, PCR gives quite low values of coefficient of determination between the response variable and the predictors.

In this work, a correlation scaled principal component regression (CSPCR) method is proposed. In this method, first we find the correlation of each predictor variable with response variable, then each predictor variable is multiplied by corresponding correlation value, then PCR is applied on varying the number of principal components, this way we construct orthogonal predictor variables having scaled by corresponding correlation with the response variable. The scaling of the predictor variable neutralizes the effect of the predictor having high deviation and low correlation. This method allows higher reduction in the number of predictors, compared to other standard methods like principal component regression (PCR) and least squares regression (LSR).

A correlation based regression (CBR) is also introduced here. It takes the first k predictors in order of non-increasing correlation with the response variable. Further the predictors are selected in such a way that the absolute correlation difference between them is greater than a threshold (say 0.02), which reduces the multicollinearity problem up to significant extent.

It is observed that PCR outperforms CBR when the response variable is led by the predictor variables with high variance, but when the response variable is highly related with the predictor with low variance, CBR outperforms PCR. The proposed method CSPCR outperforms both as it neutralizes the effect of a high variance predictor by scaling it by corresponding correlation with the response variable.

The following sections are organized as follows, Sect. 2 describes background and theory and the next two subsections are notation and algorithm for the introduced method. The Sect. 3 explains simulation and computational result. We concluded with Sect. 4.

2 Background and Theory

2.1 Notations

A linear regression line has an equation [3] of the form $Y = a + bX$ where X is the explanatory variable and Y is the dependent variable. The slope of the line is b, and a is the intercept (the value of y when $x = 0$).

$$b = \frac{\sum_{i=0}^{n} (x_i - \bar{x})(y_i - \bar{y})}{\sum_{i=0}^{n} (x_i - \bar{x})}$$

Or equivalently, $b = (X^T X)^{-1}(X^T Y)$.

And a is obtained as follows

$$a = \bar{y} - b\bar{x}$$

Where,

- y_i denotes the observed response for experimental unit i
- x_i denotes the predictor value for experimental unit i
- $\hat{y}_i$ is the predicted response (or fitted value) for i^{th} observation
- $\bar{x}$ denotes mean of X, $\bar{y}$ denotes mean of Y

The following estimates are considered to evaluate a regression model namely

Goodness of fit/correlation coefficient (R^2):$R^2 = SSM/SST$
Where
Regression sum of squared error:

$$SSM = \sum_{i=1}^{n} (\hat{y}_i - \bar{Y})^2$$

Total Sum of Squared Error:

$$SST = \sum_{i=1}^{n} (y_i - \bar{Y})^2$$

Multivariate regression is an extension of simple linear **regression**. It is used when we want to predict the value of a variable based on the value of more than one predictor variables.

Multiple regression is an extension of Multivariate and Simple linear regression. Like multivariate regression, higher degree (>1) polynomials are used in multiple regression analysis.

To perform principal components (PC) regression, we find eigen vectors and this way the independent variables transformed to their principal components. Mathematically, it is written $X'X = PDP' = W'X$ where D is a diagonal matrix of the eigenvalues of $X'X$, where X is centered by subtracting its mean. P is the eigenvector matrix of $X'X$, and W is a data matrix (similar in structure to X) made up of the principal components.

Having multicollinearity, two or more of the independent variables are highly corre-lated, such that one variable can be predicted using another variable. Consequently, the rank of $X'X$ becomes smaller than its full column rank structure. Under such situations, some of the eigenvalues of $X'X$ get very close to 0. This problem can be easily sorted out in PCR by excluding the principal components having small eigenvalues.

2.2 Description of Algorithm

The correlation based regression (CBR) method is explained as follows:

Step-I: Compute correlation of each predictor with the response variable.
Step-II: Sort the predictor variables with respect to non-increasing order of correlation with the response variable.
Step-III: Take first k predictor variables obtained from Step-II and apply multiple regression with the response variable. In our case k varies from 2 to 10 (that is less or equal to one-fourth of the total number of predictor (=40), taken in computation in this work).

The correlation scaled principal component regression (CSPCR) method is explained as follows.

Step-I: Compute correlation of each predictor with the response variable.
Step-II: Multiply each element in the predictor with the corresponding correlation obtained in step-I.
Step-III: Apply PCR. Take first k principal components as predictors. In our case k varies from 2 to 10 (that is less or equal to one-fourth of the total number of predictor (= 40)).

The computational analysis is explained in Sect. 3.

3 Simulation and Result

For simulating the computation, a synthetic dataset X is used; the data set X is generated with n = 500 samples with m = 40 predictors from specified distributions.

The data set X is generated as follows: the first 10 columns values are taken from normally distributed numbers with mean 10 and standard deviation = 5, the next 10 columns values are taken from exponentially distributed numbers with mean 5. The next 10 columns values are taken from exponentially distributed numbers with mean 25. The last 10 columns values are taken from normally distributed numbers with mean 50 and standard deviation = 25. As 10 columns are taken from the same distribution it includes a significant amount of multicollinearity. The first 20 column variables contain lesser variance than the later 20 columns. Further to randomness $X = X + \Delta$ is computed, where the matrix Δ is based on a random distribution in the range (0, 1).

Furthermore, we generate a response variable as $y = X * a + \delta$ in two different cases; Case-I: The first 20 elements of the vector 'a' are generated from a uniform distribution in the interval $[-1, 1]$, and the remaining elements of a are 0.

So, y is a linear combination of the first 20 columns of X plus an error term.

Case-II: The last 20 elements of the vector 'a' are generated from a uniform distribution in the interval $[-1, 1]$, and the remaining elements of a are 0.

So, y is a linear combination of the last 20 columns of X plus an error term.

The error term δ is obtained from the distribution N (0, 0.8).

The summary of the centered data set is tabulated in Table 1, where first column value is average of the 1st to 10th variables, and second column value is an average of the 11th–20th variables and so on.

Table 1. Summary of the data set

Var	01:10	10:20	20:30	30:40
Min.	−2.40882	−1.0954	−0.9205	−2.5838
1st	−0.67356	−0.6565	−0.6908	−0.6525
Median	0.01522	−0.2414	−0.3756	0.0712
Mean	0	0	0	0
3rd	0.76428	0.4477	0.2209	0.6592
Max.	2.02887	4.5123	2.8584	2.3883

The coefficient of determination (R^2) is used as a result of the simulation and this coefficient is compared to the different methods; namely CSPCR, PCR, and CBR. The coefficient of determination (R^2) is averaged for m = 100 iterations. The numerical values of computational results are shown in Table 2.

Table 2. Computational result of coefficient of determination and error rate

Method	#predictors	Case-I		Case-II	
	M	R^2	ErrorRate	R^2	ErrorRate
CBR	2	0.416	10.348	0.381	44.543
	4	0.514	9.574	0.573	37.526
	6	0.695	7.707	0.591	37.482
	8	0.840	5.407	0.727	31.571
	10	0.888	4.736	0.744	29.996
PCR	2	0.021	13.782	0.368	44.430
	4	0.043	13.454	0.708	28.515
	6	0.070	13.439	1.000	0.974
	8	0.558	8.785	1.000	0.851
	10	0.857	4.873	1.000	0.765
CBPCR	2	0.454	10.023	0.618	35.342
	4	0.711	6.984	0.802	25.129
	6	0.866	4.783	0.999	1.021
	8	0.941	3.074	1.000	0.719
	10	0.980	1.770	1.000	0.716

The results depicted in Figs. 1 and 2 for Case-I and Case-II respectively, show that the proposed CSPCR outperforms other standard approaches, and mainly when we select a very small number of predictor variables.

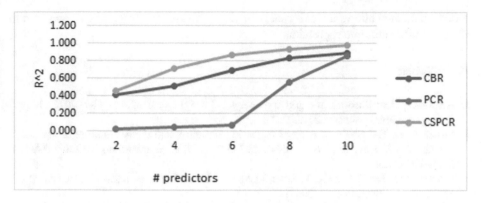

Fig. 1. Number of predictors vs. coefficient of R^2 in Case-I

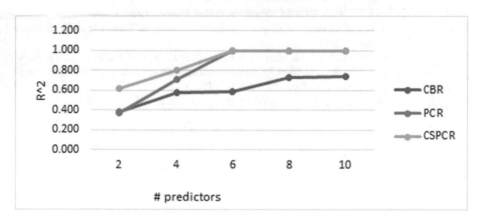

Fig. 2. Number of predictors vs. coefficient of R^2 in Case-II

Case-I: response variable (y) depends on the first twenty variables (having lower spread) of x with random multiplies in (−1, 1).

Case-II: response variable (y) depends on the first twenty variables (having higher spread) of x with random multiplies in (−1, 1).

4   Conclusion

Problems may occur in multiple linear regression if the regressor variables are highly correlated or the number of predictors are less than the number of samples. A simple way to get rid of these problems is to apply PCR because it constructs PCs (orthogonal vector) as regressors. But PCs do not take any information of the response variable into account. As it is clear from Fig. 1, when y depends on regressors with lesser variations, PCR underperforms CBR.

The proposed method (CSPCR) constructs the regressors which are scaled by corresponding correlations with the response variable. The simulation study shows that the proposed method allows a significant reduction of the predictor variables compared to PCR and other multiple regression.

References

1. Armstrong, J.S.: Illusions in regression analysis. Int. J. Forecast. **28**(3), 689 (2012). https://doi.org/10.1016/j.ijforecast.2012.02.001
2. Fisher, R.A.: The goodness of fit of regression formulae, and the distribution of regression coefficients. J. Roy. Stat. Soc. **85**(4), 597–612 (1922). https://doi.org/10.2307/2341124. JSTOR 2341124
3. Kutner, M.H., Nachtsheim, C.J., Neter, J.: Applied Linear Regression Models, 4th edn., p. 25. McGraw-Hill/Irwin, Boston (2004)
4. Filzmoser, P., Croux, C.: Dimension reduction of the explanatory variables in multiple linear regression. Pliska Stud. Math. Bulgar. **14**, 59–70 (2003)
5. Martens, H., Naes, T.: Multivariate Calibration. Wiley, London (1993)
6. Basilevsky, A.: Statistical Factor Analysis and Related Methods: Theory and Applications. Wiley & Sons, New York (1994)
7. Araújo, M.C.U., Saldanha, T.C.B., Galvão, R.K.H., Yoneyama, T., Chame, H.C., Visani, V.: The successive projections algorithm for variable selection in spectroscopic multicomponent analysis. Chemom. Intell. Lab. Syst. **57**, 65–73 (2001)
8. Ng, K.S.: A Simple Explanation of Partial Least Squares, Draft, 27 April 2013

Automated Detection of Diabetic Retinopathy Using Weighted Support Vector Machines

Soumyadeep Bhattacharjee$^{(\boxtimes)}$ and Avik Banerjee

St. Thomas' College of Engineering and Technology, Kolkata, India
soumyadeep.bh1994@gmail.com, bavik022@gmail.com

Abstract. Diabetic retinopathy is a complication of the eye caused by damage to the retinal cells due to prolonged suffering from diabetes mellitus and may lead to irreversible vision impairment in middle-age adults. The proposed algorithm detects the presence of Diabetic Retinopathy (DR) by segmentation of vital morphological features like Optic Disc, Fovea, blood vessels, and abnormalities like hemorrhages, exudates and neovascularization. The images are then classified using Support Vector Machines, based on data points in a multi-dimensional feature space. The proposed method is tested on 140 images from the Messidor database, from which 75 images are used to train an SVM model and the remaining 65 are used as inputs to the classifier.

Keywords: Diabetic retinopathy · SVM · Neovascularization
Optic disc · Fovea · Exudates

1 Introduction

According to the World Health Organization, diabetic retinopathy is a major cause of blindness, caused by prolonged damage to retinal blood vessels due to diabetes mellitus and contributing to 2.6% of global blindness [1]. Existing commercial diagnostic systems require a trained practitioner to examine the retinal fundus scan manually to detect the disease. Such a procedure is difficult in under-developed countries due to the unavailability of trained eye specialists. Computer-aided systems can assist in the early detection of the disease in such regions. Apart from this, with the rapidly growing number of diabetes patients straining the available infrastructure, detection of the disease through minimum manual intervention is the need of the hour.

Previous works reporting integrated systems for diabetic retinopathy screening included that of Singalavanija et al. [2] who used features related to vasculature, drusen, retinal hemorrhages and microaneurysms to screen input images as either normal or abnormal, thus reporting a sensitivity of 74.8% and a specificity of 82.7%. Abramoff et al. [3] prepared a mechanism for DR screening consisting of a questionnaire, visual acuity measurement and four retinal images. These images were then categorized by the software tool and obtained an inter-rater agreement of 93%. Niemeijer et al. [4] developed a DR screening system using information fusion methods that generated an ROC area of 0.881. Fleming et al. [5] developed a CAD based system to detect diabetic retinopathy using the presence of microaneurysms as a factor. The system achieved a sensitivity of 85.4% and specificity of 83.1%.

© Springer International Publishing AG, part of Springer Nature 2018
A. Abraham et al. (Eds.): ISDA 2017, AISC 736, pp. 357–367, 2018.
https://doi.org/10.1007/978-3-319-76348-4_35

Gardner et al. [6] trained a neural network to recognize features of fundus images and observed the sensitivity and specificity of recognition of these features. They examined the capability of the system to correctly predict the presence of retinopathy in fundus images. The study achieved a sensitivity of 88.4% and a specificity of 83.5%.

The proposed algorithm aims to provide an automated and integrated system that detects the risk of macular edema and the presence of diabetic retinopathy after segmentation of relevant morphological features from the retinal fundus image. The algorithm uses a number of methods to detect vasculature, exudates, hemorrhages and neo-vascularization. The procedure is applied on 140 images from the Messidor Database, a research program funded by the French Ministry of Research and Defense. Of them, 75 images are used to train a Support Vector Machine and the remaining 65 images are used as a test-set to evaluate the performance of the trained SVM. In the proposed algorithm, SVM is used over other classifiers as the feature space used is multi-dimensional and provides a higher classifying accuracy. The algorithm reports an accuracy of 95.38%, specificity of 100% and sensitivity of 94.54% in the detection of Diabetic Retinopathy. The algorithm requires 26s to perform the entire process of diagnosis on a single image.

2   Algorithm

2.1   Pre-processing

The proposed algorithm has been implemented on a series of 140 images from the Messidor database. Each image is first resized to a uniform dimension maintaining the original aspect ratio in order to enhance computational efficiency. In all the algorithms described below, the following naming convention has been followed. I_{RGB} is the original resized image, I_{CMYK} is the image converted to CMYK model, I_{YCbCr} is the image converted to YCbCr model and I_{HSV} is the image converted to the HSV color model.

2.2   Detection of Fovea (Macular Region)

In the proposed method we have utilized template matching to locate the centroid of the macular region. The diameter of the macula is estimated to be 40 pixels from existing literature [16] and hence, a 40×40 template of the fovea is extracted from a standard retinal scan. This template scans through the entire image and identifies the region having maximum similarity to it using the sum of absolute differences (SAD) measure. This region is then identified as the macular region. The procedure can be outlined as follows:

1.
$$I_{close} = (I_g \oplus S) \ominus S \tag{1}$$

where S is a disk structuring element. The closing operation eliminates blood vessels and features smaller than the fovea.

2. The contrast adjusted image I_{ad} is filtered with a disk structuring element to obtain a background separated image I_{back}. The closed image is superimposed on the background image to obtain I_1.

3.

$$I_2 = (I_1^c - adjust(I_g)) \times I_1 \tag{2}$$

isolates the macular region. Template matching is performed on I_2 and the centroid of the region in image I_2 with the minimum SAD is taken as the centroid of the fovea (x_{fovea}, y_{fovea}).

4. The image I_2 is scanned to determine and highlight the co-ordinates of each pixel (x, y) within the fovea, where each such point satisfies the equation

$$\left(\left(x_{fovea} - x \right)^2 + \left(y_{fovea} - y \right)^2 \right) \leq r_{fovea}^2 \tag{3}$$

2.3 Segmentation of Blood Vessels

Extraction of retinal blood vessels for further analysis, is a key component in detection and diagnosis of diabetic retinopathy. Vessels appear in retinal fundus images as a wire mesh or tree-like structure and describe a coarse to fine centrifugal distribution. In the proposed method, retinal vessel extraction is performed using Frangi's vesselness filter. Frangi's filter uses Eigen vectors to determine the likeliness of an image region to contain blood vessels. It treats the vessels as tubular structures and provides highly accurate results with accuracy going up to 99% with respect to ground truth images. The procedure can be outlined as follows:

1. The absolute value of the difference of the three channels, I_Y, I_{Cb}, I_{Cr} is calculated as

$$I_1 = |I_Y - I_{Cb} - I_{Cr}| \tag{4}$$

The complement of the difference of I_1 and I_H is calculated as

$$I_2 = (I_1 - I_H)^c \tag{5}$$

All negative pixels are mapped to 0 intensity.
2. Frangi's vesselness filter is applied on the complement of the difference of I_1 and I_2,

$$I_3 = (I_1 - I_2)^c \tag{6}$$

which enhances the blood vessels, generating I_4. All negative pixels are mapped to 0 intensity.
3. I_4 is binarized using Otsu's threshold to obtain image I_5, which is then skeletonized to reduce all detected vessels to single pixel width, thus generating image I_6.

4.
$$I_7 = I_6 - I_{fov} \tag{7}$$

where I_{fov} is the fovea detected image, to mask the fovea pixels thereby preventing false candidates to vessels.

5.
$$I_{vess} = I_7 \times I_{mask} \tag{8}$$

I_{vess} is the final vessel extracted image.

2.4 Detection of Optic Disc

The optic disc or the optic nerve head is one of the primary anatomical features visible in a retinal fundus image. In the macula-centered fundus images, it is located towards the left or right-hand side of the image, with a diameter about one sixth the width of the image, brighter than the neighboring region. In the proposed algorithm, the difference in intensity levels between the optic disc region and the neighboring pixels is used to detect the same. The procedure can be outlined as follows:

1.
$$I_1 = I_g \times I_r \tag{9}$$

the element wise product of the red and green channel matrices, to enhance the optic disc.

2. I_2, the contrast adjusted image is added to the vessel detected image I_{vess} as

$$I_3 = I_2 + I_{vess} \tag{10}$$

to remove blood vessels inside the optic disc. All pixels having intensity higher than the maximum intensity are reduced to the maximum intensity.

3. I_3 is smoothened to get image I_4 contains the Optic Disc as the brightest region. To enhance the computational efficiency, template matching is avoided and the co-ordinates with minimum (x_{min}, y_{min}) and maximum (x_{max}, y_{max}) magnitude having an intensity value 1 are extracted. The centroid of the Optic disc (x_{OD}, y_{OD}) is determined by taking a simple average of the two values.

4. The image I_4 is scanned to determine the co-ordinates of each optic disk pixel (x, y), where each such point satisfies the equation

$$((x_{OD} - x)^2 + (y_{OD} - y)^2) \le r_{OD}^2 \tag{11}$$

where r_{OD} is the OD radius, taken to be 45 pixels. All such pixels are raised to the maximum intensity and all other pixels are suppressed in the binary image I_{OD}, which is the final OD detected image.

2.5 Detection of Exudates

Exudates are abnormalities in the retina formed due to leakage of lipids from blood vessels. They are generally of two types, hard exudates, with distinct boundaries and soft exudates (cotton-wool spots), with less distinct, fimbriate border. Presence of hard exudates causes macular edema which in turn is a potent indicator of the presence of diabetic retinopathy. The procedure for exudate detection can be outlined as follows:

1. Equation (4) is used to remove the chroma information from the luminance channel to subdue the unwarranted illumination around vessels.
2.
$$I_2 = I_1 - I_{ad} + I_{back} \tag{12}$$

which is convoluted with a median filter to obtain I_3.
3.
$$I_4 = I_2 + I_{diff} \tag{13}$$

where I_{diff} is the contrast adjusted image $(I_2 - I_3)$. All negative intensities are mapped to 0.
4. The final image with exudates I_{Ex} is obtained by thresholding I_4 with a value equal to double the minimum intensity of I_4 in order to suppress all the soft exudates keeping only the hard exudates for macular edema detection.

2.6   Gradation of the Risk of Macular Edema

According to the specifications in the Messidor database, the risk of macular edema can have three levels of severity as mentioned in the following Table 1.

Table 1. Classification of risk of macular edema

Grade	Exudate distance from centroid of fovea	Risk
0	No Exudates present	No risk
1	≥ 1 OD Diameter	Minimal
2	< 1 OD Diameter	Maximum

The gradation process is performed as follows:

1. The minimum intensity in image I_{Ex} is determined. If it is zero (signifying no hard exudates) then $Grade_{Ex} = 0$
2. The image I_{Ex} is scanned to determine the co-ordinates of each exudate pixel $I_{Ex}(x, y)$.
3. If the following equation

$$((x_{OD} - x)^2 + (y_{OD} - y)^2) \leq D_{OD}^2 \tag{14}$$

is satisfied by at least one exudate pixel, $Grade_{Ex} = 2$. Otherwise, $Grade_{Ex} = 1$.
4. Thus the final output is a grade on the scale 0–2.

2.7   Detection of Hemorrhages

Retinal hemorrhages are dark, reddish patches in the retina indicative of bleeding. They display low intensity and amorphous, irregular shape within the retina. In the proposed algorithm, hemorrhages are detected using morphological and algebraic operations on the image after application of a decorrelation stretch on all three channels. Since SVM works best with data that are two way classified, the output of the algorithm is the hemorrhage detected image along with a grade (0 or 1), where 0 represents no hemorrhage, 1 represents the presence of hemorrhage. The procedure can be outlined as follows:

1. The image I_{RGB} is subjected to a decorrelation stretch on all three bands (R, G and B) to obtain image I_{decorr}. The green channel $I_{gdecorr}$ of I_{decorr} is extracted.
2. Equation (4) is used to remove uneven illumination, obtaining image I_1.
3. Equation (12) is applied to highlight the vessels and hemorrhages on a suppressed background owing to their similar low intensities, generating image I_2.
4. The magenta channel I_M is subjected to a top-hat transform using a disk shaped structuring element S1.

$$I_3 = (I_M \bullet S1) - I_M \tag{15}$$

5.

$$I_4 = ((I_{gdecorr} \div I_2) + I_3) \tag{16}$$

   amplifies the features other than the hemorrhages and vessels to extremely high intensities, thereby providing noticeable segmentation. $I_{gdecorr}$ and I_2 have been previously contrast adjusted.
6. I_4 is morphologically dilated using a disk shaped structuring element S_2

$$I_5 = (I_4 \oplus S_2) \tag{17}$$

7.

$$I_6 = (I_5 + I_4) \tag{18}$$

   which is binarized to obtain image I_7.
8.

$$I_{haem} = (I_7 - (I_{fov} + I_{OD})) \times I_{mask} \tag{19}$$

   segregates the hemorrhage pixels from the combined pixels of the fovea and optic disc and masks the region external to the retina.
9. The fovea and optic disc have been masked to prevent false candidature to hemorrhages, to obtain I_{haem}, the final hemorrhage detected binary image.
10. I_{haem} is scanned to detect the presence of high intensity pixels (hemorrhage). If no hemorrhages are found, the grade is marked as 0, else it is marked as 1.

2.8   Detection of Neovascularization

Neovascularization is the growth of new vessels in the retina in response to VEGF (vascular endothelial growth factor) produced by hypoxic retina. Neovascularization occurs either in the form of neovascularization on the disc (NVD) or neovascularization elsewhere (NVE). The procedure for detection of neovascularization can be outlined as follows:

1. I_g is contrast adjusted to obtain I_{g1}, which is then subjected to a bottom-hat transform using a disk shaped structuring element S_3 to suppress all features other than the vessels.

$$I_1 = \left(I_{g1} \bullet S_3\right) - I_{g1} \tag{20}$$

where $\bullet$ denotes the morphological closing operation.

2. I_2 is obtained by binarizing I_1 using Otsu's Threshold.
3. I_{skel} is obtained by skeletonization of I_2 to reduce the extracted details to single pixel width.
4.

$$I_{NV} = \left(I_{skel} - I_{vess}\right) \times I_{mask} \tag{21}$$

removes the common vessels detected by the bottom-hat and Frangi's filters. Since Frangi's filter additionally detects the finer vessels, I_{NV} provides the locations where neo-vascularization has occurred.

2.9   Classification Using SVM

Support vector machines (SVM's) are supervised learning models accompanied with learning algorithms that analyze data for classification and regression analysis. An SVM model represents the training set as points in space dividing the samples of separate categories by a discriminative hyperplane with maximum gap between the nearest training data sample from both the classes.

In our case, the feature set, which is assumed to be a p-dimensional vector, is classified by establishing a $(p-1)$-dimensional hyperplane. A decision hyperplane is defined by an intercept β and a weight vector w, perpendicular to the hyperplane. All points x on the decision hyperplane satisfy

$$w^T.x + \beta = 0 \tag{22}$$

To find the optimal hyperplane, the following objective function is to be minimized,

$$\min \frac{1}{2} \|w\|^2 + C \sum_{i=1}^{n} \xi_i \tag{23}$$

subject to,

$$y_i[w^T.x_i + \beta] \geq 1 - \xi_i \quad and$$
$$\xi_i \geq 0 \quad i = 1, 2, \ldots, n$$

Parameter 'C' is the regularization parameter to control over-fitting, whereas ξ_i is a slack parameter that provides an upper bound on the number of training errors. The dual problem of classification becomes:

$$\max \sum_{i=1}^{n} \alpha_i \quad \frac{1}{2} \sum_{i,j=1}^{n} \alpha_i \alpha_j y_i y_j K(x_i, x_j) \tag{24}$$

$$\text{Subject to,} \quad \sum_{i=1}^{n} \alpha_i y_i = 0 \quad and$$
$$0 \leq \alpha_i \leq C, \quad \forall 1 \leq i \leq n$$

where 'α_i' is a Lagrange multiplier and $K(x_i, x_j) = x_i^T.x_j$ is a kernel function corresponding to a linear kernel. The final classification function is evaluated as:

$$F(x) = sgn\left(\sum_{i=1}^{n} \alpha_i y_i K(x_i, x_j) + \beta\right) \tag{25}$$

In the proposed method, six features are used to train the SVM classifier consisting of 75 images in the training set T. Since many of the features assume different values spread across a wide range, feature normalization is performed on the dataset to enhance computational efficiency of the SVM. The normalized feature x' with the minimum and maximum feature value in the dataset x_{min} and x_{max} respectively, is evaluated as,

$$x_i' = \frac{x_i - x_{min}}{x_{max} - x_{min}} \tag{26}$$

To further enhance classification accuracy, the normalized feature x' is standardized using a standard score, so that all feature values have unit-variance and zero-mean. The weighted feature vector x_i'' with weight $W = 1/\sigma$ is evaluated as,

$$x_i'' = W.(x_i' - \overline{x'}) \tag{27}$$

where x_i' is the normalized feature, $\overline{x'}$ is the mean and σ is the standard deviation of the normalized feature vector.

Feature	Description
1	Risk of macular edema (0–2)
2	Haemorrhages (0:Absent/1:Present)
3	Area of exudates
4	Perimeter of exudates
5	Area of blood vessels
6	Area of neo-vascularization

3   Results and Discussion

Sensitivity refers to the fraction of abnormal fundus images classified correctly by the procedure. Specificity is calculated as the percentage of normal fundus images detected as normal by the procedure. The higher the sensitivity and specificity values, the better the procedure. These performance metrics can be calculated as follows:

$$Sensitivity = \frac{TP}{TP + FN} \tag{28}$$

$$Specificity = \frac{TN}{TN + FP} \tag{29}$$

TP, *TN*, *FP*, *FN* are true positives, true negatives, false positives, and false negatives, respectively. A screened fundus is considered as a true positive if the fundus is abnormal and the screening procedure also classified it as abnormal. Similarly, a true negative means that the fundus is normal and the procedure also classified it as normal. A false positive means that the fundus is normal, but the procedure classified it as abnormal. A false negative means that the procedure classified the screened fundus as normal, but it is abnormal.

The accuracy of classification is calculated as the percentage of images classified correctly as either with or without Diabetic Retinopathy.

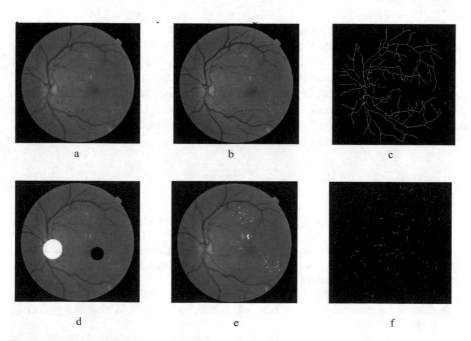

Fig. 1. (a) The input RGB image (b) The green channel extracted image (c) The vessel extracted image (d) The positions of the fovea and optic disc (e) The exudate extracted image (f) The neo-vascularization detected image

Table 2. Comparison of various DR screening methods

Authors	Sensitivity (%)	Specificity (%)	Accuracy of classification (%)
Tang et al. [7]	92.2	90	Not specified
Acharya et al. (2008) [8]	83	89	82
Acharya et al. (2009) [9]	82	86	86
Singalavanija et al. [2]	74.8	82.7	Not specified
Akram et al. [10]	99.17	97.07	98.9
Nayak et al. [11]	90	100	94
Dupas et al. [12]	83.9	72.7	Not specified
Abramoff et al. [3]	84	64	Not specified
Wong et al. [13]	91.7	100	84
Roychowdhury et al. [14]	100	54.16	Not specified
Antal et al. [15]	90	91	90
Proposed method	94.54	100	95.3

Figure 1 shows the outputs of the individual stages of the proposed procedure and Table 2 shows a comparison of the existing DR detection algorithms with the proposed algorithm.

4 Conclusion

In the proposed method we have put forward a comprehensive system for diagnosis of Diabetic Retinopathy, which automatically detects the features indicative of Diabetic Retinopathy and classifies the input retinal fundus scan based on the findings. The classifier requires 26 s on a single retinal fundus image to complete the diagnosis. The system demonstrated a classification accuracy of 95.3%, sensitivity of 94.5% and specificity of 100%. The relatively high values of sensitivity, specificity and accuracy indicate that in the absence of trained professionals, the proposed system can provide an efficient and fairly accurate screening for DR.

References

1. Prevention of blindness from Diabetes Mellitus: Report of a WHO consultation in Geneva (2005)
2. Singalavanija, A., Supokavej, J., Bamroongsuk, P., Sinthanayothin, C., Phoojaruen-chanachai, S., Kongbunkiat, V.: Feasibility study on computer-aided screening for diabetic retinopathy. Jpn. J. Ophthalmol. **50**, 361–366 (2006)
3. Abràmoff, M.D., Reinhardt, J.M., Russell, S.R., Folk, J.C., Mahajan, V.B., Niemeijer, M., Quellec, G.: Automated early detection of diabetic retinopathy. Ophthalmology **117**(6), 1147–1154 (2010)
4. Niemeijer, M., van Ginneken, B., Staal, J., Suttorp-Schulten, M.S.A., Abramoff, M.D.: Automatic detection of red lesions in digital color fundus photographs. IEEE Trans. Med. Imaging **24**(5), 584–592 (2005)

5. Fleming, A.D., Goatman, K.A., Philip, S., et al.: The role of hemorrhage and exudate detection in automated grading of diabetic retinopathy. Br. J. Ophthalmol. **94**, 706–711 (2010)
6. Gardner, G.G., Keating, D., Williamson, T.H., Elliott, A.T.: Automatic detection of diabetic retinopathy using an artificial neural network: a screening tool. Br. J. Ophthalmol. **80**(11), 940–944 (1996)
7. Tang, J., Kern, T.S.: Inflammation in diabetic retinopathy. Prog. Retinal Eye Res. **30**(5), 343–358 (2011)
8. Acharya, U.R., Chua, K.C., Ng, E.Y.K., Wei, W., Chee, C.: Application of higher order spectra for the identification of diabetes retinopathy stages. J. Med. Syst. **32**(6), 481–488 (2008)
9. Acharya, U.R., Lim, C.M., Ng, E.Y.K., Chee, C., Tamura, T.: Computer-based detection of diabetes retinopathy stages using digital fundus images. Part H J. Eng. Med. **223**(5), 545–553. Proceedings of the Institution of Mechanical Engineers (2009)
10. Usman Akram, M., Khalid, S., Tariq, A., Khan, S.A., Azam, F.: Detection and classification of retinal lesions for grading of diabetic retinopathy. Comput. Biol. Med. **45**, 161–171 (2014)
11. Nayak, J., Bhat, P.S., Acharya, U.R., et al.: J. Med. Syst. **32**, 107 (2008)
12. Dupas, B., Walter, T., Erginay, A., Ordonez, R., Deb-Joardar, N., Gain, P., Klein, J.-C., Massin, P.: Evaluation of automated fundus photograph analysis algorithms for detecting microaneurysms, haemorrhages and exudates, and of a computer-assisted diagnostic system for grading diabetic retinopathy **1698**(3), pp. 173–249 (2010)
13. Wong, L.Y., Acharya, U.R., Venkatesh, Y.V., Chee, C., Lim, C.M., Ng, E.Y.K.: Identification of different stages of diabetic retinopathy using retinal optical images. Inf. Sci. **178**(1), 106–121 (2008)
14. Roychowdhury, S., Koozekanani, D.D., Parhi, K.K.: DREAM: diabetic retinopathy analysis using machine learning. IEEE J. Biomed. Health Inform. **18**(5), 1717–1728 (2014)
15. Antal, B., Hajdu, A.: An ensemble-based system for microaneurysm detection and diabetic retinopathy grading. IEEE Trans. Biomed. Eng. **59**(6), 1720–1726 (2012)
16. Sinthanayothin, C., Boyce, J., Cook, H., Williamson, T.: Automated localisation of the optic disc, fovea, and retinal blood vessels from digital colour fundus images. Br. J. Ophthalmol. **83**(8), 902–910 (1999)

Predictive Analysis of Alertness Related Features for Driver Drowsiness Detection

Sachin Kumar[✉], Anushtha Kalia, and Arjun Sharma

Cluster Innovation Centre, University of Delhi, Delhi, India
sachin.blessed@gmail.com, anushthakalia@gmail.com,
arjunsharma147@yahoo.com

Abstract. Drowsiness during driving is a major cause of accidents of drivers which has socio-economic and psychological impact on the affected person. In Intelligent Transportation Systems (ITS), the detection of the drowsy and alert state of the driver is an interesting research problem. This paper proposed a novel method to detect the drowsy state of the driver based on three parameters, namely physiological, environmental and vehicular. The undertaken model proposes a simplistic approach and achieves comparable results to the state of the art with an ROC score of 81.28 and also elaborates on the specificity and sensitivity metrics.

Keywords: Multimodal · Drowsiness · Feature selection
Machine learning · SVM · LDA · XGBoost

1 Introduction

Intelligent transport systems (ITS) are a necessity of today, which can help tackle problems related to safety and efficiency of the transportation system. According to the EU directive, it is defined as systems in which information and communication technologies are applied in the field of road transport, including infrastructure, vehicles and users. There are several types of problems in ITS including traffic flow prediction on short or long term basis, pattern of traffic formation, prediction of new best position for cab drivers, drowsiness state of drivers and prediction etc. Drowsiness is one of the major contributors to road accidents and the associated fatalities. According to studies, more than 1.3 million people die each year on the road and 20 to 50 million people suffer non-fatal injuries due to road accidents [12]. If we take police reports into consideration, conservative estimates about the vehicles crashes are 100,000 each year that are the direct result of driver drowsiness. These crashes resulted in approximately 1,550 deaths, 71,000 injuries and 12.5 billion in monetary losses [14]. A study by the US National Sleep Foundation (NSF) in 2009 reported that 54% of adult drivers have driven a vehicle while feeling drowsy and 28% of them actually fell asleep [5]. All these studies suggest that drowsiness is indeed a grave problem and needs to be given immediate attention. Various studies have

© Springer International Publishing AG, part of Springer Nature 2018
A. Abraham et al. (Eds.): ISDA 2017, AISC 736, pp. 368–377, 2018.
https://doi.org/10.1007/978-3-319-76348-4_36

been performed in order to detect driver drowsiness based on physiological, environmental, sensorial, visual and behavioural modalities. Initial studies focused on the steering motion of vehicles as an indicator of drowsiness [3,11]. Recent approaches have tried measuring the levels of alertness of drivers using face monitoring and tracking. These methods relied on extracting facial features related to yawning [2] and eye closure [9]. Some recent approaches have also relied upon measuring physiological signals such as brain waves and heart rates and correlating them to states of drowsiness [16]. [8] developed a system consisting of an overhead capacitive sensor array which tracked the orientation of the head to detect alertness. Gundgurti et al. [6] tried to detect drowsiness effectively using extracted geometric features of the mouth in addition to features related to head movements. In order to improve on the detection rates of eye closure for people with darker skins or those wearing sunglasses, Near IR spectrum was used [18]. Vezard et al. [19] used EEG signals recorded using multiple electrodes to propose a genetic algorithm to detect alertness of individuals. We made use of the Fords dataset provided by Kaggle [4], which included features related to physiological, environmental and vehicular modalities. Our aim was to come up with an efficient and simple to deploy model based on these features to accurately differentiate between the drowsy and alert state of the driver at any point of time. Also, we tried to include minimum features from the physiological modality (so that minimum sensors are connected to the driver) while developing the model. Another important point to note is that, here we are not seeking high prediction accuracy but a high AUC score, which indicates how well our model can differentiate between drowsy and alert state. Specificity and sensitivity are two important metrics on which we evaluated the performance of our model. Abouelenien et al. [1] used a cascaded multimodal for drivers safety applications. Sigari et al. [17] used driver face monitoring system for fatigue and distraction detection. Rahman et al. [13] developed a novel haptic jacket based scheme for driver fatigue. Jo et al. [7] presented a vision-based method for driver drowsiness and distraction. Wang and Xu [21] worked on non-intrusive metrics for drowsiness. [10] recently presented review on the basis of head movement-based driver drowsiness detection: A review of state-of-art techniques. EEG and ECG signals were extracted and analyzed to detect driver drowsiness and alertness [22]. Doering et al. [15] took a sample of 60 people and recorded driving sessions in a driving simulator using a foggy highway to introduce fatigue. In order to detect drowsiness, visual and physiological measurements were extracted using a camera and multiple sensors. Wang et al. [20] related methods used to detect drivers alertness to those which were used to detect drivers state such as eyelid movements and percentage of eye closure, methods related to drivers performance such as distance between vehicles and lane tracking, and multimodal methods that combined both approaches.

2   Data Set and Environment Description

The data is available on Kaggle and was released by Ford for Stay Alert! The Ford Challenge. It consisted of a training and test set. The data consisted of results (collected from driving session and a driving simulator) for a number of trials, each representing two minutes of sequential data recorded after every 100 ms. The data set consisted of 33 columns. The first column was *TrialID*, which was the same for all the rows which were part of the same two minute observation. The second column was *ObsNum*, a value sequentially increasing by one for each recording within a *TrialID*. The third column, *IsAlert*, defined the state of the driver: 1 if the driver was alert and 0 if he/she wasn't. The challenge was to predict the *IsAlert* values for the instances in the test set, thereby making the challenge a binary classification one. The remaining 30 columns consisted of features from three categories, namely physiological (*P*), environmental (*E*) and vehicular (*V*). There were a total of 8 physiological features, 11 environmental features and 11 vehicular features. There were a total of 604,329 instances in the training set and 120,840 in the test set. Out of the 604,329 instances, 349,785 had the *IsAlert* column value as 1, representing the alert state of the driver. Table 1 gives the statistical description of the features across the three categories. It can be observed from the table that no missing values are present in the data. Categorical variables are also not present in the data set. Across the three categories, there were three features, *P8*, *V7* and *V9*, which had a value zero across all the training and test instances which can be seen in Table 1. Hence, we removed these features from both the data sets. Since the values in Table 2 have been rounded off to the first decimal, hence, value of E5 is zero across all the columns. For the purpose of this experiment, two machines were used with same configurations. The machines had Ubuntu 16.04 operating system with AMD Dual Core A6-5350M processor, 2.9 GHz clock speed and 8 GB DDR3 memory.

3   Methodology

3.1   Machine Learning Approaches

There are different classes of machine learning algorithms such as K-Nearest Neighbors (kNN), Artificial Neural Network (ANNs), Logistic Regression (LR), Naive Bayes (NB), Random Forests (RF), Decision Trees (DT) and Support Vector Machine (SVM) etc. We discuss about each algorithm briefly as follows. ANNs is a computational model that are inspired by the way biological nervous systems process information. It is based on a large collection of connected simple units called artificial neurons, loosely analogous to axons in a biological brain. Generally, neurons are connected in layers, and signals travel from the first (input), to the last (output) layer, with one or more hidden layers. They are popularly used in situations where there is dependency between various functional variables, leading to complex hypotheses. Unlike ANNs, DT, RF and XGBoost are tree based. DT, a supervised learning is used in both classification and regression problems. In DT, we split the population or sample into two or more homogeneous sets based on most significant splitter/differentiator in input variables. DT uses multiple algorithms to decide to split a node in two or more sub-nodes. We've made use of an optimised version of the CART algorithm, which is very

Table 1. Description of the statistics of the features across the three categories

		Count	Mean	Std	Min	25%	50%	75%	Max
	TrialID	604329.0	250.2	145.4	0.0	125.0	250.0	374.0	510.0
	ObsNum	604329.0	603.8	348.9	0.0	302.0	604.0	906.0	1210.0
	IsAlert	604329.0	0.6	0.5	0.0	0.0	1.0	1.0	1.0
Physiological	P1	604329.0	35.4	7.5	−22.5	31.8	34.1	37.3	101.4
	P2	604329.0	12.0	3.8	−45.6	9.9	11.4	13.6	71.2
	P3	604329.0	1026.7	309.3	504.0	792.0	1000.0	1220.0	2512.0
	P4	604329.0	64.1	19.8	23.9	49.2	60.0	75.8	119.0
	P5	604329.0	0.2	0.4	0.0	0.1	0.1	0.1	27.2
	P6	604329.0	845.4	2505.3	128.0	668.0	800.0	900.0	228812.0
	P7	604329.0	77.9	18.6	0.3	66.7	75.0	89.8	468.8
	P8	604329.0	0.0	0.0	0.0	0.0	0.0	0.0	0.0
Environmental	E1	604329.0	10.5	14.0	0.0	0.0	0.0	28.2	244.0
	E2	604329.0	102.8	127.3	0.0	0.0	0.0	211.6	360.0
	E3	604329.0	0.3	1.0	0.0	0.0	0.0	0.0	4.0
	E4	604329.0	−4.2	35.5	−250.0	−8.0	0.0	6.0	260.0
	E5	604329.0	0.0	0.0	0.0	0.0	0.0	0.0	0.0
	E6	604329.0	358.7	27.4	260.0	348.0	365.0	367.0	513.0
	E7	604329.0	1.8	2.9	0.0	0.0	1.0	2.0	25.0
	E8	604329.0	1.4	1.6	0.0	0.0	1.0	2.0	9.0
	E9	604329.0	0.9	0.3	0.0	1.0	1.0	1.0	1.0
	E10	604329.0	63.3	18.9	0.0	52.0	67.0	73.0	127.0
	E11	604329.0	1.3	5.2	0.0	0.0	0.0	0.0	52.4
Vehicular	V1	604329.0	77.0	44.4	0.0	41.9	100.4	108.5	129.7
	V2	604329.0	−0.0	0.4	−4.8	−0.2	0.0	0.1	4.0
	V3	604329.0	573.8	298.4	240.0	255.0	511.0	767.0	1023.0
	V4	604329.0	20.0	63.3	0.0	1.5	3.0	7.5	484.5
	V5	604329.0	0.2	0.4	0.0	0.0	0.0	0.0	1.0
	V6	604329.0	1715.7	618.2	0.0	1259.0	1994.0	2146.0	4892.0
	V7	604329.0	0.0	0.0	0.0	0.0	0.0	0.0	0.0
	V8	604329.0	12.7	11.5	0.0	0.0	12.8	21.9	82.1
	V9	604329.0	0.0	0.0	0.0	0.0	0.0	0.0	0.0
	V10	604329.0	3.3	1.2	1.0	3.0	4.0	4.0	7.0
	V11	604329.0	11.7	9.9	1.7	7.9	10.8	15.3	262.5

similar to C4.5. RF are an ensemble learning method that operate by constructing a multitude of decision trees at training time and outputting the class that is the mode of the classes (classification) or mean prediction (regression) of the individual trees. Each of the decision trees is fed an instance, for which these trees give their own, independent predictions. A majority voting is then performed to give the final prediction. XGBoost stands for extreme gradient boosting. It is an implementation of gradient boosted machines and focuses specifically on computational speed and model performance. Gradient boosting is an approach where the new models created predict the residuals or errors of prior models and are then added together to make the final prediction. LR is a linear classifiers is used to describe data and to explain the relationship between one dependent binary variable and one or more nominal, ordinal, interval or ratio-level independent variables. It makes use of the sigmoid function to convert a continuous result into associated probabilities. NB, a Bayesian algorithm, is a classification technique with an assumption of independence among predictors. NB assumes that the presence of a particular feature in a class is unrelated to the presence of any other feature. We used the multinomial NB. NB has a number of advantages associated as ease of predicting test data, better performance compared to other models if independence holds. SVM is a non-parametric supervised machine learning algorithm that plot each data item as a point in n-dimensional space with the value of each feature being the value of a particular coordinate. Then, we perform classification by finding the hyperplane that differentiate the two classes very well. SVM is a frontier which best segregates the two classes (hyperplane/line). KNN, an instance-based non-parametric method. An object is classified by a majority vote of its neighbors, with the object being assigned to the class most common among its k nearest neighbors (k is a positive integer, typically small). If $k = 1$, then the object is simply assigned to the class of the single nearest neighbor.

3.2   Proposed Model

Our proposed model consists of a series of steps, which are represented in Fig. 1. Our data consists of features belonging to three modalities namely, physiological, vehicular and environmental. First we took all the seven classifiers discussed in Sect. 3.1 and trained them on all the 27 raw features, as well as separately on features from individual modalities. All these seven classifiers are from different classes of machine learning algorithms, and hence, would help us identify which class represented our data set in the best way. As mentioned earlier, since the focus was not solely on the classification accuracy, the specificity and sensitivity metrics were also recorded with them which was not provided in previous studies. Specificity is the true negative rate, which tells us the accuracy of the drowsiness class while sensitivity is the true positive rate which tells us the accuracy the alert class. After training the classifiers on raw data, we performed feature selection using ensemble trees and fed the features to the classifier which performed best during our initial training process on raw data. We then performed feature generation on the selected features by computing the mean and standard deviation of each feature in the raw data for a particular *TrialID*. This was done in order to incorporate information related to a specific *TrialID* into our model. The newly generated features were then appended to the features selected via ensemble trees. This new data set was again fed into ensemble trees for feature selection, and the features obtained were finally given to our selected classifier for training and for final evaluation on the test data.

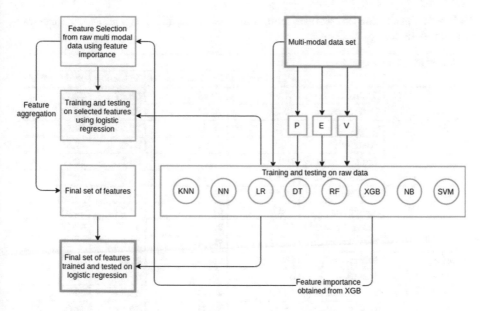

Fig. 1. Proposed model: category based feature selection method

4 Results and Analysis

4.1 Training on Raw Features

To get started our model development process, we took seven classifiers as mentioned before and trained on features from the individual modalities namely, physiological, environmental and vehicular and from the multimodal data set. We evaluated the models on accuracy, specificity and sensitivity. Note that Support Vector Machine was trained on downsampled data. The results are shown in Table 2. Multi Layer Perceptron and SVM performed poorly and were unable to differentiate between drowsy and alert state. Decision trees fared well, but overall, RF and LR performed well as compared to other classifiers. LR outperformed all other classifiers by a large margin when trained on the multimodal data set. While the performance of RF can be associated with their ability to choose the best predictor variables along with the prevention of overfitting via pruning, the higher scores of LR and the lower ones of the others represent the low signal to noise ratio or the inherent difficulty of the problem. Therefore, we inferred that ensemble trees were a good choice for feature selection and we could use the selected features on LR.

4.2 Feature Selection Using Ensemble Trees

We wanted to perform feature selection in order to extract features which were able to describe the state of the driver most appropriately. Since RF were performing fairly well across all the metrics, we decided to choose ensemble trees

Table 2. Results for the features from individual modalities as well as from the multimodal data set

ML algorithm	Metric	Data			
		Physiological	Environmental	Vehicular	Multimodal
	Train Accuracy	57.89	70.45	69.57	78.71
	Test Accuracy	72.54	79.55	73.88	85.17
LR	Sensitivity	95.44	85.89	97.24	94.08
	Specificity	2.93	60.26	2.88	57.83
	Train Accuracy	100	99.61	99.98	100
	Test Accuracy	58.65	62.66	64.06	62.77
KNN	Sensitivity	68.72	60.96	75.75	72.62
	Specificity	28.05	67.85	28.54	32.87
	Train Accuracy	57.87	72.2	57.88	53.96
	Test Accuracy	75.24	75.54	75.24	73.88
MLP	Sensitivity	100	85.01	100	98.13
	Specificity	0	46.76	0	0.14
	Train Accuracy	43.99	69.44	58.55	64.49
	Test Accuracy	26.03	83.34	40.37	47.06
NB	Sensitivity	2.36	92.88	27.24	33.96
	Specificity	97.97	54.35	80.28	86.87
	Train Accuracy	100	99.72	99.98	100
	Test Accuracy	57.95	64.03	62.38	70.99
DT	Sensitivity	66.23	62.84	72.45	74.06
	Specificity	32.87	67.68	31.75	61.63
	Train Accuracy	99.7	99.8	9.67	99.99
	Test Accuracy	58.6	65.13	65.98	79.79
RF	Sensitivity	68.79	65.13	78.81	86.99
	Specificity	28.74	68.34	27.01	57.89
	Train Accuracy	98.6	99.8	99.59	99.94
	Test Accuracy	75.02	75.34	74.85	75.24
SVM	Sensitivity	99.49	99.93	99.33	100
	Specificity	0.64	0.59	0.44	0

as the means to select features. In ensemble learning, we specifically opted for boosting techniques or Gradient Boosting Machines (GBMs) because these techniques give more importance to instances which have been classified incorrectly, which is crucial to drowsiness detection. Since the best combination of the two, ensemble trees and GBMs, is XGBoost, we trained it with the our multimodal dataset as well as with features from individual modalities, and used its feature importance attribute to choose the important features from each modality as well as from all modalities combined. Results are given in Table 3.

The XGBoost feature importance metric helped in understanding which features contributed the most in finding out the labels for the unseen data.

Table 3. Results of XGBoost

Data sets	Accuracy	Specificity	Sensitivity	Features
Physiological	65.53	10.65	83.58	P1, P5, P6
Environmental	71.05	60.82	74.42	E1, E6, E7, E8, E9, E10
Vehicular	69.59	14.41	87.6	V1, V2, V3, V5, V11
Multimodal	88.1	52.74	99.58	E7, E8, E9, V1, V11

The important features selected by XGBoost from the multimodal dataset consisted of all those on the basis of which the most number of splits were being performed for the individual modalities. On that account, we decided to select these features for further analysis. The next step that we took was feeding the features selected by XGBoost to train Logistic Regression, assuming that the features which were earlier hindering the models performance would have been removed. This model was then used to predict the test set labels, and the results achieved are given in Table 4.

Table 4. Results on features obtained from XGBoost

ML algorithm	Accuracy	Specificity	Sensitivity	AUC
LR	88.2	52.84	99.84	80.03

This model performed quite well, giving an accuracy of 88.2 on the test set with high specificity and sensitivity scores. An ROC score of 80.03 was also achieved.

4.3 Feature Generation and Final Model Training

Till here, we hadn't considered the fact that the individual instances were observations from trials, and that the observations from the same trials may have some similarity. Therefore, our next step was to aggregate the selected features corresponding to the same *TrialID*. The aggregation was performed via mean and standard deviation. A total of 10 new features, mean and standard deviation of each of the features chosen by XGBoost, was appended to the dataset. This data was then used to train XGBoost, whose feature importance metric was in turn used again to find the best performing features. The new selected features were *E9, E10, V1, V11, E7std, E7mean*. These six features were then again used to train logistic regression classifier, whose performance was measured on the test set. The results are shown in Table 5. A combined analysis of all the logistic regression models is give in Fig. 2.

Table 5. Here the results of Logistic Regression, which was evaluated on the accuracy, specificity, sensitivity and ROC score metric for the final feature set, are presented

ML Algorithm	Accuracy	Specificity	Sensitivity	AUC
LR	88.2	52.82	99.84	81.28

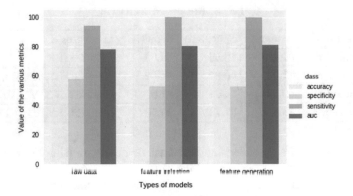

Fig. 2. Comparison of various models trained using Logistic Regression

5   Conclusion

Our model came up with features set having *E9, E10, V1, V11, E7std, E7mean*, which consisted of only environmental and vehicular modalities. All models performed well on features from the environmental modality than on features from physiological or vehicular modalities making them important for drowsiness. Our final model of based on LR which we obtained after sequential feature selection and feature generation resulted in a sensitivity of 99.8, which means our model could accurately identify the alertness class and a specificity of 52.7, meaning our model could to some extent identify drowsy class effectively. We achieved ROC score of 81% approx with simpler and efficient model. In future, Deep learning approach could be used for better feature representation and prediction.

References

1. Abouelenien, M., Burzo, M., Mihalcea, R.: Cascaded multimodal analysis of alertness related features for drivers safety applications. In: Proceedings of the 8th ACM International Conference on PErvasive Technologies Related to Assistive Environments, PETRA 2015, pp. 59:1–59:8. ACM, New York (2015)
2. Abtahi, S., Hariri, B., Shirmohammadi, S.: Driver drowsiness monitoring based on yawning detection. In: 2011 IEEE International Instrumentation and Measurement Technology Conference, pp. 1–4, May 2011
3. Barr, L., Howarth, H., Popkin, S., Carroll, R.J.: A review and evaluation of emerging driver fatigue detection measures and technologies. National Transportation Systems Center, Cambridge. US Department of Transportation, Washington (2005). Disponível em¡ http://www.ecse.rpi.edu/~qji/Fatigue/fatigue_report_dot.pdf
4. bin Tariq, T., Chen, A.: Stay alert! the ford challenge
5. Drivers Beware Getting Enough Sleep Can: Save your life this memorial day. National Sleep Foundation (NSF), Arlington (2010)

6. Gundgurti, P., Patil, B., Hemadri, V., Kulkarni, U.: Experimental study on assessment on impact of biometric parameters on drowsiness based on yawning and head movement using support vector machine. Int. J. Comput. Sci. Manag. Res. **2**(5), 2576–2580 (2013)
7. Jo, J., Lee, S.J., Jung, H.G., Park, K.R., Kim, J.: Vision-based method for detecting driver drowsiness and distraction in driver monitoring system. Opt. Eng. **50**(12), 127202 (2011)
8. Kithil, P.W., Jones, R.D., McCuish, J.: Driver alertness detection research using capacitive sensor array. Technical report, SAE Technical Paper (2001)
9. Kristjansson, S.D., Stern, J.A., Brown, T.B., Rohrbaugh, J.W.: Detecting phasic lapses in alertness using pupillometric measures. Appl. Ergon. **40**(6), 978–986 (2009)
10. Mittal, A., Kumar, K., Dhamija, S., Kaur, M.: Head movement-based driver drowsiness detection: a review of state-of-art techniques. In: 2016 IEEE International Conference on Engineering and Technology (ICETECH) (2016)
11. Omry, D.: Driver alertness indication system (daisy). Technical report (2006)
12. World Health Organization: Global status report on road safety: time for action. World Health Organization (2009)
13. Mahfujur Rahman, A.S.M., Azmi, N., Shirmohammadi, S., Saddik, A.E.: A novel haptic jacket based alerting scheme in a driver fatigue monitoring system. In: 2011 IEEE International Workshop on Haptic Audio Visual Environments and Games (HAVE), pp. 112–117. IEEE (2011)
14. Rau, P.S.: Drowsy driver detection and warning system for commercial vehicle drivers: field operational test design, data analyses, and progress. In: 19th International Conference on Enhanced Safety of Vehicles, pp. 6–9 (2005)
15. Rimini-Doering, M., Manstetten, D., Altmueller, T., Ladstaetter, U., Mahler, M.: Monitoring driver drowsiness and stress in a driving simulator. In: First International Driving Symposium on Human Factors in Driver Assessment, Training and Vehicle Design, pp. 58–63 (2001)
16. Sahayadhas, A., Sundaraj, K., Murugappan, M.: Detecting driver drowsiness based on sensors: a review. Sensors **12**(12), 16937–16953 (2012)
17. Sigari, M.-H., Fathy, M., Soryani, M.: A driver face monitoring system for fatigue and distraction detection. Int. J. Vehicular Technol. **2013**, 11 (2013)
18. Sigari, M.-H., Pourshahabi, M.-R., Soryani, M., Fathy, M.: A review on driver face monitoring systems for fatigue and distraction detection (2014)
19. Vezard, L., Chavent, M., Legrand, P., Faïta-Aïnseba, F., Trujillo, L.: Detecting mental states of alertness with genetic algorithm variable selection. In: 2013 IEEE Congress on Evolutionary Computation (CEC), pp. 1247–1254. IEEE (2013)
20. Wang, Q., Yang, J., Ren, M., Zheng, Y.: Driver fatigue detection: a survey. In: The Sixth World Congress on Intelligent Control and Automation, WCICA 2006, vol. 2, pp. 8587–8591. IEEE (2006)
21. Wang, X., Chuan, X.: Driver drowsiness detection based on non-intrusive metrics considering individual specifics. Accid. Anal. Prev. **95**, 350–357 (2016)
22. Xu, S., Zhao, X., Zhang, X., Rong, J.: A study of the identification method of driving fatigue based on physiological signals. In: ICCTP 2011: Towards Sustainable Transportation Systems, pp. 2296–2307 (2011)

Association Rules Transformation for Knowledge Integration and Warehousing

Rim Ayadi[1(✉)], Yasser Hachaichi[1,2], and Jamel Feki[1,3]

[1] Multimedia Information Systems and Advanced Computing Laboratory,
University of Sfax, Sfax, Tunisia
rim.ayadi@yahoo.fr
[2] Department of Computer Science and Quantitative Methods,
Higher Institute of Business Administration of Sfax, Sfax, Tunisia
[3] Faculty of Computing and IT, University of Jeddah, Jeddah, Saudi Arabia

Abstract. Knowledge management process is a set of procedures and tools applied to facilitate capturing, sharing and effectively using knowledge. However, knowledge collected from organizations is generally expressed in various formalisms, therefore it is heterogeneous. Thus, a *Knowledge Warehouse* (KW), which is a solution for implementing all phases of the knowledge management process, should solve this structural heterogeneity before loading and storing knowledge. In this paper, we are interested in knowledge normalization. More accurately, we firstly introduce our proposed architecture for a KW, and then we present the MOT (Modeling with Object Types) language for knowledge representation. Since our objective is to transform heterogeneous knowledge into MOT, as a pivot model, we suggest a meta-model for the MOT and another for the explicit knowledge extracted through the association rules technique. Thereafter, we define eight transformation rules and an algorithm to transform an association rules model into the MOT model.

Keywords: Knowledge Warehouse · MOT language
Transformation rules · Data mining · Heterogeneous knowledge
Knowledge Normalization

1 Introduction

For centuries, experts and decision-makers have been concerned about creating, acquiring, and sharing knowledge since it is a powerful means of success for organizations. There are several taxonomies for knowledge but the most fundamental distinction is between *explicit* and *tacit* knowledge. *Explicit* knowledge is extracted from data sources and presented in a tangible form [1]. However, tacit knowledge inhabits the minds of people and is difficult to articulate. For this reason, the need for the knowledge management process [2,3] is increased; it aims to explicate tacit knowledge and then to make it available for use by others.

© Springer International Publishing AG, part of Springer Nature 2018
A. Abraham et al. (Eds.): ISDA 2017, AISC 736, pp. 378–388, 2018.
https://doi.org/10.1007/978-3-319-76348-4_37

The obtained knowledge is generally scattered over various systems. Thus, it looks crucial to collect and store this knowledge for decision activities [2]. For this purpose, some research works introduced the concept of KW [4–6].

Actually, knowledge (i.e., *explicit* or *tacit*) is generally represented in several heterogeneous models (e.g., association rules, decision trees, neural networks, clusters [7]) [8]. Thus, as our objective is to store different types of knowledge in the KW, we need to standardize and homogenize knowledge. More specifically, we aim to transform different knowledge models into a unified model namely the MOT (Modeling with Object Types) language [9]. In this work, we focus on the transformation of the association rules knowledge model into the MOT model.

The MOT language is a semi-formal graphical language for knowledge modeling. It relies on a set of typed knowledge units and on a set of typed links allowing homogeneous representation of heterogeneous explicit knowledge models. MOT is a language based on a graphical formalism which facilitates the modeling of explicit knowledge as well as the formalization of experts' tacit knowledge.

The remainder of this paper is organized as follows: Sect. 2 introduces the context of this work and proposes the architecture of the KW. Section 3 presents the MOT language and its meta-model. In Sect. 4, we present the meta-model of the association rules data mining technique. Then, we define eight transformation rules to transform the association rules models into MOT, we develop an algorithm to apply our rules, and then we illustrate it with an example. Finally, Sect. 5 concludes this paper and projects our future work.

2   Context

After the era of information structured as databases, the era of data warehouses emerged from which knowledge is extracted by means of data mining techniques. However, knowledge is often expressed in heterogeneous formats and is generally stored in various information-processing systems. In this context, we propose the concept of KW [6] as a solution to collect, organize, homogenize and store initially heterogeneous pieces of knowledge in order to make them useful for decision-making processes. The global architecture of the suggested KW has three layers made up of seven modules [10] (see Fig. 1).

Fig. 1. Three layers architecture for a Knowledge Warehouse.

The *Knowledge Acquisition and Transformation* layer is made up of four modules: *Data Preparation, Knowledge Extraction, Tacit Knowledge Capture,* and *Knowledge Normalization format.* The main objective of this layer is to transform heterogeneous knowledge models into a single pivot model, which is the MOT in our proposal. This transformation allows the homogenization of knowledge and facilitates its integration [11]. It serves as a foundation for the *Knowledge Normalization format* module and is performed by applying rules based on the structural correspondences between the source meta-model (MM) of the knowledge model and the MOT MM [10].

The *Knowledge Storage* layer is made up of two modules: *Knowledge Unification/Integration,* and *KW Construction.* This layer consists in storing unified knowledge in KW. Actually, knowledge is firstly gathered in a global repository representing all the MOT knowledge models to be integrated via the *Knowledge Unification/Integration* module in order to remove duplicated knowledge and infer new one [11].

The *Knowledge Exploitation and Maintenance* layer is made up of the *Knowledge Exploitation* module. This layer aims to exploit and handle the KW. Decision-makers should be able to query the KW to make decisions. In addition, they inform the KW administrator to update KW when necessary.

In the remainder of this paper, we focus on the *Knowledge Normalization format* module as a keystone task for knowledge integration, and then for knowledge warehousing. For this purpose, we need a pivot language for knowledge modeling. Actually, there are several languages for knowledge representation: *Informal, formal* or *semi-formal* languages. An *Informal* language expresses knowledge as sentences; however, it is ambiguous and inaccurate. A *formal* language has a vocabulary with a rigid syntax that makes it difficult to understand. Finally, the *semi-formal* language is a knowledge expression formalism. Its rich grammar expressiveness and simplicity make it easy for use by novice people or experts. It offers many advantages compared to an informal language because it provides a representational guide that helps to structure the modeling process [12]. On the other hand, compared to a formal language, a semi-formal language extends the number of people qualified to express their knowledge with few assistance of knowledge engineers. Due to the benefits of the semi-formal language, we choose to use a semi-formal language for the knowledge graphical representation.

In what follows, we present the characteristics of for well-known semi-formal languages [9] (i.e., *semantic trees, concept maps, semantic networks* and *MOT*) to choose the most appropriate for our work. *Semantic trees* [13] provoke ambiguous interpretations because they do not clarify the direction or the nature of links between concepts. *Concept maps* [14] represent the relation between concepts using labeled but non-oriented links. The imprecise choice of labels can complicate the interpretation. *Semantic networks* [15] use oriented and labeled links between the pieces of knowledge. However, they do not have a standard terminology for the graphical modeling. Finally, the *MOT* [9] language offers graphical representations to model abstract and factual knowledge connected through oriented and typed links. Moreover, it enables a quality gain in the representation of knowledge, as it has a greater expressiveness to represent procedural and strategic knowledge in addition to declarative knowledge [12] in the same model, which is the major benefit of the MOT language.

Since the MOT offers several advantages, we elect this language as a unified language for knowledge homogenization. In the next section, we will introduce the MOT language and we develop its MM.

3 The MOT Language

The MOT, which was introduced by [9], allows a semi-formal graphical representation of knowledge. With MOT, it is possible to represent knowledge on the basis of its types (i.e., declarative, procedural, strategic). Moreover, it helps to easily represent explicit knowledge and consequently allows increasing the number of people who are ready to clarify their tacit knowledge far from the help of experts.

In the MOT language, the pieces of knowledge are represented with graphical symbols which distinguish their abstraction levels, their types and their semantics (see Fig. 2) [9]. These pieces of knowledge are connected by typed links, each of which has an appropriate semantics (see Table 1) [9,12].

Abstract Knowledge		Equivalent Factual Knowledge		Knowledge Semantics
Type	Graphic Representation	Type	Graphic Representation	
Concept		Example		Declarative knowledge representing a class of objects.
Procedure		Trace		Procedural knowledge describing the set of operations/actions to act on the objects.
Principle		Statement		Strategic knowledge, establishes causal links between objects, determines conditions to perform actions, and represents agents that act on something.

Fig. 2. Type and semantics of knowledge in the MOT language [12].

In what follows, we propose the MOT MM as described by [9]. This MM has two principal classes: *Knowledge Unit* (KU) and *Link* (see Fig. 3).

KU is labeled by its name *name_KU* which is extended by the field for multiplicity [1..*]. This extension is useful for defining our integration rules in case of synonymy between two KUs [11]. One KU can be an abstract knowledge unit AKU (*Concept, Procedure, Principle*) or a factual knowledge unit FKU (*Example, Trace, Statement*). At each AKU may correspond n (n ≥ 1) FKU.

Link can be of type {I, C, C*, P, S, I/P, R or A}; it links two KUs.

Moreover, we extend the MM by adding two new classes that we called *Position* and *PrincipleProcedure* (see Fig. 3). Indeed, the *Position* class is necessary to define the transformation rules of a decision tree source model, where each decision node is located by its position, towards the target MOT model.

Table 1. Types and semantics of the links in the MOT language [12].

Link type	Semantics associated with link
Instantiation (I)	Connects abstract knowledge to factual knowledge
Composition/multiple composition (C/C*)	Link knowledge to one of its components
Specialization (S)	Connects two abstract knowledge of the same type, one of which is one-kind-on the other
Precedence (P)	Connects two procedural or strategic knowledge, the first one must complete before the second begins
Input/Product (I/P)	Connects a procedural and conceptual knowledge to represent the input or the product of a procedural knowledge
Regulation (R)	Connects a strategic knowledge with a conceptual knowledge to specify a constraint and with a procedural or strategic knowledge to control the execution or the selection of other knowledge
Application (A)	Connects a fact to an abstract knowledge

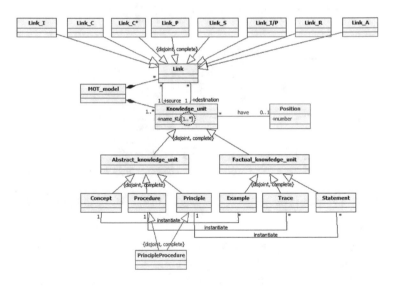

Fig. 3. UML class diagram of the extended MOT meta-model.

The *PrincipleProcedure* class is also necessary to transform an association rules model (cf., Sect. 4.2). In this way, the extended MM MOT takes into account the transformations of three knowledge models: decision tree, association rules, and clusters. We note that in order to transform other knowledge models into MOT, the extended MM may require other extensions if new concepts (i.e., attributes, classes) are to be considered.

In what follows, we deal with association rules, as a knowledge model source, and their transformation into the MOT language.

4   Transformation of Association Rules into the MOT

In this section, we are interested in the association rules data mining technique. We propose the association rules MM, then we define eight transformation rules that are applied by an algorithm allowing to transform the association rules model into the MOT target model.

4.1   Association Rules Meta-Model

An association rule is an *If-Then* rule having two parts: one antecedent (i.e., condition) and one consequent which are composed of many items.

Figure 4 proposes the MM for Association Rules Base (ARB). In this MM, the *Rule base* class has a name *RBname* and is composed of *nbAR* (nbAR ≥ 1) *association rules* each of which has a numeric identifier *num*. In an ARB, each *antecedent* implies one or more *consequent*(s). Moreover, each *item* has a name *name_I* and a type *type_I* (i.e., *Conclusion* or *Action*). One *item* can participate in both of the *antecedent* and the *consequent* of two different rules.

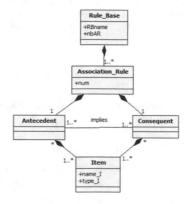

Fig. 4. Association rules meta-model.

4.2   Transformation Rules for Association Rules Base Meta-Model

In this sub-section, we define transformation rules to connect the elements of the ARB MM with those of the MOT MM.

Actually, items belonging to the antecedents of an ARB are transformed, in MOT model, into AKU of type *Principle*. However, items belonging to the consequents of an ARB are transformed into AKU of different type. Indeed, the choice of the right type depends on the semantics of the items participating in

Rules	Elements of an ARB	AKU generated by transformation rules	
RAR1	item $I1 \in$ ITA	Principle $Pr1$ with the same name as $I1$	I1
RAR2	item $I2$ (of type *Conclusion*) $\in$ ITC	Principle $Pr2$ with the same name as $I2$	I2
RAR3	item $I3$ (of type *Action*) $\in$ ITC	Procedure $Proc1$ with the same name as $I3$	I3
RAR4	item $I4$ (of type *Action*) $\in$ ITAC	PrincipleProcedure $PrProc$ with the same name as $I4$	I4
RAR5	item $I5$ (of type *Conclusion*) $\in$ ITAC	Principle $Pr3$ with the same name as $I5$	I5
RAR6	• *implies* association between an antecedent A and a consequent C of an association rule R ($R \in$ ARB B) • i : number of the rule R in B	Procedure $Proc2$ named IRi_B	IRi_B

Fig. 5. Transformation rules to generate AKU.

Rules	The obtained AKU	Linked AKU	
RAR7	• $Pr1$ or $PrProc$ or $Pr3$ issued from an association rule Ri • $Proc2$ issued from the same rule Ri	I1 —P→ IRi_B I4 —P→ IRi_B I5 —P→ IRi_B	
RAR8	• $Pr2$ or $Proc1$ or $PrProc$ or $Pr3$ issued from an association rule Ri • $Proc2$ issued from the same rule Ri	IRi_B —P→ I2 IRi_B —P→ I3 IRi_B —P→ I4 IRi_B —P→ I5	

Fig. 6. Transformation rules to link the obtained AKU.

the consequents, which are of type *Action* or *Conclusion*, and on whether these items are antecedents in other association rules or not. For this reason, we think to go through an intermediate step in which we group these items. We classify the set of items of a rule base B into three subsets:

- *ITA* the subset of items of B that participate as antecedents only.
- *ITC* the subset of items of B that participate as consequents only.
- *ITAC* the subset of items of B that participate both as antecedents and consequents.

Based on these subsets, we define, in Fig. 5, six transformation rules to generate AKU. In the fourth rule, we use the new graphic representation of an AKU of type *PrincipleProcedure* (rounded rectangle) that plays a dual role (antecedent and consequent of type *Action*). This new type is presented in the MOT MM of Fig. 3.

In Fig. 6, we define two rules that link/connect the AKU obtained by the previous six rules.

In the next section and based on the above rules, we will develop the AR2MOT algorithm that transforms a rule base into the MOT language.

4.3   The ARB2MOT Algorithm

In this section, we propose the *ARB2MOT* algorithm which helps transform an ARB into a MOT model. This algorithm shows step by step the application of our transformations rules.

Algorithm 1. *ARB2MOT*

Input: ITA,ITC, ITAC, ARB
Output: MOT model

1 Declaration: item I ∈ ARB; association rule R ∈ ARB; Type(I): method returns
   the type of I(*Action* or *Conclusion*); implies(R): method returns the association
   between an antecedent A and a consequent C in the association rule R

2 **begin**

3 **foreach** I ∈ *ITA* **do**

4 ⎿ Transform_item_to_principle (I, Pr1) /* corresponds to rule RAR1 where
       I:input and Pr1:output */

5 **foreach** I ∈ *ITC* **do**

6 ⎪ **if** *(Type(I) == Conclusion)* **then**

7 ⎪ ⎪ Transform_item_to_principle (I, Pr2)   /* rule RAR2 */

8 ⎪ **else**   /* Type(I) is Action */

9 ⎿ ⎿ Transform_item_to_procedure (I, Proc1)   /* rule RAR3 */

10 **foreach** I ∈ *ITAC* **do**

11 ⎪ **if** *(Type(I) == Action)* **then**

12 ⎪ ⎪ Transform_item_to_principleprocedure (I, PrProc)   /* rule RAR4 */

13 ⎪ **else**   /* Type(I) is Conclusion */

14 ⎿ ⎿ Transform_item_to_principle (I, Pr3)   /* rule RAR5 */

15 **foreach** R **do**

16 ⎿ Transform_implies_to_procedure (implies(R), Proc2)   /* rule RAR6 */

   /* Note that the calls for the four previous methods could be
   executed in parallel   */

17 **foreach** *AKU generated by RAR1 or RAR4 or RAR5* **do**

18 ⎿ Link_by_LinkP (AKU, Proc2) /* rule RAR7; AKU:source of the link P
      and Proc2:destination of the link P */

19 **foreach** *AKU generated by RAR2 or RAR3 or RAR4 or RAR5* **do**

20 ⎿ Link_by_LinkP (Proc2, AKU)   /* rule RAR8 */

21 **end**

4.4 Example of Transformation of an ARB into the MOT

Figure 7 shows a part of an ARB model that conforms to the MM of Fig. 4. This base is extracted from the *Bank Marketing*[1] dataset described with 17 variables: age, job, marital, default (i.e., if the client has a credit in default), housing (i.e., if the client has a housing loan), loan, contact, poutcome (i.e., Outcome of the previous marketing campaign), etc. The classification goal of this data is to predict if the client will subscribe to a term deposit (variable y).

N°	Antecedent	Consequent
1	"y=no" - "housing=no"	"default=no" - "poutcome=unknown"
2	"y=no" - "housing=no"	"poutcome=unknown"
3	"default=no" - "y=no" - "housing=no"	"poutcome=unknown"
4	"poutcome=unknown" - "housing=yes"	"y=no"
5	"default=no" - "poutcome=unknown" - "housing=yes"	"y=no"
6	"poutcome=unknown" - "marital=married"	"default=no" - "y=no"
7	"marital=married" - "housing=yes"	"default=no" - "y=no"
8	"poutcome=unknown" - "housing=yes"	"default=no" - "y=no"
9	"poutcome=unknown" - "marital=married"	"y=no"
10	"default=no" - "poutcome=unknown" - "marital=married"	"y=no"

Fig. 7. Association rules base of the bank Marketing Dataset.

The application of our *ARB2MOT* algorithm on the ARB (called *B*) of Fig. 7 generates the MOT model of Fig. 8 displayed with the G-MOT knowledge modeling editor[2].

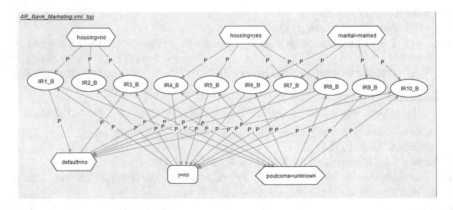

Fig. 8. MOT model of the association rules of Fig. 7.

---

[1] https://archive.ics.uci.edu/ml/datasets/bank+marketing.

[2] The editor is distributed by the LICEF research center and it is the most recent contribution comparing to editors MOT2.3 and MOTplus. It is available on http://poseidon.licef.ca/gmot/.

In the ARB of Fig. 7, the items 'default = no' and 'poutcome = unknown' belong to *ITAC*; they are of type *Conclusion*. Thus, by applying the *ARB2MOT* algorithm from line 10 to line 14, we notice that these items are transformed into AKU of type *Principle*. However, the items 'housing = no', 'housing = yes' and 'marital = married' belong to *ITA* so, by applying the two lines 3 and 4 of the *ARB2MOT* algorithm, they are transformed into AKU of type *Principle*. Finally, the item 'y = no' belongs to *ITAC* and is of type *Action*; so, by applying our algorithm from line 10 to line 14, this item is transformed into an AKU of type *PrincipleProcedure*.

On the other hand, we apply lines 15 and 16 of our algorithm to associate, for each association rule, an AKU of type *Procedure* Proc2. Indeed, this *Procedure* shows that the items of an antecedent imply items of a consequent in an association rule. Finally, we apply the *ARB2MOT* algorithm from line 17 to line 20 in order to connect, with links of type P, the antecedent and the consequent AKU (generated by previous methods) with the appropriate procedure Proc2.

5   Conclusion

Knowledge collected from organizations has different formats and origins. In practice, there are two main types of knowledge, namely *explicit* and *tacit*. Explicit knowledge is extracted from different sources using data mining techniques in accordance with different models whereas tacit knowledge is not formalized because it is kept in minds of experts. In this paper, we are interested in warehousing explicit knowledge and therefore, in how to unify the modeling so that initially heterogeneous knowledge could be used by an intelligent decision process.

For this purpose, we proposed a multi-layer architecture for a KW. In this work, we focus on the first layer *Knowledge Acquisition and Transformation* which aims to homogenize heterogeneous knowledge models. For this homogenization, we proceeded with conversion into a unified model. We elected the MOT semi-formal graphical language as a pivot language for which we designed its MM. In this paper, we focused on the transformation of the explicit knowledge model extracted through the association rules techniques. In this regard, we defined the association rules MM and then we suggested eight transformation rules. In addition, we developed an algorithm to apply our rules that allow the transformation of the association rules model into the MOT model. Finally, we illustrated this algorithm with an example.

Currently, we are working on the integration of three knowledge models extracted through decision tree, association rules and clustering techniques transformed into MOT model. We aim to gather these pieces of MOT knowledge in a global repository for more coherent decisions. In addition, we are thinking about the most adequate KW model that should facilitate the querying and handling of the MOT KW.

References

1. Nonaka, I., Takeuchi, H.: The Knowledge-Creating Company: How Japanese Companies Create the Dynamics of Innovation. Oxford University Press, New York (1995)
2. Nemati, H.R., Steiger, D.M., Iyer, L.S., Herschel, R.T.: Knowledge warehouse: an architectural integration of knowledge management, decision support, artificial intelligence and data warehousing. Decis. Support Syst. **33**(2), 143–161 (2002)
3. Liebowitz, J., Frank, M.: Knowledge Management and E-Learning. CRC Press, Boca Raton (2016)
4. Michael, Y · The knowledge warehouses reusing knowledge components. Perform. Improv. Q. **12**(3), 132–140 (1999)
5. Dymond, A.: The knowledge warehouse: the next step beyond the data warehouse. In: Data Warehousing and Enterprise Solutions, SAS Users Group International 27 (2002)
6. Ayadi, R., Hachaichi, Y., Feki, J.: Towards knowledge warehouses: definition and architecture. In: 7th Edition of the Conference on Advances in Decisional Systems, Marrakech, Morocco (2013) (In French)
7. Basciani, F., Di Rocco, J., Di Ruscio, D., Iovino, L., Pierantonio, A.: Automated Clustering of Metamodel Repositories, pp. 342–358. Springer International Publishing, Cham (2016)
8. Zaki, M.J., Meira, W.: Data Mining and Analysis: Fundamental Concepts and Algorithms. Cambridge University Press, Cambridge (2014)
9. Paquette, G.: Knowledge and Skills Modeling: A Graphical Language for Designing and Learning. University of Quebec Press, Sainte-Foy (2002). (In French)
10. Ayadi, R., Hachaichi, Y., Alshomrani, S., Feki, J.: Decision tree transformation for knowledge warehousing. In: Proceedings of the 17th International Conference on Enterprise Information Systems, ICEIS 2015, Barcelona, Spain, 27–30 April 2015, vol. 1, pp. 616–623 (2015)
11. Ayadi, R., Hachaichi, Y., Feki, J.: MOT knowledge model integration rules for knowledge warehousing. In: Knowledge-Based and Intelligent Information & Engineering Systems: Proceedings of the 21st International Conference KES-2017, Marseille, France, 6–8 September 2017, pp. 544–553 (2017)
12. Héon, M., Basque, J., Paquette, G.: Semantics validation of a semi-formal knowledge model with ontocase. In: Act of 21st Francophone Days of Knowledge Engineering, Nimes, France, pp. 55–66 (2010). (In French)
13. Dinarelli, M., Moschitti, A., Riccardi, G.: Hypotheses selection for re-ranking semantic annotations. In: 2010 IEEE Spoken Language Technology Workshop, pp. 407–411 (2010)
14. Canas, A.J., Ford, K.M., Novak, J.D., Hayes, P., Reichherzer, T.R., Suri, N.: Online concept maps: enhancing collaborative learning by using technology with concept maps. Sci. Teach. **68**, 49–51 (2001)
15. Collins, A., Quillian, M.: Retrieval time from semantic memory. J. Verbal Learn. Verbal Behav. **8**(2), 240–247 (1969)

Abnormal High-Level Event Recognition in Parking lot

Najla Bouarada Ghrab[1]([✉]), Rania Rebai Boukhriss[1], Emna Fendri[2], and Mohamed Hammami[2]

[1] MIRACL Laboratory, ISIMS, University of Sfax,
BP 242, 3021 Sakiet Ezzeit Sfax, Tunisia
najla.bouarada@yahoo.fr, rania.rebai@hotmail.fr
[2] MIRACL Laboratory, FSS, University of Sfax,
Road Sokra km 4, BP 802, 3038 Sfax, Tunisia
fendri.msf@gnet.tn, mohamed.hammami@fss.rnu.tn

Abstract. In this paper, we presented an approach to automatically detect abnormal high-level events in a parking lot. A high-level event or a scenario is a combination of simple events with spatial, temporal and logical relations. We proposed to define the simple events through a spatio-temporal analysis of features extracted from a low-level processing. The low level processing involves detecting, tracking and classifying moving objects. To naturally model the relations between simpler events, a Petri Nets model was used. The experimental results based on recorded parking video data sets and public data sets illustrate the performance of our approach.

Keywords: Object classification · Simple event · Scenario
Abnormal event

1 Introduction

High-level video event recognition has received a lot of attention recently due to its diverse application areas in video surveillance. The goal of high-level video event recognition is to describe and infer a variety of abnormal behaviors that occur in the scene. Unlike normal events, the abnormal behaviors in a scene are rare. The focus of this study was to detect abnormal high-level events that may occur in a parking lot monitored by a surveillance camera through an explicit modeling. The high-level events recognition problem is hierarchical since the high-level event or scenario consists of multiple simple events with certain spatial, temporal, and logical relations. A simple event can be an action performed by a single object or an interaction between multiple objects. For this reason, we were motivated to address the problem at different levels. Our method is based on the extraction of features from a low level processing enhanced with spatio-temporal analysis. To describe and recognize high-level events or scenarios that incorporate simple events with temporal and spatial relations, a Petri Nets model

© Springer International Publishing AG, part of Springer Nature 2018
A. Abraham et al. (Eds.): ISDA 2017, AISC 736, pp. 389–398, 2018.
https://doi.org/10.1007/978-3-319-76348-4_38

was used. Our contributions in this paper are the following: First, the objects are automatically and accurately detected and classified. Second, the simple events are defined through a spatio-temporal analysis of the features related to the detected objects in the scene. Moreover, the abnormal scenarios are represented using the Petri Nets model.

The remainder of this paper was organized as follows. Section 2 reviewed the related work to the problems addressed by the proposed approach. Section 3 described our proposed method of high-level event recognition for an abnormal event detection. Section 4 detailed the experimental results. Finally, we conclude this work in Sect. 5.

2   Related Work

The generic process for surveillance system [1, 2] consists of two processing levels: low-level and high-level. In what follows, the issues related to this system were discussed.

The low-level processing consists of three principle steps: moving object detection, tracking and classification. The moving object detection aims to isolate the foreground pixels that participate into any kind of motion observed in a given scene. The contributions reported in the literature can be classified in four main categories based either on inter-frames differences [3], background modeling [4], optical flow [5] or on a combination of two or all of these methods [6]. The next step is the objects tracking to locate the objects in time and extract their trajectories. There are two categories of tracking methods: Points based approach [7] and Model based approach [8] (silhouettes or kernel). Objects classification aims to improve the reliability of the high level processing results by classifying detected objects appearing in a given scene into classes of humans or vehicles. In the literature, object classification methods can be categorized primarily into four main approaches: Shape-based methods [9], Texture-based methods [10], Motion-based methods [11] and Hybrid methods [12]. These methods generally use a machine learning technique for the moving object classification.

The simple events can be a human action performed in a short time or an interaction between two objects. The high level processing takes into account the information obtained from the low-level processing in order to represent such simple events. These are combined afterwards to include the high-level events. In the literature, various techniques have been proposed for action recognition. These techniques can be classified as follows: (i) Human body model based methods [13], (ii) Holistic methods [14], and (iii) Local feature methods [15]. To represent the high-level events with a formal model and recognize these events as they occur in the video sequence, a great deal of work has also been proposed. Some researchers deal with state models that model the state of the video event in space and time using semantic knowledge, e.g. Finite-State Machines (FSM) [16], Bayesian Networks (BNs) [17], Hidden Markov Models (HMM) [18]... Other works model the high-level event by defining a set of semantic rules, constraints, and relations at the symbol level. Grammar based methods [19] and Petri Nets [20] are the most used in most of the studies dealing with this issue.

3   Proposed Approach

In this section we introduce our approach of high-level events recognition in a parking lot. As presented in Fig. 1, our approach consists of two levels with an increasing level of abstraction. In a low-level processing, we are interested in detecting, tracking and classifying moving objects to obtain the position and class of each object. In a high-level processing, we rely on a spatio-temporal analysis to recognize the simple events; as for the high-level events, they are identified with a Petri Nets model.

3.1   Low-Level Processing

The process of our low-level processing is composed of two phases: (1) an off-line phase adopting a data mining process to construct the appropriate prediction model for the moving object classification and (2) an on-line phase to detect, track and classify the moving objects in order to obtain the position and class of each detected moving object in the scene.

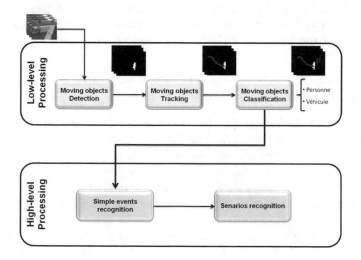

Fig. 1. The process of our approach.

Off-line Phase: The proposed method for moving objects classification operates in two steps: (a) the first step is devoted to the preparation of data for the learning phase; (b) the second step consists in finding the most appropriate prediction model using several data-mining algorithms in order to best discriminate the two classes of moving objects: Pedestrian (P) and Car (C).

(a) *Data Preparation:* In this step, we start by the construction of a large and representative learning dataset and a test dataset. Our training data is composed of 4583 moving objects of class P and 849 moving objects of class C. As for the test data set, it contains 4682 moving objects of class P and 108 moving objects of class C. Thereafter, a selection of the best features is required to ensure the best performance of the moving objects classification. In our work, we have selected seven shape based features that are Height, Width, Aspect Ratio, Area, Perimeter, Compactness and Anthropomethry; two famous texture based features that are the Histogram of Oriented Gradients (HOG) and the Local Binary Pattern (LBP) descriptors and one motion based feature that is the velocity of the object [21].

(b) *Data Mining:* The aim of this step is the construction of the appropriate prediction model. To this end, we introduce a new field of application of data mining techniques. Thus, we have selected 4 different learning algorithms: the SVM with polynomial kernel (Poly Kernel), SVM with Radial Basis Function Kernel (RBF Kernel), Multilayer Perceptron Neural Network (MLP) and the algorithm C4.5 of decision trees. The quality and the stability of the prediction models constructed by these four data mining algorithms are evaluated intensively in our experimental study to determine the most suitable prediction model for our classification task. This classification is relevant for the performance of simple event recognition.

On-line Phase: The online phase of our low-level processing is composed of three principle steps:

- *Moving Objects Detection:* To ensure better results of moving objects detection, we adopted a background modeling based method with a dynamic matrix and spatio-temporal analyses of scenes [22]. In addition, the adopted method originality lies in the integration of the principle of the methods based on inter-frame differences in the background modeling step and in the way this integration was carried out, making it more adaptable and independent of the moving objects speeds and sizes.
- *Moving Objects Tracking:* For the moving objects tracking, we adopted the method presented in [23] based on merging and splitting detection, and feature correspondence. This method can track multiple objects with long-duration even if there are partially or completely occluded.
- *Moving Object Classification:* Our moving object classification method is based on a hybrid approach combining the best features based on shape, texture and motion using a data-mining process to ensure a better classification of the moving objects. This classification is based on the prediction model constructed in the off-line phase of our approach.

3.2   High-Level Processing

Our high-level processing consists of simple events and scenarios recognition.

Simple Events Recognition. We were interested in recognizing a set of simple events based on the information extracted from the first level processing, namely the object Bounding Box, the object blob, the object center position and the object class. Thus, each moving object is characterized by the spatial features presented in Eq. 1:

$$MO = [x, y, id, R, x_{top}, y_{left}, x_{height}, y_{width}] \tag{1}$$

where (x, y) are the center coordinates, id is the class, R is the blob, x_{top}, y_{left}, x_{height}, y_{width} are the coordinates of the Bounding Box.

In our work, the simple events were defined with respect to persons and cars. A simple event can be the action performed by one object (car enters, car moves, car stops, person enters) and can also be performed between two or more objects, which is commonly referred to an interaction (person walks-towards, person joins car, person joins person, person turns car, person splits from car, person splits from person, person walks away, person runs away). We have two types of interactions: person-to-person and person-to-car. An important aspect in the abnormal event detection is to differentiate between the simple events of the person. A person running away is different from a person walking. So, we enhanced our spatio-temporal analysis with a motion based descriptor defined by Histograms of Optical Flow (HOF) [24]. This descriptor allows differentiating between walks-towards and runs-towards and between walks-away and runs-away. The HOF descriptor was computed by blocks. The number of blocks used is 3∗3 blocks in the spatial domain and 2 in the temporal domain. The responses per block were computed using Lucas-Kanade method [25]. Then for each response the magnitude was quantized in o orientations, usually $o = 8$. Finally, adjacent blocks were concatenated to form the descriptors. To determinate the action of a person, the Bounding Box extracted during the detection and classification steps were firstly resized to the maximum dimension of the Bounding Box. Then they were organized in a figure centric spatio-temporal volume for each person. The Bounding Box were segmented to cycles where a cycle begins with an initial swing of the two feet. Finally, the HOF descriptors were calculated for each cycle.

For the recognition of interactions between objects, a vector of spatio-temporal features was extracted for each moving object at each time t, as presented in Eq. 2:

$$f^t = [d_{ij}, op_{ij}, th_{ij}, se_{ij}] \tag{2}$$

where d_{ij} is the distance between the two objects, op_{ij} is the overlap between the two objects, th_{ij} is the angle between the vector that forms the objects centers and the original axis and se_{ij} is the closest segment between the objects. The extracted interactions between two objects were detailed in Table 1.

Scenarios Recognition. Scenarios were defined by a Petri Nets model which allows benefiting from the following advantages:

– It naturally models the relations between the simpler events including non sequential (and sequential) temporal relations, spatial and logical composition, hierarchy, concurrency, and partial ordering;

Table 1. Determination of interactions between two objects.

Interaction	Determination
$MoveAway$ (O_1, O_2)	$d^{t+1}(O_1, O_2) < d^t(O_1, O_2)$
$MoveFrom$ (O_1, O_2)	$d^{t+1}(O_1, O_2) > d^t(O_1, O_2)$
$Merge$ (O_1, O_2)	$op^t(O_1, O_2) = 0$ and $op^{t+1}(O_1, O_2) = 1$
$Split$ (O_1, O_2)	$op^t(O_1, O_2) = 1$ and $op^{t+1}(O_1, O_2) = 0$
$TurnB$ (O_1, O_2)	$op^t(O_1, O_2) = 1$ and $SE = $ 'b' and $th \in [-90, 90]$
$TurnL$ (O_1, O_2)	$op^t(O_1, O_2) = 1$ and $SE = $ 'l' and $th \in [-45, 135]$
$TurnR$ (O_1, O_2)	$op^t(O_1, O_2) = 1$ and $SE = $ 'r' and $th \in [-90, 0]$
$TurnF$ (O_1, O_2)	$op^t(O_1, O_2) = 1$ and $SE = $ 'f' and $th \in [90, 180]$

- It provides a nice graphical representation of the event model;
- PN event models are specified manually without any need of long video for training.

A Petri Net model is made up of set of Places, a set of Transitions, a set of Arcs between a Place and a Transition or between a Transition and a Place, and the tokens. In our work, the set of simple events are represented by the transitions and the places are the state describing a situation of the objects. Each token in a place represents a distinguished moving object which can be cars or persons. As Petri Nets model the scenarios as a combination of simple events with the spatial, temporal and logical relations, we start by presenting these relations:

Temporal Relations: Temporal relations are heavily important when describing scenarios. We adopt the Allen's relations [26] between two intervals: 'before', 'meets', 'overlaps', 'stars', 'during' and 'finishes'.

Spatial Relations: A spatial relation is a relation between spatial objects. Thus, they are only defined in their interactions.

Logical Relations: The logical relations include and, or and not. It is designed for concatenating the spatial and temporal relations.

This research study addressed two interesting scenarios of theft in a parking: theft from a car and theft from a person. The two Petri Net models are defined respectively in Fig. 2(a) and (b).

In the first scenario, we have two objects: a Car C_0 and a Person P_1. The Petri Net model combines the simple events with a sequential order. Whenever a car enters the scene, a new token is inserted in the place p_1. If the car moves, its token is moved from p_1 to p_2 and so on. Whenever a Person P_1 enters the scene, a new token is inserted in the place p_4. If the person P_1 walks towards C_0, the token is relocated to p_4. At the end, the number of token in p_7 denotes the number of objects in the scene. In the second scenario, the interactions are between two persons P_1 and P_2. Whenever a person P_1 enters and moves in the

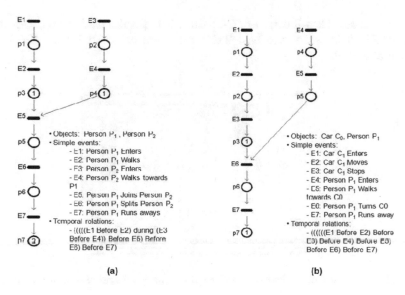

Fig. 2. Petri Net representation.

scene, a new token is inserted in the place p_1 then replaced to p_3. Whenever a person P_2 enters and moves in the scene, a new token is inserted in the place p_2 then relocated to p_4. The tokens from p_3 and p_4 are not matched in p_5 until the simple event 'joint' is detected. The same process is performed until the token is inserted into p_7.

4   Experimental Results

In this section, we presented two series of experiments: The first series detailed the experiment concerning the moving objects classification of the low level processing, whereas the second series gave the results of our high-level event recognition.

4.1   First Series of Experiments

In this series of experiments, we aimed to show the performance of the moving objects classification. The experiments were performed to select the most adequate learning technique. We used a corpus[1,2] of data recorded in different scenes and in various weather conditions. The data contain moving objects of different classes. In this study, we proceeded to select the most adequate learning technique. We used such different classifiers as the Multilayer Perceptrons (MLP), Support Vector Machines (SVM) with polynomial kernel and RBF kernel and C4.5 based decision trees. We compared the cited data mining algorithms using

---

[1] http://vcipl-okstate.org/pbvs/bench/.
[2] http://www.ino.ca/Video-Analytics-Dataset.

the Total Correct Classification (TCC) rate. The results of this experiment are shown in Fig. 3 in which the best TCC rates are achieved by the SVM with polynomial kernel.

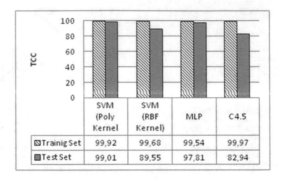

Fig. 3. Moving object classification results with different machine learning techniques.

4.2   Second Series of Experiments

Given the lack of public datasets representing abnormal scenarios in a parking lot, we applied our framework on our video stream captured from a static camera monitoring a university parking lot. In this dataset, 28 sequences are recorded where 14 participants. The dataset involves two types of thefts: theft from a car and another from a person. Our system is used to model and detect these abnormal scenarios. Once the abnormal scenario is detected, an alert is trigged by the system. For the two thefts scenarios, we have eight simple events: Enter, Join, Split, Walk-towards, Run-towards, Walk-away, Run-away, and Turn around.

Since the high-level event recognition is related to the simple events recognition, the final results are strictly bounded by the results of the simple event recognition. To validate our method in the case of a simple event performed by a single object, we extracted a model of walk and run using a popular benchmark dataset WEIZMANN [27]. The WEIZMANN dataset contains 90 videos of 10 actions performed by 9 persons. We prepared a learning data and used the SVM

Table 2. Performance of simple events recognition.

	Enters	Joins	Splits	Moves-toward	Moves-away	Turns	Walks	Runs
TP	6	5	4	134	172	601	20	6
FP	0	0	0	0	0	0	0	0
FN	0	0	0	7	0	0	6	0
Precision	1	1	1	1	1	1	1	1
Rappel	1	1	1	0.95	1	1	0.89	1

with Poly kernel to recognize the simple event. Based on this model, we achieved the results presented in Table 2 that shows the performance of the simple events defined with spatio-temporal analysis.

5 Conclusion

We have presented a new approach of high-level video recognition for an abnormal event detection in a parking lot. Our approach proceeds by detecting, tracking and classifying the moving objects in the scene first. Based on the position and the class of the detected moving object, a module of simple events recognition is then performed. The defined simple events are combined with spatial, temporal and logical relations to recognize the abnormal scenarios through a petri Nets model. The achieved results prove the efficiency of our approach.

References

1. Kim, I.S., Choi, H.S., Yi, K.M., et al.: Intelligent visual surveillance a survey. Int. J. Control Autom. Syst. **8**(5), 926–939 (2010)
2. Vishwakarma, S., Agrawal, A.: A survey on activity recognition and behavior understanding in video surveillance. Vis. Comput. **29**(10), 983–1009 (2013)
3. Cheng, Y.H., Wang, J.: A motion image detection method based on the inter-frame difference method. Appl. Mech. Mater. **490–491**, 1283–1286 (2014)
4. Jian-Ping, T., Xiao-lan, L., Jun, L.: Moving object detection and identification method based on vision. Int. J. Secur. Appl. **10**(3), 101–110 (2016)
5. Hariyono, J., Hoang, V.D., Jo, K.H.: Motion segmentation using optical flow for pedestrian detection from moving vehicle. In: Hwang, D., Jung, J.J., Nguyen, N.T. (eds.) Computational Collective Intelligence. Technologies and Applications. ICCCI 2014. LNCS, vol. 8733. Springer, Cham (2014)
6. Kushwaha, A.K.S., Srivastava, S., Srivastava, R.: Multi-view human activity recognition based on silhouette and uniform rotation invariant local binary patterns. Multimed. Syst. J. **32**(4), 451–467 (2017)
7. Gargi, P., Rajbabu, V.: Mean LBP and modified fuzzy C-means weighted hybrid feature for illumination invariant meanshift tracking. Signal Image Video Process. **11**(4), 665–672 (2017)
8. Yang, Y., Cao, Q.: A fast feature points-based object tracking method for robot grasp. Int. J. Adv. Robot. Syst. **10**(3) (2013)
9. Dedeoğlu, Y., Töreyin, B.U., Güdükbay, U., Çetin, A.E.: Silhouette-based method for object classification and human action recognition in video. In: Huang, T.S., et al. (eds.) Computer Vision in Human-Computer Interaction. ECCV 2006. LNCS, vol. 3979, pp. 64–77. Springer, Heidelberg (2006)
10. Moctezuma, D., Conde, C., Diego, I.M., Cabello, E.: Person detection in surveillance environment with HoGG: Gabor filters and histogram of oriented gradient. In: ICCV Workshops, pp. 1793–1800 (2011)
11. Wang, L., Hu, W., Tan, T.: Recent developments in human motion analysis. Pattern Recognit. **36**(3), 585–601 (2003)
12. Zhang, Z., Cai, Y., Huang, K., Tan, T.: Real-time moving object classification with automatic scene division. In: IEEE International Conference on Image Processing, pp. 149–152 (2007)

13. Chakraborty, B., Bagdanov, A.D., Gonzàlez, J., Roca, X.: Human action recognition using an ensemble of body-part detectors. Expert Syst. **30**(2), 101–114 (2012)
14. Bobick, F., Davis, W.: The recognition of human movement using temporal templates. IEEE Trans. Pattern Anal. Mach. Intell. **23**(3), 257–267 (2001)
15. Wang, H., Kläser, A., Schmid, C., Cheng-Lin, L.: Dense trajectories and motion boundary descriptors for action recognition. Int. J. Comput. Vis. **103**(1), 60–79 (2013)
16. Lv, F., Nevatia, R.: Single view human action recognition using key pose matching and Viterbi path searching. In: IEEE Conference on Computer Vision and Pattern Recognition, p. 18 (2007)
17. Wang, Y., Cao, K: A proactive complex event processing method for large-scale transportation Internet of Things. Int. J. Distrib. Sens. Netw. **10**(3) (2014)
18. Bansal, N.K., Feng, X., Zhang, W., Wei, W., Zhao, Y.: Modeling temporal pattern and event detection using Hidden Markov Model with application to a sludge bulking data. Procedia Comput. Sci. **12**, 218–223 (2012)
19. Zhang, Z., Tan, T., Huang, K.: An extended grammar system for learning and recognizing complex visual events. IEEE Trans. Pattern Anal. Mach. Intell. **33**(2), 240–255 (2011)
20. Ghanem, N., DeMenthon, D., Doermann, D., Davis, L.: Representation and recognition of events in surveillance video using Petri nets. In: Conference on Computer Vision and Pattern Recognition Workshop, pp. 112–121 (2004)
21. Boukhriss, R.R., Fendri, E., Hammami, M.: Moving object classification in infrared and visible spectra. In: International Conference on Machine Vision (2016)
22. Hammami, M., Jarraya, S.K., Ben-Abdallah, H.: On line background modeling for moving object segmentation in dynamic scene. Multim. Tools Appl. **63**(3), 899–926 (2013)
23. Yang, T., Li, S.Z., Pan, Q., Li, J.: Real-time multiple objects tracking with occlusion handling in dynamic scenes. In: IEEE Computer Society Conference on Computer Vision and Pattern Recognition (2005)
24. Uijlings, J., Duta, I.C., Sangineto, E., Sebe, N.: Video classification with densely extracted HOG/HOF/MBH features: an evaluation of the accuracy/computational efficiency trade-off. Int. J. Multimed. Inf. Retr. **4**(1), 33–44 (2014)
25. Lucas, B., Kanade, T.: An iterative image registration technique with an application to stereo vision. In: International Joint Conference on Artificial Intelligence (1981)
26. Allen, J.F., Ferguson, F.: Actions and events in interval temporal logical. J. Log. Comput. **4**(5), 531–579 (1994)
27. Gorelick, L., Blank, M., Shechtman, E., Irani, M., Basri, R.: Actions as space-time shapes. In: IEEE International Conference on Computer Vision (2005)

Optimum Feature Selection Using Firefly Algorithm for Keystroke Dynamics

Akila Muthuramalingam[3]([⊠]), Jenifa Gnanamanickam[1],
and Ramzan Muhammad[2]

[1] KPR Institute of Engineering and Technology, Coimbatore, India
jenifa.g@kpriet.ac.in
[2] Maulana Mukhtar Ahmad Nadvi Technical Campus, Malegaon, Nashik, India
ramzan145@gmail.com
[3] PGP College of Engineering and Technology, Namakkal, Tamil Nadu, India
akila@nvgroup.in

Abstract. Keystroke dynamics, an automated method and promising biometric technique, is used to recognize an individual, based on an analysis of user's typing patterns. The processing steps involved in keystroke dynamics are data collection, feature extraction and feature selection. Initially the statistical measures of feature characteristics like latency, duration and digraph are computed during feature extraction. Various advanced optimization techniques are applied by researchers to mimic the behavioral pattern of key stroke dynamics. In this study, Firefly algorithm (FA) is proposed for feature selection. The performance efficiency of FA is computed and compared with existing techniques and found that the convergence rate and iteration generations to reach the optimum solution is 41% and 18% less respectively, as compared to those by other algorithms.

Keywords: Keystroke dynamics · Duration · Latency · Digraph
Feature extraction · Feature selection · Firefly algorithm

1 Introduction

Biometrics, an automated identification system [1, 2], is based on an individual's physiological or behavioral characteristics to make positive identification. These systems identify the uniqueness in physiology and behavior attributes of an individual for biometric authentication. Biometric identifiers are distinctive in nature but physically measurable characteristics to recognize an individual. However, physiological and behavioral both can be separated as two subsets of biometric identification system. Physiological characteristics accounts recognition of natural attributes, such as fingerprint, iris, retina, face, palm veins etc. [3], whereas behavioral authentication measures habitual tendencies of individuals due to psychological and physiological differences. Writing style, typing rhythm and unconscious body movements are used in behavioral biometric system. Among multiple authentication technologies, behavioral biometric system is inherently more reliable as a consequence of natural uniqueness of each individual, psychologically as well as physiologically. Physiologically identified authentication systems are successfully implemented, though with certain recognition errors; the iris scanner, with an Equal Error Rate (EER) of 0.01%, performed the best [3].

© Springer International Publishing AG, part of Springer Nature 2018
A. Abraham et al. (Eds.): ISDA 2017, AISC 736, pp. 399–406, 2018.
https://doi.org/10.1007/978-3-319-76348-4_39

In security system, authentication is a process of verifying the digital reference identity stored in the system, matching with the identity of the individual asking for access. It is expected that, the reference psychological and behavioral characteristics may reduce the error rate in identification of an individual. These methods include keystroke dynamics [4, 5], mouse dynamics [6–8], signature verification [3] and Graphical User Interface (GUI) analysis [9].

In this study, 25 individuals' typing behavioral characteristics were studied through key stroke dynamics. Ten valid samples of each user were recorded for a common password 'welcome' and features were optimized by using Firefly Algorithm (FA). The results of FA were compared with those obtained from various advanced algorithms such as Particle Swarm Optimization (PSO), Genetic Algorithm (GA), Artificial Bee Colony (ABC) and Ant Colony Optimization (ACO). It was found that, the iterative generations and processing time required for FA is about 18% and 41% less respectively, as compared to those of other techniques.

2 Keystroke Dynamics

Keystroke dynamics (KD) [10], is a behavioral authentication tool, which extracts and analyzes how (pattern and way) an individual types rather than what the individual types. In KD, no any extra hardware is required to be added as in other biometric systems. A typical keystroke dynamic authentication system may consist of several components like data acquisition, feature extraction, feature selection, classification, decision and retraining. However, the following three major steps are involved in keystroke dynamics analysis [11]:

  i. A time vector is registered for an individual, through typing a specific password by the individual.
 ii. The specific features of the registered password are extracted.
iii. A feature selection is made by filtering redundant or irrelevant features from large scale data sets, to store the reference pattern.

In KD based authentication system, reducing computational speed and improving prediction accuracy, in feature selection, decides the accuracy and reliability of the approach.

In the following lines, the terminology and their specific significance is mentioned in accordance with KD analysis.

2.1 Data Collection

Data collection is the preliminary and essential step of keystroke dynamics [12]. It is the process of gathering and measuring information on targeted variables in an established systematic fashion. The goal of data collection is to capture quality evidence. To capture keystroke dynamics, it is necessary for users to type their own password a number of times during enrolment.

2.2   Feature Extraction

There are number of different aspects of keystroke characteristics for feature extraction that can be used for identification such as cumulative typing speed, the time elapsed between consecutive keys, the time that each key held down, the frequency of the individual in using other keys on the keyboard like the number pad or function keys, the sequence utilized by the individual when attempting to type a capital letter (for example, does the user release the shift key or the letter key first), pressure applied for a keystroke, typographical errors made etc. Initially the statistical measures of feature characteristics such as latency, duration and digraph are computed [12, 13].

- **Duration**
  Keystroke duration (Dwell/Held time) is the interval time in milliseconds (ms), between a key press and a key release of the same key. Figure 1 shows, how the duration time of the key 'T' is determined. Duration time is strictly greater than zero.
- **Latency**
  Latency is defined as the time in ms between two consecutive keystrokes. In Fig. 1, the time between the key release of 'H' and key press of 'E'. The latency times can be negative, i.e. the second key is depressed before the first key is released.
- **Digraph**
  Digraph is the time interval between the down key event of the first keystroke and the up key event of the second keystroke. In Fig. 1, the time between the key press of 'T' and key press of 'H'.

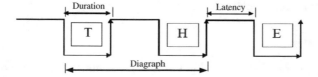

Fig. 1. Typical keystroke

The obtained values are normalized using mean (Eq. 1), standard deviation (Eq. 2) and latency (Eq. 3) as follows:

$$\mu_i = \frac{1}{N} \sum\nolimits_{j=1}^{N} x(j) \tag{1}$$

$$\sigma = \sqrt{\frac{1}{N} \sum_{i,j=1}^{N} |x(j) - \mu(i)|} \tag{2}$$

$$m = \begin{cases} N/2 & \text{When N is odd} \\ \frac{N+(N+1)}{2} & \text{When N is even} \end{cases} \tag{3}$$

where $i = 1...N$, μ is the mean, σ is the standard deviation, m is the median, j is the extracted feature of user i, x is the feature set and N is the number of letters/digits typed for each sample.

3 Firefly Algorithm

Feature extraction creates new features from functions of original features, whereas feature selection returns a subset of the features. The goal of the subset selection process is to obtain a subset of features that allows better rates of correct identification than with the entire set of features. Feature subset selection is an optimization problem, since the aim is to obtain any subsets that minimize the particular measure. An optimization problem can be solved through stochastic algorithms. In this study, for better accuracy, firefly algorithm is proposed for feature selection.

3.1 Firefly Algorithm

Firefly is an insect that mostly produces short and rhythmic flashes that produced by a process of bioluminescence. The function of the flashing light is to attract partners (communication) or attract potential prey and as a protective warning toward the predator. Thus, this intensity of light is the factor of the other fireflies to move toward the other firefly.

The light intensity is varied at the distance from the eyes of the beholder. It is safe to say that the light intensity is decreased as the distance increase. The light intensity also the influence of the air absorb by the surroundings, thus the intensity becomes less appealing as the distance increase.

3.2 Firefly Rules

Firefly algorithm is based on idealizing the flashing characteristic of fireflies [14–17]. The idealized three rules are:

  i. Fireflies are attracted toward each other regardless of gender.
 ii. The attractiveness is proportional to the brightness, and they both decrease as their distance increases. Thus for any two flashing fireflies, the less bright one will move towards the brighter one. If there is no brighter one than a particular firefly, it will move randomly.
iii. The brightness of a firefly is determined by the landscape of the objective function. For a maximization problem, the brightness is proportional to the objective function's value.

3.3 Structure of Firefly Algorithm

In firefly algorithm, there are two important variables; the light intensity and attractiveness. A firefly is attracted towards the other firefly having brighter flash than itself. The attractiveness is depended with the light intensity.

In the simplest case, for maximization problems, the brightness I of a firefly at a particular location x can be chosen as: $I(r) \propto f(x)$. However, the attractiveness β is relative; it should be seen in the eyes of the beholder or judged by the other fireflies. In the simplest form, the light intensity $I(r)$ varies with the distance r monotonically and exponentially (Eq. 4).

$$I(r) = I(0)e^{-\gamma(r*r)}. \tag{4}$$

Sometimes, we may need a function which decreases monotonically at a slower rate. In this case, we can use the approximation as follows (Eq. 5):

$$I(r) = \frac{I(0)}{\gamma(r*r)} \tag{5}$$

As firefly's attractiveness is proportional to the light intensity seen by adjacent fireflies (6), we can now define the variation of attractiveness β with the distance r by

$$\beta = \beta_0 e^{-\gamma(r*r)}. \tag{6}$$

where β_0 is the attractiveness at $r = 0$. It is worth pointing out that the exponent $\gamma(r*r)$ can be replaced by other functions such as γr^m when $m > 0$.

The distance r between any two fireflies i and j at x_i and x_j, respectively, is the Cartesian distance defined by (Eq. 7):

$$r_{ij} = ||x_i - x_j|| = \sqrt{\sum_{k=1}^{d} x(i,k) - x(j,k)}. \tag{7}$$

The movement of a firefly i attracted to another more attractive (brighter) firefly j is determined by (Eq. 8)

$$x_i^{t+1} = xi^t + \beta_0 e^{-\gamma(r(i,j)*r(i,j))}\left(x_j^t - x_i^t\right) + \alpha_t \varepsilon_i^t. \tag{8}$$

where the second term is due to the attraction, the third term is randomization with α_t being the randomization parameter, and ε_i^t is a vector of random numbers drawn from a Gaussian distribution or uniform distribution at time t. If $\beta_0 = 0$, it becomes a simple random walk. On the other hand, if $\gamma = 0$, it reduces to a variant of particle swarm optimization. Furthermore, the randomization ε_i^t can easily be extended to other distributions. The basic Firefly Algorithm is described as follows:

3.4   The Algorithm

Objective Function, which is to be maximized, $f(x)$, $x = (x_1, x_2, ..., x_d)^T$.
   Generate initial population of fireflies x_i ($i = 1, 2, ..., m$).
   Light intensity I_i at x_i is determined by $f(x_i)$.

```
if Iⱼ < Iᵢ then
  Move firefly i towards firefly j in d-dimension;
end if
Alternativeness varies with distance r via e⁻ʸʳ
for i = 1: n do
  for j =1: i do
    Define light absorption coefficient γ
    while t < Max Generation do
      Evaluate new solutions and update light intensity
  end for j
end for i
Rank the fireflies and find the current best
end while
```

The initial population is considered as mean value of the duration for individual users and objective function is the feature subsets obtained by using FA. Each mean value is chosen and compared with every other mean values in population and then the largest difference are neglected (i.e. the light intensity) to find the subset. The similar mean values are considered as single in a set. Hence the irrelevant and repeated data are removed using FA in order to find subset.

4   Results

Experiments were carried out with the key stroke data collected from 25 users, with 10 valid samples from each user, in a span of one month. Duration, latency and digraph timing were measured for each sample and their mean, median and standard deviation were stored in the reference file. A system using Visual Basic was created, to organize these measurements. The system registered each typing entry and stored the corresponding data in samples' file. Table 1 shows the measured keystroke feature values of duration timing of user-1 for the password "welcome" and the corresponding each key value obtained in milliseconds. The computations as shown in Table 1 serve as the initial reference signature or template for feature string. Extracted features are optimized using FA. Firefly was implemented using the algorithm as explained in Sect. 3.

It was observed, from Fig. 2, that the convergence rate of PSO was 58 ms and took 425 generations to reach an optimal solution and GA took 68 ms and 440 generations to converge and produced only near optimal solution. ABC was very competitive, performed relatively well and showed lower computational overhead than PSO and GA and also fewer parameters to set. However FA has proved to be most efficient algorithm in this case, with convergence rate as 40 ms and 360 generations to reach the optimum solution; which is in fact 41% and 18% less respectively as compared to those of maximum values taken by other algorithms.

Table 1. Sample key stroke timings (milliseconds) of a user

Samples	W	e	l	c	o	m	e
1	20.000	20.034	20.036	20.051	20.096	20.070	20.036
2	20.056	20.055	20.054	20.036	20.088	20.074	20.045
3	20.021	20.018	20.040	20.032	20.360	20.078	20.085
4	20.065	20.024	20.036	20.042	20.036	20.068	20.034
5	20.072	20.066	20.051	20.045	20.036	20.032	20.095
6	20.085	20.063	20.092	20.075	20.081	20.001	20.036
7	20.025	20.068	20.063	20.057	20.046	20.031	20.097
8	20.061	20.077	20.079	20.023	20.045	20.062	20.086
9	20.080	20.039	20.045	20.049	20.023	20.027	20.086
10	20.064	20.068	20.068	20.058	20.045	20.069	20.018

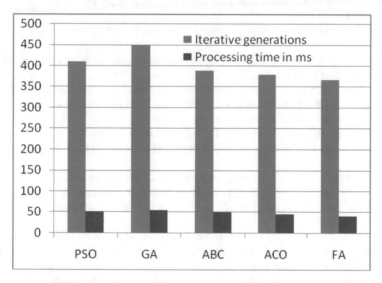

Fig. 2. Summarized results of feature selection algorithm. PSO – Particle Swarm Optimization; GA – Genetic Algorithm; ABC – Artificial Bee Colony; ACO – Ant Colony Optimization; FA – Firefly Algorithm

5 Conclusion

Keystroke dynamics present the advantage of not requiring any additional sensing device and allows on-line authentication through time. It has been demonstrated that using certain features, such as duration, latency, digraph and their combinations timing as attributes can be applied for biometric authentication. Features were collected from 25 users with 10 samples of each user in a period of one month. Using the collected samples the mean, standard deviation and median were calculated for duration, latency and digraph. Firefly algorithm was used for feature selection in keystroke dynamics. On comparing the efficiency of different algorithms, it was concluded that Firefly algorithm has the better efficiency than other algorithms.

References

1. Karnan, M., Akila, M., Krishnaraj, M.: Biometric personal authentication using keystroke dynamics: a review. App. Soft Comput. J. **11**, 1565–1573 (2010)
2. Akila, M., Suresh Kumar S.: Improving feature extraction in keystroke dynamics using optimization techniques and neural network. In: Second International Conference on Sustainable Energy and Intelligent System, SEISCON 2011, Chennai (2011)
3. Bhattacharyya, D., Ranjan, R., Alisherov, F., Choi, M.: Biometric authentic: a review. Int. J. u-&e-Serv. Sci. Technol. **2**, 13–28 (2009)
4. Gunetti, D., Picardi, C.: Keystroke analysis of free text. ACM Trans. Inf. Syst. Secur. **8**, 312–347 (2005)
5. Marsters, J.: Keystroke dynamics as a biometric [Ph.D. thesis]. University of Southampton (2009)
6. Ahmed, A., Traore, I.: A new biometric technology based on mouse dynamics. IEEE Trans. Dependable Secur. Comput. **4**, 165–179 (2007)
7. Shen, C., Guan, X., Cai, J.: A hypo-optimum feature selection strategy for mouse dynamics in continuous identity authentication and monitoring. In: IEEE International Conference on Information Theory and Information Security, pp. 349–353 (2010)
8. Zheng, N., Paloski, A., Wang, H.: An efficient user verification system via mouse movements. In: ACM Conference on Computer and Communications Security, pp. 1–12 (2011)
9. Imsand, E.: Applications of GUI usage analysis [Ph.D. thesis]. Auburn University (2008)
10. Karnan, M., Akila, M.: Personal authentication based on keystroke dynamics using soft computing techniques. In: 2nd International Conference on Communication Software and Networks, Singapore (2010)
11. Karnan, M., Akila, M.: Identity authentication based on keystroke dynamics using genetic algorithm and particle swarm optimization. In: 2nd IEEE International Conference on Computer Science and Information Technology, ICCSIT 2009, Beijing (2009)
12. Akila, M., Suresh, K., Anusheela, V., Sugumar, K.: A novel feature subset selection algorithm using artificial bee colony in keystroke dynamics. In: International Conference on Soft Computing and Problem Solving, Roorkee, India, pp. 759–766 (2011)
13. Akila, M., Suresh, K.: Performance of classification using a hybrid distance measure with artificial bee colony algorithm for feature selection in keystroke dynamics. Int. J. Compt. Intell. Stud. **2**, 187–197 (2013)
14. Farahani, Sh.M, Abshouri, A.A., Nasiri, B., Meybodi, R.: A Gaussian firefly algorithm. Int. J. Mach. Learn. Compt. **1**, 448–453 (2011)
15. Sankalap, A., Satvir, S.: The firefly optimization algorithm: convergence analysis and parameter selection. Int. J. Compt. Appl. **69**, 0975–8887 (2013)
16. Abdesslem, L., Zeyneb, B.: A novel firefly algorithm based ant colony optimization for solving combinatorial optimization problems. Int. J. Compt. Sci. Appl. **11**, 19–37 (2014)
17. Rohit, A.P., Amar, L.: Keystroke dynamics for user authentication and identification by using typing rhythm. Int. J. Compt. Appl. **144**, 975–8887 (2016)

Multi-UAV Path Planning with Multi Colony Ant Optimization

Ugur Cekmez[1,2], Mustafa Ozsiginan[2], and Ozgur Koray Sahingoz[3(✉)]

[1] Information Technologies Institute, TUBITAK BILGEM, Kocaeli, Turkey
ugur.cekmez@tubitak.gov.tr
[2] Institute of Pure and Applied Sciences, Marmara University, Istanbul, Turkey
mustafaozsiginan@gmail.com
[3] Computer Engineering Department, Istanbul Kultur University, 34158 Istanbul, Turkey
o.sahingoz@iku.edu.tr

Abstract. In the last few decades, Unmanned Aerial Vehicles (UAVs) have been widely used in different type of domains such as search and rescue missions, firefighting, farming, etc. To increase the efficiency and decrease the mission completion time, in most of these areas swarm UAVs, which consist of a team of UAVs, are preferred instead of using a single large UAV due to the decreasing the total cost and increasing the reliability of the whole system. One of the important research topics for the UAVs autonomous control system is the optimization of flight path planning, especially in complex environments. Lots of researchers get help from the evolutionary algorithms and/or swarm algorithms. However, due to the increased complexity of the problem with more control points which need to be checked and mission requirements, some additional mechanisms such as parallel programming and/or multi-core computing is needed to decrease the calculation time. In this paper, to solve the path planning problem of multi-UAVs, an enhanced version of Ant Colony Optimization (ACO) algorithm, named as multi-colony ant optimization, is proposed. To increase the speed of computing, the proposed algorithm is implemented on a parallel computing platform: CUDA. The experimental results show the efficiency and the effectiveness of the proposed approach under different scenarios.

Keywords: Multi-UAV · Multi Colony Ant Optimization · GPGPU · CUDA
Parallel evolutionary computing

1 Introduction

In recent years, due to the advances in aeronautics, electronic, and communication technologies the usage of Unmanned Aerial Vehicles (UAVs) has become increasingly attractive research area for dangerous, dull and dirty missions in which crews of manned aircraft could not provide. UAVs not only eliminate some threats to pilots' life but also they are cheaper and need less power. Therefore, they are preferred for longer missions and missions which need flexible maneuvering.

© Springer International Publishing AG, part of Springer Nature 2018
A. Abraham et al. (Eds.): ISDA 2017, AISC 736, pp. 407–417, 2018.
https://doi.org/10.1007/978-3-319-76348-4_40

The optimal path planning is one of the crucial research areas for UAVs' autonomous control system. In this autonomous structure, it is aimed to compute an optimal or near-optimal flight path between the starting point to the final point with visiting some control points by taking into account some mission specific constraints. In general, due to the physical restrictions of UAVs maintenance the starting point and the final point are accepted as the same point, and they are located in an airport. This path planning problem is an NP-Hard problem, and best path can be found by using some brute-force search algorithms. However, it needs too much time. Therefore, lots of the researchers prefer to use evolutionary algorithms and/or swarm algorithms, such as Ant Colony Optimization, Genetic Algorithms, Simulated Annealing, Particle Swarm Optimization, etc. [1, 2], which needs less time to solve the problem by finding near-optimal solutions.

Swarm intelligence (SI) is a population-based search strategy which tries to find an optimal/acceptable solution by inspiration from the behavior of animal swarms such as ants, bees, glowworm, termites, etc. Due to their swarm structure, they try to have their knowledge about the mission with other individuals directly or indirectly to make a more sophisticated intelligent solution search.

One of the mostly used SI algorithms is Ant Colony Optimization (ACO) which is inspired by the foraging behavior of real ants and it is generally used to solve large and complex combinatorial optimization problems to achieve high-quality solutions in an acceptable execution time. They are based on a colony of artificial ants to construct solutions by adding some heuristic information and artificial pheromone trails which reflect the acquired knowledge about the search problem with others in the swarm.

Due to the structure of ACO algorithm, it can be easily parallelizable by using multiple ants with different synchronization mechanism. Therefore, it is highly preferred algorithms for complex problems in real-world. However, it converges slowly to the optimal solution, and it can easily converge to local optima due to the selected parameters. This yields a stagnation which is not good to find better solutions [3].

To avoid this type stagnation and premature convergence problems of standard ACO, an alternative approach is proposed in [4]: a Multi-Colony ACO. In this approach, apart from ACO, many different ant colonies exist in the system and each of them executed separately by using a different table to maximize the search area explored. Due to its structure, different parameters can be set for each colony and therefore, each colony can produce various solutions in their iterative steps. To increase the performance of the system, each colony needs to share its knowledge with others. After a few iterations, each colony can share its valuable information with its neighbor colony/colonies such as the best solution of the colony or its pheromone table. After getting this information, each colony updates its knowledge base with incoming information.

1.1 Problem Description

The schematic diagram of the multiple UAV path planning problems is shown in Fig. 1. As can be seen from the figure there are some control points (CPs) in the mission theatre, and there are many UAVs which will be assigned to visit these control points. Additionally, there can be some threat zones such as radars or missiles, and some no-fly zones where no UAV entrance is permitted. Mainly, this problem can be treated (and formulated) as a variant of the Single Depot Multiple Vehicle Routing Problem (SDMVRP) with the following properties:

- There is no need a central control mechanism: each ant executes as a single agent. Therefore it is more robust.
- There are some UAVs, assigned as n, initially located at a single depot point.
- The main problem is to assign a sequence of CPs to each UAV such that its path contains the depot location and a subset of target CPs and return to the same depot location.
- Given a set of n UAVs such that n_i m where m is the number of CPs to be visited.
- The objective is to minimize the total distance traveled by all UAVs. Additionally, in the flying process, the multi-UAVs need to avoid the complex terrain conditions such as no-fly zones, radars, mountains, hills or buildings.

Fig. 1. A sample map in 3-D environment.

Also, to facilitate the path planning problem of multiple UAVs, we give the following assumptions:

- UAVs flight at a certain altitude and with a constant speed. Although, CPs are located on the ground with different altitude, visiting a target means going over them by getting some image from this point or receiving data from this place, which can be easily accomplished at a certain altitude. Therefore, the path planning problem is reduced to a two-dimensional planning.
- There are some threats in the mission environment. These threats can be radars, no-fly zones, mountains, hills or buildings. It is assumed that the number and location of them have already been known.

1.2 Ant Colony Optimization and Multi Colony Approach

Research on ACO algorithms showed that it produces very satisfactory solutions for many types of combinatorial and polynomial problems. It can find optimal/near optimal solution at the early stage of the search process. All ants use same parameters and information, therefore, they can quickly converge to a single, not optimal solution, which is called as search stagnation problem. To overcome this type stagnation problem, there is a need to extend the solution space for finding the global optimum solution [5] (or at least a closer solution). One proposed solution to solve this stagnation problem is the usage of Multi-Colony ACO in which each colony lays its pheromones and colonies are cooperatively working with others to solve the optimization problem. This approach is used by Sim et al. for load balancing in circuit-switched networks [6]. Due to the existence of many colonies, there is a need for information exchange. Middendorf et al. [4] proposed some mechanisms for this. According to this study, each colony sends some information to the others after executing some iterations. Experimental results showed a considerable decrease in execution time of the proposed algorithm by increasing interval between information exchanges.

The remainder of the paper is organized as follows. The details of the proposed algorithm and the experimental results are given in Sects. 2 and 3 respectively. In Sect. 4, we discuss the implications of the overall results.

2 Proposed System

In this section, the proposed algorithm and its steps are described. The multi-UAV path planning with multi-colony ant optimization algorithm has two main parts those are modeled to run with CUDA. The first part is to partition the problem space into K convenient subproblems, and the second part is to solve these subproblems separately by the multi-colony ant optimization algorithm.

The scenario requires that given set of control points to be separated into K convenient parts and each part to be assigned to a single UAV. Each part is then sent to the multi-colony ant optimization algorithm. Each UAV starts from the same depot point and then returns to that point when they visit the required control points. Multiple colonies are in charge of finding the best path for the UAVs by collaborating with each other.

Depending on the criteria of the given problem, having multiple UAVs assigned to the small parts of the problem might result better in several aspects. These can be expressed as follows: (1) If the partitioning is in a meaningful structure such as not overlapping the control points of distinct UAVs, it is satisfactory. (2) If the problem is evenly separated, and if the mission is suitable for using multiple UAVs instead of one, then the result of using multiple UAVs is satisfactory. The computation time of solving all the subproblems takes shorter comparing to solving the problem as a whole. Since the problem set for each UAV is smaller now, the probability of having the possible best solution from the algorithm is more guaranteed for each UAV. Since the whole problem takes longer to complete for a UAV in the mission, the multi-UAV problem takes as much as the biggest subproblem of the separated missions. In addition, the structure of the problem may require the UAVs to handle different conditions and different missions for the parts. So, having multiple UAVs gives an opportunity to satisfy different missions while on the go.

2.1   Parallel Implementation of Multi Colony Ant Optimization

In this study, it is aimed to solve the multi-UAV path planning problem by using Multi Colony Ant Optimization (MCAO). Ant Colony Optimization is an evolutionary optimization algorithm that is with the aim of producing feasible solutions for its intended problem domains in acceptable time. However, Ant Colony Optimization is known to be used for solving NP-hard problems where the required computational resource increase exponentially as the problem space gets bigger [7]. For this reason, standard serial applications for this optimization algorithm still stay infeasible when there are many aspects of the targeted problem. Infeasibility varies from not being able to produce the desired output in acceptable time to not having that output in any case. So this kind of algorithm needs another mechanism to yield better optimization in such conditions. In recent years, research has shown that designing a parallel approach for the Ant Colony Optimization both produce better results in feasible time and it allows the user to work on large datasets [8, 9].

Random Number Generation: The first step in Parallel MACO model is to produce reliable randomness at hand. It is known that ant colony optimization inherently uses not only the pheromone values but the probabilistic operators when determining the next control point for an ant. So it is essential to provide high-quality random numbers for that step to keep the algorithm from converging poor results. In this case, N random number generators are produced by the help of cuRAND framework defined in CUDA SDK, where N is equal to the number of ants to be used.

Distance Table Calculation: The next step is to calculate the distance table for all the control points. This table is a matrix that includes the Euclidean distance from point i to point j, where all the combinations are included. The distance matrix is created by N × N threads where each thread calculates one distance and puts the result into that matrix. The ants will later use these distances to construct their paths.

Initialization: At the beginning of constructing the paths, one step before is to initialize the pheromone tables for the control points. As known that ants make their next point selection by taking advantage of pheromone values between the points, it is essential to put a small amount of initial values between those points. The initialization process is similar to calculating distances as a model. Except, now it is known that there is more than one colony and what it means that these colonies will have their pheromone tables to consider. For this reason, the threads to put initial values in the pheromone table will take several more steps to fill in corresponding fields of other colonies.

Tour Construction: The last step for an ant is to construct its path by the help of its prior knowledge about the problem domain and heuristic approach. In this study, an ant is represented by a thread block. A thread block includes N threads where N is equal to the number of control points in the problem. Here, each thread is distributed to the control points and is required to calculate the random proportional rule of going from that point to another point.

Colony Behavior: In the colony, each ant is responsible for constructing a good path; but the resulting paths may not be good as desired. If such a case happens, there is no completed mechanism to prevent the ant from update the pheromone table. In the conventional ant colony approach, the pheromone table is filled with all the ants so that better routes influence more than the worse routes. To take it one step ahead, in this study, it is aimed to take control of baby steps of the algorithm and make sure not to converge undesirable results from the start. For this reason, a local optimization is applied to the first 200 iterations of the colonies. Experiments show that applying a local optimization yields better quality results. In this study, a 2-opt local optimization technique is adapted to the paths before influencing the pheromone table. 2-opt is a simple local search algorithm which takes the crossovered path and reorders it without crossovers to decrease the total path length.

As soon as an ant completes its path, the newly created path is then taken by a thread so that the fitness calculation is done. The fitness is calculated by summing all the Euclidean distances through the path. Once the fitness values are calculated, then it is time to update the pheromone table, the ants use to communicate. Updating the pheromones is a crucial step since all the next generation ants are fed by this table. It also helps the ants to make better selections while the tour construction is in progress. The basic workflow of ant colony optimization is depicted in Fig. 2.

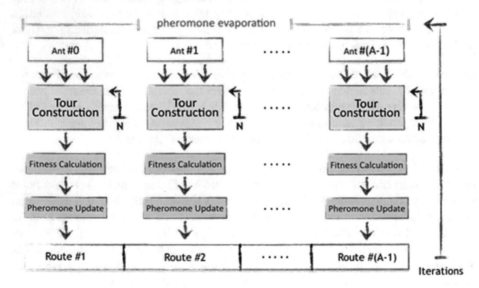

Fig. 2. Basic workflow of parallel ant colony optimization.

2.2   Knowledge Sharing Between Colonies

In the structural concept of ant colony optimization, there is a stagnation probability where the colony may converge to a suboptimal result and stuck in the same infeasible solution. This is mainly because of the pheromone values between the control points in the pheromone table. Although this problem can be reduced to a certain extent by fine-tuning the parameters, it is still not guaranteed to find better solutions.

So, a new optimization technique arises: Multi Colony Ant Optimization. In MCAO, N ACO work for the same problem set distributive to find better solutions and then they share their local knowledge among the other colonies. Each colony has a local phero-mone table for the ants to memorize their iterations. Knowledge sharing between the colonies takes place after several iterations. In this study, once a colony completes its 20th iteration, its pheromone table is shared with the next colony. The sharing is as simple as updating the next colony's pheromone table by the best 10 ants of the current colony, as expressed in Fig. 3. This process goes along all the iterations until the algo-rithm stops. By sharing the knowledge, the resulting iterations are more likely to converge to better near-optimal results. It also provides a diversity between the phero-mone values, thus yielding much more options for the ants to select.

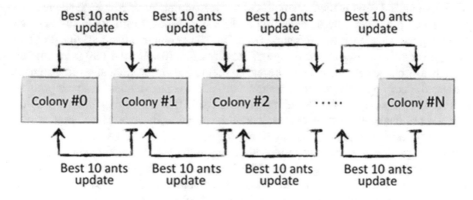

Fig. 3. Knowledge sharing between colonies.

2.3 Multi-Colony Ant Approach for Path Planning

In the proposed algorithm, the problem sets are divided into K subproblems and there are N colonies assigned to each subproblem. When each colony finishes its work, all the results of this colony's ants are sorted according to their fitness values with the Thrust sorting mechanism of CUDA SDK and then the best 10 ants are chosen to update the next colony's pheromone table as well as its own pheromone table.

The assumption here is that colonies will know that K subproblems start with the same depot point and ends with it, too. The proposed algorithm consists of the following main features:

- There exist N different colonies,
- Each colony has the same number of ants with others
- this value equals with the number of CPs in the mission area,
- Each colony is executed with the same number of iterations,
- Each colony shares the same heuristic function,
- Each colony has the same parameters (such as α, β and ρ)

- Each colony has its own knowledge base to optimize the solution
- After each iteration, the first 200 solutions of each colony are optimized by using a local optimization technique (2-opt).
- To share the valuable knowledge, after each 20 iterations, each colony shares its local best 10 ants with neighbor(s) to update the pheromone table(s).

3   Experimental Results

In this study, it is primarily aimed to combine and extend the two concepts studied before in [2] and [1]. It is to take the benefit of using multi-colony ant optimization to better optimize the path planning problem in multi-UAV problems by designing and implementing massively parallel algorithms run on the GPUs.

This study focuses on a multi-UAV path planning problem by constructing simple subproblems and solving them on multi-colony ant optimization. Since the single UAV path planning problem, by its definition and its limitations, is similar to the classical Traveling Salesman Problem, it is aimed to imitate the problem into a TSP-like problem. Thus, it is firstly focused on the design of the algorithm to solve the TSP in an efficient way, then aimed to transform it into the real concept. The virtual simulation environment is created according to the parameters shown in Table 1. In this context, the base problem sets used in this study are the TSP libraries from TSPLIB. After getting the base problem sets, a set of changes has been made to better serve the study case. There are a number of obstacles such as radars applied to the appropriate locations of the problem sets. The concept of radar represents a forbidden area where the UAVs not allowed to fly over. As applying the radars changes the known best paths of the problems from TSPLIB, there is no any fixed fitness value for the used problem sets. Thus, the multiple colony results are compared with each other by taking the 1 colony scenario result as a reference point.

Table 1.   Parameter settings used in this study

Parameters	Values	Parameters	Values
# of visiting points	53, 77, 101, 226, 440	β	5
# of UAVs	1, 2, 3, 4	ρ	0.02
# of colonies	1, 2, 4	Initial pheromone	$1/C_{NN}$
# of ants per colony	# of CPs	Iteration	500
Pheromone sharing rate	At every 10 iteration	Q	1.618
α	2	Q_{best}	2×1.618

The parallel model of K-Means and Multi-Colony Ant Optimization is implemented on NVIDIA GeForce GTX 970 graphics card. The operating system is Ubuntu 14.10 and the underlying CUDA SDK is 7.5 where the algorithms are written is C++ and CUDA C. The GPU has the Maxwell architecture with 4 GB DDR5 RAM and 1664 cores, where each core is at 1228 MHz frequency.

As in [1], it is assumed that the simulation environment provides an offline path planning with the high maneuver capability fixed wings as UAVs. As of the year 2016, small UAVs reach the capability of speed up to 60 km/h with approximately 60 to 90 min of flight time. Figure 4 is an example scenario where there are 1-2-3 and 4 UAVs for the 226-CPs problem. 1 UAV completes traversing its mission in 336 min. Using 1 UAV for this mission seems infeasible since it exceeds the flying capacity of a UAV. If there are 2 UAVs, then the mission is completed in 156 min in total. It takes 109 min for 3 UAVs and only 82 min for 4 UAVs. The traversal time is noted by the least recently completed path among the UAVs. Considering that 226-CPs problem is a large-scale space, using 4 UAVs in that kind of problem by partitioning it makes more sense if there are some available resources. Having 4 UAVs in this problem yields approximately 4x speed up. All the problem sets used in this study shows a similar speed up behavior when using multiple UAVs.

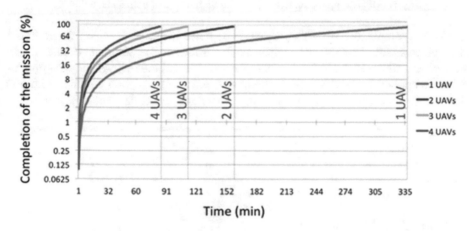

Fig. 4. Flight time comparison for 1-2-3-4 UAVs to finish the 226-CPs problem.

Using multiple UAVs also decreases the total computation time for the path planning. Since the problem space gets narrowed, the time needed to compute the same number of iterations decreases substantially. Also, it should be noted that increasing the number of colonies directly increases the time needed for computation since the colonies in this study work sequentially one after another. The overall computation time for the problem sets with different number of UAVs and corresponding colonies are shown in Fig. 5.

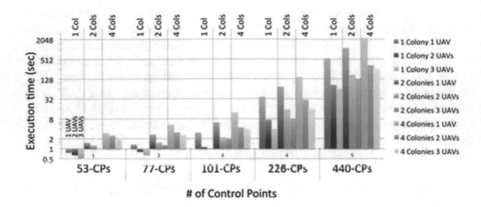

Fig. 5. GPU computation times for different number of points and colonies.

As an example, the 101-CPs problem is depicted in Fig. 6. It is partitioned into 3 subproblems with the K-Means applied. Each partition in the problem becomes a new subproblem where each UAV takes one portion of it and the multi-colony ant optimization is applied to. In this case, each subproblem is applied to 4 colonies with 500 iterations, where after each 10th iteration, each colony shares data about its stage.

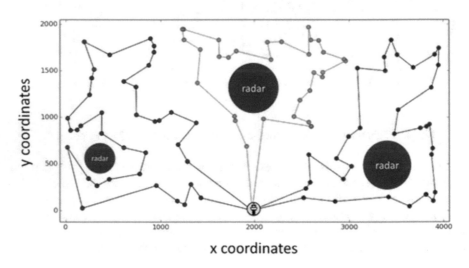

Fig. 6. Sample path for the 101-CPs problem for 3 UAVs and 4 colonies for each UAV.

4   Conclusion

The aim of this paper is to propose an efficient parallel multi-colony implementation of ACO for solving path planning problem of multiple UAVs on a parallel computing platform: CUDA. The algorithm is redesigned by making some adaptations to run on

CUDA platform to exploit data-parallel computing mechanism of NVIDIA GPUs. The proposed algorithm has the following characteristics; there is no need a central control mechanism: each ant executes as a single agent. Therefore, it is more robust. Due to the algorithm of ACO, it has a simple implementation. Therefore, it can be easily parallelized by adding some synchronization mechanism. It has a low complexity in the implementation of ants and colonies. Therefore, the calculation can be distributed to several colonies which cooperate to find better solutions complex path planning of multiple UAVs. Experimental results show that GPU-accelerated multi-colony ant optimization provides a significant increase in computing performance and better solutions for multiple UAVs path planning problems. As a future work, to find better solutions for more complex UAV path planning problems, some modifications are planned to take into account such as setting parameters and using different coordination mechanisms between colonies

References

1. Cekmez, U., Ozsiginan, M., Sahingoz, O.K.: Multi-UAV path planning with parallel genetic algorithms on CUDA architecture. In: Proceedings of the 2016 on Genetic and Evolutionary Computation Conference Companion, pp. 1079–1086. ACM (2016)
2. Cekmez, U., Ozsiginan, M., Sahingoz, O.K.: Multi colony ant optimization for UAV path planning with obstacle avoidance. In: 2016 International Conference on Unmanned Aircraft Systems (ICUAS), pp. 47–52. IEEE (2016)
3. Yueshun, H., Ping, D.: A study of a new multi-ant colony optimization algorithm. In: Advances in Information Technology and Industry Applications, pp. 155–161 (2012)
4. Middendorf, M., Reischle, F., Schmeck, H.: Multi colony ant algorithms. J. Heuristics $8(3)$, 305–320 (2002)
5. Blum, C., Dorigo, M.: Search bias in ant colony optimization: on the role of competition-balanced systems. IEEE Trans. Evol. Comput. $9(2)$, 159–174 (2005)
6. Sim, K.M., Sun, W.H.: Multiple ant colony optimization for load balancing. In: International Conference on Intelligent Data Engineering and Automated Learning, pp. 467–471. Springer, Heidelberg (2003)
7. Neumann, F., Witt, C.: Bioinspired computation in combinatorial optimization: algorithms and their computational complexity. In: Proceedings of the 15th Annual Conference Companion on Genetic and Evolutionary Computation, pp. 567–590. ACM (2013)
8. DeleVacq, A., Delisle, P., Gravel, M., Krajecki, M.: Parallel ant colony optimization on graphics processing units. J. Parallel Distrib. Comput. $73(1)$, 52–61 (2013)
9. Cekmez, U., Ozsiginan, M., Sahingoz, O.K.: A UAV path planning with parallel ACO algorithm on CUDA platform. In: 2014 International Conference in Unmanned Aircraft Systems (ICUAS), pp. 347–354. IEEE (2014)

An Efficient Method for Detecting Fraudulent Transactions Using Classification Algorithms on an Anonymized Credit Card Data Set

Sylvester Manlangit[1], Sami Azam[1(✉)], Bharanidharan Shanmugam[1],
Krishnan Kannoorpatti[1], Mirjam Jonkman[1], and Arasu Balasubramaniam[2]

[1] School of Engineering and IT, Charles Darwin University, Darwin 0909, Australia
sylvester.manlangit@students.cdu.edu.au,
{sami.azam,bharanidharan.shanmugam}@cdu.edu.au
[2] Cookie Analytix Pvt. Ltd., Bengaluru, India
arasu@cookieanalytix.com

Abstract. Credit card fraudulent transactions are causing businesses and banks to lose time and money. Detecting fraudulent transactions before a transaction is finalized will help businesses and banks to save resources. This research aims to compare the fraud detection accuracy of different sampling techniques and classification algorithms. An efficient method of detecting fraud using machine learning is proposed. Anonymized data set from Kaggle was used for detecting fraudulent transactions. Each transaction has been labeled as either a fraudulent transaction or not. The severe imbalance between fraud and non-fraudulent data caused the algorithms to under-perform. This was addressed with the application of sampling techniques. The combination of undersampling and SMOTE raised the recall accuracy of the classification algorithm. k-NN algorithm showed the highest recall accuracy compared to the other algorithms.

Keywords: Credit card · Anonymized data · Fraud detection

1 Introduction

Credit transactions have become the principal way that we make payments to buy goods and services either in person or online. As many of these services and goods can be bought with very little proof of identity, credit card usage is subjected to fraud. Billions worth of transactions are made every year using credit cards [1]. In 2014, $28.84 trillion was the size of the global card business. These have become part of the financial systems. It is now common process to pay with credit cards [3]. It can also be used in a short-term finance [2]. With the growth of credit card transactions, the fraudulent transactions have also increased. Aside from credit card fraud, identity fraud has also become one of the concerns of U.S. consumers [4]. Credit card fraud is an unauthorized account movement by a person who is not authorized to use it [5]. It can also be described as a situation where a person uses the card without the consent or knowledge of the card holder and the card issuer [5]. In the event of a fraud, the merchant loses the products sold. And they have to pay the chargeback fees, and have the risk of closing their

© Springer International Publishing AG, part of Springer Nature 2018
A. Abraham et al. (Eds.): ISDA 2017, AISC 736, pp. 418–429, 2018.
https://doi.org/10.1007/978-3-319-76348-4_41

accounts [6]. Non-traditional mode of payments was accounted for 60% of the fraud. As the mode of payments increase, new fraud patterns have emerged. This made the current fraud detection systems unsuccessful [7].

Credit card fraud detection is the use of machine learning to detect fraudulent transactions. This is one of the most explored domains. It relies on the automatic analysis of a recorded transaction and detect it if it shows any fraudulent behavior [8]. In, failing to detect a fraud, the amount is lost, while on the other, administrative costs are incurred [7]. Several factors are important in training the algorithm in fraud detection [7]. The important matter to look at in fraud detection is to let fraud investigators consider transactions with high probability of being a fraud. This will turn into creating a correct ranking of transactions based on their probability of being a fraudulent transaction [8].

2   Methodology

Initially a number of sampling techniques will be used on the data set to find the best sampling possibility of the data. After selecting the values of the different attributes in each sampling technique the data set with the applied sampling techniques will be used by four different classification algorithms. The classification algorithms used in this research are random forest, k nearest neighbor, logistic regression and naïve Bayes. The results will be evaluated by using measures which can assess the performance of a classification algorithm. Based on comparative analysis of the algorithms an efficient method of detecting fraud transactions will be proposed.

2.1   Experimental Data

This research uses an anonymized credit card data from European credit card users obtained from Kaggle. PCA has turned the original values into numbers, thereby hiding the privacy of the credit card users. It is a popular unsupervised method that reduces number of features in the original data set by using the principal direction of the data distribution [9, 10]. The data set has a total of 284, 807 transactions in a span of 2 days. Out of all the transactions, 492 are classified as fraud. This is 0.172% of all transactions. The dataset is highly unbalanced [11]. The dataset has 30 columns, features V1 to V28 are values that were PCA transformed to retain confidentiality. "Time" and "Amount" are the features which were not transformed by PCA.

2.2   Sampling Techniques

Sampling techniques are performed to address the imbalance between the class populations in the data set. Balancing techniques are done before the application of any algorithm to remove noise between two classes. Sampling techniques adjust the sizes of the training sets. Undersampling or oversampling reduces the level of class imbalance, creating a better result [8]. These techniques have shown to address the class imbalance and are very easy to implement [12].

Undersampling technique resizes the majority class by randomly removing samples until the data set has attained that each class are equal in number [8]. Another definition of the technique is getting a sub-set from the majority class to train the classifier [12]. Aside from obtaining a balanced data set, the training time becomes shorter. The problem with this technique is a large part of the majority class is excluded. There may be important information on the samples that were excluded [12, 13].

Synthetic minority oversampling technique (SMOTE) is an oversampling approach that creates synthetic samples instead of oversampling with replacements. Synthetic samples are introduced along the line segments of the original minority samples using a specified number of minority class nearest neighbors. Neighbors from the k nearest neighbors are randomly chosen depending on the specified number of synthetic samples that needs to be created [14]. These synthetic samples' effect on decision trees causes the decision trees to generalize better. These create larger and less specific regions instead of smaller and more specific ones. Algorithms can better identify the regions related to the minority class rather than overwhelmed by the majority class [14].

2.3 Supervised Methods

These methods assume that past transactions are reliable and available, but fraud patterns that have occurred are often limited [8]. The performance of these models can be checked using the model's accuracy in correctly classifying the new observations, if these are fraudulent or not [1]. These types of models use databases containing past information. When a new data comes on, these models classify a new data based on the past transaction [4, 6, 15]. These types of model needed accurate information on identifying fraudulent transactions from the past databases, since it can compare new data with the past transactions [6]. This research uses supervised methods (random forest, k nearest neighbor, logistic regression and naïve Bayes) to create fraud detection models.

2.4 Performance of the Classification Algorithms

Classification algorithms are usually assessed using the confusion matrix. Figure 1 illustrates the confusion matrix. The columns are the class prediction, while the rows are the actual class. TN denotes the number of correctly classified negative examples (True Negatives), FP denotes the number of misclassified negative examples predicted as positives (False Positives), FN denotes the number of positive examples that are misclassified as negatives (False Negatives), and TP denotes the number of correctly classified positive examples [16]. In this research, positives are referred as the fraudulent transactions, while the negatives are the non-fraudulent transactions.

	Predicted Negative	Predicted Positive
Actual Negative	TN	FP
Actual Positive	FN	TP

Fig. 1. Confusion matrix

Shown below is the formula for checking the overall predictive accuracy of a classification algorithm [16].

$$Predictive\,Accuracy = \frac{TP + TN}{TP + FP + TN + FN}\,[16]$$

Recall accuracy
This is the ratio of correctly classified positive examples to the total actual *positive examples*. This accuracy shows the ability of the classification algorithm to correctly classify the actual fraud instances.

$$Recall\,Accuracy = \frac{TP}{TP + FN}\,[16]$$

Precision accuracy
This is the ratio of correctly classified positive examples to the total predicted positive examples. This accuracy shows how much of the predicted frauds are true fraud instances.

$$Precision\,Accuracy = \frac{TP}{TP + FP}\,[16]$$

Recall accuracy will be used as a tool for performance analysis of the algorithm because failure to detect an actual fraud costs more resources and time compared to labeling a legitimate transaction as a fraud.

3   Selection of Sampling Techniques

A data set is said to be imbalanced when the size of a single class is much higher than the other class thus making it difficult for learning systems that are not designed to handle large difference of instances between each class. Relying on sampling techniques to balance the data set is one of the traditional methods in dealing with unbalanced data sets [8, 17]. When it comes to predictions, imbalanced data set needs to be handled carefully [12, 17]. Imbalance can affect the performance of the classifiers. Classifiers that cannot detect class imbalance are overwhelmed by the majority class and the minority class are ignored [12]. In fraud detection, classification algorithms are used. Cost-sensitive learning deals with class imbalance by adding cost for different classes. It is considered as one of the best ways to deal with class imbalance [12]. Another approach when dealing with imbalanced data sets is to add a bias to the classifier, this will allow the classifier to give more attention to the fraudulent class [13]. The whole data set will be classified into to majority and minority classes. Majority class refers to the class with greater population in the data set compared to the other class. Majority of the instances in the data set are non-fraudulent transactions. Thus, non-fraud class is referred as the majority class, while the fraud class is referred as the minority class. Initially sampling techniques will be applied to the data set to see if there are changes in the precision and recall accuracy of a classification algorithm. This research will use Weka, a data mining tool, to perform different sampling techniques and use classification algorithms to create a fraud detection model. Table 1 shows the classifier used to test the effects of sampling techniques. In testing the sampling techniques, the classification algorithm used will be logistic regression. The values used are the default values in the tool. These values will be explored when the best attribute values of the sampling techniques have been selected. The data set will be split into training set and test set. The training set will be used by the algorithm to identify the unique characteristic of each class.

Table 1. Settings used in testing the effects of sampling techniques

Attribute	Value
Classifier used	Logistic regression
Ridge estimator	1.00E–08
Max iterations	−1 (unlimited)
Test method	Split test
Train, Test	70%, 30%

3.1 Undersampling

Table 2 shows the ratio of the majority and minority class with the last column showing how large is the data set used compared to the original data set. The first row uses equal number of instances for each class. The fraud class has 492 transactions and this will be paired with equal amount of majority class.

Table 2. Samples used in the majority class using undersampling technique

Ratio	Majority class Non-fraud	Minority class Fraud	Total number of instances	Percentage from the total number of transaction
1 to 1	492	492	984	0.3455%
2 to 1	984	492	1476	0.5182%
3 to 1	1476	492	1968	0.6910%
5 to 1	2460	492	2952	1.0365%
10 to 1	4920	492	5412	1.9002%

Table 3 shows the results of logistic regression algorithm at different numbers of majority class. The results indicated that the more number of majority class instances are added to the data set recall accuracy decreases, while the precision accuracy increases. The number of FPs diminishes when the number of majority class instances increases, while the number of FNs increases. This research selects a balanced ratio of 1 to 1 with highest recall accuracy when using undersampling technique.

Table 3. Comparison of precision and recall accuracy between different class ratio

Confusion matrix	TP	FN	TN	FP	Precision	Recall
1 to 1	129	11	145	10	92.806%	92.143%
2 to 1	132	15	288	8	94.286%	89.796%
3 to 1	131	21	435	3	97.761%	86.184%
5 to 1	123	26	729	8	93.893%	82.550%
10 to 1	132	18	1469	5	96.350%	88.000%

3.2 Synthetic Minority Oversampling Technique (SMOTE)

Table 4 shows the value of the percentage attribute of SMOTE. The number of nearest neighbor used was 5. This value is the default value in the tool. Nearest neighbor is defined as the instance whose Euclidean distance is closest to the original minority instance.

Table 4. Comparison of synthetic samples added based on the percentage attribute in the sampling technique

Percentage of synthetic instances	Synthetic instances added	Original fraud instances	Total fraud samples	Non-fraud samples	Total number of instances	Percentage from the total number of transaction
100%	492	492	984	284315	285299	100.173%
200%	984	492	1476	284315	285791	100.345%
300%	1476	492	1968	284315	286283	100.518%
500%	2460	492	2952	284315	287267	100.864%
1000%	4920	492	5412	284315	289727	101.727%

Table 5 shows the performance of logistic regression at different numbers of synthetic instances added to the data set. Results indicated that the more synthetic instances are added to the data set it also increase the recall accuracy of the algorithm. The rise of FNs goes with the increase in synthetic instances. Although recall accuracy rises, some synthetic instances have close characteristics with majority class instances causing a misclassification. When using SMOTE sampling technique, this research will use 1000% in adding synthetic samples as it gave the high recall accuracy.

Table 5. Comparison of recall accuracy in relation to the number of synthetic samples added to the minority class

Percentage of synthetic instances	TP	FN	TN	FP	Precision	Recall
100%	217	83	85274	16	93.133%	72.333%
200%	343	88	85273	33	91.223%	79.582%
300%	462	115	85276	32	93.522%	80.069%
500%	696	184	85262	38	94.832%	79.091%
1000%	1366	304	85204	44	96.879%	81.796%

To identify the best value for k-NN attribute of SMOTE, this research used the same set-up, with the percentage of synthetic instances set to 100%. Table 6 showed that using 5 and 6 have the best recall accuracy, however using 6 neighbors resulted in better precision accuracy compared to using 5 neighbors. This research will use 6 as the number of nearest neighbors in SMOTE.

Table 6. Comparison of precision and recall accuracy between different values of k-NN used in SMOTE

Number of nearest neighbors used to create the synthetic instances	TP	FN	TN	FP	Precision	Recall
1	253	22	305	10	96.198%	92.000%
2	253	22	303	12	95.472%	92.000%
3	253	22	304	11	95.833%	92.000%
4	254	21	306	9	96.578%	92.364%
5	255	20	301	14	94.796%	92.727%
6	255	20	306	9	96.591%	92.727%
7	254	21	307	8	96.947%	92.364%
10	253	22	308	7	97.308%	92.000%

4 Classification Algorithm with Selected Sampling Techniques

Upon seeing the effects of sampling techniques on the data set. The best results will be used to test the performance of the classification algorithms. Table 7 lists the parameters and values of the data set before performing the classification algorithms. This is to show that the algorithms used a uniform data set.

Table 7. Attributes of the data set used when performing the classification algorithms

Attribute	Value
Sampling technique used	
SMOTE	
Percentage added	1000%
k-NN	6
Undersampling	
Distribution spread	1.0
Number of instances in the data set used	
Fraud (positives), Non-fraud (negatives)	5412, 5412
Split test	
Train, Test	70%, 30%

4.1 List of Classification Algorithms

Four different algorithms will be used to understand the best detection capability.

Random Forest - This classification algorithm works by creating multiple decision trees instead of a single decision tree. Random forest divides the training set into sub-sets and each sub-set has its own randomly selected sub-set of features that is going to be used to split the data. This model has performed well against other classifiers, examples of these were neural networks, discriminant analysis and support vector machines [18] The number of features used in each decision trees when value is set to 0 is 9. It is 30% of the total number of features.

k-Nearest Neighbor (k-NN) - This algorithm classifies the test data by identifying the nearest instances of the test data. A formula is used to determine the distance between the training set and the test data. And the nearest neighbors are the instances that have the shortest distance to the tested instance. Based on the specified number of nearest neighbors to be used, the number of instances are selected to classify the tested data. The classification of the test data is determined by the number of instances belonging to a class. If the set k-NN value is 5, the nearest neighbors consist of 3 fraud instances and 2 non-fraud instances. The algorithm will classify the test data as a fraud because there are more fraud neighbors than the non-fraud.

Logistic Regression - Logistic regression has been used in situation where there are 2 possible outcomes to the problem [4, 19]. The goal of logistic regression is to create an equation that will point a test data if it is a fraud or non-fraud. The co-efficient values are based on the training set [20–22]. A coefficient is a number that helps separate the two classes. This is the reason why the algorithm needs to run many iterations. It is to find the best estimated value of the coefficient of each feature. This research has set the number of iterations to unlimited and it will stop until the coefficient values have separated both classes or the coefficients cannot improve the separation of the classes in the

training set. These coefficients will then be tested using the test set. If the result probability is above 0.5, the test data will be classified as a positive class or fraud, while the
result below 0.5, the test data will be classified as a negative class or non-fraud. Ridge
values adjust the balance between each class. Raising the value sets the equation to favor
the negative class, while lowering the value leans the equation to favor the positive class.
The value used in testing the algorithm favors in detecting fraud.

Naïve Bayes - Hand et al. [23] defined Naïve Bayes as a straight forward classification
algorithm. The process starts with the estimation of probabilities that an object from
each class will fall in each category of the discrete variables. This algorithm assumes
that each feature in the data set is independent from each other. Each feature independently adds to the probability [24]. To calculate the probability of each feature, the Bayes
theorem is used. Table 8 shows short description of each attribute and its corresponding
values when using the classification algorithm.

Table 8. List of attributes that is present on each classification algorithm

Attribute name	Description	Random forest	K-NN	Logistic regression	Naïve Bayes
batchSize	Number of instances processed when testing the prediction model built by the algorithm	100	100	100	100
bagSizePercent	Size of each bag, as a percentage of the training set size	100	–	–	–
maxDepth	The maximum depth of the tree, 0 for unlimited	0	–	–	–
numFeatures	Sets the number of randomly chosen attributes	0	–	–	–
KNN	Number of neighbors used in classification decision	–	6	–	–
distance-Weighing	Allow distance between neighbor and test data to affect classification decision	–	No distance weighing	–	–
nearestNeighbor-SearchAlgorithm	Formula used to calculate distance between training data and test data	–	Euclidean distance	–	–
maxIts	Maximum number of iterations to perform	–	–	0	–
Ridge	Set the Ridge value in the log-likelihood	–	–	1.0E–8	–

5   Results

Table 9 shows the confusion matrix results of the different classification algorithms. k-NN has the lowest FNs, these are actual fraud instances that were predicted as non-fraud. On the other hand, random forest performed best in classifying the non-fraud instances. It showed that there are only 7 FPs, non-fraud instances classified as fraud, but the algorithm placed 2nd in correctly classifying the fraud data. Compared to k-NN, there were 63 FNs which is 36 misclassified instances higher. Naïve Bayes performed the least. It has the highest number of FNs, 90 instances higher than logistic regression and 198 instances higher than k-NN.

Table 9. Confusion matrix of different classification algorithms

Algorithm	TP	FN	TN	FP	Precision	Recall	Overall accuracy
Random Forest	1547	63	1630	7	99.55%	96.09%	97.84%
K-NN	1583	27	1581	56	96.58%	98.32%	97.44%
Logistic Regression	1475	135	1593	44	97.10%	91.61%	94.49%
Naïve Bayes	1385	225	1599	38	97.33%	86.02%	91.90%

Based on recall accuracy it is found that k-NN showed the highest fraud detection ability, followed by random forest, logistic regression and lastly naïve Bayes. The algorithm k-NN performed better than the rest of the algorithm because of the way SMOTE adds synthetic instances around the original fraud instances using a distance formula. This complements the way k-NN classifies the test data even though there are some fraud instances positioned in the middle of non-fraud data, if they are clustered or the calculated distance is near to each other then the algorithm will be able to classify the fraud instances correctly. Random forest was ranked next to k-NN. The algorithm did not perform better than k-NN because the features were not able to breakdown the training set into smaller groups. Increasing the number of features might allow the algorithm to perform better. However, breaking down the training set into very small end groups will lead to overfitting.

Logistic regression ranked next to random forest. The algorithm performs well when there is definite class division. Some features in the data set displayed this characteristic, but there are also features which did not have a clear class distribution. Features which did not have clear class distribution creates coefficient that does not point the probability in the right direction.

Naïve Bayes algorithm obtained the least fraud detection capability. Unlike the other algorithms that use the inter-dependence of the features, this algorithm calculates probability by each feature independently. This algorithm performs well when the data set is comprised of nominal or discrete values. Data used in this research were not transformed to discrete type to maintain the uniformity of the data set when implementing the classification algorithm.

6   Conclusion

The combination of sampling techniques, undersampling and SMOTE, showed that it increased the fraud detection capability of the classification algorithm. It has showed that the more balanced the data set is, the higher the recall accuracy will become. This research has also found out that the more synthetic minority samples are added, the better recall accuracy will be obtained. In addition the combination of both sampling techniques complements the fraud detection ability of the classification algorithm. The k-NN algorithm performed the best recall accuracy compared to random forest, logistic regression and naïve Bayes. The way K-NN classifies the test data was complemented with how SMOTE sampling technique added the synthetic minority samples and this process made it easier and efficient to detect fraud instances.

Acknowledgements. We would like to thank School of Engineering and IT, Charles Darwin University for providing funding and assistance for this research.

References

1. Jha, S., Westland, J.C.: A descriptive study of credit card fraud pattern. Glob. Bus. Rev. **14**, 373–384 (2013)
2. Liñares-Zegarra, J., Wilson, J.O.S.: Credit card interest rates and risk: new evidence from US survey data. Eur. J. Financ. **20**, 892–914 (2014)
3. Lepoivre, M.R., Avanzini, C.O., Bignon, G., Legendre, L., Piwele, A.K.: Credit card fraud detection with unsupervised algorithms (Report). J. Adv. Inf. Technol. **7**, 34 (2016)
4. Bhattacharyya, S., Jha, S., Tharakunnel, K., Westland, J.C.: Data mining for credit card fraud: a comparative study. Dec. Support Syst. **50**, 602–613 (2011)
5. Prakash, C.: A parameter optimized approach for improving credit card fraud detection. Int. J. Comput. Sci. Issues **10**, 360–366 (2013)
6. Venkata Ratnam, G., Siva Naga Prasad, M.: Credit card fraud detection using anti-k nearest neighbor algorithm. Int. J. Comput. Sci. Eng. **4**, 1035–1039 (2012)
7. Correa Bahnsen, A., Aouada, D., Stojanovic, A., Ottersten, B.: Feature engineering strategies for credit card fraud detection. Exp. Syst. Appl. **51**, 134–142 (2016)
8. Dal Pozzolo, A., Caelen, O., Le Borgne, Y.-A., Waterschoot, S., Bontempi, G.: Learned lessons in credit card fraud detection from a practitioner perspective. Exp. Syst. Appl. **41**, 4915–4928 (2014)
9. Lee, Y.J., Yeh, Y.R., Wang, Y.C.F.: Anomaly detection via online oversampling principal component analysis. IEEE Trans. Knowl. Data Eng. **25**, 1460–1470 (2013)
10. http://setosa.io/ev/principal-component-analysis/. Accessed 11 Nov 2017
11. Dal Pozzolo, A., Caelen, O., Johnson, R.A., Bontempi, G.: Calibrating probability with undersampling for unbalanced classification. In: 2015 IEEE Symposium Series on Computational Intelligence, pp. 159–166. IEEE (2015)
12. Liu, X.-Y., Wu, J., Zhou, Z.-H.: Exploratory undersampling for class-imbalance learning. IEEE Trans. Syst. Man Cybern. Part B (Cybernetics) **39**, 539–550 (2009)
13. Akbani, R., Kwek, S., Japkowicz, N.: Applying support vector machines to imbalanced datasets. In: European Conference on Machine Learning, pp. 39–50. Springer, Heidelberg (2004)

14. Chawla, N.V., Bowyer, K.W., Hall, L.O., Kegelmeyer, W.P.: SMOTE: synthetic minority over-sampling technique. J. Artif. Intell. Res. **16**, 321–357 (2002)
15. Khyati, C., Bhawna, M.: Exploration of Data mining techniques in fraud detection: credit card. Int. J. Electron. Comput. Sci. Eng. **1**, 1765–1771 (2012)
16. Chawla, N.V.: Data mining for imbalanced datasets: an overview. In: Data Mining and Knowledge Discovery Handbook, pp. 853–867. Springer, Boston (2005)
17. Nadarajan, S., Ramanujam, B.: Encountering imbalance in credit card fraud detection with metaheuristics. Adv. Nat. Appl. Sci. **10**, 33–41 (2016)
18. Liaw, A., Wiener, M.: Classification and regression by randomForest. R News **2**, 18–22 (2002)
19. Le Cessie, S., Van Houwelingen, J.C.: Ridge estimators in logistic regression. Appl. Stat. **41**, 191–201 (1992)
20. Excel Master Series. http://blog.excelmasterseries.com/2014/06/logistic-regression-performed-in-excel.html. Accessed 13 Nov 2017
21. MedCalc. https://www.medcalc.org/manual/logistic_regression.php. Accessed 13 Nov 2017
22. Analytics Vidhya. https://www.analyticsvidhya.com/blog/2015/10/basics-logistic-regression/. Accessed 15 Nov 2017
23. Hand, D.J., Mannila, H., Smyth, P.: Principles of Data Mining. MIT Press, Cambridge (2001)
24. Analytics Vidhya. https://www.analyticsvidhya.com/blog/2017/09/naive-bayes-explained/. Accessed 15 Nov 2017

A Deep Convolution Neural Network Based Model for Enhancing Text Video Frames for Detection

C. Sunil[1]([⊠]), H. K. Chethan[1], K. S. Raghunandan[2],
and G. Hemantha Kumar[2]

[1] Department of Computer Science and Engineering, Maharaja Research
Foundation, Maharaja Institute of Technology, Mysore, Karnataka, India
sunilchaluvaiah87@gmail.com, hkchethan@gmail.com
[2] Department of Studies in Computer Science, University of Mysore, Mysore,
Karnataka, India
raghu0770@gamil.com, ghk.2007@yahoo.com

Abstract. The main causes of getting poor results in video text detection is low quality of frames and which is affected by different factors like de-blurring, complex background, illumination etc. are few of the challenges encountered in image enhancement. This paper proposes a technique for enhancing image quality for better human perception along with text detection for video frames. An approach based on set of smart and effective CNN denoisers are designed and trained to denoise an image by adopting variable splitting technique, the robust denoisers are plugged into model based optimization methods with HQS framework to handle image deblurring and super resolution problems. Further, for detecting text from denoised frames, we have used state-of-art methods such as MSER (Maximally Extremal Regions) and SWT (Stroke Width Transform) and experiments are done on our database, ICDAR and YVT database to demonstrate our proposed work in terms of precision, recall and F-measure.

Keywords: Video text detection · CNN · Enhancement · Low quality images

1 Introduction

Recent survey has shown that in today's digital world, video plays a vital role due to its diverse applications. With market flooded with Hand held image devices (HHIDs) like Smartphone and cameras more and more challenges towards the field of research. Its passion or hobby to capture videos using HHIDs. These video contains vital information in the form of textual content. Automatic detection and extraction of this text from video frames is very challenging due to low resolution and distortions like complex backgrounds, font, colors, variable font sizes etc. Video OCR make an effort to create a computer system which automatically identify, extract and understand what text is embedded in images and video frames [22]. Many of the researchers wish to express the 'graphic text' for scene text and 'superimposed text' or 'artificial text' for caption. Video usually contains multilingual text which is more challenging task.

© Springer International Publishing AG, part of Springer Nature 2018
A. Abraham et al. (Eds.): ISDA 2017, AISC 736, pp. 430–441, 2018.
https://doi.org/10.1007/978-3-319-76348-4_42

Text image deblurring and denoising has gained a lot of attention due to its wide range of applications. The motivation of image restoration is to get better and important information for processing and human interpretation from degraded image. Image being degraded is due to noise, motion blur and camera mis-focus. Figure 1 shows the results of detection of the text in video frames from YVT, ICDAR and Our datasets for before enhancement of the frames and we can observe that some misclassification of text detection. This result shows that enhancement of video frames is a must for accurate detection of the text in video frames and also needs to improve the correctness of text detection.

(a) YVT   (b) ICDAR   (c) OUR

Fig. 1. Text detection results before enhancement

In recent year's deep learning techniques is the most active area of research to solve the Image Restoration which is an ill-posed inverse problem [21]. In particular to solve such an ill-posed inverse problems some efforts have been made such as model based optimization methods, discriminative learning methods and combining deep neural networks into model based optimization methods. Deep learning networks have been successfully applied for denoising the images and convolution neural networks give effective results for license plate recognition and hand written text recognition. To this end, we propose simple and most effective enhancement technique which further improves text detection results significantly.

The organization of the remaining of our paper is in the order of Sect. 2 presents literature survey of the related work, Sect. 3 describes the proposed methodology video frame enhancement for multilingual text detection Sect. 4 about database and Sect. 5 presents the experimental results and Sect. 6 provides the conclusion and future work.

2 Related Work

Enhancement technique plays an important role for generating an input for detection and recognition of text in images and video frames. Text in images and videos with uncomplicated background and high contrast can be effortlessly extracted were as low quality text in images/video is more difficult to extract. Enhancement technique for detection of text in videos and images are furthermore classified into two types single frame based and multiple frame based. Though the majority of thresholding methods are proposed for still images, this method does not perform well for video sequences. Rainer [6] a novel technique motion analysis is used to enhance the text extraction in videos based on connected components with similar size and color. Split merge algorithm is developed for segmentation of the input image and by geometric analysis non-text regions are separated. Kanade et al. [1] difficult problems in videos such as low resolution data and complex background in news video have been addressed by applying linear interpolation techniques. Incorporation of sub-pixel interpolation with independent frames and multiple frame integration and combinations of four character extraction filters have been used to obtain accurate video OCR for news video indexing. Li et al [3, 4] proposed an algorithm for detecting and tracking text in videos. A hybrid neural network based method is used to detect text and tracking is performed using sum of squared difference (SSD) to find initial position and contour module to refine the positions. Doerman et al. [2] proposed text enhancement methods for digital videos. Method is carried out in two procedures one is to enhance resolution based on Shannon interpolation and separation of text from complex backgrounds. Further, Nyquist theorem is used to determine normal text or inverse by comparing background color and global thresholding. Chen et al. [5] proposed new enhancement method based on multi-hypotheses approach and regions of text are detected based on vertical and horizontal edges. Based on base line locations text candidates are localized and then filtered by SVM.

Many approaches have been performed for image restoration based on discriminative learning and model based optimization methods. [9] proposed a novel algorithm for image restoration based on nonlocally centralized sparse representation that capitalise on the image nonlocal redundancy, which reduces the sparse coding noise. Iterative shrinkage function has been acquired for resolving l_1 regularized NCSR minimization problem. The results of image deblurring, denoising and super resolution outperforms on NSCR model compared to other existing methods. [7] proposed an idea to learn a multi-layer perceptron with noisy image patches onto clean image which is able to reduce the noise and it is implemented on large database with GPU. A super resolution convolution neural networks (SRCNN) method is used to build and attempt to learn end-to-end mapping between low and high resolution images [8]. The simplicity and robustness of the proposed structure can be also implemented on low level vision problems such as image deblurring and denoising. In [10] proposed novel deep convolution neural structure DCNN which is flexible to solve three image restoration tasks namely image deblurring, image denoising and image super resolution.

Detection of text methods is arranged into two classes one is connected component based methods [11, 12] and region based methods [13, 14]. Detection of text based on MSER (Maximally Stable Extremal Regions) uses intensity uniformity of the text strokes for detecting text and SWT (Stroke width transform) which uses the wildness of the text strokes for detecting text which shows better performance than existing methods [15, 16]. Recently deep learning techniques based convolution neural networks are developed for detecting text and non-text regions [17, 18].

3 Proposed Methodology

From the above literature survey it shows that still there is need for improvement of low level vision problems for effective detection and recognition of text in video frames. In our methodology there are two sections. In first section we describe the proposed methodology of video frame enhancement through CNN model the work presents that for an image denoising, we design and train a set of rapid and smart CNN denoisers. We effectively make use of plain convolution neural network to learn the denoisers. In the second section text detection method is applied for before enhanced video frames as well as after enhanced video frames and are evaluated in terms of recall, f-measure and precision.

3.1 Video Frame Enhancement Through CNN Model

This work describes that for an image denoising we design and train a set of rapid and constructive CNN denoisers. Here we have adopted variable splitting technique and half quadratic splitting method to solve super-resolution and image deblurring problems by plugging in learned denoiser accompanied by model based optimization method. By plug in of smart CNN denoisers we have proposed deep CNN denoisers prior based optimization method which is capable of producing good results but sophisticated image priors were as other conventional model based optimization methods tends to produce time consuming task [20]. In image denoising the context information accelerate to reconstruct the corrupted pixel. To capture the context information from CNN by making use of feed forward convolution we enlarge the respective field. Basically they are two types to enlarge respective fields by increasing the depth and increasing the filter size. As we know that dilated convolution is well known for its enlarging capability of respective fields without losing their conventional (3×3) convolution. So in this work, we make use of dilated convolution which helps to balance respective field and network depth. For dilated filter we used the dilation factor '1' and sparse filter of size $(2l + 1) \times (2l + 1)$, here we adopted a fixed locations for non zeros. Then for equivalent respective field of each layer of 3, 5, 7, 9, 5 and 3 is used. Hence proposed network obtained respective field of size 33×33.

To gain the momentum of the training we adopt batch normalization and residual learning by combing these two we get better Gaussian denoising and also helps to give fast and steady training with preferable denoising performance and one of the major advantage is that it allows to transfer faster from one model to another by taking into account various noise levels. To avoid boundary artifacts from the denoised image of CNN which is produce by some characteristics of convolution we adopt training samples with small size. Usually to handle this there are two methods one is symmetrical padding and another is zero padding. In this work we make use of zero padding which is capable to model the image boundary.

Implementation of particular denoiser model for small interval noise levels, as studies suggests iterative optimization methods make use of different denoisers models with varying noise levels.

The solutions of subproblems (i.e., Eqs. (6a) and (6b)) shows that optimization is more difficult. In other way Eq. (9) which is a denoiser gives us a different goal from conventional Gaussian denoising and one of the main aim is to retrieve the clean image in spite of various noise level of the image to be denoised.

To solve image restoration which an ill posed inverse problem, where prior is called as regularization which affect to limit the solution space, according to Bayesian statistics, the solution $\hat{l}$ is acquired to solving matrix A posterior (MAP) problem,

$$\hat{l} = \arg\max_l \log(z|l) + \log h(l) \tag{1}$$

where, $\log l$ is the log-likelihood of observation z, $\log h(l)$ delivers the prior l and is independent of z. Equation (1) can be written as,

$$\hat{l} = \arg\min_l \frac{1}{2}\|z - Cx\|^2 + \lambda\Phi(l) \tag{2}$$

where, the solution reduce an energy function composed with fidelity term $\frac{1}{2}\|z - Cx\|^2$, a regularization term $\Phi(l)$ and trade of parameter λ. model based optimization method can be directly solved by Eq. (2) and to solve discriminative learning methods we adopt to learn the prior parameter Θ and the equation is given as,

$$\min_\Theta g(\hat{l}, l) \text{ s.t. } \hat{l} = \arg\min \frac{1}{2}\|z - Cx\|^2 + \lambda\Phi(l; \Theta) \tag{3}$$

variable splitting technique is acquired to disassociate the fidelity term and regularization and plugs denoiser prior into optimization method of Eq. (2) and it can also be written as constrained optimization problem in half quadratic splitting technique by introducing an auxiliary variable m and it is reconstructed as,

$$\hat{l} = \arg\min_l \frac{1}{2}\|z - Cx\|^2 + \lambda\Phi(m) \quad \text{s.t. } m = l \tag{4}$$

half quadratic splitting method strive to shrink the cost function and it is given by,

$$L_\mu(l, m) = \frac{1}{2}\|z - Cx\|^2 + \lambda\Phi(m) + \frac{\mu}{2}\|m - l\|^2 \tag{5}$$

Where, μ is a penalty parameter and it varies iteratively in an non descending order and iterative equations are given by,

$$l_{k+1} = \arg\min_l \|z - Cx\|^2 + \mu\|l - m_k\|^2 \tag{6a}$$

$$m_{k+1} = \arg\min_l \frac{\mu}{2}\|m - l_{k+1}\|^2 + \lambda\Phi(m) \tag{6b}$$

Equation (6a) shows that fidelity term integrated with quadratic regularized least square problem to achieve fast solution for different response matrix and direct equation is shown below,

$$l_{k+1} = \left(C^T C + \mu I\right)^{-1}\left(C^T b + \mu m_k\right) \tag{7}$$

Equation (6b) is associated with regularization term and it is reformulated as

$$m_{k+1} = \arg\min_m \frac{1}{2\left(\sqrt{\lambda/\mu}\right)^2}\|l_{k+1} - m\|^2 + \Phi(m) \tag{8}$$

Equation (9) denoises the image l_{k+1} from gaussian denoiser with noise level $\sqrt{\lambda/\mu}$.

$$m_{k+1} = Denoiser\left(l_{k+1}, \sqrt{\lambda/\mu}\right) \tag{9}$$

Equations (8) and (9) can be absolutely changed by denoiser prior from the image prior $\Phi(\cdot)$.

From the above discussion we have illustrated results shows that Fig. 2(a)–(c) are before enhancement of text video frames of different datasets namely YVT, ICDAR, our multilingual respectively and also we have shown in Fig. 3(a)–(c) are after enhancement of text video frames of different datasets namely YVT, ICDAR, our multilingual respectively and Fig. 3(d)–(f) how the text edges are preserved after enhancement compared to before enhancement as shown in Fig. 2(d)–(f) of different datasets namely YVT, ICDAR, our multilingual respectively. After the enhancement, we utilized enhanced text video frames for text detection to achieve better results. In the next sub section we describe how we have used text detection method.

(a)   (b)   (c)

(d)   (e)   (f)

Fig. 2. (a) YVT dataset (b) ICDAR dataset (c) OUR dataset are video frames before enhancement and (d) YVT dataset (e) ICDAR dataset (f) OUR dataset are magnified versions from before enhanced frames.

Fig. 3. (a) YVT dataset (b) ICDAR dataset (c) OUR dataset are Video frame after enhancement and (d) YVT dataset (e) ICDAR dataset (f) OUR dataset Magnified versions from after enhanced frames.

3.2   Text Detection

In the above section we have discussed how we have enhanced video frames and in this section we utilized results of enhanced video frames to detect text. We have used modified method [19] to detect text from video frames to test before and after detection results of enhanced frames. In this method, we applied MSER (Maximally Stable Extremal Regions) with SWT (Stroke Width Transform). Firstly, we have used MSER to detect text region with fixed range of area in MSER method. Then we used SWT to remove some false alarms from the MSER method. This method adopts self-training distance metric learning algorithm which can learn distance weights and threshold simultaneously. By making use of single link algorithm the text candidates are built by clustering character candidates with learned parameters. To differentiate the text candidates and non-text candidates we make use of SVM character classifier to evaluate the posterior probability of text candidate and for remove the non-text candidate we trained SVM with non-text images. When we look into existing method, which is not efficiently remove non-text candidates because existing method trained only with text candidate images. So, compared existing method our modified method work good and improve the results.

Before enhancement proposed method gives poor result shown in the Fig. 4(a) YVT dataset, Fig. 4(b) ICDAR dataset and also Fig. 4(c) our own multilingual text in videos dataset sample. After enhancement we again tested on YVT dataset Fig. 4(d), Fig. 4(e) ICDAR dataset and Fig. 4(f) our own multilingual videos dataset. When we analyze before and after enhancement of text detection results of all the datasets, we have achieved good results for after enhancement compared to before enhancement.

From these results we conclude that before detect text in video frame enhancement is required, because video frames are affected by multiple challenges like blurriness and low quality. So, detecting text with above challenges does not give good performance.

Fig. 4. (a) YVT (b) ICDAR (c) OUR dataset are text detection results before enhancement and (d) YVT (e) ICDAR (f) OUR dataset are text detection results after enhancement.

4   Database

In this work we use standard datasets namely ICDAR (2015) in English, Spanish, Japanese, YVT (2014) are in English and French languages. When look into these datasets we notice there is no South Indian languages. So we have created our own video datasets of South Indian languages such as Kannada, English, Hindi, etc. which consists of 30 videos (Fig. 5).

5   Experimental Results

For experimentation we have considered our own mulitilingual (include Kannada, English and Hindi) dataset consists of 30 videos and standard benchmark databases, namely ICDAR-2015 of 23 videos and YVT of 30 videos. We have used key frame selection method to extract key Frames from the all dataset videos and from every video we have selected 5 key frames to evaluate text detection result. From our multilingual dataset 150 key frames, ICDAR-2015 datasets 115 key frames, and YVT dataset 150 key frames are extracted.

Fig. 5. Shows sample images of our multilingual video frames datasets.

For evaluation we used ICDAR 2013 criteria (e.g. 10, 11 and 12). Table 1 shows the evaluation of text detection results before enhancement of video frames and similarly Table 2 shows the evaluation of text detection rates after enhancement in terms of recall, precision and F-measure which is defined as,

$$Recall = \frac{Truly\ detected\ text\ block}{Actual\ text\ blocks} \tag{10}$$

$$Precision = \frac{Truly\ detected\ text\ block}{Truly\ detected\ text\ block + False\ detected\ text\ block} \tag{11}$$

$$f - measure = 2 * \frac{precision * recall}{precision + recall} \tag{12}$$

From Tables 1 and 2 we can see the difference between text detection results of before and after enhancement. Proposed method gives more false alarms in before enhancement results because that method is robust for scene images and scene images are in good quality with well-focused images so it can easily achieve good results for scene images. When we look into videos frames without enhancement it gives poor results compared to after enhancement. We have compared text detection results with state of art methods [19, 23, 24] from the Tables 1 and 2 we can see how our method is out performed compared to existing art methods. From the experimentation we conclude that we have achieved good result for video frames because of enhancement and our method is out performed compared other methods.

Table 1. Text detection results for before enhancement

ICDAR 2015 dataset			
Methods	Recall	Precision	F-measure
Proposed method	67.30	63.80	65.50
Yin et al. [19]	62.40	58.30	60.28
Epshtein et al. [23]	55.70	51.90	53.73
Chen et al. [24]	58.40	60.30	59.33
YVT dataset			
Methods	Recall	Precision	F-measure
Proposed method	74.4	70.27	72.28
Yin et al. [19]	68.57	62.38	65.33
Epshtein et al. [23]	60.85	57.26	59.00
Chen et al. [24]	60.88	65.73	63.21
Our multilingual dataset			
Methods	Recall	Precision	F-measure
Proposed method	75.16	78.34	76.72
Yin et al. [19]	71.45	68.66	70.03
Epshtein et al. [23]	65.18	58.34	61.57
Chen et al. [24]	62.36	68.71	65.38

Table 2. Text detection result for after enhancement

ICDAR 2015 dataset			
Methods	Recall	Precision	F-measure
Proposed method	73.27	68.73	70.93
Yin et al. [19]	65.13	60.40	62.68
Epshtein et al. [23]	59.61	54.78	57.09
Chen et al. [24]	60.18	63.72	61.90
YVT dataset			
Methods	Recall	Precision	F-measure
Proposed method	78.50	76.43	77.45
Yin et al. [19]	72.29	67.86	70.00
Epshtein et al. [23]	63.24	59.08	61.09
Chen et al. [24]	65.42	69.60	67.45
Our multilingual dataset			
Methods	Recall	Precision	F-measure
Proposed method	79.38	82.63	80.97
Yin et al. [19]	74.23	70.15	72.13
Epshtein et al. [23]	67.27	63.19	65.17
Chen et al. [24]	65.06	70.80	67.81

6 Conclusion and Future Work

The work in this paper contributes that for an image denoising we design and train a set of quick and constructive CNN denoisers and we have been incorporated the variable splitting technique and half quadratic splitting method to solve super-resolution and image deblurring problems by plugging in learned denoiser into model based optimization method. Then we applied text detection method on denoised image to test performance for text detection. Firstly, we have used MSER with SWT to detect text region with fixed range of area and removed false alarms with support of trained character classifier.

This method gives poor result for before enhancement and for after enhancement it gives good results. This shows that before text detection on video frames it requires image enhancement. In future we will improve the text detection accuracy and we will work on Multi-type of datasets like caption video text, poor quality scene dataset and etc.

Acknowledgment. The work carried out in this paper was supported by High Performance Computing Lab, under UPE Grant Department of Studies in Computer Science, University of Mysore, Mysore.

References

1. Sato, T., Kanade, T., Hughes, E.K., Smith, M.A.: Video OCR for digital news archive. In: Proceedings of IEEE Workshop on Content Based Access of Image and Video Databases, Bombay, India, pp. 52–60 (1998)
2. Li, H., Kia, O., Doermann, D.: Text enhancement in digital video. In: Proceedings of SPIE, Document Recognition IV, pp. 1–8 (1999)
3. Li, H., Doerman, D., Kia, O.: Automatic text detection and tracking in digital video. IEEE Trans. Image Process. **9**, 147–156 (2000)
4. Li, H., Doermann, D.: A video text detection system based on automated training. In: Proceedings of IEEE International Conference on Pattern Recognition, pp. 223–226 (2000)
5. Chen, D., Odobez, J., Bourlard, H.: Text segmentation and recognition in complex background based on Markov random field. In: Proceedings of International Conference on Pattern Recognition, Quebec, Canada, vol. 4, pp. 227–230 (2002)
6. Rainer, L., Stuber, F.: Automatic text recognition in digital videos. Technical Report, University of Mannheim (1995)
7. Burger, H.C., Schuler, C.J., Harmeling, S.: Image denoising: can plain neural networks compete with BM3D? In: IEEE Conference on Computer Vision and Pattern Recognition, pp. 2392–2399 (2012)
8. Dong, C., Loy, C.C., He, K., Tang, X.: Image super-resolution using deep convolution networks. IEEE Trans. Pattern Anal. Mach. Intell. **38**(2), 295–307 (2016)
9. Dong, W., Zhang, L., Shi, G., Li, X.: Nonlocally centralized sparse representation for image restoration. IEEE Trans. Image Process. **22**(4), 1620–1630 (2013)
10. Xu, L., Ren, J.S., Liu, C., Jia, J.: Deep convolution neural network for image deconvolution. In: Advances in Neural Information Processing Systems, pp. 1790–1798 (2014)
11. Jain, A.K., Yu, B.: Automatic text location in images and video frames. Pattern Recogn. **31**(12), 2055–2076 (1998)

12. Petter, M., Fragoso, V., Turk, M., Baur, C.: Automatic text detection for mobile augmented reality translation. In: Proceedings of the 2011 IEEE International Conference on Computer Vision Workshops (ICCV 2011), pp. 48–55 (2011)
13. Lyu, M.R., Song, J., Cai, M.: A comprehensive method for multilingual video text detection, localization, and extraction. IEEE Trans. Circ. Syst. Video Technol. **15**(2), 243–255 (2005)
14. Shivakumara, P., Phan, T.Q., Lu, S., Tan, C.L.: Gradient vector flow and grouping-based method for arbitrarily oriented scene text detection in video images. IEEE Trans. Circ. Syst. Video Technol. **23**(10), 1729–1739 (2013)
15. Epshtein, B., Ofek, E., Wexler, Y.: Detecting text in natural scenes with stroke width transform. In: Proceedings of International Conference on Computer Vision and Pattern Recognition, CVPR 2010, pp. 2963–2970 (2010)
16. Matas, J., Chum, O., Urban, M., Pajdla, T.: Robust wide baseline stereo from maximally stable extremal regions. In: Proceedings of British Machine Vision Conference, vol. 1, pp. 384–393 (2002)
17. Wang, T., Wu, D.J., Coates, A., Ng, A.Y.: End-to-end text recognition with convolution neural networks. In: Proceedings of International Conference on Pattern Recognition (ICPR 2012), pp. 3304–3308 (2012)
18. Jaderberg, M., Vedaldi, A., Zisserman, A.: Deep features for text spotting. In: Proceedings of the 13th European Conference on Computer Vision (ECCV 2014), pp. 512–528 (2014)
19. Yin, X.-C., Yin, X., Huang, K., Hao, H.-W.: Robust text detection in natural scene images. IEEE Trans. PAMI **36**(5), 970–983 (2014)
20. Zhang, K., Zuo, W., Gu, S., Zhang, L.: Learning deep CNN denoiser prior for image restoration. In: Computer Vision and Pattern Recognition, CVPR (2017)
21. Andrews, H.C., Hunt, B.R.: Digital Image Restoration. Prentice-Hall Signal Processing Series, vol. 1. Prentice-Hall, Englewood Cliffs (1977)
22. Campisi, P., Egiazarian, K.: Blind Image Deconvolution: Theory and Applications. CRC Press, New York (2016)
23. Epshtein, B., Ofek, E., Wexler, Y.: Detecting text in natural scenes with stroke width transform. In: IEEE Computer Society Conference on Computer Vision and Pattern Recognition, CVPR (2010)
24. Chen, H., Tsai, S.S., Schroth, G., Chen, D.M., Grzeszczuk, R., Girod, B: Robust text detection in natural scene images with edge-enhanced maximally stable extremal regions. In: 18th IEEE International Conference Image Processing (ICIP), pp. 2609–2612 (2011)

A Novel Approach for Steganography App in Android OS

Kushal Gurung[1], Sami Azam[1(✉)], Bharanidharan Shanmugam[1],
Krishnan Kannoorpatti[1], Mirjam Jonkman[1], and Arasu Balasubramaniam[2]

[1] School of Engineering and IT, Charles Darwin University, Darwin 0909, Australia
{sami.azam,bharanidharan.shanmugam}@cdu.edu.au
[2] Cookie Analytix Pvt. Ltd., Bengaluru, India
arasu@cookieanalytix.com

Abstract. The process of hiding information in a scientific and artistic way is known as Steganography. The information hidden cannot be easily retrieved or accessed and is unidentifiable. In this research, some of the existing methods for image steganography has been explained. These are LSB (Least Significant Bits) substitution method, DCT (Discrete Cosine Transform) and DWT (Discrete Wavelet Transform). A comparative analysis of these techniques depicted that LSB is the easiest and most efficient way of hiding information. But this technique can be easily attacked and targeted by attackers as it changes the image resolution. Using LSB technique an application was created for image steganography because it hides the secret message in binary coding. To overcome this problem a RSA algorithm was used in the least significant bits of pixels of image. Additionally, a QR code was generated in the encryption process to make it more secure and allow the quality of the image to remain as intact, as it was before the encryption. PNG and JPEG formats were used as the cover image in the app and findings also indicated the data was fully recovered.

Keywords: Image steganography · LSB (Least Significant Bits) · Wavelet
DCT (Discrete Cosine Transform) · RSA · QR code

1 Introduction

Steganography refers to hiding of the message in an artistic and scientific way. It comes from the Greek word "stegos" which means art of writing the message in a secret way [1]. By using Steganography anyone can hide a secret information in a message and send it to someone without knowing to anyone. In traditional steganography, the invisible ink, pin punctures, small holes were made on the letters and the messages were visible only in a light [2]. Due to the advancement in technologies, information management are done electronically therefore, information security has become more important. Cryptography on the other hand is applied to a digital host before sending to the network which makes the message hidden. Steganography handles the information to be passed securely which is a best approach to protect the digital rights for image, audio and video [3]. In Image steganography, image is used as a carrier file. There are different file

© Springer International Publishing AG, part of Springer Nature 2018
A. Abraham et al. (Eds.): ISDA 2017, AISC 736, pp. 442–450, 2018.
https://doi.org/10.1007/978-3-319-76348-4_43

formats as Jpeg, bmp, png for the digital images. But to compress the large digital images it is a challenge for the steganography techniques. There is a solution for this problem which is digital image compression. Two types of compression methods used in digital images; they are lossy and lossless compression. In Lossy compression, it removes the size and excessive image data by calculating the nearest value of the original image. It is usually used with 24 bit images to decrease the size of an image. Lossless compression is the process of compression where image does not lose its originality. That is the reason for most of the steganography techniques use lossless compression methods [4].

2 Research Motivation, Contributions and Novelty

In the past few decades, Image steganography has become a matter of interest. There are different file formats like Jpeg, bmp, png for the digital images are being used. Steganography refers to the technique or art that is unidentifiable to human when implemented. This point is not completely true since it removes the content of the image thinking as excess bits. This does not mean image steganography techniques are not secure to use instead it provides a gap where improvements to the existing techniques can be achieved. In this report among other techniques used in image steganography, LSB is chosen which hides the secret messages within the pixel of images. To increase the level of protection RSA encryption and QR code are used in stego image. This will make stego image more secure and robust against attackers who try to attack the image to get the information. This process is done without changing the quality of image and is unidentifiable to human eyes. The result of this report is based on JPEG (Joint Photographic Expert Group) and PNG (Portable Network Graphics) image formats. Image steganography is chosen because its embedding procedure differs according to data hiding [5]. On the other hand, the quality of an image is distorted with the presence of noise which lacks image robustness [5]. So we have proposed a new technique that provides the Image steganography more robustness and secure against attackers because RSA and QR code are implemented in this system which is one step more secure to the stego image. This technique has been implemented on the android app and web platform. This app also has sharing features enabled like email, drop box, google drive. The experiments and results also indicate that this proposed system is better than previous LSB technique. The future scope of this research is to implement the proposed technique in iOS along with enhancing existing LSB algorithm in terms of compression and speed.

The concept of developing an app might be getting matured, but the novelty lies in using QR code with LSB algorithm. The motivation was to design an app that could be very light weight that has the encryption capability on the fly. The novelty lies in combining QR code along with RSA and LSB algorithms. We were successful in our pilot study and it will be extended with several options based on the feedback received. Due to time restriction, only few images have been tested. Nevertheless, we are confident that the future tests will bring almost close to these results.

3 Literature Review

There are different techniques used in Image Steganography, some of them are described in this paper.

Blind Hide: In this technique information that is hidden can be retrieved from the embedded file without accessing original image. It can be more secure and has more applicability because carrier image is does not need to be extracted always. However, this technique is hard to implement when carrier image is considered as an image because both carrier image and secret message cannot be extracted separately at the recipient slde [6, 7].

HideSeek: In this algorithm, message is randomly distributed to the image area. It generates password when hiding the information and to decrypt the message that same password is required. It is also not the best algorithm to hide the message because it does not look in the pixel of the image [3]. It is useful to find the best possible area in the image to hide the message. In this digital world people tend to transfer files over internet via email or other ftp protocols. Hide Seek algorithm is an interesting algorithm because it is transformed in a system and not detected by visual attacks. It is very hard to detect stego image when a carrier image is embedded via Hide Seek. It provides no evidence of secret message existence [8].

Filter first: This algorithm filters the image with inbuilt filters and also it hides the highest value of pixels first. It is an upgrade of blind hide algorithm where message can be retrieved without passwords. In this algorithm MSB (Most significant bits) are filtered and LSB (least significant bits) are left to be changed. When message is hidden the area of an image is less noticeable [9]. This algorithm considers a pixel of the image and looks for surroundings neighbour pixel to determine whether it should represent or not. It simply replaces the median value of the pixel instead of replacing the mean value of the neighbouring pixel.

LSB (Least Significant Bits): LSB is a simple method to embed information in images. It is the mostly used technique in Image steganography and its embedding procedure differs according to data hiding. Some of the algorithms [12] modify the pixels areas of the image by changing the LSB of the image rather than increasing or decreasing the pixel value. This algorithm store the image bit of a carrier image file by substituting the actual bits of the image with the hidden message bits and after that it is stored into the LSB of the carrier image. The stego image produced after embedding the cover image is far better that human eye cannot detect the difference between original and stego image [5]. Therefore, slight change in the image such as; color, substitution of pixel is indistinguishable by simply looking at it. The secret message hidden in large images is most desirable it is due to the large space in the large images. The best quality of image can be produced with 24 bit pixel used in a cover image. When 24 bit image is used, each of the RGB colour (red, green and blue) components represents the components of a byte [10, 11].

DCT (Discrete Cosine Transform): DCT is applied [14] in embedding message in transform domain of cover image. The embedding process in the transform domain is performed for the robustness of the image. When data is embedded in the transform domain it spreads in the image and provides resistance to the attacks and signal processing. In a block DCT, select one or more components of the block depending on its payload requirements to create a new data which is then scrambled and process through second layer of transformations. Using several schemes and procedures modification takes place in the second layer of transformation of the image [13]. These methods have high encoding and retrieving complexity. Due to its resistance and robustness quality, these technique is generally suitable for data hiding. Most of the steganography techniques in this transform domain are inspired from this method. The signal from an image is transformed to frequency by DCT. When pixels are grouped in a 8×8 blocks and transforming those pixel block into 64 coefficients of DCT then modifying a single coefficient of DCT will affect entire pixel block (64 pixel) as shown in the figure. The compression of an image is quantized in a manner that even though a human eye can distinguish a small variation in the brightness of an image it is very hard to separate the differences in the frequency of the image brightness. Which result in the low strength of the frequency without the physical distortion of an image [12].

DWT (Discrete Wavelet Transform): Discrete Wavelet Transform (DWT) is a wavelet transformation where analogue signal (Sines and Cosines) is converted in fourier analysis to represent the signal. The simplest discrete wavelet transform is Haar. Low frequency is generated by taking an average value of the two pixels and high frequency is generated by taking difference of the same two pixels in DWT. To calculate DWT a signal is passed through the filters [13, 18]. DWT is better than DCT because it hides the secret message in different frequency sub bands and it is very difficult to distinguish the low frequency part of an image. It has very high computational speed and produce an image matrix with rows and column transformation [15].

4 System Design and Additional Technologies Used

This system supports the information security which is the most important factor in Image steganography. In this Fig. 1, at first the secret data is embedded RSA encryption (public key) and QR code [17, 19]. The resulted QR code is scanned and ciphered text is embedded in cover image. In the next phase LSB embedding algorithm is applied to generate secure stego image. RSA algorithm [16] is used in this system for data privacy and confidentiality. Plain text message (original text) is encrypted and it is embedded in either JPG or PNG format as this system supports both image formats as shown in Fig. 1.

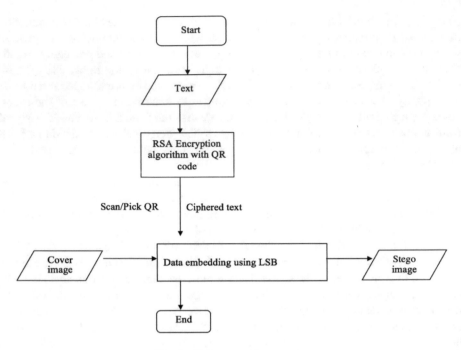

Fig. 1. Data encryption process

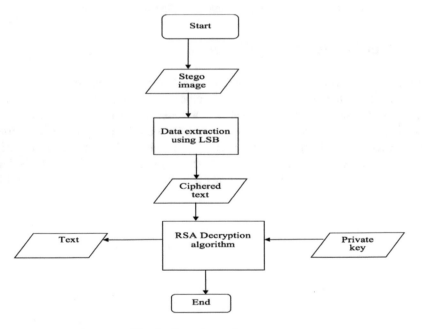

Fig. 2. Data decryption process

Stego image in taken as the input to the system. LSB decryption algorithm is applied to the stego image to get ciphered text as an output. There will be a key selected 'private key' to authenticate the ciphered text with RSA decryption algorithm. The final phase of this process is a plain text as an output which is the secret message as shown in the Fig. 2.

5   Proposed Architecture Implementation

This section describes the different parts of the experiments. First is the encryption followed by the decryption process.

Step 1: User selects the key generate button from the app.
Step 2: User will select his name to make sender easy to send message and click generate button. It provides QR code with private key and public key and saves the QR code in a jpg image format.
Step 3: The generated QR image is scanned/picked through a QR scanner. It reveals the name of the receiver and public key on his device.
Step 4: Sender starts the embedding procedure. He can either click or pick an image from his device gallery and selects the contact with receiver's name and public key. He then completes the embedding process.
Step 5: The generated image is stego image. It is then shared via email, google drive as shown in the Fig. 3.

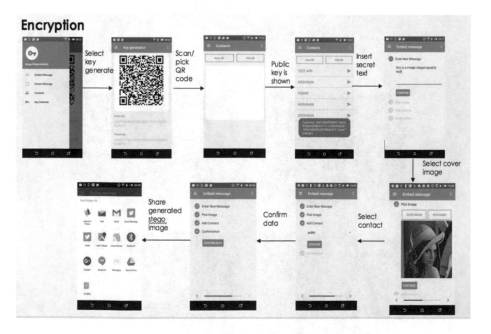

Fig. 3.   Embedding process in an app

The decryption process is as follows:

Step 1: Receiver receives the stego image. He then selects extract message from the app.
Step 2: Receiver choose pick image option from the app. After that he picks the stego image send by sender.
Step 3: If this stego image falls in wrong hand and tries to decrypt the secret message then an error message "The message is corrupted or you are not authorized person to view this".
Step 4: When a valid receiver decrypts the stego image using his authentic private keys then secret message is revealed as shown in the Fig. 4.

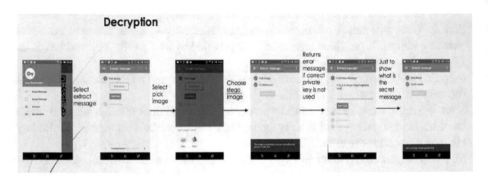

Fig. 4. Decryption process in an app

Fig. 5. Image comparison

This experiment was tested in two different Android versions (5.1 and 6.01). Tables 1 and 2 show the tested results in Android platform. In Android, two different images (Lena.jpg & Lena.png and Castle.jpg and Castle.png) as shown in Fig. 5 were compared where embedding and extraction time were calculated. Data recovery was also tested for two different versions and surprisingly 100% of data were recovered during extraction process.

Table 1. Test results on Android v5.1

Method	Format	Image size	Resolution	Android version	Embedding time	Extraction time	Recovery %
LSB	Lena.PNG	463 KB	512 × 512	5.1.0	3 s	1 s	100
LSB	Lena.PNG	468 KB	512 × 512	5.1.0	3 s	1 s	100
LSB	Castle.PNG	9 MB	5400 × 2700	5.1.0	100 s	2 s	100
LSB	Castle.PNG	10 MB	5400 × 27K	5.1.0	100 s	2 s	100

The above experiment was conducted using Android version 5.1.0 emulator with 1 GB memory.

Table 2. Test results on HTC mobile Android 6.01.

Method	Format	Image size	Resolution	Android version	Embedding time	Extraction time	Recovery %
LSB	Lena.PNG	463 KB	512 × 512	6.01	3 s	1 s	100
LSB	Lena.PNG	468 KB	512 × 512	6.01	3 s	1 s	100
LSB	Castle.PNG	9 MB	5400 × 2700	6.01	90 s	2 s	100
LSB	Castle.PNG	10 MB	5400 × 27K	6.01	120 s	2 s	100

The above experiment was conducted using HTC One M8s Android version 6.0.1 mobile with 2 GB memory.

6   Conclusion and Future Work

In this paper, RSA algorithm was used in this system for data privacy and confidentiality. Plain text message (original text) was encrypted and it was embedded in either JPG or PNG format as this system supported both image formats. LSB substitution method was used for this process. There were negligible changes in colour of the image which could not be identified by human eyes. To protect the data QR code was used in the encryption process which generated the random sequence number for the private and public keys. Then ciphered message was embedded in JPG or PNG format based on which image format was used as a cover image. The proposed system was implemented in an android app and also succeeded in web based application.

This app can be extended further with the use of different file formats other than JPEG and PNG. More improvement needs to be done in the area of capacity, embedding distortion and time complexity. This app will be made platform independent to iOS because it only works in web and android platforms. Also, further research will be done to enhance the existing LSB algorithm in terms of speed and compression. Our long term aim is to release it as open source android mobile application for the better usage.

References

1. Morkel, T., Eloff, J.H., Olivier, M.S.: An overview of image steganography. In: Information and Computer Security Architecture (ISSA), pp. 1–11 (2005)
2. Reddy, R., Ramani, R.: The process of encoding and decoding of image steganography using LSB algorithm. Int. J. Comput. Sci. Eng. Technol. **2**(11), 1488–1492 (2012)
3. Ibrahim, R., Kuan, T.S.: Steganography algorithm to hide secret message inside an image. arXiv (preprint) arXiv:1112.2809 (2011)
4. Dunbar, B.: A detailed look at steganographic techniques and their use in an open-systems environment, pp. 1–8 (2002)
5. Dabas, P., Khanna, K.: A study on spatial and transform domain watermarking techniques. Int. J. Comput. Appl. **71**(14), 38–41 (2013)
6. Muhammad, N., Bibi, N., Mahmood, Z., Kim, D.G.: Blind data hiding technique using the Fresnelet transform. SpringerPlus **4**(1), 832 (2015)
7. Umamaheswari, M., Sivasubramanian, S., Pandiarajan, S.: Analysis of different steganographic algorithms for secured data hiding. IJCSNS Int. J. Comput. Sci. Netw. Secur. **10**(8), 154–160 (2010)
8. Provos, N., Honeyman, P.: Hide and seek: an introduction to steganography. IEEE Secur. Priv. **1**(3), 32–44 (2003)
9. Krenn, R.: Steganography and steganalysis (2004). http://www.retawprojects.com/uploads/steganalysis.pdf
10. Ahuja, B., Kaur, M.: High capacity filter based steganograph. Int. J. Recent Trends Eng. **1**(1), 672–674 (2009)
11. Sharma, V.K., Shrivastava, V.: A steganography algorithm for hiding image in image by improved LSB substitution by minimize detection. J. Theor. Appl. Inf. Technol. **36**(1), 1–8 (2012)
12. Devi, G.S., Thangadurai, K.: An analysis of LSB based image steganography techniques. In: Computer Communication and Informatics (ICCCI), pp. 1–4 (2014)
13. Goel, S., Rana, A., Kaur, M.: A review of comparison techniques of image steganography. Glob. J. Comput. Sci. Technol. **13**(4), 9–14 (2013)
14. Kaur, G., Kochhar, A.: A steganography implementation based on LSB & DCT. Int. J. Sci. Emerg. Technol. Latest Trends **4**(1), 35–41 (2012)
15. Kumar, V., Kumar, D.: Performance evaluation of DWT based image steganography. In: Advance Computing Conference (IACC), vol. 6, no. 10, pp. 223–228 (2010)
16. Milanov, E.: The RSA algorithm, pp. 1–11. RSA Laboratories (2009)
17. Kak, A.: Public-key cryptography and the RSA algorithm. Lecture Notes on Computer and Network Security, Purdue University, pp. 3–7 (2015)
18. Altaay, A.A.J., Sahib, S.B., Zamani, M.: An introduction to image steganography techniques. In: 2012 International Conference on Advanced Computer Science Applications and Technologies, ACSAT 2012, pp. 122–126 (2013)
19. Association of Nova Scotia Museums 2015, QR code how-to guide. http://www.rcip-chin.gc.ca/media/pro/carrefour-du-savoir-knowledge-exchange/ansm_qr/codes_qr-qr_codes-eng.pdf

Exploring Human Movement Behaviour Based on Mobility Association Rule Mining of Trajectory Traces

Shreya Ghosh[✉] and Soumya K. Ghosh[✉]

Department of Computer Science and Engineering,
Indian Institute of Technology Kharagpur, Kharagpur, India
shreya.cst@gmail.com, skg@iitkgp.ac.in

Abstract. With the emergence of location sensing technologies there is a growing interest to explore spatio-temporal GPS (Global Positioning System) traces collected from various moving agents (ex: mobile-users, GPS-equipped vehicles etc.) to facilitate location-aware applications. This paper, therefore focuses on finding meaningful patterns from spatio-temporal data (GPS log) of human movement history and measures the interestingness of the extracted patterns. An experimental evaluation on GPS data-set of an academic campus demonstrates the efficacy of the system and its potential to extract meaningful rules from real-life dataset.

Keywords: Trajectory · Mobility · GPS traces · Association rule
Transactional database

1 Introduction

Owing to the pervasiveness of mobile phones and development of sensor-technologies, wireless networks, the availability of mobility traces including personal-GPS trajectories, taxi-traces etc. have opened up the possibility of interpreting human mobility behaviour in space and time. This myriad of mobility data fuses interesting and challenging problems namely, determining classical travel sequences and top k interesting locations, traffic monitoring, defense applications [7], mobile-user categorization [3] etc. Obviously the major challenge is capturing the inherent knowledge so that it can be used effectively in several location based services or personalized recommendation systems. The core of any mobility-behaviour analysis task is *human moves with an intent* [2] and thus people follow a highly reproducible and meaningful patterns [6] in their daily movement. Therefore, to utilize the mobility traces of people for various services, it is crucial to perceive how location, time effects their mobility patterns.

Association rule mining finds application in several domains [1] including business analysis, clinical databases, stock market analysis etc. - where inter-relation among objects contribute in the knowledge-base. With the research

© Springer International Publishing AG, part of Springer Nature 2018
A. Abraham et al. (Eds.): ISDA 2017, AISC 736, pp. 451–463, 2018.
https://doi.org/10.1007/978-3-319-76348-4_44

advancement, several new paradigms have evolved, namely *inter-sequence* patterns, *intertransactional patterns*, where association relations are discovered among attributes from different transaction records [9]. Discovering frequent or co-related patterns in heterogeneous databases is one of the challenging and most important facets in data mining research. Since the introduction of *Association Rule Mining* problem and Apriori algorithms [1], significant research efforts have been made in the direction of dynamic dataset mining, appending additional semantics, such as time, space, ontologies etc. to discover temporal or spatial association rules, sequential pattern mining, Bayesian association rule mining etc. Temporal association rule mining uncovers a wide spectrum of paradigms for knowledge extraction [10] from time series data. Sequential pattern mining is a type of temporal pattern mining which is used in web-usage mining, discovering rules from medical databases [8], classification approaches etc. A key aspect of discovering meaningful rules is interestingness measures to extract and rank patterns according to the applications and potential interest to the users [5].

Motivation and Objectives: Discovering intertransaction multidimensional (space, time) rules from human mobility traces is a novel proposition. In this work, we aim to analyze human mobility behavioral patterns from the probabilistic graphical model of historical GPS log and discover the time-featured and categorical rules based on the spatio-temporal features. For example, the proposed system should be able to extract rules like, ***R1:*** *People who visits health-care center (say, gym, playground) regularly, mostly visits point-of-interests (POIs) like hospital, medicine-shop less frequently,* or ***R2:*** *Students are more likely to visit POIs like library, academicBuilding in weekdays while in weekdends frequency of GPS footprints are higher in POIs like cafe, movieComplex etc.* Rule *R1* depicts time-feature based correlations among different movement patterns (or transaction in GPS trace database) while *R2* represents categorical movement behaviour in an acedemic region-of-interest (ROI).

Contributions: The contributions of this work can be summarized as: (i) Proposing the structure of *GPS transactional log* using *time-series discretization* and a hash-based data structure to efficiently represent the spatio-temporal attributes of movement data in different resolution; (ii) Extracting the *Time-featured* and *User-categorical* rules to analyze the interdependencies of places (or stay-points) and time features of trajectory data and user mobility behavioral rules respectively; (iii) Finding the interestingness of the extracted rules using two proposed relevance-measures.

In our previous work [2], we identify the *mobility-association rule* mining task. To the best of our knowledge, no other existing work has undertaken the association rule mining problem from human mobility traces.

The rest of the paper is organized as follows: Sect. 2 presents the proposed framework. Experimental evaluation of the framework is shown in Sect. 3 through a case-study in the academic campus. We conclude in Sect. 4 with the future directions of the work.

2   Proposed Framework

In this section, we discuss our proposed framework [Fig. 1] to discover the interesting rules from mobile-user GPS traces. The framework consists of three modules, namely, *Mobility Data Pre-processing*, *Mobility Rule Generation* and *Significant Rule Learning*. Before describing the modules, few basic concepts and the problem definitions have been presented.

2.1   Preliminaries

1. **Labelled User GPS log** (G):
   User GPS log (G) is a sequence of time-stamped geo-tagged latitude, longitude points of an individual.
   $G = (C, U, < lat_1, lon_1, p_1, t_1 >, \ldots, < lat_n, lon_n, p_n, t_n >)$, where C is the category of mobile-user (say, student, faculty of an academic ROI), U represents the unique user-id and p_i stores the geo-tagged place (say, university, cafe etc.) of ith point in the sequence.
2. **User Movement Summary (UMS):**
   UMS of an individual depicts the graphical representation of probabilistic relationship among the stay-points of the trajectory traces. $UMS = (N, \theta)$, where $N = (V, E)$ is the directed graph consisting V, stay-points and connecting edges E among the stay-points. The probability distribution among the stay-points or variables is quantified by θ. The detailed study of representing movement summary as probabilistic graphical model has been discussed in [3].
3. **Stay Point** (S) **and Point-of-interest Taxonomy** (T):
   S is defined as $S = < lat, lon, P, T_s, T_d >$ where within a radius of $d > D_{thresh}$ distance, an individual spends $T_d > T_{thresh}$ time at T_s timestamp. P represents the geo-tagging information of the stay-point. Point-of-interest taxonomy T is generated to represent the geo-tagged information in a hierarchical manner [3].
4. **GPS Transaction record** (T):
   A GPS transaction record, $T = < S, T_d, T_s, E_T, E_D >$ consists of staypoint information (S: Stay-point POI, T_d: Time-duration at S, T_s: Time-interval visited at S) along with the edge traversal information (E_T: Distance travelled from the previous stay-point, E_D: Time-duration to travel from the previous stay-point) between two consecutive stay-points.

In our proposed framework, the input dataset is a tuple of *latitude, longitude* and *timestamp* along with some application-specific information, namely, user-category, types of place visited etc. Therefore, there are several types of data available unlike transactional database. For a discrete (POIs, namely *ResidentialBuilding, Cafe* etc.) or categorical attribute (user category: *student, professor*), all the possible categories are mapped to a set of integers and continuous attribute (time) is discretized into several intervals. Further, a *day* is partitioned into 4 non-overlapping time-ranges allowing the continuous time-series to be mapped to different time-slots or buckets. Consequently, each data-item in the dataset can be represented as *attribute, integer-value*.

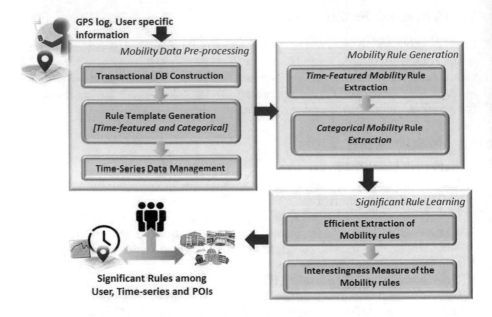

Fig. 1. Architecture of the mobility-rule mining framework

2.2   Problem Definition

In association rule mining algorithm, data record is taken as attribute-value pairs and association rules related to certain features of the attributes are extracted. In our GPS trace dataset, a number of features each having finite number of possible values are present which represent the objects or items in this domain. Example of transaction databases are shown in Figs. 2a and 3 where dimensions are user-category, place-information and time-features. We aim to extract two types of inter-transactional rules from the GPS trajectory of users.

Time-featured Mobility Rule: Given a transactional database of GPS or trajectory traces of moving agents, where each transaction consists of user-id, stay-point information (i.e., T_d: duration, T_s: Time-interval, P: place-information), edge-traversal information (i.e., E_d: distance and E_t: travel-duration to reach the stay-point), discover all the rules of the form $A \Rightarrow B$, where $A \in \{T_s, T_d, E_d\}$ and $B \in \{P, T_d, E_d, D_{AB}\}$ and $A \cap B = \phi$ and $Support(A \Rightarrow B) > thresh$. D_{AB} depicts derived attributes of the transactions within a sliding-window w. For example, D_{AB} may be frequency of visit at a stay-point or number of unique places visited within a time-interval.

Based on this definition, a rule might be "if the time of edge traversal is sufficiently large, then at a particular time-stamp, an individual's next stay-points belong to a certain set of POIs and stay-duration is likely to be more" and it can be expressed as $\Delta_{e_d} \wedge \Delta_{T_s} \Rightarrow \Delta_P \wedge \Delta_{T_d}$.

Category based Mobility Rules: Given a transactional database of summarized GPS or trajectory traces of different user-categories, where each transaction consists of user-id (u), category-id (C), stay-point information (S) along with the probability to follow the paths and time-interval (T), discover all rules of the form $A \Rightarrow B$, where $A \in \{C', T, S\}$ and $B \in \{S, T\}$, and $Support(A \Rightarrow B) \geq minSupp_{thresh}$. C' depicts derived user-categories from the transactional database. For example, 'Users visiting health-care center regularly' or 'Users having higher footprint at Library' are two examples of derived categorical attributes from the database based on movement behaviour.

2.3   Mobility Data Preparation

In this section, we briefly describe the database preparation from time-series GPS log followed by mobility rule template definitions and efficient storage of several mobility features of the GPS traces.

Generation of GPS Transactional Log from Movement Sequence: Figs. 2(a) and 3 show transactional database from typical GPS log and UMS at a particular time-instance.

Each of the stay-point related information along with the edge-traversal from the previous stay-point are recorded as a *transaction* in the database. An *item* or *literal i* in the transactional database (G_1) is an attribute value pair of the form (A_i, v), where A_i is an attribute and takes value of stay-point information (*place* and *time*) and edge-traversal information (distance covered from the previous stay-point and time duration of the travel). Any transaction in G_1 is identified by $< u_{id}, t_i >$; i.e., *user id* and the actual timestamp when the GPS point is captured.

In the next transactional database (G_2), each transaction is generated from any existing path in the UMS. Each path $< s_i, CPT_i >$ consists of a sequence of stay-points visited and the corresponding conditional probability values. Hence, there must be a corresponding entry in the transactional database to represent the movement history along the given sequence. For example, in Fig. 3, stay-points $< s_1, s_2, s_3, s_5 >$ and $< s_1, s_4, s_3, s_5 >$ depict two different sequence present in UMS and these two sequences are reflected as two transactions in G_2.

It is clear from the definitions and terminologies, some of the literals (time-duration, distance, timestamp) take continuous values while some literals (point-of-interests, user-category) form hierarchical structure or taxonomy. Figure 2(d) and 2(b) represent typical taxonomy of user-category and place-of-interests in an academic region-of-interest. In order to implement classical association rule mining techniques, we need to partition the continuous or qualitative variables into different intervals. Both of the literals *time-duration* and *distance* are partitioned into three intervals namely, *high*, *medium* and *low* and timestamp of a particular transaction falls into timeInterval-1 (0600–1200), timeInterval-2 (1200–1600), timeInterval-3 (1600–2100) and timeInterval-4 (2100-0600). We assume all the qualitative attributes belong to a non-overlapping partition or class.

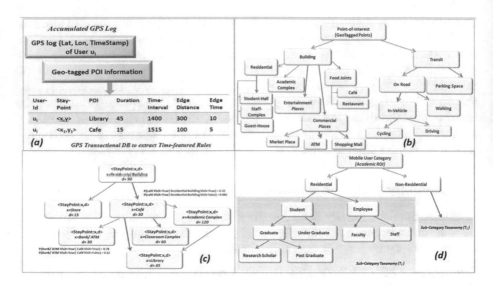

Fig. 2. (a) Sample entries on transactional DB of mobility traces from accumulated GPS log (b) Sample POI taxonomy of an academic ROI ($POI_{Taxonomy}(P)$) (c) Sample UMS (User Movement Summary): probabilistic graphical model of user's summarized mobility trace (d) User Category taxonomy (C) of an academic ROI

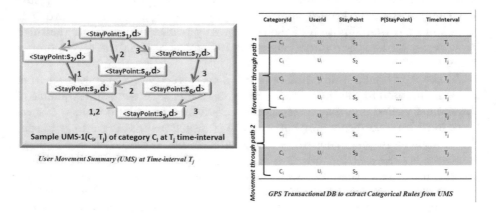

Fig. 3. Sample entries on transactional DB of categorical movement traces from UMS

Mobility Rule Template Generation: Clearly, *time* attribute is the key feature of UMS and acts as a principal factor to extract significant patterns or association rules from the movement summary. The key reason is "time has several meanings in movement data" and thus we need to combine all time-features to discover the inter-relationships between them. Examples of such interrelating time-feature rules are "if the edge-time T_e is more then next location stay-time T_s is likely to be more" or "a long-duration stay-time T_s is followed by a

Table 1. Mobility behaviour and corresponding potential rule-templates

T Id	Rule template mobility behaviour
T_1	$TimeStamp(t) \rightarrow stayPoint(f, P); f \in F, P \in POI$
	Time impacts footprint distribution at different POIs
T_2	$TimeStamp(t) \rightarrow StayPoint(d, P); d \in D, P \in POI$
	TimeStamp impacts on StayDuration at different POIs
T_3	$EdgeTraversal(dis, d) \rightarrow StayPoint(d); d \in D, dis \in L$
	Edge traversal information impacts on duration at next stay-point
T_4	$StayPoint(x, d_1) \wedge EdgeTraversal(dis, d_2) \rightarrow StayPoint(y, d_3)$
	$d_1, d_2, d_3 \in D, dis \in L, x < y \in S$
	Staypoint information and edge traversal influences consequent stay points
T_5	$StayPoint(x, d, t) \rightarrow is\_a(x, P)$
	$x \in S, d \in D, t \in T$
	Time-features of a particular stay-point implies the type of POI
T_6	$visits(x, A, f_1) \rightarrow visits(x, B, f_2)$
	$x \in U, A, B \in POI, f_1, f_2 \in F$
	Users visiting a POI with a particular frequency are likely to visit other POIs with certain frequency
T_7	$is\_a(x, c) \rightarrow visits(x, B, f_2)$
	$x \in U, c \in C, f_1 \in F, B \in POI$
	User category impacts on the visiting frequency of various POIs
T_8	$is\_a(x, c) \wedge TimeStamp(t) \rightarrow visits(x, B, f_2)$
	$x \in U, c \in C, f_1 \in F, B \in POI$
	User category and timestamp of the transaction influences the visiting frequency at different POIs

short-duration stay-time". Noticeably, all time-features of the time-series data: stay-time, duration, edge-time along need to be considered in the computation. We define potential mobility rule templates [Table 1] containing both time-featured [rule template $T_1 - T_5$] and categorical [rule template $T_6 - T_8$] mobility patterns of GPS traces. The intuition behind these rule-templates are "mobility traces (or sequence of movement) are highly dependent on timestamp values and each place-of-interest has direct correlation with several time-features which influences the movement behaviour of different categories of user."

Time-Series (GPS traces) Data Management: One of the major challenges to extract mobility rules is the huge amount of spatio-temporal traces. In order to reduce the computation load, we propose a *hash-based Place* (B_1) and *user-category* (B_2) bucket, where visit-frequency count and categorical visit-frequency count at different time-intervals are stored. A hash function is used to maintain the spatial correlation among the stay-points, i.e., two stay-points with minimum distance are stored into two consecutive buckets.

Figure 4(a) shows the visit-frequency lattice structure of the place-visit frequency where each node n_j at level j is a combination of $< H, L, M >$ or $< High, Low, Medium >$ tuple. Clearly, using a simple count function on $H(B_1, T)$ and $H(B_2, T)$, the visit-frequency lattice structure of different POIs is formed. Frequent item-set is extracted using a top-down approach on the place-frequency lattice. An item-set $< P_i, P_j >$ is *frequent* with the attribute value $< H, L >$, iff it is observed that people usually visit P_i place for a *high* stay-duration followed by *short*-duration visit at P_j.

2.4   Mobility Rule Generation

The anti-monotone property of *Apriori algorithm* presumes $\forall X, Y : (X \subseteq Y) \Rightarrow S(X) \geq S(Y)$, i.e., all superset of infrequent item-sets will be infrequent. Therefore, all the supersets of an infrequent item-set (support is less than min_{supp}) are discarded in the procedure. Our proposed algorithm utilizes the non-overlapping property of the frequency class and uses the top-down apriori algorithm. Further, we extract rules from two taxonomies, place-taxonomy and user-category taxonomy having different support values. Algorithm 1 depicts the steps of the mobility rule mining algorithm. The hash-based structures are initialized and updated when new place-visit or stay-point information is extracted from the transactional log. Using *IDDFS (iterative deepening DFS)* on the visit-frequency lattice structure, frequent item-sets are extracted. In the next run of the algorithm, derived or extended item-sets are discovered and all possible combinations of categorical and time-featured rules are extracted. Finally, all the rules having less values compared to the threshold interestingness measure value are discarded. The output of the mobility rule generation algorithm provides distinct mobility rules along with the interestingness measures.

2.5   Significant Rule Learning

We quantify the extracted rules using five measurements. While three commonly used measures *Gini Index (G)*, *Mutual Information (M)* and *J-measure (J)* along with support and confidence depict the rule-interestingness, another two proposed measures *category-relevant-measure (CRM)* and *user-relevant-measure (URM)* illustrate the applicability of a particular association rule to an user-category or group of people and an individual. In our study, we evaluate the rules using *J-measure (J)*, *Gini Index (G)* and normalized *Mutual Information (M)* [5]. All of these measures are asymmetric measure, i.e. it implies there is a difference between $X \Rightarrow Y$ and $Y \Rightarrow X$. Clearly, in our case time-featured rules maintain a sequence of stay-points visit and there is a strong need to distinguish the strength of the implication rules for both the cases.

All of these measures captures the variance of the probability of GPS footprint distribution. It is interesting to note that, although there are several existing interestingness measures of association rules in the literature, they are not

---

Algorithm 1. Extracting distinct Mobility Rules from time-series data

---

Input: The GPS Transactional log $G = (g_1, g_2, \ldots, g_l)$, $POI_{Taxonomy}(P)$, $UserCategory_{Taxonomy}(C)$, sliding window w

Output: Set of unique time-featured mobility rules (R_T) and Categorical mobility rules (R_C);

1: $R_T = \{\}; R_C = \{\}$ ▷ *Initialize Mobility Rule-set:* R_T *and* R_C

2: $L \leftarrow$ Generate *visit-frequency lattice-structure* for place-hash table

3: Iterative Deepening Depth First Search L to extract $i_m = \{< t_m, c_m, p_m > | (r_m$ *is an item − set* $) \wedge (support(< r_m >\geq supp_{thresh})|\}$ ▷ t_m: *Time-information*, p_m: *Place-information*, c_m: *category-information*

4: **while** $|i| > 0$ **do**

5:     $C' \leftarrow C, F' \leftarrow \{H, L, M\}, T \leftarrow \{1, 2, 3, 4\}$ ▷ *Derived attributes*

6:     rule $a \leftarrow NULL$

7:     **for** $c = 0; c \leq |C|; c + +$ **do** ▷ *Candidate set Generation for Categorical Mobility Rules*

8:         Generate Candidate − Set $C_i' = \{\{c_i \cup t_i\} \wedge \{c_i \cup p_i\}\}$

9:         Prune candidate items : $C_i = C_i' - \{\gamma | (\gamma \in C_k') \wedge (k \subset (i - 1)^{th}$ height of C taxonomy$) \wedge (\gamma \notin c_i)\}$

10:         Extract $a_i = \{c | (c \in C_i) \wedge (support(c) \geq supp_{thresh}(\kappa))\}$

11:         Extract pattern $a_i \in L$ with maximum confidence

12:         **if** a_i covers all instances in T **then** ▷ *Check for all time-intervals*

13:             $R_C = R_C \cup \{a_i\}; L = L - \{a\}$

14:         **end if**

15:     **end for**

16: **end while** ▷ *Repeat the same procedure to extract* R_T *using* w *time-sliding window*

17: $R = R_C \cup R_T; \kappa \leftarrow$ *Find interestingness Measures(R)* ▷ *Append all extracted mobility rules*

18: **return** $R, [\kappa]$ ▷ *Result: (Mobility Rule, Interestingness Measures)*

---

suitable for measuring associative patterns of trajectory data. There are few questions like, "Whether a rule (r_i) is prevalent in the mobility traces of an individual?" or "Whether a rule (r_j) is relevant or applicable to a particular user-category or a group of users showing higher support and confidence within the community?". Clearly, while the former rule (r_i) represents an individual's movement behaviour, the next one (r_j) is useful to capture the group-mobility behaviour. To this end, we propose two new measures, namely *category-relevant-measure (CRM)* and *user-relevant-measure (URM)*.

$$URM = 1 - \sum_{u=1}^{n} \frac{t_u(A \rightarrow B)}{\sum_{j=1}^{n} t_j(A \rightarrow B)} \log_2 \frac{t_u(A \rightarrow B)}{\sum_{j=1}^{n} t_j(A \rightarrow B)} \qquad (1)$$

where u denotes an user-id and $t_j(A \rightarrow B)$ represents count of transactions containing rule $(A \rightarrow B)$ in the mobility trace of user j. Similarly, *category-relevant-measure (CRM)* is defined on user-group or categorical mobility summaries.

$$CRM = 1 - \sum_{c=1}^{|C|} \frac{t_c(A \rightarrow B)}{|c| \times t_c} \log_2 \frac{t_c(A \rightarrow B)}{|c| \times t_c} \qquad (2)$$

Typically, CRM quantifies a rule applicability to a particular user-category, i.e., a higher value of CRM denotes only a few categories follow the mobility-rule, while a less CRM value depicts the rule is applicable for most of the user-categories. Similarly, a high value of URM represents only few individuals' mobility behaviour is represented by it.

3   Experimental Observations

DataSet: We demonstrate our approach using a real-life GPS traces of mobile-users of IIT Kharagpur Campus, an academic region. We collected 6 months GPS dataset from 56 volunteers (specifically 8 categories of users) from their mobile-GPS sensor and *GoogleMap Timeline*. Reverse geo-coding technique is used to extract the POIs of the region and each of the stay-points are geo-tagged.

Extracting Mobility Rules and Notations: Table 2 depicts few extracted association rules (TimeFeatured and Categorical) from the collected GPS traces. Rules are represented by following notations: $StayDuration(High, S, w = 1)$ implies stayduration of a movement-transaction is high at staypoint (S), where $w = 1$ and $w = T$ indicate same transaction and transactions within the same time-interval respectively. $TimeStamp(1)$ implies time-interval 1, i.e., [0600–1200]. $FootprintFrequency(f, P)$: Footprint frequency (f) at P POI(s). $countPlacePOI(a)$: The count of unique POIs(leaf nodes of $POI_{taxonomy}$) visited by the users in a time-interval is a. $User(f, P)$: Visit frequency (f) of users at P POI(s). $Category(c)$: c is a user-category of user-category taxonomy (C) [Fig. 2].

Results and Discussion: Table 2 depicts the *Support (S)*, *Confidence (C)*, *J-measure (J)*, *Mutual Information (M)*, *Category-relevant-measure (CRM)* and *User-relevant-measure (URM)* of each rule. Clearly, rules with higher URM are useful to capture individuals' mobility patterns and could be used for personalized mobility pattern mining and higher CRM-rules are useful for group-mobility pattern mining. It is worth noticing that none of the existing measures are capable to distinguish between these rules. Moreover, a simple support-pruning may eliminate personalized mobility-pattern rules as they might have a less support values. For example, rules r_4 has low support value as the number of participant of the category is less, however higher CRM value indicates meaningful patterns extracted from the GPS traces.

A rule is considered to be *interesting*, "if $A \rightarrow B$ is strong then $A \rightarrow \bar{B}$ must be a weak rule." The 'goodness of fit' between the rule hypothesis and data is measured by *J-measure*. While *mutual-information* depicts the average information shared by antecedent and consequent parts of the rule, quadratic entropy decrease is measured by *gini-index* [5]. From the experimental dataset, it is observed that rules R_1, R_2 have higher J values and average G, M values indicating higher information relative to the truth of the antecedent part.

Figure 4(b) shows a graph of running time of the algorithm (in sec) against the number of transactions. It clearly shows the improvement of running-time using the proposed data structure to capture inherent patterns. It is observed that time-featured rules are more time-intensive than categorical-rules. The reason is significant time is required to search the complete transactional log instead of only the summarized categorical patterns.

Table 2. Few examples of extracted TimeFeatured ($R_1 - R_5$) and categorical mobility-rules ($r_1 - r_5$)

Rule Id	Rule representation
R_1	EdgeDuration (High) $\wedge$ EdgeDistance (High) $\Rightarrow$ StayDuration (High,$S, w = 1$)
	S = 0.68, C = 0.85, J = 0.91, M = 0.67, G = 0.56, URM = 0.34, CRM = 0.39
R_2	StayDuration (High,S)$\wedge$ EdgeDuration (High) $\wedge$ EdgeDistance (High) $\Rightarrow$ StayDuration (Low,$S, w = T$)
	S = 0.58, C = 0.81, J = 0.92, M = 0.64, G = 0.61, URM = 0.38, CRM = 0.42
R_3	TimeStamp (1, 3) $\Rightarrow$ FootprintFrequency (High, {AcademicComplex, Transit, Department})
	S = 0.51, C = 0.75, J = 0.67, M = 0.72, G = 0.68, URM = 0.54, CRM = 0.59
R_4	TimeStamp (1, 4) $\wedge$ StayDuration (High) $\Rightarrow$ countPlacePOI (Low), countPlaceName (High)
	S = 0.52, C = 0.84, J = 0.61, M = 0.56, G = 0.52, URM = 0.56, CRM = 0.51
R_5	TimeStamp (3) $\Rightarrow$ StayDuration (Low) $\wedge$ Transit (High)
	S = 0.48, C = 0.67, J = 0.78, M = 0.66, G = 0.59, URM = 0.51, CRM = 0.56
r_1	User (High, HealthCare Center) $\Rightarrow$ User (Low, {Hospital, MedicineShop})
	S = 0.51, C = 0.87, J = 0.83, M = 0.87, G = 0.85, URM = 0.78, CRM = 0.51
r_2	User (High, {Cafe, Restaurant, Transit}) $\Rightarrow$ User (Low, {AcademicBuilding, ClassRoomComplex, Department})
	S = 0.78, C = 0.88, J = 0.78, M = 0.84, G = 0.78, URM = 0.65, CRM = 0.89
r_3	User ({High, Transit}, {Medium, Bank, ATM }) $\Rightarrow$ User (High, {MarketPlace, Store})
	S = 0.65, C = 0.84, J = 0.83, M = 0.75, G = 0.71, URM = 0.51, CRM = 0.57
r_4	TimeStamp (1,3) $\wedge$ Category (Faculty) $\Rightarrow$ User (High, {ResidentialArea, MarketPlace, Store, Transit})
	S = 0.32, C = 0.94, J = 0.51, M = 0.43, G = 0.41, URM = 0.87, CRM = 0.95
r_5	Category (Student, Residential, UnderGraduate) $\Rightarrow$ User ({High, {Library, Cafe, HealthCareCenter, AcademicComplex}}, {Medium, {MarketPlace, Store, CommercialPlace}})
	S = 0.42, C = 0.87, J = 0.79, M = 0.67, G = 0.64, URM = 0.78, CRM = 0.89

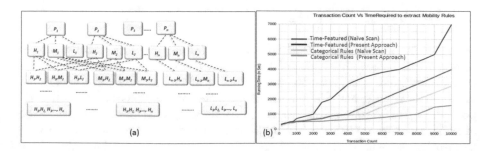

Fig. 4. (a) Lattice structure of place-visit frequency (b) Running time comparison

The experimental findings to explore human movement behaviour have following significances:

- The extracted mobility-rules (Table 2) with high *support, confidence* values demonstrate the potential of the proposed mobility rule mining framework to extract meaningful rules from accumulated GPS log. It has also been shown that although a mobility-rule may have lesser support value, it captures inherent patterns of individuals' mobility behaviour. This is a direct consequence of individuals' unique movement behaviour.
- It also turns out that mobility rules are highly dependent on timestamp values, i.e., people largely follow regular mobility patterns on the same time-intervals [Rules: R_3, r_4]. For example, footprint-frequency is high at marketplace in evening and medium in morning while cafe, restaurants are crowded at evening. These may provide important insights about aggregated footprint density on different POIs and may be useful for resource-allocation.
- People also follow regular spatio-temporal patterns in their daily movement summary. Rule R_4 indicates people generally spends a long duration at specific places (POIs) in time-interval 1 [0600–1200] and time-interval 4 [2100-0600], while at time-interval 3 [1600–2100], low stay-duration and more transit-points (i.e., travelling) have been observed [Rule R_5].
- Mining mobility traces also provide interesting movement behaviour of people. For example, R_1 indicates people generally stays a longer duration in a POI after travelling a long distance. Also, it has been observed that a long-duration stay at a place is generally followed by short-duration stop within same time-interval $w = T$ [Rule R_2]. Again *high* visit at *Cafe or Hangout-spots* may be a reason of low footprints at *Acaemic-Complex, Department etc.* [Rule r_2].

In summary, our framework is capable to extract user-specific and categorical mobility-rules depending on several key factors, namely, time-intervals, POIs, stay-point duration etc. Further, data-preparation including GPS transactional log management reduces the time complexity of the procedure which is a major challenge in spatio-temporal data mining.

4   Concluding Remarks

In this work, we proposed a novel approach for mining mobility association rules of movement behaviour of people. The proposed framework extracts movement behavioural rules from accumulated GPS log of mobile users. An experimental evaluation on a real-life dataset of an academic campus demonstrates the potential of the framework to extract meaningful rules. Discovering reasonable mobility rules from GPS log is a major contribution of this work. We strongly believe the present work will act as a foundation of association rule mining framework from GPS trajectories. In the future, we also aim to assimilate other contextual information such as, weather information, traffic information and extend the present framework to extract interesting and meaningful rules from heterogeneous spatio-temporal data-set.

References

1. Agrawal, R., Imieliński, T., Swami, A.: Mining association rules between sets of items in large databases. ACM SIGMOD Rec. **22**(2), 207–216 (1993)
2. Ghosh, S., Ghosh, S.K.: THUMP: semantic analysis on trajectory traces to explore human movement pattern. In: Proceedings of the 25th International Conference Companion on World Wide Web. International World Wide Web Conferences Steering Committee, pp. 35-36 (2016)
3. Ghosh, S., Ghosh, S.K.: Modeling of human movement behavioral knowledge from GPS traces for categorizing mobile users. In: Proceedings of the 26th International Conference on World Wide Web Companion. International World Wide Web Conferences Steering Committee, pp. 51-58 (2017)
4. Czibula, G., Marian, Z., Czibula, I.G.: Software defect prediction using relational association rule mining. Inf. Sci. **264**, 260–278 (2014)
5. Geng, L., Hamilton, H.J.: Interestingness measures for data mining: A survey. ACM Comput. Surv. (CSUR) **38**(3), 1–32 (2006). Art. No. 9
6. González, M.C., Hidalgo, C.A., Barabási, A.-L.: Understanding individual human mobility patterns. Nature **453**(7196), 779–782 (2008)
7. Zheng, Y.: Trajectory data mining: an overview. ACM Trans. Intell. Syst. Technol. (TIST) **6**(3), 1–41 (2015). Art. No. 29
8. Nahar, J., Imam, T., Tickle, K.S., Chen, Y.P.P.: Association rule mining to detect factors which contribute to heart disease in males and females. Expert Syst. Appl. **40**(4), 1086–1093 (2013)
9. Wang, C.-S., Lee, A.J.T.: Mining inter-sequence patterns. Expert Syst. Appl. **36**(4), 8649–8658 (2009)
10. Roddick, J.F., Spiliopoulou, M.: A survey of temporal knowledge discovery paradigms and methods. IEEE Trans. Knowl Data Eng. **14**(4), 750–767 (2002)
11. Tew, C., Giraud-Carrier, C., Tanner, K., Burton, S.: Behavior-based clustering and analysis of interestingness measures for association rule mining. Data Min. Knowl. Disc. **28**(4), 1004–1045 (2014)

Image Sentiment Analysis Using Convolutional Neural Network

Akshi Kumar[ID] and Arunima Jaiswal[✉]

Delhi Technological University, Delhi, India
akshikumar@dce.ac.in, arunimajaiswal@gmail.com

Abstract. Visual media is one of the most powerful channel for expressing emotions and sentiments. Social media users are gradually using multimedia like images, videos etc. for expressing their opinions, views and experiences. Sentiment analysis of this vast user generated visual content can aid in better and improved extraction of user sentiments. This motivated us to focus on determining 'image sentiment analyses'. Significant advancement has been made in this area, however, there is lot more to focus on visual sentiment analysis using deep learning techniques. In our study, we aim to design a visual sentiment framework using a convolutional neural network. For experimentation, we employ the use of Flickr images for training purposes and Twitter images for testing purposes. The results depict that the proposed 'visual sentiment framework using convolutional neural network' shows improved performance for analyzing the sentiments associated with the images.

Keywords: Sentiment analysis · Deep learning · Convolutional neural network

1 Introduction

These days due to abundant volume of opinion rich web data accessible via Internet, a large portion of recent research is going on in the area of web mining called as Sentiment Analysis [1]. It is defined as the process of computationally classifying and categorizing sentiments expressed in a piece of 'multimedia web data' (both textual or non-textual), especially in order to determine the polarity of the writer's attitude towards a particular topic, product, etc. [2] It is a way to evaluate the written or spoken language in order to determine the degree of the expression, whether it is favorable, unfavorable, or neutral [3]. Also, referred to as Opinion Mining, the ideology of sentiment analysis is to search for opinions, identify the sentiments involved in it and classify it based on the polarity. Huge measure of heterogeneous information is produced by the users of the social media via Internet which is analyzed for efficient decision making. Era of Internet has drastically modified the way of expressing of views, opinions, sentiments etc. of an individual [3]. It is essentially done via blogs, online reviews, forums, social media, feedbacks or surveys etc. These days, people are more dependent on the usage of social networking sites like Twitter, Facebook, Flickr, Instagram etc. for appropriate decision making, to share their opinions and views which in turn is generating enormous volume of 'sentiment rich data' that is often expressed in the form of texts, images, audios, videos,

© Springer International Publishing AG, part of Springer Nature 2018
A. Abraham et al. (Eds.): ISDA 2017, AISC 736, pp. 464–473, 2018.
https://doi.org/10.1007/978-3-319-76348-4_45

mixture of images and texts etc. So we can say that the masses is relying on such (online) user generated multimedia web content for the opinions etc. Images being a part of 'multimedia web data' helps to convey, express, communicate, comprehend, illustrate and carry's different level of people's opinions or sentiments to their viewers that extensively marks the increasing significance of image sentiment analysis or image sentiment prediction.

Scrutinization of this 'multimedia web data' especially sentiment analysis within textual and visual data promised to a better understanding of the human behavior as they convey the emotions and the opinions more clearly. Till date, most of the 'sentiment analysis computation' has covered the textual content only. Although, understanding the emotions and sentiments of visual media contents has attracted increasing attention in research and practical applications. However, little progress has been made in determining and estimating emotions & sentiments of 'visual user generated online content'. It is thus emerging as a recent area of research and there is a huge scope of exploring sentiments of visual multimedia content. The main motivation of this work is the explosive growth of social media and online visual content that has encouraged the research on 'large-scale social multimedia analyses'. Therefore, we aim to choose it as our problem statement for analyzing sentiments associated with the images using deep learning algorithm like convolutional neural network. Multimedia messages including videos and images encapsulating strong sentiments can strengthen the sentiment or the opinion conveyed in the content and thus influencing the audience more effectively [12]. Understanding the opinions or the sentiments expressed in visual content will significantly profit social media communication and facilitate broad applications in education, finance, advertisement, entertainment, health etc. [13] We can say that the sentiments (both textual and non-textual) of the social media is broadly influencing the thoughts and views of the public, like U.S. economy and stock market situations gets influenced by the changing sentiments of the Twitter users.

We aim to explore the application of deep learning algorithm like convolutional neural network to visual media for determining its sentiments accurately. The major problem arises in situations where we have conflicting emotions being shown by the image and the text, and thus it necessitates the requirement of a proper framework for determining the sentiments of any visual media like images etc.

Rest of the paper is organized as follows. Section 2 discusses the related work in the area of image sentiment analysis. Section 3 explains the fundamental concept of deep learning algorithm like Convolutional Neural Networks. Section 4 describe the data set collection, system architecture and the experimentation. Section 5 briefs about the experimental results. Section 6 finally concludes the paper.

2   Related Work

Most of the work in the past literature on sentiment analysis has majorly focused on text analysis [4–7]. Sentiment based models have been demonstrated to be beneficial in various analytical uses like human behavior prediction, business, and political science

[8, 22–24]. In comparison to "text-based" opinion mining or sentiment analysis, modeling of the sentiments based on the images has been much less studied.

One of the effort given by [9] suggested to design a "large scale visual sentiment ontology" based on 'Adjective Noun Pairs (ANP)'. Borth et al. [10] proposed a more tractable approach that models sentiment related visual concepts as a 'mid-level representation' to fill the gaps. These concepts include 'ANPs', such as "happy dog" & "beautiful sky", that merge the sentimental strength of "adjectives" and "detectability" of the nouns. Although such ANP concepts do not directly express emotions or sentiments, they were learned based on the strong' co-occurrence relationships with emotion tags' of web photos, and thus are valuable as effective statistical cues for detecting emotions depicted in the images.

Study by Krizhevsky et al. [11] focused on the training of a 'deep convolutional neural network' for classifying 1.2 million 'high resolution images' in the ImageNet 'LSVRC-2010 contest' and had obtained improved results. Author [12] had proposed a novel framework for visual sentiment prediction of images using Deep Convolutional Neural Networks and does the experimentation on the data obtained from two famous microblogs, Twitter and Tumblr. Jindal et al. [13] discusses about the applicability of an image sentiment prediction framework using Convolutional Neural Networks for Flickr image dataset. Author [14] had designed a framework for image sentiment analysis of Flickr images using CNN. They had also developed new strategies to tackle the noisy nature of large scale image samples taken. Their results show the improved performance of CNN. Cai et al. [15] briefs about the applicability of CNN for learning both the textual as well as visual features for determining sentiment analysis using Twitter dataset. Their study depicts that the combination of both the text and the images showed improved results. The work done by [16] shows the exploration and utilization of hyper parameters from a very deep CNN network for analyzing image sentiments drawn on Twitter dataset. The results claim that their model exhibits improved results.

3   Deep Learning Using Convolutional Neural Network

Deep learning (DL) algorithm was earlier proposed by G.E. Hinton in 2006 and is the part of machine learning process which refers to Deep Neural Network [17]. Neural network (NN) works just like our human brains, comprising of numerous neurons that make an impressive network. DL is a group of networks containing lots of other algorithms like Convolutional Neural Networks (CNN), Recurrent Neural Networks, Recursive Neural Networks, Deep Belief Networks etc. NN are very advantageous in text generation, word representation estimation, vector representation, sentence modeling, sentence classification, and feature presentation [18]. The application of DL algorithms is increasing enormously because of three prime reasons, i.e., enhanced abilities of chip processing, comprehensively lesser expenditure of hardware & noteworthy improvements in machine learning algorithms [19].

In our study, we aim to focus on the use of CNNs. A "CNN" is a type of "feed-forward" artificial NN where the individual neurons are lined in such a way that they respond to 'overlapping regions' in the 'visual field'. CNNs were inspired by 'biological processes' (the connectivity pattern between its neurons is inspired by the organization of the animal visual cortex) and are variations of multilayer perceptron (MLP), that are designed to use slight amounts of pre-processing. They comprise of several layers of receptive fields [20]. These are small neuron collections which process portions of the input image. The outputs of these collections are then lined or tiled so that their input regions overlap, in order to attain a higher-resolution representation of the original image; & this process is then repeated for every such layer [13]. Tiling permits CNNs to endure translation of the input image [12, 13].

There are 4 key operations in the ConvNet:

3.1 Convolution

ConvNets derive their name from the operator "convolution". The major motive of Convolution (of a ConvNet) is to extract features from the image taken as input.

3.2 Non Linearity (ReLU)

ReLU stands for Rectified Linear Unit and is a non-linear operation. ReLU is an element wise operation (applied per pixel) and swaps all negative pixel values in the feature map by 0. The prime intend of ReLU is to announce non-linearity in our ConvNet.

3.3 Pooling or Sub Sampling

Spatial Pooling reduces the dimensionality of each feature map but retains the most important information. Spatial Pooling can be of different types: Max, Average, Sum etc.

3.4 Classification (Fully Connected Layer)

The "Fully Connected" layer is a traditional MLP that uses a 'softmax' activation function in the output layer. The term "Fully Connected" denotes that each neuron in the previous layer is connected to every single neuron in the next layer. The output deriving from the 'convolutional & pooling layers' represents the high-level features of the input image. The prime aspect of the Fully Connected layer is to use these features in order to classify the input image into several classes based on the training dataset.

In our study, we had employed the use of CNN, where training is done by the back-propagation method. The Convolution & Pooling layers serve as Feature Extractors from the input image whereas the Fully Connected layer behaves as a classifier. Training the CNN signifies the optimization of all the weights & parameters for correctly classifying images from the training set. Whenever a new input image arrives into the CNN, it undergoes the propagation step and a probability for each of the class is obtained as output. If the training set is huge enough in size, then the network will generalize well to the new images and will eventually classify them into correct categories.

4   System Architecture and Experimentation

We had implemented the CNN using a deep learning framework called as "Caffe" [21]. It is developed by 'Berkeley AI Research (BAIR) and by community contributors'. It is under active development by the Berkeley Vision and Learning Center (BVLC).

Our proposed system intends to classify sentiment of an image based on generation of Adjective Noun Pairs (ANP's) [9, 10]. The reason why ANP's are generated for this purpose is that ANP's can easily be correlated with the sentiments. The ANP's are generated with the help of a Convolutional Neural Net that has been trained rigorously for this particular labelling task. Image recognition and labelling through CNN are topics which have been extensively researched upon. Image labelling for sentiment analysis is an approach which is in its nascent state. We aim to develop this approach into a novel method of image sentiment classification. The system can broadly be categorized into two phases (as shown in Fig. 1), the ANP generation phase and the sentiment predictor phase. The ANP generation phase is mainly comprised of the neural net while the sentiment predictor phase is comprised of a Support Vector Machine. Additional work has also been done to analyze the ANP's generated through textual sentiment analysis libraries and the results were compared with the sentiment predicted through SVM.

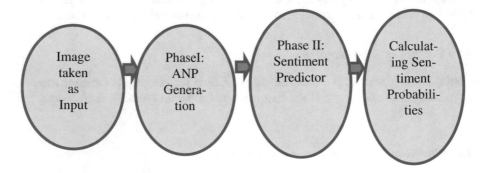

Fig. 1.   Generating sentiment probabilities

Phase I takes an image as an input and outputs the probability distribution regarding the sentiment for that image (positive, negative or neutral). The image is pre-processed to a specific resolution and is then passed through multiple layers of the neural net (ReLu, Pooling, Fully Connected and Convolutional). ANP's generated in the first phase serves as an input to the Phase II which thus outputs the sentiment probabilities. This step can be performed by two methods. The first is sentiment prediction through the use of Support Vector Machine (SVM) and the second is sentiment prediction through textual analysis of ANP's. In our study, we had implemented both the methods in order to compare them so as to know which one is best suited for the purpose of sentiment classification. Prime focus was on the generation of ANP's that closely represent the sentiment of the image. Support Vector Machine was selected for the classification task as it works well in scenarios where the number of input features are high. In our work, the features are the ANP's generated in the first phase (number of input features are

2089). The output has the sentiment probabilities. SVM performs a quite well in predicting sentiments on the basis of the ANP's. Our main aim was to generate labels or ANP's that are closely related to the sentiments. Here, a string of the top ANP's was passed through a textual sentiment analyzer and the corresponding sentiment probabilities were obtained.

Our proposed framework is divided into five modules. The first module takes input as an image from the user. The output of each of the module serves as an input for the next module. The last module outputs the desired results. Each module performs a specific task ranging from preprocessing of the dataset to the sentiment prediction. This modular approach helps in building a better system that even simplifies the system designing processing. Our proposed framework has been depicted in Fig. 2.

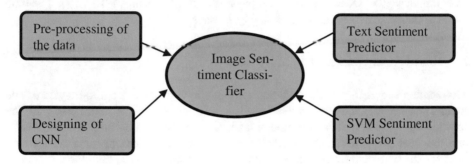

Fig. 2. Proposed visual sentiment framework

Data preprocessing involves processing of the dataset. Our system makes use of set of two databases. The Flickr database is used for training and testing the convolutional neural network or the ANP classifier. The twitter database is used for training the support vector machine and also for testing both the SVM and text sentiment predictors. It consists of 800 images and their corresponding tweets. The images have been labelled with their corresponding sentiment. For each ANP, at most 1,000 images tagged with it were downloaded that resulted into approximately one million images for 3,316 ANPs. To train the visual sentiment concept or ANP classifiers, we first filter out the ANPs associated with less than 120 images. Consequently, 2,089 ANPs with 867,919 images were left after filtration process. For each ANP, 20 images were randomly selected for testing purposes, while others were used for training purposes, ensuring at least 100 training images per ANP. The ANP tags from Flickr users were used as labels for each image. All the images, whether used for training or testing were normalized to 256×256 without keeping the aspect ratio. The 'python pandas' library was used for shaping, merging, re-shaping and slicing the datasets. The final dataset consists of approximately 800,000 images for the proposed 2089 ANP's. The CNN for ANP detection has been built through the 'Caffe' deep learning framework. We had used the python bindings of Caffe to build this network that comprises of eight main layers. Five of them were convolutional and the other three were fully connected. The output of the CNN is thus used as input for the SVM. The SVM is implemented using the 'scikit-learn python library'. The text sentiment analyzer had been implemented in python using the natural language toolkit NLTK. This toolkit is used to extract features from the text. The twitter dataset has been used for testing and training the text sentiment predictor. In the system implementation, the top ANP's generated were formed into a string. The string is constructed as follows. For all the top ANP's, the rank/probability of the ANP is normalized to an integer greater than 1. The adjective of the ANP is added to the string 'x' number of times, where 'x' is the weight of the ANP. The noun is added as it is for each ANP. The final string is passed through the text sentiment predictor and the sentiment probabilities were obtained.

All the training, testing, and experimentation was done on a Macbook (Processor – 2.7 GHz dual-core Intel Core i5, Turbo Boost up to 3.1 GHz, 3 MB shared L3 cache, 256 GB PCIe-based onboard SSD, 8 GB of 1867 MHz LPDDR3 onboard memory, Intel Iris Graphics 6100 1536 MB, Operating system – macOS Sierra v 10.12.2). The time taken to train the model was approximately 13 days.

5 Result Analysis and Conclusion

This section shows the results obtained after passing images through our system. The images have been selected to cover a diverse range of sentiments in order to push the limits of our system. It shows the input image, the mid-level representation of the ANP's and the final graph obtained on executing the system, comprising of a comparison between the two techniques used for sentiment prediction and the final label produced for the particular image. We test our system on three types of data. The first comprised only the text i.e. the tweets, the second contains only the images associated with those tweets and the third one consists of the combination of the tweets and their corresponding images. We observed that the third method gives us the most desirable results as shown in Fig. 3. We have used accuracy as a parameter for classifying images/texts. Accuracy

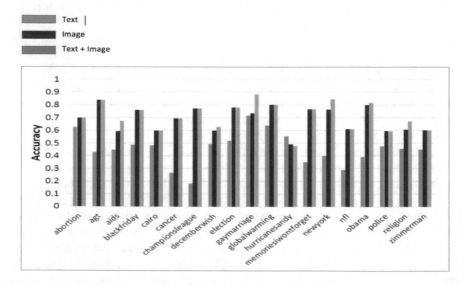

Fig. 3. Comparison between text, image, and text + image

is defined as the percentage of images/text that have been correctly labelled by the system. In our context, it is the number of images out of the 800 twitter images/text that have been correctly labelled.

In this study, we intend to propose a visual sentiment concept classification model based on deep convolutional neural networks. The deep CNNs model is trained using 'Caffe'. We did the comparative performance analysis of applying basic text analysis to that of SVM on the generated Adjective Noun Pairs (ANPs) to find out that the ANPs when passed through SVM, yields better results for visual sentiment classification. Another comparison was made pertaining to the results of textual sentiment analysis, visual sentiment analysis, and textual + visual sentiment analysis to explore the method that generates the most desirable results for correctly classifying sentiments. Performance evaluation shows that the newly trained deep CNNs model works significantly better for both annotation and retrieval in comparison to other shallow models that employed the use of an independent binary SVM classification models etc.

6 Conclusion

Visual sentiment analysis is thus a very challenging task although it is still in its infancy. Earlier works majorly focused on the textual sentiment analysis. In our study, we aim to focus on the picture sentiment analysis using convolutional neural networks. We had also used the concept of averaging for generating sentiment probabilities and it could possibly serve as a trivial approach and might not ensure most optimized and accurate results. Thus other approaches could also be used that would yield more optimized solutions for producing more accurate results. The experimental results depicts that the deep learning CNN model when properly trained can give better results and thus could be used as a variant for outperforming the challenging problem of visual multimedia mining. We can also incorporate the concept localization into the deep CNNs model, and improve network structure by leveraging concept relations. Furthermore, we can also apply other deep learning models to many other application domains using more data sources for detecting emotion analysis and predicting visual sentiment analysis.

References

1. Pang, B., Lee, L.: Opinion mining and sentiment analysis. Found. Trends Inf. Retr. 2(1–2), 1–135 (2008)
2. Liu, B.: Sentiment Analysis: Mining Opinions, Sentiments, and Emotions. Cambridge University Press, Chicago (2015)
3. Kumar, A., Teeja, M.S.: Sentiment analysis: A perspective on its past, present and future. Int. J. Intell. Syst. Appl. 4(10), 1–14 (2012)
4. Esuli, A., Sebastiani, F.: SentiWordnet: a publicly available lexical resource for opinion mining. In: Proceedings of LREC (2006)
5. Thelwall, M., Buckley, K., Paltoglou, G., Cai, D., Kappas, A.: Sentiment strength detection in short informal text. J. Am. Soc. Inf. Sci. Technol. 62(2), 419–442 (2011)

6. Vonikakis, V., Winkler, S.: Emotion-based sequence of family photos. In: Proceedings of the 20th ACM International Conference on Multimedia, pp. 1371–1372. ACM, Japan (2012)

7. Yanulevskaya, V., Uijlings, J., Bruni, E., Sartori, A., Zamboni, E., Bacci, F., Sebe, N.: In the eye of the beholder: employing statistical analysis and eye tracking for analyzing abstract paintings. In: Proceedings of the 20th ACM International Conference on Multimedia, pp. 349–358. ACM, Japan (2012)

8. Wang, X., Jia, J., Hu, P., Wu, S., Tang, J., Cai, L.: Understanding the emotional impact of images. In: Proceedings of the 20th ACM International Conference on Multimedia, pp. 1369–1370. ACM, Japan (2012)

9. Aradhye, H., Toderici, G., Yagnik, J.: Video2text: Learning to annotate video content. In: IEEE International Conference on Data Mining Workshops, ICDMW 2009, pp. 144–151. IEEE, USA (2009)

10. Borth, D., Ji, R., Chen, T., Breuel, T., Chang, S. F.: Large-scale visual sentiment ontology and detectors using adjective noun pairs. In: Proceedings of the 21st ACM International Conference on Multimedia, pp. 223–232. ACM, Spain (2013)

11. Krizhevsky, A., Sutskever, I., Hinton, G.E.: Imagenet classification with deep convolutional neural networks. In: Advances in neural information processing systems, pp. 1097–1105. ACM, USA (2012)

12. Xu, C., Cetintas, S., Lee, K. C., Li, L. J.: Visual sentiment prediction with deep convolutional neural networks. arXiv preprint arXiv:1411.5731 (2014)

13. Jindal, S., Singh, S.: Image sentiment analysis using deep convolutional neural networks with domain specific fine tuning. In: 2015 International Conference on Information Processing (ICIP), pp. 447–451. IEEE, India (2015)

14. You, Q., Luo, J., Jin, H., Yang, J.: Robust image sentiment analysis using progressively trained and domain transferred deep networks. In: AAAI, pp. 381–388. ACM, USA (2015)

15. Cai, G., Xia, B.: Convolutional neural networks for multimedia sentiment analysis. In: Natural Language Processing and Chinese Computing, pp. 159–167. Springer, Cham (2015)

16. Islam, J., Zhang, Y.: Visual Sentiment Analysis for Social Images Using Transfer Learning Approach. In: IEEE International Conferences on Big Data and Cloud Computing (BDCloud), Social Computing and Networking (SocialCom), Sustainable Computing and Communications (SustainCom) (BDCloud-SocialCom-SustainCom), pp. 124–130. IEEE, USA (2016)

17. Deng, J., Dong, W., Socher, R., Li, L. J., Li, K., Fei-Fei, L. Imagenet: A large-scale hierarchical image database. In: Computer Vision and Pattern Recognition, CVPR 2009, pp. 248–255. IEEE, USA (2009)

18. Zhang, Y., Er, M. J., Venkatesan, R., Wang, N., Pratama, M.: Sentiment classification using comprehensive attention recurrent models. In: Neural Networks (IJCNN), pp. 1562–1569. IEEE, Canada (2016)

19. Jiang, Y. G., Ye, G., Chang, S. F., Ellis, D., Loui, A. C.: Consumer video understanding: A benchmark database and an evaluation of human and machine performance. In: Proceedings of the 1st ACM International Conference on Multimedia Retrieval, p. 29. ACM, Italy (2011)

20. Ain, Q.T., Ali, M., Riaz, A., Noureen, A., Kamran, M., Hayat, B., Rehman, A.: Sentiment analysis using deep learning techniques: a review. Int. J. Adv. Comput. Sci. Appl. 8(6), 424–433 (2017)

21. Jindal, S., Singh, S.: Image sentiment analysis using deep convolutional neural networks with domain specific fine tuning. In: Information Processing (ICIP), pp. 447–451. IEEE, India (2015)

22. Kumar, A., Khorwal, R., Chaudhary, S.: A survey on sentiment analysis using swarm intelligence. Indian J. Sci. Technol. 9(39), 1–7 (2016)

23. Kumar, A., Sebastian, T.M.: Sentiment analysis on twitter. Int. J. Comput. Sci. Issues **9**(4), 372–378 (2012)

24. Kumar, A., Sebastian, T. M.: Machine learning assisted sentiment analysis. In: Proceedings of International Conference on Computer Science & Engineering (ICCSE 2012), pp. 123–130. IAENG, UAE (2012)

Cluster Based Approaches for Keyframe Selection in Natural Flower Videos

D. S. Guru[1(✉)], V. K. Jyothi[1(✉)], and Y. H. Sharath Kumar[2(✉)]

[1] Department of Studies in Computer Science, University of Mysore,
Manasagangotri, Mysore 570006, India
dsg@compsci.uni-mysore.ac.in, jyothivk.mca@gmail.com
[2] Department of Information Science, Maharaja Institute of Technology
Mysore (MITM), Mandya 571438, India
sharathyhk@gmail.com

Abstract. The selection of representative keyframes from a natural flower video is an important task in archival and retrieval of flower videos. In this paper, we propose an algorithmic model for automatic selection of keyframes from a natural flower video. The proposed model consists of two alternative methods for keyframe selection. In the first method, K-means clustering is applied to the frames of a given video using color, gradient, texture and entropy features. Then the cluster centroids are considered to be the keyframes. In the second method, the frames are initially clustered through Gaussian Mixture Model (GMM) using entropy features and the K-means clustering is applied on the resultant clusters to obtain keyframes. Among the two different sets of keyframes generated by two alternative methods, the one with a high fidelity value is chosen as the final set of keyframes for the video. Experimentation has been conducted on our own dataset. It is observed that the proposed model is efficient in generating all possible keyframes of a given flower video.

Keywords: Keyframe selection · Clustering · Gaussian Mixture Model
K-means clustering · Retrieval of flower videos

1 Introduction

Due to the ease of availability of mobile technology and storage media, users can easily capture and store a large number of videos. Designing and developing an efficient video archival and retrieval system is an important area of research as it finds tremendous applications both in general purpose and domain specific searching. Development of an efficient archival and retrieval system for flower videos is one such domain with many applications.

Applications for retrieval of flower videos can be found useful in floriculture; flower searching for patent analysis, users interest in knowing the flower names for decoration etc. With the increase in the demand for flowers, the floriculture has become one of the important commercial trades in agriculture [8]. Floriculture industry involves flower trade, nursery, seed production and extraction of oils from flowers [8]. In such cases, automation of video flower recognition and retrieval is very essential. Flower recognition is also used for searching patent flower videos to know if a flower video is already present in the database [5].

Processing all frames of a flower video incurs unnecessary computational burden. To avoid the computational burden, we need to select some of the frames containing necessary information of the video to help the users to browse and retrieve videos of their interest easily and effectively. We can call this process as a keyframe selection or video summarization [14]. Keyframe selection is one of the important areas of research in video retrieval.

Developing a system for keyframe selection from flower videos captured is a difficult task when the flower videos are captured in a real environment, that too in natural outdoor scenes, where the lighting condition varies with the weather and time, with variation in viewpoint, occlusions and in different scale of flowers in videos [9].

2 Related Works

Generally, keyframe selection techniques can be categorized into sequential and cluster based methods [13]. With sequential methods, as the name implies, consecutive frames are compared in a sequential manner [2]. In cluster-based methods, the frames are grouped into a finite set of clusters [20]. Zhou et al. [21] proposed a method for video summarization using audio-visual features. Clustering the frames can also be done by using Delaunay triangulation [16]. Keyframes in a shot are extracted using the average histogram algorithm [18]. A video is summarized based on color feature extraction from video frames and applied k-means clustering algorithm in [1]. A method for keyframe selection using dynamic Delaunay graph clustering through an iterative edge pruning strategy proposed by [17]. Loannidis et al. [12] proposed weighted multi view clustering algorithm based on convex mixture models to extract keyframes.

Keyframe selection with sequential approaches through thresholding may not generate all possible keyframes with challenges like occluded, different angled, varying appearance of flowers in frames. Clustering approaches can overcome this drawback. From literature, we found that there is no attempt on the selection of keyframes from natural flower videos. In the proposed work keyframes are selected based on the similarity of the fusion of clustering of color, texture and gradient features for the application of classification and retrieval of flower videos. The proposed methods reduce the redundancy to a greater extent. In many realistic scenarios, one may need to retrieve flower videos in addition to images when a flower video is given as a query.

The paper is organized as follows, in Sect. 3 we reveal two different methods to select keyframes with different features and clustering techniques used to design our proposed methods and also, we reveal the block diagram of the proposed work. The details of the datasets used in experimentation are given in Sect. 4. Experimentation and results are tabulated in Sect. 5 and finally the work is concluded in Sect. 6.

3   Proposed Model

We propose two different methods to illustrate the comparative study and the best method to select keyframes among two methods. Each method consists of four stages preprocessing, feature extraction, clustering and selection of keyframes. Figure 1 depicts the block diagram of the methods of our proposed work.

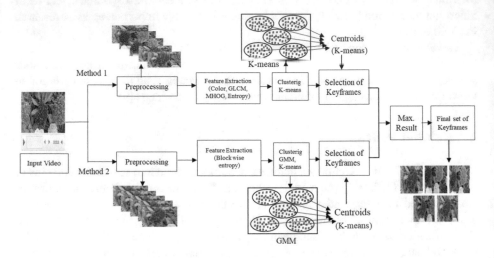

Fig. 1.   Block diagram of the proposed methods 1 and 2.

Following are the two proposed methods for keyframe selection namely,

1. Features based clustering for keyframe selection.
2. Entropy-GMM-Kmeans based keyframe selection.

3.1   Features Based Clustering for Keyframe Selection (FCK)

This method involves four stages to select keyframes, namely, preprocessing, feature extraction, clustering and selection of keyframes, the stages are explained below.

3.1.1   Preprocessing

Given is a video V_i which contains 'n' number of frames and it can be stated as,

$$V_i = \{f_1, f_2, f_3 \cdots f_n\} \tag{1}$$

Preprocessing generates on an average of 30 frames per second and resizes the frames 250×250 for further processing.

3.1.2   Feature Extraction

Different features viz., color, entropy, GLCM and Modified Histogram of Oriented Gradient (MHOG) are extracted to select most discriminating frames from a flower video V_i.

A. Color features

Color is the most expressive of all the visual features [19]. Flowers in video frames are often surrounded by greenery in the background. The proposed method extracts Red and Blue color components excluding the green color component since our concentration is on flower region. After extracting the color components Red and Blue, calculate the mean and standard deviation of histogram of red component and blue component are defined as

$$
\mu\big(H\big(F_i(R)\big)\big) \quad \text{and} \quad \sigma\big(H\big(F_i(R)\big)\big)
$$
$$
\mu\big(H\big(F_i(B)\big)\big) \quad \text{and} \quad \sigma\big(H\big(F_i(B)\big)\big) \tag{2}
$$

where $\mu\big(H\big(F_i(R)\big)\big)$, $\sigma\big(H\big(F_i(R)\big)\big)$ and $\mu\big(H\big(F_i(B)\big)\big)$, $\sigma\big(H\big(F_i(B)\big)\big)$ are the mean and standard deviation of histogram of red and blue components of ith frame respectively.

Color features are also described by HSV values of the pixels from each frame [15]. The proposed method also extracts HSV color components.

B. Gray Level Co-occurrence Matrix (GLCM)

The texture of the flower region also plays an important role in keyframe selection. In the proposed method, GLCM based texture features are extracted from each frame, as GLCM can provide useful statistical information about the texture. GLCM describes the distribution of co-occurring grey-level values within an image. Haralick et al. [10] proposed a total of 14 features based on the information contained in a GLCM to reveal the different combinations of gray level co-occur in a frame. These 14 features namely, Angular second moment, Contrast, Correlation, Sum of squares variance, Inverse difference moment, Sum average, Sum entropy, Entropy, Sum variance, Difference variance, Difference entropy, Information measure of correlation 1, Information measure of correlation 2, Maximum correlation coefficient are used as texture features for keyframe selection in the proposed work.

C. Modified Histogram of Oriented Gradients (MHOG)

Histogram of oriented gradients can be used as feature descriptors for the keyframe selection, where the occurrences of gradient orientations of frames perform an important role. HOG describes the distribution of intensity gradients, appearance and shape of the regions of the frames. The proposed method extracts HOG features with modifying the parameters such as cell size, block size and number of bins.

D. Entropy

Junding et al. [11] proposed color distribution entropy to describe the spatial information and color distribution. The method divides the frame into 'b' number of blocks and creates a vector of entropy features, then calculates the entropy at pixel $P(i, j)$ of each block, it can be defined as,

$$E_i(P_i) = \sum_{j=1}^{N} -P_{ij} log_2(P_{ij}) \tag{3}$$

3.1.3   Clustering

Create clusters with the features obtained from feature extraction phase using K-means clustering. The main reason for selecting k-means for clustering purpose is due to its ease of implementation even though fixing the parameter 'K' is challenging.

The algorithm is explained below [4, 6]

1. Given K, the number of clusters. Partition the video V_i into a 'K' nonempty subsets (clusters) of frames.
2. Compute centroids of the clusters of the current partition of V_i.
3. Create clusters with nearest frames from centroid by calculating the Euclidian distance between each pixel in each frame and the center of the cluster. Using the Euclidian distance between F_{ij} (the jth pixel of the ith frame) and the Kth cluster is defined as

$$D(F_{ij}, K_c) = \sqrt{\sum_{i,j,}^{n,m} (F_{ij} - K_c)^2} \tag{4}$$

Where $i = 1, 2, \ldots, n$, $j = 1, 2, \ldots, m$, n is the total number of frames in the cluster and m is the total number of pixels in each frame in the cluster. Then, pixel F_{ij} is assigned to its closest cluster center K_c.
4. Repeat step 3 until to create K new clusters having similar frames.

3.1.4   Selection of Keyframes

In this stage, keyframes are identified from each cluster obtained from the clustering stage. Find the minimum distanced frame from each cluster centroid 'K_c' by calculating Euclidian distance between the centroid and all the frames in the cluster as shown in the following equation.

$$D(K_c, CF_i) = \sqrt{\sum_{i=1}^{n} (K_c - Cf_i)^2} \tag{5}$$

Where $D(K_c, CF_i)$ is the distance between the cluster centroid (K_c) and all the frames in the cluster (CF_i) with 'n' number of samples, the minimum distanced frame from centroid is the keyframe. Then apply the fidelity measure and compare the result with method 2 then select keyframes obtained high fidelity value.

3.2   Entropy-GMM-Kmeans Based Keyframe Selection (EGKK)

The proposed method involves four stages to select keyframes, namely, preprocessing, feature extraction, clustering and selection of keyframes. The preprocessing explained in Sect. 3.1.1. In the feature extraction phase extract, a vector of entropy features explained in Sect. 3.1.2 then cluster similar frames through the block wise comparison between each corresponding block of consecutive frames. It can be stated as follows.

$$Sim(f_i, f_{i+1}) = Sim(f_i(E_i b_i), f_{i+1}(E_i b_i)) \tag{6}$$

Where $Sim(f_i, f_{i+1})$ Similarity of entropy E_i of block b_i i.e., $f_i(E_i b_i)$ and $f_{i+1}(E_i b_i)$ between i^{th} and $(i+1)^{th}$ frames.

Then, create 'K' clusters with Gaussian Mixture Model (GMM) to cluster similar frames and the GMM is defined below. Then select keyframes explained in Sect. 3.1.4. Then apply fidelity measure, compare the result with method1 then select keyframes obtained high fidelity value.

The difference between the first method and the second method is that the method1 extracts color, GLCM, MHOG and entropy features and creates clusters using k-means and selects keyframes using k-means. In the second method, it extracts only entropy feature to cluster the frames using GMM, then it selects keyframes using the k-means algorithm.

3.2.1   Gaussian Mixture Model

Create GMM clusters by fitting the Gaussian distribution with 'n' features of data (x) and must specify the desired 'K' number of clusters, the Gaussian function is shown below [3].

$$f(x) = \frac{1}{\sigma\sqrt{2\pi}} e^{-\frac{(x-\mu)^2}{2\sigma^2}} \tag{7}$$

Where σ is the standard deviation, μ is the mean of data (features) 'x'.

4   Datasets

Since there is no standard dataset available for flower videos, in this work we made an attempt to create flower video dataset. Our dataset consists of 225 videos with 15 different classes of flowers where in which each class contains 5 to 37 videos. The duration of the videos ranges from 2 to 60 s. It consists of challenges such as light variations, viewpoint variations, occlusion and cluttered background. We collected videos of different types of flowers that are located in southern Karnataka. Videos are captured by Samsung Galaxy Grand Prime mobile device, Sony Cyber-shot camera and Canon camera at different ecological conditions such as sunny, cloudy and rainy. Experimentation has been conducted on videos captured through Samsung Galaxy Grand Prime mobile device.

5   Experimental Results

In this section, we present the details of experimental results of the proposed methods explained in Sect. 3. Experimentation conducted on our own dataset. To evaluate the performance of the proposed methods we computed the fidelity measure and compression ratio, the results are shown in Tables 1 and 2 respectively. We have made a comparative study of the existing works of [23, 24]. Some of the sample flower video frames with different challenges mentioned above in Sect. 4 are shown in Fig. 2(a). The keyframes selected from proposed models are shown in Fig. 2(b).

Table 1. Comparison of keyframe selection methods using fidelity measure.

Classes	No. of videos	Total frames	Huayong and Huifen (2014)	Sheena and Narayan (2015)	Proposed methods	
					Method 1	Method 2
1	37	10391	0.207	0.037	0.647	**0.678**
2	34	6244	0.381	0.014	0.344	**0.487**
3	12	2140	0.274	0.008	**0.339**	0.322
4	6	1167	0.464	0.010	0.266	**0.536**
5	12	2948	0.590	0.012	0.408	**0.470**
6	12	2796	0.346	0.035	**0.696**	0.425
7	5	1515	0.895	0.009	**0.946**	0.847
8	22	21208	0.308	0.026	**0.512**	0.341
9	5	1401	0.303	0.027	0.121	**0.453**
10	20	8076	0.244	0.013	0.420	**0.576**
11	20	5794	0.238	0.014	0.207	**0.212**
12	15	5261	0.214	0.022	0.285	**0.724**
13	10	4270	0.562	0.038	**0.488**	0.289
14	10	2837	0.243	0.047	**0.378**	0.254
15	5	1283	0.202	0.020	**0.571**	0.232

Table 2. Comparison of keyframe selection methods using compression ratio.

Classes	No. of videos	Total frames	Huayong and Huifen (2014)		Sheena and Narayan (2015)		Proposed methods			
							Method 1		Method 2	
			Key-frames	CR	Key-frames	CR	Key-frames	CR	Key-frames	CR
1	37	10391	185	0.98	2859	0.72	185	**0.98**	185	**0.98**
2	34	6244	170	0.97	999	0.84	170	**0.97**	170	**0.97**
3	12	2140	60	0.97	452	0.78	60	**0.97**	60	**0.97**
4	6	1167	30	0.97	144	0.88	30	**0.97**	30	**0.97**
5	12	2948	60	0.97	548	0.81	60	**0.97**	60	**0.97**
6	12	2796	60	0.97	750	0.73	60	**0.97**	60	**0.97**
7	5	1515	25	098	372	0.75	25	**098**	25	**098**
8	22	21208	110	0.99	3374	0.84	110	**0.99**	110	**0.99**
9	5	1401	25	0.98	264	0.81	25	**0.98**	25	**0.98**
10	20	8076	100	0.98	1893	0.76	100	**0.98**	100	**0.98**
11	20	5794	100	0.98	1255	0.78	100	**0.98**	100	**0.98**
12	15	5261	75	0.98	1534	0.70	75	**0.98**	75	**0.98**
13	10	4270	50	0.98	970	0.77	50	**0.98**	50	**0.98**
14	10	2837	50	0.98	837	0.70	50	**0.98**	50	**0.98**
15	5	1283	25	0.98	107	0.91	25	**0.98**	25	**0.98**

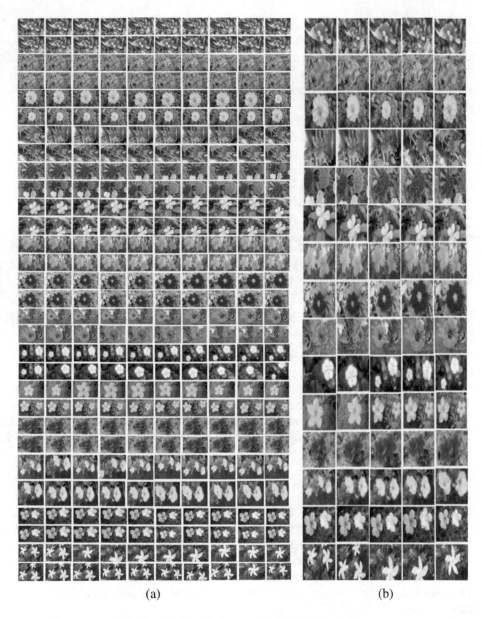

         (a)   (b)

Fig. 2. (a): Sample frames (b): Selected keyframes

Fidelity Measure

The Fidelity measure (FM) compares each keyframe with the other frames in the video sequence which is defined as a semi-Hausdorff distance [7, 22].

In Table 1, we have listed the fidelity values of the proposed method compared with existing methods [23, 24], the proposed model gives high fidelity values than the existing models. The high-fidelity values indicate that there exists a large dissimilarity (largest variation between frames) among the selected keyframes. This is due to the videos captured with different viewpoints, varying in length, pose and moments of the camera. Low-fidelity values indicate that the keyframes selected from the video sequence do not much differ each other, in that case for the videos obtained low fidelity values we can select a single keyframe to represent that video. Hence, in our work the fidelity measure depends on the videos captured.

Table 1 gives the Fidelity value of the selected keyframes of our dataset. There are 15 species of flowers which are mentioned as classes. Each class consists of 5 to 37 videos. From Table 1 it can be observed that the proposed methods provide good fidelity value compared to Sheena and Narayan (2015) for all classes and also for the majority of the classes the obtained result is better than Huayong and Huifen (2014). The result clearly shows the effectiveness of the proposed approach.

Compression Ratio
Compression ratio states that, the compactness of video summary. The compression ratio can be computed by dividing the obtained number of keyframes by a total number of frames in the video sequence. It can be defined as follows [22]

$$CR = 1 - \frac{S_{KF}}{S_F} \tag{8}$$

Where CR is the compression ratio, S_{KF} is the set of selected keyframes and S_F is the set of video frames.

Table 2 gives the compression ratio of the selected keyframes of our dataset. From the results we can observe that the proposed methods generate high compression ratio than the sequential approach of Sheena and Narayan (2015) [23]. In case of cluster based approach, the obtained result is similar to that of Huayong and Huifen (2014) [24].

6   Conclusion

In this paper, we have proposed cluster based approaches for automatic selection of keyframes from natural flower videos, which can be used for archival and retrieval of flower videos. From the experimentation, it is observed that the proposed model gives good results and generates all possible keyframes of a given flower video. As an initial attempt we have used flower data set for selection of keyframes in natural flower videos. The proposed approach is more general in nature and hence it can be applied to other datasets also.

References

1. de Avila, S.E.F., Ana, P., Antonia, D.: VSUMM: a mechanism designed to produce static video summaries and a novel evaluation method. Pattern Recogn. Lett. **32**, 56–68 (2011)
2. Chatzigiorgaki, M., Skodras, A.N.: Real-time keyframe extraction towards video content identification. IEEE, ISSN: 978-1-4244-3298 (2009)
3. Chen, W., Tian, Y., Wang, Y., Huang, T.: Fixed-point Gaussian mixture model for analysis-friendly surveillance video coding. Comput. Vis. Image Underst. **142**, 65–79 (2016)
4. Chuen-Horng, L., Chun-Chieh, C., Hsin-Lun, L., Jan-ray, L.: Fast k-means algorithm based on a level histogram for image retrieval. Expert Syst. Appl. **41**, 3276–3283 (2014)
5. Das, M., Manmatha, R., Riseman, E.M.: Indexing flower patent images using domain knowledge. IEEE Intell. Syst. **14**(5), 24–33 (1999)
6. Duda, R.O., Hart, P.E., Stork, D.G.: Unsupervised learning and clustering. Pattern Classification. Springer, New York (2001)
7. Gianluigiand, C., Raimondo, S.: An innovative algorithm for key frame extraction in video summarization. J. Real-Time Image Proc. **1**, 69–88 (2006)
8. Guru, D.S., Sharath, Y.H., Manjunath, S.: Texture features and KNN in classification of flower images. IJCA Special Issue on Recent Trends Image Process. Pattern Recogn. **1**, 21–29 (2010)
9. Guru, D.S., Sharath, Y.H., Manjunath, S.: Textural features in flower classification. Math. Comput. Model. **54**, 1030–1036 (2011)
10. Haralick, R.M., Shanmugam, K., Dinstein, I.: Textural features for image classification. IEEE Trans. Syst. Man Cybern. **SMC-3**(6), 610–621 (1973)
11. Sun, J., Zhang, X., Cui, J., Zhou, L.: Image retrieval based on color distribution entropy. Pattern Recogn. Lett. **27**, 1122–1126 (2006)
12. Loannidis, A., Chasanis, V., Likas, A.: Weighted multi-view key frame extraction. Pattern Recogn. Lett. **72**, 52–61 (2016)
13. Manjunath, S.: VARS: Video Archival and Retrieval System. Ph. D Thesis (2012)
14. Naveed, E., Tayyab, B.T., Sung, W.B.: Adaptive key frame extraction for video summarization using an aggregation mechanism. J. Vis. Commun. Image R. **23**, 1031–1040 (2012)
15. Nilsback, M.E., Zisserman, A.: A Visual vocabulary for flower classification. In: Proceedings of Computer Vision and Pattern Recognition, vol. 2, pp. 1447–1454 (2006)
16. Padmavathi, M., Yong, R., Yelena, Y.: Keyframe-based video summarization using delaunay clustering. Int. J. Digit. Librar. **6**(2), 219–232 (2006)
17. Kaunar, S.K., Panda, R., Chowdhury, A.S.: Video key frame extraction through dynamic delaunay clustering with a structural constraint. J. Vis. Commun. Image Rep. **24**, 1212–1227 (2013)
18. Song, G.H., Ji, Q.G., Lu, Z.M., Fang, Z.D., Xie, Z.H.: A novel video abstract method based on fast clustering of the region of interest in key frames. Int. J. Elec. Commun. **68**, 783–794 (2014)
19. Tremeau, A., Tominaga, S., Plataniotis, K.N.: Color in image and video processing: most recent trends and future research direction. EURASIP J. Image Video Process. **2008**(3), 1–26 (2008)
20. Zeng, X., Hu, W., Li, W., Zhang, X., Xu, B.: Key-frame extraction using dominant – set clustering. In: Proceedings of IEEE International Conference on Multimedia and Expo, Hannover, Germany, pp. 1285–1288 (2008)
21. Zhou, H., Sadka, A.H., Swash, A.H., Azizi, J., Sidiq, U.A.: Feature extraction and clustering for dynamic video summarization. Neurocomputing **73**, 1718–1729 (2010)

22. Jyothi, V.K., Sharath, Y.H., Guru, D.S.: Sequential approach for key frame selection in natural flower videos. In: 6th International Conference on Signal and Image Processing (ICSIP) (2017, accepted)
23. Sheena, C.V., Narayan, N.K.: Key-frame extraction by analysis of histograms of video frames using statistical methods. Procedia Comput. Sci. **70**, 36–40 (2015). 4th International Conference on Eco-friendly Computing and Communication Systems
24. Liu, H., Hao, H.: Key frame extraction based on improved hierarchical clustering algorithm. In: 11th International Conference on FSKD, pp. 793–797. IEEE Xplore (2014)

From Crisp to Soft Possibilistic
and Rough Meta-clustering
of Retail Datasets

Asma Ammar$^{(\boxtimes)}$ and Zied Elouedi

LARODEC, Institut Supérieur de Gestion de Tunis, Université de Tunis,
41 Avenue de la Liberté, 2000 Le Bardo, Tunisia
asmaammarbr@gmail.com, zied.elouedi@gmx.fr

Abstract. This paper investigates the problem of meta-clustering real-world retail datasets based on possibility and rough set theories. We propose a crisp then, a soft meta-clustering methods and we compare and analyze the results of both methods using real-world retail datasets. The main aim of this paper is to prove the performance gain of the soft meta-clustering method compared to the crisp one. Our novel methods combine the advantages of the meta-clustering process and the k-modes method under possibilistic and rough frameworks. Our approaches perform a double clustering (or meta-clustering) using two datasets that depend on each other consisting of the retail datasets. It uses for the meta-clustering a modified version of the k-modes method. For the new crisp meta-clustering method, we use the possibilistic k-modes (PKM) and for the soft method, the k-modes under possibilitic and rough frameworks (KM-PR) is applied.

Keywords: Meta-clustering · K-modes · Possibility theory
Rough set theory · Retail datasets

1 Introduction

Meta-clustering also known as double clustering presents a bi-clustering process where a first clustering of a first dataset is performed, then the results of this clustering are combined with a second dataset in order to make a second clustering. The meta-clustering offers the possibility to improve the results and study the relation between two datasets that depend on each other [3,4]. In the meta-clustering, we can apply a crisp or a soft method. Crisp clustering methods assign each object of the training set to exactly one cluster. However, soft methods consider that each object is a member of all available clusters with respect to their membership degrees. As a result, the result of the meta-clustering will depend on the used method i.e., crisp or soft method.

As in real-world situations, imperfection is widely and often found, we aim in this paper to study the impact of the uncertainty when applying meta-clustering

© Springer International Publishing AG, part of Springer Nature 2018
A. Abraham et al. (Eds.): ISDA 2017, AISC 736, pp. 485–495, 2018.
https://doi.org/10.1007/978-3-319-76348-4_47

on two real-world datasets using possibility and rough set theories. The possibility and rough set theories had been successfully applied in many works [4,7] using both clustering and meta-clustering processes.

In this paper, we propose to make a comparative study to analyze the impact of both crisp and soft methods on meta-clustering real-world datasets. These latter will depend on each others and they consist of the customer and product datasets. For the crisp case, we propose to apply a crisp (also known as hard) method consisting of the possibilistic k-modes (PKM) [1]. For the soft case, we use the k-modes method under possibilistic and rough frameworks (KM-PR) [2].

In fact, our new meta-clustering methods use both of the possibility and rough set theories to treat the imperfection of data and to improve the results. For the crisp and soft cases, the possibility theory handles the uncertainty when using datasets with uncertain values of attributes by presenting the uncertain values of an attribute relative to an object through a possibility distribution. For the soft case, possibility theory also represents the degree of belonging of an object to several clusters through a possibilistic membership degree.

The rest of the paper is structured as follows: Sect. 2 provides the background concerning the possibility and rough set theories. Section 3 presents an overview of the k-modes method and its improved versions. Section 4 presents with details the algorithms of our novel crisp and soft approaches applying the meta-clustering methods under possibilistic and rough frameworks. Section 5 describes and analyzes experiments. Section 6 concludes this paper.

2   Possibility and Rough Set Theories

2.1   Possibility Theory

Possibility theory is a well-known uncertainty theory devoted to handle incomplete and uncertain states of knowledge [13]. It has been developed by several researchers including Dubois and Prade [5]. One fundamental concept in possibility theory is the possibility distribution function π. Assume Ω as the universe of discourse containing several states of the world such as $\Omega = \{\omega_1, \omega_2, ..., \omega_n\}$. A possibility distribution is defined as the mapping from the universe of discourse Ω to the interval $L = [0, 1]$ which represents the possibilistic scale. The degree of uncertainty relative to ω_i is described through a possibility degree.

Several similarity and dissimilarity measures were proposed in possibility theory in order to compare pieces of uncertain information. We can mention the information affinity [6] defined by: $InfoAff(\pi_1, \pi_2) = 1 - 0.5[D(\pi_1, \pi_2) + Inc(\pi_1, \pi_2)]$.

2.2   Rough Set Theory

Rough set theory was proposed in the early 1980s by Pawlak [11,12]. In rough set theory, data are represented through an *information table*. It contains objects (in rows) and attributes (in columns). Each object is described using some attributes. A decision table contains two types of attributes mainly *condition* and *decision* attributes. In the decision table, each row represents a decision rule. An information system is defined as a system $IS = (U, A)$ where U and A are finite and nonempty sets. U is the universe containing a set of objects and A is the set of attributes. Each $a \in A$ has a domain of attribute V_a defined by an information function $f_a : U \rightarrow V_a$. The approximation of sets is a fundamental concept in rough set theory. Given an $IS = (U, A)$, $B \subseteq A$ and $Y \subseteq U$. The set Y is described through the values of attribute from B by defining as follows two sets named the B-upper $\overline{B}(Y)$ and the B-lower $\underline{B}(Y)$ approximations.

$$\overline{B}(Y) = \bigcup \{B(y) : B(y) \cap Y \neq \phi\}. \tag{1}$$

$$\underline{B}(Y) = \bigcup \{B(y) : B(y) \subseteq Y\}. \tag{2}$$

Here $y \in U$, and $B(y)$ is a subset of U such that $\forall x \in B(y)$ values of attributes in B are the same. The B-boundary region $BR(Y)$ of Y can be deduced using Eqs. (1) and (2) as follows: $BR(Y) = \overline{B}(Y) - \underline{B}(Y)$. If $BR(Y) = \phi$ is empty, then the set Y is considered as a *crisp* set otherwise, Y is a *rough* set. Besides, we have: $POS(Y) = \underline{B}(Y) \Rightarrow$ is certainly a member of Y, $NEG(Y) = U - \overline{B}(Y) \Rightarrow$ is non-member of Y, and $BR(Y) = \overline{B}(Y) - \underline{B}(Y) \Rightarrow$ is possibly a member of Y.

3 K-modes Method and Its Extensions

3.1 The K-modes Method

The k-modes method [10] has been developed to handle large categorical datasets. It is based on the k-means method [9] and uses as parameters the simple matching dissimilarity measure and a frequency based function. The simple matching is defined by $d(X_1, Y_1) = \sum_{t=1}^{m} \delta(x_{1t}, y_{1t})$. Note that $\delta(x_{1t}, y_{1t}) = 0$ if $x_{1t} = y_{1t}$ and $\delta(x_{1t}, y_{1t}) = 1$ otherwise. It is possible to find, when updating the modes, more than one possibility i.e., in some cases there is more than one mode for a cluster, as a result a random choice is used. In order to avoid the non-uniqueness of the cluster mode and to deal with uncertain datasets, improved versions of the k-modes method under possibilistic and rough frameworks are proposed. They are presented in the following subsections.

3.2 Possibilistic K-modes

The possibilistic k-modes denoted by PKM [1] is a crisp method that handles uncertain values of attributes using possibility theory. It represents the values of each attribute by a possibility distribution. The PKM uses a modified version of the information affinity measure [6] as possibilistic similarity measure. It consists of the $IA(X_1, X_2) = \frac{\sum_{j=1}^{m} InfoAff(\pi_{1j}, \pi_{2j})}{m}$, where m is the total number of attributes. The PKM uses also the mean of possibilistic distribution in order to update the clusters' modes.

3.3   K-modes Method Using Possibility and Rough Set Theories

The KM-PR that means the k-modes using possibility and rough set theories [2] is an improved method that clusters categorical datasets using the k-modes method and handles the uncertainty in the values of attributes and in the belonging of an object to several clusters using possibility theory. It also uses the rough set theory to detect the boundary regions by finding the upper and lower bounds. It uses an uncertain dataset where values of attributes are presented through possibility degrees. The KM-PR parameters are a possibilistic similarity measure $IA(X_1, X_2)$, a possibilistic membership degree $\omega_{ij} \in [0, 1]$ that presents the degree of belonging of an object i to the cluster j, and for the update of the cluster mode, it is based on ω_{ij} (it is detailed in [2]). Finally, for the detection of boundary region, it uses the lower and upper approximations. The boundary region will contains peripheral objects. We introduce a new parameter consisting of the ratio $R_{ij} = \frac{\max \omega_i}{\omega_{ij}}$. R is relative to each object is compared to a threshold $T \geq 1$ [7,8]. If $R_{ij} \leq T$, it means that the object i belongs to the upper bound of the cluster j. If an object belongs to the upper bound of exactly one cluster j, it means that it belongs to the lower bound of cluster j.

4   Meta-clustering Under Possibilistic and Rough Frameworks

In this section, we present the main algorithms of our proposals consisting of the crisp and soft uncertain meta-clustering methods (i.e. when using PKM and KM-PR). The following algorithm describes our proposals in general case when we have two datasets A and B that depend on each others.

Algorithm. *Meta-clustering algorithm in general case*

1.   *We randomly select a dataset (the dataset A) and make the clustering task by applying the KM-PR or the PKM using the static information from A.*
2.   *In this step, we can compute the dynamic representation of the dataset B using the results provided from the clustering of A.*
3.   *We cluster objects from B based on static and dynamic parts. For each object, we compute its membership degree to the cluster.*
4.   *We compute the dynamic part of the objects from A.*
5.   *We cluster objects from A using the static and dynamic parts. Then, for each object, we calculate its membership degree to all clusters.*
6.   *We repeat 3 to 5 until the dynamic parts of both datasets are stable.*
7.   *We derive the rough clustering using the ratio R and the threshold T.*

In our case, we use two retail datasets consisting of customer and product datasets where there is a relation between a customer and some products bought. The algorithm produces nk clusters of customers and mk clusters of products with possibilistic membership degrees. In case of using the PKM crisp algorithm, the $\omega_{ij} = \{0, 1\}$. Thus the resulting partitions will be crisp. However, if we use the KM-PR soft method for the meta-clustering, we will follow these steps:

Algorithm. *Meta-clustering algorithm using the KM-PR and retail datasets*

1. *At the beginning, we do not have a dynamic part neither for customers nor for products. Thus, we only use the static information. We start by clustering customers using static information from the transaction dataset based on the KM-PR [2]. As a result, we get the membership degree ω_{ij} relative to the customer c_i and the cluster cc_j.*

2. *We compute the dynamic representation of the products: $p_t\omega_j = \frac{\sum_{i=1}^{n'} \omega_{ij}}{n'}$, where $p_t\omega_j$ is the degree of belonging of the product p_t to the customer cluster cc_j, n' is the number of customers that buy p_t and ω_{ij} is obtained from 1.*

3. *We cluster products using static and dynamic parts. For each product p_t, we compute its membership degree ω_{tq} to the cluster cp_q.*

4. *We compute the dynamic part of the customers dataset: $c_i\omega_q = \frac{\sum_{t=1}^{m'} \omega_{tq}}{m'}$, where $c_i\omega_q$ is the degree of belonging of the customer c_i to the product cluster cp_q, m' is the number of products bought by c_i and ω_{tq} is obtained from 2.*

5. *We cluster customers based on the static and dynamic parts. Then, for each customer, we compute its membership degree to all clusters.*

6. *We repeat 3 to 5 until the dynamic parts are stable.*

7. *We derive the rough clustering through the ω_{ij} (similarly through ω_{tq}) for the customers (similarly for the products) by computing the ratio of each object. Finally, we assign each object to the upper or the lower bound of the cluster.*

5   Experiments

5.1   The Framework

We use an example of two databases that depend on each others. They consist of retail datasets consisting of customer and product datasets to make the meta-clustering process. They report transactions relative to a small retail store chain during a period of three years, from 2005 to 2007. The customer dataset has 15341 rows and 4 columns. It describes the behaviors of customers when buying products. The products dataset has 8987 rows and 5 columns representing different products bought by customers.

1. Customer dataset:
   - #Products: is the number of products bought by a customer. Its values are low, modest, and high.
   - Revenues: describe how much money a customer spent when visiting the store. Their values are low, modest, high, and veyhigh.
   - Profits: represent the profit from the customer to the store. Their values are low, modest, high, and veyhigh.
   - Loyalty: deduced from the number of visits by a customer to the store. Its values are low, modest, high, and veyhigh.
2. Product dataset:
   - #Customers: is the number of customers who purchased a product. Its values consist of low, modest, and high.

- Revenues: made from a product. Their values are low, modest, high, and veyhigh.
- Profits: represent to the profits made by selling the product. Their values are low, modest, and high.
- Popularity: deduced from the number of visits in which a product was bought. Its values are low, modest, and high.
- Quantity: consists of the quantity of the product sold. Its values are small, medium, and big.

As there are many customers that they do not buy often from the store and there are also different products that are rarely sold, we use the top 1000 customers and the top 500 products from the retail datasets.

5.2   Pre-treatmenet Procedure of Retail Datasets

We use the retail datasets that contain two datasets with numeric values of attributes. However, numeric value cannot provide meaningful information and hard to be analyzed. Let us take an example of a number of products (#Products) equals to 36. We cannot deduce if this number is small or large. As a result, we propose to use a discretization of the initial retail datasets by applying the *equal-frequency* function. This latter is defined in R software and it is non-sensitive to outliers. It provides intervals with the same frequency. Tables 1 and 2 provide the discretization results using this function.

Table 1. Discretization of the customer data using equal-frequency function

	Intervals			
# Products	[5, 40)	[40, 58)	[58, 546]	
	Low	Modest	High	
Revenues	[57, 712)	[712, 1041)	[1041, 1624)	[1624, 12614]
	Low	Modest	High	Veryhigh
Profits	[25, 287)	[287, 428)	[428, 657]	[657, 3484]
	Low	Modest	High	Veryhigh
Loyalty	[3, 16)	[16, 21)	[21, 29]	[29, 289]
	Low	Modest	High	Veryhigh

5.3   Artificial Creation of Uncertain Datasets

The proposed algorithm of our meta-clustering replace uncertain values by possibility degrees given by expert as follows:

1. For certain attribute values: the true value i.e., known with certainty is replaced by the possibility degree 1. Remaining values take the value of 0.
2. For uncertain attribute values: The true value also takes the degree 1. The remaining values randomly take possibilistic degrees from]0, 1].

An example of the representation of a product with the values low, low, low, low, and small by uncertain values is shown in Table 3.

Table 2. Discretization of the product data using equal-frequency function

	Intervals			
# Customers	$[7, 32)$	$[32, 48)$	$[48, 377]$	
	Low	Modest	High	
Revenues	$[98, 886)$	$[886, 1785)$	$[1785, 3305)$	$[3305, 40800]$
	Low	Modest	High	Veryhigh
Profits	$[6, 491)$	$[491, 1063)$	$[1063, 20084]$	
	Low	Modest	High	
Popularity	$[38, 58)$	$[58, 86)$	$[86, 867]$	
	Low	Modest	High	
Quantity	$[43, 73)$	$[73, 120)$	$[120, 1395]$	
	Small	Medium	Big	

Table 3. Example of the representation of two products by possibilistic values

		# Customers			Revenues			
		Low	Modest	High	Low	Modest	High	Veryhigh
Certain case	P1	1	0	0	1	0	0	0
Uncertain case	P1	1	0.16	0.56	1	0.31	0.11	0.24

		Profits			Popularity			Quantity		
		Low	Modest	High	Low	Modest	High	Small	Medium	Big
Certain case	P1	1	0	0	1	0	0	1	0	0
Uncertain case	P1	1	0.81	0.61	1	0.37	0.094	1	0.71	0.38

5.4 Experimental Results

We report and analyze the final results provided after running our meta-clustering algorithm using the PKM then, the KM-PR. We set a number of clusters equals to $k_c = 4$ for the customer dataset and equals to $k_p = 3$ for the product dataset. We can deduce 4 profiles for the customers from the clusters' modes:

- Profile 1 corresponding to cluster 1: it describes customers that bought the highest number of products. These customers are very loyal since they have very high revenue and make a very important profit to the store.
- Profile 2 corresponding to cluster 2: it is relative to customers that bought a modest number of products and have also a modest revenue. These customers are considered as loyal as they make a modest profit to the store.
- Profile 3 corresponding to cluster 3: it characterizes customers that bought a low number of products. As a result, they have a low revenues, they are not loyal to the store and they do not make profits.

– Profile 4 corresponding to cluster 4: it provides information about customers that bought a modest number of products but they have a high revenue. They are also considered as loyal since, they make a high profit to the store.

From the resulting modes relative to the products dataset, we can detect 3 profiles for the products sold:

– Profile 1 corresponding to cluster 1: it is relative to products that are bought by a small number of customers with small quantity. These products do not make profits to the store and they are not popular.
Profile 2 corresponding to cluster 2: it describes products that are bought by an important number of customers with acceptable quantity. They are considered as popular and they make a modest profits to the store.
– Profile 3 corresponding to cluster 3: it characterizes products that are the most bought by an important number of customers. They make a high profit to the store as a result, they are considered as popular.

The final possibilistic membership degrees provided after running our proposals are illustrated in Tables 4 and 5.

Table 4. Final membership degrees of the first 5 instances of the customers dataset

	KM-PR				PKM			
	C_{c1}	C_{c2}	C_{c3}	C_{c4}	C_{c1}	C_{c2}	C_{c3}	C_{c4}
C_1	0	1	0	0.25	0	1	0	0
C_2	0	1	0.5	0.5	0	1	0	0
C_3	1	0	0	0	1	0	0	0
C_4	1	0	0	0	1	0	0	0
C_5	1	0.5	0.5	0	1	0	0	0

Table 5. Final membership degrees of the first 5 instances of the products dataset

	KM-PR			PKM		
	C_{p1}	C_{p2}	C_{p3}	C_{p1}	C_{p2}	C_{p3}
P_1	0	1	0	0	1	0
P_2	1	0	0	1	0	0
P_3	1	0	0	1	0	0
P_4	1	0	0	1	0	0
P_5	0	1	1	0	0	1

After computing the final memberships of each customer to the four clusters, we calculate the ratio. Table 6 describes the ratio of the customers dataset.

Table 6. Example of ratios of the first 5 instances of customers dataset

| | Final ratios by cluster | | | | | | | |
| | KM-PR | | | | PKM | | | |
	C_{c1}	C_{c2}	C_{c3}	C_{c4}	C_{c1}	C_{c2}	C_{c3}	C_{c4}
C_1	-	1	-	4	-	1	-	-
C_2	-	1	2	2	-	1	-	-
C_3	1	-	-	-	1	-	-	-
C_4	1	-	-	-	1	-	-	-
C_5	1	2	2	-	1	-	-	-

We have tested different values of threshold then, we set $T = 2.55$ as threshold as this value provides the best separation between the rough clusters. We get the boundary regions for the customer dataset when using the KM-PR in the meta-clustering.

- Boundary region of cluster 1 C_{c1}: $C_5, C_7, C_{10}, P_{33}, P_{48}, C_{59}, C_{83}, C_{208}, C_{209}$.
- Boundary region of cluster 2 C_{c2}: $C_2, C_5, C_6, C_7, C_{10}, C_{35}, P_{48}, C_{66}, C_{176}, C_{178}$, C_{380}, C_{387}.
- Boundary region of cluster 3 C_{c3}: $C_2, C_5, C_6, C_7, C_{35}, C_{59}, P_{61}, C_{66}, C_{176}, C_{178}$, C_{208}, C_{209}.
- Boundary region of cluster 4 C_{c4}: $C_2, C_6, C_{10}, P_{33}, P_{61}, C_{66}, C_{83}, C_{208}, C_{209}$, C_{380}, C_{387}.

Similarly, we calculate the ratio of each instance for the product dataset as described in Table 7.

Table 7. Example of ratios of the first 5 instances of products dataset

| | Final ratios by cluster | | | | | |
| | KM-PR | | | PKM | | |
	C_{p1}	C_{p2}	C_{p3}	C_{p1}	C_{p2}	C_{p3}
P_1	-	1	-	-	1	-
P_2	1	-	-	1	-	-
P_3	1	-	-	1	-	-
P_4	1	-	-	1	-	-
P_5	-	1	1	-	-	1

We also get the boundary regions for the product dataset using the same threshold T when applying the KM-PR algorithm for the meta-clustering.

- Boundary region of cluster 1 C_{p1}: $P_{10}, P_{57}, P_{63}, P_{139}, P_{187}, P_{215}, P_{247}$, P_{249}, P_{420}.

- Boundary region of cluster 2 C_{p2}: $P_5, P_{57}, P_{215}, P_{247}, P_{249}, P_{420}$.
- Boundary region of cluster 3 C_{p3}: $P_5, P_{10}, P_{63}, P_{139}, P_{187}, P_{249}, P_{420}$.

Looking at Tables 6 and 7, we can observe that there are some instances of customers and products that belong to the boundary region which confirms the ability of the soft method when applied to the meta-clustering process to detect the rough clusters using the ratio.

However, when using the crisp method i.e. the PKM, we remark, for both datasets, that all objects are members of only positive regions which means that each object is certainly a member of exactly one cluster. Thus, all boundary regions are empty and we do not have any rough cluster which is not correct. Generally, the potential and ability of soft methods, such as the KM-PR, to compute the membership degrees and to keep all similarities between objects and clusters confirm their performances compared to crisp methods. These latter may make results inaccurate by ignoring similarities that can exist.

6   Conclusion

In this paper, we focused on the impact of the crisp and soft meta-clustering when being applied on real-world datasets such as retail datasets. We remarked the importance of using a soft method in a meta-clustering process to improve the results. In our case, it helped the decision maker in his analysis by providing a good description of the customers behaviors and products characteristics. The resulting profiles of customers and products can help the decision maker in the store to select, for example, the appropriate products. This comparison made in this paper proved the performance gain of the soft method when being used in a double clustering process. Besides, the use of possibility and rough set theories for handling uncertainty in the values of attributes and in the belonging of objects to different clusters and to detect boundary region improve the final results.

References

1. Ammar, A., Elouedi, Z.: A new possibilistic clustering method: the possibilistic k-modes. In: Proceedings of the 12th International Conference of the Italian Association for Artificial Intelligence. LNAI, vol. 6934, pp. 413–419 (2011)
2. Ammar, A., Elouedi, Z., Lingras, P.: The k-modes method using possibility and rough set theories. In: Proceedings of the IFSA World Congress and NAFIPS Annual Meeting, IFSA/NAFIPS, pp. 1297–1302. IEEE (2013)
3. Ammar, A., Elouedi, Z., Lingras, P.: Rough possibilistic meta-clustering of retail datasets. In: Proceedings of the 2014 International Conference on Data Science and Advanced Analytics, DSAA 2014, pp. 177–183. IEEE (2014)
4. Ammar, A., Elouedi, Z., Lingras, P.: Meta-clustering of possibilistically segmented retail datasets. Fuzzy Sets Syst. **286**, 173–196 (2016)
5. Dubois, D., Prade, H.: Possibility Theory: An Approach to Computerized Processing of Uncertainty. Plenium Press, New York (1988)

6. Jenhani, I., Ben Amor, N., Elouedi, Z., Benferhat, S., Mellouli, K.: Information affinity: a new similarity measure for possibilistic uncertain information. In: Proceedings of the 9th European Conference on Symbolic and Quantitative Approaches to Reasoning with Uncertainty, ECSQARU. LNAI, vol. 4724, pp. 840–852 (2007)
7. Joshi, M., Lingras, P., Rao, C.R.: Correlating fuzzy and rough clustering. Fundamenta Informaticae **115**, 233–246 (2012)
8. Lingras, P., Nimse, S., Darkunde, N., Muley, A.: Soft clustering from crisp clustering using granulation for mobile call mining. In: Proceedings of the GrC 2011: International Conference on Granular Computing, pp. 410–416 (2011)
9. MacQueen, J.B.: Some methods for classification and analysis of multivariate observations. In: Proceeding of the 5th Berkeley Symposium on Mathematical Statistics and Probability, pp. 281–296 (1967)
10. Huang, Z.: Extensions to the k-means algorithm for clustering large data sets with categorical values. Data Mining Knowl. Discov. **2**, 283–304 (1998)
11. Pawlak, Z.: Rough sets. Int. J. Inf. Comput. Sci. **11**, 341–356 (1982)
12. Pawlak, Z.: Rough Sets: Theoretical Aspects of Reasoning about Data. Kluwer Academic Publishers, Norwell (1992)
13. Zadeh, L.A.: Fuzzy sets as a basis for a theory of possibility. In: Fuzzy Sets and Systems, vol. 1, pp. 3–28 (1978)

Improved Symbol Segmentation for TELUGU Optical Character Recognition

Sukumar Burra[1]([✉]), Amit Patel[2], Chakravarthy Bhagvati[1], and Atul Negi[1]

[1] School of Computer and Information Sciences, University of Hyderabad,
Hyderabad 500046, Telangana, India
sukumar.leo@gmail.com, {chakcs,atulcs}@uohyd.ernet.in
[2] RGUKT IIIT Nuzvid, Krishna, Nuzvid, Andhra Pradesh, India
amtptl93@gmail.com

Abstract. In this paper, we propose two approaches to improving symbol or glyph segmentation in a Telugu OCR system. One of the critical aspects having an impact on the overall performance of a Telugu OCR system is the ability to segment or divide a scanned document image into recognizable units. In Telugu, these units are usually connected components and are called glyphs. When a document is degraded, most connected component based algorithms for segmentation fail. They give malformed glyphs that (a) are partial and are a result of breaks in the character due to uneven distribution of ink on the page or noise; and (b) are a combination of two or more glyphs because of smudging in print or noise. The former are labelled *broken* and the latter, *merged* characters. Two new techniques are proposed to handle such characters. The first idea is based on conventional machine learning approach where a Two Class SVM is used in segmenting word into valid glyps in two stages. The second idea is based on the spatial arrangement of the detected connected components. It is based on the intuition that valid characters exhibit certain clear patterns in their spatial arrangement of the bounding boxes. If rules are defined to capture such arrangements, we can design an algorithm to improve symbol segmentation. Testing is done on the Telugu corpus of about 5000 pages from nearly 30 books. Some of these books are of poor quality and provide very good test cases to our proposed approaches. The results show significant improvements over developed Telugu OCR (Drishti System) on poor-quality books that contain many ill-formed glyphs.

Keywords: Bounding boxes · Optical Character Recognition
Symbol level segmentation · Support Vector Machine · Telugu OCR

1 Introduction

Optical Character Recognition (OCR) systems deal with the recognition of printed or handwritten characters. OCR System is the one which converts a scanned document image into editable text. Any OCR system implementation

A. Abraham et al. (Eds.): ISDA 2017, AISC 736, pp. 496–507, 2018.
https://doi.org/10.1007/978-3-319-76348-4_48

involves the following basic steps: Preprocessing which involves binarization, skew detection and correction. Symbol Segmentation i.e. Line, word and character segmentation. Feature extraction of a symbol, actual Recognition using classifier and finally, Conversion of recognized symbol into text format (UNICODE). Symbol Segmentation is the process of decomposing lines into words, then words into characters or recognizable units called **Glyphs**. Glyph is a basic orthographic unit in a scripture of a language. English and other European language scriptures contain small set of alphabets which are combined to form words. So segmenting and recognizing these characters is not as big an issue as segmenting Indic script characters.

Where as an Indian languages shows huge diversity in their orthography. Especially south Indian languages like Telugu, Kannada etc. are of complex orthography with a large number of distinct character shapes which are combined to form compound characters. In order to reduce the complexity, the characters are generally broken into basic glyphs to recognize them. In most of the OCR systems this segmentation is done by extracting the connected components.

Developing the OCR systems for Indian language scripts is challenging task. Telugu is a phonetic language, written from left to right, with each character representing generally a syllable [1]. There are 52 letters in the Telugu alphabet: 16 *Achchulu* which denote basic vowel sounds, 36 *Hallulu* which represent consonants. In addition to these 52 letters, there are several semi-vowel symbols, called *Maatralu*, which are used in conjunction with *Hallulu* and, half consonants, called *Voththulu*, to form clusters of consonants.

DRISHTI is a complete OCR system for Telugu, which was developed at University of Hyderabad [2]. DRISHTI is the first comprehensive OCR system for Telugu. This system was initially described [1], and then improved subsequently [3]. It was reported [1] that DRISHTI system gives better recognition accuracy at the glyph level up to 97% on good quality Telugu document images, which are scanned at 300 dpi. It also resulted in reasonable accuracy over different kinds of inputs like novels, newspapers and laser printed text. A significant improvement were reported in [4] using inverse fringe.

DRISHTI system uses connected component extraction to segment the characters. But sometimes, imperfections in scanning, poor quality printing and binarization leads the characters to break or merge with each other. In such cases connected component extraction results in touching and broken glyphs. In this paper, we have proposed the methods to handle this situation in order to improve the symbol segmentation. We have proposed two level Segmentation in order to reduce the broken glyphs and proposed an approach to merge the broken glyphs.

The content of paper is organized as follows: apart from introduction Sect. 2 provides overview of proposed approaches. Two level Symbol Segmentation approach was described in Sect. 3. Approach to merge the broken glyphs is discussed in Sect. 4, Sect. 5 discusses about experimental results and analysis of it and Finally, Sect. 6 gives Conclusion.

2   Proposed Methods

As we discussed earlier, connected component labelling algorithm on word images may result in broken and touching glyphs, which are recognized wrongly most of the time. So there is a need to handle these touching and broken glyphs in order to get better recognition. And also we need to segment the symbols in a better way to avoid broken and touching components.

   We have proposed following approaches in order to improve the symbol segmentation:

2.1   Two Level Symbol Segmentation

In order to reduce the broken glyphs, we have proposed an approach to segment the words in two levels into connected components. If we observe the glyphs in Telugu, almost all of them are round in shape and with curves. And, there are no vertical strokes in Telugu script. So, vertical projection profile of a word will help in segmenting the connected glyphs correctly, though they may have breaks at certain locations and angles.

   Segmenting the words vertically does not result in single connected components over the words having the compound characters. So, we need to segment these compound characters into connected components in order to send it to the recognition step. For this, we have used a classifier to classify the component which is obtained after Vertical cut segmentation, into a single connected component (SCC) and Compound connected component (CCC). We will send single connected component to the recognition directly. On the Compound connected component, we apply connected component approach [10] to extract the connected components and then send to the recognition step. We have briefly discussed all the steps in Two level symbol segmentation approach in Sect. 3.

2.2   Merging Broken Glyphs Using Bounding Box Analysis

Two level symbol segmentation is able to extract most of the broken Single-CCs correctly. But, whenever broken Compound-CCs having are subjected to Connected component labelling, they result in broken glyphs.

   To handle these broken Compound-CCs, we have computed bounding boxes for each connected component resulted from the segmentation. Bounding box for a connected component is the Minimum Bounding Rectangle (MBR). Later, we analysed MBRs of CCs in different Telugu document images. This analysis helped us to identify the possible MBRs of connected components which can be generated only by valid Telugu characters.

   Along with this analysis, we have used the zone information [5] of a line for framing the rules to merge the MBRs of broken glyphs. We have described these details in Sect. 4.

3   Two Level Symbol Segmentation

As we mentioned in Sect. 2.1, in Two level symbol segmentation, first segment the words into components using Vertical Cut Segmentation algorithm, which is designed based on vertical projections. We refer these components as **Vertical Cut Components**. Later, we classify these vertical cut components into single and compound connected components using Two class SVM. Finally, connected component segmentation algorithm was applied on compound connected components to extract CCs.

3.1   Vertical Cut Segmentation

Vertical Cut Segmentation algorithm splits the word image into vertical cut components. These components are segmented by identifying the vertical boundaries for each component in a word. This is done by using vertical projection profile [6] of the word i.e. number of black pixels in each column of the word. The Vertical Cut Segmentation on the word image shown in Fig. 1(a), results in the components shown in Fig. 1(b). Connected component segmentation on this word image will definitely result in broken glyphs at third and fourth component in Fig. 1(b).

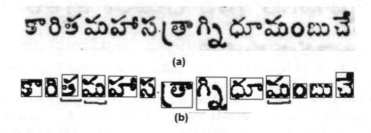

(a)

(b)

Fig. 1. Sample word image with segmented vertical cut components

3.2   Classification of Single and Compound Connected Components

The components obtained from the vertical cut segmentation algorithm are either Single connected components or Compound connected components. Only, the compound connected components are to be segmented further, in order to get Single connected components. Otherwise, the SCCs which having breaks in them will be segmented further, and results in broken glyphs. So, we have used a binary classifier i.e. Two Class SVM, in order to classify which components are to be sent directly to the recognition and which are to be segment further.

Features. We have scaled the components obtained from Vertical Cut Segmentation, into the size 32×32. We have stored pixel intensities of entire 32×32 sized components in 1024 sized array row by row. These 1024 valued array is taken as a feature vector for each component. Since the components are binarized, these 1024 values are either 0 (i.e. black pixel) or 255 (i.e. white pixel).

Classifier - (Two Class SVM). We have used **Two Class-Support Vector Machine** to classify the components into two classes. Here, we consider the first class as Single CC and the second class is Compound CC. As we know that, TC-SVM learning constructs a classification hyperplane to maximally separate the two classes in the feature space. We have used ν-Support Vector Classification (ν-SVC) implementation of Two class SVM [7].

Here, we have implemented Two Class SVM by using CvSVM [8], which is in-built OpenCV library class for Support Vector Machines. This CvSVM class implementation in OpenCV is based on LibSVM [9].

Training and Testing: We have collected 1000 samples of vertical cut components for each class i.e. 1000 Single-CCs and 1000 Compound-CCs. These 2000 samples are extracted from different the Telugu books, which include good quality printed books as well as poor quality old printed books.

As we mentioned earlier, Single-CC class samples are labelled as Class-0. Class-0 includes the single connected components which contains breaks too. These trained, broken single-CCs in this class are responsible for segmenting single-CCs correctly, though they contain breaks.

The Compound-CC class samples are labelled as Class-1. In this class, all samples are compound characters, which contain more than one CC. We have included some of basic glyphs like *sa, pa*, because they contain 2 CCs. We have also included the components having multiple basic glyphs, which we are not able to segment using our vertical cut segmentation algorithm.

Trained TC-SVM builds a model, which is able to classify the test sample i.e. a vertical cut component as either Class-0 or Class-1.

For testing, we collected 1000 samples of single-CCs and 1000 samples of Compound-CCs, other than 2000 training samples. TC-SVM is able to classify 940 samples as Class-0, out of 1000 class-0 test samples, which gives accuracy 94% for Class-0. And, it classified 898 samples as Class-1, when it is tested over 1000 Class-1 samples. The accuracy for class-1 is 89.8%.

3.3   Experiments and Observations

We have experimented Two-level Symbol Segmentation over a different type of document images which are good as well as poor quality printed.

The Vertical Cut segmentation output of a input line image in Fig. 2(a) will be shown in Fig. 2(b). Then, the Compound-CCs in Fig. 2(b) subjected to Connected component segmentation. The final output of Two level segmentation on input image i.e. Fig. 2(a) is shown in Fig. 2(c). Here, it can be observed that, current approach is able to segment broken glyphs as a combined and able to break Compound-CCs into valid glyphs.

Two level symbol segmentation is capable of handling the breaks in single-CCs and extract them as a single components. But, the broken glyphs due to the segmentation of Compound-CCs are still remain. In the next section, we have proposed an approach to merge the broken glyphs arises in Compound-CCs.

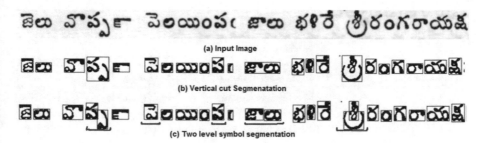

(a) Input Image

(b) Vertical cut Segmenatation

(c) Two level symbol segmentation

Fig. 2. (a) Input test line image (b) Output of vertical cut segmentation on line image (c) Output of two level segmentation on line image

4   Merging Broken Glyphs Using Bounding Box Analysis

4.1   Bounding Boxes

Bounding box of a connected component is a representation of it's maximum extents in 2D(X-Y) co-ordinate system. It's also referred as Minimum Bounding Rectangle (MBR). MBR is the rectangle formed by $min(x), max(x), min(y)$ and $max(y)$ of a connected component. Figure 3 shows the word and its bounding boxes.

Fig. 3. Word image and it's bounding boxes

4.2   Zone Information of a Line

Most of North Indian language scripts like Devanagari have Shirorekha. There is no Shirorekha in Telugu script. It is difficult to identify the top and lower zones. But, most of the Telugu base characters are in middle zone. This will be used to divide the line into three zones i.e. top, bottom and middle zone.

To divide a line into 3 zones, 4 boundaries are required. Two boundaries came to be known though the line MBR i.e. top and bottom. To identify other two boundaries, horizontal projection profile of a line is computed. Then two peaks are identified. The position at which the peak is identified in the region from top boundary to middle point of the line is considered as *shirorekha* and the region from middle point to bottom boundary is considered as *baseline*. Figure 4(a) depicts the top, middle and bottom zones along with its boundaries, of a line image shown in Fig. 4(b).

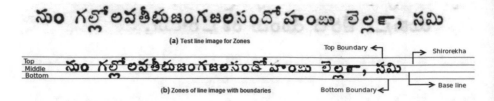

(a) Test line image for Zones

(b) Zones of line image with boundaries

Fig. 4. Test line image with zone boundaries

4.3   Rules to Merge the Broken Glyphs

In this approach, we have framed some rules, in order to combine the broken glyphs in the Compound-CCs by using their bounding boxes i.e. MBRs. We have also used Zone information of the line, word MBR and Compound-CC MBR to build these rules. The midpoints of Compound-CCs and component CCs of it are also computed. These are used for knowing to which zone of the line, the given CC belongs.

Broken Voththulu (At the Right and Left). Consider a Compound-CCs having a broken voththu which is joined at right or left to the hallu. The Connected component labelling on it leads to break that voththu. This type of breaks in voththulu are frequent and can be observed in Fig. 5.

Figure 5(A) shows a Compound-cc obtained after Vertical Cut algorithm. The connected component segmentation on it results in CCs that can be shown in Fig. 5(B). The bounding boxes of CCs in that Compound-CC, can be seen in Fig. 5(C). Figure 5(D) depicts the top, bottom boundaries, shirorekha and base line for the given Compound-CC.

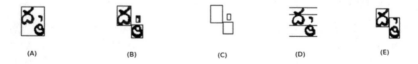

(A)   (B)   (C)   (D)   (E)

Fig. 5. Merging the broken voththu combined with hallu

To handle this type of broken Voththulu, first the broken voththu is to be identified in Compound-CC. Later they can be merged by combining their bounding boxes. This can be done by using following algorithm:

$CC \leftarrow$ Connected Component
$CCC \leftarrow$ Compound-CC
$flag \leftarrow 0$
for all $CC_i \in CCC$ **do**
   if $MBR(CC_i) \in$ Bottom Zone, and $MBR(CC_i) \in$ right part of $MBR(CCC)$
   then
      for all $CC_j \in CCC$ **do**

> **if** $MBR(CC_j) \in$ Middle Zone, and $MBR(CC_j) \in$ right part of $MBR(CCC)$
> **then**
>> **if** $distance(CC_i, CC_j) \leq Threshold$ **then**
>>> Break
>>> $flag = 1$
>> **end if**
> **end if**
> **end for**
> **if** $flag == 1$ **then**
>> Combine($MBR(CC_i)$, all $MBR(CC_j)$s for which $CC_i \in$ Middle Zone, and $MBR(cc_j) \in$ right part of $MBR(CCC)$)
> **end if**
> **end if**
> **end for**

This algorithm will combine the broken voththulu which can be possible right to the hallu. The combined MBR of broken voththu can be seen in Fig. 5(E). Here the voththu is placed at the right side of hallu. The similar rule can be written for voththulu which can be placed at the left side of hallu like ⌞ voththu.

In Top and Bottom Zones. In Telugu script, a simple or compound character never contain more than one connected component in Top zone. It implies that, if two are more bounding boxes are identified in Top zone alone, they should be broken glyphs of maatralu. So, they have to be combined.

This merging is done by using following rule:

$CC \leftarrow$ Connected Component
$CCC \leftarrow$ Compound-CC
for all $CC_i \in CCC$ **do**
> **for all** $CC_j \in CCC$ **do**
>> **if** $MBR(CC_i) \in$ Top Zone, and $MBR(CC_j) \in$ Top Zone **then**
>>> Combine($MBR(CC_i)$,$MBR(CC_j)$)
>> **end if**
> **end for**
end for

(A)   (B)   (C)   (D)   (E)

Fig. 6. Merging the broken glyphs in Top Zone

Figure 6(A) is a broken Compound-CC. The CCs can be observed in Fig. 6(B). The broken bounding boxes in Top Zone, can be seen in Fig. 6(C).

The Fig. 6(D) depicts zone boundaries. Finally, Fig. 6(E) shows the combined bounding boxes in Top zone by using the rule, which was described above.

On the other hand, in Telugu script, possibility of two more CCs in the bottom zone alone is very rare. If there are more than one CC in bottom zone, they can be combined by using the similar rule, which is framed for top zone. But, these type of broken glyphs are very rare.

Middle Zone. In Telugu script, all the base characters i.e. achchulu and hallulu will be in middle zone. There is no possibility of more than one CC in middle zone alone even in Compound CC, unless it was broken. If there are more than one CC in middle zone, they can be combined by using following rule:

$CC \leftarrow$ Connected Component
$CCC \leftarrow$ Compound-CC
for all $CC_i \in CCC$ **do**
  for all $CC_j \in CCC$ **do**
    if $MBR(CC_i) \in$ Middle Zone, and $MBR(CC_j) \in$ Middle Zone **then**
      Combine($MBR(CC_i)$,$MBR(CC_j)$)
    end if
  end for
end for

Fig. 7. Merging the broken glyphs in Middle Zone

The Compound-CC having multiple breaks in middle zone can be seen in Fig. 7(A). The CCs are shown in Fig. 7(B). The broken bounding boxes alone can be seen in Fig. 7(C). The Fig. 7(D) gives it's zone boundaries. After applying the rule mentioned above, the broken bounding boxes are combined and form a single bounding box in middle zone. This can be seen in Fig. 7.

These rules are framed by observing the frequent broken glyphs. Though, these rules doesn't handle all possible broken glyphs, but still able to merge some broken glyphs. The strength of these rules is that generalization over script i.e. rules are not framed for some specific symbols.

5   Experiments and Analysis of Results

We have done the experiments over 5000 pages of Telugu document which includes, good as well as poor quality print. Here, we have used Two level symbol segmentation to extract the valid glyphs and later, we applied the rules that are proposed in Sect. 4 to merge the broken glyphs.

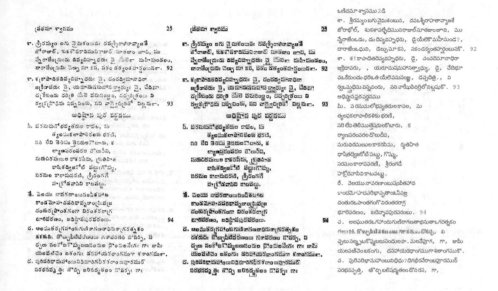

Fig. 8. Input sample image, the valid glyphs extracted from it and OCR output

Let us consider the Input image shown in Fig. 8. This is a page from the book named **Vasucharithramu**, having lots of broken glyphs.

The valid glyphs which are extracted from the input image in Fig. 8, using proposed approaches i.e. Two level symbol segmentation and Merging the broken glyphs, is shown in Fig. 8. It can be observed that, the components are segmented well, though they have breaks in them. We can also observe, some merged broken glyphs which are underlined below.

The OCR output text for the input image shown in Fig. 8, can be seen in Fig. 8. For this document image, the error rate is reduced more than 10%, when compared with the OCR output of same image by the old DRISHTI system.

The Table 1 gives the comparison of the error rates for some document images computed over the old DRISHTI system and DRISTHI with proposed approaches. It can be observed that, the error rate is reduced significantly for the documents having lots of broken glyphs.

We have done experiments over different Telugu books and novels with the proposed approaches. We have reported the average error rate for some books, using current approaches along with previous error rates (by DRISHTI with Connected component approach alone) in the Table 2.

From Table 2, it can be observed that, the error rate is reduced significantly in first two books. Because, in these books the print quality is so poor which causes lots of broken glyphs. These are segmented, merged and recognized correctly by using proposed approaches. We can also observe the reduction in error rate for next two books. Over running the proposed algorithm on 836 pages an average error rate reduced by 3.39%.

Table 1. Comparison of DRISHTI system and proposed system error rates

S.no.	Book name	Page no. (%)	DRISHTI system error rate (%)	Proposed system error rate (%)	Reduction in error rate
1	Ooragaya Navvindi	14	34.98	22.40	12.58
2	Vasucharitramu	40	38.69	26.41	12.28
3	Vasucharitramu	25	31.69	21.69	10.00
4	Sasaanka Vijayamu	3	33.33	24.34	8.99
5	Sasaanka Vijayamu	38	47.62	38.65	8.97
6	GVS Navalalu-Kathalu	1	25.45	16.87	8.58
7	Vasucharitramu	114	25.60	17.11	8.49
8	Sasaanka Vijayamu	122	28.17	19.85	8.32
9	GVS Navalalu-Kathalu	32	30.14	23.18	6.96
10	Ooragaya Navvindi	26	30.46	23.97	6.49

Table 2. Comparison of DRISHTI system and proposed system average error rate over the books

S.no.	Book name	Number of pages	DRISHTI system error rate (%)	Proposed system error rate (%)	Difference
1	Vasucharitramu	203	27.42	22.31	5.11
2	Sasaanka Vijayamu	181	30.24	26.29	3.95
3	GVS Navalalu-Kathalu	125	21.96	18.64	3.32
4	Ooragaya Navvindi	327	18.06	16.88	1.18

6   Conclusion

In this paper, we proposed two methods to improve symbol or glyph segmentation in Telugu OCR systems. The first proposed method is Two level symbol segmentation method, which is able to extract many broken characters correctly. The second method is based on bounding box analysis. From experiments which we performed over 5000 pages of Telugu documents observed that proposed methods do result in significant improvement in OCR performance on degraded pages.

References

1. Negi, A., Bhagvati, C., Krishna, B.: An OCR system for Telugu. In: ICDAR, pp. 1110–1114 (2001)
2. Vishwabharat@tdil: Journal of Language Technology, July 2003
3. Bhagvati, C., Ravi, T., Kumar, S.M., Negi, A.: On developing high accuracy OCR systems for Telugu and other Indian scripts. In: Language Engineering Conference, p. 18 (2002)
4. Patel, A., Burra, S., Bhagvati, C.: SVM with inverse fringe as feature for improving accuracy of Telugu OCR systems. In: Progress in Intelligent Computing Techniques: Theory, Practice, and Applications, p. 10 (2016)
5. Dholakia, J., Negi, A., Mohan, S.R.: Progress in Gujarati document processing and character recognition. In: Advances in Pattern Recognition, pp. 73–95 (2010)
6. Baird, H., Kahan, S., Pavlidis, T.: Components of an omnifont page reader. In: Proceedings of ICPR, pp. 34–348, October 1986
7. Schölkopf, B., Smola, A.J., Williamson, R.C., Bartlett, P.L.: New support vector algorithms. Neural Comput. **12**(5), 1207–1245 (2000)
8. OpenCV 2.4.9.0 documentation - Support Vector Machines
9. Chang, C.C., Lin, C.J.: LibSVM: a library for Support Vector Machines. ACM Trans. Intell. Syst. Technol. **2**(3), 27:1–27:27 (2011)
10. Gonzalez, R., Woods, R.: Digital Image Processing. Addison-Wesley, Reading (1993)

Semantic Attribute Classification Related to Gait

Imen Chtourou[1(✉)], Emna Fendri[2(✉)], and Mohamed Hammami[2(✉)]

[1] MIRACL Laboratory, ENIS, University of Sfax,
Road Sokra km 4, BP 1173, 3038 Sfax, Tunisia
imene.chtourou@gmail.com
[2] MIRACL Laboratory, FSS, University of Sfax,
Road Sokra km 4, BP 802, 3038 Sfax, Tunisia
fendri.msf@gnet.tn, mohamed.hammami@fss.rnu.tn

Abstract. Human gait, as a behavioral biometric, has recently gained significant attention from computer vision researchers. But there are some challenges which hamper using this biometric in real applications. Among these challenges is clothing variations and carrying objects which influence on its accuracy. In this paper, we propose a semantic classification based method in order to deal with such challenges. Different predictive models are elaborated in order to determine the most relevant model for this task. Experimental results on CASIA-B gait database show the performance of our proposed method.

Keywords: Pedestrian analysis · Semantic attributes · Classification

1 Introduction

In the two past decades, surveillance cameras are widespread in many places such as train stations, parking lots, airports, banks, etc. Therefore, surveillance video applications have drawn a large amount of research attention in the world. Several biometric identifiers are used in these applications. Biometrics identifiers used currently include signature, fingerprint, palm vein, face, iris, retina scans, etc. These identifiers are not convenient for busy environments since they all need high quality images from close distances and the subjects cooperation. Human gait as a behavioral biometric identifier has received significant interest in recent years. This is due to its unique characteristics such as unobtrusiveness, recognition from distance and no need of high quality video. However, there are difficulties that face several gait based methods caused by covariate conditions that affect the gait negatively. Examples of these covariate conditions that commonly occur in real life are changes in clothing and the carrying objects (such as carrying a bag). These covariate conditions create problems for practical gait based application and significantly deteriorate their performance. They occlude the appearance of the body shape. In addition, they have an effect on the dynamic pattern of body movements. Whenever there are occluded pixels as

A. Abraham et al. (Eds.): ISDA 2017, AISC 736, pp. 508–518, 2018.
https://doi.org/10.1007/978-3-319-76348-4_49

a result of carrying a bag or wearing a baggy coat, it is normal that the accuracy decreased. A solution for these covariates may be adopted to use parts which are unaffected by these items in re-identification application. In this paper, a new method relying on the detection of these covariates (i.e. clothing variation and carrying objects) is proposed. To this end, we have used semantic attribute as they are human-understandable properties. Semantic attribute classification related to gait may be investigated for several applications such as gait recognition or gait based person re-identification, etc. The rest of this paper is organized as follows. Section 2 summarizes some related works. Section 3 introduces the proposed method. Section 4 describes the evaluation protocol. Section 5 concludes this paper.

2   Related Work

Recently, semantic attributes, which are human-understandable properties, such as male, black eyes, long hair, etc. are gained more and more interests. This interest is due to their ability to infer high-level semantic knowledge. Vaquero et al. [7] identified people by a series of attribute detectors. They introduce an attribute-based people searching system in surveillance environments. Layne et al. [4] definite a set of human-understandable pedestrian attributes such as "longhair", "headphones", "male", "backpacks", "sunglasses", "v-necks" and "clothing" on person re-identification databases like Viper [1] and PRID [2]. The work of [8] proposed an approach to illustrate the appearance of pedestrian with several binary attributes like "is male", "has T-shirt", "glasses", "has jeans", "long hair", etc. They used a set of parts from Poselets [9] for extracting low-level features and perform subsequent attribute learning. Describing clothing appearance with semantic attributes is an appealing technique for many important applications. Yang et al. [11] proposed a clothing recognition system that identifies clothing categories such as suit, T-shirt and Jeans in a surveillance video. Their method was based essentially on color (i.e. colour histogram) and texture features (i.e. histogram of oriented gradient HOG, a bag of dense SIFT features and DCT responses). Linear Support Vector Machines (SVM) classifiers have been trained in order to learn clothing categories. Chen et al. [10] proposed a method that comprehensively describes the upper clothing appearance with sets composed of multi-class attributes and binary attributes. They consider high-level attributes such as clothing categories and deal also with some very detailed attributes like "collar presence", "neckline shape", "striped", "spotted" and "graphics". Liu et al. [3] assemble a large online shopping database and a daily photo database for the research of the cross-scenario clothing retrieval. They define a set of clothing-specific attributes. These can be summarized into three classes, i.e., global, upper-body and lower-body attributes. Lower-body attributes can be further divided into two related attributes classes. Recently, Convolutional neural networks (CNN) has been adopted also in pedestrian attribute classification [15,17,18]. Zhu et al. [17,22] applied the learned CNN for person re-identification task. They used the pedestrian attribute classification by weighted interactions from other attributes. In this method, a set of attribute was defined such as "male", "redshirt", "barelegs", "lightbottoms", "notlightdark", etc. Li et al. [16] also propose two deep learning models to learn the pedestrian attributes, one is called as deep learning based single attribute recognition model (DeepSAR) and the other is a deep learning framework which recognizes multiple attributes jointly (DeepMAR). Authors have used attributes like "Age 16–30", "Casual lower", "V-neck", "Sunglasses", "Formal lower", etc. Matsukawa and Suzuki [18] refined CNNs for attribute recognition and employ metric learning for person re-identification. They grouped attributes into 7 groups which are "Gender", "Age", "Luggage", "UpperBody Clothing", "UpperBody Color", "LowerBody Clothing", "LowerBody Color". Lin et al. [15] use person attributes as auxiliary tasks to learn more information. Attributes are composed of groups such as gender (male, female), color of shoes (dark, light), 8 colors of upper-body clothing (black, white, red purple, gray, blue, green, brown), age (child, teenager, adult, old), etc.

However, Attributes like "sunglasses", "headphones", "is male", "Age" and "neckline shape" used in the work of [3, 4, 8, 10, 11, 15, 16, 18] may not alter or affect the natural gait appearance and dynamic pattern of body motion. Therefore, we have concentrated in this work on Single Shoulder Bag, Back Pack, Hand Bag and Outerwear attributes. These attributes can influence and occlude the gait based appearance of the body shape and consequently decrease accuracy. We have considered also carrying nothing as an attribute.

3   Proposed Method

We propose to build an automatic semantic attribute classification solution based on machine learning method using a set of manually classified attributes, in order to produce a predictive model, which allow predicting the class of each semantic attribute. To classify each semantic attribute into each one class, we are based on the Knowledge Discovery in Databases (KDD) process for extracting useful knowledge from volumes data [19]. The general principle of the classification method is the following: Let S be the set of image's samples to be classified. To each sample s of S one can associate a particular class of attribute, namely, its class label C. C takes its value in the class of labels (0 for the absence of attribute, 1 for the presence of attribute).

$$C : S \rightarrow \Gamma = \{existence, nonexistence\} \tag{1}$$

$$s \in S \rightarrow C(s) \in \Gamma \tag{2}$$

Our study consists in building a model to predict the attribute class of each persons image. The total process of the detection of pedestrian semantic attributes is composed of two steps: (i) an Off-line step and (ii) On-line step. Figure 1 shows the framework of the proposed method.

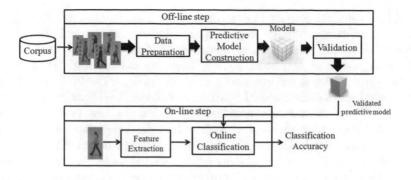

Fig. 1. The framework of the proposed method.

3.1   Off-line Step

The off-line stage involves three major sub-steps. First, we start with a data preparation of the training database related to semantic attribute. Second, a predictive model for each class of attribute is constructed. Third, a validation step is required.

Data Preparation. Given a corpus of images taken randomly, pedestrian's bounding box is detected from each image. The corpus of bounding box's images is divided in such a way that the training set contains equal number of positive and negative samples for each attribute. In this step, our goal is to construct a two-dimensional table from our training database. Each table row represents a bounding box's image and each column represents a feature. In the last column, we save the semantic attribute class denoted 1 for the presence of attribute and 0 for the absence of attribute. In our work, we are concerned with five different semantic attributes. Figure 2 gives the list adopted from CASIA-B database. It shows example of positive images for each attribute: Hand Bag, Single Shoulder Bag, Back pack, Outerwear and carrying nothing. The annotation practice was carried out automatically according to the name of images for carrying nothing (Normal), carrying Bags and wearing Coat (Outerwear) and to the image's number for each category of bags. Low-level color and texture features have

Fig. 2. Example positive images for each attribute. From left to right: Hand bag, Single Shoulder Bag, Back pack, Outerwear and Carrying nothing from CASIA-B database.

been shown their robustness in describing pedestrian images [4]. Therefore, we extracted a collection of color features (i.e. color histograms in RGB, HSV and YCbCr color spaces) and texture features (i.e. Gabor and Schmid filters) to model each semantic attribute. Several conventional methods [4,12,13] have used the same features set composed of RGB, HSV, YCbCr, Gabor, Schmid. We extracted a 2784-dimensional feature vector from each bounding box of person image. Once the data preparation step is defined, our task is to perform machine learning using different classifiers in order to prepare model for each class of semantic attribute. Further, trained classes are tested for the classification accuracy and their corresponding results are presented in the experimental result section. There are several algorithms of supervised learning in the literature. Each having its advantages and disadvantages. We used three supervised algorithms from different families like the support vector machines [20], Tree bagger based decision tree [21] and the neural networks [14]. In the end, the best performer predictive model is chosen.

Support Vector Machines (SVM)

Classification by SVM (Support Vector Machines) [5] is performed by constructing a model that iteratively separates the training data into two classes. It is defined over a vector space where categorization is achieved by linear or non-linear separating surfaces in the input space of the original data set [20].

$$\min_{w,b,\xi} \frac{1}{2} w^T w + C \sum_{i=1}^{N} \xi_i \tag{3}$$

$$Subject\ to\ \left\{ y_i(w^T \Phi(x_i) + b) \geq +1 - \xi_i,\ i = 1, ..., N \xi_i \geq 0,\ i = 1, ..., N \right. \tag{4}$$

ξ_i's are slack variables required to permit misclassifications in the set of inequalities, and $C \epsilon \mathbb{R}^+$ is a tuning hyper parameter, weighting the significance of classification errors to the margin width. Here training vectors x_i are mapped into a higher (maybe infinite) dimensional space by the function Φ. Then SVM finds a linear separating hyperplane with the maximal margin in this higher dimensional space. Furthermore, $K(X_i, X_j) = \Phi(X_i)^T \Phi(X_j)$ is called the kernel function. In our work, we use a SVM using the histogram Intersection (HI) as kernel since our feature vectors are based on histograms, as formulated below.

$$K(X_i, X_j) = \sum_{k=1}^{n} min\ \{x_i, x_j\} \tag{5}$$

where: $X_i = \left\{ x_1^i, ..., x_n^i \right\}$ and $X_j = \left\{ x_1^j, ..., x_n^j \right\}$ are two histograms with n-bins (in $\mathbb{R}^n$). HI kernel has been proved a positive result which makes it suitable as a discriminative classification kernel.

Neural Network (NN)

Neural networks [14] have been used in many image-related applications and exhibited good performances. NN is a network of simple neurons called perceptrons. The perceptron computes a single output from multiple real-valued inputs for forming a linear combination according to its input weights and then possibly putting the output through some nonlinear activation function. Mathematically this can be written as:

$$y = \varphi(\sum_{i=1}^{n} w_i x_i + b) = \varphi(wx^T + b) \tag{6}$$

Where w denotes the vector of weights, x is the vector of inputs, b is the bias and φ is the activation function.

In order to train a model based on an Artificial Neural Networks, we use a multilayer perceptron with the back propagation learning algorithm [24]. We have adopted a neural network with a three-layer architecture consisting of input, hidden and output layers for the prediction of each semantic attribute. The activation function for the neurons in the hidden layer and in the input layer is sigmoid function defined as below.

$$f(x) = \frac{1}{1 - exp(-x))} \tag{7}$$

In a back propagation network, a supervised learning algorithm controls the training phase. Then, the input and output (desired) data should be available, thus allowing the calculation of the error of the network as the difference between the calculated output and the desired vector.

Tree Bagger (FT)

Decision trees are a popular method for various machine learning tasks. Random forests is a notion of the general technique of random decision forests [21] that are an ensemble learning method for classification, regression and other tasks, that work by constructing a multitude of decision trees at training time and outputting the class that is the mode of the classes (classification) or mean prediction (regression) of the individual trees. The training algorithm for random forests assigns the general technique of bootstrap aggregating, or bagging, to tree learners. Given a training set $X = x_1, ... x_n$ with responses $Y = y_1, ... y_n$, bagging repeatedly (B times) selects a random sample with replacement of the training set and fits trees to these samples: After training, predictions for unseen samples x' can be made by averaging the predictions from all the individual regression trees on x':

$$\hat{f} = \frac{1}{B} \sum_{b=1}^{B} \hat{f}_b(x') \tag{8}$$

or by taking the majority vote in the case of decision trees.

Validation of the Predictive Model. The aim of this step is to measure the performance of the learned predictive model in order to guarantee the generality and effectiveness for future samples. Several possible metrics have been proposed in the literature to evaluate the quality of a predictive model. Among this metrics, we have opted for Correct Classification Accuracy (CCA) which denotes the ratio of correctly classified images with the total number of images. It is defined by:

$$CCA = \frac{Number\ of\ correctly\ classified\ Image}{Total\ number\ of\ images} \tag{9}$$

3.2   On-line Step

Given a new pedestrian's bounding box, we start by extracting features to represent the global information of each image. Labels are generated depending on the semantic information of the image. These feature vectors design the input of the pre-learned predictive models.

4   Experimental Results

In this section, we present two series of experiments: The first serie of experiments is realized in order to show the classification accuracy of each semantic attributes using different supervised algorithms. In this serie of experiments, we used the CASIA-B database [23]. The second serie of experiments concerns comparison of our proposed method with Layne et al. [4]. This experiment shows the independence of our proposed method regarding the database used. In this serie of experiments we have used the VIPeR database [1]. Before presenting the results of the two series of experiments, we present in the next section a description of the two databases used CASIA-B [23] and VIPeR [1].

4.1   Description of Used Databases

CASIA gait database collected by the Institute of Automation of the Chinese Academy of Sciences [23]. Database B [23] is a large multi view gait database collected indoors with 124 subjects and 13640 samples from 11 different views ranging from 0 to 180°. In our experiments, we consider only (90°). This is motivated by the fact that gait information is more significant and reliable in the side view [6]. Each person is recorded six times under normal conditions, twice under carrying bag conditions and twice under clothing variation conditions. Table 1 shows the repartition of train and test sets for the five semantic attributes from the CASIA-B database.

VIPeR viewpoint invariant pedestrian recognition (VIPeR) database [1]. This database contains 632 person image pairs taken from two non overlapping camera views (camera A and camera B). Each image is scaled to 64×128 pixels. Images appearance exhibit significant variation in pose, illumination conditions with the presence of occlusions and viewpoint.

Table 1. Train and test sets repartition of the 5 semantic attributes from CASIA-B database.

Attribute	#train	#test
Carrying nothing	1488	496
Outerwear	744	248
Single shoulder bag	522	174
Hand bag	132	44
Back pack	90	30

4.2   First Serie of Experiments: Performance Evaluation

In this section, we presented the first series of experiment. Data from CASIA-B database is divided in such a way that each attribute had an equal number of positive and negative samples. We have used three supervised algorithms namely Support Vector Machines (SVM), Tree Bagger (FT) and Neural Network (NN). For SVM classifier, we used LIBSVM [5]. For neural network, we have adopted 10 hidden neurons and back propagation algorithm is chosen to evaluate classifier. For Tree bagger, we have used 100 trees. The classification results of the three supervised algorithms are presented in Table 2. Results shows that the neural network gives better accuracy than SVM and tree bagger for 4 semantic attributes (i.e. Outerwear, Single Shoulder Bag, Back pack and Hand Bag) higher than 89%. For the semantic attribute carrying nothing, SVM shows better performance than neural network. This confirms that neural network is more precise and efficient for detecting semantic attributes that alter human shape. Thanks to its effectiveness for the majority of semantic attribute classification, neural network (NN) will be adopted for our proposed method.

Table 2. Experimental results by the three algorithms on CASIA-B database.

Attribute	Classification accuracy (%)		
	Support vector machines	Neural network	Tree bagger
Outerwear	80.645	89.9	73.4
Single shoulder bag	79.885	89.4	78.2
Hand bag	84.09	92.3	65.9
Back pack	73.333	94.4	76.7
Carrying nothing	89.314	86.2	72.0

4.3 Second Serie of Experiments: Comparison with State-of-the-Art Method

We compared our attribute classification results with the popular work related of Layne et al. [4]. Our proposed method was tested using as probe set Camera A from the database VIPeR [1]. It should be noted that images are randomly taken. This serie of experiments shows the performance of our proposed method compared to the popular method of Layne [4] and it also shows that our selected predictive model is independent of the database used. Table 3 shows that classification accuracy for carrying nothing, outerwear (coat), single shoulder bag and hand bag attributes are higher compared to [4] results. Accuracy for back pack attribute is slightly lower. This is due to occlusion that may cover back pack such as arm and may consequently alter the shape.

Table 3. Comparison with state-of-the-art method.

Attribute	Classification accuracy (%)	
	Proposed method	Layne et al. [4]
Outerwear	45.8	–
Single shoulder bag	56.9	56.0
Hand bag	76.9	54.5
Back pack	60.0	68.6
Carrying nothing	70.0	69.7

5 Conclusion

In this paper, we have investigated the classification of semantic attribute that can not only influence and occlude the appearance of the body shape, but also have an impact on the dynamic pattern of body motion and consequently on accuracy. Different supervised algorithms were proposed, we have proved that using neural network as a supervised algorithm improves the attribute classification performance compared to the state-of-the-art method. Inspired by the promising performance, we will further explore how to adopt this semantic attribute classification solution in gait based task.

References

1. Gray, D., Brennan, S., Tao, H.: Evaluating appearance models for recognition, reacquisition, and tracking. In: Proceedings of the IEEE International Workshop on Performance Evaluation for Tracking and Surveillance (PETS), vol. 3, no. 5, pp. 1–7. Citeseer, October 2007
2. Hirzer, M., Beleznai, C., Roth, P.M., Bischof, H.: Person re-identification by descriptive and discriminative classification. In: Scandinavian Conference on Image Analysis, pp. 91–102. Springer, Heidelberg, May 2011

3. Liu, S., Song, Z., Liu, G., Xu, C., Lu, H., Yan, S.: Street-to-shop: cross-scenario clothing retrieval via parts alignment and auxiliary set. In: 2012 IEEE Conference on Computer Vision and Pattern Recognition (CVPR), pp. 3330–3337. IEEE, June 2012

4. Layne, R., Hospedales, T.M., Gong, S.: Attributes-based re-identification. In: Person Re-identification, pp. 93–117. Springer, London (2014)

5. Schllkopf, B., Smola, A.J.: Learning with Kernels: Support Vector Machines, Regularization, Optimization, and Beyond. MIT Press, Cambridge (2002)

6. Bashir, K., Xiang, T., Gong, S.: Gait recognition without subject cooperation. Pattern Recognit. Lett. **31**(13), 2052–2060 (2010)

7. Vaquero, D.A., Feris, R.S., Tran, D., Brown, L., Hampapur, A., Turk, M.: Attribute-based people search in surveillance environments. In: Workshop on Applications of Computer Vision (WACV), pp. 1–8. IEEE, December 2009

8. Bourdev, L., Maji, S., Malik, J.: Describing people: a poselet-based approach to attribute classification. In: IEEE International Conference on Computer Vision (ICCV), pp. 1543–1550. IEEE, November 2011

9. Bourdev, L., Malik, J.: Poselets: body part detectors trained using 3D human pose annotations. In: IEEE 12th International Conference on Computer Vision, pp. 1365–1372. IEEE, September 2009

10. Chen, H., Gallagher, A., Girod, B.: Describing clothing by semantic attributes. In: Computer Vision – ECCV 2012, pp. 609–623 (2012)

11. Yang, M., Yu, K.: Real-time clothing recognition in surveillance videos. In: 18th IEEE International Conference on Image Processing (ICIP), pp. 2937–2940. IEEE, September 2011

12. Nguyen, N.B., Nguyen, V.H., Duc, T.N., Duong, D.A.: Using attribute relationships for person re-identification. In: Knowledge and Systems Engineering, pp. 195–207. Springer, Cham (2015)

13. Umeda, T., Sun, Y., Irie, G., Sudo, K., Kinebuchi, T.: Attribute discovery for person re-identification. In: International Conference on Multimedia Modeling, pp. 268–276. Springer, Cham, January 2016

14. Yegnanarayana, B.: Artificial Neural Networks. PHI Learning Pvt. Ltd., New Delhi (2009)

15. Lin, Y., Zheng, L., Zheng, Z., Wu, Y., Yang, Y.: Improving person re-identification by attribute and identity learning. arXiv preprint arXiv:1703.07220 (2017)

16. Li, D., Chen, X., Huang, K.: Multi-attribute learning for pedestrian attribute recognition in surveillance scenarios. In: 3rd IAPR Asian Conference on Pattern Recognition (ACPR), pp. 111–115. IEEE, November 2015

17. Zhu, J., Liao, S., Yi, D., Lei, Z., Li, S.Z.: Multi-label CNN based pedestrian attribute learning for soft biometrics. In: International Conference on Biometrics (ICB), pp. 535–540. IEEE, May 2015

18. Matsukawa, T., Suzuki, E.: Person re-identification using CNN features learned from combination of attributes. In: 23rd International Conference on Pattern Recognition (ICPR), pp. 2428–2433. IEEE, December 2016

19. Fayyad, U., Piatetsky-Shapiro, G., Smyth, P.: The KDD process for extracting useful knowledge from volumes of data. Commun. ACM **39**(11), 27–34 (1996)

20. Boser, B.E., Guyon, I.M., Vapnik, V.N.: A training algorithm for optimal margin classifiers. In: Proceedings of the Fifth Annual Workshop on Computational Learning Theory, pp. 144–152. ACM, July 1992

21. Zighed, D.A., Rakotomalala, R.: Graphes d'induction: apprentissage et data mining. Hermes, Paris (2000)

22. Zhu, J., Liao, S., Lei, Z., Li, S.Z.: Multi-label convolutional neural network based pedestrian attribute classification. Image Vis. Comput. **58**, 224–229 (2017)
23. Zheng, S.: CASIA gait database (2005). http://www.cbsr.ia.ac.cn/english/Gait %20Databases.asp
24. Shih, F.Y.: Image Processing and Pattern Recognition: Fundamentals and Techniques. Wiley, Hoboken (2010)

Classification of Dengue Gene Expression Using Entropy-Based Feature Selection and Pruning on Neural Network

Pandiselvam Pandiyarajan[1]($\boxtimes$)
and Kathirvalavakumar Thangairulappan[2]

[1] Department of Computer Science, Ayya Nadar Janaki Ammal College,
Sivakasi 626124, Tamilnadu, India
pandiselvam.pps@gmail.com
[2] Research Centre in Computer Science, V.H.N.Senthikumara Nadar College,
Virudhunagar 626001, Tamilnadu, India
kathirvalavakumar@yahoo.com

Abstract. Dengue virus is a growing problem in tropical countries. It serves diseases, especially in children. Different diagnosing methods like ELISA, Platelia, haemocytometer, RT-PCR, decision tree algorithms and Support Vector Machine algorithms are used to diagnose the dengue infection using the detection of antibodies IgG and IgM but the recognition of IgM is not possible between thirty to ninety days of dengue virus infection. These methods could not find the correct result and needs a volume of the blood. It is not possible, especially in the children. To overcome these problems, this paper proposes classification method of dengue infection based on informative and most significant genes in the gene expression of dengue patients. The proposed method needs only gene expression for a patient which is easily obtained from skin, hair and so on. The classification accuracy has been evaluated on various benchmark algorithms. It has been observed that the increase in classification accuracy for the proposed method is highly significant for dengue gene expression datasets when compared with benchmark algorithms and the standard results.

Keywords: Neural network · Feature selection · Pruning · Dengue infection
Dengue diagnosis · Classification

1 Introduction

Dengue is a man kill disease transmitted by bites of Aedes aegypti mosquito in several regions and it also spreads through tropic countries. Aedes mosquitoes breed in clean water. It serves diseases, especially in children. In Tamilnadu, deaths have been reported due to the dengue rising to 87 included three girl children and a teenage boy. Meanwhile, South Delhi Municipal Corporation (SDMA) report said, of the total 4,545 dengue cases, 2152 were residents of Delhi, while the rest were from other states. According to the SDMC, Aedes mosquito breeding has been reported from 1, 80, 687 households in Delhi [26]. There is no drug produced for the treatment of dengue diseases because Dengue infections can be difficult to differentiate from other viral infections.

© Springer International Publishing AG, part of Springer Nature 2018
A. Abraham et al. (Eds.): ISDA 2017, AISC 736, pp. 519–529, 2018.
https://doi.org/10.1007/978-3-319-76348-4_50

It can be identified as an undifferentiated fever, Dengue Fever (DF), Dengue Hemorrhagic Fever (DHF) and Dengue Shock Syndrome (DSS). Some clinical diagnostic methods are used to diagnose the later stages of Dengue infections. These methods are based on the detection of IgG and IgM antibodies in the blood. After the infection, IgM becomes unrecognized between thirty and ninety days. Among these periods, the treatment of normal viral fever is given to dengue patients. This leads to severe of dengue infections. The recognition of IgG alone is not enough to confirm the dengue infection without the presence of IgM. Recovery from infection by one creates permanent immunity against that specific serotype. Succeeding infections by other serotypes enhance the risk of increasing severe dengue. The burden of dengue in the world is to classify dengue infections. This paper suggests a useful and stable method for classifying dengue infections using dengue gene expression data.

2 Literature Review

Dengue diagnosis method had proposed by [23] using the detection of Immunoglobulin IgM and IgG antibodies. They characterized the dengue infections. The positive IgM plus negative IgG (IgM+, IgG−) values of patients had primary infections while the positive IgG plus either positive or negative IgM (IgG+, IgM+/−) values of the patients had secondary infections. Similarly, the patients with positive IgM count and positive IgG count are classified as primary and secondary infections respectively [24]. The classification and pattern recognition techniques of data mining can be used for diagnosing Arbovirus dengue [6] and classifying the patient record data of dengue [4, 14]. Rough set theory is also used for generating classifying rules of dengue [11]. Arunkumar et al. [5] have proposed a dengue disease prediction system using decision tree and Support Vector Machine (SVM). The decision tree is generated using fisher filtering method. They provide better classification result.

Pabbi [9] has provided the Fuzzy rules for classifying dengue into three classes DF, DHF and Dengue Shock Syndrome (DSS) by using the factors age, TLC, SGOT/SGPT, Platelets count and BP. Fathima et al. [8] have proposed a method for classifying different dengue serotypes. Differences between dengue serotypes are identified using SVM classifier. Shaukat et al. [12] have analyzed the attack of dengue fever in different areas of Jhelum in Pakistan using k-means, k-medoids, DB Scan and optics clustering algorithms. Singh et al. [2] have proposed recommender system for detection of dengue using fuzzy logic. They have developed an android application for detection of dengue using the factors fever, blood pressure, joint pain, skin rashes, pain behind eyes, severe headache, etc. This system analyzed the factors using fuzzy logic and identified the fever is because of dengue or not. Tanner et al. [1] have proposed a dengue diagnosis method using decision tree algorithms by analyzing the factors such as platelet count, IgM and IgG antigen count and cross over threshold values collected from real-time RT-PCR of dengue viral RNA. C 4.5 decision tree classifier has been used to differentiate dengue from non-dengue febrile illness. DHF cases were classified correctly.

The system designed by [7] have proposed three artificial neural network model for diagnosing and identifying the dengue-affected patient's data from Jalpaiguri Sadar hospital, North Bengal, India. Ibrahim et al. [16] have developed the prediction system using multilayer feed forward neural networks based on clinical symptoms and signs of Dengue Fever (DF) and Dengue Hemorrhagic Fever (DHF). Artificial Neural Network (ANN) was trained for finding the dengue patients by analyzing factors mean temperature, mean relative humidity and total rainfall. This network is trained and tested with the real data obtained from Singaporean National Environment Agency (NEA) and the city of Iloilo Philippines [15]. The warning system of dengue made to predict the future outbreaks in Srilanka [18] and Jember [10] based on risk factors. ANN-based dengue diagnosing system [13] used for identifying the severity of dengue virus in microscopic images of blood cells.

Existing methods used to diagnose and classify the later stages of dengue infection using various factors. The proposed system uses the gene expression data of dengue-infected patients because Gene expression is responsible for causing any type of viral dengue infections. In the proposed method, the informative and significance genes are identified from Gain-Ratio feature selection then Informative genes are fed into Neural Network for classifying Dengue infections as DF, DHF, Primary infection and healthy state of a man, then identified the most relevant genes by pruning neural network.

3   Proposed Method

The proposed method has three segments (i) Feature selection (ii) Feedforward Network and (iii) Network Pruning.

3.1   Feature Selection

Feature selection is used to identify informative genes. It is also used to remove irrelevant genes. Gene expression has more than ten thousands genes. Selection of informative or significance of gene is the most important task in bioinformatics. Only 40% of human genes are expressed at a specific period [3]. The entropy-based method is used to filter informative or significant genes from dengue gene expression. Entropy (H) is the measurement of information spread in gene expression dataset. Entropy is calculated for each gene using Eq. (1). These values are

$$H(\text{gene}) = -\sum_{i=1}^{n} p(i) \log p(i) \tag{1}$$

where p(i) is the probability of gene presences in gene expression, and n represents a number of genes. Entropy value of a gene is converted into binary using Eq. (2) for selecting informative or significance gene for dengue.

$$R = \begin{cases} 1 & \text{if } H(\text{gene }) > 0 \\ 0 & \text{otherwise} \end{cases} \tag{2}$$

where R is a binary value of the gene. When R = 1, the gene is informative.

3.2   Classification of Dengue Infections

Single hidden layer feedforward neural network is considered to classify the dengue infection into primary infection, DF, DHF and normal. The sigmoidal activation function is used in the hidden layer. The linear activation function is used in the output layer. X = [x1 ... xn, 1] is the input pattern to the network. W is a matrix connecting input layer and hidden layer. V is a matrix connecting hidden layer and an output layer. H and Y are vectors that represent the output of the hidden and output layer.

$$net_h = \sum XW, \qquad net_o = \sum HV \qquad (3)$$

$$H = f(net_h), \qquad Y = f(net_o) \qquad (4)$$

$$f(net) = \frac{1}{1 + e^{-net}} \qquad (5)$$

Backpropagation algorithm is used to train the single hidden layer feedforward network.

3.3   Pruning of Neural Network

Pruning is used to reduce the size of the networks. Brute force [19], sensitivity calculation [22], Penalty term method [20] and weight decay [21] are commonly used methods for pruning. Many of the methods either calculate the sensitivity of error for the removal of irrelevant elements or add terms to the error functions which favors small network [17]. In the proposed system, significance based pruning is used. Initially, the trained network has irrelevant genes in the input neurons. The pruning has to identify the significant genes from a trained network. For that, the significance measure Si and threshold value α based method [25] is used. Activation value and the initial weights of a node play a major role in the selection of the significant genes. S_i of a gene is calculated using Eq. (6).

$$S_i = \sum_{i=0}^{m} |F(t_{xip}) + w_{ij}| \qquad (6)$$

where $t_{xip} = \sum_{p=1}^{np} x_{ip}$, $F(t_{xip})$ is same as Eq. (3) and w_{ij} represents the initial weights. The threshold value α is calculated by finding the mean of the significance measure S_i. The α is calculated using Eq. (7).

$$\alpha = \sum_{i=0}^{m} \frac{S_i}{m} \qquad (7)$$

The status of each gene (Sta$_i$) is identified using Eq. (8). All the genes with insignificant status are pruned from a trained neural network.

$$Sta_i = \begin{cases} \text{significant} & \text{if}(s_i > \alpha) \\ \text{insignificant} & \text{otherwise} \end{cases} \qquad (8)$$

3.4 Evaluation Metrics

TP is the True Positive which is the infections of dengue patients correctly detected. FN is the False Negative which is the infections of dengue patients wrongly detected. The overall prediction accuracy CC is calculated using Eq. (9).

$$CC = \frac{TP \times TN - FP \times FN}{\sqrt{(TP+FP)(TP+FN)(TN+FP)(TN+FN)}} \qquad (9)$$

Sensitivity is the factor for testing of correctly detected the patients affected with the infections. Specificity is a factor for testing of correctly detected the patients not affected with infections. Sensitivity S_n and Specificity S_p are calculated using Eqs. (10) and (11).

$$S_n = \frac{TP}{TP + FN} \qquad (10)$$

$$S_p = \frac{TN}{TN + FP} \qquad (11)$$

4 Results and Discussion

Gene expression is the process of find which information from a gene is used in the protein synthesis. The source of protein is gene expression so that the proposed method uses gene expression dataset for dengue patients. Experimental results were carried out by dengue-infected patient's gene expression data collected from the National Center for Biotechnology and Information (NCBI) [3]. Single gene expression dataset contains plenty of genes with their values at specific intervals. Similarly, dengue gene expression dataset contains gene expression of 51 patients. Among this, 19 patients are affected with primary infections, 18 patients are affected with DF, 5 patients are affected with DHF and 9 patients are not affected with dengue. This gene expression contains 5100 genes for each patient. Among this, 4710 genes are selected for the feature selection task because remaining genes have some missing values.

Entropy values of genes are calculated and listed in Table 1. If the gene has positive entropy value then the R-value assigned to 1. Similarly, if the gene has negative entropy value then the R-value assigned to 0. The gene which has the positive entropy value is selected for training the neural network. Among 4710 genes, 1635 genes are selected for training the neural network. These genes are feed into the neural network for classification of dengue infections. The classification result is compared with the class labels as per the NCBI depicted in Fig. 1. According to the obtained result, primary infection, DHF and Healthy state of a man are correctly classified. Among 18 DF patients, 10 patients are correctly classified. The proposed neural network without pruning provides sensitivity as 90.4% and specificity as 9.6%. In this proposed system, some insignificant genes are feed into a network. For that, this system provides an error rate as 9.6%. For getting a better result, pruning of neural network is used. Pruning is the process of optimizing the neural network for improving performances. Most of the pruning algorithms are used for reducing the size of hidden layers in the neural network. In this system, it can be used to reduce the size of input layers as well as hidden layers. The calculated significance measure Si is listed in Table 2. By using Si value, α is calculated as 7.092669.

Table 1. Entropy and ranking values of genes

Gene Name	Patient I	Patient II	Patient III	Patient IV	Patient V	Patient VI	Patient VII	Entropy (H)	Rank (R)
DDR1	5.48026	6.88083	6.23394	5.91046	5.98107	6.23907	6.14071	0.2164677	1
RFC2	6.91033	7.13581	5.98807	6.81096	6.90024	7.31714	6.85388	0.3172565	1
HSPA6	8.89893	9.99144	8.17112	7.0995	9.98487	10.439	9.93342	0.8300973	1
PAX8	5.98468	6.15014	5.64924	6.33406	6.49025	6.81194	6.14881	0.2175674	1
GUCA1A	3.25053	2.65906	2.78711	2.79141	3.30507	2.83694	2.83561	−0.120139	0
UBA7	9.09812	8.92139	8.79076	8.98535	9.04577	9.13579	8.92389	0.6496606	1
THRA	4.0313	3.57161	4.17457	3.58135	3.75172	3.61059	4.26838	−0.005483	0
PTPN21	3.09885	3.31263	3.44781	4.26812	3.43139	3.67195	3.05206	−0.106757	0
CCL5	11.819	11.7409	11.9787	11.107	11.5155	11.9254	12.1075	1.2524969	1
CYP2E1	3.49301	3.09708	3.71753	3.34548	4.09916	3.78818	3.6564	−0.061538	0
EPHB3	3.223	3.46146	3.78034	3.87365	3.54918	3.39758	3.58659	−0.067309	0
ESRRA	7.30041	7.23027	6.47017	6.90095	6.54128	7.00392	6.6079	0.2816095	1
CYP2A6	4.79885	4.64204	4.21465	5.0826	4.63298	4.91311	4.73501	0.0432663	1
SCARB1	5.632	6.09988	5.24526	6.15725	5.72223	6.20672	5.84263	0.1767554	1
TTLL12	7.56266	7.95359	6.50853	7.36925	7.76167	7.75599	7.71392	0.4486681	1
LINC00152	5.99946	5.9313	4.82181	5.25499	5.15343	5.78826	5.87532	0.1810374	1
WFDC2	3.10097	3.55681	3.44538	3.4111	3.66418	3.95457	3.82768	−0.046819	0
MAPK1	9.34518	10.2057	8.73565	8.35432	9.24854	9.83841	9.84283	0.8134685	1
MAPK1	9.18598	9.78132	8.71371	9.0981	9.04181	9.75144	9.57217	0.764288	1
ADAM32	2.67721	2.74463	2.85403	2.49733	2.76698	2.91196	2.55405	−0.135041	0
SPATA17	3.34259	3.0677	3.3409	3.7116	3.25218	2.94436	3.11654	−0.102467	0
PRR22	3.61521	3.71602	4.09528	4.25042	4.08804	3.79467	4.02884	−0.02855	0
PRR22	4.33212	4.48863	4.70195	5.13757	4.77769	3.95291	4.67834	0.03709	1

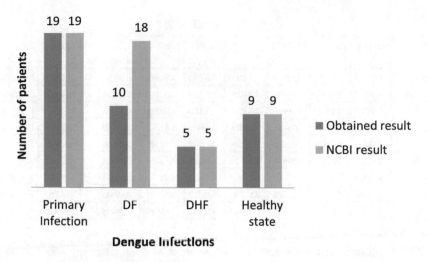

Fig. 1. Analysis of neural network classification before pruning

Table 2. Significant Measures and Threshold value of genes

Name	Patient I	Patient II	Patient III	Patient IV	Patient V	Patient VI	Significant measure (Sta)	Threshold	Status
RFC2	6.91033	7.13581	5.98807	7.31714	6.85388	6.75781	7.91033	7.092669	Significant
HSPA6	8.89893	9.99144	8.17112	10.439	9.93342	10.1633	9.89893	7.092669	Significant
UBA7	9.09812	8.92139	8.79076	9.13579	8.92389	9.10548	10.09812	7.092669	Significant
CCL5	11.819	11.7409	11.9787	11.9254	12.1075	11.2857	12.819	7.092669	Significant
ESRRA	7.30041	7.23027	6.47017	7.00392	6.6079	6.65331	8.30041	7.092669	Significant
TTLL12	7.56266	7.95359	6.50853	7.75599	7.71392	8.05993	8.56266	7.092669	Significant
MAPK1	9.34518	10.257	8.73565	9.83841	9.84283	9.69769	10.34518	7.092669	Significant
MAPK1	9.18598	9.78132	8.71371	9.75144	9.57217	9.33063	10.18598	7.092669	Significant
PXK	6.71473	7.51491	6.75391	6.74233	6.92152	7.40476	7.71473	7.092669	Significant
MSANTD3	7.78721	7.0566	7.33472	7.38274	7.66336	7.62723	8.78721	7.092669	Significant
AFG3L1P	8.01004	8.01518	7.83443	7.45653	8.01064	7.90431	9.01004	7.092669	Significant
DDR1	5.48026	6.88083	6.23394	5.91046	5.98107	6.51122	6.48026	7.092669	Insignificant
CYP2A6	4.79885	4.64204	4.21465	5.0826	4.63298	4.55945	5.79885	7.092669	Insignificant
SCARB1	5.632	6.09988	5.24526	6.15725	5.72223	5.27153	6.632	7.092669	Insignificant
PRR22	4.33212	4.48863	4.70195	5.13757	4.77769	4.81185	5.33212	7.092669	Insignificant
PXK	5.74254	6.31156	5.43929	5.43248	5.81589	5.50508	6.74254	7.092669	Insignificant
ZDHHC11	5.56816	5.69512	5.85787	6.17284	5.13606	5.07913	6.56816	7.092669	Insignificant
ATP6VE2	5.54001	5.76051	5.90709	5.60434	6.0576	5.88707	6.54001	7.092669	Insignificant
AK9	4.00129	4.99412	4.61975	5.26649	4.98231	4.29339	5.00129	7.092669	Insignificant

(*continued*)

Table 2. (*continued*)

Name	Patient I	Patient II	Patient III	Patient IV	Patient V	Patient VI	Significant measure (Sta)	Threshold	Status
TMEM106A	5.65055	6.11324	5.49945	5.5441	5.67693	5.34537	6.65055	7.092669	Insignificant
TMEM106A	4.65214	5.35579	3.93859	4.48676	4.78534	4.22718	5.65214	7.092669	Insignificant
ALG10	3.8906	4.35422	3.65073	3.83233	2.78035	3.40715	4.8906	7.092669	Insignificant
NEXN	4.11913	5.74975	4.14212	4.19027	5.69427	5.41397	5.11913	7.092669	Insignificant
MFAP3	3.99009	5.04812	3.31618	3.09144	3.21998	2.94242	4.99009	7.092669	Insignificant
CCDC65	5.71514	6.01071	5.54594	4.98224	5.13321	5.55451	6.71514	7.092669	Insignificant
CFAP53	4.61515	4.67424	7.09829	6.22021	4.72417	4.98302	5.61515	7.092669	Insignificant
DDR1	5.48026	6.88083	6.23394	5.91046	5.98107	6.51122	6.48026	7.092669	Insignificant
CYP2A6	4.79885	4.64204	4.21465	5.0826	4.63298	4.55945	5.79885	7.092669	Insignificant

The significant genes are selected based on α and S_i. The significant gene has greater S_i value than α. Among 1635 genes 686 significant genes are identified as significant genes and are selected for training. The significant and insignificant genes are represented in Fig. 5. Figure 2 depicts the obtained result and NCBI result. Figure shows that primary infection, DF and Healthy state of a man are correctly classified but among 5 DHF patients, 4 patients are correctly classified. Table 2 represents the significant and insignificant genes of dengue infections. The learning curve of a proposed method is shown in Fig. 3. The sensitivity and specificity of the proposed network are 98.0% and 2.0% respectively. Some clinical diagnostic methods take 30–60 min for identifying dengue. The proposed method takes a fraction of seconds for classifying the dengue infections. The learning curve of proposed network guarantees the convergence of training error which is shown in Fig. 3. The results of this proposed system can be used for the drug designer for dengue.

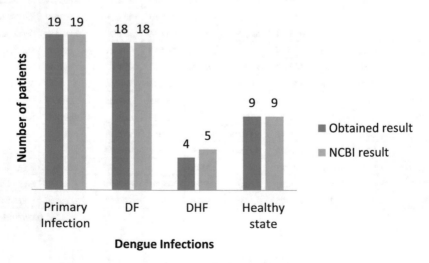

Fig. 2. Analysis of neural network classification after pruning

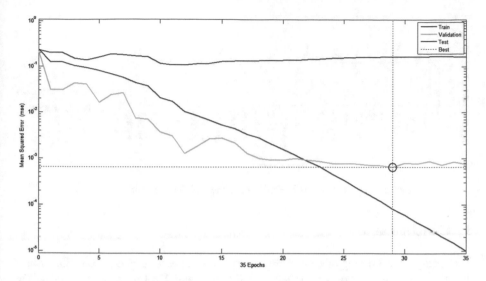

Fig. 3. Learning curve of trained network after pruning

The benchmarks classification algorithms such as Naïve Bayes, Multilayer Perceptron, Iterative Dichotomiser (ID3), Decision tree algorithm (J48) and Random Forest algorithms are evaluated by using Weka tool. The same dengue gene expression dataset is applied to these algorithms. After that sensitivity and specificity for all benchmarks algorithms are calculated using Eqs. 9–11 and are shown in Fig. 3. As per the error rates of all algorithms, the proposed method provides better performance for classifying the dengue infections using dengue gene expression data (Fig. 4).

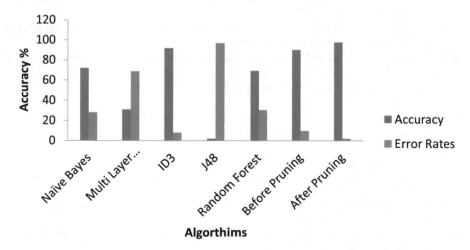

Fig. 4. Comparative analysis

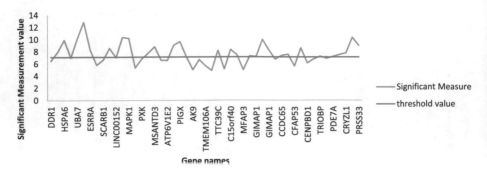

Fig. 5. Significant and insignificant genes of dengue gene expression

5 Conclusion

The entropy and pruning of gene expression play an important role for classification of dengue infections as Primary infection, DF, DHF and Healthy state of a man. The proposed method does not need to know IgG and IgM but it needs only gene expression which can be obtained from any patients at any time and this method gives correct results for classifying dengue infections of the patients which are observed from the results of the experiments.

References

1. Tanner, L., Schreiber, M., Low, J.-G.H., Ong, A., Tolfvenstam, T., Lai, Y.L., Ng, L.C., Leo, Y.S., Puong, L.T., Vasudevan, S.G., Simmons, C.P., Hibberd, M.L., Eong, E.: Decision tree algorithms predict the diagnosis and outcome of dengue fever in the early phase of illness. PLoS Negl. Trop. Dis. **3**, e196 (2008)
2. Singh, S., Singh, A., Singh, M.: Recommender system for detection of dengue using fuzzy logic. J. Comput. Eng. Technol. **7**, 44–52 (2016)
3. National Center for Biotechnology Information. https://www.ncbi.nlm.nih.gov/genomes/VirusVariation/Database/nph-select.cgi
4. Bhatt, A., Joshi, M.: Analytical study of applied data mining in health care. Int. J. Emer. Technol. **8**, 124–127 (2017)
5. Arunkumar, P.M., Chitradevi, B., Karthick, P., Ganesan, M., Madhan, A.S.: Dengue disease prediction using decision tree and support vector machine. SSRG Int. J. Comput. Sci. Eng. **1**, 60–63 (2017)
6. Fathima, S.A., Manimegalai, D., Hundewale, N.: A review of data mining classification techniques applied for diagnosis and prognosis of the arbovirus – dengue. Int. J. Comput. Sci. **6**, 322–328 (2011)
7. Saha, P., Mandal, R.: Detection of dengue disease using artificial neural networks. Int. J. Comput. Eng. **5**, 65–68 (2017)
8. Fatima, M., Pasha, M.: Survey of machine learning algorithms for disease diagnostic. J. Intell. Learn. Syst. Appl. **9**, 1–16 (2017)
9. Pabbi, V.: Fuzzy expert system for medical diagnosis. Int. J. Sci. Res. **5**, 1–7 (2017)

10. Roziqin, C.M., Basuki, A., Harsono, T.: Parameters data distribution analysis for dengue fever breaks in Jember using Monte Carlo. Int. J. Comput. Sci. Soft. Eng. **5**, 45–48 (2016)
11. Mishra, S., Mohanty, P.S., Hota, R., Badajena, J.C.: Rough set approach for generation of classification rules for dengue. Int. J. Comput. Appl. **11**, 31–35 (2015)
12. Shaukat, K., Masood, N., Shafaat, B.A., Jabbar, K., Shabbir, H., Shabbir, S.: Dengue fever in perspective of clustering algorithms. Data Min. Genomics Proteomics. **6**, 1–5 (2015)
13. Subitha, N., Padmapriya, A.: Diagnosis for dengue fever using spatial data mining. Int. J. Comput. Trends Technol. **4**, 2646–2651 (2013)
14. Tarle, B., Tajanpure, R., Jena, S.: Medical data classification using different optimization techniques: a survey. Int. J. Res. Eng. Technol. **5**, 101–108 (2016)
15. Cetiner, G.B., Sari, M., Aburas, H.M.: Recognition of dengue disease pattern using artificial neural networks. Paper Presented at Fifth International Advanced Technologies Symposium (IAST 2009) (2009)
16. Ibrahim, F., Nasir Taib, M., Wan Abas, B.A.W., Chong Guan, C., Sulaiman, S.: A novel dengue fever (DF) and dengue Haemorrhagic fever (DHF) analysis using artificial neural network (ANN). Comput. Methods Programs Biomed. **79**, 273–281 (2005)
17. Reed, R.: Pruning algorithms – a survey. IEEE Trans. Neural Netw. **4**(5), 740–747 (1993)
18. Munasinghe, A., Premaratne, H.L., Fernando, M.G.N.A.S.: Towards an early warning system to combat dengue. Int. J. Comput. Sci. Electr. Eng. **1**(2), 252–256 (2013)
19. Greer, K.: Tree pruning for new search techniques in computer games. Adv. Artif. Intell. (2013). http://dx.doi.org/10.1155/2013/357068
20. Lin, Z., Liu, R., Su, Z.: Linearized Alternating Direction Method with Adaptive Penalty for low-rank representation. In: Advances in Neural Information Processing Systems (2011)
21. Dreiseitl, S., Machado, O.L.: Logistic regression and artificial neural network classification models: a methodology review. J. Biomed. Inform. **35**(5), 352–359 (2002)
22. Zeng, X., Yeung, S.D.: Hidden neuron pruning of multilayer perceptron using a quantified sensitivity measure. Neuro Computing **69**(7–9), 825–837 (2006)
23. Lima, J.-C.R., Rouquayrol, M.Z., Callado, M.R., Guede, M.-I.F., Pessoa, C.: Interpretation of the presence of IgM and IgG antibodies in a rapid test for dengue: analysis of dengue antibody prevalence in Fortaleza City in the 20th year of the epidemic. Rev. Soc. Bras. Med. Trop. **45**(2), 163–167 (2012)
24. Vaughn, W.D., Nisalak, A., Kalayanarooj, S., Solomon, T., Dung, N.M., Cuzzubbo, A., Devine, P.: Evaluation of a rapid immunochromatographic test for diagnosis of dengue virus infection. J. Clin. Microbiol. **35**(1), 234–238 (1997)
25. Augasta, G.M., Kathirvalavakumar, T.: Pruning algorithms of neural networks – a comparative study. Cent. Eur. J. Comput. Sci. **3**(3), 105–115 (2003)
26. www.india.com/news/india/dengue-outbreak/amp

Hardware Trojan: Malware Detection Using Reverse Engineering and SVM

Girishma Jain[✉], Sandeep Raghuwanshi, and Gagan Vishwakarma

Computer Science and Engineering, Samrat Ashok Technological Institute, Vidisha, M.P., India
girishma.jain.2011@gmail.com, sraghuwanshi@gmail.com,
gagan.v.20@gmail.com

Abstract. Due to the globalization, advanced information and simplicity of computerized frameworks have left the substance of the advanced media greatly unreliable. Security concerns, particularly for integrated circuits (ICs) and systems utilized as a part of critical applications and cyber infrastructure have been encountered due to Hardware Trojan. In last decade Hardware Trojans have been investigated significantly by the research community and proposed solution using either test time analysis or run time analysis. Test time analysis uses a reverse engineering based approach to detect Trojan which, limits to the destruction of ICs in detection process.

This paper explores Hardware Trojans from the most recent decade and endeavors to catch the lessons learned to detect Hardware Trojan and proposed an innovative and powerful reverse engineering based Hardware Trojan detection method using Support Vector Machine (SVM). SVM uses benchmark golden ICs for training purpose and use them for the future detection of Trojan infected ICs. Simulation process of proposed method was carried out by utilizing state-of-art tools on openly accessible benchmark circuits ISCAS 85 and ISCAS 89 and demonstrates Hardware Trojans detection accuracy using SVM over different kernel functions. The results show that Radial kernel based SVM performs better among linear and polynomial.

Keywords: Confusion matrix · Hardware Trojan · Radial kernal
Reverse engineering · SVM

1 Introduction

The semiconductor business has warmly invited globalization and outsourcing as a philosophy for bringing down cost of Integrated Circuits (ICs). In the production process of ICs, manufacturing, bonding, and packaging are generally performed by off-shore service providers. This opens ways to various security dangers, for example, overproduction of devices, utilization of old and low-quality parts, forging, and potential modifications of the electronic circuits [1]. Hardware Trojan (maliciously and purposefully connected to the circuit) is one of the most dangerous used attacks at hardware level. Hardware Trojan attacks can enormously trade off the security and protection of hardware users, either straight forwardly or through interaction with pertinent frameworks,

© Springer International Publishing AG, part of Springer Nature 2018
A. Abraham et al. (Eds.): ISDA 2017, AISC 736, pp. 530–539, 2018.
https://doi.org/10.1007/978-3-319-76348-4_51

application software, or with information which raise the importance of its detection and handle with care. To decrease the dangers related with Trojans, researchers have proposed distinctive ways to deal with them. Many detection techniques have been proposed by researchers out of them Test-time detection approaches [2, 3] have drawn the greatest amount of consideration. In these methodologies, functional as well as side-channel behavior of suspect ICs are contrasted with a "golden model" that represents the expected behavior of Trojan free IC. If the suspect IC deviates sufficiently from the golden model, it is classified as being Trojan-infected. While these methodologies have met with some achievement, acquiring such golden models/data is for the most part an open issue.

In order to detect HTs, Reverse-engineering based approaches apply the reverse engineering (RE) process to ICs. Basically, the design/layout uncovered by RE is compared to an intended (golden) layout. Since these approaches are not only time-consuming, so frequently used. The main limitation of RE based approach that these are destructive, IC's layout is being destroyed in the detection process [4].

This paper presents a non-destructive reverse engineering based Hardware Trojan detection approach using SVM classifier. Technique simulation process investigates 5 freely accessible benchmarks, counting from 13 to more than 691 gates. The result demonstrate that our technique can recognize two various types of HTs (inserted Trojans and deleted Trojans) with high accuracy.

Organization of the paper is composed as: Sect. 2 gives a short review of hardware Trojan, reverse-engineering and its attributes and detection with SVM description. Section 3 gives a Problem formulation and examines the inspiration behind it. Section 4 clarifies our proposed methodology in detail including mathematical concept and benefits. In Sect. 5, Performance Evaluation and experimental results are presented. At last, Sect. 6 finishes and concludes the paper.

2 Background

2.1 Hardware Trojans

An intentional, malicious modification in any circuit design, when deployed results in undesired behavior of an IC are Hardware Trojan [5]. Trojan infected ICs may experience changes such as experience degraded or unreliable performance, functionality or specification, may leak sensitive information. HTs can be categorized on the basis of physical and activation characteristics [1, 6].

Physical characteristics come by addition and deletion of transistors, gates, interconnects and changes consist of thinning interconnects, weakening flip-flops or increasing susceptibility to aging [7, 8]. Whereas Activation characteristics come by triggering mechanism. Once the HT is triggered, the payload gets activated and executes the Trojans attack [8]. Figure 1 shows a typical structure of Hardware Trojan.

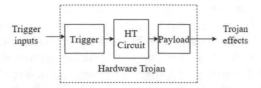

Fig. 1. Typical hardware Trojan.

2.2 Reverse Engineering

Reverse engineering is the reproduction of another manufacturer's product following detailed representations of its construction or composition at higher level of abstraction. It can also be seen as "going back through the development cycle" [9]. A common RE flow incorporates the accompanying steps [10].

(i) **Decapsulation:** From package the die is removed.
(ii) **Delayering:** While polishing the surface each layer of the die is stripped off to keep it planar one at a time, using chemical methods.
(iii) **Imaging:** Using scanning electron microscope (SEM) many high-resolution images of each exposed layer are taken. The images are stitched together using special software to obtain a complete view of layer. Multiple layers are also aligned at this step so that contacts and vias are lined up with layers above and below them.
(iv) **Annotation:** All structures in the device, such as interconnects, vias, transistors, etc., are annotated either manually or by using image recognition software [10].
(v) **Schematic creation, organization, and analysis:** A hierarchical or flat netlist is generated using the annotated images as well as public information (datasheets, papers, etc.). All of the above steps are time-consuming and error-prone.

The first 3 stages basically separate images of the structures contained in the IC. Removing and planarizing layers influence the structures in the uncovered and lower layers, resulting in additional noise in their IC structures. The last 2 steps are required for Circuit/Design extraction and a real so quite challenging. The annotation process and the schematic creation/analysis frequently require contribution from experienced experts [10].

2.3 Trojan Detection Techniques

Hardware Trojan detection techniques fall into the following three categories.

Test-time approaches are most widely method and based on post-silicon tests. These are further classified as functional testing and side-channel fingerprinting [3]. Functional testing [2, 11] aims to detect HTs that change the functionality (primary outputs) of the IC from the intended one. Side-channel fingerprinting [6, 12] is an alternative approach that measures side channel signals (timing, power, etc.) and uses them to distinguish genuine ICs from Trojan infested ones.

Run-time approaches add circuitry that monitors the behavior/state of a layout after it has been deployed. If there is any deviation from the expected golden behavior, additional circuitry detects it and can disable the layout or bypass the malicious logic before the HT can do any damage to the layout. [4, 13] provides a good survey of different run-time approaches. The major disadvantages of these approaches are their high resource overhead and assumption that the run-time circuitry is Trojan-free.

One of the important assumptions that most test-time and run-time approaches assume is that a golden model (i.e. its intended functionality and behavior) is available to compare with. They suggest Reverse-engineering be used to obtain such a golden model [4].

Reverse-engineering based approaches apply the Reverse-engineering (RE) process to ICs in order to detect HTs. Basically, the design/netlist uncovered by RE is compared to an intended (golden) netlist. These approaches are time-consuming and destroy IC layout during RE process [4]. RE based approaches can also verify golden layouts required for test time and run-time golden models.

2.4 Support Vector Machine (SVM)

Support Vector Machines (SVM) is a classification method that separates data using hyperplanes and very useful in case when data has non-regularity or data distribution is unknown [14]. SVM classifier works in two steps of learning and prediction. In learning phase, SVM identifies nearest information focuses to choose limit known as Support Vectors (SVs), use to create divisions among the classes. These SVs are utilized to foresee the class of test record in the prediction phase. Figure 2 shows SVM separating hyperplane of dataset.

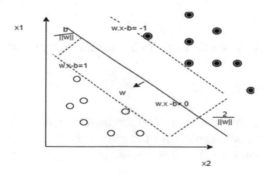

Fig. 2. SVM separating hyper plane.

SVM constructs a decision surface $W \cdot X - b = 0$ to find maximum separation between the classes and future prediction. Classification method is characterized by following equation.

$$f(x) = sgn\left(W_i \times X_i\right) + b$$

Where,

W_i is weight vector formed by using SVs,
X_i is tested record and
b is bias.

3   Problem Formulation

Problem Statement: Let a set contains ICs from at least one untrusted foundries (3PIP Vendor) and constitutes Hardware Trojan free and infected ICs. The objective is to find out Trojan infected ICs. Infected ICs are further categorized in Trajan Addition (TA) and Trojan Deletion (TD). TA includes transistors, gates, and interconnects into the original design whereas TD erases transistors, gates, and interconnects from the original design as recommended in [7].

Cases of Trojan Infected (TI) and Trojan Free (TF) are appeared in Fig. 3. The problem can be viewed as two classes classification problem. let C_0 and C_1 are the two classes denoting TI and TF ICs. Each object is within C_0 and C_1 represented by a feature vector $x = (x_1..., x_n)$ where x_i indicates the ith feature and $x_i \in R$, and n signifies the number of features. Assume objects under test are ICs layouts where TI (TA and TD both) denotes class C_0 and TF denotes class C_1.

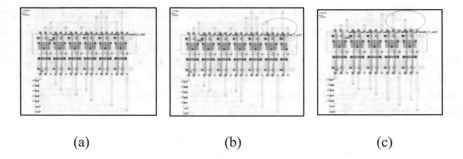

| (a) | (b) | (c) |

Fig. 3.   (a) Trojan free (b) Trojan insertion (c) Trojan deletion

4   Methodology

Proposed model uses SVM classifier for detecting TI ICs. Model works in two phases of offline learning from set of golden ICs recovered using Reverse Engineering and then apply it online for detection TI using classification. Figure 4 shows the flow of the proposed method.

- Learning flow
- Classification flow.

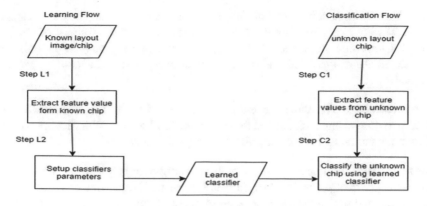

Fig. 4. Proposed model work flow.

SVM Classification

SVM [15, 16] classifier is trained to learn the feature vector help in classification of Trojan infected and Trojan free layouts. SVM get trained for binary classification with −1 and 1, where −1 denoted Trojan Infected and 1 for Trojan Free layout. SVM working can be simplified by following equation.

$$f(x_k) = sgn\left(\sum_{i=1}^{N} y_i \lambda_i K(x_i, x_k + b)\right) \tag{1}$$

Where,

N is the number of learned layouts in known Layouts,
λ_i is the Lagrange multiplier,
K (x, y) is the kernel function for vectors u and v,
b is a bias term,
sgn(u) is the sign function which returns 1 when u ≥ 0, and returns −1, otherwise.

Merits of the Proposed Approach:

- Our approach is proficient to decide the small Trojans too.
- It's accuracy to distinguish the Trojan is increased.
- Its execution time depends on the dataset, if it is large then time for execution will be less and vice-versa.
- It is capable of working on large datasets.

5 Experiments and Results

Experiment Setup Benchmarks and Results Recorded

The proposed procedure was investigated utilizing 5 freely accessible benchmarks [17] (from ISCAS85 and ISCAS89). They were altogether incorporated utilizing DSCH3.5 testing instrument to acquire input designs. Reports from DSCH3.5 device demonstrated that these benchmarks have a gate count ranging from 13 gates for C17 to 691 for C1355 (see Table 1). All specifications are run over the core i3 and all the experiments are carried over the machine using R studio. We chose 32 layouts arbitrarily from TI, TF and TD layouts and utilize them as training samples. The test samples comprise of 16 TF, 8TI, 8TD layouts. Model is trained and training time with its accuracy rate is recorded.

Implementation: The three models of SVM classification are implemented using linear, polynomial and radial kernel function over ICs Layouts. Working flow of linear, polynomial and radial is depict as following Figs. 5, 6 and 7.

Linear Kernel: In linear model, maximum-margin hyperplane is calculated that divides the group of points into two classes, 1 or −1 i.e. Trojan-free or Trojan-Infected, which is defined so that the distance between the hyperplane and the nearest point from either group is maximized and expressed as Eq. (2).

$$F(x, y) = x \cdot y \tag{2}$$

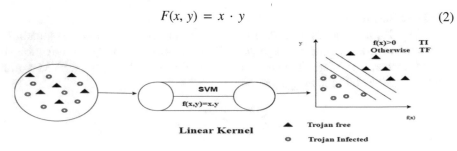

Fig. 5. SVM with linear kernel

Polynomial Kernel: In Polynomial model, the algorithm fit the maximum-margin hyperplane in a transformed feature space using Eq. (3).

$$F(x, y) = (1 + x \cdot y)d \tag{3}$$

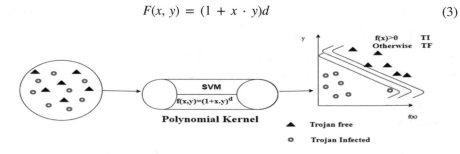

Fig. 6. SVM with polynomial kernel

Radial Kernel: In Radial model, the algorithm fit the maximum-margin hyperplane in a transformed feature space using Eq. (4).

$$F(x, y) = e^{\frac{-|x - y|^2}{2\sigma^2}} \tag{4}$$

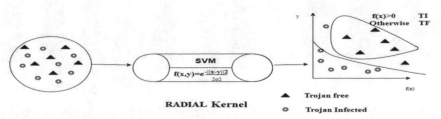

Fig. 7. SVM with radial kernel

Precision: Precision is the fraction of relevant layouts among input instances. It is also called positive predictive value or accuracy.

$$Precision = \frac{True\ positive}{True\ positive + False\ positive} \tag{5}$$

Recall: Recall is the fraction of relevant layouts that have been trained over total amount of input instances. It is also called sensitivity or completeness.

$$Recall = \frac{True\ positive}{True\ positive + False\ negative} \tag{6}$$

F-measure: The harmonic mean of precision and recall is F-measure [18].

$$F - measure = 2\left(\frac{Precision * Recall}{Precision + Recall}\right) \tag{7}$$

The values of above parameters are recorded in Table 1 and are shown in Fig. 8.

Table 1. Evaluation matrix.

Kernel	Training layouts	Recall	Precision (%)	F-measure	Training time (Sec)
Linear	32	0.89	92	0.94	0.08
Polynomial	32	0.77	85	0.87	0.07
Radial	32	1.00	100	1.00	0.06

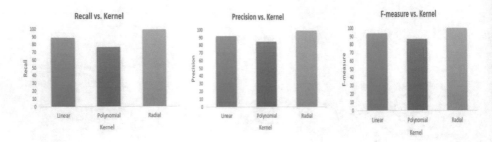

Fig. 8. (a) Detection recall (b) Precision (c) F-measure analysis SVM support vector machine

6 Conclusion

Analysis result demonstrates hardware Trojan with high accuracy and low computation effort. SVM classifier is implemented and results have been recorded. Among the models i.e. Linear, polynomial and Radial all the values of Recall, precision and F-measure are recorded and concluded that Radial Kernel gives the better result among them with less training time to detect Trojans in the IC layouts.

In our HT detection issue, we don't know which ICs are Trojan free and which are Trojan-infected, nor we realize what sort of modifications the attacker will make to those ICs. This prompts an absence of training tests from one class. In future work, we may likewise examine our SVM approach for multi classes. More sophisticated approaches can be used to detect Hardware Trojan.

References

1. Karri, R., Rajendran, J., Rosenfeld, K.: Trojan taxonomy. In: Tehranipoor, M., Wang, C. (eds.) Introduction to Hardware Security and Trust, pp. 325–338. Springer, New York (2012)
2. Zhang, X., Tehranipoor, M.: Case study: detecting hardware Trojans in third-party digital IP cores. In: 2011 IEEE International Symposium on Hardware-Oriented Security and Trust (HOST), San Diego, CA, USA, 5–6 June 2011 (2011)
3. Jin, Y., Makris, Y.: Hardware Trojans in wireless cryptographic integrated circuits. IEEE Des. Test **PP**(99), 1 (2009)
4. Narasimhan, S., Bhunia, S.: Hardware Trojan detection. In: Introduction to Hardware Security and Trust, pp. 339–364. Springer, New York (2012)
5. Tehranipoor, M., Wang, C.: Introduction to Hardware Security and Trust. Springer, New York (2012)
6. Tehranipoor, M., Koushanfar, F.: A survey of hardware Trojan taxonomy and detection. IEEE Des. Test **27**(1), 10–25 (2010)
7. Shiyanovskii, Y., Wolff, F., Rajendran, A., Papachristou, C., Weyer, D., Clay, W.: Process reliability based Trojans through NBTI and HCI effects. In: 2010 NASA/ESA Conference Adaptive Hardware and Systems (AHS), pp. 215–222. IEEE (2010)
8. Bao, C., Forte, D., Srivastava, A.: On application of one-class SVM to reverse engineering-based hardware Trojan detection. In: 15th International Symposium on Quality Electronic Design (2014)

9. Reverse Engineering. Techopedia Inc. https://www.techopedia.com/definition/3868/reverse-engineering
10. Torrance, R., James, D.: The state-of-the-art in semiconductor reverse engineering. In: DAC, pp. 333–338 (2011)
11. Chakraborty, R., Wolff, F., Paul, S., Bhunia, S.: MERO: a statistical approach for hardware Trojan detection. In: Clavier, C., Gaj, K. (eds.) Cryptographic Hardware and Embedded Systems - CHES 2009. Lecture Notes in Computer Science, vol. 5747, pp. 396–410. Springer, Heidelberg (2009)
12. Narasimhan, S., Du, D., Chakraborty, R.S., Paul, S., Wolff, F., Papachristou, C., Roy, K., Bhunia, S.: Multiple-parameter side-channel analysis: a non-invasive hardware Trojan detection approach. In: 2010 IEEE International Symposium Hardware-Oriented Security and Trust (HOST), pp. 13–18 (2010)
13. Wolff, F., Papachristou, C., Bhunia, S., Chakraborty, R.S.: Towards Trojan-free trusted ICs: problem analysis and detection scheme. In: Proceedings of the Conference on Design, Automation and Test in Europe, Munich, Germany (2008)
14. Perceptive Analytics: Machine Learning Using Support Vector Machines. R-Bloggers, 19 April 2007. https://www.r-bloggers.com/machine-learning-using-support-vector-machines/
15. Cortes, C., Vapnik, V.: Support-vector networks. Mach. Learn. 20(3), 273–297 (1995)
16. Hsu, C.-W., Chang, C.-C., Lin, C.-J.: A practical guide to support vector classification. In: National Taiwan University, Taipei 106, Taiwan (2016)
17. Jenihhin, M.: Benchmark circuits, 2 January 2007. http://www.pld.ttu.ee/~maksim/benchmarks/
18. Precision and recall (2017). https://en.wikipedia.org/wiki/Precision_and_recall

Obtaining Word Embedding
from Existing Classification Model

Martin Sustek[(⊠)] and Frantisek V. Zboril

FIT, IT4Innovations Centre of Excellence, Brno University of Technology,
Bozetechova 1/2, 612 66 Brno, Czech Republic
isustek@fit.vutbr.cz

Abstract. This paper introduces a new technique to inspect relations
between classes in a classification model. The method is built on the
assumption that it is easier to distinguish some classes than others. The
harder the distinction is, the more similar the objects are. Simple application demonstrating this approach was implemented and obtained class
representations in a vector space are discussed. Created representation
can be treated as word embedding where the words are represented by
the classes. As an addition, potential usages and characteristics are discussed including a knowledge base.

Keywords: Unsupervised learning · Artificial intelligence
Word embedding · Word2vec · CNN

1 Introduction

Word embedding can be used to represent any entity in continuous vector space
with hundreds of dimensions. The entity is typically a word obtained from a
large dataset that had been constructed from publicly available corpus (e.g.
Wikipedia). The word representations model the language and can be afterward used for different machine learning tasks such as speech recognition. As
an addition, from operation over the words represented as dots (or vectors) in
continuous vector space can be derived rich information about their relationship.
The most popular analogy is mathematically represented sex (king to queen as
man to woman) as was shown in [1]. When the corpus is used as dataset in combination with method such as *word2vec*, it creates word embedding reflecting
the given text, therefore, it might be possible to deduce the information hidden
in used text.

Even though the relations captured by this method can be considered surprisingly precise, they are based on the Distributional Hypothesis. Two words
tend to have similar meaning if they share the same context according to the
hypothesis. Some semantic information, however, is difficult to capture just with
the help of the general text corpus and the Distributional Hypothesis. Recently,
some authors [2–4] used visual information to overcome the lack of such semantic

© Springer International Publishing AG, part of Springer Nature 2018
A. Abraham et al. (Eds.): ISDA 2017, AISC 736, pp. 540–547, 2018.
https://doi.org/10.1007/978-3-319-76348-4_52

evidence in the textual data. Provided visual information is typically in the form of sentences describing pictures (e.g. "girl eats ice cream") or the system is multimodal (uses both text and pictures). The authors afterward demonstrate the improvement on the chosen dataset, therefore, these methods are aimed to bring additional semantic information into encoded word representation. Because they target on natural language processing, it is reasonable to focus on visual information in the form of sentences or build upon existing word representations obtained by text corpus.

2   Goals of the Paper

The goal is to present an approach of creating a class representation in a vector space. With this approach, we want to demonstrate the relations between classes on a simple example using convolutional neural networks (CNN). Moreover, we want to discuss important characteristics regarding the learning with the focus on unsupervised learning and the origin of the information (considered to be a knowledge). In the last part, we will propose methods benefiting from the word embedding as an potential future work.

We will present the way of obtaining word embedding from any existing classification model rather than from a corpus. We will focus on models working with pictures, because it can bring some additional benefits that will be discussed below. However, the method is capable of working with any model, not necessarily picture based. We choose convolution neural network (CNN) as an example of existing model. CNN was trained on a dataset in the form of pairs consisting of labelled class and example picture of this class.

It is typical to demonstrate that the approach is useful for some task. Moreover, obtained score can be compared with others. However, we do not present any task or provide any score since the method is intended to inspect the relations between classes in classification model and we do not demand or expect any strict relations between them.

3   Learning the Similarity

We decided to use very simple method to create word embedding by assuming that a similarity is related to an error. This means that when CNN finishes learning phase, and given a picture of dog, it is theoretically expected that the probability of seeing a dog $P(dog|picture) = 100\%$ and $P(x|picture) = 0\%$ for any class x that is not a dog. In practice, results can be typically $P(dog|picture) = 90\%$, $P(cat|picture) = 9\%$, $P(x|picture) < 1\%$ for any x that is neither dog nor cat. This can be illustrated in Fig. 1.

For this method purpose it is actually beneficial that the theoretically desired state is not reached. We suppose that similar approach is natural for humans as well since the question "Does this dog look more like a cat or like a sheep?" can be asked. The answer will probably not be indifferent.

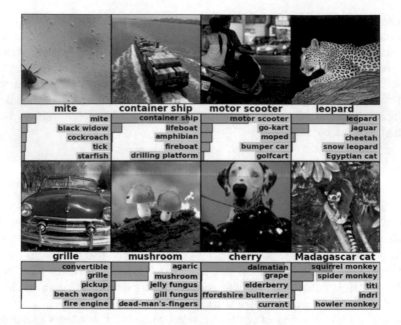

Fig. 1. An example response of existing trained CNN. Below each picture is the label of the expected class and the distribution between classes. The width of the bar corresponds to the probability that CNN assigned to the class. The fact that the probability of seeing jaguar given the picture of leopard P(*jaguar|leopard*) in the upper right picture is not negligible can be used as a witness of their similarity.

4   Experiments

Firstly, the convolutional neural networks was trained on the task to classify the image. *CIFAR-10* [6][1] was chosen as a dataset for CNN. The dataset consists of colour pictures of 32×32 size divided into following 10 categories (classes):

- airplane
- automobile
- bird
- cat
- deer
- dog
- frog
- horse
- ship
- truck

Each class includes 5000 training and 1000 testing examples. In order to train CNN, we used publicly available example[2] and about 300 epochs.

---

[1] http://www.cs.toronto.edu/~kriz/cifar.html.

[2] https://www.tensorflow.org/tutorials/deep_cnn.

After we had finished the learning phase, we used CNN as a classification model to create a new dataset containing pairs in the form of $(input, output)$. For each picture in dataset, 3 most probable classes were retrieved and pairs $(input, mp_1), (input, mp_2), (input, mp_3)$ were constructed, where $input$ is the class of the input picture and mp_k is the k-th most probable class.

A newly created dataset was afterward used as an input of the *word2vec Skip-Gram* architecture to create word embedding of used classification classes as introduced in [1]. The number of dimensions in word embedding was set to 5 since there are only 10 classes. The process is shown in Fig. 2.

5 Evaluation

Word analogy is typical measurement for the evaluation of constructed word embedding. It can be good when modelling a language, therefore, learned embedding should hold the information about the relation between words regarding the sex, geographical facts or linguistic regularities. This information can be regain by certain operations over the vector representations.

We do not think that this method is ideal for task such as language modelling at the time. It can be, however, used as an introspection tool and should provide information about *how the model intercepts classes*, there is no ground truth to be expected, therefore, we do not use any method to evaluate results. However, we as humans feel that at least organic and inorganic division can be reflected. We also know that airplane shape was inspired by the shape of a bird. Since there are just 10 classes, the cosine similarity (1) can be computed for each vector pair and visualize the similarity between classes. The cosine similarity express how similar the vectors are and ranges from -1 for opposite vectors, to 1 for vectors sharing the same direction. The cosine similarity should theoretically reach almost 1 for very similar objects. Symbol θ stands for an angle between vectors.

$$\text{similarity}(\mathbf{A}, \mathbf{B}) = \cos(\theta) = \frac{\mathbf{A} \cdot \mathbf{B}}{\|\mathbf{A}\|\|\mathbf{B}\|} \tag{1}$$

Tables 1 and 2 display 4 organic and inorganic classes respectively. Each class (column) shows the rest of the categories in descending order based on their similarity. It can be noticed that certain relations are somehow reflected through the similarity. In particular, inanimate classes (airplane, automobile, ship, truck) are the least similar objects to animate classes such as dog, cat or deer. The distinction between animate and inanimate objects remains noticeable in all categories. Cat and dog as well as automobile and truck form a strong connection. Cosine similarity between pairs of four-legged animals (cat, deer, dog, horse) is above the average value. Already mentioned similarity between airplane and bird is reflected as well.

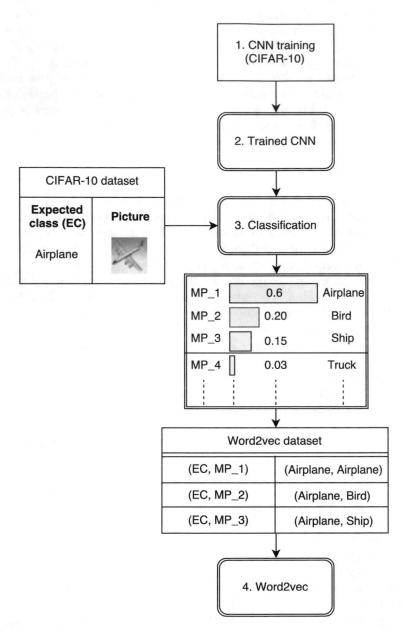

Fig. 2. The illustration of the process of creating word embedding. At first, CNN is trained. Afterward, it is used as classification model on each picture from the *CIFAR-10* dataset to construct a new dataset for word2vec method. Word2vec is learned using the pairs, where the first element is the input and the second is the output.

Table 1. Similarities between organic classes. Each column shows classification class and remaining classes in descending order sorted by the similarity.

dog		cat		deer		bird	
cat	0.9715	dog	0.9715	horse	0.6744	airplane	0.6377
horse	0.4560	frog	0.5029	bird	0.5978	deer	0.5978
frog	0.2944	horse	0.2886	frog	0.3080	frog	0.5888
deer	0.2542	bird	0.2698	dog	0.2542	cat	0.2698
bird	0.1761	deer	0.2455	cat	0.2455	horse	0.2166
ship	0.1397	ship	0.1208	airplane	0.1208	dog	0.1761
automobile	-0.0359	automobile	-0.0258	automobile	0.0726	ship	0.1719
truck	-0.1054	truck	-0.0948	truck	0.0653	truck	-0.1957
airplane	-0.1537	airplane	-0.1357	ship	-0.3449	automobile	-0.2750

Table 2. Similarities between inorganic classes. Each column shows classification class and remaining classes in descending order sorted by the similarity.

automobile		truck		ship		airplane	
truck	0.9877	automobile	0.9877	airplane	0.7608	ship	0.7608
ship	0.2725	ship	0.3723	truck	0.3723	bird	0.6377
frog	0.1371	airplane	0.2558	automobile	0.2725	truck	0.2558
airplane	0.1104	frog	0.1228	bird	0.1719	horse	0.1667
horse	0.0870	horse	0.0888	dog	0.1397	frog	0.1288
deer	0.0726	deer	0.0653	cat	0.1208	deer	0.1208
cat	-0.0258	cat	-0.0948	horse	0.0552	automobile	0.1104
dog	-0.0359	dog	-0.1054	frog	-0.0210	cat	-0.1357
bird	-0.2750	bird	-0.1957	deer	-0.3449	dog	-0.1537

5.1 Learned Intelligence

We believe that each model must have some kind of intelligence to be able to give good results. Even if a model would take into account just the colour of the object or in the corner of each picture would be written object class and the model would learn to recognize the text. Different approaches can yield different word embedding. We assume that it might not be able to strictly compare different approaches ("intelligence") in all cases but they might depend on circumstances.

6 Discussion

Learned embedding has some interesting characteristics regarding the process of obtaining the knowledge, moreover, potential knowledge rooted in the word embedding could be learned semi-automatically.

6.1 Unsupervised Learning and the Information Origin

Word2vec method is often considered to be one of unsupervised learning methods. This assumption is based on the fact that the input is in the form of unlabeled data (corpus). However, the origin of information is in the language that were created by humans, therefore the word representation is learned from humans (authors of used corpus). This is inevitable when modeling languages that were created by humans. The similarity of two word representations is then simply based on word distribution.

We consider presented method of learning relations between classes also as unsupervised, moreover, independent of humans in terms of provided information, even though the classes were assigned by humans and data were labeled. The class name is just some identifier grouping multiple objects, so no information about the relation between these classes is provided.

Building on the fact that pictures simply reflect the reality, artificial intelligence is the consequence of the CNN architecture itself. Admitting that the artificial intelligence can be the consequence of the learning, it can be suitable to deduce the knowledge from for example classification task by the model itself instead of providing the knowledge to the model to learn it. In other words, instead of forcing a model to learn to describe picture by sentences in natural language (annotations made by humans), we want it to "understand" that it is beneficial to involve it (not in natural language) automatically as it could improve some task and later discover how to acquire this information. An example annotation "cat sits on the mat" could be learned as a fact that the model is somehow able to define "cat sits" if the body position affects the task. This could satisfy the *learning to learn* condition as presented in [7].

6.2 Learning More Complex Word Embedding

Since the pictures in the dataset always show only one object (class), learned embedding mainly reflects object's visual similarity. However, embedding could contain even more information regarding the relation between objects. Learning from the real world pictures with multiple objects on single caption could be the way. Therefore, with the ability to distinguish objects and to observe, more knowledge can be potentially produced.

We supplied annotated pictures to learn to distinguish basic classes. It would be interesting to have an automatic decision system that could be based on for example clustering. Then it would be possible to decide whether a newly observed object belongs to one of the known classes or is unknown, therefore, new class is to be created. Satisfying this condition, learning would be independent of humans. Probably the most difficult task *how to extract encoded knowledge from the word embedding* was not discussed.

7   Conclusion

Even though more sophisticated methods of learning word embedding representing the classification classes could be brought, we proposed a simple method based on errors just to demonstrate the approach of representing the knowledge by the model itself. Moreover, we built strictly on information from pictures because they reflect the real world and we think that learning from the real world can be easier task than learning any language since the ability to speak in any language is only human-specific while any animal needs to classify objects.

We demonstrated that some kind of intelligence can be derived automatically from the ability to distinguish two objects or their classes. Moreover, we assume that the intelligence in some form is necessary in order to obtain acceptable results. Further experiments are needed to decide whether this approach could be sufficient as for example knowledge representation for general artificial intelligence.

Acknowledgment. This work was supported by the BUT project FIT-S-17-4014 and the IT4IXS: IT4Innovations Excellence in Science project (LQ1602).

References

1. Mikolov, T., Chen, K., Corrado, G., Dean, J.: Efficient estimation of word representations in vector space. arXiv preprint arXiv:1301.3781 (2013)
2. Kottur, S., Vedantam, R., Moura, J.M.F., Parikh, D.: Visual word2vec (vis-w2v): learning visually grounded word embeddings using abstract scenes. CoRR abs/1511.07067 (2015)
3. Xu, R., Lu, J., Xiong, C., Yang, Z., Corso, J.J.: Improving word representations via global visual context. In: NIPS Workshop on Learning Semantics (2014)
4. Lazaridou, A., Baroni, M., et al.: Combining language and vision with a multimodal skip-gram model. In: Proceedings of the 2015 Conference of the North American Chapter of the Association for Computational Linguistics: Human Language Technologies, pp. 153–163 (2015)
5. Krizhevsky, A., Sutskever, I., Hinton, G.E.: Imagenet classification with deep convolutional neural networks. In: Pereira, F., Burges, C.J.C., Bottou, L., Weinberger, K.O. (eds.) Advances in Neural Information Processing Systems 25, pp. 1097–1105. Curran Associates, Inc., New York (2012)
6. Krizhevsky, A., Hinton, G.: Learning multiple layers of features from tiny images (2009)
7. Baroni, M., Joulin, A., Jabri, A., Kruszewski, G., Lazaridou, A., Simonic, K., Mikolov, T.: Commai: evaluating the first steps towards a useful general AI. CoRR abs/1701.08954 (2017)

A Robust Static Sign Language Recognition System Based on Hand Key Points Estimation

Pengfei Sun[1], Feng Chen[1]($\boxtimes$), Guijin Wang[2], Jinsheng Ren[1], and Jianwu Dong[3]

[1] Department of Automation, Tsinghua University, Beijing 100084, China
chenfeng@mail.tsinghua.edu.cn
[2] Electronic Engineering Department, Tsinghua University, Beijing 100084, China
[3] National Computer Network Emergency Response Technical Team Coordination
Center of China (CNCERT/CC), Beijing 100029, China

Abstract. Sign language recognition is not only an essential tool between normal people and deaf, but a prospective technique in human-computer interaction (HCI). This paper proposes a robust method based on the RGB sensor and hand key points estimation. Compared with depth sensor and the wearable devices, RGB sensor has smaller size and simpler operation process. With the hand key points detection technique, the data can conquer the influence of unfavourable factors like complex background, occlusion, and different angles. During training step, 5 kinds of machine learning algorithms are used for the classification of 20 letters in alphabet, and the highest classification accuracy are realized by SVM and KNN algorithms, which are 95.54% and 97.3% respectively. Finally, a real time sign language recognition system with SVM training model is built and it's recognition accuracy can reach 97%, which confirms that our method can effectively eliminate unfavourable factors.

Keywords: Sign language · Key points estimation · SVM

1 Introduction

Sign language recognition system is of great importance in our daily life and it has wide application in both life and production, such like the communication between normal people and deaf, the command of Unmanned Aerial Vehicles and robots. The sign language in general includes the static hand gesture, dynamic hand gesture, the combined actions between face and hand, and the complex ones such as the body actions [1]. The most important component in many sign languages is the hand sign language, because it can represent hundreds of words, and most of the other sign languages like face actions and body gestures mainly play the role of auxiliary function.

Static sign language identification and recognition is composed of the hand gesture detecting and the hand gesture recognition parts [2]. The key techniques

© Springer International Publishing AG, part of Springer Nature 2018
A. Abraham et al. (Eds.): ISDA 2017, AISC 736, pp. 548–557, 2018.
https://doi.org/10.1007/978-3-319-76348-4_53

in these two parts are different according to the type of data input devices, which can be classified into the following four categories.

The first method is to let the experimenters to wear some special data acquisition devices such as the complex and heavy mechanical arm with kinds of sensors [3,4]. By using the special devices, the location of hands and the detailed figures and hands actions can be detected precisely (list the responding works). However the disadvantages of this method is also obvious, on the one hand, it will be difficult to collect the training datasets for these heavy and complex devices; on the other hand, the limit of data acquisition speed of devices is also a critical problem for this method.

The second method is based on the computer version approaches [5]. In this method, users needn't wear any special sensors devices, instead, the training dataset are acquired by RGB cameras [6]. It has the advantages of being convenient for users and simple to set up the system. However, problems are also clear in this method, firstly, hands are detected based on the skin color [7] or special color gloves [9], so various light problems like illumination, shadow and occlusion can cause pretty bad influences on the accuracy of detection. Furthermore, more images in different conditions is needed for higher accuracy, which will increase the time cost of sample training.

The third method is based on the artificial neural network (ANN). With the development of deep convolution neural network (CNN) [8], some recognition and classification problems can be transformed into how to construct a CNN structure and collect massive data [9]. However, this network cannot classify the hand's gestures correctly while it is not showed in the training sets. Besides, it needs massive data coving all conditions like the illumination and occlusion.

The last method is based on 3D model, which has become a hot spot recently [10]. The 3D model is based on depth camera such as RGB-D, which can estimates the hand parameters by comparing the 2D images and the 3D hand data. This method is very popular recently [11], for its accurate data stream and convenient application process. There are two main branches in 3D model based algorithm, which are based on volumetric and skeletal models respectively [12]. Volumetric model can use human hands' shape information to detect hand in real time, but the parameters would greatly increase the cost of computation. The skeletal model is bases on the skeletal information of human, which has less parameters but would has less accuracy either.

In this paper, we propose a robust method to realize the sign language recognition. The recognition system has two process: offline training and online recognition. The system is based on the RGB camera and the hand key points estimation. Without the complicated data acquiring equipments, like special gloves or heavy wearable devices, we can get the image easily and quickly by the RGB sensor. Besides, we have simple operations on the input data, decreasing some complex steps compared with depth sensor. The key points of hand are acquired by the machine learning method based on multi-view bootstrapping algorithm, which effectively decrease the influence of illumination and occlusion problems. After acquiring the points, several kinds of machine leaning methods for the classification of static sign language are used, and the comparison of average training time and accuracy are made between the different algorithms. And then, the algorithm which has the advantages of high operating speed and accuracy is used in our recognition system. Finally, a real time sign language recognition experiment is made, which has verified the validity of our method.

In Sect. 2 we will introduce the basic theoretical knowledge and main algorithms we adopt. In Sect. 3 we will present the main structure of the recognition task. In section, the experimental result will be gave. The conclusion part in Sect. 5 will summarizes the contribution of the paper and give our future work.

2   Methods and Algorithms

To realize sign language recognition task, the process of hand key point detection and classification are need to be done. For hand key points estimation, the multiview bootstrapping algorithm is used, while for classification, the SVM algorithm is chose for its advantages of high accuracy and efficiency.

2.1   Multiview Bootstrapping Algorithm

In multiview bootstrapping algorithm [13], a key point detector $d(\cdot)$ can map an image P to N key points, and each point has a detection confidence c_n, it can be expressed as:

$$d(P) \rightarrow (x_n, c_n) \; for \; n \in [1...N] \tag{1}$$

The important steps of multiview bootstrapping algorithm are shown in Table 1. At beginning of the algorithm, the labeled and unlabeled images as well as a simple key point detector are inputed in the network. After the training process, we can get the improved hand key points detector and the training set with more samples labeled.

Table 1. Algorithm of hand keypoint detector [13]

Method Multiview Bootstrapping
In:
• Unlabeled images: $\{\mathbf{P}_v^{frame} \text{ for } v \in \text{views}\}$
• Keypoint detector: $d_0(\mathbf{P}) \mapsto (\mathbf{x}_n, c_n) \text{ for } n \in \text{points}$
• Labeled training data: τ_0
for iteration m in 0 to M:
1. Triangulate key points from initial detector
for each frame:
(1)Run detector $d_m(\mathbf{P}_v^{frame})$ on each view v
(2)better triangulate key points
2. Get the best frames
3.train with M-best results again
$d_{m+1} \leftarrow train(\tau_0 \cup \tau_{m+1})$
Out: Excellent detector$d_M(\cdot)$ and training set τ_M

2.2   SVM Algorithm

Section 4 uses 5 kinds of machine learning algorithms. Among these methods, the SVM algorithm achieves the best recognition result while considering their performance of train time and test accuracy.

Many machine learning algorithms can classify the two samples without error, however, SVM algorithm can maximize the classify gap among all the classify schemes. By using SVM algorithm, we can get the decision function:

$$f(x) = sign(w \cdot x + b) \tag{2}$$

Where x denotes the estimation data of key points, and w, b is the weight matrix and bias respectively.

In order to maximize the classify gap, the classification objective can be transformed into a optimization problem. Its objective function is:

$$W(a) = \sum_{i=1}^{l} a_i - \frac{1}{2} \sum_{i,j=1}^{l} y_i y_j a_i a_j M(x_i, x_j). \tag{3}$$

Among the formula, the $M(x_i x_j)$ is the kernel function. The final classify function is:

$$f(x) = sign \left[\left[\sum_{i=1}^{l} a_i y_i M(x, x_i) + b \right] \right] \tag{4}$$

3   Real-Time Sign Language Recognition Framework

After the images are obtained by the RGB camera, the classifier need to be trained offline, and then a real-time online recognition can be effectively achieved. The main structure of the sign language recognition system is shown in Fig. 1.

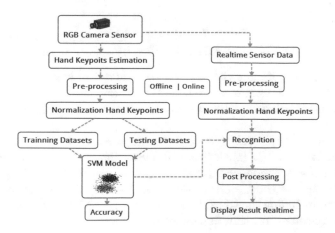

Fig. 1. Flow chart of sign language recognition system

Fig. 2. Image from the RGB sensor

The algorithm of key point estimation can be accomplished by Openpose method [14,15]. Figure 2 gives the display interface of the real time recognition system.

3.1   Offline Training of Sign Language System

During the offline training process, firstly we collect the image data by the RGB sensor. Meanwhile, we should take consider of some complex and bad conditions, like the different angles of the sign language, variable illumination and hand with shadow case. After the key points are acquired, unsuitable candidates need to be wiped out.

After acquiring the hand key points, we need to normalize samples data. During the process of sample collection, several cases such as the hand location in image, near or far distance between hand and sensor are taken into consideration. So we normalize the location and size of the data, which guarantee all samples to have consistent data format in training process. The sample data can be normalized by:

$$x_{new} = \frac{x_{old} - min\left\{x_{values}, y_{values}\right\}}{max\left\{x_{values}, y_{values}\right\} - min\left\{x_{values}, y_{values}\right\}} * 100 \tag{5}$$

Where x_{new} and x_{old} denote the position coordinates of key points in each step; the min and max functions are the minimum and maximum of the position coordinates set; the constant 100 denotes the range of normalized position data ([0, 100]).

The collected sample data has 43 dimensions, which includes the 42-D position coordinates and 1-D label. Figure 3 shows the 21 key points of hand [16] and rendered effect by connecting the key points together.

In this paper, we use 5 kinds of machine learning algorithms to classify the sign language, including the KNN, SVM, Adaboost, ANBC and BAG algorithms. This will be revisited in detail in simulation section.

Fig. 3. 21 hand keypoint and hand rendered by keypoint

3.2 Online Recongnition Process

After the training process, we can get the trained model of the recognition in sign language. In the online recognition process, the way of image data acquisition is same with the offline training process. After the hand keypoint data are normalized, the trained model can recognize the sign language in real time. During the post processing process, the classification result can be output by interactive devices.

4 Simulations and Analyses

The alphabet expressed by hand gesture is the basics of sign language. Figure 4 [17] shows the 26 letters in alphabet expressed by hand gesture. In our recognition system, we collect 20 kinds of sign language, and each one has about 400–500 image samples.

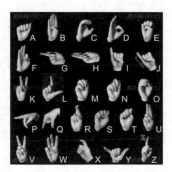

Fig. 4. Alphabet expressed by hand

The essential processes has been explained in detail in above section. Here we present some samples in data collecting process. Figure 5 shows some special but important hand samples including hand in different angles, hand in dark light condition, hand with a pen, hand having a sheltered regions.

Fig. 5. Kinds of conditions

4.1   Training Process Using Multi-machine Learning Algorithms

After finished the work of constructing the training and test dataset, we use 5 kinds of algorithms to train the samples. k-nearest neighbors algorithm (KNN), is a non-parametric method used for classification [18]. In the algorithm, the output result is one of class membership. It is classified by the vote of k neighbors. The support vector machines (SVM), which has talked about in Sect. 2, is the main algorithm adopted in this paper. The three other algorithms are Adaboost [?], Adaptive Naive Bayes algorithm (ANBC), bootstrap aggregating algorithm (BAG). The algorithms are also tested in this paper, however the accuracy are not as good as SVM or KNN algorithms.

We compare the average accuracy and cost time between the 5 algorithms, which are showed in Fig. 6. As can be seen from the graph, we can find the highest classification accuracy are realized by SVM and KNN algorithms, which are 95.54% and 97.3%.

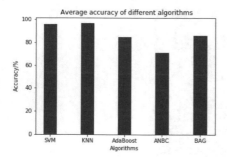 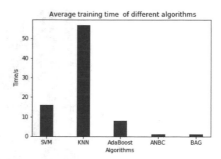

Fig. 6. Compare of different algorithms

4.2   Discussion of Two Excellent Algorithms

In order to decide which algorithm will be used in online recognition process, we perform more experiments to compare their effect. We increase the number of the sign languages (from 5 to 20 at most) to observe the changes in training time and average accuracy, which are shown in Fig. 7. The average accuracy of each class using SVM algorithm is shown Fig. 8.

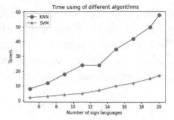
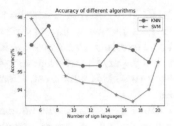

Fig. 7. Compare of algorithm

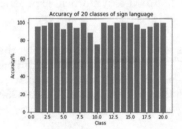

Fig. 8. Accuracy of classes by SVM

Obviously, the KNN algorithm improves average accuracy slightly but cost more time. Most important of all, we need the model to do the real time recognition work, so the time cost on calling the trained model is essential. KNN needs to find the similar data in the training data according to parameter k, which will cost more time than SVM algorithm. On the contrary, SVM has a great advantage when the amount of training samples is huge. For these reasons, we adopt the SVM algorithm in the real time recognition system. Finally in the practice test, we get 97.3% average accuracy using the SVM trained model, which is higher than the test step, 95.4%. Part of reason is that the bad samples still exist in the training data, specially in the classes that have a relatively low accuracy such as class 10 in Fig. 8.

5 Conclusion and Future Work

In this paper, we proposed an accurate and real time sign language recognition system. Using the hand's key points estimation technique, we can acquire the hand data precisely. Besides, this method has strong robustness while work against some bad recognition conditions such as illumination changes or being sheltered by other objects. When compared with the popular recognition technique based on depth camera, our method has equal or even better accuracy, while the implementary difficulty can be greatly decreased. After constructed 20 kinds of different sign languages, we trained the labeled samples by five popular machine learning algorithms, which are SVM, KNN, AdaBoost,

ANBC and BAG. Among the test, we found that the KNN and SVM methods can gain more satisfactory result than others in our method. Ultimately, the SVM algorithm was chose for its better behavior in real time recognition effect. In this paper, the average recognition accuracy of 20 kinds of sign language can attain to 95.0%, which can effectively satisfies the demand of related recognition task.

Our future work will be concentrated on two aspects, on the one hand, we need to collect some new samples to replace some bad data, which will improve the test accuracy, especially for some cannot reach to 90%. More samples will be collected and added into the training set, which will improve the accuracy of recognition; on the other hand, the dynamical sign language recognition will be studied.

Acknowledgement. This work is supported in part by the National Natural Science Foundation of China under Grant 61671266, 61327902, in part by the Research Project of Tsinghua University under Grant 20161080084, and in part by National High-tech Research and Development Plan under Grant 2015AA042306, 2015AA016304.

References

1. Cokely, D., Bakershenk, C.L.: American sign language. Language **59**(1), 119–124 (1998)
2. Chou, F.H., Su, Y.C.: An encoding and identification approach for the static sign language recognition. In: IEEE/ASME International Conference on Advanced Intelligent Mechatronics, pp. 885–889. IEEE (2012)
3. Starner, T., Pentland, A., Weaver, J.: Real-time American sign language recognition using desk and wearable computer based video. IEEE Comput. Soc. **20**, 1371–1375 (1998)
4. Wu, J., Jafari, R.: Wearable computers for sign language recognition (2017)
5. Meena, S.: A Study on Hand Gesture Recognition Technique (2015)
6. Hasan, M.M., Mishra, P.K.: HSV brightness factor matching for gesture recognition system. Int. J. Image Process. **4**(5), 456–467 (2010)
7. Yu, C., Wang, X., Huang, H., et al.: Vision-based hand gesture recognition using combinational features. In: Sixth International Conference on Intelligent Information Hiding and Multimedia Signal Processing, pp. 543–546. IEEE (2010)
8. Oz, C., Leu, M.C.: Recognition of finger spelling of American sign language with artificial neural network using position/orientation sensors and data glove. In: Advances in Neural Networks ISNN 2005, pp. 157–164. Springer, Heidelberg (2005)
9. Oz, C., Ming, C.L.: American sign language word recognition with a sensory glove using artificial neural networks. Eng. Appl. Artif. Intell. **24**(7), 1204–1213 (2011)
10. Mahajan, P.M., Chaudhari, J.S.: Review of finger spelling sign language recognition. Int. J. Eng. Trends Technol. (IJETT) **22**, 400–404 (2015)
11. Li, S.Z., Yu, B., Wu, W., et al.: Feature learning based on SAEPCA network for human gesture recognition in RGBD images. Neurocomputing **151**, 565–573 (2015)
12. Sutarman, M.M.B.A., Zain, J.B.M., et al.: Recognition of Malaysian sign language using skeleton data with neural network. In: International Conference on Science in Information Technology, pp. 231–236. IEEE (2016)

13. Simon, T., Joo, H., Matthews, I., et al.: Hand keypoint detection in single images using multiview bootstrapping (2017)
14. Cao, Z., Simon, T., Wei, S.E., et al.: Realtime multi-person 2D pose estimation using part affinity fields (2016)
15. Wei, S.E., Ramakrishna, V., Kanade, T., et al.: Convolutional pose machines, pp. 4724–4732 (2016)
16. From: https://github.com/CMU-Perceptual-Computing-Lab/openpose
17. From: http://www.photophoto.cn/tuku/sheji/104/013/312486.htm
18. Altman, N.S.: An introduction to kernel and nearest-neighbor nonparametric regression. Am. Stat. **46**(3), 175–185 (1992)
19. Cortes, C., Vapnik, V.: Support vector network. Mach. Learn. **20**(3), 273–297 (1995)

Multiobjective Genetic Algorithm for Minimum Weight Minimum Connected Dominating Set

Dinesh Rengaswamy$^{(\boxtimes)}$, Subham Datta$^{(\boxtimes)}$, and Subramanian Ramalingam$^{(\boxtimes)}$

Department of Computer Science, Pondicherry University, Puducherry, India
dineshrengaswamy@gmail.com, subhamdatta@yahoo.com,
rsmanian.csc@pondiuni.edu.in

Abstract. Connected Dominating Set (CDS) is a connected subgraph of a graph G with the property that any given node in G either belongs to the CDS or is adjacent to one of the CDS nodes. Minimum Connected Dominating Sets (MCDS), where the CDS nodes are sought to be minimized, are of special interests in various fields like Computer networks, Biological networks, Social networks, etc., since they represent a set of minimal important nodes. Similarly, Minimum Weight Connected Dominating Sets (MWCDS), where the connected weights among the CDS nodes are sought to be minimized, is also of interest in many research application. This work is based on the hypothesis that a CDS with both the properties of minimum size and minimum weight optimized would enhance performance in many applications where CDS is used. Though there are a good number of approximate and heuristic algorithms for MCDS and MWMCDS, there is no work to the best of our knowledge, that optimizes the generated CDS with respect to both the size and weight. A Multiobjective Genetic Algorithm for Minimum Weight Minimum Connected Dominating Set (MOGA-MWMCDS) is proposed. Performance analysis based on a Wireless Sensor Network (WSN) scenario indicates the efficiency of the proposed MOGA-MWMCDS and supports the advantage of MWMCDS use.

Keywords: Multiobjective Genetic Algorithm
Minimum Connected Dominating Set
Minimum Weight Connected Dominating Set
Minimum Weight Minimum Connected Dominating Set
Evolutionary algorithm · Wireless sensor networks

1 Introduction

A virtual backbone in a network corresponds to a subset of nodes that can facilitate routing effectively. It achieves this by providing reliable transmission connections between other nodes in the network. The selection of nodes that form the part of virtual backbone is a challenging task, in wireless networks especially in Wireless Sensor Network (WSN), where the topology may vary dynamically.

© Springer International Publishing AG, part of Springer Nature 2018
A. Abraham et al. (Eds.): ISDA 2017, AISC 736, pp. 558–567, 2018.
https://doi.org/10.1007/978-3-319-76348-4_54

The graph theoretical concept of Connected Dominating Set (CDS) is accepted to determine the virtual backbone nodes in WSN [1]. A Dominating Set (DS) is a subset of a graph with the property that any node from the parent graph either belongs the DS or is adjacent to any of the node in the DS. We get a Connected Dominated Set (CDS) when the nodes in the DS are themselves connected.

When considering the CDS for routing in a WSN, it is favourable that they exhibit the property of minimality in terms of number of connected components and the total weight between them. These correspond to the problem of Minimum Connected Dominating Set (MCDS) and Minimum Weight Connected Dominated Set (MWCDS) and help in reducing the energy consumption of the chosen backbone nodes as well as the network in general [2,3]. An early work that bought forward the idea of using MWCDS to form energy efficient WSN backbones is [4].

Most of the works in the literature focus on one of the either problem, as both of the problem belong to NP-Hard complexity class. But it is clear that finding a virtual network that is optimized with respect to both MCDS and MWMCDS would enhanced the network performance. Thus the requirement is that the CDS chosen should be optimal with respect to the multi objectives of minimum connectedness and minimum weight. An effective way to handle multiobjective optimization problems relating to networks is to use an evolutionary approach as in [5–7].

Hence, a Multiobjective Genetic Algorithm (MOGA) for Minimum Weight Minimum Connected Dominating Set (MWMCDS) is proposed, which aims to provide an optimized set of CDS with respect to both number of nodes and its corresponding weights.

1.1   Related Work

Genetic Algorithm (GA) was used in [8] to determine a MCDS for a given wireless network. The authors defined a probabilistic network over which MCDS of required reliability was generated using GA. While the resulting MCDS based on reliability (RMCDS) addresses the issue of lossy links in the network, energy expense of the network is not accounted. In [9], a genetic algorithm is proposed to produce energy efficient MCDS for MANETs. The input to the GA here is a CDS constructed based on reference energy. The resulting CDS is optimized with respect to this reference energy.

To optimize the load balancing in WSN using CDS, GA was applied in [10,11]. Topology changes based on CDS obtained through a Hierarchical Sub-Chromosome Genetic Algorithm was proposed in [12] to enhance performance in terms of power consumption and reliability. A hybrid genetic algorithm (HGA) is proposed in [13] to solve the MWCDS problem. The HGA functions by initiating the GA with a population primed with a defined greedy heuristic. The population based methods proposed so far generates the fitness values of chromosomes based on only one factor.

2   Fitness Model

Now the various components of the proposed fitness model, namely the multi-objective fitness, degree heuristics and the weightage assignment are defined.

2.1   Multiobjective Fitness

A multiobjective CDS problem $(\theta, F_1, F_2, \ldots, F_t)$ can be defined as

$$minF_i(C), i = 1, \ldots, t, \ subject \ to \ C \in \theta$$

where $\theta = C_1, \ldots, C_k$ is the set of possible CDS in the given network, and $F = F_1, F_2, \ldots, F_t$ is the set of different single objective functions with respect to a given CDS. The $F_i : \theta \rightarrow \mathbb{R}$ gives a measure of CDS fitness based on a particular criteria. Now based on Pareto optimality, a solution set can be obtained such that the F function vector values are optimized [14]. If C_1 and C_2 are two candidate CDS in θ, then C_1 can be said to dominate the CDS C_2 if and only if

$$\forall_i : F_i(C_1) \leq F_i(C_2) \bigwedge \exists is.t. F_i(C_1) < F_i(C_2)$$

Since the interest is in obtaining CDS that are optimized with respect to the vector of objective functions based on the application, nondominated CDS with the property of having some level of compromise between the objectives is sought and this corresponds to Pareto-Optimal solutions Π,

$$\Pi = C \in \theta : \nexists C' \in \theta \ with \ C' < C$$

2.2   Objective Functions

The objective functions considered here are minimum weight and minimum nodes in the CDS. So defining objective function Fitness-Nodes, $F_n : \theta \rightarrow [1, N]$, where θ is the set of candidate CDS and N is the number of nodes in the network. The other objective function defined is Fitness-Weight, $F_w : C \rightarrow \mathbb{N}$. Hence the MW-MCDS can be defined as a multiobjective CDS problem with $F = \{F_n, F_w\}$ and

$$min\{F_n(C) \ and \ F_w(C)\} \ subject \ to \ C \in \theta$$

2.3   Degree Heuristics

Heuristics based on the degree of the nodes is used to adjust the fitness based on the number of nodes in the CDS, F_n. Let G be a connected network and Max and $SecMax$ denoted the corresponding maximum degree and second maximum degree in G. Let $Leaf$ be the set of nodes having degree 1 in G. Then, if m is a node in C and $deg(m)$ is the degree of m, where C is set of possible nodes in the CDS, the following three heuristics are applied:

1. Reduce the fitness significantly when $deg(m) = Max$
2. Reduce the fitness partially when $deg(m) = SecMax$
3. Increase the fitness when $m \in Leaf$

It has to be noted that the modeling of MW-MCDS problem here requires the fitness to be reduced. That is, lesser the fitness value more optimal is the result.

2.4   Fitness Calculation

Let F_n and F_w be the fitness value obtained based on the number of nodes and the weights in the CDS and W_n and W_w be the respective weightage that should be assigned to the fitness values. Then the multiobjective fitness value is calculated using:

$$F = (W_n \times F_n') + (W_w \times F_w') \tag{1}$$

Where F_n' and F_w' are the normalized values of F_n and F_w, based on the total number of nodes and total weight in the network respectively.

3   Multiobjective Genetic Algorithm for MW-MCDS

The following steps are followed to design a genetic algorithm to obtain MW-MCDS. It can be noted that the GA is designed to explore the search space more by allowing the maximum possible variety.

3.1   Step 1: Representation

Let n be the number of nodes in the given connected Network G. The nodes of the G are assumed to be numbered using distinct positive integers from $[1, n]$. Then a chromosome c is represented as a list or array of size n with entries belonging to $[0, 1]$, where 0 represents a blank or no selection and 1 represents the consideration of the corresponding positioned node.

3.2   Step 2: Initial Population

Let $popsize_{ini}$ be the initial population size. Then generate randomly the set C_{ini} such that $\forall c \in C_{ini}, |c| = n$ and $|C_{ini}| = popsize_{ini}$.

3.3   Step 3: Fitness Evaluation

The Multiobjective CDS Fitness Algorithm (MCFA) described in Sect. 4 is mapping function between a chromosome from a set of population C and an integer, that is, $TFA : c -> \mathbb{N}$ where $c \in C$. The integer value mapped indicates the multiobjective fitness of the chromosome with respect to MWCDS and MCDS. Create set F such that $\forall c \in C$ corresponding $TFA(c, G) \in F$.

3.4   Step 4: Selection

Sort the chromosomes in the population C based on the corresponding values in F. Let $popsize_{select}$ be the number of chromosomes selected from the given set of C chromosomes. Let $s_{fitness}$ be the portion of $popsize_{select}$ that has to be selected based on fitness, then we have $s_{random} = 1 - s_{fitness}$ as the portion that has to be selected randomly from C. Generate set $C_{selected}$ such that $|C_{selected}| = popsize_{select}$ and $C_{selected} = [(s_{fitness} \times popsize_{select})least\ fitness\ valued\ c \in C] + [(s_{random} \times popsize_{select})randoms\ c \in C]$.

3.5   Step 5: Crossover

Divide the chromosomes in the given population C into pairs of parents. For each of the pairs p:

- Select randomly an integer value $i \in [1, n]$ that represents the crossover point
- Generate offsprings Off_1^{cross} and Off_2^{cross} by performing crossover at point i

Form $C_{crossovered}$, the population that consists of all the offsprings produced by crossover.

3.6   Step 6: Mutation

For each of the chromosome $c \in C$:

- Randomly select two integer $r1$ and $r2$ such that $r1, r2 \in [1, n]$
- Perform $r1$ mutations in c, with the mutated value as 0, to produce Off_1^{mut}
- Perform $r2$ mutations in c, with the mutated value in $[0, 1]$, to produce Off_2^{mut}

Form $C_{mutated}$, the population that consists of all the offsprings produced by mutation.

3.7   Step 7: Iteration

Form the intermediate population $C_{intermediate} = C + C_{crossovered} + C_{mutated}$. Perform steps 3 and 4 on $C_{intermediate}$ to obtain C_{new}. Continue steps 5–7 with $C = C_{new}$.

Repeat steps till termination condition is reached.

4   Multiobjective CDS Fitness Algorithm

1. Initialize three arrays A, B and Weight of length l with 1s, 0s and 0s respectively. Where l is the number of nodes in the graph and is equal to the size of the chromosome.
2. For each of the non zero entry c[i] in the chromosome do the following
   (a) Check the corresponding indexed entry in the array B, B[i], if it is 0 replace it with the index value, i
   (b) Scan the adjacent nodes of i, for each of the adjacent node j do the following
      i. Check the corresponding indexed value is in array A, A[j], if it is 1 replace it with 0
      ii. Check the corresponding indexed value in array B, B[j]
         A. If it is 0, no action
         B. If it is positive, negate it and replace the i indexed value of B, B[i], with negative of index value, −i,
            and assign Weight[j] = Distance[i, j]

      C. If it is negative, replace the i indexed value, B[i], with −i

         If Distance[i, j] is less than Weight[j], assign Weight[j] = Distance[i, j]

3. Scan the array A, if all the entries are zero continue, else assign fitness as 1000
4. Check if the sum of squares of all the non zero entry indices of the chromosome is equal to the sum of squares of the negative valued entries of array B
   (a) If yes assign fitness1 as number of non zero values in array B
   (b) If no assign fitness1 as 500, then modify it based on the Degree Heuristics by scanning each of the chromosome entry, whenever a 1 is encountered do the following
       i. If the corresponding node's degree is equal to the maximum degree, subtract 1 from the fitness1
       ii. If the corresponding node's degree is equal to the second maximum degree, subtract 0.5 from the fitness1
       iii. If the corresponding node's degree is equal to 1, add 1 to the fitness1
5. Calculate fitness2 as the sum of non zero entries in the Weight array
6. Normalize the fitness1 and fitness2 with respect to 1
7. Multiply w1 and w2 with normalized fitness1 and fitness2 respectively to get wfitness1 and wfitness2, where w1 and w2 are the required weightage of connectedness and weight of the CDS
8. Calculate and return the fitness of the CDS as the sum of wfitness1 and wfitness2

The flow of the fitness function steps is given in Fig. 1. The time complexity of the above algorithm is $O(N^2)$, where N is the number of nodes in the network.

5   Experiment and Result

5.1   Experiment Model

The goal of the experiment is limited to analyze the performance difference between a MCDS and MW-MCDS in a simplistic data transfer model. The following probabilities are defined to perform simulations:

1. The probability that a node i transfers data at a given instant, P^i_{Trans}
2. The probability that a data transfer t, is dropped or is failed in transit, P^t_{Drop}

It has to be noted that P^i_{Trans} is independent of P^t_{Drop} as the former indicates only the node i's probability to initiate a transmission, where as the later indicates the probability of the data transmission being not successful, which can be caused by various factors like congestion or link failure.

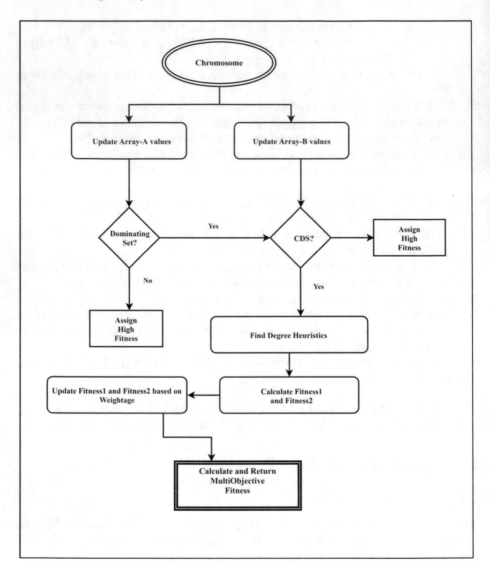

Fig. 1. Flow of fitness function

Further, let E^t be the energy expended during the transfer t. The assumption made is that:

1. In case of successful transfer $E \propto dist(t)$
2. In case of failed transfer $E \propto \frac{dist(t)}{2}$

Where $dist(t)$ is the total weight required to be covered during the transfer t. The energy for failed transfer is defined based on the assumption that a data unit would cover on average atleast half of its total distance to destination before getting lost or dropped.

5.2   Experiment Setup

The defined MOGA for MWMCDS and the energy model was coded using R language. Connected weighted networks of different sparsity was generated and the results averaged for each of the fixed node sizes. Degree heuristics and initial populations were randomly generated and fed to the MOGA-MWMCDS. The number of iterations was fixed as 100 for the MOGA. The resultant MWMCDS obtained was analyzed using the defined energy model. For comparison of the results, MCDS of the same networks was analyzed in Cytoscape software [15].

It has to be noted that [15] uses a deterministic approach to produce the MCDS. Hence a straightforward comparison with the proposed MOGA in terms of number of iterations is difficult. The energy model was defined to overcome this issue and the performance analysis is carried out based on the energy consumption and the number of nodes produced.

5.3   Result

The performance of MWMCDS with respect to energy consumption for 100 instances of data transfer is shown in Fig. 2. The energy consumption represents the total energy consumed by the network on completion of all the instances of the data transfer. Based on the model described in Sect. 5.1, this energy consumption indirectly represents the sum of distances covered by each of the data units in each transfer. It can be seen that the energy requirements of MWMCDS

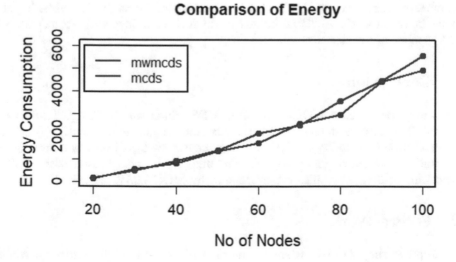

Fig. 2. MWMCDS vs MCDS - energy consumption

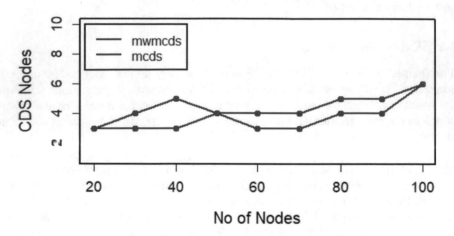

Fig. 3. MWMCDS vs MCDS - size

are lesser in most of the cases, especially when the number of nodes increases. Figure 3 shows the number of nodes produced by MWMCDS and MCDS for different networks. Here too it can be seen that MWMCDS obtained through MOGA performs better in most cases by producing CDS of minimal length. From these comparisons it can be deduced that the performance of the MWM-CDS generated by the proposed MOGA is better.

Since the time complexity of the fitness evaluation algorithm is proportional to the number of nodes, the iteration period increases with the number of nodes. Also, the performance of the proposed MOGA is limited with respect to the number of iterations set (100) and can be further increased by adjusting it. That is, the size of the CDS obtained might further reduce with an increase in the number of iterations.

6   Conclusion

This work proposed a MOGA for MWMCDS which was built upon a Fitness Model. The Fitness model defined the size and weight as the multi objectives to determine the quality of CDS. Further, heuristics based on degree was used to quicken the convergence of CDS. The analysis by comparison with MCDS algorithm indicates significant performance by MOGA-MWMCDS.

7   Future Work

To improve the CDS size when the number of nodes is high heuristic seeding techniques can be used to generate the initial populations, such as in [13].

References

1. Das, B., Sivakumar, R., Bharghavan, V.: Routing in ad hoc networks using a spine. In: Proceedings of the International Conference on Computer Communications and Networks, ICCCN, pp. 34–39 (1997)
2. Mohanty, J.P., Mandal, C., Reade, C.: Distributed construction of minimum Connected Dominating Set in wireless sensor network using two-hop information. Comput. Netw. **123**, 137–152 (2017)
3. Wang, Y., Wang, W., Li, X.-Y.: Distributed low-cost backbone formation for wireless ad hoc networks. In: Proceedings of the International Symposium on Mobile Ad Hoc Networking and Computing (MobiHoc), pp. 2–13 (2005)
4. Torkestani, J.A.: Backbone formation in wireless sensor networks. Sens. Actuators A: Phys. **185**, 117–126 (2012)
5. Lima, M.P., Alexandre, R.F., Takahashi, R.H.C., Carrano, E.G.: A comparative study of multiobjective evolutionary algorithms for wireless local area network design. In: Proceedings of the IEEE Congress on Evolutionary Computation, CEC 2017, pp. 968–975 (2017). Article no. 7969413
6. Massobrio, R., Bertinat, S., Nesmachnow, S., Toutouh, J., Alba, E.: Smart placement of RSU for vehicular networks using multiobjective evolutionary algorithms. In: Latin-America Congress on Computational Intelligence, LA-CCI 2015 (2016). Article no. 7435974
7. Gu, F., Liu, H.-L., Cheung, Y.-M., Xie, S.: Optimal WCDMA network planning by multiobjective evolutionary algorithm with problem-specific genetic operation. Knowl. Inf. Syst. **45**(3), 679–703 (2015)
8. He, J., Cai, Z., Ji, S., Beyah, R., Pan, Y.: A genetic algorithm for constructing a reliable MCDS in probabilistic wireless networks. Lecture Notes in Computer Science (including subseries Lecture Notes in Artificial Intelligence and Lecture Notes in Bioinformatics). LNCS, vol. 6843, pp. 96–107 (2011)
9. Manohari, D., Anandha Mala, G.S.: An evolutionary algorithmic approach to construct connected dominating set in MANETs. IET Semin. Digest (4) (2012)
10. He, J., Ji, S., Yan, M., Pan, Y., Li, Y.: Load-balanced CDS construction in wireless sensor networks via genetic algorithm. Int. J. Sens. Netw. **11**(3), 166–178 (2012)
11. Kumar, G., Rai, M.K.: An energy efficient and optimized load balanced localization method using CDS with one-hop neighbourhood and genetic algorithm in WSNs. J. Netw. Comput. Appl. **78**, 73–82 (2017)
12. Hosseini, E.S., Esmaeelzadeh, V., Eslami, M.: A hierarchical sub-chromosome genetic algorithm (HSC-GA) to optimize power consumption and data communications reliability in wireless sensor networks. Wirel. Pers. Commun. **80**(4), 1579–1605 (2015)
13. Dagdeviren, Z.A., Aydin, D., Cinsdikici, M.: Two population-based optimization algorithms for minimum weight connected dominating set problem. Appl. Soft Comput. J. **59**, 644–658 (2017)
14. Pizzuti, C.: A multiobjective genetic algorithm to find communities in complex networks. IEEE Trans. Evol. Comput. **16**(3), 418–430 (2012). Article no. 6045331
15. Nazarieh, M., Wiese, A., Will, T., Hamed, M., Helms, V.: Identification of key player genes in gene regulatory networks. BMC Syst. Biol. **10**(1), Article no. 88 (2016)

Modeling of a System for fECG Extraction from abdECG

Rolant Gini John[1]([✉]), Ponmozhy Deepan Chakravarthy[1], K. I. Ramachandran[2], and Pooja Anand[1]

[1] Department of Electronics and Communication Engineering, Amrita University, Coimbatore, India
j_rolantgini@cb.amrita.edu, deepanpdc@gmail.com,
anandpooj@gmail.com
[2] Center for Computational Engineering and Networking, Amrita University, Coimbatore, India
ki_ram@cb.amrita.edu

Abstract. The objective of this paper is to move a step ahead in investigation and create a feasible, cost effective fetal ECG analysis tool for clinical practice which will be easy for usage by any non-skilled personal and provide actionable medical information such as the QRS complex of fetal ECG, fetal HR etc. In this method, a composite abdominal ECG is subjected to a pre-processing stage which involves filtering and normalization, then fed into the 'thresholding and peak finding' stage to detect the maternal ECG peaks. The next stage involves construction of the MLE of maternal ECG embedded in the abdominal ECG. After this, the constructed MLE which represent the maternal ECG is subtracted from the abdominal ECG to obtain fetal ECG along with a smidgen of noise. This noise which adulterates the fetal ECG is removed by filtering, done at the post processing stage. Thresholding and peak finding is done at the post processed signal to calculate the fetal HR. This paper puts forth a promising possibility of implementing the proposed algorithm in any suitable hardware model, since an average Accuracy of 76.8% and average Sensitivity of 90.7% is attained.

Keywords: fetal ECG (fECG) · maternal ECG (mECG)
abdominal ECG (abdECG) · fetal Heart Rate (fHR)
Maximum Likelihood Estimation (MLE)

1 Introduction

Worldwide, an estimated 45% of neonatal death occurs during childbirth. The leading cause of birth defect-related death is heart defects [1]. In an official statement [2], the American College of Obstetrics and Gynecology (ACOG) admitted that no well-controlled study has yet proven that routine scanning of prenatal patients will improve the outcome of pregnancy. The Association for Improvements in the Maternity Services (AIMS), England, recorded cases of women who aborted their perfectly fit and healthy babies as a result of misinterpreted scans. An ultrasound scan may reveal if a baby suffers from Down syndrome, but it doesn't scale the seriousness of the condition, because the changes in fetal heart rate and fetal heart rate variability are not well understood.

© Springer International Publishing AG, part of Springer Nature 2018
A. Abraham et al. (Eds.): ISDA 2017, AISC 736, pp. 568–579, 2018.
https://doi.org/10.1007/978-3-319-76348-4_55

These problems being held in mind, there started to emerge techniques to estimate fetal heart rate. Low heart rate variations have been studied to identify intrauterine growth restricted fetuses (prepartum), and abnormal fHR patterns have been associated with fetal distress during delivery (intrapartum) [1]. Though there is a wide spread technology for fetal well-being, there are many cardiac defects that cannot be identified just by the estimation of fHR alone.

Hence interests for fECG extraction techniques multiplied in a visible scale. fECG monitoring can be used to study the effects of lack of oxygen in the heart and brain of the fetus. This even measures oxygen saturation in fetuses better than pulse oximetry. fECG acts as the best tool to detect fetuses which are at the risk of suffering acidosis. Once the risk is detected, the delivery is expedited before the fetus shows signs of acidosis. Moreover, this technique allows the detection of false recorded Cardiotocography (CTG) positives [3].

Generally fECG monitoring can be done using invasive or non-invasive techniques. In invasive techniques, the electrodes are placed in a way that it has direct contact with the fetus (mostly on the scalp) which may sometimes be very harmful to the fetus or to the mother as well. Moreover, this can only be done just before the time of delivery of fetus. Whereas non-invasive fetal ECG (NI-fECG) extraction techniques can be performed during any stages of pregnancy. The electrodes placed during NI-fECG will have no direct contact with the fetus and will not cause any harm to the fetus. Current non-invasive techniques [1] used are Phonocardiography, Doppler ultrasound, Cardiotocography, Fetal magnetocardiography etc. In methods such as Doppler ultrasound and fetal magnetocardiography, the heart beat can be heard without amplification till 20 weeks. After this, a highly non-conductive layer called vernix caseosa is formed around 20th week of gestation and dissolves only by 37th to 38th week in normal pregnancies [4]. This acts as a limiting factor for most of the NI-fECG techniques, especially for Cardiotocography. In commonly used techniques such as Doppler ultrasound, there exist disadvantages like unsafe radiation on the fetus since it is not passive and may pick up maternal heart rate instead of fetal heart rate (fHR) due to misorientation of ultrasound transducer.

Moreover the drawbacks of fetal magnetocardiogram are the requirement of skilled personnel, much higher SNR of maternal ECG than the fetus and no possibility of long term monitoring because of the apparatus size and cost [1].

Heart being the first organ to develop in the fetus [5], its proper functioning should be ensured by continuous or periodic monitoring to avoid upcoming anomalies. Hence fECG extraction can play a vital role to reduce mortality rate of fetus and cinch wellbeing of both mother and fetus. The framework of this paper is as follows: Literature survey is detailed in Sect. 2; Sect. 3 describes the proposed method in detail including the preprocessing stage, mECG peak identification and cancellation, and the fECG peak detection; fetal heart rate measurement is discussed subsequently. It is followed by Results and Discussions in Sect. 4. Conclusion of the work is in Sect. 5.

2 Literature Survey

Normally NI-fECG techniques use data acquired from a single channel or multiple no of channels as input. Single channel techniques do not depend on any sort of information other than one particular channel, whereas multi-channel techniques depend on more than one channel for information to process. Generally an electrode upon mother's thoracic region will be considered as part of the multi-channel technique for reference. The method which is being discussed here is a single channel technique and is capable of computing the results without any information from any other means. The main challenge with NI-fECG extraction techniques is the low signal-to-noise ratio of the fECG signal on the abdominal mixture signal which consists of a dominant maternal ECG component, fECG and noise. However the NI-fECG offers many advantages over the alternative fetal monitoring techniques. The most important advantage is the opportunity to enable morphological analysis of the fECG which is vital for determining whether an observed fHR event is normal or pathological.

Some of the common existing NI-fECG extraction techniques are Blind Source Separation (BSS) [6], Wavelet transform [7], Principal Component Analysis (PCA) [8], Independent Component Analysis (ICA) [9], Fast ICA [10], Adaptive Filtering [6, 7] etc.

In BSS, the source signal is evacuated from a mixture picked up by multiple no of channels without any earlier information on how the independent source signal got blended. Even though this method seems to be performing better than Adaptive Filtering in many aspects, Empirical Mode Decomposition (EMD) which is being used for de-noising is not effective and the signal stays noisy even after the procedure is done [11]. In addition to this, BSS is not reliable when both the mother and fetus are in movement [6].

Techniques like Principal Component Analysis [8], Independent Component Analysis [9] or Periodic Component Analysis make use of the ECG's periodicity. Generally, these methodologies are a type of Blind Source Separation [6], which plan to isolate the Independent sources into three classifications: mECG, fECG and noise. PCA [8] looks to correlate the dataset i.e. remove the second order dependencies; However in these methods picking a channel plays a noteworthy part which impacts the extraction. In PCA, it is assumed that signals originating from different sources are linearly mixed; that huge fluctuation speaks to intriguing structures and that the Components are orthogonal, where no fECG was outwardly identifiable in the time. Fast ICA [10] had been proven to be capable of extracting fECG even when it is at a very low voltage levels. It had been said that ICA outflanks majority of the other signal processing methods. Be that as it may, ICA is computationally complicated because of its utilization of higher order statistics and hence it is not appropriate either for real time applications or in lightweight and portable monitors, which aids the movement of the mother. Wavelet transform [7] is a very powerful tool for non-stationary signal analysis but the calculations used in the projection algorithm is very complex for real time applications. Hence it is majorly used in de-noising signals before or after fetal ECG extraction. The Wavelet transform is once in a while utilized along with adaptive filter for fECG extraction. In such cases, wavelet transforms are utilized for pre-processing of abdominal and thoracic signals giving out Wavelet coefficients and are delivered as inputs for adaptive filtering algorithm.

In spite of many fascinating hypothetical systems, the robustness of the most of these techniques had not been sufficiently quantitatively assessed and little advancement had been made in their utilization. This is mainly because of the upcoming three reasons: (1) the absence of good standard databases with clear Annotations; (2) the immature procedure for evaluating the calculations; (3) the absence of open source code makes the re-usage of the original algorithms inclined to mistakes, and makes objective benchmarking difficult if not inconceivable.

Certain studies on hardware were also done. Devices like Monica AN24 [12], Meridian monitor [13], Portable fECG extractor from abdECG (Raspberry Pi model) [14] are studied in order to understand how efficiently they identify fHR and helps in diagnosis of any diseases. In recent years, lot of wireless devices as the above mentioned are into existence which makes the fHR identification much easier and also provides efficient results. Drawback of the Monica AN24, Meridian monitor is that the location of electrodes happens to decide the Sensitivity rate of the output [14]. But these devices are mostly portable and can be used in any places. Most of the above mentioned machines except the portable fECG extractor from abdECG (Raspberry Pi model) are costly and are not evenly available to all segments of people which become a limiting factor in developing nations.

A simple non-invasive approach to identify the fECG of good quality and to infer the fetal heart rate from it; is being discussed here. The method yields good results, even though the algorithm involves exceptionally less number of steps. This uses a single abdominal channel and the fECG is being extracted without using any other channel information or reference channel. The concept is implemented in SIMULINK, an integrated tool in MATLAB for activities like hardware modelling, Making it readily available to apply in any affordable less costly hardware such as TMS320C5515 [15], MSC8144 [16] etc., and the real signals from daISy [17] database are used.

3   Methodology

The major component of the abdECG signal which is corrupting the fECG extraction is mECG. In this method, a composite abdECG is subjected to a pre-processing stage which involves filtering and normalization, then fed into the 'thresholding and peak finding' stage to detect the mECG peaks. The next stage involves construction of the MLE of mECG embedded in the abdECG. After this, the constructed MLE which represent the non-linearly transformed mECG is subtracted from the abdECG to obtain fECG along with a smidgen of noise. This noise which adulterates the fECG is removed by filtering done at the post processing stage. Thresholding and peak finding is done at the post processed signal to calculate the fHR. The overall flow of the proposed methodology is as shown in the Fig. 1.

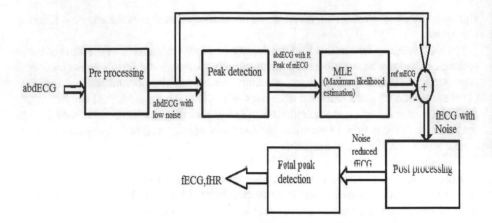

Fig. 1. Process of fECG extraction

3.1 Pre-processing Stage

This stage is totally dependent on the composite abdECG signal available in the daISy database where a single channel had been shown in the Fig. 2(A). Necessary filtering

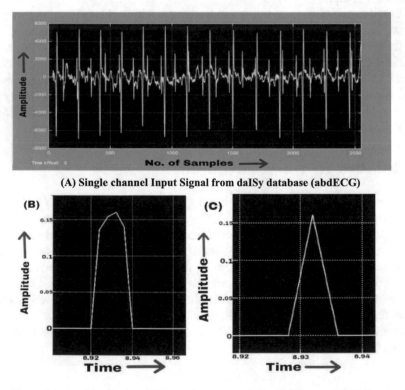

Fig. 2. Channel 3: (A) Single channel input signal from daISy Database (abdECG); (B) Output of thresholding stage of abdECG; (C) Peak finding after thresholding in abdECG

should be performed in accordance with the noise present. In common, a FIR band pass filter set at a range of 5–20 Hz has been used [18]. Hence the lower frequencies contributed by the Baseline wandering, Power-line interference and other noise sources are eliminated. The filtered signal is then normalized for further processing.

3.2 Maternal Peak Identification

The normalized abdECG serves as input for this stage. The maternal R peaks had been detected using the concepts of thresholding and local maxima. A threshold range had been set according to the nature of the signal. Based on thresholding, the region of the signal above the threshold which will contain the mECG peaks had been obtained and had been shown in the Fig. 2(B) (considered channel 3 as input). This is because the mECG peaks are of higher amplitude than fECG peaks. In this region, points of local maxima which are the R peaks of the QRS complex of mECG had been detected and had been shown in the Fig. 2(C). It should be noted that Fig. 2(C) represents one of the R peaks among the total no of peaks detected in the considered input.

3.3 Maternal ECG Reconstruction

The maternal R peaks, its location in the time domain and also the normalized abdECG are given as inputs for MLE of mECG. Generally MLE [19] is very efficient for data having large samples and it extricates all possible information from the data and does not dependent on the additional reference data. Hence MLE is considered as the right choice in case of fECG extraction. Using the identified maternal R peaks, MLE has been used to form a PQRST complex by considering the information around the R peak of every maternal beat of the considered signal. This constructs a single beat of mECG and shown in the Fig. 3(A). Based on the identified R-peak locations and repetition of this single beat, the entire mECG beats of the signal have been formed and condensed into a single reference which has been shown in the Fig. 3(B). Thus, reference PQRST complex of the mECG is created using MLE algorithm involving quite a few steps. It

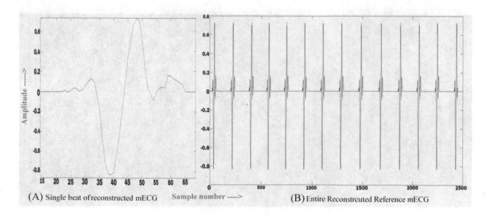

(A) Single beat of reconstructed mECG   Sample number ---->   (B) Entire Reconstructed Reference mECG

Fig. 3. Channel 2: MLE of mECG

should be noted that the repetition of reconstructed single beat of reference mECG is not blindly random or simply periodic. The time interval and the no of samples for the reconstructed mECG beats were decided in accordance with the input considered.

3.4   Fetal Peak Detection and fHR Measurement

A simple subtraction is done to obtain fECG; Normal ECG taken at the thoracic region of the mother can't be used as such to cull out mECG component embedded in abdECG, because the mECG will be non-linearly transformed on its way from heart to abdomen due to various reasons such as breathing noise etc,. Hence the reconstructed reference mECG is subtracted from the normalized abdECG to obtain fECG infused with few noise signals. These noises are removed by post processing. Normally a similar Band-pass filter, like the one used in the pre-processing stage is used but this could even be a simple high pass filter to filter out the lower frequencies which might still be present after the extraction. After this, Thresholding and Peak detection as used in mECG peak identification stage is done here to detect the fetal peaks which in-turn is used to measure fHR. Here also the threshold values for the thresholding function should be modified with respect to the fECG signal nature. The fetal heart rate is calculated as [18]

$$fHR = \frac{\text{Number of peaks detected}}{\text{Duration of the signal}} * 60 \tag{1}$$

fHR calculated and morphology of fECG give physicians a clear idea of the arrhythmias and other anomalies in the fetus [1].

4   Results and Discussion

The proposed method is implemented using SIMULINK software. The concept of FP, FN and TD can be understood clearly with the help of Fig. 4 and Channel 3 has been

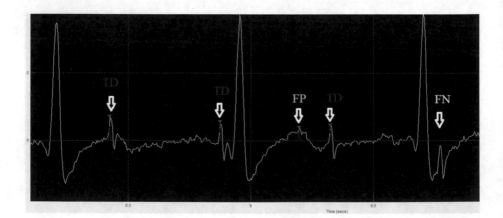

Fig. 4.   Channel 3: Illustration of FP, FN and TD

used as input for the illustration. The peaks which are present and are detected are truly diagnosed peaks (TD) and indicated in Fig. 4. Some peaks which detected when they are actually not present are categorized as false positives (FP) and the same has been marked in Fig. 4. When the actual peaks are not being detected, it is considered as false negative (FN) as indicated in Fig. 4 [18].

The accuracy and sensitivity are calculated using [18]:

$$\text{Accuracy} = \frac{TD}{(TD + FN + FP)} * 100 \tag{2}$$

$$\text{Sensitivity} = \frac{TD}{(TD + FN)} * 100 \tag{3}$$

The proposed algorithm was also tested with channel 4 as input. This would be a challenging factor to elicit good results, as channel 4 input comes from a lead placed far away from the fetus's heart; the fECG embedded in abdECG is of very low amplitude. The following figures: Fig. 5 shows the detected maternal R peaks, marked on normalized abdECG from channel 4 and it is evident that all of the maternal peaks which are originally present have been identified correctly and none is missed out; Fig. 6 shows the detected fetal R peaks, marked on normalized abdECG from channel 4. But here it can be noted that few noise signals have been misidentified as fetal R peaks (false positives) and this contributes in the reduction of accuracy and sensitivity of the proposed algorithm. Using the formulas mentioned in the Eqs. (2) and (3), the Accuracy and Sensitivity of 81.4% and 100% was obtained (for channel 4) respectively. The normal range of fHR is between 120 to 160 bpm [5]. And by using the formula mentioned in the Eq. (1), a fetal heart rate of 132 bpm (beats per minute) was obtained for channel 4.

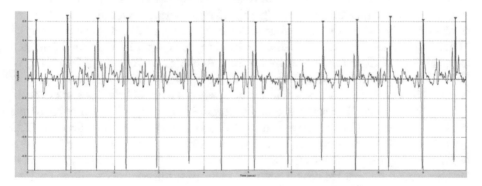

Fig. 5. Channel 4: Maternal 'R' peaks marked on normalized input signal

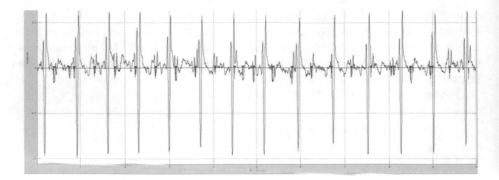

Fig. 6. Channel 4: Fetal 'R' peaks marked on normalized input signal

In daISy database, only channels 2, 3 and 4 are rich in vital information of fetus. In order to validate our method and for further assessment, the proposed approach is implemented for channels 2 and 3 as well. Overall Accuracy of 76.8% and Sensitivity of 90.7% was obtained when tested with the inputs from Channel 2, 3 and 4. The fHR obtained falls within the normal range (120–160 bpm) for all the three selected channels. Direct comparison of obtained results with those from other existing techniques hadn't been done due to unavailability of similar hardware modeling algorithms in this area. For example, in a method such as "Extraction of clinical information from non-invasive fECG [1]", "Research of fECG extraction using Wavelet Analysis and Adaptive filtering [7]", "Fast Technique for Non-invasive fECG extraction [10]" etc., real time signals, synthetic data and different database (other than daISy) has been used as inputs. Moreover those aren't dealing with hardware modeling algorithms which restricts direct comparison of obtained results. The work of "Portable fECG extractor from abdECG [14]" proposed a hardware modeling algorithm for portable system which used the same database (daISy database) achieved an average Accuracy of 87.3% and Sensitivity of 91%. The proposed hardware modeling algorithm by this paper is able to perform close to the above mentioned prototype with overall Accuracy of 76.8% and Sensitivity of 90.7%. However the following figures affirm the efficiency of this proposed method. Figure 7 shows a fetal peak being detected even when the fECG is buried completely in the negative domain. The small rise in the negative going signal which is below the marked peak denoted by the arrow, is the fECG component in the normalized abdECG from channel 3 and the same has been detected as shown in Fig. 7. Figure 8 shows a fetal peak being identified even though there is no sign of fECG. It can be noted that, in the area around the marked peak denoted by the arrow, there is no significant positive rise in the normalized abdECG which generally represents the embedded fECG component which can be observed in Fig. 8. This is because the fECG is completely overlapped with the mECG. Here also the input from channel 3 was used.

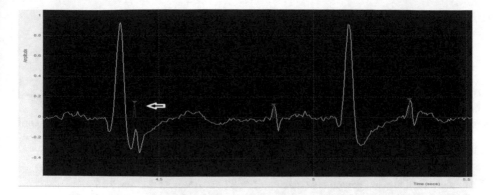

Fig. 7. Channel 3: Detection of fECG peak buried in negative domain

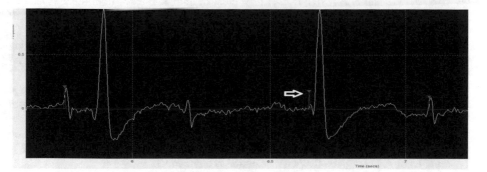

Fig. 8. Channel 3: Detection of fECG peak completely overlapped with abdECG

5 Conclusion

A simple yet effective approach is presented in this paper for fetal ECG extraction by non-invasive means. The proposed algorithm involves quite a few steps, thereby reducing the computational complexity unlike most of the existing methods.

The time taken for the entire process to execute completely is just 5 to 7 s. Even though the proposed approach might not work in case of twin fetuses, this method works effectively for single fetuses using single lead input irrespective of the electrode's placed position in the abdomen. So this reduces the hardware complexity which exists as a barricade for most of the systems. The proposed method also wields a guaranteed possibility of hardware validation using a suitable processor which can be capable of serving as a portable standalone system. Thus the research objective has been achieved as a reliable method for fECG extraction and its applicability in hardware as aimed. But the system has to be tested for other data base, real time signals and is to be clinically validated.

Conflict of Interest. The authors disclose that there is no conflict of interest present.

References

1. Behar, J.: Extraction of clinical information from the non-invasive fetal electrocardiogram (2017). Arxiv.org. https://arxiv.org/abs/1606.01093. Accessed 21 Oct 2017
2. The Risks of Ultrasound Scans for the Mother and the Unborn Child - Ener-Chi Wellness Center. In: Ener-Chi Wellness Center. http://www.ener-chi.com/the-risks-of-ultrasound-scans-for-the-mother-and-the-unborn-child. Accessed 21 Oct 2017
3. ScienceDaily: Fetal electrocardiogram helps in early detection of neonatal acidosis, Spanish researcher find. In: ScienceDaily. https://www.sciencedaily.com/releases/2011/06/110616081819.htm. Accessed 21 Oct 2017
4. Singh, G., Archana, G.: Unraveling the myotery of vernix caseosa, https://www.ncbi.nlm.nih.gov/pmc/articles/PMC2763724/. Accessed 21 Oct 2017
5. Heart development. https://en.wikipedia.org/wiki/Heart_development. Accessed 21 Oct 2017
6. Zarzoso, V., Nandi, A.: Noninvasive fetal electrocardiogram extraction: blind separation versus adaptive noise cancellation. IEEE Trans. Biomed. Eng. **48**, 12–18 (2001). https://doi.org/10.1109/10.900244
7. Wu, S., Shen, Y., Zhou, Z., Lin, L., Zeng, Y., Gao, X.: Research of fetal ECG extraction using wavelet analysis and adaptive filtering. Comput. Biol. Med. **43**, 1622–1627 (2013). https://doi.org/10.1016/j.compbiomed.2013.07.028
8. Sameni, R., Jutten, C., Shamsollahi, M.: Multichannel electrocardiogram decomposition using periodic component analysis. IEEE Trans. Biomed. Eng. **55**, 1935–1940 (2008). https://doi.org/10.1109/tbme.2008.919714
9. Yao, W., Zhao, J., Zheng, Z., Liu, T., Liu, H., Wang, J.: Fetal electrocardiogram extraction based on modified robust independent component analysis. Adv. Mater. Res. **749**, 250–253 (2013). https://doi.org/10.4028/www.scientific.net/amr.749.250
10. Martín-Clemente, R., Camargo-Olivares, J., Hornillo-Mellado, S., Elena, M., Román, I.: Fast technique for noninvasive fetal ECG extraction. IEEE Trans. Biomed. Eng. **58**, 227–230 (2011). https://doi.org/10.1109/tbme.2010.2059703
11. Luo, Z.: Fetal electrocardiogram extraction using blind source separation and empirical mode decomposition. J. Comput. Inf. Syst. **12**(2012), 4825–4833 (2012). 8th edn http://citeseerx.ist.psu.edu/viewdoc/download?doi=10.1.1.470.7411&rep=rep1&type=pdf
12. Monica Healthcare: Monica AN24 with IF24|Fetal Monitoring. In: Monica Healthcare. http://www.monicahealthcare.com/products/labour-and-delivery/monica-an24-with-if24. Accessed 21 Oct 2017
13. Mindchild: MERIDIAN MONITOR. http://www.mindchild.com/assets/22dsp011522-mindchild-meridian-monitor.pdf. Accessed 21 Oct 2017
14. Gini, J.R., Ramachandran, K.I., Nair, R.H., Anand, P.: Portable fetal ECG extractor from abdECG. In: International Conference on Communication and Signal processing, India (2016). https://doi.org/10.1109/ICCSP.2016.7754265
15. TMS320C5515 Datasheet (PDF) - Texas Instruments. Alldatasheet.com. http://www.alldatasheet.com/datasheet-pdf/pdf/349584/TI/TMS320C5515.html. Accessed 14 Nov 2017
16. MSC8144 Datasheet (PDF) - Freescale Semiconductor, Inc. Alldatasheet.com. http://www.alldatasheet.com/datasheet-pdf/pdf/183488/FREESCALE/MSC8144.html. Accessed 14 Nov 2017
17. daISy: The Datasets. Homes.esat.kuleuven.be. http://homes.esat.kuleuven.be/~smc/daisy/daisydata.html. Accessed 21 Oct 2017

18. Nair, R.H., Gini, J.R., Ramachandran, K.I.: A simplified approach to identify the fetal ECG from abdECG and to measure the fHR. In: Goh, J., Lim, C.T. (eds.) 7th WACBE world congress on bioengineering 2015. IP, vol. 52, pp. 23–26. Springer, Cham (2015). https://doi.org/10.1007/978-3-319-19452-3_7
19. Myung, I.J.: Tutorial on maximum likelihood estimation. J. Math. Psychol. **47**, 90–100 (2003). https://doi.org/10.1016/S0022-2496(02)00028-7

Supervised Learning Model
for Combating Cyberbullying: Indonesian
Capital City 2017 Governor Election Case

Putri Sanggabuana Setiawan[1], Muhammad Ikhwan Jambak[1(✉)],
and Muhammad Ihsan Jambak[2]

[1] BINUS Graduate Program-Master of Computer Science,
Computer Science Department, Bina Nusantara University,
Jakarta 11480, Indonesia
putri.setiawan@binus.ac.id , mjambak@binus.edu
[2] Department of Informatics, Faculty of Computer Science,
Universitas Sriwijaya, Palembang, Indonesia
jambak@ilkom.unsri.ac.id

Abstract. Technological growth has triggered the internet usage in
Indonesia, especially the social media. However, the cyberbullying that is
believed has a negative effect on the victims bigger than the traditional
one come as a price of this changing trend. It is not an easy task to
eliminate or reduce such bad action due to several reasons. Firstly, the
cyberbullying often appear in an informal language, slank or local languages. Secondly, the content must be understood from a specific context.
In order to have such a tool to combat the cyberbullying, such model and
tested algorithm should be made and confirmed by the linguists. This
research develops a cyberbullying corpus in order to develop a supervised
learning algorithm. The data have been crawled from the social media
during the Indonesian capital city 2017 governor election event where
the cyberbullying between the candidates supporters occurred. After the
data preprocessed and classified by the experts/linguists for the feature
space design.

Keywords: Cyberbullying · Supervised learning · Feature space design

1 Introduction

Based on *Asosiasi Penyelenggara Jasa Internet Indonesia* (APJII) data, the
number of internet user in 2016 is 132.7 million that has increased 51.8% than
number of internet user in 2014 [1]. And from that figure, on average people
spend 2 to 3 h to access social media, where the users could share their feeling,
emotion, and opinion about their personal life as well as the social, political,
financial and any other issues that concern them. Having a looser ethics in a
virtual community, such intimidation, and humiliation acts are formed relatively
easier. Cyberbullying is the use of cell phones, instant messaging, e-mail, chat

© Springer International Publishing AG, part of Springer Nature 2018
A. Abraham et al. (Eds.): ISDA 2017, AISC 736, pp. 580–588, 2018.
https://doi.org/10.1007/978-3-319-76348-4_56

rooms or social networking sites to harass, threaten or intimidate someone. The impact of cyberbullying is considered more harmful than bullying because it's content can spread out easily and can be kept for a longer period time. However, is not an easy task to eliminate or reduce such bad action due to several reasons. Firstly, the cyberbullying often appear in an informal language, slank or local languages. Secondly, the content must be understood from a specific context.

The cyberbullying was a research subject conducted by the Indonesian ministry of communication and information along with the United Nations Children's Fund (UNICEF) in a period of 2011 until 2012, which involved a representative sample of 400 kids and teenagers from urban and rural areas in 11 provinces of Indonesia [5]. According to UNICEF, cyberbullying has been considered as one of many violence experienced that often occurs at social media. The research result showed that 13% of them experiencing cyberbullying, 8.2% sent a bullying message through social media, and 4.4% sent a bullying messages through text message. In addition, according to a Global Market and Opinion Research Specialists called IPSOS, in collaboration with Reuters, Business Financial News, and US International Breaking News, they found that 74% of Indonesian respondents pointed out that social media such as Facebook and Twitter are the location of the cyberbullying [6].

Cyberbullying trend in Indonesia is gone to high and undoubtedly, an anti-cyberbullying is a necessity and Indonesia has yet to produce one but far from surcease. Beside the mentioned reasons above, a Bahasa Indonesia cyberbullying corpus where a supervised model can rely on it has yet need to be produced. This research aimed at creating a cyberbullying corpus in Bahasa Indonesia and develop a supervised model based on in. This study, taking the advantage of the cyberbullying escalation during the Indonesian 2017 capital city governor election that intensively stimulus cyberbullying between the candidate's supporters.

For the purpose of this research, similar works of Farisia, Dinakar, and Desai have been studied. Farisia [4] did the detection of cyberbullying using text mining by crawling twitter post with Hyper pipes. The text then processed with tree-based J48, and Support Vector Machine (SVM) methods using the Margono's bullying words list. However, this work did not include the accompanied punctuation and emoticons. Dinakar [3] did the cyberbullying detection by crawling the YouTube comments using Nave Bayes, Hyperpipes, J48, and SVM algorithms along with the feature space design. Nahar [7] did the cyberbullying detection by crawling the social media using SVM with linear kernel and k-10 fold cross, and feature selection. In addition, Desai [2] did the sarcasm detection using SVM with Radial Basis Function (RBF) kernel to classify sentences or phrase that are suggested containing bullying meaning.

2   Methodology

The data collection happened during the campaign period of Indonesian 2017 capital city governor election. The dataset was made up of crawled Tweeter and

Facebook users' status using the R Studio application. The dataset not only contains the texts but also includes the emoji, hashtag, punctuation marks as Dinakar [3] said that in order to detect a cyberbullying, ones should also consider the punctuation mark, emoticon, hashtag. In addition, Desai said that the cue word lists that often used in cyberbullying would also useful [2]. Filters have been set in the R studio as "Pilkada" (capital city election), "Jakarta", "debat pilkada" (candidates debate), "pemilihan" (election), "kampanye" (campaign), and the governor as well as the vice governor candidates' name. The crawling process has collected 10000 tweets in the period of February to April 2017. The number of the tweets down to 5000 after both automatic redundancy filtering and manual redundancy check by the linguist.

The dataset then assessed by the linguist expert to be classified manually into bullying or not bullying. The ones that classified as bullying then assessed further in order to design the feature space. Further, the designed feature space has been used to crawl more for a new dataset. The new dataset has been divided into two by 60 to 40% for training and testing purposes. The dataset has been preprocessed using a Python code before can be used for training or testing. The preprocessing includes removing the noise such as URL, geotag and @Mention as well as lowering the text case, tokenizing the unigram, filtering the texts with emoji, hashtag, and punctuation, encoding the emoji into emojiText, marking the word, phrase and cue word, and lastly removing the duplication.

3   Feature Space Design

In order to give a context the readers on how difficult the cyberbullying assessment, we provide one of a social media user's tweet in the following sentence: *Kesalahan terbesar Prabowo adalalah memberi tungangan kepada pecatan menteri yang cuma bermodalkan kata-kata manis dan indah hahaha!!! *ROTFL* #perudungan*. Obviously, this sentence is not easy to be assessed. The assessment must be seen from the context of sentence, and not on at the level of an individual words as they are not a dirty or rough word. In this case, there are some key words which classify it into a bullying sentence, such as: *pecatan menteri* (ex-minister), *bermodalkan kata-kata manis* (lips service), and *bermodalkan kata-kata indah* (another form of lips service). These words or phrase are called as feature space for feature extraction in machine learning for classifying the cyberbullying [2–4].

From the 5000 provided tweets, the linguist selected 250 tweets that meaningful as partly shown in Table 1. The linguist classified 200 out of 250 tweets as bullying sentence. Furthermore, the future spaces are defined as word feature space in Table 2, phrase feature space in Table 3, emoji feature space in Fig. 1, punctuation mark feature space in Table 4, hashtag feature space in Table 5, and the cue word feature space in Table 6.

Table 1. Cyberbullying manual classification by the linguist

No.	Sentence	Bullying (Y/N)
1	*Anies yang haus kekuasaan begitu dipecat karena tidak becus mengelola dana sertifikasi di Kemendiknas sampai kelebihan Rp 23 triliun meloncat ke seberang musuh politik dan bergabung dengan pecundang Pilpres 2014 Prabowo dan PKS*	N
2	*Kampanye SARA adalah satu-satunya cara Anies menuai dukungan karena kalau berkoar tentang program, maka Ahok-Djarot telah membuktikan dan pasti Anies kedodoran*	N
3	*Bagaimana dengan tatanan keberagaman dan kedamaian masyarakat kita karena kelakuan mereka yg merasa superior, merasa paling benar. Mereka merasa ada yg jadi, 'tauladan'*	N
4	*Kesalahan terbesar Prabowo adalalah memberi tunggangan kepada pecatan menteri yang cuma bermodalkan kata-kata manis dan indah hahaha!!! *ROTFL* #perudungan*	Y
5	*kalau Ahok - Djarot harus meneruskan karienya sebagai PELAYAN warga Jakarta menata kota sedangkan untuk Anies dibiarkan saja menata kata untuk beretorika dan berwacana*	N
6	*Kecoa2 radikal pada kepanasan di komen ini ckckck*	Y
7	*Pecundang itu hanya modal sembako ckckck*	Y
8	*Sesungguhnya sistem Demokrasi hanyalah berhala lainnya ckckck*	N
9	*Hebatnya Anies manfaatkan partainya Prabowo, uangnya Sandiaga Uno, Masjidnya JK, Anies kalah gak rugi, yang kalah telak itu Prabowo, Sandiaga Uno, dan JK*	N

Table 2. Word feature space

Pencitraan	Beretorika	Bodoh	Curang	Persetan	Bego	Busuk
Blak-blakan	Copas	Berengsek	Jahat	Fatamorgana	Koar	Sembako
Jadi-jadian	Kuno	Buncit	Dicap	Delusional	Keok	Chaos
Tolol	Biadab	Congek	Tipu	Kebohongan	Provokasi	Jiplak
Absurd	PHP	Congor	Fitnah	Bersandiwara	Sogokan	Diskredit
Kecoa	Bloon	Waras	Hoaker	Kecurangan	Provokator	Korban
Amplop	Sesat	Zonk	Pecatan	Kedodoran	Lawakan	Mainan
Maling	Nyinyir	Pecatan	Manja	Omdo	Siluman	Ngawur
Melawak	Otak	Retorika	Mencuri	Rasis	Topeng	Munafik
Pura-pura	Palsu	Sampah	Miskin	Penjilat	Keputusan	Primitif
Modus	Hujat	Negatif	Sirik	Kedok	Berantakan	Koruptor
Miring	Basa-basi	Becus	Ketakutan	Nyontek	Sewot	Rendahan
Dipecat	Kasar	Fasis	Jongos	Radikal	Pecundang	

Table 3. Phrase feature space

Amplop Beterbangan	Miskin Integritas	Boneka Partai	Mulut Jamban
Debat Teletubbies	Kampanye Busuk	Tidak Berbobot	Data Sesat
Haus Kekuasaan	Kampanye Hitam	Menjilat Ludah	Miskin Kreativitas
Tidak Becus	Eksportir Kebencian	Pantesan Dipecat	Duo Pecundang
Pecatan Menteri	Modal Topeng	Hanya Janji	Mulut Nyinyir
Kecoa Radikal	Pemilih Tertipu	Si Congor	Baju Pilkada
Cari Muka	Kampanye Sogokan	Hasil Jiplak	Cuman Nyinyir
Program Copas	Jongos Kroninya	Basa-basi Medsos	Corong Asap
Bermulut Manis	Manusia Sampah	Pilkada Chaos	Otak Tak Berisi
Kampanye Jahat	Tebar Pesona	Pemimpin Amatir	Janji Manis
Cuma Retorika	Mental Jiplak	Pikiran Sempit	Kekisruhan Politik
Pikiran Primitif	Bagi-bagi Sembako	Berkata Manis	Kata-kata Manis
Main Sembako	Kumpulan Penjilat	Menikmati Kehancuran	Umbar Janji
Biang Permasalahan	Janji Palsu	Tukang Berhayal	Jual Program

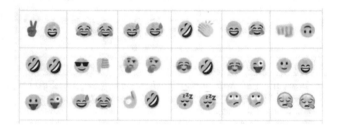

Fig. 1. Emoji feature space

4   Discussion

The feature spaces mentioned in Sect. 3 above have been used as the keywords for collecting data for training and testing dataset. From the manual classification by the linguist, five rules of how a cyberbullying is formed have been established. The feature spaces mentioned in Sect. 3 above have been used as the keywords for collecting data for training and testing dataset. From the manual classification by the linguist, five rules of how a cyberbullying is formed have been established. Firstly, if there are two or three different types of feature space appear in a sentence, then it can be classified as a cyberbullying. Consider this example: *Kesalahan terbesar Prabowo adalalah memberi tunggangan kepada pecatan menteri yang cuma bermodalkan kata-kata manis dan indah hahaha!!! *ROTFL* #perudungan* (The biggest mistake made by Prabowo is to give an ex-minister that only could do lips service a political ride - laugh out loud). This sentence is classified as bullying as it contains three types feature space, they are: *pecatan* (ex) is a word feature space, *kata-kata manis* (lips service) is

Table 4. Punctuation marks feature space

!!	!!??	???	??
!!!	(?)	????	??!!

Table 5. HashTag feature space

#ahoknyungseppp	#SalamWaras	#aniestukangnyinyir
#OTTSembakoAhok	#KamiParaMunafik	#ahokkejangkejang
#AhokPanikAhokKalah	#HaramPilihKafir	#AhoxSumberPerpecahan
#AsalBukanAhok	#ahokpastitumbang	#ahoksumbermasalah
#AHYTakutDebat	#tepokjidat	#AdaAQUA

a phrase feature space, and "hahaha" (laugh out loud) is a cue word feature space. In contras, the following example is not a bullying sentence since it contains only one feature space: *kalau Ahok - Djarot harus meneruskan karienya sebagai pelayan warga Jakarta menata kota sedangkan untuk Anies dibiarkan saja menata kata untuk beretorika dan berwacana* (Ahok - Djarot must continue their career as Jakarta people's servants, and let Anies composing his rethorical words). In this sentence, there is only one feature space that is *beretorika* (rethorical). The combination can be seen in Table 7.

The second rule is if there is only one type of feature space but appears more than one times, then it can be classified as a bullying sentence. *Pecundang itu hanya modal sembako ckckck* (The loser only have people's basic need for bribery). In this example, there is only word feature space but it appears as *Pecundang* (The loser) and *sembako* (people's basic need for bribery). More can be seen in Table 8.

The third rule is when two or more combination of emoji, punctuation and hashtag feature space are combined together then it forms a bullying sentence. And the last rule is if one of these three feature spaces appears more than one times then it forms a bullying sentence (Tables 9, 10, 11 and 12).

Table 6. Cue word feature space

ha-ha-ha	akhhhh	KOK??	LOL	he-he	hioooo????
ck-ck-ck	heuheuheu	Jlebbb!!	Haudeuhhh!	Waduuhh	Lho
wkwkwk	ciee	woyyy!!	Nyahaha	preeeettttt!	Ehemm
ow-ow-ow	hem	duuh	cihuuuyyy	ah	Hohohoho!!!
jiaaaahh	hadeeeh	haaah	oh	Wekekek	Heh!

Table 7. The feature space formed from the bullying corpus extraction for detection algorithm

Category	Object	Bullying	
		Positive	Negative
Word	*pecatan*	V	
Phrase	*pecatan menteri*	V	
Phrase	*kata-kata manis*	V	
Phrase	*jual program*	V	
Punctuation mark	!!!	V	
Emoticon	*ROTFL*	V	
#tag	#perudungan	V	
Cue word	Hahaha	V	

Table 8. Table of contribution word, phrase, and cue word by number of feature space type

Word	Phrase	Cue word	Value
0	0	0	0
0	0	1	0
0	1	0	0
0	1	1	1
1	0	0	0
1	0	1	1
1	1	0	1
1	1	1	1

Table 9. Table of contribution word, phrase, and cue word by number of keyword

Word	Phrase	Cue word	Quantity	Value
0	0	1	1	0
0	0	1	>1	1
0	1	0	1	0
0	1	0	>1	1
1	0	0	1	0
1	0	0	>1	1

Table 10. Table of contribution emoji, punctuation mark, and #Tag by number of keyword type

Emoji	Punctuation mark	#Tag	Value
0	0	0	0
0	0	1	0
0	1	0	0
0	1	1	1
1	0	0	0
1	0	1	1
1	1	0	1
1	1	1	1

Table 11. Table of contribution emoji, punctuation mark, and #Tag by number of keyword

Emoji	Punctuation mark	#Tag	Quantity	Value
0	0	1	1	0
0	0	1	>1	1
0	1	0	1	0
0	1	0	>1	1
1	0	0	1	0
1	0	0	>1	1

Table 12. Table of contribution word, phrase, cue word, emoji, punctuation mark, and #Tag by number of keyword type

Word	Phrase	Cue word	Emoji	Punctuation mark	#Tag	Value
0	0	0	0	0	0	0
0	0	1	0	0	1	1
0	0	1	0	1	0	1
0	0	1	1	0	0	1
0	1	0	0	0	1	1
0	1	0	0	1	0	1
0	1	0	1	0	0	1
1	0	0	0	0	1	1
1	0	0	0	1	0	1
1	0	0	1	0	0	1

5   Conclusion

This paper discussed the cyberbullying problems in Indonesia and the need for anti-cyberbullying means. In order to have a supervised learning based cyberbullying tool, a good corpus is needed. This paper reports the work on developing Indonesian cyberbullying corpus for a supervised learning model. A set of the cyberbullying feature space is designed and checked by a linguist. The feature space then has been used for collecting data for training and testing dataset as well as developing the rules for the supervised learning algorithm.

References

1. APJII: Buletin apjii edisi05 2016, November 2016. https://apjii.or.id/downfile/file/BULETINAPJIIEDISI05November2016.pdf. Accessed Feb 2017
2. Desai, N., Dave, A.D.: Sarcasm detection in Hindi sentences using support vector machine. Int. J. **4**(7), 8–15 (2016)
3. Dinakar, K., Reichart, R., Lieberman, H.: Modeling the detection of textual cyberbullying. Soc. Mobile Web **11**(02), 11–27 (2011)
4. Farisia, N.: Deteksi Cyberbullying pada Media Sosial di Indonesia dengan Memanfaatkan Text Mining [tesis]. Master's thesis, Magister of Information technology, Indonesia University, The Jakarta (2016)
5. Gayatri, G.: Digital citizenship safety among children and adolescents in Indonesia (2012)
6. Ipsos: Cyberbullying: Citizen in 24 countries assess bullying via information technology for a total global perspective, December 2011. https://www.ipsos.com/sites/default/files/new_and_polls/201201/5462revppt.pdf. Accessed Feb 2017
7. Nahar, V., Li, X., Pang, C.: An effective approach for cyberbullying detection. Commun. Inf. Sci. Manag. Eng. **3**(5), 238 (2013)

Improving upon Package and Food Delivery by Semi-autonomous Tag-along Vehicles

Vaclav Uhlir[✉], Frantisek Zboril, and Jaroslav Rozman

FIT, Brno University of Technology, IT4Innovations Centre of Excellence,
Bozetechova 1/2, 612 66 Brno, Czech Republic
iuhlir@fit.vutbr.cz
http://www.fit.vutbr.cz/~iuhlir/

Abstract. This paper aims to improve current last mile distribution and package delivery by introducing basic concept of delivery using Semi-autonomous Tag-along Vehicles (SaTaVs) driven by their own agency and desires. SaTaVs are introduced as vehicles capable of traveling by following leading vehicle and thus reducing requirements for their autonomy while maintaining most of the advantages. Whole system is designed as maintaining long term equilibrium with agents goal in maximizing future investment.

Keywords: Vehicle routing problem · Perishable food delivery
Intelligent transportation · Agent systems

1 Current State

Package and food delivery consists mainly of Vehicle Routing Problems (VRPs) which is NP-hard problem [1]. Distribution plan [2] is then set as plan of a fleet vehicles used to deliver (or pick-up) goods to (or from) customers with the other endpoint being warehouse. Additional complication arise when dealing with perishable goods this is usually refereed as Vehicle Routing Problem with Time Windows and Temporal Dependencies (VRPTWTD) and is comprehensively described by Hsu et al. [3].

When dealing with Point-to-Point (PtP) deliveries (like food deliveries from restaurants or delivering timely legal documents) the distribution throughout warehouses becomes impossible due to the inherent time delays. This could be solved by adaptation of methods like advanced Dial-A-Ride Paratransit (DARP) solution [4]. But in reality the cost of managing rides for food delivery exceeds the costs of self-owning of delivery vehicles and thus majority of food providers use their own delivery cars and systems of High Coverage Point-to-Point Transit (HCPPT) [5] are yet to arise. Change to this may be introduced by Shared Autonomous Vehicles (SAVs) as showed in recent studies [6,7] - one SAV can replace multiple conventional vehicles thus lowering prices of distribution.

© Springer International Publishing AG, part of Springer Nature 2018
A. Abraham et al. (Eds.): ISDA 2017, AISC 736, pp. 589–596, 2018.
https://doi.org/10.1007/978-3-319-76348-4_57

The cost benefit for packages is even greater then for person transportation due to higher potential capacity for packages. Using SAVs for package distribution could also improve Levin et al. model [8].

1.1 Last Mile Distribution

Last mile distribution represents last step in shipments when packages are delivered to end-customers of process. That usually means driving to customer address notifying him of arrival and waiting for him to sign the receive documents. Additionally when dealing with temperature sensitive cargo - the opening the cargo hold is important factor impacting inventory costs as noted in [3].

1.2 Point-to-Point Deliveries

Distribution of prepared food from restaurants is usually handled as some version of VRPTWTD where restaurants either schedule food preparation according customer location and thus creating less predicable delivery times or running distribution one item at a time which inevitably increases the prices.

2 Semi-autonomous Tag-along Vehicles - SaTaV

Our solution to this problem is introducing concept of vehicles capable of following a car or motorbike - SaTaVs. Such vehicles would need properties of maximum speed around 50 km/h (or 30 mph) - most common city speed limit. These vehicles would also of course need GPS (or other positioning) unit to monitor their real-time position, wireless communications and means of tracking leading vehicle which would provide routing.

Other features of SaTaVs would also need to be ability to travel in formation and safely manage short distance reposition on their own. Lastly they should be able to provide customers with secured access to the cargo.

Such vehicles would bring the advantage for last mile distribution in form of reduced stop time of leading vehicle – SaTaVs would wait for their customers on their own while other packages are on their way to their customers. And for PtP deliveries they would present more immediate expedition if opportunity arises (Fig. 1).

2.1 Package as Agent

This basic premise lets us intuitively treat every package as its own agent. The agent then has desire to get to the target position and constraints in form of time and cost. Considering one SaTaV carrying one package desires need to be expanded by desire of being in particular city zone and costs of travel to sender and return from customer.

2.2   Desire Zones

Every SaTaV has desire to be in the zone with lowest possible SaTaV population index i_p. For the most part this paper will present three types of zones dependent on SaTaVs population in current zone: **low** - zones where is less SaTaVs then there is expected to be demanded of: **fair** - number of SaTaVs should suffice for expected demand and **high** - number of SaTaVs is far higher then expected demand for specified zone (Table 1).

SaTaVs population index for specific zone is calculated according to minimum D_L and maximum D_H zone SaTaV expected requirement as follows:

$$x : ip_x = \frac{P_x - D_{Lx}}{D_{Hx} - D_{Lx}} \tag{1}$$

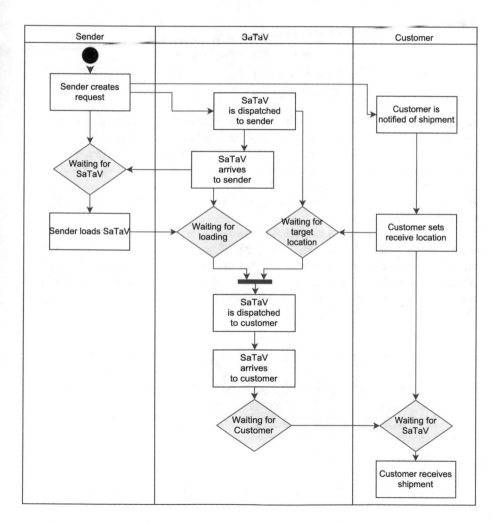

Fig. 1. Process of shipment with SaTaVs from sender and customer view

Table 1. SaTaV population zones

ip	Population	Description
$(-\infty, 0)$	low	Population less then desired
$[0, 1)$	fair	Population has reserve as expected
$[1, \infty)$	high	Too many SaTaVs in zone

Population index is then used for cost equation of SaTaV movement as from point x to y with point 0 being SaTaV initial position by:

$$C_{xy} = Ca_{0x} + Ct_{xy} + Cp_y \tag{2}$$

where:

$$Ca_{0x} = (1 - ip_0) * c_A + d_{0x} * c_d \tag{3}$$
$$Ct_{xy} = d_{xy} * c_d \tag{4}$$
$$Cp_y = ip'_y * c_P \tag{5}$$

where Ca_{0x} is allocation cost with c_A being allocation cost constant, Ct_{xy} is cost of transport with d_{xy} being distance from point x to y and c_d distance unit cost constant. Lastly Cp_y is cost ending in target zone with c_P being constant of parking cost and ip'_y is population index including traveling agents. Because ip is from interval $(-\infty, +\infty)$ part, or a whole cost equation can be negative – this is a wanted behavior for encouraging use of SaTaVs from over populated zones to underpopulated.

Let's imagine following situation - five zones with sender in zone 2 an customers in zones 4 and 5 (shown on Fig. 2). When a sender puts in a request for package transport from zone 2 to zone 4 all SaTaVs will see request and can calculate their allocation and travel expanse. For agents in zone 1 initial allocation and travel cost will be $Ca_{12} = 1 - (-0.2) * c_A + d_{12} * c_d = 1.2c_A + d_{12} * c_d$ while for agents in zone 2 (let's suppose that d_{22} is 0) allocation cost will be

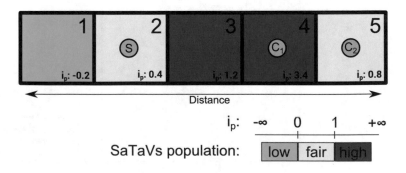

Fig. 2. Zoning example for determining SaTaVs desires.

$Ca_{22} = 1 - (0.4) * c_A = 0.6c_A$ and thus agents from zone 2 will be preferred over agents from zone 1. For the agents from zone 3 the allocation and initial cost will be $Ca_{32} = 1 - (-1.2) * c_A + d_{32} * c_d = -0.2c_A + d_{32} * c_d$. Agents from the zone 3 are by negative cost of allocation incentivized to further system zones population equilibrium. Similarly for cost of delivering packages transported to the customer $1 : d_{24} * c_d + ip_4 * c_P$ and customer $2 : d_{25} * c_d + ip_5 * c_P$ and while $d_{24} < d_{25}$ the difference in $ip_4 = 3.4$ and $ip_5 = 0.8$ depending on c_P and c_d may cause longer distance delivery to customer no. 2 to be cheaper in view of better system equilibrium reflecting agents future prospect for another delivery.

2.3   Empty Trips

SaTaVs are also capable of empty trips with cost Eq. (2) in reduced form as:

$$Ce_{0t} = Cu_{0t} + Cp_t \tag{6}$$

intended for relocation of SaTaVs in case of large system imbalance when benefits from leaving the zone with high population and benefits from arriving to low population zone cover travel expanses.

3   SaTaVs Auction System

All transactions in SaTaVs ecosystem are managed by auctions, thus allowing clients to chose their own priorities and keep prices for services in competitive perspective.

3.1   Allocation Bidding

SaTaV agents offer their services through simple auction system where sender puts request for shipment and gets bids from available units in form of price and expected arrival time as can be seen in Fig. 3.

   Client is presented with choices, where according to his priorities, he can choose allocation time versus cost ratio. After client makes his choice, SaTaV is notified of successful bid and starts his way to client.

3.2   Transportation Hiring

When SaTaVs want to get to sender or transport package, they put up bid for transport with information on origin location, target location and price willing to pay for transport. Price B is set according to priority P_B of sender or customer depending on time sensitivity of the package, where priority $P_B = 1$ represents urgency and price is equal to full price of hired personal leading vehicle. SaTaVs with less time sensitive cargo are starting bidding for transport by fraction of full price and may use increments on price initiated by customer while SaTaVs with highly time sensitive cargo are using price expected to yield transportation as soon as possible.

bid_id	zone	distance	allocation_cost ▲	approx_time
8561	D7	3.62	120	420
8565	B7	4.4	200	820
8567	D8	4.9	280	370
8563	D8	5.7	320	400
8562	C6	7.2	600	380
8566	B6	12.5	1300	840
8564	C7	0.2	1700	70

Fig. 3. Allocation table showing bids of near SaTaVs.

Prices for transport, willing to be paid by SaTaVs, are then shown in the system, where driving agents (either human drivers with cars, motorbikes or even bicycles or fully autonomous vehicles) may purchase contract to lead SaTaV from its current location to desired destination.

3.3   Empty Trips

When SaTaV is in high population zone ($ip \geq 1$) it automatically starts to scan for fair and low population zones ($ip < 1$) and calculates empty trip cost according (6). If result Ce_{0t} is negative the sum is used for transportation bid.

4   System Parameters

In above mentioned equations is a lot of parameters (like allocation constant c_A, parking constant c_P or optimal zone population limits) without defining their value or relationship. This is intentional because this constants have highly specific nature depending on factors such as geography, traffic, zone input or output of packages and lot more.

This can be seen on specific example in Fig. 4 showing disproportioned package flow in Brno city caused by high concentration of small businesses in city center in proportion to high consumption by residential areas on edges of city. This creates high demand for SaTaVs on one side and unwanted overpopulation on other. Such effects are countered by system constants causing SaTaVs to engage in empty trip migration from edges back to center funded by high parking costs on edges and high allocation costs in center.

Net package flow

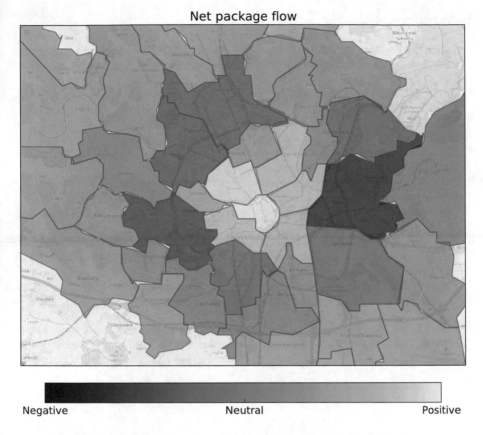

Negative   Neutral   Positive

Fig. 4. Simulated package flow based on district type in Brno.

5 Conclusion

This paper presented basic overview of delivery with Semi-autonomous Tag-along Vehicles representing alternative to now used package delivery systems. System is dependent on future technology of economically viable SaTaVs, but principle can be implemented on packages delivered by human drivers directly yet economical viability is depended on low time cost autonomous systems.

Future work should be directed to establishing parameters automatically based on feedback from simulation of particular zones using real-time traffic data and creating comparative studies to current distribution systems.

Acknowledgements. This work was supported by the BUT project FIT-S-17-4014 and the IT4IXS: IT4Innovations Excellence in Science project (LQ1602).

References

1. Paolo, T., Daniele, V.: The vehicle routing problem. SIAM Monographs on Discrete Mathematics and Applications. Society for Industrial and Applied Mathematics, Philadelphia (2002)
2. Toth, P., Vigo, D.: Vehicle Routing: Problems, Methods, and Applications. SIAM, Philadelphia (2014)
3. Hsu, C.I., Hung, S.F., Li, H.C.: Vehicle routing problem with time-windows for perishable food delivery. J. Food Eng. **80**(2), 465–475 (2007)
4. Fu, L.: Scheduling Dial-A-Ride Paratransit under time-varying, stochastic congestion. Transp. Res. Part B Methodol. **36**(6), 485–506 (2002)
5. Jung, J., Jayakrishnan, R., Nam, D.: High coverage point-to-point transit: local vehicle routing problem with genetic algorithms. In: 2011 14th International IEEE Conference on Intelligent Transportation Systems (ITSC), pp. 1285–1290. IEEE (2011)
6. Fagnant, D.J., Kockelman, K.M., Bansal, P.: Operations of shared autonomous vehicle fleet for austin, texas, market. Transp. Res. Rec. J. Transp. Res. Board **2536**, 98–106 (2015)
7. Fagnant, D.J., Kockelman, K.M.: Dynamic ride-sharing and fleet sizing for a system of shared autonomous vehicles in Austin. Texas. Transportation **45**(1), 143–158 (2016). https://doi.org/10.1007/s11116-016-9729-z
8. Levin, M.W., Kockelman, K.M., Boyles, S.D., Li, T.: A general framework for modeling shared autonomous vehicles with dynamic network-loading and dynamic ride-sharing application. Comput. Environ. Urban Syst. **64**, 373–383 (2017)

A Novel Multi-party Key Exchange Protocol

Swapnil Paliwal$^{(\boxtimes)}$ and Ch. Aswani Kumar

Vellore Institute of Technology, Vellore, Tamil Nadu, India
swapnil.paliwal18@gmail.com, cherukuri@acm.org

Abstract. A key exchange protocols are generally used in order to exchange a cryptographic session key, such that no one except the communicating party must be able to deduce the keys lifetime. In this paper we propose a protocol which can be used for multi-party key exchange amongst the entities with less number of iterations, and also the same protocol can be used for sharing the messages amongst the entities, the key advantage of this protocol is when we are using this protocol for the multi-party key exchange as it requires very less result when compared with conventional two move Diffie-Hellman key exchange protocol.

Keywords: Key exchange protocol · Safe multi-party key exchange protocol
Message sharing protocol · Protocol based on DDH assumption

1 Introduction

A key exchange protocol is the one where two or more entities communicate over a network to attain the common secret key (which is usually called a session key), Many key exchange protocols have been proposed in the past based on DDH, the drawback is that in many of these protocols which are based on Decisional Diffie-Hellman assumption usually involve entities having to work on a common base (a common primitive root of a prime modulo) or the generator 'g', and security of the protocol lies in the difficulty to solve the discrete logarithm problem, the problem arises when we have multi-parties communicating for attaining the same session key here all of these entities are required to perform computation over the same generator 'g', and many of these protocols cannot be used for message sharing thus these protocols are confined to key exchange. Moreover even in the key exchange protocol the each entity P_i must satisfy the following pair (Pi, Pj, s, role) uniquely where p_i is an entity which is trying to communicate with entity p_j in the session s (where s is the id) where the entity p_i has a specific role of being either an initiator (begins the communication) or a responder, thus both of these pair must match for communication to begin i.e. (Pi, Pj, s, initiator) and (Pj, Pi, s, responder), this pair is said to be matching as one of the entity is initiator and other is responder (the pair must match for the communication to begin), usually these protocols are called by some higher level protocol, where these higher level protocols also ensure that the session id is not similar for any other communication which they are initiating currently or in some later time in the future. The most common key exchange protocol is Diffie-Hellman key exchange protocol, it is used for exchanging session key and also helps in creating a secure channel but Diffie-Hellman

© Springer International Publishing AG, part of Springer Nature 2018
A. Abraham et al. (Eds.): ISDA 2017, AISC 736, pp. 597–607, 2018.
https://doi.org/10.1007/978-3-319-76348-4_58

key exchange protocol is vulnerable as per recent studies i.e. As per the logjam attack (which allows the man in the middle attacker to downgrade the communication to 512 bit export grade cryptography) we come to a conclusion that if 1024 most frequently used generator 'g' is broken then we have about 17.9% of the top one million domains [2] which will become vulnerable, thus allowing a passive eavesdropping over these websites (a nation-state can break a prime modulo of most commonly used 1024 bit prime using number field-sieve which is the fastest way of breaking these public-key protocols). As per the same study it was concluded that one must allow a prime generator 'g' of 2048 bit, and discard the primes less than 1024 bits, thus in a way 1024 bit generator 'g' can also be used provided it is not commonly used one, another drawback of this protocol is that there exists no way of preventing weak keys from being exchanged. In this paper we design protocols which can be used to overcome the drawbacks of the above protocol, i.e. the protocol will allow the entities to use multiple generators 'x', 'y' and this protocol has major advantage when compared with many multi-party key exchange protocols in terms of number of iterations, and also this protocol can be used for sharing message as well.

2   Protocol over the Authenticated Model

Assuming that we are communicating over an authenticated model (where the communication is done over a network which has been authenticated), Which states that the entity A or B wishes to communicate with other entity using some protocol 'π' where entities are ensured as of whom they are communicating with (entities are aware as of who the designated sender is and who is the designated receiver). We first establish our protocol in the authenticated model then extend this to the unauthenticated model, Let us first establish the protocol and prove the correctness of the protocol. Figure 1 below depicts the key exchange protocol, where we are assuming that entity A is the initiator and entity B is the receiver.

2.1   Brief Description of the Protocol

Here q is a prime modulo of the order 1024–2048 bit, and x is a primitive root (which generates all the values ranging between $[1, (q-1)]$) of the modulo q, here both the values i.e. x, q are announced publically or sent to communicating network by some means. 'A' in the meantime while waiting for 'B' to send its intermediate value computes the session key $x^r \bmod (q)$ in advance (main motive for 'A' is to generate a desired key or rather eliminate possibility of weak key being exchanged same has been communicated as of how it is achieved in the next section) this is what is happening at STEP 1, now 'B' generates a secret number p such that $p^{-1} \bmod (\phi(q))$ exists, where ϕ is the Euler Totient function, now since B has successfully generated p it can compute $x^p \bmod (q)$ and once this is done, B sends the value of this to A (B's secret number 'p' is safe as discrete logarithm has to be computed in order to attain 'p' value, and the protocol is making a DDH assumption i.e. decisional Diffie-Hellman assumption that it is tough to solve the discrete log problem), since only 'A' currently is in possession of the session key $x^r \bmod (q)$ and now if it embeds the value of r in $x^p \bmod (q)$ and if B

Entities: A, B

Public Values: $x, \mathrm{mod}(q), x^p \bmod(q), x^{pr} \bmod(q)$

Private Values: $p, r, x^r \bmod(q)$

STEP 1: A chooses a prime modulo ' q ', a corresponding primitive root ' x ' and a secret number ' p ' using which the entity computes $x^p \bmod(q)$. Upon completion it sends, $x^p \bmod(q), x, \mathrm{mod}(q)$ to entity B.

STEP 2: B chooses it's secret number ' r ' and computes the session key which is to be shared i.e it computes $x^r \bmod(q)$ now the entity B also computes $[x^p \bmod(q)]^r \bmod(q) \equiv x^{pr} \bmod(q)$ so that A can attain the shared session key as well. Upon completion it sends $x^{pr} \bmod(q)$ to A.

STEP 3: A is already in possession of ' p ' it can compute $p^{-1} \bmod(\phi(q))$, now it receives $x^{pr} \bmod(q)$ from B it can attain the session key by performing exponentiation of the received value with $p^{-1} \bmod(\phi(q))$ and thereby attaining $x^r \bmod(q)$

Fig. 1. Key exchange protocol

nullifies the effect of p then even 'B' shall have the session key, thus A embeds the value of r by performing $x^{pr} \bmod (q)$ as seen in step 3, and sends it to B. Now B proceeds to nullify the effect of p as $p^{-1} \bmod (\phi(q))$ exists it raises the value sent by A with $p^{-1} \bmod (\phi(q))$ value and thereby resulting in $x^r \bmod (q)$, which is our resulting session key which was initialized by A.

It is clear from STEP 2 that entity B will be in possession of the key i.e. the receiver is in possession of the key, now it has been established that A also gets the key in STEP 3.

2.2 Correctness

From the famous Fermat-Euler theorem we have the following $a^{\phi(q)} \equiv 1 \bmod (q)$, where let ϕ be the Euler totient function. We take a value p, k such that $p * k = 1 \bmod (\phi(q))$, where $k = p^{-1} \bmod (\phi(q))$. As in STEP 3 A is computing equivalent to $x^{prk} \bmod (q)$ now we have to Prove that $x^{prk} \equiv x^r \bmod (q)$. As we are making and assumption that $x^{pk} \equiv x \bmod (\phi(q))$.

Proof: On the basis of above discussion made, We have $x^{prk} \equiv (x^r)^{pk-1} * x^r \bmod (q)$, and also $x^{\phi(q)} \equiv 1 \bmod (q)$, $pk = 1 \bmod (\phi(q))$, Now since we are in possession of p, k we can extend the above shown relation as follows $p * k \equiv 1 \bmod (\phi(q)) \Rightarrow pk - 1 \equiv 0 \bmod (\phi(q))$. Thus on basis of the above steps we can attain and have the following relations $(x^r)^{pk-1} * x^r \bmod (q) \Rightarrow (x^{pk-1})^r * x^r \bmod (q)$ since we know that $pk - 1 \equiv 0 \bmod (\phi(q))$, we have $(x^r)^{pk-1} * x^r \bmod (q) \equiv (x^{pk-1})^r * x^r \bmod (q)$. Since $pk - 1 \bmod (\phi(q)) \equiv 0$ this implies that $pk - 1$ is divisible by $q - 1$. Thus we can on the basis of the results attained above that $(x^{pk-1})^r * x^r \bmod (q) \equiv (x^{c*\phi(q)})^r * x^r \equiv 1^r * x^r \bmod (q)$. Both the Parties now have the shared secret key $x^r \bmod (q)$. In the next section we discuss as of how this protocol is efficient for multi-party communication, but before that we prove as of how weak keys and repeated key (sharing same key over and over again) sharing can be avoided.

2.3   Elimination of Weak Keys

Since 'A' is generating the session key value it has high control over the values it can generate since there exists no dependency on the other entity unlike that of Diffie-Hellman key exchange where the resulting value depends totally on the basis of their (entities) secret values. Now in above protocol and the protocols described below this is how 'A' can eliminate weak keys, initially it makes an array where all the possible weak values are stored and then it sorts the array in the ascending order this is the initial process of initialization where all the weak keys are identified and now it (the main entity) generate values of 'r' such that the resulting value $(x^T mod(q))$ is not in the above established array (here again sorting helps as now binary searches can be applied to the search values efficiently) and if the resulting value lies in the array then discard the value of 'r' (and generate new value of 'r' whose result do not coincide with the array), again another observation is that in any symmetric encryption algorithm weak keys are algorithm dependent but in general values having high number of either '0' or '1' are considered to be weak, and there are few weak keys in the domain of the session keys which can be generated and used thus this occupies a very less space when compared with total key space. Again similar approach can be used to eliminate use of same session key for multiple sessions as follows, the entity might generate another array which stores all the values of 'r' that are in use (if communicating with multiple parties) or that have been used in the recent past and then generate value of 'r' which are not similar to that of values stored in the array, this array needs to be periodically updated in order to facilitate keys which were used initially (as is no longer a recently used key). In the next session we discuss as of how our protocol which is discussed above; can be extended to multi-party communication and also how this can also be used for multi-party communication effectively when compared with Diffie-Hellman protocol.

3   Multi-party Communication over the Authenticated Links

The main advantage of our algorithm over Diffie-Hellman key exchange protocol is during multi-party key exchange, in terms of iterations required for exchanging the keys. The table below depicts how many iterations are required in Diffie-Hellman as the number of entities increase. Again here the communication is done over an authenticated model. The assumptions made are 1. The role of attacker in the authenticated network is just to observe the communication made and all the others discussed in [1]. We discuss the extension of these protocols over the unauthenticated network in the later phases.

Here is how we obtained the above results, first assuming that four entities (A, B, C, D) are communicating in order to exchange the keys, thus each entity must generate a secret number and all of them must agree on a base (generator or primitive root of the modulo which they will choose) and a modulo, now all of them compute g^a mod $(q), g^b$ mod $(q), g^c$ mod $(q), g^d$ mod (q) where a, b, c, d are their individual secret key. Now for entity D to attain the secret key it must receive g^{abc} mod (q) so that it can perform $(g^{abc}$ mod $(q))^d$ mod (q) and there by attain the session key g^{abcd} mod (q), similarly for entity C to attain the secret key it must receive g^{abd} mod (q), now from above it is clear that the iterations which will be required to compute the same shared keys are $n(n-1)$ (the same results are obtained using Newton's Forward Difference method), where n (which is always greater than or equal to 2) is the number of entities participating. By using Newton's Forward Difference method [6] we obtain the following equation (which tells us as of how many iterations are required when the number of entities are known) $2 + 4(n-2) + (n-2)(n-3)$ where n is the number of entities participating, in other words $n(n-1) = 2 + 4(n-2) + (n-2)(n-3)$ (which has been derived using Newton's Forward Difference Method using Table 1) number of iterations are required. Following are the observations were made **1.** If there are 'n' participating entities then $n-1$ times each entity's involvement is required in order for all the entities to obtain the same final key. **2.** Each entity initialize the communication once (once all the public values of the entities are known one of the entity will perform the operation using it's secret value and then send it to the other entity, let us assume three parties are communicating (using same base and the modulo) and all of their public values are known then first entity will perform operation on second entity's value and then send it to the third entity so that third entity attains the shared secret, now third party will perform operation on any other entity's public value assuming first entity's value and send it to second entity and now second entity can obtain the shared secret key similarly second entity will make use of third entity's public value and send it to first entity thereby first party can attain the shared secret key) Another approach (as discussed in [5]) can also be used but here we are assuming that instead of disclosing the public value of A to everyone we only disclose it to the entity which requests it. Our method requires less number of iterations when compared with multi-party Diffie-Hellman key exchange protocol. Now coming to the protocol, the main entity α generates a base (primitive root) and a modulo and announces it publically to the participating entities in the mean while it also generates a secret number m and generates the resulting session key k^m mod (q) which is known only to α at this time. As seen in Fig. 1 we extend the same method to multiple parties this is how it works assume there are three entities willing to exchange or generate a same net resulting key, now all the entities are in possession of k, mod(q) as these are public values provided by entity α, now each of the two entity generates their respective private number u, v and perform k^u mod $(q), k^v$ mod (q) and send it to the entity α, now the entity computes $(k^u)^m$ mod $(q) \equiv k^{um}$ mod (q) and send it to the entity from which it received k^u mod (q), and similarly sends $(k^v)^m$ mod $(q) \equiv k^{vm}$ mod (q) to the entity from which it received k^v mod (q), now similar to protocol discussed in Fig. 1. Once the two parties received these values, both of them are in possession of either 'u' or 'v' which are their secret key and now these entities can attain the inverse of these values by using the same approach described in Fig. 1. And can attain k^m mod (q). Same approach is applied when multiple parties exist in the communication. The table below depicts the number of entities involved and the iterations required to attain the same shared keys (Table 2).

Here is how we attained the above results; it is obvious from the Fig. 1. That when two entities are involved two iterations or steps are required in order to attain the shared key. From above explanation of three entities it is clear that total of two iterations were

Table 1. Depicts total number of iterations required as the entities increase in Diffie-Hellman protocol.

Number of entities	Iterations
2	2
3	6
4	12
5	20
6	30
7	42
8	56
9	72

Table 2. Depicts total number of iterations required as the entities increase in our proposed protocol.

Number of entities	Iterations
2	2
3	4
4	6
5	8
6	10
7	12
8	14
9	16

required for each performing entity other than that of α as first, the entities sent $k^u \bmod (q), k^v \bmod (q)$ to entity α and it replies with $(k^u)^m \bmod (q) \equiv k^{um} \bmod (q), (k^v)^m \bmod (q) \equiv k^{vm} \bmod (q)$ and at the end both the entities compute the shared session key i.e. $k^m \bmod (q)$. As seen when three entities were involved a total of four iterations were made, similarly it can be easily seen that if n entities are involved then $2(n-1)$ iterations are required in total. Same has been proved using Newton's Forward Difference method which yields $2 + 2(n-2)$ which equals $2(n-1)$. Following observations were made on this method they are [6]:

1. The main initiating party is in possession of the session key and can eliminate the weak keys by choosing the secret number such that a weak key is not attained, as main entity chooses its own secret number which generates the secret keys as $k^m \bmod (q)$.
2. If there are n entities then each entities involvement is required only twice (once for initiating and once for receiving), excluding the main entity α. There exists a major computational and performance task on the main entity (the entity which is

generating this session key) as it has to receive the participating entities public key and compute the pre-final key (as seen in step 2 of the protocol) and send it to the communicating entities.

3. The ratio of our algorithm with Diffie-Hellman algorithm in terms iterations is $\frac{n}{2}$, as our algorithm requires $2(n-1)$ iterations and Diffie-Hellman algorithm require $n(n-1)$ iterations for sharing of keys. Thus by dividing results in the ratio of $\frac{n}{2}$.

4 Employing Message Sharing

The above approach can be used for message sharing as well, in the above protocol there were no restrictions on the value of 'r' unlike that of 'p' where it was required that p is such that $p^{-1} \bmod (\phi(q))$ exists; thus total possible values of p were those values which were relatively prime to $(q-1)$, in this protocol we make the similar assumption i.e. $r^{-1} \bmod (\phi(q))$ exists, the main focus here is that instead of sending $x, \bmod(q)$ like which was done in the above protocol, in this protocol it is desired to send $x^r \bmod (q)$ from the entity A side (in this protocol there is a difference from above protocol i.e. here x is the message whereas it was a primitive root in the above protocol, and also there is a restriction on 'r' which did not exist on the above protocol) now since 'p' has an inverse which can be used to nullify the effect and attain 'x' thus now

---

Entities: A, B

Public: $x^r \bmod(q), \bmod(q), x^{rp} \bmod(q), x^p \bmod(q)$

Private: $x(message), r, p$

Exchanged value: x

STEP 1: 'A' initially generates a message x which it wants to send to 'B', and then it also generates a secret number r such that $r^{-1} \bmod(\phi(q))$ exists and then computes $x^r \bmod(q)$ and sends it to 'B' along with $\bmod(q)$ via a network which is presumably insecure.

STEP 2: 'B' also generates a secret number p such that $p^{-1} \bmod(\phi(q))$ exists, and computes $x^{rp} \bmod(q)$ and sends it back to entity 'A'.

STEP3: 'A' nullifies the effect of **'p'** by computing $[x^{pr} \bmod(q)]^{p^{-1}\bmod(\phi(q))} \bmod(q)$ and thereby resulting in $x^r \bmod(q)$, which is sent to B.

STEP4: 'B' nullifies the effect of **'r'** by computing $[x^r \bmod(q)]^{r^{-1}\bmod(\phi(q))} \bmod(q)$ which results in the value 'x', which was the message generated by A.

Fig. 2. Protocol for encryption and decryption of the messages

entity B computes x^{pr} mod (q) here this the same step as STEP 3 in the first protocol but it is step 2 in our protocol. Now A nullifies the effect of 'r' and sends it back to B and now B can nullify the effect of 'p' and thereby attaining the message value from A's side. See the protocol below in Fig. 2. For more clarity. The next section introduces a protocol which can work where both the entities wish to be initiator and not be concerned regarding matching of the pairs i.e. (Pi, Pj, s, initiator) and (Pj, Pi, s, responder).

5   Extending Protocol to Multiple Bases

The main advantage of this protocol is that both the entities are free to generate their own session key, and not be dependent on other entity to generate a session key for them. Thus one entity might not wait for the entity to first exchange key and then start encrypting here the entities can encrypt and later share the key required to attain the message (by decrypting it). This plays a very vital role as it is time saving. Here there

---

Entities: A,B

Key Info: Both the entities agree on a prime modulo q. Again it is possible that both the entities generate their own modulo and then send it along with the values of x,y.

Public: $x, y, \text{mod}(q), x^t \text{mod}(q), y^s \text{mod}(q), x^{st} \text{mod}(q), y^{st} \text{mod}(q)$

Private: s, t

Exchanged: $x^s \text{mod}(q), y^t \text{mod}(q)$

STEP 1: 'A' generates a primitive root x and a prime modulo q; and sends these values to entity 'B'.

STEP 2: 'B' generates another primitve root y of the order q and a secret number t which it sends along with the value of $x^t \text{mod}(q)$ to Entity A.

STEP 3: 'A' computes $[x^t \text{mod}(q)]^s \text{mod}(q) \equiv x^{st} \text{mod}(q)$ and sends it back to entity A along with the value of $y^s \text{mod}(q)$.

STEP 4: 'B' computes $[x^{st} \text{mod}(q)]^{t^{-1} \text{mod}(\phi(q-1))} \text{mod}(q)$ and attains the secret session key which A has generated i.e $[x^s \text{mod}(q)]$, and computes $[y^s \text{mod}(q)]^t \text{mod}(q)$ and sends it back to entity 'A'.

STEP 5: 'A' nullifies the effect of s and attains $[y^t \text{mod}(q)]$, which is the secret session key generated by entity B.

Fig. 3. Protocol using multiple bases

are less possible values of {s, t} as these must be relatively prime to q − 1, but in protocol 1 there are a total of q − 1 possible values. Thus a modulo must be chosen whose $\phi(q - 1)$ is large. The approach of this protocol is similar to that of first protocol, it is an extension of the first protocol itself, although in the Fig. 3 it is shown that both the entities agree on the same modulo; again it is not a necessary that both of these entities agree on the modulo, each of the two entity can generate its own session key. Another advantage of using this approach is that the attacker will have to computed discrete logarithm twice in order to break the communication from both the ends. It is essential to discuss the security aspect of the protocol, in order to prove its security against various attacks.

6 Security Discussion

This section discusses the security aspect of the protocols, thus proving as of how secure each of these protocols actually are, first type of attack is Guessing attack in this attack the attacker proceeds by guessing the potential session key again this attack is not an optimal attack, as there are a total of q − 1 potential value for protocol 1 as the value of 'r' will still be accepted if they are not invertible, and a total of $\phi(q - 1)$ for both protocol 2 and protocol 3 as it is expected that these secret values which are generated be invertible, thus it is impractical (if the order of prime is greater than or equal to 1024 bits) to just make an assumption regarding the value of the session key. As a result protocol 1 is more secure in terms of possible values of the key, our next attack is Trivial attack in this attack the attacker may proceed by directly computing the session key based on the communications made between the entities since it is difficult to compute discrete logarithm based on CDH assumption (which states it is difficult to compute discrete logarithm), same holds for all of the above protocols, next attack is Known Message/Key attack here we assume that the attacker is in possession of the session key or the message which is sent to the entity, now the aim of the attacker is to attain the secret values or long term value of entity A or B, here the resulting value though known will not be useful as discrete logarithm needs to be computed for all of the protocols, These protocols ensure perfect forward secrecy i.e. even if the attacker has compromised the current session key or message which was sent then also the attacker will not be able to jeopardize either previous or past communications as ephemerals are generated, thus distinct values of {p, r} will be generated each time and will be random and used for exchanging the values of the session keys or messages (in this case the value of x must also be changed). Thus the above discussed protocols are secure over the authenticated network, in the next section we show as of how we can secure the protocol over an unauthenticated network and prevent man-in-the-middle attack.

7   Employing These Protocols over an Unauthenticated Network

The above protocols when employed with the help of either MAC (message authentication code) codes where it is assumed that both the entities are in possession of the key in advance, thus each of the entities can proceed by encrypting these values before sending over the network since the observer is not in possession of this key thus cannot attain the actual value which is sent but rather just attains the encrypted values. Another approach is to use hashing thus the entities can take up a message hash it and then send it by encrypting using a pre-existing (same will also work if the entity can encrypt using other entities public key, since the other entity is in possession of the private key it can decrypt it and attain the message) secret session key, this ensures that the content of the message has not be altered or tempered with. Similarly there are many techniques which can be employed; here the basic idea is that the values must not be sent as is; if the communication is being made over an unauthenticated network, but rather be encrypted or hashed in such a way that attacker cannot derive meaning out of it, or cannot tamper it. These types of applications can prevent reply attacks where the attacker assumes to be either of entities and tries to attain the session key, here the attacker is incapable of inverting these encrypted values as it is not in possession of the secret key which it can use to decrypt it. Again man-in-the-middle attack can also be prevented using the above approach discussed here, the basic logic is similar to as discussed in [3, 9, 10] which states that it is not practical for the attacker to derive meaning if we encrypt the values and then send it.

8   Conclusion

The introduced protocol is highly efficient for the multi-party communication and its variants can be used for making the communication over multiple bases and also proposed an interesting approach as of how this protocol can be used for message encryption as well, for message encryption it is required that the value of the message be such that it is invertible, moreover the credibility of the entire protocol lies on the assumption that it is difficult to solve the discrete logarithm problem.

References

1. Canetti, R., Krawczyk, H.: Analysis of key-exchange protocols and their use for building secure channels. In: EUROCRYPT 2001, pp. 451–469 (2001)
2. Adrian, D., Bhargavan, K., Durumeric, Z., Gaudry, P., Green, M., Halderman, J.A., Heninger, N., Springall, D., Thomé, E., Valenta, L., VanderSloot, B., Wustrow, E., Zanella-Béguelink, S., Zimmermann, P.: In: 2nd ACM Conference on Computer and Communications Security (CCS 2015), Denver, CO, pp. 1–5, October 2015
3. Sauerbier, C.: Computing a Discrete Logarithm in $O(n^3)$. https://arxiv.org/ftp/arxiv/papers/0912/0912.2269.pdf. Accessed 17 Sept 2017
4. Lu, R., Cao, Z.: Simple three-party key exchange protocol. Comput. Secur. **26**, 94–97 (2007)

5. University of Texas. https://www.cs.utexas.edu/~shmat/courses/cs395t_fall04/gupta_gdh.pdf. Accessed 24 Sept 2017
6. AL-Sammarraie, O.A., Bashir, M.A.: Generalization of Newton's forward interpolation formula. Int. J. Sci. Res. Publ. **5**(3), 1–5 (2015)
7. Stallings, W.: Cryptography and Network Security Principles and Practices, 4th edn. Prentice Hall, Upper Saddle River (2006)
8. Ahmed, M., Sanjabi, B., Aldiaz, D., Rezaei, A., Omotunde, H.: Diffie-Hellman and its application in security protocols. Int. J. Eng. Sci. Innov. Technol. (IJESIT) **1**(2), 69–71 (2012)
9. Khader, A.S., Lai, D.: Preventing man-in-the-middle attack in Diffie-Hellman key exchange protocol. In: 22nd International Conference on Telecommunications (ICT 2015), pp. 204–208 (2015)
10. McCurley, K.S.: Proceedings of Symposia in Applied Mathematics, vol. 42, pp. 49–69, 8–9 August 1989

NLP Based Phishing Attack
Detection from URLs

Ebubekir Buber[1], Banu Diri[1], and Ozgur Koray Sahingoz[2(✉)]

[1] Computer Engineering Department, Yildiz Techical University, Istanbul, Turkey
ebubekirbbr@gmail.com, banu@ce.yildiz.edu.tr
[2] Computer Engineering Department, Istanbul Kultur University, 34158 Istanbul, Turkey
o.sahingoz@iku.edu.tr

Abstract. In recent years, phishing has become an increasing threat in the cyber-space, especially with the increasingly use of messaging and social networks. In traditional phishing attack, users are motivated to visit a bogus website which is carefully designed to look like exactly to a famous banking, e-commerce, social networks, etc., site for getting some personal information such as credit card numbers, usernames, passwords, and even money. Lots of the phishers usually make their attacks with the help of emails by forwarding to the target website. Inexperienced users (even the experienced ones) can visit these fake websites and share their sensitive information. In a phishing attack analysis of 45 countries in the last quarter of 2016, China, Turkey and Taiwan are mostly plagued by malware with the rate of 47.09%, 42.88% and 38.98%. Detection of a phishing attack is a challenging problem, because, this type of attacks is considered as semantics-based attacks, which mainly exploit the computer user's vulnerabilities. In this paper, a phishing detection system which can detect this type of attacks by using some machine learning algorithms and detecting some visual similarities with the help of some natural language processing techniques. Many tests have been applied on the proposed system and experimental results showed that Random Forest algorithm has a very good performance with a success rate of 97.2%.

Keywords: Machine learning · Phishing attack · Random Forest Algorithm
Cyber attack detection · Cyber security

1 Introduction

In the last few decades, Internet has become the most preferred medium between the users for transferring necessary information between them. This constructs a cyberspace in the digital world, which facilitates and speeds up human life such as electronic banking, social networks, e-mail usage and electronic commerce, etc. In addition to these advantages of the Internet, it has an open anonymous infrastructure for cyber-attacks which presents serious security vulnerabilities for the users especially inexperienced ones. Cyber-attacks are causing massive loss of personal and sensitive information and even money to companies and individuals every day. The material losses are at the level of billions of dollars annually. Phishing Attacks are one of the attack types that cause

© Springer International Publishing AG, part of Springer Nature 2018
A. Abraham et al. (Eds.): ISDA 2017, AISC 736, pp. 608–618, 2018.
https://doi.org/10.1007/978-3-319-76348-4_59

the most material damage. In this attack method, the attacker (also named as phisher) opens a site which is similar to a known legal site in the Internet through his domain and tries to capture some personal and sensitive information such as username, password, bank information, personal information of the victim. This type of attack especially started with fake emails and in a study conducted by the Anti-Phishing Working Group (APWG) on 45 countries around the world. The countries which are most plagued by malware is firstly China with the rate of 47.09% (of machines are infected) and then followed by Turkey (42.88%) and Taiwan (38.98%) [1].

When users connect to this site, which is believed to be legitimate, they enter the desired information without knowing that the site is fraudulent. The attacker usually directs the user to his or her site via email. An example scenario is shown in Fig. 1. The web page visited when the link in the e-mail is actually a hidden link is shown in it. When this page is entered, it appears that the original web page is designed almost the same. Although the difference in URL address is noteworthy, for an inexperienced user, this page is a good trap.

Fig. 1. Example of a mail and a web page used in phishing attack

There are many studies in the literature to detect Phishing Attacks [3]. These efforts are mainly focused on Natural Language Processing [4], Image Processing [5, 6], Machine Learning [7], Rule [8], White List - Black List [9], etc. Detection of an attack from a URL address can be a trivial task even for many experts, because an attacker can even deceive knowledgeable users by using different techniques. Simplest one is the use of rules as mentioned in [10] with similar rule format. However, rule based techniques are not satisfactory and there are many additional techniques which can be classified as depicted in Fig. 2 [2]. For this reason, it is important to detect these attacks with software support and a single approach cannot be sufficient for a good mechanism, some hybrid approaches can be preferred.

In this paper, domain names used in punctuation attacks were tried to be detected by using Natural Language Processing (NLP) techniques. We also identified some supporting features for the detection of such attacks and evaluated the effect on the performance of the proper use of these features. Specified features alone are not enough to decide whether a domain name is a cheat, but it will provide very useful information in order to make a right decision. The results obtained for this purpose should be evaluated as a result of a decision support system and should be evaluated as a parameter in the final decision.

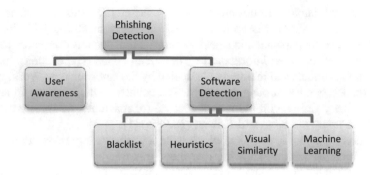

Fig. 2. An overview of phishing detection approaches

The rest of the report is organized as follows. Factors that make it difficult to identify the domain names used in Phishing attacks in the next section. Explanations about the data set used and pre-operations applied to the data set to extract the features to be used for the detection of Phishing attacks are explained in Sect. 3 and evaluations of the performed test results in Sect. 4. Finally, Conclusions are depicted.

2   URL and Attackers' Techniques

Attackers use different number of techniques to ensure that phishing attacks cannot be detected by either safety mechanisms or user. In this section these techniques and topics will be mentioned. The components in the URL (Uniform Resource Locator) structure need to be known to understand the attackers' techniques. Figure 3 shows the components found in a standard URL.

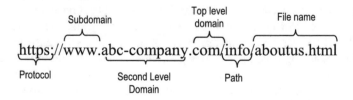

Fig. 3. URL components

The URL begins with a protocol used to access the page. The second level domain name (SLD) identifies the server hosting the web page and top level domain name specifies the domain name extension.

An attacker can purchase and use any SLD name that is not currently in use for attack purposes. This part of the URL can only be set once. An attacker can generate an unlimited number of URLs that can be used in phishing attacks by making the desired change in the Free URL field.

Because the unique part of each URL is the domain name part, cyber security companies make a lot of effort to identify this domain. When a domain name used in a phishing attack is detected, it is easy to block access to that domain and access to an unlimited number of URLs under that domain name can be blocked.

An attacker who wishes to perform a phishing attack uses the following basic methods to increase attack performance and steal more user information: typosquatting, cybersquatting, combined word usage, and use of random characters.

3   Data Sets and Preprocessing

In order to develop a system for detecting phishing attacks, two classes of URLs are needed. These classes are Malicious URL and Legal URL. In order to detect the URLs used in phishing attacks, it is necessary to extract distinguished the features. Malicious URLs may contain known brand names or some key words. In order to be able to detect a usage in this way, brand names and keywords lists are needed. In this study, these lists were created using different sources.

The malicious URLs analyzed in this study are provided from PhishTank [11]. Yandex Search API [12] has been used to collect legal URLs. Because the malicious URLs are not allowed to have high ranking in search engines, the URLs that are ranked high obtained by sending the query words to Yandex Search API are regarded as legal URLs.

The components within a URL are separated from each other using some special characters. For example, the Domain Name and the TLD are separated from each other by a dot mark. Likewise, the domain and the subdomain are also separated by a dot. Separating within the file path is done according to the "/" sign.

Each element within the URL may contain a separation mark of its own within it. For example, "-" in the Domain in the URL "abc-company.com" made a separation. Similar usage can be done with "=, ?, &" characters in File Path Area. Before starting the URL preprocessing, each word is added to the list of words to be analyzed by extracting words that are separated from each other by a special character. The flow diagram for the data preprocessing is given in Fig. 4a.

The objectives for data pre-processing can be summarized as follows:

1. To detect words with a known brand name or something similar in the URL
2. To detect keywords in the URL or something similar
3. To detect words created with random characters.

In the flowchart shown in Fig. 4b, the brand name or key words are first extracted in the words to be analyzed and the remaining words to be analyzed are given to the Random Word Detection Module.

The Random Word Detection Module (RWDM) checks whether a word is composed of random characters or not. The detected words are saved in a random word list and removed from the list of words to be analyzed.

When the words used in the phishing attacks were examined, it was determined that the length of the words with the contiguous writing was excessive. Therefore, the heuristic based threshold is determined to detect adjacent words. Words with more than 7 characters in length have been tried in Word Decomposer Module (WDM) to separate

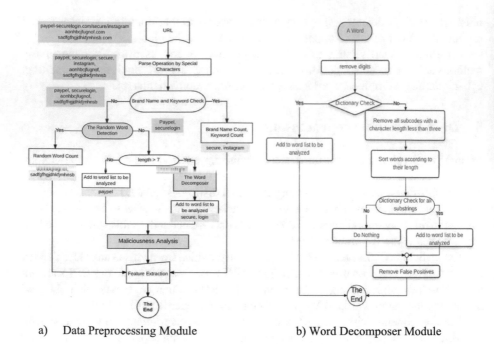

a)   Data Preprocessing Module   b) Word Decomposer Module

Fig. 4. The flow diagram for the data preprocessing and word decomposer [16]

the words they contain. If the word to be analyzed has two or more words, WDM separate words. If the word has not two or more words within it, WDM return raw the word.

The words with less than 7 characters in length and the words obtained from the Word Decomposition Module were analyzed and given to the Maliciousness Analysis Module (MAM). Finally, some features related to the words analyzed have been extracted.

3.1   The Word Decomposer Module

Word Decomposer Module (WDM) is a module that returns words as separate objects if there is more than one written word in a given character string. If no words are found in the analyzed word, the main word is returned. The flow diagram for performing the decomposition process is given in Fig. 3. First, in order to start the process of separating a character string, it was checked whether the related string was found in the dictionary. A word found in the dictionary has been added to the Word List without being parsed. The designed module can distinguish the English words which are written contiguously. The Enchant [13] package, developed in Python language, is used to check if a word is found in the English dictionary. All possible substrings are extracted for the word being parsed. Main string is divided into consecutive characters. An example of the process of splitting main word into substrings is given in Fig. 5.

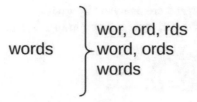

Fig. 5. Substring extraction

Extracted substrings are sorted by character length. The string with the highest character length is checked first. It is checked whether each substring is in the dictionary or not. The substrings in the dictionary are added to the word list.

There may be some false positive words in the word list obtained. For example, the "secure" word is in the list while the "cure" word is in the list. Both words are in the English dictionary and both words are in substring list. In this case, some words need to be eliminated. If the longer words contain smaller words in substring list, smaller words are eliminated on the word list to eliminate the false positive words.

It is not the case that any brand name or key word with a character length greater than 7 is given to the word parser module because the keywords are cleaned from the words to be analyzed with the control in the first stage. However, dictionary words that have a character length greater than 7 are not on the keyword list enter the WDM. These words are returned as a single word without attempting to be parsed.

3.2   The Random Word Detection Module (RWDM)

In URLs used in phishing attacks, words formed from random characters can be used. A project in Github [14] was used to identify random words. The Markov Chain Model was used in the study. First of all, the system is trained in a given English text. The probability of finding two successive consecutive characters in the training phase is calculated. Characters can only be letters or space characters found in the dictionary. The probabilities of the characters other than these are not extracted.

In order to be able to understand whether a word is random, consecutive letter in the word extracted during the test phase. For the extracted letter pairs, the probabilities of the letter pairs learned during training are found and the probabilities for all letter pairs are multiplied. If the analyzed word looks like a real word, then the multiplicative result is a high value, otherwise the multiplicative result is a low value.

A threshold value is determined so that it can be decided that a word is random. If the multiplication of the likelihoods of the letter pairs is below this threshold value, the word is classified as random.

3.3   Maliciousness Analysis Module (MAM)

In this study, MAM has been developed to detect whether the words used in phishing attacks are used for fraudulent purposes or not. In this module words written with Typosqautting can be detected.

The words to be analyzed are analyzed by giving this module. The flow diagram showing the operation of this module is given in Fig. 6 [16].

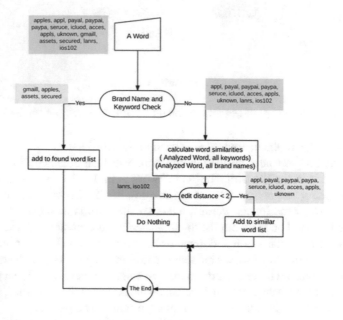

Fig. 6. MAM flow diagram

The words gathered from WDM may include a brand name or keyword. Therefore, the words to be analyzed are checked by brand name and keyword lists in this module again. Then the similarity is calculated by comparing the analyzed word with all the words in the brand name and keyword lists. Levenshtein Distance (Edit Distance) algorithm is used in the calculation of similarity. Then the similarity is calculated by comparing the analyzed word with all the words in the brand name and keyword lists. Levenshtein Distance (Edit Distance) algorithm is used in the calculation of similarity.

The Levenshtein Distance algorithm calculates how many moves are needed to convert an analyzed word to target word. The number of moves required for the conversion process gives the distance of the two words. Transactions used for conversion are; adding characters, deleting characters, and changing characters.

4   Experimental Results

The tests carried out in this study are grouped under 3 main titles.

- *Testing with Natural Language Processing (NLP) Features:* Tests with NLP-based features determined after the data preprocessing steps described in Sect. 4.
- *Test with Word Vectors:* Tests made with the features obtained after the vectorization of the words in the URL.

- *Hybrid Tests:* Tests where the features obtained with the word vectors and the NLP features are used together.

Within the scope of the tests performed, 3 different algorithms have been tried. Algorithms tested; Random Forest (RF) is a tree based algorithm, Sequential Minimal Optimization (SMO) is a kernel based algorithm, and Naive Bayes (NB) is a statistical based algorithm.

During the tests, the data set collected as described in Sect. 3. In the collected data set, there are 73,575 URLs including 37,175 malicious URLs and 36,400 legal URLs. Tests processed on a MacBook Pro device with 8 GB of 1867 MHz DDR3 RAM and 2.7 GHz Intel Core i5 processor. Weka [15] was used for testing. 10-fold Cross Validation and the default parameter values of all algorithms were used during the tests.

Tests were performed on a 10% sub-sample set instead of testing all samples because of lack of test device capacity. 3,717 malicious URLs and 3.640 legal URLs was ware used during the tests.

4.1   Testing with NLP Features

Data preprocessing steps which are explained in Sect. 4 has facilitated the extraction of some distinctive features. In this section, the extracted features and the obtained success rates are evaluated.

In the implemented system, firstly a URL was parsed according to some special characters such as ("=", ".", "/", "&", "?"). The obtained raw word list was processed to extract distinguishing features for malicious URL detection. The number of extracted distinguished features is 40. Detailed information on these features is given in Table 1.

Table 1. NLP features

The name of feature			
Raw word count (1)	www, com (2)	Alexa check (2)	Known TLD (1)
Punny code (1)	Separated word count (1)	Target keyword count (1)	Average adjacent word length (1)
Average word length (1)	Keyword count (1)	Other words count (1)	Brand check for domain (1)
Longest word length (1)	Brand name count (1)	Digit count (3)	Target brand name count (1)
Shortest word length (1)	Similar keyword count (1)	Subdomain count (1)	Consecutive character repeat (1)
Standard deviation (1)	Special character (8)	Random domain (1)	Similar brand name count (1)
Adjacent word count (1)	Random word count (1)	Length (3)	

4.2   Test with Word Vectors

Word Vectorization is one of the main approaches which are used for machine learning based text processing. In this approach, the words found in the text are used as features instead of the features that are manually extracted in the text based data.

For each URL, the words obtained after applying the preprocessing step given in Fig. 3 have been converted into vectors. The *StringToWordVector* module in Weka is used for the vectorization process. The generated word vectors are processed by machine learning algorithms.

For the 7,357 URLs used for the test, 10,572 features were extracted after the vectorization process. Then Feature Selection was made to reduce the number of features. The CfsSubsetEval algorithm is used for feature selection. The algorithm has been run Forward with the BestFirst search method. With feature selection algorithm execution, the number of features has dropped from 10,572 to 238.

4.3   Hybrid Test

The Word Vectorization and Feature selection steps were repeated for Hybrid Tests. After the implementation of the feature selection step, 238 word features are obtained. Then 40 NLP features were added to the features obtained. Classification algorithms have been tested with a total of 278 features.

4.4   Test Results

40 features are used for Testing NLP Features, 238 features are used for Testing Word Vectors and 278 features are used for Hybrid Tests. The F-measurement values for the achievement values obtained as a result of the tests performed are given in Table 2.

Table 2. Test Results

Algorithm	NLP features	Word vector	Hybrid
Random Forest	0.966	0.868	0.972
SMO	0.941	0.865	0.964
Naïve Bayes	0.655	0.817	0.755

According to the experimental results given in Table 2, it is seen that the system formed by the NLP features is more successful than the system formed by the Word Vector features. Hybrid approach using NLP Features in conjunction with Word Vector Features has been observed to have a boosting effect for Random Forest and Sequential Minimal Optimization algorithms.

In the previous work [16] of this study, we constructed our system with 17 NLP based features and 209 Word Vector features (and totally 226 features). The enhancement in this study increased the success of our prediction about 7%.

5 Conclusion

In this paper, a system has been developed to detect URLs which are used in Phishing Attacks. In the proposed system some features have been taken out by using NLP techniques. The extracted features are evaluated in two different groups. The first one is a person-determined attribute that is thought to be distinctive to malicious URLs and legal URLs. The second group focuses on the usage of the words in the URL without performing any other operations by applying only the vectorization process. Experimental study is constructed over three different test scenarios, including tests for NLP-based features, tests for Word Vectors, and Hybrid approach tests for both of these features.

During the tests; Random Forest which is a tree based algorithm, Sequential Minimal Optimization which is a kernel based algorithm and Naive Bayes algorithm which is a statistical based algorithm are used. According to the results obtained, the tests made for the hybrid approach were more successful than the other tests. In the hybrid approach, the Random Forest Algorithm was observed to be more successful than the other algorithms tested with 97.2% success rate.

Acknowledgement. Thanks to Normshield Inc., BGA Security, SinaraLabs and Roksit for contributing to the development of this work.

References

1. Anti-Phishing Working Group (APWG): Phishing activity trends report—last quarter (2016). http://docs.apwg.org/reports/apwg_trends_report_q4_2016.pdf
2. Khonji, M., Iraqi, Y., Jones, A.: Phishing detection: a literature survey. IEEE Commun. Surv. Tutor. **15**(4), 2091–2121 (2013)
3. Garera, S., Provos, N., Chew, M., Rubin, A.D.: A framework for detection and measurement of phishing attacks. In: Proceedings of the 2007 ACM Workshop on Recurring Malcode, pp. 1–8. ACM, November 2007
4. Stone, A.: Natural-language processing for intrusion detection. Computer **40**(12), 103–105 (2007)
5. Fu, A.Y., Wenyin, L., Deng, X.: Detecting phishing web pages with visual similarity assessment based on earth mover's distance (EMD). IEEE Trans. Dependable Secur. Comput. **3**(4), 301–311 (2006)
6. Toolan, F., Carthy, J.: Phishing detection using classifier ensembles. In: 2009 eCrime Researchers Summit, eCRIME 2009, pp. 1–9 (2009)
7. Abu-Nimeh, S., Nappa, D., Wang, X., Nair, S.: A comparison of machine learning techniques for phishing detection. In: Proceedings of the Anti-Phishing Working Groups 2nd Annual eCrime Researchers Summit, eCrime 2007, pp. 60–69. ACM, New York (2007)
8. Cook, D.L., Gurbani, V.K., Daniluk, M.: Phishwish: a stateless phishing filter using minimal rules. In: Financial Cryptography and Data Security, pp. 182–186. Springer (2008)
9. Cao, Y., Han, W., Le, Y.: Anti-phishing based on automated individual white-list. In: DIM 2008: 4th ACM Workshop on Digital Identity Management, New York, pp. 51–60 (2008)

10. Sahingoz, O.K., Erdogan, N.: RUBDES: a rule based distributed event system. In: 18th International Symposium on Computer and Information Sciences - ISCIS 2003, Antalya, Turkey, pp. 284–291 (2003)

11. Phistank: join the fight against phishing. https://www.phishtank.com/developer_info.php. Accessed Oct 2017

12. Yandex account: Yandex Technologies. https://tech.yandex.com.tr/xml/. Accessed Oct 2017

13. PyEnchant—PyEnchant v1.6.6 documentation. http://pyenchant.readthedocs.io/en/latest/index.html. Accessed Oct 2017

14. A small program to detect gibberish using a Markov chain. https://github.com/rrenaud/Gibberish-Detector. Accessed Oct 2017

15. Weka 3: data mining software in Java. http://www.cs.waikato.ac.nz/ml/weka/. Accessed Oct 2017

16. Buber, E., Diri, B., Sahingoz, O.K.: Detecting phishing attacks from URL by using NLP techniques. In: 2017 International Conference on Computer Science and Engineering (UBMK), Antalya, Turkey, pp. 337–342 (2017)

Hand Pose Estimation System Based on a Cascade Approach for Mobile Devices

Houssem Lahiani[1,3,4(✉)], Monji Kherallah[2], and Mahmoud Neji[3,4]

[1] National School of Electronics and Telecommunications, University of Sfax, Sfax, Tunisia
lahianihoussem@gmail.com
[2] Faculty of Sciences, University of Sfax, Sfax, Tunisia
monji.kherallah@gmail.com
[3] Faculty of Economics and Management, University of Sfax, Sfax, Tunisia
mahmoud.neji@gmail.com
[4] Multimedia Information Systems and Advanced Computing Laboratory, Sfax, Tunisia

Abstract. The rise in the use of mobile devices requires finding new ways to interact with this type of devices. Gestures are an effective way to interact with the mobile device and to place order to it. However, gesture recognition in this context constitute a challenging task due the limited computational capacities of this type of devices. In this work, we present a hand pose estimation system for mobile device. The gesture is recognized by using a boosting algorithm and Haar-like features. The system is designed for Android devices. The method used consists of capturing gestures by a smartphone's camera to recognize the hand sign. It presents a system based on a real-time hand posture recognition algorithm for mobile devices. The aim of this system is to allow the mobile device interpreting hand signs made by users without the need to touch the screen.

Keywords: Hand gesture recognition · Android · Haar-like features · AdaBoost HCI

1 Introduction

Since the down of time, gestures have an important role in communication between humans, and now they constitute an interface between humans and machines. Currently, many technologies of Human Machine Interaction are being developed to deliver user's command to mobile devices like smartphones. Some of them are based on interaction with machines through head, hand, voice or touch. In this paper we aim to describe a system based on hand gestures to control smartphones by using Computer Vision. Thus, vision-based hand gesture recognition is being a challenging task in this context due to the computational limits of mobile devices and diversity of gestures. In this paper a simple and fast algorithm for hand pose estimation for mobile device is described. The hand pose is recognized by using Adaboost classifier. The system is developed on an Adroid OS platform. Hand pose estimation can be very important in different scenarios. Training some system to understand and react to different hand gesture in a complex background using smartphones is an exciting idea. It enables the Smartphone to perform

© Springer International Publishing AG, part of Springer Nature 2018
A. Abraham et al. (Eds.): ISDA 2017, AISC 736, pp. 619–629, 2018.
https://doi.org/10.1007/978-3-319-76348-4_60

all necessary steps to recognize gestures without the need to connect to a distant device. Android platform was choose for technical and non-technical reasons. Among the non-technical reason, most of mobile device users use Android OS [1]. According to Stats-Counter, an independent web analytics company, Android is the most popular operating system and reached a very big milestone in March 2017, surpassing Windows OS as the world's most popular operating system in terms of total Internet usage on the tablet, Portable, desktop, and mobile handset [2]. In addition, Android ecosystem allows integrating libraries for image processing like Open CV library (Open Computer Vision library) [3], a free graphics library for image processing. In addition, according to a comparative study for image processing between Android and iOS [4], better performance was achieved with Android platform. Mobile devices are computationally limited and vision based tasks are computationally expensive which make developing a hand gesture recognition system for mobile devices a challenging task. Several approaches for hand pose estimation for mobile devices has been developed. Among those approaches, we find glove-based approach [5, 6], which is annoying for users. Vision-based approaches were used too [10]. However, even some vision-based methods are suitable for mobile device since they are based on color, they are still non-robust against various kinds of lighting. Moreover, Shape based approaches could achieve good results for rigid objects, but not for articulated objects; in addition, most of them are computationally demanding and not suitable for mobile devices. Our proposed system is based on a cascade architecture using Boosting algorithm. This architecture was firstly introduced by Viola and Jones for face recognition [11] and it was widely used for object detection due to its robustness against illumination changes and its fast detection. We propose a robust and real-time system for hand gesture recognition, which is based on the (GAB) Gentle AdaBoost and Haar-like features.

2   Related Works

Many systems have been made in hand gesture recognition field using mobile devices. Due to their availability, mobile devices have motivated researchers to use them in hand gesture recognition. Many approaches have been used in this field. Vision-based approaches were widely used in recognizing gestures by mobile devices due to inexpensive color vision. However, those systems are vulnerable against luminance changes and the difference in skin color, which risks producing a low accuracy compared to glove-based approaches. Thus, we propose a cascade-based approach that is efficient against illumination changes. In addition, it is more computationally inexpensive which makes it suitable for mobile device. Haar features and boosting algorithm has been widely used in object detection and recognition because its performance does not depends on lightning conditions.

Lahiani et al. [7, 8] developed a real time hand gesture recognition system based on skin color and SVM classifier. The mobile did all steps and recognition rate of this system was about 93%.

Lahiani et al. [9] proposed a hand pose estimation system to control the smartphone, the system used the Viola-Jones Algorithm to detect the palm and SVM classifier to recognize gesture. The mobile device does all steps and recognition rate of this system was about 91%.

In [12] Song et al. developed a gestures recognition system for smartphones, smartwatches and Google Glasses. Their system uses the RGB camera integrated in mobile devices. They used the skin color detection for the preprocessing step. They experimented skin detection with HSV and GMM-based adaptive thresholding.

In [13] Jin et al. makes Sign Language Translator for mobile devices. They used the Canny edge detection to detect the hand and SURF algorithm for the feature extraction step and Support Vector Machine for classification. Overall accuracy of the system was about 97.13%.

In [14] Guerra-Casanova et al. proposed a gesture recognition system to authenticating a person. To accomplish this purpose, the user is prompted to be recognized by a gesture he performs moving his hand while holding the mobile device with an embedded accelerometer. The robustness of this biometric system has been tested within 2 different tests analyzing a database of 100 users with real falsifications. Equal Error Rates of 2.01 and 4.82% have been achieved in a zero effort and an active impostor attack respectively.

In [15, 16] Pouke et al. developed a multi-modal system for 3D interaction with portable device. The developed system is constituted by gaze-tracking system based on corneal reflection approach and hands gestures based on accelerometer and machine learning algorithms. The gaze tracking was used for object selection and hand gestures for manipulation. This interaction is touchless without the need to any connected device like mouse or keyboard.

In [17] Prasuhn et al. developed a static hand gesture recognition system for the "ASL" American Sign Language for mobile device. They make a system based on Histogram of Oriented Gradients (HOG) features due to its invariance against rotation. The sensitivity of HOG is treated by using a database of Histogram of Oriented Gradients descriptors corresponding to generated hand images.

3 Proposed System

In this work, we propose a system for human-mobile interaction based on hand gesture. This system aims to design a natural and intuitive interaction with mobiles.

In this paper, we present a system running on an android OS platform. The captured hand pose by the camera of the device constitute the input of our system. We used approaches that are computationally inexpensive to compensate the weak processing capability of mobile devices. The proposed system consists of four main steps: Hand detection, features extraction, gesture classification and recognition (Fig. 1).

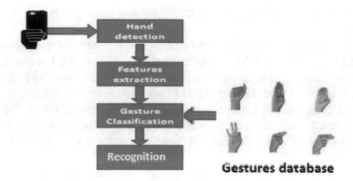

Fig. 1. System design

3.1 Hand Detection

Hand Gesture Recognition is a computer science field which have to interpret signs made by humans using mathematical algorithms. It represents a way for computers to understand the humans' intentions, creating an interface between them and computers. An input device like smartphone in our case, detect frames. They are then interpreted using algorithms based on artificial intelligence algorithm.

The first step in any vision based hand gesture recognition system is to extract Hand from the input image. This step is more difficult when the acquired image contains several objects and non-uniform background that creates a texture disturbing the correct segmentation of the hand.

3.2 Feature Extraction: Haar Features

The values of a pixel inform us only about the luminance and the color of a point given. It is therefore more sensible to find detectors based on more global features of the object. This is the case of Haar descriptors. The Haar descriptors are functions for knowing the difference in contrast between several adjacent rectangular regions in an image. In this way, existing contrasts in an object and spatial relationships are coded (Fig. 2) [18]. Indeed, these descriptors make it possible to calculate the difference between the sum of the pixels in the white zones and the sum of the black zones. The descriptor value is calculated by:

$$f_i = Sum(r_{i,white}) - Sum(r_{i,black}) \tag{1}$$

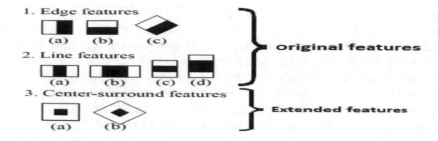

Fig. 2. Haar descriptor

These descriptors are calculated in a fixed size window (24 × 24 pixels). Generally, they are classified in 3 kinds: 2-rectangles, 3-rectangles and 4 rectangles descriptors. The 2-rectangles descriptors are used horizontally and vertically. White regions have positive weights and black regions have negative weights.

A Haar descriptor is characterized by:

- the number of rectangles (2, 3 or 4)
- the position (the top left vertex) (x, y) of each rectangle
- the width w and the height h of each rectangle with $0 < x$, $x + w < W$; $0 < y$, $y + h < H$
- the positive or negative weights of each rectangle.

Many works based on Haar-like features were done in the field of object detection with mobile devices and show that the Haar-like feature provides similar performance to traditional approaches while being more suitable for smartphones and mobile device [19].

3.3 Gesture Classification: AdaBoost

The popularity of Boosted cascades is due to its efficiency in detecting a large number of objects. Success is conducted by the effectiveness and ability of this approach to recognize objects in real time and under different conditions and on low-cost ARM architectures like mobile devices and embedded devices. One of the most used algorithms in boosting is called AdaBoost, an abbreviation for adaptive boosting. Adaboost relies on the iterative selection of weak classifier according to a distribution of the learning examples. Each example is weighted according to its difficulty with the current classifier.

AdaBoost uses several weak classifiers over several cycles, selecting the best weak classifier in each cycle and combining the best classifier to generate a strong classifier.

The proposed architecture is based on a cascade architecture using AdaBoost algorithm. Viola and Jones firstly introduced this architecture. The Viola-Jones Algorithm has four stages: Haar Feature Selection, creation of an Integral Image, Adaboost Training and Cascading Classifiers. There are many variant of AdaBoost like gentle AdaBoost, Logit Adaboost and Real AdaBoost. In our work, we used "GAB" gentle AdaBoost. It is a variant of the powerful boosting learning technique [20]. The different AdaBoost algorithms differ in the weights update scheme. In [21] according to Lienhart et al. the gentle AdaBoost (GAB) is the most successful learning method tested for face detection applications. Gentle AdaBoost doesn't require the calculation of log-ratios that can be numerically unstable. The Experimental results on benchmark data demonstrates that the conservative Gentle AdaBoost shows similar performance to Real AdaBoost and Logit AdaBoost, and in majority of cases outperforms these other two variants [22]. Consequently, we chose to use the Gentle Adaboost for our system.

The Haar-based features cited above are used in the framework of AdaBoost learning for detecting Hand gestures. In this paper, Gentle AdaBoost is proposed as a classifier. A set of positive and negative samples were used for training purposes. During the training process, only weak classifiers made out of Haar features that would be able to improve the prediction are selected. For every weak classifier AdaBoost chose an acceptance threshold, but a single weak classifier is not able to classify wanted hand gesture with low error rate. In every iteration, weights are normalized and the best weak classifier is chose in function of the weighted error value. In the next stage, the weight values are updated and it could be decided if an example is correctly classified or not. After having traversed all the iterations, a set of weak classifiers characterized by a specified error rate is selected and we obtain a resulted strong-trained classifier. The classification is done iteratively and the number of learning examples affects the effectiveness of the classification process [23].

4   Experiments

4.1   Training Data

To train our system, we used The NUS hand gesture database I [24]. This database contain 240 images containing hand gestures. It has 10 classes of postures and 24 sample images per class that have been captured by varying the position and size of the hand within the frame. And to train classifier we used those images to create samples that show the hand postures we want to detect positive sample) and even more images without the hand gesture which represents negative samples.

Training an accurate classifier takes a lot of time and requires a huge number of samples. In our case, due to the small number of images in each class in the NUS dataset we enhanced it by adding to each class more than 50 images representing the gesture. Images were taken with different persons and in different lightening conditions and backgrounds. We obtained after that an enhanced NUS dataset, which has 75 images for each class.

For each hand posture, we choose the 75 sample images representing the hand gestures as positive samples to train the classifier. We cropped positive images to obtain only the wanted area that represents the hand gesture. The ratios of the cropped images should not differ that much. The best results come from positive images that look exactly like the ones we want to detect (Fig. 3).

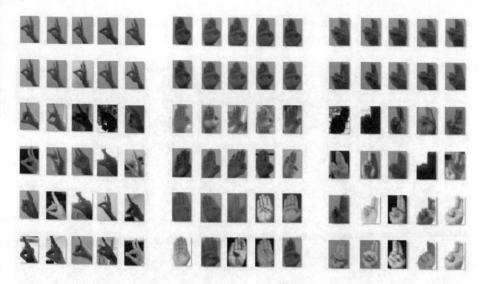

Fig. 3. Some gestures of the enhanced NUS dataset

To increase the number of positive samples, we used the opencv_createsamples utility to obtain 1500 samples for each class of gestures. Thus, a large set of positive images is generated from the set of positive data having the background from the set of negative images. If we want to train a highly accurate classifier, we must have many negative images that do not contain the hand gesture we want to detect. We used more than 4000 negative images containing images that do not contain gestures like backgrounds, walls, faces, etc. To recognize those gestures, we assign to each one of them a letter or word (Fig. 4).

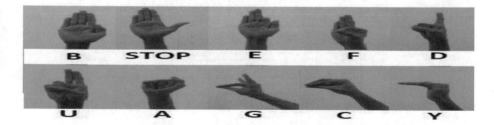

Fig. 4. Grammar to interpret gestures

4.2 Evaluation

To test the system and to know how well the recognition rate is, we tested hand poses made by different persons in different lightning condition and different places. For each posture, fifty frames were captured in different backgrounds and lightening condition. The Fig. 5 presents some gestures recognized correctly made by different persons in different lightening conditions and with different backgrounds. We can see the name that corresponds to each sign on the top of the green box in the capture.

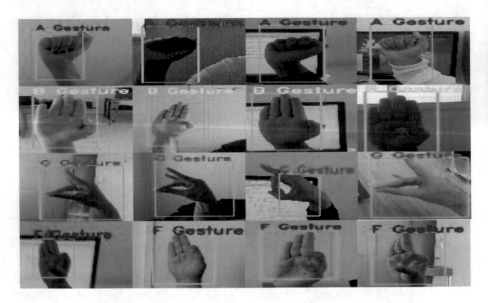

Fig. 5. Different poses in different luminance conditions and backgrounds

We tested the system by inviting 50 different people to make the gestures in different lightning condition and different backgrounds in front of the camera of the device. Results are shown respectively in Table 1.

Table 1. Recognition rate of different hand poses.

Distance ≤ 75 cm			
Sign	Number of capture	Recognized	Recognition rate
B	50	46	92%
STOP	50	45	90%
E	50	44	88%
F	50	46	92%
D	50	44	88%
U	50	45	90%
A	50	45	90%
G	50	43	86%
C	50	44	88%
Y	50	43	86%
Average			89%

The Table 1 gives summary of the hand pose recognition systems for mobile devices. It summarize the comparison result between different gestures and the average recognition rate.

According to the obtained result, detection of different hand poses would be optimal if done at a distance less than or equals to 75 cm.

Here, we have used Open CV library and Tegra Android Development Pack (TADP). Hardware used is Intel® Core i7® CPU, Windows 10 (64 bit), 8 GB RAM and a 2 Smartphones: the first one uses a 13 MP camera, Android 5.1 Os (Lollipop) version and the other uses a 4 MP camera, Android 4.4 KitKat.

The system was deployed on those different mobile device models. The average execution time of detecting poses was about 68 ms in a smartphone which has 1.5 GB RAM and a 1.4 GHz Octo-core processor and it was about 90 ms in a smartphone equipped with a 1.2 GHz Quad-Core processor and 1 GB RAM. The performance of the system essentially relies on the training data and the quality of positive images. In other terms, if the training step is well done, the probability that the hand gesture could be detected is high. A similar system based on LBP features was developed using the same dataset and was trained under same conditions [24]. The recognition rate for the system based on Haar features was better than the other based on LBP. In contrast, the detection speed was better with LBP. At the end, we can say that the recognition system with both Haar and LBP provides a satisfactory result since it achieve a good recognition rate and at the same time, it is not vulnerable against illumination changes like systems cited at the related work section.

5   Conclusion

In this paper, we proposed a new system of human-mobile interaction based on hand gestures. We used a system of hand gesture recognition that detected static hand poses. The experiment results proved the robustness of the system in despite of some limits.

The NUS hand poses database was used for training purposes. Due to its small numbers of samples, we enhanced it with a bigger number of sample for each gestures. We have choose to work with Haar approach that demonstrate a good performance in the detection rate. However, training time has taken many days and compared to LBP approach, it was long enough since the LBP approach takes just some hours for training. The cascade architecture demonstrated that it is suitable for mobile device as it is computationally inexpensive. In a future work, to improve the training step, we will try to enhance the NUS hand poses database with more and more samples. We will try also to enhance the experimentation by using PR curve and by inviting more and more people to test our application in real time.

References

1. Manjoo, F.: A murky road ahead for android, despite market dominance. The New York Times (2015). ISSN 0362-4331. Accessed 27 May 2015
2. Statcounter Company. http://gs.statcounter.com/press/android-overtakes-windows-for-first-time. Accessed 3 Apr 2017
3. OpenCV. http://opencv.org/platforms/android.html
4. Cobârzan, C., Hudelist, M.A., Schoeffmann, K., Primus, M.J.: Mobile image analysis: android vs. iOS. In: 21st International Conference on MultiMedia Modelling (MMM), pp. 99–110 2015
5. Seymour, M., Tšoeu, M., A Mobile application for South African Sign Language (SASL) recognition. In: IEEE AFRICON 2015, pp. 281–285 (2015)
6. Xie, C., Luan, S., Wang, H., Zhang, B.: Gesture recognition benchmark based on mobile phone. In: You, Z., Zhou, J., Wang, Y., Sun, Z., Shan, S., Zheng, W., Feng, J., Zhao, Q. (eds.) CCBR 2016. LNCS, vol. 9967, pp. 432–440. Springer, Cham (2016). https://doi.org/10.1007/978-3-319-46654-5_48
7. Lahiani, H., Elleuch, M., Kherallah, M.: Real time hand gesture recognition system for android devices. In: 15th International Conference on Intelligent Systems Design and Applications (ISDA), pp. 592–597 (2015)
8. Lahiani, H., Elleuch, M., Kherallah, M.: Real time static hand gesture recognition system for mobile devices. J. Inf. Assur. Secur. **11**, 067–076 (2016). ISSN 1554-1010
9. Lahiani, H., Kherallah, M., Neji, M.: Hand pose estimation system based on Viola-Jones algorithm for android devices. In: 13th ACS/IEEE International Conference on Computer Systems and Applications, (AICCSA) (2016)
10. Lahiani, H., Kherallah, M., Neji, M.: Vision based hand gesture recognition for mobile devices: a review. In: Abraham, A., Haqiq, A., Alimi, Adel M., Mezzour, G., Rokbani, N., Muda, A.K. (eds.) HIS 2016. AISC, vol. 552, pp. 308–318. Springer, Cham (2017). https://doi.org/10.1007/978-3-319-52941-7_31
11. Viola, P., Jones, M.: Rapid objet detection using a boosted cascade of simple features. In: Proceedings of IEEE Conference on Computer Vision and Pattern Recognition, pp. 511–518 (2001)
12. Song, J., Sörös, G., Pece, F., Fanello, S.R., Izadi, S., Keskin, C., Hilliges, O.: In-air gestures around unmodified mobile devices. In: 27TH ACM User Interface Software and Technology Symposium (UIST 2014), pp. 319–129 (2014)
13. Jin, C., Omar, Z., Jaward, M.H.: A mobile application of american sign language translation via image processing algorithms. In: 2016 IEEE Region 10 Symposium (TENSYMP), pp. 104–109 (2016)
14. Guerra-Casanova, J., Sánchez-Ávila, C., Bailador, G., de Santos Sierra, A.: Authentication in mobile devices through hand gesture recognition. Int. J. Inf. Secur. **11**(2), 65–83 (2012)
15. Pouke, M., Karhu, A., Hickey, S., Arhippainen, L.: Gaze tracking and non-touch gesture based interaction method for mobile 3D virtual spaces. In: Proceedings of OzCHI, pp. 505–512 (2012)
16. Prasuhn, L., Oyamada, Y., Mochizuki, Y., Ishikawa, H.: A HOG-based hand gesture recognition system on a mobile device. In: IEEE International Conference on Image Processing (ICIP), pp. 3973–3977
17. Harwood, D., Ojala, T., Pietikäinen, M., Kelman, S., Davis, S.: Texture classification by center-symmetric auto-correlation, using Kullback discrimination of distributions. Technical report CAR-TR-678, Computer Vision Laboratory, Center for Automation Research, University of Maryland, College Park, Maryland (1993)

18. Dixit, V., Agrawal, A.: Real time hand detection & tracking for dynamic gesture recognition. Int. J. Intell. Syst. Appl. **08**, 38–44 (2015)
19. Tresadern, P.A., Ionita, M.C., Cootes, T.F.: Real-time facial feature tracking on a mobile device. Int. J. Comput. Vis. **96**(3), 280–289 (2012)
20. Lienhart, R., Maydt, J.: An extended set of Haar-like features for rapid object detection. In: Proceedings of IEEE Conference on Image Processing (ICIP 2002), New York, USA, pp. 155–162, September 2002
21. Ferreira, A., Figueiredo, M.: Boosting algorithms: a review of methods, theory, and applications. In: Zhang, C., Ma, Y. (eds.) Ensemble Machine Learning: Methods and Applications, pp. 35–85. Springer, New York (2012). https://doi.org/10.1007/978-1-4419-9326-7_2
22. Frejlichowski, D., Gościewska, K., Forczmański, P., Nowosielski, A., Hofman, R.: Applying image features and AdaBoost classification for vehicle detection in the 'SM4Public' system. In: Choraś, Ryszard S. (ed.) Image Processing and Communications Challenges 7. AISC, vol. 389, pp. 81–88. Springer, Cham (2016). https://doi.org/10.1007/978-3-319-23814-2_10
23. The NUS Hand Posture Datasets I. https://www.ece.nus.edu.sg/stfpage/elepv/NUS-HandSet/
24. Lahiani, H., Kherallah, M., Neji, M.: Hand gesture recognition system based on Local Binary Pattern approach for mobile devices. In: 17th International Conference on Intelligent Systems Design and Applications (ISDA) (2017)

HMI Fuzzy Assessment of Complex Systems Usability

Ilhem Kallel[1,2]([✉]), Mohamed Jouili[3], and Houcine Ezzedine[4]

[1] ISIMS: Higher Institute of Computer Science and Multimedia of Sfax,
Route de Tunis Km 10 BP 242, 3021 Sfax, Tunisia
ilhem.kallel@ieee.org
[2] REGIM Lab: Research Groups in Intelligent Machines, University of Sfax,
ENIS, BP 1173, 3038 Sfax, Tunisia
[3] ISIMG, University of Gabes, Gabes, Tunisia
Jouilim8@gmail.com
[4] UVHC, LAMIH Le Mont-Houy, 59313 Valenciennes cedex, France
Houcine.Ezzedine@univ-valenciennes.fr

Abstract. Testing and assessing are core activities in the development cycle of software applications, dedicated to evaluating interactive products in order to improve their quality by identifying various usability problems and defects. For complex system such as multiagent ones, usability evaluation is still an issue, and requires new test techniques to assess autonomous and interactive behaviors. This paper deals with investigation about the evaluation of the Human-Machine Interaction (HMI) in Complex Systems: A review on evaluation methods and introduce domain-specific requirements is presented; A mechanism for the evaluation of HMI is proposed. Also, an implementation of an automatic tool dedicated to the assessment of complex interactive systems based on fuzzy logic approaches is explained; A solution to automate the evaluation of the HMI that reduces the need for expert assessment and fully integrates end-users into the HMI evaluation loop is suggested. These are assessed in the urban transit control room in the city of Valenciennes, France. The comparative study deals with acceptance, motivation and perceived happiness.

1 Introduction

Currently, evaluation tools are trying to integrate end-users into their evaluation procedures (i.e. interviews, survey, and questionnaires). Therefore, it is important to ensure the production of interactive systems with a respectable degree of usability as well as expecting better evaluation results. Under these circumstences, it is recommended to completely automate the usability evaluation of the Human Machine Interaction (HMI). Thus, the majority of UI evaluation tools do not allow the interpretation of the metrics used in the evaluation (e.g. the relationships between the different usability criteria used in the evaluation of UI). To overcome this problem, we propose to adopt universal and standard

© Springer International Publishing AG, part of Springer Nature 2018
A. Abraham et al. (Eds.): ISDA 2017, AISC 736, pp. 630–639, 2018.
https://doi.org/10.1007/978-3-319-76348-4_61

measures, and develop a new method to automatically assess the quality of interactive complex systems. Our Proposed method aims to reduce the need for an appraisal, evaluate the ergonomic aspect of the UI as well as the usability, integrate end-users into the evaluation loop, develop an automated evaluation interface that integrates the ISO/IEC 25066 (2016) standards and design an evaluation model that allows the interpretation of HMI evaluation data, as well as the production of proportional relationships between the usability criteria used in the evaluation. This paper is organized as follows; in Sect. 1, we present a review on subjective and objective methods for evaluating UI and some recent works on automatic HMI evaluation. In Sect. 2, we describe in detail our proposed method dedicated to the automatic evaluation of UI (based on the process ISO/IEC 25066 (2016))[1]. In Sect. 3, we present our related software implementation about an automatic questionnaire. Finally, we detail the process of analysis applied to the control system of the urban transport network (bus/tramway) Valenciennes (France) and discuss the obtained results.

2   Review Attempt on HMI Evaluation

The usability evaluation of a UI can be processed in two formats: (1) Subjectively according to the opinions of users and experts in order to determine the quality of an interactive system; (2) objectively using definite rules and parameters to infer the quality of an interactive system [1]. Users' opinions and data used by the rules to evaluate the system can be collected manually or automatically. In a recent study [2], the authors propose a classification of evaluation methods in two main classes: (1) Subjective methods (ThinkAloud protocol, heuristics and surveys/questionnaires). (2) Objective methods (electronic cookie, ergonomic guides, etc.).

2.1   Subjective Evaluation

Subjective evaluation of UI is carried out through the analysis of end-users' opinions in order to give meaningful assessment as a result [1]. Heuristic assessment of UI is a subjective evaluation as a usability inspection method handling a number of experienced UI evaluators who give their appreciations according to defined guidelines (heuristics) [1] and standards [3]. Another type of subjective evaluation based on the usability scale method [2], uses standardized questionnaires, filled out by users of targeted UI, about ergonomics and effectiveness of the Interactive System. The results (data sets) have to be analyzed in order to evaluate the UI according to the experience of these users.

In [4], the authors performed a heuristic evaluation of a website interfaces with a team of experts and then compared the results with those obtained from a

---

[1] Systems and Software Engineering - Quality and System and Software Evaluation Requirements (SQuaRE) - Common industry format for use - Evaluation report. Consulted in: https://www.iso.org/fr/standard/63831.html (Accessed date: 02/01/2017).

usability test performed with a group of typical users. Their results showed that typical users identified 90% of user interface problems which were already identified by the team of experts in the heuristic evaluation. Conversely, they also showed that only 20% of problems scored as critical by experts in heuristic evaluation were relevant to typical users. This study demonstrates the importance of the participation of system users in the evaluation process.

In [2], the authors showed that the classic heuristics of Nielsen [3] fail at evaluating interactive and transactional websites, so they proposed a set of 15 heuristics to cover all types of web sites. They conducted a comparative study by asking a group of students to perform the heuristic evaluation of a transactional website using both heuristics sets after being trained. Then, they analyzed student' surveys after the evaluation and the results showed that their proposed heuristics surpassed the traditional ones in terms of "perceived ease of use", "perceived utility" and "intent use". However, they stated that the difference is not significant enough to revise these lowscore heuristics.

In [5], the authors proposed a generic questionnaire tool for the evaluation of Web interfaces. It is based on the ISO/IEC 14598-5 standard evaluation process. The proposed questionnaire collects and analyzes the views of end-users and focuses on three phases: the preparation of the evaluation (generation of the software tool), the evaluation (presentation of questions to users), and the evaluation control (analyzing opinions and presenting the evaluation report). However, the proposed tool does not provide any interpretation of the evaluation criteria involved in the UI assessment process.

Some researchers are adopting the usability satisfaction questionnaires to evaluate UI. In [6], they used the Usability Satisfaction Questionnaires CUSQ [5], which are developed according to psychometric factors to evaluate an online product catalog website interface. They hired a group of users and gave each one some search tasks to conduct on this website. Next, users were asked to give their opinions on the website interface via a questionnaire. Accordingly, they recommended a catalog type where product selection is done by user navigation in the website instead of entering specific attributes and then getting the necessary recommendations as evaluation results. The major disadvantage of this approach is its slowness.

2.2   Objective Evaluation

Observation of the user's behavior while interacting with a UI may reflect the user's current feedback and provide an objective assessment of the UI. If the number of users who are involved in this evaluation is sufficient, an index can be generated from their experiences to compare the different user interfaces. In [7], researchers showed that an evaluation can be used as an indicator of the level of machine intelligence since it tests usability in terms of interactions with end-users. They compared the results obtained from the evaluation of the interactions with the users and the results of the survey based evaluation. They collected data from users with similar backgrounds and experience while the system tested performed the same tasks. Then, they used a fuzzy logic system to infer evaluation. Accordingly, an objective index for each user interface was obtained. These indexes were compared with the results of the survey given to the same users involved in the first test. The findings showed that their approach match with the survey in 70% of the cases. They corroborate that these results can be improved if registration of user interactions is done automatically, and more users are involved in the process.

A group of researchers showed in [8] that evaluation of UI can be used as an intermediate step for design improvement. They used the heuristic evaluation by teachers familiar with usability in order to improve their interface design of digital textbooks platform for schools. Then, they used the recommendations obtained from the "Heuristic evaluation" step in order to modify their prototype. They used the surveys and student log files that interact with their modified version of the interface, and made an interactive changes to their prototype. After two test levels, students admitted that the modified interface is more usable and user friendly, proving that both subjective and objective evaluation methods can be combined to improve the design of the UI and subsequently its quality.

2.3   Automatic Evaluation

Even using the same evaluation technique, the results of the usability assessment performed by many end users could be different. In other words, the number of users participating in the evaluation should be increased in order to reach a good evaluation. Therefore, the evaluation process should be automated. Capture of usability data can also be automated to highlight important advantages such as [9]:

- Reducing the cost of evaluation: automating the evaluation reduces the time of the process and the number of participants.
- Improving the consistency: automation allows detecting all system errors.
- Improving the quality of evaluation: automation increases the amount of evaluation data, which contribute at increasing the quality and coverage of evaluation.
- Reducing the number of expert evaluation: the need for expert evaluator may be reduced by automation because of the increase in the amount of evaluation data.
- Possibility to compare different alternatives: automation can allow comparison by analytical modeling.

In order to identify websites usability, researchers in [10] proposed "USEful framework" to automate certain aspects of the usability evaluation and follows a set of proposed guidelines. Results showed that the proposed system detected more usability violations than the heuristic evaluation of Nielsen [3]. However, it failed in covering the specific recommendations of the website which were included by Nielsen. Automatic capture of user behaviors can be done using intelligent agents and these data can be analyzed to assess the UI. In [11], authors proposed an evolutionary multi-agent system for semi-automated study of websites usability. It aims to detect usability problems by simulating the browsing process of users and analyzing the usability of web pages. For this reason, they proposed a system based on two types of specialist agents named "HTML Analyser Agent" and "User Agent". They tested the prototype of the proposed system on one of the oldest and best known internet sites. Results showed that the system succeeded to automatically draws conclusions and suggestions to improve the usability of the website and to facilitate the human expert task. However, since the proposed system focuses only on HTML codes, it is unable to analyze the possible impact that server-side code might have on usability.

2.4 Synthesis

Table 1 presents a comparative survey that analyzes and summarizes some features of the previously studied works. This table is mainly based on a set of criteria such as: (1) Users integration into the evaluation loop. (2) Expert participation in the evaluation loop. (3) Interpretation of the metrics involved in the evaluation process (4) evaluation location in the development cycle (5) Evaluation type and (6) the nature of the data used in the evaluation loop. Despite the deversity of these approaches, all of them are not interpretable (see column 3). This encourages us to investigate and propose a new approach of automatic and interpretable evaluation.

Table 1. Comparative analysis of the mentioned evaluation approaches

Criteria						
(1)	(2)	(3)	(4)	(5)	(6)	Ref.
Yes	Yes	No	Execution	Static	Qualitative	[4]
Yes	Yes	No	Execution	Static	Qualitative	[2]
Yes	Yes	No	Execution	Dynamic	Qualitative	[5]
Yes	Yes	No	Execution	Static	Qualitative	[6]
No	Yes	No	Execution	Static	Quantitative	[7]
No	Yes	No	Design	Static	Quantitative	[8]
Yes	No	No	Execution	Dynamic, Automatic	Qualitative	[11]
No	No	No	Execution	Dynamic, Semi-automatic	Qualitative	[10]

3 Automatic Tool for HMI Evaluation in Interactive Systems

To evaluate the HMI of an interactive system, it is important to adopt among others, the criteria of ergonomics. In fact, the usability of interactive systems was widely defined in literature and norms. For instance, ISO/IEC 25010[2] norm define three main usability criteria: fiability, efficiency and satisfaction.

---

[2] Systems and software engineering – Systems and software Quality Requirements and Evaluation (SQuaRE) – System and software quality models, 2011. International

3.1   Adopted Process for HMI Evaluation: ISO/IEC 25066 (2016)

The purpose of our method is to avoid the intervention of experts and to provide a full participation of the user in the evaluation process. Figure 1 presents an overview of our automated process to obtain a subjective assessment using the questionnaire tool. It is based on the improved standard version ISO/IEC 25066 witch integrates ISO/IEC 25040 (2011). Our method consist of three main phases, Input Data phase (collection of information), Analysis phase which consist of five steps and Generation of evaluation report phase. Based on the standardized questionnaires [12], the criteria adopted for our approach to evaluate HMI in the system of supervision of a complex process related to urban transport of Valenciennes are the following: Satisfaction, Learnability, ergonomics of user interface, and the accessibility.

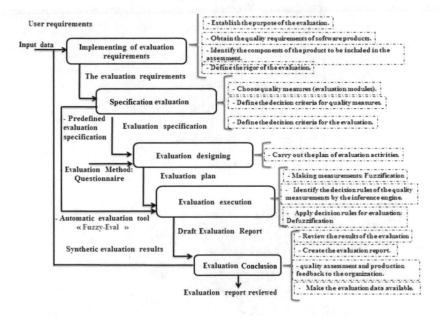

Fig. 1. Evaluation process proposed on the basis of ISO/IEC 25040 (2011) (integrated in the standard ISO/IEC 25066 (2016))

3.2   Fuzzy-Eval: A Qualitative HMI Fuzzy Evaluation

As illustrated in Fig. 2, users of the Information Assistance System (IAS) [13] answer our automated questionnaire to stimulate the HMI evaluation of this system. All responses are automatically stored in the dashboard of our questionnaire application in form of an EXCEL file. These responses follow an automatic evaluation process in order to assess the usability of an interactive system by communicating a usability report to the designers in order to improve its ergonomic quality.

Organization for Standardization, Geneva, Switzerland. Disponible dans: https://www.iso.org/standard/23779.html (consulté le 09/06/2016).

636   I. Kallel et al.

Since the usability evaluation is a naturally uncertain measure, we apply the
fuzzy reasonning in order to infer a value within the closed interval [0,1] in one
of the ranges: very low, low, medium, high and very high. We define a question
for each usability criterion, and thus a membership function in the evaluation
system. For example for user satisfaction: "What do you think about the information quality that the system gives to you?" To collect users' opinions on this
ergonomic criterion, we have defined a membership function "Satisfaction" of
three linguistic variables: Low, Medium, and High (see Eq. 1x). In addition, the
usability of IAS can be evaluated with a value between 0 and 1 in one of the
ranges, very low, low, medium, high and very high (see Eq. 2).

$$
\mu_i(x) = \begin{cases} \dfrac{x}{0.4}, 0 \le x \le 0.4 \\ \dfrac{0.65 - x}{0.25}, 0.3 \le x \le 0.65 \\ \dfrac{1 - x}{0.42}, 0.58 \le x \le 1 \end{cases} \tag{1}
$$

$$
\mu_0(x) = \begin{cases} \dfrac{x}{0.25}, 0 \le x \le 0.25 \\ \dfrac{0.45 - x}{0.20}, 0.15 \le x \le 0.45 \\ \dfrac{0.68 - x}{0.30}, 0.38 \le x \le 0.68 \\ \dfrac{0.82 - x}{0.42}, 0.58 \le x \le 0.82 \\ \dfrac{1 - x}{0.25}, 0.75 \le x \le 1 \end{cases} \tag{2}
$$

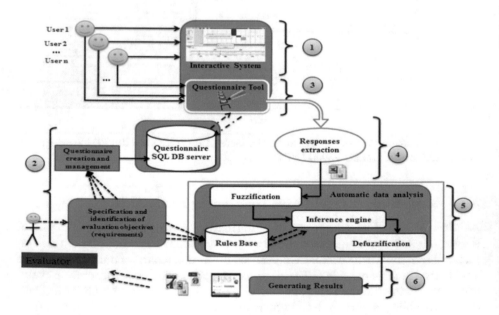

Fig. 2. The proposed qualitative HMI fuzzy evaluation method Fuzzy-Eval

4   Results and Discussion

Figure 3 illustrates the results of the simulation of different rule surfaces. It shows the relationship between usability criteria (Satisfaction, Learnability, etc.) and their impact on the final value of usability (i.e.: If (Satisfaction is high) & (Learnability is high) & (Accessibility is high) & (Aesthetics is high) then (Usability is high)). For example, from the first simulation, we can see that when the Satisfaction and Learnability values between 0 and 0.5, the usability value will be less than 0.35. However, when they are greater than 0.5, the final value of usability closer to 0.6.

In order to test and validate the automatic HMI evaluation tool, we have integrated our assessment system in the control system of the urban transport network (bus/tramway) Valenciennes (France).

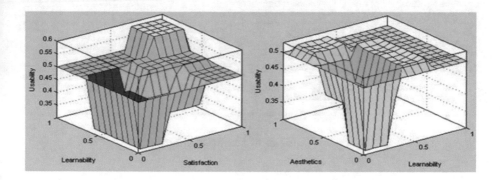

Fig. 3. Simulation of fuzzy rules surfaces

As presented in Fig. 4, Fuzzy-Eval allows the generation of three types of evaluation results: graphical result, textual result (table form), and arithmetic result. We can say that "Fuzzy-Eval" is quite promising in the field of evaluation of interactive complex systems and goes further than measuring usability. The proposed approach allows to evaluate the interactive systems. So, Fuzzy-Eval is Objective and systematic subjective. This hybridization makes the evaluation of fuzzy-Eval less time-consuming. Also, it is possible detect defects in usability and it is easy to translate the results into recommendations for design. However, it makes it possible to detect defects in usability and is more flexible, easy to administer than interviews, it is easy to translate the results into recommendations for design, and it is also adapted to the a priori evaluations of Usability, so our method is objective and systematic. However, the proposed tool may have some disadvantages, among which we can cite that a disgusted user can answer the questionnaire in a random manner without any reflection from where the survey loses its credibility (Table 2). The table below represents the position of our approach according to an analysis grid defined previously in Sect. 2 (see Table 1).

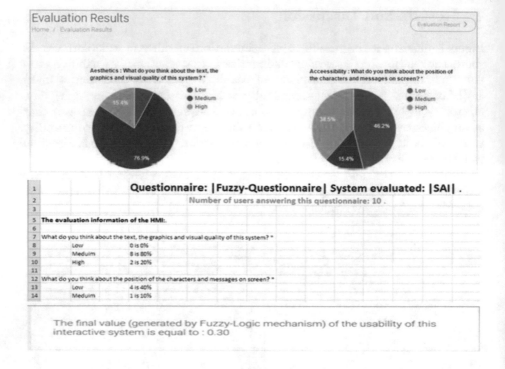

Fig. 4. Evaluation results interfaces

Table 2. Categorization of Fuzzy-Eval according the literature synthesis in Table 1

	(1)	(2)	(3)	(4)	(5)	(6)
Fuzzy-Eval	Yes	No	Yes	Execution	Dynamic/Automatic	Qualitative

5   Conclusion and Future Works

In this paper, we have proposed an automatic tool, called Fuzzy-Eval, to assess the usability of interactive complex systems applied to the control system of the urban transport network (bus/tramway) Valenciennes (France). Users interact with the interactive system by making decisions that have a direct impact on the overall regulation of the public transport network. Generally, the complexity of the architectures of supervisory systems makes their evaluation more complicated. Therefore, we proposed to hyridize a fuzzy evaluation with a questionnaire tool to determine the usability of a distributed system. This hybridization makes the evaluation of fuzzy-Eval less time-consuming. Also, it is possible detect defects in usability and it is easy to translate the results into recommendations for design. As perspectives, we intend to use ontology in order to integrate a context-based questionnaire tool.

Acknowledgment. The first author acknowledges the fruitful discussion with Rim Rekik and Abir Abid to improve the results. The authors acknowledge also the financial support of this work by grants from Tunisian General Direction of Scientific Research (DGRST) under the ARUB program.

References

1. Paz, F., Paz, F.A., Villanueva, D., Pow-Sang, J.A.: Heuristic evaluation as a complement to usability testing: a case study in web domain. In: 2015 12th International Conference on Information Technology-New Generations (ITNG), pp. 546–551. IEEE (2015)
2. Paz, F., Paz, F.A., Pow-Sang, J.A.: Evaluation of usability heuristics for transactional web sites: a comparative study. In: Information Technology: New Generations, pp. 1063–1073. Springer, Cham (2016)
3. Nielsen, J.: The use and misuse of focus groups. IEEE Softw. **14**(1), 94–95 (1997)
4. Paz, F., Pow-Sang, J.A.: Usability evaluation methods for software development: a systematic mapping review. In: 2015 8th International Conference on Advanced Software Engineering & Its Applications (ASEA), pp. 1–4. IEEE (2015)
5. Assila, A., Ezzedine, H., de Oliveira, K.M., Bouhlel, M.S.: Towards improving the subjective quality evaluation of human computer interfaces using a questionnaire tool. In: 2013 International Conference on Advanced Logistics and Transport (ICALT), pp. 275–283. IEEE (2013)
6. Callahan, E., Koenemann, J.: A comparative usability evaluation of user interfaces for online product catalog. In: Proceedings of 2nd ACM conference on Electronic commerce, pp. 197–206. ACM (2000)
7. Al Zarqa, A., Ozkul, T., Al-Ali, A.: A study toward development of an assessment method for measuring computational intelligence of smart device interfaces. Int. J. Math. Comput. Simul. **8**, 87–93 (2014)
8. Lim, C., Song, H.-D., Lee, Y.: Improving the usability of the user interface for a digital textbook platform for elementary-school students. Educ. Technol. Res. Dev. **60**(1), 159–173 (2012)
9. Campos, J.C., Fayollas, C., Martinie, C., Navarre, D., Palanque, P., Pinto, M.: Systematic automation of scenario-based testing of user interfaces. In: Proceedings of 8th ACM SIGCHI Symposium on Engineering Interactive Computing Systems, pp. 138–148. ACM (2016)
10. Dingli, A., Mifsud, J.: Useful: a framework to mainstream web site usability through automated evaluation. Int. J. Hum. Comput. Interact. (IJHCI) **2**(1), 10 (2011)
11. Mosqueira-Rey, E., Alonso-Ríos, D., Vázquez-García, A., del Río, B., Moret-Bonillo, V.: A multi-agent system based on evolutionary learning for the usability analysis of websites. In: Nguyen, N.T., Jain, L.C. (eds.) Intelligent Agents in the Evolution of Web and Applications, pp. 11–34. Springer, Berlin (2009). https://doi.org/10.1007/978-3-540-88071-4_2
12. Berkman, M.I., Karahoca, D.: Re-assessing the usability metric for user experience (UMUX) scale. J. Usabil. Stud. **11**(3), 89–109 (2016)
13. Ezzedine, H., Trabelsi, A., Kolski, C.: Modelling of an interactive system with an agent-based architecture using Petri nets, application of the method to the supervision of a transport system. Math. Comput. Simul. **70**(5), 358–376 (2006)

A Novel Hybrid GA for the Assignment of Jobs to Machines in a Complex Hybrid Flow Shop Problem

Houda Harbaoui[1,3]([✉]), Soulef Khalfallah[2], and Odile Bellenguez-Morineau[3]

[1] Institut Supérieur d'Informatique et des Techniques de Communication,
University of Sousse, Rue G.P.1, 4011 Hammam Sousse, Tunisia
houda.harbaoui@ymail.com
[2] Institut Supérieur de Gestion de Sousse, University of Sousse,
rue Abdlaaziz il Behi, BP763, 4000 Sousse, Tunisia
soulefk2004@yahoo.fr
[3] Institut Mines Telecom Atlantique, LS2N, UMR CNRS 6004, 4 rue Alfred Kastler,
La Chantrerie, BP20722, 44307 Nantes Cedex 3, France
odile.bellenguez@imt-atlantique.fr

Abstract. This paper, investigates a complex manufacturing production system encountered in the food industry. We consider a two stage hybrid flow shop with two dedicated machines at stage1, and several identical parallel machines at stage 2. We consider two simultaneous constraints: the sequence dependent family setup times and time lags. The optimization criterion considered is the minimization of makespan. Given the complexity of problem, an hybrid genetic algorithms (HGA) based on an improving heuristic is presented. We experimented a new heuristic to assign jobs on the second stage. The proposed HGA is compared against a lower bound (LB), and against a mixed integer programming model (MIP). The results indicate that the proposed hybrid GA is effective and can produce near-optimal solutions in a reasonable amount of time.

Keywords: Hybrid flow shop · Sequence dependent family setup
Hybrid genetic algorithm · Time lag · New heuristic
Dedicated machine

1 Introduction

This paper is concerned with a manufacturing configuration encountered in the food industry. The problem is to minimize total completion time in a two-stage hybrid flow shop scheduling. There are two dedicated machines at stage 1: generally production machines, and several parallel identical machines at stage 2; generally drying cabins. This problem was for the first time addressed in IEEE conference, by [1]. It could be defined as follows: at time zero, a set J of n jobs; $J = 1, 2, \ldots, n$ is waiting to be processed. They are first processed on two dedicated machines at stage 1 called m_1^1 and m_2^1. After completion time at stage 1,

© Springer International Publishing AG, part of Springer Nature 2018
A. Abraham et al. (Eds.): ISDA 2017, AISC 736, pp. 640–649, 2018.
https://doi.org/10.1007/978-3-319-76348-4_62

a job is processed on one of the m2 parallel machines at stage 2. Because of the dedicated nature of the machines at the first stage, the n jobs are regrouped on two disjoint families of jobs. Each family of jobs contains several groups. Furthermore, each group of jobs needs a setup time at stage 1. This configuration is known in the literature as sequence dependent family setup. Indeed, there are sequence-dependent setup times between groups of jobs, especially when the job processing is based on group technology (GT) where two successive jobs do not belong to the same group, will need a setup time. In this paper the above mentioned condition is considered, which means sequence dependent setup times with respect to processing jobs is considered on a GT basis.

After completion time at stage 1, certain jobs are allowed to wait for an interval of time before processing on the second stage. In this case, a maximal time lag is considered. Other jobs are "No-wait", and must begin processing at stage 2 immediately when they leave stage 1. In this case the maximum time lag will be null. Finally no job pre-emption is allowed.

The rest of this paper is organized as follows: In Sect. 2, a literature review is provided. Section 3 presents a mixed integer model for the problem. Section 4 is devoted to the solution methodology. Computational analysis follows in Sect. 5. Finally, some concluding remarks are presented.

2   Related Works

Using $\alpha/\beta/\gamma$ notation, our problem could be noted as $F2$ $(PD_m, P_m)/ST_{sd}$, max timelag$/C_{max}$ where ST_{sd} represents sequence dependent setup time. To our knowledge, this problem with the configuration described above, was considered in the literature by [1]. Thus, in this review we will concentrate on related problems. First, it is worth mentioning that the two stage hybrid flow shop was proved to be NP-hard by [2]. Since our problem is a HFS, with additional constraints, it is also NP-hard. In the following, we first address the literature related to the hybrid flow-shop with dedicated machines, and then we go through the literature in relation with the hybrid flow shop with sequence dependent setup and/or time lag.

1. **Hybrid flow shop with dedicated machines**

   The two stage hybrid flow shop with two dedicated machines at stage 1 was studied by [3]. The authors proposed two integer programming formulations and presented lower bounds for the objective function. They also presented two heuristics to find approximate solutions for large-size problems. A hybrid approach combining a heuristic approach with $(B\&B)$ was proposed by [4], in order to minimize the makespan in a two-stage hybrid flow shop scheduling problem with dedicated machines at the second stage. They derived several lower bounds and used four constructive heuristics to obtain initial upper bounds, and then they enhanced the performance of the proposed heuristic method by using three dominance properties.

   More recently, Lin [5] studied a two stage hybrid flow shop with a common machine at stage one and two parallel dedicated machines at stage two. They first developed a linear-time algorithm to solve the case where two sequences of the two job types are given a priori. The authors aimed to minimize the maximum lateness.

2. **Flow shop with sequence dependent setup and/or time lag**
   Several researchers studied the sequence dependent setup (SD_{ST}) and/or time lag for the hybrid flow shop. Zandieh et al. [6] presented an immune algorithm and they showed that their algorithm outperforms the random keys genetic algorithm of [7]. Ruiz et al. [8,9] considered a complex $SDST$ flexible flow shop problem, where the machines in each stage are unrelated, and some machines are not eligible to perform some jobs. Also, Ebrahimi et al. [10] considered the sequence dependent family setup time in a k stage hybrid flow shop configuration. They proposed two procedures based on GA in order to minimize the makespan and the total tardiness simultaneously. They also assumed that the due date is uncertain. Several authors used GA for SD_{ST} hybrid flow shop with or without time lags, we could cite Naderi et al. [11], Ziaeifar et al. [12] and Ruiz and Maroto [8]). Other researchers, such as Wang et al. [13], Huang et al. [14] and [15] worked on the no-wait two-stage hybrid flow shop. Recently, Pan et al. [16] worked on SD_{ST} hybrid flow shop and they proposed a total of nine algorithms to minimize the makespan.

3   A Mixed Integer Model for the Problem

In this section, we propose a new mixed-integer program model (MIP) for the considered problem. To evaluate the quality of the solution, we describe 4 lower bounds, as presented next. The tighter LB will be used to evaluate the proposed MIP. This new model is more effective from the one proposed in [1]. In fact, we used fewer constraints and fewer binary variables than model proposed in [1].

3.1   Notations

- **Sets**
  The following notations are used in this paper:
  J set of jobs, where $J = \{1 \ldots n\}$
  J_1 set of jobs with first job as dummy $\{0\} \cup J$
  J_2 set of jobs with last job as dummy W where $J_2 = n + 1$
  n number of jobs to be scheduled ($i = 1, 2, \ldots n$)
  K set of stages where $K = \{1, 2\}$
  m_k number of parallel machines at stage k
  $n1$ number of jobs to be scheduled on the first dedicated machine
  $n2$ number of jobs to be scheduled on the second dedicated machine
- **Parameters**
  $P_{i,k}$ processing time of job i on stage k
  $S_{i,k}$ set up time of job i on stage k
  $L_{i,k}$ time lag of job i from stage k to k + 1
  $Y_i^{m,k}$ Binary matrix of pre-assignment of job i to machine m at stage k

$B_{i,j}$ 1 if job i belongs to same group as job j
ST_G setup of group G
G_1:   number of group on machine 1
G_2:   number of group on machine 2.

3.2   Decision Variables

$t_{i,k}$: the starting time of job i at stage k

C_{max}: Makespan or completion time of all jobs

$$x_{ij}^{m,k} = \begin{cases} 1 \text{ if job } j \text{ is excecuted exactly after job } i \text{ on machine m at stage k} \\ 0 \text{ otherwise} \end{cases}$$

3.3   Scheduling Model

$$Min \quad C_{max}$$

$$\sum_{m=1}^{M_k} x_{i0}^{m,k} \leq 1, \qquad i = 1 \ldots n, k \in K \tag{1}$$

$$\sum_{m=1}^{M_k} x_{0i}^{m,k} \leq 1, \qquad i = 1 \ldots n, k \in K \tag{2}$$

$$\sum_{i=0}^{n} \sum_{m=1}^{M_k} x_{ij}^{m,k} . Y_j^{m,k} = 1, \qquad j = 1 \ldots n, i \neq j, k \in K \tag{3}$$

$$\sum_{j=0}^{n} \sum_{m=1}^{M_k} x_{ij}^{m,k} . Y_i^{m,k} = 1, \qquad i = 1 \ldots n, i \neq j, k \in K \tag{4}$$

$$\sum_{\substack{i=0 \\ i \neq j}}^{n} x_{ij}^{m,k} . Y_j^{m,k} - \sum_{\substack{i=0 \\ i \neq j}}^{n} x_{ji}^{m,k} . Y_i^{m,k} = 0, \qquad j = 1 \ldots n, m = 1 \ldots M_k, k \in K \tag{5}$$

$$t_{j,k} \geq t_{i,k} + P_{i,k} - M(1 - x_{ij}^{mk}) + S_{j,k} . B_{ij}, \quad i = 0 \ldots n, j = 1 \ldots n, m = 1, \ldots, m_k, k \in K, i \neq j \tag{6}$$

$$t_{i,2} \leq t_{i,1} + P_{i,1} + L_{i,1}, \qquad i = 1 \ldots n \tag{7}$$

$$t_{i,2} \geq t_{i,1} + P_{i,1}, \qquad i = 1 \ldots n \tag{8}$$

$$C_{max} \geq t_{i,2} + P_{i,2}, \qquad i = 1 \ldots n \tag{9}$$

$$x_{ij}^{mk} \in \{0,1\}, m = 1, \ldots, m_k, i = 1 \ldots n, j = 1 \ldots n, k \in K, i \neq j \tag{10}$$

$$C_{max}, t_{i,k} \geqslant 0, \ i = 1 \ldots n, k \in K \tag{11}$$

M: a big number

Constraint (1) and (2) ensure that the fictitious job 0 has only one successor and one predecessor. **Constraint** (3) **and** (4) state that all jobs must be processed exactly once on each machine at each stage. **Constraint** (5) ensures that each job i has at most one predecessor and one successor. **constraint** (6) expresses the disjunctive constraints on each machine. **Constraint** (7) **and** (8) ensure that no job can be processed to the second. stage unless its execution on the first stage is completed plus a maximal time lag. **Constraint** (9) implies that the completion time of the last job must be greater or equal to the end of processing of any job. **Constraint** (10) declares x_{ij}^{mk} binary variable. **Constraint** (11) specifies the non negativity of $C_{max}, t_{l,k}$

4   Solution Methodology

The solution methodology is based on hybridisation on genetic algorithm with local search heuristic. Since the problem is NP-Hard; several hybrid procedures based on a local search and GA are proposed. In order to evaluate the quality of the solution, and since no benchmark for the problem is available, we recall the lower bound in [1].

4.1   Proposed Genetic Algorithms

Genetic algorithms proved their efficiency in solving optimization problems. They are inspired from the Darwinian natural selection theory. The basic concepts are described by [17]. We propose a new version of GA: Hybrid Genetic Algorithm (HGA).

Job	4	7	1	3	2	8	6	5	10	9
Machine (S1)	1	1	2	2	2	1	2	1	1	2

Machine (S2)	1	2	1	3	2	1	1	2	3	3

Fig. 1. Solution encoding

1. **Encoding solution and initialization of the population**
   For HGA, the solution is represented by $2*n$ matrices on stage 1, and a permutation on stage 2 which represents a sequence of machines on stage 2. In our case the assignment of jobs to machines at stage 2 is done simultaneously with generation of jobs in each permutation. As illustrated in the following example: *10 jobs, 2 machines on stage 1, and 3 machines on stage 2*. Jobs *4,7,8,5 and 10* are processed on *M1 of S1* (Stage 1) and the other are processed on *M2 of S1*. To obtain sequences on parallel machine on stage 2, we proposed a heuristic called HAPM() (heuristic of assignment to parallel machines). There are 2 phases to follow, at phase I, 50% of jobs are assigned

randomly to parallel machines under the assumption that no machine receives more than $2 * (1/m)\%$ of total processing time (20% in our case). In phase II, assign the rest of jobs to machines according to the least total processing time. The HAPM() is presented below (Algorithm 1). About initial population, HGA generate an initial population by generating 2 sub-populations: first sub-population generated by dispatching rules, and second sub-population generated randomly. For each individual, we call HAPM() 3 times and we keep the best sequence of parallel machine (Fig. 1).

Input: π: permutation stage 1
Result: sequence on each parallel machine
`/* `TPT_m`: total processing time on parallel machine m           */`
i, m, n, job, TPT, TPT_m: Integer
$i, j \longleftarrow 0$
Phase I
while $n \leq \frac{n}{2}$ **do**
$\quad$ job$\longleftarrow \pi[i]$
$\quad$ **repeat**
$\quad\quad$ | generate randomly m (number of parallel machine);
$\quad$ **until** $TPT_m \leq 0.2 * TPT$;
$\quad$ assign job i to machine number m;
$\quad$ $i{+}{+}$
end
Phase II
for $(i = (\frac{n}{2} + 1); i \leq n; i{+}{+})$ **do**
$\quad$ | assign job to the machine with the lowest TPT ;
end

$\quad\quad$ **Algorithm 1.** HAPM: Heuristic to assign job to parallel machine

---

2. Selection, crossover and mutation operators

In this work, we experiment well-known genetic operators. The roulette wheel method is used to select two parents from the current population for the crossover. Then, the elitism selection is applied to choose the best individuals among all individuals. In this work, we tested the operators RMPX proposed by [11] of parallel machines.

3. Fitness evaluation

In this phase, the fitness of each individual of HGA is based on recursive inequalities (linear program (PL)), taken from the proposed MIP cited above:

$$t_{j,k} \geq t_{i,k} + P_{i,k} + S_{j,k}.B_{ij}, i = 1 \ldots n, j = 1 \ldots n, k = \{1, 2\}, i \neq j,$$

$\forall j$ immediate successor of i, read directly from the solution

$$t_{i,2} \leq t_{i,1} + P_{i,1} + L_{i,1}, \qquad i = 1 \ldots n$$
$$t_{i,2} \geq t_{i,1} + P_{i,1}, \qquad i = 1 \ldots n$$
$$C_{max} \geq t_{i,2} + P_{i,2}, \qquad i = 1 \ldots n$$

The PL will browse sub-sequences of jobs immediately successor on each machine, and then, calculate starting processing date of each job, by respecting setup time and time lag between two stages.

4. **Local search phase**

In this phase, we hybrid The GA with an iterative local search (ILS) algorithm. We aim to improve the population by escaping from local optima. ILS is applied to 25% of the last generation.

5   Computational Analysis

In this section, we present the computational experiments. we will start with a presentation of the instances, and then we discuss the results generated by the proposed HGA. The solutions are compared to the lower bound and to the MIP output.

5.1   Data Generation and Settings

Since there is no benchmark available for this problem, we generated randomly 30 instances which describe various conditions. These instances depend on the number of jobs, the number of machines at stage 2, the distribution of processing times, the distribution of sequence-dependent family setup, and the distribution of time lag's. The following parameters are considered to construct these instances:

Number of stages $|K| = 2$
Number of jobs n: $(n = 10, 20, 50,$ and $100)$
Distribution of the *setup times*: *uniform* $[5 \ldots 20]$
Distribution of the *time lags*: *uniform* $[0 \ldots 30]$
Number of dedicated machines on stage 1: 2
Number of parallel machines on stage 2: 10
Briefly, there are two classes of problems. The processing times at the first stage and at second stage are as following:
Class 1: $[1, 40]$ at stage 1 and $[5, 200]$ at stage 2
Class 2: $[40, 80]$ at stage 1 and $[200, 400]$ at stage 2
These instance have been randomly generated using a uniform distribution.

5.2   Parametrization of the GA

To calibrate the GA, we chose the other parameters experimentally as following:

- Number of generation $= 200$
- Probability of crossover $p_c = 0.9$
- Probability of mutation $p_m = 0.2$
- Population size $= 100$
- Stopping criterion: maximum CPU time $(300 \,\mathrm{s})$, maximum number of generations

Given the probabilistic aspect of the algorithm, we lunch five runs for each instance.

5.3   Results and Discussion

In this section, we aim to study the performances of the MIP and the GA. At first, we run the mathematical model for 1 hour using CPLEX and we compared the generated results to the lower bound LB proposed by [1]. Secondly, we analysed the performance of GA by studying the effect of the local search on it. Furthermore, we compared the best solution to LB. In the following, we define the relative deviation RD, RD1, RD2 and AT as follows:

- RD: the relative deviation between CPLEX solution $CPLEX_{sol}$ and the lower bound LB:

$$RD = \frac{CPLEX_{sol} - LB}{LB} \times 100$$

- $RD1$: the relative deviation between best solution $best_{solGA}$ found by GA without ILS and LB:

$$RD1 = \frac{best_{solGA-LB}}{LB} \times 100$$

- $RD2$: the relative deviation between best solution $best_{solHGA}$ found by GA with ILS and LB:

$$RD2 = \frac{best_{solHGA} - LB}{LB} \times 100$$

- AT: the average run time of the algorithm

Firstly, we present the results of the MIP (Table 1).

Table 1. Average of the deviation (%) according to the number of jobs

Instance	Class 1		Class 2	
	CPLEX	AVG (time)	CPLEX	AVG (time)
10	0.0	7.1	0.0	21.2
20	13.8	3608.8	3.7	3600.0

We can see that CPLEX can find optimal solution of small instances (problems of 10 jobs), so the average of RD (ARD) is null. For the problems of 20 jobs, ARD is high only for the first class (14%). In addition, ARD is very low in regards to class 2 (3.7%) However, the solver can't find solution for the instances with big size in 2 classes.

In Table 2, we compare the results obtained by GA without ILS and GA with ILS for the different families of instances in the 2 classes. The first column represents $|J|$, the second column represents average of RD1 (ARD1) found by GA without ILS, the third column represents AT and the fourth column represents average of RD2 (ARD2) found by GA with ILS.

Table 2. Average value of the deviation (%) according to the number of jobs in 2 classes

Instance	Class 1				Class 2			
	GA		GA+ILS		GA		GA+ILS	
	ARD1	AT	ARD2	AT	ARD1	AT	ARD2	AT
10	0.3	0.0	0.0	0.0	0.2	0.0	0.0	0.0
20	10.85	1.0	8.27	1.0	2.69	2.0	1.71	3.0
50	10.67	3.0	7.22	5.0	3.98	5.0	1.87	15.0
100	12.13	8.0	8.12	76.0	7.86	12.0	2.75	75.0

As we can see in Table 2, ARD1 did not exceed 12% in the 2 classes. This gap confirms the effectiveness of the applied operators and the fixed parameters. The application of ILS improved ARD1. This shows that the GA with ILS improved the quality of the solutions found for all instances of jobs in the 2 data sets. Besides, the low AT shows that HGA converged quickly to the best solutions. Indeed, AT is null for instances of 10 jobs in the 2 classes. In addition, it did not exceed 15 s for instances of 20 and 50 jobs. For instances of 100 jobs, AT does not exceed 80 s.

ARD2 found is null for all the instances of 10 jobs, this means that HGA found optimal solutions. Also, ARD2 is very small for instances of 20, 50 and 100 jobs in class 2. In this case ARD generated solutions near to optimum (maximum ARD2 is 2.75%). However, ARD remains slightly low for instances of the first class (8%). This result is explained by the fact that the heuristic used to assign jobs HAPM() is limited in the first class while it is effective in the second class.

5.4   Conclusion

In this paper, we presented a new mathematical model and a new version of hybrid genetic algorithm for the two stage hybrid flow shop with dedicated machines. We considered two additional constraints: sequence dependent family setup and time lag. Experimental results showed the efficiency of the proposed approach. The results are encouraging to develop our investigation to other evaluation criteria or other methods.

References

1. Harbaoui, H., Bellenguez-Morineau, O., Khalfallah, S.: Scheduling a two-stage hybrid flow shop with dedicated machines, time lags and sequence-dependent family setup times. In: 2016 IEEE International Conference on Systems, Man, and Cybernetics (SMC), pp. 002990–002995 (2016)
2. Gupta, J.N.D.: Two-stage hybrid flow shop scheduling problem. J. Oper. Res. Soc. **39**, 359–364 (1988)

3. Chikhi, N., Benmansoury, R., Bekrarz, A., Hanafiy, A., Abbas, M.: Makespan minimization for two-stage hybrid flow shop with dedicated machines and additional constraints. In: 9th International Conference of Modeling, Optimization and Simulation - MOSIM 2012, 6–8 June 2012, Bordeaux, France (2012). https://hal.archives-ouvertes.fr/hal-00728687
4. Wang, S., Liu, M.: A heuristic method for two-stage hybrid flow shop with dedicated machines. Comput. Oper. Res. **40**, 438–450 (2013)
5. Lin, B.M.T.: Two-stage flow shop scheduling with dedicated machines. Int. J. Prod. Res. **53**(4), 1094–1097 (2015)
6. Zandieh, M., Fatemi Ghomi, S.M.T., Moattar Husseini, S.M.: An immune algorithm approach to hybrid flow shops scheduling with sequence-dependent setup times. Appl. Math. Comput. **180**, 111–127 (2006)
7. Kurz, M.E., Askin, R.G.: Scheduling flexible flow lines with sequence-dependent setup times. Eur. J. Oper. Res. **159**, 66–82 (2004)
8. Ruiz, R., Maroto, C.: A genetic algorithm for hybrid flow shops with sequence dependent setup times and machine eligibility. Eur. J. Oper. Res. **169**, 781–800 (2006)
9. Ruiz, R., Maroto, C., Alcaraz, J.: Solving the flow shop scheduling problem with sequence dependent setup times using advanced metaheuristics. Eur. J. Oper. Res. **165**, 34–54 (2005)
10. Ebrahimi, M., Fatemi Ghomi, S.M.T., Karimi, B.: Hybrid flow shop scheduling with sequence dependent family setup time and uncertain due dates. Appl. Math. Model. **38**, 2490–2504 (2014)
11. Naderi, B., Ruiz, R., Zandieh, M.: A two stage flow shop with parallel dedicated machines. In: 8th International Conference of Modeling and Simulation, MOSIM 2010, Hammamet, Tunisia. IEEE Explore (2010)
12. Ziaeifar, A., Moghaddam, R.T., Pichka, K.: Solving a new mathematical model for a hybrid flow shop scheduling problem with a processor assignment by a genetic algorithm. Int. J. Adv. Manuf. Technol. **61**, 339–349 (2012)
13. Wang, S., Liu, M.: A genetic algorithm for two-stage no-wait hybrid flow shop scheduling problem. Comput. Oper. Res. **40**, 1064–1075 (2013)
14. Huang, R.H., Yang, C.L., Huang, Y.C.: No-wait two-stage multi processor flowshop scheduling with unit setup. Int. J. Adv. Manuf. Technol. **44**, 921–927 (2009)
15. Sioud, A., Gravel, M., Gagné, M.: A genetic algorithm for solving a hybrid flexible flowshop with sequence dependent setup times. In: IEEE Congress on Evolutionary Computation, Cancan, Mexico. IEEE Explore (2013)
16. Pan, Q.-K., Gao, L., Li, X.-Y., Gao, K.-Z.: Effective metaheuristics for scheduling a hybrid flowshop with sequence-dependent setup times. Appl. Math. Comput. **303**, 89–112 (2017)
17. Holland, J.: Adaptation in Natural and Artificial Systems. University of Michigan Press, Ann Arbor (1975)

Selecting Relevant Educational Attributes for Predicting Students' Academic Performance

Abir Abid[1(✉)], Ilhem Kallel[1,2], Ignacio J. Blanco[3], and Mounir Benayed[1,4]

[1] REGIM-Lab: Research Groups in Intelligent Machines, ENIS, University of Sfax, BP 1173, 3038 Sfax, Tunisia
{abir.abid,ilhem.kallel,mounir.benayed}@ieee.org
[2] ISIMS: Higher Institute of Computer Science and Multimedia of Sfax, Route de Tunis Km 10 BP 242, 3021 Sfax, Tunisia
[3] University of Granada, 18071 Granada, Spain
iblanco@decsai.ugr.es
[4] Computer Science and Communications Department, Faculty of Sciences of Sfax, University of Sfax, Route Sokra Km 3.5 BP 1171, 3000 Sfax, Tunisia

Abstract. Predicting students' academic performance is one of the oldest and most popular applications of educational data mining. It helps to estimate the unknown evaluation of a student's performance. However, a huge amount of data with different formats and from multiple sources may contain a large number of features supposed as not-relevant that could influence the prediction results. The main objective of this paper is to improve the effectiveness of a predictive model for students' academic performance. For this purpose, we propose a methodology to carry out a comparative study for evaluating the influence of feature selection techniques on the prediction of students' academic performance. In our study, F-measure parameter is used to evaluate the effectiveness of the selected techniques. Two real data sources are used in this work, Mathematics and language courses. The outcomes are compared and discussed in order to identify the technique that has the best influence for an accurate predictive model.

1 Introduction

Applying data mining techniques on educational data has led to the apparition of the new emerging interdisciplinary research field called Educational Data Mining (EDM) [1]. These techniques provide the capability to explore, visualize and analyze large amounts of data in order to reveal valuable patterns in students' performance. One of the most popular applications of EDM is predicting students' academic performance [2]. It helps to improve the efficiency of learning systems. However, a huge amount of data with different formats may contain a large number of irrelevant features that could influence the prediction results. Furthermore, attributes within a dataset can be correlated, redundant or with no predictive information. Consequently, the data must be preprocessed to select an appropriate subset of attributes and ignore not-relevant and redundant ones.

This paper is organized as follows: The second section propose a review about feature selection techniques and their use in some different domains. We mainly focus on the use of these algorithms and their influence on the accuracy of the classification model as well as on the predictive model. Next, in the third section, we introduce the methodology adopted in this study as well as a description of the used dataset. Finally, we carried out a comparative study on four well known feature selection techniques.

2   Overview on Known Feature Selection Processes

Feature Selection is one of the prominent preprocessing steps in many research area such as pattern recognition, machine learning and data mining communities [3,4]. In fact, these techniques select the relevant features from the original feature set according to an evaluation criterion. It aims to reduce dimensionality, remove not-relevant or redundant data, increase learning accuracy, etc.

2.1   Presentation of the Selected Selection Techniques

Among several available feature selection techniques, we select to present in this paper four techniques, namely:

Correlation Based Feature Selection (CFS): [5] is a heuristic for evaluating the merit of features subset. Its main idea is selecting feature subsets including attributes that are highly correlated to the class, yet uncorrelated with each other. In other word, the heuristic handles not-relevant features as they will be poor predictors of the class. In addition, the redundant features will be excluded as they will be highly correlated with one or more of the remaining features. Equation 1 formalises the heuristic:

$$Merit_s = \frac{k\overline{r_{ca}}}{\sqrt{k + k(k-1)\overline{r_{aa}}}} \tag{1}$$

Where $Merit_s$ models the correlation between the summed attributes and the class variable. s is an attribute subset that has k attributes. $\overline{r_{ca}}$ is the average of correlation between the attributes and the class variable. Finally, $\overline{r_{ca}}$ is the average inter-correlation between attributes.

ReliefF (RF): is an instance-based attribute ranking. It was introduced by Kira and Rendell [6] and later enhanced by Kononenko [7]. This algorithm evaluates the worth of an attribute by ranking it with a value between -1 and 1, more positive weights indicates more the predictively of the attribute. As shown in the pseudo code (Fig. 1), for each instances m, Relief randomly sampling an instance from the dataset. Then, it locates its nearest neighbor (H_j) from the same and opposite class (M_j). Finally, it updates the relevance scores for each attribute by comparing the weight of the attributes of the nearest neighbors and the sampled instance.

```
Set all Weights W[A] = 0.0;
for i=1 to m do
    Begin ;
    Randomly select an instance R ;
    Find K nearest hits Hⱼ ;
    foreach class C ≠ class(R) do
    |  find K nearest misses Mⱼ(C)
    end
    for A=1 to ♯attributes do
```
$$W[A] = W[A] - \sum_{j=1}^{k} diff(A, R, H_j)/(m \times k) +$$
$$\sum_{c \neq class(R)} [\frac{P(C)}{1 - P(class(R))} \sum_{j=1}^{k} diff(A, R, M_j(C))/(m \times k)$$
```
    end
end
```

Fig. 1. A pseudo code of Relief feature selection [7]

Information Gain (IG): In probability and information theory, IG is a measure of the difference between two probability distributions. This method evaluates an attributes by measuring its information gained with respect to the class. In other word, IG is an amount by which the entropy of the class decreases reflects the additional information about the class provided by the attribute [8].

If A is an attribute and C is the class, the entropy of the class before and after observing the attribute is presented in Eqs. 2 and 3.

$$H(C) = - \sum_{c \in C} p(c) log_2 p(c) \tag{2}$$

$$H(C \mid A) = - \sum_{a \in A} p(a) \sum_{c \in C} p(c \mid a) log_2 p(c \mid a) \tag{3}$$

For each attribute A_i, a score is assigned based on the information gain between itself and the class (See Eq. 4).

$$\begin{aligned} IG_i \quad &= H(C) - H(C \mid A_i) \\ &= H(A_i) - H(A_i \mid C) \\ &= H(A_i) + H(C) - H(A_i \mid C) \end{aligned} \tag{4}$$

Symmetrical Uncertainty (SU): [9] uses an information theoretic measure to evaluate the rank of an attribute by Eq. 5. This method is symmetric in nature, $SU(A, C)$ is same as that of $SU(A, C)$, which help at reducing the number of comparisons required especially when A and C are two features.

$$SU(A, C) = 2[\frac{IG(A \mid C)}{H(A) + H(C)}] \tag{5}$$

Where $IG(A|C)$ is the information gain of feature A, that is an independent attribute and C is the class attribute. $H(A)$ is the entropy of attribute A and $H(C)$ is the entropy of feature/Class C. It's worth to mention that SU is not influenced by multivalued attributes as is the case for IG.

2.2   Literature Review of Some Feature Selection Methods

Feature selection process can play an important role in improving the performance of the learning algorithm. Several researchers have performed these techniques before applying classification algorithms in many application domains such as Education, Security, etc.

In [10] the authors carried out a comparative study of six well-known filter feature section algorithms in order to improve the effectiveness of a students' model that could predict their academic performance. They were able to define the best method and reach an optimal dimensionality of the feature subset. The finding of their work show a reduction in computational time and constructional cost in both training and classification phases of the student performance model and finally they deduced that "Information-Gain Attribute evaluation" is the best feature selection technique for their predictive model. Similarly, the main purpose presented in [11] is to enhance the results of predicting students' performance in the final semester examination. To achieve their goal, the authors analyzed the impact of feature selection techniques on the classification task in order to identify the best one. They compared four feature selection algorithms, and based on the results, they selected "CFS Subset Evaluator" as the best feature selection algorithms.

Authors in [12] carried out a comparative study on the effectiveness of four feature selection techniques to early predict students likely to fail in introductory programming courses. For this purpose, they applied the four techniques on two different and independent data sources on introductory programming courses. The outcome of this comparative study selected "Information Gain algorithm" since it presented the best results in both data sources.

In the work presented in [13], seven filters were applied on 11 synthetic dataset with a not-relevant features, redundancy and interdependency between attributes. The authors review the efficiency of feature selection methods and finally they were able to select "ReliefF algorithms" as the best method. Authors in [14] proposed a new feature selection technique based on random forest. Their aim was to select the most relevent and non-redundant features. They used the random forest to measure the relevance value attributes and the correlation coefficient to calculate the value of redundancy. The proposed approach has been tested and validated on nine different databases.

As a summary, from the previous works, we may notice that there is not a one common feature selection method which can be accurate for all dataset even for the same domain. For this reason, we select a set of four feature selection techniques in order to identify the best attribute selection algorithm for our predictive students' performance model.

3   A Feature Selection Methodology in Educational Dataset

The main purpose of this paper is to analyze the impact of feature selection techniques on the classification task in order to improve the performance of the predictive students' performance model. Accordingly, we consider four feature selection techniques, introduced in the previous section, to reduce the dimensionality of data are as follows: Correlation Based Feature Selection (CFS), Information Gain (IG), ReliefF (RF) and Symmetrical Uncertainty (SU). These algorithms are combined with five classifiers in order to evaluate their performance. Therefore, a classification model is built using RandomForest [15], REPTree [16], LogitBoost [17], JRip [16] and J48 [18].

3.1   Description of the Proposed Methodology

The methodology adopted for this comparative study is displayed in Fig. 2. The first step is transforming the available dataset into data format of the destination data mining system, which in this case from Excel to Attribute-Relation File Format (ARFF)[1] format of Weka [19]. As we suspected that the first and second grades of learners would have a high impact on the final results, we decided to divide the dataset into social and academic data. Then, by applying the feature selection techniques on the transformed data, a set of five features sets are the results of this step. Next, balanced data step is needed in order to deal with one of the important problem that may exist in educational data, which is unbalanced data [20]. This problem may occur when one class has a number of instances larger than the one in other classes. Thus, prediction algorithms may focus on learning from them [21]. After preprocessing data and applying the four feature selection techniques, the collection of five dataset are now ready to be the input of the classification model. Finally, the collected results are used to analyze the performance of each classification algorithms in order to select the best features.

3.2   Description of the Dataset

In this paper, two real datasets are used [22]. Each one has 33 attributes presenting students' academic and social data for a specific core classes: Mathematics (395 instances) and Portuguese language (649 instances). During the school year, students are evaluated in three periods and the last one corresponds to the final grade. A 20-point grading scale is used to evaluate student academic performance in each period, where 0 is the lowest grade and 20 is the perfect score. As the main purpose of this study is to select the best feature selection technique using a classification model. According to the Erasmus[2] grade conversion system, the final grade can be transformed into European Credit Transfer System (ECTS) grades. The student's final grade is discretized into five categorical classes consisting of Excellent, Good, Satisfactory, Sufficient and Fail.

---

[1] https://www.cs.waikato.ac.nz/ml/weka/arff.html.

[2] European exchange programme that enables student exchange in 31 countries.

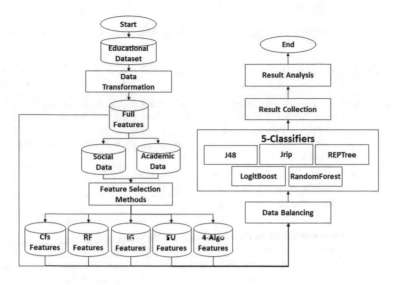

Fig. 2. Flow chart of the proposed methodology

4   Results and Discussions

The performance of a classification model highly depends on the dataset's features. Accordingly, in this section, we are going to apply the different steps of the methodology presented above in order to select the best feature selection technique based on the F-measure parameter [23] of each classifier. By applying the four selected feature selection techniques on both datasets, we collect four sets (Cfs features, Rf features, IG features, SU features) where each of them is the outcome of each techniques. Using these results, a "4-Algo features" set were generated. Each attribute within the full dataset that was selected by more than two feature selection techniques is a part of "4-Algo features". Table 1 (respectively Table 2), shows the final selected features for each set as well as the number of the selected attributes of Mathematics course dataset (respectively Language course dataset) the description of the features are detailed in [22]. According to these tables, each feature selection technique gives different results than each other, which make the decision of selecting one of them as the main feature selection technique for the predictive model harder. Thus, five well known classification algorithms were used to evaluate the obtained features. In this experiment, we used the data mining tool Weka [19] with the testing option "cross validatio" using ten-folds. Tables 3 and 4 show the performance of each feature selection method with each classification techniques by evaluating the "F-measure" for each classifier. We can notice that none of the algorithms shows exceptional results, despite of improving the performance of the five classifiers by applying features selection methods, as shown in Tables 3 and 4. These findings confirm the fact that reducing the number of redundant or irrelevant features improve the performance of the classifiers.

Table 1. Selected attributes using feature selection techniques for "Mathematics course dataset"

Feature subset	Attributes	No of features
Full features set	School, sex, age, address, famsize, Pstatus, Medu, Fedu, Mjob, Fjob, reason, guardian, traveltime, studytime, failures, schoolsup, famsup paid, activities, nursery, higher, famrel, freetime, internet, romantic, gout, Dalc, Walc, health, absences, G1, G2, G3	33
4-Algo features	School, Sex, Address, Mjob, famsize, paid, Failures, Schoolsup, Higher, Internet, famsup, Pstatus, G1, G2, G3	15
Cfs features	Sex, failures, G1, G2, G3	5
Rf features	G2, G1, Mjob, sex, Walc, Medu, paid, failures, studytime, schoolsup famsup, Dalc, address, Pstatus, higher, famsize, internet, health, absences, Fedu, age, school, goout, G3	24
IG features	G2, G1, failures, Mjob, schoolsup, Fjob, higher, reason, guardian, paid, romantic, address, sex, internet, famsize, nursery, school, Pstatus, activities, famsup, G3	21
SU features	G2, G1, failures, higher, schoolsup, Mjob, Fjob, reason, paid, guardian, romantic, address, sex, internet, famsize, nursery, school, Pstatus, activities, famsup, G3	21

Table 2. Selected attributes using feature selection techniques for "Language course dataset"

Feature subset	Attributes	No of features
Full features Set	School, sex, age, address, famsize, Pstatus, Medu, Fedu, Mjob, Fjob, reason, guardian, traveltime, studytime, failures, schoolsup, famsup paid, activities, nursery, higher, internet, romantic, famrel, freetime gout, Dalc, Walc, health, absences, G1, G2, G3	33
4-Algo features	G2, G1, failures, higher, school, studytime, schoolsup, internet, paid, activities, famsup, Sex, address, Medu, Mjob, famsize, Pstatus, G3	18
Cfs features	Studytime, failures, schoolsup, paid, activities, internet, G1, G2, G3	9
Rf features	G2, G1, Mjob, sex, Walc, Medu, paid, failures, studytime, address, schoolsup, famsup, Dalc, Pstatus, higher, famsize, internet, health, absences, Fedu, age, school, goout, G3	24
IG features	G2, G1, failures, higher, school, Mjob, Medu, studytime, reason, Fjob, Fedu, schoolsup, address, internet, guardian, sex, activities, paid, nursery, romantic, famsup, famsize, Pstatus, G3	24
SU features	G2, G1, failures, higher, school, Medu, studytime, Mjob, address, internet, guardian, sex, paid, activities, nursery, romantic, famsup famsize, Pstatus, G3	20

Table 3. Classification results for "Mathematics course dataset"

Feature subset	F-measure				
	Jrip	Random forest	J48	REP tree	Logit boost
Full features	0.745	0.697	0.716	0.739	0.716
4-Algo features	0.745	0.673	0.736	0.742	0.730
Cfs features	0.731	0.665	0.736	0.759	0.734
Rf features	**0.748**	**0.734**	**0.737**	**0.762**	0.704
IG features	0.747	0.703	0.727	0.727	**0.736**
SU features	0.741	0.692	0.727	0.731	**0.736**

Table 4. Classification results for "Language course dataset"

Feature subset	F-measure				
	Jrip	Random forest	J48	REP tree	Logit boost
Full features	0.708	0.708	0.674	0.696	0.723
4-Algo features	0.713	0.710	0.705	0.707	0.705
Cfs features	0.719	0.690	**0.712**	**0.727**	0.726
Rf features	**0.728**	**0.717**	0.699	0.698	**0.739**
G features	0.717	0.707	0.678	0.701	0.713
SU features	0.725	0.709	0.691	0.708	0.715

Figure 3 shows the comparison of the performance between the different feature selection techniques. For each technique, the classifier with the best performance was selected. Therefore, the fact that stands out from Fig. 3 is that Rf features has outperformed other techniques using 24 selected features for both datasets. It gives the best result F-measure 0.762 combined with REP Tree classifier using the "Mathematics course dataset". Rf features combined with Logit Boost classifier has also produced better result 0.739 using the "Language course dataset". Further analysis is necessary to better understand these results. Figure 4 attests the best performance of each classifier. Indeed, the feature selection technique with the best F-measure was selected for each classification algorithm. Rf features technique has the highest influence for most classifiers. According to the "Mathematics course dataset", this technique was selected by four classifiers. Jrip, RandomForest, J48 and REPTree give the best performance combined with Rf features. Similarly, Rf features was selected as the best technique by three classifiers using the "Language course dataset".

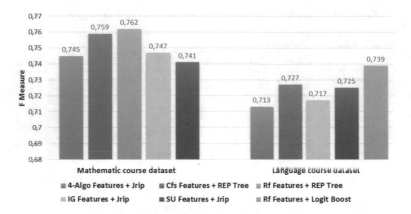

Fig. 3. Graphical comparison of the performance of the feature selection techniques using the selected classification algorithms

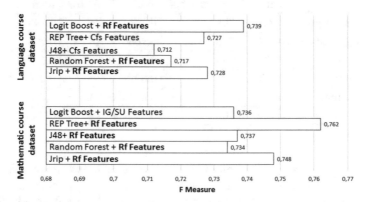

Fig. 4. Graphical comparison of the performance of the selected classification algorithms

5   Conclusion

The purpose of this paper is to analyze the impact of feature selection techniques on the classification model. Accordingly, we carried out a comparative study in order to identify the best predictive technique for students' academic performance. In this paper, we presented the methodology adopted for this study, where four feature selection techniques combined with five classification algorithms were compared in order to identify the best feature selection technique. Based on F-measure parameter, the experimental results shows that Rf Feature is the best feature selection technique compared to other standard techniques. For future work, we intend to integrate our findings into a distance learning system [24] for grouping learners by adopting bioinspired behaviors applied in previous works to collaborative multirobot systems [25] and taking into account the emotional aspect of learners [26] in order to improve their performance.

Acknowledgment. The authors express thanks to the Erasmus+ project for funding the research reported under the Grant Agreement number 2015-1-ES01-K107-015469.

References

1. Romero, C., Ventura, S.: Data mining in education. Wiley Interdisc. Rev. Data Min. Knowl. Discovery **3**(1), 12–27 (2013)
2. Abid, A., Kallel, I., BenAyed, M.: Teamwork construction in e-learning system: a systematic literature review. In: 2016 15th International Conference on Information Technology Based Higher Education and Training (ITHET). IEEE, pp. 1–7 (2016)
3. Mitra, P., Murthy, C., Pal, S.K.: Unsupervised feature selection using feature similarity. IEEE Trans. Pattern Anal. Mach. Intell. **24**(3), 301–312 (2002)
4. Miller, A.: Subset Selection in Regression. CRC Press, Boca Raton (2002)
5. Hall, M.A.: Correlation based feature selection for machine learning (1999)
6. Kira, K., Rendell, L.A.: A practical approach to feature selection. In: Proceedings of the Ninth International Workshop on Machine Learning, pp. 249–256 (1992)
7. Kononenko, I.: Estimating attributes: analysis and extensions of RELIEF. In: European conference on machine learning, pp. 171–182. Springer (1994)
8. Quinlan, J.R.: C4.5: Programs for Machine Learning. Elsevier, Amsterdam (2014)
9. Press, W.H., Teukolsky, S.A., Vetterling, W.T., Flannery, B.P.: Numerical Recipes in C, vol. 2. Cambridge University Press, Cambridge (1996)
10. Ramaswami, M., Bhaskaran, R.: A study on feature selection techniques in educational data mining. arXiv preprint arXiv:0912.3924 (2009)
11. Velmurugan, T., Anuradha, C.: Performance evaluation of feature selection algorithms in educational data mining. Perform. Eval. **5**(02) (2016)
12. Costa, E.B., Fonseca, B., Santana, M.A., de Araújo, F.F., Rego, J.: Evaluating the effectiveness of educational data mining techniques for early prediction of students' academic failure in introductory programming courses. Comput. Hum. Behav. **73**, 247–256 (2017)
13. Bolón-Canedo, V., Sánchez-Maroño, N., Alonso-Betanzos, A.: A review of feature selection methods on synthetic data. Knowl. Inf. Syst. **34**(3), 483–519 (2013)
14. Noura, A., Shili, H., Romdhane, L.B.: Reliable attribute selection based on random forest (RASER). In: International Conference on Intelligent Systems Design and Applications, pp. 11–24. Springer (2017)
15. Breiman, L.: Random forests. Mach. Learn. **45**(1), 5–32 (2001)
16. Cohen, W.W.: Fast effective rule induction. In: Proceedings of the Twelfth International Conference on Machine Learning, pp. 115–123 (1995)
17. Friedman, J., Hastie, T., Tibshirani, R., et al.: Additive logistic regression: a statistical view of boosting (with discussion and a rejoinder by the authors). Ann. Stat. **28**(2), 337–407 (2000)
18. Quinlan, J.R.: C4.5: Programming for Machine Learning, vol. 38. Morgan Kauffmann, Burlington (1993)
19. Smith, T.C., Frank, E.: Introducing machine learning concepts with WEKA. In: Statistical Genomics: Methods and Protocols, pp. 353–378 (2016)
20. Márquez-Vera, C., Morales, C.R., Soto, S.V.: Predicting school failure and dropout by using data mining techniques. IEEE Revista Iberoamericana de Tecnologias del Aprendizaje **8**(1), 7–14 (2013)
21. Gu, Q., Cai, Z., Zhu, L., Huang, B.: Data mining on imbalanced data sets. In: IEEE 2008 International Conference on Advanced Computer Theory and Engineering, ICACTE 2008, pp. 1020–1024 (2008)

22. Cortez, P., Silva, A.M.G.: Using data mining to predict secondary school student performance (2008)
23. Han, J., Pei, J., Kamber, M.: Data Mining: Concepts and Techniques. Elsevier, Amsterdam (2011)
24. Volungevičienė, A., Daukšienė, E., Caldirola, E., Blanco, I.J.: Success factors for virtual mobility exchange on open educational resources (2014)
25. Chatty, A., Kallel, I., Alimi, A.M.: Counter-ant algorithm for evolving multirobot collaboration. In: Proceedings of the 5th International Conference on Soft Computing as Transdisciplinary Science and Technology. ACM, pp. 84–89 (2008)
26. Abdelkefi, M., Kallel, I.: Towards a fuzzy multiagent tutoring system for M-learners' emotion regulation. In: 2017 16th International Conference on Information Technology Based Higher Education and Training (ITHET). IEEE, pp. 1–6 (2017)

Detection and Localization of Duplicated Frames in Doctored Video

Vivek Kumar Singh[✉], Pavan Chakraborty,
and Ramesh Chandra Tripathi

Indian Institute of Information Technology, Allahabad, India
vivekkr.singh@hotmail.com

Abstract. With the advent of high-quality digital video cameras and sophisticated video editing software, it is becoming easier to tamper with digital videos. A common form of manipulation is to clone or duplicate frames or parts of a frame to remove people or objects from a video. We describe a computationally efficient technique for detecting this form of tampering. After detection of frame duplication localization of frame sequence has also been done. Detection as well as localization of forgery are found better than other techniques.

Keywords: Correlation · Frame duplication · Sub blocking · RMSE

1 Introduction

Twenty first century is no doubt an expectancy of digital world. Each individual is drowned into the sea of digital multimedia. Knowingly or unknowingly people come across thousands of pictures and videos in their daily life routines. These contents get instilled somewhere in the human mind. Images and videos are powerful channel to convey a wealth of information as well as emotional impact. Therefore, response of viewer depends on what he/she perceives from the color, texture, shape and moving images or a video. Taking advantage of this fact, these major features (color, texture and motion) are sometimes twisted by some cheat professionals to leave a long lasting emotional impression on the viewers though later on the truth prevails but when harm has already been done. This results in serious consequences, increasing apathy and creating fallacy in many daily life applications and interferences.

With the advent of high-quality digital video cameras and sophisticated video editing software, it is becoming easier to tamper with digital videos. A common form of manipulation is to clone or duplicate frames or parts of a frame to remove people or objects from a video. In this paper, we describe a computationally efficient technique for detecting this form of tampering.

The main objectives of frame duplication are:

- to hide a sequence of real incidence and/or
- to repeat any incidence so that the viewer gets misleading interference from the scene.

© Springer International Publishing AG, part of Springer Nature 2018
A. Abraham et al. (Eds.): ISDA 2017, AISC 736, pp. 661–669, 2018.
https://doi.org/10.1007/978-3-319-76348-4_64

In order to hide a sequence, a group of frames has to be replaced by other group of frames preferably from the same video. Whereas to repeat an event, a group of frames containing desired event is inserted somewhere else in the same video sequence. However, to defy the detection in both the cases, group of frames is duplicated preferably from the same video.

A lot of work has been done to detect video related forgery. Wang and Farid [1] have introduced solution for duplication type of forgery. They have used very obvious method to compare correlation of the two consecutive frames. This technique is computationally high. To overcome with complexity problem, They have used group of frames to overcome complexity. Results are good up to 90% of accuracy.

In another approach, Kobayashi et al. [2] proposed used consistency of noise level function in each frames is sufficient to differentiate between attacked and original region. However, this method is good only for static camera and performance dramatically decreases as the compression is performed on the videos. Thus this method is not acceptable to find such forgeries.

Hsu et al. [3] used GMM but still problem exists with accuracy and complexity. Whereas Lin et al. [4] used histogram difference and this reduces time complexity and found more accurate than previous methods.

Bestagini et al. [5] propose a method to detect spatio-temporal forgeries by analyzing left footprints. But results was not that prominent on all type of videos. Mondaini et al. [6] proposed a method to detect whole frame duplication and for object insertion. Authors used camera sensor pattern noise as a characteristics of a camera. Presented method also claim to be some level of invariance to MPEG compression.

Singh et al. [7] used nine features to match suspected frames these nine features are also considered in this paper but the sub blocking method is changed.

Various challenges are encountered while detecting above duplication type of forgery. Being very expensive, trivial methods have limited reach because of the involved computational cost. Let us assume, targeted video sequence is of length L. Comparing each frame with all other frames on the basis of intensity values is computational big task and computation cost will be proportional to L^2 which cannot be considered as a good solution. Another predominant challenge is to reduce the number of false positives in case of detection. For a practical solution some of the assumptions made while detecting frame duplications are following:

1. Duplication of a single frame in a video sequence does not make any perceivable forgery. Therefore, duplication over one second i.e. at least of 30 frames (in case of a video with 30 fps rate) is regarded as forged frames.
2. The technique should be based on fact that each video capturing device inserts its own some hardware noise while recording.

Assumption: It has been assumed that video acquisition devices insert some noise while capturing videos. This has been examined with the help of various test videos. A graph as shown in Fig. 1. It depicts that each consecutive frames are different even for the static scene (Only 6 consecutive frames have been considered).

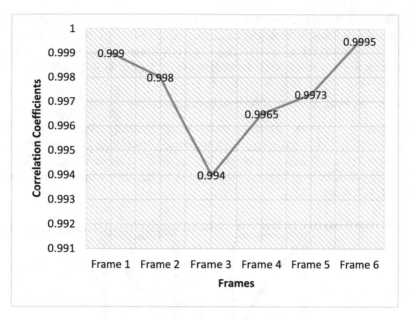

Fig. 1. Correlation of two consecutive frames in a static scene captured by fixed camera.

2   The Proposed Methodology

In case of frame duplication type of forgery, targeted video is first divided into frame by frames. Since our objective is to detect similar frames hence some suitable frame matching algorithm needs to be applied. The method which is used for said purpose is discussed as below:-

2.1   Conversion of Video Sequence in Frames and then to the Gray Scale

Video sequence is a group of pictures called as frames. Frames of targeted videos are extracted. Each of the frames have three different channels R (Red), G (Green) and B (Blue). To reduce computation, RGB frames are converted into their equivalent gray scale frames using Eq. 1 conversion technique.

$$I_t(x, y) = 0.2989 \times R_t(x, y) + 0.5870 \times G_t(x, y) + 0.1140 \times B_t \qquad (1)$$

Here, I_t is the particular frame at time t and $R_t(x, y), G_t(x, y), B_t(x, y)$ are the intensity values at location (x, y) for corresponding Red, Green and Blue channel.

2.2   Suspected Frame Duplication Detection

To extract similar frames, correlations between frames are computed. All frames with high correlation are considered to be 'Key frames' and kept as suspected frames. For N number of frames, N^2 computations are required for comparison of each frame with each other. Trivial method of exact matching of frames requires high amount of computation and hence is regarded as inefficient for larger video size. As a practical solution, to reduce number of matching, only key frames are chosen for further processing. Key frames would be those frames where forgery is suspected.

2.3   Sub Blocking, Feature Selection and Extraction

After getting key frames, the important part of matching algorithm is to choose appropriate features. Various researchers have used correlation as a matching feature.

In this work, we divided each frames in four parts as shown in Fig 2. This type of sub blocking allowed mostly pixel contribution in feature making. Features are then calculated as follows [7].

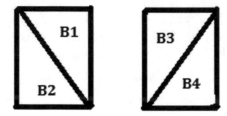

Fig. 2.   A single frame division

(i)   Mean (μ): First feature is taken as the mean of overall frame intensity values.
(ii)   Ratio of each sub division with mean of frame.

$$r_i = Normalized\left(\frac{B_i}{B}\right) for\ i = 1, 2, 3, 4$$

(iii)   Residual of each sub division from mean of frame

$$e_i = Normalized(B - B_i)\ for\ i = 1, 2, 3, 4$$

Hence, we get nine features for each frame F_t. For each frame, we maintain a feature vector $f_t = \{\mu, r_1, r_2, r_3, r_4, e_1, e_2, e_3, e_4\}$, where t is the number of that frame in video sequence.

At the end of this step, once we calculated the feature vector for all the frames we got the feature matrix with the dimension of $N \times 9$. Here N is the number of frames in video sequence.

2.4   Suspected Frame Selection

After getting all feature vectors for N number of frames, root mean square error is calculated between each feature vectors of two consecutive frames i and j, using the formulae as below:-

$$RMSE(i,j) = \sqrt{Sqr\left(\vec{f_i} - \vec{f_j}\right)}$$

Pair of frames with RMSE zero or nearly zero could be considered as suspected frames. These frames are marked as suspected frames for further processing. Since there may be high number of false positives, therefore to reduce false positives further processing is required.

Looking into the final event of only human interpretation, duplication with at least one or two seconds is taken for consideration. As discussed earlier, human perception is an important aspect in deciding whether a forgery has been attempted or not. Let us assume a video with 30 fps rate. For one second, at least 30 consecutive frames would be duplicated. To reduce number of comparisons between frames following assumption can be made.

It is obvious that frame i, is highly correlated to frame $i+1, i+2$....and so on. Whenever we are performing frame matching than these nearby frames also show duplication case. To avoid these false matching we perform a jump while performing matching. Frame 'i' is not matched with adjacent frames within the frame length $(1 \ll N)$.

2.5   Localization of Forgery

After retrieval of the duplicated clips successfully, it is imperative to locate the original and duplicated frames in the video sequence so as to enable any one get convinced that frame duplication type of forgery has really been crafted and the result found is a sequence of frames which is similar to another sequence of frames. Let us assume these two sequences are located as $P(i, i+1, i+2, \ldots i+n)$ and $Q(j, j+1, j+2, j+3 \ldots j+n)$. In these two frame sequences, one sequence is original and another is duplicated. To localize duplicated clip, it is assumed that consecutive frames show a high correlation value between such pairs.

Hence if sequence starting from ith frame is similar to jth frame, then for both frames correlation with the previous frame is examined. Whichever of these two frames show higher correlation with its previous frame, such clips containing frames are regarded as the case of duplication. It can be summarized as below:

$$\text{If}(Corr(i, i-1) > Corr(j, j-1))$$
$$\text{Clip Q is duplicated;}$$
$$\text{Else}$$
$$\text{Clip P is duplicated;}$$

3   Result and Discussion

To examine the result of the proposed method, number of experiments have been performed. A total number of 50 videos have been tested for this experiment. Most of the videos are self-made. Some videos are taken from SULFA (Surrey University Library for Forensic Analysis: http://sulfa.cs.surrey.ac.uk/videos.php).

Dataset of the videos contains a wide variety of videos including static scenes to very highly activity intensive videos. Description of some of videos are given as below:

Properties → Test videos ↓	Total frames (numbers)	Resolution (pixels)	Time length (seconds)
Video 1	465	1080 × 1920	43
Video 2	414	1080 × 1920	36
Video 3	353	1080 × 1920	30
Video 4	469	240 × 320	17
Video 5	274	240 × 320	8

Following are the some snap shots of considered videos (Fig. 3).

Fig. 3. Snapshot of the considered videos.

Graphically, the results can be seen as in Fig. 4 for a duplicated video. Here X and Y axis show the number of frames starting from 1. Each of the matched frames is denoted by red color. Initially, it could be seen that frame duplication detection results in various false matches. Spread red patches shows a number of false positives. Whereas after double checking as in proposed methodology, red marks settle at a straight line. This confirms that a sequence of frames have been duplicated in some other area. Graph is plotted for upper half only because lower half will be just a reflection along y = x.

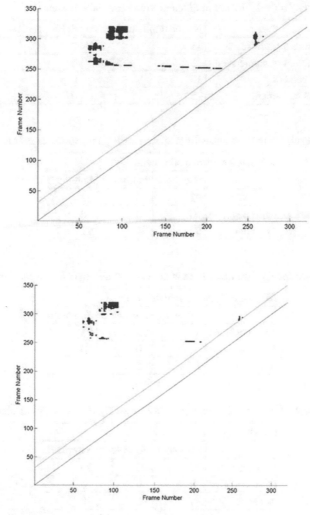

Fig. 4. Graphs showing matched pairs after various steps of methods for two different videos (forged i.e. frames duplicated).

Table 1 is showing the false positive found with static and dynamic cameras with the help of our proposed algorithm.

A prominent results can be seen through Tables 2, 3 and 4 that our experiment is good for each scenario even with moving as well as static camera.

The achievement of our proposed method is localization of duplicated frames. Uniquely, our proposed method localizes the duplicated frames. This feature of localization is advantageous over all other methods.

Table 1. Result in different video acquisition scenario:

	No. of duplicated video	False positives found
Static camera static scene	15	2
Static camera dynamic scene	15	2
Moving camera	10	0
Surveillance camera	5	0

Table 2. Result for three samples from static camera and static scene from database.

(A) Static camera static scene			
	Sample 1	Sample 2	Sample 3
True positive	92	94	92
False negative	0	0	0
Accuracy	94	94	92

Table 3. Result for three samples from static camera and dynamic scene from database.

(B) Static camera dynamic scene			
	Sample 4	Sample 5	Sample 6
True positive	94	95	96
False negative	0	0	0
Accuracy	95	95	95

Table 4. Result for three samples from moving camera from database.

(c) Moving camera			
	Sample 7	Sample 8	Sample 9
True positive	95	95	96
False negative	0	0	0
Accuracy	95	95	95

4   Conclusion

We have proposed a sub-blocking algorithm to detect frame duplication on the basis of similarity analysis between frames. We extracted nine symmetrical features of each frames and implemented the Euclidean distance method. Hence suspected duplication is found. Such key frames found are then further processed in second step to confirm the duplication in videos. The advantage of our proposed method is localization of duplication.

References

1. Wang, W., Farid, H.: Exposing digital forgeries in video by detecting duplication. In: Proceedings of ACM 9th Workshop on Multimedia and Security, pp. 35–42 (2007)
2. Kobayashi, M., Okabe, T., Sato, Y.: Detecting video forgeries based on noise characteristics. In: Pacific-Rim Symposium on Image and Video Technology, pp. 306–317 (2009)
3. Hsu, C.-C., Hung, T.-Y., Lin, C.-W., Hsu, C.-T.: Video forgery detection using correlation of noise residue. In: Proceedings of the IEEE 10th Workshop on Multimedia Signal Processing, pp. 170–174 (2008)
4. Lin, G.S., Chang, J.F., Chuang, C.H.: Detecting frame duplication based on spatial and temporal analyses. In: Proceedings of the IEEE Conference on Computer Science and Education, pp. 1396–1399 (2011)
5. Bestagini, P., Milani, S., Tagliasacchi, M., Tubaro, S.: Local tampering detection in video sequences. In: IEEE International Workshop on Multimedia Signal Processing, pp. 488–493 (2013)
6. Mondaini, N., Caldelli, R., Piva, A., Barni, M., Cappellini, V.: Detection of malevolent changes in digital video for forensic applications. In: Proceedings of SPIE, Security, Steganography, and Watermarking of Multimedia Contents IX, vol. 6505 (2007)
7. Singh, V.K., Pant, P., Tripathi, R.C.: Detection of frame duplication type of forgery in digital video using sub-block based features. In: James, J., Breitinger, F. (eds.) Digital Forensics and Cyber Crime. Lecture Notes of the Institute for Computer Sciences, Social Informatics and Telecommunications Engineering, vol 157. Springer, Cham (2015)

A Novel Approach for Approximate Spatio-Textual Skyline Queries

Seyyed Hamid Aboutorabi, Nasser Ghadiri$^{(\boxtimes)}$, and Mohammad Khodizadeh Nahari

Department of Electrical and Computer Engineering, Isfahan University of Technology,
84156-83111 Isfahan, Iran
{h.aboutorabi,m.khodizadeh}@ec.iut.ac.ir,
nghadiri@cc.iut.ac.ir

Abstract. The highly generation of spatio-textual data and the ever-increasing development of spatio-textual-based services have attracted the attention of researchers to retrieve desired points among the data. With the ability of returning all desired points which are not dominated by other points, skyline queries prune input data and make it easy to the user to make the final decision. A point will dominate another point if it is as good as the point in all dimensions and is better than it at least in one dimension. This type of query is very costly in terms of computation. Therefore, this paper provides an approximate method to solve spatio-textual skyline problem. It provides a trade-off between runtime and accuracy and improves the efficiency of the query. Experiment results show the acceptable accuracy and efficiency of the proposed method.

Keywords: Skyline query · Spatio-textual data · Approximated method

1 Introduction

The increased number of location-enabled devices such as smartphones and the development of social networks, including Twitter, Flickr, and Foursquare, have generated a massive amount of spatio-textual data in recent years. Also, location-based search services that use such data have been rapidly developed [1]. According to the statistics from a report published in 2015, about 20% of daily Google queries and 53% of smartphone users' searches are associated with spatial content [2, 3]. An accurate response to such queries, demands models that make the best possible trade-off between different features of data (here spatial and textual features) and facilitate decision-making process through dataset purification.

The skyline query was developed to cover such problems. It was first introduced by Börzsönyi et al. [4] in the database research field, and then received the attention of computer science researchers so that they offered methods to solve it [5, 6]. Consequently, its initial definition was gradually expanded and various issues were put forward based on the initial skyline. The spatio-textual skyline query was first introduced by Shi et al. [7]. It is only one of the various types of skyline queries with a set of data points and a set of query points; each point has spatial data and a set of textual descriptions. The spatio-textual distance between the data points and query points is measured using the integration of spatial distance and the textual relevance between these points. As the

A. Abraham et al. (Eds.): ISDA 2017, AISC 736, pp. 670–682, 2018.
https://doi.org/10.1007/978-3-319-76348-4_65

output of the query, the skyline set returns all data points that are not spatio-textually dominated by other data points. A data point (p2) is spatio-textually dominated by another data point (p1), if spatio-textual distance of p2 to all query points is not better than p1 and spatio-textual distance of p2 to at least one query point is worse than p1.

Spatio-textual skyline queries are among the newest and the most challenging queries of skyline problems. This is because skyline query has rarely been studied on spatio-textual data. Although there are studies on spatial skyline queries, the textual dimension is a new area of focus, which has been less studied. It should be noted that the top-k queries have previously been studied on spatio-textual data. However, they cannot be adopted in some applications because of the unknown weights of the different dimensions of data. On the other hand, the skyline queries are very costly from both processing time, and disk usage points of view [8, 9] and the addition of spatio-textual features to the queries will increase these costs. These challenges demand a new efficient approach. The primary aim of this research is to introduce an efficient approximate approach for spatio-textual skyline computation with acceptable accuracy and efficiency.

Our new approach does not necessitate the investigation of whole data space, and it shows, in many cases, a better performance than other approaches by making a trade-off between accuracy and efficiency. This is the main advantage of the proposed approximate method.

This paper has been organized as follows. In Sect. 2 we present related work. Section 3 discusses the proposed approach to executing spatio-textual skyline queries. In Sect. 4, the obtained results are analyzed and Sect. 5 concludes the paper and presents recommendations for future studies.

2 Background

From the time of its introduction by Börzsönyi et al., the skyline family of database queries has widely been studied. According to the classification shown in Fig. 1, three areas serve as the source of the diversity of the skyline challenges: data diversity, different execution environments, and different computation methods. Data diversity is the first influential factor. Handling deterministic data are the simplest case of this class where each data point has a set of features which are complete, fixed and reliable from the beginning of algorithm, and lacks any spatio-textual features. Some algorithms, including divide and conquer and bitmap skylines, are executed on such data [4, 10]. Although these algorithms are the basis of other methods, they have received less attention of researchers due to introducing new requirements.

A challenging class of skyline query focuses on spatial data. In this category, the Euclidean distance-based algorithms were first introduced in Sharifzadeh and Shahabi's work to solve spatial skyline problems [11]. The difference between the spatial skyline and spatio-textual skyline queries is that in the former, the data points and query points, lack textual descriptions. Lee et al. proposed a set of accurate and approximate approaches to solve this type of queries [12].

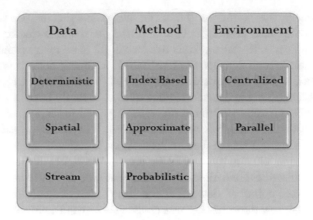

Fig. 1. Source of the diversity of the skyline challenges

Stream data are another type of data which is not available at the beginning of algorithm entirely. They are incomplete data; therefore, prior approaches are not applicable to them. There are different papers about spatial skyline query [13, 14] and this study concentrates on spatio-textual data.

Selection of the solution approach is another influential factor. It depends on the attitude of the researcher towards the skyline problem and his/her aim of solving the problem. The problem-solving methods can be divided into three categories: index-based, approximate and probabilistic methods. In the index-based methods, different index structures including B-tree, R-tree, ZB-tree, and Zinc are used to increase the efficiency of algorithms [15–17]. Although index-based methods do not necessitate the scanning of entire data space, they add additional storage overloads. Our proposed method uses simple database-ready built-in indexes to cope with the problem. In the approximate methods, such as those proposed by Chen and Lee [18], all or parts of skyline set along with the set of points which are closer to the skyline points may be returned to the user to satisfy the user or to increase algorithm efficiency. Our developed method is an approximate method that will be explained in the next section. Probabilistic methods are another class of skyline processing methods where the probability may be associated with data and users' preferences. Sacharidis et al. first presented an algorithm to introduce this method [19] followed by several algorithms. In such algorithms, the user defines a probability threshold and an error value. All responses returned as the skyline set should meet this threshold considering the defined error value. Our work considers non-probabilistic preferences.

The execution environment and hardware platform is another influential factor on which the problem is introduced and solved. There have been studies on the quality of the execution of skyline query on mobile-based networks, cloud platform, peer to peer networks, ad-hoc networks and wireless sensor networks [20]. These studies can be divided into two classes: centralized and parallel. Certain requirements should be taken into account in the design of the algorithms for each architecture. In the centralized methods, memory restriction is the only limitation of algorithm while parallel methods go with several challenges including concurrency, load balancing on CPU cores, resource management, software and hardware failures and task scheduling. Different methods have been introduced to develop skyline parallel algorithms [21, 22]. Our hardware structure is a conventional centralized and standalone computer.

3   Method

3.1   Spatio-Textual Skyline Query Model

A spatio-textual skyline query gets a set of data points and query points as inputs, and returns those data points which are not spatio-textually dominated by other data points considering the location and preferences of query points. The spatio-textual skyline query contains two data sets: a set of data points (P) on which the query should be executed, and a set of query points from which the query is issued. Each member of the mentioned sets has a geographical coordinate shown as $p.\gamma$ for data points and as $q.\gamma$ for query points. In addition, each data point has a textual description shown as a set of keywords $p.\psi$. All query points have the same textual description indicating users' preferences and shown as $Q.\psi$. According to [7], a data point (p_i) will spatio-textually dominate another data point (p_j) if Eq. (1) holds:

$$\forall q_k \in Q : st(q_k, p_i) \leq st(q_k, p_j), \exists q_k \in Q : st(q_k, p_i) < st(q_k, p_j) \tag{1}$$

In Eq. (1), $st(q_k, p_i)$ is a model for integrating spatial distance and the textual relevance of data points and query points. The relationship between spatio-textual distance with spatial distance and textual relevance is direct and inverse, respectively:

$$st(q_k, p_i) = \frac{d(q_k, p_i)}{w(Q.\psi, p_i.\psi)} \tag{2}$$

In Eq. (2), $st(q_k, p_i)$ shows spatio-textual distance. The term $d(q_k, p_i)$ shows the spatial distance between data points and query points which is computed using Euclidean distance on the geographical latitude and longitude of the points. In addition, $w(Q.\psi, p_i.\psi)$ shows the textual relevance between the keywords of a data point and the keywords of query points. This relevance is calculated as shown in Eq. (3):

$$w(Q.\psi, p_i.\psi) = \prod_{t_j \in Q.\psi} (\hat{w}(t_j, p_i.\psi)) \tag{3}$$

In Eq. (3), $\hat{w}(t_j, p_i.\psi)$ indicates the weight of the keywords t_j of data point p_i. Equation (3) is interpreted as follows: To compute the relevance between the keywords set of query points $Q.\psi$ and the keywords set of data points $p_i.\psi$, the corresponding weights should be defined for each keyword Q available in $p_i.\psi$ and the product of the weights should be considered as the textual relevance between them. If only a few weights are zero, a reduction coefficient is considered for them, instead of considering their value as zero. This will prevent the value of the term to become zero. However, if all weights are zero, no textual relevance is considered between the points. $\hat{w}(t_j, p_i.\psi)$ is computed using *tf-idf* method [23] as follows:

$$\hat{w}(t_j, p_i.\psi) = \frac{1}{|p_i.\psi|} \times TF \times \log \frac{N}{\left|\{p_i \in P : t_j \in p_i.\psi\}\right|} \tag{4}$$

In Eq. (4), TF is the total frequency of a keyword in the textual description of a data point and N is the total number of data points, (|P|). The whole term is divided by $|p_i.\psi|$ in order to give more value to low frequent words.

The corresponding weight of the keywords of data points are computed before running the algorithm, and the outputs are stored in a MongoDB database. The MongoDB database is used because the corresponding weights of the keywords of data points are non-structured data and MongoDB is an optimized tool for managing such structure.

3.2   Propose Method

Initial assessments of spatio-textual skyline query results show that in most cases, the ideal points appear around the convex hull of query points. Theoretically, this may root in two reasons. The first one is that those data points which are in the vicinity of the convex hull of query points are more likely to be selected because their distance from query points is lower than that of other data points. The second reason is the accumulation of data points around the convex hull of query points. In this case, there are sufficient data points with a high textual relevance with query points. These points dominate other data points and are introduced as skyline points. The black solid line pentagon shown in Fig. 2 shows the convex hull of query points. As the area of the convex hull decreases/increases (dash line pentagons), algorithm accuracy decreases/increases. Also, the problem space becomes smaller/larger proportional to the changes. This significantly influences algorithm efficiency. If the area of whole data space is S and the area of circumscribed quadrilateral of the convex hull of query points is S_q, ΔS_q will be defined as follows:

$$\Delta S_q = S_{q_2} - S_{q_1} \tag{5}$$

In Eq. (5), S_{q_1} and S_{q_2} respectively are the area of the circumscribed quadrilateral of the convex hull of query point before and after its expansion, respectively.

Then, the α coefficient is defined as follows:

$$\alpha = \Delta S_q / S \tag{6}$$

Based on Eq. (6), α is the ratio of the changes in the area of the circumscribed quadrilateral of the convex hull of query points to the total area of data space. It belongs to [0,1) interval and varies by the variation of S_q. The optimal value of α is obtained by time-accuracy trade off. The effect of α on the accuracy and efficiency of the proposed method will be discussed in the experiments section.

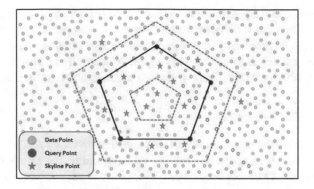

Fig. 2. The main idea of the proposed algorithm

Algorithm 1 shows the pseudo code of Approximate Centralized Method (ACM). It receives a set of data points and a set of query points as inputs. Then, it computes spatio-textual skyline for this query and returns the set of data points with the maximum match with users' requests. The convex hull of query points is first created (line 1), then the optimal value of alpha is computed based on the created convex (line 2) and accuracy required. The optimal value of alpha is predetermined with respect to query points and required accuracy. Then convex is expanded considering alpha (line 3).

Algorithm 1. Approximate spatio-textual skyline

Approximate_Spatio_Textual_Skyline (DataPoint P[ ], QueryPoint Q[ ])

---

1. CH = Create convex hull of Q
2. α = Optimal Value of α based on created convex(CH) and accuracy required
3. CHE = Expand (CH, α)
4. **for** each $p_i \in CHE$
5.   TF = Read corresponding weights of p's keywords from MongoDB Database
6.   TR = Compute textual relevance between $p_i.\psi$ and $Q.\psi$ using TF
7.   **if** (TR$\neq$ 0)
8.   **for** each $q_j \in$ Q
9. SD_j = Compute spatial distance between $p_i.\gamma$ and $q_j.\gamma$
10. $STD_j = SD_j$ / TR
11.   skyline_set = Skyline ($p_i.id$, STD[ ])
12. **return** skyline_set

Then, the spatio-textual distance of each data point from query points is computed, and it is investigated that whether they can be present in the final skyline set. In line 5, the corresponding weights of the keywords of data points are retrieved from the MongoDB database. In Line 6, the textual relevance between the keywords of data points and query points is computed. If the relevance is zero, the corresponding data point is pruned without additional consideration because the denominator becomes zero in Eq. 2, making spatio-textual distance reach infinity. If the relevance is not zero, the spatial distance between query points and data points is computed (line 8 to 10).

Line 11 calls skyline algorithm and compares current data point with current members of skyline set. Then, the variable of *skyline_set* changes depending on the comparison result so that some members may be eliminated, some new members may be added, or the skyline set may remain unchanged. Following the evaluation of all current points of data set, *skyline_set*, which contains the final skyline set, is returned as the output.

3.3 Experimental Evaluation

This section assesses the proposed algorithm regarding accuracy and efficiency. Instead of monitoring whole data space, this approximate algorithm assesses only points in the vicinity of the convex hull of query points. Therefore, in some cases, it may fail to find all skyline points. The accuracy of this algorithm depends on alpha to a large extent so that by selecting a suitable alpha it can achieve an accuracy of 95%, and even higher. The next section measures the effect of alpha on the accuracy using sensitivity and specificity criteria:

$$Sensitivity = \frac{TP}{TP + FN} \tag{7}$$

$$Specificity = \frac{TN}{TN + FP} \tag{8}$$

In Eq. (7), TP is skyline points truly marked by the algorithm. FN is skyline points falsely marked by the algorithm as non-skyline points. In relation (8), TN is non-skyline points truly marked by the algorithm as non-skyline points, and FP is non-skyline points falsely marked by the algorithm as skyline points. Relying on more than thousands of tests, it can be argued that the specificity is always higher than 99.9%. Therefore, it is neglected and the accuracy of approximate algorithms is measured and declared only based on sensitivity.

The proposed algorithm is not a recursive algorithm, and the number of data points and query points is not endless. It is, therefore, naturally proven that the proposed algorithms are terminable. To assess the efficiency of proposed algorithms, we designed tests by changing the number of data points, the number of query points and the number of the keywords of query points, and measured algorithm runtime in all cases. The next section provides the results.

4 Experiments

In this section, we report on the results of the experiments. First we describe the evaluation platform and the datasets used, followed by results and their interpretation.

4.1 Test Specifications-Methods and Data Set

The tests were conducted on a system with Intel core i5 4460 CPU, 8 GB of RAM, and 1 TB hard disk. We developed the code for Approximate Centralized Method (ACM) and Exact Centralized Method (ECM) to provide accurate assessments. They differ in that ECM evaluates all data points, instead of evaluating only data points in the vicinity of the convex hull of skyline points, to raise accuracy to 100%. The methods mentioned were compared with three methods presented in Shi et al. paper [7] i.e., ASTD, BSTD and BSTD-IR. Amongst them, ASTD shows a better efficiency. The comparison was made to show the improvements in the proposed methods.

The datasets were created in accordance with the standards indicated in Börzsönyi et al. paper [4] to evaluate different methods. The number of the keywords of each data point was variable and was 10 keywords on average. The variable parameters of the problem were alpha, number of data points (default = 400,000 points), number of query points (default = 10 points) and number of the keywords of query points (default = 10 points). The effect of these variables on the efficiency and accuracy of the studied methods were evaluated by conducting separate tests. The execution of the algorithm was measured from the moment that algorithm starts running until the final output is recorded. Each query was executed 100 times to avoid any deviation in results, and the average values of the obtained results were depicted.

4.2 Effect of Alpha on ACM Accuracy

This experiment was made to find the optimal alpha of ACM by which an acceptable trade-off is provided between the execution time and accuracy. To this end, separate tests were conducted with different S_q (from 0.1% to 12.8% of total data space) and different values of alpha (0% to 0.1% with an interval of 0.02). Figure 3 shows the results where the vertical and horizontal axes show the accuracy and runtime, respectively. Different values of S_q are shown by distinct colors. In addition, each group of bars shows results associated with a given alpha. For example, the farthest left group bar shows values associated with $\alpha = 0$ and, the farthest right one shows values associated with $\alpha = 0.1$. Observing the results of this test reveals that as alpha rises from zero to 0.1, the accuracy and execution time of the algorithm will increase. Surprisingly, as alpha exceeds a given value, the accuracy remains almost constant. This threshold value is defined as the optimal alpha. For example, the optimal alpha of $S_q = 0.01\%$ is defined to be 0.04. This figure shows another important point. As S_q rises, the optimal alpha increases proportionally. For example, for $S_q = 0.08$, the optimal alpha is defined to be 0.06.

In summary, for any given accuracy and S_q the optimal alpha can be determined using this figure. In the next experiments, the value of alpha is selected in a manner that the accuracy of algorithm always remained about 95%.

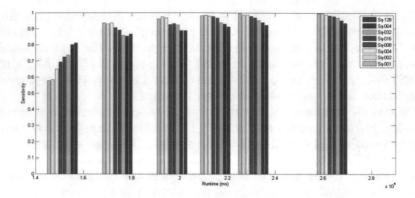

Fig. 3. The effect of the area of the convex hull of query points on the accuracy and efficiency of the studied approximate algorithm

4.3 Effect of the Number of Data Points on Efficiency

In this set of experiments, the number of data points, $|P|$ varied from 100000 to 1500000 and other parameters remained unchanged and were set at their default values. Figure 4 shows the execution time of three competitor algorithms and our two algorithms i.e. ECM (red curve) and ACM (green curve) with an accuracy of about 95%.

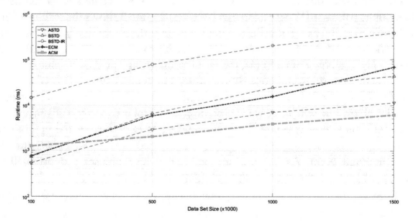

Fig. 4. The effect of *data points* on execution times of different algorithms

Deliberating in the results of this test reveals three crucial points. The first point is that as the number of data points increases, the execution time of all algorithm increases due to the enlargement of problem space. The algorithms differ with each other in their raising speed, i.e., their slopes. The second point is that the slope of ACM curve is lower than that of other methods. This implies the higher scalability of ACM compared with competitor algorithms. The third point is that in the test with 100000 data points, two competitor algorithms showed lower execution time compared to our ACM algorithm.

This may root in two reasons. First, they use modern index structures like IR-Tree which are optimized for such data. Apparently, the smaller the problem space is, the lower is search structures depth, and lower nodes should be explored to obtain target nodes. This, in turn, shortens algorithm execution time. However, the advantage of our algorithms is well demonstrated in the larger search space.

The second reason is that we tried to keep the accuracy of our algorithm in a constant level in all tests. This demanded accuracy vs execution trade-off. If we ignore the accuracy of 95% for ACM, better execution times could be achieved.

4.4 Effect of the Changes to Query Parameters on Efficiency

Two experiments were designed to show the non-significant effect of changes to query parameters on the proposed algorithm efficiency. In the first experiment, the number of query points $|Q|$ varied from 1 to 40 and in the second one, the number of the keywords of query points, $|Q.\psi|$, varied from 1 to 20. Other parameters were set in their default value and remained unchanged in all tests. According to Figs. 5 and 6, the changes of ACM resemble the changes of ASTD while the changes of ECM resemble the changes of BSTD. In addition, ACM is faster than ECM by 2 to 4 times.

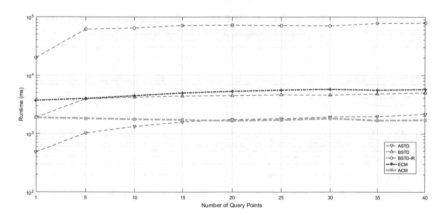

Fig. 5. The effect of *query points* on execution times of different algorithms

Focusing on the results of the tests reveals two crucial points. The first one is that the efficiency of our proposed methods is as good as that of competitors. The second point is that the execution time of algorithms arguably remained almost unchanged despite making a change in parameters. Therefore, any change in parameters has no significant effect on the efficiency of the proposed algorithms.

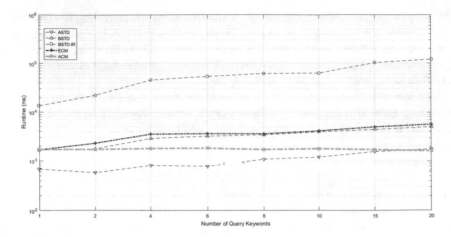

Fig. 6. Effect of the number of the keywords of query points on algorithms runtime

5   Conclusion

Spatio-textual skyline query returns all data points which are not spatio-textually dominated by other points. This paper provided an approximate method for spatio-textual
skyline query, assessed its efficiency and compared it with exact version of the method
and other competitor algorithms. According to the results of many experiments, the
proposed method generally shows a higher efficiency than other methods of this field.
If users do not need the accuracy of 100%, they can provide a trade-off between runtime
and accuracy in order to neglect the accuracy of final result, to some extent, and get the
result in a shorter time, instead. However, this method, and other competitor methods,
fails to show good efficiency on large-scale input data. In future, we will provide a
method to increase the scalability of spatio-textual skyline query using distributed
methods. The implementation of spatio-textual index structures and consolidating them
with ACM is another approach for improving the efficiency of the proposed method.

References

1. Bao, J., Mokbel, M.: GeoRank: an efficient location-aware news feed ranking system. In:
   Proceedings of the 21st ACM SIGSPATIAL International Conference on Advances in
   Geographic Information Systems, Orlando, Florida, USA, pp. 184–193. ACM, New York
   (2013)
2. Statistic Brain (2016). http://www.statisticbrain.com/google-searches. Accessed 10 Nov
   2016
3. Chen, L., Cong, G., Cao, X.: An efficient query indexing mechanism for filtering geo-textual
   data. In: Proceedings of the 2013 ACM SIGMOD International Conference on Management
   of Data, New York, USA, pp. 749–760. ACM, New York (2013)

4. Börzsönyi, S., Kossmann, D., Stocker K.: The skyline operator. In: Proceedings of the 17th International Conference on Data Engineering, Heidelberg, Germany, pp. 421–430. IEEE Computer Society (2001)
5. Kossmann, D., Ramsak, F., Rost, S.: Shooting stars in the sky: an online algorithm for skyline queries. In: Proceedings of the 28th International Conference on Very Large Databases, Hong Kong, China, pp. 275–286. VLDB Endowment (2002)
6. Chomicki, J., Godfrey, P., Gryz, J., Liang, D.: Skyline with presorting. In: Proceedings of the 19th International Conference on Data Engineering, Bangalore, India, pp. 717–719. IEEE Computer Society (2003)
7. Shi, J., Wu, D., Mamoulis, N.: Textually relevant spatial skylines. IEEE Trans. Knowl. Data Eng. **28**, 224–237 (2016)
8. Mullesgaard, K., Pedersen, J., Lu, H., Zhou, Y.: Efficient skyline computation in MapReduce. In: Proceedings of the 17th International Conference on Extending Database Technology (EDBT), Athens, Greece, pp. 37–48. OpenProceedings.org (2014)
9. Woods, L., Alonso, G., Teubner, J.: Parallel computation of skyline queries. In: IEEE 21st Annual International Symposium on Field-Programmable Custom Computing Machines (FCCM), Seattle, Washington, USA, pp. 1–8. IEEE Press (2013)
10. Tan, K., Eng, P., Ooi, B.: Efficient progressive skyline computation. In: Proceedings of the 27th International Conference on Very Large Databases, Rome, Italy, pp. 301–310. Morgan Kaufmann Publishers Inc. (2001)
11. Sharifzadeh, M., Shahabi, C.: The spatial skyline queries. In: Proceedings of the 32nd International Conference on Very large Databases, Seoul, South Korea, pp. 751–762. VLDB Endowment (2006)
12. Lee, M., Son, W., Ahn, H., Hwang, S.: Spatial skyline queries: exact and approximation algorithms. GeoInformatica **15**, 665–697 (2011)
13. Lin, Q., Zhang, Y., Zhang, W., Li, A.: General spatial skyline operator. In: 17th International Conference Database Systems for Advanced Applications: DASFAA, Busan, South Korea, pp. 494–508. Springer (2012)
14. You, G., Lee, M., Im, H., Hwang, S.: The farthest spatial skyline queries. Inf. Syst. **38**, 286–301 (2013)
15. Lee, K., Zheng, B., Li, H., Lee, W.: Approaching the skyline in Z order. In: Proceedings of the 33rd International Conference on Very large Databases, Vienna, Austria, pp. 279–290. VLDB Endowment (2007)
16. Lee, K., Lee, W., Zheng, B., Li, H., Tian, Y.: Z-SKY: an efficient skyline query processing framework based on Z-order. The VLDB J. **19**, 333–362 (2010)
17. Liu, B., Chan, C.: ZINC: efficient indexing for skyline computation. Proc. VLDB Endow. **4**, 197–207 (2010)
18. Chen, Y., Lee, C.: The σ-neighborhood skyline queries. Inf. Sci. **322**, 92–114 (2015)
19. Sacharidis, D., Arvanitis, A., Sellis, T.: Probabilistic contextual skylines. In: IEEE 26th International Conference on Data Engineering (ICDE), Long Beach, CA, USA, pp. 273–284. IEEE Press (2010)
20. Hose, K., Vlachou, A.: A survey of skyline processing in highly distributed environments. VLDB J. **21**, 359–384 (2012)
21. De Matteis, T., Di Girolamo, S., Mencagli, G.: A multicore parallelization of continuous skyline queries on data streams, In: Euro-Par 2015: Parallel Processing: 21st International Conference on Parallel and Distributed Computing, Vienna, Austria, pp. 402–413. Springer (2015)

22. Liou, M., Shu, Y., Chen, W.: Parallel skyline queries on multi-core systems. In: Proceedings of the 14th International Conference on Parallel and Distributed Computing, Applications and Technologies, Taipei, Taiwan, pp. 287–292. IEEE Press (2013)
23. Manning, C.D., Raghavan, P., Schütze, H.: Introduction to Information Retrieval. Cambridge University Press, New York (2008)

SMI-Based Opinion Analysis of Cloud Services from Online Reviews

Emna Ben-Abdallah$^{(\boxtimes)}$, Khouloud Boukadi, and Mohamed Hammami

Mir@cl Laboratory, Sfax University, Sfax, Tunisia
emnabenabdallah@ymail.com

Abstract. Nowadays, online reviews have become one of the most useful sources upon which cloud users can rely on to construct their purchasing decisions. This widespread adoption of online reviews results in nourishing the online opinion-based information. Analyzing and studying this kind of information help considerably in the fastidious task of cloud user decision-making. The contribution of this volume can be divided into two major parts: (i) a new opinion mining based cloud service analysis approach for the cloud service selection purpose. The proposed approach extracts and classifies user opinions from online reviews according to each cloud service property and (ii) an Opinion based cloud service ontology to effectively detect cloud service properties from reviews based on Service Measurement Index (SMI) metrics. To illustrate the proposed approach, we develop some experiments and we present the relevant results.

Keywords: Cloud computing · Opinion mining · Social analytic
SMI · Service property · Ontology

1 Introduction

Over the past few years, online reviews have played an important role in providing useful product/service insights. There are several domains in which online reviews are requisite, such as hotels, restaurants, retail outlets, airlines, and other Internet related services. Similar to shopping a cloud user finds other users' comments in online reviews as the most helpful to take a certain decision when it comes to purchase a cloud service [1]. However, browsing through the extensive collection of reviews to search for useful information might be a time-consuming and a tedious task. Hence, opinion mining, sentiment analysis and feature analysis are used in the present paper to analyze and understand the emotional tone of the shared reviews.

Sentiment analysis entails several interesting and challenging tasks. One traditional and fundamental task is polarity classification, which determines the overall polarity (e.g., positive or negative) of a sentence or a document [2]. However, judging the whole sentence or the document won't give a credible insight about the latter. In other words, in many cases, users give their opinions about cloud services in terms of the quality of the service (QoS) properties such as

© Springer International Publishing AG, part of Springer Nature 2018
A. Abraham et al. (Eds.): ISDA 2017, AISC 736, pp. 683–692, 2018.
https://doi.org/10.1007/978-3-319-76348-4_66

availability, security, etc., for example, "price is cheap and availability is worse". On the other hand, the service property based sentiment analysis faces different challenges such as the heterogeneity of the service and QoS property descriptions within the user reviews. Opinion mining based on a dedicated cloud service ontology can be used to overcome the lack of common understanding. By doing so, the cloud service ontology will be the building block for an easy cloud service property detection from user reviews and an effective storage for user opinion scores. The primary purpose of this paper is to evaluate QoS properties for each cloud service based on feature-level sentiment analysis from online reviews that are available on well known cloud platforms such as cloudReviews and TrustRadius. Thus, we propose an Opinion mining Based Cloud Service Ontology (OBCSO) that covers the cloud service types and accounts for a set of service properties as well as their associated opinion scores. The main contributions of this paper are summarized as the following:

1. An Opinion Based Cloud Service Ontology (OBCSO) that models QoS cloud service properties based on cloud and CSMIC-SMI (Cloud Service Measurement Index Consortium - Service Measurement Index) [3] standards to effectively extract properties and opinions from online reviews.
2. A cloud service opinion analytic approach based on sentiment analysis from online reviews.

The rest of this paper is organized as follows. Section 2 describes related works that are concerned with online reviews and sentiment analysis. Section 3 defines the OBCSO ontology that will be used to analyze cloud services based on opinion mining in Sect. 4. Section 5 presents the qualitative and quantitative evaluations of the proposed approach. Finally, the paper is enclosed with a summary of our contributions and their future potential extensions.

2   Related Work

A multitude of theories in literature are meant to evaluate and analyze the consumer's views on products or services using other users' insights through online reviews in several domains. For instance, Ye [4] has proved that online users reviews have an important impact on online hotel bookings. In addition, the work of Han [1] has studied the factors that influence users' perceptions of online reviews' usefulness of experiencing goods. To analyze and understand the large number of online reviews, many studies have adopted the sentiment analysis techniques. For instance, Fang and Zhan [2] have proposed a sentiment analysis system using product review data collected from Amazon.com. They have applied their experiments on the sentence and document levels. The authors in [5] have proposed a feature based approach for review mining using appraisal words. They have presented a new framework for review classification that uses appraisal words lexicon and product feature extraction for review categorization. In the cloud computing field, Alkalbani et al. [6] have analyzed, in their paper, around 6000 cloud users reviews to identify the attitude (positive or negative) of each review. The authors have performed a document level sentiment analysis based on four models using four supervised machine learning algorithms: K-Nearest Neighbor (K-NN), Naive Bayes, Random Tree and Random Forest. Most works that are based on sentiment analysis or opinion mining involve text or document categorization into positive or negative. Existing applications of sentiment analysis include stock market analysis, foretelling election results, movie reviews, etc. Sentence level or document level work is done predominantly instead of feature level sentiment analysis. Nevertheless, the document or the sentence level cannot highlight what the user precisely liked or not. In reality, in many cases consumers give their opinions about a cloud service/product in terms of features/service properties. Indeed, We believe that feature-level [7] is the most appropriate solution to extract users' opinions about a cloud service properties. Existing studies have shown that the sentiment analysis techniques based on machine learning are more likely suitable to be used in document-level sentiment analysis, while the lexicon-based sentiment analysis techniques are well suited for the feature-level sentiment analysis. Besides, the role of domain ontology is meagerly or not used in sentiment analysis works. To the best of our knowledge, less work is done on this research area using semantic web ontologies like OWL, RDF, SPARQL for sentiment analysis, where phrases or features are considered.

3   OBCSO: Opinion-Based Cloud Service Ontology

During the past few years, ontology has gained a lot of attention from both industrial and research communities belonging to different areas, such as cloud service selection, opinion mining and information extraction [8,9]. To enhance the pertinence and relevancy of the discovered service properties and their related opinions, we propose a dedicated domain ontology. It is worth mentioning that the use of ontology in the search process from user reviews is of a prominent importance as it allows for a better information discovery and a well knowledge extraction. The proposed ontology called Opinion-Based Cloud Service Ontology (OBCSO) is constructed while analyzing standards and proposals from both the industry and the literature, such as NIST[1], OCCI[2], CIMI[3] and metric standard such as CSMIC-SMI [3]. The proposed ontology encompasses three interrelated levels (see Fig. 1): service, property and opinion levels.

Service level: The service level, built upon our previous ontology called CSO (Cloud Service Ontology) [9], covers the three layers of cloud models, namely IaaS, PaaS and SaaS. CSO ontology relies on the cloud standards such as NIST, OCCI, CIMI, etc.

---

[1] http://www.nist.gov/itl/cloud.

[2] http://occi-wg.org.

[3] https://www.dmtf.org/standards/cloud.

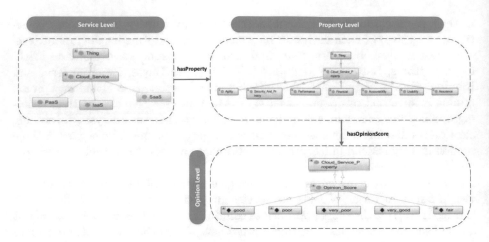

Fig. 1. OBCSO ontology representation

Property level: The property level contains all sub-classes needed to describe properties of cloud services. QoS property classes are defined based on CSMIC-SMI [3] metric standard to assess different cloud services based on user requirements. CSMIC-SMI was launched by Carnegie Mellon University as a standard measure to evaluate any service based on the user's requirements. The major categories of the SMI metrics are discussed as follows: (1) Accountability which is evaluated through different metrics such as auditability, compliance, data ownership, provider ethicality, sustainability, etc.; (2) The agility which depends on adaptability, elasticity, portability, etc. which can be measured in terms of how quickly new features can be incorporated into an IT infrastructure; (3) Performance which can be evaluated based on suitability, interoperability, accuracy and so forth; (4) Assurance which can be defined as the likelihood of the expected performance in the SLA[4] and the original performance. To evaluate assurance, SMI considers reliability, resiliency and service stability; (5) Security and Privacy which includes confidentiality, data integrity, access control, etc.; and (6) Usability which can be assessed with respect to accessibility, installability, learnability and operability.

Opinion level: In this level, we define Opinion_Score class to describe the different opinions related to cloud user feelings about a service property. To associate an opinion to each service property, we define five Opinion_Score individuals: very poor, poor, fair, good and very good. The association of each property to an opinion_Score is described in Sect. 4.

---

[4] Service Level Agreement.

4   Opinion-Mining Based Cloud Service Analysis

The principal objective of the proposed approach is to analyze cloud service properties based on cloud user opinions from online reviews by employing OBCSO ontology in the selection of features and SentiwordNet in the classification of the opinions. In order to achieve these goals, we present a framework composed of four main modules (see Fig. 2), namely, the data collection and Natural Language Processing (NLP), the OBCSO-based feature identification, the opinion mining and OBCSO interrogation modules. A detailed description of these components is provided below.

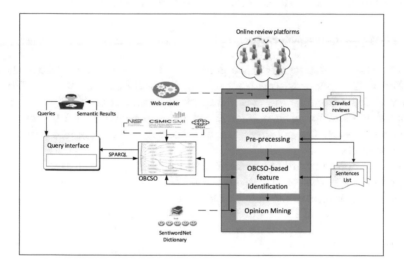

Fig. 2. Opinion-mining based cloud service analysis approach

4.1   Data Collection and Pre-processing Module

Data collection refers to the collection of cloud service reviews from the well known review platforms using the Web crawler named Crawler4j[5]. The latter is an open source Java crawler which provides a simple interface for crawling the Web. The crawling is carried out periodically in order to be update with the new reviews and the new cloud service releases. Secondly, NLP pre-processing procedures such as tokenization, Part-Of-Speech (POS) tagging, stop word removal and stemming are used to pre-process user reviews to eliminate needless information, such as dates, user's name, and tags and to increase the accuracy of the information search. Moreover, we split compound reviews to simple sentences [7].

---

[5] https://github.com/yasserg/crawler4j.

4.2   OBCSO-Based Feature Identification Module

To identify cloud service properties mentioned in a sentence, we use the proposed OBCSO ontology for mapping sentence onto property classes. This module is divided into two basic steps: The first step corresponds to the extraction of features from sentences. We assume each noun phrase as a feature. For example, the *performance* noun in the sentence 'performance is exceptional' is extracted as a candidate feature. The second step consists in mapping each candidate feature with the OBCSO ontology. In another word, each extracted feature from a sentence is compared with OBCSO property classes. Once the extracted feature from a sentence matches with a service property or it synonym, the next phase is proceeded to analyze the opinion related to this property.

4.3   Opinion Mining Module

Based on the set of matched features identified in the previous phase, the main objective of this phase is to extract service property opinion words. The adjectives, verbs and adverbs are considered as opinion words. Then, we determine the polarity score for each opinion word using SentiWorNet 3.0 dictionary as well as we compute the polarity score and the normalized opinion score for each service property.

Given an opinion word ow, SentiwordNet provides 3 measures for determining the opinion score: $Pos(ow)$: Positive score, $Neg(ow)$: Negative score for ow and $Neu(ow)$: Neutral score for ow. The more neutral the opinion word is, the less opinionated it is i.e. if this value is 1, it implies that the Pos(ow) and Neg(ow) are zero. So, given the positive and negative scores, the neutral score can be computed as follows:

$$Neu(ow) = 1 - (Pos(ow) + Neg(ow)) \tag{1}$$

Then, the opinion score for each service property is determined as follows:

$$opinion(sp) = \frac{\sum SentiScore(ow_i)}{N} \tag{2}$$

Where sp is a service property, ow_i is an opinion word and N is the number of opinion word concerning the service property sp.

Normalizing opinion score for a service property sp is determined as follows:

- $op\_score(sp) = verypoor$ if $opinion(sp) \leq -0.5$;
- $op\_score(sp) = poor$ if $-5 < opinion(sp) < 0$;
- $op\_score(sp) = fair$ if $opinion(sp) = 0$;
- $op\_score(sp) = good$ if $0 < opinion(sp) < 0.5$;
- $op\_score(sp) = verygood$ if $opinion(sp) \geq 0.5$;

4.4   OBCSO Interrogation Module

In order to apply the proposed approach in an adequate cloud services analysis, we implement an Opinion Based Cloud Service Interrogation Module (OBCSO-IP). The OBCSO-IP can be adopted by users who are generally unfamiliar with the SPARQL language. Indeed, this module offers an adequate interface for the user to express his query in natural language and then applies NLP rules and feature mapping as defined in Sects. 4.1 and 4.2 to generate the equivalent SPARQL query. To retrieve results from the OBCSO, the OBCSO-IP relies on Jena API [10], which is a Java API that provides classes and interfaces for the management of OWL based ontologies. The running of this module is shown in Fig. 3 which depicts the results of two distinct queries from two cloud users. The first user wants to evaluate virtual machines belonging to different providers while referring to their scalability as a QoS. The second requests for virtual machine with a high availability. The result of the first query is presented in Fig. 3(a), in which the scalability opinion score for each virtual machine is showed. While, the second query result is displayed in Fig. 3(b).

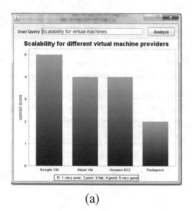   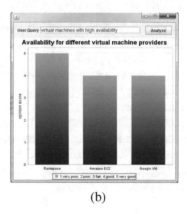

(a)   (b)

Fig. 3. Cloud user interface examples

5   Experiments

To conduct the experiments, we adopted real-life cloud service reviews datasets from CloudReviews, Trustradius, G2CROWD and Clutch. These reviews were published between January 2015 and August 2017 and mainly concern the IaaS (especially the VMs). In particular we choose to analyze the most recognized cloud service providers such as: Amazon Elastic Compute Cloud (VM1), Michrosoft Azure VM (VM2), Google Compute Engine VM Servers (VM3) and Rackspace Virtual Cloud (VM4). The proposed approach is developed using the Java language under the Eclipse environment[6]. Moreover, a set of components

---

[6] https://eclipse.org/ide/.

are exploited to better achieve its functionality. The construction and the visualization of the proposed ontology is achieved by the Protégé. The Stanford CoreNLP parser[7] is applied to POS-Tag reviews and the GATE (Generalized Architecture Text Engineering) [11] for stemming them. Two series of experiments were conducted in order to validate our approach: the first one evaluates the quality of the approach, while the second compares the number of reviewers considered for property ratings in our approach to the ones used in existing reviews platforms (Table 1).

Table 1. Dataset description

Dataset	VM1	VM2	VM3	VM4	All
#Reviews	1532	1638	973	654	4797
#Sentences	4986	4253	1853	1571	12663

5.1   Experiment1: Qualitative Evaluation

This experiment is conducted in conjunction with cloud instructors from the IT department of our university (considered as expert) and aims at evaluating the quality of our approach in terms of performance. For this purpose, 10 queries have been randomly generated, each with various QoS requirements. The number of requirements is increased in each query and the performance is measured. To meet each query, the cloud instructors relied on cloud service benchmarking tools, such as (cloudHarmony[8], Cloudlook[9] and Cloudorado[10]). Then, we evaluate our approach with regard to the selection decisions taken by the cloud instructors. The metrics used in this comparison are Precision (P) and Recall (R) [12]. Besides, we compare our approach to the ratings provided by cloud service review platforms.

$$P = \frac{Number\,of\,Relevant\,services \cap Number\,of\,Retrieved\,services}{Number\,of\,Retrieved\,services} \quad (3)$$

$$R = \frac{Number\,of\,Relevant\,services \cap Number\,of\,Retrieved\,services}{Number\,of\,Relevant\,services} \quad (4)$$

Figure 4 summarizes the results of our approach precision and recall compared to TrustRadius, G2CROWD and Clutch as review platforms. From the results in Fig. 4(a), we notice that our approach provides a high precision in most cases compared to platforms' reviews. That is, in the majority of the queries, our approach achieves the desired results as outlined by the experts. For the approach sensitivity, Fig. 4(b) demonstrates the result pertinence of our approach.

---

[7] http://nlp.stanford.edu/software/tagger.shtml.
[8] http://cloudharmony.com/.
[9] http://www.cloudlook.com/.
[10] www.cloudorado.com.

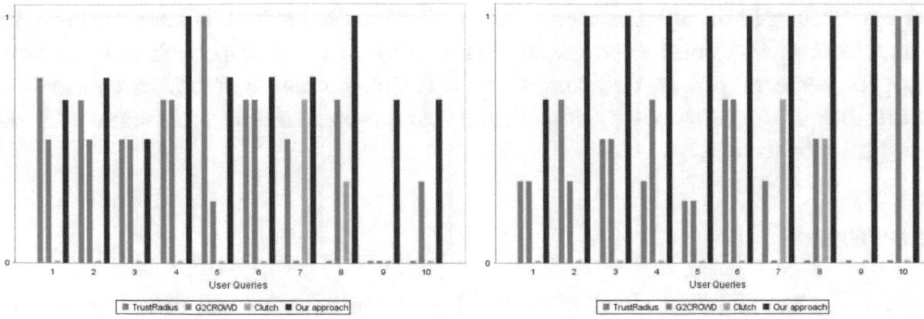

(a) (b)

Fig. 4. Performance results

Table 2. Reviewers number for service property rating

Service property	TrustRadius				G2CROWD				Clutch				Our approach			
	VM1	VM2	VM3	VM4	VM1	VM2	VM3	VM4	VM1	VM2	VM3	VM4	VM1	VM2	VM3	VM4
Usability	3	2	1	-	76	86	35	17	175	158	161	30	**826**	**628**	**512**	**105**
Availability	1	2	-	-	-				-				**752**	**584**	**474**	**261**
Support	3	2	-	-	97	104	27	24	175	158	161	30	**713**	**943**	**260**	**123**

5.2 Experiment2: Quantitative Evaluation

Table 2 presents the number of reviewers considered in service properties' ratings in TrustRadius, G2CROWD, Clutch as well as in our approach. As shown in this table, there is a huge difference between the number of reviewers present in the review platforms in comparison to our approach. This could be explained by the fact that, in the review platforms, the service property ratings are directly given by the reviewers. And, the majority of consumers prefer to just write the reviews. Contrariwise, our approach deduces the property ratings from the contents of the reviews through the OBCSO-based feature identification and the opinion mining modules. Indeed, the ratings of all service providers are unbiased since they are based on user's volunteer reviews.

6 Conclusion

Reviews, in online platforms, can be considered as one of the most useful kind of information for cloud users since it enlightens them for a better services' selection. Opinion mining, for instance, paves the way for automatic analysis of reviews and the extraction of what is relevant to users. Linking opinions to the cloud service properties is, however, a difficult task because of the heterogeneity of service property descriptions within users reviews. To tackle these issues, we proposed OBCSO ontology that defines cloud service properties based on CSMIC-SMI standard metrics to ease the property detection in reviews and to link these properties to an opinion score. Moreover, we proposed a new approach that extracts opinion words for each cloud service property and then deduces its opinion score. The conducted experiments evaluate our approach and demonstrate its performance and efficiency. As a future endeavor, we plan to increase the number of user reviews by considering other social media sites such as Facebook and Twitter.

References

1. Han, A., Hao, L., Jifan, R.: An empirical study on inline impact factors of reviews usefulness based on movie reviews. In: 2016 13th International Conference on Service Systems and Service Management (ICSSSM), pp. 1–5, June 2016
2. Fang, X., Zhan, J.: Sentiment analysis using product review data. J. Big Data 2(1), 5 (2015)
3. Cloud Service Measurement Index Consortium (CSMIC): SMI Framework. http://beta-www.cloudcommons.com/servicemeasurementindex
4. Ye, Q., Law, R., Gu, B.: The impact of online user reviews on hotel room sales. Int. J. Hosp. Manag. 28(1), 180–182 (2009)
5. Chaudhari, D.D., Deshmukh, R.A., Bagwan, A.B., Deshmukh, P.K.: Feature based approach for review mining using appraisal words. In: 2013 International Conference on Emerging Trends in Communication, Control, Signal Processing and Computing Applications (C2SPCA), pp. 1–5, October 2013
6. Alkalbani, A.M., Gadhvi, L., Patel, B., Hussain, F.K., Ghamry, A.M., Hussain, O.K.: Analysing cloud services reviews using opining mining. In: 2017 IEEE 31st International Conference on Advanced Information Networking and Applications (AINA), pp. 1124–1129, March 2017
7. Jeong, H., Shin, D., Choi, J.: FEROM: feature extraction and refinement for opinion mining. ETRI J. 33, 720–730 (2011)
8. Ali, F., Kwak, K.S., Kim, Y.G.: Opinion mining based on fuzzy domain ontology and support vector machine: a proposal to automate online review classification. Appl. Soft Comput. 47, 235–250 (2016)
9. Rekik, M., Boukadi, K., Ben-Abdallah, H.: Cloud description ontology for service discovery and selection. In: Proceedings of the 10th International Conference on Software Engineering and Applications ICSOFT-EA, vol. 1, pp. 26–36 (2015)
10. Carroll, J.J., Dickinson, I., Dollin, C., Reynolds, D., Seaborne, A., Wilkinson, K.: Jena: implementing the semantic web recommendations. In: Proceedings of the 13th International World Wide Web Conference on Alternate Track Papers & Amp; Posters. WWW Alt. 2004. ACM, New York, pp. 74–83 (2004)
11. Cunningham, H.: GATE, a general architecture for text engineering. Comput. Humanit. 36(2), 223–254 (2002)
12. Manning, C.D., Raghavan, P., Schütze, H.: Introduction to Information Retrieval. Cambridge University Press, New York (2008)

Heuristics for the Hybrid Flow Shop Scheduling Problem with Parallel Machines at the First Stage and Two Dedicated Machines at the Second Stage

Zouhour Nabli[1(✉)], Soulef Khalfallah[2], and Ouajdi Korbaa[1]

[1] MARS (Modeling of Automated Reasoning Systems) Research Lab LR17ES05,
Higher Institute of Computer Sciences and Communication Technologies (ISITCom),
University of Sousse, Sousse, Tunisia
nablizouhour@yahoo.fr
[2] Higher Institute of Management, University of Sousse, Sousse, Tunisia

Abstract. In this article, we present four heuristics for solving the problem of the hybrid flow shop scheduling problem denoted HFS with parallel machines at the first stage and two dedicated machines at the second stage. We compare heuristics with lower bounds and upper bounds generated by a mathematical model.

Keywords: Flow shop · Hybrid · Dedicated machines
Parallel machines · Heuristics

1 Introduction

The scheduling problems in a production workshop system can be classified according to the number of machines and their order of use to manufacture a product. In addition to that, it depends on job characteristics and the evaluation criteria. In this paper we consider the hybrid flow shop problem defined as follow: a set of n jobs j= 1, .., n must be processed on two stages in series e = 1, 2. The first stage contains m identical parallel machines and the second stage contains two dedicated machines. Each job must be processed on one machine for every stage and without interruption; each machine can process one O_{ij} operation at a time. For stage two, each job is pre-assigned to one machine. We assume that the storage space between machines is sufficient. The objective is to minimize the completion time of the last job on stage 2, called makespan and denoted Cmax. This kind of configuration is encountered in the food industry where stage 1 corresponds to drying cabins and stage 2 to packing machines. There are many types of configurations for the two-stage hybrid flow shop problem with dedicated machines that we can mention; Riane et al. [1] treat the n-job, two-stage hybrid flow shop problem with one machine in the first stage and two dedicated machines in the second stage. The objective is to minimize the makespan.

© Springer International Publishing AG, part of Springer Nature 2018
A. Abraham et al. (Eds.): ISDA 2017, AISC 736, pp. 693–701, 2018.
https://doi.org/10.1007/978-3-319-76348-4_67

All the jobs are initially processed at the first stage. Then the jobs are transferred to the second stage, which is composed of two machines that are dedicated to two different classes of products in the sense that the set of jobs is partitioned into two subsets and each subset goes to only one of these two dedicated machines. They present an exact algorithm based on a dynamic program (DP) and three heuristics. Lin and Liao [2] consider a two-stage hybrid flow shop with the characteristics of a single high speed machine and sequence-dependent setup time at stage 1, dedicated machines at stage 2, and two due dates. They proposed a heuristic to find the near-optimal schedule for the problem. The performance of the heuristic was evaluated by comparing its solution with both the optimal solution obtained by a branch and bound algorithm for small-sized problems, and the solution obtained by the scheduling method currently used in the shop. Wang and Liu [3] consider a two-stage hybrid flow shop scheduling problem with dedicated machines, in which the first stage contains a single common critical machine, and the second stage contains several dedicated machines. To solve the problem, they proposed a heuristic method based on branch and bound (B&B) algorithm. They derived several lower bounds and they used four constructive heuristics to obtain initial upper bounds. Lin [4] developed a linear-time algorithm for the problem of two-stage flow shop scheduling with a common machine at stage one and two parallel dedicated machines at stage two. Harbaoui et al. [5] consider the opposite of our problem that can be defined as a two-stage hybrid flow shop scheduling problem with dedicated machines at the first stage and parallel machines at the second stage. They proposed a mathematical model for the problem and developed an upper bound using genetic algorithm and a lower bound. Yang [6] consider the same problem studied in this paper; he has established some simple lower bounds and examines three heuristics. In [7] we proposed two mathematical models and lower bounds for the same problem. In this paper, we will present four heuristics and a new lower bound that we will compare with the mathematical model and the best lower bound that we proposed in [7].

2   Lower Bounds

In this section, we present two lower bounds. We use the following notation,

- P_{ej}: Processing time of job j at stage e,
- P_1^1 and P_1^2 are the processing time of jobs with smallest and second smallest processing time on stage 1.
- P_2^1: processing time of job P_1^1 at stage 2.
- N1: set of jobs dedicated to machine 1 at stage 2.
- N2: set of jobs dedicated to machine 2 at stage 2.
- P1: The sum of processing times of jobs at stage 2 that are dedicated to machine 1:

$$P1 = \sum_{j \in N1} P_2 j$$

- P2: The sum of processing times of jobs at stage 2 that are dedicated to machine 2:

$$P2 = \sum_{j \in N2} P_{2j}$$

In [7] we found that the LB_1 was the best bound compared to the other lower bound (see Eqs. (1)–(3)). LB_{11} (LB_{12}) is equal to the period of inactivity of machine 1 (machine 2) at stage 2 plus the sum of the processing times of jobs dedicated to machine 1 (machine 2) at stage 2 minus processing time at stage 2 of job with minimum processing time at stage 1.

$$LB_{11} = \max_{j \in N1}\{P_1^1 + P_1^2, P_2^1\} + P1 - P_1^2 \tag{1}$$

$$LB_{12} = \max_{j \in N2}\{P_1^1 + P_1^2, P_2^1\} + P2 - P_1^2 \tag{2}$$

$$LB_1 = \max\{LB_{11}, LB_{12}\} \tag{3}$$

When the processing times in stage 2 are smaller than the processing times in stage 1 the performance of LB_1 decreases. To solve this problem we have defined a new lower bound LB_2 as follow:

$$LB_2 = \left\{ \frac{\sum_{j=1}^{n} P_{1j}}{m} + \min_{j \in N}\{P_{2j}\} \right\} \tag{4}$$

LB_2 is equal to the sum of the processing times of jobs on stage 1 divided by the number of parallel machine at stage 1 plus the minimum processing time at stage 2. This lower bound takes into consideration the case where the processing time at stage 1 is strictly greater than the processing time at stage 2.

3   Heuristics

In this section we present a local search heuristic and three known priority rules that we adapted to our problem.

3.1   Local Search

The principle of this heuristic is to generate random permutations and apply the local search method on each permutation. In each permutation, we compare the best result obtained with the current solution, the better one become current. This heuristic allows us to visit a large number of possible solutions and improve them. We get a solution closer to the optimal solution. The generation of permutations concerns only stage 1, for stage 2 the jobs will be sorted according to their completion time at stage 1. The different steps of this heuristic are as follows:

1. Generate a set of permutations of jobs randomly on Stage 1.
   For each permutation:

2. Process the jobs one by one on the first available machine on stage 1 (non increasing order of starting time).
3. Process the jobs on stage 2 according to the first completion time on Stage 1.
4. Make swap permutations on the sequence of stage 1 and retain the permutation that improved result. Stop the swap permutation when we have not had any improvements in the iteration.

3.2   LPT (Longest Processing Time) on Stage 2

The principle of this houristic is to order the jobs according to their processing time on stage 2. So, jobs that have the maximum processing time on stage 2 will be processed first. The objective of this priority rule is to occupy the machines on stages 2 as much as possible at the beginning until the jobs in stage 1 are processed which minimizes the waiting time on stage 2. The different steps of this heuristic are as follows:

1. Group the jobs on stage 1 according to the machine on which they will be process on stage 2 alternately.
2. Sort each job group (sort descending) on stage 1 according to the processing time of the jobs on stage 2.
3. Sort the jobs on stage 1 in a table by taking a job from group 1 and a job from group 2.
4. Execute the jobs one by one on the first available machine on stage 1.
5. Process the jobs on stage 2 according to the first completion time on stage 1.

3.3   SPT (Smallest Processing Time) on Stage 1

The principle of this heuristic is to order the jobs according to their processing times on stage 1. Which allows us to process jobs that have the minimum processing time on stage 1 first. In this way, the machines on stage 2 begin the execution as soon as possible. The different steps of this heuristic are as follows:

1. Sort the jobs (sort ascending) on stage 1 according to their processing times on stage 1.
2. Process the jobs one by one on the first available machine on stage 1.
3. Process the jobs on stage 2 according to the first completion time on stage 1.

3.4   STPT (Shortest Total Processing Time)

The principle of this heuristic is to order the jobs according to their processing times on both stages. The goal is to process jobs that take less processing time on both stages first. The different steps of this heuristic are as follows:

1. Sort the jobs on stage 1 according to their total processing time.
2. Process the jobs one by one on the first available machine on stage 1.
3. Process the jobs on stage 2 according to the First Completion Time on stage 1 recording to FIFO.

4 Computational Analysis

4.1 Instances Generation

In this work, regarding processing times, we generate data based on [6]. We searched in the literature for benchmark problems (input) for the HFS with this configuration, but all articles do not include their input data, they only explain how they generate their processing times. We will test problems under different conditions of n, m, $n1$, $n2$ that we call problem family where n is the number of jobs, $n1$ number of jobs processed on machine 1 at stage 2, n2 number of jobs processed on machine 2 at stage 2, and m the number of identical parallel machines at stage 1. There are 45 families of problems. For each family, the processing times are generated randomly based on a uniform distribution; $P_{ej} \in$ [PLB, PUB] where PLB is a lower limit of the processing time and PUB is an upper limit. To test the effects of the number of jobs, three different values of n are considered: 10, 50, and 100 as in [7]. To determine the effect of job assignment (The number of jobs dedicated to the first and second machine at stage 2), we consider three different combinations of n is as follows for each value of n: $n1 = 0.5 * n$, $n1 = 0.6 * n$ and $n1 = 0.7 * n$. We consider three different distributions for processing times:

$P_{ej} :\in [1, 99]$, with standard deviation 29.42.
$P_{ej} :\in [25, 75]$, with standard deviation 14.93.
$P_{ej} :\in [40, 60]$, with standard deviation 6.17.

We added two other distributions for which the processing time of stage 1 and stage 2 are different:

$P_{1j} :\in [10 - 49], P_{2j}[50 - 99]$, with standard deviation 25.58.
$P_{1j} :\in [50 - 99], P_{2j}[10 - 49]$, with standard deviation 25.64 (Closer to real settings).

We set the number of machines at stage 1 to 2. Finally, to test the effect of numbers of parallel machines on stage 1, three values are considered: 2, 5, and 10. In these cases, we fixed n to 50 jobs. For each family of problems where $n \leq 50$, we generated 5 instances. To evaluate the performance of the heuristics we measure the relative deviation $RD = (HE - MB)/MB$, where HE is the result obtained by the heuristics (LS: Local Search, SPT_E1: Smallest Processing Time on stage 1, LPT_E2: longest processing time on stage 2, STPT: Shortest total processing time). MB is equal to the solution obtained by the mathematical model if it is optimal else MB is equal to the value of the best lower bound LB. The average relative deviation ARD is calculated as follows:

$$\frac{\sum_{i \in instances} (RD_i)}{number of instances}$$

4.2   Comparison of Results

In this section we will compare the results obtained by the LS heuristic and the priority rules (SPT_E1, LPT_E2 and STPT) with the available bound (optimal solution or best lower bound). We tested the heuristics over 5 instances; Each instance is composed of 15 problems for 5 different distributions for P_{ej}.

Problem of 10 jobs and 2 parallel machines. In Table 1; Where n = 10, the mathematical model is solved to optimality. Indeed, we compared the heuristics with the optimal solution. The results obtained show that the heuristic LS gives results that are very close or equal to the optimal for most of the instances with a maximum mean deviation of 1.2% and a computational time (CT) lower than 1 s. In the second place, the heuristic LPT_E2 gave a maximum average deviation of 8.8% with computational time less than that of the heuristic LS. The performance of the SPT_E1 rule and the STPT rule are poor compared to LS heuristic and the LPT_E2 rule.

Table 1. Comparisons of heuristics with the best bound for 10 jobs

			10 jobs and $m_1=2$														
			LS				SPT_E1				LPT_E2				STPT			
inst	PT	(n_1,n_2)	ARD	Max RD	Min RD	CT (s)	ARD	Max RD	Min RD	CT (s)	ARD	Max RD	Min RD	CT (s)	ARD	Max RD	Min RD	CT (s)
pb1		(5,5)	0%	0%	0%	1.0	18,4%	41,9%	0%	0	8,0%	14,4%	0%	0	23,2%	34,3%	2,5%	0
pb2	[1,99]	(6,4)	0,2%	0,8%	0%	0.9	16,8%	41,9%	0%	0	4,7%	10,7%	0%	0	20,6%	34,3%	0%	0
pb3		(7,3)	0%	0%	0%	0.9	13,3%	35,1%	0%	0	2,9%	10,1%	0%	0	12,1%	22,9%	0%	0
pb4		(5,5)	1,2%	4,1%	0%	0.9	7,3%	17,6%	0%	0	8,7%	13,8%	1,9%	0	11,4%	19,7%	5,4%	0
pb5	[25,75]	(6,4)	0%	0%	0%	0.9	4,8%	12,8%	0%	0	8,8%	12,5%	0%	0	3,7%	12,0%	0%	0
pb6		(7,3)	0%	0%	0%	0.9	0,6%	2,4%	0%	0	7,5%	10,8%	0%	0	1,0%	2,4%	0%	0
pb7		(5,5)	0,1%	0,3%	0%	0.9	17,0%	26,0%	6,6%	0	5,8%	18,8%	0%	0	18,2%	20,2%	15,6%	0
pb8	[40,60]	(6,4)	0%	0%	0%	0.9	6,2%	18,1%	0%	0	3,8%	12,4%	0%	0	3,7%	7,0%	0%	0
pb9		(7,3)	0%	0%	0%	0.9	3,2%	15,7%	0%	0	0,3%	1,6%	0%	0	0,7%	1,6%	0%	0
pb10		(5,5)	0%	0%	0%	0.9	3,2%	12,8%	0%	0	3,7%	9,4%	0%	0	3,2%	7,3%	0%	0
pb11	[10,49]	(6,4)	0%	0%	0%	0.9	0,7%	3,7%	0%	0	3,1%	8,0%	0%	0	1,3%	2,8%	0%	0
pb12	[50,99]	(7,3)	0%	0%	0%	0.9	0,6%	3,0%	0%	0	3,1%	6,8%	0%	0	0,9%	1,5%	0%	0
pb13		(5,5)	0%	0%	0%	0.9	7,7%	13,8%	1,1%	0	4,5%	6,5%	3,2%	0	15,1%	18,5%	9,2%	0
pb14	[50,99]	(6,4)	0%	0%	0%	0.9	7,9%	13,8%	1,1%	0	3,7%	7,0%	2,1%	0	15,2%	18,5%	9,8%	0
pb15	[10,49]	(7,3)	0%	0%	0%	1.0	7,9%	13,8%	0,8%	0	3,3%	7,0%	0,8%	0	14,9%	17,6%	10,1%	0

Problem of 50 jobs and 2 parallel machines. In Table 2; Where n = 50, for this size, the mathematical model was not able to give the optimal solution for most of the instances in acceptable time. So we compared the heuristic and the priority rules with the best lower bound. The results obtained show that the LS heuristic generates results equal or very close to the lower bound for most instances with a maximum ARD of 0.5% and maximum computational time of 19 s. The ARD of the priority rules LPT_E2, SPT_E1 and STPT is high. On the other hand, we note that even when n = 50, these heuristics do not exceed 1 s of computational time. In fact, priority rules have $O(n \log n)$ complexity.

Table 2. Comparisons of heuristics with the best bound for 50 jobs

inst	PT	(n₁,n₂)	LS				SPT_E1				LPT_E2				STPT			
			ARD	Max RD	Min RD	CT (s)	ARD	Max RD	Min RD	CT (s)	ARD	Max RD	Min RD	CT (s)	ARD	Max RD	Min RD	CT (s)
pb1		(25,25)	0,3%	1,5%	0%	18,7	2,0%	6,1%	0%	0	3,6%	4,9%	0,5%	0	13,6%	18,4%	1,2%	0
pb2	[1,99]	(30,20)	0%	0%	0%	18,3	0,1%	0,3%	0%	0	3,1%	4,6%	0,4%	0	7,6%	12,9%	0,8%	0
pb3		(35,15)	0%	0%	0%	18,1	0%	0,2%	0%	0	2,7%	4,0%	0,3%	0	4,0%	9,6%	0%	0
pb4		(25,25)	0%	0%	0%	18,0	2,3%	4,0%	0%	0	1,5%	3,0%	0,3%	0	5,0%	11,4%	1,6%	0
pb5	[25,75]	(30,20)	0%	0%	0%	18,0	0%	0%	0%	0	0,8%	1,7%	0,2%	0	0,8%	2,1%	0%	0
pb6		(35,15)	0%	0%	0%	18,3	0%	0%	0%	0	0,9%	2,5%	0,2%	0	0,4%	1,6%	0%	0
pb7		(25,25)	0,5%	1,3%	0%	18,6	6,8%	9,5%	4,4%	0	0,4%	0,8%	0,1%	0	10,8%	16,5%	6,6%	0
pb8	[40,60]	(30,20)	0%	0%	0%	18,2	1,6%	4,8%	0%	0	0,3%	0,5%	0%	0	0,3%	0,6%	0%	0
pb9		(35,15)	0%	0%	0%	18,2	0,9%	2,3%	0%	0	0,5%	0,9%	0,2%	0	0,1%	0,3%	0%	0
pb10		(25,25)	0%	0%	0%	18,2	0,1%	0,6%	0%	0	1,3%	1,8%	0,5%	0	1,2%	2,3%	0%	0
pb11	[10,49]	(30,20)	0%	0%	0%	18,5	0,2%	0,5%	0%	0	1,4%	1,7%	0,9%	0	0,2%	0,7%	0%	0
pb12	[50,99]	(35,15)	0%	0%	0%	18,4	0,1%	0,4%	0%	0	1,0%	1,4%	0,6%	0	0,2%	0,6%	0%	0
pb13		(25,25)	0,1%	0,2%	0%	18,6	1,9%	3,0%	0,8%	0	0,9%	1,5%	0,4%	0	3,3%	4,5%	1,8%	0
pb14	[50,99]	(30,20)	0,1%	0,2%	0%	19,0	1,9%	3,0%	0,8%	0	1,4%	1,8%	0,8%	0	3,6%	4,5%	1,8%	0
pb15	[10,49]	(35,15)	0,1%	0,2%	0%	18,2	1,9%	3,0%	0,8%	0	1,4%	2,6%	0,4%	0	3,6%	4,5%	1,8%	0

Problem of 100 jobs and 2 parallel machines. In Table 3; Where n = 100, the mathematical model was not able to give the optimal solution for most of the instances. We compared the heuristics with the best lower bound $LB = max(LB_1, LB_2)$. LS heuristic give results equal or very close to the lower bound for most instances with a maximum ARD of 0.8% and maximum computational time of 76.7 s. In the second place, LPT_E2 rule gave a maximum ARD of 1.4% in less time. The mean deviation of the heuristics SPT_E1 and STPT is very high compared to the heuristics LS and LPT_E2.

Table 3. Comparisons of heuristics with the best bound for 100 jobs

inst	PT	(n₁,n₂)	LS				SPT_E1				LPT_E2				STPT			
			ARD	Max RD	Min RD	CT (s)	ARD	Max RD	Min RD	CT (s)	ARD	Max RD	Min RD	CT (s)	ARD	Max RD	Min RD	CT (s)
pb1		(50,50)	0,2%	0,7%	0%	74,2	1,1%	3,9%	0%	0	1,1%	2,1%	0,2%	0	6,2%	9,4%	2,7%	0
pb2	[1,99]	(60,40)	0%	0%	0%	72,9	0,1%	0,2%	0%	0	1,6%	2,3%	0,7%	0	2,7%	7,6%	0,2%	0
pb3		(70,30)	0%	0%	0%	74,1	0%	0%	0%	0	1,1%	2,0%	0,5%	0	1,2%	2,6%	0,2%	0
pb4		(50,50)	0,3%	0,8%	0%	73,5	1,7%	3,7%	0%	0	1,4%	3,7%	0,1%	0	7,2%	8,6%	3,9%	0
pb5	[25,75]	(60,40)	0%	0%	0%	74,3	0,3%	0,8%	0%	0	0,6%	1,6%	0%	0	3,1%	6,6%	0,3%	0
pb6		(70,30)	0%	0%	0%	76,7	0,1%	0,7%	0%	0	0,4%	1,2%	0%	0	0,7%	1,9%	0,1%	0
pb7		(50,50)	0,8%	1,4%	0%	72,9	4,0%	6,7%	2,3%	0	0,8%	2,3%	0%	0	5,6%	8,8%	2,9%	0
pb8	[40,60]	(60,40)	0%	0%	0%	75,0	1,4%	3,4%	0%	0	0,2%	0,5%	0%	0	1,4%	3,6%	0%	0
pb9		(70,30)	0%	0%	0%	74,0	1,0%	2,9%	0%	0	0,2%	0,4%	0%	0	0,3%	1,3%	0%	0
pb10		(50,50)	0%	0%	0%	73,0	0,1%	0,3%	0%	0	0,6%	1,0%	0,1%	0	0,2%	0,5%	0%	0
pb11	[10,49]	(60,40)	0%	0%	0%	74,3	0%	0%	0%	0	0,5%	0,8%	0,1%	0	0,1%	0,2%	0%	0
pb12	[50,99]	(70,30)	0%	0%	0%	74,1	0%	0%	0%	0	0,4%	0,7%	0,1%	0	0,1%	0,2%	0%	0
pb13		(50,50)	0,1%	0,1%	0%	72,6	0,8%	1,1%	0,5%	0	0,4%	0,9%	0%	0	1,9%	2,2%	1,5%	0
pb14	[50,99]	(60,40)	0%	0,1%	0%	72,8	0,7%	0,9%	0,5%	0	0,4%	0,9%	0,2%	0	1,9%	2,2%	1,5%	0
pb15	[10,49]	(70,30)	0%	0,1%	0%	74,9	0,9%	1,4%	0,5%	0	0,5%	0,9%	0,2%	0	1,9%	2,2%	1,6%	0

Problem of 50 jobs and m parallel machines. We set n to 50 jobs for different number of parallel machines at stage 1; m = 2, 5 and 10. We tested two cases: the first when the processing times $P_{ej} : \in [1, 99]$ and the second when $P_{1j} \in [50, 99]$ and $P_{2j} \in [10, 49]$.

Case 1 problems. In Table 4; Where n = 50 and m = 2, 5 and 10, we set the processing times between 1 and 99. We compared the heuristics with the best lower bound. The results obtained show that the LS give the same results as the lower bound with a maximum ARD of 1.2% and computational time that does not exceed 20 s. In the second place, the SPT_E1 rule with maximum ARD of 2%.

Table 4. 50 jobs and m parallel machine at stage 1 $P_{e}j : \in [1, 99]$

inst	m_1	(n_1,n_2)	LS				SPT_E1				LPT_E2				STPT			
			ARD	Max RD	Min RD	CT (s)	ARD	Max RD	Min RD	CT (s)	ARD	Max RD	Min RD	CT (s)	ARD	Max RD	Min RD	CT (s)
pb1		(25,25)	0,3%	1,5%	0%	18,7	2,0%	6,1%	0%	0	3,6%	4,9%	0,5%	0	13,6%	18,4%	1,2%	0
pb2	2	(30,20)	0%	0%	0%	18,3	0,1%	0,3%	0%	0	3,1%	4,6%	0,4%	0	7,6%	12,9%	0,8%	0
pb3		(35,15)	0%	0%	0%	18,1	0%	0,2%	0%	0	2,7%	4,0%	0,3%	0	4,0%	9,6%	0%	0
pb4		(25,25)	0%	0%	0%	19,5	0%	0,2%	0%	0	3,1%	4,9%	0,5%	0	0,9%	2,9%	0%	0
pb5	5	(30,20)	0%	0,1%	0%	19,2	0%	0,1%	0%	0	1,5%	2,7%	0,4%	0	0,7%	2,4%	0%	0
pb6		(35,15)	0%	0,1%	0%	20,2	0%	0,1%	0%	0	1,3%	2,3%	0,3%	0	0,3%	0,9%	0%	0
pb7		(25,25)	0%	0%	0%	20,1	1,4%	2,7%	0%	0	0,2%	0,7%	0%	0	0%	0%	0%	0
pb8	10	(30,20)	0,1%	0%	0%	20,0	0%	0,1%	0%	0	0,8%	1,7%	0%	0	0,2%	0,5%	0%	0
pb9		(35,15)	0%	0,1%	0%	20,2	0%	0,1%	0%	0	0,9%	1,5%	0,3%	0	0,1%	0,2%	0%	0

Case 2 problems. In Table 5; we compared the heuristics with the best lower bound. The results confirm the performance of heuristic LS with a maximum ARD of 0.7% and computational time that does not exceed 20 s. In the second place, LPT_E2 rule with a maximum mean deviation of 1.7%.

Table 5. 50 jobs and m parallel machine at stage 1 $P_{1}j \in [50,99]$ and $P_{2}j \in [10,49]$

inst	m_1	(n_1,n_2)	LS				SPT_E1				LPT_E2				STPT			
			ARD	Max RD	Min RD	CT (s)	ARD	Max RD	Min RD	CT (s)	ARD	Max RD	Min RD	CT (s)	ARD	Max RD	Min RD	CT (s)
pb1		(25,25)	0,1%	0,2%	0,1%	18,6	1,9%	3,0%	0,8%	0	1,3%	1,8%	0,7%	0	3,6%	4,5%	1,8%	0,8
pb2	2	(30,20)	0,1%	0,2%	0,1%	19,0	1,9%	3,0%	0,8%	0	1,5%	2,6%	0,7%	0	3,6%	4,5%	1,8%	0,4
pb3		(35,15)	0,1%	0,2%	0,1%	18,2	4,2%	7,3%	0%	0	1,7%	5,2%	0%	0	13,2%	21,5%	4,5%	0,2
pb4		(25,25)	0,7%	3,6%	0%	19,7	0,4%	2,2%	0%	0	1,0%	2,2%	0%	0	4,0%	6,9%	1,0%	0,4
pb5	5	(30,20)	0%	0%	0%	19,2	0%	0%	0%	0	0,9%	1,9%	0%	0	1,3%	3,9%	0%	0,4
pb6		(35,15)	0%	0%	0%	19,4	0%	0%	0%	0	0,7%	1,8%	0%	0	1,6%	4,8%	0,1%	0,5
pb7		(25,25)	0%	0%	0%	19,7	0%	0%	0%	0	0,6%	2,2%	0%	0	0,2%	0,7%	0%	0,4
pb8	10	(30,20)	0%	0%	0%	20,1	0%	0%	0%	0	0,7%	1,9%	0%	0	0,2%	0,6%	0%	0,4
pb9		(35,15)	0%	0%	0%	20,1	0%	0%	0%	0	0,1%	0%	0%	0	0%	0,1%	0%	0,1

5 Conclusion

In this paper we proposed a local search heuristic and adapted 3 priority rules for hybrid flow shop scheduling with parallel machines at the First stage and two dedicated machines at the second Stage. The results obtained are compared with a mathematical model and lower bounds. We demonstrated that, the local search heuristic proves its efficiency for different problem instances.

References

1. Riane, F., Artiba, A., Elmaghraby, S.E.: Sequencing a hybrid two-stage flowshop with dedicated machines. Int. J. Prod. Res. **40**(17), 4353–4380 (2002)
2. Lin, H.T., Liao, C.J.: A case study in a two-stage hybrid flow shop with setup time and dedicated machines. Int. J. Prod. Econ. **86**(2), 133–143 (2003)
3. Wang, S., Liu, M.: A heuristic method for two-stage hybrid flow shop with dedicated machines. Comput. Oper. Res. **40**(1), 438–450 (2013)
4. Lin, B.M.: Two-stage flow shop scheduling with dedicated machines. Int. J. Prod. Res. **53**(4), 1094–1097 (2015)
5. Harbaoui, H., Bellenguez-Morineau, O., Khalfallah, S.: Scheduling a two-stage hybrid flow shop with dedicated machines, time lags and sequence-dependent family setup times. In: IEEE International Conference on Systems, Man, and Cybernetics (SMC), Budapest, Hongrie, October 2016
6. Yang, J.: Hybrid flow shop with parallel machines at the first stage and dedicated machines at the second stage. Ind. Eng. Manag. Syst. **14**(1), 22–31 (2015)
7. Nabli, Z., Korbaa, O., Khalfallah S.: Mathematical programming formulations for hybrid flow shop scheduling with parallel machines at the first stage and two dedicated machines at the second stage. In: IEEE International Conference on Systems, Man, and Cybernetics (SMC), Budapest, Hongrie, October 2016

Breast Density Classification for Cancer Detection Using DCT-PCA Feature Extraction and Classifier Ensemble

Md Sarwar Morshedul Haque[1(✉)], Md Rafiul Hassan[1], G. M. BinMakhashen[1],
A. H. Owaidh[1], and Joarder Kamruzzaman[2,3]

[1] King Fahd University of Petroleum and Minerals, Dhahran 31261,
Kingdom of Saudi Arabia
{smhaque,mrhassan}@kfupm.edu.sa, sarwar.haque@gmail.com
[2] Federation University Australia, Ballarat, Australia
[3] Monash University, Melbourne, Australia

Abstract. It is well known that breast density in mammograms may
hinder the accuracy of diagnosis of breast cancer. Although the dense
breasts should be processed in a special manner, most of the research
has treated dense breast almost the same as fatty. Consequently, the
dense tissues in the breast are diagnosed as a developed cancer. In contrast, dense-fatty should be clearly distinguished before the diagnosis
of cancerous or not cancerous breast. In this paper, we develop such a
system that will automatically analyze mammograms and identify significant features. For feature extraction, we develop a novel system by
combining a two-dimensional discrete cosine transform (2D-DCT) and a
principal component analysis (PCA) to extract a minimal feature set of
mammograms to differentiate breast density. These features are fed to
three classifiers: Backpropagation Multilayer Perceptron (MLP), Support
Vector Machine (SVM) and K Nearest Neighbour (KNN). A majority
voting on the outputs of different machine learning tools is also investigated to enhance the classification performance. The results show that
features extracted using a combination of DCT-PCA provide a very high
classification performance while using a majority voting of classifiers outputs from MLP, SVM, and KNN.

Keywords: Breast cancer · Breast Dense and Fatty · DCT, PCA
Machine learning tools · Pattern Recognition

1 Introduction

According to National Cancer Institute (NCI), the cancer is a term used to
describe abnormal cells division without control that has the power to spread to
other tissues via blood and lymph system [1]. Breast cancer is one of the most
dangerous and lethal diseases that found to be common among females. However,
early breast cancer detection leads to high chances of survival. Several studies

© Springer International Publishing AG, part of Springer Nature 2018
A. Abraham et al. (Eds.): ISDA 2017, AISC 736, pp. 702–711, 2018.
https://doi.org/10.1007/978-3-319-76348-4_68

have pointed out the importance of the breast density as a mammographic risk indicator, since dense breasts can influence the interpretation of the mammogram compared to fatty ones. Therefore, automatic assessment of breast density will be highly beneficial for breast cancer screening.

Regardless of the amount of research conducted on breast cancer early screening, mortality rate from breast cancer remains high. Moreover, it was found that within the last three years 75–80% who came late and diagnosed with advanced stages resulted in a degraded successful treatment [2]. Therefore, an early breast cancer screening is recommended to be conducted from the age of 40 [3]. It is not known until now the exact relationship between breast density and cancer, but as per current understanding, it is most probable that the increased density turn to cancers tissue [4]. In this study, we aim to enhance the breast density detection using a two dimensional discrete cosine transform with Principle Component Analysis (2D DCT-PCA) as a feature extraction technique on mammogram images for accurate classification of dense-fatty breast types.

A typical 2D-DCT (for short DCT) technique transforms an image based on the frequency domain information. This information is divided into three parts (low, moderate and high bands). It has been found that the sensitivity to variations can be noticed by the human visual system in low-frequency band [5]. Therefore, this allows the important information to be concentrated within small area of the DCT domain. DCT has been widely used as feature extractor to detect micro calcification in the breast tissues. For instance, Farag and Moshali have adopted the DCT to locate the ROI that contains the micro-calcifications [3]. Prathibha and Sadasivam [6] have combined DCT and DWT to tackle benign-malignant breast tissue using Nearest Neighbor (KNN) classifier. Interestingly, to our knowledge there is no research that has adopted DCT to distinguish breast density from mammogram images.

In this paper, we propose and develop a combination of 2D-DCT feature extraction method with principal component analysis to identify the most influential features from mammogram images. These features are then fed to a number of well known classifiers, e.g. Multilayer Perceptron (MLP), Support Vector Machine (SVM) and K-nearest neighbor (KNN) to classify the images as either fatty or dense-glandular. This sort of classification is important in the early diagnosis of breast cancer. It has been noticed that none of the classifier can achieve very high classification accuracy. Hence, a majority voting approach is adopted to decide about the class of each mammogram image, which achieves acceptably high beast density detection accuracy.

2 Related Works

As separating dense breast tissues from the fatty ones in a mammogram image can be treated as a classification problem, it needs a set of good features to characterize inter and intra variations. Texture based analysis is one big field of feature extraction techniques that has been considered deeply to represent breast images with a set of statistical measurements of the textures [2]. Another work by Mudigonda et al. [7] extracted a gray-level co-occurrence matrices (GCM) features. Their feature set was based on a polygonal modeling of boundaries to excerpt a ribbon of pixels across the mass margin. Tan et al. [8] has used texture based image features to establish association between changes in mammographic image features and risk for breast cancer development. Gabor filter bank was also presented in terms of extracting texture representation of the mammograms. Hussain [9] has adopted a bank of Gabor filters that constituted of different scales (5) and angles (8) to represent micro-patterns. Muthukarthigadevi and Anand in [10] have used wavelet transform to analyze the portrait, angular spread of power and fractal analysis as a features to determine the existence of calcification cells.

Transformation based feature extraction techniques could lead to a curse of dimensionality such as wavelet transform, Gabor filter bank, Discrete Cosine Transform (DCT). These techniques have a strong way to represent the visual characteristics of the raw image. However, the resultant feature set may have high dimensionality and this will lead to expensive computations and besides redundant information. Therefore, a feature selection step may be necessary to reduce the features dimensionality and remove the redundant information.

Abnormality (e.g. dense Vs fatty) detection could be very hard from mammogram images. Therefore, some efforts have been devoted to develop automatic classification systems for breast density. For example, Oliver et al. [11] explored the applicability of morphological features for density classification. They reported an 81% classification accuracy using MIAS database. A recent review by Ganesan et al. [12] on computer aided breast cancer detection using mammogram suggests that K-NN, neural network and support vector classifiers are along the most widely used classifiers in this domain.

3   System Overview

In this work, we propose a novel method for breast density classification using a combination of DCT, PCA and majority voting approach of outputs from a number of classifiers. The method consists of four phases as shown in Fig. 1. The phases are:

- Data acquisition
- Preprocessing
- Feature extraction
- Classification

In the data acquisition, the system uses hardware (medical sensory) such as an Ultrasound or MRI for capturing some raw representation of the patients breast. After that, these representations (images) are preprocessed to concentrate only on specific location called region of interest (ROI). Furthermore, the raw images may need some enhancement such as intensity adjustment (contrast enhancement) [13]. Once the ROI is determined, in the phase of feature extraction, important features are extracted such that the features are able to distinguish among various classes (fatty/dense breast). In case of a very large

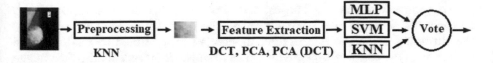

Fig. 1. System overview

number of features, the system may adopt a feature selection technique to filter out noisy or less significant features. Finally, the system classifies (recognize or detect) the mammograms through using the selected features.

4   Methodology

4.1   Preprocessing

To identify the region of interest (ROI) from a mammogram image, K-means algorithm has been adopted. K-means would group the image based on pixel gray level intensity [14]. An analysis of the available mammogram images revealed that there exist three main groups of pixels: black, white and other. Therefore, the well known clustering technique like K-means has been adopted. A search window of 8 by 8 cells has been used to find the cluster label for the center area of the mammogram to avoid small black background. Figure 2 illustrates an example of the input images in (A), the clustered version of this input is in (B), and (C) is the results after a searching window. The clustered points are obtained by minimizing the objective function:

$$O_d = n \sum_{j=1}^{k} \sum_{x_j \in I_j} (x_j - C_i) \tag{1}$$

where $I_j = 1, 2, \ldots, n$, C_i is the centroid of the cluster, is the data points.

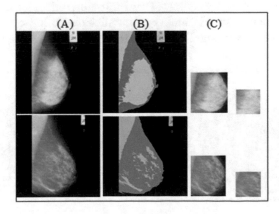

Fig. 2. Two examples of right breast clustering results, (A) original image, (B) after clustering representations, (c) using the window to select the cluster label and segment the ROI.

To remove noise in the images, the clustering technique is applied twice on each of the images, i.e. having received the resultant image after executing the clustering technique the resultant image is further tuned by executing the clustering. This removes background noise from the images. Finally, each image is re-sized within a fixed frame of width and height as 300 X 300.

4.2   Feature Extraction

In our proposed method we adopted the DCT to characterize the dense-fatty domain. Since DCT has been proved to be able to aggregate the important information into a small location in a number of studies [15], we used DCT in this study. Figure 3 shows how application of DCT can transform the image of breast mammograms. To find the DCT of image (for our case N = 300), the DCT can be computed as follows [3]:

$$C_{pq} = \alpha_p\alpha_q \sum_{x=1}^{N} \sum_{y=1}^{N} I(x,y)cos(\frac{(2x+1)\pi p}{2N})cos(\frac{(2y+1)\pi q}{2N}) \tag{2}$$

where (x, y) is the spatial coordinates of the image, p and q are the frequency coordinates $0 \leq p, q \leq N$, and α_p, α_q are DCT coefficients of the two dimensions that are computed as follows:

$$\alpha_{p,q} = \begin{cases} \frac{1}{\sqrt{N}}, & p = q = 0 \\ \sqrt{\frac{2}{N}}, & 1 \leq p, q \leq N \end{cases} \tag{3}$$

Breast density-fatty problem rely very much on the pixel level intensity. Therefore, every single pixel is important to be considered as a feature point. However, for a ROI of size 300×300, there are 90000 features which is huge. Therefore, we apply the following two strategies to reduce the number of features without losing any significant information.

Strategy 1: Since DCT concentrates the most significant information in the lower-band frequency, a small number of coefficients in zigzag manner starting from the upper left part of the DCT domain are selected as significant features.

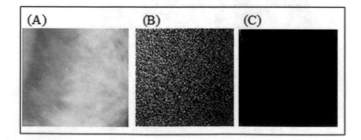

Fig. 3. Mammograms transformed using DCT, (A) is the ROI, (B) is the DCT Coefficients, (C) DCT transformed back into gray level image.

Strategy 2: A PCA is applied to reduce and select the best DCT coefficients as significant features.

4.3   Classification and Majority Voting

In the process of breast density-fatty classification, three well known classifiers are used: (1) Multilayer perceptron (MLP), (2) Support Vector Machine (SVM) and (3) K nearest neighbor (KNN).

MLP: MLP has been a popular tool to automate classification problems for the last two decades. The underlying learning blocks, i.e., artificial neurons of MLP are similar to a biological neuron. Being capable of dealing with nonlinear problems efficiently, MLP has been widely used to solve complex classification problems. In this paper, MLP has been used to classify breast density-fatty groups while the inputs are extracted features as discussed in Sect. 4.2.

SVM: SVM is another very popular classifier with strong generalization ability for nonlinear tasks using a kernel trick. Nevertheless, it may consume a very long time to find an optimal discriminant decision boundary. To reduce such computations the sequential minimal optimization (SMO) has been adopted. Details about SVM and SMO can be found in [16].

KNN: Compared to MLP and SVM the underlying methodology of KNN is very simple and straightforward. Given a target point in KNN algorithm distance be-tween the target point and each of the points in the training data is computed. Then, final decision is taken based on the nearest distance to the target point. Provided that each classifier has its own characteristics and capabilities, these classifiers may show a negative correlation in which the integration of these can drastically enhances the overall system performance [17]. Therefore, a majority voting technique has been adopted to combine the three classifiers predictions to a final decision. Figure 4 illustrates the combination of classifiers.

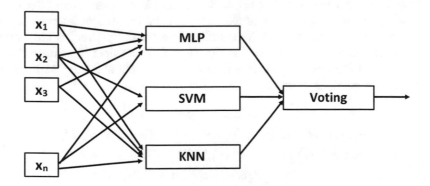

Fig. 4. Integrating classifiers using majority voting.

5   Experiments and Discussion

5.1   Mammography Dataset

To validate our proposed method, we used the Mammographic Image Analysis Society (MIAS) database that is publicly available at [18]. It contains 322 mammograms of left and right breasts. In this paper our aim is accurately classify fatty and dense breasts and hence we considered only those mammograms that belong to either fatty or dense-glandular in our experiments. The total number of mammogram images for fatty type is 115 and 105 for a dense glandular type.

5.2   Experimental Setup and Performance Metrics

The mammograms were initially preprocessed following methods described in Sect. 4.1. Then features were extracted as:

* Scenario 1: Select features from the preprocessed mammogram images by using PCA. In the subsequent part of this paper, we refer to this scenario as PCA.
* Scenario 2: Select 1000 features following the method as described in Sect. 4.2 (Strategy 1). Then select the most significant features out of these 1000 features by using PCA. In the subsequent part of this paper, we refer to this scenario as DCT(10^3)-PCA.
* Scenario 3: Select the features simply applying PCA to the DCT domain (i.e., DCT transformed images). Following this only 7 features were selected. In the subsequent part of this paper, we refer to this scenario as Direct DCTPCA.

A 10-fold cross validation scheme [19] was used to test the efficacy of our proposed method. The performance of classification is measured based on the following metrics (given the confusion matrix as Table 1):

The sensitivity (Sen) is the fraction of dense cases that the classifier predicted as dense. Sen is calculated using Eq. 4. Sen is also known as Recall (Rec). The specificity (Spc) is computed using Eq. 5. Spc is the fraction of fatty cases that the classifier expected as fatty. The classification accuracy (Acc) is the correct prediction of dense and fatty over the number of all considered examples as in Eq. 6. The precision (Prc) measures how precise is the classification with respect to the positive class. Equation 7 is used to compute Prc.

$$Sensitivity(Sen) = TP/(TP + FN) \qquad (4)$$

$$Specificity \ or \ Recall(Spc \ or \ Rec) = TN/(TN + FP) \qquad (5)$$

$$Accuracy(Acc) = (TP + TN)/(TP + FN + TN + FP) \qquad (6)$$

$$Precision(Prc) = TP/(TP + FP) \qquad (7)$$

$$FScore = 2 \times Prc \times Rec/(Prc + Rec) \qquad (8)$$

In our experiment, we used a three layer MLP (One input layer, one hidden layer and one output layer). The activation function for input and hidden layer was chosen as hyperbolic tangent function while that of output layer was chosen as

Table 1. Confusion matrix

Classes	Positive	Negative
Positive	True Positive (TP)	False Positive(FP)
Negative	False Negative (FN)	True Negative (TN)

pure linear. The number of neurons in the hidden layer was chosen empirically. The conjugate gradient backpropagation with Fletcher-Reeves algorithm [20] was used to train the MLP. We used an SVM with radial basis kernel. The parameters of SVM (i.e., sigma and scaling factor) were chosen empirically. Finally, for KNN, K was chosen as 3.

5.3   Results and Discussion

We observe from Tables (2, 3, and 4) that SVM has better accuracy compared to other classifiers (consider performance of individual classifier only) for the cases where data features are collected using PCA, and data features are collected using DCT (10^3)-PCA.

SVM can recognize the dense breasts over fatty better compared with the other classifier which is revealed through a better sensitivity (see Table 3). On the other hand MLP and KNN provide a better recognition to the fatty breast type over the dense type by considering specificity. It is interest-ing to note that, KNN achieves a very high recognition rate for the fatty type using Direct DCTPCA (97.18% Specificity). The precision of KNN and MLP is higher than SVM because these classifiers have suffered misclassification of dense breast types. The opposite is true with SVM classifier. However, the tradeoff the Rec and Prc can be witnessed from the FScore metric. Therefore, SVM shows a good Fscore over the other two classifiers for all feature types. According to the behavior of these classifiers, we can observe the strength and weaknesses of the classifiers over different breast types (dense or fatty). Therefore, we cannot depend merely on a signal classifier for such classification problem. As seen in Table 5 the majority voting technique obtains better recognition in terms of accuracy compared to the others for cases: DCT-PCA(103) and Direct DCTPCA. This is due to the complementary of the three classifiers. It considers the best results from different classifiers instead of depending on one, and then vote between these results to obtain final decision. Taking into consideration all the performance metrics, the majority voting technique has a balanced degree of performance over all classifiers.

Table 2. MLP classifier performance

Method	Accuracy	Sensitivity	Specificity	Precision	FScore
PCA	88.35	86.82	90.45	91.35	88.29
DCT(10^3)-PCA	89.81	90	89.73	91.20	90.14
Direct DCTPCA	88.79	86.36	91.36	91.84	88.75

Table 3. SVM classifier performance

Method	Accuracy	Sensitivity	Specificity	Precision	FScore
PCA	**91.69**	93.64	**89.64**	**91.43**	**92.06**
DCT(10^3)-PCA	91.17	94.55	87.73	89.50	91.62
Direct DCTPCA	87.90	**96.36**	79.18	83.59	88.75

Table 4. KNN classifier performance

Method	Accuracy	Sensitivity	Specificity	Precision	FScore
PCA	**87.99**	**81.82**	94.18	94.49	**86.92**
DCT(10^3)-PCA	86.08	80.91	91.55	91.32	85.36
Direct DCTPCA	87.94	79.09	**97.18**	**97.22**	86.5

Table 5. Vote (MLP, SVM, KNN)

Method	Accuracy	Sensitivity	Specificity	Precision	FScore
PCA	**92.12**	90.91	93.27	94	92.04
DCT(10^3)-PCA	**92.55**	**93.64**	91.55	92.40	**92.64**
Direct DCTPCA	91.58	89.09	**94.27**	**94.41**	91.35

6   Conclusion

In this paper, we have investigated the breast density classification using a combination of three classifiers over DCT-PCA feature sets. The majority voting of these classifiers reveals an enhanced performance to classify fatty and dense breast types. In conclusion, the combination of DCT-PCA as a feature extraction methodology along with majority voting could be a good choice to classify fatty and dense breast from breast mammogram images.

References

1. NIH: comprehensive cancer information - national cancer institute. http://www.cancer.gov/. Accessed 12 June 2017
2. Silva, W., Menotti, D.: Classification of mammograms by the breast composition. In: Proceedings of the International Conference on Image Processing, Computer Vision, and Pattern Recognition (IPCV), WorldComp, pp. 1–6 (2012)
3. Farag, A., Mashali, S.: DCT based features for the detection of microcalcificationsin digital mammograms. In: 2003 IEEE 46th Midwest Symposium on Circuits and Systems, vol. 1., pp. 352–355. IEEE (2003)
4. Komen, S.G.: Understanding breast cancer. http://ww5.komen.org/BreastCancer/HighBreastDensityonMammogram.htm. Accessed 12 June 2017

5. Chen, W., Er, M.J., Wu, S.: PCA and LDA in DCT domain. Pattern Recogn. Lett. **26**(15), 2474–2482 (2005)
6. Prathibha, B., Sadasivam, V.: An analysis on breast tissue characterization in combined transform domain using nearest neighbor classifiers. In: 2011 International Conference on Computer, Communication and Electrical Technology (ICCCET), pp. 50–54. IEEE (2011)
7. Mudigonda, N.R., Rangayyan, R., Desautels, J.L.: Gradient and texture analysis for the classification of mammographic masses. IEEE Trans. Med. Imaging **19**(10), 1032–1043 (2000)
8. Tan, M., Zheng, B., Leader, J.K., Gur, D.: Association between changes in mammographic image features and risk for near-term breast cancer development. IEEE Trans. Med. Imaging **35**(7), 1719–1728 (2016)
9. Hussain, M.: False positive reduction using Gabor feature subset selection. In: 2013 International Conference on Information Science and Applications (ICISA), pp. 1–5. IEEE (2013)
10. Muthukarthigadevi, R., Anand, S.: Detection of architectural distortion in mammogram image using wavelet transform. In: 2013 International Conference on Information Communication and Embedded Systems (ICICES), pp. 638–643. IEEE (2013)
11. Oliver, A., Freixenet, J., Marti, R., Pont, J., Pérez, E., Denton, E.R., Zwiggelaar, R.: A novel breast tissue density classification methodology. IEEE Trans. Inf. Technol. Biomed. **12**(1), 55–65 (2008)
12. Ganesan, K., Acharya, U.R., Chua, C.K., Min, L.C., Abraham, K.T., Ng, K.H.: Computer-aided breast cancer detection using mammograms: a review. IEEE Rev. Biomed. Eng. **6**, 77–98 (2013)
13. Basheer, N.M., Mohammed, M.H.: Segmentation of breast masses in digital mammograms using adaptive median filtering and texture analysis. Int. J. Recent Technol. Eng. (IJRTE) **2**(1), 39–43 (2013)
14. Liu, H., Guo, Q., Xu, M., Shen, I.F.: Fast image segmentation using region merging with a k-nearest neighbor graph. In: 2008 IEEE Conference on Cybernetics and Intelligent Systems, pp. 179–184. IEEE (2008)
15. El-Alfy, E.S.M., BinMakhashen, G.M.: Improved personal identification using face and hand geometry fusion and support vector machines. Networked Digit. Technol. **294**, 253–261 (2012)
16. Platt, J.: Fast training of support vector machines using sequential minimal optimization. In: Scholkopf, B., Burges, C., Smola, A. (eds.) Advances in Kernel Methods-support Vector Learning (1998)
17. Kuncheva, L.I., Whitaker, C.J., Shipp, C.A., Duin, R.P.: Limits on the majority vote accuracy in classifier fusion. Pattern Anal. Appl. **6**(1), 22–31 (2003)
18. Suckling, J.: Mammographic image analysis society (2017). http://peipa.essex.ac. uk/info/mias.html. Accessed 12 June 2017
19. Kohavi, R., et al.: A study of cross-validation and bootstrap for accuracy estimation and model selection. In: IJCAI, vol. 14, pp. 1137–1145, Stanford, CA (1995)
20. Charalambous, C.: Conjugate gradient algorithm for efficient training of artificial neural networks. IEE Proc. G (Circ. Devices Syst.) **139**(3), 301–310 (1992)

Scheduling Analysis and Correction of Periodic Real Time Systems with Tasks Migration

Faten Mrabet[1]([✉]), Walid Karamti[2]([✉]), and Adel Mahfoudhi[3]([✉])

[1] Faculty of Economics and Management, University of Sfax, Sfax, Tunisia
mrabet.faten@gmail.com
[2] Higher Institute of Computer Science and Mathematics,
University of Monastir, Monastir, Tunisia
walid.karamti@yahoo.fr
[3] College of Computers and Information Technology, Taif University,
Ta'if, Saudi Arabia
a.mahfoudhi@tu.edu.sa

Abstract. Satisfying a real-time system (RTS) timing constraints is a serious deal for the prevention or the decreasing of the human and economic possible problems. In this vein, multiprocessor RTS scheduling is increasingly difficult, especially since it became more complex and dynamic. In this context, scheduling correction turns out to be very necessary for a better schedulability. In fact once done at an early stage of the scheduling cycle, it can effectively reduce the potential temporal faults. It helps then avoiding the risk of a very expensive complete reiteration of the scheduling cycle, thus reduce the temporal cost.

The present paper focuses on providing a scheduling correction solution that may be integrated into the scheduling analysis. The proposed correction method is based on rectifying the partitioning if the RTS is detected non-schedulable. This method is based on the migration of some tasks from the non-schedulable overloaded partitions toward the not loaded ones.

Keywords: Real time systems · Scheduling analysis
Scheduling correction · Tasks migration

1 Introduction

The real-time systems (RTS) continue to spread in our daily lives in the definition of services and applications. These applications have an important role in our society and are manifested in several areas such as:air traffic, robotics, multimedia and others [19]. However, they are characterized by a growing complexity [2]. Therefore, using powerful architectures, such as multiprocessor and multicores [3,9], is highly recommended. In fact, satisfying the timing constraints is

© Springer International Publishing AG, part of Springer Nature 2018
A. Abraham et al. (Eds.): ISDA 2017, AISC 736, pp. 712–723, 2018.
https://doi.org/10.1007/978-3-319-76348-4_69

the most crucial part when projecting a RTS application on a specific architecture. Two main steps exist for multiprocessor scheduling, the partitionning and the scheduling analysis [5]. The first defines the spatial organization of tasks on processor. While the second, defines their temporal organization. Indeed, the scheduling analysis must be carried out at an early stage during the development life cycle to predict and validate the system temporal behavior. Thus, the possible eventual risks can be distinguished and may be corrected.

Two main families are widely used for the multiprocessor real-time scheduling [20] the global family [7] and the partitioned family [10]. Indeed, contrary to the partitioned family, the global one allows the tasks migration from one processor to another. For this latter, only near optimal solutions exist [12]. This is due to the additional migration costs [8] which can alter the scheduling results in practice [9,30]. Nevertheless, in the partitioned family each partition are treated separately using the one processor scheduling algorithms. Thus, optimal solutions were proved [13]. However, the major problem in this family is the risk of non-schedulable partitions and the time needed for its correction. Actually, the correction of the partitionning (know as a NP-Hard problem [22]) is based on the regeneration of a new partitioning solution, and this can lead to a very important time cost.

For several years, RTS have been continuously growing and complex, the time needed for the detection of a feasible solution is consequently very important. It seems crucial then to propose a new partitionning and analyzing method, to correct the non-schedulable partitions and to avoid the regeneration cost of new ones. In this paper, we focus on reducing the correction cost while analyzing in order to attempt a valid solution for a feasible RTS.

The rest of this paper is organized as follows. Section 2 focuses on exposing the state of the art related to the classical partitioned RTS scheduling. In this section the problems of the partitioning step and scheduling analysis step are exposed. Section 3 describes the proposed approach to solve the problems of non schedulability. In this section, we present a flow chart of an algorithm dedicated to the correction of non-schedulable partition based on the principle of targeted migration. Section 4 is devoted to discuss results and evaluation. Section 5 draws conclusions.

2   Classical Partitioned RTS Scheduling

An RTS is defined by its formal 4-tuplet scheduling specification:

$$RTS = \langle Task, Proc, Alloc, Prec \rangle \qquad (1)$$

with:

- $Task : \{T_1, T_2, \cdots, T_n\}$, is a finite set of tasks, where $(n > 0)$ is the number of tasks with each $Task_i \in Task$ is defined by:

$$T_i = \langle R_i, P_i, C_i, D_i \rangle \qquad (2)$$

with:

- R_i: the date of the first activation.
- P_i: the period associated with the task.
- C_i: the execution period of the task for the P_i period.
- D_i: the deadline of the task. (In this paper $D_i = P_i$, deadline on request)

- $Proc : \{P_1, \cdots, P_m\}$ is a finite set of tasks, where $m > 0$ is the number of processors.
- $Alloc : Task \to proc$, a surjective function which allocates a task to a processor. In fact a processor is allocated to at least one task, but a task must be assigned to only one processor.
- $-Prec : Task \times Task \to \{0, 1\}$ a function which initializes precedence relations between tasks referring to the dependency between them.

The key problems in the classical partitioned multiprocessor RTS scheduling is the projection of a software application on hardware architecture while preserving the schedulability. This projection requires two main steps [28], a partitioning step and a scheduling analysis step. The first consists on distributing tasks into partitions(processors). While, the second aim to give and verify the schedule of the tasks on processors. For the partitioned family, each task is not allowed to migrate from its initial partition towards another during scheduling. In fact, the tasks affected within the same processor can be scheduled by the well known optimal single-processor algorithms such as: Rate Monotonic (RM), Earliest Deadline First (EDF) [26], etc.

Unfortunately, despite the amount of research in NP-hard scheduling problems, there still common problems that always occur. The most common is the inability to give a reliable partitioning solution from the beginning, which increases the risk of non schedulability [24]. In fact, in order to correct non schedulability, the regenerations of an exponential number of partitioning solutions is needed. Thus a complete costing repetition of the scheduling cycle is caused [1,11]. Indeed, browsing exhaustively these solutions generates the risk of obtaining non-optimal ones with a computation time that remains substantially high. Indeed, for such algorithms that depend on several essays it is usually impossible to guarantee that the solution found is the best even after a significant number of regenerations [29] and a great waste of time. That's why, most of the works promotes solutions where they prefer having relatively a good solution quickly rather than having the optimal solution indefinitely. Indeed, for such problem where there is no exact method or where the solution is unknown [21], the most common is the use of "heuristics" [1,28]. Among the main and simple heuristics used, first fit decreasing (FFD), best fit decreasing (BFD) [18], etc. Actually, the combination of First fit, Next Fit, Best Fit, Worst-Fit, heuristics associated with the various conditions of single processor schedulability, has given rise to many extension of these algorithms [14]. However, these algorithms generally allow obtaining approximate results, close to the optimal solution [16]. As some existing works focused on the partitioning quality [25,27], as the only corrective action to reduce non schedulability. Other works, focused on the scheduling analysis quality [23] to save time and effort.

In fact as said above, multiprocessor scheduling analysis in partitioned family [13] consists on analyzing each partition using the well-known mono-processor scheduling analysis methods. However, if a non-schedulable partition is detected, partitioning a task across multiple processors in order to schedule the task system is sometimes needed. But, unfortunately this family does not allow such partitioning.

In the literature, several techniques exist for the scheduling analysis, such as analytical techniques [28], simulation techniques [28], and other formal techniques in the process of progression based on "the proof of theorem" [15] and the "model-checking" [23]. The diversity of these techniques explains the effort to improve the quality of analysis. However, improving the quality, is limited to improving the feedback on system schedulability [23]. In fact, it's only about well specifying the non-schedulable partitions without proposing a concrete method for scheduling correction. Indeed, when detecting "non schedulable" configurations the only corrective action is limited to calling the partitionning tool for a new (NP-hard)time costing regeneration, without the proposal of possible corrections [4, 6]. Consequently such regeneration can lead to a very important effort and time cost.

So, despite the diversity of scheduling analysis techniques, they remain insufficient. They are generally, not autonomous and require other means (tools) to assist them in correcting the system eventual faults. However, only few of them are assisted by a tool due to their delicacy and complexity [31]. So, it is very interesting to integrate into them, methods for correction to ensure autonomy and to save time and effort.

The purpose of this paper is to add a method of correction in the scheduling analysis process. In fact, on the basis of previous studies, we assume that if we focus on minimizing the back and forth between the scheduling analysis and the partitioning tool by the way of scheduling correction we can afford time and complexity. Therefore, it will be interesting to propose a method enabling both the analysis and correction of a RTS in the case of non-schedulability. This method aims to ensure that the correction would be faster and less expensive than other configurations generation.

3 Proposed Approach

Contrary to the classical partitioned scheduling, our approach focuses on correcting the non-schedulable partitions [23] detected after launching the scheduling analysis step. This correction step is far from being trivial. In fact, it allows to reduce the costly returns to the partitioning tool for another regeneration [22] and another launch of the analysis step. Indeed, a correction step leads whether to a valid solution where the system becomes schedulable, or a non valid solution where the system remains non-schedulable and require regenerations. Figure 1 outlines the proposed approach.

Previous studies indicate that schedulability tests based on system use may fail when the partition uses is very different. More precisely when the processor

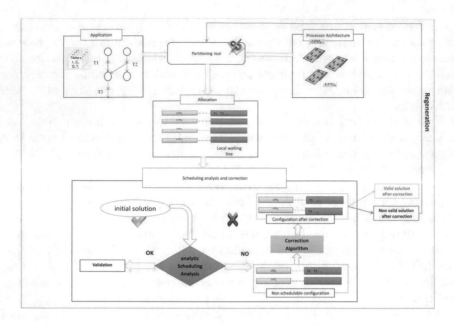

Fig. 1. Proposed approach

overload is the cause of system faults [31]. In the current paper we base our approach on calculating analytically the load on each processor, then on correcting shedulability by migrating some tasks from these overloaded partitions toward the non-loaded ones. In fact, this adds a degree of system flexibility and enables load balancing [17]. However, on the best of our knowledge this balancing does not prevent system faults from occurring and was never considered for the correction of non-schedulable configurations. In this paper, the main concern is about correcting non schedulability using the migration principle. It is about a repartitionning of tasks on processor keeping the schedulability.

The proposed approach for resolving the previously identified problems and to attain the described objectives will be concretized by an algorithm based on targeted migrations of some tasks dedicated for correction. In the following, a global pattern of the proposed approach.

3.1 Algorithm Description

In the current manuscript, we propose an algorithm that is dedicated for both of the scheduling analysis and the correction of a periodic RTS. The proposed solution works on a set of independent, periodic, and deadline on request ($P_i = D_i$) Real-Time tasks, running synchronously ($r_i = 0$) on a multiprocessor identical architecture [3,9]. The priority of the tasks is monitored by a dynamic priority-driven scheduling algorithm Earliest Deadline First (EDF).

In the first place, a partitioning algorithm allocates the tasks over the processors to propose a feasible solution [26]. In the second place, based on the system

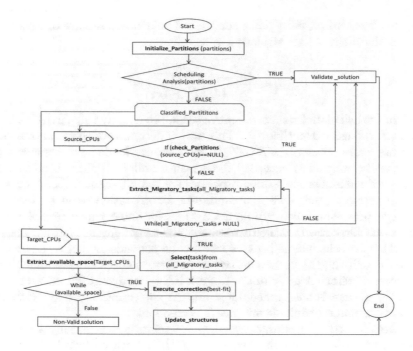

Fig. 2. Algorithm-flowchart

load calculation, the schedulability is analyzed. For this reason, an analytical technique is needed to detect the partitions cause of temporal faults if exists. In the third place, it is interesting to identify the tasks that can be migrated to correct the partitions failure. In the last place, the migration of the detected tasks to other partitions able to receive them viewpoint processor load, can correct the system schedulability. This algorithm gives a generic solution to correct all the generated partitions given by a non efficient partitionning algorithm. In fact, correcting based on task migration level, allows a repartitioning of tasks and minimizes then the number of returns for regenerations.

Figure 2 illustrates the flowchart of the algorithm based on structured programming concepts. It gives an overall view of the main part of the algorithm.

From Fig. 2, it can be seen that, we start by initializing the input partitions of the system **"initialize_Partitions"**. Then, we move to check the schedulability of these partitions **"scheduling_analysis (partitions)"**. To do so, we need first to calculate the load on each partition **"calculate_load"**.

The load or the use U of a processor $proc_i$, $U_{(proc_i)}$ is equal to the sum of the execution time (C_i) of each task on the processor $proc_i$ multiplied by the number of times each task can execute over its hyper period (divider), that is to say $LCM_{proc_i}(P_i)$, divided by P_i, $\sum_{proc_i} C_i \cdot \frac{LCM_{proc_i}(P_i)}{P_i}$, the result is divided by $LCM_{proc_i}(P_i)$, where $LCM_{proc_i}(P_i)$ defines the lower common multiplier period of a set of independent tasks existing on this processor $proc_i$. The final result calculates the load on each processor that should not normally exceed 1. The equation that defines the calculation is as follows:

$$U_{proc_i} = \frac{\sum_{proc_i} C_i \cdot \frac{LCM_{proc_i}(P_i)}{P_i}}{LCM_{proc_i}(P_i)} \tag{3}$$

Based on the calculated results, we move second to classify the partitions in separate lists **"classify_Partitions"**. This action decomposes them on overloaded processors, called **"source_CPUs"**. In fact, an overloaded processor is a processor that has obviously exceeded some task deadlines (temporal faults), so it is non-schedulable. On not loaded processors called **"Target_CPUs"** on which they can support other tasks, so candidate for migration. And on load limit or charged processor called **"bloked_CPUs"**, for those which cannot support other tasks. They are then neither source nor target for migration. They are those which are schedulable because there is no dependency between the tasks.

After classifying the partitions, we move then to check the schedulability of these latter's **"check_Partitions"**. In fact, if the **"source_CPUs"** structure is empty, the system is then schedulable and no correction is needed. Otherwise the system is non_schedulable and needs correction.

In fact, in order to correct, we need to migrate sequentially some tasks **"migratory_tasks"** from the **"source_CPUs"**, toward their best-fit in the **"Target_CPUs"**. To do so, we start by scrolling the structure **source_CPUs**, one by one until it becomes empty to **Extract_migratory_tasks**. In fact on a processor, we can have one or more migrating tasks.

The Migratory_tasks structure does not necessarily contain tasks that have exceeded deadlines. It is about finding the tasks that their migration can correct the schedule on each processor. In fact, to extract the migratory tasks on a source processor, we need to calculate the task-load. The task-load is the result of the division of its execution time (C_i) by its period (P_i). It can be computed by the following equation: $\frac{C_i}{P_i}$. The migratory tasks are those having the minimum task-load till the processor becomes not loaded or on load limit. That's to say the task with the smallest (C_i) on the largest (P_i) from the processor having exceeded deadlines. In fact these tasks have more instances in the LCM of all the periods (P_i) of the tasks existing on overloaded processors, which generate more preemption. So, moving those gives more chance to other task to be schedulable so gives more chance for the correction. "all_migratory_tasks" is a structure containing all the migratory tasks in the system.

Once "all_migratory_tasks" are detected, we move on to extract the places of tasks to be affected in the target processors **Extract_available_space**, the available space on a target processor is the free capacity or the load on this partition. Since the strategy of migration to come is best-fit, we need then to order the processor by their ascendant available space, so that the best-fit takes less space and leaves more room for others. In fact, if there is available spaces, we can proceed to execute the correction, else the system is not schedulable.

We select then the tasks one by one and we search its best fit from the available_spaces. Ones the task and the available space is ready we move on to

execute_correction(best_fit). Indeed affecting an execution time in nearly the exact space, increase the probability, for other tasks to be affected, and then the probability of correction.

To guarantee an efficient correction during the migration process and after being conducted, =an update series of the structures used are necessary "update_structures".

In fact before every task migration, =it is crucial to check whether there is a place for this migration viewpoint processor load "**Update_load_target**". That is to say does not become overloaded or loaded. The equation can be written as:

$$U_{proc_i} = \frac{\sum_{proc_i} C_i + C_{i\_mig} \cdot \frac{LCM_{proc_i}(P_i, P_{i\_mig})}{P_i}}{LCM_{proc_i}(P_i, P_{i\_mig})} \tag{4}$$

Where, C_{i_mig} is the execution time of the new migrating task is and P_{i_mig} his period which will be integrated into the CPU_load Eq. 1.

The correction on a source processor stops when it becomes on load limit or not loaded "Update_load_source".

The equation can be written as:

$$U_{proc_i} = \frac{\sum_{proc_i} C_i - C_{i\_source} \cdot \frac{LCM_{proc_i}(P_i)/P_{i\_source}}{P_i}}{LCM_{proc_i}(P_i,) \ P_i/_{source}} \tag{5}$$

Where, C_{i_source} is the execution time of the migrating task which will be liberated and P_{i_source} his period which will be subtracted from the CPU_load Eq. 1.

If the structure "all_migratory_tasks" became empty, the correction finish and the system becomes schedulable.

4   Results and Evaluation

To illustrate the result, a simulation of a scheduling analysis and correction tool called "SAC-tool" was performed.

Figure 3 represents two steps performed in SAC-tool, the configuration step and the scheduling analysis and correction step. The first step allows inserting a new RTS input configuration, where we can set the number of tasks "tasks_number" and processors "procs_number" as shown in Fig. 3(1). Once validated, the temporal characteristics of the input tasks and their partitionning to the available processor must be done, see Fig. 3(2), the configuration step is then finished.

The second step allows performing two actions. The first one is always active and dedicated for analyzing the systems as can be seen in Fig. 3(3). However, the second is active only if the system is detected non-schedulable and it is dedicated for the scheduling correction as shown in Fig. 3(4).

In Fig. 3(3), a summary interface of the system configuration. It contains the processors and the assigned tasks to each one, and in front of each one the load calculated. During analysis and according to the processor load calculated,

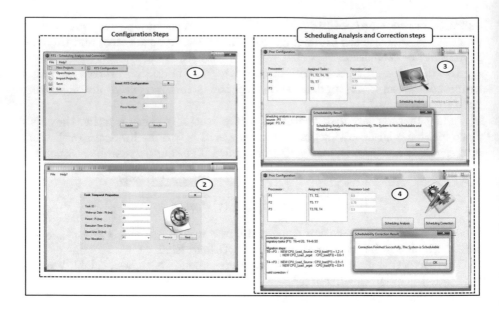

Fig. 3. SAC-tool

it displays the analysis process. In this part, information about the source and the target processor is presented. In fact, if there is at least one source processor, then the system is non-schedulable.

Once the analysis process is complete, two scenarios can be resulted. Or, it is finished with success and the system is schedulable. Or it is finished with failure, the system is then non-schedulable and needs correction.

When activating the correction, the interface displays the correction process Fig. 3(4). In this part, informations about the migratory tasks of each source processor, the migration steps and the new cpu_loads. In fact in every migration step, the new load on the source and the target processor is calculated. The processors Load are then dynamic. Once the new load on the source and the target processor becomes inferior to one, the system is then schedulable. Else the system remains non-schedulable.

After correction, the load on each processor becomes stable, and the initial partitionning "assigned tasks" in front of the source and the target processors changes.

At the end of the correction process, two scenarios can be resulted. Either the correction is finished successfully and the system becomes schedulable or the correction is finished with failure and the system remains non-schedulable. The first scenario displays the same page with the changes previously reported in the interface. The second gives the hand to exit correction and return to the initial configuration step.

Due to space limitation, we detail in the following a pedagogical case study. Indeed to clarify the sequences done, we present the RTS definition, the initial and final calculated loads, the source and target cpus, the migration steps needed, and the result of migration.

In the following a synchronous real-time task system ($r_i = 0$), where the tasks are independent ($ec = \emptyset$) and where the scheduling between competing tasks is managed by the dynamic priority scheduling algorithm EDF Earliest dealline first.

$RTS = \langle Task, Proc, Alloc \rangle$ with:

- $Task_i : T_1, T_2, T_3, T_4, T_5, T_6, T_7$
- $Proc_i : P_1, P_2, P_3$
- $Alloc : \langle T_1, P_1 \rangle, \langle T_2, P_1 \rangle, \langle T_4, P_1 \rangle, \langle T_6, P_1 \rangle, \langle T_5, P_2 \rangle, \langle T_7, P_2 \rangle, \langle T_3, P_3 \rangle$
- $T_i = \langle r_i, P_i = D_i, C_i \rangle$
- $T_1 = (0, 20, 8)$, $T_2 = (0, 30, 15)$, $T_4 = (0, 20, 6)$, $T_6 = (0, 20, 4)$, $T_5 = (0, 40, 15)$, $T_7 = (0, 40, 15)$, $T_3 = (0, 20, 8)$.

In the analysis step, the P_1 is detected as overloaded processor (source_CPUs), because the calculated CPU_load was found superior to one (CPU_load(P_1) = $1, 4 > 1$), while P_2 (CPU_load(P_2) = $0, 75 < 1$) and P_3 (CPU_load(P_3) = $0, 4 < 1$) was detected as not overloaded ones (target_CPUs). Thus The system is at the beginning not schedulable and needs correction.

From the source processor P1 two migrant tasks was distinguished. These tasks are respectively T_6 and T_4. That's why, two migration steps were needed to correct the system schedulability.

In the first migration of T_6 on the processor P_3, the schedulability was not corrected (CPU_load(P_1) = $1, 2 > 1$ (overloaded processor), CPU_load(P_3) = $0, 6 < 1$), because the cpu_load on the source processor P_1 ramined superior to one. Moving to the next migration step of the task T_4 to P_3, the loads on the source and the target becomes inferior to one (CPU_load(P_1) = $0, 9 < 1$), (CPU_load(P_3) = $0, 9 < 1$). After two migration step, the correction is then done.

5   Conclusion

Despite the number of studies in scheduling analysis techniques, there are very few studies in the best of our knowledge that integrates scheduling correction methods toward it. In fact, these techniques remain only with detection techniques. Indeed, they allow only the detection of non-schedulable configurations without the proposal of possible corrections. Thus, these techniques are not autonomous and require other means (tools) to assist them to correct the system faults.

The present paper focuses on providing scheduling correction solutions that can be integrated into the scheduling analysis-tool. The proposed correction method is based on migration of some tasks to the most relaxed processors. Ones the correction is done, the initial partitioning solution will be then changed without recurring to regenerate a new configuration.

In our future research we intend to concentrate on improving furthermore the schedulability. In fact, sometimes when the task-level migration does not correct the schedulability, it is possible to correct with the migration in job-level. The next stage of our research will be then including the scheduling correction in different migration level.

References

1. Agrawal, S., Gupta, R.K.: Data-flow assisted behavioral partitioning for embedded systems. In: Proceedings of the 34th Annual Design Automation Conference, pp. 709–712. ACM (1997)
2. Alur, R., Henzinger, T.A.: Real-time logics: complexity and expressiveness. Inf. Comput. **104**(1), 35–77 (1993)
3. Åsberg, M., Nolte, T., Kato, S.: Towards partitioned hierarchical real-time scheduling on multi-core processors. ACM SIGBED Rev. **11**(2), 13–18 (2014)
4. Baker, T.P.: An analysis of EDF schedulability on a multiprocessor. IEEE Trans. Parallel Distrib. Syst. **16**(8), 760–768 (2005). https://doi.org/10.1109/TPDS.2005.88
5. Baker, T.P.: Comparison of empirical success rates of global vs. partitioned fixed-priority and EDF scheduling for hard real time. Citeseer (2005)
6. Baker, T.P.: A comparison of global and partitioned EDF schedulability tests for multiprocessors. Technical report, in International Conference on Real-Time and Network Systems (2005)
7. Baruah, S.: Techniques for multiprocessor global schedulability analysis. In: 28th IEEE International Real-Time Systems Symposium, RTSS 2007, pp. 119–128. IEEE (2007)
8. Bertogna, M., Baruah, S.: Tests for global EDF schedulability analysis. J. Syst. Architect. **57**(5), 487–497 (2011)
9. Bertozzi, S., Acquaviva, A., Bertozzi, D., Poggiali, A.: Supporting task migration in multi-processor systems-on-chip: a feasibility study. In: Proceedings of the Conference on Design, Automation and Test in Europe: Proceedings, European Design and Automation Association, pp. 15–20 (2006)
10. Burchard, A., Liebeherr, J., Oh, Y., Son, S.H.: New strategies for assigning real-time tasks to multiprocessor systems. IEEE Trans. Comput. **44**(12), 1429–1442 (1995)
11. Chehida, K.B.: Méthodologie de partitionnement logiciel/matériel pour plate-formes reconfigurables dynamiquement. Ph.D. thesis, Université Nice Sophia Antipolis (2004)
12. Chéramy, M., Déplanche, A.M., Hladik, P.E.: Ordonnancement temps réel: des politiques monoprocesseurs aux politiques multiprocesseurs. HALUNIV-NANTESFR (2012)
13. Chéramy, M., Hladik, P.E., Déplanche, A.M.: Algorithmes pour lordonnancement temps réel multiprocesseur. J. Européen des Systèmes Automatisés (JESA) **48**(7–8), 613–639 (2015)
14. Davis, R.I., Burns, A.: A survey of hard real-time scheduling for multiprocessor systems. ACM Comput. Surv. (CSUR) **43**(4), 35 (2011)
15. De Rauglaudre, D.: Vérification formelle de conditions d'ordonnançabilité de tâches temps réel périodiques strictes. In: JFLA-Journées Francophones des Langages Applicatifs-2012 (2012)

16. Dorin, F., Yomsi, P.M., Goossens, J., Richard, P.: Semi-partitioned hard real-time scheduling with restricted migrations upon identical multiprocessor platforms. arXiv preprint arXiv:10062637 (2010)
17. Funk, S., Baruah, S.: Restricting EDF migration on uniform multiprocessors. In: Proceedings of the 12th International Conference on Real-Time Systems (2004)
18. Gen, M., Cheng, R.: Genetic Algorithms and Engineering Optimization, vol. 7. Wiley, Hoboken (2000)
19. Goossens, J., Richard, P.: Ordonnancement temps réel multiprocesseur. In: État de l'art–ETR 2013 (2013)
20. Gracioli, G., Fröhlich, A.A., Pellizzoni, R., Fischmeister, S.: Implementation and evaluation of global and partitioned scheduling in a real-time OS. Real-Time Syst. **49**(6), 669–714 (2013)
21. Houbad, Y., Souier, M., Hassam, A., Sari, Z.: Ordonnancement en temps réel dun jobshop par métaheuristique hybride: étude comparative. dspaceuniv-tlemcendz (2011)
22. Johnson, D.S.: Fast algorithms for bin packing. J. Comput. Syst. Sci. **8**(3), 272–314 (1974)
23. Karamti, W., Mahfoudhi, A.: Scheduling analysis based on model checking for multiprocessor real-time systems. J. Supercomput. **68**(3), 1604–1629 (2014)
24. Khardon, R., Pinter, S.S.: Partitioning and scheduling to counteract overhead. Parallel Comput. **22**(4), 555–593 (1996)
25. Korf, R.E.: A new algorithm for optimal bin packing. In: AAAI/IAAI, pp. 731–736 (2002)
26. Liu, C.L., Layland, J.W.: Scheduling algorithms for multiprogramming in a hard-real-time environment. J. ACM (JACM) **20**(1), 46–61 (1973)
27. Martello, S., Pisinger, D., Vigo, D.: The three-dimensional bin packing problem. Oper. Res. **48**(2), 256–267 (2000)
28. Ndoye, F.: Ordonnancement temps réel préemptif multiprocesseur avec prise en compte du coût du système d'exploitation. Ph.D. thesis, Université Paris Sud-Paris XI (2014)
29. Niemann, R., Marwedel, P.: Hardware/software partitioning using integer programming. In: Proceedings of the 1996 European conference on Design and Test, p. 473. IEEE Computer Society (1996)
30. Shekhar, M., Ramaprasad, H., Sarkar, A., Mueller, F.: Architecture aware semi partitioned real-time scheduling on multicore platforms. Real-Time Syst. **51**(3), 274–313 (2015)
31. Silva, F.A., Maciel, P., Matos, R.: SmartRank: a smart scheduling tool for mobile cloud computing. J. Supercomput. **71**(8), 2985–3008 (2015)

Generating Semantic and Logic Meaning Representations When Analyzing the Arabic Natural Questions

Wided Bakari[1,3]([✉]), Patrice Bellot[2,4]([✉]), and Mahmoud Neji[1,3]([✉])

[1] Faculty of Economics and Management, 3018 Sfax, Tunisia
wided.bakkari@isegs.rnu.tn
[2] Aix-Marseille University, University of Toulon,
CNRS, ENSAM, Marseille, France
[3] MIR@CL, Sfax, Tunisia
[4] LSIS, Marseille, France

Abstract. In this paper, we provide a performance analysis of the question and describe their different tasks in Arabic language. Regardless of the approaches being studied this language, the first step is to analyze the question for extracting all the information exploited by the processes of searching for documents and selecting relevant passages. Question analysis shows that there are few studies provide semantic and logic-inference based-approaches in the Arabic. After extracting the keywords, determining the declarative form, generating the focus and the expected answer type, we transform the questions into semantic representations via the conceptual graph formalism and into logic representations using a transformation algorithm. This analysis is classified into tree modules: The first one emphasizes a preprocessing of the question; the second one generates a question transformation; and the third one provides a linguistic analysis. The goal of the first module is to extract the main features from each question (list of keywords, focus and expected answer type). The focus and the keywords are identified to retrieve the short and relevant answers located in small passages containing the accurate answer. The second module allows transforming the question into a declarative form. The third module makes some linguistic analyses that are used in the graph construction and logic representation phases. Lastly, we assess the process of analysis with examples of 5 types of questions collected in our corpus.

Keywords: Arabic question-answering · Question analysis
Declarative from · Conceptual graph · Logic representation
Focus · Expected answer type · Keywords

1 Introduction

Question-answering is divided into three modules namely. Question processing module, document processing module, and answer extraction module [1]. The problem inherent in such a system lies in the difficulty of finding answers based on a question. Indeed, a question in natural language is the most natural way to express a need for

© Springer International Publishing AG, part of Springer Nature 2018
A. Abraham et al. (Eds.): ISDA 2017, AISC 736, pp. 724–734, 2018.
https://doi.org/10.1007/978-3-319-76348-4_70

information. In addition, the language variability and the fact that there are multiple ways of expressing the same information are relatively rapid. In fact, question analysis is a fundamental task in question-answering systems. Indeed, in order to be able to answer a question correctly, it must first be analyzed.

The limiting aspect of the results retrieved by the search engines is particularly concerned with the lack of precision of the search. The problem of question-answering fascinates researchers since Turing proposed to consider the following question "Can machine think?". Such a system takes as input a question in textual form like "إخترع الحاسوب الألي من؟" and returns the reply "تشارلز بابيج" as well as the passage from which it was extracted and which is supposed to justify it "اخترع الحاسوب شارلز بابيج العالم الأول الذي". The first step is to take into account the analysis stage of the questions. The objective of this step is to obtain characteristics of the question that could be useful in the following steps. Subsequently, by interrogating a search engine, we consider looking for documents that can answer the question after having analyzed it. This step calls for more complicated treatment; the selection of passages may contain an answer. It is also necessary to return a certain number of documents, of which the sentences, called candidates, are likely to contain the answer to the question asked.

Question analysis is a fundamental task in question-answering systems because in order to be able to correctly answer a question, it must first be analyzed. Indeed, one of the main challenges of question-answering systems is to analyze the question asked by the user. In our proposal, we developed and evaluated a prototype for analyzing the question, which the goal is to extract the keywords of the question, to determine its declarative form, to generate focus and to determine the expected answer type. In our case, we proposed two sub-tasks related to the question analysis. The first is for question preprocessing that begins with a factual question and attempts to determine some elements (keywords focus and expected answer type); these elements are later used by other modules for generating the accurate answer. The second one is for the question transformation. It allows building a transformation of this question in declarative form by eliminating the question mark and the interrogative particle. This must be used shortly for generating logical forms. This representation is very promising because it helps us later in the selection of the justifiable answer.

In this study, we explore a method for analyzing the question. We propose to transform each question in a logic representation in order to determine textual implication between the question and the passage answers it. In addition, this work begins by presenting precisely the outline of the question-answering and question analysis tasks. It raises the contribution of such question analysis for the generation of an accurate answer. It also presents some proposals for the question analysis in Arabic investigations. Finally, it presents its implementation; a conclusion shall close this paper.

2   Question Analysis

The problem we tackle in this study is analysis of questions. The question-answering is a multidisciplinary field in which a question-answering system often incorporates techniques and potential resources, including information retrieval (IR), Automatic Natural Language Processing (ANLP), information extraction (IE), machine learning (ML) [2], etc. in order to achieve the above mentioned objective, we have added various techniques to improve the whole process such as, techniques of artificial intelligence, logical reasoning and recognizing textual entailment (RTE) techniques. Indeed, our objective is to analyze a question, to search for a passage of text, to understand this given passage, then to select the precise answer. To do this, we propose an approach with which we can integrate logic in this field of research by transforming Arabic statements (question + passage) into logical representations. Our approach is based on the recognition of textual implication.

A question is defined as a natural language sentence, which usually begins with an interrogative word and expresses a need for information from the user. Sometimes, a question has a form of imperative construction and begins with a verb. In such a case, the request for information is called a declaration [3]. In fact, in all question-answering systems, the generation of a precise answer to a natural question involves, necessary, a step of analyzing this question. Moreover, this step is an important and even necessary task not only for documents retrieval but also for the extraction of a justifiable and a precise answer.

A question analysis is a primordial step in the processing chain of a question-answering system. Several studies emphasize that the task of extracting an answer to a given question, essentially, requires a deep analysis of the issue [4]. This analysis extracted the key indicators of the question, namely, the expected answer type, the object of the question (focus), the terms that will be used later in the search of the documents that could be prove the answers [7]. These features could be also useful in the next steps when searching the accurate answers [5]. A further essential and complementary purpose for this step is to identify the named entities in the input questions and address the relationships that link those entities.

Our approach is different from other ones proposed in Arabic question-answering systems in that we use the logic representation to analyze the Arabic statements and generate the accurate answer. In addition, we present in our survey [6] a performance analysis of the different investigations in Arabic. In fact, we explore an analysis of main question answering tasks (question analysis, passages retrieval, and answer extraction). To analyze a given question, we suggest that the most studies in this language are focused on question classification. However, in our research, in one hand, the step of analyzing the question determines the expected answer type, the focus and the keywords of the question. In other hand, we propose to formulate and generate the declarative form of the question in order to generate logic representation. This transformation could help us to extract the accurate answer.

3   The Contribution of Question Analysis When Extracting the Answer

The role of the question analysis is the subject of numerous studies [7, 8]. Indeed, the main purpose of such an analysis is to define the information in question which can guide or facilitate the task of generating answers in natural language. In this spirit, Zweigenbaum and his associates have shown in their work that recourse to an effective analysis of the question is increasingly common to find the precise answer. Indeed, the analysis of the question includes the ability to determine the form of the question. When the answer is factual, it is to specify the type of expected answer [7].

Other investigators recommend that how some elements of question analysis, namely its syntactic structure, the lexicalization of concepts or the type of expected answer, may contribute to the generation of good qualities of linguistic answers [8].

In addition, other researchers like [9] suggest the contribution of the different components to overall accuracy of the question-answering system. This is, also, the component that identifies what filters to apply over the harvested n-grams, along with the actual regular expression filters themselves, contributes the most to overall performance.

4   A Proposed Method for Analyzing the Questions

In this section, we present in detail the step of question analysis while presenting their various characteristics, which can help us to select the precise answer. Indeed, although the techniques differ from one system to another, most question-answering systems are based on a step of question analysis. This step is defined as a preliminary step in the process of finding precise answers to questions in natural language. In our context, this phase is a succession of four stages; the result of each step will be used by the next. In general, the characteristics extracted from this step, in particular, the keywords, the focus, the expected answer type facilitate the extraction of the precise answer. And then, all those characteristics are given in the following steps of the system [5]. In Arabic, the majority of studies focus on the keyword extraction and the recognition of named entities. In our case, we add the reformulation of the question in a declarative form. This is used soon to generate logical forms.

In order to answer a question in natural language, several characteristics of the question were at least highlighted and/or used in our study on real questions. We focus on a particular type of questions, namely the factual questions. Indeed, the step of analyzing the question consists of different sub-steps such as, the pre-processing of the question which has as results the extraction of the keywords, the expected answer type, and the focus of the question and the transformation of the question that derives the declarative form of the question. Our analysis module assumes each question to be a simple declarative sentence that is composed of a sequence of words and looks for the focus of each sentence as a useful proof to extract the precise answer. We describe the analysis process with examples of 5 types of questions collected in our corpus.

We focus on a particular type of questions, such as the factual question. Given a concrete example of a question analysis and the treatments involved. Consider the following question: « ‏ما هو أكبر تنظيم اقتصادي في العالم؟‏ »„ «What is the largest economic organization in the world?», the pre-treatments are performed by generating key features. Then, the declarative form of the question (new form) is constructed by eliminating the particle and the question mark. Then, the new form of the question is annotated by a tool of recognition of named entities. Finally, the identification of the grammatical categories is performed by a morphological analyzer. The Fig. 1 shows the output of each step; all of this information will be used in the subsequent retrieval process in the document collection.

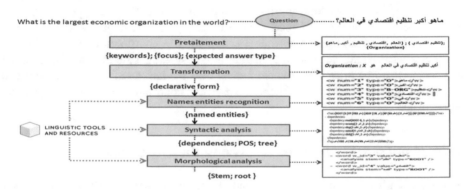

Fig. 1. The steps of analyzing the question

The following steps are performed to analyze a given question by specifying their main characteristics that are more commonly used for the documents/passages retrieval process, their analysis, and the accurate answer selection.

4.1   Pretreatment

The first step of analyzing the question undergoes pre-processing tasks. This step tries to determine the principal features (i.e., keywords, focus, and expected answer type) for any factual question. These features can help us later to select the relevant text passages from the Web. They are also used by other modules to generate the exact answer (selection of the precise answer). The treatments, which we propose to solve the problem of finding answers fit well within the framework of analyzing the question and having what it wants to find. Even in the other Arabic question-answering studies, in the question analysis phase, pre-processing tasks were applied to eliminate irrelevant data (stop words are eliminated and query particle is deleted).

More precisely, pre-processing produces these main characteristics before the transformation of the question. In this context, a question is analyzed in order to extract these characteristics, which will be transmitted to the other modules. The extracted features are typically basic information's of the question, such as, the keywords, the focus and the expected answer type:

- The keywords: It is necessary to analyze the questions beyond the division into key words: for the search of documents, for example, it is necessary to know the most relevant words in order to give different weights to the words of the query. These keywords are obtained by removing the symbols and empty words from the user's question.
- The focus is a word or sequence of words that defines the question and indicates what the question seeks to examine [10]. Thus, focus is the primordial element and the object about which information is requested. This is in fact the part of the question that is a reference to the answer [11].
- The expected answer type: This type corresponds to the named entity expected to answer. According to the answers, there are five categories of questions such as, place, person, date, organization, and numeric expression. To find this type, the analysis takes into account the structure of the question and the list of interrogative pronouns. In this respect, the recognition of named entities makes it possible in particular to obtain the type of answer to questions. For example, if the interrogative pronoun is "Who" the question waits a "person" as the answer.

Each question-answering system extracts information according to its approach. For example, in the following question: « ماهو أكبر تنظيم اقتصادي في العالم؟ », «What is the largest economic organization in the world?», the pre-processing step in our analysis module of the question gives the following characteristics:

- Keywords: العالم ,اقتصادي ,تنظيم ,أكبر
- Focus: تنظيم اقتصادي
- Expected answer type: Organization

4.2   Transformation

To increase the capacity of other steps of the answer processing to identify a specific answer, the analysis of the question is mainly focused on possible reformulation. Indeed, for any question, we have generated a rewriting, which is a representation in its declarative form. This transformation will be useful in the logical representation module. Pre-processed and transformed questions in natural language are analyzed to give the information that can help us to locate the correct answer. In particular, we eliminated the special characters and the interrogative particle just to get the textual content of the questions (for example, remove the particle and the question mark). Deleted information's are considered non-important information. They are removed in order to obtain more significant results. A question is formulated and given to the logical representation module. More precisely, the declarative form extracted from each question can help us to construct the semantic and the logical representation. In particular, to produce a "reformulated" version of the input question so that it can be used directly and efficiently by the logical representation step.

Question : «‏ما هو أكبر تنظيم اقتصادي في العالم؟‏», « What is the largest economic organization in the world? »

Déclaration de la Question : Organisation : X هو أكبر تنظيم اقتصادي في العالم

4.3 Linguistic Analysis

This step is divided into three sub-steps: recognition of named entities, syntactic analysis and morphological analysis. The new form of the question is annotated by a tool for recognizing named entities. Then, each sentence requires a syntactic analysis. Finally, the identification of the stem (stem) is performed by a morphological analyzer.

- Named entities recognition

    The NER is therefore used to distinguish the types of entities appearing in a text. As a result, the ability to find accurate answers relies heavily on the quality of named entity recognition performed on phrases that are relevant to the user's question. For example, Fig. 2 shows the text annotation structure that gives the extraction of the named entities found in our sample question: «‏ما هو أكبر تنظيم اقتصادي في العالم؟‏», «What is the largest economic organization in the world?». Specifically, this step receives the text of the question and provides an XML file that contains all the named entities.

```
<?xml version="1.0" encoding="UTF-8" ?>
- <text>
  - <p>
    - <s num="1">
        <w num="1" type="O">ماهو</w>
        <w num="2" type="O">أكبر</w>
        <w num="3" type="B-ORG">تنظيم</w>
        <w num="4" type="O">اقتصادي</w>
        <w num="5" type="O">في</w>
        <w num="6" type="O">العالم</w>
      </s>
    </p>
  </text>
```

Fig. 2. Name entities generated by ArNER [12]

- Syntactic analysis

    The analysis process begins by identifying the constituents of the text with thematic roles and producing the syntactic patterns of the sentences. Indeed, word tags help to minimize the synsets from Arabic Wordnet by extracting only the synsets having the same tag as the word. Thus, the dependencies and the tags help us to find the relations between the words and to construct the conceptual graph correspondent. Among the information that we will use, that are extracted by Stanford parser[1] for the example of

---

[1] https://nlp.stanford.edu/software/lex-parser.shtml.

```
<?xml version="1.0" encoding="UTF-8" standalone="no" ?>
- <QuestionSyntacticAnalysis>
  - <Question>
      <Tree>(ROOT (S (VP (VBD سام) (ADJP (JJR اغر) (NP (NN نظم) (JJ تصفي))) (NP (DTNN العلم)))))</Tree>
    - <Dependencies>
        <Dependency>root(ROOT-0, 1-سام)</Dependency>
        <Dependency>xcomp(2-اغر, 1-سام)</Dependency>
        <Dependency>dep(3-نظم, 2-اغر)</Dependency>
        <Dependency>amod(4-تصفي, 3-نظم)</Dependency>
        <Dependency>dobj(5-العلم, 1-سام)</Dependency>
      </Dependencies>
      <Tag>سام/VBD اغر/JJR نظم/NN تصفي/JJ العلم/DTNN</Tag>
    </Question>
  </QuestionSyntacticAnalysis>
```

Fig. 3. Information extracted by Stanford

the previous question are, the dependency relations, tags and sentence structure trees; this information is mentioned in Fig. 3.

- Morphological analysis

In this framework, a morphological analysis of the words of the input question is carried out using Khoja Stemmer [13]. In fact, this tool removes the longest suffix and prefix. It then corresponds to the remaining word with verbal and nominal models to extract the root. It uses several linguistic data files such as a list of all diacritical characters, punctuation characters, defined articles, and 168 stop words. For example, the words in the previous question and their stems or roots are saved in an XML file, as shown in the following figure (Fig. 4):

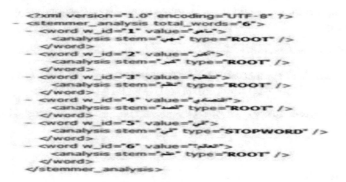

Fig. 4. Morphological analysis with Khoja Stemmer

5 Semantic and Logic Meaning Representation of the Questions

In order to determine which logical representation is best suited to each specific type of the question, we aim to represent the question in a conceptual graph formalism which can be transformed later into logical representations via the operator Φ of Sowa [14]. Then, we ensure the validity of the logical implication between these representations.

We present a method for constructing a conceptual graph from an Arabic text of the question. In fact, a conceptual modeling in natural language processing is a way of modeling the semantics. So, the semantic of the texts is transformed into semantics of conceptual models at a high level of abstraction, in terms of concepts and relationships.

After that, a conceptual graph can also be constructed as a logical structure of a certain type. Such a conceptual modeling of textual content facilitates their transformation into logical forms. Indeed, Sowa [14] proposed several fundamental rules allowing the coherent manipulation of conceptual graphs. In particular, an important characteristic of their model is that the reasoning's presented on the conceptual graphs can be made while maintaining a link with the first order logic (Fig. 5).

```
[كبير]-
  -(is)->[Organisation:OX]
[تنظيم]-
  -(attributeOf)->[كبير]
  -(propertyOf)->[اقتصادي]
[العالم]-
  -(is)->[Organisation:OX]
```

Fig. 5. A conceptual graph corresponding to a question

Thus, we describe an algorithm for the transformation of conceptual graphs based on the principle of the operator Φ proposed in [14] in order to obtain a first-order logic representation of the information extracted from these graphs of the question user. Using this algorithm, we obtain for each given graph its corresponding first-order logic representation:

∃X ∃Y ∃Z ∃W : Organisation(OX) Λ كبير(X) Λ is(X,OX) Λ تنظيم(Y) Λ attributeOf(Y,X) Λ اقتصادي(Z) Λ propertyOf(Y,Z) Λ العالم(W) Λ is(W,OX)

6 Evaluation

In this section, we present the empirical evaluation results to evaluate the number of questions that are correctly analyzed and translated into the conceptual graphs and then into the logical forms. Typically, the performance of a question transformed into semantic and logical form is measured by calculating the accuracy. This evaluation measures the transformation of questions into their corresponding conceptual graph that is then translated into a logical form. In fact, the transformation is correctly ensured when the question is correctly analyzed. Accuracy would be the number of correctly processed questions **CT**, divided by the total number of questions collected **TQ** (correctly translated and false translated). This measure is defined as follows:

$$\mathbf{Accuracy} = \frac{\mathbf{CT}}{\mathbf{TQ}} = \frac{185}{250} = 0.74 \quad (1)$$

It should be noted that our module for analysis of the questions is tested on a total of 250 questions. We obtained 185 correct semantic and logical transformations of the 250 analyzed questions and 65 false ones (28 false transformations are mainly due to errors in the recognition of the named entity, 25 are mainly due to errors of the morphological and syntactic analyzes. For 12 of the 250 questions, we found difficulties in their transformations into conceptual graphs, which meant that we could not identify their logical representations. Thus, the evaluation gives an accuracy value of 0.74.

7 Conclusion

The current presented research discusses the first stage for generating an answer to a question in Arabic language. In addition, we have presented our proposed question analysis in order to retrieve passage texts from the web of our collected questions. We carefully analyzed each question, as well as we extracted the features of those questions. Then, we have implemented this task. In order to obtain passages from the web to questions in natural language, we collected 250 factual questions and analyzed them. For each question, some features are automatically, extracted such as the keywords, the focus, the expected answer type and the declarative form. Some of those features are used to retrieve passages when the answer is located. Furthermore, the prototype presented in this paper tries to face up question analysis in order to facilitate the extraction of relevant passages from the web were the accurate answer is located. Our work uses a new point of view that is the logic representation of Arabic statements.

References

1. Tellex, S.: Pauchok: a modular framework for question answering. Master thesis Submitted to the Department of Electrical Engineering and Computer Science, Maccachusetts Institute of Technology, June 2003
2. Lee, C.-W., Shih, C.-W., Day, M.-Y., Tsai, T.-H., Jiang, T.-J., Wu, C.-W., Sung, C.-L., Chen, Y.-R., Wu, S.-H., Hsu, W.-L.: Perspectives on Chinese question answering systems. In: Proceedings of the Workshop on the Sciences of the Artificial (WSA 2005), Hualien, Taiwan (2005)
3. Kolomiyets, O., Moens, M.F.: A survey on question answering technology from an information retrieval perspective. Inf. Sci. **181**(24), 5412–5434 (2011)
4. Embarek, M.: Un système de question-réponse dans le domaine médical Le système Esculape. Ph.D. thesis, Université de Paris-Est, Juillet 2008
5. Rodrigo, Á., Perez-Iglesias, J., Peñas, A., Garrido, G., Araujo, L.: A question answering system based on information retrieval and validation. In: InCLEF (Notebook Papers/LABs/Workshops) (2010)
6. Bakari, W., Bellot, P., Trigui, O., Neji, M.: Towards logical inference for Arabic question-answering. Res. Comput. Sci. **90**, 87–99 (2015). rec. 2015-01-31; acc. 2015-02-25
7. Zweigenbaum, P., Grau, B., Ligozat, A.L., Robba, I., Rosset, S., Tannier, X., Bellot, P.: Apports de la linguistique dans les systèmes de recherche d'informations précises (2008)
8. Mendes, S., Véronique, M.: L'analyse des questions: intérêt pour la génération des réponses. In: Workshop Question-Réponse (2004)

9. Brill, E., Dumais, S., Banko, M.: An analysis of the AskMSR question-answering system. In: Proceedings of the ACL 2002 Conference on Empirical Methods in Natural Language Processing, vol. 10, pp. 257–264. Association for Computational Linguistics, July (2002)
10. Damljanovic, D., Agatonovic, M., Cunningham, H.: Identification of the question focus: combining syntactic analysis and ontology-based lookup through the user interaction. In: LREC, May 2010
11. Lally, A., Prager, J.M., McCord, M.C., Boguraev, B.K., Patwardhan, S., Fan, J., Chu-Curroll, J.: Question analysis: how Watson reads a clue. IBM J. Res. Dev. **56**(34), 2:1 (2012)
12. Zribi, I., Hammami, S.M., Belguith, L.H.: L'apport d'une approche hybride pour la reconnaissance des entités nommées en langue arabe. In: TALN 2010, Montréal, 19–23 juillet 2010, pp. 19–23 (2010)
13. Larkey, L., Connell, M.E.: Arabic information retrieval at UMass in TREC-10. In: Proceedings of TREC. NIST, Gaithersburg (2001). http://dx.doi.org/10.1.1.14.9079
14. Sowa, J.F.: Conceptual Structures: Information Processing in Mind and Machine. Addison-Wesley Company, Boston (1984)

An Arabic Question-Answering System Combining a Semantic and Logical Representation of Texts

Mabrouka Ben-Sghaier[1](✉), Wided Bakari[1,2](✉),
and Mahmoud Neji[1,2]

[1] Faculty of Economics and Management, 3018 Sfax, Tunisia
mabrouka.bensghaier@gmail.com,
widod.bakkari@fsegs.rnu.tn
[2] MIR@CL, Sfax, Tunisia

Abstract. In this paper, we present the overall structure of our specific system for generating answers of questions in Arabic, named NArQAS (New Arabic Question Answering System). This system aims to develop and evaluate the contribution of the use of reasoning procedures, natural language processing techniques and the recognizing textual entailment technology to develop precise answers to natural language questions. We also detail its operating architecture. In particular, our system is seen as a contribution, rather than a rival, to traditional systems focused on approaches extensively used information retrieval and natural language processing techniques. Thus, we present the evaluation of the outputs of each of these components based on a collection of questions and texts retrieved from the Web. NArQAS system was built and experiments showed good results with an accuracy of 68% for answering factual questions from the Web.

Keywords: NArQAS · Question-answering system · Arabic · Architecture
Logic representation · Implementation · Evaluation

1 Introduction

The interest of our approach [1] was illustrated through a question-answering system, called "NArQAS". This system has been evaluated to show that taking into account new types of approaches, in particular, those based on logic and semantic improves the Arabic question-answering and improves the results of different modules of the answer processing. In addition, a large number of systems, including their proposed approaches in Arabic are well presented in [3, 4]. Therefore, the performance of these systems is limited by the difficulty of processing the Arabic language and the widespread lack of effective natural language processing (NLP) tools that support the Arabic language [7, 10].

It is a hybrid system combining a semantic analyzer with logical reasoning; it mainly involves five steps, such as the analysis of the question, the passage retrieval, the logical representation. Indeed, NArQAS is a complete system ranging from the analysis of the question to the generation of the answer in natural language. In fact, the logic representation of Arabic statements into logic forms are presented in [5];

© Springer International Publishing AG, part of Springer Nature 2018
A. Abraham et al. (Eds.): ISDA 2017, AISC 736, pp. 735–744, 2018.
https://doi.org/10.1007/978-3-319-76348-4_71

the detection of the textual implications between the question and the passage of text that can answer it and the extraction of the answer. In general, each step has a particular need for input information and performs actions on the extracted information to produce results.

Therefore, finding an answer to a question in a large collection of question documents (instead of keyword search) answers (instead of documents) [9] is a major challenge. The main characteristic of our system is that the sentences of the text and the question are converted into existentially closed logical formulas which encode the essential semantic relations between the words of the sentences. As it presented in Fig. 1 below, this system is composed of several modules and integrates a module of logical representation of Arabic statements (question + passage containing the answer). NArQAS is a modular system which each of these phases plays a crucial role in the overall performance of Arabic question-answering systems.

This paper is organized as follows: Sect. 2 is devoted for presenting the NArQAS system. Section 3 presents the operating architecture of NArQAS. The operating view is indicated in Sect. 4; Sect. 5 presents the semantic and logical representation of Arabic texts and Sect. 6 indicates the implementation. Before concluding this paper and giving the future work, the evaluation of this system is discussed in Sect. 7.

2   System Description

We present NArQAS, a question-answering system for Arabic that allows searching for answers from the Web to factual questions. We have combined Artificial Intelligence, Information Retrieval, NLP and Automatic Reasoning techniques to improve the performance of our system by taking into account the limitations of previous systems, especially in Arabic, such as the reading comprehension for answering questions, the integration of logic and inference in the Arabic language, the use of the Recognizing Textual Entailment (RTE) technique to find the exact answer among many others candidates. Indeed, the use of these last two in question-answering systems has been largely demonstrated in particular by applications in English.

Our idea, in our work, is to design and develop a system that takes as input a natural language question and returns its output answer. Treatment is based on NLP tools and information retrieval techniques. Although the existing tools were mainly designed to improve the performance of traditional information retrieval technologies, we find that their performance is strongly influenced by those of the natural language automation techniques used.

Although the existing tools were mainly designed to improve the performance of traditional information retrieval technologies, we find that their performance is strongly influenced by those techniques of natural language processing used. Therefore, the process of generating the precise answer is based on a logical transformation step. In this context, the objective of our system is to address the following preoccupations:

- Analyze the collected questions.
- Transform each question into its declarative form.
- Query a search engine to search for the relevant document.

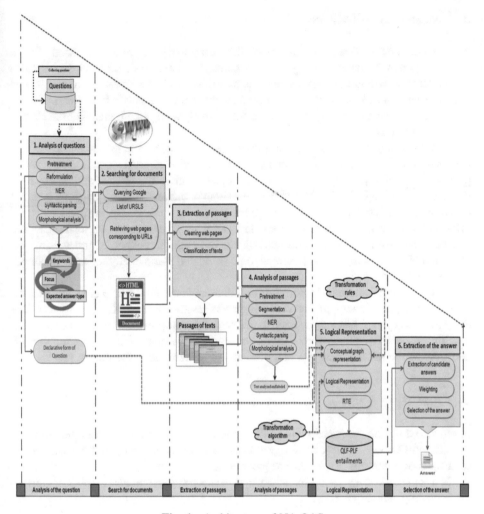

Fig. 1. Architecture of NArQAS

- Retrieve the text passages containing the answers to these questions.
- Perform morphosyntactic analyzes to determine the grammatical categories of the verbal and nominal sentences.
- Construct semantic representations of the question and the passages with a conceptual graph formalism.
- Conduct logical representations of the conceptual graph representations of the questions and the sentences constituting each passage of text.
- Integrate reasoning procedures with knowledge from Web pages and a knowledge base to produce intelligent answers in natural language.
- Apply a RTE technique between logic representation of each question and the correspond sentences of the passage answer.
- Extract the answer.

3 System Architecture

To design our system, we adopted the commonly architecture used for a question-response system. Our system is located downstream of the analysis modules of the question and the text answering this question. First, the research process begins with the analysis of the asked question until the precise answer is reached. Nevertheless, if the elements of the question are not correctly recognized, there is little chance of finding the answer.

Schematically, the design of our system generally uses pipeline architecture that chains group together six main modules namely: question analysis, document retrieval, passage extraction, passage analysis, logical representation and answer extraction; each module refers to a particular discipline. For example, the analysis of questions is based on techniques related to automatic language processing; document retrieval relies on information retrieval techniques to locate the relevant documents in relation to the question and information extraction techniques to extract the expected precise answer, etc. Each of these components deserves to be evaluated intrinsically, or their assembly is evaluated as a whole.

Our proposed system has a complex architecture and support of more sophisticated search techniques namely, logical reasoning and recognition of textual entailments. The design of this system has largely contributed to the development of question-answering systems, especially for Arabic. Therefore, more information's on the number of questions and each question type adopted in this system are presented in [2].

4 Operating View

In order to describe the functioning of the various components and to find the accurate and coherent answer to this question, we propose to follow those operations. Therefore, each operation can contain several sequential actions.

In a schematic way, we can consider that the search for a precise answer comprises the following phases (Fig. 2):

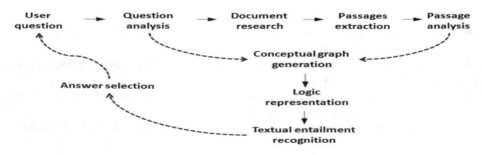

Fig. 2. Schematic of the NArQAS system

- The user writes the question in Arabic.
- The analyzer of the question must analyze it. In particular, it is responsible for identifying all the keywords in the question; the focus; the type of question and its expected answer type. It also generates the corresponding declarative form for each question.
- The document search module uses Google that accepts in input the keywords of each question and produces documents that are closely related to those keywords. In this context, the search for documents is based on a match between the terms of the question and those of the documents. Research is not limited to finding resources referenced by keywords, but attempts to identify relevant passages that contain answers to the questions.
- The extractor of the relevant passages must identify the relevant passages containing the correct answers to the questions.
- In the passage analysis module: the system divides the passages into sentences and labels the named entities therein. The named entities designate names of persons, places, organization, dates or monetary units. These are very important elements for retrieving answers, especially when the question waits as answer a named entity.
- The logical form generator takes into consideration the logical representation of these texts by finding argument-predicate structures from conceptual graphs.
- Finally, the extractor of the answer is responsible for selecting the most consistent and exact answer among other candidates.

5   Semantic and Logical Representation of Arabic Texts

In our work, the main challenge is how to convert the information found in the text into the language with which a machine can think and make decisions, and correctly answer a user's natural language questions. Indeed, our objective is to allow an in-depth analysis of Arabic documents (texts and questions) which can go as far as a semantic representation. To do this, we propose to use the conceptual graph model, which allows a richer representation of the textual content. In contrast, text analysis does not pretend to replace the interpretation of the meaning of texts; it is to extract contents or a structure to answer specific questions. This graphical conceptual model represents the sentence of the text and the question with a structure formed by vertices and edges in the graph [12]. It provides a higher level of understanding of the text by capturing the semantics in the text.

Indeed, the conceptual graph is a connected, finite bipartite graph presented by Sowa in 1984 [12]. In addition, this formalism is used as an intermediate language to interpret object-oriented formalism and natural language. The graph being constituted by a set of nodes connected by links; the nodes of the graph are either concepts (denoted by rectangles), or conceptual relations (noted by ovals). Conceptual relationship nodes indicate a relationship involving one or more concepts. Concepts consist of a type of concept and a referent (instantiation of the type of concept); Relationships consist of a type of relationship. Conceptual modeling in the processing of natural language is a way of modeling semantics. The semantics of texts turns into semantics of conceptual models at a high level of abstraction, in terms of concepts.

With regard to the Arabic language, the representation of texts in conceptual graphs was already the subject of some research. However, this research has not yet reached the same level of advancement as that of the Latin languages. Thus, the representation of semantically texts using conceptual graphs is one of the recent techniques that facilitate the manipulation process of different NLP applications, such as information retrieval, question-answering, machine translation, Automatic summary, reasoning, etc.

In reality, the transformation of the text (passage, question) into a conceptual graph is carried out by 4 phases. We first introduce the list of terms. Then, we detail the step of extracting concepts associated with these terms. Then we extract the list of relations between these concepts. Finally, we construct the graph corresponding to these concepts and relations. More precisely, the construction process begins with an initial sub-step that extracts the set of terms that have meanings in these documents. Then, Arabic Wordnet is used to associate these terms with concepts. Then, the relations between these concepts are also identified using the syntactic relations between the words and the rule-based technique proposed by [11].

An example of the conceptual graph corresponding to the following passage is illustrated in Fig. 3.

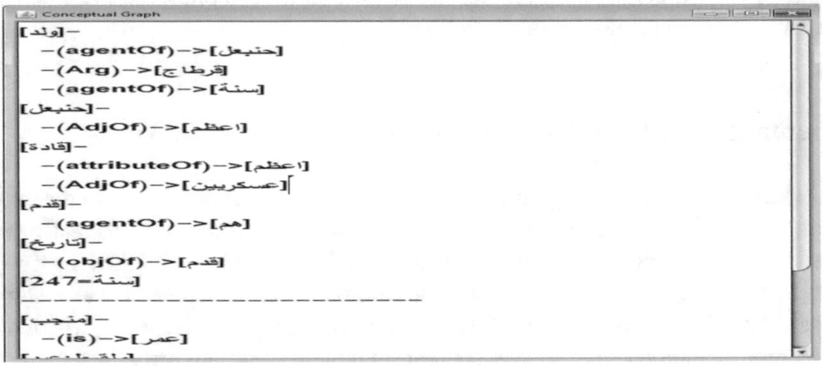

Fig. 3. A conceptual graph corresponds to the previous passage

Our implementation of document analysis (text and question) consists in representing them in conceptual graphs to determine the predicate-arguments structures. This task is complex, identifying the events and actors involved are one of the main objectives in information retrieval. Predicate-argument structures give a semantic interpretation of sentences by determining who has done what, to whom, where, when, how and why.

A conceptual graph can also be constructed as a logical structure of a certain type. Indeed, such a conceptual modeling of textual content (text or question) facilitates their transformation into logical form. In our work, we propose an approach that not only builds conceptual graphs from Arabic documents (question, text) but is also effective for the logical representation of these documents by finding predictive arguments structures from graphs conceptual. These graphs have been translated into first-order logic by associating a new variable with each generic concept vertex, while all the individual concept vertices with the same marker are associated with the same constant.

Indeed, Sowa [12] proposed several fundamental rules allowing manipulating conceptual graphs in a coherent way. In particular, an important feature of their model is that the reasoning's presented on the graphs can be made while maintaining a link with first-order logic. However, the idea of using formalized graphs to construct a logical formula is not new. Currently, there are several works based on the transformation of conceptual graphs into logical forms being used to solve a particular problem. To do this, we describe a conceptual graph transformation algorithm based on the principle of the operator Φ proposed in [12] in order to obtain a first order logic representation of the information extracted from these graphs of the question of the user and of the passage answer.

The transformation of a conceptual graph into a formula of first-order logic is done as follows. At each node of the graph corresponds a quantified variable to which is applied a unary predicate having for name the word associated with the node. The arcs introduce a binary prediction with the label of the arc and for arguments the variables associated with its source and destination nodes. The final formula is the conjunction of the predications thus obtained. Indeed, the identification of such a predicate depends on its context. The predictive structure is a graph of predicate-argument relations. In the verbal sentence, to assume the logical representations of the Arabic states, we treat the two cases of verbs such as transitive and intransitive. However, in the nominal case, we deal with the common name and the proper name.

The transformation of a sentence into a natural language in the logic of predicates serves to write the predicate first, then the arguments. Indeed, before generating its logical form of this sentence, one must first determine which ones are used as predicate and as arguments.

In addition, logical representation has a long history in natural language; it is an intermediate step between deep syntactic and semantic analysis [13]. It has made considerable progress in key areas of natural language processing; such an application is the question-answer where the problem is to find exact answers to questions expressed in natural language by searching for a large collection of documents [14]. On the other hand, this type of research by integrating logical reasoning into Arabic language processing applications has not yet reached an advanced stage. This is due, on the one hand, to the complexity of this language and, on the other, to the inadequacy of research on this language. Thus, advanced question-answering requires sophisticated text-processing tools based on NLP and logical reasoning methods [13].

Thus, for the graph of Fig. 3, we have the logical formula illustrated below.

∃X ∃Y ∃Z ∃W ∃T ∃E ∃F ∃G ∃H ∃I ∃J : ولد(X) Λ حنبعل(Y) Λ agentOf(X,Y) Λ
اعظم(Z) Λ AdjOf(Y,Z) Λ قائد(W) Λ attributeOf(W,Z) Λ عسكريين(T) Λ AdjOf(W,T) Λ
قدم(E) Λ هم(F) Λ agentOf(E,F) Λ تاريخ(G) Λ objOf(G,E) Λ قرطاج(H) Λ Arg(X,H) Λ
سنة(I) Λ agentOf(X,I) Λ 247(J) Λ isEqual(I,J)

∃X ∃Y ∃Z ∃W ∃T ∃E : منجب(X) Λ عمر(Y) Λ is(X,Y) Λ ملقرط(Z) Λ عبد(W) Λ
is(Z,W) Λ is(X,Z) Λ رافق(T) Λ حملة(E) Λ Arg(T,E)

---

6 Implementation

After a detailed description of this system, its architecture and its operation, we present
the procedures of its implementation. More specifically, we will see how we implement
the different modules of this system. Indeed, each of these modules plays a crucial role
in the overall performance of our question-answering system. The programming language we have chosen for the implementation is JAVA. Indeed, we have developed an
interface in Java that manages upstream all the processing required for the tasks
requested by the user in order to obtain a precise answer to a question in Arabic. The
implementation of the proposed approach in the framework of a system enriched sake
of better performance of the Arabic question-answering.

The actual implementation of such a system in Arabic with the end users has never
been carried out, except for demonstration purposes. Their development, particularly at
the level of the Web, is encountered with sizeable locks (scale-up, noise management,
identification of duplicates…), and the founded solutions require significant implementation efforts. This interface includes the following processes: analysis of the
question, document search, selection of relevant passages, analysis of passages, logical
representation of questions and passages and detection of textual implication between
them; And finally, the extraction of the precise answer.

7 Evaluation of NArQAS

In this section, we present the evaluations carried out by our system. Indeed, to evaluate
finely the contribution of the components of a system assumes to measure the contribution of each module compared to the overall results obtained by the system. It is
then necessary to successfully measure the relevance of the strategies developed as well
as the operation of the components. In fact, the evaluation is generally based on the
validity of an individual answer supported by a candidate passage. We have chosen to
evaluate not only NArQAS but also its main constituents, especially those that allowed
the passage of natural language statements into logical forms in order to have a detailed

view of the origin of its performances. Besides, for our experiments, we built a corpus of questions-texts using 250 questions collected from TREC, CLEF, FAQ and forums [2]. On their side [8], the authors suggest that the evaluation of question-answering systems is an important area of research that needs more attention.

The evaluation of a question-answering system is still a very open question. Indeed, the overall evaluation is carried out on the number of correct answers returned by the system. Thus, the performance at each stage affects the end result. In this framework, the answers to be evaluated are provided by a question-answering system that seeks an answer to the various questions. The questions used in the evaluation process correspond of five types; they are collected from different sources and used to build our corpus presented in [6]. From these 250 questions, 170 questions have correct answers. Generally, the performance of a question answering system can be measured by calculating the accuracy of the number of correct answers on a set of candidate answers. Therefore, the accuracy would be the number of correctly answered questions, divided by the total number of answered questions; this measure is defined as follows:

$$\mathbf{Accuracy} = \frac{\text{Number of correctly answered questions}}{\text{Total number of answered questions}} = \frac{170}{250} = 0.68$$

8 Discussion

The possibility of choosing the appropriate technique for each type of question makes it possible to achieve performances close to the desired ones, such as obtaining the answer in real time. In our research work, we seek to consider the influence of logical representation and the recognition of textual entailment on the process of searching for the precise answer. We present the results of empirical evaluation to count the number of questions that are correctly analyzed and translated into the logical form to seek the right answer. The results obtained during our experiments are explained by the decisions taken during the preparation of the data and the realization of the proposed approach. We have carried out the steps of this approach in an experimental setting with the aim of implementing a complete system of answering a question in Arabic. The results of each module provide evidence of problem.

9 Conclusion

In this paper, we presented the interest of our approach which was illustrated through a question-answering system for the Arabic language "NArQAS". More specifically, we have described the architecture of this system, as well as the principles of the various modules that compose this architecture and the functioning of our system. This system has a modular architecture and is based on the combination of information retrieval, information extraction, automatic language processing and automatic reasoning. Thus, it will allow users to find a precise answer to a factual question. Future work may complete the implementation of our system; use the conceptual graphs to logic representation and further exploring other questions types.

References

1. Bakari, W., Trigui, O., Neji, M.: Logic-based approach for improving Arabic question answering. In: 2014 IEEE International Conference on Computational Intelligence and Computing Research (ICCIC), pp. 1–6. IEEE, December 2014
2. Bakari, W., Bellot, P., Neji, M.: AQA-WebCorp: web-based factual questions for Arabic. Procedia Comput. Sci. **96**, 275–284 (2016)
3. Bakarı, W., Bellot, P., Neji, M.: Literature review of Arabic question-answering: modeling, generation, experimentation and performance analysis. In: FQAS, pp. 321–334 (2015)
4. Bakari, W., Bellot, P., Neji, M.: Researches and reviews in Arabic question answering: principal approaches and systems with classification (2016)
5. Bakari, W., Bellot, P., Trigui, O., Neji, M.: Towards logical inference for the Arabic question-answering (2015)
6. Bakari, W., Bellot, P., Neji, M.: A preliminary study for building an Arabic corpus of pair questions-texts from the web: AQA-WebCorp. iJES **4**(2), 38–45 (2016)
7. Al-Khalifa, H., Al-Wabil, A.: The Arabic language and the semantic web: challenges and opportunities. In: The 1st International Symposium on Computer and Arabic Language (2007)
8. Olvera-Lobo, M.D., Gutiérrez-Artacho, J.: Question answering track evaluation in TREC, CLEF and NTCIR. In: New Contributions in Information Systems and Technologies, pp. 13–22. Springer (2015)
9. Chavan, G., Gore, S.: Design of the effective question answering system by performing question analysis using the classifier. Int. J. Comput. Appl. **139**(14), 1–3 (2016)
10. Al Agha, I., Abu-Taha, A.: AR2SPARQL: an Arabic natural language interface for the semantic web. Int. J. Comput. Appl. **125**(6), 19–27 (2015)
11. Cheddadi, A.: Three-levels Approach for Arabic Question Answering Systems. Diss. Ecole Mohammadia d'Ingénieurs (2014)
12. Sowa, J.F.: Conceptual Structures: Information Processing in Mind and Machine. Addison-Wesley, Reading (1984)
13. Moldovan, D., Harabagiu, S., Girju, R., Morarescu, P., Novischi, A., Lacatusu, F., Badulescu, A., Bolohan, O.: LCC tools for question answering. In: Voorhees, E., Buckland, L. (eds.) Proceedings of TREC 2002 (2002)
14. Voorhees, E.: Overview of the TREC 2002 question answering track. In: TREC 2002 (2002). http://trec.nist.gov

Algorithms for Finding Maximal and Maximum Cliques: A Survey

Faten Fakhfakh[1]([✉]), Mohamed Tounsi[1], Mohamed Mosbah[2],
and Ahmed Hadj Kacem[1]

[1] ReDCAD Laboratory, University of Sfax, Sfax, Tunisia
{faten.fakhfakh,mohamed.tounsi}@redcad.org, ahmed.hadjkacem@fsegs.rnu.tn
[2] LaBRI Laboratory, Bordeaux INP, University of Bordeaux, Bordeaux, France
mohamed.mosbah@u-bordeaux.fr

Abstract. Finding maximal and maximum cliques are well-known problems in the graph theory. They have different applications in several fields such as the analysis of social network, bioinformatics and graph coloring. They have attracted the interest of the research community. The main goal of this paper is to present a comprehensive review of the existing approaches for finding maximal and maximum cliques. It presents a comparative study of the existing algorithms based on some criteria and identifies the critical challenges. Then, it aims to motivate the future development of more efficient algorithms.

Keywords: Maximal and maximum cliques · Algorithms
Comparative study · Challenges

1 Introduction

The detection of densely connected groups, called communities, is a well known problem in complex networks. It plays a vital role in several domains such as computational biology [1], emergent pattern detection in terrorist networks [2], data mining [3] and system analysis [4]. For example, it is possible to identify closely people in a social network, build the tree of life, etc.

In order to model communities in a rigorous way, we consider maximal and maximum cliques. A clique is a fully connected (or complete) subgraph of the graph. The maximal clique problem consists in finding a clique that is not a subset of any larger clique in the same graph. A maximum clique represents a clique which has the maximum cardinality.

Given a graph G with six vertices in the Fig. 1. In this graph, there are two maximal cliques ({e, g, h} and {b, c, f, e}) and one maximum clique {b, c, f, e}.

Various algorithms are available in the literature to deal with the problem of enumerating maximal and maximum cliques [5–11]. They are based on different search strategies. However, this problem still a challenging research task.

The remainder of this paper is structured as follows: In Sect. 2, we introduce the research methodology that we have applied in this paper. Section 3 provides

© Springer International Publishing AG, part of Springer Nature 2018
A. Abraham et al. (Eds.): ISDA 2017, AISC 736, pp. 745–754, 2018.
https://doi.org/10.1007/978-3-319-76348-4_72

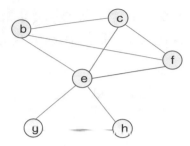

Fig. 1. Example of a graph

a detailed review of the existing algorithms for finding maximal and maximum cliques. Afterwards, we present the taxonomy that we have adopted to compare the existing algorithms as well as a rich discussion. Finally, we conclude the paper and highlights a research roadmap for our future work.

2   Research Methodology

In order to have a clear picture of the proposed algorithms for finding maximal and maximum cliques, a systematic literature review (SLR) which presents an exhaustive summary of the literature is introduced, while following a rigorous methodology. According to Kitchenham [12], an SLR is composed of three phases: planning the review, conducting the review and reporting the review. In the rest of this section, we describe each phase.

2.1   Phase 1: Planning the Review

The purpose of this phase is to investigate the studies that have proposed algorithms for finding maximal and maximum cliques. To this end, we will examine the reviewed publications using some research questions (RQs):

RQ 1. Which are the goals and the characteristics of each proposed algorithm?
RQ 2. What taxonomy of existing algorithm can be used?
RQ 3. What are the evaluating methods adopted to determine the efficiency of each algorithm?

- **Search strategy**
  To conduct the review, we are interested in searching IEEE Xplore, ACM Digital Library, Elsevier Scopus, Google Scholar, arXiv.org and Web of Science. We have combined the following search terms in order to find relevant studies: "detect" and "maximal clique"; "find" and "maximal clique"; "enumerate" and "maximal clique", "detect" and "maximum clique"; "find" and "maximum clique"; "enumerate" and "maximum clique".

- **Inclusion and exclusion criteria**
  To evaluate each primary study, we need to determine the inclusion and exclusion criteria. We had selected only papers published in English language from 2009 to 2017. Also, we had considered papers published in journals, conferences or workshops and we had discarded other publications such as summaries and presented slides.

2.2   Phase 2: Conducting the Review

In this second phase, we follow these two steps:

- **Search for studies**
  Through the search strategy that we have already defined, we had found a total of 120 papers. However, some of these publications were redundant because they exist in many databases. After filtering them, we selected only 70 publications for the next step.

- **Study selection**
  In this step, we determine the papers that meet some inclusion criteria by analyzing their abstracts, keywords and titles. We had retrieved 55 papers that refer to our defined inclusion criteria. After a full reading, only 32 papers were selected. Then, we had analyzed each one to extract the pertinent information while answering the three research questions.

2.3   Phase 3: Reporting the Review

In this phase, we use the research questions that we have already presented to evaluate each solution. RQ1 is very useful question as it allows us to resume the selected papers. To respond to the second question RQ2, we classify the proposed algorithms based on different taxonomy criteria. Finally, the answer of RQ3 determines the method adopted to evaluate each algorithm.

3   Overview on the Existing Algorithms for Finding Maximal and Maximum Cliques

Several works have addressed the problem of finding maximal and maximum cliques with different goals and assumptions. The proposed algorithms can be classified into two types according to the studied graph. In fact, we distinguish algorithms that are devoted to static graphs and others to dynamic graphs whose topology can change over time.

3.1   Algorithms Dedicated to a Static Graph

Segundo et al. [13] introduced an efficient exact maximum clique algorithm, which consists in finding the largest possible clique for large and massive sparse graphs. This algorithm is based on a prior bit-parallel algorithm *BBMC (BB-MaxClique)* [14] and it uses some optimization techniques according to the application domain. The efficiency of this algorithm is evaluated using a repository of real graphs.

In [15], the authors proposed a parallel and scalable algorithm for maximal clique enumeration (MCE). Its complexity is linear and it is able to handle the data-intensive nature of the MCE problem for large graphs.

Conte et al. [16] present a new technique for detecting maximal cliques in large networks. This technique consists of a two-level decomposition process. The goal of the first level is to recursively identify tractable portions of the network. The second level decomposes the tractable portions into small blocks. The authors formally proved the correctness of their proposed solution based on manual proofs.

Another approach cited in [17] provides a distributed algorithm based on MapReduce platform for enumerating maximal cliques. This algorithm is based on two phases. Firstly, a method which generates small subgraphs from input graph is proposed. Secondly, it applies a strategy for enumerating maximal based on subgraphs. It uses pruning technique which can eliminate duplicate cliques and improve the efficiency.

Additionally, the approach cited in [18] presents a new algorithm called MULE (Maximal Uncertain cLique Enumeration) which enumerates all maximal cliques. This work is focused on uncertain graph which represents a set of vertices that has a high probability of being a completely connected subgraph. Uncertain graphs have been extensively used in modeling biological systems, social networks, etc. The authors compare their algorithm with others in terms of different quality of service (QoS) parameters.

More recently, Sun et al. [5] have studied the issue of finding maximal cliques in a dynamic graph. The major problem is how to maintain maximal cliques taking into consideration graph changes. To this end, the authors have proposed two algorithms to handle edge insertion and edge deletion. These algorithms are evaluated in terms of different criteria including time performance and memory consumption.

Furthermore, Rezvanian and Meybodi [6] have proposed some algorithms based on learning automata for solving maximum clique in a stochastic graph. In such graph, the probability distribution functions of weight corresponding to each edge is unknown. Each vertex is equipped with a learning automaton whose actions correspond to selecting the edges of the associated vertex. To evaluate the efficiency of these algorithms, many experiments have been presented using a set of stochastic graphs. They aim to study the convergence behavior of the algorithms, the impact of learning rate on the accuracy, etc.

In [7], Jiang et al. have developed an exact branch-and-bound algorithm to solve maximum weight clique problem (MWCL). MWCL consists of two phases. The first one aims to reduce the size of the graph using a preprocessing step. In the second phase, another algorithm is used to decrease the number of branches in the search space. Different experiments are conducted using real-world graphs. The results show the efficiency of MWCL algorithm compared to others algorithms.

A new exact algorithm for extracting maximum cliques is proposed by Shimizu et al. [19]. It is composed of two steps. The first step needs several procedures. It consists in computing the weights of maximum cliques and generating the optimal tables. In the second phase, each problem is divided into some subproblems and resolved recursively. This algorithm is evaluated by generating random graphs having different number of vertices and edges.

3.2 Algorithms Dedicated to a Dynamic Graph

The major contribution of Luo et al. [20] is a distributed algorithm called MC (Maximal Clique). It allows to find all maximal cliques in a graph with linear complexity. Then, the authors used the MCP (Maximum Clique Problem) algorithm to get all maximal cliques with the same number of memberships and they computed a unique maximum clique by UMCP (Unique MCP) algorithm.

Xu et al. [21] have studied two major problems. The first one consists in computing the set of maximal cliques in a graph and the second aims to manage these cliques. In order to resolve the first problem, the authors have proposed an efficient algorithm to distribute data and enumerate maximal cliques. When the underlying graph is updated, an algorithm which updates the set of maximal cliques is used. Two update operations are considered: edge deletion and edge insertion. To evaluate the efficiency of the proposed algorithms, a set of real-world graphs from different application domains are used.

Das et al. [22] have treated the problem of constructing and maintaining maximal cliques in a dynamic graph. In such graph, edges can be added or deleted at any time. The authors have studied the case of adding a new edge in a graph. Then, they have enumerated the new maximal cliques that are built after change. The experimental results show the efficiency of this solution in terms of memory consumption and computation time by comparing it with other algorithms.

Cheng et al. [23] have developed an algorithm that recursively enumerate maximal cliques in a large graph. This algorithm consists in constructing special subgraphs (called B*-graph) in order to locally search maximal cliques while ensuring the correctness of the result. When some topological changes are applied, another version of B*-graph is used to update the maximal cliques and compute the maximum clique. The authors have evaluated the performance of their algorithm using real large networks. Furthermore, they have presented some proofs in order to verify the correctness of their results.

A recent algorithm has been presented in [8]. It aims to enumerate maximal cliques based on the operation of binary graph partitioning. This operation consists in dividing a graph until each task becomes sufficiently small in order to be processed in parallel. In addition, the authors introduced a hybrid algorithm for maximal cliques enumeration, which benefits from the advantages of the algorithm and an existing one called BK [24]. Moreover, they evaluated the performance of the proposed solutions based on a wide variety of graph data available in an open source.

4   Comparative Study and Discussion

Throughout this survey paper, we provide an overview of the major efforts of solving the maximal and maximum cliques problems. In Table 1, we present an attempt to compare existing research approaches based on some criteria as illustrated in Fig. 2.

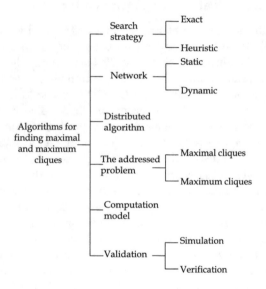

Fig. 2. Taxonomy of different criteria of comparing existing algorithms

- **Search strategy**
  It represents the strategy which has been adopted to enumerate maximal or maximum cliques. In fact, some approaches are based on exact algorithms which ensure optimum solutions. Others algorithms use some heuristics which determine a solution after a certain number of iterations.

- **Network**
  This criterion denotes whether the proposed approach considers a static or dynamic network.

Table 1. A comparative table of some approaches for finding maximal and maximum cliques

Approaches	Criteria								
	Type of search strategy	Network		Distributed algorithm	Addressed problem		Computation model	Validation	
		Static	Dynamic		Maximal cliques	Maximum clique		Simulation	Verification
Schmidt et al. [15] (2009)	heuristic	✓			✓			✓	
Segundo et al. [13] (2016)	exact	✓				✓		✓	
Hou et al. [8] (2016)	heuristic	✓			✓			✓	
Luo et al. [20] (2015)	heuristic	✓		✓		✓		✓	
Conte et al. [16] (2016)	heuristic	✓		✓	✓				✓
Wu et al. [17] (2009)	heuristic	✓		✓	✓			✓	
Xu et al. [21] (2016)	heuristic	✓		✓	✓			✓	
Mukherjee et al. [18] (2017)	heuristic	✓			✓			✓	
Sun et al. [5] (2017)	heuristic	✓			✓			✓	
Rezvanian et al. [6] (2015)	heuristic	✓				✓		✓	
Jiang et al. [7] (2017)	exact	✓				✓		✓	
Shimizu et al. [19] (2017)	exact	✓				✓		✓	
Das et al. [22] (2016)	heuristic		✓		✓			✓	
Cheng et al. [23] (2011)	heuristic		✓		✓	✓			✓

- **Distributed algorithm**
  This attribute determines if the adopted algorithm is distributed. In contrast to centralized algorithms, a distributed algorithm is designed to run on interconnected computing entities in order to realize a common task [25].

- **The addressed problem**
  It tells us whether the proposed solution allows to address the problem of enumerating maximal or maximum cliques.

- **Computation model**
  This criterion represents the model used to design the proposed algorithms. This model describes the behavior of the different steps of the algorithm.

- **Validation**
  This criterion indicates the way used to validate the proposed algorithms. In fact, we can distinguish two types of validation: simulation and verification. The simulation consists in using simulators to evaluate the performance of algorithms based on some metrics. However, the verification provides an efficient way for the designer to evaluate the behavior of a system and to detect possible errors. It is characterized by their ability to reason rigorously on programs in order to demonstrate their validity in relation to a certain specification.

Based on the above observation, we note that most of the proposed works use centralized algorithms [8,13–15] and only few works [16,20] introduced distributed algorithms for finding maximal or maximum cliques. Moreover, we identify that only some solutions have addressed this problem in dynamic networks. These networks are characterized by frequent topology changes due to unpredictable appearance and disappearance of mobile devices and/or communication links.

Despite the complexity of algorithms and especially in dynamic networks, we notice that the proposed solutions in the literature do not consider any model to design their algorithms.

Another metric that has been considered in our comparison is the validation method adopted for each algorithm. We notice that the majority of the research studies [8,13,18] rely on simulation to improve the performance of these solutions. While simulation represents the most commonly applied approaches, it suffers from a strong limit and particularly in the context of the distributed systems. In fact, these approaches are therefore insufficient to ensure an exhaustive verification.

Some works [16,21] have formally proved the correctness of their algorithms. However, the proofs which have been introduced are performed manually. These proofs are long and tedious specially in the case of complex algorithms and a minor error can have serious consequences on the system operation. Then, ensuring the correctness of the proposed algorithms becomes crucial because it gives us confidence that systems perform as designed and do not behave harmfully.

5   Conclusion and Research Challenges

To conclude, we have introduced in this paper a detailed review of the existing algorithms for finding maximal and maximum cliques. After that, we have provided a comparative study of these algorithms based on some criteria. Finally, we have pointed out some research challenges.

In our ongoing work, we intend to propose a distributed algorithms for identifying maximal and maximum cliques in dynamic networks. We aim to use local computation model to design our algorithms [26]. The latter allows us to encode algorithms at a high level of abstraction independently from the network topology. In fact, an algorithm is simply given by a set of relabeling rules which are locally executed. These rules, which are closely related to mathematical and logic formulas, are able to derive the correctness of distributed algorithms.

To specify the abstraction provided by local computations, we plan to use a formal method. In fact, formal methods provide a real help for ensuring correctness while respecting safety properties in the design of distributed algorithms. Particularly, the correct-by-construction approach [27] provides a simple way to construct and prove algorithms. The main idea relies upon the development of distributed algorithms following a top/down approach controlled by the refinement of models. The purpose of this approach is to simplify the formal proofs and also to validate the incorporation of requirements.

References

1. Abu-Khzam, F.N., Baldwin, N.E., Langston, M.A., Samatova, N.F.: On the relative efficiency of maximal clique enumeration algorithms, with applications to high-throughput computational biology. In: Research Trends in Science and Technology (2005)
2. Berry, N., Ko, T., Moy, T., Smrcka, J., Turnley, J., Wu, B.: Emergent clique formation in terrorist recruitment. In: Agent Organizations: Theory and Practice (2004)
3. Matsunaga, T., Yonemori, C., Tomita, E., Muramatsu, M.: Clique-based data mining for related genes in a biomedical database. BMC Bioinform. **10**(1), 205 (2009)
4. Zhang, H., Zhao, H., Cai, W., Liu, J., Zhou, W.: Using the k-core decomposition to analyze the static structure of large-scale software systems. J. Supercomput. **53**(2), 352–369 (2010)
5. Sun, S., Wang, Y., Liao, W., Wang, W.: Mining maximal cliques on dynamic graphs efficiently by local strategies. In: Proceedings of the 33rd International Conference on Data Engineering (ICDE), pp. 115–118. IEEE (2017)
6. Rezvanian, A., Meybodi, M.R.: Finding maximum clique in stochastic graphs using distributed learning automata. Int. J. Uncertain. Fuzziness Knowl. Based Syst. **23**(01), 1–31 (2015)
7. Jiang, H., Li, C.M., Manya, F.: An exact algorithm for the maximum weight clique problem in large graphs. In: Proceedings of the Thirty-First AAAI Conference on Artificial Intelligence (AAAI), pp. 830–838 (2017)
8. Hou, B., Wang, Z., Chen, Q., Suo, B., Fang, C., Li, Z., Ives, Z.G.: Efficient maximal clique enumeration over graph data. Data Sci. Eng. **1**(4), 219–230 (2016)

9. Fazlali, M., Zakerolhosseini, A., Gaydadjiev, G.: Efficient datapath merging for the overhead reduction of run-time reconfigurable systems. J. Supercomput. **59**(2), 636–657 (2012)

10. Kuz, A., Falco, M., Giandini, R.: Social network analysis: a practical case study. Computación y Sistemas **20**(1), 89–106 (2016)

11. Eppstein, D., Löffler, M., Strash, D.: Listing all maximal cliques in sparse graphs in near-optimal time. In: Proceedings of the International Symposium on Algorithms and Computation (ISAAC), Jeju Island, Korea, 15–17 December 2010, pp. 403–414. Springer, Heidelberg

12. Kitchenham, B.: Procedures for performing systematic reviews. Keele, UK, Keele University **33**(2004), 1–26 (2004)

13. Segundo, P.S., Lopez, A., Pardalos, P.M.: A new exact maximum clique algorithm for large and massive sparse graphs. Comput. Oper. Res. **66**, 81–94 (2016)

14. Segundo, P.S., Rodríguez-Losada, D., Jiménez, A.: An exact bit-parallel algorithm for the maximum clique problem. Comput. Oper. Res. **38**(2), 571–581 (2011)

15. Schmidt, M.C., Samatova, N.F., Thomas, K., Park, B.H.: A scalable, parallel algorithm for maximal clique enumeration. J. Parallel Distrib. Comput. **69**(4), 417–428 (2009)

16. Conte, A., Virgilio, R.D., Maccioni, A., Patrignani, M., Torlone, R.: Finding all maximal cliques in very large social networks. In: Proceedings of the 19th International Conference on Extending Database Technology (EDBT), pp. 173–184 (2016)

17. Wu, B., Yang, S., Zhao, H., Wang, B.: A distributed algorithm to enumerate all maximal cliques in mapreduce. In: Proceedings of the Fourth International Conference on Frontier of Computer Science and Technology (FCST), pp. 45–51. IEEE (2009)

18. Mukherjee, A.P., Xu, P., Tirthapura, S.: Enumeration of maximal cliques from an uncertain graph. IEEE Trans. Knowl. Data Eng. **29**(3), 543–555 (2017)

19. Shimizu, S., Yamaguchi, K., Saitoh, T., Masuda, S.: Fast maximum weight clique extraction algorithm: optimal tables for branch-and-bound. Discret. Appl. Math. **223**, 120–134 (2017)

20. Luo, C., Yu, J., Yu, D., Cheng, X.: Distributed algorithms for maximum clique in wireless networks. In: Proceedings of the 11th International Conference on Mobile Ad-hoc and Sensor Networks (MSN), pp. 222–226 (2015)

21. Xu, Y., Cheng, J., Fu, A.W.C.: Distributed maximal clique computation and management. IEEE Trans. Serv. Comput. **9**(1), 110–122 (2016)

22. Das, A., Svendsen, M., Tirthapura, S.: Change-sensitive algorithms for maintaining maximal cliques in a dynamic graph. arXiv preprint arXiv:1601.06311 (2016)

23. Cheng, J., Ke, Y., Fu, A.W.C., Yu, J.X., Zhu, L.: Finding maximal cliques in massive networks. ACM Trans. Database Syst. (TODS) **36**(4), 21 (2011)

24. Bron, C., Kerbosch, J.: Algorithm 457: finding all cliques of an undirected graph. Commun. ACM **16**(9), 575–577 (1973)

25. Tel, G.: Introduction to Distributed Algorithms. Cambridge University Press, Cambridge (2000)

26. Ehrig, H., Rozenberg, G., Kreowski, H.J.: Handbook of Graph Grammars and Computing by Graph Transformation, vol. 3. World Scientific, River Edge (1999)

27. Leavens, G.T., Abrial, J.R., Batory, D., Butler, M., Coglio, A., Fisler, K., Hehner, E., Jones, C., Miller, D., Peyton-Jones, S., Sitaraman, M., Smith, D.R., Stump, A.: Roadmap for enhanced languages and methods to aid verification. In: Proceedings of the 5th International Conference on Generative Programming and Component Engineering (GPCE), pp. 221–236. ACM (2006)

K4BPMN Modeler: An Extension of BPMN2 Modeler with the Knowledge Dimension Based on Core Ontologies

Molka Keskes, Mariam Ben Hassen$^{(\boxtimes)}$, and Mohamed Turki

MIRACL Laboratory, ISIMS, University of Sfax,
BP 242, 3021 Sakiet Ezzeit, Sfax, Tunisia
keskesmolka92@gmail.com, {mariem.benhassen,
mohamed.turki}@isims.usf.tn

Abstract. In this paper, we aim at enriching the graphical representation of sensitive business processes (SBPs) in order to identify and localize the crucial knowledge mobilized and created by these processes. Therefore, we develop a specific Eclipse plug-in, called «K4BPMN Modeler: Knowledge for Business Process Modeling Notation Modeler», implementing and supporting the BPMN extension «BPMN4KM». This extension was designed in a previous research project based on core ontologies in order to integrate all relevant aspects related to the knowledge dimension in sensitive business process models. Besides, we illustrated the application of some extended concepts on a model of medical care process.

Keywords: Knowledge management · Sensitive business process modeling
BPM4KI · BPMN 2.0 · BPMN4KM · BPMN2 Modeler

1 Introduction

Modeling Sensitive Business Processes (SBP) is not an easy task due to its essential characteristics [1]. A SBP usually comprises critical activities based on acquisition, sharing, storage, (re)use of knowledge, and collaboration/social interaction among participants, so that the amount of value added to the organization depends on the knowledge of the process agents. They deal with unpredictable decisions, creativity-oriented tasks, and dynamic execution that evolves based on the experience acquired by the agents/experts. Research on SBP points out its essential characteristics [1–4]. All those issues hinder their representation, and make them subject to different interpretations [2]. However, it is difficult to find out approaches and formalisms that address all or at least most of these characteristics in the representation of their processes, mainly due to the lack of proper modeling strategies, as discussed in [2]. Moreover, most approaches and formalisms (like UML 2.0 activity diagram [5], BPMN 2.0 [6], KMDL [7], PROMOTE [8] and NKIP [9]) do not provide special attention to the notation applied in SBP representation [2].

The Business Process Meta-Model for Knowledge Identification (BPM4KI) [4, 10] was developed, from an extensive literature review, with the target to comprise all the

© Springer International Publishing AG, part of Springer Nature 2018
A. Abraham et al. (Eds.): ISDA 2017, AISC 736, pp. 755–770, 2018.
https://doi.org/10.1007/978-3-319-76348-4_73

elements to make a SBP explicit. This meta-model is semantically rich and well-based on core domain ontologies [11–13], which are based on top of the DOLCE foundational ontology [14]. BPM4KI comprises concepts from several perspectives/dimensions that are crucial for a complete understanding and representation of a SBP, namely: the Functional perspective, the Organizational perspective, the Behavioral perspective, the Informational perspective, the Intentional perspective and the Knowledge perspectives.

In this paper, we focus more on the representation of «Knowledge Perspective» which represents the most relevant aspect of SBP modeling. This dimension has not yet fully supported, integrated and implemented within business process (BP) models and business process modeling (BPM) approaches and formalisms (the BPM and knowledge modeling formalisms) [2, 10, 15]. To address this research gap, enrich and improve the SBP modeling, Ben Hassen et al. [10, 15] have proposed a valid BPMN 2.0 extension for integrating knowledge dimension in SBP models. The proposed extension, entitled «BPMN4KM» is developed using the extensibility mechanisms of BPMN [6]. In fact, BPMN 2.0 is a standard for BP modeling that is very common in professional practice due to its expressiveness, the well defined meta model and the possibility of workflow integration. It was selected as the most suitable BPM notations for SBP representation, because addresses the highest representation coverage of the set of BPM4KI concepts and incorporates requirements for SBP modeling better than other formalisms [2].

The objective of this research work is the development of a specific plug-in based on the Eclipse platform, called «K4BPMN: Knowledge for Business Process Modeling Notation», implementing and supporting BPMN4KM [10, 15] to explicitly incorporate all relevant aspects related to knowledge management (KM) within BPs models, and on the other hand, to enrich the graphical representation of SBPs and improve the localization and identification of crucial knowledge mobilized and created by these processes. K4BPMN is an extension of the already existing Eclipse BPMN2 Modeler plug-in [16]: it completes this later by integrating new attributes, properties, elements and specific icons for introduce new semantics.

The paper is organized as follows. Section 2 presents the proposed approach for extending BPMN 2.0 with the knowledge dimension. Section 3 presents the concrete syntax of the proposed BPMN4KM extension and its implementation. Section 4 illustrates the application and the relevance of some BPMN4KM concepts, based on a real case study. Finally, Sect. 5 concludes the paper and underlines some future research topics.

2   BPMN4KM: A BPMN Extension for Integrating the Knowledge Dimension in SBP Models

To date, to the best of our knowledge, there is a lack of works providing systematic approaches for the development of extensions to the BPMN 2.0 meta-model to consider the knowledge aspect in BPM [17, 18]. However, there are previous works providing approaches to extend BPMN 2.0 to represent their domain specific requirements. Some interesting extension proposals are presented in [17–21]. The differences between the different research works unveil the need for a unified method for the conceptual modeling of extensions and their representation in terms of the BPMN extension mechanism. However, none of the proposals adequately and fully support and represent all relevant aspects of knowledge dimension within BPs models (e.g., differentiation between tacit and explicit knowledge, the different types of knowledge conversion, the dynamic aspects of knowledge, the different sources of knowledge, etc.).

To address this research gap between BPM and KM, Ben Hassen et al. [10, 15] proposed in a previous research work a rigorous scientific approach to extend BPMN 2.0 for KM, using the extensibility mechanisms of BPMN [6]. This extension, called «BPMN4KM», is semantically rich and well-based on core domain ontologies [11–13]. It considers and incorporates all relevant aspects of the knowledge dimension in SBP modeling to improve the localization and identification of crucial knowledge mobilized and created by these processes.

The method for the development of BPMN4KM [10, 15] consists of five steps. First a requirements analysis based on literature review should be performed in order to define the according domain requirements. Thereafter, an in-depth analysis of the SBP domain is conducted in order to identify concepts, properties, rules and constraints of the domain. Then, a core domain-ontology has to be created which offers a referential of generic and central concepts and semantic relationships relevant to the BPM-KM domain for conceptualizing SBPs in various contexts. In this step, the «Knowledge Perspective» is modeled as an Ontological Design Patterns (ODP) [22] represented as a UML class diagram [10]. It is semantically rich and well-based on «core» domain ontologies [11–13] which are based on top of the DOLCE foundational ontology [14]. Precisely, the extended ODP is based on the reuse and the specialization of central generic concepts (and the relationships between them) defined in different ontological modules of the global and consistent ontology OntoSpec [11–13]. The Knowledge ODP offers a referential of generic concepts and relationships characterising the BPM-KM domain. It represents an overview of the organizational and individual knowledge mobilized and created by an BP/organization. Besides, it describes the different knowledge sources, the knowledge flow and the dynamics of acquisition, preservation, conversion, transfer, sharing, development, and (re)use of knowledge within and between organizations. The authors [10, 15] proposed a general characterization of the knowledge concept: A Knowledge is the Capacity (or Disposition) to carryOut (and affects) a type of Action (e.g. an Organizational Action) aiming to achieve an objective. It isBornedBy an Agentive Entity (which can be a Human, an Expert, a Collective, or an Organization). Moreover, several features of Knowledge have been further described in [10, 15] classifying them according to some dimensions: *affiliation of knowledge, source of knowledge, organizational coverage of knowledge/sharing scope, nature of knowledge, cruciality* and the *strategic* dimensions. For example, Knowledge is divided into Individual Knowledge and Collective Knowledge according to the *organizational coverage of knowledge* dimension. With respect to the limited space of this paper, a comprehensive description of the different knowledge concepts cannot be presented. Subsequently, the described elements of the core domain ontology have to be examined regarding their matching to BPMN elements. In the third step an equivalence check is performed. The check aims on the identification of BPMN extension elements and their extension requirements. Step 4 refers to the method of

Stroppi et al. [19] and were used to create a valid BPMN extension model (BPMN+X model) with an abstract syntax. In the fifth step the concrete syntax is implemented and graphical notations have to be defined for the extension elements.

Figure 1 presents an extract of the resulting extended BPMN meta-model (BPMN4KM). In this figure only the relevant standard BPMN classes are shown in white. The BPMN4KM concepts are shown in grey. The semantics and the abstract syntax of the BPMN4KM elements are based on the specification of the BPMN extension mechanism [6]. *ExtensionModel* is the topmost container of all the elements defining a BPMN extension. *BPMNElement* allows representing an original element of the BPMN meta-model. *ExtensionElement* allows representing a new element in the extension model which is not defined in the BPMN meta-model (such as Knowledge, InternalKnowledge, TacitKnowledge, ExplicitedKnowledge, Explicitable Knowledge, ProceduralKnowledge, ExternalKnowlede, InternalKnowledge, PhysicalKnowledgeSupport, Information, DistalIntention, Combination, Socialization, Internalization, Externalization and Explicitation). *ExtensionDefinition* allows specifying a named group of attributes which are jointly added to the original BPMN elements (such as KnowledgeFlow, Experiencer, Collective, KnowledgeConversionAction, Knowledge Intensive Activity, Critical Organizational Activity, Collaborative Organizational Activity, and Sensitive Business Process). *ExtensionDefinition* has the same meaning than the ExtensionDefinition element of the BPMN metamodel. The semantics defined by the *ExtensionAttributeDefinition* element of the BPMN meta-model is captured by the Property metaclass of the UML metamodel. Thus, ExtensionAttributeDefinition is represented in BPMN4KM models by UML properties, either owned by the ExtensionDefinition elements or navigable from them through associations. The properties of *ExtensionDefinition* and *ExtensionElement* elements can be typed as a *BPMNElement*, *ExtensionElement*, *BPMNEnum*, *ExtensionEnum* or UML primitive type. Finally, *ExtensionRelationship* specifies a conceptual link between a *BPMNElement* and an *ExtensionDefinition* element aimed to extend it. The BPMN extension mechanism cannot express the BPMN element to be extended by an extension definition. Thus, the definition of an *ExtensionRelationship* does not produce any effect in the resulting BPMN extension. *ExtensionRelationship* is provided to help conceptualizing extensions since extensions are generally defined to customize certain elements of the BPMN meta-model [10]. With respect to the limited space of this paper, the application of each applied transformation rule cannot be presented.

In the following, we will present the implementation of the concrete syntax of the extended BPMN4KM meta-model as well as the graphical notations for the extension elements.

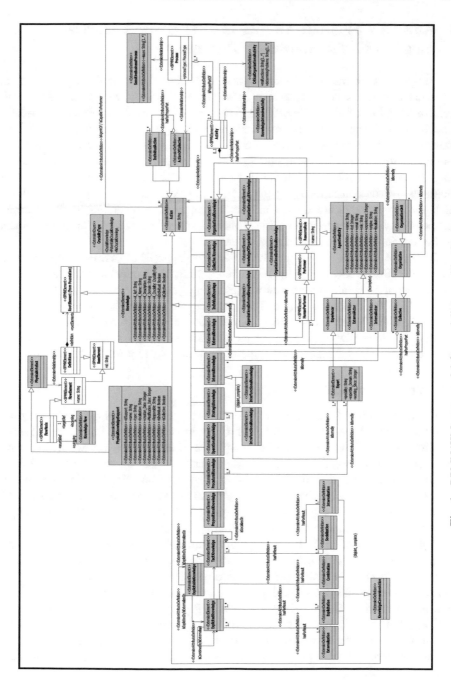

Fig. 1. BPMN4KM meta-model for modeling sensitive business process

3   K4BPMN Modeler: An Extension of BPMN2 Modeler Plug-in with the Knowledge Dimension

We proposed an advanced concrete syntax that defines new and specific graphical representation for the new concepts of BPMN4KM. Precisely, we developed a specific Eclipse plug-in, entitled «K4BPMN: Knowledge for Business Process Modeling Notation», to integrate and represent all relevant aspects related to the knowledge dimension in SBP models (to improve the localization of crucial knowledge that is mobilized and created by these processes). This plug-in extends an existing Eclipse plug-in, named BPMN2 Modeler [16] dedicated to the definition of different BPMN diagrams. Let us first remind that the BPMN 2.0 extended meta-model (BPMN4KM Meta-model) presents the BPMN elements allowing the modeling of SBPs from a perspective of knowledge localization. This meta-model contains the original elements of the notation as well as the extensions made.

Therefore, to represent graphically SBPs according to BPMN4KM, we opt to use the Eclipse BPMN2 Modeler plug-in [16]. Indeed, this plug-in offers a graphical modeling tool allowing to create business processes using graphical elements of the notation BPMN 2.0. Admittedly this plug-in does not take into account the proposed extensions on the BPMN 2.0 meta-model. Hence our mission is to extend the functionality of the BPMN2 Modeler plugin to consider these extensions. This section briefly presents Eclipse Extensibility Principle and explains how we extend it for SBP modeling.

3.1   Eclipse Extensibility Principle

The Eclipse platform framework consists of a set of extensible plug-ins allowing new plug-ins to extend them through the PDE (Plug-in Development Environment) tool.

The extensibility of the whole Eclipse plug-in is concretized via two main notions namely: extension points and extensions. It must declare an extension point in its plugin.xml file that is described by an XML Schema (xsd) file, defining the possible extension's grammar from that point. Any plug-in that wants to extend the functionality of an extensible plug-in must declare extensions in their plugin.xml file conforming to the grammars associated with the corresponding extension points.

Figure 2 graphically illustrates the Eclipse extensibility principle. It indicates that the plugin B is extensible because it declares the three points of extensions: Point 1, Point 2 and Point 3. The grammars associated with these three extensions are described respectively in the files point1.exsd, point2.exsd and point3.exsd. These are specified in their plugin.xml file. The plug-in A extends the B plug-in by declaring two extensions: Ext1 and Ext2 which correspond to the points of extension Point1 and Point2 of the plug-in B. These extensions are defined in the plugin.xml file of the plug-in A. Thus any extension must be plugged into an extension point which in turn can be connected to several extensions.

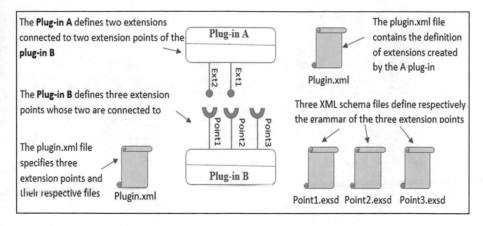

Fig. 2. Extensibility principle of Eclipse

3.2 K4BPMN Modeler Plug-in

K4BPMN Modeler is a dedicated tool for modelling SBP as instances of the BPMN4KM meta-model. It is an extension of the already existing Eclipse BPMN2 Modeller plug-in: it extends this later by integrating (i) new elements with their own properties (property tabs) and forms and, (ii) new icons for representing a new semantics.

These extensions are realized by the implementation of several extensions associated with an extension point defined by the BPMN2 Modeler plug-in. In the following we detail the different elements of the extension point and the extensions used in the development of the K4BPMN Modeler plug-in.

(a) ***BPMN Modeler plug-in extension point.*** BPMN 2.0 Modeler [16] is built on the architecture of the Eclipse Plug-in, it is an extensible plug-in that provides a set of extension points. Indeed, considering the extensions we add to the BPMN notation, we propose to add *custom Task* corresponding to the new elements such as the element `Tacit Knowledge, Explicited Knowledge, Explicitable Knowledge, Information`, etc. and *Property tabs* containing the properties of these elements.

Thus, we propose to decorate the shapes of the basic elements `Lane` and `Participant` with specific icons. To do this we have used the BPMN2 Modeler *RuntimeSpecialization* extension point, which is identified by org.eclipse.bpmn2.modeler. core.org.eclipse.bpmn2.modeler.runtime and is used to customize and extend the BPMN2 Modeler user interface as well as the BPMN2 model itself. The file org. eclipse.bpmn2.modeler.runtime.exsd contains the grammar associated with this extension point. Figure 3 shows an extract of this file.

From this figure, we see that an extension associated with this extension point must include the Extension element, and the following extension elements: *runtime, model, propertyTab, customTask, modelExtension, modelEnable, propertyExtension, featureContainer, Style, toolPalette, dataType, typeLanguage, expressionLanguage,*

```
<element name="extension">
    <annotation>
        <appinfo>
            <meta.element />
        </appinfo>
    </annotation>
    <complexType>
        <choice minOccurs="1" maxOccurs="unbounded">
            <element ref="runtime"/>
            <element ref="model"/>
            <element ref="propertyTab" minOccurs="0" maxOccurs="unbounded"/>
            <element ref="customTask" minOccurs="0" maxOccurs="unbounded"/>
            <element ref="modelExtension" minOccurs="0" maxOccurs="unbounded"/>
            <element ref="modelEnablement" minOccurs="0" maxOccurs="unbounded"/>
            <element ref="propertyExtension" minOccurs="0" maxOccurs="unbounded"/>
            <element ref="featureContainer" minOccurs="0" maxOccurs="unbounded"/>
            <element ref="style" minOccurs="0" maxOccurs="unbounded"/>
            <element ref="toolPalette" minOccurs="0" maxOccurs="unbounded"/>
            <element ref="dataType" minOccurs="0" maxOccurs="unbounded"/>
            <element ref="typeLanguage" minOccurs="0" maxOccurs="unbounded"/>
            <element ref="expressionLanguage" minOccurs="0" maxOccurs="unbounded"/>
            <element ref="serviceImplementation"/>
        </choice>
        <attribute name="point" type="string" use="required">
            <annotation>
                <documentation>

                </documentation>
            </annotation>
        </attribute>
        <attribute name="name" type="string">
            <annotation>
                <documentation>

                </documentation>
                <appinfo>
                    <meta.attribute translatable="true"/>
                </appinfo>
            </annotation>
        </attribute>
    </complexType>
</element>
```

Fig. 3. Extract from the file including the grammar of the extension point BPMN2 Modeler runtime

serviceImplementation. Among these elements, we used mainly the following five elements:

- *Runtime*: This element allows to define a namespace for a specific execution engine, through attributes described in the contract of this extension point.
- *CustomTask*: is used to define extended BPMN2 model elements that can be made available as Creation Tools in the Tool Palette.
- *PropertyTab*: is used to define additional Property Sheet tabs, or replacements for existing default tabs provided by the editor.
- *ModelExtension*: is used to dynamically extend BPMN2 elements. The extension features and classes are defined by property elements contained in this modelExtension.
- *FeatureContainer*: This element allows the extension plug-in to define custom behavior for graphical components managed by the editor to replace their creative and drawing behaviors on the editing canvas.

(b) **Extensions to the K4BPMN Modeler plug-in.** The K4BPMN Modeler plug-in defines a set of extension elements associated with the extension point of the BPMN2 Modeler plug-in, allowing the new and specific graphical representation for the new concepts of BPMN4KM as illustrated in Table 1. For instance, the Participant element (represented in the form of LaneSet) is specified by new markers for representing Internal Actor, External Actor, Organization, Organization Unit, and Informal Group. Furthermore, we have incorporated new notational elements with specific properties for Knowledge typologies (individual/collective dimension, propositional/procedural dimension, etc.), Information, Information Medium, Knowledge Conversion Actions (Socialization, Internalization, etc.), Knowledge Flows (between knowledge, activities and agentive entities), Physical Knowledge Supports, Agentive Entities (Expert, Collective, Human, etc.)

Table 1. Main objects of the K4BPMN

BPMN4KM Elements	Modeling Notation	BPMN4KM Elements	Modeling Notation
Tacit Knowledge (Individual/Collective Crucial, Internal/External, Strategic/Operational Procedural/Propositional)	Individual Tacit Knowledge · Collective Tacit Knowledge · Crucial Collective Tacit Knowledge · Internal Collective Tacit Knowledge · External Collective Tacit Knowledge · Propositional Collective Tacit Knowledge · Procedural Collective Tacit Knowledge · Strategic Collective Tacit Knowledge · Operational Collective Tacit Knowledge	Critical Organizational Activity (Individual/ Collective)	Individual Critical Task · Collective Critical Task · Collective Critical Sub Process
Explicitable Knowledge (Individual/Collective, Crucial, Internal/External, Operational/Strategic, Procedural/Propositional)	Individual Explicitable Knowledge · Crucial Collective Explicitable Knowledge · Strategic Collective Explicitable Knowledge · Internal Individual Explicitable Knowledge · Propositional Collective Explicitable Knowledge · Procedural Collective Explicitable Knowledge	Knowledge Flow	– · – · – · →
Explicited Knowledge (Individual/Collective, Crucial, Internal/External, Operational/Strategic, Procedural/Propositional)	Individual Explicited Knowledge · Collective Explicited Knowledge · Crucial Collective Explicited Knowledge · External Collective Explicited Knowledge · Propositional Collective Explicited Knowledge · Strategic Collective Explicited Knowledge	Expert (Internal, External)	Expert · Internal Expert · External Expert
Information (Individual/Collective, Procedural /Propositional)	Collective Information · Individual Internal Propositional Information · Collective External Procedural Information	Human (Internal, External)	Human · External Human · Internal Human
Physical Knowledge Support (Individual/Collective)	Physical Knowledge Support · Individual Physical Knowledge Support · Collective Physical Knowledge Support	Collective (Internal, External)	Collective · Internal Collective · External Collective
Organization	(As an Agentive Entity) [External Organization] (As a BPMN Participant)	Knowledge conversion modes	Knowledge conversion Action
		Internalization (Red)	**Internalization** ·········►
Organization Unit	(As an Agentive Entity) [Internal Organization Unit] (As a BPMN Participant)	Combination (Blue)	**Combination** ·········►
		Socialization (Green)	**Socialization** ·········►
		Externalization (Gray)	**Externalization** ·········►
		Explicitation (Orange)	**Explicitation** ·········►

and Critical Organizational Activities. The concrete syntax of the extended Knowledge concept is depicted in Table 1. Knowledge elements are marked with source, nature, affiliation, strategic and organizational coverage information according to the five knowledge dimensions [10, 15] as introduced in Sect. 2.

In the following, we illustrate the addition of the Collective Tacit Knowledge element as an example of adding a type of knowledge. To add the latter, we define the *CustomTask* extension element. The XML code for this extension is shown in Fig. 4.

```
<customTask
       category="Knowledge"
       featureContainer="mynewexemple.ColKnowledgeFeatureContainer"
       icon="int.PNG"
       id="mynewexemple.knowledge"
       name="Collective Tacit Knowledge"
       runtimeId="mynewexemple.runtime"
       type="Task">
  <property
         name="type"
         value="Collective Tacit Knowledge">
  </property>
  <property
         name="name"
         type="EString"
         value="Collective Tacit Knowledge">
  </property>
  <property
         label="Is General"
         name="IsGeneral"
         type="EBoolean"
         value="false">
  </property>
  <property
         label="Domain"
         name="Isdomain-specifc"
         type="EString">
  </property>
  <property
         name="Internal"
         type="EBoolean"
         value="false">
  </property>
  <property
         name="External"
         type="EBoolean"
         value="false">
  </property>
  <property
         name="Procedural"
         type="EBoolean"
         value="false">
  </property>
  <property
         name="Knowledge Of Organization"
         type="EBoolean"
         value="false">
  </property>
  <property
         name="Organizational Unit Knowledge"
         type="EBoolean"
         value="false">
  </property>
  <property
         name="Propositional"
         type="EBoolean"
         value="false">
  </property>
  <property
         name="Strategic"
         type="EBoolean"
         value="false">
  </property>
  <property
         name="Operational"
         type="EBoolean"
         value="false">
  </property>
  <property
         label="Is Crucial"
         name="Cruciality"
         type="EBoolean"
         value="false">
  </property>
  <property
         label="Ref"
         name="Num"
         type="EString">
  </property>
  <property
         label="Description"
         name="documentation"
         type="EString">
  </property>
  <property
         name="Organizational Informal Group Knowledge"
         type="EBoolean"
         value="false">
  </property>
</customTask>
```

Creating Collective Tacit Knowledge Properties

Fig. 4. The CustomTask extension element associated with the Collective Tacit Knowledge

```
public class CollectiveTacitKnowledgeFeatureContainer extends CustomShapeFeatureContainer {
    private final static String TYPE_VALUE = "Collective Tacit Knowledge";
    private final static String KNOWLEDGE_ID = "mynewexemple.knowledge";
    public final static String IMAGE_ID_PREFIX =
            CollectiveTacitKnowledgeFeatureContainer.class.getPackage().getName() + ".";
    public final static String ICONS_FOLDER = "icons/";
        @Override
        public String getId(EObject object) {
        EStructuralFeature f = ModelDecorator.getAnyAttribute(object, "type");
        if (f!=null) {
        Object id = object.eGet(f);
        if (TYPE_VALUE.equals(id))
        return KNOWLEDGE_ID;}
        return null;}
    @Override
        protected IShapeFeatureContainer createFeatureContainer(IFeatureProvider fp) {
        return new TaskFeatureContainer(){
            @Override
            public IAddFeature getAddFeature(IFeatureProvider fp){
                return new AddTaskFeature(fp){
                    @Override
                    protected void decorateShape(IAddContext context, ContainerShape containerShape,
                            Task businessObject) {
                        IGaService gaService = Graphiti.getGaService();
                        IPeService peService = Graphiti.getPeService();
                        Rectangle selectionRect = (Rectangle)containerShape.getGraphicsAlgorithm();
                        int width = 140;
                        int height = 60;
                        selectionRect.setWidth(width);
                        selectionRect.setHeight(height);
                        peService.deletePictogramElement(containerShape.getChildren().get(0));
                        Shape rectShape = peService.createShape(containerShape, false);
                        peService.sendToBack(rectShape);
                        org.eclipse.graphiti.mm.algorithms.Ellipse roundedRect = gaService.createEllipse(rectShape);
                        StyleUtil.applyStyle(roundedRect, businessObject);
                        gaService.setLocationAndSize(roundedRect, 0, 0, width, height);
                        setFillColor(containerShape);
                        setImage(containerShape);
                        Image img = ImageProviderRole.createImage(roundedRect, customTaskDescriptor, 38, 38);
                        Graphiti.getGaService().setLocation(img, 2, 2);
                        for (PictogramElement pe : containerShape.getChildren()) {
                            GraphicsAlgorithm ga = pe.getGraphicsAlgorithm();
                            if (ga instanceof MultiText) {
                                Graphiti.getGaService().setLocationAndSize(ga, 40, 2, width-42, height-2);}}}};}
    private void setImage(ContainerShape containerShape) {
        Task ta = BusinessObjectUtil.getFirstElementOfType(containerShape, Task.class);
        if (ta!=null) {
            ExtendedPropertiesAdapter adapter = ExtendedPropertiesAdapter.adapt(ta);
            Boolean attributeValue = (Boolean)adapter.getFeatureDescriptor("External").getValue();
            Boolean attributeValueinternal = (Boolean)adapter.getFeatureDescriptor("Internal").getValue();
            Boolean attributeValueprocedural = (Boolean)adapter.getFeatureDescriptor("Procedural").getValue();
            Boolean attributeValuepropositional= (Boolean)adapter.getFeatureDescriptor("Propositional").getValue();
            Boolean attributeValuestrategic = (Boolean)adapter.getFeatureDescriptor("Strategic").getValue();
            Boolean attributeValueoperational= (Boolean)adapter.getFeatureDescriptor("Operational").getValue();
            Boolean attributeValuecruciality= (Boolean)adapter.getFeatureDescriptor("Cruciality").getValue();
            Shape shape = containerShape.getChildren().get(0);
            ShapeStyle ss = new ShapeStyle();
            shape.getGraphicsAlgorithm().getGraphicsAlgorithmChildren().clear();
            IGaService gaService = Graphiti.getGaService();
            String imageId = getImageId2();
            String internalimageId = getImageId3();
            String proceduralimageId = getImageId4();
            String collectiveimageId = getImageId5();
            String propositionalimageId = getImageId7();
            String strategicimageId = getImageId9();
            String operationalimageId = getImageId10();
            String crucialityimageId = getImageId11();
            String imageId8 = getImageId8();
            String propertyValue;
            ss.setDefaultColors(IColorConstant.YELLOW);
            if (Boolean.TRUE.equals(attributeValueinternal)) {
                propertyValue = Boolean.TRUE.toString();
                GraphicsAlgorithmContainer ga = containerShape.getChildren().get(0).getGraphicsAlgorithm();
                Image img = gaService.createImage(ga, internalimageId.trim());
                gaService.setLocationAndSize(img, 5,35, 24, 24);}
            else { EList<GraphicsAlgorithm> elts = shape.getGraphicsAlgorithm().getGraphicsAlgorithmChildren();
                for (GraphicsAlgorithm graphicsAlgorithm : elts) {
                    if(graphicsAlgorithm instanceof Image)
                    {graphicsAlgorithm.getGraphicsAlgorithmChildren().clear();}}
                propertyValue = Boolean.FALSE.toString();}
            StyleUtil.applyStyle(shape.getGraphicsAlgorithm(), ta, ss);
            FeatureSupport.setPropertyValue(containerShape, "Internal.property", propertyValue);
        if (Boolean.TRUE.equals(attributeValuecruciality)) {
            propertyValue = Boolean.TRUE.toString();
            GraphicsAlgorithmContainer ga = containerShape.getChildren().get(0).getGraphicsAlgorithm();
            Image img = gaService.createImage(ga, crucialityimageId.trim());
            gaService.setLocationAndSize(img, 20,0, 24, 24);}
        else {
            EList<GraphicsAlgorithm> elts2 = shape.getGraphicsAlgorithm().getGraphicsAlgorithmChildren();
            for (GraphicsAlgorithm graphicsAlgorithm : elts2) {
                if(graphicsAlgorithm instanceof Image)
                {
                graphicsAlgorithm.getGraphicsAlgorithmChildren().clear();}}
            propertyValue = Boolean.FALSE.toString();}
        StyleUtil.applyStyle(shape.getGraphicsAlgorithm(), ta, ss);
        FeatureSupport.setPropertyValue(containerShape, "Cruciality.property", propertyValue);}}

    public String getImageId3() {
    return ImageProvider.IMG_16_INTERNAL;}
```

Fig. 5. An extract of the methods of the CollectiveTacitKnowledgeFeatureContainer class

This element implements the CollectiveTacitKnowledgeFeatureContainer class to (i) create the forms of the `Collective Tacit Knowledge` element (ii) to add the icons associated with the `Propositional, Procedural, Strategic, Operational, Internal, External` type of knowledge. Figure 5 illustrates an extract of the decorateShape and SetImage methods of the CollectiveTacitKnowledgeFeatureContainer class.

Also a *PropertyTab* extension element implements the CollectiveTacitKnowledgePropertySection class to (i) create labels, text boxes, and Boolean attributes that exist in the property tab through the createSectionRoot method, and (ii) display the properties of the `Collective Tacit Knowledge` selected through the resourceSetchanged method, Fig. 6 shows an extract of these methods. This element allows to

```java
public class CollectiveTacitKnowledgePropertySection extends DefaultPropertySection {
    @Override
    protected AbstractDetailComposite createSectionRoot() {
        DefaultDetailComposite composite = new KnowledgeDetailComposite(this);
        setProperties(composite,new String[] {"Num","name", "Cruciality","IsGeneral","Isdomain-specifc","Internal","External",
        "Procedural","Propositional","Strategic","Operational","Knowledge Of Organization","Organizational Unit Knowledge","Organizational Informal Group Knowledge"});
        return composite;}
    @Override
    public AbstractDetailComposite createSectionRoot(Composite parent, int style) {
        DefaultDetailComposite composite = new KnowledgeDetailComposite(parent,style);
        setProperties(composite,new String[] {"Num","name", "Cruciality","IsGeneral","Isdomain-specifc","Internal","External",
        "Procedural","Propositional","Strategic","Operational","Knowledge Of Organization","Organizational Unit Knowledge","Organizational Informal Group Knowledge"});
        return composite;}
    public class KnowledgeDetailComposite extends DefaultDetailComposite {
    public KnowledgeDetailComposite(AbstractBpmn2PropertySection section) {
        super(section);}
    public KnowledgeDetailComposite(Composite parent, int style) {
        super(parent, style);}
    @Override
    protected void cleanBindings() {
        super.cleanBindings();
        descriptionText = null;}
    @Override
    public void createBindings(EObject be) {
        bindAttribute(be,"Num"); //$NON-NLS-1$
        bindAttribute(be,"name"); //$NON-NLS-1$
        bindAttribute(be,"Cruciality"); //$NON-NLS-1$
        bindAttribute(be,"IsGeneral"); //$NON-NLS-1$
        bindAttribute(be,"Isdomain-specifc"); //$NON-NLS-1$
        bindAttribute(be,"Internal"); //$NON-NLS-1$
        bindAttribute(be,"External"); //$NON-NLS-1$
        bindAttribute(be,"Procedural"); //$NON-NLS-1$
        bindAttribute(be,"Propositional"); //$NON-NLS-1$
        bindAttribute(be,"Strategic"); //$NON-NLS-1$
        bindAttribute(be,"Operational"); //$NON-NLS-1$
        bindAttribute(be,"Knowledge Of Organization"); //$NON-NLS-1$
        bindAttribute(be,"Organizational Unit Knowledge"); //$NON-NLS-1$
        bindAttribute(be,"Organizational Informal Group Knowledge"); //$NON-NLS-1$
        bindList(be, "documentation"); //$NON-NLS-1$
        }
    protected boolean isModelObjectEnabled(String className, String featureName) {
        if (featureName!=null && "name".equals(featureName)) //$NON-NLS-1$
            return true;
        return super.isModelObjectEnabled(className,featureName);}
    @Override
    public void resourceSetChanged(ResourceSetChangeEvent event) {
        super.resourceSetChanged(event);
            final Task ta = (Task)getBusinessObject();
            final EStructuralFeature nameAttribute = ModelDecorator.getAnyAttribute(ta, "name");
            final TransactionalEditingDomain editingDomain = getDiagramEditor().getEditingDomain();
            for (Notification n : event.getNotifications()) {
                int et = n.getEventType();
                if (et==Notification.SET) {
                    if (n.getFeature()==nameAttribute) {
                        Display.getDefault().asyncExec( new Runnable() {
                            public void run() {
                                editingDomain.getCommandStack().execute(new RecordingCommand(editingDomain) {
                                    @Override
                                    protected void doExecute() {
                                        String text = (String)ta.eGet(nameAttribute);
                                        text += "\n";}});});}}}}}
```

Fig. 6. An extract of the methods of the CollectiveTacitKnowledgePropertySection class

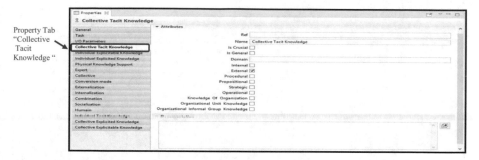

Fig. 7. The property tab `Collective Tacit Knowledge`

```
<propertyTab
        class="mynewexemple.CollectiveTacitKnowledgePropertySection"
        id="mynewexemple.KnowledgepropertyTab1"
        image="knowledge.png"
        label="Collective Tacit Knowledge"
        runtimeId="mynewexemple.runtime">
</propertyTab>
```

Fig. 8. PropertyTab extension element associated with the `Collective TacitKnowledge` element

specify properties related to this knowledge, that is, when the user selects a `Collective Tacit Knowledge`, their properties are displayed in the corresponding fields created by the createSectionRoot method (see Fig. 7). The XML code for this is shown in Fig. 8.

The following section illustrates the practical applicability of the proposed extension considering a real SBP scenario from medical domain.

4   Case Study

This section illustrates a SBP model using the extended K4BPMN Modeler plug-in which is based on BPMN4KM meta-model [10, 15]. We aim to evaluate their potential in providing an adequate and expressive representation of SBPs in order to improve the localization and the identification of crucial medical knowledge. This experimentation was carried out in the context of the Association of Protection of the Motor-disabled of Sfax-Tunisia (ASHMS). A depth description of the case study has been presented in [2, 10]. Figure 9 illustrates the evolved BPMN extension by presenting a simplified SBP model of the initial evaluation of a child with cerebral palsy enriched with some extensions according to BPMN4KM [10]. During our experimentation, we identified different types and modalities of medical knowledge mobilized and produced by the SBP critical activities. For example, the knowledge A_3K_{p1} related to «Evaluation results of the motor acquisition capacities development of the disabled children and its disorders» is produced by the critical activity A_3 «Evaluation of functional and psyco-motor acquisition capacities». It is an Explicited Propositional Knowledge which is created as a result of the activity execution by the medical community, during which

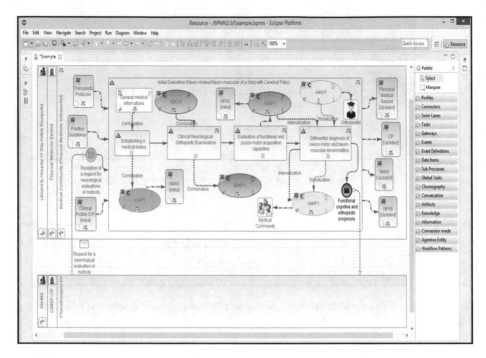

Fig. 9. Instance of the extended BPMN demonstrating a part of a (simplified) SBP model related to the initial neuro-motor evaluation of a child with CP using K4BPMN Modeler

they interact with the information (source of knowledge) related to the disabled child. It is also a Collective Procedural Knowledge which can be borne by the Collective Physical Knowledge Supports: the neurological and neuro-motor assessment sheets (NMAS) in the medical record of the children with cerebral palsy. These physical supports are located internally within the Physical Medicine service in the University Hospital Habib Bourguiba of Sfax-Tunisia, precisely in the various archives drawers or patients'directories. A_3K_{p1} is of a scientific nature which is related to patients.

5 Conclusions

In this research work, we presented K4BPMN Modeler, an extension of the plug-in BPMN2 Modeler with the knowledge dimension for modeling SBPs. This plug-in extension is designed methodically by application of the BPMN extension «BPMN4KM» which was designed in a previous research project, based on core ontologies. We aim at enriching the graphical representation of SBPs in order to identify and localize the crucial knowledge mobilized and created by these processes. The applicability of the proposed K4BPMN Modeler extension is demonstrated by a medical process. Our current research activities focus on achieving the implementation of the different SBP modeling aspects (i.e. the functional, the behavioral, the informational, the intentional and the knowledge perspectives). Future work includes the development of a methodology for mapping, analyzing and representing SBP.

References

1. Ben Hassen, M., Turki, M., Gargouri, F.: Towards extending business process modeling formalisms with information and knowledge dimensions. In: 30th International Conference on Industrial and Engineering Applications of Artificial Intelligence and Expert Systems (IEA/AIE 2017), Arras, France (2017)
2. Ben Hassen, M., Turki, M., Gargouri, F.: Sensitive business processes representation: a multi-dimensional comparative analysis of business process modeling formalisms. In: International Symposium on Business Modeling and Software Design. Revised Selected Papers. LNBIP, vol. 257, pp. 83–118. Springer, Cham (2017)
3. Ben Hassen, M., Turki, M., Gargouri, F.: Modeling dynamic aspects of sensitive business processes for knowledge localization. In: International Conference on Knowledge Based and Intelligent Information and Engineering Systems, KES 2017, Marseille, France (2017)
4. Ben Hassen, M., Turki, M., Gargouri, F.: Extending sensitive business process modeling with functional dimension for knowledge identification. In: International Conference on e-Business (ICE-B 2017), Madrid, Spain (2017)
5. OMG: Unified Modeling Language (UML). Version 2.0 (2007). http://www.uml.org/
6. OMG: Business Process Model and Notation (BPMN), Version 2.0.2 (2013). http://www. omg.org/spec/BPMN/2.0.2/pdf/
7. Arbeitsbericht.: KMDL® v2.2 (2009). http://www.kmdl.de/
8. Woitsch, R., Karagiannis, D.: Process oriented knowledge management: a service based approach. J. Univ. Comput. Sci. 11(4), 565–588 (2005)
9. Netto, J.M, Franca, J.B.S., Baião, F.A., Santoro, F.M.: A notation for knowledge-intensive processes. In: IEEE 17th International Conference on Computer Supported Cooperative Work in Design, vol. 1, pp. 1–6 (2013)
10. Ben Hassen, M., Turki, M., Gargouri, F.: Using core ontologies for extending sensitive business process modeling with the knowledge perspective. In: Proceedings of the Fifth European Conference on the Engineering of Computer-Based Systems (ECBS 2017), p. 2. ACM (2017)
11. Kassel, G.: Integration of the DOLCE top-level ontology into the OntoSpec methodology (2005)
12. Kassel, G., Turki, M., Saad, I., Gargouri, F.: From collective actions to actions of organizations: an ontological analysis. In: Symposium Understanding and Modelling Collective Phenomena (UMoCop), England (2012)
13. Turki, M., Kassel, G., Saad, I., Gargouri, F.: A core ontology of business processes based on DOLCE. J. Data Semant. 5(3), 165–177 (2016)
14. Masolo, C., Vieu, L., Bottazzi, E., Catenacci, C., Ferrario, R., Gangemi, A., Guarino, N.: Social roles and their descriptions. In: Dubois, D., Welty, C. (eds.) Proceedings of the Ninth International Conference on the Principles of Knowledge Representation and Reasoning, pp. 267–277 (2004)
15. Ben Hassen, M., Turki, M., Gargouri, F.: A BPMN extension for integrating knowledge dimension in sensitive business process models. In: European, Mediterranean, and Middle Eastern Conference on Information Systems, pp. 559–578. Springer, Cham (2017)
16. BPMN2 Modeler. http://www.eclipse.org/bpmn2-modeler/

17. Supulniece, I., Businska, L., Kirikova, M.: Towards extending BPMN with the knowledge dimension. In: Bider, I., Halpin, T., Krogstie, J., Nurcan, S., Proper, Ukor, R. (eds.) EMMSAD 2010. LNBIP, vol. 50, pp. 69–81. Springer, Heidelberg (2010)
18. Ammann, E.M.: Modeling of knowledge-intensive business processes. World Acad. Sci. Eng. Technol. Int. J. Soc. Behav. Educ. Econ. Bus. Ind. Eng. **6**(11), 3144–3150 (2012)
19. Stroppi, L.J.R., Chiotti, O., Villarreal, P.D.: Extending BPMN 2.0: Method and tool support. In: Dijkman, R., Hofstetter, J., Koehler, J. (eds.) BPMN 2011. LNBIP, vol. 95, pp. 59–73. Springer, Heidelberg (2011)
20. Jankovic, M., Ljubicic, M., Anicic, N., Marjanovic, Z.: Enhancing BPMN 2.0 informational perspective to support interoperability for cross-organizational business processes. Comput. Sci. Inf. Syst. **12**(3), 1101–1120 (2015)
21. Braun, R., Schlieter, H., Burwitz, M., Esswein, W.: Extending a business process modeling language for domain-specific adaptation in healthcare. In: Wirtschaftsinformatik, pp. 468–481 (2015)
22. Gangemi, A.: Ontology Design Patterns: A primer, with applications and perspectives. Tutorial on ODP, Laboratory for Applied Ontology Institute of Cognitive Sciences and Technology CNR, Rome, Italy (2006)

Exploring the Integration of Business Process with Nosql Databases in the Context of BPM

Asma Hassani[1](✉) and Sonia Ayachi Ghannouchi[2]

[1] Laboratory RIADI-GDL, ENSI, Mannouba, Tunisia
Asmahassani08@yahoo.fr
[2] High Institute on Management of Sousse, Sousse, Tunisia
s.ayachi@coselearn.org

Abstract. Business process is defined as a set of interrelated tasks or activities which allows the fulfillment of one of the organization's objectives. Modeling business process can be applied in several domains such as healthcare, business, education, etc. Modeling such process allows to facilitate and understand the functioning of corresponding systems. Steps in the process need input data and generate new output data. Business Process Management Systems (BPMS) play the role to model, configure and execute business processes. These latters are facing new challenges toward big data area. Data in business process originate from multiple sources with a variety of formats and are generated in a high speed and hence need in one hand, a storage infrastructure gathering all data types and forms. And on the other hand, analytics infrastructure that makes those data ready for analysis is needed. Therefore, regarding the flexibility and the dynamics of the execution of learning process, Not Only SQL (NoSQL) databases should be taken into consideration. So, the idea of combining business process and NoSQL databases becomes one merging and critical research area. In this paper, we propose the adoption of a Nosql database schema with MongoDB to model learning data in the context of MOOCs. Then, we explore the idea of integrating such database with the designed and configured massive learning process.

Keywords: Business process · SQL · NoSQL · MongoDB · Learning process BPMS · Data storage

1 Introduction

A business process is a collection of related, structured activities or tasks that produce a specific service or a particular goal for a particular person(s). It can often be viewed as a sequence of activities with decision points based on data in the process. Business processes are inseparable from execution data, artifacts and data generated or exchanged during the execution of the process. Van der Aalst and Ter Hofstede [1] have determined four perspectives considered as suitable aspects over business process modeling (control flow, data flow, resource and operational perspective). The data flow perspective defines the data manipulated by activities, their structures, their sources, their destinations, and their transformation rules if they are exported to the information system or the invoked applications [1]. In this context, business process models are

© Springer International Publishing AG, part of Springer Nature 2018
A. Abraham et al. (Eds.): ISDA 2017, AISC 736, pp. 771–784, 2018.
https://doi.org/10.1007/978-3-319-76348-4_74

integrated with external information systems to ensure their execution by involving actors in the realization of their tasks. Furthermore, business processes become part of a complex area where storing and integrating data are essential steps for future analytics and decision applications. But it's difficult to relational database management systems (RDBMS) to carry out data from highly distributed environments under a variety of heterogeneous systems and at very high speed [2]. Thereby, the amount of digital data is becoming massive and expanding. This technology, in full evolution, has been called big data and is adopted in various fields such as marketing, E-Commerce, E-Heath, E-Learning, E-Government... To define and characterize this buzzword, many authors used volume, velocity and variety known as 3Vs. Volume refers to the size of the data set, velocity describes the speed of data, and variety indicates the range of data types and sources [3]. According to Gao [4] this new data phenomenon, will lead to knowledge revolution in all sectors, including Business Process Management (BPM). The big data perspective on BPM approach shows new challenges and opportunities in managing a large number of process data instances, and thus can bring huge value for process decision makers and process actors. Here Business Process Management Systems (BPMS) must be integrated with new big data infrastructure and have the ability to capture, generate and store vast amounts of data.

Otherwise, storage and data processing technologies have known a remarkable and important evolution. Relational databases have limits in their ability to process very large volumes of data, data with complex structure or without structure, or data generated and/or received at very high speed. Thereby, another trend in the era of big data is the increasing use of Not Only SQL (NoSQL) databases as the suitable method for storing and retrieving information. Recently, this paradigm started to attract more attention with models such as key-store, column-oriented, document-based stores and graph oriented. The causes of such raise in interest are better performance, capacity of handling unstructured data and suitability for distributed environments [5]. Thus, combining BPMS with big data storage server is the first step for future process and analysis data towards future improvement based on data process analytics.

Among the domains in which a huge amount of data is produced, we find the contemporary domain of MOOCs. MOOC has been revolutionizing the pedagogical methods and the e-learning technologies by offering to a large number of participants (massive), for free (open) and entirely over the web (online) a structured content (course). In an E-Learning scenario, activities can be seen as processes. It represents a series of tasks or activities to be carried out to achieve individual or collective learning objectives. Thus, adopting the BPM approach in the MOOC framework, allows us to have an overall view of all the process and the corresponding interactions in order to optimize and automate the operations in general as well as improve the process.

The idea of this paper is to propose a Nosql database schema with MongoDB which will be designed and configured for the execution of MOOC's launching process.

The remainder of the paper is organized as follows. Firstly, a background section is devoted to present the main related information to our research topic. Secondly, we will uncover related works and we give our research issue and related research questions. Third, experimental results are presented. Finally, the paper closes with discussion and conclusions.

2 Background

2.1 Business Process Management

Business Process management (BPM) includes methods, techniques, and software to design, enact, control and analyze operational processes [6]. This approach has received considerable attention in recent years due to its potential for significantly increasing productivity and reducing costs. BPM can be seen as successive steps that form the lifecycle of such approach. BPM lifecycle model systemizes the steps and activities that should be followed for conducting a BPM project. According to Netjes et al. [7] BPM lifecycle is composed of five steps: design, configuration, execution, control and diagnosis.

The notion of process model is fundamental for BPM. A process model aims to understand and capture the different manners in which a case can be treated and processed. A range of notations exists to model business process (Petri nets, UML, BPMN, EPC). These notations have a common point that is the fact that processes are described in terms of activities. The order of these activities is modeled by describing causal dependencies. Further, the process model may also demonstrate temporal conditions and indicate the creation and use of data. Indeed, Van der Aalst and Ter Hofstede [1] has delimited four perspectives considered as pertinent aspects during business process modeling. The data flow perspective defines the data handled by activities, their structures, their sources, their destinations and their transformation rules if they are exported to the information system or the invoked applications.

Thus, modeling a business process and analyzing it is usually based on the use of management tools and specially Business Process Management Systems (BPMS) that help process decision makers to improve part of or the overall process [16]. Basically, the enactment is based on data which are collected, captured, processed and treated to offer suitable and new information allowing future decision making. Otherwise, BPMS is associated with other software to manage, control and support data process. During the process execution, process instance interact with database applications to use, capture, manipulate and involve execution data. So, BPMS must be configured and implemented with database systems which present new technical challenges to deal with varied, voluminous and volatile data, which are known as big data.

So we can consider that the key point of the success of BPM projects is on one hand, the modeling of the process and its configuration with external systems, and on another hand, the corresponding data which requires be capturing, storing and manipulating. As more and more business process are managed, data modeling and analyzing are becoming increasingly important components of a BPM project in the age of big data.

2.2 Big Data

Over the last 20 years, data has increased in a large scale in various fields. Hundreds of Petabytes are generated every second by industries, Internet companies, media and social networks. To cope with the explosion of data volume, a new technological field was born: the Big data.

Big data is defined through the 3Vs which are presented by Laney [8] in 2001 as "high-volume, high velocity and high variety information assets that demand cost-effective, innovative forms of information handling for improved insight and decision making". In 2012, Gartner [9] updated the definition as follows: "Big data is high volume, high velocity, and/or high variety information assets that require new forms of processing to enable enhanced decision making, insight discovery and process optimization". Other authors, researchers and engineers have extended the 3Vs to 4Vs and 5Vs by adding Value and Veracity to the definition of the big data definition. These concepts can be briefly described as follows [10]:

- Volume: refers to large amounts of any kind of data from any different sources.
- Variety: refers to different types of data collected via sensors, smartphones or social networks, such as videos, images, text, audio, and so on. Moreover, these data can be in structured or unstructured formats.
- Velocity: refers to the speed of data transfers. Data content is constantly changing.

We remark that in the concept of big data it is not really the volume of data that makes the novelty but rather the combination of the 3Vs. In addition, the big data domain is evolving rapidly, while placing great importance on the storage and processing of large data sets. Big data can be identified by three essential criteria: volume, velocity and variety. New ways of collecting, processing and analyzing large volumes of data have been created and adopted.

2.3   NoSQL Databases

The main aim of this part is to give an overview of NoSQL databases, about how they reduce the dominance of SQL, with its main characteristics. Later, a comparison between Nosql and SQL is given in order to highlight advantages and disadvantages of these two fields. Relational databases have limits concerning their ability to process very large volumes of data, data with complex structure or without structure, or data generated or captured at very high speed. ACID indicates Atomicity, Consistency, Isolation and Durability, adopted for relational databases. Otherwise, this theorem will no more be suitable for distributed environment and creates problems to Nosql.

Nosql databases follow three principles known as CAP (Consistency, Availability, Partition Tolerance) [11]:

- Consistency: All the servers in the system will have the same data so anyone using the system will get the same copy regardless of which server answers their request.
- Availability: data must be available permanently and should be accessible at any time.
- Partition tolerance: during server failure or any faults in the system databases must continue to work fine without stopping their work.

On the basis of CAP theorem, NoSQL databases are classified into four new different types of data stores. In the following table (Table 1), we present the different types of Nosql databases and we give for each category the major examples of servers.

Table 1. NoSQL databases

Type	Presentation	Server example
Key-value databases	A key value data store allows the user to store data in a schema-less manner. The data consists of two parts, a string which represents the key and the actual data which is referred to as value thus creating a "key-value" pair	Amazon DynamoDB, RIAK, Redis…
Column-oriented databases	Rather than store sets of information in a heavily structured table of columns and rows with uniform sized fields for each record, as is the case with relational databases, with this model, the number of columns can vary from one record to another, which avoids finding columns with null values	Bigtable (Google), Cassandra, simpleDB (Amazon), DynamoDB,…
Document-stores Databases	Document Stores Databases refer to databases that store their data in the form of documents. Document stores offer great performance and horizontal scalability options. Documents inside a document-oriented database are somewhat similar to records in relational databases, but they are much more flexible since they are schema-less. Documents in the database are addressed using a unique key that represents that document. These keys may be a simple string or a string that refers to a URI or a path	CouchDB, MongoDB,…
Graph databases	Graph databases are databases which store data in the form of a graph. The graph consists of nodes and edges, where nodes act as the objects and edges act as the relationship between the objects. The graph also consists of properties related to nodes	Neo4j, InfoGrid,…

Nosql databases have advantages as well as disadvantages compared to relational databases. In this paragraph, we provide some points of comparison between these two fields. The advantages of Nosql compared to SQL are summarized as follows:

- It provides a wide range of data models
- It's easy scalable
- It's faster, more efficient and flexible
- It has evolved at a very high pace

Despite the importance of the new models of Nosql databases, they still have problems such as:

- No standard query language
- No standard interface
- Maintenance is difficult

In this part we have presented Nosql databases, the different models as well as advantages and disadvantages of this new buzzword. This new solution aims to provide the robustness, scalability and reliability.

3 Related Works, Problematic and Research Questions

With the increasing amount of data, several works carried out in recent years are interested in collecting, storing, processing and visualizing big data. In this part, we will introduce different works dealing with business process data storage.

Meyer et al. [12] confirm that process and data are both substantial fields for business process management. The BPMN notation offers capabilities to model data. BPMN allows explicit specification of data objects by associating them with process tasks. A data object can also be associated with a sequence flow to visualize data moving. BPMN generally provides a high data consciousness in terms of data object specifications and implicit data dependencies. Wang et al. [13] observe that in recent years, process data (especially event log) is more and more increasing and becoming a typical type of big data. Behavioral analysis on big process data is in urgent need. Thus, efficiently managing a large number of models and instances is challenging and can bring huge value. In fact, the author used MongoDB as data storage layer in the developed process space project.

In recent years, big data has modified the way that we adopt in dealing with businesses, managements as well as researches. Data-intensive science especially in data-intensive computing aims to provide the tools that we need to handle the big data problems [3]. Yoo et al. [14] propose a BPM-based analysis modeling system which consists of 3 layers. The data storage layer is based on hadoop storage space. It integrates a system to provide the input data of analysis or store data.

The storage, management and analysis of big data generated from big data-intensive process are becoming one of the most important topics in BPM research area. The literature underlines the importance of modeling data in Business process models and BPMN notation is a suitable graphical method for that. Furthermore, all studies focus and put a great attention on how big data will be captured, stored and analyzed. For this purpose, new big data tools must be adopted and combined with business process management tools to achieve BPM projects.

In this paper, we seek to answer the following research questions:

- Which challenges are met by e-learning, process management and big data?
- How to identify data-intensive processes for which big data is appropriate?
- Is MongoDB a well adapted system for such processes?

4 Modeling of Learning Process

A learning process represents a series of tasks or activities executed to achieve learning goals. BPM provides tools to support e-learning in the modeling, configuration, deploying, management and enactment steps of learning processes which involve various interactions between external systems (basically data store systems).

In BPM, a business process is defined using a graphical notation such as Petri networks, YAWL (Yet Another Workflow Language), UML (Unified Modeling Language) and BPMN (Business Process Model and Notation). BPMN is a comprehensible graphical notation for defining a learning process including learning activities, stakeholders, their roles and their interactions. Hence, BPMN allows the representation of activities, actors and their roles in a learning situation by the intermediate of lanes and pools. BPMN defines business process diagrams, which can be used to create graphical learning process models especially learning process models [16].

In a previous work, we were interested in modeling massive learning process with the BPMN notation and we focused on the design phase of the BPM cycle [16]. We can consider that a MOOC is a suitable form of massive learning which is mainly based on interaction between multiple participants, through various applications which are involved. In fact, our focus is on massive learning processes because, on the one hand, we observe nowadays an expansion of the use of MOOCs and some issues appear with this growing trend. On the other hand, a BPM approach seems very useful for a continuous improvement of such learning processes.

Being a new trend of distance learning, MOOCs are becoming more and more important. So it is primordial to understand the structuring of such process of learning. We tried to understand the learning activities insured by learners and tutors and modelled them with the BPMN notation.

Figure 1 presents the general model of massive learning process. We already consider that it is composed of the following sub-processes namely:

- Course's launching process
- Course's running process
- Evaluation process

Fig. 1. General model of learning process [15]

In this paper we will focus on the first sub-process which model is shown in Fig. 2. This model consists in putting online details of a MOOC (objectives, target audience, prerequisites, timetable and pace, the teaching staff, dates and deadlines, etc.). Internet users consult the published details and if they are interested, they sign up. To log in, users must have an account on the platform hosting the MOOC, otherwise they must create one.

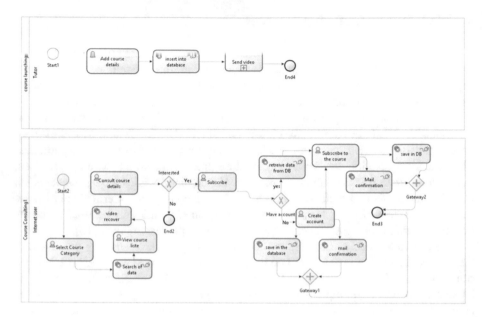

Fig. 2. Course's launching process model

Automatic activities in the process include connectors to store data of process execution. The generated data must be stored in order to be able to be exploited and analyzed for future process improvement.

5 Process and Nosql Environments

As presented and described in above sections, there are several types of Nosql databases which focus on storing and retrieving large amounts of unstructured, semi-structured or even structured data. In our work, we choose to adopt the MongoDB system to store process data instances for several reasons. According to the data production speed, variety and volume, RDBMS fails to deal with data with those characteristics. So MongoDB can be the suitable system to store and manage big volumes of data. Also, MongoDB allows to store unstructured data (video course, images, comments…). As well, it does not need a predefined schema of data in tables and with MongoDB it is easy to represent open structured schema and store flexible data structure. Moreover, document-oriented database is known by its scalability and high performance to treat big data.

We propose in Fig. 3 an architecture presenting the way to link the process model and MongoDB database.

To make a BPM project successful, several systems need to be brought together. First, the considered Business process must be modeled to describe clearly the scenario of learning. It leads people and systems to accomplish activities. These latters require input data and result from other output data. We have adopted BonitaBPM as the BPM

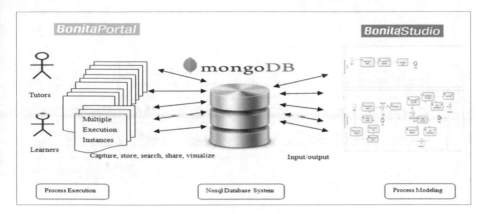

Fig. 3. Overview of the architecture

solution for creating, configuring and executing our process based learning application. BonitaStudio aims to model processes with the BPMN notation. Actors (tutors and learners) interact with the Bonita Web-portal to accomplish their activities and manage their assigned tasks.

The database model (MongoDB) is responsible for collecting data from multiple instances of processes and storing them in the appropriate place. The Business Process Management System (BPMS) connects the process model with the database model via connectors. The BPMS and the Nosql system work together to accomplish the complete business process. As a result, big data technologies and BPMS systems open up new opportunities for interacting, communicating and relying on components of the learning environment. So neither the database systems nor the processes are isolated from each other.

As described above, the sub-process of launching the course consists of publishing details of a MOOC. The following figure (Fig. 4) shows the interface for adding the course by the tutor.

Internet user will consult the details of the proposed course. Once interested, he/she will pass to create an account if he/she does not have one, as presented in Fig. 5.

For several reasons, we choose to use MongoDB instead of SQL Systems to store process data. In fact, a major feature of a database on MongoDB is the missing join aspect as not the case with relational databases. Moreover, data (document in the context of MongoDB) are modeled in a single collection where they are embedded. Thus, one of the great benefits of this document-oriented database is the ability to get rid of joins that end up in SQL. MongoDB has a flexible schema, which means that we can add several documents with different structures in the same collection. A revolutionary aspect is the absence of the notion of foreign key. Documents are in BSON format (Binary JavaScript Object Notation). Also, MongoDB offers a clear and powerful API (Application Programming Interface) to be used and integrated into other applications. To ensure CRUD (Create, Read, Update, Delete) operations, MongoDB has four main commands: insert(), find(), update() and remove().

Fig. 4. Add course interface

Fig. 5. Create account interface

In our use case study, we have a many to many relationship between course and participant. The is-registered table is the junction table. Using the schema-less nature of MongoDB, we can store an array of the ObjectId's from the course collection in participant collection to identify what courses a participant is registered in. Likewise, we can store an array of ObjectID's from the participant collection in the course document to identify which participants are registered in a course.

Figure 6 presents how one participant documents would look like in MongoDB. The ObjectID is generated automatically by MongoDB server.

```
db.participant.insert({
   "firstName": "walter",
   "lastName": "Bates",
   "gender": "male",
   "date-of-birth":05-10-2001,
   "level": "master",
   "city": "UK",
   "town": "UK",
   "username": "Walter.bates",
   "email": "Walter.bates@gmail.com",
   "password" : "***",
   "course": [
     ObjectId ("58ea40fb4296651fa8321507"),
     ObjectId ("58ea540af1a7234e852490dd")
   ]
});
```

Fig. 6. Participant documents in MongoDB

Figure 7 presents how some course documents would look in MongoDB.

To summarize, the oriented document modeling offered by MongoDB adapts perfectly to our case study, which is massive pedagogy and e-learning. Performance, solidity, schema-less and high availability are the criteria that underpin our choice for this Nosql database. As work's extension, MongoDB also allows to communicate with MapReduce and Hadoop ecosystem for massive learning process data computing and analysis.

6 Discussion

Nowadays learning environments especially MOOCs, have known a rapid evolution and are more and more complex and produced increasingly vast volumes of high velocity data in various formats. Modeling such processes allows, on one hand to understand the learning scenario of this new form of learning and on another hand to analyze data in order to bring process's improvement. BPMN is the notation adopted for modeling the massive learning process. The main purpose of BPMN is to provide

```
db.course.insert({
  "title": "initiation à la conception orientée objet",
  "category". "Computing",
  "details": "Les concepts et les technologies introduites par l'approche orientée objet se sont imposés en quelques années dans tous les domaines du génie
  logiciel : spécification, conception, programmation, test, base de données",
  "prerequisites": "Ce MOOC ne suppose aucune connaissance spécifique préalable, mais s'adresse cependant à un public composés de professionnels,
  de stagiaires ou d'étudiants du domaine informatique (niveau IUT, licence, Mastere ou études d'ingénieurs) recherchant à découvrir ou à mieux maîtriser
  le standard de modélisation orientée objet UML",
  "end-of-registration": "01-01-2017",
  "start course": "01-01-2017",
  "end course": "15-02-2017",
  "effort": "04:00h/semaine",
  "language": "fr",
  "university" : "university of Sousse",
  "plan": "Semaine 1 : Introduction à UML et DCU Démarrage du MOOC ICOO Leçon 1.0 : Du génie logiciel vers les méthodes de conception
  Leçon 1.1 : Introduction à UML Leçon 1.2 : Diagramme de cas d'utilisation (DCU) Semaine 2 : DCL, DOB, DCP et DDP
  Leçon 2.1 : Diagramme de Classe (DCL) et Diagramme d'objets (DOB) Leçon 2.2 : Diagramme de composants (DCP) et diagramme de déploiement (DDP)
  Semaine 3 : DET et DAC Leçon 3.1 : Diagramme d'état-transition (DET) Leçon 3.2 : Diagramme d'activités (DAC) Semaine 4 : DSE et DCO
  Leçon 4.1 : Diagramme de séquences (DSE) Leçon 4.2 : Diagramme de collaboration (DCO)",
  "educational team": "SONIA AYACHI GHANNOUCHI ,LILIA CHENITI, MAHA KHEMAJA, EMNA SOUISSI",
  "course video URL": "https://www.youtube.com/watch?v=AaD9FWciqi4",
  "participant": [
    ObjectId("59031c9b55f82f3f79b98da9"),
    ObjectId("58ebd81af485dd099c37a870"),
    ObjectId("58ebd808f485dd099c37a86f"),
    ObjectId("58ebd7f4f485dd099c37a86e")
  ]});
```

Fig. 7. Course documents in MongoDB

understandable models by all stakeholders. Another feature of BPMN is its ability of being executable. BPMN also allows to associate process and data flow.

On the other hand, the success of BPM projects is based on one hand, on the modeling of the process and its configuration with external systems, and on another hand, on data requiring capturing, storing and manipulating. As more and more business processes are managed, data modeling and analyzing are becoming an increasingly important component of a BPM project in the age of big data.

New big data solutions must be integrated with BPMS in order to store, manage and analyze big data-intensive processes. Known by its performance, solidity, schema-less and high availability, we choose MongoDB to store instance's process data.

In the current state of our work, the course launching process model is designed, implemented and connected to a MongoDB database. In the next steps, we will collect data from our database through concrete executions of our process and then analyze them.

7 Conclusion

In this paper, we have shown the importance of the combination of BPM and big data in the context of massive learning. Big data feeds the BPM approach to provide in first step an abstraction, description and definition of business processes as well as important capabilities to manage and control those processes. In our work we have concentrated on the integration of our proposed process model with the MongoDB Data store. A vast amount of structured and unstructured data generated by the execution of the course's launching process are collected and stored in a less-schema database.

Being able to explore data to better understand the learning process is a great challenge to extend our future work and carry out analytics on big data for BPM approach.

References

1. Van der Aalst, W.M., Ter Hofstede, A.H.: YAWL yet another workflow language. Inf. Syst. **30**(4), 245–275 (2005)
2. Vera-Baquero, A., Colomo-Palacios, R., Molloy, O.: Business process analytics using a big data approach. IT Prof. **15**(6), 29–35 (2013)
3. Chen, C.P., Zhang, C.Y.: Data-intensive applications, challenges, techniques and technologies: a survey on big data. Inf. Sci. **275**, 314–347 (2014)
4. Gao, X.: Towards the next generation intelligent BPM–in the era of big data. In: Business Process Management, pp. 4–9 (2013)
5. Assunção, M.D., Calheiros, R.N., Bianchi, S., Netto, M.A., Buyya, R.: Big data computing and clouds: trends and future directions. J. Parallel Distrib. Comput. **79**, 3–15 (2015)
6. Van der Aalst, W.M., Ter Hofstede, A.H., Weske, M.: Business process management: a survey. In: Business Process Management, vol. 3, pp. 1–12. Springer, Heidelberg (2003)
7. Netjes, M., Reijers, H., Van der Aalst, W.M.: Supporting the BPM life-cycle with FileNet. In: Proceedings of the Workshop on Exploring Modeling Methods for Systems Analysis and Design (EMMSAD), pp. 497–508. Namur University, Namur (2006)
8. Laney, D.: 3D data management: controlling data volume, velocity, and variety, Technical report (2001). http://blogs.gartner.com/doug-laney/files/2012/01/ad949-3D-Data-Management-Controlling-Data-Volume-Velocity-andVariety.pdf
9. Beyer, M.A., Laney, D.: The Importance of 'Big Data': A Definition. Gartner, Stamford, CT (2012)
10. Bello-Orgaz, G., Jung, J.J., Camacho, D.: Social big data: recent achievements and new challenges. Inf. Fusion **28**, 45–59 (2016)
11. Sharma, V., Dave, M.: SQL and NoSQL databases. Int. J. Adv. Res. Comput. Sci. Softw. Eng. **2**(8), 20–27 (2012)
12. Meyer, A., Smirnov, S., Weske, M.: Data in business processes. No. 50. Universitätsverlag Potsdam (2011)
13. Wang, S., Lv, C., Wen, L., Wang, J.: Managing massive business process models and instances with process space. In: BPM (Demos), p. 91 (2014)

14. Yoo, Y.S., Yu, J., Bang, H.C., Park, C.H.: A study on data analysis process management system in MapReduce using BPM. In: Proceedings of the 4th International Conference on Security-Enriched Urban Computing and Smart Grid (SUComS) pp. 7–12 (2013)
15. Hassani, A., Ghanouchi, S.A.: Modeling of a collaborative learning process in the context of MOOCs. In: International Conference on Systems of Collaboration (SysCo), pp. 1–6 (2016)
16. Kahloun, F., Ayachi, S.A.: Evaluating the quality of business process models based on measures and criteria in higher education developing a framework for continuous quality improvement. In: ISDA Conference (2016)

An Effective Heuristic Algorithm for the Double Vehicle Routing Problem with Multiple Stack and Heterogeneous Demand

Jonatas B. C. Chagas[1]([✉]) and André G. Santos[2]

[1] Departamento de Computação,
Universidade Federal de Ouro Preto, Ouro Preto, Brazil
jonatas.chagas@iceb.ufop.br
[2] Departamento de Informática, Universidade Federal de Viçosa, Viçosa, Brazil
andre@dpi.ufv.br

Abstract. In this work, we address the Double Routing Vehicle Problem with Multiple Stacks and Heterogeneous Demand, a pickup and delivery problem. The objective of the problem is to determine a set routes for a fleet of vehicles in order to meet the demand of a set of customers so that the distance travelled by the vehicles is the minimum possible, while ensuring the feasibility of the loading plan. We propose a heuristic approach based on the Simulated Annealing for solving the problem. The computational results show that our heuristic is able to find high quality solution in short computational time when compared to the exact methods found in the literature.

Keywords: Vehicle routing · Pickup and delivery
Loading constraints · Heterogeneous demand
Simulated Annealing · Meta-heuristic

1 Introduction

The Double Vehicle Routing Problem with Multiple Stacks and Heterogeneous Demand (DVRPMSHD) is a realistic generalization of the Double Vehicle Routing Problem with Multiple Stacks (DVRPMS), which in turn is a generalization of the well-known and well-studied Double Travelling Salesman Problem with Multiple Stacks (DTSPMS).

The DTSPMS was proposed by Petersen and Madsen [1]. This problem arises in a real world application where a single vehicle must transport items from a pickup region to a delivery region. These regions are widely separated from each other so that the vehicle must load all items in the pickup region before unload them in the delivery region. Each item is associated with a pickup customer and a delivery customer. The loading compartment of the vehicle (container) is divided into stacks of fixed height. The vehicle starts its tour at the pickup depot,

© Springer International Publishing AG, part of Springer Nature 2018
A. Abraham et al. (Eds.): ISDA 2017, AISC 736, pp. 785–796, 2018.
https://doi.org/10.1007/978-3-319-76348-4_75

collects all the assigned items in the pickup region loading them in horizontal stacks, and then returns to the pickup depot. After checking-in in the pickup depot, the vehicle travels to the delivery depot. The items cannot be rearranged in container, i.e., once that a particular item is stored in some stack it must remain unmoved up to the unloading. In the delivery region the vehicle starts its tour at the delivery depot, delivers all the items and then returns to the delivery depot. The items must be unloaded and delivered respecting the last-in-first-out (LIFO) policy, i.e., the last collected and stored item in a specific stack must be the first one to be delivered from that stack. The objective is to find two tours, one for each region, of minimum total cost while ensuring the feasibility of the loading plan.

In 2015, Iori and Riera-Ledesma [2] proposed the Double Vehicle Routing Problem with Multiple Stacks (DVRPMS). According to Iori and Riera-Ledesma, the DVRPMS is a generalization of the DTSPMS, motivated by the fact that not always a single vehicle is enough to transport all items, thus more than one vehicle is needed. The DVRPMS has the same characteristics and restrictions as the DTSPMS, except that now there is a fleet of vehicles available to meet the demand of the customers. Some studies have already addressed the DVRPMS: Iori and Riera-Ledesma [2] proposed three exact algorithms, while Silveira et al. [3] and Chagas et al. [4] proposed heuristic algorithms.

Recently, the DVRPMSHD was introduced by Chagas and Santos [5], who presented a mathematical formulation and a branch-and-price algorithm to solve the problem. According to the authors, the DVRPMSHD is a more realistic generalization of the DVRPMS, which considers that customers have multiple and heterogeneous demand and all demand of each customer must be served by a single vehicle.

To the best of our knowledge no previous work has approached the DVRPMSHD by heuristic methods. Therefore, in this work we describe a heuristic algorithm to find high quality solutions for the problem and we also compare its results with the results obtained by the exact methods proposed by Chagas and Santos [5].

The remaining of this paper is organized as follows. The next section provides a formal definition of the DVRPMSHD. In Sect. 3 we present a heuristic approach for the problem, based on the Simulated Annealing meta-heuristic. In Sect. 4 we show the results obtained by our proposed algorithm, as well as a comparative analysis with the results already present in the literature. Finally, Sect. 5 presents the conclusions of this work.

2   Formal Definition

As was formally introduced by Chagas and Santos [5], the Double Vehicle Routing Problem with Multiple Stacks and Heterogeneous Demand (DVRPMSHD) can be defined on two complete directed graphs, $G^P = (V^P,\ E^P)$ and $G^D = (V^D,\ E^D)$, where $V^P = V_c^P \cup \{0^P\}$ and $V^D = V_c^D \cup \{0^D\}$ represents, respectively, the vertices in the pickup and delivery regions, being 0^P and 0^D the depots on each region. Let $I = \{1, 2, ..., n\}$ be the set of n customer requests, and let also $V_c^P = \{1^P, 2^P, ..., n^P\}$ be the set of vertices that represent the locations of the pickup customers and $V_c^D = \{1^D, 2^D, ..., n^D\}$ representing the locations of the delivery customers. Each request $i \in I$ is associated with a pickup vertex i^P, a delivery vertex i^D, and a value d_i representing the request demand, i.e., the number of items that has to be loaded at the vertex i^P, transported and unloaded at the vertex i^D.

The set of edges in pickup and delivery regions are defined by $E^P = \{(i^P, j^P, c_{ij}^P) \ \forall i^P, j^P \in V^P \mid i^P \neq j^P\}$ and $E^D = \{(i^D, j^D, c_{ij}^D) \ \forall i^D, j^D \in V^D \mid i^D \neq j^D\}$, where c_{ij}^P and c_{ij}^D are, respectively, the routing costs associated to edges $\{i^P, j^P\}$ and $\{i^D, j^D\}$.

The vehicles of set K depart always from the pickup and delivery depots. The loading compartment of each vehicle $k \in K$ is divided into R_k stacks of height L_k. The vehicles must collect all items from customers located in the pickup region, storing them in their containers and then deliver them in the delivery region. The routes of each vehicle must satisfy the LIFO policy in all stacks of all vehicles, i.e., if a pickup vertex i^P is visited before the pickup vertex j^P, and some item of the request i and some item of the request j are loaded into the same stack, then the delivery vertex j^D must be visited before the delivery vertex i^D.

The aim of the DVRPMSHD is to find optimal routes for the fleet of $|K|$ vehicles so that all n customers' requests are attended while ensuring the feasibility of the loading plans.

3   Heuristic Approach

As showed by Chagas and Santos [5], the DVRPMSHD presents a high complexity and solving it by exact methods requires excessive computational time. Therefore, in this section we describe a Simulated Annealing meta-heuristic followed by an Improvement Phase approach in order to obtain high quality solutions in short computational time.

3.1   Solution Representation and Loading Heuristic

The representation of the DVRPMSHD solutions was designed with the purpose of facilitating the exploration of the solution space. Initially, we intended to use the representation used by Chagas et al. [4] for the DVRPMS solutions, which represents the loading plan for each vehicle. However, the fact that more than one item may be associated with the same customer request makes it difficult to maintain solutions in the feasible solution space of the problem, since neighborhood structures must be designed to ensure that the loading plan of a given container is viable, i.e., ensure that there is at least one route in the pickup region and another in the delivery region that respects the LIFO policy of the loading plan. Figure 1 illustrates an infeasible loading plan for a container (2×3). Note that one cannot construct routes in the pickup and in the delivery

8	4
4	9
12	4

Fig. 1. Infeasible loading plan for a container (2×3).

region, since the items of the requests 4 and 9 generate dependencies between each other (deadlock).

Therefore, in order to avoid infeasible solutions (containers), we represent a solution of the DVRPMSHD through a matrix of $|K|$ rows, where each row informs which customer requests are assigned to each vehicle (k-th row refers to the k-th vehicle). Then, from this representation we create a loading plan for each vehicle through a simple loading heuristic.

In Fig. 2 we present an example of a representation of a solution, as well as the containers formed after the application of the loading heuristic. This example involves 9 customer requests that are assigned, as showed in the figure, to three vehicles. For the first vehicle (first row of the matrix) are assigned the requests 2, 4 and 3, while for the second and third vehicles (second and third rows, respectively, of the matrix) are assigned the requests 5, 1, 7 and 8, and the requests 6 and 9, respectively. The demand value of each request is indicated by the number in the upper right corner of each request. The loading heuristic is applied independently for each row of the matrix, i.e., for each vehicle. Analyzing the requests from left to right, considering the order they appear in the solution representation, for each request i assigned to the vehicle k, the loading heuristic seeks to allocate all d_i items in the stack R less filled so far. In this way, the loading plan is constructed iteratively in an attempt to decrease the number of LIFO constraints. If it is not possible to allocate all the items of a request in the same stack, then the maximum number of possible items are allocated, and the others ones are allocated in other stack, following the same strategy of trying to allocate them together in one stack. Consider for example the vehicle 2 (second row of the matrix and second container). Firstly the items of the request 5 are

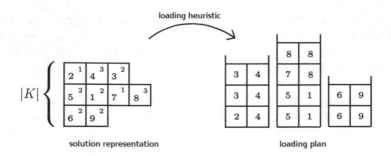

Fig. 2. Solution representation and loading heuristic.

stored in the first stack, since both stacks are empty. Then, the items of the request 1 are stored in the second stack (least filled so far) of the vehicle, the only item of the request 7 is stored in the first stack (lexicographically smaller), and, finally, two items of the request 8 are stored in the remainder of the least filled stack (second stack) and the third is then stored in another stack (first stack).

3.2 Evaluation Function

The solution representation and the loading heuristic define only the assignment of requests and the loading plan for each vehicle, respectively. In order to evaluate a solution of the DVRPMSHD, i.e., get the pickup and delivery routes for each vehicle, as well as their costs, we use the two evaluation strategies proposed by Chagas et al. [4] to evaluate the solutions of the DVRPMS. These two evaluation strategies are briefly presented below:

- Greedy evaluation: a greedy strategy, where the pickup (and delivery) route is constructed iteratively choosing among the requests at the bottom (or top) of each stack, the request that has the least distance on the pickup (or delivery) route costs is chosen. This strategy is $O(R^2 L)$ [4].
- Optimum evaluation: among all possible routes, the pickup and delivery routes having the smallest distance are chosen, i.e., the optimal pickup and delivery routes are selected, given the loading plan. This strategy is $O((RL)^R)$ [4].

The optimal evaluation strategy turns out to be computationally impracticable to be used in all evaluations due to its exponential complexity. Therefore, we use both strategies described above, but the optimal one is applied only in the final iterations of the algorithm.

3.3 Neighborhood Structure

The neighborhood structure defined to explore the solution space of the DVRPMSHD consists of exchanging the position of two or more customer requests from their current designations. A neighbor solution s' is then obtained from a solution s through the following steps: select a vehicle $k \in K$ and select a request i assigned to this vehicle; select a vehicle $k' \in K$ and select a subset of requests q assigned to this vehicle; exchange the position of request i with all requests of the subset q, if the capacity of the vehicles are not extrapolated after the exchange.

3.4 Initial Solution

The initial solution of our heuristic approach is created randomly, where a random subset of customer requests is assigned to each vehicle in such way that the demand does not exceed the capacity of each vehicle and that each request is to be served by a single vehicle.

3.5   Simulated Annealing

Proposed by Kirkpatrick et al. [6], Simulated Annealing (SA) is a probabilistic meta-heuristic inspired by the annealing process of steel thermodynamic optimization. It starts with a high temperature that is slowly reduced until the atoms of steel reach equilibrium. Analogously, a solution follows the behavior of the atoms, initially the atoms are in great agitation (great perturbations in the solution), but as the temperature decreases, the atoms decrease the agitation (small perturbations in the solution) until a state of equilibrium (final solution) is reached. In order to escape from local optimal solutions, the SA meta-heuristic allows to accept some worse solutions compared to the current one. Worse solutions are accepted according to a probability given in function of the temperature and the worsening of the new solution respect to the current one. The higher the temperature and the lower the worsening, the greater the chance of accepting a worse solution.

Algorithm 1 describes the method proposed to solve the DVRPMSHD based on the Simulated Annealing technique. Note that in this algorithm all solution evaluations are performed using the greedy evaluation, previously mentioned and referenced by $f_{greedy}(.)$.

Algorithm 1. Simulated Annealing (SA)

---
1   $s \leftarrow$ Generate a random solution
2   $BestSolution \leftarrow s$
3   $CurrentTemp \leftarrow InitialTemp$
4   **while** $CurrentTemp \geq FinalTemp$ **do**
5       $iter \leftarrow 0$
6   **while** $iter < NumIterSA$ **do**
7   Select a random solution $s' \in N(s)$
8           $\Delta \leftarrow f_{greedy}(s') - f_{greedy}(s)$
9   **if** $\Delta < 0$ *or* $rand(0,\ 1) < e^{-\Delta/CurrentTemp}$ **then**
10              $s \leftarrow s'$
11   **if** $f_{greedy}(s') < f_{greedy}(BestSolution)$ **then**
12                  $BestSolution \leftarrow s'$
13          $iter \leftarrow iter + 1$
14      $CurrentTemp \leftarrow CurrentTemp \times (1.0 - \alpha)$
15 **return** $BestSolution$

The initial solution of the algorithm is generated randomly and it is considered as the best solution so far. While the stop criteria is not met, the algorithm randomly selects a solution nearby the current solution and accepts it according to a probability function given by the method, selecting it as the best solution so far if it is better than the last best solution. After a given number of iterations, the temperature is updated by a cooling rate. The algorithm ends when the minimum temperature is reached.

3.6   Improvement Phase

In order to reach a local minimum that may not have been found previously due to the possibility of not obtaining the best routes of each vehicle, the best solution found by the Algorithm 1 is then submitted to the procedure described by the Algorithm 2, which performs a local search with a limited number of iterations. Note that all solution evaluations performed in this procedure use the optimum evaluation described before and referenced by $f_{opt}(.)$.

Algorithm 2. Improvement Phase (IP)

---

1 $s \leftarrow$ Get the best solution found by the SA algorithm (Algorithm 1)
2 $iter \leftarrow 0$
3 **while** $iter < NumIterIP$ **do**
4 $\quad$ Select a random solution $s' \in N(s)$
5 $\quad$ **if** $f_{opt}(s') - f_{opt}(s) < 0$ **then**
6 $\quad\quad$ $s \leftarrow s'$
7 $\quad$ $iter \leftarrow iter + 1$
8 **return** s

Our approach to solve the DVRPMSHD, named by SAIP, consists to apply the SA-based heuristic (Algorithm 1) and subsequently apply the IP procedure (Algorithm 2).

4   Computational Experiments

The proposed algorithm (SAIP) have been implemented in C/C++ language and run sequentially on an Intel Core i7-4790K CPU @ 4.00 GHz $\times$ 8 desktop computer with 32 GB RAM, Ubuntu 14.04 LTS 64 bits.

All tests have been performed on instances from the set of 90 randomly generated instances from Chagas and Santos, which are described in details in [5].

4.1   Parameter Tuning

We use the I/F-Race implementation provided in the Irace package [7] to find the best parameter values for our algorithm, which has five parameters: initial temperature ($InitialTemp$), final temperature ($FinalTemp$), number of iterations per temperature ($NumIterSA$), cooling factor (α) and the number of iterations performed in the improvement phase ($NumIterIP$). We tested the following values: $InitialTemp \in \{15, 20, 50, 100\}$, $FinalTemp \in \{0.1, 1.0, 5.0, 10.0\}$, $NumIterSA \in \{|K||I|, 100, |I|^2, 2|I|^2, |K||I|^2\}$, $\alpha \in \{0.001, 0.002, 0.005, 0.01, 0.05\}$ and $NumIterIP \in \{|K||I|, 100, |I|^2, 2|I|^2, |K||I|^2\}$. After calibration the following parameter values were determined: $InitialTemp = 20$, $FinalTemp = 1.0$, $NumIterSA = |K||I|^2$, $\alpha = 0.002$ and $NumIterIP = |I|^2$.

4.2   Experimental Results

In this section we present the results obtained by our SAIP algorithm described in Sect. 3, and we also make a comparative analysis between our results and the ones obtained by the methods proposed by Chagas and Santos [5].

Tables 1 and 2 report the results on the 45 smallest instances and the 45 largest ones, respectively. Each table has four blocks, the first corresponding to the instances and the others corresponding to each algorithm indicated in the first row. In both tables the four first columns describe the instances, being each instance identified by a number (column ID), a name indicating the pickup and delivery regions (column R), a type of vehicle feet (column F) (see Table 1 in [5]) and a demand distribution (column D) (see Table 2 in [5]). The results obtained by the Integer Linear Programming (ILP) formulation and the branch-and-price (B&P) algorithm, both proposed by Chagas and Santos [5], are described in columns UB, Opt and $t(s)$ that inform, respectively, the best solution value obtained at the end of the computation, if each instance was solved to optimality (indicated by an asterisk) and total processing time in seconds. The results obtained by the SAIP heuristic proposed here are presented in the last four columns. The columns Avg, $Best$ inform, respectively, the value of the average and best solution obtained after 10 runs of the SAIP algorithm, while the column $t(s)$ reports the total processing time for these 10 runs, and the column $Better$ inform by an asterisk the instances that the SAIP found solutions equal or better than the ones found by the ILP formulation and B&P algorithm. Notice that it is possible for SAIP to obtain a better solution because due to their high asymptotic complexity, the exact strategies had a limited execution time.

As concluded by Chagas and Santos [5] the B&P algorithm solved more instances than the ILP formulation, even considering that for some instances B&P algorithm was not able to reach any solution within the time limit of 1 h. In Table 1, the exact methods found the optimal solution for all 45 instances up to 12 customer requests, while our SAIP algorithm obtained the optimal solution for 39 of these instances. Analyzing the 6 instances (instances with ID 16, 18, 33, 34, 35 and 39) in which the SAIP algorithm obtained worse solutions than the methods ILP and B&P, we can note that the relative difference between the optimal solutions and ones found by the algorithm SAIP has a value close to zero, the largest relative difference (1.32%) occurring on the instance with ID 34.

Still regarding the Table 1, analyzing the columns Avg and $Best$, referring to the SAIP algorithm, we observe that our algorithm has a good convergence, since the difference between the value of the average and best solution is equal to zero or close to zero for all instances. On average (last row of Table 1), the SAIP algorithm was able to find solutions with a relative difference of only 0.1% in comparison to the optimal solutions. Besides that, the SAIP algorithm was more efficient among all the methods already proposed, performing 13.5 times faster than the ILP formulation and approximately 2.7 times faster than the B&P algorithm.

As seen from Table 2, for the 45 largest instances, the SAIP found significantly better solutions for many instances, specially for instances with ID 76 to 90

Table 1. Comparative analysis of algorithms proposed in the literature and the SAIP algorithm on instances with $6 \leq n \leq 12$.

Instance				ILP [5]			B&P [5]			SAIP			
ID	R	F	D	UB	Opt	t(s)	UB	Opt	t(s)	Avg	Best	Better	t(s)
01	R00	(a)	(i)	555	*	0.1	555	*	0.1	555.0	555	*	1.1
02	R01			636	*	0.2	636	*	0.2	636.0	636	*	1.1
03	R02			614	*	0.1	614	*	0.2	614.0	614	*	1.1
04	R03			507	*	0.2	507	*	0.2	507.0	507	*	1.1
05	R04			573	*	0.2	573	*	0.3	573.0	573	*	1.2
06	R00	(b)		572	*	0.2	572	*	0.5	572.0	572	*	1.1
07	R01			589	*	0.1	589	*	1.2	590.0	589	*	1.1
08	R02			590	*	0.1	590	*	0.9	590.0	590	*	1.1
09	R03			480	*	0.1	480	*	0.6	480.0	480	*	1.1
10	R04			559	*	0.2	559	*	0.8	559.0	559	*	1.1
11	R00	(c)		699	*	0.2	699	*	0.2	699.0	699	*	1.7
12	R01			783	*	0.3	783	*	0.1	783.0	783	*	1.8
13	R02			746	*	0.2	746	*	0.1	746.0	746	*	1.8
14	R03			616	*	0.3	616	*	0.1	616.0	616	*	1.8
15	R04			678	*	0.3	678	*	0.2	678.0	678	*	1.7
16	R00	(d)	(ii)	725	*	1.2	725	*	4.1	728.0	728		3.3
17	R01			741	*	1.5	741	*	5.8	741.0	741	*	3.5
18	R02			776	*	4.9	776	*	5.7	785.0	785		3.4
19	R03			693	*	1.5	693	*	4.5	693.0	693	*	3.4
20	R04			679	*	1.2	679	*	7.3	679.0	679	*	3.6
21	R00	(e)		887	*	3.5	887	*	1.8	887.0	887	*	4.6
22	R01			932	*	3.7	932	*	2.6	932.0	932	*	4.5
23	R02			995	*	3.3	995	*	8.4	995.0	995	*	5.3
24	R03			870	*	22.1	870	*	10.7	870.0	870	*	4.6
25	R04			863	*	20.4	863	*	3.3	875.0	863	*	4.8
26	R00	(f)		1013	*	3.5	1013	*	9.4	1013.0	1013	*	6.2
27	R01			1054	*	4.5	1054	*	1.5	1054.0	1054	*	6.2
28	R02			1077	*	4.2	1077	*	1.2	1077.0	1077	*	6.1
29	R03			985	*	33.9	985	*	1.6	990.0	985	*	6.4
30	R04			977	*	16.9	977	*	2.0	995.0	977	*	6.3
31	R00	(g)	(iii)	844	*	87.0	844	*	33.3	844.0	844	*	7.5
32	R01			852	*	42.0	852	*	13.6	856.4	852	*	7.5
33	R02			813	*	151.0	813	*	78.6	819.0	819		7.6
34	R03			836	*	133.2	836	*	54.2	847.8	847		7.8
35	R04			763	*	58.8	763	*	85.9	765.4	765		7.8
36	R00	(h)		959	*	80.5	959	*	7.2	959.0	959	*	9.8

(continued)

Table 1. (*continued*)

Instance				ILP [5]			B&P [5]			SAIP			
ID	R	F	D	UB	Opt	t(s)	UB	Opt	t(s)	Avg	Best	Better	t(s)
37	R01			1032	*	67.2	1032	*	8.3	1032.0	1032	*	9.7
38	R02			944	*	178.3	944	*	13.6	944.0	944	*	9.9
39	R03			988	*	231.5	988	*	17.1	992.0	992		9.8
40	R04			842	*	59.6	842	*	17.2	842.0	842	*	9.8
41	R00	(i)		1162	*	575.6	1162	*	7.5	1162.0	1162	*	13.6
42	R01			1176	*	224.7	1176	*	5.2	1176.0	1176	*	11.8
43	R02			1129	*	384.2	1129	*	75.0	1129.0	1129	*	13.5
44	R03			1095	*	510.8	1095	*	159.7	1097.0	1095	*	12.9
45	R04			992	*	366.8	992	*	6.9	992.0	992	*	12.6
Average				819.8	$\frac{45}{45}$	72.9	819.8	$\frac{45}{45}$	14.6	821.5	820.6	$\frac{39}{45}$	5.4

Table 2. Comparative analysis of algorithms proposed in the literature and the SAIP algorithm on instances with $15 \leq n \leq 24$.

Instance				ILP [5]			B&P [5]			SAIP			
ID	R	F	D	UB	Opt	t(s)	UB	Opt	t(s)	Avg	Best	Better	t(s)
46	R00	(j)	(iv)	909	*	2740.5	909	*	1873.4	920.6	917		17.3
47	R01			935	*	703.0	935	*	338.5	935.6	935	*	17.2
48	R02			822	*	864.8	822	*	1701.3	822.0	822	*	15.8
49	R03			938	*	608.3	938	*	751.9	950.6	949		18.1
50	R04			855	*	376.4	855	*	277.2	856.3	855	*	15.8
51	R00	(k)		1094		1 h	1094	*	1017.4	1103.0	1103		18.3
52	R01			1088	*	3302.0	1088	*	79.0	1097.0	1097		17.7
53	R02			1010		1 h	1006	*	163.8	1006.0	1006	*	17.8
54	R03			1115		1 h	1115	*	752.2	1118.0	1118		19.0
55	R04			1008	*	1863.1	1008	*	92.5	1008.0	1008	*	18.2
56	R00	(l)		1162		1 h	1151	*	44.2	1158.4	1158		23.3
57	R01			1243		1 h	1243	*	45.3	1249.0	1249		21.8
58	R02			1129		1 h	1126	*	63.3	1127.2	1126	*	22.4
59	R03			1227		1 h	1208	*	118.5	1247.0	1247		22.2
60	R04			1098		1 h	1098	*	57.9	1104.3	1098	*	22.0
61	R00	(m)	(v)	1005		1 h	1118		1 h	1016.3	1011		59.4
62	R01			985		1 h	–		1 h	953.8	951	*	43.9
63	R02			884		1 h	1026		1 h	869.4	867	*	40.3
64	R03			1081		1 h	1062		1 h	1023.1	1017	*	74.4
65	R04			930		1 h	1108		1 h	924.0	917	*	44.6

(*continued*)

Table 2. (*continued*)

Instance				ILP [5]			B&P [5]			SAIP			
ID	R	F	D	UB	Opt	t(s)	UB	Opt	t(s)	Avg	Best	Better	t(s)
66	R00	(n)		1196		1 h	–		1 h	1139.6	1139	*	32.3
67	R01			1148		1 h	1113	*	930.3	1113.6	1113	*	31.7
68	R02			1103		1 h	1154		1 h	1060.0	1060	*	31.2
69	R03			1234		1 h	1102	*	2779.9	1169.7	1169		32.7
70	R04			1151		1 h	1101		1 h	1079.0	1079	*	32.2
71	R00	(o)		1265		1 h	1255	*	441.2	1256.8	1255	*	39.3
72	R01			1431		1 h	1254	*	595.6	1254.0	1254	*	38.1
73	R02			1266		1 h	1182	*	3058.0	1187.3	1187		37.5
74	R03			1385		1 h	1337		1 h	1326.8	1326	*	38.5
75	R04			1271		1 h	1191	*	432.8	1195.0	1195		37.1
76	R00	(p)	(vi)	2645		1 h	2072		1 h	1124.9	1105	*	648.7
77	R01			3047		1 h	–		1 h	1165.8	1154	*	648.0
78	R02			2619		1 h	1642		1 h	1095.6	1063	*	651.3
79	R03			2783		1 h	1793		1 h	1190.0	1166	*	648.9
80	R04			1716		1 h	–		1 h	1054.3	1041	*	646.8
81	R00	(q)		1624		1 h	–		1 h	1211.3	1205	*	108.2
82	R01			1854		1 h	–		1 h	1276.5	1240	*	102.5
83	R02			1725		1 h	–		1 h	1209.4	1197	*	97.9
84	R03			1646		1 h	–		1 h	1345.1	1334	*	102.2
85	R04			1348		1 h	–		1 h	1165.0	1152	*	104.7
86	R00	(r)		1802		1 h	–		1 h	1402.3	1401	*	86.5
87	R01			1954		1 h	–		1 h	1469.8	1467	*	84.4
88	R02			1709		1 h	–		1 h	1340.0	1340	*	85.1
89	R03			1842		1 h	–		1 h	1509.0	1508	*	84.2
90	R04			1706		1 h	–		1 h	1301.2	1301	*	86.5
Average				1399.7	$\frac{7}{45}$	≈1 h	–	$\frac{16}{40}$	2267.0	1136.3	1131.2	$\frac{34}{45}$	113.0

when compared to the solutions obtained by the ILP and B&P exact methods. The SAIP algorithm found solutions of same or higher quality for 34 out of 45 instances reported in Table 2. Regarding the computational time, the SAIP algorithm was more efficient on all instances reducing one order in magnitude in average.

5 Conclusions and Future Work

In this paper we proposed a heuristic algorithm for the Double Vehicle Routing Problem with Multiple Stacks and Heterogeneous Demand. The proposed approach was tested using instances available in the literature, and the results found show a significant superiority when compared to the results already presented in previous work that addressed the problem, mainly, for the large instances.

Acknowledgments. The authors thank Coordenação de Aperfeiçoamento de Pessoal de Nível Superior (CAPES) and Fundação de Amparo à Pesquisa do Estado de Minas Gerais (FAPEMIG) for the financial support of this project.

References

1. Petersen, H.L., Madsen, O.B.: The double travelling salesman problem with multiple stacks - formulation and heuristic solution approaches. Eur. J. Oper. Res. **198**, 139–147 (2009)
2. Iori, M., Riera-Ledesma, J.: Exact algorithms for the double vehicle routing problem with multiple stacks. Comput. Oper. Res. **63**, 83–101 (2015)
3. Silveira, U.E.F., Benedito, M.P.L., Santos, A.G.: Heuristic approaches to double vehicle routing problem with multiple stacks. In: 15th IEEE International Conference on Intelligent Systems Design and Applications, pp. 231–236. IEEE Press, Marrakesh (2015)
4. Chagas, J.B.C., Silveira, U.E.F., Benedito, M.P.L., Santos, A.G.: Simulated annealing metaheuristic for the double vehicle routing problem with multiple stacks. In: 19th IEEE International Conference on Intelligent Transportation Systems, pp. 1311–1316. IEEE Press, Rio de Janeiro (2016)
5. Chagas, J.B.C., Santos, A.G.: A branch-and-price algorithm for the double vehicle routing problem with multiple stacks and heterogeneous demand. In: International Conference on Intelligent Systems Design and Applications, pp. 921–934. Springer, Cham (2016)
6. Kirkpatrick, S., Gelatt, C.D., Vecchi, M.P.: Optimization by simulated annealing. Science **220**(4598), 671–680 (1983)
7. López-Ibáñez, M., Dubois-Lacoste, J., Cáceres, L.P., Birattari, M., Stützle, T.: The Irace package: iterated racing for automatic algorithm configuration. Oper. Res. Perspect. **3**, 43–58 (2016)

Named Entity Recognition from Gujarati Text Using Rule-Based Approach

Dikshan N. Shah[1](✉) and Harshad B. Bhadka[2]

[1] Faculty of Computer Applications, S S Agrawal Institute of Computer Science,
Navsari, India
dikshan817@gmail.com
[2] Faculty of Computer Science, C U Shah University, Wadhwan, India
harshad.bhadka@yahoo.com

Abstract. NER which is known as Named Entity Recognition is an application of Natural Language Processing (NLP). NER is an activity of Information Extraction. NER is a task used for automated text processing for various industries, a key concept for academics, artificial intelligence, robotics, Bioinformatics and much more. NER is always an essential activity when dealing with chief NLP activity such as machine translation, question-answering, document summarization etc. Most NER work has been done for other European languages. NER work has been done in few Indian constitutional languages. Not enough work is possible due to some challenges such as lack of resources, ambiguity in language, morphologically rich and much more. In this paper, to identify various named entities from a text document, rules are defined using Rule-based approach. Based on defined rules, three different test cases computed on the training dataset and achieved 70% of accuracy.

Keywords: NER · Rule-based approach · Constitutional languages
Tagset · Tithi

1 Introduction

The phrase Named Entity (NE) was coined during the 6th Message Understanding Conference (MUC-6) in 1995. Many NER systems were developed after that. Foremost work has been done in European languages and all systems were highly precise [6]. Named Entity is the structured information mentioning to predefined proper names like persons, locations, and organizations, year, date, month, monetary amounts, percentages as well as temporal and numeric expressions from text [2].

Named Entity Recognition (NER) systems proved to be very significant for many tasks in Natural Language Processing (NLP) such as information retrieval, machine translation, information extraction, question answering systems. The objectives of NER is to classify each word of a document into predefined target named entities classes.

© Springer International Publishing AG, part of Springer Nature 2018
A. Abraham et al. (Eds.): ISDA 2017, AISC 736, pp. 797–805, 2018.
https://doi.org/10.1007/978-3-319-76348-4_76

1.1   Existing NER Approaches

Present NER systems have been built using mainly knowledge-based or linguistic, and machine learning approach.

1.1.1   Rule-Based Approach

The linguistic approach or Knowledge-based approach is basically called as a rule-based approach which uses a set of hand-crafted rules deliberate and described by human experts, especially linguists. This approach considers a set of patterns containing grammatical, syntactic, linguistic and orthographic features in a grouping with dictionaries. It is a prerequisite to have a thorough knowledge of target language as it is a time-consuming task to develop such kind of system.

1.1.2   Statistical Approach

Machine Learning or a Statistical approach is a swift way to build an NER system which fundamentally supports rule-based systems or use sequence labeling algorithms to collect knowledge from a collection of training examples. The accuracy of this approach is purely dependent upon the training dataset. Various Machine Learning models used for NER systems like Hidden Markov Model, Conditional Random Field, and Maximum Entropy.

1.1.3   Hybrid System

Use of Statistical tools as well as linguistic rules and combinations of both approaches make a system more precise and effective.

1.2   About the Gujarati Language

Basically, Among the Indo-European language family, Gujarati is well-known Indo-Aryan language and it was tailored from the Devanagari script. Alphabets of this language mainly include 34 consonants and 14 vowels. [1] A language is very widespread and spoken by more than 50 million people across the India. It is the official language of the Gujarat state of India.

2   Related Work on Different Indian Languages in NER

Among the constitutional Indian languages, NER work has been done in some languages. NER approaches used in various Indian languages with their accuracies are mentioned in Table 1 as follows:

Table 1. Different approaches used for various Indian Languages according to their accuracies

Author	Language	Method/Approach	Precision	Recall	F-Measure	Accuracy
[1]	Gujarati	Inflectional stemmer	–	–	–	90.7%
[1]	Gujarati	Derivational stemmer	–	–	–	70.7%
[2]	Hindi	Rule based	75.86%	79.17%	77.48%	–
[3]	Kannada	Rule based	78.6%	77.22%	77.2%	–
[4]	Malay	Rule based	85%	94.44%	89.47%	–
[5]	Kannada	Hybrid	–	–	–	94.85%
[8]	Hindi	Rule based	–	–	–	79.06%
[9]	Dogri	Rule based	–	–	–	90%
[10]	Hindi	HMM	–	–	–	98.37%
[11]	Tamil	CRF	–	–	–	87.20%
[11]	Tamil	SVM	–	–	–	86.06%
[12]	Hindi	Hybrid	–	–	–	95.77%
[13]	Arabic	Rule based	92.25%	91.25%	91.71%	–
[14]	English	Rule based	–	–	–	88.19%

3   Rule-Based Approach

A morphological analyzer for the Hindi language analyze Hindi sentences and produce its features with its root words. [7] As Rule-based approach is a domain specific, rules define for one language will not apply for other languages. Some Rules used to identify different tags in the Gujarati language are as follows:

3.1   Date and Time

This Rule is applied on given input which contains the various date and Time formats. Regular Expressions are used to identify these kinds of tags [15]. Following are date and time tagset examples:

3.1.1   Year વિક્રમસંવત, ઇસવિસન, વર્ષ, સાલ

- 'સંવત' *(Samvat)* refers to the epoch of the several Hindu calendar systems in India and also in Nepal. There are three most significant 'સંવત' *(Samvat)*: Vikrama era, Old Shaka era and Shaka era of 78 AD [15].

3.1.2   Month Names

- જાન્યુઆરી ફેબ્રુઆરી માર્ચ એપ્રિલ મે જૂન જુલાઈ ઓગસ્ટ સપ્ટેમ્બર ઓક્ટોબર, નવેમ્બર, ડિસેમ્બર

- The names of the Indian months diverge by region. Hindu calendars are based on lunar cycle and usually phonetic variants of each other.
- કારતક, માગશર, પોષ, મહા, ફાગણ, ચૈતર, વૈશાખ, જેઠ, અષાડ, શ્રાવણ, ભાદરપો, આસો   [15].

3.1.3   Days

- સોમવાર, મંગળવાર, બુધવાર, ગુરુવાર, શુક્રવાર, શનિવાર, રવિવાર

The Hindu calendar has two measures of a day, one based on the lunar movement and the other on solar. The solar day or civil day is called divas (દિવસ), and the lunar day is called tithhi (તિથિ). A lunar month has 30 tithhi. Lunar month starts with Kartak (કારતક).

પ્રતિપદા, દ્વિતીયા, તૃતીયા, ચતુર્થી, પંચમી, શ્રષ્ટિ, સપ્તમા, અષ્ટમી, નવમી, દસમી, એકાદસી, દ્વાદસી, ત્રાયોદશી, ચતુર્દસી, પૂર્ણિમા, અમાવસ્યા, એકમ, બીજ, ત્રિજ, ચોથ, પંચમ, છઠ, સાતમ, આઠમ, નોમ, દસમ, અગિયારસ, બારસ, તેરસ, ચૌદસ, અમાસ, પૂનમ [16].

3.2   Location

Suffix matching is used for types of location names and terms. Different suffix makes different location names of Indian States and Cities are as follows: [17]

- Location names that end with 'pure' (પુર) i.e. - રામપુર, સુંદરપુર, જયપુર, ઉદયપુર
- Location names that end with 'Ghar' (ગઢ) i.e. - રાયગઢ, ચંદીગઢ, સોનગઢ
- Location names that end with 'stan' (સ્તાન) i.e. - હિન્દુસ્તાન, પાકિસ્તાન, અફઘાનિસ્તાન
- Location names that end with 'bad' (બાદ) i.e. – અમદાવાદ, ફરિદાબાદ, હૈદરાબાદ
- Location names that end with 'Nagar' (નગર) i.e. – શ્રીનગર, ગાંધીનગર, ગંગાનગર
- Location names that end with 'pat' (પત) i.e. – પાણીપત, સોનિપત
- Location names that end with 'nath' (નાથ) i.e. - કેદારનાથ, સોમનાથ, બદ્રીનાથ
- Location names that end with 'mer' (મેર) i.e. – જેસલમેર, બારમેલ
- Location names that end with 'kot' (કોટ) i.e. - રાજકોટ, પઠાણકોટ, સિયાલકોટ
- Location names that end with 'Ishwar' (ઇશ્વર) i.e. - અંકલેશ્વર, મહાબળેશ્વર, ભુવનેશ્વર
- Location names that end with 'Wada' (વાડા) i.e. – બામનવાડા, ભિલવારા, તેલવાડા
- Location names that end with 'giri' (ગિરિ) i.e. – રત્નાગિરિ, ચંદ્રગીરી, ધૌલગીરી
- Location names that end with 'Puram' (પુરમ) i.e. – તિરુવનંતપુરમ, મલ્લાઇપુરમ
- Location names that end with 'uru' (ઉરૂ) i.e. – બેંગલુરૂ, મેંગલુરૂ
- Location names that end with 'patnam' (પટનમ) i.e. – વિશાખાપટ્ટનમ, માસુલિપટનમ
- Location names that end with 'guri' (ગુડી) i.e. – સિલિગુડી, જલપાઇગુડી, મેનાગુરી
- Location names that end with 'tal' (તાલ) i.e. – નૈનિતાલ
- Location names that end with 'Dwar' (દ્વાર) i.e. – હરિદ્વાર, કોટવાર
- Location names that end with 'Wada' (વાડા) i.e. – ભીલવાડા, બામનવાડા

- Location names that end with 'Palli' (પલ્લી) i.e. – તિરુચિરાપલ્લી, જલાહલ્લી
- Location names that end with 'Malai' (માલાઇ) i.e. – કોલામાલાઇ, અન્નામલાઇ,, સબરીમાલાઇ

3.3 To Identify Some Abbreviations in Date and Time Tag Entities

Abbreviations point to an original name. Some words used in their abbreviated form for a date, month and year entities. Examples: ਫ਼ਸ , વિ સ , તા., જાન્યુ, ફ્રેબુ., એપ્રિ., જૂ., ઓગ., સપ્ટે.,ઓક્ટો.,નવે.,ડિસે., સોમ, મંગળ, બુધ, ગુરુ, શુક, શનિ, રવિ [15].

3.4 For Numerals

There is a difference between mentioning numbers. Two types of number system we used: Hindu Arabic Numerals and Gujarati Numerals. Numbers have different number names in different languages. Number 0 to 100 written in both format is different and their Gujarati names also [18] (Table 2).

Table 2. Number Names of Hindu Arabic Numerals in Gujarati

Hindu-Arabic Numeral	Gujarati numeral	Gujarati name	Hindu-Arabic numeral	Gujarati numeral	Gujarati name
0	૦	શૂન્ય	31	૩૧	એકત્રીસ
1	૧	એક	32	૩૨	બત્રીસ
2	૨	બે	33	૩૩	તેત્રીસ
3	૩	ત્રણ	34	૩૪	ચોત્રીસ
4	૪	ચાર	35	૩૫	પાંત્રીસ
5	૫	પાંચ	36	૩૬	છત્રીસ
6	૬	છ	37	૩૭	સડત્રીસ
7	૭	સાત	38	૩૮	અડત્રીસ
8	૮	આઠ	39	૩૯	ઓગણચાલીસ
9	૯	નવ	40	૪૦	ચાલીસ
10	૧૦	દસ	41	૪૧	એકતાલીસ
11	૧૧	અગિયાર	42	૪૨	બેતાલીસ
12	૧૨	બાર	43	૪૩	ત્રેતાલીસ
13	૧૩	તેર	44	૪૪	ચુંમાલીસ
14	૧૪	ચૌદ	45	૪૫	પિસ્તાલીસ
15	૧૫	પંદર	46	૪૬	છેતાલીસ
16	૧૬	સોળ	47	૪૭	સુડતાલીસ
17	૧૭	સત્તર	48	૪૮	અડતાલીસ
18	૧૮	અઢાર	49	૪૯	ઓગણપચાસ

19	૧૯	ઓગણિસ	50	૫૦	પચાસ
20	૨૦	વીસ	51	૫૧	એકાવન
21	૨૧	એકવીસ	52	૫૨	બાવન
22	૨૨	બાવીસ	53	૫૩	ત્રેપન
23	૨૩	તેવીસ	54	૫૪	ચોપન
24	૨૪	ચોવીસ	55	૫૫	પંચાવન
25	૨૫	પચ્ચીસ	56	૫૬	છપ્પન
26	૨૬	છવીસ	57	૫૭	સત્તાવન
27	૨૭	સત્તાવીસ	58	૫૮	અઠ્ઠાવન
28	૨૮	અઠ્ઠાવીસ	59	૫૯	ઓગણસાઠ
29	૨૯	ઓગણત્રીસ	60	૬૦	સાઈઠ
30	૩૦	ત્રીસ	81	૮૧	એક્યાસી
61	૬૧	એકસઠ	82	૮૨	બ્યાસી
62	૬૨	બાસઠ	83	૮૩	ત્યાસી
63	૬૩	ત્રેસઠ	84	૮૪	ચોર્યાસી
64	૬૪	ચોસઠ	85	૮૫	પંચાસી
65	૬૫	પાંસઠ	86	૮૬	છ્યાસી
66	૬૬	છાસઠ	87	૮૭	સિત્યાસી
67	૬૭	સડસઠ	88	૮૮	ઈઠ્યાસી
68	૬૮	અડસઠ	89	૮૯	નેવ્યાસી
69	૬૯	અગણોસિત્તેર	90	૯૦	નેવું
70	૭૦	સિત્તેર	91	૯૧	એકાણું
71	૭૧	એકોતેર	92	૯૨	બાણું
72	૭૨	બોતેર	93	૯૩	ત્રાણું
73	૭૩	તોતેર	94	૯૪	ચોરાણું
74	૭૪	ચુમોતેર	95	૯૫	પંચાણું
75	૭૫	પંચોતેર	96	૯૬	છ્નું
76	૭૬	છોતેર	97	૯૭	સત્તાણું
77	૭૭	સિત્યોતેર	98	૯૮	અઠ્ઠાણું
78	૭૮	ઇઠ્યોતેર	99	૯૯	નવ્વાણું
79	૭૯	ઓગણાએંસી	100	૧૦૦	સો
80	૮૦	એંસી	1000	૧૦૦૦	હજાર

4   Research Methodology

We have developed various rules using a rule-based approach which helps to recognize various named entities. For Identification of Named Entity, we have collected document in the Gujarati language as a corpus from E-newspaper '**Gujarat Samachar**'. There are various categories of news as Entertainment, Sports, Religious and much more. Among them, we have gathered 100 sports category documents to identify various Named Entity tagset.

A. Preparation of Database

Based on various categories of tagset, following dictionaries are created.

Date Dictionary: Date tagset contains Day, Month number and name, and Year. The day is also categorized based on Hindu calendar and Panchang (પંચાગ). Tithis (તિથિ) and days (દિવસો) stored in gazetteer list.

Location Dictionary: Here Location names are only within a limited range of area or for a specific country. 21 Suffix stripping rules are created for Location Names as City or State or Village names of India.

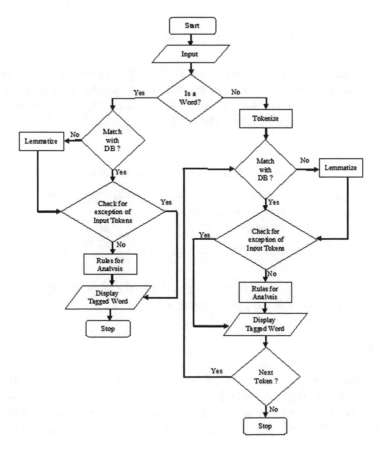

Fig. 1. Flowchart for Rule based NER for Gujarati language

Abbreviation Dictionary: Various abbreviations of date, month, day are listed in it.

Number-names dictionary: Based on Hindu-Arabic numerals, Gujarati number names listed for 0 to 100 digits (Fig. 1).

B. **Architecture of System**

Step – 1 Input text - Through file upload, upload a file which is a text file comprising raw data in the Gujarati language.

Step – 2 Preprocessing – Prepare Gujarati text document for preprocessing.

Step – 3 Tokenization - Input text tokenized word by word for pattern matching.

Step – 4 Entity detection - Detection of Date, Time and Location entities based on created rules and if any rules matched go to step 7.

Step – 5 Detection of **Abbreviation Names**, if matches are found and compared with gazetteer list, go to step 7.

Step – 6 Detection of **Numbers and its Number names** from Non-numerals practice and if found any matches then go to step 7.

Step – 7 Display **tagged output** generated by the system with the untagged result to the user.

Step – 8 End

5   Experimental Result and Analysis

The core objectives of such experiment are to identify the kinds of patterns of named entities by the proposed NER algorithm. We have collected documents of Sports category to recognize various Named Entities such as date, Day names, Month Names, Tithhi (રિથિ), Location and numerals (Table 3).

Table 3. Apply various test cases on dataset

Accuracy	Test No	Tagset	No. of Words	Correctly observed tag
	Test 1	Date, Days, Tithi	147	97
	Test 2	Months	129	68
	Test 3	Locations	87	29

Among the given 363 words, 194 entities are correctly identified by applying various test cases and achieved 70% of accuracy.

6   Conclusions

An innovative technique can be build up to develop the performance of NER in the Gujarati language. We have developed rules to identify various named entities which is a very beneficial in many significant applications. We have studied various existing approaches of NER and analyzed that among the various constitutional Indian languages, lots of scopes is for NER in Indian languages. By implementing various rules on given dataset we attained 70% of accuracy. As a future work, we can build more precise rules for much more named entities to achieve good accuracy.

References

1. Athavale, V., Bharadwaj, S., Pamecha, M., Prabhu, A., Shrivastava, M.: Towards Deep Learning in Hindi NER: An approach to tackle the Labelled Data Scarcity (2016)
2. Jiandani, K.S.D., Bhattacharyya, P.: Hybrid inflectional stemmer and rule-based derivational stemmer for Gujarati. In: Proceedings of the 2nd Workshop on South and Southeast Asian Natural Language Processing (WSSANLP 2011), November 2011
3. Amarappa, S., Sathyanarayana, S.V.: Kannada named entity recognition and classification (nerc) based on multinomial naïve Bayes (MNB) classifier. Int. J. Nat. Lang. Comput. (IJNLC) **4**, 39–52 (2015)
4. Alfred, R., Leong, L.C., On, C.K., Anthony, P.: Malay named entity recognition based on rule-based approach. Int. J. Mach. Learn. Comput. **4**(3), 300–306 (2014)
5. Sathyanarayana, S.A.: A hybrid approach for named entity recognition, classification and extraction (NERCE) in Kannada documents. In: Proceedings of International Conference on Multimedia Processing, Communication, and Info. Tech., MPCIT (2013)
6. Singh, A.K.: Named entity recognition for south and south east asian languages: taking stock. In: Proceedings of the IJCNLP Workshop on NER for South and South East Asian Languages, pp 5–16 (2008)
7. Agarwal, A., Singh, S.P., Kumar, A., Darbari, H.: Morphological analyser for hindi-a rule-based implementation. Int. J. Adv. Comput. Res. **4**(1), 19 (2014)
8. Sharma, L.K., Mittal, N.: Named entity based answer extraction from hindi text corpus using n-grams. In: 11th International Conference on Natural Language Processing, p. 362, December 2014
9. Sasan, T.S., Jamwal, S.S.: Transliteration of name entities using rule-based approach. Int. J. Adv. Res. Comput. Sci. Soft. Eng., **6**(6) (2016)
10. Jahan, N., Morwal, S., Chopra, D.: Named entity recognition in Indian languages using gazetteer method and hidden Markov model: a hybrid approach. IJCSET, March 2012
11. Abinaya, N., Kumar, M.A., Soman, K.P.: Randomized kernel approach for named entity recognition in Tamil. Indian J. Sci. Technol. **8**(24), 1–7 (2015)
12. Kaur, Y., Kaur, E.: Named Entity Recognition system for Hindi Language using a combination of rule-based approach and list lookup approach. Int. J. Sci. Res. Manag. (IJSRM) **3**(3), 2300–2306 (2015)
13. Aboaoga, M., Ab Aziz, M.J.: Arabic person names recognition by using a rule-based approach. J. Comput. Sci. **9**(7), 922 (2013)
14. Bhalla, D., Joshi, N., Mathur, I.: Rule-based transliteration scheme for English to Punjabi (2013)
15. To download. Guj-Ind-StyleGuide. http://download.microsoft.com/download/7/2/0/720b015e-94f9-4b6e-911f-539f38c60774/guj-ind-styleguide.pdf
16. Tithi (Internet). https://en.wikipedia.org/wiki/Tithi
17. Indian Place Names (Internet). http://www.irfca.org/docs/place-names.html
18. Gujarati Number names for Digits (Internet). https://www.omniglot.com/language/numbers/gujarati.htm

A Meta-modeling Approach to Create
a Multidimensional Business Knowledge
Model Based on BPMN

Sonya Ouali[1,2](✉), Mohamed Mhiri[1,2], and Faiez Gargouri[1,2]

[1] University of Sfax, Sfax, Tunisia
sonyawali@hotmail.fr
[2] MIR@CL Laboratory, Technopark of Sfax,
Tunis Road Km 10, BP. 242, 3021 Sfax, Tunisia

Abstract. Business processes are everywhere. To be more efficient, organizations look for a good business process modeling. In this way, to model a business process, the are a lot of knowledge which are responsible to present a part of the process in a good level of understanding. However, the issue to be addressed in this paper is how to formalize implicit pieces of knowledge figured in the business process so as to construct a business knowledge model which will be treat in a high understanding level. This paper contributes with a meta-modeling approach that the principle is to transform a business process model to a business knowledge model. The purpose of such an approach is to provide a way to automatically build a business ontology based on easy processing of business knowledge dimensions.

Keywords: Knowledge · Multi-dimension
Business process modeling · Meta-models · M2M · ATL
Business ontology

1 Introduction

To date, organizations made more efforts to efficiently manage their business process. Successful business process management is closely underpinned to a well understanding of manner in which inputs are used to perform outputs (the various activities and the used links). The business process models are a good way to ensure this understanding due to a graphical presentation which mainly focus on the logical relationships between the business activities.

Business Process Management (BPM) is the discipline that allows the modeling, the implementation, and the control of all activities taking place in an organization to perform a product or a service. Our work subscribes in this discipline, exactly in the business process modeling. Despite the efforts made in this field, there is yet a gap between business view and IT. In reality, sometimes, the designed process are completely different than the modelled processes [1]. Several researchers focus in the quality-improving direction, exist in the literature

© Springer International Publishing AG, part of Springer Nature 2018
A. Abraham et al. (Eds.): ISDA 2017, AISC 736, pp. 806–815, 2018.
https://doi.org/10.1007/978-3-319-76348-4_77

to obtain a clear business process model [2]. This objective constitutes the most important prerequisite in this respect. For that, it is necessary to well explicit the understanding view of the business process. Undoubtedly, it is the crucial role of the semantic level. First, such a level allows to present a business process at a high understanding level by give a meaning to its components. Second, it helps to eliminate the ambiguities in modeling the business processes.

Recently the focus moved to integrate the Knowledge Management to express the semantic of the business process [3]. In fact, knowledge has many definitions in the literature. As an instance, we can cite that of [4–6]. Accordingly, projecting knowledge in the context of business modeling, gives birth to the notion of business knowledge (BK). In this regard, we have closely followed those definitions and we describe business knowledge as "*a common path between the business process model and the knowledge management domain*". Indeed, we have defined it as "*a set of knowledge dimensions that serve to better understand processes and especially to facilitate modeling tasks for designers*". Therefore, we have defined seven dimension which are organized around the business process main perspectives [7].

The main problem has been addressed in this paper is "*how to formalize Business Knowledge Dimensions from BP Model?*".

Owing to the wide issue, we proposed as a solution to opt to a meta-modeling level. First, it is defined as the model of models. Second, it identifies more precisely the model elements in ensuring a high abstract definitions [8].

This paper is outlined as follows. Section 2 describes briefly the related work in terms of knowledge using in the business process. Our meta-modeling approach is detailed in Sect. 3. It first gives an overview of the Multidimensional Business Knowledge approach followed by a BPMN (Business Process Modeling Notation) and MBKM (Multidimensional Business Knowledge Model) meta-models description. Section 4 highlights the BPMN and MBKM mapping and gives some implementation details. Finally Sect. 5 draws conclusions and suggests future directions.

2   Related Work

Different users, such as information systems developers, process designers and participants may have inconsistent, or even conflicting, understanding of the same process. Indeed, the lack of understanding business process can give many risks. For this regard, knowledge plays a famous role to improve the business process modeling. In leafing the studies treating both business process and knowledge, we remark that knowledge is taken as a whole and unable to be divided or separated. As an illustration, we can cite [9] where the authors used the knowledge as a whole in order to improve the semantic completeness and expressiveness of business process models according to domain knowledge contained in the ontologies. Hence, they tried to build an uniform description of both process models and ontology. Similarly, [10,11] used the knowledge to detect naming conflicts during the business process modeling. In this regard, knowledge plays a crucial role so as to assist the process designers in the labeling of process elements.

In prior research, a few of them consider the knowledge as a set of divided dimension. Examples to mention are [12–14], that are especially focused to the functional, informational and behavioral dimensions of knowledge. An other example this of [15], the authors used knowledge as a set of dimension which are the organizational, operational, decisional and cooperative. The idea is to examine these dimensions to treat the Information system of sensitive situation.

To summarize, our analysis of the state of-the-art led us to argue that our approach is different from the others for the simple reason that, first, we treat knowledge as a set of dimension which are organized around the three-key perspective of the business process [7]. Second, our idea is to integrate these dimensions a step forward the business process-lifecycle phases to respond to the designers' needs in making available exactly the needed piece (dimension) of knowledge [3]. Nevertheless, the questions that might be posed at this level is how formalize these dimensions? which support can be used to create them? And, what techniques will be used to express them? The answers of these questions are the topic of the followed section.

3   How to formalize Business Knowledge Dimensions from BP Model?

3.1   An Overview of the Multidimensional Business Knowledge Approach

When treating the different phases of the business process lifecycle, designers must deal with many pieces of information to ensure a good and efficient business process model. In this way, problems of understanding are provided because designers have not the same visions and ideas when designing business processes. Simply, they have different skills, manner of understanding and explaining given that this is graved in their memory and work habits. In fact, this diversity of skills and manner can product lots of problems in an understanding level. Hence, our approach is considered as a solution to limit modeling problems. By the way, the proposed solution does not contribute to the business process lifecycle. It is subscribed in a pre-modeling level. Indeed, the objective of our proposed solution is to facilitate the designers' tasks. Ontologies provide a much appreciated semantic signification for specific domain, thanks to their ability to share and reuse pieces of knowledge. Business domain is such a delicate domain [16]. For that, to build a business ontology from scratch by hand is not such an easy task to do. To resolve this issue, we propose, as a solution, to construct a business ontology from existing business process models. Figure 1 demonstrates the approach taken to construct the business ontology which is based on the MBKM.

Eventually, the business ontology is the result of a formalization step. Hence, it could be worthwhile to use specialized techniques. In this context, we have

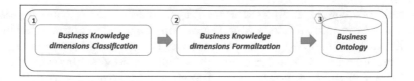

Fig. 1. Steps of multidimensional business knowledge construction

chosen to opt to a meta-modeling approach. Since to present something in a high-level of abstraction, this requires the use of a meta-modeling presentation. In this regard, there are lots of business process meta-models. Some of them are related to the business process itself such as [2, 17]. Some others are related to the languages supporting the business process modeling like the BPMN notation, the UML language through the activity diagram, the Petri Net and the EPC (Event driven Process Chain) [18, 19].

3.2 BPMN and MBKM Meta-models

In follows, we have briefly described the BMPN and MBKM meta-models.

Business Process Meta-Model. Among the graphical modeling language and notations cited in the Sect. 3, we are particularly interested to BPMN for the simple reasons that, on the one hand, BPMN was presented as a standard by the OMG (Object Management Group). On the other hand, BPMN provides a readily-understood notation by many business actors (designers, analysts and developers) [20].

In short, as it is graphically presented in Fig. 2(a)(2), a BPMN meta model is composed of a set of graphical object which can be connecting objects, artifacts, swimlanes or flow objects. A connecting object can be a sequence flow, a message flow or an association. An association represents the link between two concepts. The sequence flow defines the order of activities execution. A message flow represents the information exchanged between two-process participant. An artifact represents the resource consumed by the activities throughout the process. it can be text annotation, group or data object. A swimlane represents process participant. It can be a pool which represents a participant in a collaboration or a lane that represent the performer of an activity. A flow object can be an event (initial, intermediate or final), an activity (task or subProcess), or a gateway. A gateway is responsible to manage the convergence or divergence of activities flows.

Multidimensional Business Knowledge Meta-Model. In this way, as it is well presented in the Fig. 2(a)(1), a Process is composed of activities, which can be atomic or composite. The atomic activity is presented as a task and the composed one is presented as a sub_process. This fragment represents the functional dimension of knowledge. Activities are linked together using control patterns which can be conditional or non conditional. This fragment represents the behavioral dimension of knowledge. Once the activity is composite, its components are the operations what is classified as an operational knowledge dimension. To perform the activities, it is necessary to apply an actor. An actor is belonged to an organizational unit and he can be internal, external, human or not human. This fragment represents the organizational dimension of business knowledge. An actor performs an activity only if he played a special role which represents the skill. Finally, an activity produced or consumes an informational resource that can be process data, system data or application data. This fragment represents the informational resource knowledge.

4   BPMN and MBKM Mapping

4.1   Description

Recall that a standard BPMN provides business with the capability of understanding their internal business procedures in a graphical notation and gives organizations the ability to communicate these procedures in a standard manner [21], we have chosen the BPMN elements since they are considered the best way to construct our MBK concepts. In follows, we point the correspondence between the BPMN elements and the MBK main concepts.

Figure 2(a) resumes the main links. A process P in a BPMN model is transformed to a functional knowledge dimension. In this way, a concept named "*Process*" is created in the hierarchic of the MBK model. The *Process* instances are automatically created. The *Process* has a name and an identifier. Their values are respectively the P_name and the P_id. Similarly, *Artifact* in a BPMN model is transformed to an informational resource knowledge. The related concepts are *Process Data*, *System Data* and *Application Data*. Additionally, a *Lane* is transformed to an organizational knowledge and the related concept is *Actor*. An actor can be internal or external, human or not human. The *Pool* is also transformed to an organizational knowledge. Here the related concept is the *Organizational Unit*. To additional signification we have created an additional knowledge dimension which is *Skill*. Indeed, the corresponding concept is *Role*. This concept is a derived one, it is related to the actor concept of the MBK model and the lane element of the BPMN model. An *activity* is transformed to a functional knowledge and the relevant concept is activity. Hence, an activity can be a subprocess or a task. In this regard, if it is a subprocess then the corresponding concept is *composite activity*. Else, the corresponding concept is *atomic activity* which is the equivalence of the *operation* concept. Once it is a task, it will be transformed to an operational knowledge. As a behavioral knowledge, on the one hand, *ConnectingObjcet* allows to create a *NonConditionalCoordinationPattern* concept. On the other hand, *Gateway* allows to create a *ConditionalCoordinationPattern* concept since it represents a conditional transaction. Finally, an Event is transformed to a conditional (Contextual) knowledge. The corresponding concept is *Condition*.

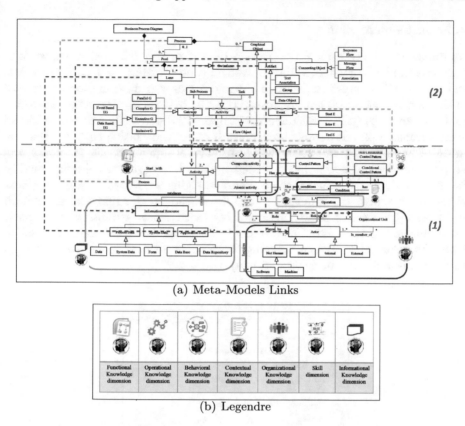

(a) Meta-Models Links

(b) Legendre

Fig. 2. Links between the BP and the BPMN in a meta-modeling level

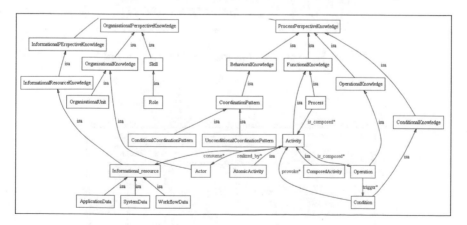

Fig. 3. Fragment of the proposed business ontology skeleton [3]

These correspondences allow to create the skeleton of our Business Ontology as it is presented in Fig. 3.

4.2   Implementation

To implement the transformation rules proposed in Sect. 3.2, since we have elaborated a model to model (M2M) transformation starting from BPMN models
to create a MBK target model. One of the most used transformation techniques
is the *meta-modeling Approach* throughout the MDA [22]. The Atlas Transformation Language (ATL) was defined to cope with operations referred to model
transformation [23]. It allows the definition of transformation rules for the creation of one or more output models from several input models. Moreover, the
scope of model transformation provided by this language is wide being quality
improvement, model refinement or model merging examples of some applications. In this way, the principle directives of the MDA M2M transformation
is to start by a set of BPMN models (*ls*) (with *.bpmn* extension) that must
be conformed to the BPMN meta-model. Those models have received a set of
transformation rules (*Tr*). In applying these rules, a target model (*lt*) conforms
to its meta-model will be generated (Fig. 4).

```
trans.atl
     -- @path BPMN=/eclipse.emf.businessknowledge/model/MyBPMN.ecore
     -- @path BK=/eclipser.emf.businessknowledge/model/businessknowledge.ecore

   module trans;
   create OUT : BK from IN : BPMN;

   rule Process2Process{
       from s: BPMN!Process(s.oclIsTypeOf(BPMN!Process))
       to t: BK!Process {
           id <- s.id,
           name <- s.name}
   }
   rule Activity2Activity {
       from s: BPMN!Activity (s.oclIsTypeOf(BPMN!Activity))
       to t: BK!Activity {
           name <- s.name}
   }
   rule Lane2Actor2Role {
       from s: BPMN!Lane (s.oclIsTypeOf(BPMN!Activity))
       to t: BK!Actor {
           name <- s.name},
       tt: BK!Role {
           name <- s.name
       }
   }
   rule SequenceFlow2UnconditionalCoordinationPattern{
       from s: BPMN!SequenceFlow
       to u: BK!UnconditionalCoordinationPattern {
           name: "is followed by",
           conceptIn <- s.sourceRef, -- to define the source of the created behavioral knowled
           conceptOut <- s.targetRef,-- to define the target of the created behavioral knowled
       }
   }
```

Fig. 4. Some transformation rules written in *.atl*

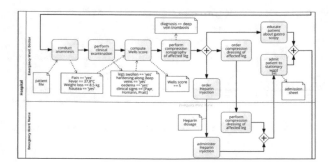

Fig. 5. A partial cancer treatment model expressed in BPMN

(a) MBK with .*bk* format   (b) MBK with.*xmi* format

Fig. 6. The MBK generated model

To execute the M2M transformation, we have chosen as an example, the treatment scenarios. These scenarios are critical and complex, mainly those that lead to oncology care (cancer) care that involve not only a fairly large number of specialists but also many coordinated steps. Figure 5 shows a partial model of an example of a cancer treatment process [24]. In follows, we give some rules written in atl language. In applying the transformations rules expressed in the ATL language we obtain the generated model: a MBK model. The Fig. 6 shows the related model expressed with .*bk* (*businessknowledge*).

5   Conclusion and Future Work

Talking about a high level of abstraction, is talking about a meta-modeling level. Such a level provides a clear representation of concepts used independently of all languages and notations helping in the understandability of models. This paper deals with the meta-modeling approach that aims at creating a MBKM from existing business process models based on a M2M transformation using the ATL. In fact, the key idea of this transformation is to automatically generate a business ontology skeleton and to instantiate its concepts. The result of this transformation is a MBKM which is conformed to the proposed business knowledge meta model.

In future endeavours, on the one hand, it will be envisaged to integrate other transformation rules to automatically explore the semantic relationships which play a very important role to semantically interrelate the business concepts. On the other hand, based on the generated MBKM that is expressed in a bk extension (.*bk*), we envisage to obtain an *owl (Web Ontology Language)* code since it is the most used ontology notation. To do this, we plan to use a code generator so as to perform a model to text (code) transformation.

814   S. Ouali et al.

Acknowledgements. This manuscript is in the memory of Mr. Lotfi Bouzguenda who provided insight and expertise which greatly assisted the research for two years but he suddenly passed away.

References

1. van der Aalst, W.M.: Trends in business process analysis. In: Proceedings of the 9th International Conference on Enterprise Information Systems (ICEIS), pp. 12–22 (2007)
2. Cherfi, S.S.S., Ayad, S., Comyn-Wattiau, I.: Improving business process model quality using domain ontologies. JoDS J. Data Semant. **2**, 75–87 (2013). Special issue on Evolution and Versioning in Semantic Data Integration Systems
3. Ouali, S., Mhiri, M.B.A., Gargouri, F.: Knowledge engineering for business process modeling, pp. 81–90 (2017)
4. Nonaka, I., Takeuchi, H.: The Knowledge-Creating Company: How Japanese Companies Create the Dynamics of Innovation. Oxford University Press, New York (1995)
5. Jablonski, S., Bussler, C.: Workflow management: modeling concepts, architecture and implementation (1996)
6. Sveiby, K.E.: The New Organizational Wealth: Managing & Measuring Knowledge-Based Assets. Berrett-Koehler Publishers, San Francisco (1997)
7. Ouali, S., Mhiri, M., Bouzguenda, L.: A multidimensional knowledge model for business process modeling. Procedia Comput. Sci. **96**, 654–663 (2016)
8. Engels, G., Heckel, R., Küster, J.M.: Rule-based specification of behavioral consistency based on the UML meta-model. In: International Conference on the Unified Modeling Language, pp. 272–286. Springer (2001)
9. Benjamins, V.R., Radoff, M., Davis, M., Greaves, M., Lockwood, R., Contreras, J.: Semantic technology adoption: a business perspective. In: Handbook of Semantic Web Technologies, pp. 619–657. Springer (2011)
10. Havel, J.M., Steinhorst, M., Dietrich, H.A., Delfmann, P.: Supporting terminological standardization in conceptual models-a plugin for a meta-modelling tool (2014)
11. Leopold, H., Eid-Sabbagh, R.H., Mendling, J., Azevedo, L.G., Baião, F.A.: Detection of naming convention violations in process models for different languages. Decis. Support Syst. **56**, 310–325 (2013)
12. Barba, I., Weber, B., Del Valle, C., Jiménez-Ramírez, A.: User recommendations for the optimized execution of business processes. Data Knowl. Eng. **86**, 61–84 (2013)
13. Born, M., Brelage, C., Markovic, I., Pfeiffer, D., Weber, I.: Auto-completion for executable business process models. In: International Conference on Business Process Management, pp. 510–515. Springer (2008)
14. Smith, F., Bianchini, D.: Selection, ranking and composition of semantically enriched business processes. Comput. Ind. **65**(9), 1253–1263 (2014)
15. Zahaf, S., Gargouri, F.: The urbanized bid process information system. Procedia Comput. Sci. **112**, 874–885 (2017)
16. Antoniou, G., Franconi, E., Van Harmelen, F.: Introduction to semantic web ontology languages. In: Reasoning Web, pp. 1–21. Springer (2005)
17. Heidari, F., Loucopoulos, P., Kedad, Z.: A quality-oriented business process meta-model. In: Enterprise and Organizational Modeling and Simulation, pp. 85–99. Springer (2011)

18. Model, B.P.: Notation (BPMN) version 2.0. OMG Specification, Object Management Group (2011)
19. La Rosa, M., Dumas, M., Uba, R., Dijkman, R.: Business process model merging: an approach to business process consolidation. ACM Trans. Softw. Eng. Methodol. (TOSEM) **22**(2), 11 (2013)
20. Chhun, S., Cherifi, C., Moalla, N., Ouzrout, Y.: Business process implementation using an ontology-driven web service selection algorithm. In: 5th Journées Francophones sur les Ontologies (JFO 2014) (2014)
21. Gao, X., Liu, Z., Zhao, Y., Chen, Z., Li, A., Xu, B., Shen, B.: BPM-driven educational informationization technology. (2015)
22. Bazoun, H., Zacharewicz, G., Ducq, Y., Boye, H.: Transformation of extended actigram star to BPMN2. 0 and simulation model in the frame of model driven service engineering architecture. In: Proceedings of the Symposium on Theory of Modeling & Simulation-DEVS Integrative M&S Symposium, Society for Computer Simulation International, p. 20 (2013)
23. Jouault, F., Allilaire, F., Bézivin, J., Kurtev, I.: ATL: a model transformation tool. Sci. Comput. Program. **72**(1), 31–39 (2008)
24. Hewelt, M., Kunde, A., Weske, M., Meinel, C.: Recommendations for medical treatment processes: the PIGS approach. In: International Conference on Business Process Management, pp. 16–27. Springer (2014)

Toward a MapReduce-Based K-Means Method for Multi-dimensional Time Serial Data Clustering

Yongzheng Lin[1,2], Kun Ma[1,2], Runyuan Sun[1,2(⊠)], and Ajith Abraham[3]

[1] School of Information Science and Engineering,
University of Jinan, Jinan 250022, China
{ise_linyz,ise_mak,sunry}@ujn.edu.cn
[2] Shandong Provincial Key Laboratory of Network Based Intelligent Computing,
University of Jinan, Jinan 250022, China
[3] Machines Intelligence Research Labs (MIR Labs),
Scientific Network for Innovation and Research Excellence,
Auburn, WA 98071, USA
ajith.abraham@ieee.org

Abstract. Time series data is a sequence of real numbers that represent the measurements of a real variable at equal time intervals. There are some bottlenecks to process large scale data. In this paper, we firstly propose a K-means method for multi-dimensional time serial data clustering. As an improvement, MapReduce framework is used to implement this method in parallel. Different versions of k-means for several distance measures are compared, and the experiments show that MapReduce-based K-means has better speedup when the scale of data is larger.

Keywords: Clustering · Time serial data · K-means · MapReduce

1 Introduction

Unlike static data, time series data is a sequence of real numbers that represent the measurements of a real variable at equal time intervals. The values of time series data are changed with time. A data stream is an ordered sequence of points $x_1, ..., x_n$. These data can be read or accessed only once or a small numbers, each number indicating a value at a time point. Data flows continuously from a data stream at high speed, producing more examples over time in recent real world applications. Most of the time series encountered in cluster analysis are discrete time series. Multi-dimensional time series are an extension and generalization of regular time series. Regarding the expansion of the amount of the data, it is natural to consider parallelism in a distributed computational environment. Currently, MapReduce framework is a classical method to process large-scale datasets by the map and reduce paradigm [2]. It is appreciable that several researchers use MapReduce for big data clustering [1,13]. In this paper,

© Springer International Publishing AG, part of Springer Nature 2018
A. Abraham et al. (Eds.): ISDA 2017, AISC 736, pp. 816–825, 2018.
https://doi.org/10.1007/978-3-319-76348-4_78

we attempt to investigate the behavior of the k-means algorithm for several multi-dimensional time series data. As an improvement, we use MapReduce framework to implement this method in parallel. Different versions of k-means for several distance measures are compared.

The paper is organized as follows. Related work are given in Sect. 2, Sect. 3 presents the K-means for multi-dimensional time series data clustering and the distance measures that are used, Sect. 4 gives the implementation of the parallel K-means method for multi-dimensional time serial data clustering with MapReduce framework, Sect. 5 contains the experiments and comparisons and Sect. 6 presents the conclusions and future.

2   Related Work

In the past few years, there are three kinds of time serial data clustering methods depending upon whether they work directly with raw data (raw-data-based), indirectly with features extracted from the raw data (feature-based), or indirectly with models built from the raw data (model-based).

2.1   Raw-Data-Based Approaches

Raw-data-based methods work with raw data either in the frequency or time domain. The two time series that are compared are normally sampled at the same interval, but their length might not be the same.

Table 1. Raw-data-based time series clustering algorithms.

Ref.	Distance measure	Clustering algorithm	Application
[5]	Gaussian models of data errors	Agglomerative hierarchical	Seasonality pattern in retails
[7]	Short time series	Modified fuzzy c-means	DNA microarray
[9]	Kullback Leibler discrimination information measures	Agglomerative hierarchical	Earthquakes and mining explosions
[11]	N/A	Neural network	Functional MRI brain activity mapping

Table 1 summarizes several raw-data-based time series clustering algorithms.

2.2   Feature-Based Approaches

Raw-data-based clustering methods working with high dimensional space, especially for data collected at fast sampling rates. But it is not applicable when the raw data that are highly noisy. Therefore, several feature-based clustering methods have been proposed to address these issues. Though most feature extraction methods are generic in nature, the extracted features are usually application dependent. Several recent works even take another feature selection step to further reduce the number of feature dimensions after feature extraction.

Table 2. Feature-based time series clustering algorithms.

Ref.	Features	Distance measure	Clustering algorithm	Application
[3]	Perceptually important points	Sum of the mean squared distance along the vertical and horizontal scales	Modified SOM	Stock market
[8]	Timefrequency representation of the transient region	Euclidean	Modified k-means	Tool condition monitoring
[10]	Haar wavelet transform	Euclidean	Modified k-means	Non-specific

Table 2 summarizes major components used in each feature-based clustering algorithm. They all can handle series with unequal length because the feature extraction operation takes care of the issue. For a multivariate time series, features extracted can simply be put together or go through some fusion operation to reduce the dimension and improve the quality of the clustering results, as in classification studies.

2.3   Model-Based Approaches

Model-based approaches consider that each time series are generated by some kind of model or by a mixture of under lying probability distributions. Time series are considered similar when the models characterizing individual series or the remaining residuals after fitting the model are similar [5].

Table 3. Model-based time series clustering algorithms.

Ref.	Model	Distance measure	Clustering algorithm	Application
[4]	AR	Euclidean	Public data	
[6]	Discrete HMM	Log-likelihood	EM learning	Tool condition monitoring
[12]	ARMA mixture	EM learning	Public data	

Table 3 summarizes several model-based clustering algorithm. Model-based methods can handle series with unequal length as well as feature-based methods through the modeling operation. Most those methods use log-likelihood as the distance measure.

3 K-Means for Multi-dimensional Time Serial Data Clustering

The similarity between two time series is usually calculated using a distance or a similarity measure. In this section, we consider the difference between each time series (of a multidimensional time series instance) as an objective function which has to be minimized. Thus, the goal is to compare how similar two object X and Y are, where X and Y are given by Eq. 1.

$$X = \begin{pmatrix} X_1 \\ X_2 \\ ... \\ X_n \end{pmatrix}, Y = \begin{pmatrix} Y_1 \\ Y_2 \\ ... \\ Y_n \end{pmatrix} \tag{1}$$

3.1 Similarity Measures

$$F = \begin{pmatrix} f_1 = d(X_1, Y_1) \\ f_2 = d(X_2, Y_2) \\ ... \\ f_n = d(X_n, Y_n) \end{pmatrix} \tag{2}$$

We define an N dimensional objective function $F = (f_1, f_2, ..., f_N)$ as Eq. 2, where d defines a similarity measure.

$$d_{sim} = \sum_{i-1}^{N} w_i * f_i \tag{3}$$

We use traditional K-means for clustering multi-dimensional time series data. In our case, each item is assigned to a cluster based on the values of the F function. We consider a weighted combination of all f_i, $1 \leq i \leq N$ as a result of the similarity and denote this by d_{sim} in Eq. 3, where w is a vector of weights denoting the importance of that particular time series in the clustering. For our experiments we considered all time series as having equal importance and in this case $w_i = 1$, $1 \leq i \leq N$.

We implemented four different distances d: Euclidian distance, Manhattan distance, Maximum distance, and Average distance.

3.2   Learning the k Value

One of the four distance measures (Euclidian distance, Manhattan distance, Maximum distance, Average distance) is selected from the main menu, and sent as parameter for the algorithm to use while computing. Also a Maximum Dis tance Percent can be introduced before running the algorithm; the default value for this variable is 0.6 in our experiments.

The algorithm starts with a large k (equal to the no. of items to cluster) which is decreased step-by-step (by moving data, if convenient, from initial clusters - containing only one item from the data set - to new clusters - containing similar items) until it reaches a value that satisfies the stability of each cluster (small distance between data belonging to same cluster, large distance between data belonging to distinct clusters).

4   MapReduce-Based K-Means for Multi-dimensional Time Serial Data Clustering

In this section, we use MapReduce batch computing framework to implement K-means in parallel. The architecture is shown in Fig. 1. MapReduce job will read HDFS where the large-scale datasets are stored, and then they will be shuffled over the whole clusters each time. Finally, they are reduced to several K/V values. The map function performs the procedure of assigning each sample to the closest center, and the reduce function performs the procedure of updating the new centers. All the intermediate values are stored on HDFS.

The input is a sequence file of <key, value> pairs, each of which represents a record in the dataset. The output of map function is several K/V (<key', $value'$>) pairs, where the key' is the index of the closest center point and <$value'$ is a string comprise of sample information. The key is the offset in bytes of this record to the start point of the data file, and the value is a string of the content of this record. The pseudocode of map function is shown in Algorithm 1.

The input of the reduce function is the data obtained from the combine function of each host. The output is a set of (<key', $value'$>) pairs, , where the key' is the index of the cluster, $value'$ is a string comprised of sum of the samples in the same cluster and the sample number. Then, the new centers which are used for next iteration are got. The pseudocode for reduce function is shown in Algorithm 2.

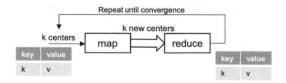

Fig. 1. K-means clustering with MapReduce.

Algorithm 1. map function of K-means *map*

Input:
    Variable *centers*, the sample *value*, and the offset *key*;
Output:
    <*key'*, *value'*> pair;
 1: Construct the sample instance from value;
 2: minDis = Double.MAX VALUE;
 3: index = -1;
 4: **for** each *center* ∈ *center* **do**
 5:     dis= ComputeDist(instance, center);
 6:     **if** dis ¡ minDis **then**
 7:         minDis = dis;
 8:         index = i;
 9:     **end if**
10: **end for**
11: Take index as key';
12: Construct value' as a string comprise of the values of different dimensions;
13: output <*key'*, *value'*> pair;

Algorithm 2. reduce function of K-means *reduce*

Input:
    keyistheindexofthecluster, *V* is the list of the partial sums from different host
Output:
    <*key'*, *value'*> pair
 1: Initialize one array record, e.g. the samples in the list *V*;
 2: Initialize a counter *NUM*;
 3: **for** each *instance* ∈ *V* **do**
 4:     Construct the sample instance;
 5:     Add the values of different dimensions of instance to the array
 6:     NUM += num;
 7: **end for**
 8: Divide the entries of the array by *NUM* to get the new center's coordinates;
 9: Take key as *key'*;
10: Construct *value'* as a string comprise of the center's coordinates;
11: output <*key'*, *value'*> pair;

5 Experiment Results

We perform experiments considering three datasets from various domains. Silhouette coefficient is used to compare the performance of k-means for various distance measures.

5.1 Weather Data

This dataset contains data about countries with respect to temperature, precipitation level, atmospheric pressure and humidity. The countries have to be clustered based on the records over time for all these parameters together.

These are the details of the dataset:

- 14 Countries;
- No. of parameters: 5 (Precipitation Level (L/m^2), Wind Speed (m/s), Temperature (°C), Atmospheric Pressure (mmHg), and Humidity (RH%));
- No. of time points: 77.

Table 4. K-means results for the first dataset.

Algorithm	Number of clusters	Silhouette coefficient
k-means with Euclidian distance	8	0.2105
k-means with Manhattan distance	10	0.1574
k-means with Maximum distance	12	0.0200
k-means with Average distance	8	0.2105

The results obtained by k-means are presented in Table 4.
From the experiments we observe that:

- Best average silhouette coefficient: Euclidian distance and Average distance;
- Better average silhouette coefficient for cluster 0 is obtained using Manhattan distance (0.745) not Euclidian/Average distance (0.181) or Maximum distance (0.240);
- Better average silhouette coefficient for cluster 1 is obtained using Euclidian distance or Average distance (0.674);
- Best average silhouette coefficient obtained for a cluster is 0.828 using Euclidian, Average or Manhattan distance.

5.2   Machine Learning Repository Data

This dataset if from the UCI Machine Learning Repository. The files contains 19 activities (like sitting, lying on back and on right side, ascending and descending stairs, running on a treadmill with a speed of 8 km/h, etc.). Data is acquired from one of the sensors (T_xacc) of one of the units (T) over a period of 5 sec, for each subject and for each of the activities.

Table 5. K-means results for the second dataset.

Algorithm	Number of clusters	Silhouette coefficient
Max Distance Percent = 0.6		
k-means with Euclidian distance	18	0.028
k-means with Manhattan distance	18	0.028
k-means with Maximum distance	17	0.028
k-means with Average distance	18	0.028
Max Distance Percent = 0.9		
k-means with Euclidian distance	17	0.028
k-means with Manhattan distance	16	0.038
k-means with Maximum distance	17	0.028
k-means with Average distance	18	0.028

Results obtained by k-means are presented in Table 5. In this case we tested the algorithm with two values for the maximum Distance Percent parameter (used to decide which k (number of clusters) is best): 0.6 and 0.9.

We observed that:

- Best average silhouette coefficient: Manhattan distance using Max Distance Percent 0.9;
- The same average silhouette coefficient for cluster 1 is obtained using Manhattan distance, Euclidian distance or Average distance and the default Max Distance Percent (0.6) or Average distance and a Max Distance Percent = 0.9 (0.521);
- The same average silhouette coefficient for cluster 0 is obtained using Maximum distance and the default Max Distance Percent or Euclidian distance or Average distance and a Max Distance Percent = 0.9 (0.491);
- Best average silhouette coefficient obtained for a cluster is 0.613 using Manhattan distance and Max Distance Percent 0.9;
- For Max Distance Percent lower than default (0.6) worse clustering results have been obtained.

5.3   KEGG Data

The third dataset if from the KEGG database and is not a time series dataset. We wanted to test the algorithm for this kind of data as well in order to validate the findings. The data is a Metabolic Relation Network (Directed) Data Set. It has 8 attributes such as: Nodes (min:2, max:116), Edges (min:1, max:606), Connected Components (min:1, max:13), Network Diameter (min:1, max:30), Network Radius (min:1, max:2), Shortest Path (min:1, max:3277), Characteristic Path Length (min:1), Average number of Neighbors (min:1)). The data set has 1,000 instances.

Table 6. K-means results for the third dataset.

Algorithm	Number of clusters	Silhouette coefficient
k-means with Average distance	18	0.028
k-means with Euclidian distance	618	0.0014
k-means with Manhattan distance	618	0.0014
k-means with Maximum distance	618	0.0014
k-means with Average distance	618	0.0014

The results obtained by k-means are given in Table 6.
We can observe that:

- The same average silhouette coefficient is obtained for all distance measures (0.0014);

- Using different values for Max Distance Percent (0.2, 0.6, 0.9) has not improved the results;
- The best average silhouette coefficient obtained for a cluster is 0.920.

5.4 MapReduce Speedup

In this section, we evaluate the performance of our proposed MapReduce-based K-means algorithm with respect to speedup. Experiments are run on a cluster of computers, where each node has two 2.8 GHz cores and 8 GB memory. Hadoop version 2.7.4 and Java 1.8 are used as the MapReduce system for all experiments. Linear speedup is difficult to achieve because the communication cost increases with the number of clusters becomes large. The ratio of map and reduce jobs is 2:1.

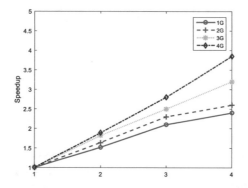

Fig. 2. Speedup of MapReduce-based K-means.

The speedup evaluation on datasets is performed on datasets with different sizes and systems. The number of computing nodes varied from 1 to 4. The size of the dataset increases from 1 GB to 8 GB. Figure 2 shows the speedup result for different datasets. Our MapReduce-based K-means has a good speedup performance. Speedup performance is better especially when the data size is large.

6 Conclusions

The paper investigates the role of various distance measures in k-means algorithm for clustering multidimensional time series data. Euclidian distance is the most frequent used and most common measure. Then, a MapReduce-based parallel K-means method is proposed as an improvement. On one hand, our experiments reveal that Manhattan distances (and sometimes the average distance) are better candidates for similarity between two multidimensional time series instances. On the other hand, MapReduce-based K-means has linear speedup with the number of map and reduce jobs.

Acknowledgments. This work was supported by the Science and Technology Program of University of Jinan (XKY1623 & XKY1734), the National Natural Science Foundation of China (61772231), the Shandong Provincial Natural Science Foundation (ZR2017MF025), the Shandong Provincial Key R&D Program of China (2015GGX106007), and the Project of Shandong Province Higher Educational Science and Technology Program (J16LN13).

References

1. Cui, X., Zhu, P., Yang, X., Li, K., Ji, C.: Optimized big data k-means clustering using mapreduce. J. Supercomput. **70**(3), 1249–1259 (2014)
2. Dean, J., Ghemawat, S.: Mapreduce: simplified data processing on large clusters. Commun. ACM **51**(1), 107–113 (2008)
3. Fu, T.C., Chung, F.L., Ng, V., Luk, R.: Pattern discovery from stock time series using self-organizing maps. In: Workshop Notes of KDD2001 Workshop on Temporal Data Mining, pp. 26–29 (2001)
4. Kalpakis, K., Gada, D., Puttagunta, V.: Distance measures for effective clustering of ARIMA time-series. In: Proceedings IEEE International Conference on Data Mining 2001, ICDM 2001, pp. 273–280. IEEE (2001)
5. Kumar, M., Patel, N.R., Woo, J.: Clustering seasonality patterns in the presence of errors. In: Proceedings of the Eighth ACM SIGKDD International Conference on Knowledge Discovery and Data Mining, pp. 557–563. ACM (2002)
6. Mehrabi, M.G., Kannatey-Asibu Jr., E.: Hidden Markov model-based tool wear monitoring in turning. J. Manufact. Sci. Eng. **124**(3), 651–658 (2002)
7. Möller-Levet, C.S., Klawonn, F., Cho, K.H., Wolkenhauer, O.: Fuzzy clustering of short time-series and unevenly distributed sampling points. In: International Symposium on Intelligent Data Analysis, pp. 330–340. Springer (2003)
8. Owsley, L.M., Atlas, L.E., Bernard, G.D.: Self-organizing feature maps and hidden Markov models for machine-tool monitoring. IEEE Trans. Signal Process. **45**(11), 2787–2798 (1997)
9. Shumway, R.H.: Time-frequency clustering and discriminant analysis. Stat. Probab. Lett. **63**(3), 307–314 (2003)
10. Vlachos, M., Lin, J., Keogh, E., Gunopulos, D.: A wavelet-based anytime algorithm for k-means clustering of time series. In: Proceedings of Workshop on Clustering High Dimensionality Data and its Applications, Citeseer (2003)
11. Wismüller, A., Lange, O., Dersch, D.R., Leinsinger, G.L., Hahn, K., Pütz, B., Auer, D.: Cluster analysis of biomedical image time-series. Int. J. Comput. Vis. **46**(2), 103–128 (2002)
12. Xiong, Y., Yeung, D.Y.: Mixtures of ARMA models for model-based time series clustering. In: 2002 IEEE International Conference on Data Mining 2002, ICDM 2003. Proceedings, pp. 717–720. IEEE (2002)
13. Zhao, W., Ma, H., He, Q.: Parallel k-means clustering based on mapreduce. In: IEEE International Conference on Cloud Computing, pp. 674–679. Springer (2009)

Mining Communities in Directed Networks: A Game Theoretic Approach

Annapurna Jonnalagadda$^{(\boxtimes)}$ and Lakshmanan Kuppusamy

School of Computer Science and Engineering, VIT University, Vellore 632014, India
{annapurna.j,klakshma}@vit.ac.in

Abstract. Detecting the communities in directed networks is a challenging task. Many of the existing community detection algorithm are designed to disclose the community structure for undirected networks. These algorithms can be applied to directed networks by transforming the directed networks to undirected. However, ignoring the direction of the links loses the information concealed along the link and end-up with imprecise community structure. In this paper, we retain the direction of the graph and propose a cooperative game in order to capture the interactions among the nodes of the network. We develop a greedy community detection algorithm to disclose the overlapping communities of the given directed network. Experimental evaluation on synthetic networks illustrates that the algorithm is able to disclose the correct number of communities with good community structure.

Keywords: Cooperative game · Directed networks · Game theory
Utility of node

1 Introduction

Many complex systems in nature can be visualized in the form of a network to understand and analyze the topology as well as functionality [4,10,20]. These real world networks follow power-law distribution, small world phenomenon and smaller average path length [20]. The interactions among the nodes of the network result in denser subgroups of nodes which are called communities. Even though detecting the communities is a challenging and complex problem, received attention from researchers from multiple disciplines due to its impact applications [2,4,10].

Most of the existing algorithms are able to disclose the community structure in the undirected networks [2,4]. One can extend these algorithms to directed networks either by ignoring the direction of the edge or by adding weight to the edge according to the prominence of direction of links [1,7–9,11,13–16,18]. The existing algorithms for community detection in directed networks looks for some predefined patterns such as cycles or partite graphs (bi/tri) or flow of information etc. But ignoring the direction of the link loses very important communication across the nodes of the network. For example, consider a network having 8 nodes

© Springer International Publishing AG, part of Springer Nature 2018
A. Abraham et al. (Eds.): ISDA 2017, AISC 736, pp. 826–835, 2018.
https://doi.org/10.1007/978-3-319-76348-4_79

as shown in Fig. 1a. There are two cycles {1, 2, 3, 4} and {5, 6, 7, 8} which can be claimed as communities as shown in Fig. 1b. But if the direction of the links is ignored, the corresponding undirected network is shown in Fig. 1c. This representation results in a grand coalition as shown in Fig. 1d. In spite of this, determining the overlapping communities is still an attractive challenge as the nodes of overlap plays an important role in promoting the communication across the communities [3,5,19,21].

The existing algorithm tries to optimize objective functions like modularity, density or demands the pre-requisite information on size of communities. In real world networks, no apriori information is available on community structure of the network. Hence, there is an urge for the community detection algorithms that discloses the overlapping structure of the network with out demanding for any prior information.

Assuming the nodes of the network as selfish players, Game theory can model the scenario of cooperative and competitive players [12,17]. In this paper, we propose a cooperative game called Directed Weighted Graph Game (DWGG) to model the interaction among the nodes of the network. We compute the weights of link using the common neighborhood information, popularity and reachability of source and target nodes. In due course of game play, the nodes forms the coalitions to maximize their utility. The proposed utility function captures the interest of the nodes in the given coalition. The stable coalitions are determined, which can are used as seed for disclosing the community structure of the network.

To the best of our knowledge, this is the first co-operative game framework to disclose the overlapping communities of the given directed network.

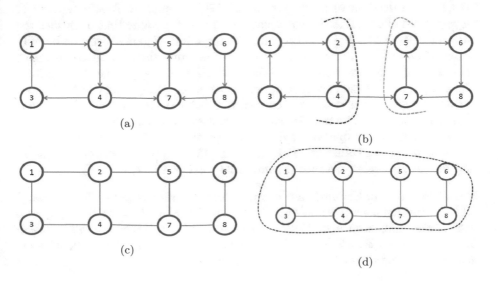

Fig. 1. Example network

2   Notations and Preliminaries

In this section, we discuss different notations used and the required preliminaries to understand the concepts.

Notations:

$d[i]$: *Degree of node i*

$d_i(s)$: *Degree of node i inweighted subgraph $S \in G_w$*

$d_i(G_w)$: *Degree of node i in weighted graph G_w*

$outd[i]$: *Out degree of node i*

$ind[i]$: *In degree of node i*

$N_{in}[i]$: *In neighbors of node i*

$N_{out}[i]$: *Out neighbors of node of out neighbors of nodes of c*

W_{ij} : *Weight of the link between i and j*

$wdg[i]$: *Number of non zero weighted edges incident on node i*

$c_i^{p\_merge}$: *Possible coalition that can be merged*

c^{best_poss} : *Best possible coalition that can be merged*

c_i^{merged} : *Merged coalition for c_i*

C^{poss} : *Stable coalition or possiblecommunity*

$G_{C^{poss}}$: *Subgraph formed by the nodes of C^{poss}*

$e_{i,j}$: *Edge between the pair of nodes i and j*

$COMM$: *Set of all communities*

Preliminaries: Game theory is an abstract mathematical framework that focuses on modelling the situations, in which the decision of one player can influence the decision of other players [12,17]. A game involves a number of players, a set of strategies for each player and a utility value that quantifies the outcome of each play of the game [12]. The discipline of game theory is broadly classified in to two categories: non-cooperative game theory and co-operative game theory. Non-cooperative game theory deals with players primitive actions. The well known solution concept of non-cooperative game theory is Nash equilibrium [12,17]. The primitives of cooperative game theory are the joint actions of group of players. The popular solution concepts of cooperative game theory are core, nucleolus, stable-sets and bargaining sets.

As we propose a transferable utility game (TU game), we restrict our discussion to TU games. For detailed preliminaries refer [4,12,17].

Definition 1. *Coalitional game with transferable utility (TU games)* *[12]: A coalitional form game on a finite set of players $N = \{1, 2, \ldots, n\}$ is a characteristic function v that assigns a number $v(S)$ to every possible coalition S. i.e., $v : 2^n \rightarrow \Re$. vS is called the worth of coalition S. $v(\varphi) = 0$, where φ denotes the empty coalition.*

Definition 2. *Payoff allocation*: *A payoff allocation is any vector $u = (u_i)_{i \in N}$ in $\Re^N$, where u_i represents the utility payoff to player i.*

Definition 3. Feasible allocation: *An allocation x is said to be a* **feasible allocation** *for a coalition S iff $\sum_{i \in S} x_i \leq v(S)$. A coalition S is said to improve on an allocation y iff $v(S) \geq \sum_{i \in S} y_i$.*

Definition 4. Core: *An allocation x is in Core of v iff x is feasible and no coalition can improve on x. i.e., x is in the core iff $\sum_{i \in N} x_i = v(N)$ and $\sum_{i \in S} x_i \geq v(S), \forall S \subseteq N$.*

Thus, if the feasible coalition is not in the core, then the coalition S may be inconsistent. The core of a coalition game is not guaranteed, it may be empty or quite large.

Definition 5. Monotonic game: *A game is said to be Monotone if $\forall S \subseteq T \subseteq N$, $v(S) \leq v(T)$.*

That is, the worth of a larger coalition is always greater than the worth of its subset coalitions.

For TU games, the notion of convex games introduced by Shapley provides a natural way to formalize the distribution of payoff's among the players [12,17]. Convex games capture the intuition that the benefit for joining the coalition increases as the coalition grows.

Definition 6. Convex game: *A game is Convex if $\forall S, T \subseteq N : v(S \cup T) + v(S \cap T) \geq v(S) + v(T)$.*

Equivalently, a game is convex if $v(S \cup \{i\}) - v(S) \leq v(T \cap \{i\}) - v(T)$ whenever $S \subseteq T$ and $i \notin T$. In other words, the marginal value that the player i adds to a coalition S is no greater than the marginal value i adds to a coalition $T \supseteq S$. The following section discusses the proposed game framework and the greedy algorithm to reveal the underlying community structure of the network.

3 Proposed Work

Let $G = (v, E)$ be the given directed network, where V denote the nodes of the network and E represents the edges/links between the nodes of the network. Let $A = a_{ij}$ be the adjacency matrix of G, where $a_{ij} = 1$ if there is an edge between the pair of nodes i and j otherwise $a_{ij} = 0$. As G is the directed network, the adjacency matrix need not be a symmetric matrix. Let $d[i]$ be the total degree of the node i which is the sum of the in degree ($ind[i]$) and out degree ($out[i]$) of node i. Let W_{ij} denote the weight of the edge between a pair of nodes i and j. We compute the distinct edge weights depending on the connectivity of neighbors, in degree, out degree and reachability of the pair of nodes using the Eq. 1.

$$W_{ij} = \begin{cases} 0 & \text{If } d[i] = 0 \text{ or } d[j] = 0 \\ \dfrac{|N_{in}[i] \cap N_{out}[i]|}{|N_{in}[i] \cup N_{out}[i]|} & \text{If } j \notin N_{out}[i] \\ \dfrac{|N_{in}[i] \cap N_{out}[i]|}{|N_{in}[i] \cup N_{out}[i]|} + \dfrac{1}{outd[i]} + \dfrac{1}{ind[j]} & \text{if there is a directed edge} \\ & \text{between node i and node j} \end{cases} \quad (1)$$

The function to compute edge weights (Eq. 1 assigns different weights to the edges between the pair of nodes i and j. If i and j are isolated nodes, then the weight of the edge will be 0. According to Eq. 1, the pair of nodes that are connected through a directed edge has been given the utmost weight. Even though node i and node j are not directly connected, we augment an edge whose weight is computed based on their number of common neighbors. Let G_w be the augmented weighted directed graph. Now, we define a cooperative game called Directed weighted Graph Game (DWGG) to model the interactions among the nodes of the given social network.

Definition 7. *Directed weighted Graph Game (DWGG):* *A cooperative game defined over weighted directed graph $G_w = (N, \Upsilon)$ is said to be is said to be Directed Weighted Graph Game, where N represents the nodes of the given social network and Υ is the characteristic function which specifies the worth of any coalition $S \subset N$.*

The characteristic function for any coalition $S \subset N$ is defined as follows:

$$\Upsilon(S) = \begin{cases} 0 & \text{if } |S| < 2 \\ \displaystyle\sum_{i,j \in S} W_{i,j} & \text{if } |S| \geq 2 \end{cases} \tag{2}$$

The nodes of the social network are allowed to play the cooperative game DWGG. In due course of their play, the nodes tend to form coalitions. The utility of a node i in a coalition S is given as $\psi_i(S) = \displaystyle\sum_{j \in S-i} W_{ij}$. The proposed game satisfies monotonicity and super additive properties. Hence, it turns to be a convex game. When the game reaches the state of equilibrium, we call the set of coalitions as stable coalitions or possible communities. We define these stable coalitions or possible communities as follows.

Definition 8. *Stable coalitions or possible communities:* *A coalition $S \subset N$ is said to be possible community if all the members of coalition S are fully interested to be members of the coalition S. i.e., all the members are satisfied with their received utility. $\forall i \in S, \psi_i(S) \geq \delta * \psi_{max}(i)$, where $\delta = \dfrac{d_i(S)}{d_i(G_w)}$.*

The set of possible communities undergoes a merging procedure in order to derive the community structure of the given social network. we define the community as follows:

Definition 9. *Community:* *A community S is a stable coalition, if S cannot be merged with any other stable coalition S'.*

Here, we propose a greedy algorithm called Directed Community Detection Algorithm (DCDA) that discloses the underlying community structure of the given network. The algorithm works as follows: (1) The *seedlist* is generated by computing the maximum possible coalition for each node $i \in N$. (2) The *seedlist*

Algorithm 1. DCDA_Community_Detection

1: **procedure** COMMUNITY˙DETECTION(G') ▷ The weighted graph
2:    $for\ each i \in N$
3:        $seedlist \leftarrow \{c = i, j | c\ commits\ maximum\ worth\ for\ i\}$
4:    **while** $seedlist is not empty$ **do**
                                              ▷ repeat the following till the *seedlist* is not empty

5:
6:        $\{$
7:        $choose\ c \in seedlist\ such\ that\ c\ has\ maximum\ worth$
8:        $repeat$
9:        $c^{expand} = c \cup N_{out}[c]$
10:        $\forall i \in c^{expand}$
11:        **if** $\psi_i(expand) \geq \delta * \psi_{max}[i]$ **then**
12:            $C^{poss} = c^{poss} \cup c^{expand}$
13:        **else**
14:            $repeat$
15:            $\{$
16:            $C^{uIn} = \{set\ of\ all\ nodes\ \psi_i(expand) < \delta * \psi_{max}[i]$
17:            $C^{uIn} = \{set\ of\ all\ nodes\ \psi_i(expand) < \delta * \psi_{max}[i]$
18:            $c^{poss} = c^{expand} - c^{uIn}$
19:            $\}\ until\ no\ node\ has\ utility\ value\ less than\ \delta * \psi_{max}[i]$
20:        $seedlist = seedlist - set of edges of C^{poss}$
21:        $\forall e_{i,j} \in G_{C^{poss}}, W_{ij} = 0$
22:        $\}$
23:    $repeat$
24:    $\{$
25:    **for** $c_i \in C^{poss}$ **do**
26:        $\{$
27:        **for** $c_j \in C^{poss} - c_i$ **do**
28:            $\{$
29:            **if** $\frac{|c_i \cap c_j|}{min(|c_i|,|c_j|)} \geq 0.3$ **then**
30:                $c_i^{p\_merge} = c_i \cup c_j$

31:            $\}$
32:            $c^{best\_poss} = \{c_i^{p\_merge} | max.\ no.\ of\ nodes\ in\ c_i, c_j\ improve\ their\ utility\ value\}$
33:        $\}$
34:        $c_i^{merged} = c_i \cup c_i^{bestposs}$
35:        $COMM = COMM \cup \{c_i^{merged}\}$

36:    $\}$
37:    $until\ no\ two\ coalitions\ can\ be\ merged$
38:    **return** $COMM$ ▷ returns the communities of the network

---

is sorted according to the worth of coalitions in ascending order. (3) choose the maximum worth coalition from the *seedlist* and expand it by adding its neighbors. (4) The expanded coalition is checked for the interest of the individuals in the coalition. (5) If no node is willing to deviate from the coalition in order to improve its utility, then the coalition is listed as stable coalition otherwise the non-interested nodes are removed from the coalition and the resultant coalition again checked for its stability. (6) After finding the stable coalition, the *seedlist* is updated by removing edges belonging to the coalition and also the weighted graph is updated by assigning weights of edges as 0.6. Then, the next maximum worth coalition is chosen and the process of expansion and update is repeated.

This procedure is repeated until the *seedlist* becomes empty. (7) The set of all the stable coalitions undergoes a voting based merge procedure. (8) When no more merges are possible, we disclose the final community structure of the network. The detailed algorithm is depicted in Algorithm 1.

The following section illustrates the performance of the proposed algorithm on LFR benchmark networks.

4   Results and Discussion

In this section, we analyze the performance of DCDA on LFR benchmark networks [6]. we generate the LFR networks with the following parameter settings: Number of nodes $N = 1000$, Average in-degree $k = 15$, maximum in-degree $maxk = 50$, minus exponent for degree distribution $\tau_1 = 2$, minus exponent for community size distribution $\tau_2 = 1$, overlapping membership $om = 2$, fraction of overlapping nodes $on = 0, 0.05, 0.1, 0.15, 0.2, 0.25, 0.3$ and the mixing parameter $\mu = o.1, 0.3$. The minimum $(minc)$ and maximum $maxc$ community sizes are taken as follows: (minc, maxc) = (5, 30), (10, 50), (20, 50), (20, 100). The metrics such as Normalized Mutual Information (NMI) and Omega index are used to measure the quality of communities detected by DCDA on the LFR networks with ground truth communities.

Figure 2a and b depicts the NMI obtained for different fraction of overlapping nodes of LFR networks with $\mu = 0.1$ and $\mu = 0.3$ respectively. The X-axis represents the fraction of overlapping nodes and the Y-axis represents the corresponding NMI. Similarly, Fig. 3a and b depicts the Omega index obtained for different fraction of overlapping nodes of LFR networks with $\mu = 0.1$ and $\mu = 0.3$ respectively. For $\mu = 0.1$, the average value of NMI and Omega index over different fraction of overlapping nodes for different (minc, maxc) are 98%, 96%, 95% and 93% respectively. For $\mu = 0.1$, the average value of NMI and Omega index for different (minc, maxc) are 93%, 89%, 87% and 80%. Even though DCDA is able to detect good community structure for both $\mu = 0.1$ and $\mu = 0.3$, the

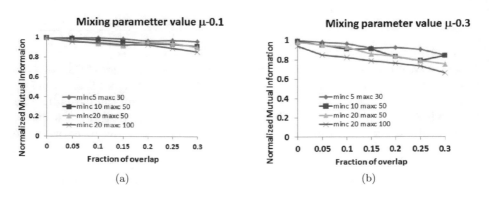

Fig. 2. Performance of DWGCG

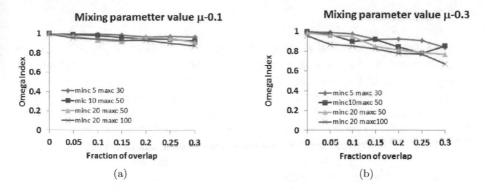

(a)   (b)

Fig. 3. Performance of DWGCG

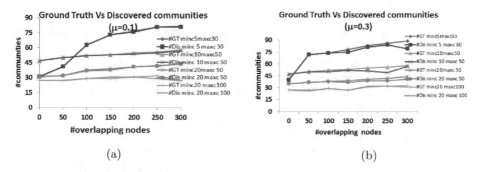

(a)   (b)

Fig. 4. Performance of DWGCG in disclosing the number of communities

algorithm is performing very well for $\mu = 0.1$. DCDA is able to detect the good community structure even for the large community sizes.

According to our claim, DCDA does not require any prior information on number of communities. To justify our claim, we further analyze the number of communities detected by DCDA and number of ground truth communities. Figure 4a and b illustrates the comparison of actual number of communities and the number of communities detected by DCDA over different values of (minc, maxc) for $\mu = 0.1$ and $\mu = 0.3$ respectively. The x-axis represents the number of overlapping nodes and the Y-axis represents the number of communities. Figure 4a demonstrates that DCDA is able to detect the exact number of communities of the given network for $\mu = 0.1$. Even for $\mu = 0.3$, DCDA is able to detect the correct number of communities till the number of overlapping nodes is 200. When the number of overlapping nodes is between 200 and 300, there is slight variation in the number of detected communities. This analysis evidences that DCDA mechanism is able detect the correct number of communities with out any apriori information.

5 Conclusions

In this paper, we proposed Directed Weighted Graph Game (DWGG) in order to capture the interactions among the nodes of the network. We developed a greedy community detection algorithm (DCDA) in order to disclose the underlying community structure of the given network. The evaluation on LFR benchmark networks demonstrates that the algorithm is able to detect the good overlapping community structure of the given network. In addition to this, DCDA is able disclose the correct number of communities with out providing any apriori information. One future work is to analyze the performance of DCDA on large synthetic and real world networks.

References

1. Dugué, N., Perez, A.: Directed Louvain: maximizing modularity in directed networks. Ph.D. thesis, Université d'Orléans (2015)
2. Fortunato, S.: Community detection in graphs. Phys. Rep. **486**(3), 75–174 (2010)
3. Havens, T.C., Bezdek, J.C., Leckie, C., Ramamohanarao, K., Palaniswami, M.: A soft modularity function for detecting fuzzy communities in social networks. IEEE Trans. Fuzzy Syst. **21**(6), 1170–1175 (2013)
4. Jonnalagadda, A., Kuppusamy, L.: A survey on game theoretic models for community detection in social networks. Soc. Netw. Anal. Min. **6**(1), 83 (2016)
5. Jonnalagadda, A., Kuppusamy, L.: A cooperative game framework for detecting overlapping communities in social networks. Phys. A Stat. Mech. Appl. (2017)
6. Lancichinetti, A., Fortunato, S.: Benchmarks for testing community detection algorithms on directed and weighted graphs with overlapping communities. Phys. Rev. E **80**(1), 016118 (2009)
7. Leicht, E.A., Newman, M.E.: Community structure in directed networks. Phys. Rev. Lett. **100**(11), 118703 (2008)
8. Levorato, V., Petermann, C.: Detection of communities in directed networks based on strongly p-connected components. In: 2011 International Conference on Computational Aspects of Social Networks (CASoN), pp. 211–216. IEEE (2011)
9. Long, H., Li, B.: Overlapping community identification algorithm in directed network. Procedia Comput. Sci. **107**, 527–532 (2017)
10. Malliaros, F.D., Vazirgiannis, M.: Clustering and community detection in directed networks: a survey. Phys. Rep. **533**(4), 95–142 (2013)
11. Mathias, S.B., Rosset, V., Nascimento, M.C.: Community detection by consensus genetic-based algorithm for directed networks. Procedia Comput. Sci. **96**, 90–99 (2016)
12. Myerson, R.B.: Game Theory. Harvard University Press, Cambridge (2013)
13. Nicosia, V., Mangioni, G., Carchiolo, V., Malgeri, M.: Extending the definition of modularity to directed graphs with overlapping communities. J. Stat. Mech. Theory Exp. **2009**(03), P03024 (2009)
14. Ning, X., Liu, Z., Zhang, S.: Local community extraction in directed networks. Phys. A Stat. Mech. Appl. **452**, 258–265 (2016)
15. Palla, G., Ábel, D., Farkas, I.J., Pollner, P., Derényi, I., Vicsek, T.: k-clique percolation and clustering. In: Handbook of Large-Scale Random Networks, pp. 369–408. Springer (2008)

16. Santos, C.P., Carvalho, D.M., Nascimento, M.C.: A consensus graph clustering algorithm for directed networks. Expert Syst. Appl. **54**, 121–135 (2016)
17. Shoham, Y., Leyton-Brown, K.: Multiagent Systems: Algorithmic, Game-Theoretic, and Logical Foundations. Cambridge University Press, Cambridge (2008)
18. Sun, P.G., Gao, L.: A framework of mapping undirected to directed graphs for community detection. Inf. Sci. **298**, 330–343 (2015)
19. Wang, Q., Fleury, E.: Overlapping community structure and modular overlaps in complex networks. In: Mining Social Networks and Security Informatics, pp. 15–40. Springer (2013)
20. Wasserman, S., Faust, K.: Social network analysis: methods and applications (1995)
21. Zhou, L., Lü, K., Yang, P., Wang, L., Kong, B.: An approach for overlapping and hierarchical community detection in social networks based on coalition formation game theory. Expert Syst. Appl. **42**(24), 9634–9646 (2015)

A Support Vector Machine Based Approach to Real Time Fault Signal Classification for High Speed BLDC Motor

Tribeni Prasad Banerjee[1] and Ajith Abraham[2(⊠)]

[1] B. C. Roy Engineering College, Durgapur 713209, India
tribeniprasad.banerjee@bcrec.ac.in
[2] Machine Intelligence Research Labs (MIR Labs), Scientific Network
for Innovation and Research Excellence, Auburn, Washington 98071, USA
ajith.abraham@ieee.org

Abstract. In this paper we propose a new methodology for designing an intelligent incipient fault signal classifier. This classifier can classify the fault signal. The design has been validated to a sate observer which indicates the valve controller output signal and communicate the health status of the embedded processor based valve controller in right time without any false alert signal to the actuator through FPGA processor. This has been achieved by using an SVM-based classifier and time duration based state machine modeling. The design methodology of a fault aware controller using one against all strategy is selected for classification tool due to good generalization properties. Performance of the proposed system is validated by applying the system to induction motor faults diagnosis. Experimental result for BLDC motor (which is mostly used for aircraft) valve controller, and computer simulations indicate that the proposed scheme for intelligent control based on signal classification is simple and robust, with good accuracy.

Keywords: Cyber physical system · Short Term Fourier Transform (STFT)
Support Vector Machine (SVM) · Intelligent control · Fault signal classifier

1 Introduction

Induction motor is a critical component in many industrial processes and it is very important part to support industry in producing the product. It is also frequently integrated with any commercially available equipment and the process itself. Fault diagnosis of induction motor is very important for save operation and preventing rescue [1]. It is preferable to find fault before complete motor failure. This is called incipient fault detection. Often the motor can run with incipient fault but it will lead to motor catastrophic failure causing downtime and large losses.

Fault diagnostics is a kind of fault classification as far as its essence is concerned. Support Vector Machines (SVM) has a good generalization capability even in the small-sample cases of classification and has been successfully applied in fault detection and diagnosis [2–4, 9]. Support vector machines (SVMs) [5–7] provide efficient and powerful classification algorithms that are capable of dealing with high-dimensional

© Springer International Publishing AG, part of Springer Nature 2018
A. Abraham et al. (Eds.): ISDA 2017, AISC 736, pp. 836–845, 2018.
https://doi.org/10.1007/978-3-319-76348-4_80

input features and with theoretical bounds on the generalization error and sparseness of the solution provided by statistical learning theory [5, 8]. Classifiers based on SVMs have few free parameters requiring tuning, are simple to implement, and are trained through optimization of a convex quadratic cost function, which ensures the uniqueness of the SVM solution. Furthermore, SVM-based solutions are sparse in the training data and are defined only by the most "informative" training points.

In control and automata theory, a state is defined as a condition that characterizes a prescribed relationship of input, output and input to the next state. Thus knowledge of the state of a system gives better understanding through observation and we can subsequently take necessary steps for controlling the system, for example, stabilizing a system using state feedback. This implies that the current state of a system determines the future plan of action. So it becomes extremely important to devise some methodology by which the state of a system can be observed correctly within a stipulated time frame. For example a particular state of a feedback control system cannot be observed directly due to its own compensator which is one of the limitations of the feedback controller. A separate hardware with very low observation latency period [17] is necessary to observe this unobserved state which is called a sub state or internal state directly related to the health of the system. A sub state observer generally models a real time system in order to estimate its internal state, from the input and output words of the system. The disadvantage of the microcontroller based stitch controlling mechanism is that it does not take into account the possibility of fault under degraded condition of a component (qualified through its sub state) during its operation i.e. it is not fault aware although it is flexible in nature. In order to make system fault aware a sub state observer is conceived whose execution time of the event sequences (necessary for sub state observation) is much faster and due to this high speed execution and subsequent scheduling technique, it can observe the sub state behavior of a system and thus differentiate between two almost similar states with different end effects.

In this paper, we proposed a incipient signal classification which is optimally separated by SVM and after classification of the signal or a optimal threshold value a controller strategy is implanted in a embedded processor which takes the finite or real time decision make it hard and real time responsive or responsive system.

2 Principal of Operation Signal Classification

The basic scheme of Signal classification has been shown in Fig. 1(a) and (b).

2.1 Hard Real Time Digital Signal Processing

Hard real time system is that system which has a critical time limit [20]. Within that time duration the system has to response otherwise the system is going to a catastrophic loss or high probability for human death. The real time systems are basically two types one is reactive system and another is embedded system. A reactive system is always react with the environment (online aircraft valve signal monitoring from a actuator signal) and another is embedded system which is used to control specialized hardware that is installed within a larger system (such as a microprocessor that controls anti-lock brakes

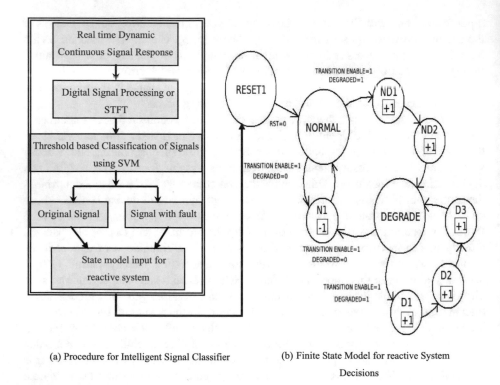

(a) Procedure for Intelligent Signal Classifier

(b) Finite State Model for reactive System

Decisions

Fig. 1. A generic scheme for signal classification and reactive system and their consequences

in an automobile) In our system is more reactive with the environments. Here in our proposed system the online dynamic continuous signals is comes and we capture the signal and transform it and pass through into our classifier and take the decision. The Extended Finite State Machine (EFSM) model [17] does the real time operation within a few micro seconds. This makes the system first responsive according to the decision of the output and the reactive system makes a more intelligent and adaptable system.

2.2   Short Time Fourier Transform (STFT)

Signal classification is a major research area in the real time DSP, because of the real time operation the classification of signal is very crucial issue. Even the signal classification has lots of problem, so we prefer the Short-Time Fourier Transform (STFT). Here we briefly review the chaos detector based on the STFT proposed in [31]. The STFT is defined as [32] The STFT, or alternatively short-term Fourier transform, is a Fourier-related transform used to determine the sinusoidal frequency and phase content of local sections of a signal as it changes over time.

$$STFT(t,f) = \int_{-\infty}^{\infty} x(t+\tau)w(\tau)e^{-j2\pi f\tau}d\tau \tag{1}$$

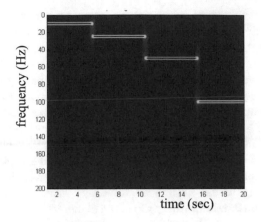

Fig. 2. Original signal after the STFT

Where, x(t) is signal of interest (in this paper, it is a voltage or a current from the BLDC motor), and $w(\tau)$ is the window function, where $w(\tau) = 0$ for $|\tau| \succ T/2$ and T is window width. In order to avoid complex-valued STFT, we use its squared magnitude, i.e., the spectrogram $SPEC(t,f) = |STFT(t,f)|^2$.

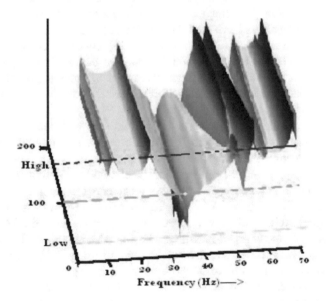

Fig. 3. Original Signal and the fault signal are characterized using the threshold based time duration model

2.3 Classification of Signal Using SVM

Support vector machines (SVM) [11] have been widely used in the fault classification and diagnosis of machines in the past few years, especially in the fault diagnosis [18], for dynamic procedures such as the starting, the stopping, and the changing of working mode. SVM certainly outperform some other artificial intelligence methods because it is able to maximize the generalization performance of a trained classifier. This may not be easily achieved by using HMM or ANN, because ANN are prone to having a higher specificity and a lower sensitivity and HMM are prone to having a higher sensitivity and a lower specificity. SVM aims at the optimal solution in the available information rather than just the optimal solution when the sample number tends towards infinitely large. It has a good generalization when the samples are few, so it is especially fit for classification, forecasting and estimation in small-sample cases such as fault diagnosis, in most cases whose bottleneck problem is the lack of fault samples.

Once the frequency is separated then the signals are taken into the classification stage, here according to the maximized threshold value are calculated by SVM. Because that optimum threshold value is to given to the state machine for taken into the real time decision processes. Depending upon the magnitude value the threshold values are set to a specific signal and the different signal magnitude are set to a different class. A signal which is exceed a specific threshold value it is classified as high risk and other which are not exceed is fallen into other class. This threshold value is set dynamically and depending upon the timed model components is already classified form the inputs.

Support vector machine (SVM), developed by Vapnik [11, 21], as well as neural networks, have been extensively employed to solve classification problems due to the excellent generalization performance on a wide range problem. In machine fault diagnosis, some researchers have employed SVM as a tool for classification of faults. For example, ball bearing faults [12], gear faults [13], condition classification of small reciprocating compressor [14], cavitation detection of butterfly valve [15] and induction motor [16]. In this paper, we proposed a incipient faults signal classification in dynamic signal diagnosis system of induction motor based on start-up transient current signal using component analysis and SVM.

In SVM, the original input space is mapped into a high-dimensional dot product space called a feature space, and in the feature space, the optimal hyper plane is determined to maximize the generalization ability of the classifier. The optimal hyper plane is found by exploiting the optimization theory, and respecting insights provided by the statistical learning theory. SVMs have the potential to handle very large feature spaces, because the training of SVM is carried out so that the dimension of classified vectors does not have as a distinct influence on the performance of SVM as it has on the performance of conventional classifiers. That is why it is noticed to be especially efficient in large classification problems. This will also benefit in fault classification, because the number of features to be the basis of fault diagnosis may not have to be limited. Also, SVM-based classifiers are claimed to have good generalization properties compared to conventional classifiers, because in training the SVM classifier, the so-called structural misclassification risk is to be minimized, whereas traditional classifiers are usually trained so that the empirical risk is minimized. When a signal falls outside the clusters, it is tagged as a potential motor failure. Since a fault condition is not a spurious event it can degrade the system, the postprocessor alarms the user only after multiple indications of a potential failure have occurred. In this way, the time

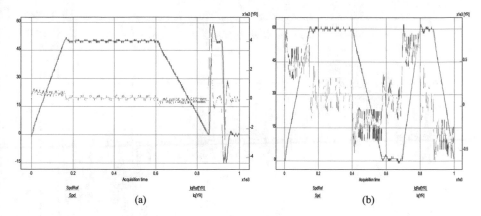

Fig. 4. Mixed perturbed signal output due to injected fault signal with the original signal. In (a) and (b) blue signal is the motor trajectory path signal red is the noise

duration modeling of the State machine is incorporated into the monitoring system and protects the reactive system by alarming on random signals that have been identified (Fig. 4).

2.4 Finite State Model for Reactive System

The implementation of the concept was performed on an Altera Max II CPLD [19] based board. The design part consist of two stages, one that qualifies the stitch i.e. implementation of the timed safety automata and second that takes the actual decision based on the parameter generated using the algorithm developed and generates proper alarm signal. The CPLD based implementation of the design has been named as Faultdec. The first stage uses an EFSM to implement the timed safety automata. The threshold for the signal classifier has taken from the SVM output, because the SVM output gives us the optimized threshold value which is can be e found out. The output was optical interrupter used as the trigger. The output of this stage is a signal that indicates whether the state is normal or degraded described as the count enable signal for the counter used in the first stage. The second stage implements the degradation number generation algorithm. This stage also uses an EFSM (Fig. 6) which has the count enable signal from the first stage as trigger. The output of this stage is the state ratio number that is used to generate the alarm by comparing it with the critical value.

3 Experimental Set up and System Description

3.1 Hardware Design Environment of the Signal Classifier

The data acquisition and the simulation we use standard PC with XP, Pentium I3 Processor, 4 Gb RAM, and integrated with LabVIEW software of National Instrument and there data and signal acquisition system coupled with Texas Instrument Embedded DSP kit TMS320F2812. TMS320F2812 is the 32 bit Fixed-point DSP of Harvard

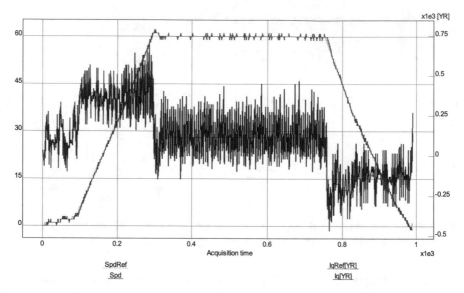

Fig. 5. Original Signal and the fault signal are classified by our proposed STFT and SVM based system

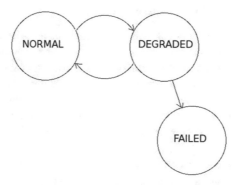

Fig. 6. State model of signal condition monitoring

Bus Architecture which has separated Data Bus and Program Bus. It also supports atomic operations, 32×32 bit MAC operations, fast interrupt response. It is able to control the motor easily with TMS320F2812 because it has two event managers which include capture function, PWM function, and QEP (Quadrature-Encoder Pulse) function Since it also supports variable serial port peripherals such as CAN, SCI, and SPI [26, 27], it is possible to communicate with external device and exchange control signal and data. The prototype is designed with a control circuit based on the "MSK2812 Kit C Pro VS" (from Technosoft S.A.) [22]. Operating at a 150-MHz frequency, with double event-manager signals, increased internal memory, etc., this new kit offers all features needed for an advanced digital control implementation. All communication between the PC and the DSP board is done through the RS-232 interface using a real-time serial communication monitor with download/upload functions, debug and inspect facilities [23]. PROCEV28x, a graphical evaluation/analysis software for the specific peripherals embedded in the TMS320F2812 DSP, is provided to rapidly familiarize the designer with these functions (Figs. 7 and 8).

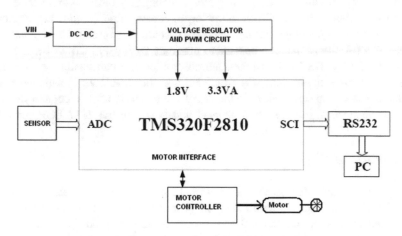

Fig. 7. Hardware configuration BLDC motor with Embedded DSP

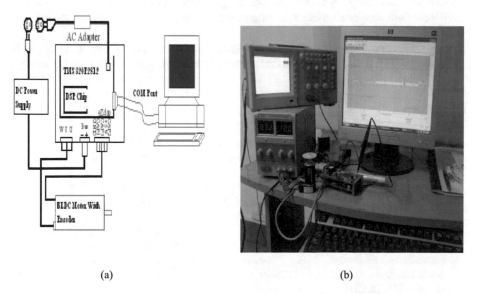

(a)   (b)

Fig. 8. (a) Hardware configuration in block diagram with the Texas DSP chip with BLDC motor. (b) Actual hardware configuration in with the Texas DSP chip with BLDC motor

3.2   Software Design of the Incipient Signal Classifier

After the classifier the motion controller gives the actual output to the system which makes the systems response smooth and Intelligent. The state model of Incipient fault classifier is show in DMCD28x Lite, the Digital Motion Control Development platform offered by Technosoft S.A. integrates a debugger, basic assembler, linker, C compiler and many other facilities, allowing to create, modify and test applications within a highly organized project management system. It is offered also a collection of simulink control blocks that can be used to program Technosoft Control Kits based on the TMS320F2812 DSP [24–26]. So, the control system can easily be simulated in simulink, and once the system was simulated and the developer is satisfied with the expected behavior, he can proceed to the next level generate C/C++ code or the control blocks of the system, in order to implement and test it on the 2812 DSP controller.

4   Conclusion

The experiments are done in the laboratory. In our proposed system the intelligent signal classifier classifies the required signal (blue colored) from the signal as received from the motor sensory data (red colored) shown in the Fig. 5. At the initial state of the system the classifier separate the signal by STFT methods, and the duration based model as shown in Figs. 2 and 3 after that the SVM classified the signal extremely smooth and accurately. We assumed that the proposed system has possible to serve an intelligent fault diagnosis system in other hard real time system applications also.

Acknowledgement. The author is thankful to Embedded System Lab of CMERI, Durgapur for giving support to his work in experimental setup and funding support to continue his research.

References

1. Yongming, Y., Bin, W.: A review on induction motor online fault diagnosis. In: The 3rd International Power Electronic and Motion Control Conference 2000 (IEEE IPEMC), pp. 1353–1358 (2000)
2. Matthias, P., Stefan, O., Manfred, G.: Support vector approaches for engine knock detection. In: International Joint Conference on Neural Networks. IEEE Press, Washington, pp. 969–974 (1999)
3. Chiang, L.H., Kotanchek, M.E., Kordon, A.K.: Fault diagnosis based on fisher discriminant analysis and support vector machines. Comput. Chem. Eng. **28**, 1389–1401 (2004)
4. Samanta, B., Al-Balushi, K.R., Al-Araimi, S.A.: Artificial neural networks and support vector machines with genetic algorithm for bearing fault detection. Eng. Appl. Artif. Intell. **16**, 657–665 (2003)
5. Schölkopf, B., Smola, A.: Learning With Kernels. MIT Press, Cambridge, MA (2002)
6. Herbrich, R.: Learning Kernel Classifiers: Theory and Algorithms. MIT Press, Cambridge, MA (2002)
7. Cristianini, N., Shawe-Taylor, J.: An Introduction to Support Vector Machines. Cambridge University Press, Cambridge, UK (2000)

8. Sebald, D.J., Bucklew, J.A.: Support vector machine techniques for nonlinear equalization. IEEE Trans. Signal Process. **48**(11), 3217–3226 (2000)

9. Jack, L.B., Nandi, A.K.: Support vector machines for detection and characterization of rolling element bearing faults. J. Mech. Eng. Sci. **9**, 1065–1074 (2001)

10. Bengtsson, J., Yi, W.: Timed automata: semantics, algorithms and tools. In: Lectures on Concurrency and Petri Nets, pp. 87–124. Springer, Heidelberg (2004).

11. Vapnik, V.N.: The Nature of Statistical Learning Theory. Springer, New York (1999)

12. Jack, L.B., Nandi, A.K.: Fault detection using support vector machines and artificial neural networks, augmented by genetic algorithms. Mech. Syst. Signal Process. **16**(2), 373–390 (2002)

13. Samanta, B.: Gear fault detection using artificial neural networks and support vector machines with genetic algorithms. Mech. Syst. Signal Process. **18**(3), 625–644 (2004)

14. Yang, B.-S., et al.: Condition classification of small reciprocating compressor for refrigerators using artificial neural networks and support vector machines. Mech. Syst. Signal Process. **19**(2), 371–390 (2005)

15. Yang, B.-S., et al.: Cavitation detection of butterfly valve using support vector machines. J. Sound vib. **287**(1), 25–43 (2005)

16. Yang, B.-S., Han, T., Hwang, W.-W.: Fault diagnosis of rotating machinery based on multi-class support vector machines. J. Mech. Sci. Technol. **19**(3), 846–859 (2005)

17. Alur, R., Dill, D.: The Theory of Timed Automata. Theoret. Comput. Sci. **126**, 183–235 (1994)

18. Ma, X.-x., Huang, X.-y., Chai, Y.: 2PTMC classification algorithm based on support vector machines and its application to fault diagnosis. Control Decis. **18**(3), 272–276 (2003)

19. Getting Started User Guide, MAX II Development Kit, Altera, USA, 2005. https://www.altera.com/content/dam/altera-www/global/en_US/pdfs/literature/ug/ug_max_ii_devel_kit.pdf

20. Mall, R.: Real Time Systems Theory and Practice. Pearson Publication, India (2007)

21. Matthias, P., Stefan, O., Manfred, G.: Support vector approaches for engine knock detection, In: International Joint Conference on Neural Networks. IEEE Press, Washington, pp. 969–974 (1999)

22. Texas Instruments: TMS320F2812 Digital Signal Processors Data Manual (2005)

23. Texas Instruments: TMS320F28x DSP System Control and Interrupts Reference Guide (2005)

24. Texas Instruments: TMS320F28x DSP Event Manager (EV) Reference Guide (2004)

25. Texas Instruments: TMS320F28x DSP External Interface (XINTF) Reference Guide (2004)

26. Texas Instruments: Running an Application from Internal Flash Memory on the TMS320F28xx DSP (2005)

27. Texas Instruments: TMS320F28x DSP Serial Communication Interface (SCI) Reference Guide (2004)

28. Texas Instruments: TMS320F28x DSP Enhanced Controller Area Network (eCAN) Reference Guide (2005)

29. Texas Instruments: TMS320F28x Analog-to-Digital Converter (ADC) Reference Guide (2002)

30. Simeu-Abazi, Z., Bouredji, Z.: Monitoring and predictive maintenance: Modeling and analyse of fault latency. Comput. Ind. **57**(6), 504–515 (2006)

31. Bouyer, P.: Timed Automata-From Theory to Implementation LSV- CNRS & ENS de Cachan France (2003)

32. Rubežić, V., Djurović, I., Daković, M.: Time–frequency representations-based detector of chaos in oscillatory circuits. Signal Process. **86**(9), 2255–2270 (2006)

33. Boashash, B. (ed.): Time frequency Signal Analysis and Applications. Elsevier, Amsterdam (2003)

Automatic Identification of Malaria Using Image Processing and Artificial Neural Network

Mahendra Kanojia[1]([⊠]), Niketa Gandhi[2], Leisa J. Armstrong[3], and Pranali Pednekar[4]

[1] JJT University, Jhunjhunu, Rajasthan, India
kgkmahendra@gmail.com
[2] Machine Intelligence Research Labs (MIR Labs), Auburn, Washington, USA
niketa@gmail.com
[3] Edith Cowan University, Perth, Western Australia, Australia
l.armstrong@ecu.edu.au
[4] Sathaye College, University of Mumbai, Mumbai, Maharashtra, India
panupednekar@gmail.com

Abstract. Malaria is a mosquito-borne infectious disease, which is diagnosed by visual microscopic assessment of Giemsa stained blood smears. Manual detection of malaria is very time consuming and inefficient. The automation of the detection of malarial cells would be very beneficial in the treatment of patients. This paper investigates the possibility of developing automatic malarial diagnosis process through the development of a Graphical User Interface (GUI) based detection system. The detection system carries out segmentation of red blood cells (RBC) and creates a database of these RBC sample images. The GUI based system extracts features from smear image which were used to execute a segmentation method for a particular blood smear image. The segmentation technique proposed in this paper is based on the processing of a threshold binary image. Watershed threshold transformation was used as a principal method to separate cell compounds. The approach described in this study was found to give satisfactory results for smear images with various qualitative characteristics. Some problems were noted with the segmentation process with some smear images showing over or under segmentation of cells. The paper also describes the feature extraction technique that was used to determine the important features from the RBC smear images. These features were used to differentiate between malaria infected and normal red blood cells. A set of features were proposed based on shape, intensity, contrast and texture. These features were used for input to a neural network for identification. The results from the study concluded that some features could be successfully used for the malaria detection.

Keywords: Feature extraction · Image processing · Malaria
Microscope image analysis · Matlab program · Plasmodium · Red blood cell
Segmentation

© Springer International Publishing AG, part of Springer Nature 2018
A. Abraham et al. (Eds.): ISDA 2017, AISC 736, pp. 846–857, 2018.
https://doi.org/10.1007/978-3-319-76348-4_81

1 Introduction

Malaria is caused by protozoan parasites of the genus Plasmodium [1]. It is a serious global disease and a leading cause of morbidity and mortality in tropical and sub-tropical countries. As per Centers for Disease Control and Prevention, it affects between 350–500 million people and causes more than 1 million deaths every year [2]. Yet, malaria is both preventable and curable. Rapid and correct diagnosis is an essential requirement to control the disease. There are four species of Plasmodium that infects humans and result in four types of malarial fever: P. falciparum, P. vivax, P. ovale, and P. malariae. P. vivax shows the widest distribution and is characterized by reappearances of symptoms after a latent period of up to five years. With the similar characteristics, P. ovale appears mainly in tropical Africa. P. falciparum is most common in tropical and subtropical areas [1]. It causes the most dangerous and malignant kind of malaria and contributes to the majority of deaths associated with the Malaria disease. P. malariae is also widely distributed but much less than P. vivax or P. falciparum [1, 3, 8] (Fig. 1).

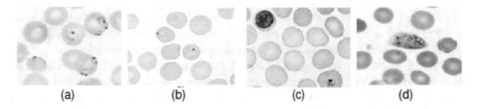

(a)   (b)   (c)   (d)

Fig. 1. (a) P. Falciparum (b) P. Vivax (c) P. Malariae (d) P. Ovale [3]

Two kinds of blood film are used in malaria microscopy. The thick film is always used to search for malaria parasites. The thin film is used to confirm the malaria parasite species. The most widely used technique for determining the development stage of the malaria disease is visual microscopical evaluation of Giemsa stained blood films [4, 8, 15]. This process consists of manually counting the infected red blood cells against the number of red blood cells in a slide prepared. The manual analysis of slides is time consuming and requires a trained operator. The accuracy of the final diagnosis also depends on the skill and experience of the technician and the time spent studying each slide. It has been observed that the agreement rates among the clinical experts for the diagnosis are surprisingly low.

The objective of this paper was to improve malaria microscopy diagnosis by removing the dependency on the performance of a human operator for the accuracy of diagnosis. This paper proposes a system to segment red blood cells, extract image features and classify the images into infected or uninfected red blood cells using thin blood smear film images and identify the type of malaria parasite using thick blood smear film images. Image processing techniques, Levenberg – Marquardt Backpropagation Neural Network and Euclidean distance measure were used for feature extraction, classification and parasite identification respectively.

2 Related Work

The majority of the research carried out so far for automated detection of malaria either includes image processing or artificial neural network techniques. There are very few instances where both the techniques are used. Further some researchers have used thick or thin blood simmers film images for the detection purpose whereas such methods cannot be termed as complete.

A study by [4] proposed a thin smear blood image classification technique using local image histogram equalization and adaptive thresholding and gradient edge detection technique. The proposed work was semiautomatic and claimed to reduce human error in detection. Another study by [3] used median filtering thresholding and morphological operations for image processing. The study applied the clump splitting algorithm for object identification and reported a sensitivity of 85.5%. Another study by [5] reported on the color normalization with K-nearest neighbor and Bayesian decision algorithm. The study used a stained pixel classifier which resulted in 88.5%, 5.6% true and false detection, parasite/non-parasite classifier achieves 74% sensitivity, 98% specificity, 88% positive prediction, 95% negative prediction. Another study by [6] reviewed the use of minimum watershed transformation and color normalization image processing methods for segmentation, whereas Bayesian pixel classifier with double threshold is reported as preferred classifier with accuracy of 96.72%.

Other studies have investigated the possibility of rapid and accurate automated diagnosis of red blood cell disorders and described a method to detect and classify malarial parasites in blood sample images [7]. The study applied image processing techniques including morphological operations and thresholding with feed forward back propagation neural network as soft computing methodology. A study by [8] adopted frequency domain discrete fourier transform to convert the images to log-polar coordinates and neural network support vector machine and reported accuracy of 96.72%. Similarly, [9] uses images taken from leishman-stained blood smears. The computing technique used are Zack's thresholding, euclidian distance and clustering. Another similar study conducted by [10] used highest quality oil immersion views (10 × 1000), of Giemsa stained blood films images. Two stage tree classifiers were used to achieve a sensitivity of 99%. Other studies have implemented support vector machine neural network with image processing techniques which resulted in sensitivity of 93.12% [11]. Another similar study which used only image processing techniques achieve a accuracy of 73.75% [12]. A study by [13] used machine learning model based on convolution neural network to automatically classify single cells in thin blood smears and reported accuracy of 97.37%. Work done by [14] include an automatic microscopic image acquisition system equipped with GUI based image processing modeled software to quantify the number of red blood cells. Several features grouped under shape features, intensity features, and texture features are classified and the result is presented to physician for further inspection. Review on computer vision malaria diagnostic technologies by [15] reports the challenges in commercial malaria diagnosis system such as availability of malaria parasite database to train the algorithm and disability of computer vision system to catch basic hematologic abnormalities.

3   Research Methodology

This section outlines the methods and phases used to develop a system for the accurate identification of malaria cells using either the microscopic examination using either thin or thick blood smears. A description of the techniques used for image features extraction from both the types of images for the diagnosis of malaria and malaria parasite are provided. The features extracted using thin blood smear images were simulated using Levenberg – Marquardt backpropagation neural network for diagnosis of malaria. The euler thin smear images were used to diagnose malaria whereas the thick smear images were used to identify the type of parasite.

3.1   Malaria Detection Using Thin Blood Smears Images

The input thin blood smear images are Giemsa stained. The images are processed through a number of stages as shown in Fig. 2. Details of the stages are provided below.

Image Enhancement: Image processing techniques were used to enhance the required features in the image [1, 2, 9, 10].

Color to Gray Scale Conversion: Thin smear blood images are in Red Green Blue (RGB) color space. Applying gray scale conversion gives darker shade to red blood cells and the background gets lighter shades.

Median Filter: The noise in the gray scale image was removed using nonlinear median filter [11]. The 5×5 kernel is identified to give best noise reduction results. Noise reduction was needed in order to remove false object detection and improve the results in the later stages.

Histogram Equalization for Contrast Enhancement: Filtering was used to smoothen the edges with the noise. Enhancement of the edges of the red blood cells and the contrast of the image is achieved using adaptive histogram equalization (AHE).

Image Segmentation: Ostu global thresholding and watershed algorithm was used for image segmentation. The goal of image segmentation was to segment individual red blood cells from the background [5, 9, 10, 14].

Ostu's Global Thresholding: Ostu's method [11] is the clustering based global thresholding technique. It minimizes within class variances and auto calculates optimum threshold to separate the foreground (red blood cells) and background pixels.

Flood-Fill Algorithm: Red blood cells in some blood smear images have ring-like shapes. The corresponding binary mask obtained by thresholding of such an image can result in holes in the centre of the cells. These holes need to be filled to obtain correct masks and to allow the subsequent methods to work properly. This operation was used starting with the border pixels of the background to fill the background area connected to the edges and identify the holes in the image as background pixels.

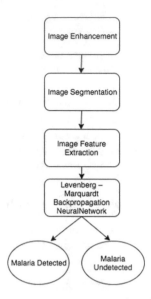

Fig. 2. Malaria detection using thin smear blood film images.

Marker Controlled Watershed Segmentation: The final stage of segmenting all the individual red blood cells from the background was achieved by using marker controlled watershed transformation [11]. This allowed the overlapping red blood cells to be uniquely identified.

Image Feature Extraction: To train the proposed neural network to diagnose malaria, all the principal image component features corresponding to red blood cells were extracted. The extracted feature vector space has large between-class distance and small within-class variance. The set of features that extracted were used to discriminate between infected and uninfected red blood. A feature based dataset was created to train the neural network [2, 5, 6, 8, 11, 13].

Shape Descriptor Features: This feature was used to identify the roundness of the identified objects.

First Order Statistics Features: The set of features such as mean, standard deviation, variance, skewness, kurtosis, fifth and sixth moment (Higher moments) and entropy under this category were extracted.

Second Order Statistics Features (GLCM): Gray-level co-occurrence matrices, contrast, correlation, energy and homogeneity are all higher order statistical features which covers the texture based information.

Levenberg – Marquardt Backpropagation Neural Network [18]. The Levenberg-Marquardt algorithm is the damped least-squares method. The loss function is the sum of squared error. It computes the gradient vector and the Jacobian matrix without computing exact Hessian matrix. This process was based on the following algorithm

1. Compute the Jacobian matrix J
2. Compute the error gradient

$$g = J^T E \text{ where E is vector of all error}$$

3. Approximate the Hessian using the cross-product Jacobian

$$H = J^T J$$

4. Solve $(H + \lambda I)\delta = g$ to find δ where I is identity matrix
5. Update the network weights w using δ
6. Recalculate the sum of squared errors using the updated weights
7. If the sum of squared errors has not decreased.
8. Discard the new weights, increase λ using v and go to step iv.
9. Else decrease λ using v and stop.

3.2   Malaria Detection Using Thin Blood Smears Images

The thick blood smear image analysis process was used to count the number of malaria parasites which existed in each digitized red blood specimen image. Thick blood smear examination was considered a necessary part for rapid screening of malaria parasite. The proposed system carried out malaria detection based on analysis of colors. The detection of malaria parasites from thick smear images thresholding used segmentation techniques, which separates foreground (parasite) from background. The images were pre-processed to enhance the image details followed by application of thresholding technique. The stages for automatic diagnosis of malaria using thin smear blood images is shown in Fig. 3.

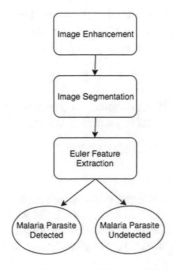

Fig. 3. Malaria detection using thick smear blood film images.

Image Enhancement: The red, green and blue planes of input color image was separated using the adaptive histogram equalization applied to each plane. This method found to be suitable for improving the local contrast and enhancing the definitions of edges in each red blood cells of an image [3] (Fig. 4).

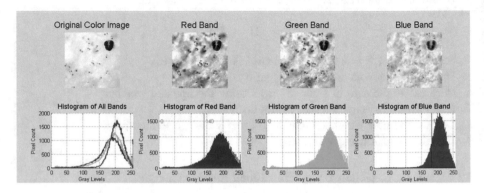

Fig. 4. Separated RGB planes of input image

Image Segmentation: Low and high threshold values for each red, green and blue planes were selected manually which provided a better segmentation result by extracting out parasites from background of blood images. The binary image produced after thresholding of all three planes were concatenated for generating parasite mask image. The borders were smoothened using morphological closing operation. Connected components where then counted for determining the number of malaria parasites present in thick blood smear images [3–5, 8–10].

Feature Extraction: The Euler number [19] for malaria thick blood smear images is the total number of parasites in the image minus the total number of holes in those images. Objects were connected sets of pixels, that is, pixels having a value of 1.

$$E = C - H$$

Where E is Euler Number, C is Connected components and H is the holes in Object. If Euler value is less than zero than the counted objects are infer as malaria parasite.

4 Proposed System

The proposed system was tested using microscopic blood images of both thick and thin smear which were obtained from an open source library CDC DPDx - Malaria image library [16]. DPDx is an education resource designed for health professionals and laboratory scientists. CDC's DPDx Parasite Image Library where parasites and parasitic diseases are listed alphabetically and are cross-referenced [17]. The image processing techniques were applied to the data set of 134 downloaded images which included 76 thick and thin smear images respectively. System testing was carried out on 20 images collected from a pathology lab in India. The sample included 15 images which were infected and 5 images were uninfected. All the images were based on 300 X 300 in .jpg format with resolution of 72 dpi.

The MatLab software tools were used to implement image processing techniques and Levenberg – Marquardt backpropagation neural network. The GUI was designed to execute all the image processing functionality, feature extraction and saving the malaria cell identification results (Fig. 5).

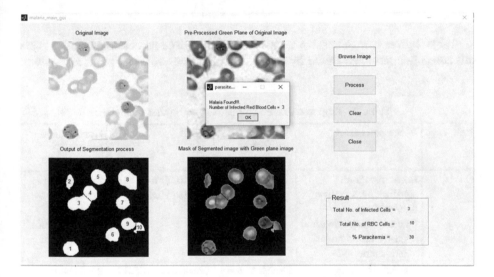

Fig. 5. GUI for thin smear blood cell image processing

5   Result

Table 1 provides results of the accuracy of identification of the malaria testing dataset. The system tested 20 images of which 5 images where from blood infected with malaria. Extracted features were simulated using SVM algorithm where the results found were less than 50% accuracy, hence Levenberg - Marquardt Backpropagation Neural Network was selected owing to better accuracy.

The proposed system correctly identified malaria cells with an accuracy of 78%. Overall performance of the system is found to be 80%.

Table 1. Result comparison (Infected by malaria)

Image No.	Actual parasite count	System count
1	1	3
?	1	1
3	3	6
4	1	1
5	1	1
6	2	2
7	2	2
8	4	3
9	2	2
10	2	1

Table 2 gives comprehensive comparison of the image processing techniques and soft computing techniques used by the related work done with the proposed system.

Table 2. Proposed system comparison with other methods

Reference	Image processing	Soft computing technique	Result	Comparative remark
[1]	Local histogram equalization adaptive thresholding Image segmentation and intensity based malaria identification	Not used	Localization of infected RBC	Feature extraction not done, soft computing not implemented
[2]	Median filtering thresholding and morphological operations	Decision tree and back propagation feed forward algorithm	Sensitivity = 85.5%. Positive predictive value = 81%	Analysis of overlapping cells is absent
[3]	Color normalization	Bayesian classifier, KNN algorithm	Sensitivity = 98%. Positive prediction = 88%,	Data set used is very less to develop reliability
[6]	Discrete fourier transform	Support vector machine	Accuracy: Neural network = 7853% SVM = 96.72%	GUI and complete automation missing
[7]	HSV color space, sequential edge linking algorithm and clustering	Not used	Localization of infected RBC	Less image database, soft computing not implemented

(*continued*)

Table 2. (*continued*)

Reference	Image processing	Soft computing technique	Result	Comparative remark
	based on Euclidian distance			
[13]	Not used	Convolutional Neural Network (CNN)	Accuracy of 97.37%	Image processing not implemented
[14]	RGB to HSV, adaptive estimated threshold, Ostu's thresholding	Not used	Test results are compared with human experts	No statistical results reported
Proposed system	Color to gray scale conversion, median filter, histogram equalization for contrast enhancement, Ostu's global thresholding, flood-fill algorithm, marker controlled watershed segmentation	Levenberg – Marquardt Back Propagation Neural Network	Accuracy = 72%	GUI implemented, 134 image dataset, real time implementation

6 Conclusion

This paper presents a proposed method for segmentation of red blood cells in microscopic blood smear images infected with malaria cells. The paper also described the design of a graphical user interface for automation of this process. The main purpose of the GUI was to execute the segmentation method, label the samples and save the information about individual segmented cells. The GUI based system was used to create a database of labeled red blood cells containing non-infected cells and cells infected by malaria parasites is created. A set of features was extracted from the area of a red blood cell in order to detect malaria.

The segmentation method presented a relatively simple solution of separating red blood cells from the background and isolating overlapping or occluded cell. The method was based on processing of a binary image obtained by thresholding and utilizes the watershed transformation for separating cell compounds. The method produced good results on all input images with only occasional over-segmented and under-segmented cells.

The designed GUI provided the tools necessary for segmenting cells, labeling of samples, manipulating with files, and saving the results. As some future enhancement certain improvements could be made in the GUI including the implementation of some semi-automatic methods for the correction of cell contours, a set of shortcut keys, the use of mouse wheel for zooming, etc. Some of these improvements would be dependent on the programing environment. Future studies will be devoted to enhance the accuracy of the proposed system using larger data set.

There are many possible classification scenarios that can exist with the malaria cell diagnosis. This paper has described a study which has focused on distinguishing between infected and non-infected red blood cells. The findings from the study indicated the potential to use this system to identify malaria cells from images taken from both thin and thick blood smear.

References

1. World Health Organization. Malaria, Fact sheet no. 094 (2017)
2. Centers for Disease Control and Prevention. CDC Health Information for International Travel 2010. Atlanta, Mosby, pp. 128–159 (2009)
3. Sudhakar, P.: Detection of malarial parasite in blood using image processing. Int. J. Eng. Innov. Technol. 2(10), 124–126 (2013)
4. Khan, M., Acharya, B., Singh, B., Soni, J.: Content based image retrieval approaches for detection of malarial parasite in blood images. Int. J. Biom. Bioinform. 2, 97–110 (2011)
5. Tek, B., Dempster, A., Kale, I.: Malaria parasite detection in peripheral blood images. In: BMVC, pp. 347–356 (2006)
6. Tek, B., Dempster, A., Kale, I.: Computer vision for microscopy diagnosis of malaria. Bio-Med Central Malar. J. 8(1), 153 (2009)
7. Ahirwar, N., Pattnaik, S., Acharya, B.: Advanced image analysis based system for automation detection and classification of malarial parasite in blood images. Int. J. Inf. Technol. Knowl. Manag. 5(1), 59–64 (2012)
8. Annaldas, S., Shirgan, S.: Automatic diagnosis of malaria parasites using neural network and support vector machine. Int. J. Adv. Found. Res. Comput. (IJAFRC) 2, 60–66 (2015)
9. Damahe, L., Krrishna, K., Janwe, J., Nileshsingh, T.: Segmentation based approach to detect parasites and RBCs in blood cell images. Int. J. Comput. Sci. Appl. 4(2), 71–81 (2011)
10. Soni, J., Mishra, N.: Advanced image analysis based system for automatic detection of malarial parasite in blood images. In: International Conference on Advanced Computing Communication and Networks, pp. 59–64 (2011)
11. Savkare, S., Narote, P.: Automatic detection of malaria parasites for estimating parasitemia. Int. J. Comput. Sci. Secur. (IJCSS) 5(3), 310–315 (2011)
12. Ghate, D., Jadhav, C., Rani, U.: Automatic detection of malaria parasite from blood images. Int. J. Adv. Comput. Technol. 4(1), 129–132 (2012)
13. Liang, Z., Powell, A., Ersoy, I., Poostchi, M., Silamut, K., Palaniappan, K., Guo, P., Hossain, M., Sameer, A., Maude, R., Huang, J., Jaeger, S., Thoma, G.: CNN-based image analysis for malaria diagnosis. In: IEEE International Conference on Bioinformatics and Biomedicine (BIBM), Shenzhen, pp. 493–496 (2016)
14. Mehrjou, A., Abbasian, T., Izadi, M.: Automatic malaria diagnosis system. In: First RSI/ISM International Conference on Robotics and Mechatronics (ICRoM), pp. 205–211 (2013)
15. Pollak, J., Houri-Yafin, A., Salpeter, S.: Computer vision malaria diagnostic systems— progress and prospects. Front. Public Health 5, 1–5 (2017)
16. Centers for Disease Control and Prevention (CDC) Laboratory Identification of Parasitic Diseases of Public Health Concern (DPDx) - Malaria image library. http://www.cdc.gov/dpdx/malaria/gallery.html. Accessed 04 May 2017

17. Centers for Disease Control and Prevention (CDC) Laboratory Identification of Parasitic Diseases of Public Health Concern (DPDx) - Malaria image library listed alphabetically and cross referenced. http://www.cdc.gov/dpdx/az.html. Accessed 04 May 2017
18. Ahiwar, N., Patnaik, S.: Texture and intensity based classification of malaria parasite in blood images using Levenberg-Marquardt (LM) algorithm. Int. J. Comput. Intell. Inf. Secur. **2**(12), 4–9 (2011)
19. Chakrabortya, K., Chattopadhyayb, A., Chakrabarti, A., Acharyad, T.: A combined algorithm for malaria detection from thick smear blood slides. Health Med. Inf. **6**(1), 2–6 (2015)

Comparative Analysis of Adaptive Filters for Predicting Wind-Power Generation (SLMS, NLMS, SGDLMS, WLMS, RLMS)

Ashima Arora$^{(\boxtimes)}$ and Rajesh Wadhvani

Computer Science and Engineering,
Maulana Azad National Institute of Technology, Bhopal 462003, M.P., India
arora.ashima1993@gmail.com

Abstract. Adaptive filters play an important role in prediction. This ability of adaptive filters have been successfully used in prediction of wind-power generation. This paper focuses on the comparison between adaptive filtering algorithms in order to determine which filter produces least error for predicting wind-power generation. Algorithms such as Standard least mean square (SLMS), Normalized least mean square (NLMS), Weighted least mean square (WLMS), Stochastic Gradient Descent least mean square (SGDLMS), Recursive least Square (RLS) are implemented. The performance of the filters is evaluated using actual operational power data of a wind farm in America. Four performance criteria are used in the study of these algorithms: Mean Absolute Error, R-squared value, Computational Complexity, and Stability of the system.

Keywords: Adaptive filtering algorithms · Adaptive filter
Computational complexity · Least mean square
Mean absolute error · R-squared · Wind power generation

1 Introduction

Wind energy has proved itself as an alternative energy resource for adding new capacity to the power grid in markets around the globe. Wind energy has become a viable resource for electricity generation. A high share of electricity is produced in the energy sector. So, the performance of wind power generators is of prime concern. Generated output power has a lot of uncertainties and deviations which causes serious issues in the energy management systems (EMS) and impact the reliability of the power grid [1]. Predicting wind-power energy enable better functioning in power grids. Manufacturers often assume ideal meteorological and geomorphological conditions and supply their theoretical power curves. In the real world, wind turbines never get these ideal conditions and therefore, the factual power curves could be substantially different from the theoretical ones. This deviation is due to various factors like wind direction, temperature,

© Springer International Publishing AG, part of Springer Nature 2018
A. Abraham et al. (Eds.): ISDA 2017, AISC 736, pp. 858–867, 2018.
https://doi.org/10.1007/978-3-319-76348-4_82

humidity, pressure, precipitation, wind velocity distribution, altitudes, lightning directly or indirectly, independently or in combination with others affects wind energy generation [2].

Practically, there is no one filter that could produce the least error and over-rule all others upon all possible observations which are obtained from different wind turbines. It is observed that on a particular data set, a specific adaptive filter might work best, but on other data sets, other filters might be more applicable [3]. An adaptive filter is a time-variant filter which adjusts its coefficients in order to optimize a cost function in accordance with the unknown environment. These filters process the input signals and are concerned with following three major parameters: the design, analysis, and implementation of systems whose structure changes in response to the incoming data [4]. It works in online mode i.e. input data arrive continuously with respect to time. The main characteristic of the adaptive filter is that it can automatically adapt itself in following scenarios: When there is a change in the environment or when requirements of the existing system change with respect to time. Moreover, these filters can be trained to perform specific filtering tasks or decision-making tasks based on some updating equations or some set of training rules [5]. Such abilities make it a powerful device for signal processing and control operations. These filters have been successfully applied in various fields like seismology, radar, sonar, mobile phones and other communication devices, camcorders and digital cameras, and medical monitoring equipment.

In this paper, we focus on five different adaptive filters for predicting the power of the wind turbine. These approaches include- Standard Least Mean Square Algorithm (SLMS), Normalized Least Mean Square Algorithm (NLMS), Weighted Least Mean Square Algorithm (WLMS), Stochastic Gradient Descent Least Mean Square Algorithm (SGDLMS) and Recursive Least Square Algorithm (RLS).

2   Adaptive Filters for Wind Power Estimation

An adaptive filter is a computational device that enables to find the relationship between two signals in real time in an iterative manner. Adaptive filters can be one of the following: a set of program instructions or set of logic operations which run on arithmetical processing devices. Adaptive filters can also be implemented on a custom VLSI integrated circuit or field-programmable gate array (FPGA). Purpose of the adaptive filter is to estimate the future values of a signal with respect to the past values of the input signal [6]. Four basic aspects define an adaptive filter. The signal that is fed to the filtering structure for producing the output signal. The filtering structure that defines how the output signal of the filter is computed with the help of its input signal; the parameters which change with every iteration to alter the filter's input-output relationship; the adaptive algorithm that describes how the parameters are adjusted or updated from one time instant to the next. Each adaptive filter has a different set of parameters. When a particular adaptive filter structure is selected, the number and type of parameters associated with that can be adjusted. The adaptive algorithm is used as a form of the optimizer that minimizes the error of a particular dataset [7].

In this paper, we present the general adaptive filtering structure and introduce their implementation for updating the weights after every iteration and then, determining the error form of the adaptive filter. The basic structure of an adaptive filter is shown in Fig. 1.

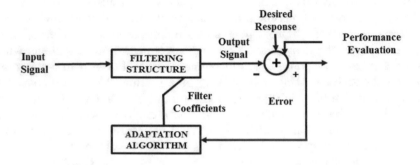

Fig. 1. Basic structure of adaptive filter

Each adaptive filter has different filtering structure and adaptive algorithm according to the nature of its adaptivity. Adaptive filters based on various algorithms are discussed in this paper which can be used to predict future values of the wind power turbine.

2.1   Standard Least Mean Square Algorithm (SLMS)

The standard LMS algorithms are a class of adaptive filters. These filters are utilized to produce the desired filter after it evaluates the least mean square of the error signal. The desired filter is produced by finding the filter coefficients that identify even the slightest mean square error signal. The filter coefficients of the algorithm are adjusted so as to obtain minimum cost function. In this algorithm, the filter is only adapted based on the error that occurs at current time [8]. For a given input signal $\widehat{u}(k) = [u(k), u(k-1), \ldots, u(k-p+1)]^T$, where p is the order of the filter; SLMS performs the following operations in order to update the coefficients of an adaptive filter: Firstly, the output signal $y(k)$ from the adaptive filter is calculated. Using the output signal, the error signal $e(k)$ is calculated by using the following equation for $k = 0, 1, 2, \ldots, k_{max}$:

$$e(k) = d(k) - y(k). \tag{1}$$

where, $y(k) = \widehat{w}^T(k).\widehat{u}(k)$ and $d(k)$ is the desired signal. Then, following equation is used to update the filter coefficients over iteration:

$$\widehat{w}(k+1) = \widehat{w}(k) + \mu \cdot e(k) \cdot \widehat{u}(k). \tag{2}$$

where, $\widehat{w}(k+1)$ is equal to the weights vector at instance $k+1$, $\widehat{w}(k)$ is the filter coefficients vector at the instant k, and $\widehat{u}(k)$ is the filter input vector which is stored in the filter delayed line, $e(k)$ corresponds to the filters error signal, μ is the step size of the adaptive filter (also known as convergence factor). The convergence factor μ determines the minimum square average error and the convergence speed [8].

As observed, exact values of estimation cannot be used while applying LMS algorithms. That is why optimal weights are not achievable in an absolute sense. However, LMS algorithms converge to mean value. This property helps to achieve optimal weights, even-though, the weights may change by small amounts. When the change in weights is large, the problem of variance arises. Then convergence towards mean value can be misleading. So, proper selection of variable parameter is necessary. An upper bound is given to step-size μ, defined as follows:

$$0 < \mu < \frac{2}{\lambda_{max}}. \tag{3}$$

where, λ_{max} is an autocorrelation matrix with non-negative eigenvalues. The algorithm becomes unstable if μ does not satisfy the relation in (3). Also, by analyzing the relation we can conclude that the convergence of the algorithm is inversely proportional to the eigenvalue spread of the correlation matrix. When the eigenvalues of the correlation matrix are widespread, then algorithm converges slowly. This spreading of the eigenvalue is calculated by taking ratio of the largest eigenvalue to the smallest eigenvalue. Convergence speed of the algorithm is dependent on the choice of μ. When μ is small then convergence takes place slowly. When the value of μ is large then, a fast convergence of the algorithm is observed. However, upper bound of the μ should always be taken into consideration as the algorithm may tend to move to an unstable state if μ is high. To improvise the stability of the algorithm, max achievable convergence speed is given by:

$$\mu = \frac{2}{(\lambda_{max} + \lambda_{min})}. \tag{4}$$

where, λ_{min} and λ_{max} are the smallest and largest eigenvalues respectively. So, faster convergence is achieved when λ_{max} is close to or equal to λ_{min}. Thus, maximum convergence speed is solely dependent on the spread of the eigenvalues.

2.2 Normalised Least Mean Square Algorithm (NLMS)

Normalized LMS algorithm is introduced to overcome the limitations of standard LMS algorithm. Standard LMS algorithm is sensitive to the scaling of its input which causes difficulty in choosing appropriate learning rate μ. NLMS is the modified form of the standard LMS algorithm [9]. For a given input signal $\widehat{u}(k) = [u(k), u(k-1), \ldots, u(k-p+1)]^T$, where p is the order of the filter; this algorithm updates the coefficients of an adaptive filter by using the following equation:

$$\widehat{w}(k+1) = \widehat{w}(k) + \mu.e(k)\frac{\widehat{u}(k)}{||\widehat{u}(k)||^2}. \tag{5}$$

Substituting $\mu(k) = \frac{\mu}{||u(k)||^2}$ in (5), the above equation becomes:

$$\widehat{w}(k+1) = \widehat{w}(k) + \mu(k).e(k).\widehat{u}(k). \tag{6}$$

Here, (6) illustrates that the NLMS algorithm becomes the same as the SLMS algorithm. The only differential factor here is time-varying step size $\mu(k)$. When parameters like convergence speed and mean square error are considered, NLMS shows better performance than SLMS i.e. NLMS gives faster convergence and less mean square error.

2.3 Stochastic Gradient Descent Least Mean Square Algorithm (SGDLMS)

SGDLMS algorithm is very popular and because of its robustness and low computational complexity, it is used worldwide. In this filter, a vector named as Gradient $P(k)$ is used for updating the weights of the filter. The expression is defined as follows:

$$\widehat{w}(k+1) = \widehat{w}(k) + \frac{1}{2}\mu[-\nabla P(k)]. \tag{7}$$

where, μ is the step size, $\widehat{w}(k+1)$ is equal to the weights vector at instance $k+1$, $\widehat{w}(k)$ is the filter coefficients vector at the instant k. If step size parameter (μ) is suitably selected, this filter enables to compute tap weight vector $\widehat{w}(k)$. Here, matrices like correlation matrix R of the tap input vector and the cross-correlation vector T between the tap inputs and the desired response are used to optimally estimate the gradient. Formulation of the gradient vector is given below:

$$\nabla P(k) = -2t + 2r\widehat{w}(k). \tag{8}$$

Instantaneous estimates for auto-correlation function r and the cross-correlation function t are defined in (9) and (10) respectively:

$$\widehat{r}(k) = u(k) \cdot u^H(k). \tag{9}$$

$$\widehat{t}(k) = u(k) \cdot d * (k). \tag{10}$$

A hat is used in representation of $\widehat{r}(k)$ and $\widehat{t}(k)$ which intend to signify that these quantities are just "estimates". To include a non-stationary environment, (9) and (10) are introduced such that all signals and desired responses can vary with respect to time. These estimates are collaborated by using different discrete magnitude values of the tap input vector and necessary response. Using the values of r and t from (9) and (10), the instantaneous value of gradient vector is defined as:

$$\nabla P(k) = -2u(k) \cdot d * (k) + 2u(k) \cdot u^H(k) \cdot \widehat{w}(k). \tag{11}$$

Substituting the estimate of the gradient vector from (11) in the steepest descent algorithm, following relation is obtained:

$$\widehat{w}(k+1) = \widehat{w}(k) + \mu \cdot u(k)[d * (k) - u^H(k)\widehat{w}(n)]. \tag{12}$$

In (12), we see that weight is updated for instance $k+1$ with the help of tap input vector and necessary response. This approach allows the adaptive filter to perform robustly and update the weight for k^{th} instance. This filter simply minimizes the instantaneous error squared which results in reducing the storage requirement to the minimum. It is also compatible for use in non-stationary environments as well as online settings [10]. Its faster computation gives reasonable results. Moreover, it can be used for more complex models where there are more error fluctuations.

2.4   Weighted Least Square Algorithm (WLMS)

When adaptive filter is produced using least mean squares algorithm, an assumption follows that there is constant variance in errors. This is known as homoscedasticity. In other words, each value in the data-set is treated with equal weight. Despite the fact that some values of variables may be more likely accountable while calculating error. So, weighted least square algorithm is used when the least square algorithm assumption of constant variance in the errors is violated. This is known as heteroscedasticity. That is, few values in the dataset has more weightage with respect to other values in the same data-set [11].

A normally distributed weight matrix P with mean vector 0 and non-constant variance-co-variance is shown below. The reciprocal of each variance, σ_i^2 in P is defined as the weight, $w_i = 1/\sigma_i^2$, then let matrix W be a diagonal matrix containing these weights, as shown below:

$$P = \begin{bmatrix} \sigma_1^2 & 0 & \cdots & 0 \\ 0 & \sigma_2^2 & \cdots & 0 \\ \vdots & \vdots & \ddots & \vdots \\ 0 & 0 & \cdots & \sigma_n^2 \end{bmatrix} ; W = \begin{bmatrix} w_1 & 0 & \cdots & 0 \\ 0 & w_2 & \cdots & 0 \\ \vdots & \vdots & \ddots & \vdots \\ 0 & 0 & \cdots & w_n \end{bmatrix}$$

Now, the equation of weighted least square is given below:

$$Y = \mu \cdot X + \varepsilon. \tag{13}$$

where, ε is assumed to be (multivariate) normally distributed with mean vector 0 and non-constant variance-co-variance matrix, is the weighted least square estimate which is evaluated by minimizing the error ε, Y is the responses that reflect in the data set. To evaluate the weighted least square estimate minimize the error in a matrix form in following way:

$$(Y - \mu \cdot X)^T W (Y - \mu \cdot X). \tag{14}$$

where, W is a diagonal matrix. Now, taking the derivative with respect to μ, the solution is:

$$\widehat{\mu} = (X^T W X)^{-1} X^T W Y. \tag{15}$$

As per observation, each weight is inversely proportional to the error variance as it reflects the information in that observation. Therefore, an observation with small error variance has a large weight since it contains relatively more information than an observation with large error variance (small weight). In small datasets, WLS works best in retrieving maximum information. It is the only method that can be applied to datasets where data points are of varying quality. However, application of this requires the exact knowledge of the weights, which is not always feasible [12]. Estimating the weights may give unpredictable results, especially in small data sets. So, this technique should only be used when the weights estimates are accurate and precise.

2.5   Recursive Least Squares Algorithm (RLS)

RLS is also a class of adaptive filter. When filters are implemented using RLS algorithm, filters recursively find the coefficients relating to the input signals. The goal of this algorithm is to minimize the cost function. This is done by appropriately selecting the filter coefficients and updating the filter as new data arrives. In order to update the old estimates, computation begins with prescribed initial conditions and the information that is contained in new data samples. The RLS algorithms work best in time-varying environments but the cost of computational complexity increases and some stability problems arises.

In RLS, the input vector is simultaneously given to traversal filter unit as well as to the adaptive weight control mechanism unit. Once the output is generated, it is matched with the desired response to evaluate the error. This error is then fed into adaptive weight control mechanism unit where it is recursively called till stability state is achieved [13]. There are two main reasons to use RLS: (a) when the number of variables in the linear system exceeds the number of observations, under these circumstances the ordinary least-squares problem becomes ill-posed which makes it impossible to fit as infinitely many solutions are obtained for optimization. RLS algorithm enables to overcome this limitation by introducing further constraints that uniquely determine the solution. (b) Secondly, RLS is used when the learned model suffers from poor generalization (even when the number of variables does not exceed the number of observations). In such cases, RLS improves the generalizability of the model by putting some constraints at training time [13]. To solve RLS, a weighting factor is introduced to the sum of-errors-squares definition:

$$\varepsilon(n) = \sum_{i=1}^{k} \beta(k,i)|e(i)|^2. \tag{16}$$

where, weighting factor has the property $0 < \beta \leq 1$ for $i = 1, 2, \cdots, k$. This weighting factor is used to ensure that less weight is given to older error samples such that statistical variations in the data-set can be observed more easily when the filter operates in the non-stationary environment [14]. A special form of the weighting factor that is commonly used is the exponentially weighting factor or forgetting factor. It is defined as $\beta(k,i) = \lambda^{k-i}$ for $i = 1, 2, \cdots, k$ where, λ is a positive constant, $0 < \lambda < 1$. In this algorithm the filter tap weight vector is updated using following equations:

$$\widehat{w}(k) = \widehat{w}(k-1) + K(k) \cdot \widehat{e}_{k-1}(k). \tag{17}$$

$$K(k) = \frac{u(k)}{\lambda + X^T(k) \cdot u(k)}. \tag{18}$$

$$u(k) = \widehat{w}_\lambda^{-1}(k-1) \cdot X(k). \tag{19}$$

where, λ is a small positive constant whose value is close to 1 but not equal to 1. Equations (18) and (19) represents the intermediate gain vector that are used to compute tap weights. By using the filter tap weights of above iteration and the current input vector, the filter output is calculated as in (20) and (21).

$$\widehat{y}_{k-1}(k) = \widehat{w}^T(k-1) \cdot X(k). \tag{20}$$

$$\widehat{e}_{k-1}(k) = d(k) - \widehat{y}_{k-1}(k). \tag{21}$$

In the RLS Algorithm, the estimate of previous samples of output signal, error signal and filter weight is required that leads to higher memory requirements.

3   Simulation and Results

3.1   Description Data-Sets

The data-sets for the experiment is taken from a site in America. This data-set is collected by National Renewable Energy Laboratory (NREL), located in Golden, Colorado, which specializes in renewable energy and energy efficiency research and development. The data-set has approx. hundred thousand entries of wind-power generation upon which training and testing are performed.

3.2   Parameters for Error Evaluation

In order to compare the algorithms discussed in Sect. 2, few measures are used to evaluate error. These expressions will evaluate net error and will help to verify the accuracy of the filter based on a particular algorithm. Expressions for evaluating error is as follow:

$$MAE = \frac{1}{k} \sum_{t=1}^{k} |y_t - f_t|^2, \quad R^2 = 1 - \frac{\sum_{t=1}^{k}(y_t - f_t)^2}{\sum_{t=1}^{k}(y_t - y_{tm})^2}. \tag{22}$$

where, MAE is mean absolute error, R^2 is R-squared value, y_t is the actual wind power at time t, f_t is the estimated value of the power at time t and y_{tm} is the mean of actual wind power. Anyone of these evaluation expressions in (22) can be used to estimate the error of the adaptive filter. MAE is directly proportional to the error. The lesser value of MAE corresponds to less error. On the other hand, R^2 is inversely proportional to error i.e. larger value of R^2 means lesser error. Value of R^2 score lies between 0 and 1. When the value of R^2 tends to 1, we can conclude that adaptive filter has least error while training and testing.

3.3   Experiment and Results

In the first simulation, each adaptive filter is optimized for its variable parameter. Here, step-size (μ) is the variable parameter in all the adaptive filters. In this simulation, an important issue is finding μ for which least error is obtained. Therefore, the value of μ is chosen by k-fold cross validation [15], where the value of k = 10. In the beginning, value of μ is initialized for range 0.001 to 1. With this range-variation, error of the data set is checked with respect to μ. The optimized value of μ is selected for which the filter produces least error. Once μ is final for a filter, its value is fixed and is used to evaluate mean absolute error and R^2. This process is repeated for each filter and with its respective optimized μ. Mean absolute error (MAE), R-squared value (R^2), computational complexity and stability of each filter is computed and performance is monitored as shown in Table 1.

Table 1. Performance comparison of adaptive algorithms

S. No.	Algorithm	MAE	R^2	Complexity	Stability
1	SLMS	0.28	0.990	$2N+1$	Highly stable
2	NLMS	1.90	0.375	$3N+1$	Less stable
3	SGDLMS	0.88	0.935	$2N$	Less stable
4	WLMS	0.44	0.987	$2N+N^2$	Stable
5	RLS	0.22	0.996	$4N^2$	Highly stable

From the Table 1, we can observe that for the selected data set, the performance of RLS adaptive algorithm is high as compared to other algorithms as it has least mean absolute error and highest R-squared value. Also, RLS produces highly stable state for the system. On the other hand, poor performance is shown when NLMS is applied over the same data set. Moreover, we can observe that RLS and SLMS show comparable results as their MAE values and R-squared values are very close to each other. Computational complexity for each filter is computed which helps to determine the performance of the algorithm in a unit time.

4   Conclusion

Here, the comparison between different adaptive algorithms is successfully completed. From Table 1, we can conclude that RLS algorithm is the most suitable algorithm for predicting wind-power generation for the selected data set. NLMS being an extension of the SLMS algorithm, fail to produce least error for the dataset. On the contrary, SLMS is found to be comparable to RLS and can be used in similar data-sets. SDGLMS and WLMS show average performance in terms of error as well as in terms of computational complexity and stability. RLS shows high computational complexity whereas SLMS and NLMS give less complexity results. In terms of stability, RLS and SLMS have proved to be best among other algorithms. Performance can be further improved but at the cost of increasing the computational complexity of the algorithm. Purpose of comparing various adaptive filters was to find out which algorithm takes the unknown system to a stable state. Other algorithms too can be implemented to form adaptive filters which may show better performance in terms of error, complexity, and stability.

References

1. Ipakchi, A., Albuyeh, F.: Grid of the future. IEEE Power Energy Mag. **7**(2), 52–62 (2009)
2. Shokrzadeh, S., Jozani, M.J.: Wind turbine power curve modeling using advanced parametric & nonparametric methods. IEEE Trans. Sustain. Energy **5**(4), 827–835 (2014)
3. Gasch, R., Twele, J.: Wind Power Plants: Fundamentals, Design, Construction and Operation, pp. 46–113. Springer, Berlin (2012)
4. Madisetti, V.K., Douglas, B.W.: Introduction to Adaptive Filters Digital Signal Processing Handbook, 2nd edn. CRC Press LLC, Boca Raton (2009). Chap. 18
5. Macchi, O.: Adaptive Processing: The Least Mean Squares Approach with Applications in Transmission, vol. 23, no. 11, pp. 45–78. Wiley, Chichester (1995)
6. Patil, A.P., Patil, M.R.: Computational complexity of adaptive algorithms in echo cancellation. SSRG Int. J. Electron. Commun. Eng. (SSRG-IJECE) **2**(7), 16 (2015)
7. Clarkson, P.M.: Optimal and Adaptive Signal Processing. CRC Press, Boca Raton (1993)
8. Dhiman, J., Ahmad, S., Gulia, K.: Comparison between adaptive filter algorithms (LMS, NLMS and RLS). Int. J. Sci. Eng. Technol. Res. (IJSETR) **2**(5), 1100–1103 (2013)
9. Sharma, A., Juneja, Y.: Acoustic echo cancellation of from the signal using NLMS algorithm. Int. J. Res. Advent Technol. **2**(6) (2014)
10. Shoval, D.J., Snelgrove, W.: Comparison of DC offset effects in four LMS adaptive algorithms. IEEE Trans. Circ. Syst. II Analog Digit. Sig. Process. **42**(3), 176–185 (1995)
11. Zaknich, A.: Principles of Adaptive Filters and Self-learning Systems. Springer, London (2005)
12. Haykin, S.S.: Adaptive Filter Theory. Prentice Hall, Upper Saddle River (1996)
13. Sayed, A.H., Kailath, T.: Recursive Least-Squares Adaptive Filters. Wiley, Los Angeles (2003)
14. Marshall, D.F., Jenkins, W.K., Murphy, J.J.: The use of orthogonal transforms for improving performance of adaptive filters. IEEE Trans. Circ. Syst. **36**, 474–485 (1989)
15. Fushiki, T.: Estimation of prediction error by using K-fold cross-validation. Stat. Comput. **21**(2), 137–146 (2011)

Blind Write Protocol

Khairil Anshar[✉], Nanna Suryana, and Noraswaliza Binti Abdullah

Universiti Teknikal Malaysia Melaka, Alor Gajah, Malaysia
p031420004@student.utem.edu.my,
{nsuryana, noraswaliza}@utem.edu.my

Abstract. The current approach to handle interleaved write operation and preserve consistency in relational database system relies on locking protocol. The application system has no other option to deal with interleaved write operation. In other hand, allowing more write operations to be interleaved will increase the throughput of database. Since each application system has its own consistency requirement then database system should provide another protocol to allow more write operation to be interleaved. Therefore, this paper proposes blind write protocol as a complement to the current concurrency control.

Keywords: Concurrency control · Interleaved transaction · Locking
Consistency · Availability · Deadlock · Blind write

1 Introduction

Current implementation of concurrency control in Database Management System (DBMS) [1] is handling the interleaved write operation at system level. Eswaran et al. described in [2], when someone is transferring money from one bank account to another, then there will be one or more entity in the temporary inconsistent state. It is due to all operations in one transaction are executed one by one. As a result, it makes the balance in one account has been deducted and another account not yet added. If this happens, then there should no other transaction are able to access those two bank accounts in order to preserve the consistency. To achieve this, Eswaran et al. proposed Locking Protocol. This locking operations is handled by Transaction Manager at system level.

Another concurrency control is utilizing an entity version [3] and local copies [6] to handle the temporary inconsistent state of entity. Similar with the locking protocol, this validation is performed at the system level. Thus, application system has no option to deal with the interleaved write operations. In other hand, each application has different consistency requirement which is transformed into read and write operation in the application code. Therefore, the application system has the knowledge on every operation. Hence, this paper proposes blind write protocol as a complement of current concurrency control to give more option to the application to deal with interleaved write operation.

There are also some discussions which aims to allow more operation to be interleaved such as in [4, 11]. They discussed about read committed and snapshot isolation. Kemme et al. explained in [13] that snapshot isolation with First Committer Wins

© Springer International Publishing AG, part of Springer Nature 2018
A. Abraham et al. (Eds.): ISDA 2017, AISC 736, pp. 868–879, 2018.
https://doi.org/10.1007/978-3-319-76348-4_83

(FCW) feature can prevent dirty read, lost update, nonrepeatable read, and read skew but it still allows write skew concurrency anomalies. Therefore, it still relies on the locking protocol to preserve consistency or to make interleaved operations are serializable [13].

The latest discussion on the concurrency control is trying to make the snapshot isolation is able to prevent write skew concurrency anomaly and makes the interleaved transactions become serializable [16, 17, 20]. The discussion on making the interleaved transactions in the read committed isolation become serializable is started in [21]. Their approaches are similar with [3, 6], i.e. the system has to abort or restart one of the interleaved transactions if conflict pattern called dangerous structure appears [22].

This discussion above has same objective i.e. to preserve consistency at any cost and trade off and it is performed at system level. While, the blind write protocol is proposed to give more option to the application to allow more write operation to be interleaved with the consequence that preserving consistency becomes application system responsibility. This approach can be applied in the application code level.

We found several discussions on the blind write but no one discuss it in detail. On 1981, Stearns et al. explained in [7] *"We make the assumption, called the no blind writes assumption, that a process does not issue a write request on a particular entity without first issuing a read request on that entity."* On 1994, Mendonca et al. explained in [10] *"In this paper we present a new replica control protocol that logically imposes a hierarchy onto the set of copies and introduces the blind write as another operation. During a blind write operation, copies are modified regardless of their previous values; such situation occurs, for instance, in initializations."* On 1997, Burger et al. explained in [12] *"One of the significant differences between our work and the works reviewed above is that we have simulated a write as a blind write (a read is not performed before the data item is written)."*

This paper is discussing the blind write in detail to understand more on how the blind write protocol is able to achieve or deliver the consistency required by the application system. Terry explained in [18], *high availability is not sufficient for most application system, but strong consistency is not needed either.* Vogels argued in [15], *there is a range of applications that can handle slightly stale data, and they are served well under this model.* In other hand, Bernstein argued in [19] that the high availability increases the application complexity to handle inconsistent data. The discussion is started with the concurrency control. Then, we describe about blind write protocol and its implementation in next section. The last section concludes the topic.

2 Concurrency Control

The discussion on the concurrency control aims to preserve the consistency. The more transactions are being processed then it will increase the throughput of accesses to the database [6], but it can result to an inconsistent database state [5]. Therefore, database system requires a concurrency control to handle two or more transactions, that access the same entity, in order to preserve the consistency. In the absence of concurrency control, any two or more transactions will have concurrency anomaly. Gray et al. in [4], Bernstein et al. [8], Berenson et al. [11] and Kemme et al. [13] discussed about uncommitted (dirty) read, lost update, inconsistent retrieval, non-repeatable read, read skew, write skew anomaly.

The discussion in [11, 13, 14] is focusing on allowing more operation to be interleaved because of strict 2-phase-locking protocol, all locking is released after the transaction perform commit or rollback [9]. Streans et al. [7] argue that since the new value of T_1 has not been committed yet, then it is natural choice to make the T_2 wait until T_1 is committed. However, Streans explained further, since the database system also has the initial value before any other transaction commit their changes, then giving the initial value to T_2 can improve the concurrency [7]. The uncommitted read, committed read, snapshot isolation level and read lock were proposed to improve the concurrency.

Uncommitted read allows the read operation to get uncommitted entity value of other transaction. But DBMS will become inconsistent if the uncommitted transaction is rollback. Committed read allows the read operation to get committed entity value only. But two read committed operations to the same entity in one transaction may return different value. It is due to in between two operations there could be a write operation by other transaction. It is known as non-repeatable read anomaly.

Read lock, and snapshot isolation guarantee the two read operations in one transaction get same entity value. Read lock should wait until the lock is released by other transaction or preempt. Snapshot isolation was proposed to avoid read lock [11]. Snapshot isolation reads the entity value from the log, but it still relies on the locking protocol to prevent write skew [13].

These concurrencies control with different protocols and their limitation give base knowledge to the blind write protocol discussion. It is an important information to develop an algorithm that can handle concurrency anomaly and preserve consistency in blind write protocol.

3   Blind Write Protocol

The blind write protocol is proposed as a complement to allow more write operation to be interleaved and no transaction should be restarted. Since the blind write protocol is a complement then the application system has another option to perform write operation. If application system does not want to create their own specific approach to achieve consistency, then it can use normal write protocol. Moreover, since these two write operation, i.e. normal and blind, can be used together, then the blind write protocol should be able to make them work together.

There will be three combinations if two write operations, i.e.:

1. both are normal write protocols,
2. one transaction is normal and another one is blind write protocols,
3. both are blind write protocols.

Point no. 1 above is clear. Since both are using locking protocol, then one transaction should wait or preempt. Before we discuss point no. 2, let discuss point no. 3 first. Because, we should know how to develop the approach to prevent the lost update and write skew anomaly when two blind write operations are executed at the same time.

3.1   Two Blind Write Operation

To begin with, let start with making proper definition and its principal of blind write protocol. This definition is related to DBMS discussed in [1, 3, 6] which refers to [1]. Interaction between client end point with DBMS is known as transaction. This interaction consists of one or more operations. The operation can be read or write. Write operation is an action to create new entity, modify or delete any entity. Read operation is an action to get entity value.

Blind Write Definition

Before we discuss more detail on how to handle 2 or more interleaved transactions that use blind write protocol, we need to give proper definition on database system. We define database system as D which consists of n number of entity.

$$D = \{e_1, e_2, e_3, \ldots, e_n\}$$

These entities can be either tables, rows, or columns. This paper is focusing on the Data Manipulation Language (DML) protocol, which create, modify or delete a row into, in, or from a table. The Data Definition Language (DDL) is not part of our paper scope. We also consider that modifying a current value of one column as modifying a row. Therefore, the write operation is action to assign a value to the entity. We use $\leftarrow$ notation as assigning a value on the right to entity on the left as discussed on Sect. 2.

Create operation is considered as assigning any value, v, to new entity, e_{n+1},

$$e_{n+1} \leftarrow v \text{ where } v \text{ is not NULL.}$$

Delete operation is considered as assigning NULL to existing entity, e_i,

$$e_i \leftarrow NULL; \text{ where } 1 < i < n.$$

Modify/update operation is considered as assigning a value, v, to existing entity, e_i,

$$e_i \leftarrow v; \text{ where } v \text{ is not NULL and } 1 < i < n.$$

The value of v above can be defined as:

1. function of any entity, e_j. Therefore, it is known as normal write operation.

$$e_{n+1} \leftarrow f(e_j); \text{ where } 1 \leq j \leq n,$$
$$e_i \leftarrow f(e_j); \text{ where } 1 \leq i \leq n, \text{ and } 1 \leq j \leq n.$$

If i = j then it means the new value depends on the initial value of entity.
2. Constant or Fixed value, e.g. 'APPROVED', '536980 MALAYSIA', '+6012345678', 20, etc. Therefore, it is known as blind write operation.

$$e_{n+1} \leftarrow c; \text{ where c is fixed value and c is not NULL.}$$

$$e_i \leftarrow c; \text{ where } 1 < i < n \text{ and c is fixed value and c is not NULL.}$$

Since delete operation is considered as assigning NULL to the entity, then there is no different between normal and blind write protocol. The main different between them is that blind write protocol will not apply any locking to any entity. Based on Bernstein argument in [20] that *the high availability increases the application complexity to handle inconsistent data*. One concrete example is handling lost update and write skew anomaly.

Achieving the Consistency using Blind Write Protocol

The example of lost update and write skew anomaly can be seen in Sect. 2. In that example, it is utilizing one entity only to handle and maintain the operation. The entity in that example is considered as a table. To give more explanation please see Table 1 below. It is a **balance table** consist of one entity, in this case the entity is a row, with 4 columns i.e. *account_id*, *account_number*, *balance_amount* and *last_updated_date*.

Table 1. Balance table

account_id	account_number	balance_amount	last_updated_date
1	1234-567-890	1000	10-Jan-1980 00:00:01

As explained above, the value of blind write operation should be a fixed value, c. Therefore, one table is not enough to preserve the consistency using blind write protocol. To achieve that, then it required at least one table to handle and maintain historical write operation as can be seen on Table 2.

Table 2. History table

history_id	account_id	transaction_amount	transaction_date	status
0	1	1000	10-Jan-1980 00:00:01	Approved
1	1	100	20-Jan-1980 00:00:01	Approved
2	1	300	20-Jan-1980 00:00:01	Approved

The **history table** has foreign key of **balance table**, i.e. *account_id*. For deposit operation then the *transaction_amount* should be greater than 0. For withdraw operation, the *transaction_amount* should be less than 0. The *transaction_date* is used to record the timing when the operation is committed. The *status* is used to differentiate whether the operation is approved or rejected. The *status* will be set to rejected if the operation for particular *account_id* do not meet with specific constrain. The *history_id* is primary key of the **history table**, it is a running number generated from sequence object. Two or more entity (row) may have same *transaction_date* but they should have unique *history_id* value.

Using **history table**, there will be no aborted transaction. All the operation from all transactions will be recorded in this table as one entity (record). The *balance_amount* of particular *account_number* in the **balance table** is aggregation of *transaction_amount* from **history table** which has same *account_id* and the *status* should be approved.

To achieve the consistency using blind write operation, then we need to discuss the possibility of interleaved write operation combination, i.e.:

1. both operations are deposit
2. both operations are withdrawal
3. one operation is withdrawal and other one is deposit.

Let say the entity of **balance** and **history table** is e_b and e_t respectively. It may consist of many *account_id*. To indicate *account_id* $= 1$, we use e_{b1} and $e_t [1]$. The update operation value of e_b is (*balance_amount, last_updated_date*) and for insert operation of $e_t [1]$ is (*history_id, account_id, transaction_amount, transaction_date, status*). If two transactions, T_1 and T_2, are using blind write protocol and executed at the same time follow the same step, as can be seen on Fig. 1, then the result can be same if they are executed one by one, either T_1 first or T_2.

Both Operations are Deposit.

Seq.	Initial State $e_{b1} = 1000$;	
	T_1	T_2
1	begin	begin
2	seq_id ← history_seq.nextval;	seq_id ← history_seq.nextval;
3	$e_t[1]_{n+1}$ ← (seq_id,1,100,sysdate,'approved');	$e_t[1]_{n+1}$ ← (seq_id,1,300,sysdate,'approved');
4	e_{b1} ← ($\{\sum_{j=1}^{n} e_t[1]_j$(transaction_amount); where status='approved'}, sysdate);	e_{b1} ← ($\{\sum_{j=1}^{n} e_t[1]_j$(transaction_amount); where status='approved'}, sysdate);
5	end;	end;
	Final State $e_{b1} = 50$;	

Fig. 1. The aggregation

To achieve this of course there are some conditions need to be applied as follows:

1. The read operation should use read committed isolation level.
2. It should apply auto commit on each write operation to prevent the lost update anomaly.

The first condition is clear. It was explained on the previous section. To show that condition no. 2 is required then let say there are two commit operations. The first commit is between seq. no. 3 and 4 and the second one is between seq. no 4 and 5. The sequence of operation is as follow:

$T_1[\text{seq. no. 1}] \rightarrow T_1[\text{seq. no. 2}] \rightarrow T_1[\text{seq. no. 3}] \rightarrow T_1[\text{commit}] \rightarrow T_1[\text{seq. no. 4}] \rightarrow$
$T_2[\text{seq. no. 1}] \rightarrow T_2[\text{seq. no. 2}] \rightarrow T_2[\text{seq. no. 3}] \rightarrow T_2[\text{commit}] \rightarrow T_2[\text{seq. no. 4}] \rightarrow$
$T_2[\text{commit}] \rightarrow T_1[\text{commit}]$

Since the $T_1[\text{seq. no. 4}]$ has not been committed then the $T_2[\text{seq. no. 4}]$ and the second $T_2[\text{commit}]$ will be overwritten by the second $T_1[\text{commit}]$ which will eventually experience the lost update anomaly. Therefore, to prevent this the blind write protocol should apply auto commit. Since it is using auto commit and balance_amount of e_{b1} is aggregation of transaction_amount of e_t [1], then it always gives the latest result, regardless T_1 is executed first or T_2.

Both Operations are Withdrawal

From previous section, we find that the blind write protocol can handle lost update anomaly without lock any entity. The example above is involving deposit operation only. But how if both operations are withdrawal and it must be in accordance with certain rules as follow:

1. the balance_amount should not be minus
2. the operation should not be rejected if the balance_amount is greater or equal than absolute(transaction_amount). The withdrawal amount is always less than 0.

To discuss this, let say the current balance amount $e_{b1} = 1000$ as shown in Table 1 above. We set two interleaved transactions, T_1 and T_2, and execute at the same time. These transactions are performing withdrawal operation respectively with different scenarios as follows:

1. −100 and −300. Since 1000–100–300 > 0 then both should not be rejected.
2. −900 and −500. Since 1000–900–500 < 0 and 1000–900 > 0 and 1000–500 > 0 then one of them should be rejected and the other one should be approved.
3. −1100 and −900. Since 1000–1100–900 < 0 and 1000–1100 < 0 and 1000–900 > 0 then T1 should be rejected and T2 should be approved.
4. −1100 and −1200. Since 1000–1100–900 < 0 and 1000–1100 < 0 and 1000–1200 < 0 then both transactions should be rejected.

To handle all the scenarios above, we introduce 2 functions. The first function is simple function used to get *account_id* for specific account number from the balance table. The second function has 2 input arguments, i.e. account id and transaction amount. It has one output either true or false. This discussion is focusing more on the second function, we do not explain the first function in detail.

Let name the second function as transact. We modify the steps in Fig. 1 above to implement both functions as shown in Fig. 2 below. The transact function is shown in Fig. 3.

Seq.	Initial State $e_{b1} = 1000$; account_no = '1234-567-890';	
	T_1	T_2
1	begin	begin
2	v_acct_id <-- *getAccountId*(account_no)	v_acct_id <-- *getAccountId*(account_no)
3	if (*transact*(v_acct_id, -100)) then	if (*transact*(v_acct_id, -300)) then
4		
	$e_{b1} \leftarrow (\{\sum_{j=1}^{n} e_t[1]_j(\text{transaction\_amount});$	$e_{b1} \leftarrow (\{\sum_{j=1}^{n} e_t[1]_j(\text{transaction\_amount});$
5	where status='approved'}, sysdate);	where status='approved'}, sysdate);
6	end if;	end if;
	end;	end;
	Final State $e_{b1} = 600$;	

Fig. 2. The aggregation with transac function

Seq.	Transact Function
1	Function *transact* (**a_account_id** number, **a_transaction_amount** number) return boolean
2	begin
3	seq_history_id ← history_seq.nextval;
4	v_status ← 'approved';
5	if (**a_transaction_amount** <0) v_status ← 'not approved';
6	$et[1]_{n+1}$ ← (seq_history_id, **a_account_id**, **a_transaction_amount**, sysdate, v_status);
7	v_sysdate ← get transaction_date from history table where history_id = seq_history_id;
8	if (**a_transaction_amount** >=0) then return true;
9	else
10	v_array ← get $\{\sum_{j=1}^{n} e_t[1]_j$(transaction_amount); where status='approved'} union $[e_t[1]_k$; where $1 \leq k \leq n$ and $e_t[1]_k$(status) ='not approved' and sysdate <= v_sysdate] order by transaction_date and history_id;
11	v_balance ← v_array[0](amount);
12	v_success ← false;
13	for (i=1; i<length(v_array); i++) then
14	v_amount ← v_array[i] (amount);
15	v_history_id ← v_array[i] (history_id);
16	if (v_history_id = seq_history_id) then
17	if (v_balance + v_amount >=0) then
18	v_balance ← v_balance + v_amount;
19	v_success ← true;
20	$et[1]_{seq\_history\_id}$ (status) ← 'approved';
21	else
22	$et[1]_{seq\_history\_id}$ (status) ← 'rejected';
23	end if;
24	else
25	if (v_balance + v_amount >=0) then
26	v_balance ← v_balance + v_amount;
27	end if;
28	end if;
29	end for;
30	return v_success;
31	end if;
32	end;

Fig. 3. Transact function

The explanation of transact function is as follows:

- Seq. no. 1 defines function name, its input argument and output.
- Seq. no. 2 begins the function.
- Seq. no. 3 gets history id from sequence object and put into seq_history_id. It is running number.
- Seq. no. 4 sets default value of v_status to 'approved'. If both operations are Deposit, there is no any validation required since it will not make the balance amount become negative. Hence, the status should always be approved.

- Seq. no. 5 assigns v_status value to 'not approved' if a_transaction_amount is minus (withdrawal operation).
- Seq. no. 6 inserts new record to History table with history_id value is seq_history_id. The operations from seq. no. 3 until 6 can be executed as one statement by utilizing output in insert statement and decode clause. So, it can be treated as one operation. The example of DML statement for these operations is:

insert into history values (history_seq.nextval, a_account_id, a_transaction_amount, sysdate, decode((a_transaction_amount/abs(a_transaction_amount)), 1,'approve','not approve')) returning history_id into seq_history_id;

The explanation for the DML statement above as follows:
The returning *returning history_id into seq_history_id* is used for seq. no. 3.
The *decode((a_transaction_amount/abs(a_transaction_amount)), 1,'approve','not approve')* is used for seq. no. 4 and 5.

- Seq. no. 7 gets transaction_date from the history table where history_id is equal to seq_history_id. The example of DML statement for this operation is:

select transaction_date into v_ sysdate from history where history_id = seq_history_id;

- Seq. no. 8 determines whether the operation is deposit or withdrawal. If a_transaction_amount > 0 then end the function and return true. Otherwise, then it continues to Seq no. 9. It means for deposit operation, it does not need any further validation.
- Seq. no. 9 else condition.
- Seq. no. 10 gets collection of history records for specific account_id. The example of DML statement for this operation is:

select min (transaction_date), −1 history_id, sum(transaction_amount) transaction_amount from history where account_id = a_account_id and status='approved' union
select transaction_date, history_id, transaction_amount from history where account_id = a_account_id and status='not approved' and transaction_-date <=v_sysdate order by transaction_date, history_id

This DML statement is utilizing 'union' that will be executed as one operation. If it does not use 'union' in the statement above, then the DML will become two statements (operations) as follows.

DML statement 1:
select min (transaction_date), −1 history_id, sum(transaction_amount) transaction_amount from history where account_id = a_account_id and status='approved';

DML statement 2:
select transaction_date, history_id, transaction_amount from history where account_id = a_account_id and status='not approved' and history_id <=seq_history_id;

Moreover, if there is blind write operation, which update the status from 'not approved' to either 'approved' or 'rejected', between these two DML statements then it will affect sum(transaction_amount) in DML statement 1 and the collection of records for DML statement 2. This will end with lost update anomaly. Therefore, the third condition required by blind write protocol is:

Since blind write protocol will not lock any entity, then the transaction should request one read operation to be used in validation to prevent write skew anomaly.

To prove this, let execute T_1 and T_2 at the same time. T_1 is a transaction with history_id = 1 and T2 is a transaction with history_id = 2 Table 3. T_1 has executed DML statement 1 and it returns 1000. Then T_2 is executing seq. no. 11 and 20 (it updates and auto commits T_2 status become 'approved' see Fig. 3). If T_1 continue to execute DML statement 2 then it will return one record only since T_2 status has become 'approved'. Thus, T_1 status will be updated and auto committed to 'approved' also because the validation $1000 - 900 > 0$. Now, update Balance table as shown in Fig. 2 seq. no. 4. It shows that the balance amount will become $1000 - 900 - 500 = -400$ since T_1 and T_2 was updated as 'approved'.

Table 3. History table for withdrawal operation

history_id	account_id	transaction_amount	transaction_date	status
0	1	1000	10-Jan-1980 00:00:01	Approved
1	1	−900	20-Jan-1980 00:00:01	Not approved
2	1	−500	20-Jan-1980 00:00:01	Not approved

- Seq. no. 11 assigns sum(transaction_amount) value of approved status to v_balance.
- Seq. no. 12 sets default value of v_success to false.
- Seq. no. 13 until 29 validates the transaction amount with balance amount.
- Seq. no. 32 returns the validation result. If the history status is updated and committed to 'rejected' then it returns false, otherwise it returns true.

Combination of Deposit and Withdrawal Operation

There are no significant obstacles with the deposit operation in this combination. Likewise, with the withdrawal operation. The main obstacle with this combination is about the timing.

As explained above that the seq. no. 10 operation is fetching collection of history record which ordered by *transaction_date* and *history_id*. If two transactions have same *transaction_date* then it will be ordered by *history_id* which is unique for each history record.

3.2   Combination of Normal and Blind Protocols

Until this section, we have already shown that blind write protocol can preserve the consistency in different approach with the normal write protocol. For two or more transactions that use different protocol, then they are two options. First option is the blind write protocol should wait until the locked entity is released. The second option is the blind write protocol should not wait other transaction to release the lock on the entity. These options provide more choice to the application to determine which one suits with the business requirements. This wait and no wait option should be applied in the DML statement along with blind write option.

The wait option will work for blind write protocol to wait until the locked entity is released. Since the blind write protocol will not lock any entity then the normal write protocol can start to perform any operation including lock any entity at any time. Once the entity is locked then any write operation that wants to access the locked entity, including blind write with wait option, should wait or preempt.

4 Summary

This paper proposes blind write protocol as a complement of current concurrency control to give more option to the application on dealing with the interleaved write operation. The blind protocol provides more option besides wait or preempt. The blind write protocol also can be used together with normal write operation with wait or no wait option.

Since, the blind write operation does not use locking protocol, then the database system will experience a lost update and write skew anomaly. Therefore, the blind write protocol should apply their own approach to prevent these anomalies. To achieve this, there are some conditions need to be applied in the transaction as follows:

1. The read operation should use read committed isolation level.
2. It should apply auto commit on each write operation to prevent the lost update anomaly.
3. The transaction should request one read operation to be used in validation to prevent write skew anomaly.

References

1. Codd, E.F.: A relational model of data for large shared data banks. Commun. ACM **13**, 377–387 (1970)
2. Eswaran, K.P., Gray, J.N., Lorie, R.A., Traiger, I.L.: The notions of consistency and predicate lock in a database system. ACM Comput. Surv. **19**, 624–633 (1976)
3. Stearns, R.E., Lewis, P.M., Rosenkrantz, D.J.: Concurrency control for database systems. In: Proceedings of 7th Symposium on Foundations of Computer Science, pp. 19–32 (1976)
4. Gray, J., Lorie, R.A., Putzolu, G.R., Traiger, I.L.: Granularity of locks and degrees of consistency in a shared data base. In: IFIP Working Conference on Modelling in Data Base Management Systems, pp. 365–394 (1976)
5. Bernstein, P.A., Shipman, D.W., Wong, W.S.: Formal aspects of serializability in database concurrency control. In: IEEE, vol SE-5, no. 3 (1979)
6. Kung, H.T., Androbinson, J.T.: An optimistic methods for concurrency control. ACM Trans. Database Syst. **6**(2), 213–226 (1981)

7. Stearns, R.E., Rosenkrantz D.J.: Distributed database concurrency controls using before-values. In: Proceedings of the 1981 ACM SIGMOD International Conference on Management of Data (1981)

8. Bernstein, P.A., Goodman, N.: Concurrency control in distributed database systems. ACM Comput. Surv. (CSUR) **13**(2), 185–221 (1981)

9. Bernstein, P.A., Hadzilacos, V., Goodman, N.: Concurrency Control and Recovery in Database Systems. Addison-Wesley Longman Publishing Co., Inc., Boston (1986)

10. das Chagas Mendonca, N., de Oliveira Anido, R.: Using extended hierarchical quorum consensus to control replicated data: from traditional voting to logical structures. In: Proceedings of the Twenty-Seventh Hawaii International Conference on System Sciences, Wailea, HI, USA, pp. 303–312 (1994)

11. Berenson, H., Bernstein, P., Gray, J., Melton, J., O'Neil, E., O'Neil, P.: A critique of ANSI SQL isolation levels. In: Carey, M., Schneider, D. (eds.) Proceedings of the 1995 ACM SIGMOD International Conference on Management of Data (SIGMOD 1995), pp. 1–10. ACM, New York (1995)

12. Burger, A., Kumar, V., Hines, M.L.: Performance of multiversion and distributed two-phase locking concurrency control mechanisms in distributed databases. Inf. Sci. **96**(1–2), 129–152 (1997)

13. Kemme, B., Alonso, G.: A new approach to developing and implementing eager database replication protocols. ACM Trans. Database Syst. **25**, 333–379 (2000)

14. Balling, D.J., Lentz, A., Zawodny, J.D., Tkachenko, V., Zaitsev, P., Schwartz, B.: High Performance MySQL, 2nd edn. O'Reilly Media, Sebastopol (2008)

15. Vogels, W.: Eventually consistent. Commun. ACM **52**, 40–44 (2009)

16. Alomari, M., Fekete, A., Röhm, U.: A robust technique to ensure serializable executions with snapshot isolation DBMS. In: IEEE 25th International Conference on Data Engineering, Shanghai, pp. 341–352 (2009)

17. Cahill, J., Rohm, U., Fekete, A.D.: Serializable isolation for snapshot databases. ACM Trans. Database Syst. **34**(4), 20 (2009)

18. Terry, D.: Replicated data consistency explained through baseball. Commun. ACM **56**(12), 82–89 (2013)

19. Bernstein, P.A., Das, S.: Rethinking eventual consistency. In: Proceedings of the 2013 ACM SIGMOD International Conference on Management of Data (SIGMOD 2013), pp. 923–928. ACM, New York (2013)

20. Zhou, X., Yu, Z., Tan, K.L.: Posterior snapshot isolation. In: 2017 IEEE 33rd International Conference on Data Engineering (ICDE), San Diego, CA, pp. 797–808 (2017)

21. Alomari, M., Fekete, A.: Serializable use of read committed isolation level. In: 2015 IEEE/ACS 12th International Conference of Computer Systems and Applications (AICCSA), Marrakech, pp. 1–8 (2015)

22. Zendaoui, F., Hidouci, W.K.: Performance evaluation of serializable snapshot isolation in PostgreSQL. In: 2015 12th International Symposium on Programming and Systems (ISPS), Algiers, pp. 1–11 (2015)

Ontology Visualization: An Overview

Nassira Achich[1(✉)], Bassem Bouaziz[1], Alsayed Algergawy[2],
and Faiez Gargouri[1]

[1] Higher Institute of computer science and Multimedia,
University of Sfax, Sfax, Tunisia
achichnassira@gmail.com, {bassem.bouaziz,faiez.gargouri}@isims.usf.tn
[2] Friedrich-Schiller-Universität Jena, Jena, Germany
alsayed.algergawy@uni-jena.de

Abstract. As the main way for knowledge representation for the purpose of completely machine understanding, ontologies are widely used in different application domains. This full machine understanding makes them harder to be easily understood by a human. This necessitates the need to develop ontology visualization tools, which results in the existence of a large number of approaches and visualization tools. Along with this development direction, the number of published research papers related to ontology visualization is largely increasing. To this end, in this paper, we introduce a systemic review on different directions related to ontology visualization. In particular, we start by describing different application domains that make use of ontology visualization. Then, we propose a generic visualization pipeline that incorporates main steps in ontology visualization that could be later used as main criteria during comparing and discussing different visualization tools. By this review, we aim to introduce a general visualization pipeline that is useful when comparing ontology visualization tools and when developing a new visualization technique. Finally, the paper moves into the description of future trends and research issues that still need to be addressed.

1 Introduction

Ontologies are the basic components of the semantic Web, where underlying data are well structured for the purpose of full machine understanding. As defined by Gruber, "an ontology is a formal, explicit specification of a shared conceptualization". It consists of a set of concepts (classes), a set of attributes (data type properties), relationships (object properties), and constraints to abstractly represent a specific event [20,36]. An important aspect is how to facilitate the process of design, manage, and browsing such kind of complex structure. This results in a growing needs for ontology visualization tools that simplify the user involvement in these ontology-based management processes [24].

As a main way of knowledge representation, they have been becoming more and more largely used in different application domains. Even its importance in different application domains, ontology visualization is not a simple task. Since

© Springer International Publishing AG, part of Springer Nature 2018
A. Abraham et al. (Eds.): ISDA 2017, AISC 736, pp. 880–891, 2018.
https://doi.org/10.1007/978-3-319-76348-4_84

an ontology is more than a hierarchy of concepts. It further models role relations among concepts, and each concept has a set of properties attached to it. Therefore, a large number of ontology visualization approaches have been proposed and a set of tools have been developed [11,24,31,33,36]. Ontology visualizations is an important step for working with ontologies and it should provide a multi-level view in order to handle even entities, classes, properties, and blocks. To be useful, ontology graphical representation tools should encourage domain experts in the ontology creation and manipulating processes. Consequently, a smart visualization tool will enable the direct input from domain experts and reduce the dependency on knowledge engineers at every step of ontology development.

For example, by reducing the semantic gap between different overlapping representations of the same domain ontologies, visualizations techniques should provide support for ontology matching techniques. The process and the result of ontology matching need to be visualized as well, to get contextual feed back from user and produce better alignment results [2,11]. Ontologies visualization tools like *CODEX* [22], *REX* [10] and *OntoVIEW* [34] focus on ontology evolution, aiming at tracking and marking the changes happened between versions of ontologies. In this field, visualization tools should provide distinguish representation symbols, color or pattern to differentiate the old ontologies content (concepts, links, entities, etc.) from the newest.

Motivated by these challenges, in this paper, we introduce a systematic review of existing ontology visualization approaches and tools in order to draw a road map of using ontology visualization implementations in ontology-based management systems. In particular, we start by motivating this review by presenting different application domains that make use of ontology visualization. We further introduce a generic ontology visualization pipeline to be used as a basis for discussing and comparing existing tools. After that, we elaborate a set of visualization tools. The paper also includes a discussion on the challenges and benefits that the field of ontology visualization brings forward. It is hoped that the survey would be helpful both to developers and to users. The rest of the paper is organized as follows: a set of application domains that benefit from ontology visualization is presented in the next section. We then introduce a generic pipeline that guides the ontology visualization in Sect. 3. In Sect. 4, we review ontology visualization tools and techniques. In Sect. 5, we discuss current state of the art solutions and enumerate most common challenges to be solved in our future work.

2   Application Domains

As mentioned before, ontologies are being widely used in different domains, and visualizing the ontology becomes an important step in the whole pipeline within these domains. To motivate the importance of ontology visualization, we summarize its use in some of these application domains.

- **Ontology creation and development** - Developing ontologies is an important aspect of the Semantic Web [19]. For this reason, there exist a number of tools for ontology development [19], such as Protege [26], SWOOP [32], OntoLingua [15] and OBO-edit [9]. Ontology development tools allow the user to create and/or to modify the ontology by adding new concepts, relations and instances. Those tools may also contain many features like graph visulizor and search and constraint checking capabilities [19].

– **Ontology browsers** - Several tools were developed to browse ontologies. Some of them are dedicated to analyse specific data sets such as agriGO [12], some others are for ontology exploration in general, such as Amigo [7], OLSViz [39] and FLEXViz [13]. Text-based ontology viewer, like Amigo [7], use a folder/subfolder-interface to explore hierarchies [39]. OLSViz [39] and FLEXViz [13], were created to improve the exploration of bio-ontologies. However, OLSViz [39] makes more efficient and intuitive use of the available screen space than FLEXViz [13].

– **Ontology matching** - is the process that identifies and discover corresponding concepts across different ontologies [35]. It plays a crucial role in different shared-data applications, such as data and ontology integration [30]. Due to its importance, many matching algorithms have been proposed and a myriad of matching tools have been developed. Most of these tools provide a way to visualize the whole matching process for different goals [28,37]. One is to enhance and support visual analytic in a semi-automatic matching process [28]. For example, *ENViz* is an approach to integrate data that performs joins enrichment analysis of two type matched datasets. In that tool, the role of ontology visualization is obvious to support user analysis during matching different datasets [28]. Another dimension is to involve the user in the matching process to validate the automatically generated alignment [11]. To sum up, there are different ways of user intervention in the ontology matching process: to select base matchers, to adjust similarity weights, or to validate matching result. All these kinds require an effective way to visualize ontologies to support user intervention.

– **Ontology evolution** - is the process that timely adapts of the ontology due to the arisen changes and the consistent propagation of these changes to dependent artifacts [38]. For example, and based on statistics on BioPortal[1], the gene ontology in one month has about 17 different versions, which indicates a high rate of expanding the ontology. This necessitates the need for tools that support users during the ontology evolution process [27]. An ontology evolution system should have a set of functionalities: among them, showing ontology version, compare different versions, identify conflicts, showing conflicts, etc. Ontology visualization techniques could be used to support the desired functionality [5,27]. For example, *REX* is a tool for discovering evolution in ontology regions by providing an interactive and user-friendly visualization to determine (un)stable regions in large life science ontologies [10]. *CODEX* is a tool that allows identifying semantic changes between two versions of an ontology which users can interactively analyze in multiple ways [22].

---

[1] http://bioportal.bioontology.org/ontologies/GO.

– **OBDA** - Ontology-based data access (OBDA) is an elegant approach to improve data accessibility in which one ontology (or a set of ontologies) can be used as a mediator between data users and data sources [6,18,25,36]. In the context of *OBDA* systems, visualization can be used in different scenarios. The development of the conceptual layer ontology is a clear scenario where ontology visualization is needed, where the ontology engineer can manage the ontology development and evolving process. Another dimension is the development of an ontology-based visual query system as an extension for the stream temporal one [36].

3   Generic Visualization Pipeline

In order to conduct a good survey and to construct a fair basis for comparing existing ontology visualization tools, a high-level architecture for a generic pipeline for ontology visualization is proposed. Figure ?? depicts the pipeline with three basic steps: ontology parsing, processing, and visual representation.

3.1   Ontology Representation

As mentioned before an ontology is a formal representation of a set of concepts within a domain and their relationships. In designing an ontology language, a tradeoff between the power of expressiveness and the efficiency of reasoning should be considered. As a result, there exist variant ontology languages based on these two criteria [1,21]. For example, the Ontology Web Language (*OWL*) is currently the standard language for the semantic Web and it is compatible with early ontology languages, such as *SHOE* (Simple HTML Ontology Extension), *DAIM + OIL* (DARPA Agent Markup Language + Ontology Interchange Language). However, it is not feasible to satisfy the tradeoff between the expressiveness and the efficiency of reasoning, *OWL* comes with three various formats:

OWL Lite adds the possibility to express definitions and axioms, together with a limited use of properties to define classes. *OWL DL* supports those users who want the maximum expressiveness while retaining good computational properties. *OWL Full* is meant for users who want maximum expressiveness with no computational guarantees.

After reading and parsing an input ontology, the next step is how to internally represent it. Two common representation scheme could be used: *tree-based* or *graph-based* scheme. The first one is used to illustrate super/subclass relationships. Several tools such as *SQuaRE* [3] and *CODEX* [22] adopted this technique. The second one is the node-link diagrams (i.e., graphs), which represents ontologies as a set of interconnected nodes through edges that illustrate ontological entities and the relationships that exist among them. A number of visualization techniques have been used over the tools, such as tree-maps, tree-layouts, fisheye views [16], birdeye views, hyperbolic and 3D hyperbolic layouts. *REX* [10], the ontology evolution tool, as an example, used the fisheye view, since it makes the selected concept in the center, surrounded with its subconcepts. It's useful when it's the case of displaying a large structures. An ontology visualization tool should be generic and has the ability to handle ontology in different languages and formats. Therefore, a first step in the visualization is *ontology representation*, how the tool reads the input ontology and how it internally represents the ontology to capture its content. To simplify reading input ontologies, several APIs and frameworks have been proposed. Two commons are Apache Jena API[2] and the OWL API[3].

3.2   Processing

As part of the whole ontology visualization pipeline, processing has to support not only the presentation and of ontology components which are: classes (or entity types), relations, instances, and properties (or slots), but also the coarse and fine ontology manipulation. Those operations are fundamentals when dealing with domain expert ontology creation, editing and validating.

Basically, processing step might include zoom-in and zoom-out for locating specific nodes and provide a comprehensive view of the hierarchy level the user is zoomed in or out. Rotation and moving nodes are also the basic operation that should be integrated. However, an ontology visualization tool as we assume, should also contain some deep operations, like ontology alignment and merging. Consequently, matching two ontologies, for example, can be achieved by automatically discovering correspondences between nodes and by graphical mapping interfaces that might assist the process of refining these correspondences. Additionally, merging two ontologies in a new one, translating a query addressed to a source ontology into a query addressed to a target ontology should be mapped into the user interface and then handled graphically. As fine operation, merging two concepts in ontologies means that the user has to determine the scope of overlapping and decide if two concepts or more that are similar.

Verbalization is a part of processing step that should be integrated into the visualization tool. It gives a textual description or details about the selected ontology element from the graphical representation. In order to facilitate the access to ontology content, querying and search related operations should allow highlighting a specific element that user is looking for. An interpreter that parse textual or visual query is part of the processing step. As for the evolution related operations, history browsing, saving and loading of customized ontologies views is required. Consequently, a toolbox managing ontology version should be considered. Hence, our goal is to give an easy graphical way to deal with ontology processing.

3.3   Visual Representation

An ontology is composed of several elements. Visual representation of the ontology is to represent those elements visually and conceptually, in a way that the

---

[2] https://jena.apache.org/.

[3] http://owlapi.sourceforge.net/.

user can understand the whole hierarchy of the ontology [24]. The visualization method should display all the ontology classes, providing at least their name, in an intelligible manner. Also, it should display the data associated with the ontology which called the instances. The presentation of the "Isa relations" on which the ontology is based is also essential for understanding the inheritance relations between classes. For this reason, the system should at least provide a holistic view of this taxonomy, in a hierarchical representation. Finally, the properties associated with an entity are also very important and a complete visualization should include their representation, either on the main ontology visualization or within separate space.

4   Visualization Tools

This section is devoted to present current ontology visualization tools directed by generic steps in the proposed pipeline.

- **OWLGred**[4] - is implemented as a web based application for ontology visualization [29]. After parsing input data using OWL API, the processing step includes data transformation aiming to retrieve information necessary for visualization. The analysis of axioms (*TBox* and *ABox* elements) aims at constructing UML-class diagram for the representation since most OWL features are mapped to UML concepts (OWL classes to UML classes, datatype properties to class attributes). As for verbalization process, authors of the tool used a Grammatical framework to facilitate the implementation of the CNLs. The user can then select an element from the ontology to generate a CNL verbalization of the corresponding axioms in ACE (Attempto Controlled English). Visual representation of this tool is provided through UML-notations. Generated graphs use an orthogonal layout where the inheritance-defining relations are presented in a hierarchical layout [29] and all other relations "flow" in the direction perpendicular to it. It is based on HTML5 canvas[5] element using KineticJS library where visualization is drawn according to the graph structure, element coordinates, and styles contained in its JSON structure.
- **ProtgVOWL**[6] - are plugins for the ontology editor Protege [31], aiming to give a visual language dedicated not only for human expert, but also for users not familiar with ontologies. After parsing by OWL API, the processing step allows to transforming internal representation of parsed ontologies into the data model required by the Prefuse visualization toolkit[7] [23]. For visual representation, protégéVOWL is based on VOWL specifications which provide a visual language that can be understood by users less familiar with ontologies. Graphical primitives forming the alphabet of the visual language contains circles to depict classes, lines representing to represent property relations, while

---

[4] http://owlgred.lumii.lv/.

[5] https://developer.mozilla.org/en-US/docs/Web/API/Canvas_API.

[6] http://vowl.visualdataweb.org/webvowl.html.

[7] http://prefuse.org/.

property labels and datatypes are shown in rectangles. VOWL defines also a color scheme for a better distinction between the different elements. Interface of VOWL contains three main layouts which are: VOWL Viewer for displaying the ontology, VOWL Sidebar for giving details about a selected element from the ontology and VOWL Controls for adapting the force directed graph layout.

– **MEMO GRAPH** [17] - is an ontology visualization tool that follows the design-for-all philosophy. A preprocessing step is based on the parsing of OWL/RDF using OWL API. For the processing step, authors integrate a force-directed algorithm to display visualization and provide zoom in-out on nodes of ontologies. Besides, the tool supports the search functionality by keywords based search according to modalities typing and dictation. It supports also node selection to look for information expressed in easy to understand wording rather than the ontology jargon. Enriched by delivering auditory information, visual representation in *MEMO GRAPH* covers all the key elements of the ontology which are classes, instances, data and object properties. Relations between related nodes are represented using labeled links. Nodes of the ontology generated graph are identified by using pictures and labels. The interface of MEMO GRAPH is divided into three parts: The "MEMO GRAPH Viewer" displaying the ontology visualization as a graph, the "MEMO GRAPH details" listing details about a selected node, and the "MEMO GRAPH search" providing a key word search option.

– **OntoSphere**[8] - is a 3D ontology visualization tool developed in Java as a stand-alone application [4]. For the preprocessing step Jena API is chosen, consequently the tool can allow to easily load and manage ontologies and taxonomies written either in RDF, RDFs, DAML, OWL or N-triple. The processing step has many manipulation features provided for users like the rotation, zoom, object selection, etc. It relies on browsing ontology as well as updating it by adding new concepts and new relations. The visual representation is based mainly on a 3D representation on which authors worked on increasing the number of dimensions (colors, shapes, transparency, etc.). It exploits different scene managers as *RootFocus*, *TreeFocus* and *ConceptFocus* that present and organize the information on the screen. The RootFocus Scene presents a big earth-like sphere bearing on its surface a collection of concepts represented as small spheres. The TreeFocus Scene shows the sub-tree displaying the hierarchical structure as well as semantic relations between classes. The ConceptFocus Scene depicts all the available information about a single concept, at the highest possible level of detail.

– **REX**[9] - Region Evolution Explorer is based on a region discovery algorithm used mainly to determine differently changing regions for periodically updated ontologies and to interactively explore the changing intensity of those regions [10]. The tool handle supports the import of ontologies in different

---

[8] http://ontosphere3d.sourceforge.net/userGuide.html.
[9] https://dbs.uni-leipzig.de/de/research/projects/evolution_of_ontologies_and_mappings.

formats such as OWL and OBO. Based on the resulted parsed ontologies, the processing step aims at first computing differences between two versions to determine changes. It then propagates change costs within the is-a hierarchy of the ontology and transfers these costs from the first to the last considered version. Based on computed change intensities differently evolving ontology regions are discovered. Visual representation of ontology is a graph where nodes represent the concepts and edges represent the relations. The color feature was used in this tool to describe the concepts change intensity. Quantitative changes are also visualized in a graphical way as to curve of frequencies changes.

– **CODEX** - is an ontology evolution tool, used to obtain knowledge about the evolving ontology [22]. It provides support for determining complex changes between two versions of ontologies. The tool deals with OBO and OWL ontologies and flat files formats in the preprocessing step. The processing offered by the tool, supports changes such merging, splitting, adding and moving subgraph. Others functions are implemented to facilitate exploration of changes. The user interface contains multiple views. A high-level view provides statistics about the number of relations and concepts in two ontology versions as well as the number of changes between the versions. After selecting a change, the changes can be explored in a tree-like manner. Hence, customizable overview statistics such frequent updates nodes, Tag clouds to visualize changes and modified content Tree-based change explorer and Impact analysis.

– **Other tools** - *OWLeasyViz* [8] is based on an approach that combines a textual and graphical representation of OWL ontologies, where the textual representation presents class, data properties and object properties in a three-column table and the graphical representation which contains the graphical representation of the ontology, where we find child nodes which are visualized inside their parents, with smaller size. Also, nodes are represented by many shapes according to their hierarchy. Leaf nodes are visualized as rounded rectangles, while parent nodes as elliptical shapes. Also in this tool, they used zoomable technique, and also Searching and filtering mechanisms. They exploit the visualization strategies used in Grokker, which is a generic system for displaying of knowledge maps.

SQuaRE [3] is a query and R2RML mappings environment. It provides a visual editor for creating and managing R2RML mappings as well as for creating and executing SPARQL queries. It contains two sides, the client side, and the server side. The client side consists of modules working on a client's machine in a user's web browser and provides a presentation layer. The server side contains the data source, DBMS manager, ontology handler for OWL ontology processing, and SPARQL query executor. SQuaRE applies a set of tools to handle OWL ontologies, relational data and SPARQL queries. For the visualization of ontologies, it uses OWL-API and javascript libraries like *AngularJS*[10], *jQuery*[11], *Cytoscape.js* with *CoSE Bilkent* layout, *jsPlumb*[12] and *jsTree*[13].

ViZiQuer is a tool for data analysis query definition and translation into SPARQL [40]. The ViziQuer notation is based on UML class diagrams. A query in the ViziQuer notation is a graph of class boxes connected with association links. Visualizer tool for the Agreement Maker system[14] [14] focuses on ontology matching visualization. They add to Agreement Maker tool, which is an ontology matching tool, the capability to visualize the results of matching large ontologies with a user interface that supports navigation and search of the ontologies.

5   Discussion and Future Directions

Keeping in mind the pipeline that we described for the ontology visualization tools, the overview of most existing works shows that the preprocessing step is fundamentals in the whole system and consists on primitives designed for parsing ontologies even in OWL or RDF based on JENA or OWL APIs. This preprocess in sometimes extended by some mapping primitives to generate intermediate data representations aiming at facilitating the processing step.

Actually, the main operations enabled in most of the current systems are limited to some transformation functions aiming at generating graphs except a few number of tools focusing on ontologies evolution. In that case, most functions are related to matching between two versions of ontology and aiming at highlighting the change in general by intensity color and distinguished links. Some quantitative measures to visualize the frequency of changes are implemented in these tools. Search functions are also provided based on keyword textual query.

Within the last step of the pipeline, the visual representation is handled in most of the existing tools by a hierarchical representation based on graphs (circle, line, color, etc.) and less on such UML language notation. A particular attention is given to 3D visualization based tools which add a new dimension to get more flexibility for ontology content visualization. The main objective is to facilitate navigation in the ontology and providing multi level view options.

Although the interest given to ontology visualization expressed in recent publications and already developed tools, we think that the goal of building an ontology tool for unfamiliar user is far to be reached. The challenges arisen can be summarized in the following points:

- Dealing with very large ontologies since most of the tools suffer from overlapped links and labels due to lack of compact visualizations widgets.

---

[10] https://angularjs.org/.
[11] https://jquery.com/.
[12] https://jsplumbtoolkit.com/.
[13] https://www.jstree.com/.
[14] https://github.com/AgreementMakerLight/AML-Jar.

- Looking for optimization techniques aiming at accelerating the process of large ontology even for parsing and visualization.
- Most tools are designed only for ontology experts and lack for generality.
- Few represent all key elements of the ontology (i.e., classes, instances, datatype properties and object properties) and show data properties as labeled links.
- Basics processing integrated into most of the tools need to be extended by more advanced operations such as merging nodes according their similarities (matching). Those primitives when implemented can provide a real support for domain expert construction and editing ontologies.
- Implement visual language representation for ontology based on standards is a step for building common useful tools.
- Integrate natural language processing techniques to the verbalization process. This should provide the semantic description of ontology elements and make ontology content common for expert and non-expert users.
- Working on dynamic visualization approach that could be adapted to the content and size of input ontology.
- Including ontology evolution related primitives to manage changes and version.

As for continuity of our work on ontology matching, alignment, and merging, we are currently working on an approach that provides at first a guideline for the development of ontology visualization tool. This guideline considers all the above challenges and aims at designing and implementing tools that can easily deal with large ontologies and generalize its use by nonexpert users.

Acknowledgments. This work was supported by the DAAD funding through the BioDialog project.

References

1. Antoniou, G., Franconi, E., van Harmelen, F.: Introduction to semantic web ontology languages. In: First International Summer School on Reasoning Web, Tutorial Lectures, pp. 1–21 (2005)
2. Aurisano, J., Nanavaty, A., Cruz, I.F.: Visual analytics for ontology matching using multi-linked views. In: Proceedings of the International Workshop on Visualizations and User Interfaces for Ontologies and Linked Data Co-located with 14th International Semantic Web Conference (ISWC 2015), p. 25 (2015)
3. Blinkiewicz, M., Bąk, J.: SQuaRE: a visual approach for ontology-based data access. In: LNCS (including subseries LNAI and LNBI), vol. 10055, pp. 47–55. Springer, Cham (2016)
4. Bosca, A., Bonino, D.: OntoSphere3D: a multidimensional visualization tool for ontologies. In: 17th International Workshop on Database and Expert Systems Applications (DEXA 2006), pp. 339–343 (2006)
5. Burch, M., Lohmann, S.: Visualizing the evolution of ontologies: a dynamic graph perspective. In: International Workshop on Visualizations and User Interfaces for Ontologies and Linked Data Co-located with ISWC 2015, p. 69 (2015)

6. Calvanese, D., Cogrel, B., Komla-Ebri, S., Kontchakov, R., Lanti, D., Rezk, M., Rodriguez-Muro, M., Xiao, G.: Ontop: answering SPARQL queries over relational databases. Semant. Web **8**(3), 471–487 (2017)
7. Carbon, S., Ireland, A., Mungall, C.J., Shu, S., Marshall, B., Lewis, S.: AmiGO: online access to ontology and annotation data. Bioinformatics **25**(2), 288–289 (2009)
8. Catenazzi, N., Sommaruga, L.: Generic environments for knowledge management and visualization. J. Ambient Intell. Hum. Comput. **4**(1), 99–108 (2013)
9. Catenazzi, N., Sommaruga, L., Mazza, R.: User-friendly ontology editing and visualization tools: the OWLeasyViz approach. In: 13th International Conference Information Visualisation, pp. 283–288 (2009)
10. Christen, V., Hartung, M., Groß, A.: Region evolution explorer - a tool for discovering evolution trends in ontology regions. J. Biomed. Semant. **6**, 26 (2015)
11. Dragisic, Z., Ivanova, V., Lambrix, P., Faria, D., Jiménez-Ruiz, E., Pesquita, C.: User validation in ontology alignment. In: 15th International Semantic Web Conference (ISWC 2016), pp. 200–217 (2016)
12. Du, Z., Zhou, X., Ling, Y., Zhang, Z., Su, Z.: agriGO: a GO analysis toolkit for the agricultural community. Nucleic Acids Res. **38**(suppl 2), W64–W70 (2010)
13. Falconer, S.M., Callendar, C., Storey, M.-A.: FLEXVIZ: visualizing biomedical ontologies on the web. In: International Conference on Biomedical Ontology, Software Demonstration, Buffalo, p. 1 (2009)
14. Faria, D., Martins, C., Nanavaty, A., Taheri, A., Pesquita, C., Santos, E., Cruz, I.F., Couto, F.M.: Agreementmakerlight results for OAEI 2014. In: Proceedings of the 9th International Workshop on Ontology Matching Collocated with the 13th International Semantic Web Conference (ISWC 2014), Riva del Garda, pp. 105–112, 20 October 2014
15. Farquhar, A., Fikes, R., Rice, J.: The Ontolingua server a tool for collaborative ontology construction. Int. J. Hum. Comput. Stud. **46**(6), 707–727 (1997)
16. Furnas, G.W.: Generalized fisheye views. In: Proceedings of the SIGCHI Conference on Human Factors in Computing Systems, pp. 16–23, April 1986
17. Ghorbel, F., Ellouze, N., Métais, E., Hamdi, F., Gargouri, F., Herradi, N.: MEMO GRAPH: an ontology visualization tool for everyone. In: 20th International Conference on Knowledge-Based and Intelligent Information & Engineering Systems (KES-2016), pp. 265–274 (2016)
18. Giese, M., Soylu, A., Vega-Gorgojo, G., Waaler, A., Haase, P., Jiménez-Ruiz, E., Lanti, D., Rezk, M., Xiao, G., Özçep, Ö.L., Rosati, R.: Optique: zooming in on big data. IEEE Comput. **48**(3), 60–67 (2015)
19. Grover, P., Chawla, S.: Ontology creation and development model. **6**(1991), 3878–3883 (2014). ijircce.com
20. Guarino, N., Oberle, D., Staab, S.: What is an ontology? In: Handbook on Ontologies, pp. 1–17 (2009)
21. Gutierrez-Pulido, J.R., Ruiz, M.A.G., Herrera, R., Cabello, E., Legrand, S., Elliman, D.: Ontology languages for the semantic web: a never completely updated review. Knowl. Based Syst. **19**(7), 489–497 (2006)
22. Hartung, M., Groß, A., Rahm, E.: CODEX: exploration of semantic changes between ontology versions. Bioinformatics **28**(6), 895–896 (2012)
23. Heer, J., Card, S.K., Landay, J.A.: prefuse: a toolkit for interactive information visualization. In: Proceedings of the 2005 Conference on Human Factors in Computing Systems (CHI 2005), pp. 421–430 (2005)
24. Katifori, A., Halatsis, C., Lepouras, G., Vassilakis, C., Giannopoulou, E.G.: Ontology visualization methods - a survey. ACM Comput. Surv. **39**(4), 10 (2007)

25. Kharlamov, E., Hovland, D., Jiménez-Ruiz, E., Lanti, D., Lie, H., Pinkel, C., Rezk, M., Skjæveland, M.G., Thorstensen, E., Xiao, G., Zheleznyakov, D., Horrocks, I.: Ontology based access to exploration data at statoil. In: 14th International Semantic Web Conference (ISWC 2015), pp. 93–112 (2015)

26. Klein, M., Fensel, D., Kiryakov, A., Ognyanov, D.: Ontology versioning and change detection on the web. In: Knowledge Engineering and Knowledge Management: Ontologies and the Semantic Web, pp. 197–212 (2002)

27. Lambrix, P., Dragisic, Z., Ivanova, V., Anslow, C.: Visualization for ontology evolution. In: Second International Workshop on Visualization and Interaction for Ontologies and Linked Data co-located with the 15th International Semantic Web Conference, VOILA@ISWC 2016, pp. 54–67 (2016)

28. Li, Y., Stroe, C., Cruz, I.F.: Interactive visualization of large ontology matching results. In: International Workshop on Visualizations and User Interfaces for Ontologies and Linked Data Co-located with ISWC 2015, p. 37 (2015)

29. Liepins, R., Grasmanis, M., Bojars, U.: Owlgred ontology visualizer. In: Proceedings of the ISWC Developers Workshop 2014, pp. 37–42 (2014)

30. Livingston, K.M., Bada, M., Baumgartner Jr., W.A., Hunter, L.E.: KaBOB: ontology-based semantic integration of biomedical databases. BMC Bioinform. **16**, 126:1–126:21 (2015)

31. Lohmann, S., Negru, S., Haag, F., Ertl, T.: Visualizing ontologies with VOWL. Semant. Web **7**(4), 399–419 (2016)

32. Malik, S.K., Rizvi, S.: Ontology design towards web intelligence: a sports complex ontology case study. In: Fourth International Conference on Computational Aspects of Social Networks (CASoN), pp. 366–371 (2012)

33. Negru, S., Lohmann, S.: A visual notation for the integrated representation of OWL ontologies. In: 9th International Conference on Web Information Systems and Technologies (WEBIST 2013), pp. 308–315 (2013)

34. Noy, N.F., Klein, M.C.A.: Ontology evolution: not the same as schema evolution. Knowl. Inf. Syst. **6**(4), 428–440 (2004)

35. Shvaiko, P., Euzenat, J.: Ontology matching: state of the art and future challenges. IEEE Trans. Knowl. Data Eng. **25**(1), 158–176 (2013)

36. Soylu, A., Giese, M., Jiménez-Ruiz, E., Kharlamov, E., Zheleznyakov, D., Horrocks, I.: Ontology-based end-user visual query formulation: why, what, who, how, and which? Univ. Access Inf. Soc. **16**(2), 435–467 (2017)

37. Steinfeld, I., Navon, R., Creech, M.L., Yakhini, Z., Tsalenko, A.: ENViz: a cytoscape app for integrated statistical analysis and visualization of sample-matched data with multiple data types. Bioinformatics **31**(10), 1683–1685 (2015)

38. Stojanovic, L.: Methods and tools for ontology evolution. Ph.D. thesis, Karlsruhe Institute of Technology, Germany (2004)

39. Vercruysse, S., Venkatesan, A., Kuiper, M.: OLSVis: an animated, interactive visual browser for bio-ontologies. BMC Bioinform. **13**(1), 116 (2012)

40. Zviedris, M., Barzdins, G.: Viziquer: a tool to explore and query SPARQL endpoints. In: 8th Extended Semantic Web Conference (ESWC), pp. 441–445 (2011)

Towards a Contextual and Semantic Information Retrieval System Based on Non-negative Matrix Factorization Technique

Nesrine Ksentini[✉], Mohamed Tmar, and Faïez Gargouri

MIRACL Laboratory, University of Sfax,
ISIMS BP 242, 3021 Sakiet Ezzeit, Sfax, Tunisia
ksentini.nesrine@ieee.org, mohamedtmar@yahoo.fr,
faiez.gargouri@isims.usf.tn

Abstract. With the fast speed of technological evolution, information retrieval systems are trying to confront the large amount of textual data in order to retrieve pertinent information to meet users needs or queries. Information retrieval systems depend also on the user's query who often finds difficulty to express his need.

To resolve these problems, we propose, in this paper, a new approach that provides a contextual and semantic information retrieval system.

Our proposed system is based firstly on NNMF (Non-negative Matrix Factorization) technique for data analysis in order to present textual data with new and small representations and to organize this data into categories. Secondly, our system try to how ameliorate the user's need with new semantic keywords that keeping the same context of the original query, by exploiting obtained results by the used data analysis technique and the *LSM* method that defines semantic relationships between terms.

Experimental results performed on the ClefEhealth-2014 database demonstrate the performance of our proposed approach on large scale text collections.

Keywords: Semantic search system · Data analysis
Information retrieval · Text representation · Semantic relationships

1 Introduction

The constantly growing amount of data stored and treated on the web becomes a major problem for web users who want to find pertinent results when searching their needs. Hence, information retrieval systems (IRS) were appeared to help web users to find speedily their needs, but the quality of the search results depends firstly on the manner how data are represented in the text based systems and secondly in the quality of the query issued by the user.

IRS often represent text data by a set of terms and they adopt the vector space model (VSM) [1] to represent the defined keywords and to organize texts.

© Springer International Publishing AG, part of Springer Nature 2018
A. Abraham et al. (Eds.): ISDA 2017, AISC 736, pp. 892–902, 2018.
https://doi.org/10.1007/978-3-319-76348-4_85

This model is frequently used in IRS because of its simplicity when calculating the similarity between the query vector and the document vectors using in general the cosine measure similarity. When the volume of the data collection is small, this simplicity is assured. But for large data collection, the VSM model generates high-dimensional and sparse vectors whose several values will be close to zero and the calculus of similarity will be expensive in terms of time and memory [2].

From where, IRS lose their effectiveness and efficiency and the quality of returned results will be degraded.

At the same time, users are finding it difficult to express their needs. Usually, they use queries that consist of maximum three terms and contain little contextual information.

Therefore, we notice the incentive to develop, in the first time, a new representation of data which becomes a major task of data analysis specially for large data collections [2,3]. In the second time, we try to ameliorate the user's query using query expansion via pseudo relevance feedback techniques for boosting the global performance of the IRS and to improve the quality of the returned results.

We propose in this paper, a new representation of textual data based on dimensionality reduction technique using non-negative matrix factorization technique (NNMF) and we study how to improve the original user's query by adding semantic and related terms and exploiting obtained results by the used data analysis technique and the LSM method which defines semantic relationships between terms.

The outline of the remainder of this paper is as follows. Firstly, in the next section, we present a literature review of the different techniques of dimensionality reduction of the data. The third section is enshrined to describe details of our proposed approach. In section four, the evaluation process is presented and discussed. Finally, we draw the conclusion and outline future works.

2   Related Works

In recent years, a variety of dimensionality reduction techniques have been proposed to reduce and to handle data which usually has a high dimensionality.

Dimensionality reduction technique is the transformation of high-dimensional data into a representation of reduced dimensionality. In other words, dimensionality reduction of data is a technique that allows to cut down vectors from high dimension to small and meaningful dimensions that describes multidimensional data sample with a small sample's coordinates in a new feature space [2,3].

Dimensionality reduction is important in many domains, since it mitigates the problem of dimensionality. In information retrieval area, this technique is frequently achieved by finding factorizations of the terms $\times$ documents matrix representing the whole dataset.

Until now, it has been proposed several techniques of factorizations which describe the data in feature space instead of in keywords space. Most widely-known methods are latent semantic indexing (LSI) which based on Singular

Value Decomposition (SVD) technique, Principal Component Analysis (PCA) and recently defined Non-negative Matrix Factorization ($NNMF$) which is an approach also has been successfully applied in pattern recognition [3,4]. In this section, we present an overview to explain the functionality of these different proposed techniques.

2.1   Latent Semantic Indexing

LSI is an information retrieval method put by S.T. Dumais and one of the most widespread methods to resolve the classical problem of polysemy and synonymy [5].

The main purpose of this technique is to discover the latent semantic relationship between keywords and documents in the corpus and to eliminate the usage of different words to express the same meaning. In other way, it tries to represent documents (and queries) not by terms, but by the underlying concepts (latent, hidden) referred to by the terms.

The famous method implementing LSI is SVD that generate as a result the best rank (k) approximation of the terms × documents matrix. In fact, it tries to project the vectors of documents into an approximate and low-dimensional subspace [2,5,6].

This low-dimensional space is generated by the eigenvectors that correspond to the few largest eigenvalues, thus, to the few most striking correlations between terms.

Given as an example a terms-documents matrix X and suppose that it has as rank r, LSI decomposes X using SVD technique as follows:

$$X = TSD^T \tag{1}$$

Where:

- X is a m * n matrix.
- $S = diag(\sigma_1, \sigma_2, \cdots, \sigma_r)$ and $\sigma_1 > \sigma_2 > \cdots > \sigma_r > 0$ present the singular values of X.
- T presents a m * r matrix and the left singular vector.
- D presents a r * n matrix and the right singular vector.

At the end, LSI selects the best k vectors in T as the transformation and the dimensionality reduction of matrix X in order to project the documents in the dataset into a k dimensional space.

This method presents some shortcomings like the considerable and huge computation specially when we treat a large database. Another gap which can reveal, is existing some negative values in the approximate result matrix which cannot give a logical explanation [6].

2.2   Principal Component Analysis

PCA is a data analysis method used successfully in a large number of domains such as face recognition, image compression. It is an unsupervised and linear technique for finding patterns in high dimensional data.

Its purpose is to extract the important information to represent it as a set of new orthogonal variables called principal components, and to display the pattern of similarity of both the observations and the variables as points in maps [7].

Indeed, it constructs a low-dimensional representation of the data that describing the variance in the data as possible. It involves transforming variables linked together in new variables unrelated to each other. These new variables are named "main components" or main axis. It allows to reduce the number of variables and make the least redundant information.

PCA can be done by singular value decomposition of a data matrix or by eigenvalue decomposition of a covariance matrix. The result of a PCA is usually discussed in terms of component scores called factor scores.

The main drawback of PCA is that the size of the covariance matrix is proportional to the dimensionality of the data. As a result, the computation of the eigenvectors and eigenvalues might be impractical for high-dimensional data.

2.3   Non-negative Matrix Factorization

NNMF has been a recently defined approach that organizes text as matrix and condenses matrix by factorization, also successfully applied in several areas such as pattern recognition, speech recognition, information retrieval, natural language processing [3,4].

NNMF is a mighty and an unsupervised learning method that also reduces the dimensionality of the data and decomposes a high-dimensional nonnegative matrix (X) into the product of two nonnegative matrices $(W,$ and $H)$ as follows:

$$X \approx WH \tag{2}$$

Where X represents the $(m \times n)$ data Matrix or the terms $\times$ documents matrix, W presents the $m \times k$ matrix of basis feature vectors and H is a $k \times n$ matrix that represents the coordinates of the samples in the new feature space. The rank k of this factorization is chosen such that [4]:

$$k < \frac{m \times n}{m + n} \tag{3}$$

The inspired definition of $NNMF$ is to find two matrices whose product approximates a the original data matrix.

3   The Proposed Approach

After setting out techniques of dimensionality reduction and reveal problems met by web users in information retrieval systems, we present in this section our

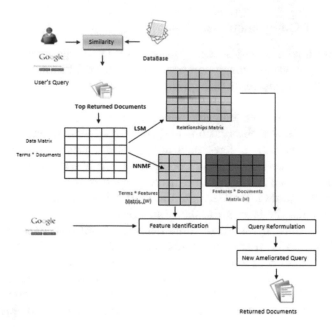

Fig. 1. Process of the proposed approach

proposed approach to define a contextual and semantic information retrieval system based firstly on dimensionality reduction using non-negative matrix factorization technique and secondly on how to ameliorate the original user's query by adding semantic and related terms exploiting the obtained results by the used data analysis technique and the obtained results by our LSM method [9, 11, 12].

The goal is to ameliorate at the end the quality of text retrieval system. In Fig. 1 we illustrate the process of the proposed approach.

We will choose to adopt the non-negative matrix factorization (NNMF) technique to represent the top returned documents in a feature space. This unsupervised learning method can be used for high dimensional matrix and put the condition that the data matrix must contain only nonnegative values, which is often the case, which facilitate the interpretation of given results.

Indeed, the terms × documents matrix that presents the top returned documents contain the values of terms weights in documents which are always positive.

Considering a collection of documents which is summarized as an m × n matrix (X) which the rows denote the terms or the bag-of-words existing in the documents, after removing the stop-words and getting only the stems, and the columns represent the documents. The values of X present the frequencies of terms in each document. We adopt to calculate frequencies using the TF-IDF measure illustrated in Eq. 4:

$$TF - IDF_{term_i,d} = nbOcc \times \log \frac{\mid D \mid}{\mid d_j : term_i \in d_j \mid} \tag{4}$$

Where:

$nbOcc$: number of times that $term_i$ occurs in document d.

$\mid D \mid$: total number of documents in the collection.

$\mid d_j : term_i \in d_j \mid$: number of documents where the $term_i$ appears.

Our goal is to find features that group related terms and to present documents within the defined features space. This is accomplished by using $NNMF$ method that identifies two matrix W and H where W size $m \times k$ and each column presents a feature, H size $k \times n$ and presents the feature frequency for each document.

The favored choice of k is to verify the condition mentioned in Eq. 3 to give better results of dimension reduction of the initial matrix and features identification.

This proposed work will be exploited in order to ameliorate user's query and to facilitate search system to identify pertinent results.

Indeed, we try to search the topic (set of features) of user's original query then to add new terms which express the same topic as the original query and linked together semantically by adopting our LSM method described in [9,11,12].

This step of query amelioration contains two sub-steps:

- Feature Identification.
- Query Reformulation.

Feature Identification

When the user issued his query, our proposed search system tries to seek the context (topic) of this query by searching for each word in what feature it belongs. The context will be the set of defined features.

Query Reformulation

The idea in this sub-step is to use the pseudo relevance feedback technique (blind relevance feedback) in order to automatically expand the original query without any user interaction [8].

In this method based on local analysis, the top q documents returned by the search system are assumed as pertinent (relevant). Then we calculate semantic relationships between terms in the query and terms of selected documents with referring to our statistical method called least square method (LSM) [9,11,12].

Indeed, we calculate for each term (t_i) its relation with other terms in a linear way (see Eq. 5).

$$t_i \approx \alpha_1 t_1 + \alpha_2 t_2 + \cdots + \alpha_{i-1} t_{i-1} + \alpha_{i+1} t_{i+1} + \cdots + \alpha_n t_n + \epsilon \qquad (5)$$

Where:

- α present real values of the model and degrees of relationships between terms.
- ϵ represents the minimum associated error of the relation.

In fact, we add to the original query only terms which have $\alpha > 0$ and belong in features of the original query.

In this section, we have presented our proposed approach which its purpose is to make a semantic information retrieval system based firstly on new representation of the data using NNMF method and secondly on semantic relationships between terms by adopting our LSM method [9,11,12].

4   Experiments and Evaluation

In this section, we carried out experiments using the NNMF technique to define the main context of the user's query (set of features) and semantic relationships between terms defined by our LSM method in order to ameliorate the original query [9,11,12,14].

4.1   The Dataset

Our evaluation was performed on clefehealth database 2014. The dataset for the ehealth information retrieval task in the clef competition, consists of a set of medical documents, around 1 *million* of English documents, provided by the Khresmoi project [10]. This collection does not contain patient information and contains documents covering a broad set of medical topics. The documents in the collection come from several online sources, including the Health On the Net organisation certified websites, as well as well-known medical sites and databases (e.g. Genetics Home Reference, ClinicalTrial.gov, Diagnosia) [13,14].

The topics or queries set consists of 50 medical professional queries containing the following fields [10]

- Title: text of the query.
- Description: longer description of what the query means.
- Narrative: expected content of the relevant documents.
- Profile: main information on the patient (age, gender, condition).
- Discharge summary: ID of the matching discharge summary provided by Ehealth Information Extraction task.

We have only use the title and the description in our experiments.

4.2   Experiment Setup

Our proposed system identifies the most relevant documents for each provided query. Then, we prepare the TF-IDF matrix of the top returned documents. After that, we apply firstly, the NNMF technique to achieve as a result the two matrix W, H and to express terms and documents in the feature space.

Secondly, we apply our LSM method to define semantic relationships between query terms and terms in the returned documents [9,11,12,14].

Once the relationships and the feature space are defined, we seek in the first time the main context of the query. Indeed, we examine the query terms to which features they belong. In the second time, we study the defined relationships in order to expand the original query with semantically related terms and which talk about the same context of the original query.

4.3   Results

In this sub-section, we present our obtained results that bring to attention trends. We present some examples of ameliorated queries with different values of the top returned documents, the number of features k and the weights of relationships between terms.

– Example1: Query number 17

$< id >$ qtest2014.17 $<$ /id $>$
$< title >$ chronic duodenal ulcer $<$ /title $>$
$< desc >$ How common is it that the ulcer starts to bleed again $<$ /desc $>$

We assume that the top 50 returned documents are relevant. Then we apply the NNMF technique to the defined terms $\times$ documents matrix, we choose the number of feature equal to 10 which verify the condition mentioned in the Eq. 3. After that, we determine the context of the original query and we calculate the relations between terms in the query with this set of terms. Then we expanded the original query by adding terms which have values of $\alpha > 0$ (positive relationships) and which are related to all terms in the initial query and express the same topic as the query. We have obtained this new query:

$< newQuery >$
chronic duoden ulcer common ulcer start bleed **gastrointest aortoenter antral ectasia**
$<$ /newQuery $>$

Terms that are in bold are the added terms to the original query and they express the same context of the original query. Indeed, the term **antral** is the antral gastritis is an autoimmune problem that can cause a lot of discomfort to the sufferer. Often people who have this condition are not aware of the symptoms that accompany it. There are also many kinds of gastritis that are treated with medications as well as acid suppressants to deal with the root cause. Most types of gastritis cause an inflammation of the entire gastric mucosa but antral gastritis affects only one region of the stomach [15].

The term **ectasia** expresses the gastric antral vascular ectasia (GAVE) is an uncommon cause of chronic gastrointestinal bleeding or iron deficiency anemia. The condition is associated with dilated small blood vessels in the antrum, or the last part of the stomach [16].

We notice that added terms talk very well on the context of the original query.

Take another example of a query with the same chosen values.

– Example2: Query number 19

$< id >$ qtest2014.19 $< /id >$
$< title >$ common carotid aneurysm $< /title >$
$< desc >$ What is common carotid aneurysm $< /desc >$

We have obtained the following new query which express the same context of the original query:

$< newQuery >$
common carotid aneurysm common carotid **aneurysm flow atherosclerosi balloon**
$< /newQuery >$

– Example3: Query number 44

$< id >$ qtest2014.44 $< /id >$
$< title >$ how can I avoid small bowel obstruction $< /title >$
$< desc >$ What are the causes of small bowel obstruction and how could the disease be avoided $< /desc >$

In this example, we have obtained a new query based also on the top 50 returned documents and the number of features equal to 5 illustrated as below:

$< newQuery >$
can avoid small bowel obstruct caus small bowel obstruct diseas avoid **pouchiti dilat gastroenterol ileocolon fazio dozoi**
$< /newQuery >$

In Table 1, we show our obtained results using NNMF technique and LSM method in query expansion. In fact, we have obtained with our basic system (without query expansion) $MAP = 0.16$, $P@10 = 0.55$ and $Recall = 0.59$. We notice that our proposed approch can ameliorate the performance of our system by retrieving more pertinents documents (the rate of recall has increased ($Recall = 0.70$)).

Table 1. Obtained results using NNMF and LSM in query expansion

k	$P@5$	$P@10$	MAP	$Recall$
2	0.5520	0.5520	**0.30**	0.63
3	0.5640	0.5580	**0.30**	0.63
4	0.5840	0.5780	0.26	0.65
5	0.5560	0.5620	0.25	**0.70**

5   Conclusion and Future Works

We present in this paper a literature review of the different techniques of dimensionality reduction of data. Then we present the process of our proposed approach. The goal is to make an appropriate contextual and semantic information retrieval system for large scale collection.

This approach contains two steps; the first step is dedicated to reducing dimensionality of the data using the NNMF technique and to present the top returned documents in features space. The second step is to ameliorate the user's query automatically by using the results obtained from the first step and the semantic relationships defined by our LSM method.

We have obtained performed results and meaningful ameliorated queries.

For future work, we will study the impact of the new defined queries on the information retrieval system and we will study which optimum values representing the number of features and the top returned documents will be used. Then, we will test this approach with another large collection such as CLEFeHealth-2015 provided by the user-centred health information retrieval task at the CLEFeHealth-2015.

References

1. Salton, G., Wong, A., Yang, C.-S.: A vector space model for automatic indexing. Commun. ACM **18**(11), 613–620 (1975)
2. Liu, M., Wu, C., Chen, L.: A vector reconstruction based clustering algorithm particularly for large-scale text collection. Neural Netw. **63**, 141–155 (2015)
3. Zurada, J.M., et al.: Nonnegative matrix Factorization and its application to pattern analysis and text mining. In: Federated Conference on Computer Science and Information Systems (FedCSIS), pp. 11–16. IEEE (2013)
4. Devarajan, K., Wang, G., Ebrahini, N.: A unified statistical approach to nonnegative matrix factorization and probabilistic latent semantic indexing. Mach. Learn. **99**(1), 137–163 (2015)
5. Ye, Y.Q.: Comparing matrix methods in text-based information retrieval. School of Mathematical Sciences, Peking University, Technical report (2000)
6. Zhang, W., Yoshida, T., Tang, X.: A comparative study of TF * IDF, LSI and multi-words for text classification. Exp. Syst. Appl. **38**(3), 2758–2765 (2011)
7. Abdi, H., Williams, L.J.: Principal component analysis. Wiley Interdisc. Rev. Comput. Stat. **2**(4), 433–459 (2010)
8. Carpineto, C., Romano, G.: A survey of automatic query expansion in information retrieval. ACM Comput. Surv. (CSUR) **44**(1), 1 (2012)
9. Ksentini, N., Tmar, M., Gargouri, F.: Detection of semantic relationships between terms with a new statistical method. In: WEBIST Conference (2014)
10. Goeuriot, L., et al.: Share/clef ehealth evaluation lab 2014, task 3: user-centred health information retrieval (2014)
11. Ksentini, N., Tmar, M., Gargouri, F.: Controlled automatic query expansion based on a new method arisen in machine learning for detection of semantic relationships between terms. In: 15th IEEE International Conference on Intelligent Systems Design and Applications (ISDA), pp. 134–139 (2015)

12. Ksentini, N., Tmar, M., Gargouri, F.: The impact of term statistical relationships on Rocchio's model parameters for pseudo relevance feedback. Int. J. Comput. Inf. Syst. Ind. Manage. Appl. **8**, 135–144 (2016)
13. Ksentini, N., Tmar, M., Gargouri, F.: Miracl at CLEF 2014: eHealth information retrieval task. In: Proceedings of the ShARe/CLEF eHealth Evaluation Lab (2014)
14. Ksentini, N., Tmar, M., Gargouri, F.: Towards automatic improvement of patient queries in health retrieval systems. Appl. Med. Inf. **38**(2), 73–80 (2016)
15. DietHealthClub.   http://www.diethealthclub.com/health-issues-and-diet/antral-gastritis.html
16. Wikipedia. https://en.wikipedia.org/wiki/Gastric-antral-vascular-ectasia

Design and Simulation of Multi-band M-shaped Vivaldi Antenna

Jalal J. Hamad Ameen$^{(\boxtimes)}$

College of Engineering Electrical Engineering Department,
University of Salahaddin, Erbil, Iraq
Jalal.hamadameen@su.edu.krd

Abstract. Vivaldi antenna is a co-planar broadband antenna which is made from a dielectric plate metalized on both sides. Double – sided printed circuit board used in the design of this type of antenna makes it cost effective at microwave frequencies exceeding 1 GHz. This type of antenna used in broadband of frequency specially ultra-wide band because its manufacturing is easy for which PCB production is used. In this paper, a new model of Vivaldi antenna been designed based on the dual Vivaldi with M-shaped, the design based on the standards and parameters for the antenna, the dimensions and size of the designed antenna given depending on the wavelength. After the design process, experimental setup done to obtain the practical parameters for the antenna. Optimization for the best VSWR versus operating frequencies, the design and simulation results presented like radiation pattern and the VSWR for many operating frequencies, comparison with the other Vivaldi antenna presented to show the improvements in the proposed and designed antenna, finally, some conclusions presented.

Keywords: Antennas · Antenna parameters · Vivaldi antenna
M-shaped Vivaldi antenna

1 Introduction

Antenna is that transducer device which receives/radiates radio waves and some times called aerial as a means for radiating or receiving radio waves and used when the transmission channel. There are many types of antennas [2] for example wire antenna, aperture antenna, micro strip antenna, array antenna and reflector antenna. The main antenna parameters are voltage standing wave ratio (VSWR), gain, radiation pattern, return loss. Vivaldi antenna is one of the special antennas invented by Peter Gibson and supposedly chose to the name after Antonio Vivaldi. In this paper, after a proposed new model of Vivaldi antenna is presented, for designed proposed Vivaldi antenna, the main parameters calculated in the design simulation like the radiation pattern, impedance, reflection coefficients, directivity, efficiency and the gain.

© Springer International Publishing AG, part of Springer Nature 2018
A. Abraham et al. (Eds.): ISDA 2017, AISC 736, pp. 903–912, 2018.
https://doi.org/10.1007/978-3-319-76348-4_86

2 Vivaldi Antenna

The most popular directive antenna is the Vivaldi antenna which is used for commercial UWB applications due to its simple structure and small size, it is widely used in different applications such as microwave imaging, wireless communications and ground penetrating radars [1]. Vivaldi antenna always needs a large size to achieve good performance. It is widely studied in UWB antenna research owing to the merits of low profile, wide impedance bandwidth, moderately high gain, good directivity, benign time domain characteristics, and symmetric beam both in E-plane and H-plane. The Vivaldi (Fig. 1) is a member of a class of aperiodic continuously scaled traveling-wave antenna structures [7]. Finding suitable feeding techniques for the Vivaldi is the important step. Understanding the characteristics of the Vivaldi is fundamental and would help a great deal in designing the antenna.

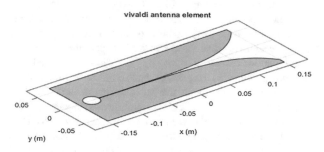

Fig. 1. Vivaldi antenna

Figure 1 is the Vivaldi antenna, Fig. 2 is the radiation pattern for the antenna given in Fig. 1 with the operating frequency of 3.5 GHz, all the data about the dimensions for

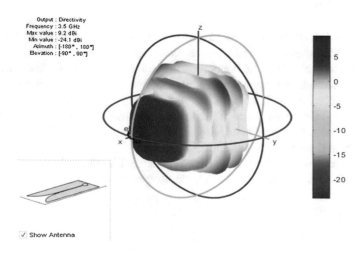

Fig. 2. Radiation pattern for Vivaldi antenna

the antenna is given after the Figs. 2 and 3 is the designed Vivaldi antenna for the dimensions given after Fig. 2.

vi = Vivaldi with properties in (mm): Taper Length: 0.2430, Aperture Width: 0.1050, Opening Rate: 0.2500, Slot Line Width: 5.0000e-04, Cavity Diameter: 0.0240, Cavity To Taper Spacing: 0.0230 Ground Plane Length: 0.3000, Ground Plane Width: 0.1250, Feed Offset: −0.1045, Tilt: 0, Tilt Axis: [1 0 0].

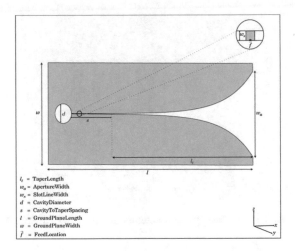

l_t = TaperLength
w_a = ApertureWidth
w_s = SlotLineWidth
d = CavityDiameter
s = CavityToTaperSpacing
l = GroundPlaneLength
w = GroundPlaneWidth
f = FeedLocation

Fig. 3. Vivaldi antenna dimensions parameters

3 Proposed Design Methodology

In this paper a dual Vivaldi antenna has been designed, using the design dimensions in accordance to the antenna parameters and using the HFSS software as shown in Fig. 4 below.

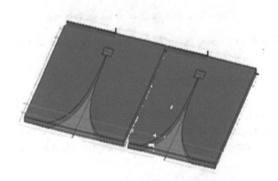

Fig. 4. Dual Vivaldi antenna

First, using the HFSS ANSOFT, the proposed design of the dual Vivaldi antenna has been designed. Figure 5 is the proposed design in this paper, the dimensions for the antenna are given in Fig. 6, as shown, the width and the length of the substrate are 20 cm and 13.5 cm respectively, the other antenna dimensions are as follows (in centimeters): Taper Length: 14, Aperture Width for each taper: 6, Cavity Diameter: 2, Cavity To Taper Spacing: 1, Ground Plane Length: 20, Ground Plane Width: 13.5.

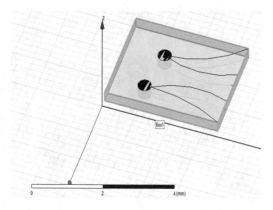

Fig. 5. The proposed Vivaldi antenna design using ANSoft HFSS

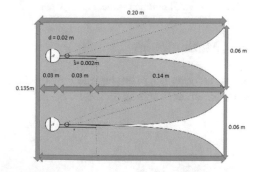

Fig. 6. The proposed Vivaldi antenna design dimensions

4 Design and Simulation Results

To design Vivaldi antenna, some conditions related with it is parameters must be satisfied, the highest frequency of operation f_H, the width d + 2w of the antenna should satisfy Eq. (1) to circumvent any grating lobes for the Vivaldi array [10]:

$$(d + 2 * w) < \frac{c}{fHsqrt(\varepsilon_e)} \tag{1}$$

where ε_e is the effective dielectric constant, c is the velocity of light = 3×10^8 m/s. For the proposed antenna in this paper as given in Fig. 6:
w = 0.005 m, d = 0.06 m, Eq. (1) will be:

$$(2 * d + 3 * w) < \frac{c}{fH \, sqrt \, (\varepsilon_e)} \qquad (2)$$

This is because the proposed antenna is dual Vivaldi.
Therefore, 2 * d + 3w = 2 * 0.06 + 3 * 0.005 = 0.135 m,
The stripline to slot line transition design starts with the choice of substrate material and thickness. Stripline width is calculated using the stripline characteristic impedance formulas of Eq. (3) [11]):

$$Zo = \frac{60}{sqrt(\varepsilon_e)} In \left(\frac{4H}{0.67\pi(T + 0.8)} \right) \quad \Omega \qquad (3)$$

Where,
Zo is characteristic impedance of the stripline in Ω
ε_e *is the relative permitivity of the dielectric,*
H is the dielectric substrate thickness in mm,
T is the stripline thickness in mm,
W is the stripline width in mm
According to the dimensions of the taper length and the thickness in addition with the parameters in Eqs. (3) and (4), the effective dielectric constant calculated will be 0.065, therefore, the highest frequency of operation f_H using Eq. (1) will be: f_H < 8.7 GHz.

Therefore, maximum operation frequency for the proposed antenna is less than 8.7 GHz. The characteristic impedance will be 53.57 Ω.

In this paper, the design simulation will be for some operation frequencies not exceeding 8.7 GHz. After the procedure of the proposed design using ANSoft HFSS software shown in Fig. 5 with the given dimensions and parameters, the proposed antenna has been designed practically. A signal generator and transmitter with different frequencies, a receiver and a motor shaft circulator to obtain the data for radiation pattern, and antenna software simulator used in the design and simulation. The experimental setup to obtain the radiation pattern and the antenna parameters shown in Fig. 7 (for the vertical polarization) and Fig. 8 (for the horizontal polarization) shown respectively done and completed.

During the practical setup, the operating frequencies 800 MHz, 1 GHz and 1.3 GHz has been used because the maximum frequency for the transmitter was 1.3 GHz. Figure 9 is the radiation pattern for the proposed antenna for the frequency 1.3 GHz horizontal polarization given in Fig. 8, while Fig. 10 is the radiation pattern for the same frequency vertical polarization. Figure 11 is the radiation pattern for the frequency 1 GHz horizontal and vertical polarization, Fig. 12 is radiation pattern for

Fig. 7. Experimental setup of the work (vertical polarization)

Fig. 8. Experimental setup of the work (horizontal polarization)

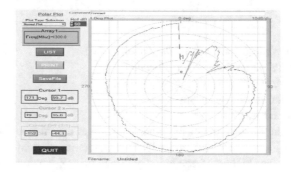

Fig. 9. Radiation pattern for the designed Vivaldi antenna at 1300 MHz (Horizontal polarization)

frequency 800 MHz, Finally, Fig. 13 is the received signal power spectrum for the frequency 800 MHz.

It is shown from Fig. 9 that at 1.3 GHz, horizontal polarization has greater main lobe than at vertical as shown in Fig. 10, beam width angle is about 150°, this means that the proposed antenna is better to use in horizontal polarization.

As shown in Figs. 12 and 13, lower frequency causes less in main lobe radiation pattern which is for 1 GHz, at the same time horizontally is better.

From Fig. 13 which is for 800 MHz, it is shown that lower frequency lower main lobe and radiation pattern.

About VSWR, Figs. 14, 15 and 16 are VSWR versus frequency, it is shown that from Fig. 14 for the range 790 MHz–810 MHz, best VSWR is at the frequencies 800 MHz and 802 MHz which is about 1.2 according to the reading in the right hand scale vertically of the analyzer because the left hand scale is the voltage and the right hand scale is the VSWR.

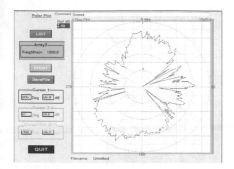

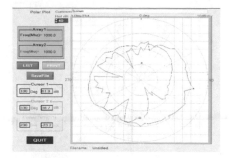

Fig. 10. At 1300 MHz (Vertical polarization)

Fig. 11. At 1000 MHz (horizontal and vertical polarization)

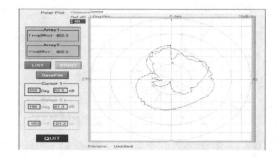

Fig. 12. At 800 MHz (horizontal and vertical polarization)

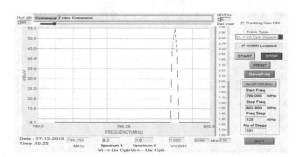

Fig. 13. Received signal using spectrum analyzer

For the range 990 MHz – 1009 MHz as shown from Fig. 15, best VSWR is at 992 MHz an equals 1.2.

For the range 1190 MHz – 1209 MHz as shown in Fig. 16, best VSWR is about 1.2 at 1202 MHz.

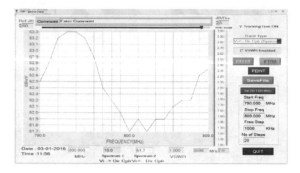

Fig. 14. VSWR vs. frequency for frequency range 790 MHz – 810 MHz

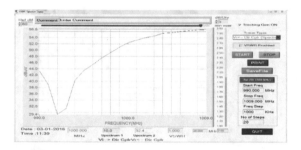

Fig. 15. VSWR vs. frequency for frequency range 990 MHz – 1009 MHz

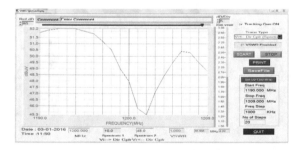

Fig. 16. VSWR vs. frequency for frequency range 1190 MHz – 1209 MHz

5 Conclusions

In this paper, a new model of Vivaldi has been designed, the designed antenna is M-shaped dual Vivaldi for Ultra-wide band, the dimensions of the antenna chosen depends on the frequency and the parameters given in this paper' equations, first, using HFSS AnSoft used to design the antenna, then practically, it is designed and tested using spectrum analyzer, micro-wave signal transmitter, receiver and a motor shaft to calculate the radiation pattern with the use of the Signet software, from the results the radiation pattern for the antenna shown with different frequencies, also, the VSWR has been determined practically, it was about 1.2 which is good value because near the ideal value 1, from the results shown that this type of antenna better for vertical polarization than horizontal, the beam-width is between 120 – to – 160 °, the designed antenna can be used for frequencies from 100 MHz to about 7 GHz as given in the design results.

References

1. Vignesh, N., Sathish Kumar, G.A., Brindha, R.: Design and development of a tapered slot Vivaldi antenna for ultra-wide band application. Int. J. Adv. Res. Comput. Sci. Soft. Eng. **4** (5), 174–178 (2014)
2. Constantine, A.B.: Antenna Theory Analysis and Design, 3rd edn. A John Wiley & Sons, Inc., Publication, New Jersey (2005)
3. Ma, K., Zhao, Z., Wu, J., Ellis, S.M., Nie, Z.P.: A printed Vivaldi antenna with improved radiation patterns by using two pairs of eye-shaped slots for UWB applications. Prog. Electromagnet. Res. **148**, 63–71 (2014)
4. Rajaraman, R.: Design of a wideband Vivaldi antenna array for the snow radar, M.Sc. thesis B.E. (Electronics & Communications Engineering), Coimbatore Institute of Technology, India (2001)
5. Monti, G., Congedo, F., Tarricone, L.: Novel planar antenna with a broadside radiation. Prog. Electromagnet. Res. Lett. **38**, 45–53 (2013)
6. Hamzah, N., Othman, K.A.: Designing Vivaldi antenna with various sizes using CST software. In: Proceedings of the World Congree on Engineering 2011, vol. II. WCE 2011, London, UK, 6–8 July 2011
7. Sarkar, H.: Some parametric studies on Vivaldi antenna. Int. J. u- e-Ser. Sci. Technol. **7**(4), 323–328 (2014)
8. Fang, G., Sato, M.: Optimization of Vivaldi antenna for demining by GPR, Center for Northeast Asian Studies Tohoku University, Kawauchi, Sendai, Japan
9. Alshamaileh, K.A., Almalkawi, M.J., Devabhaktuni, V.K.: Dual band-notched microstrip-fed Vivaldi antenna utilizing compact EBG structures. Int. J. Antennas Propag. **2015**, 1–7 (2015)
10. Pandey, G.K., Singh, H.S., Bharti, P.K., Pandey, A., Meshram, M.K.: High gain Vivaldi antenna for radar and microwave imaging applications. Int. J. Signal Process. Syst. 3(1), 35–39 (2015)
11. Yang, Y., Wang, Y., Fathy, A.E.: Design of compact Vivaldi antenna arrays for UWB see through wall applications. Prog. Electromagnet. Res. PIER **82**, 401–418 (2008)
12. Erdoan, Y.: Parametric study and design of Vivaldi antenas and arrays, M.Sc. thesis, The graduate school of natural and applied sciences of middle east technical university, March 2009
13. Safatly, L., Al-Husseini, M., El-Hajj, A., Kabalan, K.Y.: A reduced-size antipodal Vivaldi antenna with a reconfigurable band notch. In: PIERS Proceedings, Moscow, Russia, pp. 220–224, 19–23 August 2012
14. Wang, Y., Fathy, A.E.: Design of a Compact Tapered Slot Vivaldi Antenna Array for See Through Concrete Wall UWB Applications, EECS Department, University of Tennessee, Knoxville, USA

15. Alsulaiman, K.A., Al alShaykh, M., Alsulaiman, S., Elshafiey, I.: Design of ultra wideband balanced antipodal Vivaldi antenna for hyperthermia treatment. In: PIERS Proceedings, Progress in Electromagnetics Research Symposium Proceedings, Stockholm, Sweden, pp. 1752 1755, 12–15 August 2013
16. John, M., Ammann, M., McEvoy, P.: UWB Vivaldi Antenna Based on a Spline Geometry with Frequency Band-notch. Dublin Institute of Technology, AAROW@DIT, 01 January 2008, IEEE (2008)
17. Zhou, B., Li, H., Zou, X., Cui, T.J.: Broadband and high-gain planar Vivaldi antennas based on inhomogeneous anisotropic zero-index metamaterials. Prog. Electromagnet. Res. **120**, 235–247 (2011)

Performance Evaluation of Openflow SDN Controllers

Sangeeta Mittal[✉] 🄳

Jaypee Institute of Information Technology, Noida, India
sangeeta.mittal@jiit.ac.in

Abstract. Software Defined Networks (SDN) is the recent networking paradigm being adopted by stakeholders in big way. The concept works towards dramatically reducing network deployment and management costs. Controllers, also known as Network Operating System of SDN, are critical to success of SDNs. Many open source controllers are available for use. In this paper, performance of four popular Openflow based controllers has been evaluated on various metrics. Latency, Bandwidth utilization, Packet Transmission rate, Jitter and Packet loss has been calculated for TCP and UDP traffic on varying network sizes, topologies and Controllers. Floodlight is one of the best performing as compared to reference controller.

Keywords: Software Defined Networks · Mininet · Floodlight · POX
Open daylight · Bandwidth utilization · Latency

1 Introduction

With the advent of technology, cost of storage and processing power is decreasing sharply with every passing year. The cost of networks that is not in tandem with this downfall has become the bottleneck of overall cost. Software Defined Network (SDN) is the novel solution to cost effective scalability problems of traditional networking. The paradigm is about centralized control of the network implemented by decoupling control logic from forwarding devices. A simplistic architecture of SDN is shown in Fig. 1. Apart from usual networking devices, additional terminology in the architecture is SDN Controller, SDN applications, Northbound Interfaces and Southbound Interfaces [1].

SDN Controller is the centralized entity to control the network. All network orchestration functions like how switches should handle network traffic, topology management etc. are implemented in it.

Northbound Interfaces: The APIs here are defined to communicate with user domain network administration applications to help network engineers in programmatically deploying network services, implementing business logic and keep track of network health.

Southbound Interfaces: The set of APIs in these interfaces are used for communication between forwarding devices such as hubs and switches and controlling devices. Openflow has become the de facto Southbound API.

© Springer International Publishing AG, part of Springer Nature 2018
A. Abraham et al. (Eds.): ISDA 2017, AISC 736, pp. 913–923, 2018.
https://doi.org/10.1007/978-3-319-76348-4_87

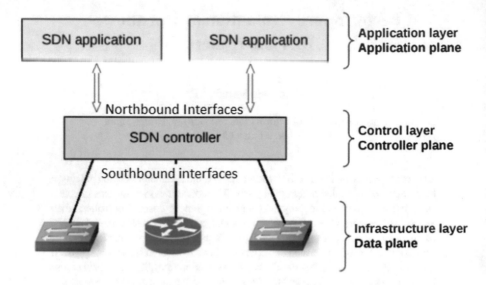

Fig. 1. Simplified SDN architecture [1]

SDN concept of networking offers numerous advantages [2–4]. A directly programmable network using open source tools has become possible. Centralized control of network algorithms by controllers gives a global view of the network for making it easy to design, implement and scale networks. Requirement of purchasing custom Built proprietary hardware solutions is reduced. Unlike earlier, now network can also be implemented as Network-as-a-Service. As a result, Implementers of SDN will eventually get an agile and flexible network. Ease of usage can also lead to more frequent innovations in network design. Moreover, there is huge potential of network cost savings due to reduced time to launch and expand a network. Enhanced interoperability of networking device will; encourage an open and more competitive market for networking devices.

Besides cost, core network functions of traffic management, guaranteed delivery of Quality of Service (QoS) can also be easily implemented in SDN. The centralized view will also enable easy debugging in case of failures.

All these advantages are possible due centralized control implemented by SDN Controllers with most common protocol, Openflow [3]. In this paper, it is intended to bring out the effect of using different SDN Openflow controllers on network performance.

Rest of the paper is organized as follows. Some work already done towards this problem has been cited in Sect. 2. In Sect. 3, various SDN controllers selected for evaluation and network emulation software have been described. Performance metrics, results obtained and their discussion & analysis has been discussed in Sect. 4. The paper is concluded in Sect. 5.

2 Related Work

There have been several attempts at evaluating the performance of SDN Controllers. Authors in [5] have evaluated five open-source SDN controllers. Experiments have been done on benchmark tests as well as other system settings. They have claimed that Beacon controller is most scalable on increasing number of switches. In [6], three existing controllers have been compared against a custom defined controllers defined in the paper. However, in both of these papers, controllers haven't been evaluated on various common topologies of network. [7] presents an insight into detailed evaluation of open-daylight controller system messages and protocol setup. This paper, gave a detailed knowledge of working of ODL and highlights the degraded performance of the controller on increasing scale. Impact of ODL on increasing network scale however, has been omitted in this work too. Authors in [8] report their experiences in doing performance comparison of POX and Floodlight Controller with different inbuilt and custom defined topologies. The work done in this paper is similar to it, but with increased number of controllers and network settings. Authors in [9] provide a feature wise comparison of various SDN Controllers, their Mininet compatibility and visualization capabilities. Mininet is one of the most popular lightweight network emulator. The tool has been introduced with examples and some basic performance studies in [10].

Overall, after going through the literature, a need for more detailed study of effects of choice of SDN controller on network emulation was found. Same has been done in subsequent sections of this paper.

3 Implementation of Software Defined Network

Implementation of SDN requires a network emulation tool compatible to the SDN controllers. In this section, the mininet tool used for network emulation and chosen SDN controllers has been described.

3.1 SDN Emulation in Mininet

Mininet is a network emulation tool [10]. Mininet provides functionality for easy creation of various network components listed here.

(a) End Hosts: - End user machines implemented to run user applications. Hosts are assigned easy to use MAC and IP addresses.
(b) Open Virtual Switches: - The hosts are connected to one or more switches. These are implemented in mininet as open virtual switches that don't have any logic for forwarding and it depends on controller for routing any traffic.
(c) Links between Hosts and Switches: - In mininet links with desired characteristics of delay and bandwidth can be created easily to design any custom topology. Easy configuration of links and their characteristics is a main feature of mininet. Predefined topologies available in mininet are star, linear and tree topology.

The topologies tested in this paper are listed in Table 1.

Table 1. Topologies tested with each controller.

Name	#of Switches	#of Hosts
Single	1	n
Linear	n	n
Tree	$2^{depth} - 1$	fanout * fanout

3.2 SDN Controllers

Various Openflow based controllers evaluated in this work are listed in Table 2. First one is the OpenFlow reference controller implemented as bridging switch. It is also called OVS-controller or test-controller and is currently the default controllers for Mininet. This controller has been used as a baseline controller to compare with others.

Table 2. Topologies tested with each controller.

Name	Developed and maintained by	Development language	Location
OVS-controller	Default controller	Python	Reference controller
POX controller	open source development platform	Python	Internal controller
Floodlight controller	Big switch networks	Java	External controller
OpenDayLight controller	Cisco and opendaylight	Java	External controller

POX controller is an open source controller available for download on github [11]. Like OVS-controller it is also written in python and comes bundled with mininet virtual machine. POX has been tested on localhost by running into a SSH login of mininet VM.

The next controller chosen for testing is floodlight of big switch networks. It is written in Java and has been designed for actual heavy traffic of real data centers. The project is available for free download from its project site at [12]. Apart from supporting several switches, floodlight also has a collection of applications for traffic engineering. For testing floodlight, Floodlight VM was downloaded from the project site and implemented in virtual box as an ubuntu OS based system.

The last controller chosen here is the Opendaylight controller. It is a popular controller and is supported by vendors like Cisco. It is a java based project initiated in 2013 and continuing to evolve uptill now. In this work, the most recent Carbon distribution of the controller has been used. Opendaylight has certain prerequisites of exclusive java version. Distribution package is unzipped in an Ubuntu machine and Opendaylight is initiated by running ./bin/karaf file in it. Opendaylight has an easy to use frontend called DLux. It enables us to view the topologies in a browser based window. Figure 2(a)–(c) shows tested topologies with 16 hosts each as seen in the DLux visualizer. Linear topology doesn't show all the hosts as the hosts failed to ping in this case.

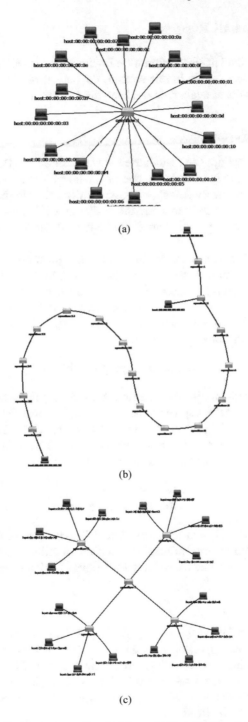

Fig. 2. Arrangement of 16 nodes in (a) Star (b) Linear (c) Tree (depth = 2, fan-out = 4) Topologies

4   Experiments and Results

The topologies described in last section were implemented in mininet by connecting to different controller every time. In this section, the performance metrics used to draw comparisons and results of evaluation has been discussed.

4.1   Performance Parameters

(1) Bandwidth Utilization: Bandwidth used was measured for TCP traffic using Iperf tool [14]. For using iperf in SDN, one of the hosts is implemented as client and another one as server. By default, only the bandwidth from the client to the server is measured. If you want to measure the bi-directional bandwidth simultaneously, use –d option. For all experiments the default TCP window size set in mininet as 85.3 Kbytes has been used.

(2) Delay: Round Trip Transmission time (RTT) has been used to calculate delay or latency of packet transmission between two nodes. In SDN, the first pings always take more time than the subsequent ones due to route queries to controller in first attempt. Therefore, to get fair idea of delay, two RTTs has been considered namely first_RTT and Second_RTT. These are computed by taking maximum RTT of two back to back execution of following command.

Mininet > h1 ping −c5 h16

(3) Jitter: It is the variation in response time between a pair of hosts. Jitter has been calculated using iperf's udp performance testing options. For example following command will test the jitter between host h1 and host h16. UDP packets will be sent for 10 s and report of jitter after every 1 s has been generated. However, in evaluation average jitter of 10 s has been used.

h1 −s −u −i1&
h16iperf −ch1 −u −d

(4) Datagram loss: UDP is non-reliable protocol. The number of datagrams that were lost during UDP based transmission is a measure of reliability of the link in routing UDP traffic. It can also be measured with an Iperf UDP test described in jitter.

4.2   Results

Experiments were conducted on an Intel CORE i3 Laptop with 6 GB RAM. Each controller was tested on supporting all three types of inbuilt topologies with number of end hosts starting from 16 and doubling subsequently upto 1024. Beyond that number, mininet wasn't able to create the network due to configuration limitations of host computer. All Controllers and Mininet were implemented as Virtual Machines in Oracle's Virtual Box Virtualization software.

Figure 3(a–c) shows the graph of RTT statistics on increasing counts of ICMP packets from 5 to 100. This study was done on reference controller with 16 hosts in each topology. It can be seen that the RTT stabilizes after 5 packets. Hence in later experiments, this has been considered as idle for measuring first and second RTTs. As expected by the number of links, RTT is maximum in linear topology followed by tree.

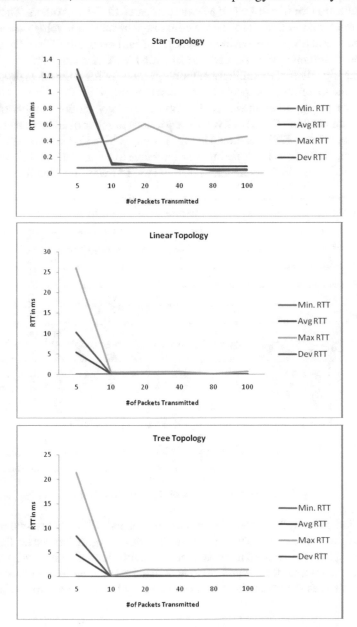

Fig. 3. RTT of three topologies on increasing transmitted packets in reference controller

Detailed perfomance study of first_RTT, Second_RTT, Bandwidth Utilization and tranferred data at client and server has been shown in Table 3. The table is with respect to star topology with increasing number of nodes. POX controller is the work performing here, as utilization for it is in several MBs as compared to GBs for others. The controller also performs worst in first_RTT, hence citing high latency in topology establishment. Floodlight is second worse in terms of latency, but has best bandwidth utilization. Jitter, which is an important QoS parameter, is also worst in POX controller.

Interestingly, given the simplicity of topology, none of the controllers drop any UDP datagrams and therefore, loss percentage is zero for all.

The nest topology to be evaluated was linear topology. In this number of switches in equal to the number of hosts. Given the enormous number of links to be established in this topology, First_RTTs as shown in Fig. 4, are very high in this case. As the network scales in size, POX controller is taking very long time to eastablish the first connections. In these experiments, POX Controllers, switch based forwarding module was used.

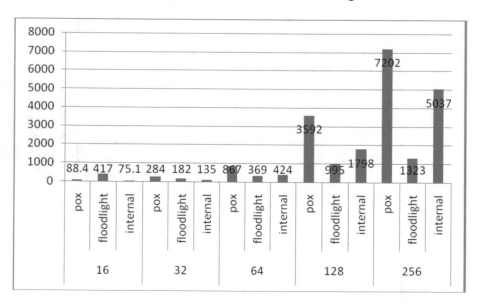

Fig. 4. Latency in linear topology in terms of First_RTT

Figure 5, demonstrates latency experienced for second_RTT, that is time when route should have been pre established. The numbers in the figure represent significantly reduced delay, except in case of internal reference controller which showed unusual high latency even on second_RTT round. Also note that the study omits results of Opendaylighht controller, as the controller didn't run correctly for any of the topologies except for 16 nodes.

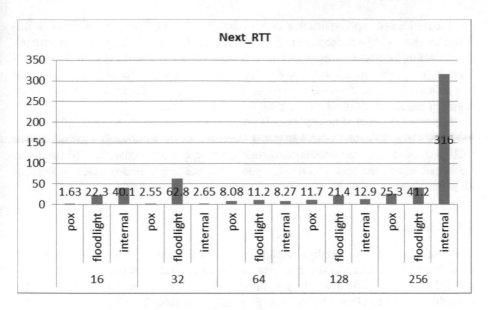

Fig. 5. Latency in linear topology in terms of Second_RTT

As shown in Fig. 6, interestingly, in linear topology each controller was dropping some UDP datagrams and hence loss percentage was non-zero. This number is very small for all controllers except the POX controller that dropped high number of packets for larger networks with 64 - 1024 nodes.

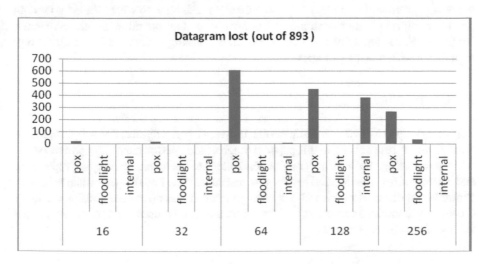

Fig. 6. Controller wise number of UDP datagram lost in linear topology

For floodlight and reference controller, this number reminaed less than 10 for all configurations.

The next set of experiments was done on tree topologies. There are several possibiliities for generating a particular number of end hosts in tree topology by varying depth and fanout of the tree. However, here all tree topologies are generated with depth 2 and doubling the fanout in each run. This resulted in 4 equivalent sized networks as compared to earlier studies. These sizes are 16, 64, 256 and 1024 nodes. Therefore, results for tree topology are shown only all three topologies of these sizes.

The detailed reults of this stdy has been presented in Table 4. Interestingly, in tree topology Opendaylight Controller performed worst. Its bandwidth utilization was multiple times lesser than other controllers. However, in jitter and latency this controller performed one of the best. Floodlight again was medicore in performance in terms of all parameters.

4.3 Discussion

The results depict that floodlight controller has an edge over all other controllers in almost all metrics. This was laso experienced during experimentation that this controller worked smoothly with all topologies and all sizes. On the other hand both POX and Openflow Controller were stressed out when either the number of hosts or the number of links (linear topology) were increased. From the current set of controllers thus, floodlight is an ideal choice for designing

SDN. The controllers can further be tested with benchmark tests to establish the relevance of this fact more robustly. In this paper, the motive was to give baseline results for a beginner researcher starting to program the SDNs in mininet. The obtained results tally with those obtained in earlier works discussed in related work section. However, the results given in this paper, gives more detailed insights into where things go wrong for other controllers. For example, it has been clear that openflow contollers cannot handle the linear topology and POX controllers learning switch module suffers from high data loss in linear topologies.

5 Conclusion

Software Defined Networks is an upcoming networking paradigm and is a topic of recent interest among researchers. In this paper, a much needed performance comparison study of open source openflow based popular controller has been done on varying topologies and network sizes. The detailed results give useful insights into the performance of each controller on differnet metrics. It has been concluded that out of the considered set of controllers, floodlight controller performs best and can be used for further design and testing of new SDN protocols.

References

1. Network Virtualization Report: SDN Controllers, Cloud Networking and More - Online Edition (2017). https://www.sdxcentral.com/sdn
2. Xia, W., et al.: A survey on software-defined networking. IEEE Commun. Surv. Tutor. **17**(1), 27–51 (2015)
3. Kreutz, D., et al.: Software-defined networking: a comprehensive survey. In: Proceedings of the IEEE 103.1, pp. 14–76 (2015)
4. Zhao, Y., Iannone, L., Riguidel, M.: On the performance of SDN controllers: a reality check. In: 2015 IEEE Conference on Network Function Virtualization and Software Defined Network (NFV-SDN), IEEE (2015)
5. Tootoonchian, A., et al.: On Controller Performance in Software-Defined Networks. Hot-ICE **12**, 1–6 (2012)
6. Zara, S., et al.: Performance debugging in SDN controllers—a case study (short paper). In: 2016 5th IEEE International Conference on Cloud Networking (Cloudnet), IEEE (2016)
7. Bholebawa, I.Z., Jha, R.K., Dalal, U.D.: Performance analysis of proposed openflow-based network architecture using mininet. Wirel. Pers. Commun. **86**(2), 943–958 (2016). https://doi.org/10.1007/s11277-015-2963-4
8. Cassongo, A.B.: The comparison of network simulators for SDN. Новітні інформаційні системи та технології-Modern information system and technologies 5 (2016)
9. Kaur, K., Singh, J., Ghumman, N.S.: Mininet as software defined networking testing platform. In: International Conference on Communication, Computing & Systems (ICCCS) (2014)
10. https://github.com/noxrepo/pox
11. https://www.opendaylight.org/
12. http://www.projectfloodlight.org/floodlight/
13. Megyesi, P., et al.: Available bandwidth measurement in software defined networks. In: Proceedings of the 31st Annual ACM Symposium on Applied Computing, ACM (2016)

Monitoring Chili Crop and Gray Mould Disease Analysis Through Wireless Sensor Network

Sana Shaikh[1]([✉]), Amiya Kumar Tripathy[1,2], Gurleen Gill[1],
Anjali Gupta[1], and Riya Hegde[1]

[1] Department of Computer Engineering,
Don Bosco Institute of Technology, Mumbai, India
{sana,amiya}@dbit.in, gurleen0694@gmail.com,
anjaligupta1307@gmail.com, riya.hegde26@gmail.com
[2] School of Science, Edith Cowan University, Perth, Australia

Abstract. The purpose of this work is to design and develop an agricultural monitoring system using Wireless Sensor Network (WSN) to increase productivity and quality of chili farming remotely. Temperature and humidity levels are the most important factors for productivity, growth, and quality of chili plant in agriculture. It is necessary that these are to be observed all the time in real time mode. The farmers or the agriculture experts can observe the measurements through the website or an android app simultaneously. The system will be immediately intimated to the farmer in detection of any critical changes occurs in one of the measurements. Which would helps the farmer to know about the possible disease range. With the continuous monitoring of many environmental parameters, the grower can analyze an optimal environmental conditions to achieve maximum crop productiveness and to save remarkable energy.

Keywords: Agriculture · Wireless sensor network · Environmental parameters
IOT

1   Introduction

Chili is one of the most valuable crops of India and its economy. Therefore, producing a better yield of the crop is considered as an important goal to be work out.

This crop is sensitive to environmental conditions, for example, excess water, humidity, hot weather conditions. It is necessary to monitor the crops periodically so as to avoid such conditions which could lead the crop to attracting diseases [3, 10]. As WSN in the field of agriculture helps in remote data collection and monitoring assistance in harsh environments, there is a great need to modernize the conventional agricultural practices for the better productivity. Due to undesigned and unplanned use of water, the water level is decreasing continuously and solubility of the fertilizers in water used for irrigation contains various chemical constituents some of which may interact with dissolved fertilizers with undesired effects which may leads to inferior quality of Chili crop production. WSN technology is one of the most effective technologies used for continuous monitoring of environmental parameters for real-time

© Springer International Publishing AG, part of Springer Nature 2018
A. Abraham et al. (Eds.): ISDA 2017, AISC 736, pp. 924–931, 2018.
https://doi.org/10.1007/978-3-319-76348-4_88

detection of specific phenomena. The major goal of WSN technology is to sense agriculture characteristics and advice farmers to properly grow and treat chili crops [5]. The aim of the system is to construct a crop monitoring system to observe soil moisture, humidity, temperature and water content for chili crop [6]. It is based on WSN for precise decision making to produce profuse chili crop production while diminishing cost and assisting farmers in real time data gathering.

Agriculture is the backbone of most developing countries like India and it provides food for humans, directly and indirectly. It is also one of the strongholds of the Indian economy and accounts for 14.6% of the country's gross domestic product (GDP) in 2009–10, and 10.23% of the total exports. Furthermore, the sector provided employment to 58.2% of the work force [1]. Chili crop is of great importance to Indian Economy as World chili production is primarily concentrated in South Asian countries to an extent of about 55% of total world production. India is the single largest producer contributing for about 38% followed by neighbors China with 7%, Pakistan and Bangladesh contributing for about 5% each [2].

In an agricultural field, the farmers or the agriculture expert can observe the measurements periodically like temperature and humidity levels inside the fields. Thus by this way, the agriculture expert can interfere in such a needy situation at an earliest time possible and may be able to prevent possible damaging effects of those changes. Applying cabling for monitoring would make the measurement system expensive, vulnerable, and also the measurement points are difficult to change or relocate. Due to these facts, a WSN has been designed which consists of small-size wireless sensor nodes with sensors. This option is more cost efficient to build the required parameters measurement system [7].

A WSN consists of a large number of nodes which are deployed densely in an agriculture land for various physical parameters to be measured or monitored as shown in Figs. 3 and 4. Each of these nodes gathers the data and route this information. The system is intended to monitor the environment of chili crop to help the farming community in understanding the micro level change in the field condition and hence can able to take decisions.

2 Materials and Methods

In order to study the crop-weather-disease interactions, a test bed for WSN experiment was chosen at Don Bosco Institute of technology, Kurla (W), Mumbai, which has a tropical wet and dry climate. Long term weather-based experiments are being carried out on 2 test beds of chili crop.

2.1 Standard Experimental Setup

Studies on crop-weather-disease interaction were carried out consecutively on slots of three months (slot 1: 20/1/2016 – 19/2/2016, slot 2: 20/2/2016 – 19/3/2016, slot 3: 20/3/2016 – 19/4/2016). A standard field experiment design was laid out in the test bed of two plots each of 1 m^2, where P1 is plot 1 and P2 is plot 2. Both plots are under normal circumstances faced in a farmer's field. Apart from this, to have uniform and unbiased observation, surveillance data has been collected from each plot in randomly selected one square meter area locations of the plot. The five points under surveillance in the first plot P1 are named as (X1, X2, X3, X4, X5) that is (P1X1, P1X2, P1X3, P1X4, P1X5) and the five points under surveillance in the second plot P2 are named as (Y1, Y2, Y3, Y4, Y5) that is (P2Y1, P2Y2, P2Y3, P2Y4, P2Y5) [8]. From flowering stage to harvest stage, the crop's phenological stages as shown in Fig. 1.

Fig. 1. Standard experimental layout   **Fig. 2.** Surveillance data collection

2.2   Sensory Data Collection

The surveillance data has been collected daily and stored in the MySQL database. The data were obtained at every week from flowering to reproductive phenological stages, where majority of disease incidences occurs at various locations in the experimental site. The weekly average of surveillance data was used for disease identification as there will not be any significant visible changes in disease incidences. A total of 48 observations (12 weeks * 4 readings) were made. Along with this, the chilli crop age (that is at which stage of the crop the disease attack takes place and their dynamics trends) also recorded week-wise to understand the infection dynamics of gray mould. Five sticks were placed in each one square meter and numbered 1, 2, 3, 4 and 5 (Fig. 2).

Subsequently, data has been collected manually from the plants adjacent to each stick. The surveillance data includes number of leaf-lets have infected, how many plants infected in the one square meter area, date of flowering, date of recording the surveillance data, etc. The plants have been numbered in the form of ID like P1X1, P1X2 etc. so as to retrieve their respective data easily from the database. (P1X1 indicates plot 1 and plant number 1 and P1X2 indicates plot 1 and plant number 2. Similarly, P1Y1 indicates plot 2 and plant number 1 and P1Y2 indicates plot 2 and plant number 2.) This is the one of the standard design practices that has been used in the long term experiment in the test bed [11]. Correlation index of disease intensity was recorded on at least 20 plants each plot from in the test bed. The system architecture can be seen in Fig. 5 as shown below.

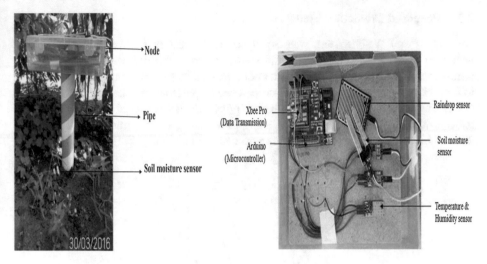

Fig. 3. Network deployment in the field   **Fig. 4.** Components present in the deployed box

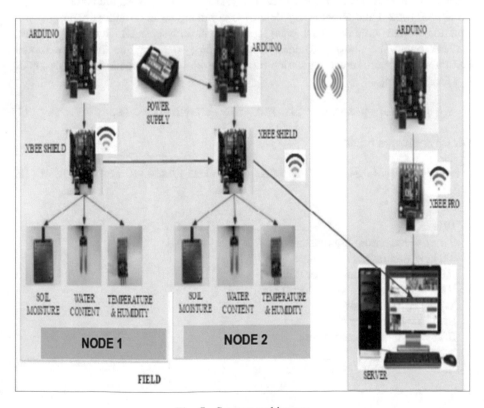

Fig. 5. System architecture

2.3   Proposed Prediction Model

An end to end WSN system was deployed in the test bed to monitor micro level weather parameters. A set of battery powered nodes embedded with wireless sensors (temperature and humidity sensor, water content sensor, soil moisture sensor), was used to continuously monitor sensory soil-crop-weather parameters. The data from the two nodes are collected and sent to the base station. Each node is able to transmit data packets to base station in every 3 h using sleep alert, over a transmission range of 200 m. Linear Regression function in matrix notation [4, 12]:

$$
\begin{bmatrix} y_1 \\ y_2 \\ \vdots \\ y_n \end{bmatrix} = \begin{bmatrix} 1 & x_1 \\ 1 & x_2 \\ \vdots & \vdots \\ 1 & x_n \end{bmatrix} \begin{bmatrix} \beta_0 \\ \beta_1 \end{bmatrix} + \begin{bmatrix} \varepsilon_1 \\ \varepsilon_2 \\ \vdots \\ \varepsilon_n \end{bmatrix}
\tag{1}
$$

$$Y = X\beta + \varepsilon$$

The regression algorithm estimates the value of the target as a function of the predictors for each case of the test data. The relationships between predictors (example Tmax, RH1, etc.) and target (Gray mould) is summarized in a multivariate regression equation, which could then be applied to a different data set in which the target values are unknown. The above technique was used in generating the prediction model. Following is a multivariate regression models developed/tailormade for Gray mould. For Gray mould,

$$Y = 0.94 * \text{Tmax} + 0.32 * \text{Tmin} - 0.13 * \text{RH1} - 0.63 * \text{RH2} + 0.85 * \text{WC} + 0.38 * \text{SM} + \text{SSE} \tag{1}$$

For Phytophthora blight,

$$Y = 0.89 * \text{Tmax} + 0.34 * \text{Tmin} - 0.7 * \text{RH1} - 0.68 * \text{RH2} + 0.41 * \text{WC} + 0.4 * \text{SM} + \text{SSE} \tag{2}$$

Where,
Tmax = maximum temperature
Tmin = minimum temperature
RH1 = relative humidity at 7:30 am
RH2 = relative humidity at 3:00 pm
WC = water content
SM = soil moisture
SSE = sum of squared errors.

3 Results and Discussions

Raw sensory data have been converted to real data at server end. Correlation studies of disease–weather-environment crop interaction were carried out and their hidden correlations were discovered and quantified with tailor-made mining techniques in the test bed. Subsequently, the developed multivariate regression mining techniques were applied for disease modeling and predictions.

The correlation values (both positive and negative) of predictor (example Tmax, RH1, etc.) versus target (Gray mould) infection index were obtained from various datasets (sensors, weather station and surveillance) during flowering to harvesting stages and are depicted in Fig. 6. Based on concept of infection index with respect to pest/disease risk model, correlation index greater than 0.5 in the scale of −1 to +1 has been considered as strong positive, whereas the value of 0.5 or more considered as strong negative.

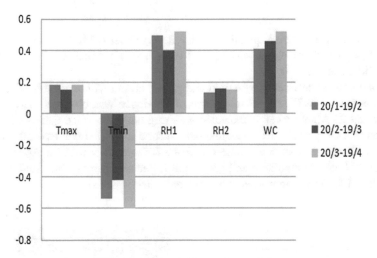

Fig. 6. Gray mould weather statistics

It was found that RH1 have strong positive correlation and Tmin has strong negative correlation with Gray mould. There is an increase in disease intensity with the period of high humidity as the latter favors abundant sporulation and germination. Since the optimum temperature for gray mould ranges from 15 °C to 25 °C naturally occurring temperature during the rainy season are not likely to limit the process. The SSE value is 1.861021.

3.1 Gray Mould and Phytophthora Blight Dynamics

Gray mould incidence in chili crop has been observed round the year with prevailing weather conditions as seen in Fig. 7. By means of the developed model, with standard ground level surveillance observation and weather parameters from sensory as well as

Fig. 7. Gray mould incidence as seen in the test bed

Fig. 8. Phytophthora incidence as seen in the test bed

nearby weather station were quantified [9]. . It was observed that symptoms of initial stages of phytophthora blight was noticed as it favors similar weather condition as Graymould as seen in Fig. 8.

The correlation between gray mould and phytophthora with respect to all the sensor readings is shown in Fig. 9. Visible range of incidence was observed between 0.18–0.45. During 23 March week unexpected watering had been done due to which it has been seen that Gray mould has been reappeared for a week and decrease after that. The data from all the sensors were collected from 20th January to 13th April. It shows indication of both phytophthora and gray mould. Reading from sensors and their status are displayed for the farmer to enable him make better decisions.

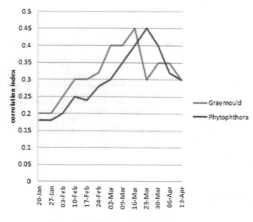

Fig. 9. Correlation between gray mould and phytophthora

Fig. 10. Screenshot of android app

4 Conclusion

Gray mould and Phytophthora has been observed in the field and its prediction model has been developed. This will help the farmer to take control of the situation. The system further has four parameters monitoring the environment in different areas in different states and countries. This system have considered parameters which are most sensitive to the crop; it has a scope of monitoring other parameters which affects the crop. The system is intended to monitor the environment of Chilli crop, other crops can also be monitored so as to produce a better yield. It can be implemented so as to facilitate automated irrigation, control over temperature around the crop, and automated pest supply when the crop is prone to diseases.

The system informs the farmer about the crop environment and its status as shown in Fig. 10. If any measured parameters are found to be extreme, the farmer is provided with a possible solution which could implement in order to save the crop.

Acknowledgments. This project was sponsored by the Don Bosco Institute of Technology, Mumbai, India. The authors are grateful to the Don Bosco Institute of Technology for providing all kinds of resources for this study.

References

1. The Indian Agricultural Scenario. http://appscmaterial.blogspot.in/2010/08/indian-economic-scene.html
2. Chilli Climate and Cultivation. http://www.commoditiescontrol.com/eagritrader/staticpages/index.php?id=67
3. Rangan, K., Vigneswaran, T.: An embedded systems approach to monitor green house. In: Research Advances in Space Technology Services and Climate Change, pp. 61–65. IEEE, November 2010
4. Tan, P.N., Steinbach, M., Kumar, V.: Data Mining and Regression, pp. 729–733. Pearson Addison Wesley (2006)
5. Balaji Bhanu, B., Raghava Rao, K., Ramesh, J.V.N., Hussain, M.A.: Agriculture field monitoring and analysis using wireless sensor networks for improving crop production. In: IEEE Wireless and Optical Communications Network, pp. 1–7, September 2014
6. Patil, S.S., Davande, V.M., Mulani, J.J.: Smart wireless sensor network for monitoring an agricultural environment. Int. J. Comput. Sci. Inf. Technol. **5**(3) (2014)
7. Cárdenas Tamayo, R.A., Lugo Ibarra, M.G., Macias, J.A.G.: Better crop management with decision support systems based on wireless sensor networks. Computer Science Department Research Center, Ensenada, México, pp. 412–417, September 2010
8. Kumar, J.P., Umar, S., Nagasai, C.S.H.B.: Implementing intelligent monitoring techniques in agriculture using wireless sensor networks. Int. J. Comput. Sci. Inf. Technol. **5**(4), 5797–5800 (2014)
9. Royle, D.J., Burr, D.J.: The place of multiple regression analysis in modern approaches to disease control. EPPO Bull. **9**(3), 155–163 (2008)
10. Agrawal, R., Mehta, S.C.: Weather based forecasting of crop yields, pests and diseases – IASRI models. J. Indian Soc. Agric. Stat. **61**(2), 255–263 (2007)
11. Entomology Work 2008–09: AgroMet-Cell. Agriculture Research Institute, ANGRAU, Hyderabad, Annual Report, pp. 57–69 (2009)
12. Gupta, M.R., Chen, Y.: Theory and use of the EM algorithm. Found. Trends Sign. Process. **4**(3), 223–296 (2010)

Intelligent AgriTrade to Abet Indian Farming

Kalpita Wagaskar[1](✉), Nilakshi Joshi[1], Amiya Kumar Tripathy[1,2](✉),
Gauri Datar[1], Suraj Singhvi[1], and Rohan Paul[1]

[1] Department of Computer Engineering,
Don Bosco Institute of Technology, Mumbai, India
{kalpita,nilakshi,amiya}@dbit.in,
gauri.datar@gmail.com, surajsinghvi11@gmail.com,
rohan.thekanath93@gmail.com
[2] School of Science, Edith Cowan University, Perth, Australia

Abstract. The present Indian agricultural system is embedded with advanced
services like GPS, weather sensors, etc. which facilitate in communicating with
each other, and analysis of the ground level real-time or near real data. Information and Communication Technology (ICT) provides services in the form of
cloud to agricultural systems. Agriculture-Cloud and ICT offers knowledge
features to farmers regarding ultra-modern farming, pricing, fertilizers,
pest/disease management, etc. Scientists/Experts working at Agriculture
research stations and extensions can add their findings, recommendations
regarding up-to-date practices for cultivation, and related practices. In this work
an attempt has been made to design and implement a simple Cloud based
application on Agriculture System which is based on AgriTrade on cloud that
will enrich agriculture production and also boost the accessibility of data related
to field level investigation and also in laboratory. The impact of undertaking
such a tool would cut the cost, time and will increase the agricultural production
in a relatively much faster and easier way. The system is intelligent to tell the
environment statistics to the farmer for improved approach towards agriculture.

Keywords: Cloud computing · ICT in agriculture · AgriTrade
Service oriented architecture

1 Introduction

Indian culture can be experienced through its villages and small towns. Rural India has
been ignored for many years and the Intelligent AgriTrade cloud technology will bring
the change that is required to bridge the gap between rural India and urban India, and
will improve the Indian rural economy. The principal source of income of India is
agriculture. The development of Indian farming is basically focused on the Indian
agriculture sector. Cloud computing is a general term used to describe a new class of
network based computing that take place over the internet. Latest technological
development has brought about a dramatic change in every field, and agriculture is no
exception to it. Intelligent cloud computing technology will have a positive impact in
the agricultural field which can also progress to other traditional approaches being
moved to cloud.

© Springer International Publishing AG, part of Springer Nature 2018
A. Abraham et al. (Eds.): ISDA 2017, AISC 736, pp. 932–941, 2018.
https://doi.org/10.1007/978-3-319-76348-4_89

Nowadays farmers are not able to cultivate the amount of crops needed because of new types of diseases and the unpredictable weather environment. The farmers also do not get paid fairly for their hard work because they sell their vegetables and crops at a very cheap price to the vendor and then the vendor sells it ahead to the customer at a very high price. Therefore the middle agent earns the profit completely, whereas the farmer faces a huge loss. This financial loop hole results in loans not being paid by the farmer as he is in huge dept, which results in suicides and the economic status of the farmer halts at being always very low with no improvement in the society. By implementing this tool AgriTrade we are trying to help the farmers in every possible way and reduce the number of farmer suicides by giving them the income they are actually worth of getting for his hardwork.

In this work an attempt has been made to make use of services such as real-time computation, data access without the need to know the physical location and config-uration of the system that delivers the services. If we need to improve the economic condition of India as a developing nation then the only way to do that is to improve the Indian agricultural sectors. The proposed solution of Intelligent AgriTrade in this paper is an attempt to make improvement by using the technology in this sector.

The problems that the farmers are currently facing is lack of knowledge about crop related problems, weather, fertilizers to be used, actual selling price of the crops.

Due to the unpredictable weather the farmers are not aware of which crop to cultivate in the favourable season. This results in total wastage of the crops. Sometimes when the crops are affected by some disease, the farmer is unaware of what has caused it and they use any kind of low quality fertilizers which destroys the crops. The farmers are also not aware of the actual city prices of the yields. Although many technologies are available, these have not helped our farmers to use it for their crops. Hence they do not get the intended profit. These are the problems the farmers are currently facing.

A solution to the mentioned problems can be AgriTrade. It offers expertise service to farmers regarding cultivation of crops, pricing, fertilizers, and disease related help, and also teaches the farmers how to use this new tool. By implementing this system it would cut the cost of farming, time of cultivation due appropriate forecast of crop type based on climate and make the life of the farmers much easier.

The Cloud allows information technology to be infused into the smallest hamlet of India. This makes information available to the poorest of the community so they can aim for a better life with the knowledge obtained through laptops or mobile phone connected to the cloud.

Here, the farmers with the help of such a system can learn about new farming techniques, learn more about the quality of the fertilizers used, know more about the weather, sell their crops at a better rate to get more profit and also solve their doubts related to crop problems. As this system will be on the cloud and available on android every farmer from any part of India can get access to it.

2 Literature Survey

In India, many attempts are made to improve the condition of Indian agriculture by the youth with technological variations and expert advice. This would help improve the condition of farmers and farming. Studies are being carried out to implement

E-Farming using Cloud Computing [4] that is selling of products for the farmers to different customers at different locations. It is proposed that there would a web portal to assist the farmers with the selling of their products. This could be done if the farmers have a knowledge about computers or else the company professionals would provide help to interested farmers to register and utilize these services. E-Farming doesn't comprise of total electronic farming but just helps the farmers to sell their valuable products by their own means through a cloud and web portal platform so as to reach out to the customers efficiently.

Another study is formulated to improve the Indian Agriculture through an Application of Cloud Computing in Agricultural Development of Rural India [1, 4, 7, 8] which emphasizes on the architecture of the usage of cloud computing with the devices and the user Interface for any particular web portal used to interface cloud with agriculture. Here the system is designed to perform database utilities and connection with the cloud to read the data so as to improve the local and global communication between the rural and urban India. This would indirectly help to improve the agricultural sector of the nation through reduction in man power, infrastructure etc. and will show progress in the economic condition of India.

Attempts have been made to incur technology in the field of agriculture through Agro Mobile: A Cloud-Based Framework for Agriculturists on Mobile Platform [2], an application on mobile phones with cameras to recognize and prevent the crops from the diseases. In this system the farmers will be educated about everything from weather forecast to the crop analysis and prevention by just a click on the mobile phones.

The Application of Cloud Computing to Agriculture and Prospects in Other Field [3] is developed and utilized by the farmers of Japan to keep a track of all the data regarding themselves as well as the cultivable land. Also they use this system to patrol their farms through sensors and camera on a hardware implementation which allows to analyze the growth in the agriculture and maintain graphs and records for the same.

These different forms of works carried out throughout India and outside India are portraying a part of the whole technological way in which agriculture could be improved. Thus all of these studies are very helpful to increase the Gross Domestic Product (GDP) of our country.

In our system we are providing E-commodity exchange where the farmer has access to all the vendors in the city to directly sell the goods, the E-farming service where all the information regarding the farming is provided and telemedicine in which assistance is given incase of infected crops. This tool will also provide real time data of the weather to cultivate crops accordingly. We will be implementing cloud based service which will be accessible through web or mobile app. The system will also help general population to purchase food grains and vegetables much cheaper as the actual price of the crops will be now directly awarded to the crops.

3 Proposed System

The proposed system aims to provide solution to the farmer in yielding productive crops, reduce cost of farming and increase farmers earning. In the process we too will pay less for the goods bought and will make farmer to earn more and boost the economy of the country.

The whole system consist of four modules Real-time computation, E-Farming, telemedicine for crops and E-Commodity Exchange. Real-time computation which gathers real-time data from various sources and database like soil database, weather database, farmer database which computes the required information according to the required real time data. Example, weather forecasting for a particular region for coming weeks [10, 11].

E-Farming aims to promote organic farming, provide assistance using new technologies in farming that yield more crops, provide tips and solution to general problems. Telemedicine for crops solves the crop related problems through expert advise who would help them with some solution.

E-Commodity Exchange enables the farmer to sell the products to the consumers directly at a reasonable price and the farmer will be able to sell the products to the customer reducing the involvement of middle agent and will help the farmer to earn maximum profits through an e-commerce website or an Android application.

The data required for this system such as crops, fertilizers, and their nutrient content etc. was obtained from government websites like farmer portal, agricoop and Mahaagri [5, 6]. The data obtained was either in word or pdf formats. Data collection was a very important part of this system to test and give the required results.

The approach applied here is called intelligent 'AgriTrade' because based on the results that we have obtained using R-programming the current status of the climate and environment is known to the farmer and the crops that can survive in that environment are encouraged to grow for better yield of the farmers. They also get information about current market condition due to the experts sharing their advise on the portal and can also get help about crop diseases through the portal by just uploading the picture of the infected crop.

3.1 System Architecture

The system consist of many different databases implemented in MongoDB and computation is done over cloud on Amazon AWS service for faster query response. The whole system is divided into four different modules, Real-time computation, E-Farming, Tele-medicine for crops and E-Commodity Exchange. All these modules can be accessed via Graphical User Interface through devices like laptops, desktops, and smartphones. End users of this system includes farmers, agricultural experts, administrators, and can be used by government services to know any kind of information anytime.

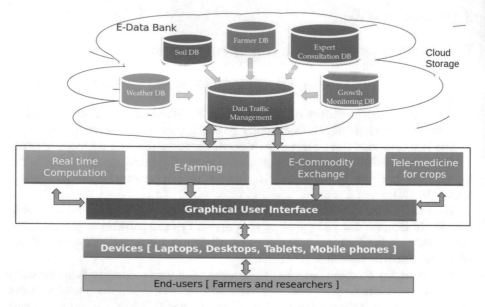

Fig. 1. System architecture

4 Implementation

The web portal depicts the exact analysis and view of the whole system through which all the services would be available to the farmers as well as the few authorities in charge of the whole system. This system is basically going to help the farmers to get all required informations related to farming under one roof. The proposed system provides a user interface along with a database which is used to store all the registered and available input data [9].

Real time computation can be administered and can monitor various databases such as farmer, soil etc., weather forecast information along with growth monitoring of particular sector. This makes the system intelligent. E-Farming provides learning tutorials to the farmers to improve their agricultural skills efficiently and Telemedicine helps the farmers to cure their affected crops and help protect those crops from diseases through fruitful guidance from experts. Next service most needed for the farmers is the E-Commodity Exchange which is an E-Commerce web page helping the farmer to sell their products to the customers at different locations by means of technology. Basically the farmer would have to register with the portal to use the services provided. Figure 2 shows the screen shot of the main page and services page of AgriTrade.

There would be computer professional to help the farmer out with the registration till they receive at least the first payment from the yield.

A Warehouse would be provided to the farmer to keep the yield so as to submit the product to the authorities for sale. Here the system helps the farmers to improve themselves as well their agricultural condition and reduce the number of suicides due to lack of crop or due to crop damage by diseases. As India has maximum of his economy

Fig. 2. Main web page and the service page for E-exchange of crops, E-farming, telemedicine inquiries

dependent on agricultural sector this will indirectly improve the conditions and livelihood of farmers as well as the economy of India to a greater extend.

The implementation of the system starts with the user interface development for the web portal which is implemented using Bootstrap which includes HTML and CSS libraries. The pages for the services are designed and would create a proper user interface for the farmers to use it easily which will include regional languages too.

The second section of work is with the database which is implemented using MongoDB as a database and integrating it with the web portal through PHP script.

5 Results

The results are performed by doing live analysis using R-programming.

The result in Fig. 3 that is generated using this software gives the graphical interpretation of the weather attributes such as Temperature, dew point etc.

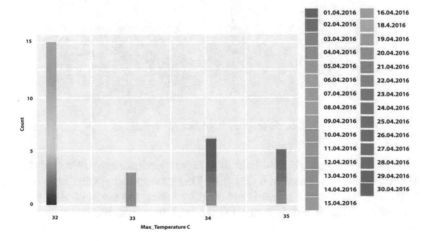

Fig. 3. R-programming temperature analysis

This graph shows the month of April along with its maximum temperature.

The result in Fig. 4 is generated by using dew point as a parameter and the graphs shows the number of days with its respective dew point. The different colors indicate the different days of the month. Figure 5 is a similar graph but is generated by taking humidity as a parameter.

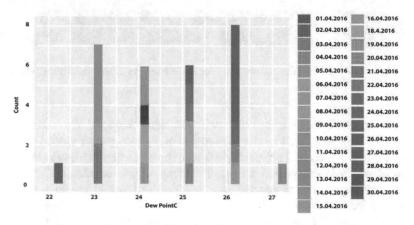

Fig. 4. R-programming dew point analysis

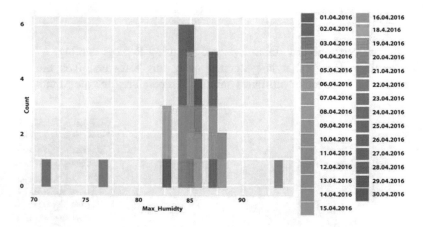

Fig. 5. R-programming humidity analysis

The data is collected with the help of a pre-defined package: Weather data. The dataset once received is stored in a CSV file. This process is repeated for a number of past years. The forecasted weather is obtained by taking the mean of all the years for every parameter. This dataset is then imported back into R-studio and then analysis is done on it.

Entire web portal is language friendly as the language could be changed into different Indian Languages with respect to the farmer's requirement. This will help the farmers to access the web portal in his regional language.

Along with Real Time Computation there are three more services such as E-Farming which provides learning tutorials to the farmers to improve their agricultural skills efficiently through textual help and videos which teach farming. Telemedicine helps the farmers to cure their affected crops and help protect those crops from diseases through fruitful guidance from experts.

Here the farmers have to upload an image of their damaged crop and then check for the experts answers to their queries and also check for the one which have been already asked by other farmers, last but the most needed service for the farmers is the E-Commodity Exchange which is an E-Commerce web page helping the farmer to sell their products to the customers at different locations by means of technology. The customer is able to buy all the commodities sold out by the farmer and the farmer is also benefited as he earns profit by directly selling the products rather than going through a middleman. Basically the farmer would have to register with the portal to use the services provided.

All the three services E-Farming, E-Commodity Exchange and Telemedicine would be available to the farmer to utilize it to the fullest limit. If the farmer does not have any knowledge about computers there would be a computer professional to help them out with the registration till the receivable of payment from the customers for their yield. Along with the farmer the customer can also use the services if customer wants to learn new techniques regarding farming and more importantly he can buy commodities fresh and at very less price.

The tutorial page will help the farmers learn new techniques in farming and would help them improve their way of farming and achieve better outcome. With farmers producing better yields their economic condition is improved. The textual and video are as per the regional languages of the state, for now we have taken only languages of India but it can expanded to all the languages of the world.

The system helps the farmer to earn more profit through the web page as the main disadvantage of middleman is reduced and thus the farmer can sell his yields to the customers directly through the web page and the customers can also access it to buy the sold out yields helping the farmers directly.

The reason why cloud computing is included in this system is that the data processing and storage needed is dynamic and will keep on increasing with time. Therefore the system should have enough storage and processing capabilities.

6   Discussions

The developed system needs to be deployed in the farms. At initial stages there will be an agent who will be helping the farmers to use the system. He will train them as well as help them in selling their crops online and show them how their queries can be answered online. Farmers will be trained to know that he is not alone and there are many people out to help him out. The farmer will get web as well as mobile app training till he becomes fluent to use the AgriTrade tool.

7 Conclusion

This system has been completely developed and all the connections of the website, database and the cloud has been established. All the mentioned modules have been developed and tested as per the requirements. As of now we have only considered Maharashtra for this project but the future scope can include all the states of India. In the e-commodity exchange module online payments methods such as credit/debit can also be included. As our requirements were less we have used Amazon's EC2 layer which can also be changed when there is more data and processing to be done. When this system will be available to the farmers it will help solve many of their problems and make their lives easier. This Intelligent approach presented in this paper help the farmers better visualize their yield and make them aware of the current market condition. The approach makes the farmer more knowledgeable, which can help them bring improvement in their production and financial condition.

Acknowledgement. This work is fully supported by Don Bosco Institute of Technology, Mumbai, India.

References

1. Patel, R., Patel, M.: Application of cloud computing in agricultural development of rural India. Int. J. Comput. Sci. Inf. Technol. **4**, 922–926 (2013)
2. Prasad, S., Peddoju, S.K., Ghosh, D.: AgroMobile: a cloud-based framework for agriculturists on mobile platform. Int. J. Adv. Sci. Technol. **59**, 41–52 (2013)
3. Hori, M., Kawashima, E., Yamazaki, T.: Application of cloud computing to agriculture and prospects in other fields. Fujitsu Sci. Tech. J. **46**(4), 446–454 (2010)
4. Pandey, A.: E-farming using cloud computing. Int. J. Cloud Comput. Serv. Sci. **2**(5), 345–350 (2013)
5. National Informatics Center (NIC). http://www.mahaagri.gov.in/. Accessed 16 Oct 2017
6. Department of Agriculture & Cooperation and Farmer Welfare, Ministry of Agriculture and Farmers Welfare, Government of India. http://www.agricoop.nic.in/. Accessed 17 Oct 2017
7. Agarwal, P.: AGROCLOUD - open surveillance of Indian agriculture via cloud. In: 2016 International Conference on Information Technology - The Next Generation IT Summit on the Theme - Internet of Things: Connect Your Worlds, Noida, 6–7 October 2016. https://doi.org/10.1109/incite.2016.7857613
8. Ahmad, T., Ahmad, S., Jamshed, M.: A knowledge based Indian agriculture: with cloud ERP arrangement. In: 2015 International Conference on Green Computing and Internet of Things, Noida, 8–10 October 2015. https://doi.org/10.1109/icgciot.2015.7380484
9. Hu, H., Chen, T.: Design and implementation of agricultural production and market information recommendation system based on cloud computing. In: 2015 8th International Conference on Intelligent Computation Technology and Automation, Nanchang, China, 14–15 June 2015. https://doi.org/10.1109/icicta.2015.99

10. Khattab, A., Abdelgawad, A., Yelmarthi, K.: Design and implementation of a cloud-based IoT scheme for precision agriculture. In: ICM 2016, Giza, Egypt, 17–20 December 2016. https://doi.org/10.1109/icm.2016.7847850
11. Kruize, J.W., Wolfert, S., Goense, D., Scholten, H., Beulens, A., Veenstra, T.: Integrating ICT applications for farm business collaboration processes using FI space. In: 2014 SRII Global Conference, San Jose, CA, USA, 23–25 April 2014. https://doi.org/10.1109/srii. 2014.41

Evaluating the Efficiency of Higher Secondary Education State Boards in India: A DEA-ANN Approach

Natthan Singh$^{(\boxtimes)}$ and Millie Pant

Indian Institute of Technology Roorkee, Roorkee 247667, Utttrakhand, India
nsingh.iet@gmail.com, millidma@gmail.com

Abstract. This study proposes the integration of two nonparametric methodologies - Data Envelopment Analysis and Artificial Neural Network for efficiency evaluation. The paper initially outlines the research work conducted in the education sector using DEA and ANN. Furthermore, the case study for the paper is conducted on various State Boards (which are used as DMU's) in Indian Higher Secondary Education System for efficiency evaluation using DEA which is integrated with soft computing technique ANN in order to increase discriminatory power, ranking and future prediction. The above two methods are compared on their practical use as a performance measurement tool on a set of Indian State Boards in Indian Higher Secondary Education System with multiple inputs and outputs criteria. The results demonstrate that ANN-DEA Integration optimizes the performance and increases the discriminatory power and ranking of the decision making units.

Keywords: Data Envelopment Analysis · Artificial Neural Network
Indian Higher Secondary Education

1 Introduction

Efficiency in very simple terms is defined as the ratio of output to input. It is a means to avoid wastage of energy, materials, resources and time in order to achieve desired output. It is a key concept for any organization. It is an important concept that is being incorporated into different sectors of the society to evaluate and analyze the respective efficiency and performance of different units, departments or individuals. The more efficient process is more productive which means more output can be achieved with less input. This is why efficiency evaluation is very important. This evaluation is done on basis of selected criteria which have maximum impact on the organization. In our present work, our focus is on Data Envelopment Analysis (DEA), is a performance measurement tool that calculates the efficiency irrespective of a number of inputs and outputs [1]. DEA is being used by various researchers for a long time in various fields of their interest. A little peek is given about the work done in this area later in this paper.

    The core of this research is to analyze how one can improvise the structure of DEA by combining it with Artificial Neural Network (ANN), a soft computing technique. Results indicate that an integration of ANN with DEA increases the discriminatory

© Springer International Publishing AG, part of Springer Nature 2018
A. Abraham et al. (Eds.): ISDA 2017, AISC 736, pp. 942–951, 2018.
https://doi.org/10.1007/978-3-319-76348-4_90

power of DEA and provides a better ranking of the decision making units for better efficiency and performance evaluation.

A small case study taken from the education sector is considered to validate the results of the integrated DEA and ANN method. Education is the process of gaining knowledge, values, skill. This is mainly done in some school or an institute. Education is indispensable for any individual in order to live a better life. It provides us the knowledge and insight so that we can form our own opinion and have a point of view in life.

Education System is divided into three sectors namely primary, secondary and higher education. Primary education includes classes from Class I to Class VIII. Secondary education includes classes from Class IX to Class XII. Higher education includes the graduation, post graduation or diploma courses in various fields like technology, science, medicine, commerce, etc. We can also say secondary education in India comprises of High School and Intermediate. Here in our study higher secondary education, that is, intermediate is considered.

The paper discusses a case study of evaluating the efficiency of various states Boards in India in higher secondary education system using DEA-ANN approach. Here various state boards are taken as DMU's and DEA-ANN is applied on it to evaluate the performance and efficiency of various state boards which thereby evaluates the performance of various states of India in higher secondary education. The result of this case study clearly reveals the efficient and inefficient state boards in India and some very interesting facts about the higher secondary education system.

This paper is divided into 5 parts including introduction in Sect. 1. Section 2 contains a brief literature review. Section 3 of the paper comprises of the methodology used for evaluating efficiency using DEA and ANN. In Sect. 4, our case study which is Indian Higher Secondary education System is discussed and its efficiency is evaluated. Section 5 contains the conclusion.

2 Literature Review

DEA was developed by Charnes et al. [1] and was extended by Banker et al. [2]. The computation of efficiency is the main objective of the DEA. This is done mainly in cases with multiple output and input criterion and when it is not feasible to turn these into a single aggregate output or an input parameter. This methodology is especially undistinguished to evaluate the efficiency of non-profit entities that run outside the market because their performance is not satisfactory.

Data Envelopment Analysis is basically a linear programming based non parametric technique. Here, the relative efficiency of a homogenous set of decision making units (DMUs) is measured. It optimized each individual observation with an objective of calculating a discrete piecewise frontier, determined by the set of efficient DMUs. This method works when DMU's are homogeneous or assumed to be homogeneous in the aspects of their input and output and industry and DMU's must be minimizing inputs and maximizing outputs.

The relative importance of input and output and the assumption regarding the functional form is not required by DEA beforehand. It calculates a maximal performance measure for each decision making units relative to all other decision making units in the population. Here, the most important necessity is that each decision making unit should lie below or on the frontier. A review of DEA being implied in the education sector is presented in the following paragraph.

Bradley et al. [3] calculate the technical efficiencies, based on the multiple outputs (school exam performance and attendance rates) of secondary schools in England and they conclude that greater the degree of competition between schools the more efficient they are. Afonso and Aubyn [4] addresses the efficiency of expenditure in education and health sectors for a sample of OECD countries using the nonparametric methods DEA and FDH and authors gives a comparison across the two sectors, education and health, to see whether efficiency and inefficiency are country specific. Johnes and Li [5] use DEA to examine the relative efficiency in the production of research of some Chinese regular universities. In this study output variables measure the impact and productivity of research; input variables reflect staff, students, capital and resources. The authors find the significant differences between HEIs is associated with either geographical location, source of funding or type of university produces some interesting results. Kuah and Wong [6] presented a DEA model which consisting multiple inputs and outputs for the measurement of the efficiencies of universities based on their teaching and research activities. Agasisti [7] used DEA to compute efficiency scores for a sample of Italian schools. The author show that at least one indicator of competition is statistically associated with higher performances of schools. Sibel and Sibel [8] used two- stage DEA to determine factors on the efficiency of universities in Turkey. Author finds that the number of female student has a positive and important effect on relative efficiency of universities in Turkey. Sagarra et al. [9] used three - way scaling model (INDSCAL) to analyze the data using scaling techniques, in order to visualize the results and make them accessible to policy makers.

An artificial neural network (ANN) is an information processing paradigm that is inspired by biological neural networks. These are parallel computing systems and they are highly interconnected with a huge number of processors. These network draws analogy to the human neural network. Hence they are called artificial neural networks (ANNs) and their neurons are called as artificial neurons. ANN is used for all kinds of work, be it to conduct the study in machines and animals or for engineering motives like data compression, forecasting and pattern recognition. Also, a review of ANN being implied in various different sectors is presented in the following paragraph.

Oladokun et al. [15] used ANN for the prediction of academic performance in the University education. Chen et al. [16] conducted a study on weather giving the classification of weather type for forecasting the solar power online. Cheh et al. [17] presented a study on stock market and conclude that very successful prediction rate of stocks is exhibited. Chhachhiya et al. [18] used ANN for university education system and swarm optimization technique is used to find the optimal architecture of a neural network model.

3   DEA-ANN Approach

Here the nonparametric technique Data Envelopment Analysis is integrated with the soft computing technique Artificial Neural Network (ANN) in order to improve the discriminatory power. This promotes best performance modeling and benchmarking. The efficiency scores are initially evaluated by DEA. Then, the new data set obtained after applying DEA is used to train the ANN. The output and input parameters of the problem act as an input layer and efficiency obtained from DEA act as an output layer. Then, this trained network is used to predict the future. Now, both the efficiencies obtained from DEA and ANN is compared.

Literature shows encouraging but rare outcomes of the combined approach of DEA and ANN with most of the studies focused on predicting DEA efficiency as an indirect performance measure. Researchers have extended pilot studies conducted by Athanassopoulos and Curram [10] through comparative and complementary studies using DEA-ANN and reported promising potential of the combined approach [11–13]. However, very limited empirical studies have been conducted in the past and more promising successes yet to be made in this research stream. Recent literature reports the approach possibility of extending this combined model to the application of best performance benchmarking through prediction of optimal outputs beyond efficiency scores [14]. Moreover, this study presents practical implementation of the method as a decision support tool with the capacity to test what-if scenarios. Exploring an innovative performance measurement and prediction framework using DEA-ANN, this study fulfills a practical need and improves benchmarking and decision-making processes.

3.1   Methodology

In this section, our proposed DEA-ANN methodology is introduced in detail through step by step procedure in order to evaluate efficiency. This section comprises of four steps that are performed accordingly: In Fig. 1, the steps of the algorithm are shown.

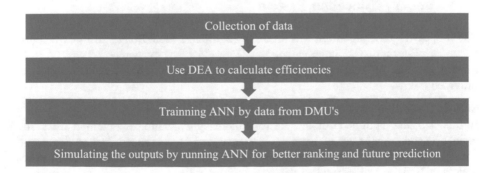

Fig. 1. An integrated DEA–ANN algorithm

Step 1: Collection of Data

In this step, data is collected in order to perform various techniques on it. Generally, the data is sourced from the government organizations which can be trusted. The data comprises of input and output parameters.

Step 2: Using DEA to Calculate Efficiencies

In this step, efficiency is calculated using either CCR model or BCC model of the DEA. After correctly applying output and input parameters and running one of the models, the efficiency for each DMU is received.

Step 3: Training ANN by Different Architectures

In this step, the output criteria and input criteria obtained from data are used as the input layer and the output parameters of DEA, that is, efficiency scores act as the output layer for training the ANN. Therefore, by training the Artificial Neural Network, we can discover the relations among these parameters. Output criteria are considered as the targets and in the training phase, ANN is designed to close up the outputs to the targets. Finally, the relations are estimated in the process of training.

In order to find a suitable architecture or the network, the transfer function for each layer, the random number of neurons in the hidden layer, which is determined by trial and error and learning algorithm, is also specified. Also, the performance function for calculating the error in each step of the algorithm between actual and estimated outputs is chosen.

Step 4: Simulating the Outputs by Running ANN for Future Prediction and Ranking

The DEA-ANN model is used to provide a better ranking of the different decision making units which makes it easy to differentiate efficient DMU's from inefficient ones. It improves the discriminatory power of the DEA. Also, simulation helps in the future prediction of the efficiencies and performance of the upcoming years through the neural network provided we have input parameters.

4   The Case Study: Efficiency Evaluation in Indian Higher Secondary Education System

The proposed hybrid algorithm was used for 22 different State Boards of Indian Higher Secondary Education (i.e. Intermediate level) for efficiency evaluation of various State Boards in India by altering the input parameters. The details for all the steps are provided below.

Step 1: Collection of Data

In this step, the input and output data for efficiency evaluation is obtained from Statistical Report of Ministry of Human Resources and Development, India. The report comprises of results of Higher Secondary Examination of the year 2013. Here 22 states boards of India are considered as decision making units.

The input and output parameters are mentioned below in Table 1. And also Output and Input for DEA efficiency calculation are mentioned in Table 2:-

Table 1. Input and Output criteria

Input criteria	Output criteria
I_1: Number of boys appeared	O_1: Number of boys passed
I_2: Number of girls appeared	O_2: Number of girls passed
I_3: Total number of students appeared	O_3: Total number of students passed
	O_4: Percentage of students passed
	O_5: Percentage of students scoring 60% and above

Table 2. Input and Output for DEA efficiency calculation

Characteristic	Input			Output				
	I_1	I_2	I_3	O_1	O_2	O_3	O_4	O_5
Max	1385804	1096036	2481840	1246843	1054941	2301784	93.0	35.56
Min	5177	5015	10192	3051	3075	6126	46.3	3.24
Avg/Mean	257387	202518	459905	194054	167286	361340	74.3	16.26

Step 2: Using DEA to Calculate Efficiency Scores

The input oriented CCR model is used for evaluating the efficiency of DMUs and based on the efficiency scores; the efficient and inefficient units are identified. The efficiency scores obtained in this step and the input-output specifications from step 1 are used for the next step. Here we consider the 22 DMU's, that is, State Boards and their technical efficiency (TE) and pure technical efficiency (PTE). Both the TE and PTE are calculated using DEAP software. Technical efficiency is CRS (constant return to scale value) and pure technical efficiency is VRS (variable return to scale value).

Step 3: Training ANN by Different Architectures

In this step, the Artificial Neural Network is trained. The 3 inputs and 5 outputs parameters of this higher secondary education problem are considered as input layer and the efficiency scores obtained from DEA in step 2 is used as the output/target layer. The ANN also comprises of the hidden layer and the number of neurons which are determined by hit and trial method.

Step 4: Simulating the Outputs by Running ANN for Future Prediction and Better Ranking

In this step, better ranking is obtained. The DEA provides a ranking of various DMU's. But in DEA ranking, many DMU's are of the same rank. So, the problem is faced to know which DMU is more efficient and which is inefficient. The ANN model solves this problem. It increases the discriminatory power and provides distinguished rankings for all the DMU's so that we can know which DMU's are efficient and which are inefficient. The Fig. 2 given below clearly depicts the error in training, validation and the test data sets. And Fig. 3 shows the correlations coefficient among outputs and targets in validation, training and test data sets. Both of these figures are obtained from implementing the network in MATLAB software. Also, Table 3 depicts the comparison of TE, PTE and ANN efficiencies and their comparative ranking and Fig. 4 shows the comparison in the bar chart form. It can be very clearly seen that the discriminatory power of ANN is very much improved after applying ANN. Now we can discriminate the efficient state board from the inefficient ones. Hence this result can be analyzed that Uttar Pradesh Board of High School & Intermediate Education has come up as the most efficient state board in India and as the State Board directly depicts the performance and efficiency of students in that state. So, it can be an inference that Uttar Pradesh State is the most efficient state in India in Higher Secondary Education System (i.e., Inter-mediate level). Similarly, J.K State Board of School Education with the last ranking is the most inefficient board. Also, it is very interesting to note that the initial data collected from MHRD portal showed West Bengal Council of Higher Education, Kolkata is having maximum number (93%) of passed students but still it is not the most efficient state. Rather, it is ranked 7th in ANN efficiency ranking.

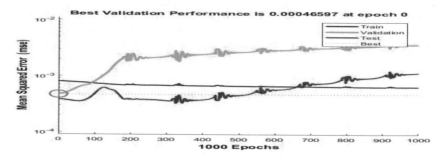

Fig. 2. Error in training, validation and the test data sets.

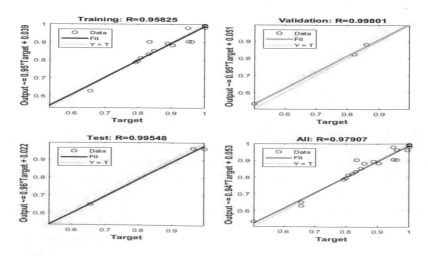

Fig. 3. Regression analysis

Table 3. Comparison of the TE, PTE and ANN efficiencies and their ranking

S.No	DMU (state boards)	TE	Rank	PTE	Rank	ANN	Rank
1	Board of Intermediate Education, Andhra Pradesh	0.655	17	0.830	1	0.649	20
2	Telangana State Board of Intermediate Education-Hyderabad	0.655	17	0.675	3	0.628	21
3	Assam Higher Secondary Education Council	0.799	15	0.841	1	0.795	18
4	Bihar Intermediate Education Council	1	1	1	1	0.992	2
5	Chhattisgarh Board of Secondary Education	0.859	9	0.859	5	0.882	13
6	Goa Board of Secondary & Higher Secondary Education	1	1	1	1	0.981	4
7	Gujarat Secondary & Higher Secondary Education Board	0.949	6	0.949	1	0.98	5
8	Board of School Education Haryana, Bhiwani	0.821	13	0.908	10	0.822	16
9	H.P. Board of School Education	0.888	8	0.927	8	0.894	11
10	J.K State Board of School Education	0.504	18	0.506	1	0.533	22
11	Jharkhand Academic Council, Ranchi	0.808	14	0.809	12	0.812	17
12	Department of Pre-University Education, Karnataka	0.792	16	0.826	13	0.787	19
13	Kerala Board of Higher Secondary Examination	0.961	3	1	15	0.963	6
14	Maharashtra State Board of Secondary & Higher Secondary Education	0.848	10	0.848	4	0.853	14
15	Board of Secondary Education, Madhya Pradesh	0.832	12	0.833	6	0.834	15
16	Council of Higher Secondary Education, Imphal, Manipur	1	1	1	2	0.990	3
17	Mizoram Board of School Education	0.951	5	1	1	0.905	8
18	Nagaland Board of School Education	0.834	11	.922	7	0.903	10
19	Tripura Board of Secondary Education	0.960	4	1	1	0.904	9
20	Uttar Pradesh Board of High School & Intermediate Education	1	1	1	9	0.994	1
21	Board of School Education Uttarakhand	0.904	7	0.936	14	0.883	12
22	West Bengal Council of Higher Education, Kolkata	0.994	2	1	11	0.96	7

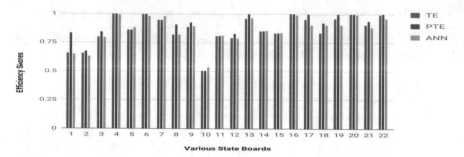

Fig. 4. Comparison between TE, PTE and ANN efficiency scores of various State Boards.

5 Conclusion

A highly flexible ANN algorithm was proposed to measure the DMU's efficiency score. The algorithm is flexible due to the nonlinearity of neural network. Also, it can handle outliers, corrupted and noisy data much better than conventional econometric techniques. The performance and efficiency evaluation of various state boards in Indian higher Secondary education is a challenge considering the multiple input and output criteria. Various researches conducted in the education sector, show that the concept of performance and efficiency and ranking of various state boards in Indian higher Secondary education is still under development. In this paper, these factors are evaluated using the discriminatory power of DEA-ANN model. This model provides the better and distinguished ranking to all the DMU's which thereby evaluates the efficiency and performance. From the result analysis, it can be said that DEA-ANN approach is more robust than the conventional DEA approach as it clearly gives the proper ranking structure amongst the data sets that turn out to be efficient as per DEA model. This is an improvement in the discriminatory power of DEA.

Also, this ANN approach can also be used for prediction purpose. ANN can be used to predict the future data, of several different years, by looking at the current trends, which is extremely beneficial to the education sector growth. These soft computing methods like ANN are an improvement to conventional DEA method for measuring performances. Not only do they provide a correct result, but also improves the discrimination power among efficiency scores which is a big limitation of the DEA technique. Since the research is an iterative and everlasting process, the future work can be carried on social concerns and sensitivity analysis in this education sector.

References

1. Charnes, A., Cooper, W.W., Rhodes, E.: Measuring the efficiency of decision making units. Eur. J. Oper. Res. **2**(6), 429–444 (1978)
2. Banker, R.D., Charnes, A., Cooper, W.W.: Some models for estimating technical and scale inefficiencies in data envelopment analysis. Manage. Sci. **30**(9), 1078–1092 (1984)
3. Bradley, S., Johnes, G., Millington, J.: The effect of competition on the efficiency of secondary schools in England. Eur. J. Oper. Res. **135**(3), 545–568 (2001)

4. Afonso, A., St Aubyn, M.: Non-parametric approaches to education and health efficiency in OECD countries. J. Appl. Econ. **8**(2), 227–246 (2005)
5. Johnes, J., Li, Y.U.: Measuring the research performance of Chinese higher education institutions using data envelopment analysis. China Econ. Rev. **19**(4), 679–696 (2008)
6. Kuah, C.T., Wong, K.Y.: Efficiency assessment of universities through data envelopment analysis. Procedia Comput. Sci. **3**, 499–506 (2011)
7. Agasisti, T.: The efficiency of Italian secondary schools and the potential role of competition: a data envelopment analysis using OECD-PISA2006 data. Educ., Econ. Taylor Francis J. **21**(5), 520–544 (2013)
8. Selim, S., Bursalıoğlu, S.A.: Efficiency of higher education in turkey: a bootstrapped two-stage DEA approach. Int. J. Stat. Appl. **5**(2), 55–67 (2015)
9. Sagarra, M., et al.: Exploring the efficiency of Mexican universities: integrating data envelopment analysis and multidimensional scaling. Omega **63**, 123–133 (2017)
10. Athanassopoulos, A.D., Curram, S.P.: A comparison of data envelopment analysis and artificial neural networks as tools for assessing the efficiency of decision making units. J. Oper. Res. Soc. **47**(8), 1000–1016 (1996)
11. Emrouznejad, A., Shale, E.: A combined neural network and DEA for measuring efficiency of large scale datasets. Comput. Ind. Eng. **56**(1), 249–254 (2009)
12. Liu, H.-H., Chen, T.-Y., Chiu, Y.-H., Kuo, F.-H.: A comparison of three-stage DEA and artificial neural network on the operational efficiency of semi-conductor firms in Taiwan. Mod. Econ. **4**(1), 20–31 (2013)
13. Kuo, R.J., Wang, Y.C., Tien, F.C.: Integration of artificial neural network and MADA methods for green supplier selection. J. Clean. Prod. **18**(12), 1161–1170 (2010)
14. Kwon, H.: Performance modeling of mobile phone providers: a DEA-ANN combined approach. Benchmarking Int. J. **22**(6), 1120–1144 (2014)
15. Oladokun, V.O., Adebanjo, A.T., Charles-Owaba, O.E.: Predicting students' academic performance using artificial neural network: a case study of an engineering course. Pac. J. Sci. Technol. **9**(1), 72–79 (2008)
16. Chen, C., et al.: Online 24-h solar power forecasting based on weather type classification using artificial neural network. Sol. Energy **85**(11), 2856–2870 (2011)
17. Cheh, J.J., Weinberg, R.S., Yook, K.C.: An application of an artificial neural network investment system to predict takeover targets. J. Appl. Bus. Res. (JABR) **15**(4), 33–46 (2013)
18. Chhachhiya, D., Sharma, A., Gupta, M.: Designing optimal architecture of neural network with particle swarm optimization techniques specifically for educational dataset. In: 7th International Conference on Cloud Computing, Data Science & Engineering-Confluence, pp 52–57. IEEE, Noida (2017)

Design of Millimeter-Wave Microstrip Antenna Array for 5G Communications – A Comparative Study

Saswati Ghosh[(⊠)] and Debarati Sen

G. S. Sanyal School of Telecommunications, Indian Institute of Technology,
Kharagpur, Kharagpur, West Bengal, India
saswatikgp@gmail.com, debarati@gssst.iitkgp.ernet.in

Abstract. Millimeter wave communication is found as a suitable technology for future 5G communications. The beamforming antenna is chosen to increase the link capacity considering the atmospheric losses at millimeter wave frequencies. This work compares the performance of different microstrip antenna arrays in terms of the return loss bandwidth, gain, half power beamwidth, side lobe level etc. The results are simulated using commercial electromagnetic software. A suitable array structure is suggested from the study.

Keywords: 5G communications · Millimeter wave · Microstrip antenna array

1 Introduction

With the ever increasing demand of wireless data traffic, the millimeter wave communication has been found as an attractive technology for the 5G mobile communication systems. In the millimeter wave frequency range from 30–300 GHz, especially the unlicensed 57–64 GHz band is chosen for 5G communication due to the increase in the allowed maximum outdoor emission power level by the Federal Communications Commission (FCC) in the United States [1]. However, there are challenges in the design of the communication system at millimeter wave frequencies due to higher path and penetration losses. The antenna beamforming technology employing multiple antennas is found as a viable solution to increase the link capacity by allowing directional communication [2]. The design of suitable millimeter wave antenna array has become an interesting research area. The millimeter wave technologies are applied to automotive radar systems, high bit-rate wireless communication systems for HD video transmissions standardized by Wireless HD, IEEE802.15.3c and IEEE802.11ad [3–5].The commonly used mm-wave antennas are classified in different groups e.g. leaky-wave antennas, integrated and microstrip antennas [6]. The microstrip antennas are preferred due to the attractive features e.g. light weight, low profile, low production cost, ease of integration with other components etc. [7]. However, they have several disadvantages also e.g. narrow impedance bandwidth, low gain, substrate loss etc.

This paper presents a brief literature survey and also the detailed analysis of performance of some recent interesting array structures in terms of gain, impedance bandwidth, size, beam scanning etc. using efficient electromagnetic software

© Springer International Publishing AG, part of Springer Nature 2018
A. Abraham et al. (Eds.): ISDA 2017, AISC 736, pp. 952–960, 2018.
https://doi.org/10.1007/978-3-319-76348-4_91

e.g. Computer Simulation Technology (CST). The novelty of this work lies in finding out the gaps and challenges in the design of microstrip arrays and suggesting a suitable antenna structure for millimeter wave communication from the comparative study of performance.

2 Literature Review

The common mm-wave systems require antennas of high directive gain with small beamwidth, which has reasonable size at this high frequency ranges [8]. Researches on the mm-wave antennas are continuously developing. The feasibility study of the application of microstrip antenna for mm-wave frequencies is performed in [9]. A 4×4 microstrip element array was designed for 60 GHz communication in this work. The high efficiency of the array proved that the application of the microstrip approach to antenna systems did not deteriorate in the millimeter wave range [9]. The advancement in the design of mmwave microstrip antennas was presented in [6]. The broadband microstrip antennas e.g. bow tie, vivaldi and spiral antenna were compared for mm-wave operation in [10]. The vivaldi antenna was found most suitable among these three antennas considering the required characteristics e.g. bandwidth, gain, ease of design and tuning ability. Also the design of broadband printed yagi antenna array operating in the 60 GHz unlicensed band was presented in [11]. The array antenna with normalized gain of 10.3 dB was found suitable for V-band short range wireless application. The feeding network plays an important role in the performance of microstrip array antenna. The optimization of the microstrip antenna array and feed network is necessary to minimize the loss and increase the antenna efficiency [7]. The aperture efficiency of microstrip array antenna reduces due to the feed-line losses. An interesting microstrip comb line array configuration was designed in [12, 13] with the rectangular radiating elements connected directly to the straight feed-line. This resulted in travelling wave operation of the array. Also the aperture efficiency increased to 53% with gain of 22.4 dBi in mm-wave frequencies [12]. Further, the reflection cancelling slit structure was inserted on the feed-line around each radiating element to reduce the reflection [14]. Later to improve the return loss characteristics of the array, the matching-circuit integrated to the radiating elements were used [15]. Several microstrip comb-line arrays with different beam directions were used for the design of mmwave beam switching applications [16]. The other commonly used feed techniques for array elements are the linear and corporate feed network [17, 18]. A combination of linear and corporate feeding technique may be simultaneously applied for designing a high gain microstrip patch array [19]. Works are in progress on the design of high gain and broadband microstrip antenna [18, 20, 21]. The shape of the array configuration requires to be properly chosen to avoid the fluctuation in gain in the main lobe of the radiation pattern with varying angles [22].

3   Different Types of Microstrip Array Antennas

Initially, different types of microstrip antenna elements e.g. rectangular patch, wideband patch antenna, elliptical dipole antenna are studied. All the structures are simulated and optimized using electromagnetic software CST Microwave Studio. The most commonly used microstrip antenna for array design is the probe-fed rectangular patch antenna. The basic formulations for the design of rectangular patch antenna depending on the operating frequency are available in the literature [23]. The RT Duroid 5880 substrate with relative permittivity $\varepsilon_r = 2.2$ and thickness of 20 mil (0.508 mm) is chosen for the desired 57–64 GHz frequency band. For microstrip lines, the effective permittivity ε_{reff} is evaluated to take into account the fringing effect of the substrate material. The ε_{reff} is expressed in terms of ε_r, width of the patch (w) and height of the substrate (h) as follows [23]:

$$\frac{w}{h} \geq 1$$

$$\varepsilon_{reff} = \frac{\varepsilon_r + 1}{2} + \frac{\varepsilon_r - 1}{2} \times \left[1 + 12\frac{h}{w}\right]^{-1/2} \tag{1}$$

Considering the fringing effect, the effective length of the patch is increased by an amount ΔL.

$$\Delta L = 0.412 \times h \times \frac{\left(\varepsilon_{reff} + 0.3\right)\left(\frac{w}{h} + 0.264\right)}{\left(\varepsilon_{reff} - 0.258\right)\left(\frac{w}{h} + 0.8\right)} \tag{2}$$

$$L_{eff} = \frac{c}{2f_r \times \sqrt{\varepsilon_{reff}}} - 2\Delta L \tag{3}$$

Initially the dimension of the patch and feed location is determined considering the resonance frequency as 60 GHz. The problem with the patch antenna is its narrow bandwidth. The impedance bandwidth of the patch is increased by using U-shaped slot of suitable dimension on the patch surface [21]. The arms of the U-slot work as separate and compactly coupled resonators. The mutual coupling between the resonators moves the resonances toward higher and lower end frequencies and thus increases the overall impedance bandwidth [21]. The average lengths of current paths of the first and second modes are evaluated using the following formulations [24]:

$$L_1 = \frac{c}{2\sqrt{\varepsilon_{eff}}\, f_1} \tag{4}$$

$$L_2 = \frac{c}{2\sqrt{\varepsilon_{eff}}\, f_2} \tag{5}$$

Here f_1 and f_2 are the resonant frequencies.

The broadband microstrip patch antenna is designed considering the same substrate. Also compared to the classical patch structures, the microstrip dipole has wider bandwidth [25]. The impedance bandwidth is further increased by using elliptical dipole of suitable dimension. By adjusting the length of the two axes, a good impedance bandwidth can be achieved.

The antenna structures are shown in Fig. 1. The simulated S_{11} data for all the antennas are presented in Fig. 2 for comparison. It is noticed that the microstrip patch antenna with slot cut in the patch shows wider 10 dB return loss bandwidth (over 55.5–65.4 GHz) compared to the simple microstrip patch and elliptical dipole antenna.

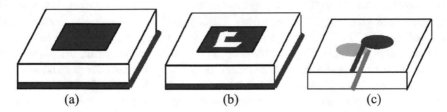

(a)   (b)   (c)

Fig. 1. Different types of microstrip antenna elements – (a) rectangular patch; (b) broadband patch antenna; (c) elliptical dipole.

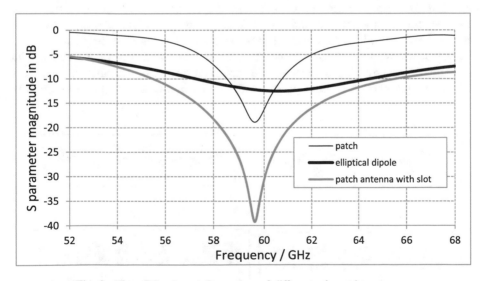

Fig. 2. Plot of S_{11} versus frequency of different microstrip antennas.

Next, the performances of different microstrip antenna arrays are studied in terms of gain, half-power beamwidth, side lobe level etc. All the antenna structures are simulated using the electromagnetic software CST Microwave Studio. The arrays are printed on the RT Duroid 5880 substrate with relative permittivity $\varepsilon r = 2.2$ and thickness of 20 mil (0.508 mm). The microstrip comb line antennas are used for its high efficiency because of relatively low feeding loss (Fig. 3(a)) [26]. The radiation and also the direction of the beam is controlled by changing width of the radiating elements, feedline length etc. To reduce the side lobe level the Taylor tapering in the element width is used (b). The linear series-fed microstrip patch antennas have high directivity and small size [27]. To reduce the side lobe level, the tapering in the width of the rectangular patch is used for non uniform excitation in different patches (Fig. 3(c)). The array structures shown in Fig. 3(a)–(c) are designed following the literatures and then optimized for the desired frequency and substrate material. Later, 10 element array consisting of conventional microstrip patch antenna (Fig. 3(d)) and microstrip patch with slot (Fig. 3(e)) are designed. The numbers of elements are decided from the desired gain (~ 15 dBi) of the array calculated from the link budget [28]. The link budget was calculated for LOS propagation of mm-wave signal for 10 m distance with average effective isotropic radiated power (EIRP) combined with receiver antenna gain as 40 dBm [28]. The simulated gain of the antenna arrays in terms of dBi for various theta and constant phi ($0°$) is presented in Fig. 4 for comparison. Also the other parameters for the performance evaluation of different antenna array structures e.g. size, half power beamwidth, sidelobe level etc. are presented in Table 1. The simulated data show that the 10-element linear microstrip patch antenna array with slot satisfies all the desired conditions e.g. wide impedance bandwidth over 55.5–65.4 GHz, high gain ~ 15.3 dBi, narrow half-power beamwidth $\sim 10.2°$ and low sidelobe level (~ 13.6 dB down) etc. The results for some interesting array structures available in literature are also presented in Table 1 for comparison. The microstrip comb array antenna shows sufficient gain and narrow half power beamwidth. The array of comb antenna with specific beam direction may be used for beam switching applications. However, for continuous beam scanning antennas the microstrip patch antenna with slot array is preferred. Figure 5 shows the scan performance of the same antenna array for different excitation phases applied to the antenna. Good scan performance with

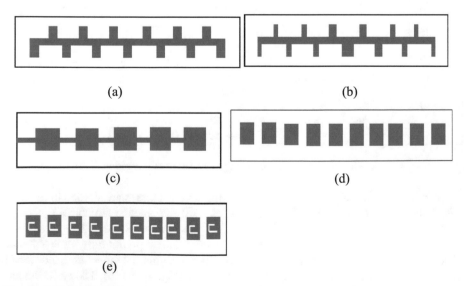

(a)   (b)

(c)   (d)

(e)

Fig. 3. Different types of microstrip array antennas – (a) microstrip combline array, (b) microstrip combline array with Taylor distribution, (c) linear series fed patch array, (d) microstrip probe-fed patch antenna array, (e) Microstrip patch with slot array.

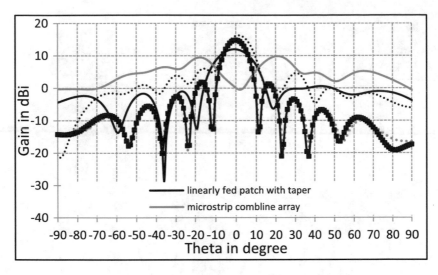

Fig. 4. Plot of gain in dBi versus theta in degree of different microstrip array antennas for phi = 0°.

Table 1. Comparison of the performance of different microstrip antenna arrays

Antenna type/Reference	Surface area (mm²)	Gain (dBi)	Half-power beamwidth (in °)	Side lobe level
Linearly-fed tapered patch antenna	5 × 20	12.0	17.1	−11.4
Microstrip comb array antenna (probe-fed)	4 × 21	9.9	17.0	−4.6
Microstrip comb array antenna with Taylor distribution (probe-fed)	5 × 38	16.3	11.6	−10.3
Microstrip patch antenna array (1 × 10)	5 × 27.5	15.2	10.3	−13.2
Microstrip patch antenna with slot on patch (1 × 10)	5 × 27.5	15.3	10.2	−13.6
Elliptical dipole with reflect plane [25]	20 × 20	19.0	–	–
Co-planar waveguide-fed wideband microstrip patch antenna array [18]	17 × 17	14.5	–	–
Microstrip comb line array antenna [27]	–	11.4	–	−16.5

desired sidelobe level and beamwidth is noticed upto ±60° scan angle in the broadside direction for constant phi (0°). However, the performance can be further improved by applying tapering in the magnitude of excitation and reducing the sidelobe level.

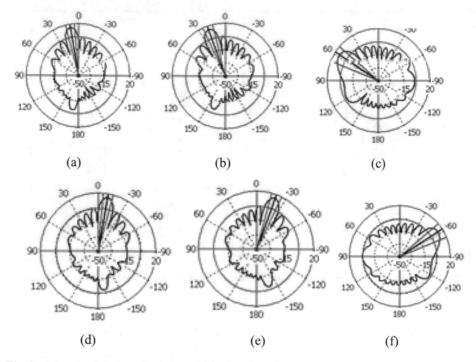

Fig. 5. Plot of gain pattern in the broadside direction phi = 0° for various main beam directions (a) theta = 10°; (b) theta = 20°; (c) theta = 57°; (d) theta = −10°; (e) theta = −20°; (f) theta = −56°.

4   Gaps and Challenges

The above study on different microstrip array antennas working in the millimeter-wave frequency range shows that though works have been started on the relevant areas there are still scopes of improvement of the performance. The design of RF front ends specially the array antenna working over the 57–64 GHz band is a big challenge. Though the microstrip antennas are preferred due to their low profiles, low cost and other advantages, the gain of microstrip antennas are quite low. Also the mutual coupling between the array elements needs to be reduced to avoid the degradation in array performance. For future application in 5G communications, the microstrip array antenna combined with appropriate beam forming architecture may be used for designing suitable phased array antenna.

5   Conclusion

The existing microstrip antennas for millimeter-wave communications are discussed and compared here. From the comparative studies of the performance of the arrays, it is noticed that the array of microstrip patch antenna with slot satisfies the broadband as well as high gain characteristics simultaneously. The fabrication of the array structures is underway for testing and verification with the simulated results. Later the array with other circuit components may be used for designing of phased array antenna for commercial telecommunication in millimeter wave frequencies.

References

1. FCC Document: Revision of part 15 of the commission's rules. In: ET docket No. 07-113 regarding operation in the 57–64 GHz band, RM- 11104-report and order, 9 August 2013
2. Kutty, S., Sen, D.: Beamforming for millimeter wave communications: an inclusive survey. IEEE Commun. Surv. Tutor. **18**(2), 949–973 (2016)
3. Kawakubo, A., Tokoro, S., Yamada, Y. et al.: Electronically scanning millimeter wave radar for forward objects detection. Soc. Automotive Eng., Warrendale, PA, SAE Technical Paper 2004-01-1122 (2004)
4. Singh, H., Jisung, O., et al.: A 60 GHz wireless network for enabling uncompressed video communication. IEEE Commun. Mag. **46**(12), 71–78 (2008)
5. Gilbert, J.M., Doan, C.H., et al.: A 4-Gbps uncompressed wireless HD A/V transceiver chipset. IEEE Micro **28**(2), 56–64 (2008)
6. Schwering, F.K.: Millimeter wave antennas. Proc. IEEE **80**(1), 92–102 (1992)
7. Albert, S.: Applications of MM wave microstrip antenna arrays. In: Proceedings on International Symposium on Signals, Systems and Electronics, ISSSE 2007, pp. 109–122. IEEE, Montreal (2007)
8. Schwering, F.K., Oliner, A.A.: Millimeter wave antennas. In: Lo, Y.T., Lee, S.W. (eds.) Antenna Handbook. Springer Science+Business Media, New York (1988)
9. Weiss, M.A.: Microstrip antennas for millimeter waves. IEEE Trans. Antennas Propag. **29** (1), 171–174 (1981)
10. Pitra, K., Raida, Z.: Planar millimeter-wave antennas: a comparative study. Radioengineering **20**(1), 263–269 (2011)
11. Briqech, Z., Sebak, A.: Low cost 60 GHz printed yagi antenna array. In: IEEE International Symposium on Antennas and Propagation Society (APSURSI), Chicago, USA (2012)
12. Iizuka, H., Watanabe, T., et al.: Millimeter-wave microstrip array antenna for automotive radar. IEICE Trans. Commun. **E86-B**(9), 2728–2738 (2003)
13. Alavi, S. E., Soltanian, M. R. K. et al.: Towards 5G: a photonic based millimeter wave signal generation for applying in 5G access fraunthaul. Sci. Rep. **6**, 19891 (2016). Nature Publishing Group
14. Hayashi, Y., Sakakibara, K., et al.: Millimeter-wave microstrip comb-line antenna using reflection-canceling slit structure. IEEE Trans. Antennas Propag. **59**(2), 398–406 (2011)
15. Sakakibara, K., Sugawa, S. et al.: Millimeter-wave microstrip array antenna with matching-circuit-integrated radiating-elements for travelling-wave excitation. In: Proceedings of the Fourth European Conference on Antennas and Propagation (EuCAP), pp. 1–5. IEEE, Barcelona (2010)
16. Sakakibara, K., Hayashi, Y. et al.: Two-dimensional array design techniques of millimeter-wave microstrip comb-line antenna array. Radio Sci. **43**(RS4S25) (2008). https://doi.org/10.1029/2007rs003801
17. Tamijani, A.A., Sarabandi, K.: An affordable millimeter-wave beam-steerable antenna using interleaved planar subarrays. IEEE Trans. Antennas Propag. **51**(9), 2193–2202 (2003)
18. Mingjian, L., Luk, K.M.: Low-cost wideband microstrip antenna array for 60-GHz applications. IEEE Trans. Antennas Propag. **62**(6), 3012–3018 (2014)

19. Rida, A., Tentzeris, M. et al.: Design of low cost microstrip antenna arrays for mm-wave applications. In: IEEE International Symposium on Antennas and Propagation (APSURSI), Spokane, WA, USA, pp. 2071–2073 (2011)
20. Hu, C.N., Chang, D.C. et al.: Millimeter wave microstrip antenna array design and an adaptive algorithm for future 5G wireless communication systems. Int. J. Antennas Propag. Hindawi Publ. Corp., 1–10 (2016). Article ID 7202143
21. Alam, M.S., Islam, M.T., et al.: A wideband microstrip patch antenna for 60 GHz wireless applications. Elektron. Elektrotech. 19(9), 65–70 (2013)
22. Zhang, J., Qiang, X., et al.: 5G millimeter-wave antenna array: design and challenges. IEEE Wirel. Commun. 24(2), 106–112 (2017)
23. Balanis, C.A.: Antenna Theory – Analysis and Design, 3rd edn. Wiley, Hoboken (2005)
24. Ghalibafan, J., Attari, A.R., Kashani, F.H.: A new dual-band microstrip antenna with U-shaped slot. Prog. Electromagn. Res. C 12, 215–223 (2010)
25. Xu, J., Wang, W.: A low cost elliptical dipole antenna array for 60 GHz applications. In: PIERS Proceedings, Taipei, pp. 1044–1047 (2013)
26. Wu, D., Tong, Z. et al.: A 76.5 GHz microstrip comb-line antenna array for automotive radar system. In: Proceedings of the Ninth European Conference on Antennas and Propagation (EuCAP), pp. 1–5. IEEE, Lisbon (2015)
27. Sengupta, S., Jackson, D.R., et al.: A method for analyzing a linear series-fed rectangular microstrip antenna array. IEEE Trans. Antennas Propag. 63(8), 3731–3736 (2015)
28. Rakesh, R.T., Chowdhary, A. et al.: A scalable subband subsampled radio architecture for millimeter wave communications. In: Proceedings of the 26th International Symposium on Personal, Indoor and Mobile Radio Communications - (PIMRC): Fundamentals and PHY. pp. 309–314. IEEE, Montreal (2015)

Simulation Design of Aircraft CFD Based on High Performance Parallel Computation

Yinfen Xie[✉]

School of Information Science and Engineering,
Linyi University, Linyi, Shandong, China
xyf.ly@163.com

Abstract. This paper introduces the application of high performance parallel computing in CFD numerical simulation. Architecture and implementation method of the aircraft CFD simulation based on large scale and high performance are raised after analyzing the complex flow field of multi area parallel computing research.

Keywords: CFD · High performance computation · Parallel computation
Aircraft

1 Introduction

CFD (Computational Fluid Dynamics) was formed in 1960s. It is a special subject that uses computer and numerical methods to solve the hydrodynamic equations to obtain the flow rules and solve the flow problem. With the continuous progress of numerical methods and the rapid development of computer and computing technology, the role and superiority of CFD in the field of aerospace are becoming more and more obvious. Now, it has become one of the three main research methods together with paratactic theoretical analysis and wind tunnel experiment and plays a more and more important role in the optimization design and aerodynamic characteristics analysis.

In the 21st Century, along with the further development of computer and CFD technology, CFD will bring a revolution in aerodynamic design. The performance of future aircraft will depend on the "virtual wind tunnel (CFD)" data based on the "virtual flight". The computing power of CFD is becoming more and more powerful, and the demand for high performance computing is becoming more and more urgent, and the computational performance needs to reach 10 billion, 100 billion, or even tens of millions of secondary computing power per second. Due to the limit of the components and the process level, the single processor cannot meet the large-scale flow field calculation and multi-processor parallel computing is an effective way to solve large-scale engineering problems. The rapid developments of large scale computer technology greatly improve the accuracy and resolution of the solution. CFD revolutionized the design process of the aircraft for it not only improves and expand the capacity of the aircraft design to make the existing tunnel is more effective, but also makes people can also be carried out in future aircraft design for those in the specific flight range not yet tested, which opened up a new world of aircraft design. This paper

© Springer International Publishing AG, part of Springer Nature 2018
A. Abraham et al. (Eds.): ISDA 2017, AISC 736, pp. 961–968, 2018.
https://doi.org/10.1007/978-3-319-76348-4_92

mainly introduces the application of high performance parallel computing in CFD simulation of aircraft design.

2   CFD Parallel Computing Technology

2.1   CFD Parallel Computing Algorithm

CFD parallel computing is an effective way to expand the scale of computation and improve the computing speed. The most common method of CFD parallel is the zone divided parallel computing method.

CFD parallel computing is to decompose the whole flow field into several sub regions, each sub region is assigned to the node calculation and single node or single CPU serial iteration is carried out in each sub region, next step begins after every step of the calculation is completed and data exchanged between adjacent sub regions, according to the topology, so the cycle until convergence. The advantage of this parallel algorithm is that each CPU is independently assigned to the interior points of the computational region when the interior points in the flow field calculating, so parallelism is very high. It is only necessary to exchange data over the network when calculation in zone boundary, for the calculation of three-dimensional flow field, the data exchange occurs only in a few two-dimensional zone boundaries, and the data exchange is very small, so the parallel efficiency is high. In each iteration, the computation flow in each sub region is exactly equivalent to the single zone serial program, and the data between interface and line is passed and exchanged through the parallel statement in the boundary processing. This kind of parallel strategy can easily transplant the single zone program originally debugged in parallel framework, which can be used to realize the parallel function quickly. The whole process is shown in Fig. 1.

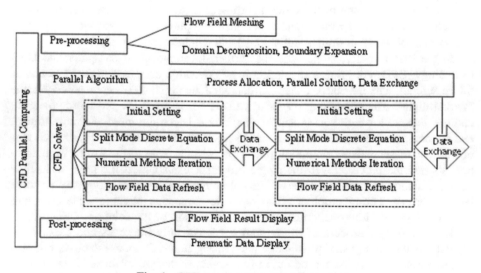

Fig. 1.   CFD parallel computing flow chart

2.2 High Performance Parallel Computing Environment

The hardware and software configuration information of the computer system of a supercomputer center is shown in Table 1.

Table 1. Hardware and software of super computer system

Name	Configuration
Computing Node	8704
Processor	64 Bit 16 Core 1.0–1.1 GHz
Memory	170TB
MIC Processor	700Blade Node, Each 2 Inter Xeon X5657 6 Core 3.06 GHz
OS	RaiseOS 2.0.5
I/O Aggregate Bandwidth	200 GB/s
MIC Drive	MPSS Version 3.4.3-1

2.3 Numerical Method of CFD Flow Field Simulation

CFD can be regarded as the numerical solution of the flow basic equations such as mass conservation equation, momentum conservation equation and energy conservation equation by using the discretization method such as the finite volume method, then numerical simulation and analysis of these equations are carried out under the control of the inviscid flow and viscous flow, as well as the related physical phenomena. At present, the commercial CFD software based on the existing theory is developing rapidly, so other professional researchers can also conduct fluid numerical calculations and be liberated from the complex, repetitive programming to make more efforts to study the physical nature of the problem, the boundary conditions and the rationality of the calculation results and other important aspects. give full play to the intellectual superiority of commercial CFD software, to open up the road to solve practical engineering problems. Superiority of commercial CFD is given full play.

The typical control equations of continuum fluid mechanics are N-S (Navier-Stokes) equation set. The equation set can accurately descript flow phenomenon encountered in nature and Engineering. Under appropriate initial and boundary conditions, the flow field result can be obtained by the numerical solution of the equations.

The governing equations used in this paper are the Reynolds averaged N-S equations in the form of three-dimensional integrals. The equations include mass conservation, momentum conservation and energy conservation equations. In the Cartesian coordinate system, the three-dimensional conservative NS equation is as follows:

$$\frac{\partial Q}{\partial t} + \frac{\partial E}{\partial x} + \frac{\partial F}{\partial y} + \frac{\partial G}{\partial z} - \left(\frac{\partial E_v}{\partial x} + \frac{\partial F_v}{\partial y} + \frac{\partial G_v}{\partial z} \right) = 0 \qquad (1)$$

After decades of continuous development, CFD has formed a complete set of numerical solutions and methods. Before the start of the calculation, the flow field is divided into the mesh, the decomposition region, the input and output parameters; then the flow field is initialized before the calculation of the mesh geometry is carried out. In the calculation, the discretization of the NS equation is mainly carried out in the whole flow field, including the use of the appropriate difference scheme and the numerical method for the iterative step, until computation convergence.

Post-processing after the finish computation is to display calculated flow field, to solve the aerodynamic and thermal data, to discuss the flow structure and flow mechanism, so as to provide support CFD value for aerodynamic characteristics analysis and optimization design. All the links in the calculation, such as difference scheme, numerical method, are constantly enriched, developed and perfected.

3   Overall Architecture

The system consists of three components: the workbench, the solver, and the front/back processing interface (see Fig. 2).

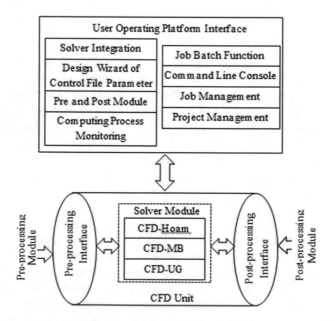

Fig. 2. Architecture of CFD software

3.1   Operating Platform Component

This component is a graphical interface for user to operate the CFD system, mainly to provide project management, job management, batch job processing, control file parameter setting, the pre and post command line console modules and etc. Here the scheme refers to the user to create, for the numerical simulation problems. The project here refers to the numerical simulation problem created by the user.

3.2 Solver Component

After the user creates and submits the job on the operating platform, CFD calls the solver component and uses the parallel numerical simulation method to solve the computational fluid dynamics problem. The component currently integrates 3 solvers:

CFD-Hoam

N-S solver of compressible and incompressible flow in finite difference method for single block structured grids.

CFD-MB

N-S solver of unstructured hybrid meshes finite volume method for compressible flow for.

CFD-UG

N-S solver of multi block structured grid finite volume method for compressible flow.

3.3 Pre-post Processing Interface

Through the communication and interaction with the pre-post processing tools, the pre-post processing tools of numerical simulation are integrated into the CFD system. The pre-processing mainly refers to the grid data generation of numerical simulation, and the post-processing mainly refers to the visualization of the results.

3.4 Center Control Cluster

Center control cluster is mainly used to manage and optimize event model, which comprises of event detection optimization module and event models module. Event format module is composed of two parts, Complex Event Format Parser and Primitive Event Format Parser, the former is used to process event queries provided by the application layer and the latter to process the real-time data stream provided by the sensor layer, both of their data formats are diverse. In order to improve the subsequent detection and communication efficiency, all data of ESCEP use a unified event model to express, in JSON format.

Event Detection Optimization Module is used to optimize event query structure and event query deployment, which consists of three parts: event query hash parser, event Query Generator and event Query Distributor. Event query hash parser hashes complex events into several intermediate things based on linear sequence relation. Event Query Generator creates a new event detection model based event sharing cost algorithm. Event Query Distributor deploys the event query to the working node based on event communication cost model.

4 Unique Feature of the System Function

4.1 High-Resolution Turbulence Numerical Simulation

CFD uses the comparative analysis of various computational models and the method to improve the resolution of the computational grid to obtain a high precision simulation results and a more detailed flow field model. Increasing drag reduction, noise reduction, vibration and other design goals of large aircraft are related to the turbulence problem. By using this CFD software, the user can carry out large eddy simulation of the complex turbulence problem of key area of large aircraft wing and fuselage, also the direct flow simulation of key local flow on supercomputers.

4.2   Supporting Scalable Parallel Computing with One Ten Thousand Cores

Because the traditional parallel computing model does not adapt well to the structural characteristics of ten thousand cores supercomputer, it has been improved in many aspects in CFD.

Parallel Programming Mode
According to the characters of ten thousand cores grade supercomputer, new parallel programming models are put to use, such as flow model, VP (virtual processor) model and the asymptotic model.

Overlap Technology of Computation and Communication
Overlap technology of communication and computation is realized using non-blocking communication mode to reducing the extra overhead caused by data communications.

Load Balance
The fine-grained tasks generated at the partitioning phase are combined into coarse grained tasks, with each tasks allocated for one processor cores using domain decomposition technique.

Efficient Parallel Computing
The key technologies such as efficient and accurate calculation method, turbulence model, the reasonable division of the computational grid and the selection of the spatial and temporal discretization schemes are adopted.

4.3   Convenient Function

When the user creates job attributes can be specified for the operation of the batch jobs. Then, the user specifies the starting value, termination value and interval size for the specific operation parameters. The system automatically generates a set of parameter values for each individual parameter values according to the user. For each of the individual parameter values, the system automatically generates a new sub job.

5   Calculation and Analysis of Flow Field Simulation

Parallel performance analysis of CFD on ten thousand cores computer based on large scale numerical simulation of turbulent channel with grid about 13×10^8 ($1280 \times 1025 \times 1024$), the Mach number 1.5, the Reynolds number 4.5×10^4. The operating environment is Sunway BlueLight MPP of the National Supercomputing Center in Jinan. For the same work 128, 512, 1024, 2048, 4096 and 8192 processors

Table 2. Parallel performance comparison of numerical simulation

Process number	Solution time/h	Speed-up ratio
128	164.8	1.00
512	18.9	8.78
1024	7.51	21.6
2048	4.8	34.2
4096	3.19	51.61
8192	2.88	57.52

are used to do parallel computing. Table 2 shows comparison of computing time and speedup under different parallel scale. It can be seen from the table with the parallel scale from 128, increased to nearly ten thousand cores (8192), the running time of the job is stabilized and the speedup is increased steadily, compared with the parallel scale of 128 cores.

The speed-up ratio in 8192 accelerated cores scale is 57.52, compared with the growth of parallel scale, there is super-linear characteristic of computational speedup when computing in 512, 1024, 2048 and 4096 (see Fig. 3). The main reason for the occurrence of super-linear is that the data of the calculation process has some locality in the above parallel scale, and then communication cost between processes is reduced. It is obvious that the parallel performance of CFD in the scale of ten thousand cores is relatively ideal.

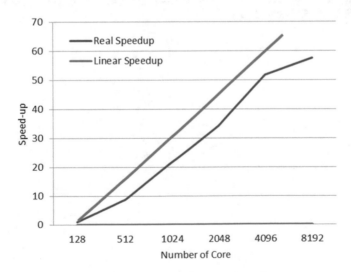

Fig. 3. Speed-up radio on each node

6 Summary

This paper introduces simulation design of CFD based on large scale and high performance parallel. CFD can support the numerical simulation and high resolution numerical simulation of complex flow field in large aircraft design.

Being aimed at the structural characteristics of super computer in thousands of core grade, some kinds of high efficiency parallel computing technology and methods are achieved, and a user-friendly interface is also provided. The test proved that the CFD has good reliability and high resolution numerical calculation in complex flow field, can solve the engineering problems in the design of large aircraft.

References

1. Fan, X.H., Wang, K.Y., Xiao, S.F., et al.: Some progress on parallel modal and vibration analysis using the JAUMIN framework. Math. Probl. Eng. **2015**, 1–9 (2015). (S1024-123X)
2. Pan, S., Fan, X., Li, X.: A parallel computation method based on decomposition for hypersonic. e-Science **5**, 61–68 (2010)
3. Chen, G., Wang, L., Lu, Z.: Development of ten-thousand-core parallel software CCFD for aircraft aerodynamics simulation. J. Huazhong Univ. Sci. Tech. (Nat. Sci. Educ.), June 2011
4. Chetlur, S., Woolley, C., Vandermersch, P., et al.: cuDNN: efficient primitives for deep learning. Eprint Arxiv (2014)
5. Liu, B., Clapworthy, G.J., Dong, F.: Fast isosurface rendering on a GPU by cell rasterization. Comput. Graph. Forum **28**(8), 2151–2164 (2009)
6. Kamakoti, R., Shyy, W.: Fluid-structure interaction for aeroelastic applications. Prog. Aerosp. Sci. **40**, 535–558 (2004)

Determining the Optimum Release Policy Through Differential Evolution: A Case Study of Mula Irrigation Project

Bilal[1(✉)], Millie Pant[1], and Deepti Rani[2]

[1] Department of Applied Science and Engineering,
Indian Institute of Technology, Roorkee 247667, India
bilal25iitr@gmail.com, millidma@gmail.com
[2] National Institute of Hydrology, Roorkee, India
deeptinatyan@yahoo.com

Abstract. The present study shows an implementation of Differential Evolution (DE) algorithm for determining the optimum flow policy for the reservoir operation. The case study is done for Mula Major Irrigation Project for river Mula (Godavari basin), Ahmednagar district, Maharashtra. The problem is formulated in terms of an unconstrained optimization model having 12 variables as the data collected is for one year.

Keywords: Differential Evolution · Reservoir operation · Release Storage

1 Introduction and Literature Review

Differential Evolution(DE) is most popular algorithm based on evolutionary algorithm which is developed by Price and Storn (1) for solving complex real life problem over different domains like continuous, discrete and combinatorial optimization problems. DE is easy to use, has a simple structure and has been applied to a wide variety of problems occurring in different domains. In the present study, DE is employed for determining the optimum release policy for reservoir operation. The case study is done for Mula river project, described briefly in Sect. 2. Some past instances of application of DE and its variants to reservoir operation problems are as follows:

The present study is divided into four sections first we introduce Differential Evolution and give the review of DE in reservoir problem, in second section we discuss the project of Mula river and In section third we discuss the results and lastly in section four we conclude our work.

1.1 Working of Differential Evolution

DE is works in two phases, first phase is initialization and the second phase is evolution. In first phase population is generated randomly and in the second phase, which is evolution, the generated population goes through mutation, crossover and selection process and the process is repeat until termination criteria is met.

Table 1. DE and its variants for reservoir problems.

Description	Reference
Short-term scheduling of hydrothermal power system with cascaded reservoirs by using modified differential evolution	Lakshminarasimman et al. (2)
Solving a reservoir problem using multi-objective differential Evolution	Reddy et al. (3)
Application of differential evolution for irrigation planning	Vasan et al. (4)
Evolving strategies for crop planning and operation of irrigation reservoir system using multi-objective differential evolution	Reddy et al. (5)
Short-term combined economic emission scheduling of hydrothermal power systems with cascaded reservoirs using differential evolution	Mandal and Chakraborty (6)
Multi-objective cultured differential evolution for generating optimal trade-offs in reservoir flood control operation	Qin et al. (7)
Differential evolution algorithm for optimal design of water distribution networks	Suribabu (8)
Optimization of water distribution network design using differential evolution	Vasan and Simonovic (10)
Differential evolution algorithm for solving multi-objective crop planning model.	Josiah and Otieno (9)
Differential evolution algorithm with application to optimal operation of multipurpose reservoir	Regulwar *et al.* (11)
Reservoir operation using multi-objective evolutionary algorithms-a review	Adeyemo (12)
Extraction of multi-crop planning rules in a reservoir system: application of evolutionary algorithms	Fallah-Mehdipour et al. (13)
Reservoir optimization in water resources: a review	Ahmad et al. (14)
Coupled self-adaptive multi-objective differential evolution and network flow algorithm approach for optimal reservoir operation	Schardong and Simonovic (15)
Fuzzy differential evolution method with dynamic parameter adaptation using type-2 fuzzy logic	Ochoa et al. (16)
A new approach for dynamic mutation parameter in the differential evolution algorithm using fuzzy logic	Ochoa et al. (17)
Differential evolution using fuzzy logic and a comparative study with other metaheuristics	Ochoa et al. (18)

(a) **Initialization:** During initialization, a set of uniformly distributed population is generated as follows:

Let $S^G = \left\{ X_j^G : j = 1, 2, \ldots, NP \right\}$ be the population at any generation G, NP denotes the size of population. Here, X_j^G denotes a D-dimensional vector as $X_j^G = \left\{ x_{1,j}^G, x_{2,j}^G, \ldots, x_{D,j}^G \right\}$. X_j^G is generated using uniformly distributed random number $rand(0, 1)$

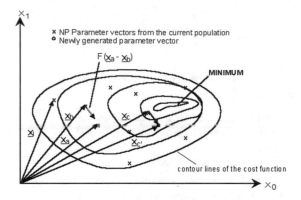

Fig. 1. Working strategy of differential evolution

$$X_j^G = X_{low} + \left(X_{upp} - X_{low}\right) * rand(0,1) \tag{1}$$

Where X_{low}, X_{upp} are lower and upper bounds of search space S^G.
Once the initial population is generated, the next phase of evolution is activated.

(b) **Evolution:** This is the second phase where mutation, crossover and selection operations are performed.

Mutation: In mutation we generate a mutant vector V_j^G for each target vector X_j^G at generation G as

$$V_j^G = X_{r1}^G + F * \left(X_{r2}^G - X_{r3}^G\right) \tag{2}$$

Where F is the scaling factor and value of F is vary from 1 to 0 and $r1, r2, r3 \in \{1, 2, \ldots, NP\}$ are mutually different, randomly chosen vectors.

Crossover: After mutation, crossover is done to generate a new vector called trial vector denoted as $U_j^G = \left\{u_{1,j}^G, u_{2,j}^G, \ldots, u_{D,j}^G\right\}$. Crossover is performed between target vector $X_j^G = \left\{x_{1,j}^G, x_{2,j}^G, \ldots, x_{D,j}^G\right\}$ and mutant vector $V_j^G = \left\{v_{1,j}^G, v_{2,j}^G, \ldots, v_{D,j}^G\right\}$ using a crossover probability Cr whose value is between $0\ to\ 1$. U_j^G is generated as

$$u_{i,j}^G = \begin{cases} v_{i,j}^G & if\ rand_j \leq Cr \\ x_{i,j}^G & otherwise \end{cases} \tag{3}$$

Where $i \in \{1, 2, \ldots, D\}$ $and\ Cr \in [0,1]$(Figs. 1, 2, 4, 5 and Table 1).

Selection: In this operation, a comparison is done between the target vector and trial vector according to their fitness value. The one having better fitness survives to the next generation. This operation is perform as:

$$X_j^{G+1} = \begin{cases} U_j^G & if f\left(U_j^G\right) \le f\left(X_j^G\right) \\ X_j^G & otherwise \end{cases} \tag{4}$$

Mutation, Crossover and selection of evolution phase are repeated till a predefined termination criteria is satisfied.

Pseudo code of Differential evolution:
Generate initial population of size NP
Do while
For each individual j in the population
Generate three random numbers r1 ≠ r2 ≠ r3 ≠ j
For each parameter j, in the population
$$X'_{i,j} = \begin{cases} X_{i,r3} + F*\left(X_{i,r2} - X_{i,r1}\right) & if\ rand_j \le Cr \\ X_{i,j} & otherwise \end{cases}$$
End for
Replace X_J with new particle X'_j, if X'_j is better
End For
Until the termination condition is satisfied

2   Case Study

The area of study selected is 'Mula project', a major irrigation project on the river Mula, tributary of Pravara, sub-tributary of Godavari. This multipurpose project provides irrigation water supply to Ahmednagar city and fulfils the water supply needs of industries and villages. It has two canals, right bank canal and left bank canal and distribution system to create an irrigation potential of 80,810 ha. Both the canals are perennial taking off from the reservoir. Average rainfall in this area is about 500 mm. in

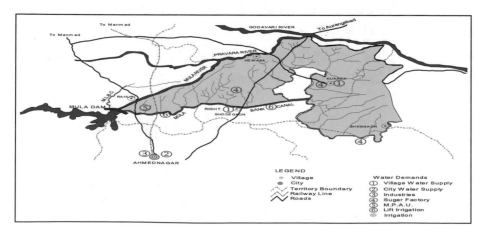

Fig. 2. Map of Mula project

lower catchment (but in upper catchment it is 5080 mm. which is sufficient). Major portion of irrigated land is black cotton soil. Evaporation is moderate.

2.1   Development of Reservoir Operation Models

The purpose of developing an optimal reservoir operation is to obtain a policy for regulating the water in a reservoir while satisfying the desired objectives. Here we assume that operating policy is composed of decision variable which is the release from the reservoir at each time period. The objective is to minimize the squared deviation of release from the target demands. Figure 3 shows a single reservoir system and the variables associated with a reservoir operation problem.

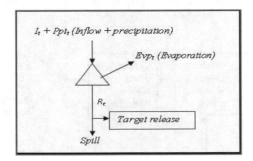

Fig. 3.   Variables associated with a reservoir problem

There are upper and lower limits for releases and storages. These limitations form the constraints of the problem. Another constraint of the problem is due to continuity equation which is to be satisfied for each time period.

Mathematical model: In general a reservoir operation optimization problem may be expressed as

The objective function is:

Minimization of squared deviation from target demands

$$g_t(R_t) = Min \sum_{t=1}^{N} (Q_t - D_t)^2 \qquad (5)$$

Where $g_t(R_t)$ is function of release at time period t. Q_t is the release for period t *and* D_t is target demand for time period t.

Satisfy the continuity equation, which is stated as:

$$S_{t+1} = S_t + I_t - Q_t - Evp_t \; \forall t = 1, \ldots, N \qquad (6)$$

where S_t, I_t and R_t are the storage, inflow and releases for the given reservoir at time period t and N is the time horizon for the problem under consideration. Evp_t is evaporation from reservoir surface during time period t, respectively.

Limits on storage impose constraints of the form

$$S_{min} \leq S_t \leq S_{max} \forall t = 1, \ldots, N \qquad (7)$$

This ensure that the storage (S_t) will remain within specified minimum and maximum storages.

Limits on release

$$Q_{min} \leq Q_t \leq Q_{max} \forall t = 1, \ldots, N \qquad (8)$$

Where release (Q_t) should be within specified minimum and maximum range.

Releases are the decision variables in the problem. Constraints of releases are identified during generation of initial population and as a matter of fact they are

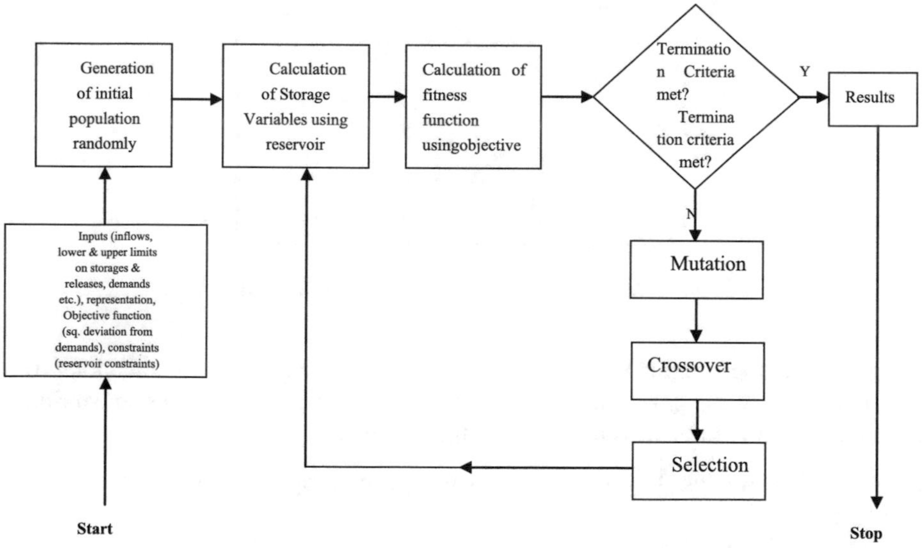

Fig. 4. Flowchart showing the steps used in reservoir operation optimization using DE

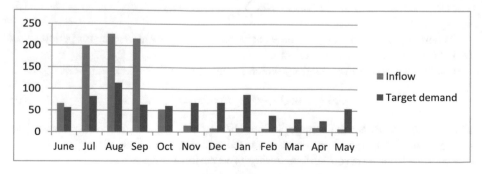

Fig. 5. Monthly inflow vs demand

satisfied. Continuity equation is readily satisfied since the storages are computed by making use of continuity equation given in Eq. (6).

3 Experimental Settings

The present study is done for 12 months. There are two constraints in this problem on storage and release. Programming is done in MATLAB. The MATLAB program is run for $NP = 20\ and\ Max\_Gen = 500$. The program is run on computer with configuration of 16 GB RAM Intel(R) Xeon(R) CPU 3.50 GHz. The search area is shown in Table 2 and Fig. 6. The release of the problem is shown in Table 3 and Fig. 7

Table 2. Monthly release limits

Monthly release search limits		
Months	Lower limit	Upper limit
June	0	56.45
Jul	0	82.33
Aug	0	113.5
Sep	0	63.33
Oct	0	60.83
Nov	0	68.39
Dec	0	68.39
Jan	0	86.6
Feb	0	38.21
Mar	0	30.45
Apr	0	25.72
May	0	54.39

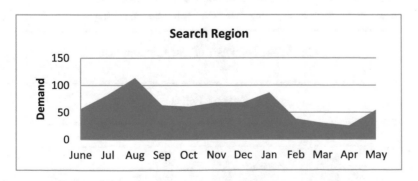

Fig. 6. Monthly demand

Results of Reservoir Problem using Differential Evolution
No. Of Generation = 500 Population size = 20
   Objective value = 7.7017e-04 Time elapsed = 0.527234 s
   Release policies and Objective values of GA and PSO is shown in
   Table 4 and Comparison between DE, PSO and GA is shown in Fig. 8.

Table 3. Monthly release and storage using DE

Months	Initial storage	Inflow	Target demand	Release	Final storage	Evaporation	Spill	Irrigation deficit
June	0	65.8	56.45	56.44322	9.356778	0	0	0
Jul	9.356778	199.67	82.33	82.32011	122.9949	3.711747	0	0
Aug	122.9949	226.52	113.5	113.4864	231.7062	4.322333	0	0
Sep	231.7062	216.18	63.33	63.3224	377.8509	6.71292	0	0
Oct	377.8509	52.5	60.83	60.8227	363.7737	5.754469	0	0
Nov	363.7737	14.87	68.39	68.38179	303.3442	6.917752	0	0
Dec	303.3442	8.21	68.39	68.38179	237.512	5.660423	0	0
Jan	237.512	8.43	86.6	86.5896	154.7306	4.621793	0	0
Feb	154.7306	7.5	38.21	38.20541	119.8609	4.164317	0	0
Mar	119.8609	8.48	30.45	30.44634	93.65207	4.242443	0	0
Apr	93.65207	9.7	25.72	25.71691	72.10274	5.532417	0	0
May	72.10274	7.32	54.39	54.38347	19.03401	6.005257	0	0

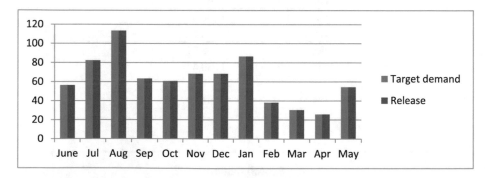

Fig. 7. Release policy of reservoir using DE

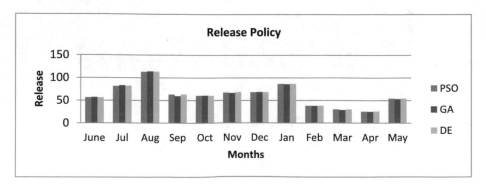

Fig. 8. Comparison on release policy of DE, PSO and GA

Table 4. Release policies of GA and PSO

Release policy of GA and PSO		
Objective value	24.1421	3.88e-25
Time	3–4 s	3–4 s
Months	GA	PSO
June	57.15	56.43
Jul	83.28	82.32
Aug	114.18	113.3
Sep	59.64	62.96
Oct	60.74	60.73
Nov	66.6	68.39
Dec	68.67	68.34
Jan	85.75	86.5
Feb	37.87	38.19
Mar	28.75	30.46
Apr	25.03	25.7
May	53.32	54.35

4 Conclusion

This paper presents an application of DE for finding the optimum release policy for Mula river reservoir. The numerical results obtained by DE are compared with GA and PSO. It is observed that the release policy obtained by the three algorithms is more or less similar to each other but in terms of objective function value PSO and DE give much better results in comparison to GA. While in terms of computational time, DE gave the best results in comparison to the other two algorithms. In this paper the case study is done for a single year but as we increase the size of the problem, the complexity will also increase. Therefore presently the authors are working on modifying the DE/PSO algorithms for solving large scale reservoir problems.

References

1. Storn, R., Price, K.: Differential evolution–a simple and efficient heuristic for global optimization over continuous spaces. J. Glob. Optim. **11**(4), 341–359 (1997)
2. Lakshminarasimman, L., Subramanian, S.: Short-term scheduling of hydrothermal power system with cascaded reservoirs by using modified differential evolution. IEE Proceedings-Generation, Transmission and Distribution **153**(6), 693–700 (2006)
3. Reddy, M.J., Kumar, D.N.: Multiobjective differential evolution with application to reservoir system optimization. J. Comput. Civil Eng. **21**(2), 136–146 (2007)
4. Vasan, A., Raju, K.S.: Application of differential evolution for irrigation planning: an Indian case study. Water Resour. Manag. **21**(8), 1393 (2007)
5. Reddy, M.J., Kumar, D.N.: Evolving strategies for crop planning and operation of irrigation reservoir system using multi-objective differential evolution. Irrig. Sci. **26**(2), 177–190 (2008)

6. Mandal, K.K., Chakraborty, N.: Short-term combined economic emission scheduling of hydrothermal power systems with cascaded reservoirs using differential evolution. Energy Convers. Manag. **50**(1), 97–104 (2009)

7. Qin, H., et al.: Multi objective cultured differential evolution for generating optimal trade-offs in reservoir flood control operation. Water Resour. Manag. **24**(11), 2611–2632 (2010)

8. Suribabu, C.R.: Differential evolution algorithm for optimal design of water distribution networks. J. Hydroinform. **12**(1), 66–82 (2010)

9. Adeyemo, J., Otieno, F.: Differential evolution algorithm for solving multi-objective crop planning model. Agric. Water Manag. **97**(6), 848–856 (2010)

10. Vasan, A., Simonovic, S.P.: Optimization of water distribution network design using differential evolution. J. Water Resour. Plan. Manag. **136**(2), 279–287 (2010)

11. Regulwar, D.G., Choudhari, S.A., Raj, P.A.: Differential evolution algorithm with application to optimal operation of multipurpose reservoir. J. Water Resour. Prot. **2**(06), 560 (2010)

12. Adeyemo, J.A.: Reservoir operation using multi-objective evolutionary algorithms-a review. Asian J. Sci. Res. **4**(1), 16–27 (2011)

13. Fallah-Mehdipour, E., Bozorg Haddad, O., Mariño, M.A.: Extraction of multicrop planning rules in a reservoir system: application of evolutionary algorithms. J. Irrig. Drain. Eng. **139** (6), 490–498 (2012)

14. Ahmad, A., et al.: Reservoir optimization in water resources: a review. Water Resour. Manag. **28**(11), 3391–3405 (2014)

15. Schardong, A., Simonovic, S.P.: Coupled self-adaptive multiobjective differential evolution and network flow algorithm approach for optimal reservoir operation. J. Water Resour. Plan. Manag. **141**(10), 04015015 (2015)

16. Ochoa, P., Castillo, O., Soria, J.: Fuzzy differential evolution method with dynamic parameter adaptation using type-2 fuzzy logic. In: 2016 IEEE 8th International Conference on Intelligent Systems (IS). IEEE (2016)

17. Ochoa, P., Castillo, O., Soria, J.: A new approach for dynamic mutation parameter in the differential evolution algorithm using fuzzy logic. In: North American Fuzzy Information Processing Society Annual Conference. Springer, Cham (2017)

18. Ochoa, P., Castillo, O., Soria, J.: Differential evolution using fuzzy logic and a comparative study with other metaheuristics. In: Nature-Inspired Design of Hybrid Intelligent Systems, pp. 257–268. Springer International Publishing, Cham (2017)

Characterising the Impact of Drought on Jowar (Sorghum spp) Crop Yield Using Bayesian Networks

Shubhangi S. Wankhede[1(✉)] and Leisa J. Armstrong[1,2]

[1] Department of Computer Science, University of Mumbai, Mumbai, India
shubh.4688@gmail.com
[2] School of Science, Edith Cowan University, Joondalup, WA, Australia
l.armstrong@ecu.edu.au

Abstract. Drought is a complex, natural hazard that affects the agricultural sector on a large scale. Although the prediction of drought can be a difficult task, understanding the patterns of drought at temporal and spatial level can help farmers to make better decisions concerning the growth of their crops and the impact of different levels of drought. This paper studied the use of Bayesian networks to characterise the impact of drought on jowar (Sorghum spp) crop in Maharashtra state on India. The study area was 25 districts on Maharashtra which were selected on the basis of data availability. Parameters such as rainfall, minimum, maximum and average temperature, potential evapotranspiration, reference crop evapotranspiration and crop yield data was obtained for the period from year 1983 to 2015. Bayes Net and Naïve Bayes classifiers were applied on the datasets using Weka analysis tool. The results obtained showed that the accuracy of Bayes net was more than the accuracy obtained by Naive Bayes method. This probabilistic model can be further used to manage and mitigate the drought conditions and hence will be useful to farmers in order to plan their cropping activities.

Keywords: Bayesian networks · Classification · Drought · Jowar crop
Prediction

1 Introduction

Drought is a complex, natural hazard that has a major effect on many industrial sectors including agriculture. It can have a long lasting social and environmental impact which can extend for long periods of time even after the period of drought [1]. Drought occurs if there is low or no rainfall for prolonged time that may result in hydrological imbalance in the environment [1]. India Meteorological Department has defined the drought condition as a time period in which the noted rainfall in a particular area is less that 25% of average rainfall [2]. Drought can be further classified as (a) meteorological: deficiency of precipitation from normal level, (b) hydrological: reduction in the volume of water supply and (c) agricultural: deficiency of soil moisture and rainfall which affects crop growth [3]. Around 70% regions of the total land area ns of India are prone to drought conditions [4]. India has regularly faced number of drought periods [5].

© Springer International Publishing AG, part of Springer Nature 2018
A. Abraham et al. (Eds.): ISDA 2017, AISC 736, pp. 979–987, 2018.
https://doi.org/10.1007/978-3-319-76348-4_94

Along with rainfall, there are other climatic parameters such as temperature, evapotranspiration, soil moisture and vegetative indices which are responsible for drought conditions. These parameters have also been used to predict the impact of periods of drought [6]. Drought is inevitable global but region specific event that requires to be predicted [7]. There are number of drought indices which have been developed for drought characterization which comprise of Standardized Precipitation Index (SPI), Palmer Drought Severity Index (PDSI), Crop Moisture Index (CMI), Surface Water Supply Index (SWSI) and Normalized Difference Vegetation Index (NDVI) and many more [6].

Drought forecasting is important for proper resource management and agricultural planning with respect to societal and agricultural perspective [8]. It has been reported that 70% of Indian population depends on agricultural sector directly or indirectly for their livelihood [9]. This sector is the major one to be affected by drought events. Frequent drought occurrences leads to low crop production which eventually affects farmers profit and results in food security of the economy [9].

This paper examined different agricultural and climatic parameters in order to characterise the drought impact at the division level in Maharashtra. This study applies a probabilistic approach to integrate the climate and cropping dataset. A total of 25 districts of Maharashtra were selected for the application of Bayesian Network techniques on 33 (1983 to 2015) years of agricultural and climatic parameters. The parameters included rainfall, minimum temperature, maximum temperature, average temperature, potential evapotranspiration, reference crop evapotranspiration and jowar crop yield. The observed impact and probabilistic approach can be used to create a drought decision support system for farmers in Maharashtra state, India.

The paper will outline related studies and provide a review of literature followed by details of the study area. This will be followed by an outline of the research methods used for the study including details of data set and methods used for collate and analysis of the data. The following section outlines the evaluation metrics used to evaluation the classifier performance. The paper concludes by reporting the results from the study and conclusions.

2 Related Work

Various studies have reported on the application of traditional statistical and computational techniques for drought forecasting and prediction. Other techniques such as data mining, neural networks are now being considered as a viable method for drought prediction. A study by [10] reported that data mining tools can prove helpful for drought modeling and understanding drought characteristics with the help of large data sets. In order to obtain the spatial and temporal patterns of drought, another study by [11] reported that data mining was a viable technique to interrogate large amount of agricultural data in order to understand complex spatial and temporal characteristics of drought. This research made use of association rule mining and decision tree classification.

A similar study [12] used association rule mining to help decision makers to improve their understanding of causes of drought, its prediction and impacts. In one study by [13], Artificial Neural Network was used to predict drought. This study found that Dynamic structures of ANN including Recurrent Network (RN) and Time Lag Recurrent Network (TLRN) were most effective. It was found that ANN is an efficient tool to model and predict drought events. Another study by [14] reported the comparison of the effectiveness of three data-driven models based on artificial neural networks (ANNs), support vector regression (SVR), and wavelet neural networks (WN) for forecasting drought conditions. It was found that ANN is more effective in forecasting the drought conditions. Recursive Multi-Step Neural Network (RMSNN) and Direct Multi-Step Neural Network (DMSNN) have also been used to forecast SPI values across the river basin [15]. This study observed that RMSNN performed better in predicting droughts. Other research reported by [16] used hybrid model of artificial neural network (ANN and Standardized precipitation index) to predict droughts. In the study by [17], statistical forecast of drought was done using logistic regression along with cross validation. The method shown good performance for high-amplitude flash drought events.

Supervised learning and Naïve Bayesian classification techniques have also been applied to evaluate the drought parameters [18]. Bayesian networks are able to model the probabilistic relations among variables. It creates a probabilistic model which can be used to query possible outcomes. It assigns probable values to the set of inputs based on its analysis. Bayesian network have unique advantages over other machine learning techniques like artificial neural network, decision trees, support vector machines. Bayesian networks can be used when the data set contains missing values and can also be queried [19]. The Bayesian forecast model has been used to apply multivariate distribution functions to forecast drought conditions when the historical drought status is provided. The forecast model can be used to develop conditional probabilities of a given forecast variable, and returns the highest probable forecast value along with an assessment of the uncertainty around that value [8].

3   Study Area

The study area for this research is Maharashtra state in India. Over 68–70% of total sown area in India is vulnerable to drought [20]. About 23 districts are drought prone and drought events are frequent [21]. Maharashtra is the third largest state in India. The north latitude is 15° 40′ and 22° 00′ & east longitude is 72° 30′ and 80° 30′. Maharashtra occupies the western and central part of the Indian sub-continent. It has a coastline of nearly 720 km along Arabian Sea. The total geographical area of the state is 3,07,713 Km^2. Maharashtra has 35 districts divided into seven revenue divisions namely Konkan, Pune, Nasik, Aurangabad, Amravati, Nagpur, Kolhapur and Latur. The state experiences tropical climatic conditions. The maximum and minimum temperature varies between 27 °C and 40 °C & 14 °C and 27 °C respectively. The rainfall period starts from first week of June and the wettest period is in the month of July and it varies from region to region [21]. Agriculture is the main occupation of rural population of Maharashtra. Rice, Jowar, Bajra, Wheat, Tur, Moong, Urad and other pulses are the major crops cultivated in Maharashtra [21]. The state is prone to disasters like drought, floods, cyclones, earthquakes [21]. Figure 1 shows the Maharashtra map, highlighting the seven divisions which are considered for this study. This map was created in QGIS 2.8.1.

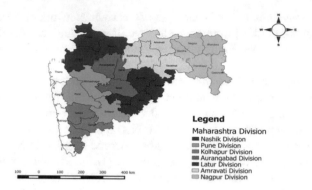

Fig. 1. Map seven divisions and twenty five districts selected as study area in Maharashtra [22, 23].

4   Materials and Methods

4.1   Data

The datasets for this study were collected from publically available government climate and agriculture datasets. The agricultural and climatic parameters datasets was sourced from Commissioner of Agriculture, Government of Maharashtra, Pune and online sources. Parameters including annual seasonal rainfall, minimum, maximum and average temperature, potential evapotranspiration, reference crop evapotranspiration and jowar (Sorghum spp) crop yield for 33 years from 1983 to 2015. The data sets were taken from 25 districts of Maharashtra, India. The districts considered for the study are: Nasik, Dhule, Jalgaon, Ahmednagar, Pune, Solapur, Satara, Sangli, Kolhapur, Aurangabad, Jalna, Beed, Latur, Osmanabad, Nanded, Parbhani, Buldhana, Akola, Amravati, Yavatmal, Wardha, Nagpur, Bhandara, Chandrapur and Gadchiroli which were divided into 7 divisions.

The data was collated and processed using MS Excel spreadsheet. The monthly total rainfall from June to October was summed up to calculate the seasonal rainfall of 33 years. The jowar crop yield data was calculated as tonnes per hectare for Kharif season. The average of minimum, maximum and average temperature, potential evapotranspiration and reference crop evapotranspiration were calculated for the Kharif season for each year and thus for 33 years.

4.2   Methodology

Data Collection. The data from different sources was obtained and was stored in excel spread sheets. The data sets for rainfall (mm), minimum, maximum and average temperature (°C), potential evapotranspiration (mm), reference crop evapotranspiration (mm) and jowar crop yield (t/h) were stored in numeric format. All data sets were collected for 25 districts and for 33 years (1983 to 2015). The next step after the collection was preprocessing of the data sets according to the requirement for further use.

Data Preprocessing. Datasets were preprocessing in MS Excel datasheets before undergoing further data analysis. The following steps were performed:

• Data gaps in rainfall, minimum, maximum and average temperature, potential evapotranspiration and reference crop evapotranspiration were filled by taking average of nearest neighbor values from the respective data values from the data sets.

• The monthly rainfall data was summed for the Kharif season (June to October) to obtained the total annual rainfall of each year for 33 years.

• Minimum, maximum and average temperature were averaged to get monthly mean values for Kharif months and potential and reference crop evapotranspiration data was averaged for the Kharif season.

• Jowar crop yield production (tonnes per hectare) and area (hectares) was available for each of the 25 districts.

• Final yield was calculated in tonnes per hectare for each district based on area and production for each year.

• The total seasonal rainfall in millimeters, minimum, maximum and average temperatures in degree Celsius, potential and reference crop evapotranspiration in millimeters and annual jowar crop yield in tonnes per hectare data sets for 33 years (1983 to 2015) was then integrated in the Excel sheet.

• A total of 825 data records were finally available after preprocessing for the further analysis.

• The general features of the data sets were visualised using graphs for each division. From these graphs, trends were identified for each division and district.

• The production and productivity data of jowar crop yield was used to identify the yield behavior.

• From these combinations and differences in production and productivity, the impact class was generated in nominal form. The classes for impact were high, medium and low.

• The final sheet for analysis included the columns as: rainfall, min temperature, max temperature, avg, temperature, potential evapotranspiration, reference crop evapotranspiration, jowar yield and impact.

• Bayesian network was applied to characterise and predict the future impact values. Bayes Net and Naïve Bayes algorithms were applied using Weka tool.

Bayesian Networks for Prediction Model Using Weka. Bayesian network classifiers are based on Bayes theorem. These are statistical classifiers that predict class membership probabilities. It is the probability of the given data instance that belongs to the specific class [24]. The naïve bayes concept is based on Bayes theorem with independence amongst predictors. It is a supervised classification technique which is used effectively to model a predictive problem probabilistically. In this technique the overall probabilities of attributes belonging to a class are calculated by pre-assuming that the likelihood of an attribute on a given class value is independent on the other attributes. This pre-assumption leads this classifier to obtain better results. This is called conditional independence. The Naïve bayes theorem is represented as:

$$P(c|X) = P(x|c)P(c)/P(x) \qquad (1)$$

The posterior probability, $P(c|x)$ can be calculated, from the Class Prior Probability $P(C)$, the Predictor Prior Probability $P(x)$ and the Likelihood $P(x|c)$ [25].

Bayes net and Naïve Bayes classifiers were applied to the prepared datasets using the Weka software. The Weka software tool is able to perform the data mining tasks like pre processing, classification, visualisation and feature selection.

The data analysis of the prepared datasets included a training and testing phase. Training and testing sets were created using resampling technique. The data was split as train and test sets. The training data was used to construct the Bayesian network. Training a Bayesian network involved creating the network structure and then creating the probabilities between the networks' nodes [19]. Climate parameters as used as the input data for the study. The impact classes were based on jowar yield production and productivity and were categorized as high, medium and low. The Bayes Net and Naïve Bayes classifiers were run on train data set and the model generated was saved and further applied on the test set. The performances of both the classifiers were compared in order to infer the best classifier for current data set.

5   Classifier Performance Evaluation Techniques

The accuracy, sensitivity and specificity are the major performance evaluators of the two classifiers (BayesNet and Naïve Bayes) used in this study. Sensitivity and specificity are statistical measures of the performance of a classification test. Sensitivity (also called the true positive rate, the recall) measures the proportion of positives that are correctly identified as such. Specificity (also called the true negative rate) measures the proportion of negatives that are correctly identified as such. Accuracy is total correctly classified instances. True Positive (TP): Correctly classified instances as positive. False Positive (FP): Incorrectly classified instances as positive. False Negative (FN): Incorrectly classified instances as negative. True Negative (TN): Correctly classified instances as negative. The calculations of accuracy, specificity and sensitivity are as given below:

$$\text{Accuracy} = (TP + TN)/(TP + FP + FN + TN) \qquad (2)$$

$$\text{Specificity} = TP/(TP + FP) \qquad (3)$$

$$\text{Sensitivity} = TP/(TP + FN) \qquad (4)$$

The other performance evaluators considered included mean absolute error (MAE) which is a measure of difference between two continuous variables; root mean squared error (RMSE) which is a measure of the differences between values predicted by a model; relative absolute error (RAE) which is the magnitude of difference between the actual and the individual values; root relative squared error (RRSE) which is the square root of the relative squared error one reduces the error; F1-score which is a weighted average of the precision and recall; and ROC area which is the accuracy measured by the area under the ROC curve.

6   Results

6.1   Bayes Net Classifier

The Bayes Net classifier in Weka was used on the dataset BayesNet classifier parameters were set at: debug: False, estimator chosen was SimpleEstimator with alpha value as 0.5. The probabilities were estimated directly from data. The K2 Bayes Network search algorithm was applied to the dataset. This algorithm is a hill climbing algorithm which is restricted by an order on the variables. The final value, useADtree was set to false. This classifier achieved an accuracy of 98.3871%. The sensitivity of the classifier was 98.4% and specificity was 99.1%. It gave a mean absolute error of 0.022, root mean squared error was 0.096, relative absolute error was 4.135% and root relative squared error of 19.30%.

6.2   Naïve Bayes Classifier

Naïve Bayes classifier was selected in Weka with parameters as: debug, display ModeInOldFormat, useKernelEstimator and useSupervisedDiscretization all set to false. The algorithm had an accuracy of 83.87%, sensitivity as 83.9% and specificity was 92%. It gave a mean absolute error of 0.13, root mean squared error was 0.27, relative absolute error was 30.59% and root relative squared error of 57.96%.

6.3   Comparison of Results

A comparison of the performance of the two Bayesian algorithms was made to assess which algorithm could be used for future drought prediction studies. Table 1 shows the comparative results for accuracy, sensitivity and specificity. It was observed that the Bayes Net algorithm provided performance. Tables 2 and 3 show a comparison other performance indicators of including MAE, RMSE, RAE, RRSE, F1-score and ROC. The Bayes Net also displayed greater performance for these indicators compared to the Naïve Bayes algorithm.

Table 1. Comparison of classifier performance

Classifier name	Accuracy	Sensitivity	Specificity
Bayes Net	98.38%	98.4%	99.1%
Naïve Bayes	83.87%	83.9%	92%

Table 2. Performance evaluation.

Classifier name	MAE	RMSE	RAE	RRSE
Bayes Net	0.022	0.096	4.135%	19.30%
Naïve Bayes	0.134	0.277	30.59%	57.96%

Table 3. F1-score and ROC.

Classifier name	F measure	ROC
Bayes Net	0.984	0.998
Naïve Bayes	0.838	0.95

7 Conclusions

The state of Maharashtra, India has experienced a number of periods of drought which has cause economic and social problems for local farming communities. Developing systems that can help predict droughts will provide management strategies for these farmers.

A number of meteorological parameters like rainfall, minimum, maximum and average temperature and potential and reference crop evapotranspiration influence on jowar crop yield in Maharashtra, India. Historical patterns of these parameters can be used to predict and characterize drought impact. The results presented in this study have shown how Bayesian networks can be used as one technique for understanding the effects of climate parameters on predicting periods of drought. The performance of Bayes Net was found to be higher than that of Naïve Bayes classifier. These results can be incorporated into drought monitoring systems for farmers in Maharashtra as well as other states in India. This could help in developing early warning systems and drought monitoring systems which could assist farmers in making better agricultural decisions. These decision support systems could help farmers and other stakeholders to combat risks associated with drought and improve the profitability of agricultural sectors of Maharashtra and India.

Acknowledgement. The Centre for Environmental Management and School of Science provided funding for the attendance at WICT 2017 conference.

References

1. Gupta, A., Tyagi, P., Sehgal, V.: Drought disaster challenges and mitigation in India: strategic appraisal. Curr. Sci. **100**, 1795–1806 (2011)
2. Shewale, M., Kumar, S.: Climatological features and drought incidences in India. Meteorological Monograph, Climatology, India Meteorological Department, 21 (2005)
3. Nandakumar., T.: Manual for drought management, Department of Agriculture and Cooperation, Ministry of Agriculture, Government of India, New Delhi (2009)
4. Roy., P., Dwiwedi., R., Vijayan,. D.: Remote sensing applications. National Remote Sensing Center, ISRO, Hyderabad (2011)
5. Singh., N., Saini., R.: Contingency and compensatory agriculture plans for drought and floods in India. National Rainfed Area Authority. New Delhi (2013)
6. Keyantash, J., Dracup, J.: The quantification of drought: an evaluation of drought indices. Am. Meteorol. Soc. **83**(8), 1167–1180 (2002)
7. Sadoddin, A., Shahabi, M., Sheikh, V.: A Bayesian decision model for drought management in rainfed wheat farms of North East Iran. Int. J. Plant Prod. **10**(4), 527–542 (2016)

8. Moradkhani, H.: Statistical-dynamical drought forecast within Bayesian networks and data assimilation: how to quantify drought recovery. Geophysical Research Abstracts 17 (2015)

9. Wankhede, S., Gandhi, N., Armstrong, L.: Role of ICTs in improving drought scenario management in India. In: Proceedings of the 9th Asian Federation for Information Technologies in Agriculture, (AFITA), Perth, pp. 521–530 (2014)

10. Tadesse, T.: Discovering associations between climatic and oceanic parameters to monitor drought in Nebraska using data-mining techniques. Am. Meteorol. Soc. **18**(10), 1541–1550 (2005)

11. Rajput, A.: Impact of data mining in drought monitoring. Int. J. Comput. Sci. Issues **8**(2), 309–313 (2011)

12. Dhanya, C., Kumar, D.: Data mining for evolution of association rules for droughts and floods in India using climate inputs. J. Geophys. Res. **114**, 193–209 (2009)

13. Dastorani, M., Afkhami, H.: Application of artificial neural networks on drought prediction in Yazd (Central Iran). Desert **16**, 39–48 (2011)

14. Belayneh, A., Adamowski, J.: Standard precipitation index drought forecasting using neural networks, wavelet neural networks, and support vector regression. Appl. Comput. Intell. Soft Comput. **2012**, 1–13 (2012)

15. Wambua, R.: Drought forecasting using indices and artificial neural networks for Upper Tana river basin, Kenya-a review concept. J. Civil Environ. Eng. **4**(4), 1 (2014)

16. Boudad, B.: Using a model hybrid based on ANN-MLP and the SPI index for drought prediction case of Inaouen basin (Northern Morocco). Int. J. Sci. Eng. Technol. **2**(6), 1301–1309 (2014)

17. Lorenz, D., Otkin, J., Svoboda, M., Hain, C.: Predicting the US drought monitor using precipitation, soil moisture and evapotranspiration anomalies. Part II: intraseasonal drought intensification forecasts. Am. Meteorol. Soc. **18**(7), 1963–1981 (2017)

18. Sriram, K., Suresh, K.: Machine learning perspective for predicting agricultural droughts using Naive Bayes algorithm. IIECS **24**, 178–184 (2016)

19. Heaton, J.: Bayesian networks for predictive modeling. Forecast. Futurism 6–10 (2013)

20. Choudhary, K., Tahlani, P., Bisen, P., Saxena, R., Ray, S.: Assessment of drought indicators. Technical report, New Delhi (2017)

21. NIDM: Maharashtra. In: National Disaster Risk Reduction Portal, Maharashtra, pp. 1–26 (2012)

22. Shapefile India. http://www.gadm.org/country. Accessed 2 Nov 2016

23. Shapefile Indian Village Boundaries. https://github.com/datameet. Accessed 2 Nov 2016

24. Gayathri, A.: A survey on weather forecasting by data mining. IJARCCE **5**(2), 298–300 (2016)

25. Netti, K., Radhika, Y.: Minimizing loss of accuracy for seismic hazard prediction using Naive Bayes classifier. IRJET **3**(4), 75–77 (2016)

Linear Programming Based Optimum Crop Mix for Crop Cultivation in Assam State of India

Rajni Jain[1]([⊠]), Kingsly Immaneulraj[1], Lungkudailiu Malangmeih[1],
Nivedita Deka[2], S. S. Raju[3], S. K. Srivastava[1], J. P. Hazarika[2],
Amrit Pal Kaur[4], and Jaspal Singh[4]

[1] ICAR-National Institute of Agricultural Economics
and Policy Research, New Delhi, India
Rajni.jain@icar.gov.in
[2] Assam Agricultural University, Jorhat, Assam, India
[3] ICAR-Central Marine Fisheries Research Institute,
Vishakhapatnam, Andhra Pradesh, India
[4] NITI Aayog, New Delhi, India

Abstract. The Assam, the gate way to the north east is an agrarian state with rice as the staple crop. Nearly 89% of the state's population of about 3 crores lives in rural area. The productivity of the major crops like rice, pulses and oilseeds is yet to reach acceptable level despite various efforts in the past. Optimum crop mix for Assam has been developed using linear programming to maximize the net returns ensuring the best use of land and other natural resources for the state. Attempts have been made to obtain crop combination under rainfed and irrigated area separately in both kharif and Rabi season. The result from the linear programming based model indicates higher returns than the existing crop plan (Rs. 39.30 hundred crore in optimum plan over 34.16 hundred crore in existing plan at market price). Sensitivity analysis also indicated that the profit can be increased substantially by bringing the fallow land during Rabi season under irrigation. This implies that there is potential in bringing more land under double or triple cropping considering vast water availability in the state.

Keywords: Linear programming · Optimum crop plan · Net returns
Sensitivity analysis · Assam agriculture

1 Introduction

Assam is the largest state among 'the seven sisters' in the North-East region of India. The economy of the state is largely rural and agrarian. Agriculture is still the principal occupation of the majority of the rural population in the state in terms of employment and livelihood. About 98.4% of the total land mass of the state is rural. The net cultivated area of the state is 28.11 lakh hectares (2011–2012) which is about 87.38% of the total land available for agricultural cultivation [1]. The Gross cropped area (GCA) of the state has witnessed increasing trends over the past few decades resulting

© Springer International Publishing AG, part of Springer Nature 2018
A. Abraham et al. (Eds.): ISDA 2017, AISC 736, pp. 988–997, 2018.
https://doi.org/10.1007/978-3-319-76348-4_95

in increase in Cropping Intensity from 141–148% during the same period [1] but have not been translated into the growth of output. This is evident from the fact that cropping intensity of the state has been consistently higher than the national average but crop productivity has been consistently lower than the national average over the years. More than 85% of the farmers belong to the small and marginal category with the average land holding of only 0.63 ha [1].

The contribution of agriculture sector to state GDP declined from 21.70% in 2004–2005 to 16.52% in 2014–2015 at 2004–2005 prices. However, Agriculture sector continues to support more than 75% population of the state directly or indirectly providing employment to about 50% of the total workforce [2].

The state has its climatic and physiographic features favourable for rice cultivation and the crop is grown in a wide range of agro-ecological situations occupying around 77 to 80% of GCA. The state has a wide variety of crops in different seasons such as summer (April–September), locally known as kharif season, and winter (December–March), locally known as Rabi season.

Only about 6% of the Net Sown Area (NSA) is irrigated and that portion of irrigated area is mainly occupied by rice, wheat, maize and sugarcane [2]. However, contrastingly the state is still not self-sufficient in food produce and there is an untapped potential in the state in the agricultural sector. Alternatives combinations of crops give different outputs. The utilization of land for appropriate crops is the key problem for crop planning in Assam. Therefore a linear programming model has been developed and solved to provide an annual crop production plan that determines the area to be used for different crops for the maximum possible contribution and self-sufficiency in food having satisfied the constraints arising out of food demand, capital, and land. However, factors like rainfall, weather conditions, floods, cyclones, and other natural calamities are difficult to predict [3] and have not been considered.

2   Review of Literature

Linear programming is a tool applicable to cropping planning and optimization of resources allocation, such as land, water, and labor, taking into account constraints on those resources availability and production [4]. Some studies developing optimal crop plan in India are mentioned here. The linear programming model formulated for maximization of annual net return with optimal water and cropping pattern allocation considering the saline and non-saline soil type, rainfed and irrigated agriculture and the monsoon and winter seasons crops are found to be an effective tool for land and water resources allocation [5]. Authors developed two models viz. groundwater balance model and optimum cropping and groundwater management model to determine optimum cropping pattern and groundwater allocation from private and government tube wells for coastal river basin in Orissa State, India. Linear Programming with multiple objective functions was used by Rani and Rao for Crop Planning strategies which increases the productivity with minimum input cost with the constraints of available resources like water usage and also labour, fertilizers, seeds, etc., and ultimately getting maximum net benefits [6]. In another study, Linear programming technique was applied to determine the optimum land allocation to 10 major crops of the saline track of rainfed zone of

Maharashtra using agriculture data, with respect to various factors viz. cost of seeds, cost of fertilizers and pesticides, yield of crops, daily wages of labour and machine charges, selling base price of commodities [7]. They found that the Linear Programming approach was appropriate for finding the optimal land allocation to the major crops in the region. The optimal cropping pattern for maximizing net returns and ensuring significant savings of groundwater with the aim of sustaining groundwater use in the Punjab agriculture was formulated using Linear Programming Approach considering the constraints such as land, labour, working capital and irrigation water with maximizing net returns with judicious use of water (see [8, 9]).

Except for Punjab, few statewide crop planning has been carried out in India. Linear programming approach to obtain optimal crop plan for the Assam state as a whole has not been reported anywhere to the best of our knowledge probably because of lack of data availability. This paper attempts optimum crop planning for Assam using Linear Programming approach in GAMS software (GAMS win64 24.7.4).

3 Data and Methodology

To develop a multi-crop model for the state of Assam, maximization of net returns based on land as a constraint and minimizing the cost of other available resources such as irrigation, seeds, fertilizers, human labour, pesticides, capital etc. has been used. For estimation of technical coefficients, the plot level data collected under "Comprehensive Scheme for Cost of Cultivation (CCS) of Principal Crops of Directorate of Economics and Statistics, Ministry of Agriculture has been used. In this study, the data pertaining to the year 2009–2010 and 2010–2011 have been used. Secondary data sources were also used in the study for few indicators like subsidy rate on fertilizers. For the development of optimum crop model, cost and returns based on market prices were estimated by subtracting variable costs at market price from gross returns at market price (see [8, 10]). Mathematically, the model specification for Assam is presented by Eqs. 1–5 as follows.

$$Max\,Z = \sum\nolimits_{c=1}^{n} (Y_c P_c - C_c)A_c \tag{1}$$

$$\sum\nolimits_{t=jan}^{Dec} \sum\nolimits_{c=1}^{n} a_{tc}A_c \leq NS_t - OA_t \tag{2}$$

$$A_c \geq Amin_c \tag{3}$$

$$A_c \leq Amax_c \tag{4}$$

$$A_c \geq 0 \tag{5}$$

In the above equations, Y_c = yield of a crop c per hectare of land, P_C = unit price received for crop C, C_c = the cost incurred to cultivate crop c in one hectare of land considering all inputs like seeds, irrigation, fertilizers, human labour, pesticide and capital, A_c = area under cultivation of crop c as allocated by the model, and a_{tc} = coefficient of crop calendar month for month t and crop c. The value of a_{tc} is 1 if a crop c needs area allocations during month t. The objective function is to maximize the net revenue (Z) based on the optimum crop plan (Eq. 1). Crop calendar for Assam is given in last column of Table 1.

Table 1. Existing cropping patterns and technical coefficients for rainfed conditions

Crops	Existing area (000 ha)	MinA (000 ha)	MaxA (000 ha)	NRmp (Rs)	Crop calendar
AutumnRice	316.90	235.09	391.81	7698	Feb–Jun
WinterRice	1816.15	1368.00	2280.00	7462	Jul–Nov
SummerRice	261.65	196.24	327.06	13344	Jan–Apr
Wheat	52.11	38.99	64.99	7420	Dec–Mar
Maize	19.42	14.63	24.38	2227	Mar–Jun
Jute	63.47	47.63	79.38	6774	Mar–Jun
Mustard	244.89	183.67	306.11	3630	Nov–Feb
Sesamum	11.40	9.00	15.00	5314	Aug–Oct
Black gram	43.70	34.50	57.50	6283	Sep–Nov
Green gram	8.08	6.38	10.63	7421	Sep–Nov
Pea	20.90	15.68	26.13	2988	Nov–Mar
Lentil	22.33	16.74	27.91	21873	Oct–Feb
Chillies	17.10	12.83	21.38	26080	Dec–May
Potato	79.80	59.85	99.75	12871	Oct–Jan
Sugarcane	28.48	20.96	34.93	30282	Annual
Cabbage	28.85	21.64	36.06	52695	Oct–Jan
Cauliflower	19.74	14.81	24.68	28291	Oct–Jan
Brinjal	15.35	11.51	19.19	35850	Oct–Dec
Tomato	15.65	11.73	19.56	66456	Sep–Dec
Okra	10.47	8.27	13.78	49260	Jun–Aug
Onion	6.51	4.88	8.13	8596	Oct–Jan
Ginger	15.20	12.00	20.00	114890	Mar–Dec
Turmeric	13.78	10.88	18.13	75411	Apr–Dec
GCArf	**3131.91**	**2356**	**3926**		

Notes: GCArf- Total cropped area under rainfed crops, MinA-Minimum area, MaxA-Maximum area, NRmp- Net Returns at market price (estimated by authors based on methodology explained in [5]

Optimum use of land for each month is required. Equation 2 represents 12 land constraint equation one each for January to December for a year. This helps to have separately sown area for each month and ensures that total cultivated area under selected crops in each month should be less than net sown area (NSt) minus area under orchard (OAt) crops.

In some modelling solutions, some major crops may drastically lose their relevance and the corresponding area allocations may become negligible. Then, even though estimates are robust and mathematically proven, such allocations may not be desirable and practically possible from the view point of food security of the country and livelihood security of the farmer because appropriate changes are required in policy framework of the country to adopt the optimum sustainable model. Similarly, area allocations for some minor crops may be over estimated ignoring the demand. Such an area allocation is again undesirable as it may lead to a glut in the market. To avoid such undesirable over-estimation or under-estimation, assigning values to minimum and maximum area of the selected crops become essential in the model. To eliminate such practically undesirable solutions, the concept of min, max constraints is used in the model as specified by the Eqs. 3 and 4. Equation 5 refers to non-negativity constraint the Linear Programming model.

4   Optimal Crop Model Development for Assam

In the case of Assam, the model was developed with the plot level data collected under "Comprehensive Scheme for Cost of Cultivation (CCS) of Principal Crops of Directorate of Economics and Statistics, Ministry of Agriculture for the year ending 2010–2011 and also based on primary survey of some crops. A diverse range of crops is grown in the state out of which rice occupied a major share of total cropped area. Attempts have been made to incorporate the maximum number of crops grown in the state but due to the limitation of coefficients for some minor crops, ultimately 21 crops have been selected for obtaining optimum crop mix for the state. A separate model was framed for rainfed and irrigated crops as there was a difference in the returns in each scenario. It was observed that above 70% of the cropped area are occupied in kharif season and most of the cultivable land is left fallow in Rabi season. For right hand side of land constraint, total net sown area adjusted after deducting the orchard and plantation crops (2503.33 thousand hectares) has been considered for the rainfed model. As only 6% of the net sown area has been reported with assured irrigation (according to secondary data and the publication by the state's department), 150 thousand hectares of area has been considered in the case of irrigated model. Based on this scenario, sensitivity analysis has been carried out by gradually increasing irrigated area by increasing from 6% to 12% i.e. double the existing area under irrigation to find out the incremental benefit if irrigation structures are developed and fallow land can be utilized and double cropping could be achieved. The lower and upper limit of the area under each crop in the season has been taken in such a manner that an economical requirement of farmers as well as the food requirement of the society is satisfied. This has also been decided based on the importance of crops and the inputs from a team of experts from the state and the same is displayed in Tables 1 and 2 for rainfed and irrigated model respectively.

Table 2. Existing cropping patterns and Technical Coefficients* used for Irrigated conditions

Crops	Existing area (000 ha)	MinA (000 ha)	MaxA (000 ha)	NRmp (Rs)
AutumnRice	12.60	9.45	15.75	8160
WinterRice	7.85	5.89	9.81	7462
SummerRice	134.85	134.85	202.28	14059
Wheat	0.40	0.30	0.49	14348
Maize	0.08	0.06	0.10	2227
Jute	0.03	0.02	0.04	6774
Mustard	1.61	1.21	2.01	11943
Sesamum	0.60	0.45	0.75	5314
Black gram	2.30	1.73	2.88	6283
Green gram	0.43	0.32	0.53	7421
Pea	1.10	0.83	1.38	6261
Lentil	1.18	0.88	1.47	25207
Chillies	0.90	0.68	1.13	31321
Potato	4.20	3.15	5.25	19547
Sugarcane	0.02	0.01	0.02	37880
Cabbage	1.52	1.14	1.90	63995
Cauliflower	1.04	0.78	1.30	36931
Brinjal	0.81	0.61	1.01	42889
Tomato	0.82	0.62	1.03	22593
Okra	0.55	0.41	0.69	49260
Onion	0.34	0.26	0.43	11276
Ginger	0.80	0.60	1.00	114890
Turmeric	0.73	0.54	0.91	75411
GCAirri	174.74	164.77	252.13	

Notes:
1. For those coefficients which were not available for irrigated crops, rainfed crops coefficients have been used
2. GCAirri- Total cropped area under irrigated conditions
3. MinA-Minimum area, MaxA-Maximum area
4. NRmp- Net Returns at market price which was estimated by authors based on methodology explained in [8]

5   Results and Discussion

In the study, the gross cropped area (GCA) and the net sown area (NSA) have been adjusted after deducting the area under plantation crops and orchards to arrive at the area utilized fully by the crops considered for the model. About 87% (3306.65 thousand) of the gross cropped area and 100% Net Sown Area (NSA) has been used. Among the kharif crops, major ones such as winter rice, autumn rice, jute, okra, chilies, tomato, sesame, and maize were considered. Rabi crops include summer rice, wheat, brinjal, mustard, black gram, green gram, pea, lentil, onion, potato, cabbage, and cauliflower. In Assam, different varieties of Paddy are grown in different seasons.

Three important annual crops such as sugarcane, ginger, and turmeric were selected. Land availability for cultivation was taken as an important resource constraint. Water availability is ignored because Assam has plenty of water and its water potential is underutilized. However, irrigation facilities for the Rabi crops have been considered by using separate coefficients for the rainfed and irrigated crops. Based on these specifications Linear Programming model based on GAMS code [10] was used to develop optimum crop plan. Further, irrigated area is gradually increased in the model from 6 to 12%. Analysis of the results shows that Gross Cropped Area increased from 3306.65 thousand hectares under existing pattern to 3616.68 thousand hectares under the optimum condition for the state as a whole and for both the rainfed and irrigated crops taken together. The comparison of the existing cropping plan with the optimum plan is given in Table 3.

Table 3. Optimum crop allocations based on Linear Programming approach

Crops	Rainfed	Irrigated	Pooled	Crops	Rainfed	Irrigated	Pooled
Kharif				Rabi			
W.Rice	1955	9.81	1964	S.Rice	327.06	134.85^a	461.91
A.Rice	391.8	10.79^b	403	Wheat	64.99	0.30^b	65.28
Jute	79.38	0.04	79.41	Brinjal	19.19	1.01	20.20
Okra	13.78	0.69	14.46	Mustard	183.7^b	1.21^b	184.9^b
Chillies	21.38	1.13	22.50	Black gram	34.50^b	2.88	37.38^b
Tomato	19.56	1.03	20.59	Green gram	6.38^b	0.53	6.91^b
Sesame	15.00	0.75	15.75	Pea	15.68^b	0.83^b	16.50^b
Maize	24.38	0.10	24.48	Lentil	27.91	0.88^b	28.79
Annual	**Rainfed**	**Irrigated**	**Pooled**	Onion	8.13	0.43	8.56
Sugarcane	34.93	0.02^a	34.95	Potato	99.75	3.39^b	103.14
Ginger	20.00	1.00	21.00	Cabbage	36.06	1.90	37.96
Turmeric	18.13	0.91	19.03	Cauliflower	24.68	1.30	25.98

Note: W.Rice-Winter rice; A.Rice: Autumn rice; S.Rice-Summer rice; 'a' denotes no change in area, 'b' denotes the decrease in area under optimum plan

Thus cropping intensity of 132% under existing cropping pattern increased to 144% in the optimum pattern. The current crop production is predominantly rainfed and only some area of each crop is irrigated as displayed in the table. It can be seen that for the rainfed and irrigated area, all the crops acreage except summer rice, mustard, black gram, green gram, pea, and sugarcane have been allotted more area in all the three optimum plans than the existing area. It is observed that during kharif, autumn rice and tomato crops area decreased while other crops area increased in the irrigated model. In rainfed model during the same season, acreage under all the crops has increased. In Rabi season, the area under mustard has decreased both in rainfed and irrigated condition. This could be due to the low net returns and higher costs involved in the mustard production. Rainfed area of black gram decreased while irrigated area increased. In the case of green gram, the area under rainfed decreased while irrigated area increased. The area under both rainfed and irrigated condition for pea as well as potato decreased. Among annual crops, sugarcane area under irrigated remains the same with the existing area. Area under all other crops in both kharif and Rabi season under both rainfed as well as irrigated conditions have increased. The comparison in profit under existing and optimum plans at different price scenario is displayed in Table 4.

Table 4. Gains due to optimum crop model over existing scenario

Parameters	Rainfed	Irrigated	Pooled
Net cropped area in optimum model (000 ha)	3441.09	175.59	3616.68
Change in GCA (%)	9.87	0.49	9.38
Existing revenue (Rs. Crores)	3153.95	261.72	3415.67
Optimum net returns (Rs. Crores)	3661.00	268.82	3929.82
Change in farmers' revenue (Rs. Crores)	507.05	7.10	514.15
Change in revenue in optimal plan (%)	16	3	15

The formulated optimum model increased the Gross cropped area by 9.87% in rainfed crops and 0.49% increase in irrigated crops contributing to 9.38% increase in gross cropped area when rainfed and irrigated area are combined. The comparison in profit showed an increased Net Return of 16% over existing pattern at market price. Similarly, in irrigated area, the increase in profit at market price was found to be 3%. Rainfed and irrigated model results are combined to obtain the pooled results for all the cropping seasons of the state. The increase in net return in optimum plan at market price is 15% over the existing net returns. In absolute term, the change in farmer's revenue (at market price) due to the optimum plan is observed to be rupees 507 crores from rainfed and Rupees 7.10 crores from irrigated crops giving a total increase of Rupees 514.15 crores from both rainfed and irrigated crops taken together.

Existing irrigated area is 6% (150 thousand hectares) of the net sown area of the state. There is a potential to increase the net revenue from crop sector by bringing more area under irrigation. Therefore, sensitivity analysis was carried out by gradually increasing the net irrigated area by one per cent in each subsequent iteration (i.e., 175 (7%), 200 (8%), 225 (9%), 250 (10%), 275 (11%) and 300 (12%) thousand hectares). This is because there are untapped water resources that can be utilized if the state provides the required infrastructures for the farmers. Further, vast area is left fallow during Rabi season due to lack of irrigation facilities. During the sensitivity analysis, the lower and upper bound of the area for each crops used were existing area under irrigation as minimum area and twice the irrigated area under the crop as maximum area. The increase in net returns due to increase in percentage of irrigated area is shown in Fig. 1. It is observed that the state can be benefited greatly by bringing more and more of the fallow land during Rabi season under irrigation. At market price, the net returns increased to 40.1 hundred crore rupees at 7% irrigated area (per cent of NSA) from 39.3 hundred crore rupees at 6%. When the net irrigate area became double of the existing one, i.e., at 12%, the return at market price is computed as 41.8 hundred crore rupees using Linear Programming based model. An increase in the irrigated land area

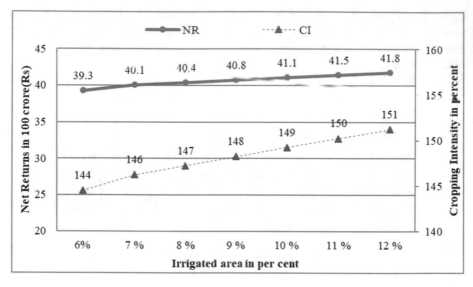

Fig. 1. Increase in objective function with percentage increase in irrigated area

can be made possible in the next few years by educating farmers, increasing credit facilities for irrigation, improving technological options and market infrastructures etc.

6 Conclusion and Policy Implication

Agriculture sector supports more than 75% population of Assam directly or indirectly providing employment to about 50% of the total workforce. So, the overall prosperity of the state and the nation is impossible without the improvement of the socio-economic condition of the agricultural people. It cannot be improved unless the proper return from their investment is ensured. This can be done by improving productivity by removing various production constraints, ensuring proper and stable market prices, and by proper production planning. However at existing level of resources, the result from the Linear Programming model for optimum crop planning indicates higher returns than the existing plan (Rupees 39.30 hundred crore in optimum plan over 34.16 hundred crore in existing plan at market price). Sensitivity analysis also indicated that the profit can be increased substantially by bringing the fallow land during Rabi season under irrigation. This implies bringing more land under double or triple cropping considering the available water in the state which is still untapped. As discussed earlier, Assam is faced with errant floods during rainy season and droughts on the other hand in some other parts of the state especially in hilly zone. The state can make sincere efforts by diverting the flood water into use by creating storage structures which can prevent the crop damage due to flood on one hand and make the water available for agriculture on the other hand. This can be achieved by proper investment and implementation from various departments.

Acknowledgement. This work was supported by Indian Council of Agricultural Research (ICAR) under the ICAR Social Science Network project 'Regional Crop Planning for Improving Resource Use Efficiency and Sustainability'. The authors also acknowledge Assam Agricultural University, Jorhat, Assam, nodal agency of cost of cultivation scheme for sharing unit-level data for the project.

References

1. Economic Survey: Directorate of Economics and Statistics, Government of Assam (2015)
2. Statistical abstracts of Assam, various issues. http://ecostatassam.nic.in/. Accessed Jan 2017
3. Sarker, R.A., Talukdar, S., Haque, A.F.M.A.: Determination of optimum crop mix for crop cultivation in Bangladesh. Appl. Math. Model. **21**(10), 621–632 (1997)
4. Borges Junior, J.C.F., Ferreira, P.A., Andrade, C.L.T., Hedden-Dunkhorst, B.: Computational modeling for irrigated agriculture planning. Part I: general description and linear programming. Engenharia Agricola **28**(3), 471–482 (2008)
5. Nath, S., Mal, B.C.: Optimal crop planning and conjunctive use of water resources in a coastal river basin. Water Res. Manag. **16**, 145–169 (2002)
6. Rani, Y.R., Rao, P.T.: Multi objective crop planning for optimal benefits. Int. J. Eng. Res. Appl. **2**(5), 279–287 (2012)
7. Wankhade, M.O., Lunge, H.S.: Allocation of agricultural land to the major crops of saline track by linear programming approach: a Case Study. Int. J. Sci. Technol. Res. **1**(9), 21–25 (2012)
8. Jain, R., Kingsly, I., Chand, R., Kaur, A., Raju, S.S., Srivastava, S.K., Singh, J.: Farmers and social perspective on optimal crop planning for ground water sustainability: a case of Punjab state in India. J. Indian Soc. Agric. Stat. **71**(1), 75–88 (2017)
9. Kaur, B., Sidhu, R.S., Vatta, K.: Optimal crop plans for sustainable water use in Punjab. Agric. Econ. Res. Rev. **23**, 273–284 (2010)
10. Jain, R., Kingsly I., Raju S.S., Srivastava, S.K., Kaur A., Singh, J.: Manual on methodology for regional crop planning and resource use efficiency, published by ICAR-NIAP (2015). www.ncap.res.in

eDWaaS: A Scalable Educational Data Warehouse as a Service

Anupam Khan[✉], Sourav Ghosh, and Soumya K. Ghosh

Department of Computer Science and Engineering,
Indian Institute of Technology Kharagpur, Kharagpur 721302, India
{anupamkh,sourav.ghosh,skg}@iitkgp.ac.in

Abstract. The university management is perpetually in the process of innovating policies to improve the quality of service. Intellectual growth of the students, the popularity of university are some of the major areas that management strives to improve upon. Relevant historical data is needed in support of taking any decision. Furthermore, providing data to various university ranking frameworks is a frequent activity in recent years. The format of such requirement changes frequently which requires efficient manual effort. Maintaining a data warehouse can be a solution to this problem. However, both in-house and outsourced implementation of a dedicated data warehouse may not be a cost-effective and smart solution. This work proposes an educational data warehouse as a service (eDWaaS) model to store historical data for multiple universities. The proposed multi-tenant schema facilitates the universities to maintain their data warehouse in a cost-effective solution. It also addresses the scalability issues in implementing such data warehouse as a service model.

Keywords: Educational data management
Data warehouse · Service-oriented architecture · Scalability

1   Introduction

Higher education is a public welfare provided by nonprofit organisations with a societal mission. However, nowadays it is becoming a global service in an ever-more complex and competitive knowledge marketplace. To survive, the higher educational universities started prioritising revenue earning and promoting the university at global level. In addition to that, sustaining reputation has become the primary focus of the management. Competition in academia is the driving force that influences the universities to always innovate policy for enhancing the quality of service [9]. In this age of information, several ranking framework help a university to maintain a better visibility globally. In addition to global reputation, management also tries to increase popularity of the departments, specialisation, courses etc.

The decision makers of a university often need historical data and its' analysis for framing future policies and providing information to several ranking

© Springer International Publishing AG, part of Springer Nature 2018
A. Abraham et al. (Eds.): ISDA 2017, AISC 736, pp. 998–1007, 2018.
https://doi.org/10.1007/978-3-319-76348-4_96

frameworks. Unfortunately, these required data, especially the historical information, are not easily available. A data warehouse can be an effective solution to store the required historical data after extract-transform-load (ETL) operation of information from heterogeneous sources. The online analytical processing (OLAP) of queries help in generating report easily in desired format. In fact, there is a pressing demand of designing a data warehouse for the universities [2]. Moreover, it can facilitate the educational data mining, which is becoming popular nowadays [7]. However, it is a costly affair to maintain such data warehouse in-house. Similarly, outsourcing the dedicated data warehouse maintenance to third-party organisation is not a smart solution. A dedicated data warehouse usually requires high-end hardware for complex information retrieval but may not yield healthy utilisation out of it.

The data warehouse as a service (DWaaS) model has emerged as a growing market in recent years, but there is little significant advancement in this field so far. To best of our knowledge, the term DWaaS was first coined in 2012 [6]. Though, the performance issues due to large data volume, heavy duty ETL process are still matter of concern [1]. Some researchers have even tried to implement ETL and OLAP processes on cloud platform [3,10]. However, the scalability issues are not clearly addressed by them. Some researchers have even proposed various data marts suitable in education domain [2,4,5,8]. However, a robust implementation of DWaaS in education domain is not clearly identifiable in literature. Therefore, the aim of this work is as follows: (i) designing a multi-tenant schema for universities in cloud environment, and (ii) implementing a scalable ETL and OLAP process.

This study proposes an educational data warehouse as a service (eDWaaS) model offered in multi-tenant environment. It helps the universities to maintain their historical statistics in a cost effective environment. The subscriber gains by offering the service to multiple universities; thus increasing the utilisation of resources. The multi-tenant architecture reduces the operating cost for the service consumer. The proposed approach to ETL process and report generation using OLAP query handles the scalability issues in cloud data warehouse implementation.

Section 2 describes the proposed eDWaaS model. Section 3 presents the analysis on the scalability of the proposed model. Finally, Sect. 4 summarises this work and highlights the future direction.

2 eDWaaS Model

This section elaborates the proposed eDWaaS model for building up the academic data repository. Figure 1 presents the architecture of multi-tenant eDWaaS model that enable multiple universities to subscribe the services provided by it. The multi-tenant schema on Hive is the central component of eDWaaS. The ETL service felicitates the university to upload the data in widely accepted comma separated value (CSV) format. The OLAP service produces report from the multi-tenant schema.

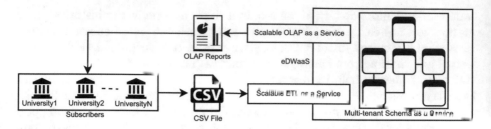

Fig. 1. Architecture of eDWaaS

2.1   Schema as a Service

As a core of the eDWaaS model, it provides a multi-tenant schema of academic data mart. It is important to mention that the proposed schema does not consider all aspects of academic data mart. Rather, it is only a representative one which considers the following few aspects among them: (i) student performance (ii) teaching quality, and (iii) student counts.

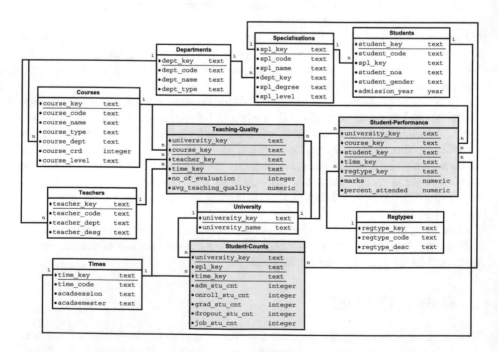

Fig. 2. Multi-tenant schema for academic data mart

The snowflake schema of this data mart is designed in such a way that it can efficiently manage information for multiple universities. The proposed schema, containing dimensions and fact tables, represents a multi-tenant model which ensures the data abstraction between multiple universities. The dimension of academic data mart enables the consumer to categorise various interesting measures. The proposed schema contains eight dimensions and three fact tables. Figure 2 presents these fact tables in grey background and dimension tables in white background.

In the proposed snowflake schema, unification of records in dimension table is handled by prefixing university key with the dimension key. For example, the dimension key of *student1* of *university1* in *Students* table is *university1_student1*. The fact table maintains quantifiable measure of certain attributes based on different dimensions. In addition to that, the proposed schema maintains the university key as a separate attribute in each fact table to ease out the process of filtering data during OLAP queries. For example, the *university_key* in *Student-Performance* table helps in this filtration process.

2.2 ETL as a Service

In this multi-tenant model, growth of data and increasing amount of simultaneous requests are the two challenging issues for the DWSP. From the perspective of a university, performance of the data warehouse should be independent of the activities of other tenants. Therefore, the next concern is to design the data warehouse related processes in a manner which would scale well with increasing data and request. Transferring the data to the multi-tenant cloud data warehouse through ETL extract-transform process is a difficult task. The simplicity of this process can make the eDWaaS model successful. In this approach, the universities upload the student record using the ETL process offered as a service. The ETL service converts the uploaded data to make it suitable for storing in the multi-tenant schema. An overview of the ETL process is presented in Fig. 3 and discussed in the following part of this section.

The universities upload the dimension and fact data in CSV format from their tenant login. Complex operations are handled by the data transformation process. One of such complex operation is the dimension key generation. Let us consider the example of *student_key* in *Dim-Students* table. In this implementation, the subscribed university uploads any of their specific student identifier like roll number, enrolment id, registration number etc. The ETL process generates the *student_key* by concatenating *university_key* with uploaded student identifier. This helps the processing to be simple. It also helps the DWSP to maintain identity of the records in multi-tenant schema. The process involved in transferring data from university sources to the multi-tenant schema is collectively referred to as ETL as a service. In order to make these process scalable, this study has exploited Map-Reduce framework on a Hadoop cluster for extract and transform process.

Extract Process. This study uses a predefined structure of CSV data which the university has to follow. The system detects any deviation from such format by comparing the CSV structure with the acceptable input format. The process

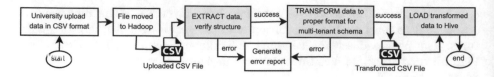

Fig. 3. Overview of ETL process

aborts once in case of any deviation. In such cases, the system generates a report mentioning the line in the CSV file where the error has been observed. Otherwise, the process continues, parsing lines from the CSV file one at a time, and storing the data in an intermediate file. In this implementation, several mappers consume the lines from the CSV file and verify the structure of the data. The proposed model left the data cleansing, the process of detecting and rectifying erroneous and/or inconsistent records, as a manual process to the subscriber.

Transform Process. The transformation process converts the extracted data in the proposed schema acceptable format. Since the data from multiple universities are stored in a multi-tenant schema, the transformation process generates dimension keys by concatenating the *university_key* with the key provided by the university in the CSV input. An important concern here is whether to store only the transformed key as a single attribute; or to store the tenant provided key and the *university_key* as separate attributes in dimension tables. The former approach allows faster join operations in reporting. However, it requires some post-processing overhead in report generation. Similarly, the second approach implies a concatenation overhead in OLAP queries. Thus, the former approach achieves performance benefits during query execution at the cost of overhead in report generation. The second approach does the opposite.

This work adapts a third approach in this implementation, where the uploaded key of a dimension is stored along with the transformed key value. This sacrifices linear fraction of storage space but decreases computation time for both OLAP query execution and report generation. In this approach, the transformation program scans each record and appends the *university_key* with all attributes which are either a key or a reference to some dimension table. In this implementation, mappers transform the data immediately after extracting from input file and store the transformed output in an intermediate file of Hadoop file system.

Load Process. The final step in ETL involves loading the transformed data in the multi-tenant schema. The primary requirement of ETL process is to ensure that the data warehouse is in a consistent state at any point of time. A secondary requirement is to conveniently generate a report of erroneous data or inconsistencies observed at this stage. The proposed ETL process discards the entire input file in case of error in data. The generated report enlists the line numbers, and tenant provided unique key values, and the nature of the error.

However, in case of no such error, the load process moves the transformed data to the Hive tables in Hadoop file system (HDFS).

Directly loading the transformed file using Hive queries would lead to faster load operation at the cost of occasional inconsistent or duplicate data. However, these can be discarded easily during report generation without much overhead. Thus, this implementation uses a simple LOAD DATA query to load the intermediate file of transform phase to the corresponding schema in HDFS. It is important to mention that if the HDFS and Hive are residing on the same Hadoop cluster, then this operation would involve merely moving the file from one location in HDFS to another. Hence, it requires only constant time regardless of the input file size.

2.3   OLAP as a Service

In the proposed eDWaaS model, the subscriber initiates the report generation through an online interface provided to them. The subscribers gain access to this interface by authenticating themselves with login identifier and password provided by the DWSP. It is important to mention here that the present eDWaaS model deals with some pre-defined reports only. This study facilitates the subscriber to generate those reports by processing the pre-defined data cubes at backend. The system, in turn, executes the OLAP query on the cube to extract the data required for a report. This mechanism of report generation is termed as OLAP as a service. In the proposed eDWaaS model, the logical separation of data is achieved by defining scope of the OLAP queries. The *university_key* attribute in the fact table ensures the scope to be limited to the data concerned. This study has filtered the data by subscribers' *university_key* available in the authenticated session of the web-based application software. The system then generates the report based on the filtered data only.

Data Cube Generation. The process of data cube generation takes a considerable time in any data warehouse implementation. In order to reduce this time, this work takes an approach where the data are processed to generate the pre-defined cubes offline. These cubes are generated from the fact tables and related dimensions. The system updates the content of the cubes at regular intervals. As a result, after some data is uploaded through ETL service, the system requires a minimum time to make it available in the report.

This study creates the data cube using queries of the form CREATE TABLE ... WITH CUBE. The Hive, in turn, creates a table where redundant data are stored from the fact and related dimensions tables. The data cube takes up more space than the collective space taken up by the related fact and the dimension tables. However, data cube enables the parallel processing of OLAP queries by the Hadoop nodes in less time with respect to the non-cube queries on the dimensions and fact tables. Joining of relevant tables is the major bottleneck of the parallel processing in the case of non-cube queries. In such cases, the non-cube query first fetches the filtered data from these individual tables and then joins them as per requirement. This join operation needs to wait until data from all individual tables have been fetched. However, in OLAP cube, parallel system can filter records by reading them independently. Thus, the query executes faster as there is no waiting time for joining. One of such data cube is presented below.

```
CREATE TABLE cube_student_performance
AS SELECT grouping_id, F.university_key, C.course_code, T.time_code,
  R.regtype_code, AVG(marks) AS avg_marks, AVG(percent_attended) AS avg_per_att
FROM Student-Performance F JOIN Courses C ON C.course_key = F.course_key JOIN
  Times T ON T.time_key = F.time_key JOIN Regtypes R
  ON R.regtype_key = F.regtype_key
GROUP BY F.university_key, C.course_code, T.time_code, R.regtype_code
WITH CUBE;
```

The system adds the `grouping_id` during the offline data cube generation. The attributes of data cube contains `NULL` value in many records. If an attribute has a `NULL` value, the corresponding bit in `grouping_id` is 0; otherwise it is 1. It is important to mention here that the most significant bit is from the reverse order of attributes in the `CREATE` statement. For example, a record in the above cube with `regtype_code` and `course_code` only as `NULL` value, the corresponding `grouping_id` is 0101. Similarly, a record with only `time_code` as `NULL`, the corresponding `grouping_id` is 1011.

OLAP Query. The system executes an OLAP query on the cube to get the final data for reporting. One of such query based on the `cube_student_performance` is presented below.

```
SELECT time_code, regtype_code, avg_marks
FROM cube_student_performance
WHERE(grouping_id = conv("010", 2, 10) OR grouping_id = conv("110", 2, 10))
  AND university_key = 'University1' AND time_code = '2016-17-SPR';
```

In this query, the `university_key` =‘University1’ clause filters all data of `University1` from the data cube. This clause provides the logical abstraction of data in multi-tenant environment. Furthermore, the `time_code` =‘2016-17-SPR’ statement filters all data of 2016-17-SPR. The `conv` is a Hive function that converts a number in a specified base to another. For example, `conv("010", 2, 10)` converts 010 from base 2 to base 10. Therefore, the above OLAP query extracts the pre-computed aggregated values for all possibile `regtype_code` form the OLAP cube.

3   Scalability of eDWaaS

This study proposes a multi-tenant schema as a service model. It also presents the service model of two major operations in a data warehouse: ETL and OLAP. However, a cloud implementation of a data warehouse can not succeed if these operations are not scalable. The size of data warehouse grows with increasing subscription from multiple universities. The service model may fail if the ETL or OLAP process takes considerable amount of time in such situation. Therefore, this work analyses the scalability aspects of the eDWaaS. The scalability analysis is laid out in two broad directions: (i) assessing the scalability of the ETL process and evaluating the benefits over a conventional system, and (ii) analysing the performance for OLAP queries over the conventional data warehouse.

3.1 Scalability of ETL Process

In order to verify the scalability of the designed process, this work have configured the ETL process in two ways. *Case 1*: It allocates only two mappers regardless of the input file size. *Case 2*: Secondly, it allocates number of mappers which is proportional to the size of the input file. Say, S_{ip} is the input file size and S_b is the size of each block required to store files in the HDFS. The S_{min} and S_{max} are configurable parameters for minimum and maximum split size on the input file. The size of the split files (S_{split}) can be calculated using Eq. 1.

$$S_{split} = max(S_{min},\ min(S_{max},\ S_b)) \tag{1}$$

Now, by configuring S_{min} and S_{max} to use a constant value, this work has made the S_{split} independent of S_b. This is achieved by taking the following approach. The S_{min} and S_{max} are two constant parameters. For this experiments, $S_{min} = S_{max} = c$, where c is a constant. Therefore, the Eq. 1 can be re-written as Eq. 2.

$$S_{split} = max(c,\ min(c,\ S_b)) \tag{2}$$

When $(c \leq S_b)$, then $S_{split} = max(c,\ c) = c$. Again, if $(c > S_b)$, then $S_{split} = max(c,\ S_b) = c$. Therefore, S_{split} is constant irrespective of the value of S_b. The number of mappers (n_m) can be calculated using Eq. 3.

$$n_m = \frac{S_{ip}}{S_{split}} \tag{3}$$

Therefore, the number of mappers is directly proportional to the size of input file, if S_{split} is a constant. For *Case 1* of this analysis, this study have set S_{split} equal to the half of the S_{ip}. This ensures the number of mappers to be 2. This work have prepared eleven datasets comprising data of various size like 2 MB, 4 MB, 8 MB, 16 MB, ..., 1 GB, 2 GB. Thereafter, it has performed the proposed ETL process multiple times on these datasets and recorded the time taken to perform this operation. For each of the eleven datasets, the ETL process is performed for 200 times through an automated program. Out of these 200 observations in each case, this study have removed the outliers first by dropping data points beyond the inter-quantile range of $25\%^{ile}$ - $75\%^{ile}$ in each identical input file size. Furthermore, this study has also removed data points that lie outside the inclusive range of $\mu \pm (1.5)\sigma$. Here μ is the mean time taken and σ is the standard deviation for a certain input file. Once the outliers have been removed, this work calculates the mean time for each input file size. The same process is repeated for both *Case 1* and *Case 2*. Figure 4 plots two line graph with the observed data for *Case 1* and *Case 2*. The observed result ensures that the time taken by ETL process increases almost linearly with input file size in

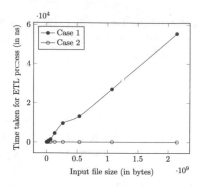

Fig. 4. Comparison of ETL process   **Fig. 5.** Comparison of OLAP query

Case 1. However, in *Case 2*, the time is relatively constant. In *Case 2*, this study has set the number of mappers to be directly proportional to the input file size. The number of mapper increases proportionately with growing file size. It is achieved by fixing S_{split} to a constant value. The processing time indicates that the proposed ETL process is scalable, subject to the fact that there are sufficient number of mapper available in the system.

3.2 Scalability of OLAP Query

In order to check the scalability of the OLAP query process, this study has analysed the proposed OLAP as a service model on the data cube mentioned in Subsect. 2.3. The analysis considers ten data cube with varied data size starting from 7.5 to 42 million approximately. This work has executed the OLAP query over the each such cube 200 times and measured the mean effective time taken to fetch the results. It also records the mean cumulative time taken by all the nodes. The mean cumulative time indicates the effective time where the number of processing nodes is constant. The observed mean effective and cumulative time taken for the OLAP query are presented in Fig. 5.

The plot in Fig. 5 indicates that the cumulative query execution time increases with growing data size in OLAP cube. However, the effective query execution time remains almost constant. Therefore, it can maintain the effective OLAP query execution time constant by increasing number of mappers.

4   Conclusion

The proposed eDWaaS model plays an important role in retrieving aggregate data efficiently from large dataset. In the present context, the proposed eDWaaS model helps the university management to generate report on student and teacher performance, various statistics on placement activity, performance of department and specialisation etc. This study presents a scalable data warehouse as a service model which can cater to the needs of multiple universities.

The proposed model involves a data warehouse service provider which improves the economy of scale. This study has presented a multi-tenant schema to facilitate the service of data aggregation and analysis. In addition to this, it also proposes the scalable service model for uploading data in multi-tenant schema and extracting reports from it. The scalability of the ETL process and the performance of OLAP query have been verified with theoretical justification.

The proposed model uses a pre-defined schema which may not be suitable for all universities. They may need a customized schema that serves other individual requirements. Therefore, further research can be carried out to design a generalized schema which can accommodate variation in data structure. Data security in this multi-tenant schema is also an interesting area which can be explored further.

References

1. Agrawal Sr, M., Joshi Sr, A.S., Velez, A.F.: Best Practices in Data Management for Analytics Projects (2017)
2. Aziz, A.A., Jusoh, J.A., Hassan, H., Idris, W., Rizhan, W.M., Zulkifli, M., Putra, A., Yusof, M., Anuwar, S.: A framework for educational data warehouse (EDW) architecture using business intelligence (BI) technologies. J. Theor. Appl. Inf. Technol. **69**(1), 50–58 (2014)
3. Cuzzocrea, A., Moussa, R.: A cloud-based framework for supporting effective and efficient OLAP in big data environments. In: Proceedings - 14th IEEE/ACM International Symposium on Cluster, Cloud, and Grid Computing, CCGrid 2014, pp. 680–684 (2014)
4. Dell'Aquila, C., Di Tria, F., Lefons, E., Tangorra, F.: An academic data warehouse. In: Proceedings of the 7th Conference on 7th WSEAS International Conference on Applied Informatics and Communications, pp. 229–235 (2007)
5. Di Tria, F., Lefons, E., Tangorra, F.: Academic data warehouse design using a hybrid methodology. Comput. Sci. Inf. Syst. **12**(1), 135–160 (2015). https://doi.org/10.2298/CSIS140325087D
6. Kaur, H., Agrawal, P., Dhiman, A.: Visualizing clouds on different stages of DWH-an introduction to data warehouse as a service. In: 2012 International Conference on Computing Sciences (ICCS), pp. 356–359. IEEE (2012)
7. Khan, A., Ghosh, S.K.: Analysing the impact of poor teaching on student performance. In: 2016 IEEE International Conference on Teaching, Assessment and Learning for Engineering (TALE), IEEE (2016)
8. Kurniawan, Y., Halim, E.: Use data warehouse and data mining to predict student academic performance in schools: A case study (perspective application and benefits). In: Proceedings of 2013 IEEE International Conference on Teaching, Assessment and Learning for Engineering, TALE 2013, pp. 98–103 (2013)
9. Pucciarelli, F., Kaplan, A.: Competition and strategy in higher education: managing complexity and uncertainty. Bus. Horizons **59**(3), 311–320 (2016)
10. Saada, A.I., El Khayat, G.A., Guirguis, S.K.: Cloud computing based ETL technique using warehouse intermediate agents. In: Proceedings - ICCES 2011: 2011 International Conference on Computer Engineering and Systems, pp. 301–306 (2011)

Online Academic Social Networking Sites (ASNSs) Selection Through AHP for Placement of Advertisement of E-Learning Website

Meenu Singh[1($\boxtimes$)], Millie Pant[1], Arshia Kaul[2], and P. C. Jha[2]

[1] Department of Applied Science and Engineering, Indian Institute of Technology (IIT), Saharanpur Campus, Roorkee, India
msingh.dase.iitr@gmail.com, millifpt@iitr.ernet.in
[2] Department of Operational Research, University of Delhi, Delhi, India
arshia.kaul@gmail.com, jhapc@yahoo.com

Abstract. Over the years the use of Social networking sites (SNSs) has grown tremendously. This popularity has led to the incessant need for marketers to concentrate advertising through these websites. The need of the hour is therefore to develop advertising strategies which are effective and efficient. With new SNS being launched each day, the important decision is to determine which SNS is the most appropriate for advertising for a firm. There are many forms and options of SNSs available with different purposes which allow the people to interact with each other such as social connection (facebook, google+, etc.), multimedia sharing (youtube, flickr, etc.), professional (linkedln, classroom 2.0, etc.), hobbies (on my bloom, pinterest, etc.), academic (researchgate, academic.edu, etc.), etc. Due to increase in the types of SNSs, the evaluation and selection of right SNS have become a complex problem for advertisers. In this research, the selection of Academic Social networking sites (ASNSs) is considered as a Multi Attribute Decision Making (MADM) problem for advertising of E-learning website. Analytical Hierarchy Process (AHP) methodology for the selection of ASNSs has been adopted. A real life case study is also presented to show the applicability of the proposed methodology.

Keywords: Social Networking Sites · Selection · AHP

1 Introduction

With the improvement and advancement in technology, Internet has emerged as a new advertising medium which reaches and delivers the advertising messages to the online audience [1]. It provides a plethora of opportunities to a firm as compared to the traditional media and has changed the way firms consider advertising. It includes many ways to advertise such as web advertising, search engine advertising, email advertising, mobile advertising, etc. but in this research we are concentrating on web advertising available on sites through personal computers. The web advertising delivers the advertising messages visually to the online audience on different websites. While there are many different types of websites available but here we are specifically concentrating on

© Springer International Publishing AG, part of Springer Nature 2018
A. Abraham et al. (Eds.): ISDA 2017, AISC 736, pp. 1008–1017, 2018.
https://doi.org/10.1007/978-3-319-76348-4_97

Social Networking site (SNS) as it has gained a lot of popularity over the past decade among all the age groups [2]. Nowadays SNS is the easiest and the fastest way of interacting and communicating with other users. Thus, it can be considered as a powerful advertising tool that not only provides a particular platform to the users to connect with the other users but also encourages online participation, communication and sharing of their product content globally, between consumer-to-consumer and between firm-to-consumer. There are various types of SNSs based on people's interest and hobbies which have introduced many opportunities for advertiser to easily reach their target audience online such as: social connection (facebook, google+, etc.), video sharing (snapchat, youtube, etc.), image sharing (pinterest, instagram, etc.), music sharing (soundcloud, last.fm, etc.), professional (linkedln, classroom 2.0, etc.), informational (Super Green Me, Do-It-Yourself Community, etc.), hobbies (on my bloom, pinterest, etc.), academic (researchgate, academic.edu, etc.), bookmarking (Digg, delicious, etc.), etc. [2–4]. Due to the increasingly competitive business environment, it has become a very important requirement for advertisers to select the most appropriate SNSs for advertising. So, an advertiser should be clear about choosing the right type and right alternatives for its placement of advertisements. But selecting the right SNSs have become a daunting task as these problems have complex and conflicting criteria.

Therefore, this paper discusses the optimal selection of academic SNSs for placement of advertisements for a well-known e-learning website, Coursera, which is a great source of online education. Coursera provides online universal access to many of the top universities and educational institutions that cover a large gamut of courses. Its goal is to attract all academicians from all over the world in order to increase their user registration. Therefore, Academic Social Networking sites (ASNSs) have been chosen for advertising as it will help in covering all types of academicians working in different fields and areas. ASNSs also allows the academicians and researchers to upload and share their research papers, journal articles, conference papers, posters, data and code to an online repository [5, 6]. But the decision of choosing the most appropriate ASNSs for advertising is a challenging task. Many authors have carried out research on selection of websites but to the best of our knowledge there is limited research on the selection of ASNSs. Typically, the selection of ASNSs is a complex decision with multiple conflicting criteria that influence the decision-making process and involves subjectivity. Such problems can be solved by using MADM methods which are well-known for their ability to solve the multi-criteria problems [7]. Each MADM method helps the Decision Makers (DMs) to judge the best among the different selected alternatives. Therefore, such problems can be considered as a Multi Attribute Decision Making (MADM) problem and we have utilised Analytical Hierarchy Process (AHP) methodology for the decision making.

The rest of the paper is organized as follows. In Sect. 2, a literature review on selection of websites as a MADM problem as considered by authors. In Sect. 3, detail of the methodology is proposed. In Sect. 4, a real life case study is presented to show the applicability of the proposed methodology. In Sect. 5, the results and analysis are presented. In Sect. 6, conclusions and future research directions are presented.

2   Literature Review

Selection of Websites is a MADM problem as there are many criteria on which they are based, such as cost criteria, service criteria, reliability, security, response time, design, compatibility navigation, quality etc. Many researchers have carried out research to define the criteria and sub-criteria and have employed different MADM methods as per their problem to select the optimal website. In [8], the best website for online advertising is selected based on five different criteria using Analytic Hierarchy Process(AHP). In [9], authors have developed a new model for selecting the Internet advertising networks by evaluating the relative weights by AHP and then ranking the alternatives by using Grey Relational Analysis (GRA). [10] evaluated the quality of e-commerce websites by Fuzzy Analytical Hierarchy Process (FAHP) for website development. [11] evaluated and selected the website for m-commerce through Fuzzy Set Theory and Technique to Order Preference by Similarity to Ideal Solution (TOPSIS). In [12], selection of best website that offers the similar services or facilities is conducted by AHP and COmplex PRoportional ASsessment of alternatives with Grey relations (COPRAS-G) method to measure the usage of each website. In [13], the selection of social media platforms was performed by integrating the Analytical Network Process (ANP) and COPRAS-G method for the advertising of Trans-Gulf Airlines. In [14], the selection of potential websites is performed by Delphi, AHP and Cuckoo search for promoting the internet based advertising campaigns. [15] has differentiated the gender for comparative evaluation of e-commerce website by the hybrid method of FAHP and TOPSIS. [16] examined the criteria and sub-criteria that affects the consumer's online purchasing decision with the help of AHP. In [17], evaluation and selection of popular e-learning website used in Turkish is determined through Weighted Distance Based Approximation (WDBA). The focus of [18] is to measure the usability of academic websites of higher education by fuzzy AHP and Entropy method. The selection of e-learning website related to C programming language for software development is carried out by using Fuzzy Multi Attribute Decision Making (FMADM) i.e., the weights are evaluated using FAHP and then ranking is performed by COPRAS, Visekriterijumsko Kompromisno Rangiranje (VIKOR) and WDBA [19].

3   Methodology

Analytical Hierarchy Process (AHP) was proposed by Saaty (1977 and 1980) [20]. It is one of the extensively used MADM methods for its ability to integrate both subjective and objective data that helps the DMs to resolve a complex problem having multiple conflicting criteria. It decomposes the problem into a hierarchical structure and easily express one's opinion over the two alternatives and simultaneously on all the alternatives based on pair-wise comparison judgments. Due to its simplicity and flexibility, it is being widely used and accepted by many corporations, organization and country all over the world. Its application can be seen in different areas like resource allocation, planning, selection, ranking, performance measurement, optimization, etc. in the fields of management, engineering, manufacturing, education, medical, health care, technology, social, political, industry, etc. [21, 22]. The various steps needed to be performed in AHP are summarized below:

3.1 Step 1: Establish a Hierarchical Model

The criteria, sub-criteria, and alternatives of the problem are identified to formulate the problem into elements. Then these elements are decomposed into a hierarchical structure of different levels for providing better focus on each criterion and sub-criteria while allocating the weights. It is an important step because the different structure can result in the different ranking of alternatives.

3.2 Step 2: Establish the Pairwise Comparison Matrix and Compute Local Priority

The elements of a particular level are compared pair-wise with the specific element of the immediate level. The comparison is a relative value or a quotient of two quantities measured on the 1–9 ratio scale, as given in Table 1. These values are recorded in a positive reciprocal matrix, called judgmental matrix, denoted by A.

$$A = \begin{bmatrix} 1 & a_{12} & \cdots & a_{1n} \\ a_{21} & \cdots & a_{ij} & \cdots \\ \cdots & a_{ji} = 1//a_{ij} & 1 & \cdots \\ a_{n1} & \cdots & \cdots & 1 \end{bmatrix}, \text{ where } a_{ij} \text{ is the comparison between } i \text{ and } j. \tag{1}$$

Table 1. The 1–9 ratio scale used in AHP.

Intensity of importance	Definition
1	Equal importance
3	Moderate importance
5	Strong importance
7	Very strong importance
9	Extreme importance
2, 4, 6, 8	Intermediate values between two adjacent judgments

The local priority of element of each row is calculated using geometric mean. The geometric mean of rows and columns provides the same ranking.

$$W = \sqrt[n]{\prod_{j=1}^{n} a_{ij}} \Big/ \sum_{i=1}^{n} \sqrt[n]{\prod_{j=1}^{n} a_{ij}}, \quad \forall\, i, j = 1, 2, \ldots.n \tag{2}$$

3.3   Step 3: Checking Consistency of Comparison Matrix

It is important to check the consistency of each judgmental matrix by Consistency Ratio (CR), which is defined as the ratio of the consistency index (CI) and random index (RI) shown in Table 2.

$$CR = CI/RI \tag{3}$$

where $CI = \left(\lambda_{max} - n\right)/(n - 1)$

$$\lambda_{max} = \text{maximal eigen value} = \sum_{j=1}^{n} a_{ij}\left(w_j/w_i\right) \tag{4}$$

n = dimension of the matrix

RI = Random Index is the average consistencies of 500 randomly generated reciprocal matrices.

If the $CR < 0.10$, then the matrix is considered to be consistent. Otherwise, the elements are needed to be revised in the judgmental matrix until they are consistent.

Table 2. Random index values

Matrix rank	1	2	3	4	5	6	7	8	9	10	11	12	13	14	15
RI	0.00	0. 00	0.58	0.90	1.12	1.24	1.32	1.41	1.45	1.49	1.51	1.48	1.56	1.57	1.59

3.4   Step 4: Aggregation of Local Priorities

The local priorities of elements of different level are aggregated to obtain the final priority. This final priority obtained represents the ranking of the alternative with respect to the objective of the problem.

$$Final\, Priority = \sum_i \left[\begin{array}{l} Local\, priority\, of\, criteria\, with\, respect\, to\, goal\, X \\ Local\, priority\, of\, alternative\, with\, respect\, to\, criteria \end{array} \right] \tag{5}$$

4   Problem Statement

We are considering the advertising for an electronic learning (e-learning) website that imparts the education and knowledge online to all the academicians. Many institutions are adopting this new system as most of the academicians wants to gain information online. Coursera itself works with many universities and institutions to offer multiple online courses and programs. It aims to draw attention of online academicians to its website for increasing its user registrations and course subscriptions. For this advertising of the website is required. The question in front of us is that which platform is best suited for advertising the e-learning website. Coursera uses the MCDM methodology for selecting the appropriate ASNSs for its advertising out of the many which are available. ASNSs encompass the academicians to create, upload and share their notes, networks, and information with the other academicians. With a view to advertise online, DM has considered fourteen criteria and summarized them into three main criteria and eleven sub-criteria and has selected four ASNSs as alternatives: Academia.edu (A1), Research-Gate(A2), Slideshare (A3) & LabRoots (A4) for the evaluation process [9, 10, 12–14, 17, 19]. AHP methodology is used for deriving the weights of these alternatives according to the importance of each criterion and modeling the problem into a four-level hierarchy structure as presented in Fig. 1.

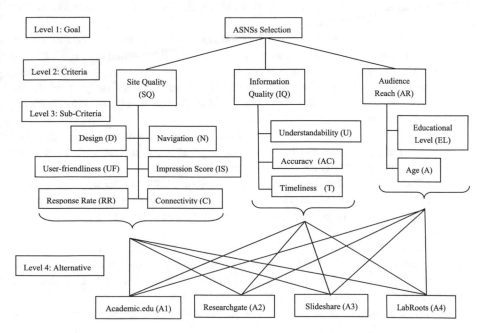

Fig. 1. The four-level AHP hierarchy structure model for ASNSs selection.

4.1 Criteria for Selection of Academic Social Networking Sites (ASNSs)

The three main criteria are site quality, information quality and audience reach. The site quality indicates the functional features of ASNSs in relation to provide convenience to the users, enhance operational performances and reduce browse time. Site quality includes the following features:

Design - It is the appearance of the site which should be attractive and well-organized in order to generate a positive impression so that it can hold the user's sight once they arrive on the site.

Navigation - It provides the directions for accessing the relevant information to the user. An easy navigation system plays a crucial role in reducing search time and increasing the satisfaction level of a user.

Response Rate - It refers to the time taken for loading the information requested by the user. If the site takes too much time to download the information, the user is likely to switch to the other website.

Impression Score - It is measured by counting the number of times an advertisement appeared on the webpage or displayed advertisement is viewed by the user.

User-friendliness - It signifies the simplicity of the site that provides the ease or understandability to the user for operating the site effectively. It enables the users to easily access the site, thus increasing their satisfaction levels.

Connectivity - It broadly describes the social connections of a site with the other available Social Networking sites (SNs) such as Facebook, Twitter, youtube etc. Their link allows the user to make connection worldwide which is likely to increase the number of potential users on the site.

The information refers to the quality of the information present on the site for the users. The quality of information can be measured by different elements but here, we will measure through the following elements:

Understandability - It refers to the way the relevant information (needed by the user) is presented on the site. A complete set of information enables the user to understand the content better.

Accuracy - It refers to the correctness of the information delivered to the user. The accurate content resists the uncertainty coming in the user's mind and leads to increase the reliability of user and the easy acceptance of the site.

Timeliness - It indicates the amount of information present on the site is up to date. A user starts losing interest if the content is static and is not updated. Thus, updating the content regularly not only influences the user to continue the site but also improves its quality.

The audience reach defines how ASNSs delivers the product or service to the potential users based on the advertiser's defined profile. Every advertiser has their specific audience and thus defines their profiles depending on different elements. The advertiser here has defined the target profile based on two elements:

Educational Level - It includes the users who are pursuing higher education or conducting research on any topic.

Age - It includes the users of both the gender above the age of 22 yrs.

5   Results and Analysis

The pairwise comparison was performed by a committee of 5 experts and then the relative weights of each pairwise matrix were aggregated by using Eqs. (2), (3) and (4). The relative weights of criteria w.r.t objective and sub-criteria w.r.t main criteria are presented in Table 3 and for alternatives w.r.t sub-criteria are presented in Table 4.

Table 3. Relative weights of three criteria and eleven sub-criteria

Criteria	Weights of criteria	Sub-criteria	Weights of the sub-criteria
Site quality (SQ)	0.223	Design	0.102
		Navigation	0.158
		Response rate	0.204
		Impression score	0.219
		User-friendliness	0.184
		Connectivity	0.134
Information quality (IQ)	0.481	Understandability	0.312
		Accuracy	0.411
		Timeliness	0.278
Audience reach (AR)	0.296	Educational level	0.561
		Age	0.439

Table 4. Relative weights of all four alternatives

Alternatives	Weights w.r.t. sub-criteria											Weights w.r.t. criteria		
	D	N	RR	IS	UF	C	U	AC	T	EL	A	SQ	IQ	AR
A1	0.104	0.153	0.144	0.309	0.177	0.216	0.258	0.189	0.153	0.280	0.243	0.193	0.201	0.264
A2	0.139	0.301	0.614	0.087	0.203	0.073	0.418	0.396	0.173	0.329	0.217	0.253	0.341	0.280
A3	0.345	0.193	0.302	0.058	0.225	0.293	0.193	0.130	0.187	0.128	0.338	0.220	0.165	0.220
A4	0.412	0.353	0.109	0.546	0.395	0.418	0.130	0.285	0.487	0.263	0.201	0.368	0.293	0.236

From the above Table, it can be seen that information quality is the most important criteria for an e-learning website. In its sub-criteria, 'accuracy' has been given the most priority. This shows that educators are more conscious about the information to be accurate on any e-learning website.

After all the comparisons and weighting process, the overall or global weights of each criterion are evaluated. The results are then presented in Table 5 which shows that Researchgate (A2) is the best ASNS for advertisement with 30.3% weight followed by the LabRoots (A4) as the second suitable ASNS for advertisement with 29.3% weight, i.e., A2 > A4 > A1 > A3. Academic.edu (A1) and Slideshare (A3) have 21.8% and 19.4% weights respectively. This analysis concludes that Coursera wants to advertise on the famous ASNS which has a large number of users with heavy traffic flow. Thus, audience reach is an important concern. Secondly, the information quality is the most influential factor which can create the interest of the user if provided with correct and accurate information. Thirdly, connectivity should be increased by linking other social networking sites in order to attract more number of users. Therefore, Coursera can choose Researchgate (A2) and Labroots (A4) for the placement of advertisements.

Table 5. Final weights of all four alternatives

Alternatives	Weights of alternatives w.r.t. criteria (W1)			Weights of criteria w.r.t. objective (W2)	Final priority (W1 × W2)	Ranking
	SQ	IQ	AR			
A1	0.193	0.201	0.264	0.223	0.218	III
A2	0.253	0.341	0.280	0.481	0.303	I
A3	0.220	0.165	0.220	0.296	0.194	IV
A4	0.368	0.293	0.236		0.293	II

6   Conclusion and Future Directions

Social Networking Sites have gained a lot of attention within the ambit of internet advertising due to its two-way communication and mass coverage feature. With the increase in competition it has become very important for advertisers to select the most appropriate SNSs for advertising as these problems have complex and conflicting criteria. So, advertisers need to be more cautious while selecting the right SNSs for the placements of advertisements. Thus, in this paper a real world problem has been introduced where an optimal selection of ASNSs has been made for the placement of advertisement for a well-known e-learning website, Coursera. The ranking of the alternatives is performed by employing AHP, which is one of the widely used MADM methods for measuring both quantitative and qualitative data. The result obtained from this study is practical for comparing the criteria and for ranking the alternatives. This study can be extended to find the interrelationship between the criteria and sub-criteria and the results of this present study can be compared with the other MADM methods or a hybrid approach can be developed to solve this problem.

References

1. Winer, R.S.: New communications approaches in marketing: issues and research directions. J. Interact. Mark. **23**(2), 108–117 (2009)
2. Mangold, W.G., Faulds, D.J.: Social media: the new hybrid element of the promotion mix. Bus. Horiz. **52**(4), 357–365 (2009)
3. Jayaprakash, M., Baranidharan, K.: Advertising in social media: an overview. Int. J. Logist. Supply Chain Manag. Perspect. **2**(4), 452 (2013)
4. THE World University Ranking, 10 October 2017. https://www.timeshighereducation.com/a-z-social-media
5. Madhusudhan, M.: Use of social networking sites by research scholars of the University of Delhi: a study. Int. Inf. Libr. Rev. **44**(2), 100–113 (2012)
6. Shanks, J., Arlitsch, K.: Making sense of researcher services. J. Libr. Adm. **56**(3), 295–316 (2016)
7. Tzeng, G.H., Huang, J.J.: Multiple attribute decision making: methods and applications. CRC Press, Boca Raton (2011)

8. Ngai, E.W.T.: Selection of web sites for online advertising using the AHP. Inf. Manag. **40**(4), 233–242 (2003)
9. Lin, C.T., Hsu, P.F.: Selection of internet advertising networks using an analytic hierarchy process and grey relational analysis. Int. J. Inf. Manag. Sci. **14**(2), 1–16 (2003)
10. Liu, Y.W., Kwon, Y.J.: A fuzzy AHP approach to evaluating e-commerce websites. In: 5th ACIS International Conference on Software Engineering Research, Management & Applications, SERA 2007, pp. 114–124. IEEE (2007)
11. Zhang, H.M.: Mobile commerce website selection base on fuzzy set theory and TOPSIS. In: 15th Annual Conference Proceedings of International Conference on Management Science and Engineering, ICMSE 2008, pp. 72–77. IEEE (2008)
12. Madhuri, B.C., Chandulal, A.J., Padmaja, M.: Selection of best web site by applying COPRAS-G method. Int. J. Comput. Sci. Inf. Technol. **1**(2), 138–146 (2010)
13. Tavana, M., Momeni, E., Rezaeiniya, N., Mirhedayatian, S.M., Rezaeiniya, H.: A novel hybrid social media platform selection model using fuzzy ANP and COPRAS-G. Expert Syst. Appl. **40**(14), 5694–5702 (2013)
14. Kar, A.K.: A decision support system for website selection for internet based advertising and promotions. In: Emerging Trends in Computing and Communication, pp. 453–457. Springer, New Delhi (2014)
15. Anand, O., Srivastava, P.R.: A comparative gender based evaluation of e-commerce website: a hybrid MCDM approach. In: 2015 Eighth International Conference on Contemporary Computing (IC3), pp. 279–284. IEEE (2015)
16. Singh, D.K., Kumar, A., Dash, M.K.: Using analytic hierarchy process to develop hierarchy structural model of consumer decision making in digital market. Asian Acad. Manag. J. **21**(1), 111–136 (2016)
17. Jain, D., Garg, R., Bansal, A., Saini, K.K.: Selection and ranking of E-learning websites using weighted distance-based approximation. J. Comput. Educ. **3**(2), 193–207 (2016)
18. Nagpal, R., Mehrotra, D., Bhatia, P.K.: Usability evaluation of website using combined weighted method: fuzzy AHP and entropy approach. Int. J. Syst. Assur. Eng. Manag. **7**(4), 408–417 (2016)
19. Garg, R., Jain, D.: Fuzzy multi-attribute decision making evaluation of e-learning websites using FAHP, COPRAS, VIKOR, WDBA. Decis. Sci. Lett. **6**(4), 351–364 (2017)
20. Saaty, T.L.: A scaling method for priorities in hierarchical structures. J. Math. Psychol. **15**(3), 234–281 (1977)
21. Vaidya, O.S., Kumar, S.: Analytic hierarchy process: an overview of applications. Eur. J. Oper. Res. **169**(1), 1–29 (2006)
22. Vargas, L.G.: An overview of the analytic hierarchy process and its applications. Eur. J. Oper. Res. **48**(1), 2–8 (1990)

Fingerprint Based Gender Identification Using Digital Image Processing and Artificial Neural Network

Mahendra Kanojia[1]([✉]), Niketa Gandhi[2], Leisa J. Armstrong[3], and Chetna Suthar[4]

[1] JJT University, Jhunjhunu, Rajasthan, India
kgkmahendra@gmail.com
[2] Machine Intelligence Research Labs (MIR Labs), Auburn, Washington, USA
niketa@gmail.com
[3] Edith Cowan University, Perth, WA, Australia
l.armstrong@ecu.edu.au
[4] Sathaye College, University of Mumbai, Mumbai, Maharashtra, India
chetnasuthar30@gmail.com

Abstract. Every person has a unique fingerprint which can be used for identification. Fingerprints are widely used for forensic cases in criminal investigations. It would be useful to be able to distinguish fingerprints samples based on gender to reduce number of persons of interest in a criminal investigation. This paper discusses a system implemented for identification of gender based on fingerprints. Digital Image Processing and Artificial Neural Network (ANN) techniques were used to implement the gender identification system. Various preprocessing techniques such as cropping, resizing and thresholding were carried out on each image. Feature extraction was carried on each pre-processed image using Discrete Wavelet Transform (DWT) at 6 levels of decomposition. The extracted features were used for implementing ANN based on Back Propagation algorithm. Fingerprint images were sourced from publicly available online datasets. A data set of 200 images of left thumbprint, which included 100 male and female fingerprint images. A training set of 100 images was used for testing purpose, which included 50 male and female images respectively. The accuracy achieved for identifying fingerprints was found to be 78% and 82% for male and female samples respectively.

Keywords: Artificial Neural Network · ANN · Back propagation
Discrete wavelet transform · Fingerprint

1 Introduction

Every time a person touches a surface with their fingers such as handling a coffee cup or a car door handle, a computer keyboard, mobile phone or pen they leave a unique signature which is referred to as 'fingerprint'. Every person in the world has unique fingerprints. Even identical twins with the same deoxyribonucleic acid (DNA) are known to differ in fingerprint patterns. This uniqueness in fingerprint patterns has been

© Springer International Publishing AG, part of Springer Nature 2018
A. Abraham et al. (Eds.): ISDA 2017, AISC 736, pp. 1018–1027, 2018.
https://doi.org/10.1007/978-3-319-76348-4_98

used for biometric security, mass disaster identification and to identify persons of interest at crime scenes. The classification of fingerprint samples based on the gender of the individual will assist in reducing the time to eliminate suspects in criminal cases.

Images taken from fingerprints have been found to display a number of unique features. These include patterns of ridges and valleys of lines, various shapes, which are classified as loops, whorls and arches. Figure 1 provides images of selected finger prints which show different patterns of loops, whorls and arches. Various combinations of these features result in a unique fingerprint pattern for each finger. Studies have shown that male and female fingerprints have different characteristics. Studies that have compared male and female fingerprints have found that female fingerprints have more loops. The frequency of whorls is more prevalent in males compared to females [1]. Other studies have reported that female fingerprints normally have greater ridge thickness and valley thickness ratios compared to male fingerprints [2].

Fig. 1. Sample of fingerprint images showing A: Loops B: Whorls C: Arches [3]

The aim of this paper is to examine whether various classification and neural network techniques can be used to accurately distinguish the gender of fingerprint images. The paper reports on a study which used freely available fingerprint images from male and females. The aim of the study was to develop a method that could extract features from fingerprint images using frequency domain analysis approach. The images were collated and preprocessed and features extracted based on cluster based image thresholding and DWT techniques. A back propagation Neural Network was also used to improve the accuracy of classification.

2   Related Work

A study by [1] investigated the relationship between gender, blood groups and fingerprint patterns. A sample of 200 images was analyzed and common patterns identified based on gender and blood groups. This study found that loops were a more common pattern in images compared to arches. Loops were identified in samples with blood group A, B, AB and O in both Rh positive and negative whereas whirls were more common in O negative. The study concluded that there is an association between distribution of fingerprint patterns, blood group and gender and thus prediction of gender and blood group of a person is possible based on the fingerprint pattern.

Another similar study by [4] proposed a method of classifying gender based on fingerprint images. Preprocessing techniques such as resizing, filtering and thresholding are used on all the images. After this Block Based Discrete Cosine Transform (BBDCT) was used, in which all the images were divided into specific block and coefficient feature is collected from each block. This feature set was used for classification based on K-Nearest algorithm.

Frequency domain analysis for identifying gender using fingerprint has been reported in another [5]. This study used 220 images collected from different college students using biometric fingerprint sensor. The sample data set consisted of equal number of male and female fingerprints. In the proposed paper three different frequency domain transformation were used i.e. Fast Fourier Transform (FFT), Discrete Cosine Transform (DCT) and Power Spectral Density (PSD). A threshold value was set for each transformation in order to classify each fingerprint based on gender. A Rule was developed that tested if the fundamental frequency was greater than the threshold then the sample was deemed to be female. The accuracy of this system was found to be 90% for female and 70.09% for male.

Another similar study by [6] where frequency and spatial domain image processing techniques were used to analyze finger print images for gender classification. This study combined features such as FFT, eccentricity and major axis length to perform gender classification. The data set consisted of 1000 images (450 males and 550 females) of left thumb. The proposed system was reported to have an accuracy of up to 80% for male and 78% for female.

Another study by [7] also proposed gender identification based on fingerprint using frequency domain analysis. The study reported on the use of 2D DWT and Principal Component Analysis (PCA) for feature extraction and the maximum distance method used for classification of a dataset of 400 fingerprint images. The study reported that an accuracy of 70% was achieved.

Differences in epidermal ridge density features of different male and female fingerprint images have also been used to distinguish genders from different community populations. Research by [8] on dataset of Indian Sikh Jat and Banian communities found variations between the different community groups in relation to gender differences in fingerprint patterns. This study found that 93% Sikh Jat females had a higher mean ridge density, compared to 76% of males. The study also found that samples from the Banian community were different in that 100% of females and 80% of males had a greater mean ridge density. This provided interesting insight into the variations between communities and gender of two similar ethnic populations.

Inconsistent results have been reported in terms of statistical significance in the classification of fingerprint images based on gender. The accuracy of the determination of gender for features such as mean ridge height is considered to not be satisfactory. The application of other techniques such as Bayesian networks has been used to determine that ridge density <13 ridges/25 mm^2 is most likely to be of male origin and those having ridge count >14 ridges/25 mm^2 are most likely to be of female origin.

Another study by [9] proposed a method of gender identification based on fingerprints using DWT and Singular Value Decomposition (SVD). Bands of DWT were calculated and combined with the non-zero singular values which were obtained by SVD of fingerprint images. The study classified 3750 (1980 male and 1590 female) fingerprint images using K Nearest Neighbor (KNN) Classification. The overall accuracy of the system was found to be 88.28%.

Another study by [10] classified an image dataset consisting of 550 images (275 female and 275 male) to identify the gender. Two different methods were used for the implementation of the system including DWT based on 5 level of decomposition and a back propagation based ANN. This study reported an accuracy of more that 91%.

3 Proposed System

This research implemented a system for the identification of gender based on finger-prints. Digital image processing and ANN techniques were used to implement the gender identification system. The extracted features were used for implementing ANN based on Back Propagation algorithm.

This section outlines the research methods used for this study including a description of processes for the acquisition and collation of images, feature extraction and transformation of the images. This section also describes the neural network methods which were used on the fingerprint image dataset. The study was carried out in a number of phases which included dataset acquisition, image preprocessing, image transformation, feature selection, feature   extraction and classification. Figure 2 summarizes the flow of gender identification using image processing and ANN.

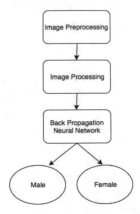

Fig. 2. Flow graph for gender identification using image processing and ANN

3.1   Image Acquisition and Preprocessing

The fingerprint images were sourced from freely available dataset provided by the University of Ilorin, Nigeria [3]. The data set was of 200 images of left thumbprint, which included 100 male and female fingerprint images each. A training set of 100 images was used for testing purpose, which included 50 male and female images each. The background enhancement, cropping, thresholding and image resizing was done on the images collected using image preprocessing techniques. All the fingerprint images are resized to 512 × 512 pixels and converted to binary image from gray scale image using Ostu's Method.

The input RGB image was first converted to the gray level representation. The input gray scale image was then converted into binary image. The binary mask is obtained by thresholding the gray-scale image with an automatically estimated threshold using Otsu's method [11]. This method is used to automatically perform clustering-based image thresholding. The algorithm used for this method assumes that the image contains two classes of pixels following bi-modal histogram (foreground pixels and background pixels), it then calculates the optimum threshold which separates the two classes so that their combined spread is minimal and their inter-class variance is maximal. It is one-dimensional Fisher's Linear Discriminant

Ostu's Thresholding method reduces the intra class variance.

$$V_w(Th) = P_0(Th)V_0(Th) + P_1(Th)V_1(Th) \tag{1}$$

The probabilities of the two classes are weights P_0 and P_1, separated by a threshold Th and V_0 and V_1 are variances of these two classes. The class probability $P_0(Th)$ and $P_1(Th)$ are computed from the histograms of an image (L):

$$P_0(Th) = \sum_{i=0}^{T-1} p(i) \tag{2}$$

$$P_1(Th) = \sum_{i=T}^{L-1} p(i) \tag{3}$$

3.2   Image Processing

After pre-processing DWT was used on all the fingerprint images up to six level of decomposition. Since most of the energy concentrated on low frequency, the sub band LL was further decomposed. Six levels of decomposition and 19 sub bands were obtained. Energy of each sub band is calculated and this feature set was later used for training the neural network.

Feature extraction was carried on each preprocessed image using DWT at 6 levels of decomposition. The fingerprint images were transformed to frequency domain using DWT. Approximation and details components are extracted from the frequency domain images. Approximations are high scale and low frequency components of image. Details are low scale and high frequency components of image. Multi-resolution representation of the finger print image is produced with minimum coefficients and losing important image information.

The two-dimensional wavelet image results in low-low, low-high, high-low, and high-high sub-band images. [12].

Different image properties are represented by each sub bands. Low-low sub band represents concentrated energy in the image and the low-low sub band decomposition is known as dyadic decomposition. (3 * x) + 1 sub bands are created for x levels in DW. Sub band energy feature vector E_n is calculated using Eq. 4.

The formula for calculating energy of each sub band is as follows:

$$E_n = \frac{1}{AB} \sum_{i=1}^{A} \sum_{j=1}^{B} |a_n(i,j)| \tag{4}$$

Where $a_n(i, j)$ is the pixel value of nth sub-band; A is width and B is height of the sub-band. Sub band energy feature vector E_n for all the images are stored in csv format.

3.3 Back Propagation Neural Network

An ANN are based on the structure of biological nervous systems in order to solve a specific problem. It processes the input information and provides relevant output. Each ANN is organized based on various layers. Each layer has nodes with signal activation function [13]. Features vectors are provided to the network through input layer, which are computed in one or more intermediate hidden layers. In a network each connection between nodes have associated weight. The Levenberg-Marquardt Back Propagation algorithm (LMA) was used for training ANN.

Non-linear least squares curve problems can be solved using LMA. It is also known as the damped least-squares (DLS) method [10]. The LMA finds local minima whereas it's not necessarily being global minima. The LMA interpolates between the Gauss–Newton algorithm (GNA) and the Gradient descent which makes LMA more reliable than GNA. LMA can find the solution though it may start far from the final minima (Figs. 3 and 4).

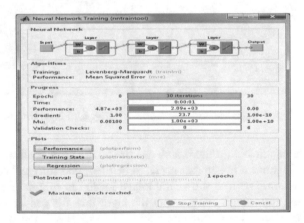

Fig. 3. Neural Network set up after applying Levenberg–Marquardt algorithm

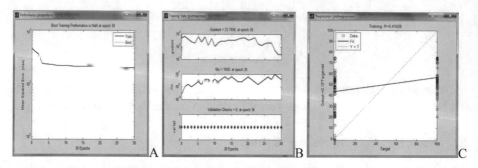

Fig. 4. Result of Neural Network set up A: Performance B: Training state C: Regression

4 Results

A simple Matlab Graphical User Interface (GUI) was created for the purpose of testing the gender identification system. The steps followed for gender identification are described below and shown in the series of Figs. 5, 6, 7, 8, 9 and 10.

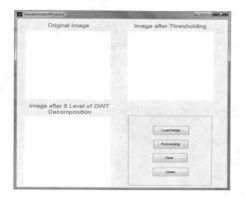

Fig. 5. GUI of the application

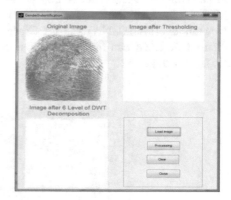

Fig. 6. Fingerprint image loaded

Step 1. Load the application. Figure 5 shows the GUI of the application.

Step 2. Load a fingerprint image that needs to be identified for gender using the load image button. Figure 6 shows the image loaded in original image axes.

Step 3. Click on processing button to run the algorithm for gender identification. It will decompose fingerprint image in 6 level of DWT and provide each sub band image to neural network for identification. Figure 7 shows image after thresholding, Figs. 8 and 9 shows image after 1 level of DWT and 6 level of DWT decomposition respectively.

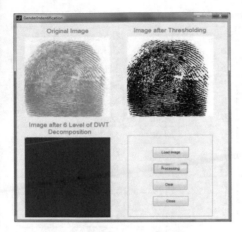

Fig. 7. Image after thresholding

Fig. 8. Image after 1 level of DWT

Fig. 9. Image after 6 level of DWT

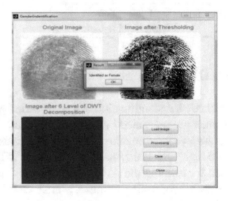

Fig. 10. Result after gender identification

Step 4. Once the processing is completed the result will be shown whether the image considered for identification was of male or female. Figure 10 shows the result on the sample image selected as female.

Table 1 provides results of the accuracy of identification of gender for the fingerprint testing dataset. The system tested 50 fingerprint images. The system correctly identified female fingerprints for 39 out of 50 images (78%) and 41 out of 50 images (82%) for male fingerprints. Overall performance of the system is found to be 80%.

Table 1. Results obtained by the system

Image no.	Gender	System identified output as	Image no.	Gender	System identified output as
1	Female	Female	16	Male	Male
2	Female	Female	17	Male	Male
3	Female	Female	18	Male	Male
4	Female	Female	19	Male	Male
5	Female	Female	20	Male	Male
6	Female	Female	21	Male	Female
7	Female	Female	22	Male	Female
8	Female	Male	23	Male	Male
9	Female	Male	24	Male	Female
10	Female	Female	25	Male	Male
11	Female	Female	26	Male	Male
12	Female	Female	27	Male	Female
13	Female	Female	28	Male	Female
14	Female	Female	29	Male	Male
15	Female	Female	30	Male	Male

5   Discussion and Conclusion

Methods for identification of gender based on fingerprint images collected by biometric scanner are presented in this paper. Instead of using ridge and valley ratio, ridge density frequency domain analysis is used. A dataset of 200 fingerprint images of left thumb (100 male and 100 female) undergoes DWT for feature extraction. These features were used for training the ANN. A total of 100 fingerprint images (50 male and 50 female) were used for testing purpose. Matlab GUI interface and database were created for male and female fingerprint images. A set of features were extracted from DWT sub band images. The method was based on processing of a binary image obtained by thresholding and DWT. The method was found to provide satisfactory results (80%).

The designed GUI system was used to carry out pre-processing, DWT feature extraction, manipulating with files, and display and saving of analysis results. Some improvements could be made in the GUI for further enhancement of using semi-automatic methods and the use of mouse wheel for zooming of images.

This study has shown the potential for using different fingerprint features to discriminate between male and female fingerprints. This method can be used for forensic anthropology to reduce the criminal suspects list and can limit gender of the person of interest.

References

1. Rastogi, P., Pillai, K.R.: A study of fingerprints in relation to gender and blood group. J. Indian Acad. Forensic Med. **32**(1), 11–14 (2010)
2. Omidiora, E.O., Ojo, O., Yekini, N.A., Tubi, T.O.: Analysis, design and implementation of human fingerprint patterns system towards age & gender determination, ridge thickness to valley thickness ratio (RTVTR) & ridge count on gender detection. Int. J. Adv. Res. Artif. Intell. **1**(2), 57–63 (2012)
3. University of Ilorin, Nigeria Archives. http://www.unilorin.edu.ng/step-b/biometrics. Accessed 21 Oct 2016
4. Anjikar, A., Tarare, S., Goswami, M.M.: Fingerprint based gender classification using block-based DCT. Int. J. Innov. Res. Comput. Commun. Eng. **3**(3), 1611–1618 (2015)
5. Kaur, R., Mazumdar, S.G.: Fingerprint based gender identification using frequency domain analysis. Int. J. Adv. Eng. Technol. **3**(1), 295–299 (2012)
6. Gornale, S.S., Geetha, C.D., Kruthi, R.: Analysis of fingerprint image for gender classification using spatial and frequency domain analysis. Am. Int. J. Res. Sci. Technol. Eng. Math. **1**(1), 24–50 (2013)
7. Tom, R.J., Arulkumaran, T.: Fingerprint based gender classification using 2D discrete wavelet transforms and principal component analysis. Int. J. Eng. Trends Technol. **4**(2), 199–203 (2013)
8. Kaur, R., Garg, R.K.: Determination of gender differences from fingerprint ridge density in two Northern Indian populations. Probl. Forensic Sci. **85**, 5–10 (2011)
9. Gnanasivam, P., Muttan, S.: Fingerprint gender classification using wavelet transform and singular value decomposition. Eur. J. Sci. Res. **59**(2), 191–199 (2011)
10. Gupta, S., Rao, A.P.: Fingerprint based gender classification using discrete wavelet transform & artificial neural network. Int. J. Comput. Sci. Mob. Comput. **3**(4), 1289–1296 (2014)
11. Liu, D., Yu, J.: Otsu method and K-means. In: Ninth International Conference on Hybrid Intelligent Systems, vol. 1, pp. 344–349 (2009)
12. Burrus, C.S., Gopinath, A.R., Guo, H.: Introduction to Wavelets and Wavelet Transforms: A Primer. Prentice Hall, Upper Saddle River (1997)
13. Leshno, M., Lin, V.Y., Pinkus, A., Schocken, S.: Multilayer feedforward networks with a nonpolynomial activation function can approximate any function. Neural Netw. **6**(6), 861–867 (1993)

Indian Mobile Agricultural Services Using Big Data and Internet of Things (IoT)

Pallavi Chatuphale[1(✉)] and Leisa Armstrong[1,2]

[1] Department of Computer Science,
University of Mumbai, Mumbai, Maharashtra, India
chatuphale2003@yahoo.co.in, l.armstrong@ecu.edu.au
[2] School of Science, Edith Cowan University, Perth, Australia

Abstract. Mobile services are now widely available to Indian farmers and other agriculture stakeholders. These services are providing a range of functionality from access to market prices and climate services. This paper presents an overview of various services available as mobile apps which are using Big Data and IoT technologies. This paper provides findings from a survey of mobile application services which are available to India agricultural sector stakeholders. The mobile applications where categorized in various service types and assessments made of the type of functionality provided and proportion of apps providing each service category. The paper also addresses the strengths and weaknesses of these services and draw conclusion sand the future trends in the availability of these mobile app services.

Keywords: Agriculture · Big Data · Internet of Things (IoT)
Mobile applications

1 Introduction

Farmers face many risks from climate change, economic and market forces. Rises is temperature and increasing erratic rainfall can have detrimental effects on crop production [1]. Famers have to adapt to these changing environments. Information and communication technology (ICT) through various mediums like radio, TV, online newspapers can provide valuable knowledge to farmers to assist them with decision making. Various agricultural advisories are available which gives suggestions at each stage of crop life cycle for betterment of Agricultural field and farmers [2]. These websites and portals work on changing patterns of weather conditions and accordingly broadcasting advices.

Technologies like Internet of Things (IoT), big data, cloud computing, android application development are helping to deliver supporting messages at real time as well as independently providing guidelines from farm management to delivery of agricultural produce to market or at the door of customer [3].

This paper provides findings from a survey of mobile application services, which are available to India agricultural sector stakeholders. The survey examined services, which has used big data sources or IoT. The mobile applications were categorized in various service types and assessments made of the type of functionality provided and

© Springer International Publishing AG, part of Springer Nature 2018
A. Abraham et al. (Eds.): ISDA 2017, AISC 736, pp. 1028–1037, 2018.
https://doi.org/10.1007/978-3-319-76348-4_99

proportion of apps providing each service category. The paper will also establish the strengths and weaknesses of these services and draw conclusion to the future trends in the availability of these mobile app services.

2 Review of Literature

Big data is term that refers to large databases, which are very complex in nature. The data cloud is of three types - public, private and hybrid. To extract meaningful data from cloud, various cloud computing techniques need to be applied [4]. The Internet of Things (IoT) is the concept of connecting to services through the Internet without the need for human intervention. The concept of "Things" refers to real world events data, properties or sensor data directly connected to Internet [5]. Other technologies such as GPS and sensor data and RFID technologies are also used to improve these services by identifying locations of animals and precision agricultural services. IoT is opening new horizons of applications development [6].

Big Data has a new generation demand. Big data provides the ways for processing information, storage of data in volumes, in real time make it available as well as different kinds of data sources including unstructured, structured, semi-structured. It can be used with all types of data such as machine and sensor data, geospatial, image, application, transaction, including human generated data [7]. Some of the advantages of big data are that it provides veracity, value, variety and velocity. Veracity treats uncertain data values, its source and accordingly inform to the user. Variety concerns with various kinds of data. Velocity is the uncommon speed of data generation and receiving. These characteristics attract the developers to create handy applications [8].

IoT services have been found to have some barriers to its uptake. These include issues of limited storage, security, performance, privacy and reliability. Cloud services can be used to overcome some of these issues because it has unlimited storage capacity, processing power [9]. The merging of both technologies gives us new views of future generation software applications. Various examples exist of cloud services for smart and mobile applications including SaaS, SEssS, DBaaS [10].

Changing climatic and soil factors have great influence agricultural production [11]. Big data and IoT with cloud computing techniques can assist the farmers to face such challenging weather conditions by assisting with decision making about agronomic, fertilizer and pest management decisions. Various android applications have been released in the Indian market through government, private sector and NGOs [12]. The following section provides a survey of mobile applications currently available for the Indian agricultural sector.

3 Survey of Mobile Applications

Mobile applications are developed as per the need of local farmer or the business objective of software developing company or market demand [13]. A survey of 40 mobile applications services available in India was carried out to assess the range of services during the period from July to October 2017. These services were categorized in to the following categories Weather forecast, Market prices/transactions, Kisan call center, Govt. schemes, Soil, Agricultural news, Crop life cycle, Insurance, Agro advisory, Technology sharing, Location based, Audio-Video learning, Multilingual, Organic farming and Special Services. Mobile applications have been divided into following categories.

3.1   Weather Forecast

Some Mobile app users provide services for farmers to receive weather forecast and other climatic details. Example services included KisanSuvidha [14], Kisan Market [15], MyAgriGuru [16], IFFCO Kisan [17], Agricultural Rural Technology [18] and RainbowAgri [19].

3.2   Market Prices/Transaction

Some applications are providing the rates of agricultural goods in local districts as well as over all country rates. These services provide the farmer with this information to assist them to make the best decisions about buying agricultural produce. Such applications act as an online advertisement for farmers, traders, experts and agents. Some applications provide single window service networked with APMC mandis. These applications also provide details about stock availability. Example services included SmartAgri [20], Kisan Market [15], AGRISCIENCE [21], eNAM [22], Indian AgroPrice [23], Agri.Live [24], Nafed [25], Agribuzz-AgriMarket [26], RainbowAgri [19] and Digital Green [27].

3.3   Kisan Call Center

The Kissan call center uses SMS services to provide information related to crop disease and market prices. Example services included KisanSuvidha [14], Gramseva: Kisan (Mandi Prices) [28], KisanMandi [29] and KISAAN USSD [30].

3.4   Govt Schemes

Various Government schemes are delivered through apps like PradhanMantri KrishiSinchaiYojana, Soil Health Cards, related to organic farming. Examples services included MyAgriGuru [16], PusaKrishi [31] and RML Farmer-KrishiMitr [32].

3.5   Soil

Farmers can receive information about soil, land, water, irrigation and rain fed areas through Soil Health Cards provided by government services. Some examples of these services include soil testing services [33] and soil health cards with Mahadhan program [34].

3.6   Agri News

All agricultural news related to government policies, mandi, latest technology, sale/buy product is available via the KRISHI SAMACHAR, online news service. Example services included RML Farmer – KrishiMitr [32], AGRISCIENCE [21], MyAgriGuru [16], AgriApp [35], Agro India [36].

3.7   Crop Life Cycle

Some applications provide information about crop cycle stages using online and real time technologies through SMS services. The services provide information from seed sowing period to delivery of farm produce. Various requirements of crop at different stages like, fertilizer dose, insecticides and pesticides information is made available on request with video demos. Example services included AGRISCIENCE [21], MyAgriGuru [16], KISAN NETWORK [37], Farm-o-pedia [38], AgriApp [35], Agricultural Rural Technology [18], Insecticides [39] and Mahadhan [34].

3.8   Insurance

Crop Insurance apps helps to calculate insurance premiums based on specific regions. The ISRO has developed an app to collect hailstorm data which will help in calculating crop insurance claim. Example services included Crop Insurance [40] and Bhuvan Hailstorm [41].

3.9   Agro Advisory

Advisories are available to give query based solutions related with plant protection, Govt. schemes and policies, agricultural services, selection of machines, crop specific advisory for various agro climatic zones, insecticide detailed information, expert and scientific advices to improve quality and quantity of farm produce. Farm management advisory is provided by Cropin [42]. Example services included KisanSuvidha [14], mKisan Application [43], KISSAN USSD [30], Digital Mandi Application for Indian Kisan [44], IFFCO Kisan [17], RML Farmer – KrishiMitr [32], Agricultural Rural Technology [18], Insecticides [39], Cropin [42], eKrishi [45], SmartAgri [20] and Agri.Live [24].

3.10   Technology Sharing

The service eNAM [22] is a single window service connected with APMC mandis. Registered users can access web portal information directly. Another service PusaKrishi [31] provides license of software to public and private sectors. Example services included KrishiGyan [46] and the Sikkim Food Security and Agriculture Department [47].

3.11   Location Based

Applications like Insurance, Hailstorm require real time field data is provided through GPS technology to the server for further processing. Market related applications require location details to search for nearest market place for business or for sending text or voice alert messages of nearest mandi to the registered users. Farmers' land records of Gujarat and Madhya Pradesh are available on 7/12 Ultra. Example services included Crop Insurance [40], Agri Market [48], DigitalMandi Application for Indian Kisan [44], Bhuvan Hailstorm [11], AgroBuzz-AgriMarket [26], RainbowAgri [19], Mandi Trades [49], Jayalaxmi Agro Tech [50], Kissan Market [15], AGRISCIENCE [21], MyAgriGuru [16] and Nafed [25].

3.12   Audio-Video Learning

Free to air Television services such as KISAN SAFAL GATHA on AGRISCIENCE [21] provides contact details and photos and inspirational stories of successful farmers. AGRISCIENCE TV shows informative short videos. The services promote organic farming and provide detail informative videos. Audio-video visuals services provide illiterate farmers an opportunity to receive proper knowledge about their queries. Example services included Jayalaxmi Agro Tech [50], SmartAgri [20], Organic farming [51], AgriApp [35], KrishiGyan [46], Digital Mandi Application for Indian Kisan [44] and Gramseva: Kisan (Mandi Prices) [28].

3.13   Multilingual

Most of the android applications provide information in both English and local language state-wise. Some services also provide in Hindi language. Example services included Farm-o-pedia [38], KhetiBadi [52], RainbowAgri [19], Jayalaxmi Agro Tech [50], Kissan Market [15], Indian AgroPrice [23], Agri.Live [24] and AGRISCIENCE [21].

3.14   Organic Farming

Some services are providing information on organic farming for the effective culti-vation of vegetables and medicinal plants. Examples of applications that provide information about organic farming including government schemes and policies advi-sory included MyAgriGuru [16], KrishiApp [53], Organic farming [51], eKrishi [45] and KhetiBadi [52].

3.15   Special Services

Applications, like Cropin provides additional features ranking of registered farmers. Other platforms are providing information on cattle management and harvesting storage methods. These services included Farm-o-pedia [38], AgriApp [35] and Cropin [42].

3.16   Analysis of Services

A summary of the fifteen agricultural services types of mobile apps from the 40 example android applications surveyed is shown in Fig. 1. The greatest percentage of services was for Market prices transactions, Crop Lifecycle, Ago advisory and Agricultural news. Services such as natural resources affecting by climate change, insurance claim, weather forecast and soil related applications were lower in percentage of services.

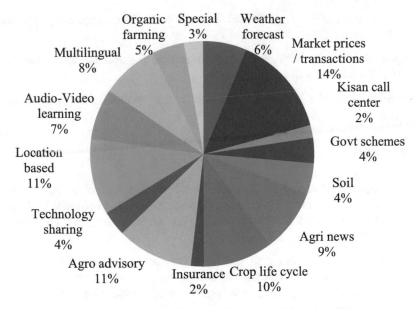

Fig. 1. Indian agricultural services (%) as android based mobile apps.

4   Weaknesses and Strengths of Big Data and IoT for Agricultural Mobile Apps

A number of weakness and strengths were identified for each of the surveyed mobile application services. These are listed below.

4.1   Weaknesses

Technology Ignorance. The majority cultivators are small farm holders. However, these farmers are not aware of this progressive technology or how to use it for their farming. Few farmers or agricultural stakeholders have an understanding of what Bigdata and IoT technologies. This understanding is limited to application developers and academicians [54].

Data Security. Emergence in the cloud computing technology increases need of analytics for data generated from IoT and Big data. The data collections from identical devices need to be separated for security reasons [55].

Data Access Permissions. In most of the applications, restricted access is given to data. This creates a divide between the needs of the user and information which is publicly available [55].

Device Privacy and Regulations. Integrating devices like sensors and phone or smart TV requires backend support for processing transaction by the third party. In such cases regulatory and legal information has to be focus for privacy law and protection.

Integrated devices required to share data. In many cases device regulatory environment also needs to be set [56].

Development Technical skill, market readiness and policy fixation are the requirements of IoT application development [36].

Device Standards. For networking, proper device configurations should have to incorporate in design [56].

4.2   Strengths

Data sharing. Cloud can share data received form IoT and collected in terms of Bigdata. These services provide the capabilities to analyze data collectively to produce desired results efficiently. Data is accessible for users, developers, experts, stakeholders, and all involved agencies [57].

Rapid Growth. Both private and public sectors are contributing knowledge and technologies to develop applications for the productivity of agricultural sector. The rapid increase in technology services will assist in economic development [58].

Supply Chain Pathway. Big data, IoT and cloud computing creates a smooth pathway for digital communication between stakeholders of mobile app users for market transactions by providing data sharing feature and computing analytics [59].

Easy-Handy Network Computing. Android apps has the features of connecting any other handheld device including sensors, etc. with any type of internet connection available to share, upload, download data. IoT provides backend facilities for easy-handy network computing [60].

5   Conclusions

The survey of mobile application services in India has indicated that there is a wide range of services that are now available to farmers. These provide various climatic cropping news and markets/pricing information. The access to Big Data and IoT services has allowed farmers real time access to information and expertise to assist with making important decisions in their daily activities. These mobile apps do provide some strength such as access and sharing of data and the ability for farmers to be responsive. A number of weaknesses were also observed with the mobile application services related to issues of understanding of the technology and data security, standards and regulations, which need to be addressed. The findings from this survey of mobile application services provide some insight into the range of current services and the lack of services in some categories.

It is recommended that agricultural government agencies and private service providers will need to continue develop new applications that are responsive to emerging technologies that are being adopted by the Indian agricultural sector. Training for technology awareness may be required to ensure that the benefits of these services are fully realized.

Acknowledgement. The authors acknowledge the Centre for Environmental Management, and the School of Science at Edith Cowan University for partial funding for Dr. Leisa Armstrong to attend the WICT 2017 conference.

References

1. Asseng, S.: Rising temperature reduce global wheat production. Nat. Clim. Change **5**(2), 143–147 (2015)
2. Dhaka, B.: Farmers' experience with ICTs on transfer of technology in changing agri-rural environment. Indian Res. J. Ext. Educ. **10**(3), 114–118 (2016)
3. Han, Q.: Mobile cloud sensing, Big Data and 5G networks make an intelligent and smart world. IEEE Netw. **29**(2), 40–45 (2015)
4. Bahrami, M.: The role of cloud computing architecture in Big Data. In: Pedrycz, W., Chen, S.M. (eds.) Information Granularity, Big Data, and Computational Intelligence, Studies in Big Data, vol. 8, pp. 275–295. Springer International Publishing, Cham (2015)
5. Botta, A.: On the integration of cloud computing and Internet of Things. In: International Conference on Future Internet of Things and Cloud (FiCloud), Barcelona, Spain, pp. 23–30. IEEE (2014)
6. Gil, D.: Internet of Things: a review of surveys based on context aware intelligent services. Sensors **16**(7), 1069 (2016)
7. Protopop, I.: Big Data and smallholder farmers: Big Data applications in the agri-food supply chain in developing countries. Int. Food Agribus. Manag. Rev., Spec. Iss. **19**(A), 173–190 (2016)
8. Yogaraj, G., Arun, A.: Mining high dimensional data sets using Big Data. Int. J. Adv. Res. Comput. Sci. Softw. Eng. **5**(2), 970–974 (2015)
9. Botta, A.: Integration of cloud computing and Internet of Things: a survey. Future Gener. Comput. Syst. **56**, 684–700 (2016)
10. Khan, I.: A review on integration of cloud computing and Internet of Things. Int. J. Adv. Res. Comput. Commun. Eng. **5**(4), 1046–1050 (2016)
11. Chandel, S.: Overview of the initiatives in renewable energy sector under the national action plan on climate change in India. Renew. Sustain. Energy Rev. **54**, 866–873 (2016)
12. Krintz, C.: SmartFarm: improving agriculture sustainability using modern information technology. In: KDD Workshop on Data Science for Food, Energy, and Water. ACM (2016). ISBN 978-1-4503-2138-9
13. Das, A.: Accessing agricultural information through mobile phone: lessons of IKSL services in West Bengal. Indian Res. J. Ext. Educ. **12**(3), 102–107 (2016)
14. KisanSuvidha: A Smart Mobile App for Farmers. Accessed 10 Oct 2017
15. AppAgg Rikwaa Tech Lab. https://appagg.com/developer/rikwaa-tech-lab/. Accessed 16 Oct 2017
16. MyAgriGuru. http://www.myagriguru.com/. Accessed 17 Oct 2017
17. IFFCO Kisan. http://www.iffcokisan.com/. Accessed 12 Oct 2017
18. Agriculture Rural Technology. http://oer.nios.ac.in/wiki/index.php/Rural-Technology. Accessed 13 Oct 2017
19. RainbowAgri. http://www.rainbowagri.com/. Accessed 14 Oct 2017
20. SmartAgri Advisory Services. http://smartagri.in/. Accessed 15 Oct 2017
21. Agri Science. http://agriscienceindia.com/. Accessed 17 Oct 2017
22. National Agricultural Market. http://www.enam.gov.in/NAM/home/about_nam.html#. Accessed 18 Oct 2017

23. Indian AgroPrice (Live Mandi). https://www.socialapphub.com/app/indian-agri-price-live-mandi. Accessed 17 Oct 2017
24. Agri.Live. http://www.agri.live/index.php. Accessed 17 Oct 2017
25. Indian Cooperative. http://www.indiancooperative.com/. Accessed 18 Oct 2017
26. Agribuz. http://www.agribuz.com/#/home. Accessed 13 Oct 2017
27. Digital Green. https://www.digitalgreen.org. Accessed 15 Oct 2017
28. Featured Community Apps/Gramseva: Kisan (Mandi Prices). https://data.gov.in/community-application/gramseva-kisan-mandi-prices-android. Accessed 12 Oct 2017
29. KisanMandi.Com Retail & Wholesale. http://www.kisanmandi.com. Accessed 12 Oct 2017
30. KISAAN USSD APK. https://www.apkmonk.com/app/in.gov.mgov.kisaanUssd/. Accessed 12 Oct 2017
31. PusaKrishi. https://www.9apps.com/android-apps/Pusa-Krishi. Accessed 10 Oct 2017
32. RML Farmer–KrishiMitr APK. https://www.apkmonk.com/app/com.rml.Activities/. Accessed 12 Oct 2017
33. Application for Soil Testing. https://apps.mgov.gov.in/descp.do?appid=386. Accessed 11 Oct 2017
34. Mahadhan. https://mahadhan.co.in/. Accessed 16 Oct 2017
35. AgriApp Connecting Farmers. http://agriapp.co.in/. Accessed 13 Oct 2017
36. Agro India. http://www.agroindia.net/. Accessed 15 Oct 2017
37. Kisan Network. http://kisannetwork.com/. Accessed 11 Oct 2017
38. Farm-o-pedia: A multilingual android application. http://gyti.techpedia.in/project-detail/farm-o-pedia-a-multilingual-android-application/2599. Accessed 12 Oct 2017
39. IndiaAgroNet.Com. https://www.indiaagronet.com/Agriculture-Information/agriculture-information-Insecticides-mobile-app.html. Accessed 13 Oct 2017
40. Crop Insurance. http://agri-insurance.gov.in/CCEAppVersions.aspx. Accessed 10 Oct 2017
41. NRSC Indian Geo-Platform of ISRO, Ministry of Agriculture and Farmer Welfare, Hailstorm. http://bhuvan.nrsc.gov.in/governance/moafw_hailstorm. Accessed 13 Oct 2017
42. Cropin. http://cropin.co.in/2017/04/03/cropin-improves-farmers-lives-data-analytics-digital-apps-ashwani-mishra-etcio-com/. Accessed 13 Oct 2017
43. mKisan. http://mkisan.gov.in/aboutmobileapps.aspx. Accessed 11 Oct 2017
44. Digital Mandi Application for Indian Kisan. http://www.tcoe.in/?q=content/digital-mandi-application-indian-kisan. Accessed 12 Oct 2017
45. Center for Sustainable Agriculture eKrishi Mobile Application. http://csa-india.org/ekrishi-app/. Accessed 15 Oct 2017
46. KrishiGyan. http://app.appsgeyser.com/1464980/krishi%20gyan. Accessed 12 Oct 2017
47. Government of Sikkim Food Security and Agriculture Development. https://www.sikkim.gov.in/portal/StatePortal/Department/-FoodSecurityAndAgricultureDevelopment. Accessed 11 Oct 2017
48. Agri Market. http://agmarknet.gov.in. Accessed 10 Oct 2017
49. Mandi Trades. https://itunes.apple.com/in/app/mandi-trades. Accessed 14 Oct 2017
50. Jayalaxmi Agro Tech. http://www.jayalaxmiagrotech.com/. Accessed 16 Oct 2017
51. Organic Farming. http://www.xenonnation.com/. Accessed 15 Oct 2017
52. KhetiBadi. https://shop.kheti-badi.com/. Accessed 13 Oct 2017
53. Krishi App. http://www.philosan.com/about.html#mobile. Accessed 15 Oct 2017
54. Jin, X.: Significance and challenges of Big Data research. Big Data Res. 2(2), 59–64 (2015)
55. Shang, W.: Challenges in IoT networking via TCP/IP architecture. NDN Project, Technical rep. NDN-0038 (2016)
56. Lee, I.: The Internet of Things (IoT): applications, investments, and challenges for enterprises. Bus. Horiz. 58(4), 431–440 (2015)

57. Bello-Orgaz, G.: Social Big Data: recent achievements and new challenges. Inf. Fusion **28**, 45–59 (2016)
58. Yang, C.: Big Data and cloud computing: innovation opportunities and challenges. Int. J. Digit. Earth **10**(1), 13–53 (2017)
59. Hashem, I.: The rise of "Big Data" on cloud computing: review and open research issues. Inf. Syst. **47**, 98–115 (2015)
60. Sun, Y.: Internet of Things and Big Data analytics for smart and connected communities. IEEE Access **4**, 766–773 (2016)

A Study of the Privacy Attitudes of the Users of the Social Network(ing) Sites and Their Expectations from the Law in India

Sandeep Mittal[1]([:envelope:]) and Priyanka Sharma[2]

[1] NICFS (MHA), New Delhi, India
sandeep.mittal@nic.in
[2] Raksha Shakti University, Ahmedabad, India

Abstract. In an era of information revolution and Web 2.0 technologies, the Social Network(ing) Sites (SNSs) have become a popular medium for the freedom of expression, the networking and maintaining the networks with the strangers and known others. A large amount of personal data is disclosed by the users intentionally or unknowingly on these social networking sites. The protection of this data at residence and in motion and its further processing by SNSs and their third parties is a cause of concern. In the present study, the attitude of Indian users of SNSs towards data privacy and their expectations from law in India have been explored and analyzed to validate the need for creation of a data privacy law in India. This study would provide timely guidance for policy makers who are currently engaged in framing a data protection framework on the directions of Supreme Court of India by following due process of law.

Keywords: Information privacy · Data privacy attitude · Data privacy law

1 Introduction

The general privacy beliefs are results of complex interaction of social norms and moral value beliefs often mediated in space and time by a number of social variables at individual or collective levels. In real-life social interactions, the individuals have a control over the personal information shared amongst each other. The personal information thus shared in physical world has a limited and slow flow to others and generally dissipates with time with no trace after a relatively reasonable timespan. Its impact on a person's reputation is also relatively limited to a close social-circle.

The rise of the Internet, Web 2.0 and easy availability of smart devices has resulted in an era of privacy development where the use of social networking sites (SNSs) like Facebook, LinkedIn, and Twitter etc. for exchanging information in virtual space has become the norm. The personal information exchanged over such SNSs generically differ from that in real world in that the persons exchanging information are not face to face with each other thus compromising the real world controls on the information, travels fast and far beyond the control of anyone and has perpetual availability on internet. The general privacy, initially defined either by value-based approach or

© Springer International Publishing AG, part of Springer Nature 2018
A. Abraham et al. (Eds.): ISDA 2017, AISC 736, pp. 1038–1051, 2018.
https://doi.org/10.1007/978-3-319-76348-4_100

cognate-based approach, gradually shifted in present information era to 'privacy as a right' concept to "control physical space and information" [1]. The protection of privacy and confidentiality of this personal data at residence and in motion within and across the borders is a cause of concern.

In India, until recently even the right to privacy was not even recognized as a fundamental right and a data privacy legal framework is still lacking. The recent judgement in the landmark case on Fundamental Right to Privacy by the 'Nine Judges Constitutional Bench' of Hon'ble Supreme Court of India [2] has recognized right to privacy as a fundamental constitutional right in India and has directed Government of India to put in place, a robust data privacy regime expeditiously for which Government of India have constituted a Committee called 'Justice B. N. Srikrishna Committee'. As the current process of drafting a data privacy framework in India has commenced, the present study is scoped to understand the privacy attitudes of the Indian users of the SNSs and their expectations on the thought process of law on data privacy.

2 The Literature Review

2.1 The Definition of Privacy

A perusal of the scholarly reviews on privacy reveals mainly two approaches to defining the general privacy, viz., value-based and cognate-based, the former being more prevalent in legal, sociological and political studies while the latter being more explored in psychological studies. In the present study a mix of these two approaches is used to explore the cognitive aspect (attitudes towards privacy) and the right-based aspect (expectations from law to protect privacy).

Let us briefly look at these two approaches to definitions of privacy. As cognate-state approach, the general privacy is defined as "a state of limited access to a person" which narrowed down to Information systems broadly translates to "a state of limited access to information" [1]. As cognate-control approach the general privacy is defined as "the selective control of access to the self" [3] and as "control of transactions between person(s) and other(s), the ultimate aim of which is to enhance autonomy or/and to minimize vulnerability" [4]. As a right-based approach, the general privacy is treated differently in different parts of the world, e.g., in the EU, privacy is seen as a fundamental human right; while in the U.S., privacy is seen as a commodity subject to the market and is cast in economic terms.

2.2 The Social Network(ing) Sites (SNSs)

While the two terms have been used interchangeably, the scholars have distinguished between "social network site" and "social networking site", the former being "unique (in) not that they allow individuals to meet strangers, but rather that they enable users to articulate and make visible their social networks" resulting in frequent connections between individuals ("latent ties") who share some offline connection, while the latter "emphasizes relationship initiation, often between strangers" [5]. However with the advent of Web 2.0 after this distinction was made, it has almost faded away and for the purpose of this study we use the term "Social Network(ing) Site" (SNSs) to mean and include features of the two. The basic definition of SNSs includes the following structural elements [6],

(a) A feature to build friends list and restrict access to users' personal data to this group or its subset.
(b) Pre-structured 'user profiles' disclosing personal information.
(c) Disclosure of some data in free form.
(d) A 'timeline' amalgamating personal data disclosed by different users at different times in different context to the site.
(e) Non-explicit disclosure of personal data.

While these features are the *raison d'être* for social network(ing) sites, at the same time are the very cause for most of the users' privacy concerns.

2.3 The Privacy and the Social Network(ing) Sites

In course of social interactions in the physical world, while an individual uses his physical senses to perceive and manage threats to his privacy, he has no such social and cultural cues to evaluate the target of self-disclosures in a visually anonymous online space of SNSs. Therefore, while the cognitive management of protection of privacy in offline world is performed unconsciously and effortlessly, deliberate actions are required for effective self-protection are required on SNSs [7]. These deliberative actions can be understood in terms of the "Theory of Planned Behavior" (TPB) [8] which stipulates that "an individual's intention is a key factor in predicting his or her behavior.

2.4 Understanding the Attitudes Towards Privacy on SNSs

Several theoretical and empirical studies across disciplines have been conducted to understand the attitudes on privacy and data privacy protection laws in jurisdictions worldwide. A few findings relevant to the present work are enumerated here,

(a) An information disclosure by SNSs' users is associated with their level of concern for privacy [9].
(b) SNSs' users are aware of privacy setting and change default settings as per their need [10, 12].
(c) Perception of trust by SNSs' users improves with greater information disclosure by SNSs [11].
(d) Privacy Policies of SNSs help in protecting privacy of SNSs' users [12].
(e) Disclosure of personal information on SNSs is a bargaining process where perceived benefits and gratifications of networking outweigh the privacy [13].
(f) More knowledge and experience of using the Internet improves privacy concern of SNSs' users [14].
(g) Demographic factors influence SNSs' user's privacy behavior [15].

2.5 Thought Process on Data Privacy Law

The right to privacy has been recently recognized as a constitutional fundamental right in India by the landmark judgement in the case on Fundamental Right to Privacy by the 'Nine Judges Constitutional Bench' of Hon'ble Supreme Court of India. The information (data) privacy was also pondered upon during the hearings and Government of India was directed to put in place a robust data privacy regime maintaining a careful and sensitive balance between individual interests and legitimate concerns of the State. During the course of Court hearings, several government and non-governmental stake-holders argued their view-points on privacy and information privacy giving an insight into the thought process on this issue as depicted info-graphically (Fig. 1).

Fig. 1. Infographic on the thought process on data privacy debate in India [16]

These propositions of Indian thought process and some others based on the thought processes in other parts of the world, e.g., GDPR etc. are also explored in the present study.

3   The Research Methodology

The population for the present study is the users of the SNSs in India grouped into five strata, namely, Law Enforcement Officers, Judicial and Legal Professionals, Academicians, Information Assurance and Privacy Experts and the Internet Users (other than listed in strata above) in India adopting disproportionate, stratified, purposive, convenience mixed sampling technique, and a statistically adequate sample size of 385 having 95% Confidence Level, 5% Margin of Error (Confidence Interval), 0.5 Standard Deviation and 1.96 Z-score was calculated.

A questionnaire was designed for this study by incorporating modified questions based on the Eurobarometer [17] and modified in Indian context and limited to the objectives of the present study. The variables included in the tool can be categorized as nominal and ordinal variables. A pilot study was conducted and reliability of instrument was checked by running reliability analysis which returned a Cronbach Alpha value of 0.700, and modified to adjust the scale and a Cronbach Alpha value of 0.795 was obtained which is well within the acceptable norms (>0.700) [18]. All the 401 respondents gave their informed explicit consent signifying their willing participation in this study. The data was collected during the month of August, 2017. The data was analyzed in SPSS for statistically significant trends regarding high privacy concern and its association for thought process on the expectations from law between variables by applying Pearson's Chi-Square (χ^2) Test of Independence with significance levels of 1% or 5% ($p < 0.01$ or $p < 0.05$) to test Null Hypotheses. The post-hoc analysis was done to determine the strength of the effect size of the association by calculating the Cramer's V values (ϕ').

As the study relied upon disproportionate, stratified, purposive, Convenience Sampling, the study may have limitation of non-generalization to wider population, and not taking into account the children presumptively below 18 years of age using the SNSs with fake accounts.

4   The Results, Data Analysis and Discussion

4.1   The Socio-Demographic Profile of Respondents

Out of 401 respondents, the majority was between 28 years to 45 year of age (42%), while the age group above 60 years has the minimum respondents (8.0%). Out of total population, 74.8% are males and 25.2% are females. The educational level of respondents were spread across categories with majority of the respondents being postgraduate (61%), followed by graduate (31%), Ph.D. (7%) and a small proportion (1%) below graduate level. The distribution of respondents across professional groups is multimodal, i.e., Judiciary and Legal profession (10%), Law Enforcement (24%), Information Assurance and Privacy Experts (17%), Academic (20%) and other users of Internet (28.9%). However, this ensures that all stakeholders involved in policy making for data privacy in India are accounted for. About 97% of respondents are users of SNSs (e.g., Facebook, Twitter, LinkedIn, etc.). Majority of the respondents (46%) spent less than one hour on Internet followed by 29% of respondents spending between one to two hours on Internet. 83% of respondents have high level of online privacy literacy and 17% of respondents had low level of Online Privacy Literacy.

4.2   The Attitudes: A Descriptive Analysis

This study explores the attitude of the Indian users of SNSs and the responses to the variables are grouped in various constructs to describe and understand the attitude of the population. The following trends on data privacy attitudes were discerned in the present study.

(a) The three most important personal information identified are financial information (59%), Aadhaar, Passport, License (56%), and Biometrics (46%). The friends' list (4%), work history (5%) and medical information (12%) are not considered as important personal information.

(b) The most important reason for information disclosure was denial of access by SNSs if the information is not disclosed (63%) followed by urge to connect with others (31%).

(c) The three most important risks of information disclosure identified are data sharing without consent (86%), identity theft (75%) and frauds (72%).

(d) The monitoring and recording of behavior is a major privacy concern (87%).

(e) The acts of users to protect their identity in daily life and on SNSs are similar

(f) While majority of respondents read privacy policy (53%) only 18% understand the same. The majority of respondents failed to adopt a change after reading privacy policy (53%). The lengthy and complex text was a major reason for ignoring the privacy policy (59%).

(g) The majority perceived only partial control over disclosed information (52%).

(h) The majority was not comfortable with SNSs collecting their personal data for commercial advertisements (55%).

(i) The majority of respondents (55%) wanted their explicit consent to be always taken.

(j) The majority did not trust default settings (43%).

(k) While 43% of respondents perceive a part role for all stakeholders including Government, 19% envisaged full responsibility of Government to protect their information.

5   The Hypotheses Testing, Analysis and Discussion Regarding Thought Process on Privacy Law in India

As the major objective of the study is to understand the thought process on privacy law in India, let us analyze the data to understand various aspects of this thought process. For this purpose, we would test the following hypotheses,

5.1   Hypothesis-1: High Concern for Privacy is Associated with Perception of Privacy Embedded in the Honor and the Dignity of a Person:

The perception that the privacy is embedded in the honor and dignity of a person is not found to be statistically significantly independent of, (a) perceived responsibility for safety of users' personal information at 5% level of significance, (b) perceived requirement of users' consent for sharing their data at 1% level of significance, (c) perceived concern for users' behavior monitoring by SNSs at 1% level of significance. The association is not only significant but has a medium effect size. As these three factors are indicative of high concern for privacy, it is reasonable to conclude that the high concern for privacy significantly effects the perception that the privacy is embedded in honor and dignity of a person (Table 1).

Table 1. Hypothesis-1: High concern for privacy is associated with perception that privacy is embedded in honour and dignity of a person

Chi-square (χ^2) statistic

S. no.	Null Hypotheses (H_0)	χ^2	df	p	Remarks	Effect size (ϕ')
1.1	Perception of privacy embedded in honour and dignity of person is significantly independent of perceived responsibility for safety of users' personal information	23.51	12	0.024	Significant (p < .05)	Medium 0.121 (r = 5)
1.2	Perception of privacy embedded in honour and dignity of person is significantly independent of perceived requirement of users' explicit consent for sharing their data	21.81	6	0.001	Significant (p < .01)	Medium 0.164 (r = 5)
1.3	Perception of privacy embedded in honour and dignity of person is significantly independent of perceived concern for users' behaviour monitoring by SNSs	52.34	12	0.000	Significant (p < .01)	Medium 0.180 (r = 5)

5.2 Hypothesis-2: High Concern for Privacy is Associated with the Perception that Privacy Should Be a Constitutional Fundamental Right:

The perception that the privacy should be a constitutional fundamental right is found not to be statistically significantly independent of, (a) Perceived responsibility for safety of users' personal information at 5% level of significance, (b) perceived concern for users' behavior monitoring by SNSs at 1% level of significance. The association is not only significant but has a medium effect size. However perceived requirement of users' consent for sharing their data did not have a statistically significant effect on the perceived need for privacy as a constitutional fundamental right. As these two factors are indicative of high concern for privacy, it is reasonable to conclude that the high concern for privacy significantly affects the perceived need for privacy as a constitutional fundamental right (Table 2).

Table 2. Hypothesis-2: High concern for privacy is associated with perceived need for privacy as a constitutional fundamental right

Chi-square (χ^2) statistic

S. no.	Null Hypotheses (H_0)	χ^2	df	p	Remarks	Effect size (ϕ')
2.1	Perceived need of privacy as a constitutional fundamental right is significantly independent of perceived responsibility for safety of users' personal information	30.25	16	0.017	Significant (p < .05)	Medium 0.137 (r = 5)
2.2	Perceived need of privacy as a constitutional fundamental right is significantly independent of perceived requirement of users' explicit consent for sharing their data	14.83	8	0.062	Insignificant	0.136 (r = 2)
2.3	Perceived need of privacy as a constitutional fundamental right is significantly independent of perceived concern for users' behaviour monitoring by SNSs	43.21	16	0.000	Significant (p < .01)	Medium 0.164 (r = 5)

5.3 Hypothesis-3: High Concern for Privacy is Associated with the Perception that Data Privacy Should Be Part of the Constitutional Fundamental Right of Privacy:

The perception that the data privacy should be part of the Constitutional fundamental right of privacy is found not to be statistically significantly independent of the perceived concern for users' behavior monitoring by SNSs at 1% level of significance. The association is not only significant but has a medium effect size. As the high concern for behavior monitoring is indicative of high concern for privacy, it is reasonable to conclude that the high concern for privacy significantly affects the perceived need for Data Privacy as a part of the Constitutional Fundamental Right of Privacy (Tables 3 and 4).

Table 3. Hypothesis-3: High concern for privacy is associated with perceived need for data privacy as a part of the constitutional fundamental right of privacy

Chi-square (χ^2) statistic

S. no.	Null Hypotheses (H_0)	χ^2	df	p	Remarks	Effect size (ϕ')
3.1	Perceived need of data privacy as a part of constitutional fundamental right is significantly independent of perceived responsibility for safety of users' personal information	19.62	16	0.238	Insignificant	0.110 (r = 2)
3.2	Perceived need of data privacy as a part of constitutional fundamental right is significantly independent of perceived requirement of users' explicit consent for sharing their data	14.06	8	0.080	Insignificant	0.132 (r = 2)
3.3	Perceived need of data privacy as a part of constitutional fundamental right is significantly independent of perceived concern for users' behaviour monitoring by SNSs	47.34	16	0.000	Significant (P < .01)	Medium 0.171 (r = 5)

Table 4. Hypothesis-4: High concern for privacy is associated with perceived need for the right to be forgotten as a fundamental right

Chi-square (χ^2) statistic

S. no.	Null Hypotheses (H_0)	χ^2	df	p	Remarks	Effect size (ϕ')
4.1	Perceived need of right to be forgotten is significantly independent of perceived responsibility for safety of users' personal information	12.35	16	0.499	Insignificant	0.097 (r = 2)
4.2	Perceived need of right to be forgotten is significantly independent of perceived requirement of users' explicit consent for sharing their data	6.11	8	0.635	Insignificant	0.087 (r = 2)
4.3	Perceived need of right to be forgotten is significantly independent of perceived concern for users' behaviour monitoring by SNSs	49.24	16	0.000	Significant (P < .01)	Medium 0.175 (r = 5)

5.4 Hypothesis-4: High Concern for Users' Behavior Monitoring by SNSs is Associated with Perceived Need for the Right to Be Forgotten as a Fundamental Right:

The perception that the right to be forgotten should be a fundamental right is found not to be statistically significantly independent of the perceived concern for users' behavior monitoring by SNSs at 1% level of significance. The association is not only significant but has a medium effect size. As the high concern for behavior monitoring is indicative of high concern for privacy, it is reasonable to conclude that the high concern for privacy significantly affects the perceived need for the Right to be forgotten as a Fundamental Right.

5.5 Hypothesis-5: High Concern for Privacy is Associated with the Perception that Right to Data Privacy is Difficult to Enforce on SNSs:

The perception that the right to data Privacy is difficult to enforce on SNSs is found not to be statistically significantly independent of the perceived concern for users' behavior monitoring by SNSs at 5% level of significance. The association is not only significant but has a medium effect size. However, no statistically significant effect was found to be exerted by perceived responsibility for safety of users' personal information and perceived requirement of users' consent for sharing their data on the perception that the

Table 5. Hypothesis-5: High concern for privacy is associated *with perception that right to data privacy is difficult to enforce on SNSs*

Chi-square (χ^2) statistic						
S. no.	Null Hypotheses (H_0)	χ^2	df	p	Remarks	Effect size (ϕ')
5.1	Perceived difficulty in enforcement of data privacy on SNSs is significantly independent of perceived responsibility for safety of users' personal information	14.95	16	0.528	Insignificant	0.124 (r= 2)
5.2	Perceived difficulty in enforcement of data privacy on SNSs is significantly independent of perceived requirement of users' explicit consent for sharing their data	12.51	8	0.130	Insignificant	0.124 (r = 2)
5.3	Perceived difficulty in enforcement of data privacy on SNSs is significantly independent of perceived concern for users' behaviour monitoring by SNSs	27.24	16	0.039	Significant (p < .05)	Medium 0.130 (r = 5)

right to data Privacy is difficult to enforce on SNSs. As the high concern for behavior monitoring is indicative of high concern for privacy, it is reasonable to conclude that the high concern for privacy significantly effects the perception that Right to Data Privacy is Difficult to Enforce on SNSs (Table 5).

5.6   Hypothesis-6: High Concern for Privacy is Associated with the Perception that Stratified Right to Data Privacy is Difficult to Implement:

The Chi-square statistic and post-hoc analysis by calculating Cramer' s V values and interpretation of the same indicates that the perception that the stratified right to data privacy is difficult to implement is found not to be statistically significantly independent of, (a) Perceived requirement of users' consent for sharing their data at 5% level of significance, (b) perceived concern for users' behavior monitoring by SNSs at 1% level of significance. The association is not only significant but has a medium effect size. However, no statistically significant effect was found to be exerted by perceived responsibility for safety of users' personal information on the perception that the stratified right to data privacy is difficult to implement. As these two factors are indicative of high concern for privacy, it is reasonable to conclude that the high concern for privacy significantly effects the perception that stratified right to data privacy is difficult to implement (Table 6).

Table 6. Hypothesis-6: High concern for privacy is associated *with perception that stratified right to data privacy is difficult to enforce*

Chi-square (χ^2) statistic						
S. no.	Null Hypotheses (H$_0$)	χ^2	df	p	Remarks	Effect size (ϕ')
6.1	Perceived difficulty in enforcement of stratified right of data privacy is significantly independent of perceived responsibility for safety of users' personal information	23.05	16	0.112	Insignificant	0.119 (r = 2)
6.2	Perceived difficulty in enforcement of stratified right of data privacy is significantly independent of perceived requirement of users' explicit consent for sharing their data	18.42	8	0.018	Significant (p < .05)	Medium 0.151 (r = 5)
6.3	Perceived difficulty in enforcement of stratified right of data privacy is not significantly dependent on perceived concern for users' behaviour monitoring by SNSs	43.19	16	0.000	Significant (p < .01)	Medium 0.164 (r = 5)

5.7   The Perception of SNSs Users Regarding Regulation of Privacy in India is Summarized as Follows (Table 7).

Table 7. Perception on privacy regulation in India

Perception on privacy regulation in India	% agreed
Personal data should not be kept in a form by which the data subject can be identified	88.1
Personal data stored may be processed solely in public interest, for scientific, historical or statistical research to safeguard the rights and freedom of the data subject	84.4
Personal data should be processed securely deploying appropriate technical measures	90.5
Personal data to be processed only if for the performance of a task in the public interest or in the exercise of lawful authority by a law enforcement agency	87.1
Right to withdraw the consent given to Service Providers at any time	89.7
Provision of a service not dependent on the condition of affirmative consent of data subject	82.9
For child, the collection and processing of data with explicit and verified parental consent	89.0
A written and clear communication to data subject in plain language for data processing	88.3
Right to obtain, from Service Providers without undue delay, rectification of inaccurate data	84.7
If the personal data are no longer right to erase the personal data without undue delay	88.8
Right of data portability from one service provider to other service provider	77.4
Automated processing of personal data should not be permitted	79.8
Law should restrict processing of personal data if processing prevents investigation, detection or prosecution of criminal offences or the execution of criminal penalties	75.3
Any international transfer of personal data only subject to the explicit consent of 'data subject' and provisions of law of jurisdiction where data subject ordinarily resides	84.5
The personal data should be transferred to a third country only if the Service Providers has provided appropriate safeguards and on condition that enforceable 'data subject' rights and effective remedy for 'data subject' is easily available and accessible at place where data subject ordinarily resides	75.7
In case infringement of the privacy right, a remedy for compensation in time-bound manner through an easily available mechanism of a Public Authority at his place of ordinary residence	85.9
The Personal data held by a public authority or a public body or a private body be disclosed in public interest accordance with law of jurisdiction where the data subject ordinarily resides	71.3

6 Conclusions

In the present study, we defined the problem of protection of personal data on SNSs. We began with the understanding of concept of privacy in general as 'cognate-based' and 'right-based' approach, and its expansion in an online environment of SNSs followed by adoption of a theoretical framework, viz., theory of planned behavior to understand the behavior of users on SNSs as a result of the intention which is shaped by attitude, subjective norm and perceived behavioral control. A review of various empirical studies on data privacy attitudes was followed by underlining the thought process on the current data privacy debate in India. The present study is a maiden attempt in India to understand the privacy behavior of users of SNSs and their expectations from Law on data privacy and presents a reasonable scientific and validated insight for the policy makers to strike a balance between the concerns of the State and the individual while framing data privacy law in India.

References

1. Smith, H.J., Dinev, T., Xu, H.: Information privacy research: an interdisciplinary review. MIS Q. **35**, 989–1016 (2011)
2. "Justice K S Puttaswamy v Union of India," Supreme Court of India (Nine Judges Constitutional Bench) (2017). http://supremecourtofindia.nic.in/pdf/LU/ALL%20WP(C)%20No.494%20of%202012%20Right%20to%20Privacy.pdf
3. Altman, I.: The Environment and Social Behavior: Privacy, Personal Space, Territory, and Crowding (1975)
4. Margulis, S.T.: Conceptions of privacy: current status and next steps. J. Social Issues **33**, 5–21 (1977)
5. Ellison, N.B.: Social network sites: definition, history, and scholarship. J. Comput. Mediated Commun. **13**, 210–230 (2007)
6. Edwards, L.: Privacy, law, code and social networking sites. In: Research Handbook on Governance of the Internet, p. 309 (2013)
7. Yao, M.Z.: Self-protection of online privacy: a behavioral approach. In: Privacy Online, pp. 111–125. Springer, Heidelberg (2011)
8. Ajzen, I., Fishbein, M.: The influence of attitudes on behavior. In: The Handbook of Attitudes, vol. 173, p. 31 (2005)
9. Gross, R., Acquisti, A.: Information revelation and privacy in online social networks. In: Proceedings of the 2005 ACM Workshop on Privacy in the Electronic Society, pp. 71–80 (2005)
10. Strater, K., Richter, H.: Examining privacy and disclosure in a social networking community. In: Proceedings of the 3rd Symposium on Usable Privacy and Security, pp. 157–158 (2007)
11. Dwyer, C., Hiltz, S., Passerini, K.: Trust and privacy concern within social networking sites: a comparison of Facebook and MySpace. In: AMCIS 2007 Proceedings, p. 339 (2007)
12. Debatin, B., Lovejoy, J.P., Horn, A.-K., Hughes, B.N.: Facebook and online privacy: attitudes, behaviors, and unintended consequences. J. Comput. Mediated Commun. **15**, 83–108 (2009)
13. Ellison, N.B., Vitak, J., Steinfield, C., Gray, R., Lampe, C.: Negotiating privacy concerns and social capital needs in a social media environment. In: Trepte, S., Reinecke, L. (eds.) Privacy Online, pp. 19–32. Springer, Heidelberg (2011)

14. LaRose, R., Mastro, D., Eastin, M.S.: Understanding Internet usage: a social-cognitive approach to uses and gratifications. Soc. Sci. Comput. Rev. **19**, 395–413 (2001)
15. Tufekci, Z.: Grooming, gossip, Facebook and MySpace: what can we learn about these sites from those who won't assimilate? Inf. Commun. Soc. **11**, 544–564 (2008)
16. Mahapatra, D.: Supreme Court for 3-tier right to privacy: intimate, private and public. The Times of India edition, New Delhi (2017)
17. Special Eurobarometer 359: Attitudes on Data Protection and Electronic Identity in the European Union (2011). http://ec.europa.eu/commfrontoffice/publicopinion/archives/ebs/ebs_359_en.pdf
18. Serbetar, I., Sedlar, I.: Assessing reliability of a multi-dimensional scale by coefficient alpha. Revija za Elementarno Izobrazevanje **9**, 189 (2016)

Author Index

© Springer International Publishing AG, part of Springer Nature 2018
A. Abraham et al. (Eds.): ISDA 2017, AISC 736, pp. 1053–1056, 2018.
https://doi.org/10.1007/978-3-319-76348-4